AF572507

ORGANIC REACTION MECHANISMS · 1973

ORGANIC REACTION MECHANISMS · 1973

An annual survey covering the literature dated December 1972 through November 1973

Edited by

A. R. BUTLER, University of St. Andrews

M. J. PERKINS, Chelsea College, University of London

An Interscience® Publication

JOHN WILEY & SONS

London · New York · Sydney · Toronto

An Interscience® Publication

Library of Congress Catalog Card Number 66-23143

ISBN 0 471 12690 X

Printed in Great Britain by
William Clowes & Sons Limited
London, Colchester and Beccles

Contributors

R. BAKER	Department of Chemistry, The University, Southampton
T. W. BENTLEY	Department of Chemistry, University College of Swansea
R. B. BOAR	Department of Chemistry, Chelsea College, University of London
A. R. BUTLER	Department of Chemistry, The University, St. Andrews
B. CAPON	Department of Chemistry, Glasgow University
M. R. CRAMPTON	Department of Chemistry, Durham University
T. L. GILCHRIST	Department of Organic Chemistry, University of Liverpool
A. C. KNIPE	Department of Chemistry, The New University of Ulster
D. C. NONHEBEL	Department of Pure and Applied Chemistry, University of Strathclyde
M. I. PAGE	Department of Chemistry, The Polytechnic, Huddersfield
F. L. SCOTT	Pennwalt Chemical Corporation, Pharmaceuticals Division, P.O. Box 1710, Rochester, N.Y. 14603, U.S.A.
B. V. SMITH	Department of Chemistry, Chelsea College, University of London
I. D. R. STEVENS	Chemistry Department, The University, Southampton
R. C. STORR	Department of Organic Chemistry, University of Liverpool
J. C. WALTON	Department of Chemistry, St. Salvator's College, University of St. Andrews

Preface

This year's edition of *Organic Reaction Mechanisms*, the eighth in the series, does not differ greatly from the highly successful pattern developed by the previous editors. The literature dated December 1972 to November 1973 has been covered and the aim has been, as in past years, to make the coverage comprehensive. Over 5000 papers have been reported and many of the important ones discussed in some detail.

One modification which will be immediately apparent to those familiar with the series is the changed order of the Chapters. This change has been made to facilitate separate publication of selected sections of the book at prices more likely to be within the reach of the individual researcher. It should be stressed that not all of this book will be available in sectional form; however, if it is successful, sectional publication may be extended with the 1974 Volume. We have also decided to omit the chapter on Photochemistry. This topic is covered comprehensively in a Specialist Report published annually by the Chemical Society.

A further change, made with great reluctance, but as an economy measure necessitated by spiralling publishing costs, has been the relegation of references to the ends of the chapters. If this innovation or any other aspect of the book's format is found irksome, please do not hesitate to let us know. Indeed, it is our intention to offer *Organic Reaction Mechanisms* as a service, and we shall always be pleased to consider suggestions for its improvement.

We are grateful to the established contributors for their continued support, and welcome those who have joined this year. The British office of John Wiley and Sons have given us every possible help and have made the rapid publication of this volume possible. We sincerely hope that *Organic Reaction Mechanisms 1973* will be of as great a value to organic chemists as its predecessors.

August 1974

A. R. B.
M. J. P.

Contents

CHAPTER 1

Reactions of Aldehydes and Ketones and their Derivatives

B. CAPON

Chemistry Department, Glasgow University

Formation and Reactions of Acetals and Ketals[1]

3-(2-Hydroxyphenoxy)phthalide (**2**) is an intermediate in the hydrolysis of *O,O'*-(2-carboxybenzylidene)catechol (**1**) in aqueous buffers of pH 2.9–5. The sponaneous reaction is about 30 times faster than that of *O,O'*-(4-carboxybenzylidene)catechol and may involve nucleophilic catalysis by the carboxylate group. The rate constants for catalysis by acetic and formic acid are similar for the two compounds. The pH–rate profile for the conversion of 2,3-(4-carboxybenzylidenedioxy)benzoic acid (**3**) into the phthalide

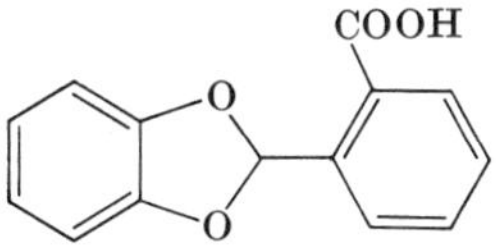

COOH O O

(**1**)

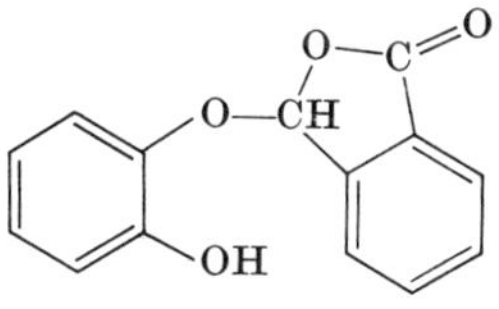

O—C=O O—CH OH

(**2**)

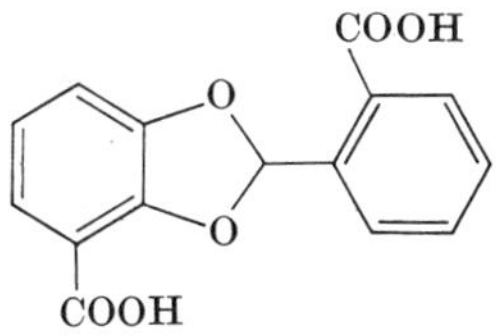

COOH O O COOH

(**3**)

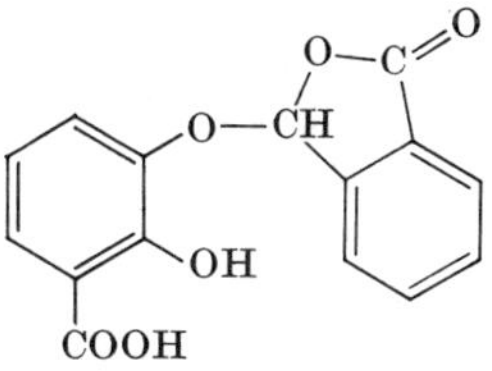

O—C=O O—CH OH COOH

(**4**)

(**4**) is bell-shaped, but this reaction occurs only about three times faster than the hydrolysis of the acid (**3**) and hence there can be little nucleophilic assistance.[2]

Details of Anderson and Fife's work on the hydrolysis of benzaldehyde disalicylyl acetal have been published.[3]

The kinetics of hydrolysis of a series of 3-alkyl-3-methoxyphthalides have been measured and interpreted in terms of the mechanism of equation (1), which was preferred to the *A*2 mechanism previously proposed.[4]

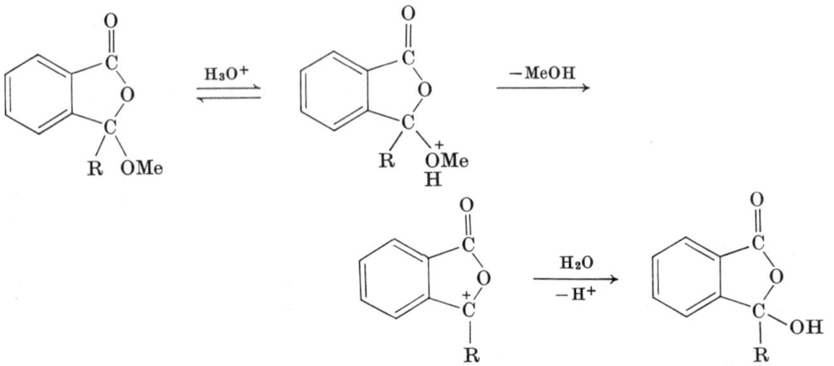

...(1)

The relative rates of hydrolysis of compounds (**5**),(**6**), and (**7**) are 1:2.3:120. Therefore, if the faster rates for compounds (**6**) and (**7**) are the result of participation by the double bond and the cyclopropane ring, the rate enhancements are much less than found with the analogous *p*-nitrobenzoates and arenesulphonates.[5] This result is not surprising in view of known reduction of anchimeric assistance brought about by an α-carbonium ion stabilizing group.[6]

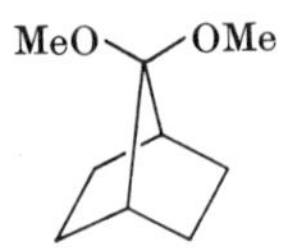

(**5**) 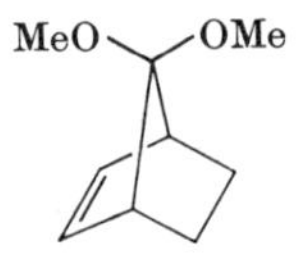 (**6**) 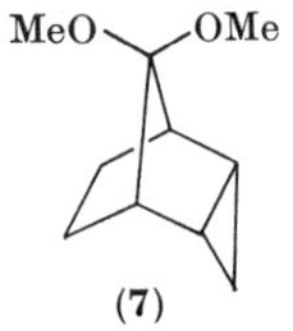 (**7**)

The kinetics of alcohol-exchange of acetals and exchange of the α-hydrogen atom have been studied. It was thought that the latter reaction proceeded by way of the vinyl ethers, and the equilibrium constants for the interconversion of the acetals and the vinyl ethers were evaluated from these results together with the rate constants for the addition of alcohols to the vinyl ethers.[7]

Tracer studies have shown that the formation of the formal from benzyl alcohol and paraformaldehyde in aqueous sulphuric acid involves aldehyde—oxygen fission, not benzyl—oxygen fission.[8]

Other related reactions studied include that between acetylferrocene and triethyl orthoformate,[9] formation of cyclic acetals from enones,[10] formation of 2-furyl-1,3-dioxans catalysed by ion-exchange resins,[11] and ketone-exchange of 2,3:4,5-di-isopropylidene-α-L-sorbofuranose with butan-2-one.[12]

There have been numerous investigations of the conformations of cyclic acetals. The classes of compounds studied include 1,3-dioxans,[13–15] 1,3-dioxolans,[16] spirodioxolans,[17,18] bicyclic 1,3-dioxans,[19] 1,3-oxathians,[20a] and 1,3-dioxacycloheptanes.[20b]

It has been shown that independent estimates of hydrogen-ion activity in moderately concentrated acids are in good agreement. This is of considerable potential use for studies of reaction mechanism.[21]

Hydrolysis and Formation of Glycosides

Non-enzymic Reactions

A reinvestigation of the hydrolysis of ferrocenylmethyl β-D-glucopyranoside has shown that it proceeds with ferrocenylmethyl—oxygen fission, not glucosyl—oxygen fission as previously supposed. Acid-catalysed methanolysis gave glucose and methoxymethylferrocene.[22]

Details of Sinnott and his co-workers' investigation of the hydrolysis of 1-adamantyl glycosides[23] and of Clark and Hay's investigation of the metal-ion catalysis of 8-quinolyl β-D-glucoside[24] have been published. It was suggested that the acid-catalysed hydrolysis of 8-quinolyl β-D-glucoside proceeds by an *A*2 pathway.[24]

Other glycosides whose acid-catalysed hydrolysis has been studied include alkyl[25] and aryl[26] β-D-galactopyranosides, methyl 4-*O*-alkyl-α- and -β-D-galactopyranosides,[27] β-D-fructofuranosides,[28a] the methyl glycoside of *O*-acetylated *N*-acetylneuramic acid,[28b] sucrose,[29] and cellulose.[30] The alkaline degradation of sucrose has also been studied.[31]

There has been an investigation of the Fischer glycoside synthesis with glucose and galactose.[32]

Brown and Bruice[33] have studied the hydrolysis of 1-β-D-glucopyranosyl benzoate, 2,3-di-*O*-methyl-1-β-D-glucopyranosyl benzoate and a mixture of α- and β-2,3,4,6-tetra-*O*-methyl-D-glucopyranosyl benzoates. These reactions show acid-catalysed and neutral components. The acid-catalysed hydrolysis is 10–100 times faster than that of methyl or phenyl β-D-glucoside and, on the basis of $\Delta S^{\ddagger}$ and the solvent isotope effect, it appears to follow an *A*1 mechanism with glycosyl—oxygen fission. The neutral hydrolysis also appears to be a unimolecular process with glycosyl—oxygen fission, as benzoate esters are not normally hydrolysed under these conditions. In alkaline solution the hydrolysis of 1-β-D-glucopyranosyl benzoate shows complex kinetics and must occur with benzoyl migrations. Since complex kinetics are also found with the 2,3-di-*O*-methyl compound, migration to the 6- as well as to the 2-hydroxy-group must occur.[33]

Enzymic Reactions[34]

(a) *Lysozymes.* A detailed kinetic investigation of the hydrolysis of hexa-*N*-acetylchitohexaoside (NAG-6) catalysed by lysozyme from hens' egg-white has been reported. From the pH-dependence of k_{cat} and $k_{\text{cat}}/K_{\text{m}}$ it was concluded that the pK_{a} of glutamic acid[35] is 6.1 in the free enzyme and 6.7 in the non-productive complex of the enzyme and NAG-6, and that the pK_{a} of aspartic acid 52 is 3.4–3.7 in the free enzyme and 3.8 in the complex.[35]

The α-tritium isotope effect on the lysozyme-catalysed hydrolysis of NAG-3 to NAG-2 and NAG is $k_{\text{H}}/k_{\text{T}} = 1.19$, corresponding to an α-deuterium isotope effect $k_{\text{H}}/k_{\text{D}} = 1.14$, from which it was concluded that the rate-limiting step is an $S_{\text{N}}1$-type ionization. Unfortunately it is difficult to judge this conclusion since there are few if any models for the possible nucleophilically assisted processes for this system, i.e. reactions at a glycosyl-carbon atom where the nucleophile is an amido or carboxylate group.[36]

The enthalpy of binding of NAG-4 to lysozyme is 2.8 kcal/mol less negative than that for the binding of NAG-3, but the free energies of binding differ only by 0.3 kcal/mol at 25°. It was suggested that two complexes are formed between lysozyme and NAG-4, one similar to that formed by NAG-3 and the other in which the terminal residue partly fills site D.[37]

The fluorescence of human lysozyme has been studied. NAG-3 and NAG-4 are bound less strongly to human lysozyme than to lysozyme from hens' egg white.[38]

The binding of aryl di-*N*-acetylchitobiosides to lysozyme appears to be more complex than hitherto thought. Whereas at 32° only the signal in the PMR spectrum of the acetamido-group proximal to the aglycone of *p*-methoxyphenyl di-*N*-acetyl-β-chitobioside is shifted in the presence of lysozyme, at 65° the signals of both acetamido-groups are shifted so that they coalesce. It was suggested that there was a temperature-dependent transition of the interaction between lysozyme and the aryl di-*N*-acetylchitobioside.[39]

An X-ray study has confirmed that when 2′,3′-epoxypropyl di-*N*-acetylchitobioside reacts with lysozyme it becomes attached to the carboxyl group of aspartic acid 52.[40] Further work on the reaction of iodine with lysozyme has been reported.[41]

A polypeptide with lysozyme activity has been synthesized.[42]

There have been numerous other investigations of lysozyme from hens' egg white[43] and also of the lysozymes from human leukocytes,[44] baboon milk,[45] bacteriophage T4,[46] turkeys' egg white,[47] and geese's egg white.[48]

(b) *β-Galactosidases.* The α-deuterium isotope effect for the hydrolysis of some β-galactosides catalysed by β-galactosidase from *E. coli* has been determined. It was suggested that a conformational change of the enzyme is the rate-limiting step in the reactions of the more reactive galactosides.[49]

Nucleophilic competition by alcohols in the hydrolysis of galactosides catalysed by the β-galactosidase from *E. coli* has been extensively re-investigated. It now seems that the levelling off of the rate at high methanol concentrations arises from factors other than a change in rate-limiting step from degalactosylation to galactosylation.[50,51]

The β-D-galactopyranosylpyridinium cation is a poor substrate for β-galactosidase from *E. coli*. As electrophilic catalysis is not possible with this substrate, substrate distortion and electrostatic or nucleophilic catalysis must be responsible for the enzymic catalysis.[52] The binding of β-D-galactopyranosyltrimethylammonium bromide to β-galactosidase has been studied,[53a]

Although the acid-catalysed hydrolysis of *tert*-butyl, 1,1-diethylpropyl and diphenylmethyl β-D-galactopyranoside proceed partly with alkyl—oxygen fission, the hydrolyses catalysed by β-galactosidase proceed wholly with galactosyl—oxygen fission.[53b]

Compound (**8**) has been used as a spin-labelled substrate for β-galactosidase.[54]

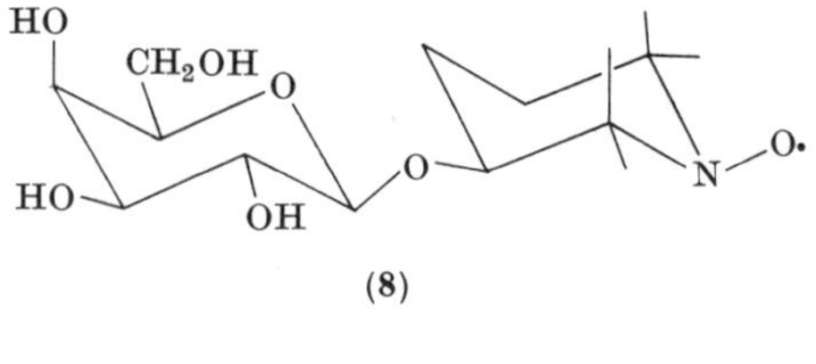

HO
CH_2OH O
O N—O•
HO OH

(**8**)

The β-galactosidase from bovine testes has been studied.[55]

(c) *Other glycosidases.* Evidence for the presence of 2–3 carboxyl groups at the active site of glucoamylase I from *A. niger* has been presented. When the enzyme reacts with glycine methyl ester and a water-soluble carbodiimide, maltose prevents the reaction of between two and three carboxyl groups per molecule of enzyme.[56]

Inhibition of the α-glucosidase from yeast and of the β-glucosidase from almond emulsin by α- and β-D-glucosylimidazole,[57] and of the glycosidases from almond emulsin by *C*-glycosides,[58] has been studied.

Other glycosidases that have been investigated include α-amylases,[59] β-amylases,[60] α-mannosidase from bakers' yeast,[61] α-glucosidase from *Mucor javanicas*[62a] and rabbit muscle,[62b] β-L-arabinosidase from *Cajanus indicus*,[63] *N*-acetyl-β-D-glucosaminidase from *A. oryzae*,[64a] and the β-D-glucuronidases from bovine liver and *E. coli*.[64b]

Hydration of Aldehydes and Ketones and Related Reactions

Equilibrium constants for the addition of hydroxyl ion to a series of substituted benzaldehydes have been measured and the results used to establish an acidity-function scale.[65]

The kinetics of dehydration of glycolaldehyde hydrate have been measured by determining the rate of scavenging of free aldehyde with semicarbazide and with sulphite. The reaction is general-acid and general-base catalysed.[66]

The kinetics of hydration of isobutyraldehyde have been studied by temperature-jump and NMR spectroscopy,[67] and the hydration of acetaldehyde has also been studied by NMR spectroscopy.[68]

The effect of pressure on the equilibrium constants for the hydration of aliphatic aldehydes has been determined.[69]

Hydration and polymerization of succinaldehyde, glutaraldehyde, and adipaldehyde,[70] and the oligomerization reaction of formaldehyde,[71] have been studied. An X-ray structure determination of the dimer of DL-glyceraldehyde has been reported.[72]

There have been investigations of the chemistry of hemiacetals[73] and of ring–chain tautomerism of 8-acyl-1-naphthoic acids.[74]

Diphenyl phosphate, phenylphosphinic acid, and trichloroacetic acid appear to act as tautomeric catalysts for the mutarotation of tetra-*O*-methylglucose in benzene as they are more effective catalysts than expected from a Brønsted plot for catalysis by phenols.[75]

The mutarotation of glucose in mixtures of water with dimethyl sulphoxide (DMSO) has been studied. The rate of the acid-catalysed reaction varies only slightly with solvent composition, but that for the water-catalysed reaction decreases strongly with increasing DMSO concentration in the range $\chi_{DMSO} = 0\text{–}0.3$ and remains approximately constant at higher concentrations. It was proposed that the acid-catalysed reaction proceeds by a stepwise mechanism without the intervention of a water molecule and that the water-catalysed reaction involves three water molecules.[76]

Glucose shows a complex mutarotation in aqueous dimethylformamide (DMF) at high DMF concentrations. This has been attributed to formation of a furanose form and one has been detected in solutions of glucose in pure DMF by gas chromatography plus mass spectrometry of the trimethylsilyl ethers.[77]

The mutarotation of β-L-arabinopyranose yields furanose forms as well as the α-pyranose form, but the *aldehydo*-form does not appear to cyclize more rapidly to furanose forms than to pyranose as reported for 2-deoxyribose.[78]

The equilibrium composition of fructose 6-phosphate has been shown by ^{13}C-NMR to be 19% of α- and 81% of β-furanose with less than 1.5% of the keto-form.[79]

The mutarotatase from bovine kidney cortex has been investigated.[80]

Reactions with Nitrogen Bases

Schiff Bases

On the basis of X-ray crystal-structure determination of compounds that have a carbonyl group close to an amino-group it has been suggested that the preferred angle of attack by nitrogen nucleophiles is not perpendicular to the carbonyl group but at an angle of 107°.[81]

Imine formation from acetone and the monoprotonated forms of 2-(dimethylamino)-ethylamine and *trans*-(2-dimethylaminomethyl)cyclopentylamine is, respectively, 1000- and 60-fold faster than expected on the basis of a Brønsted plot for the reactions with other amines. Smaller rate enhancements are found with the monoprotonated forms of 3-(dimethylamino)propylamine, 4-(dimethylamino)butylamine, and 5-(dimethylamino)-pentylamine. It seems likely that intramolecular catalysis of dehydration of the carbinolamine intermediate is occurring, as symbolized by (**9**).[82]

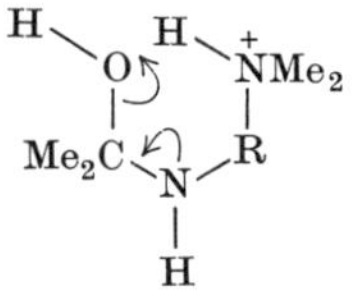

H O H $\overset{+}{N}Me_2$ Me_2C N R H

(**9**)

The condensation of isobutyraldehyde with diamines to form imidazolines has been studied.[83]

The rate constants for the uncatalysed reaction of piperazine (pK_a 9.97) and piperazine monocation (pK_a 5.80) with pyridine-4-aldehyde to form the carbinolamine are 2.3×10^5 and 65 l mol^{-1} s^{-1}, respectively. The reaction of piperazine shows general-base catalysis which was thought to involve catalysis of the proton-transfer step, i.e. conversion of the zwitterionic carbinolamine into the anionic form.[84]

The hydrolyses of 2-*tert*-butyl-3-phenyloxaziridine (**10**) and *N*-benzylidene-*tert*-butylamine *N*-oxide (**12**) are thought to proceed through the same ion (**11**). The effect of micelles on the rates of these reactions was studied.[85]

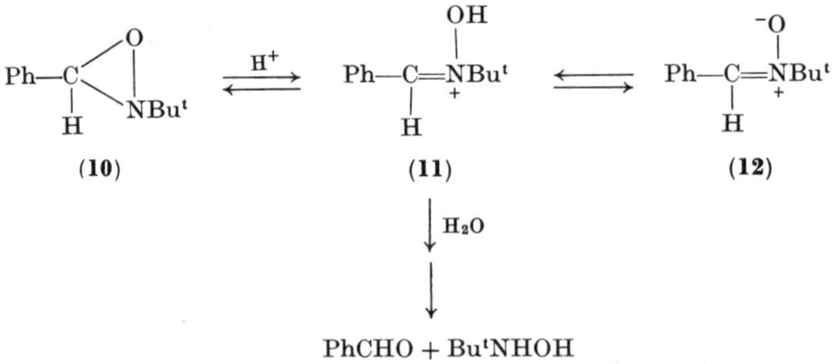

Ph—C(H)(O)NBu^t (**10**) $\overset{H^+}{\rightleftarrows}$ Ph—C(H)=$\overset{+}{N}$(OH)Bu^t (**11**) $\rightleftarrows$ Ph—C(H)=$\overset{+}{N}$(–O)Bu^t (**12**)

(**11**) $\xrightarrow{H_2O}$ PhCHO + Bu^{t}NHOH

The hydrolysis and synthesis of benzylideneaniline in the presence of micelles of sodium dodecyl sulphate[86a] and the hydrolysis of Schiff bases derived from ferrocenyl aldehyde and ketones[86b] have been studied.

The kinetics of hydrolysis of salicylideneanilines in acidic solutions with H_0 down to –4 have also been studied.[87] So have amine exchange reactions of substituted benzylideneanilines.[88, 89]

The reaction of glycine with benzaldehyde in ethanol in the presence of potassium hydroxide to form phenylserine is thought to involve formation of the Schiff base which forms a carbanion by loss of the α-proton. Reaction of this carbanion with another molecule of benzaldehyde would then lead to the product.[90]

The reaction of substituted benzylideneanilines with diethylmagnesium has been studied. Electron-withdrawing substituents in either benzene ring increase the rate of reaction and electron-releasing substituents decrease it.[91]

Dehydration of the β-ketol 9-hydroxy-10-methyl-*cis*-2-decalone by secondary and primary amines proceeds via iminium ions.[92]

There have been PMR studies[93] and MO calculations[94] on the protonated imine group.

The reaction of the following compounds have been studied: glycine with glyoxal,[95] imidazole with formaldehyde,[96] isoxazol-5-ones with Schiff bases,[97] aldehydes with ammonia,[98,99] and Schiff bases with nitrosonium borofluoride.[100]

Ring–chain tautomerism of *o*-benzoylbenzamides has been investigated.[101]

Transamination

The interconversion of *N*-(α-methyl-4-methoxybenzylidene)-1-phenylethylamine (**13**) and *N*-(α-methylbenzylidene)-1-(4-methoxyphenyl)ethylamine (**14**) catalysed by potassium *tert*-butoxide in *tert*-butyl alcohol occurs with about 50% of intramolecular proton-transfer and about 60% of racemization. The starting material also undergoes substantial racemization. It was suggested that reaction proceeds through twisted carbanions.[102a]

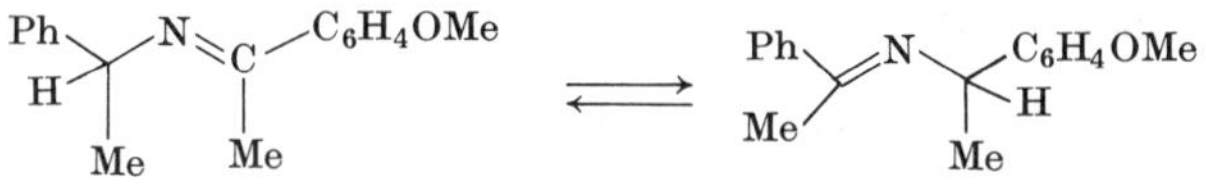

Ph, H, Me, N, C, C_6H_4OMe, Me ⇌ Ph, Me, N, C_6H_4OMe, H, Me

(**13**) (**14**)

Transamination of the imine derived from 3-hydroxypyridine-4-carboxaldehyde and alanine is 2–3 times faster than racemization in aqueous buffers, the exact value depending on the catalyst and the prototropic state of the imine. This indicates that the intermediate carbanion is protonated with approximately equal facility at the two sites.[102b]

Other workers on transamination are cited in reference 103.

Enamines

The stereochemistry of the acidic hydrolysis of 2,6-disubstituted cyclohexanone enamines has been studied,[104] as also has the formation of dienamines.[105]

Hydrazones, Oximes, Semicarbazones and Related Compounds[106]

The pH–rate profile for the reaction of methoxyamine with *p*-chlorobenzaldehyde shows two breaks. That at low pH (ca. 1) was interpreted as arising from a change in rate-determining step from attack of the methoxyamine below this pH to proton-transfer to the zwitterionic form of the carbinolamine above it.[107] Details of Sayer and Jencks's investigation of the reaction of 2-methylthiosemicarbazide with *p*-chlorobenzaldehyde have been reported;[108] and there has been a review of "Encounter-Limited Rate-determining Steps in Carbonyl and Acyl Group Reactions".[109]

The rate of the base-catalysed dehydration of carbinolamines formed from substituted hydrazines and *p*-chlorobenzaldehyde is enhanced by electron-withdrawing substituents in the hydrazine. A transition state (**15**) involving a large degree of negative charge on nitrogen and little double-bond formation was proposed. The reaction is also general-acid catalysed and the rate of that reaction is also decreased by electron-withdrawing substituents in the hydrazine, a transition state (**16**) being here proposed.[110]

Details of the investigation by Cordes and his co-workers on the α-deuterium isotope effect for addition of semicarbazide and phenylhydrazine to benzaldehyde have been published.[111]

Other work on hydrazone formation is cited in reference 112. There have been several investigations of the formation of osazones.[113] Molecular-orbital calculations for the products of reaction of glyoxal with phenylhydrazine have been reported.[114] Formaldehyde phenylhydrazone reacts with an excess of formaldehyde to yield (**17**).[115] The reaction of butane-2,3-dione dihydrazone with formaldehyde in the presence of metal ions has been studied.[116]

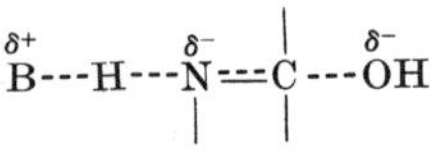

$\overset{\delta+}{B}$---H---$\overset{\delta-}{N}$=--C---$\overset{\delta-}{OH}$ (**15**)

>$\overset{\delta+}{N}$=--C---O---$\overset{\delta+}{H}$---B (with H on O) (**16**)

Ph—N, N—Ph (**17**)

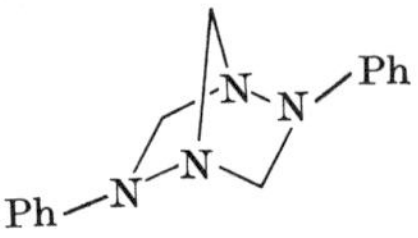

On investigation of the reaction of mesityl oxide with hydroxylamine to form 3,5,5-trimethyl-2-isoxazoline (**20**) it was shown that (**18**) is not an intermediate as it is not converted into (**20**) under the reaction conditions, but an NMR investigation of the reaction solution showed that (**19**) is an intermediate.[117]

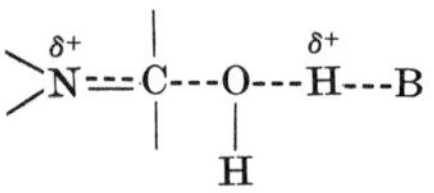

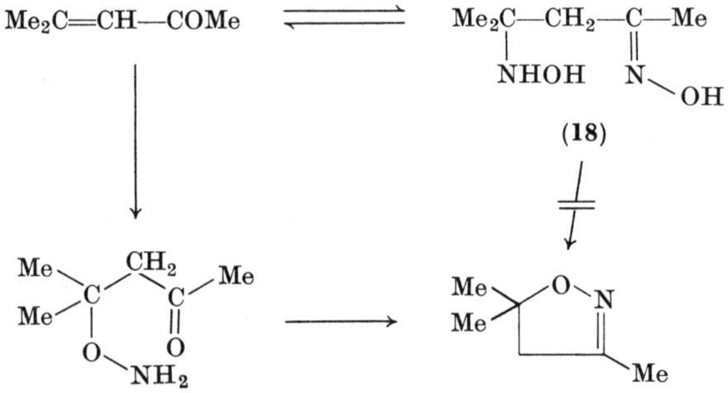

$Me_2C{=}CH{-}COMe \rightleftharpoons Me_2C(NHOH){-}CH_2{-}C(=N{-}OH){-}Me$ (**18**)

(**18**) ↛ (**20**)

Mesityl oxide → (**19**) → (**20**)

(**19**): Me, Me, CH_2, Me, O, O, NH_2

(**20**): Me, Me, O, N, Me

There have been several investigations of the hydrolysis of oximes.[118]

The reaction of *N*-arylhydroxylamines with nitrosoarenes appears not to be a simple reaction analogous to oxime formation but to be a radical process involving transfer of hydrogen.[119]

Hydrolysis of Enol Ethers

The rate constants for the hydrolysis of ethyl vinyl ether catalysed by positively charged amino-acids gives a Brønsted plot that is almost parallel to that for neutral carboxylic acids but lying about 0.3 logarithm unit below it. On the other hand, the points for negatively charged acids fall above the line for neutral carboxylic acids. An electrostatic interaction in the transition state was proposed as an explanation of these observations.[120]

The hydrolysis of ethoxypropadiene (equation 2) is general-acid catalysed with $\alpha = 0.62$. The rate constant for hydronium-ion catalysis shows a negative deviation from the Brønsted plot. The isotope effect $k(H_3O^+)/k(D_3O^+) = 3.05$, $\Delta S^\ddagger = -10$ cal mol^{-1} K^{-1}, and Bunnett's w parameter is 0.36. A mechanism similar to that for the hydrolysis of ordinary enol ethers seems likely.[121]

$$H_2C{=}C{=}CHOEt + H_2O \rightarrow H_2C{=}CHCHO + EtOH \quad (2)$$

The hydration of the β-oxy-α,β-unsaturated ketones (**21**) and (**22**) is specific-acid catalysed with $k(D_3O^+)/k(H_3O^+) =$ ca. 2.6. The ρ^+-values are -0.18 and $+0.26$, respectively. These results are clearly inconsistent with the normal mechanism of vinyl ether hydrolysis and that shown in equation (3) was proposed.[122]

The alkylation of vinyl ethers has also been studied.[123]

ArOCH=CHCOMe (**21**)   MeOCH=CHCOAr′ (**22**)

$$ROCH{=}CHCOR' \rightleftharpoons ROCH{=}CH{-}\overset{+OH}{\overset{\|}{C}}R' \longrightarrow \underset{+OH_2}{ROCH}{-}CH{=}\overset{OH}{C}R'$$

$$\downarrow \quad \ldots(3)$$

$$O{=}CH{-}CH_2{-}\overset{O}{\overset{\|}{C}}{-}R' \longleftarrow \underset{OH}{ROCH}{-}CH_2{-}\overset{O}{\overset{\|}{C}}R'$$

Enolization and Related Reactions*

A detailed investigation of the halogenation of butan-1-one has shown that the previously reported anomalies can be explained in terms of the heterogeneity of the system, reaction of the halogen with DMF (for reactions in aqueous DMF), formation of iodate, and nucleophilic displacement of the halide by hydroxide ion.[124]

The enolization of 2-oxobicyclo[2.2.2]octane-1-carboxylic acid (RCO_2H) follows a rate law of the form:

$$\text{Rate} = \{k_H[H^+] + k_0\}[RCO_2H]$$

k_H is similar to that for the methyl ester. k_0 could arise from the acid-catalysed enolization of the carboxylate form and if this were so the catalytic constant would be 230 times greater than k_H for enolization of the methyl ester. This rate enhancement was attributed to hydrogen bonding or electrostatic stabilization as symbolized by (**23**) or (**24**).[125] More work on the stereochemistry of hydrogen-exchange of bicyclic ketones has also been reported.[126]

The rates of enolization (measured as iodination) of 4-(diethylamino)butan-2-one and 5-(diethylamino)pentan-2-one are much greater than normally found for aliphatic ketones, which was attributed to intramolecular catalysis as symbolized by (**25**) and (**26**).[127]

* See also Chapter 10.

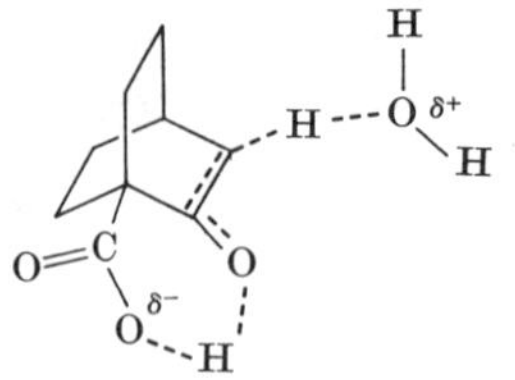

(**23**) 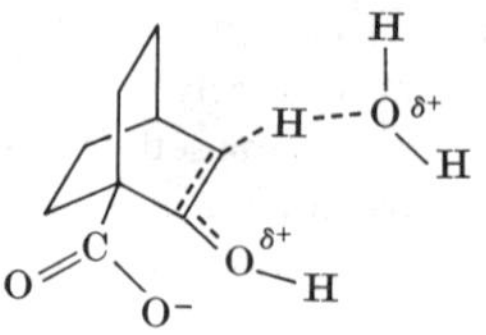 (**24**)

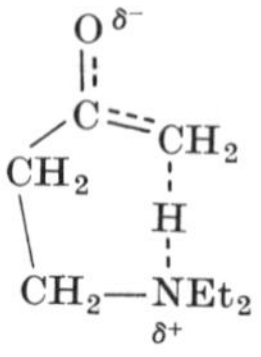

(**25**) 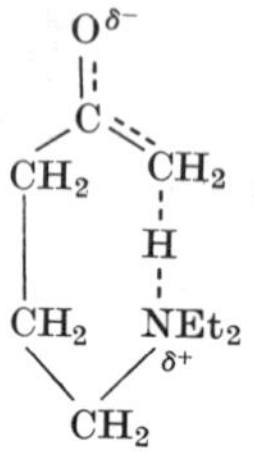 (**26**)

Diamines of type $Me_2N(CH_2)_nNH_2$, where $n = 2$, 3, 4 or 5, are not better catalysts than monoamines for the dedeuteriation of [2-^{2}H]isobutyraldehyde, showing the absence of any intramolecularly catalysed reaction of the Schiff base. This was thought to be because the Schiff bases have the *anti*-structure, but even the diamines undecane-1,11-diamine and dodecane-1,12-diamine, which have sufficiently long chains to permit catalysis even with this structure, showed no catalysis. However, evidence for catalysis was found with N^1,N^1-dimethyloct-2-yne-1,8-diamine.[128] Catalysis of dedeuteriation of [2-^{2}H]isobutyraldehyde by polyethyleneimines is also thought to involve reversible transformation of the aldehyde group into an iminium ion, followed by intramolecular removal of a deuteron by another amino-group.[129]

Ethylenediamine causes an increase in the rate of deuterium exchange of [2-^{2}H]-isobutyraldehyde (0.06M) up to a concentration of 0.03M and then a decrease at higher concentration. This decrease arises from conversion of the aldehyde into 2-isopropylimidazoline. At concentrations of diamine above 1M the rate increases again owing to catalysis by the imidazoline.[130]

The kinetics of iodination of acetone have a term that is of the second order in amine when this is 2-aminomethylpyridine but not when it is 2-aminoethylpyridine. It was thought that this term arose from a rate-limiting dehydration of a carbinolamine with another molecule of the 2-aminomethylpyridine acting as a bifunctional catalyst. This would yield an imine which should be iodinated rapidly.[131]

The Brønsted β-coefficient for the enolization of 3-methylacetylacetone (measured as bromination) is 0.49. The kinetics were also measured in H_2O–D_2O mixtures and the results analysed in terms of the Gross–Butler equation.[132]

The hydroxide-ion-catalysed enolization of 2,4-dimethylpentan-3-one and 3,5-dimethylheptan-4-one show primary deuterium isotope effects $k_H/k_D = 6.09$ and 3.67, respectively, at 25°, but no primary tritium isotope effect. Possible explanations of this surprising observation were discussed.[133]

On treatment with chlorine in carbon tetrachloride 4-*tert*-butylcyclohexanone yields 4-*tert*-butyl-*cis-trans*-2,6-dichlorocyclohexanone under conditions where 4-*tert*-butyl-*cis*- and -*trans*-2-chlorocyclohexanone are not further chlorinated. It was suggested that the dichloro-ketone was formed by the mechanism of equation (4). A process analogous to the conversion of (**27**) into (**28**) was demonstrated by treating the enol acetate (**29**) with chlorine in carbon tetrachloride to yield (**30**). It seems that, in carbon tetrachloride, C—H bond-cleavage of the enol (**27**) is competitive with O—H bond-cleavage.[134] Other work on the halogenation of ketones in carbon tetrachloride solution is reported in reference 135.

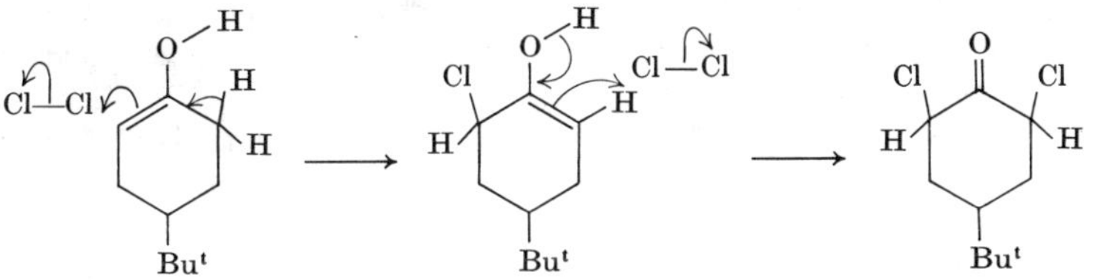

...(4)

(**27**) (**28**)

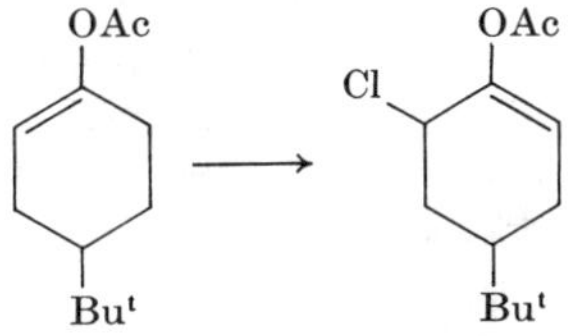

(**29**) (**30**)

The protonation and epimerization of 4-*tert*-butyl-2-halocyclohexanones and of 1-halo-2-decalones in strong acids has been studied. The epimerization was thought to proceed via the enol,[136] as is that of 1,1-trichloro-2-hydroxy-3-methylhexan-4-one in glacial acetic acid.[137]

The rates of enolization of the ketones (**31**), (**32**) and (**33**) at the α-carbon atom vary only slightly with the size of the fused ring but the rates of enolization at the α'-carbon decrease strongly as the size of the fused ring is reduced.[138]

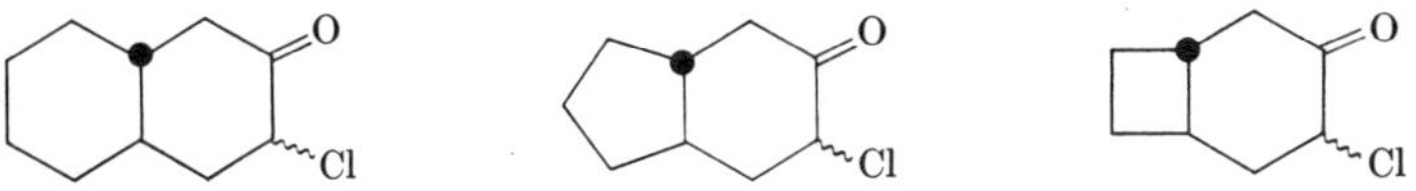

(**31**) (**32**) (**33**)

The rate constants for the acid-catalysed bromination, chlorination and iodination of acetone under conditions where formation of the enol is not rate-limiting are approximately equal. It seems that the equilibrium constant for enolization is 1.5×10^{-5}, which is considerably smaller than hitherto thought, and that the reactions of the halogens with the enol are diffusion-controlled with rate constant ca. 10^9 mol^{-1} s^{-1}.[139]

The kinetic solvent isotope effects for the enolization of ketones in alcohols ROH and ROD have been measured.[140]

Magnesium ions catalyse the exchange of hydrogen atoms of the methylene group of acetonyl phosphate in alkaline solutions, the reaction being thought to involve a chelate as shown in equation (5).[141]

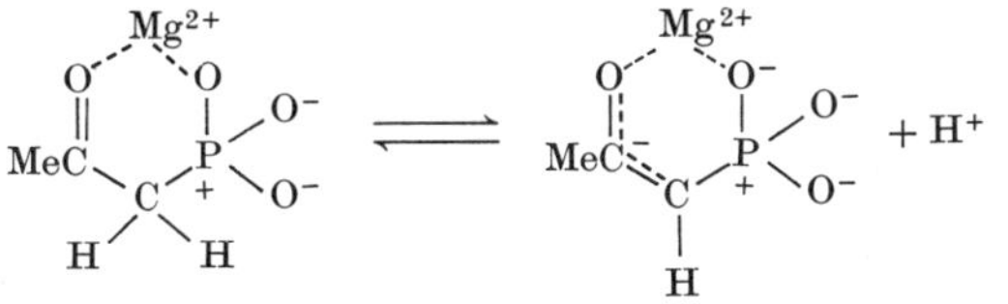

...(5)

The protolysis and iodination of the copper derivatives of the enolates of α,β-unsaturated carbonyl compounds proceeds with retention of configuration, but the lithium enolates react with racemization. It seems that the latter have the α-carbon sp-hybridized and that the direction of attack depends on steric factors.[142]

Metal-ion catalysis of the Lobry de Bruyn–Alberda van Ekstein rearrangement has been studied.[143]

More work on the keto–enol tautomerization of 3-hydroxy-2,4-dimethylcyclobutenone has been reported.[144]

Other reactions which have been studied include enolization of tetrahydro-2,4-dioxothiophen[145] and isomerization of protonated cyclohepta-3,5-dienones.[146]

There have been several investigations of keto–enol equilibria.[147]

Phosphoglucose isomerase produces and uses both the α- and β-form of D-fructofuranose 6-phosphate but with a preference for the α-form. The α-anomer of D-glucopyranose 6-phosphate is the preferred substrate, but the β-form also reacts. The mechanism outlined in Scheme 1 was preferred. It seems that the enzyme catalyses the ring-opening processes as well as the interconversion of the ring-opened forms via the enediol.[148] The carboxyl group of a glutamic acid residue has been implicated in the active site of this enzyme by allowing it to react with 1,2-anhydromannitol 6-phosphate, which causes rapid inactivation.[149]

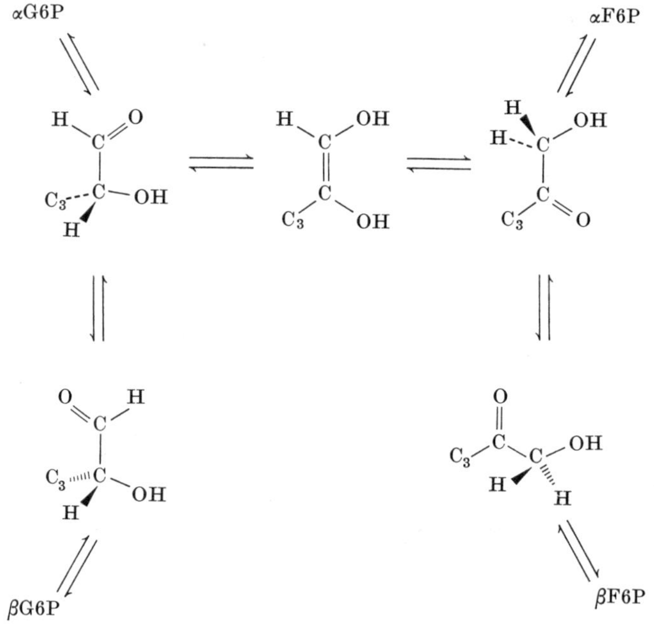

SCHEME 1.

Phosphomannose isomerase uses only the β-anomer of mannose 6-phosphate. This enzyme (or phosphoglucose isomerase) catalyses the anomerization of mannose 6-phosphate.[150]

There have also been investigations of enolase[151] and triose phosphate isomerase.[152]

Aldol and Related Reactions

On treatment with methylsulphinyl carbanion, cyclopentanone undergoes an intriguing reaction to form 5-(cyclopent-1-enyl)hex-5-enoic acid. The pathway shown in Scheme 2 was proposed.[153]

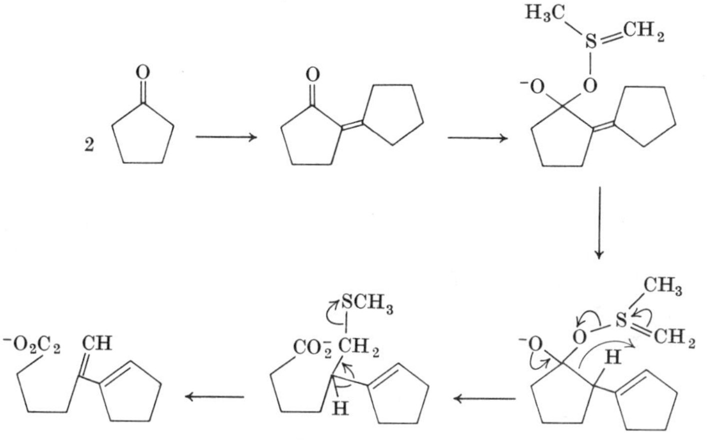

Scheme 2.

Dodecyldimethylsulphonium salts react with carbonyl compounds in a mixture of benzene and aqueous sodium hydroxide to form epoxides, presumably by the pathway shown in equation (6).[154]

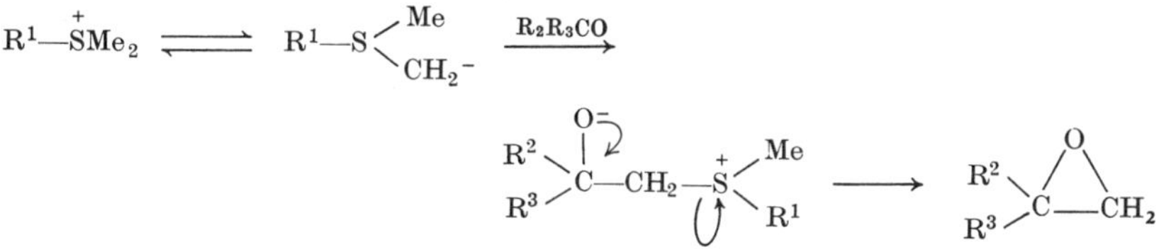

$$R^1\!-\!\overset{+}{S}Me_2 \rightleftharpoons R^1\!-\!S\!<^{Me}_{CH_2^-} \xrightarrow{R_2R_3CO} R^2R^3C(O^-)\!-\!CH_2\!-\!\overset{+}{S}(Me)R^1 \longrightarrow R^2R^3C\overset{O}{-}CH_2 \quad \ldots(6)$$

There has also been work on the aldol reaction,[155] the retro-aldol reaction,[156] the Claisen–Schmidt reaction,[157] the Darzens condensation,[158] the Knoevenagel condensation,[159] and the condensation of carbonyl compounds with 3,5-dihydroxy-4-phenylisoxazole.[160] The condensation of acetylenes with aldehydes and ketones has been reviewed[161] (in Russian), and the stereochemistry of additions to carbonyl compounds has been discussed.[162]

Pyridoxal phosphate has been shown to form a Schiff base with the amino-group of a lysine residue of rabbit muscle aldolase contained in the sequence Gly—Gly—Val—Val—Gly—Ile—Lys*—Val—Asp—Lys.[163] Other work on aldolase is reported in reference 164.

Other Reactions

The ^{31}P-NMR spectrum of an intermediate in the Wittig reaction of the ylide from ethyltriphenylphosphonium bromide and cyclohexanone has shown that it has the oxaphosphetane structure (**34**).[165] The kinetics of the Wittig reaction of $Ph_3P{=}CEtCO_2Et$

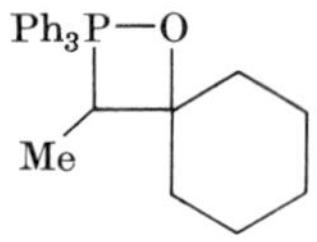

Ph3P—O, Me, cyclohexane spiro (four-membered ring)

(34)

with substituted benzaldehydes yields a ρ-value of 2.89. The rate of reaction of the compounds $Ph_3P{=}CRCO_2Et$ with benzaldehyde decreases as R is changed along the series Me > H > Et > Bu^n.[166] The rate of the Wittig reaction of $Ph_3P{=}CHCOPh$ with *p*-nitrobenzaldehyde in hydrocarbon solvents decreases with increasing dielectric constant of the solvent.[167] Other work on the Wittig reaction is noted in reference 168.

p-Nitrobenzaldehyde does not undergo the Cannizzaro reaction in aqueous potassium hydroxide but instead yields *p*-nitrosophenol.[169] An investigation of the Cannizzaro reaction is reported in reference 170.

The Schmidt reaction of hydrazoic acid with ketones PhCOR gave a ρ-value of 2.28. The rate-limiting step is addition of hydrazoic acid to the protonated ketone.[171]

Acetophenone reacts with thionyl chloride in pyridine to form *N*-styrylpyridinium chloride.[172] The mechanism of Scheme 3 was proposed.[173]

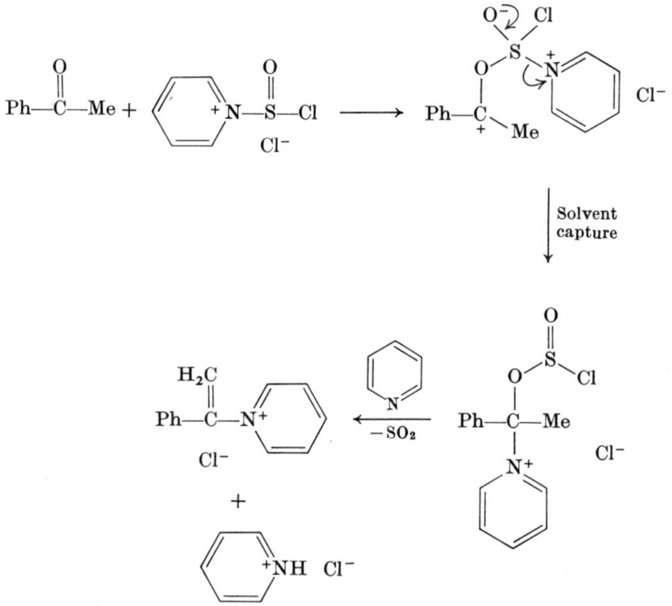

Ph—C(=O)—Me + $C_5H_5N^+$—S(=O)—Cl Cl^- ⟶ Ph—C^+(Me)—O—S(=O)(Cl)—$N^+C_5H_5$ Cl^-

Solvent capture ↓

Ph—C(Me)(O—S(=O)—Cl)—$N^+C_5H_5$ Cl^- ⟶ (pyridine, $-SO_2$) Ph—C(=CH_2)—$N^+C_5H_5$ Cl^- + $C_5H_5N^+H$ Cl^-

SCHEME 3.

Other reactions that have been studied include decarbonylation of tricyclo[2.1.0.0^{2,5}]-penten-3-one catalysed by transition metals,[174] formation of 1,2-dithiolylium tri-iodides from β-diketones, hydrogen sulphide and iodine,[175] and the reaction of disiamyl(3-phenylbut-2-enyl)borane with aldehydes.[176]

There have been several investigations of the protonation of carbonyl compounds.[177]

References

[1] B. M. Dunn and T. C. Bruice, *Adv. Enzymology*, **37**, 1 (1973).
[2] B. Capon and M. I. Page, *J.C.S. Perkin II*, **1972**, 2057.
[3] E. Anderson and T. H. Fife, *J. Am. Chem. Soc.*, **95**, 6437 (1973); cf. *Org. Reaction Mech.*, **1971**, 394–395.
[4] D. P. Weeks and J. P. Crane, *J. Org. Chem.*, **38**, 3375 (1973); see *Org. Reaction Mech.*, **1968**, 349.
[5] G. Lamaty, A. Malaval, J.-P. Roque and P. Geneste, *Bull. Soc. Chim. France*, **1972**, 4563, 4567.
[6] See *Org. Reaction Mech.*, **1968**, 30; **1969**, 33; **1970**, 39.
[7] A. Kankaanperä, P. Salomaa, P. Juhala, R. Aaltonen and M. Mattsen, *J. Am. Chem. Soc.*, **95**, 3618 (1973).
[8] N. K. Moshchinskaya and N. I. Parnyuk, *Zh. Org. Khim.*, **9**, 998 (1973); *Chem. Abs.*, **79**, 41624 (1973).
[9] Y. Sasaki and C. U. Pittman, *J. Org. Chem.*, **38**, 3723 (1973).
[10] J. W. de Leeuw, E. R. de Waard, T. Beetz and H. O. Huisman, *Rec. Trav. chim.*, **92**, 1047 (1973).
[11] Z. I. Zelikman and V. G. Kul'nevich, *Geterogen. Kataliz Reakts. Poluch. Prevrashch. Geterotsikl. Soedin.*, **1971**, 93; *Chem. Abs.*, **78**, 15216 (1973).
[12] R. S. Glass, S. Kwoh and E. P. Oliveto, *Carbohydrate Res.*, **26**, 181 (1973).
[13] M. Anteunis, D. Tavernier and G. Swaelens, *Rec. Trav. chim.*, **92**, 531 (1973).
[14] D. Tavernier and M. Anteunis, *Bull. Soc. Chim. Belges*, **82**, 405 (1973); M. Anteunis and M. Corynon, *ibid.*, p. 413.
[15] P. Ayras, *Suomen Kem.*, B, **46**, 151 (1973).
[16] M. Anteunis, R. Van Cauwenbergde and C. Becu, *Bull. Soc. Chim. Belges*, **82**, 591 (1973).
[17] R. A. Y. Jones, A. R. Katritzky, D. L. Nicol and R. Scattergood, *J.C.S. Perkin II*, **1973**, 337.
[18] D. Tavernier, M. Anteunis and N. Hosten, *Tetrahedron Letters*, **1973**, 75.
[19] A. K. Bhatti and M. Anteunis, *Tetrahedron Letters*, **1973**, 71.
[20a] P. Pasanen, *Suomen Kem.*, *B*, **45**, 363 (1972).
[20b] M. H. Gianni, J. Saavedra and J. Savoy, *J. Org. Chem.*, **38**, 3971 (1973).
[21] K. Yates and R. A. McClelland, *J. Am. Chem. Soc.*, **95**, 3055 (1973).
[22] P. M. Collins, W. G. Overend and B. A. Rayner, *J.C.S. Perkin II*, **1973**, 310.
[23] D. Cocker, L. E. Jukes, and M. L. Sinnott, *J.C.S. Perkin II*, **1973**, 190; see *Org. Reaction Mech.*, **1972**, 394.
[24] C. R. Clark and R. W. Hay, *J.C.S. Perkin II*, **1973**, 1943; see *Org. Reaction Mech.*, **1970**, 423.
[25] C. K. De Bruyne and G. van der Groen, *Carbohydrate Res.*, **25**, 59 (1972).
[26] C. K. De Bruyne, J. Wouters-Leysen and M. Yde, *Carbohydrate Res.*, **29**, 387 (1973).
[27] J. N. BeMiller and E. R. Doyle, *Carbohydrate Res.*, **25**, 429 (1972).
[28a] J. Szejtli, *Stärke*, **42**, 321 (1972).
[28b] A. Neuberger and W. A. Ratcliffe, *Biochem. J.* **133**, 623 (1973).
[29] A. D. Pethybridge, *J.C.S. Perkin II*, **1973**, 102.
[30] A. Mohn-Wehner, H. K. Rouette and H. Zollinger, *Helv. Chim. Acta*, **56**, 323 (1973).
[31] G. W. O'Donnell and G. N. Richards, *Austral. J. Chem.*, **26**, 2041 (1973).
[32] R. H. Pater, R. A. Coelho and D. F. Mowery, *J. Org. Chem.*, **38**, 3272 (1973).
[33] A. Brown and T. C. Bruice, *J. Am. Chem. Soc.*, **95**, 1593 (1973).
[34] J. F. Kirsch, "Mechanism of Enzyme Action", *Ann. Rev. Biochem.*, **42**, 205 (1973).
[35] S. K. Banerjee, I. Kregar, V. Turk and J. A. Rupley, *J. Biol. Chem.*, **248**, 4786 (1973).
[36] L. E. H. Smith, L. H. Mohr and M. A. Raftery, *J. Am. Chem. Soc.*, **95**, 7497 (1973).
[37] S. K. Banerjee and J. A. Rupley, *Arch. Biochem. Biophys.*, **155**, 19 (1973); see also *idem.*, *J. Biol. Chem.* **248**, 2117 (1973); C. Bjurulf and I. Wadso, *Eur. J. Biochem.*, **31**, 95 (1972); S. Kuramitsu, K. Ikeda and K. Hamaguchi, *J. Biochem.* (*Tokyo*), **74**, 143 (1973).
[38] R. S. Mulvey, R. J. Gaultieri and S. Beychok, *Biochemistry*, **12**, 2683 (1973); V. I. Teichberg, T. Plesse, S. Sorell and N. Sharon, *Biochem. Biophys. Acta*, **278**, 250 (1973); S. Kuramitsu, K. Ikeda, K. Hamaguchi, S. Miwa and T. Nishina, *J. Biochem.* (*Tokyo*), **72**, 1109 (1972).
[39] C. S. Tsai, *Biochem. Biophys. Res. Comm.*, **55**, 205 (1973); see also R. Otson, C. Reyes-Zamora, J. Y. Tong and C. S. Tsai, *Can. J. Biochem.*, **51**, 1 (1973).
[40] T. Moult, Y. Eshdat and N. Sharon, *J. Mol. Biol.*, **75**, 1 (1973); see *Org. Reaction Mech.*, **1970**, 426.
[41] T. Imoto, F. J. Hartdegen and J. A. Rupley, *J. Mol. Biol.*, **80**, 637 (1973); F. J. Hartdegen and J. A. Rupley, *ibid.*, p. 649; T. Imoto and J. A. Rupley, *ibid.*, p. 657; see also *Org. Reaction Mech.*, **1970**; 427–428.
[42] J. J. Sharp, A. B. Robinson and M. D. Kamen, *J. Am. Chem. Soc.*, **95**, 6097 (1973).

[43] N. Hiramatsu, N. Okabe and K. Tomita, *J. Biochem. (Tokyo)*, **73**, 971 (1973); T. Monodane, S. Hora and Y. Matsushima, *ibid.*, **72**, 1175 (1972); Y. Nakae, K. Ikeda, T. Azuma and K. Hamaguchi, *ibid.*, **72**; 1155 (1972); H. Tada and I. Kakitani, *Bull. Chem. Soc. Japan*, **46**, 1226 (1973); N.A. Kravchenko and L. A. Chistyakova, *Biokhimiya*, **37**, 1126 (1972); S. Lapanje and J. A. Rupley, *Biochemistry*, **12**, 2370 (1973); C.-L. Lee and M. Z. Atassi, *ibid.*, p. 2690; N.-I. Yu and B. H. Jo, *ibid.*, **156**, 469; K. J. Kramer and J. A. Rupley, *ibid.*, **156**, 414 (1973); **158**, 566 (1973); K. Y. Chang and C. W. Carr, *Biochem. Biophys. Acta*, **285**, 377 (1972); A. A. Aboderin, E. Boedefeld and P. L. Luisi, *ibid.*, **328**, 20 (1973); V. K. Srivastava and C. C. Bigelow, *ibid.*, **285**, 373 (1972); H. Hayashi, K.-I. Amano, Y. Araki and E. Ito, *Biochem. Biophys. Res. Comm.*, **50**; 641 (1973); M. Kugimiya and C. C. Bigelow, *Can. J. Biochem.*, **51**, 581 (1973).

[44] J. P. Penn and P. Jolles, *Clin. Chim. Acta*, **42**, 77 (1972); *Chem. Abs.*, **78**, 54675 (1973).

[45] J. Hermann, J. Jolles, D. H. Buss and P. Jolles, *J. Mol. Biol.*, **79**, 587 (1973).

[46] B. W. Matthews, F. W. Dahlquist and A. Y. Maynard, *J. Mol. Biol.*, **78**, 575 (1973).

[47] W. L. Riggle, J. A. Long and C. L. Borders, *Can. J. Biochem.*, **51**, 1433 (1973).

[48] N. Arnheim, M. Inouye, L. Law and A. Laudin, *J. Biol. Chem.*, **248**, 233 (1973).

[49] M. L. Sinnott and I. J. L. Souchard, *Biochem. J.*, **133**, 89 (1973).

[50] M. L. Sinnott and O. Viratelle, *Biochem. J.*, **133**, 81 (1973); O. M. Viratelle and J. M. Yon, *Eur. J. Biochem.*, **33**, 110 (1973); G. van der Groen, J. Wouters-Leysen, M. Yde and C. K. De Bruyne, *ibid.*, **38**, 122 (1973).

[51] Cf. *Org. Reaction Mech.*, **1970**, 424–425.

[52] M. L. Sinnott, *Chem. Comm.*, **1973**, 535.

[53a] G. S. Case, M. L. Sinnott and J.-P. Tenu, *Biochem. J.*, **133**, 99 (1973).

[53b] M. Maybury and M. Sinnott, *J.C.S. Perkin II*, **1973**, 300.

[54] W. G. Struve and H. M. McConnell, *Biochem. Biophys. Res. Comm.*, **49**, 1631 (1972).

[55] J. J. Distler and G. W. Jourdian, *J. Biol. Chem.*, **248**, 6772 (1973).

[56] C. J. Gray and M. E. Jolley, *FEBS Letters*, **29**, 197 (1973).

[57] E. J. Bourne, P. Finch and A. G. Nagpurkar, *Carbohydrate Res.*, **29**, 492 (1973).

[58] Y. A. Zhdanov and R. M. Kessler, *Dokl. Akad. Nauk SSSR*, **207**, 607 (1972); *Chem. Abs.*, **78**, 68657 (1973).

[59] D. W. Darnall and E. R. Birnbaum, *Biochemistry*, **12**, 3489 (1973); C. E. Weill and J. Guerrera, *Carbohydrate Res.*, **27**, 451 (1973); K. Omechi, T. Ikenaka and Y. Matsushima, *J. Biochem. (Tokyo)*, **73**, 491 (1973); D. J. Stiefel and P. J. Keller, *Biochem. Biophys. Acta*, **302**, 345 (1973); J. Robyt and R. J. Aikerman, *Arch. Biochem. Biophys.*, **155**, 445 (1973).

[60] E. Laszlo, J. Hollo and B. Banky, *Carbohydrate Res.*, **25**, 355 (1972); J. J. Marshall, *Eur. J. Biochem.*, **33**, 494 (1973).

[61] T. Kaya, M. Shiband and T. Kutsumi, *J. Biochem. (Tokyo)*, **73**, 181 (1972).

[62a] Y. Yamasaki, T. Miyake and Y. Suzuki, *Agr. Biol. Chem.*, **37**, 251 (1973).

[62b] J. Carter and E. E. Smith, *Arch. Biochem. Biophys.*, **155**, 82 (1973).

[63] P. M. Dey, *Biochem. Biophys. Acta*, **302**, 393 (1973).

[64a] T. Mega, T. Ikenaka, H. Arita and K. Fukukawa, *J. Biochem. (Tokyo)*, **73**, 55 (1973); T. Mega, T. Ikenaka and Y. Matsushima, *ibid.*, **72**, 1391 (1972); K. Yamamoto, *ibid.*, **73**, 631, 749 (1973).

[64b] J. Tomasic and D. Keylevic, *Biochem. J.*, **133**, 789 (1973).

[65] W. J. Bover and P. Zuman, *J. Am. Chem. Soc.*, **95**, 2531 (1973); *J.C.S. Perkin II*, **1973**, 786; P. Greenzaid, *J. Org. Chem.*, **38**, 3164 (1973).

[66] P. E. Sørensen, *Acta Chem. Scand.*, **26**, 3357 (1972).

[67] L. R. Green and J. Hine, *J. Org. Chem.*, **38**, 2801 (1973).

[68] W. E. Hull, B. D. Sykes and B. M. Babior, *J. Org. Chem.*, **38**, 2931 (1973).

[69] C. A. Lewis and R. Wolfenden, *J. Am. Chem. Soc.*, **95**, 6685 (1973).

[70] P. M. Hardy, A. C. Nicholls and H. N. Rydon, *J.C.S. Perkin II*, **1972**, 2270.

[71] T. M. Gorrie, S. K. Raman, H. K. Rouette and H. Zollinger, *Helv. Chim. Acta*, **56**, 175 (1973).

[72] M. Senma, Z. Tara, K. Osaki and T. Taga, *Chem. Comm.*, **1973**, 880.

[73] G. Siegemund, *Chem. Ber.*, **106**, 2960 (1973); M. Tsuge, T. Miyabayashi and S. Tanaka, *Nippon Kagaku Kaishi*, **1973**, 1016; *Chem. Abs.*, **79**, 31136 (1973).

[74] K. Bowden and A. M. Last, *J.C.S. Perkin II*, **1973**, 1144.

[75] P. R. Rony and R. O. Neff, *J. Am. Chem. Soc.*, **95**, 2896 (1973).

[76] N. M. Ballash and E. B. Robertson, *Can. J. Chem.*, **51**, 556 (1973).

[77] J. A. Hveding, O. Kjølberg and A. Reine, *Acta Chem. Scand.*, **27**, 1427 (1973).

[78] A. H. Conner and L. Anderson, *Carbohydrate Res.*, **25**, 107 (1972); cf. *Org. Reaction Mech.*, **1971**, 403.

[79] T. A. W. Koerner, L. W. Cary, N. S. Bhacca and E. S. Younathan, *Biochem. Biophys. Res. Comm.*, **51**, 543 (1973).
[80] P. H. Fishman, P. G. Pentchev and J. M. Bailey, *Biochemistry*, **12**, 2490 (1973).
[81] H. D. Bürgi, J. D. Dunitz and E. Shefter, *J. Am. Chem. Soc.*, **95**, 5065 (1973).
[82] J. Hine, M. S. Cholod and W. K. Chess, Jr., *J. Am. Chem. Soc.*, **95**, 4270 (1973); see also L. S. Mushketik, N. V. Volkova, E. A. Shilov and A. A. Yasnikov, *Ukr. Khim. Zh.* (Russian Ed.), **38**, 1259 (1972); *Chem. Abs.*, **78**, 96785 (1973); I. E. Kalinichenko and N. A. Emtsova, *Zh. Obshch. Khim.*, **42**, 2223 (1972); *Chem. Abs.*, **78**, 83685 (1973); R. W. Green, P. W. Alexander and R. J. Sleet, *Austral. J. Chem.*, **26**, 1653 (1973).
[83] J. Hine and K. W. Narducy, *J. Am. Chem. Soc.*, **95**, 3362 (1973); cf. ref. 130.
[84] H. Diebler and R. N. F. Thorneley, *J. Am. Chem. Soc.*, **95**, 896 (1973).
[85] C. J. O'Connor, E. J. Fendler and J. H. Fendler, *J.C.S. Perkin II*, **1973**, 1744, 1900.
[86a] K. Martinek, A. K. Yatsimirski, A. P. Osipov and I. V. Berezin, *Tetrahedron*, **29**, 963 (1973).
[86b] A. Mesli and J. Tirouflet, *Compt. Rend.*, *C*, **275**, 1541 (1972).
[87] J. Hoffmann and V. Štěrba, *Coll. Czech. Chem. Comm.*, **38**, 2091, 3146 (1973).
[88] P. Nagy, *Szegedi Tanarkepzo Foiskola Tud. Kozlem.*, **1971**, 147; *Chem. Abs.*, **78**, 110162 (1973).
[89] P. Nagy, *Magy. Kem. Folyoirat*, **79**, 145 (1973); *Chem. Abs.*, **78**, 158599 (1973).
[90] Y. Ogata, A. Kawasaki, H. Suzuki and H. Kojoh, *J. Org. Chem.*, **38**, 3031 (1973).
[91] J. Thomas, *Bull. Soc. Chim. France*, **1973**, 1296, 1300.
[92] D. J. Hupe, M. C. R. Kendall, G. T. Sinner and T. A. Spencer, *J. Am. Chem. Soc.*, **95**, 2260 (1973); D. J. Hupe, M. C. R. Kendall and T. A. Spencer, *ibid.*, p. 2271.
[93] G. M. Sharma and O. A. Roels, *J. Org. Chem.*, **38**, 3648 (1973).
[94] P. A. Kollman, W. F. Trager, S. Rothenberg and J. E. Williams, *J. Am. Chem. Soc.*, **95**, 458 (1973).
[95] J. Velisek, J. Davidek, J. Pokorny, K. Grundova and G. Janicek, *Z. Lebensm.-Unters. Forsch.*, **149**, 323 (1972); *Chem. Abs.*, **78**, 70969 (1973).
[96] P. Dunlop, M. A. Marini, H. M. Fales, E. Sokoloski and C. J. Martin, *Bio-Organic Chemistry*, **2**, 235 (1973).
[97] A. M. Knowles and A. Lawson, *J.C.S. Perkin I*, **1973**, 537.
[98] A. T. Nielsen, R. L. Atkins, D. W. Moore, R. Scott, D. Mallory and J. M. LaBerge, *J. Org. Chem.*, **38**, 3288 (1973).
[99] W. E. Hull, B. D. Sykes and B. M. Babior, *J. Org. Chem.*, **38**, 2931 (1973).
[100] M. P. Doyle, M. A. Zaleta, J. E. De Boer and W. Wierenga, *J. Org. Chem.*, **38**, 1663 (1973).
[101] M. V. Bhatt and M. Ravindranathan, *J.C.S. Perkin II*, **1973**, 1160.
[102a] R. D. Guthrie and J. L. Hedrick, *J. Am. Chem. Soc.*, **95**, 2971 (1973).
[102b] J. E. Dixon and T. C. Bruice, *Biochemistry*, **12**, 4762 (1973).
[103] K. Harada and T. Yoshida, *J. Org. Chem.*, **37**, 4366 (1972); A. E. Kitahchi, A. H. Nathans and C. L. Kitchell, *J. Biol. Chem.*, **248**, 841 (1973); H. J. Rhodes, R. B. Kluza and M. I. Blake, *J. Pharm. Sci.*, **62**, 59 (1973).
[104] S. Fatutta, A. Risaliti, C. Russo and E. Valentin, *Gazz. Chim. Ital.*, **102**, 1008 (1972).
[105] F. Weisbuch, C. Yamagami and G. Dana, *J. Org. Chem.*, **37**, 4334 (1972).
[106] J. P. Freeman, "Less Familiar Reactions of Oximes", *Chem. Rev.*, **73**, 283 (1973).
[107] S. M. Silver and J. M. Sayer, *J. Am. Chem. Soc.*, **95**, 5073 (1973).
[108] J. M. Sayer and W. P. Jencks, *J. Am. Chem. Soc.*, **95**, 5637 (1973); see *Org. Reaction Mech.*, **1972**, 405–406.
[109] R. E. Barnett, *Accounts Chem. Research*, **6**, 41 (1973).
[110] J. M. Sayer, M. Peskin and W. P. Jencks, *J. Am. Chem. Soc.*, **95**, 4277 (1973).
[111] L. do Amaral, M. P. Bastos, H. G. Bull and E. H. Cordes, *J. Am. Chem. Soc.*, **95**, 7369 (1973); see *Org. Reaction Mech.*, **1972**, 405.
[112] Ya. I. Turyan and O. A. Tohstilcova, *Organic Reactivity* (*Tartu*), **9**, 794, 804 (1972).
[113] V. A. Afanas'ev and I. F. Strel'tsova, *Zh. Fiz. Khim.*, **46**, 2545 (1972); *Chem. Abs.*, **78**, 58688 (1973); I. Dyong and H. P. Bertram, *Chem. Ber.*, **106**, 1743, 2654 (1973); A. J. Fatiadi, *Chem. Ind.* (*London*), **1973**, 38.
[114] J. Arriau, J. P. Campillo and J. Deschamps, *Bull. Soc. Chim. France*, **1973**, 1398, 1403.
[115] S. R. Johns, J. A. Lamberton and E. R. Nelson, *Austral. J. Chem.*, **26**, 1297 (1973).
[116] V. L. Goedken and S.-M. Peng, *Chem. Comm.*, **1973**, 62.
[117] A. Belly, F. Petrus and J. Verducci, *Bull. Soc. Chim. France*, **1973**, 1395.
[118] I. Christenson, *Acta Pharm. Suec.*, **9**, 309, 343, (1972); *Chem. Abs.*, **78**, 3450, 3442 (1973); T. Balakrishnan and M. Santappa, *Curr. Sci.*, **42**, 418 (1973); *Chem. Abs.*, **79**, 52522 (1973).

119 G. T. Knight and B. Saville, *J.C.S. Perkin II*, **1973**, 1550.
120 A. J. Kresge and Y. Chiang, *J. Am. Chem. Soc.*, **95**, 803 (1973).
121 I. de Jonge and W. Drenth, *Rec. Trav. chim.*, **92**, 420 (1972).
122 L. R. Fedor, N. C. De and S. K. Gurawa, *J. Am. Chem. Soc.*, **95**, 2905 (1973); cf. *Org. Reaction Mech.*, **1969**, 419; **1972**, 407.
123 G. Stork and R. L. Danheiser, *J. Org. Chem.*, **38**, 1775 (1973).
124 A. C. Knipe and B. G. Cox, *J.C.S. Perkin II*, **1973**, 1391; *J. Org. Chem.*, **38**, 3429 (1973); cf. *Org. Reaction Mech.*, **1972**, 408.
125 R. P. Bell and M. I. Page, *J.C.S. Perkin II*, **1973**, 1681.
126 G. A. Abad, S. P. Jindal and T. T. Tidwell, *J. Am. Chem. Soc.*, **95**, 6326 (1973); cf. *Org. Reaction Mech.*, **1971**, 410–411.
127 R. P. Bell and B. A. Timini, *J.C.S. Perkin II*, **1973**, 1518.
128 J. Hine, J. L. Lynn, J. H. Jensen, and F. C. Schmalstieg, *J. Am. Chem. Soc.*, **95**, 1577 (1973); cf. *Org. Reaction Mech.*, **1971**, 409–410.
129 J. Hine, E. F. Glod, R. E. Notari, F. E. Rogers and F. C. Schmalstieg, *J. Am. Chem. Soc.*, **95**, 2537 (1973).
130 J. Hine, K. W. Narducy, J. Mulders, F. E. Rogers and N. W. Flachskam, *J. Org. Chem.*, **38**, 1636 (1973); cf. ref. 83.
131 L. P. Koshechkina, I. V. Mel'nichenko and A. A. Yasnikov, *Ukr. Khim. Zh.* (Russian Ed.), **38**, 1248 (1972); *Chem. Abs.*, **78**, 96784 (1973).
132 D. B. Dahlberg and F. A. Long, *J. Am. Chem. Soc.*, **95**, 3825 (1973).
133 R. A. Lynch, S. P. Vincenti, Y. T. Lin, L. D. Smucker and S. C. S. Rao, *J. Am. Chem. Soc.*, **94**, 8351 (1972).
134 K. E. Teo and E. W. Warnhoff, *J. Am. Chem. Soc.*, **95**, 2728 (1973).
135 Y. Jasor, M. Gaudry, A. Marquet and M. Bettahar, *Bull. Soc. Chim. France*, **1973**, 2732, 2735.
136 P. Moreau, A. Casadevall and E. Casadevall, *Bull. Soc. Chim. France*, **1973**, 1014.
137 E. Kiehlmann, F. Masaro and F. J. Slawson, *Can. J. Chem.*, **51**, 3182 (1973).
138 P. Metzger, A. Casadevall and E. Casadevall, *Tetrahedron Letters*, **1973**, 2027.
139 J. Toullec and J. E. Dubois, *Tetrahedron*, **29**, 2851, 2859 (1973).
140 J.-O. Levin, *Chem. Scripta*, **4**, 85 (1973).
141 R. Kluger and P. Wasserstein, *J. Am. Chem. Soc.*, **95**, 1071 (1973).
142 J. Klein and R. Levene, *J.C.S. Perkin II*, **1973**, 1971.
143 B. E. Tilley, D. W. Porter and R. W. Gracy, *Carbohydrate Res.*, **27**, 289 (1973).
144 J. S. Chickos and R. E. K. Winter, *J. Am. Chem. Soc.*, **95**, 506 (1973); cf. *Org. Reaction Mech.*, **1972**, 411.
145 B. Macierewicz, *Rocz. Chem.*, **47**, 1735 (1973).
146 K. E. Hine and R. F. Childs, *J. Am. Chem. Soc.*, **95**, 3289 (1973).
147 H. Koshimura, J. Saito and T. Okubo, *Bull. Chem. Soc. Japan*. **46**, 632 (1973); M. C. Fernandez and E. Melendez, *An. Quím.*, **68**, 1035 (1972); A. D. Taneja and K. P. Srivastava, *J. Indian Chem. Soc.*, **49**, 889 (1972); A. D. Taneja, K. P. Srivastava and M. C. Gupta, *J. Chin. Chem. Soc.* (*Tapei*), **19**, 115 (1972); *Chem. Abs.*, **77**, 163839 (1972).
148 K. J. Schray, S. J. Benkovic, P. A. Benkovic and I. A. Rose, *J. Biol. Chem.*, **248**, 2219 (1973); B. Wurster and B. Hess, *Z. Physiol. Chem.*, **354**, 407 (1973).
149 E. L. O'Connell and I. A. Rose, *J. Biol. Chem.*, **248**, 2225 (1973).
150 I. A. Rose, E. L. O'Connell and K. J. Schray, *J. Biol. Chem.*, **248**, 2232 (1973).
151 T. Y. S. Shen and E. W. Westhead, *Biochemistry*, **12**, 3333 (1973); S.-K. Oh and J. M. Brewer, *Arch. Biochem. Biophys.*, **157**, 491 (1973); T. Nowak, A. S. Mildvan and G. L. Kenyon, *Biochemistry*, **12**, 1690 (1973).
152 S. G. Waley, *Biochem. J.*, **135**, 165 (1973); K. J. Schray, E. L. O'Connell and I. A. Rose, *J. Biol. Chem.*, **248**, 2214 (1973).
153 W. T. Comer and D. L. Temple, *J. Org. Chem.*, **38**, 2121 (1973).
154 Y. Yano, T. Okonogi, M. Sunaga and W. Tagaki, *Chem. Comm.*, **1973**, 527.
155 W. L. Dilling, N. B. Tefertiller and J. P. Heeschen, *J. Org. Chem.*, **37**, 4159 (1972); L. P. Koshechkina and A. A. Yasnikov, *Ukr. Khim. Zh.* (Russian ed.), **38**, 786 (1972); *Chem. Abs.*, **78**, 3518 (1973); T. Mukaiyama, K. Narasaka and K. Banno, *Chem. Letters* (*Tokyo*), **1973**, 1011; A. A. Gallo and H. Z. Sable. *Biochem. Biophys. Acta*, **302**, 443 (1973); J. E. Dubois and P. Fellmann, *Tetrahedron Letters*, **1972**, 5085; J.-E. Vik, *Acta Chem. Scand.*, **26**, 3165 (1972); T. J. Clark, *J. Org. Chem.*, **38**, 1749 (1973).

[156] M. Calmon and J.-P. Calmon, *Compt. Rend., C.*, **276**, 197 (1973); M. Calas, B. Calas and L. Giral, *Bull. Soc. Chim. France*, **1973**, 2079.
[157] S. A. Fine and P. D. Pulaski, *J. Org. Chem.*, **38**, 1747 (1973).
[158] A. Takeda, S. Tsuboi and T. Hongo, *Bull. Chem. Soc. Japan*, **46**, 1844 (1973); E. Elkik and C. Francesch, *Bull. Soc. Chim. France*, **1973**, 1277, 1281.
[159] T. I. Crowell and T.-R. Kim, *J. Am. Chem. Soc.*, **95**, 6781 (1973); F. S. Prout, U. D. Beaucaire, G. R. Dyrkarz, W. M. Koppes, R. E. Kuznicki, T. A. Mattewski, J. J. Pienkowski and J. M. Puda, *J. Org. Chem.*, **38**, 1512 (1973).
[160] G. Zvilichovsky and U. Fotadar, *J. Org. Chem.*, **38**, 1782 (1973).
[161] A. V. Shchelkunov, *Vest. Akad. Nauk Kaz. SSR*, **29**, 17 (1973); *Chem. Abs.*, **79**, 104364 (1973).
[162] F. Fernandez-Gonzal and R. Perez-Ossorco, *An. Quím.*, **68**, 1411 (1972).
[163] M. Anai, C. Y. Lai and B. L. Horecker, *Arch. Biochem. Biophys.*, **156**, 712 (1973); **158**, 449 (1973).
[164] A. L. Crowder, C. A. Swenson and R. Barker, *Biochemistry*, **12**, 2852 (1973); E. Grazi, C. Sivieri-Pecovari, R. Gayliano and G. Trombetta, *ibid.*, p. 2583.
[165] E. Vedejs and K. A. J. Snoble, *J. Am. Chem. Soc.*, **95**, 5778 (1973).
[166] M. Kuchar, B. Kakac, O. Nemecek, E. Kraus and J. Holubek, *Coll. Czech. Chem. Comm.*, **38**, 447 (1973).
[167] G. Aksnes and R. Y. Khalil, *Phosphorus*, **2**, 105 (1972); *Chem. Abs.*, **78**, 70865 (1973).
[168] N. A. Nesmeyanov, E. V. Vishtok and O. A. Reutov, *Dokl. Akad. Nauk SSSR*, **210**, 1102 (1973); *Chem. Abs.*, **79**, 91259 (1973); D. W. Allen, B. G. Hutley and T. C. Rich, *J.C.S. Perkin II*, **1973**, 820.
[169] I. S. Kislina and M. I. Vinnik, *Organic Reactivity* (*Tartu*), **9**, 299 (1972); I. S. Kislina and M. I. Vinnik, *Dokl. Akad., Nauk SSSR*, **206**, 1118 (1972); *Chem. Abs.*, **78**, 71090 (1973).
[170] I. V. Andreeva, M. M. Koton, A. I. Turbina and N. V. Kukarkina, *Zh. Org. Khim.*, **9**, 562 (1973); *Chem. Abs.*, **78**, 147018 (1973).
[171] V. A. Ostrovskii, A. S. Enin and G. I. Koldobskii, *Zh. Org. Khim.*, **9**, 802 (1973); *Chem. Abs.*, **79**, 4770 (1973).
[172] R. Varma and B. French, *Carbohydrate Res.*, **71**, (1972).
[173] H. M. Relles, *J. Org. Chem.*, **38**, 1570 (1973).
[174] H. Ona, M. Sakai, M. Suda and S. Masarume, *Chem. Comm.*, **1973**, 45.
[175] J.-P. Geumas and H. Quiniou, *Bull. Soc. Chim. France*, **1973**, 592.
[176] I. Mehrota and D. Devaprabhakara, *J. Organometal. Chem.*, **51**, 93 (1973).
[177] P. Talts and U. Haldna, *Organic Reactivity* (*Tartu*), **10**, 118 (1973); S. V. Zukerman, C. P. Pivovarevich, L. A. Kutulya, N. S. Pionenko and V. F. Laorushin, *Organic Reactivity* (*Tartu*), **9**, 130 (1971); K. L. Cook and A. J. Waring, *J.C.S. Perkin II*, **1973**, 84; D. M. Brouwer and A. A. Kippen, *Rec. Trav. chim.*, **92**, 906 (1973).

CHAPTER 2

Reactions of Acids and their Derivatives

M. I. PAGE

Department of Chemistry, The Polytechnic, Huddersfield

CARBOXYLIC ACIDS [1a, 1b]

Tetrahedral Intermediates

New examples of stable tetrahedral intermediates have been reported. When the imidazole derivative (**1**) is treated with acetyl chloride in pyrimidine, IR and NMR studies suggest that it is converted into the acyl tetrahedral intermediate (**2**). The latter is capable of acyl migration and in water it rearranges quantitatively to (**3**) by a first-order process.[2]

Pinacol monotrifluoroacetate (**4**) in various non-aqueous solvents exists predominantly as the cyclic diester of the orthoacid (**5**), i.e. as a tetrahedral intermediate.[3]

(1) **(2)** **(3)**

$Me_2C—OH$
$Me_2C—OCOCF_3$

(4)

(5)

The methyl basicity of alcohols, *gem*-diols and hemiacetals, as defined by the free-engergy change of:

$$R^1R^2R^3COH + MeOH \rightleftharpoons R^1R^2R^3COMe + H_2O$$

(where R is H, Me, OH or OMe), has only a small dependence upon the nature of the substituents. Extrapolation to orthoesters permitted calculation of the free-energies of hydration of orthoacids and their methyl esters, i.e. tetrahedral intermediates (**6**).

$$RCO_2Me + H_2O \rightleftharpoons RC(OH)_2OMe$$
(6)

If $k_{hydrolysis}/k_{exchange}$ is assumed to be 13, a free-energy profile, in kcal mol^{-1}, such as that shown for methyl acetate in Figure 1, is then obtainable.[4]

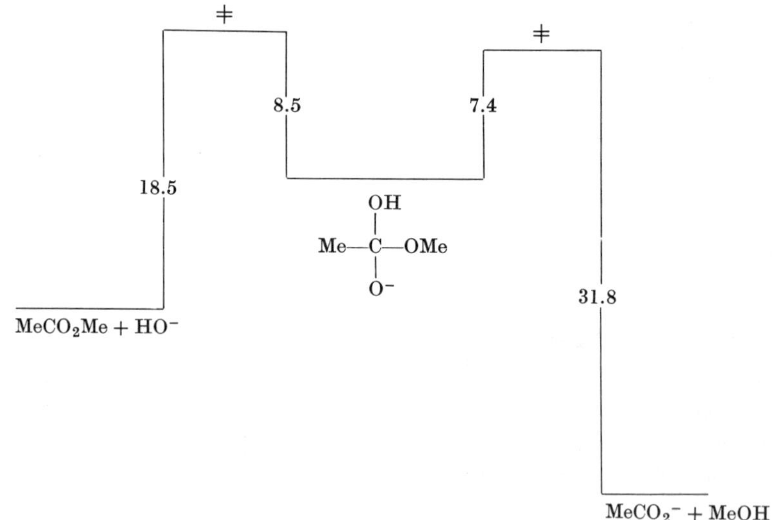

FIG. 1. Free-energy profile for methyl acetate.

It has been suggested that the conformation of the tetrahedral intermediate formed from hydroxide ion and imidate salts depends on whether the *syn*- or the *anti*-isomer (**7**) is attacked. If the conformation formed (**8**) then breaks down stereospecifically by having a lone-pair antiperiplanar to the leaving group, it determines whether the product is ester and amine (**9**) or amide plus alcohol.[5]

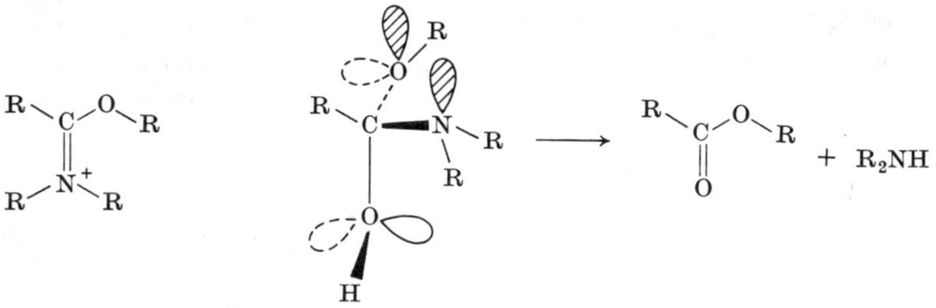

(**7**) (**8**) (**9**)

The criteria for a reaction to exhibit a rate-limiting step that is diffusion-controlled, and the experimental methods available to test such suggestions, have been described.[6] The Brønsted plot for the general-base catalysed hydrazinolysis of acetylimidazole is non-linear, with limiting slopes of $\beta \leqslant 0.2$ and $\beta > 0.7$ for strong and weak bases, respectively. This indicates a change in the nature of the rate-limiting step with changing pK of the catalyst and thus invokes an intermediate. For strongly basic catalysts, B, the rate-limiting step is diffusion-controlled encounter (k_1) with the tetrahedral intermediate in Scheme 1. For weakly basic catalysts either a "concerted" general-base catalysed

$RNH_2 + MeCOIm$

$\rightleftharpoons$

$R\overset{+}{N}H_2—C(O^-)(Me)—Im$ ($T^{\pm}$) $\underset{}{\overset{k_1[B]}{\rightleftharpoons}}$ $B.T^{\pm}$ $\xrightarrow{k_2}$ Products

$\rightleftharpoons$

$B\overset{+}{H}.T^-$ $\xrightarrow{k_4}$ Products

$\rightleftharpoons$ k_3

$BH^+ + T^-$ $\longrightarrow$ Products

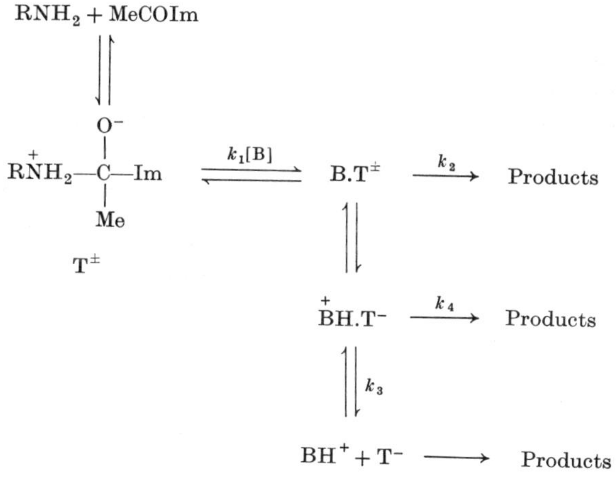

Scheme 1.

decomposition of $T^{\pm}$ (k_2), or a stepwise proton-transfer mechanism, k_3 or k_4, was suggested to be rate-limiting. Such mechanisms and their relationship to the life-time of the intermediate and the position of the break in the Brønsted plot were discussed.[7]

The decomposition of the tetrahedral intermediate formed by hydroxide-ion attack on *N*-methyltrifluoroacetanilide is general-acid catalysed and shows a Brønsted exponent α of 1.0. These results were interpreted in terms of complete proton-transfer, the rate-limiting step being either diffusion away of the conjugate base of the catalyst or cleavage of the C—N bond in the presence of the conjugate base within the solvent cage.[8] The dependence of the second-order rate constants for methoxide ion-catalysed methanolysis of ring-substituted *N*-methyltrifluoroacetylanilides, in CH_3OD–CH_3OH, upon the mole fraction of CH_3OD has been analysed in terms of an isotopic fractionation factor for the bridging proton, ϕ, = 0.39 and $k_H/k_D = 2.6$ when the substituent is *m*-nitro, $\phi = 0.29$ and $k_H/k_D = 3.5$ when it is *p*-chloro, and $\phi = 0.13$ and $k_H/k_D = 7.4$ when it is *p*-methoxy. It was suggested that this is evidence for proton-transfer catalysis with the C—N bond not breaking in the transition state (**10**) in the case of a *p*-methoxy substituent, but C—N bond fission occurring with hydrogen-bonding (**11**) with a *m*-nitro-substituent.[9] The

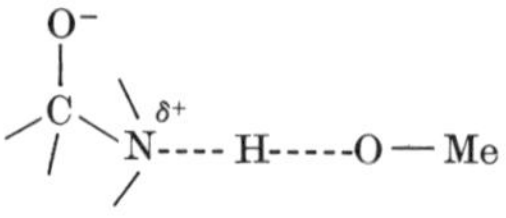

O⁻–C(–)–N$^{\delta+}$---- H-----O—Me

(**10**)

O=C(δ⁻)···N····H···O—Me

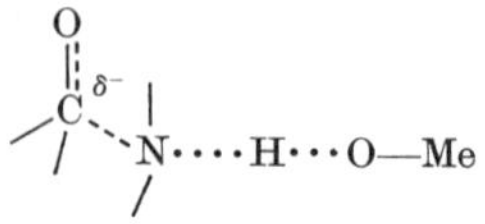

(**11**)

hydrolysis of *N*-(trifluoroacetyl)-*p*-nitroaniline is of the first order in hydroxide ion with no detectable buffer catalysis at pH 9; presumably formation of the tetrahedral intermediate is rate-limiting. At lower pH the kinetic order in hydroxide ion is greater than first and there is catalysis by the free-base form of Tris buffers. This was interpreted as a dual pathway for the breakdown of the intermediate, uncatalysed breakdown of the dianion and general-acid catalysed breakdown of the monoanion.[10a]

It is estimated that the ratio for breakdown of the tetrahedral intermediate, formed by hydroxide-ion attack upon *N*-(trifluoroacetyl)-2′,6′-dimethylaniline, k_2/k_{-1}, is 0.0008 compared with 0.025 for trifluoroacetanilide. This is attributed to steric hindrance by the methyl groups, preventing protonation of the anilide-nitrogen atom.[10b]

The Brønsted plot for the reaction of substituted pyridines with methyl chloroformate is curved, suggesting a change in rate-limiting step from formation to breakdown of a tetrahedral intermediate.[11]

The rate constant for the tetra-*n*-hexylammonium benzoate hemihydrate-catalysed reaction between *p*-nitrophenyl acetate and imidazole in toluene has a non-linear dependence upon the catalyst concentration, which was taken to indicate the presence of an intermediate and a change in the rate-limiting step.[12]

The rate of hydrolysis of strongly acyl-activated thiol esters, RCOSR′ ($R = CF_3$, CCl_3 or CF_3CF_2), is inhibited by acid at pH < 2, but weakly acyl-activated thiol esters ($R = CH_2Cl$ or CH_2NO_2) undergo acid-catalysed hydrolysis. A general mechanism as shown in Scheme 2 was considered to rationalize the observed pH–rate profiles. For

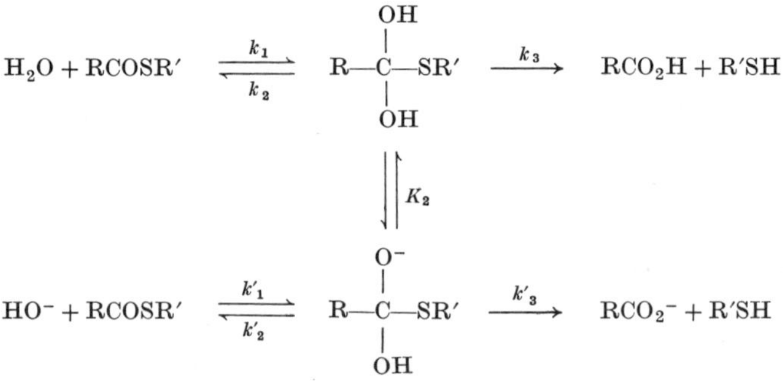

$$H_2O + RCOSR' \underset{k_2}{\overset{k_1}{\rightleftharpoons}} R\text{—}C(OH)_2\text{—}SR' \xrightarrow{k_3} RCO_2H + R'SH$$

$$\updownarrow K_2$$

$$HO^- + RCOSR' \underset{k'_2}{\overset{k'_1}{\rightleftharpoons}} R\text{—}C(O^-)(OH)\text{—}SR' \xrightarrow{k'_3} RCO_2^- + R'SH$$

SCHEME 2.

strongly acyl-activated thiol esters formation of the intermediate is rate-limiting at pH >2, but at lower pH the slow step is the breakdown of an anionic or neutral intermediate. With less electron-withdrawal in the acyl group the transition at pH ca. 1 is only partial, i.e. at pH < 1 both formation and decomposition of the intermediates are partially rate-limiting.[13]

The products of the hydrolysis of the protonated imino-ether (**12**) are pH-dependent, indicating an intermediate. At pH > 8 the product is mainly the amide (**13**) but at low pH it is the amine and phthalide (**14**). Thus, at high pH $k_2 \gg k_3$ and the rate-limiting step in the lactonization of (**13**) is k_3 (breakdown of the tetrahedral intermediate[14]), and not

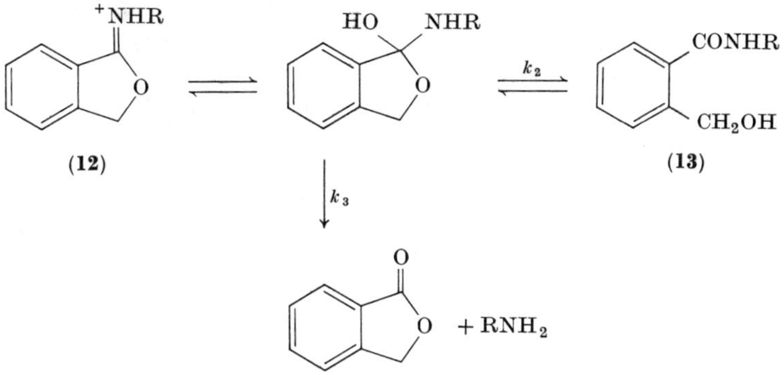

$^{+}$NHR HO NHR k_2 CONHR CH_2OH

(**12**) (**13**)

k_3

O + RNH_2

(**14**)

its formation as suggested earlier.[15] The yield of amine from the hydrolysis of ethyl *N*-arylformimidates (**15**) derived from amines of pK_a > 0 decreases as the pH is increased. Imidate esters derived from very weakly basic amines (pK_a < –6) produce increasing yields of amine as the pH is increased. The yield of amine from the hydrolysis of the imidate ester of 2,4-dinitroaniline passes through a minimum at pH 4.5. These observations were interpreted in terms of an equilibrium between cationic, neutral (**16**) and

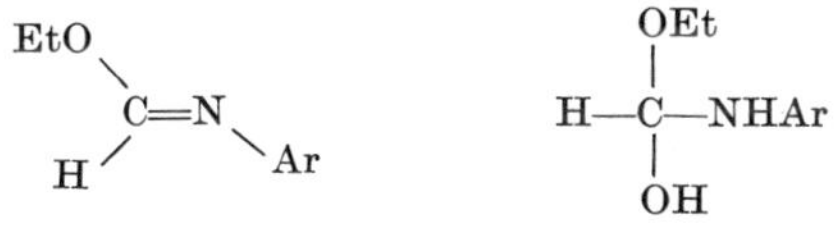

EtO C=N Ar H   OEt H—C—NHAr OH

(**15**) (**16**)

anionic tetrahedral intermediates, product distribution being controlled by their relative abundance at a given pH and the pathway of breakdown of each species.[16] The yield of amine also depends upon the nature of the general acid–base catalyst present. The phosphate monoanion favours amine formation compared with imidazole, and this was attributed to bifunctional catalysis facilitating breakdown of the tetrahedral intermediate. This effect decreases with increasing electron-withdrawal in the amine. All buffers decrease the yield of amine from the hydrolysis of ethyl *N*-(2,4-dinitrophenyl)formimidate, presumably because of the low basicity of the nitrogen atom in the tetrahedral intermediate.[17]

The products of hydrolysis of benzoylthionocholine are pH-dependent,[18] benzoic acid being obtained below pH 9.5 and thiobenzoic acid above pH 11.[19]

It has been suggested that there is a better correlation of the rate constants for nucleophilic attack on phenyl esters with σ° than there is with σ^- and that therefore the rate-limiting step is formation of the tetrahedral intermediate. However, intramolecular nucleophilic attack on phenyl esters correlates best with σ^-, and therefore breakdown of the intermediate is rate-limiting.[20]

The hydrolysis of 1-acetyl- and 1-benzoyl-pyrrole is of the second order in hydroxide ion at low base concentration and of the first order in hydroxide ion at high base concentration. There is no general base-catalysis and $k_{H_2O}/k_{D_2O} = 0.59$. This was taken to indicate formation of an intermediate followed by an equilibrium ionization to the dianion which collapses to products.[21] The hydrolysis of 1-salicyloylpyrrole exhibits a sigmoid pH–rate profile and is pH-independent at pH > 11. The point of inflexion occurs 1.4 pH units above the pK_a of the substrate. It was claimed that this was the result of ionization of the tetrahedral intermediate (Scheme 3),[21] but it has been pointed out that this interpretation is not required by the kinetics of the reaction.[22]

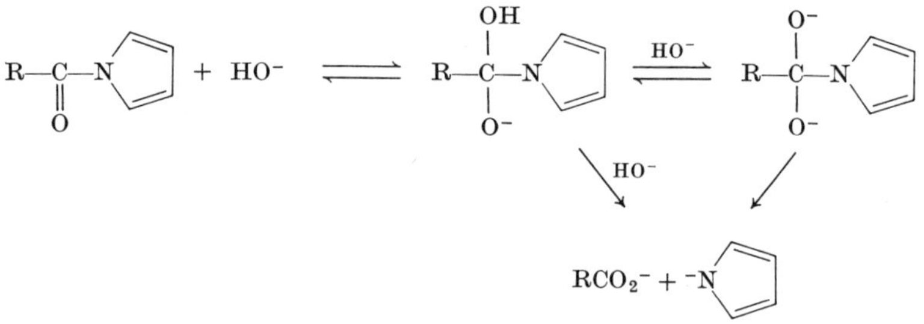

SCHEME 3.

The observed rate constants for the hydrolysis of perbenzoic acids have a non-linear dependence on the concentration of acetate buffer, indicating the presence of an intermediate and a change in the rate-limiting step.[23]

The effect of dimethyl sulphoxide (DMSO) on the alkaline hydrolysis of several acetanilides has been studied and interpreted in terms of the effect upon the rates of formation and breakdown of the tetrahedral intermediate.[24] However, the basic methanolysis of anilides in DMSO-methanol has been examined as a function of substituents in the aniline residue and of solvent composition and here, also, there is a marked dependence of the solvent effect on the nature of the substituent. This was rationalized in terms of a spectrum of transition states, but all involving rate-limiting breakdown of the intermediate. It was suggested that this explanation is also satisfactory for the hydrolysis.[25]

The volumes of aldehyde hydration are comparable in magnitude with the volumes of activation for the hydrolysis of esters and amides.[26]

The effect of successive methylation of methyl benzimidate upon the rate constants for hydrolysis and the activation parameters does not parallel that found with the corresponding *N*-methylated benzamides. It was claimed that this is evidence that acid-catalysed amide hydrolysis does not involve *O*-protonation and formation of a tetrahedral intermediate.[27]

Adenosine deaminase has been found to catalyse the hydration of pteridine, owing, it was suggested, to stabilization of a tetrahedral intermediate.[28]

Intermolecular Catalysis and Reactions

Reactions in Hydroxylic Solvents[29]

Rules to distinguish between concerted and step-wise mechanisms for general-acid or general-base catalysis have been suggested. Concerted mechanisms can occur only (*a*) at sites that undergo a large change in pK in the course of the reaction and (*b*) when the pK of the catalyst is intermediate between the initial and final pK values at the substrate site.[30] Similar conclusions have been reached by considering the energies required to form the intermediates of a step-wise reaction.[31]

The hydrolysis of D-glucono-δ-lactone (**17**) is subject to both intermolecular general-acid and general-base catalysis with Brønsted exponents α and β of about 0.4. The authors[32] prefer a cyclic concerted mechanism similar to the hydration of acetaldehyde and the mutarotation of glucose.

The reaction between aliphatic amines and 1-(*p*-nitrophenyl)azetidin-2-one (**18**) is general-base catalysed, and hydroxide ion catalysis becomes predominant at high pH. The reaction of piperidine with the β-lactam (**18**) is at least 10^3 times that with the corresponding γ-lactam. The Brønsted coefficient of 1.05 for the *uncatalysed* reactions of (**18**) with alkylamines is indicative of a transition state of intermediate type (**19**).

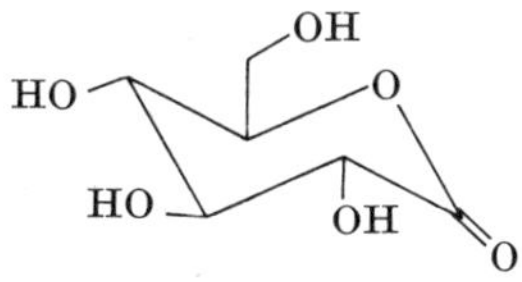

(**17**) 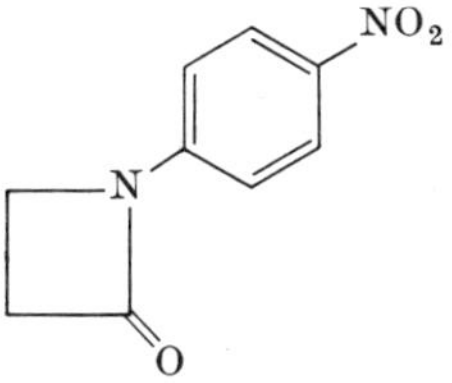 (**18**) 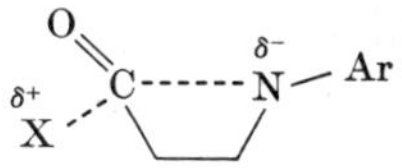 (**19**)

This is unexpected for a poor leaving group and is probably the result of release of strain in the β-lactam ring.[33] The hydrolysis of the β-lactam in 6-aminopenicillanic acid is general-acid and general-base catalysed.[34]

The hydrolysis of a variety of isoimides to the amide-acid is catalysed by hydronium and hydroxide ions and general bases. For the last reaction the Brønsted β-value is 0.47, and the detection of intermediates indicates a nucleophilic pathway for some of the bases.[35]

The acid-catalysed hydrolyses of orthoesters, $RC(OEt)_2OCH_2CCl_3$ in 65/35 w/w dioxan–water show solvent isotope effects $k_{D_3O^+}/k_{H_3O^+}$ of 0.98–1.83. General-acid catalysis is observed for R = phenyl in 2:1 v/v DMSO–water. Variation of R yields a ρ^*-value of −2.2, which is much less than that observed in acetal hydrolysis (−3.8).[36]

Rate constants for the reactions of 1-acetoxy-4-methoxypyridinium perchlorate, acetylpyridinium ions and 2,4-dinitrophenyl acetate[37] show a good unit-slope correlation with N^+ values which are obtained from reactions of nucleophiles with cations and are believed to be a measure of the energy required to desolvate the nucleophile.[38]

The catalytic constants for decomposition of benzisoxazoles catalysed by tertiary amines depend linearly on the pK's of the amine and phenolic product. Despite a total variation in rate of 10^{11} there is no evidence of non-linearity in the Brønsted plot. It is suggested that most proton-transfer reactions have transition states characterized by a sensitivity to substituent change that is invariant over appreciable changes in transition-state energy. This argument has been generalized to other reactions, and the usually assumed selectivity–reactivity relationship has been seriously questioned.[39]

The generally accepted significance of Hammett ρ values is contrary to the Hammond postulate, and such interpretations of *linear* free-energy relations are, by definition, incompatible with structure–reactivity correlations invoking variable transition states.[40] The requirements for a linear dependence of the Brønsted exponent α upon the energy of reaction has been evaluated and the relationship with Marcus rate theory has been discussed.[41] Other aspects of linear free-energy relationships are reported in ref. 42.

Activation parameters for the alkaline hydrolysis of *o*-substituted benzoylcholine esters have been interpreted on the basis of the Hepler isoergonic relationship.[43]

The α-effect has been reviewed[44] and suggested to be due to a disappearing lone-pair[45] and the involvement of aromatic-like transition-states.[46]

The hydrolysis of benzoyl chloride catalysed by tertiary amines has been investigated.[47]

The methanolysis of a series of piperidin-1-yl benzoates (**20**) involves cleavage of the N—O bond to generate a nitrenium ion (**21**) and shows a ρ-value of 0.68, which may be

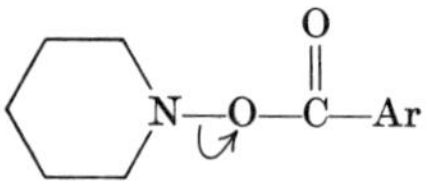

(**20**) 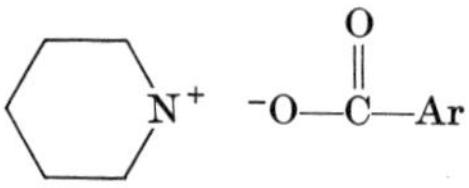 (**21**)

compared with the value of 1.34 for the methanolysis of 1-phenylcyclohexyl benzoates. This indicates that the transition state for cleavage of an N—O bond to generate nitrenium ions occurs earlier than that for C—O bond cleavage to form a tertiary carbonium ion.[48]

The values of $k_{^{16}O}/k_{^{18}O}$ for the alkaline-, acid- and general-base catalysed hydrolysis of methyl formate having ^{18}O in the methoxyl group are 1.0091, 1.0009 and 1.0115, respectively, indicating early transition states involving little C—OMe bond loosening in the rate-limiting step. The kinetic isotope effects for the hydrazinolysis reaction are 1.0621 at low pH and 1.0048 at high pH, compatible with rate-limiting breakdown and formation, respectively, of the tetrahedral intermediate.[48a]

The alkaline hydrolysis of a series of methyl pseudo-8-acyl-1-naphthoates (**22**) has a small dependence upon the nature of R. Hydrolysis in ^{18}O-enriched water results in incorporation of label in both the keto-carbonyl and the carboxyl group. This is consistent with rate-limiting attack on the carbonyl group as in (**22**).[49]

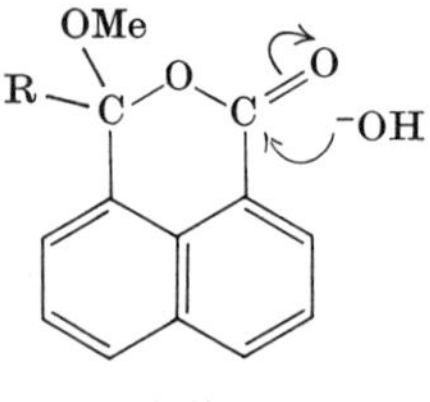

(**22**)

The hydrolysis of 3-methoxy 3-substituted phthalides probably proceeds via an A1 mechanism.[50]

The alkaline hydrolysis of alkyl *N*-phenylcarbamates has a Brønsted β-value of –1.15, whereas that for the *N*-methyl-*N*-phenylcarbamates is –0.25. An *E*1c*B* mechanism (equation 1), similar to that suggested for phenolic leaving groups,[51] was suggested.

$$HO^- + PhNHCOX \rightleftharpoons H_2O + Ph\bar{N}COX \longrightarrow Ph{-}N{=}C{=}O + X^- \quad (1)$$

$$\downarrow H_2O$$

Products

It was estimated that the change-over from elimination *E*1*cB* to substitution occurs when the pK_a of the leaving group exceeds ca. 17.[52]

The alkaline hydrolysis of aryl *N*-carbamates has a ρ-value for the *O*-aryl ring of 3.17[51] and the value for substituents in the *N*-aryl ring is 0.64, indicating an *E*1*cB* mechanism which is supported by trapping of the isocyanate intermediate.[53]

Other work on carbamate hydrolysis has also been reported,[54] and the hydrolysis of thiocarbamates has been investigated.[55]

Quaternary ammonium salts are formed as intermediates in the triethylenediamine-catalysed aminolysis of aryl *N*-arylcarbamates with aniline in dioxan.[56]

The uncatalysed and amine-catalysed elimination of HCl from *N*-arylcarbamoyl chlorides to give aryl isocyanates has been investigated.[57] The position of bond cleavage in the reaction of nucleophiles with 1-(chloroalkyl)carbamoyl chlorides depends on the nucleophile.[58]

The effect of substituents on the acid-catalysed hydrolysis of aliphatic hydroxamic acids has been studied.[59]

Protonation equilibria of benzamides and phenylureas follow the H_A acidity function,[60] and those for anilides and benzanilides have ρ-values of 1.35 and 1.63, respectively.[61]

The rate constants for the acid-catalysed hydrolysis of 4-chlorobenzamides do not correlate with the Bunnett or Bunnett–Olsen criteria. It is claimed, on the basis of a correlation with an empirical two-parameter rate equation, that hydrolysis proceeds by two distinct pathways, possibly involving *O*- and *N*-protonated transition states.[62] This equation has also been used to suggest an *A*1 mechanism for the acid-catalysed hydrolysis of phenylureas.[63]

The acid-catalysed hydrolysis of acetylglycine and glycyltyrosine occurs by an *A*2 mechanism based on w, w^* and ϕ values.[64]

The acid-catalysed hydrolysis of *N*-acetylsulphanilic acid changes from *A*2 to *A*1 in concentrated sulphuric acid, as indicated by Bunnett and Bunnett–Olsen criteria.[65] In >80% w/w H_2SO_4 acetanilide undergoes sulphonation in preference to hydrolysis; the observation of a rate minimum as a function of H_2SO_4 concentration is due to sulphonation rather than a change in mechanism from *A*2 to *A*1.[66]

Other work on the rates of hydrolysis of amides as a function of acid concentration has been described,[67] and the basicity of anilides determined.[68]

The position of bond cleavage in the neutral and base-catalysed hydrolyses of *N*-acylphthalimides as a function of the acyl group has been determined.[69]

It is claimed, on the basis of Yates' r and Bunnett–Olsen's ϕ parameters, entropies of activation, ρ–σ and ρ–σ^+ relationships, and steric effects, that the acid-catalysed hydrolysis of 1-acyl-2-thiohydantoins changes from an *A*2 to an *A*1 mechanism in strongly acidic solutions. Since the pK_a's of these compounds are ca. 7 and the rates of hydrolysis between pH 11 and 14 are of the first-order in hydroxide ion, the latter probably attacks the conjugate base. Secondary deuterium isotope effects for these reactions as a function of temperature were also determined.[70]

Reactions in Aprotic Solvents

The benzoylation of *m*-chloroaniline by benzoyl chloride in benzene proceeds by three pathways: uncatalysed, and catalysed by *N*-methylimidazolium chloride and by *N*-methylimidazole. The last compound is suggested to act as a nucleophilic catalyst and is more than 100-fold more effective than pyridine as a catalyst.[71]

The ρ-value for the second-order reaction of substituted anilines with 3-thenoyl chloride in benzene is –3.21[72] and that for substituted 2-thenoyl chlorides with aniline is 1.79.[73] Activation parameters for these reactions have been determined.[74]

Other articles on the acylation of amines with acyl halides are listed in ref. 75.

The ρ-values for formation of sulphamides from toluene-*p*-sulphonyl bromide and *m*-chloroaniline catalysed by substituted pyridines and pyridine *N*-oxides are –3.33 and –2.37, respectively. The mechanism is believed to involve nucleophilic catalysis.[76]

The acylation of ethylene oxide with benzoyl chloride in aprotic solvents gives 2-chloroethyl benzoate and is catalysed by acids and bases,[77] and the kinetics of benzoylation of substituted phenols in the presence of triethylamine has also been investigated.[77a]

The pyridine-catalysed acylation of substituted phenols by acetic anhydride in carbon tetrachloride has a ρ-value of 1.52. Substituted pyridines yielded a Brønsted β-value of 0.9, and it was claimed these observations indicate general-base rather than nucleophilic catalysis by pyridine.[78]

Investigations of the kinetics, activation parameters and ρ-values have been reported for various acylation reactions of amines with acid anhydrides.[79]

Cyclic transition states are suggested to occur in the acylation reactions of arylamines with both acyl halides and acid anhydrides.[80]

Tetra-*n*-hexylammonium benzoate hemihydrate is a 10^3 times better catalyst than piperidine for the reaction between piperidine and *p*-nitrophenyl acetate in toluene, despite a 10^7 lower basicity in water.[12]

The rate of acylation of aniline by phenyl thioacetate and its 2,4-dinitro-derivative in benzene is accelerated more by acetic acid than by 3,4-dinitrophenol. This is attributed to bifunctional catalysis by acetic acid.[81]

Other work on acylation reactions is cited in ref. 82.

The free-radical mechanism of the reaction of picoline *N*-oxides with carboxylic acid anhydrides has been further investigated.[83]

The kinetics and the catalytic effect of carboxylic acids on peptide linkage formation in non-aqueous solvents have been investigated.[83a]

Intramolecular Catalysis and Neighbouring-group Participation

The entropic and potential-energy contributions to intramolecular reactions have been reviewed.[84] Orbital steering has received further criticism and it has been estimated[85] from distance distribution functions that the maximum acceleration due to proximity and orientation in intramolecular reactions is 5×10^7 mol l^{-1}, in agreement with entropy arguments[86] which have been extended to aqueous solutions.[87] The experimental evidence of the rates of lactonization of various hydroxy-carboxylic acids used to form the basis of orbital steering[88] is questionable since the structures of some of the acids were incorrectly stated.[89] It is claimed that entropy calculations support orbital steering[90] and that thermodynamic parameters of reactions should be different in the gas and the liquid phase.[91]

The pH–rate profiles for the hydrolysis of phenyl *N*-(*o*-carboxyphenyl)carbamate (**23**) and its *N*-methyl derivative show a pH-independent region between pH 4 and 10. The presence of the *o*-carboxy-group enhances phenol release by ca. 10^6. The solvent isotope effect k_{H_2O}/k_{D_2O} is 1.2 and the intermediate anhydride (**24**) may be detected, indicating a nucleophilic mechanism rather than intramolecular general-base catalysis to form an isocyanate.[92]

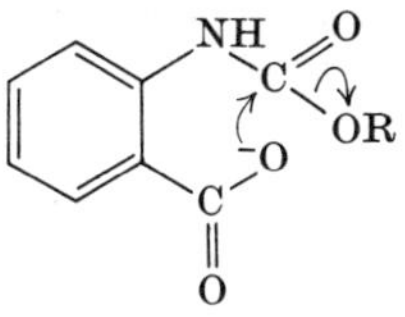

(**23**) 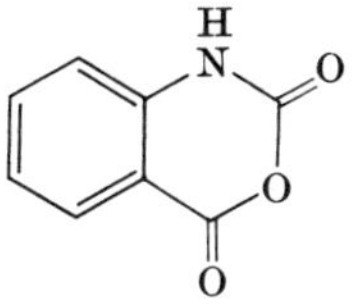 (**24**)

The first-order rate constant for the hydrolysis of (dichloroacetyl)salicylic acid anion is a linear function of the atom fraction of deuterium in mixtures of H_2O and D_2O. It is claimed that this is evidence for transfer of one proton in the transition state,[93] i.e. the generally accepted intramolecular general-base catalysed mechanism,[94] although such an interpretation may be criticized.[95]

The acid-catalysed hydrolysis of *N*-phenylphthalanilic acid proceeds via phthalic anhydride.[96]

The effect of ring substituents on the rates of proton transfer from arylethylmalonate monoanions to hydroxide ion yields a $\rho°$-value of 0.38.[97]

The mechanism of carbonyl group participation in the hydrolysis of esters has been further investigated. The hydroxide ion-catalysed hydrolyses of methyl 8-acyl-1-naphthoates (**25**) occur with participation of the carbonyl hydrate monoanions (**26**),

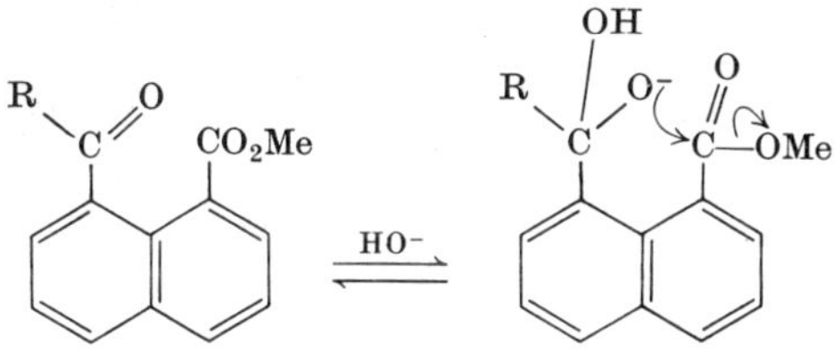

(**25**) (**26**)

as indicated by ^{18}O exchange and relative rates.[98] However, when the acyl group contains an ionizable hydrogen, as in (**27**), the dione is formed via the carbanion (**28**) and

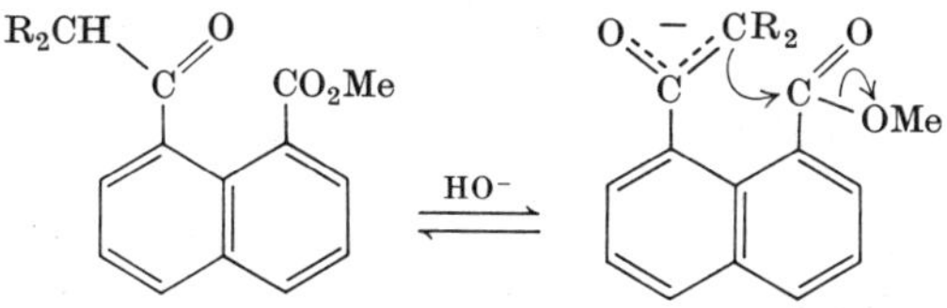

(**27**) (**28**)

rate-limiting base-catalysed enolization, $k_H/k_D = 5–6$.[99] This situation is in contrast with methyl *o*-acetylbenzoate which hydrolyses with *O*-participation.[100]

Reaction of methyl *o*-formylbenzoate with secondary amines gives the lactone (**29**) and variation of the amine basicity yields a Brønsted β-value of 0.33.[101]

Details of Capon's investigation of hydroxy-group participation in ester hydrolysis has been published.[102] The catalytic constant for the acetate-catalysed lactonization of phenyl 4-hydroxybutyrate is 50–100 times greater than that for the hydrolysis of phenyl acetate and may involve general-base catalysis (**30**). The hydrolyses of several other lactones were also studied.[102] The imidazole general-base catalysed lactonization of ethyl *o*-(hydroxymethyl)benzoate is also thought to proceed as in (**31**), on the basis of a solvent isotope effect of $k_{Im}^{H_2O}/k_{Im}^{D_2O}$ of 3.46 and a Brønsted β-value for various catalytic bases of 0.87.[103]

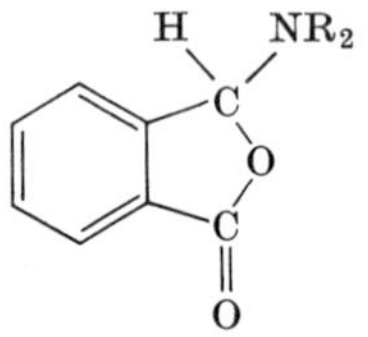

(**29**) 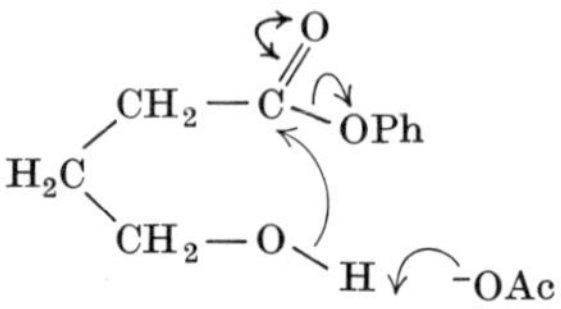 (**30**) 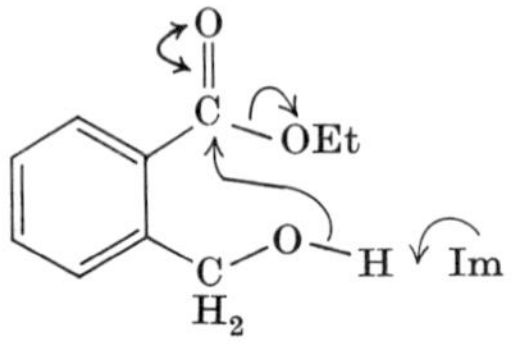 (**31**)

Nucleophilic participation of the phenoxide group is evident in the cyclization of phenyl *N*-(*o*-hydroxyphenyl)carbamate (**32**; R = H) and its *N*-methyl derivative which proceed at least 50 times faster than the *E*1*cB* hydrolysis of phenyl *N*-phenylcarbamate at pH 10.[104] The rates of hydroxide-ion catalysed ring closure of ethyl and *p*-nitrophenyl esters (**33**) of 2-hydroxymethyl-*N*-methylcarbanilic acid are ca. 10^5 greater than those for the hydrolysis of the corresponding compounds without the hydroxymethyl group, suggesting nucleophilic participation as in (**33**). The *p*-nitrophenyl ester of the non-methylated derivative cyclizes via the isocyanate.[105]

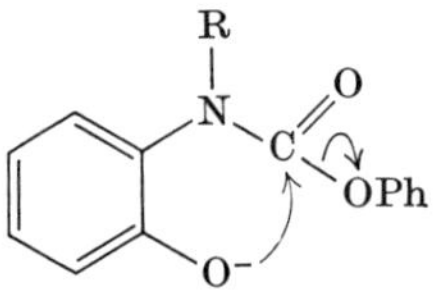

(**32**) 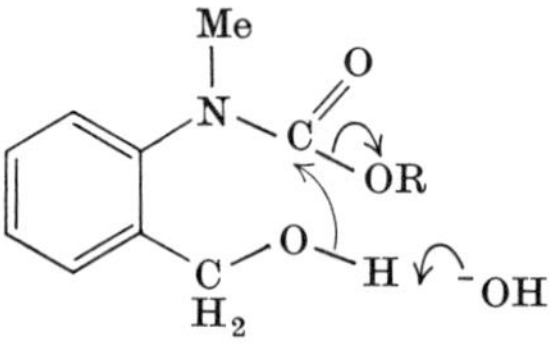 (**33**)

The lactonization of (*o*-hydroxyphenyl)acetamide and 3-(*o*-hydroxyphenyl)propionamide is general-acid catalysed and exhibits a Brønsted α-value of 0.15. The pH–rate profile shows a pH-independent region between pH 4 and 8, and at pH 6 the rates of lactonization are ca. 10^5 times faster than the rate of combined specific-acid and base-catalysed hydrolysis of phenylacetamide. It is suggested that the rate-limiting step is breakdown of the tetrahedral intermediate.[106]

The lactonization of *o*-(hydroxymethyl)benzoic acid in water is general-acid and general-base catalysed,[107] confirming earlier results.[108]

The pH–rate profile for lactonization of substituted coumarinic acids (**34**) depends markedly on the nature of the substituent. The results are interpreted in terms of Scheme 4. At low pH the formation of the intermediates via acid-catalysed and uncatalysed pathways is rate-limiting. At higher pH there is a change in rate-limiting step to breakdown of the intermediates by acid-, water- and base-catalysis. The pH-values at

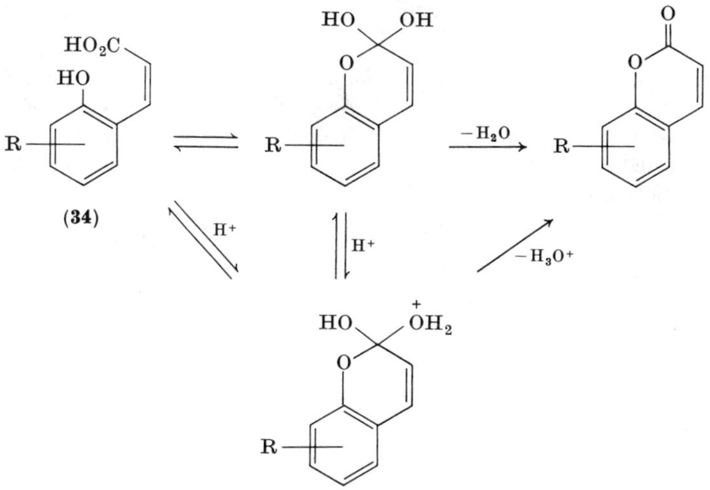

SCHEME 4.

which the change in rate-limiting step takes place and the partitioning of the intermediates between reactants and products are strongly influenced by substituents. A comparison with the lactonization of saturated phenolic acids was made.[109]

Details of the enormous rate increases in the lactonization of 3-(*o*-hydroxyphenyl)-propionic acid brought about by alkyl substitution in the aromatic ring and in the side chain have been published.[110] This effect has also been investigated in intramolecular displacement and addition reactions.[111] There appears to be some relief of non-bonded interaction upon lactonization, as indicated by the X-ray-determined structure of the methyl-substituted lactone and an analogue of the corresponding hydroxy-acid.[112]

The reaction of (tertiary-amino)-alcohols ($R^1R^2NCH_2CH_2OH$) with phenyl and *p*-nitrophenyl esters results in *O*-acylation. Direct nucleophilic attack by the hydroxyl group is favoured over *N*-attack followed by a rapid N → O shift. It was suggested that the amino-nitrogen acts as a general-base, removing the hydroxyl hydrogen. In some systems, catalysed hydrolysis is observed since the amino-ester formed undergoes deacylation.[113]

The adduct (**35**) from the reaction of cyanide-ion and pyridoxal is unstable, being converted at pH 6 into the cyclic imine (**36**); presumably this involves neighbouring-group participation of the phenolic hydroxyl group.[114]

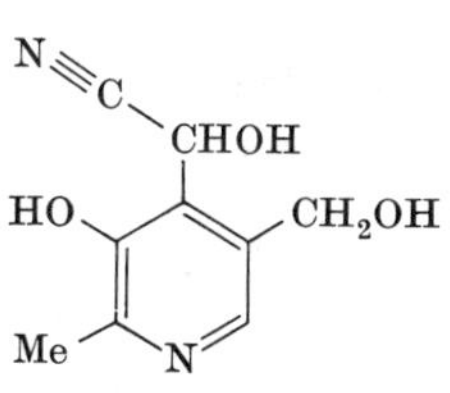

(**35**) 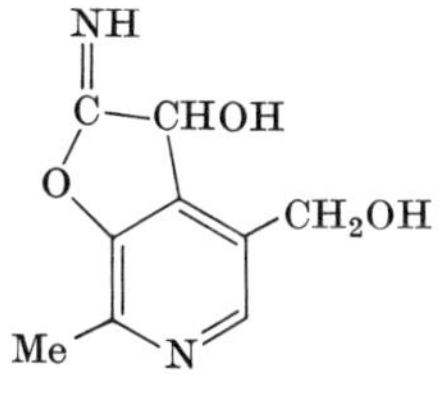 (**36**)

Other papers on the participation of hydroxyl groups are listed in ref. 115.

The rates of aminolysis of acetylimidazole (**37**) by diamines are 20–200 times faster than those by amines of comparable basicity. This is interpreted as evidence for intramolecular general-base catalysis by the second nitrogen atom in these diamines. Comparison of the second-order rate constant for diamines with the third-order rate constant for intermolecular catalysis of aminolysis gives a low effective molarity of ca. 1M. The rate acceleration has a very small dependence upon the structure of the diamine. These observations indicate a loose transition-state for the proton transfer (**38**), perhaps involving rate-limiting internal rotation of the diamine chain.[116]

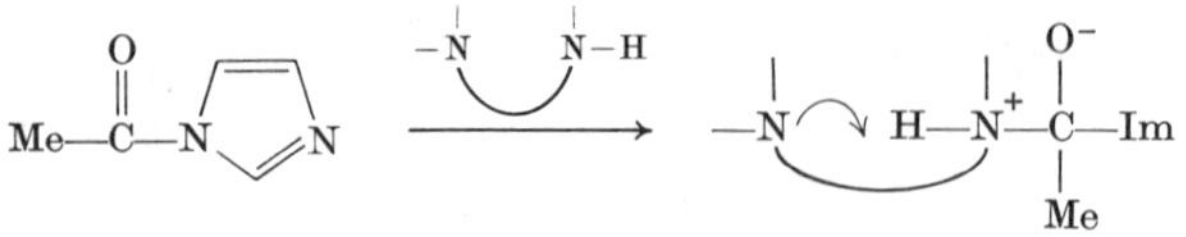

(**37**) (**38**)

The cyclization of methyl esters of various dipeptides to piperazine-2,5-diones is self-catalysed, and other amines also act as catalysts.[117]

Intramolecular nucleophilic attack of a sidechain amino-group upon the β-lactam carbonyl group in 6-epiampicillin occurs to give a diketopiperazine.[118]

The kinetics of cyclization of $H_2NCH_2CHRCH_2SC(:NH)NH_2 \cdot HBr$ has been investigated.[119]

Other reactions involving neighbouring-group participation are listed in ref. 120.

Association-prefaced Catalysis

The aggregation of *p*-nitrophenyl alkanoates occurs at extremely low concentrations in aqueous solution.[121,122] The dodecanoate ester is monomeric only at concentrations below 10^{-7}M, and addition of 10% methanol causes only a small increase in the critical micelle concentration. The evaluated rate constants for the hydrolysis of the longer-chain esters are larger than those previously determined, presumably because the earlier work was done under conditions where the esters are aggregated.[121] A water-soluble steroidal imidazole catalyses the hydrolysis of esters and the rates correlate with the Hansch hydrophobic π-value for the acyl substituent. Although rate enhancements of only 10-fold were observed it was claimed that this is an unambiguous example of favourable hydrophobic interactions.[123]

The third-order rate constant for the aminolysis of *p*-nitrophenyl dodecanoate by dodecylamine catalysed by another molecule of the amine is ca. 10^7 times greater than that for the ethylamine-catalysed reaction between ethylamine and *p*-nitrophenyl acetate. The rate enhancement of similar reactions depends on the chain lengths of both the amine and the ester, which is attributed to hydrophobic interaction although, peculiarly, there is no rate enhancement of the *uncatalysed* reaction between long-chain amines and esters.[124]

"Saturation" kinetics are observed in the poly-(4-vinylimidazole)-catalysed hydrolysis of long-chain fatty acid esters. Values of k_{cat} for this process are 2–1000 times greater than for the imidazole-catalysed reaction and the rate enhancement is chain-length- and solvent-dependent.[125] Deviations from first-order behaviour are observed such that the apparent rate constant increases with time, and this is attributed to a long-lived acyl intermediate that is more hydrophobic than the initial catalyst.[126]

The cyclodecapeptide (His–Glu–Cys–D-Phe–Gly)$_2$-catalysed hydrolysis of *p*-nitrophenyl acetate obeys Michaelis–Menten kinetics, shows a pH optimum at 7.6 and is a three-fold better catalyst than the linear decapeptide.[127]

Hydrolyses of diaryl carbonates and diaryl methylphosphonates are catalysed by cycloamyloses, and Michaelis–Menten kinetics are observed. It has been suggested that the rate-limiting step is nucleophilic attack by an ionized hydroxy-group in the cycloamylose; this is followed by attack of a second hydroxy-group to form a cyclic cycloamylose carbonate or methylphosphonate as shown in Scheme 5.[128]

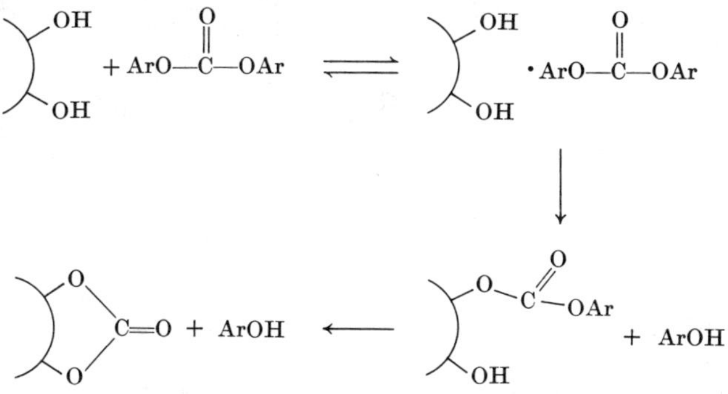

$$\text{(OH)}_2 + \text{ArO—C(=O)—OAr} \rightleftharpoons \text{(OH)}_2\cdot\text{ArO—C(=O)—OAr} \rightarrow \text{(OH)O—C(=O)—OAr} + \text{ArOH} \rightarrow \text{(O)}_2\text{C=O} + \text{ArOH}$$

SCHEME 5.

Isomerization of 2-(hydroxymethyl)-4-nitrophenyl trimethylacetate to 2-hydroxy-5-nitrobenzyl trimethylacetate is accelerated six-fold by cyclohexa-amylose and decelerated five-fold by cyclohepta-amylose.[129]

For alkaline hydrolysis of *p*-nitrophenyl esters of carboxylic acids in the presence of polyelectrolytes, results have been interpreted in terms of electrostatic and hydrophobic interactions.[130]

The problem of pH in micellar-catalysed reactions has been discussed.[131]

Various organic solutes exert a specific effect upon the structure of cetyltrimethylammonium bromide micelles.[132]

The imidazole-catalysed hydrolysis of *p*-nitrophenyl acetate in water pools dissolved in octane ("micellar water"), which is solubilized by di-(2-ethylhexyl) sodium sulphosuccinate, has been investigated.[133]

Metal-ion Catalysis

The aminolysis of the Co(III)-chelated amino-acid ester (**39**) by glycine ethyl ester in dimethyl sulphoxide occurs in two steps. Formation of the tetrahedral intermediate (**40**) is an uncatalysed pathway, but its breakdown is catalysed by another molecule of amine. Quenching and IR studies also indicate the presence of a chelated tetrahedral intermediate (**40**).[134]

The hydroxide ion-catalysed hydrolysis of glycylglycine methyl ester occurs 10^3 times faster in the presence of Cu(II). There is no metal-ion catalysis when the hydrogen on the

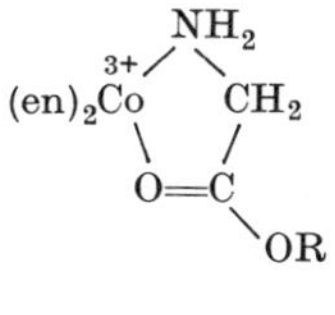

$$(\text{en})_2\text{Co}^{3+}(\text{NH}_2\text{CH}_2\text{C(OR)=O})$$

(**39**)

$$(\text{en})_2\text{Co}^{2+}(\text{NH}_2\text{CH}_2\text{C(OR)(O—)}\overset{+}{\text{N}}\text{H}_2\text{CH}_2\text{CO}_2\text{Et})$$

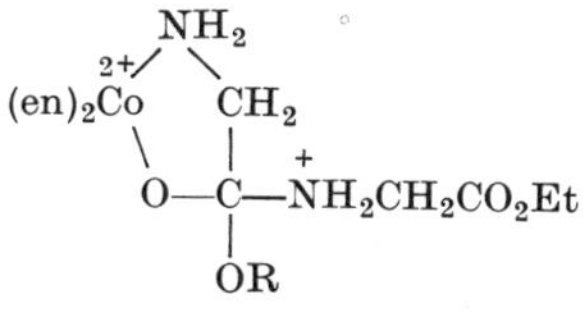

(**40**)

amide-nitrogen is replaced by methyl. It is suggested that the amide-hydrogen in the dipeptide ester is ionized and that the conjugate base co-ordinates with the metal ion, facilitating co-ordination of the ester group (see **41**).[135]

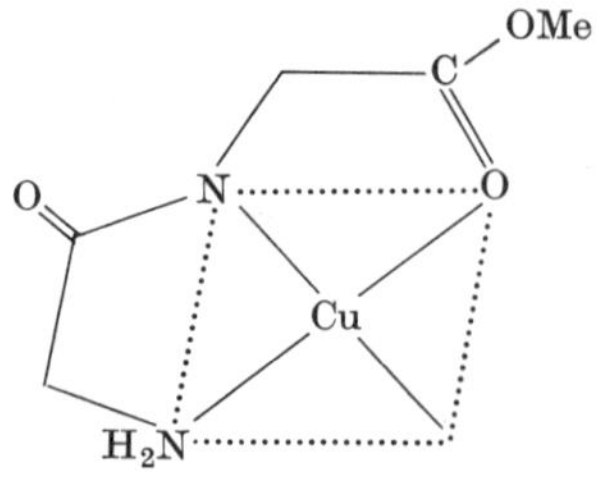

(41)

Hydroxide ion-catalysed hydrolysis of the ester function in bis-complexes of methyl DL-2,3-diaminopropionate with Cu(II) and Hg(II) proceeds 10^2–10^3 times faster than that of the free ester. This relatively small increase is taken to indicate little direct interaction between the carbonyl group and the metal ion.[136]

The rate of base-catalysed hydrolysis of acetonitrile to acetamide is increased by a factor of ca. 10^6 when the nitrile is co-ordinated to pentaaminecobalt(III). Direct attack of hydroxide ion on the nitrile-carbon atom to give the acetamido-complex, as shown in equation (2), is the suggested mechanism. During hydrolysis, exchange of about half of

$$(H_3N)_5\overset{3+}{Co}N{\equiv}C{-}Me + HO^- \longrightarrow (H_3N)_5\overset{2+}{Co}{-}NH{-}\overset{\overset{O}{\|}}{C}{-}Me \tag{2}$$

the methyl protons occurs, but it is not known if this is relevant to the hydrolysis mechanism.[137]

O-Acylation of esters by triethanolamine (TEA) is catalysed by bivalent metal ions with rate enhancements of 10^2–10^3. The predominant active species is the $Hg(TEA)_2$ complex, the metal ion being chelated by the amine-nitrogen atom and the alkoxide anion as in (**42a**).[138]

Peptide-bond formation from amino-acid esters in anhydrous alcoholic solvents is catalysed by $CuCl_2$. It is claimed that the mechanism involves the co-ordination of the amino-anion to Cu(II) and that this species attacks a non-activated carbonyl group of another ester molecule co-ordinated to the same Cu(II) ion (cf. **42b**).[139]

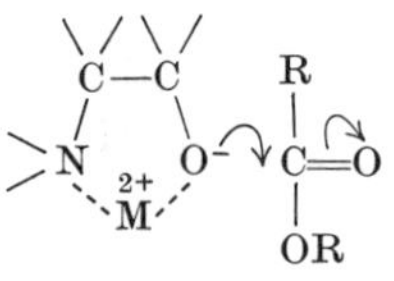

(42a) 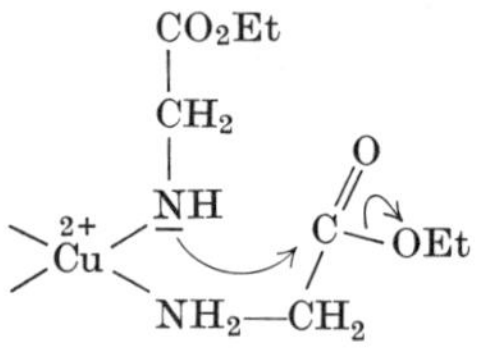 **(42b)**

The catalytic activity of a (poly-L-lysine)–Cu(II) complex in ester hydrolysis and the structure of hemin—(poly-L-lysine) as a haeme protein model has been investigated.[140]

Hydrolysis of diethyl L-aspartate and various peptides is promoted by Co(III).[141]

The rate of aquation of the tris(oxalato)chromate(III) anion is increased more than

10^6 times by surfactant-solubilized water in benzene, compared with the rate in water.[142]

The hydroxide ion-catalysed interconversion of mono- to bis-(triglycinato)cuprate(II) shows a preference for *cis*- over *trans*-geometry.[143]

The following reactions have also been studied: transition-metal-catalysed isomerization of β-lactones to α,β-unsaturated acids,[144] zinc octanoate-catalysed[145] and copper-chelate-catalysed[146] urethane formation, carbonyl metal-catalysed reaction of ethers and acid halides,[147] carbonyl metal-catalysed formation of imides from anhydrides and isocyanates,[148] copper-catalysed substitutions of aryl and vinyl halides by phthalimide ion,[149] reaction between acetylferrocene and triethyl orthoformate,[150] palladium-catalysed exchange of allylic groups of esters with carboxylic anhydrides[151] and alkylation of acid halides by alkylrhodium(I) complexes.[152]

Enzymic Catalysis[153,154]

The importance of binding energy and its consequences in enzymic catalysis have been examined.[155,156] It is claimed that the efficiency of enzymic catalysis is a result of the enzyme's effect on the time required for coupling of vibrational motions.[157]

Serine Proteinases

An anomalous pH-dependence of k_{cat}/K_m in enzyme reactions may occur if, at the optimum pH for the reaction, the rate-limiting process in this term is the association of enzyme and substrate. As the pH is changed, chemical steps may become rate-limiting and hence cause changes in the slow step with pH. This is suggested to be the case for the δ-chymotrypsin-catalysed hydrolysis of acetyl-L-tryptophan *p*-nitrophenyl ester.[158]

It has been suggested that anilide substrates of α-chymotrypsin may bind in a non-productive mode with the anilide ring in the substrate binding pocket.[159] Hydrophobic substituents such as Cl and NO_2 groups cause non-productive binding to be 50–100 times stronger than productive binding, and the tight binding of these substrates is not due to the accumulation of a covalent intermediate. Similarly, the lowered k_{cat} values are not due to electronic effects on the chemical reaction but are a consequence of non-productive binding. Both productive and non-productive binding correlate with Hansch's hydrophobic π-constants. There is no simple relationship between σ^- and the observed values of k_{cat} and K_m and there is now no experimental evidence of pre-transition-state protonation.[160] Spectroscopic studies revealed no accumulation of a tetrahedral intermediate as had been suggested for anilide hydrolysis.[161] When competition between productive and non-productive binding is pH-dependent the pH-dependence of k_{cat} is controlled by a pK_a different from that of the catalytic group in the productive complex.[159]

The variation of k_{cat}, K_m and k_{cat}/K_m with pH for the α-chymotrypsin-catalysed hydrolysis of *N*-formyl-L-phenylalanine hydrazide shows their dependence upon ionizations with pK_a values of 6.0, 6.1 and 6.7, respectively. The mechanism suggested, shown in Scheme 6, in which a tetrahedral intermediate (TI) is formed and accumulates

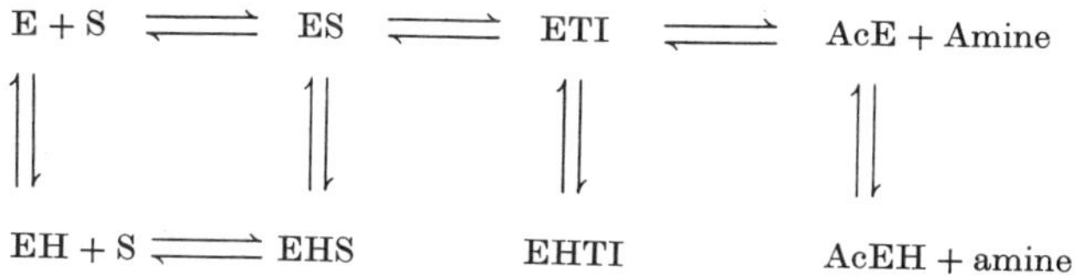

$$
\begin{array}{ccccccc}
E + S & \rightleftharpoons & ES & \rightleftharpoons & ETI & \rightleftharpoons & AcE + \text{Amine} \\
\upharpoonleft\downharpoonright & & \upharpoonleft\downharpoonright & & \upharpoonleft\downharpoonright & & \upharpoonleft\downharpoonright \\
EH + S & \rightleftharpoons & EHS & & EHTI & & AcEH + \text{amine}
\end{array}
$$

Scheme 6.

in a pre-equilibrium reaction before the acyl-enzyme (AcE) formation—the site of proton addition (H) is His-57—requires the equivalence of the pK_a values influencing k_{cat} and K_m.[162] These results do not agree with those reported previously,[163] but they do not exclude a mechanism involving general-acid catalysed breakdown of EHTI.[163] The reverse of this mechanism would involve, kinetically, the imidazolium ion-catalysed attack of formylhydrazine on a formylphenylalanyl enzyme, but there is no evidence of this pathway.[164] The partitioning of the acyl-enzyme between amine and water at high and low pH does not show an increased yield of amide at lower pH.[164] This also appears to be inconsistent with the previously proposed change in rate-limiting step.[163]

The ratio of AcPhe to AcPhe–AlaNH_2 produced from the α-chymotrypsin-catalysed hydrolysis of AcPhe–OMe and AcPhe-anilides in the presence of 1M-AlaNH_2 shows that the latter is 44 times more reactive than 55M water for both substrates. The decrease in k_{cat}/K_m for the hydrolysis of AcPhe–AlaNH_2 in the presence of AlaNH_2 is also accounted for by AlaNH_2 being 44 times more reactive. This suggests a common intermediate in the hydrolysis of ester, anilide, and peptide substrates which is the acyl-enzyme, as shown by the agreement between k_{cat}/K_m calculated for the hydrolysis of AcPhe–AlaNH_2 and the directly measured value.[165]

The rate constants for the attack of amines on various specific and non-specific acyl-chymotrypsins show that the acylation of chymotrypsin by amides and peptides involves a specificity of the leaving group.[166]

Details of Bender's investigation of α_1-chymotrypsin have been published, which suggest that deprotonation of Ala-149 is responsible for the inability of α-chymotrypsin to bind substrates at high pH.[167]

The rate of deacetylation of acetyl-α-chymotrypsin decreases linearly with the atom fraction of deuterium in mixtures of H_2O and D_2O. It is claimed that this is evidence for transfer of only *one* proton in the transition state,[168] thus being inconsistent with the "charge-relay" mechanism involving an additional proton transfer between Asp-102 and His-57.[169] Such an interpretation is, however, subject to criticism.[95]

The deacylation of acetyltyrosyl-α-chymotrypsin, a specific acyl-enzyme, by various amines is general-base catalysed. For substituted anilines a Brønsted β-value of 0.52 is obtained, although for non-aromatic amines the rate varied by a factor of only four over a basicity range of 10^8. Together with a solvent isotope effect k_{H_2O}/k_{D_2O} of ca. 3, this was interpreted in terms of a concerted proton-transfer in preference to a stepwise mechanism.[170]

It has been suggested that the deacylation of acyl-α-chymotrypsin formed from di-(*p*-nitrophenyl) carbonate occurs by the imidazole group of His-57 acting as both a general-base and a nucleophile, as shown by reaction products, pH-dependence and solvent isotope effects.[171]

The inhibition of α-chymotrypsin by long-chain anionic and cationic species is a slow, uncompetitive deactivation process which, it is suggested, is controlled by a hydrophobic site separate from the active site. Hydrophobic interaction at these two sites may account for the parabolic structure–reactivity relationship often found for a homologous series of substrates.[172]

The variation in the rates of the α-chymotrypsin- and elastase-catalysed hydrolysis of a series of normal fatty acid *p*-nitrophenyl esters with chain length shows up mainly in the enthalpy of activation. This was interpreted in terms of a conformational change in the enzyme–substrate complex or differences in solvation.[173]

The rates of acylation and deacylation of the α-chymotrypsin-catalysed hydrolysis of various acyl-substituted *p*-nitrophenyl esters have been analysed by using the Taft–

Ingold relationship to separate polar, steric and special effects. Deacylation is favoured when the acyl-enzyme has an acyl chain-length of 10–12 Å and by the hydrophobic character of the chain end.[174] With *p*-nitrophenyl esters of *N*-acylamino-acids as substrates, acylation specificity increases linearly with the Hansch hydrophobic π-value of the amino-acid side-chain.[174]

Alkyl isocyanates inhibit α-chymotrypsin and other serine proteinases with a degree of specificity.[175]

The kinetics of the reaction of a series of peptide chloromethyl ketones with chymotrypsin A have been interpreted in relation to the crystallographic model.[176]

At subzero temperatures, increasing DMSO concentration decreases k_{cat} but increases K_m for the α-chymotrypsin-catalysed hydrolysis of *N*-acetyl-L-tryptophan *p*-nitrophenyl ester. Acylation is still rapid at low temperatures and acyl-enzyme may be isolated.[177]

The effect of oxidation of Met-192 on the conformation of α-chymotrypsin has been studied.[178]

Other investigations of α-chymotrypsin include: changes in tertiary structure with changing pH,[179,180] the effect of sodium dodecyl sulphate on its circular dichroism,[181] and dimerization at low pH.[182]

The use of α-chymotrypsin for the selective removal of aromatic acyl groups has been further investigated.[183]

The work on the spin-labelled di-isopropyl fluorophosphate analogues of trypsin[184] has been re-examined and it is concluded that the enzyme is not irreversibly locked into a conformation dependent on its pH of inhibition.[185]

Bovine trypsin has been immobilized on an insoluble support in the presence and absence of a specific substrate, *N*-α-benzoyl-L-arginine ethyl ester. The enzyme is apparently "frozen" in a conformation which differs from that of the enzyme bound in the absence of substrate.[186]

Other papers on trypsin are cited in ref. 187.

Bacterial proteinase subtilisin catalyses the hydrolysis of the non-specific substrates, 2,2,2-trifluoro-*p*-nitroacetanilide and its *N*-methyl derivative, but not that of *p*-nitroacetanilide. Inhibited subtilisin experiments indicated a dependence on the active-site serine-hydroxyl group. The results are inconsistent with the pre-transition-state theory of chymotryptic anilide hydrolysis,[160] but are compatible with the intervention of a tetrahedral intermediate.[188]

Peptide chloromethyl ketones react with elastase by alkylation of a histidine residue. Peptide aldehydes bind to the enzyme 5000 times more tightly than analogous amide substrates, and the complexes are taken as transition-state analogues. Subsite, S_3, of elastase[189] cannot bind proline residues of bound peptides. k_{cat}/K_m for peptide amide hydrolysis increases with increasing chain-length and this is attributed to an increased rate of acylation resulting from a conformational change in the enzyme. Kinetic constants for elastase-catalysed hydrolysis of a series of tri- and tetra-peptides vary markedly with the nature of the group three amino-acid residues separated from, and *N*-terminal to, the bond undergoing cleavage.[190]

Thiol Proteinases

The rate of deacylation in the papain-catalysed hydrolysis of *N*-benzoyl-L-arginine *p*-nitroanilide is 25 times slower than that for the corresponding ethyl ester. It is suggested that the papain-catalysed hydrolysis of amides may not proceed via an acyl-enzyme intermediate.[191]

The pH–rate profile for the sulphonylation of the thiol group at the active site of papain by an aromatic six-membered sultone (**43**), to give an enzyme-thiosulphonate species (**44**), has been interpreted[192] in terms of general-base catalysis by a group in the enzyme with pK_a ca. 4. Intramolecular attack on the phenolic hydroxyl group in (**44**) regenerates the enzyme and substrate and is competitive with the hydrolysis of the intermediate (**44**).

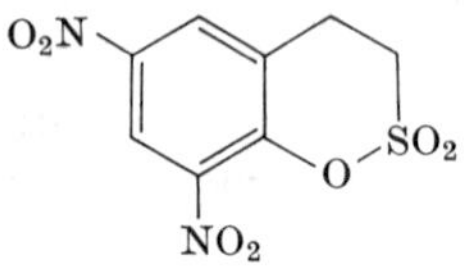

O_2N SO_2 O NO_2 O_2N SO_2—SCH_2 ~~Papain OH NO_2

(**43**) 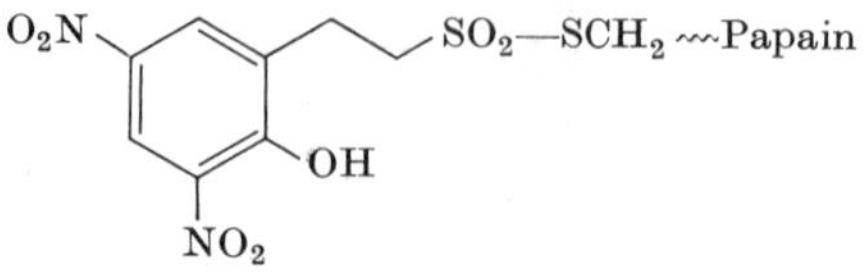 (**44**)

The pH–rate profile for the alkylation of papain by α-bromo-4-hydroxy-3-nitroacetophenone indicates the participation of two enzymic groups. Nucleophilic attack by the active-site mercapto group (p$K = 9.06$) is presumed to be assisted by a general-base (p$K = 3.08$).[193]

Proflavine is a non-competitive activator of papain.[194]

There is kinetic evidence for an acyl-enzyme intermediate in thiol streptococcal proteinase-catalysed reactions. The effect of pH and organic solvents has been determined. The solvent isotope effect k_{H_2O}/k_{D_2O} is 2–3. The kinetic properties have been compared with those of papain and ficin. Deacylation shows a dependence on a basic group of pK 4.1 but there is apparently no dependence of k_{cat}/K_m on a group of pK 8–9 (thiol group).[195]

Other work on thiol proteinases is reported in ref. 196.

Acid Proteinases.

The mechanism of action of pepsin has been reviewed.[197]

Metallo-proteinases

The inhibition constants of carboxylate anions in reactions catalysed by carboxypeptidase A are correlated by Hansch's hydrophobic π-values with the large slope of 1.77, which is interpreted in terms of an inhibitor-dependent conformational change.[198]

Other work on carboxypeptidase A is reported in ref. 199.

Zinc-free leucine aminopeptidase, with no enzyme activity, is re-activated upon addition of Zn^{2+}. The enzyme has two metal binding sites; one must be occupied by Zn^{2+} and the other, an activation site, may be occupied by Zn^{2+}, Mg^{2+} or Mn^{2+}. The nature of these ions effects mainly V_{max}.[200]

Esterases

Work on acetylcholinesterases[201] and butylcholinesterase[202] has been reported.

The kinetics of the beef-liver esterase-catalysed hydrolysis of many substrates have been determined. For a series of ethyl esters, K_m decreases with increasing acid chain-length, and k_{cat} has a maximum value with pentanoate. Alkyl acetates are better substrates as the alkyl moiety is elongated, but branching decreases reactivity. Esters of dicarboxylic acids are good substrates but only one ester group is cleaved.[203]

The hydrolysis of glycerol tripropionate by pancreatic lipase is enhanced by absorption on to siliconized glass beads, and this is ascribed to an increased local concentration of the substrate at the liquid–solid interface.[204a] The enzymic reaction of porcine pancreatic lipase with insoluble monolayers of glycerol trioctanoate has been determined. At pH 7.6 the second-order rate constant for the hydrolysis of the primary ester function is 8.5×10^5 mol^{-1} sec^{-1}.[204b]

Other Enzymes

Rate and inhibition constants for the binding of two sulphonamides to carbonic anhydrase are consistent with the ionized sulphonamides reacting with the form of the enzyme in which the water ligand bound to the active-site Zn(II) is un-ionized.[205]

Human and bovine carbonic anhydrases react with bromoacetazolamide at histidine-64.[206]

The carbonic anhydrase-catalysed hydrolysis of 3-pyridyl and nitro-3-pyridyl acetates has been studied[207] and solvent isotope effects have been determined for the reaction of the enzyme with phenyl *N*-methylacetimidate.[208]

The complete primary structure of carbonic anhydrase B has been determined.[209]

The temperature-dependent changes in spin state and the co-ordination geometry of Co(II) carbonic anhydrase have been examined.[210]

Other enzymes studied include choline acetyltransferase,[211] urea carboxylase,[212] acetyl-CoA carboxylase[213] and uricase.[214]

Decarboxylation*

The Schiff bases (**45**) and (**46**) have been synthesized; in non-polar solvents they undergo decarboxylation ca. 10^6 times as fast as do the corresponding β-keto-acids. This is consistent with earlier interpretations of amine-catalysed decarboxylations.[215]

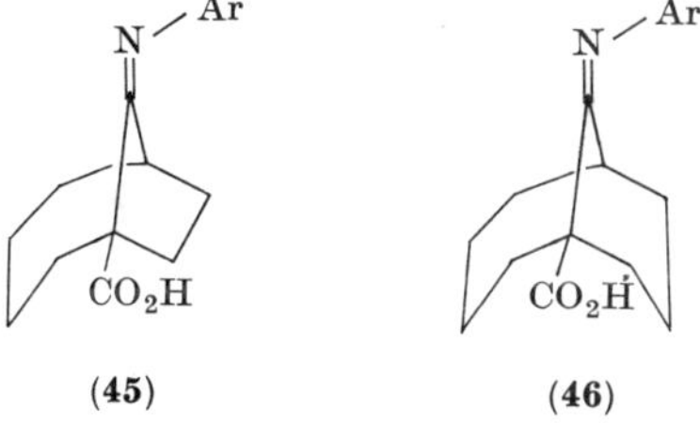

(**45**) (**46**)

Kinetic analysis of the aminoacetonitrile-catalysed decarboxylation of acetoacetic acid suggests that the reaction involves partial rate-limiting formation of the imine, which decarboxylates ca. 10^5 times faster than the acid itself.[216]

The rate of the *p*-*n*-butylaniline-catalysed decarboxylation of oxalacetic acid is increased 45-fold in the presence of cationic micelles of cetyltrimethylammonium bromide.[217]

The unimolecular decarboxylation of 6-nitrobenzisoxazole-3-carboxylate (**47**) is strongly catalysed by micelles of cationic and anionic surfactants. Micelle–solute interactions were studied by various physical techniques.[218]

The decarboxylations of substituted phenylcyanoacetate anions, benzoyl acetic acids and acetate anions are catalysed by cyclohepta-amylose. The acceleration is

* See also Chapter 7.

suggested to arise from solvation changes upon binding and, in the case of the acids, from loss of internal rotation.[219]

The mole fractions of neutral and zwitterionic forms of 2- and 4-pyridylacetic acid in aqueous propanol have been evaluated from macroscopic and model pK_a values. The calculated rate constants for decarboxylation of the zwitterions vary smoothly with solvent composition. This is claimed to indicate that decarboxylation occurs through the zwitterion (**48**) and is held also to be the mechanism of decarboxylation of β-keto-acids.[220]

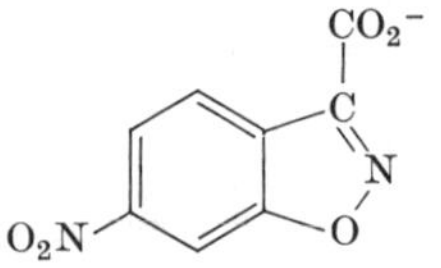

CO_2^- C N O O_2N $\overset{+}{N}$ CH_2 H ^-O C O

(**47**) 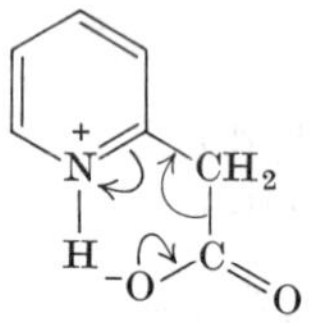 (**48**)

Oxidative decarboxylation of 2-hydroxy-carboxylic acids by lead tetra-acetate in anhydrous acetic acid follows second-order kinetics. A small solvent deuterium isotope effect, the entropy of activation, lack of sensitivity to solvent polarity and substituent effects, and unsuccessful radical-trapping experiments were taken to indicate a rate-limiting step involving the decomposition of a Pb(IV)–hydroxy acid cyclic intermediate.[221]

Other papers on decarboxylations are listed in ref. 222.

Other Reactions

Acyl transfer from the *S*-acetate of 4-mercaptoacetanilide to *n*-butylamine and imidazole has been studied.[223]

Hammett ρ-values have been obtained for the base-catalysed hydrolysis of phenyl benzoates as a function of temperature.[224]

There have been investigations of the base-catalysed hydrolysis of the following compounds: ethyl acetate,[225] alkyl benzoates,[226] γ-substituted propargyl benzoates,[227] substituted phenyl acetates and ethyl benzoates,[228] *p*-isopropenylphenyl acetate,[229] ethyl acrylates,[230] alkyl(2-methoxyethyl)malonic esters[231a] and α-chlorobenzyl esters.[231b]

The relative rates of the alkaline hydrolysis of the seven stereoisomers of esters of decahydro-1,2-dimethylquinolin-4-ol have been rationalized in terms of steric effects.[232]

The rate equations describing the alkaline hydrolysis of triethyl citrate have been solved and used to determine the rate constants in dioxan–water mixtures of various compositions.[233]

The kinetics of the alkaline hydrolysis of 32 cyclic diesters in 50% aqueous methanol have been determined.[234]

Substituent effects and activation parameters have been determined for the neutral hydrolysis of esters of trichloroacetic and trifluoroacetic acids.[235]

There have been several reports of the acid-catalysed hydrolysis of esters.[236]

Solvent effects in ester hydrolysis have been studied.[237]

The mechanism of ultrasonically induced hydrolysis of aliphatic esters is said to be identical to that of normal hydrolysis.[238]

Investigations of the reactions and hydrolysis of amides,[239a] substituted phthalimides[239b] and phenobarbitol[239c] have been reported.

The hydroxide ion-catalysed hydrolysis of benzoylglucose acylals is competitive with $O \rightarrow O$-benzoyl migration.[240] The acid-catalysed hydrolysis has also been studied.[241]

Rate constants and activation parameters have been determined for the degenerate migration of acyl and aroyls groups from O-1 to O-2 in *O*-substituted *cis*-enolates (**49**) of acetylacetone; the rates are correlated with σ^+ by a ρ-value of 0.81.[242]

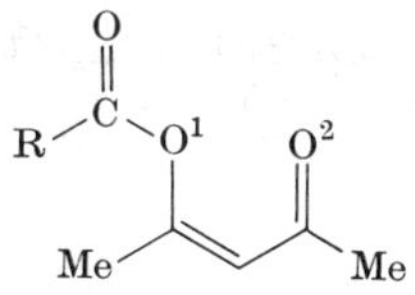

O
‖
R–C–O¹   O²
Me   Me

(**49**)

Papers dealing with other rearrangement reactions are listed in ref. 243.

Ring-opening reactions of lactones have been studied.[244]

The reactions of monomeric carbonic acid derivatives have been reviewed.[245]

The rates of decomposition of *p*-substituted *tert*-butylperoxy α-phenylisobutyrates in benzene correlate with σ^+, giving a ρ-value of -0.75.[246]

The hydrolysis of xanthates has been studied.[247]

The kinetics of the hydrolysis of urea in the presence of hydrazine and semicarbazide hydrochlorides[248] and of the reaction between iodine and thiourea[249] have been reported.

Other reactions of urea derivatives have also been investigated.[250]

The kinetics of various esterification reactions catalysed by acids[251] and by cation-exchange resins[252] have been studied.

The hydrolysis of acetic anhydride has been re-examined.[253]

β-Lactones may be formed in the Perkin reaction with quinones and the mechanism is thought to involve formation of a ketene from the carboxylic acid anhydride followed by addition of the ketene to the carbonyl group.[254]

The rates of the acid-catalysed hydrolysis of benzonitrile give a linear plot against H_0. Application of the Bunnett and Bunnett–Olsen criteria and inverse solvent deuterium isotope effects of k_{D^+}/k_{H^+} of 1.6–2 indicate a mechanism involving *N*-protonation followed by rate-limiting attack of water on carbon.[255]

Other reactions of nitriles have also been discussed.[256]

The rates of the reaction between imidazole and alkyl isocyanates have been determined.[257]

The solvolysis of 4,5-diphenylisosydnone has been studied.[258]

Bifunctional catalysis has been suggested to occur in the racemization of 2-phenylbutyric acid in trifluoroacetic acid–trifluoroacetic anhydride.[259] Other racemization studies are reported in ref. 260.

H-Exchange in amides has been investigated.[261]

Other reactions studied include: hydrolysis of 1-acyl-3,5-dimethylpyrazoles,[262] protonation of isoxazole derivatives,[263] reaction of aryl thiocyanates with *p*-nitrothiophenoxide,[264] iodolactonization of unsaturated acids,[265] reaction of cystine and 3-sulphoalanine with halogens,[266] ethyl acetate formation from an ether condensate of acetaldehyde,[267] reactions of alkenyl orthoesters with carboxylic acids,[268] dimerization of *o*-benzoylbenzoyl chlorides with iodide ion,[269] reduction of methyl acetate with trichlorosilane[270] and the formation and hydrolysis of methylthiomethyl esters.[271]

Solid-state transcarboxylations[272] and the reaction between aspirin and homatropine,[273] and the fusion reactions of chloropurines and 1,2,3,5-tetra-*O*-acetyl-D-ribofuranoses[274] have been investigated.

The following conformational studies have been made: esters by IR spectroscopy,[275] ^{13}C-labelled phenylsuccinic acid by NMR spectroscopy,[276] diacylsulphides, sulphonic anhydrides and thioanhydrides by dipole-moment measurements[277] and (–)-menthone lactam by the Cotton effect.[278]

NON-CARBOXYLIC ACIDS*

Phosphorus-containing Acids[279]

Non-enzymic Reactions

The hydrolysis of *O*-phenyl(carboxylatomethylamino)phosphoramidate (**50**) is pH-independent at pH > 6 and in this region occurs with P—O bond fission with a rate enhancement of ca. 10^4 over *O*-phenyl phosphoramidate, which hydrolyses by P—N bond cleavage. Together with a solvent isotope effect k_{H_2O}/k_{D_2O} of 1.2, this indicates nucleophilic participation by the neighbouring carboxy-group to give a cyclic acyl intermediate (**51**) which may be partially trapped with hydroxylamine. The intermediate (**51**) is hydrolysed by attack of water on phosphorus.[280]

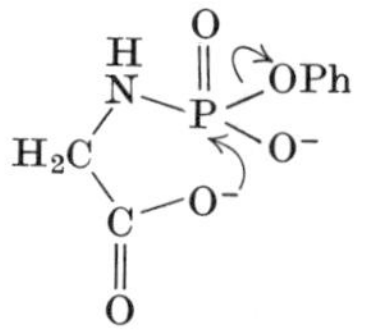

(**50**) 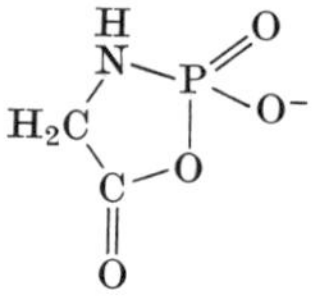 (**51**)

The phosphate diesters (**52**) and (**53**) are also hydrolysed with intramolecular nucleophilic participation. At pH > 6, exclusive expulsion of phenoxide occurs and the rate is pH-independent and ca. 10^7-fold greater than that of hydrolysis of diphenyl phosphate.[281] At saturating concentrations of Zn^{2+} ion the rates of hydrolysis of (**50**), (**52**) and (**53**) are increased 269-, 304- and 74-fold, respectively. The observed kinetics can be described in terms of a pre-equilibrium formation of a reactive 1:1 ester–metal ion complex. Rather than that simple electrostatic neutralization should promote carboxylate attack, it is suggested that the metal ion interacts with the pentacovalent intermediate (**54**) and may facilitate its breakdown by complex-formation with the leaving group.[282]

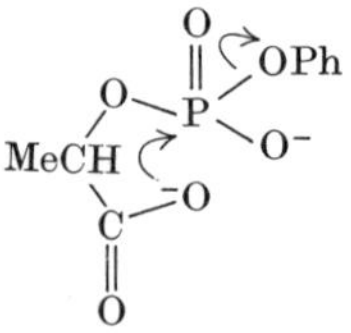

(**52**) 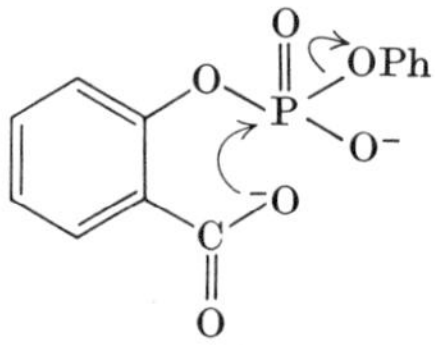 (**53**) 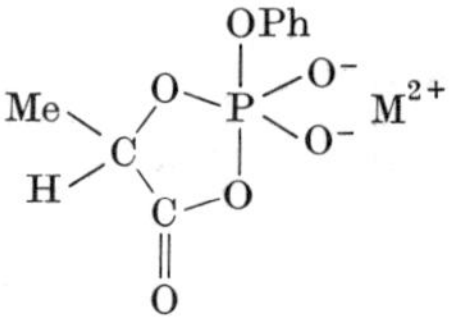 (**54**)

* See also Chapter 9.

It is suggested that the alkaline hydrolysis of α-hydroxyimino-*p*-nitrobenzyl phosphates, phosphonates and phosphinates (**55**) (X is alkyl or alkoxy) proceeds by intramolecular attack by the oximinate function on phosphorus, to give a pentacovalent intermediate. Fission of the latter and a subsequent Lossen rearrangement gives *p*-nitroaniline and the parent phosphorus acid. Unlike intermolecular nucleophilic attack, these reactions appear not to be subject to steric hindrance.[283]

The pH–rate profiles for hydrolysis of 2-pyridylethyl and 2-pyridylpropyl phosphate are bell-shaped with maxima at pH 2–3. This is attributed to intramolecular general-acid catalysis, shown in (**56**).[284] Further support comes from the deviations these compounds show in various linear free-energy relationships.[285]

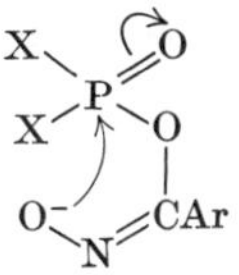

(**55**) 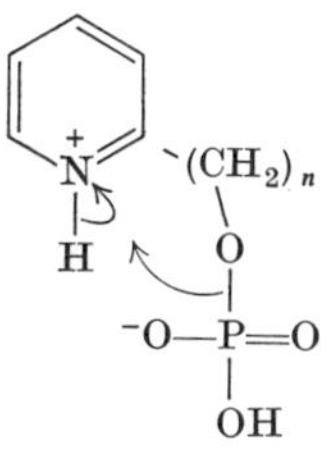 (**56**)

The hydrolysis of bis-(2,6-dimethoxyphenyl) phosphate in 50% aqueous acetic acid proceeds mainly via the anion, and the undissociated species is unreactive. This is the opposite of the behaviour usually shown by diesters.[286]

Acid-catalysed hydrolyses of diallyl phosphate,[287] tris-(2,6-dimethoxyphenyl) phosphate[288] and vinylic phosphates[289] have been investigated.

The reaction of diethyl 1-phenylvinyl phosphate with either HCl or bromine in aqueous dioxan proceeds with predominant C—O bond fission.[290]

The hydrolysis of glyceraldehyde dihydrogen phosphate is catalysed by aromatic amines[291] and various diamines.[292]

Micelles of (2-hydroxyethyl)-*N*,*N*-dimethylalkylammonium bromides are good catalysts for the hydrolysis of di-(*p*-nitrophenyl) phosphate and ethyl *p*-nitrophenyl phosphate monoanions in the presence of hydroxide ions. Rate enhancements are rationalized in terms of nucleophilic participation of the alkoxide ion of the micelle. Electrophilic catalysis is thought not to be important.[293]

Other reactions studied include: the polyamine-catalysed hydrolysis of adenosine 5′-triphosphate[294] and the decomposition of *p*-nitrophenyl phosphate in solutions containing polyethyleneimine or poly(benzyltrimethylammonium) chloride.[295]

The oxidative hydrolysis of *p*-hydroxyphenyl phosphate with periodic acid has been investigated.[296]

The following reactions have been studied: magnesium ion-catalysed hydrolysis of phenylphosphosulphate,[297a] the transfer of the terminal phosphate of adenosine triphosphate to water and inorganic phosphate,[297b] alkaline hydrolysis of trisubstituted phosphate esters,[298] reaction of amidophosphites and acetyl phosphites with acetic acid, alcohols and phenol,[299a] exchange of bound water in an Mn(II)–adenosine triphosphate complex[299b] and the hydrolysis of cyclophosphamide.[300a]

Details of the hydrolysis of phosphoroguanidines have been published and the results related to the phosphorylation of ADP by phosphocreatine.[300b]

The equilibrium constant for the hydrolysis of diethyl phosphorofluoridate[301] is

1.1×10^6. Heat capacities of activation and solvent isotope effects have been used to distinguish between S_N1 and S_N2 mechanisms for the hydrolysis of disubstituted alkyl phosphorochloridates,[302] and the reactions of phosphorus acid chlorides with nucleophiles have been studied.[303]

o-Substituted pentaphenoxyphosphoranes hydrolyse much more slowly than the corresponding *p*-derivatives (e.g. a factor of 10^3 for cresol derivatives). The reaction shows a Grunwald–Winstein *m*-value of only 0.3, no salt effect, and a solvent isotope effect k_{H_2O}/k_{D_2O} of 3. These observations indicate a displacement mechanism involving a hexaco-ordinated phosphorus intermediate or transition state (see equation 3).[304]

$$(ArO)_5P + H_2O \longrightarrow H^+ + (ArO)_5\bar{P}OH \longrightarrow (ArO)_3P{=}O + ArO^- + ArOH \quad (3)$$

The rate constants for hydrolysis of substituted phenyl *N*,*N'*-diphenylphosphorothiodiamidates (**57**) show a non-linear dependence upon hydroxide ion concentration. The mechanism shown in Scheme 7 was suggested. The elimination of phenol from the conjugate base of the ester (k_1) to form an unstable thiophosphorothioimidamide (**58**)

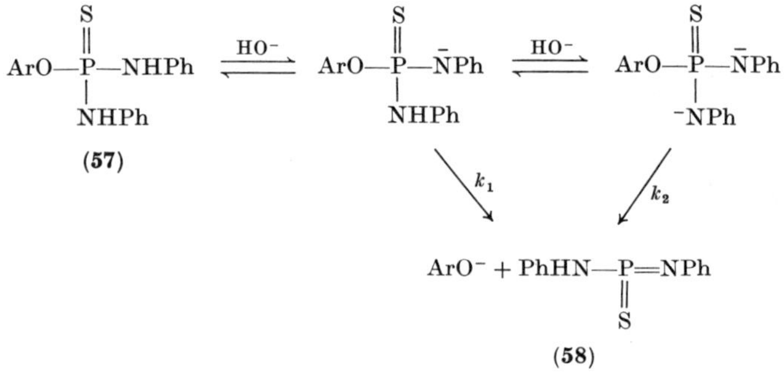

$$ArO{-}P(=S)(NHPh){-}NHPh \underset{}{\overset{HO^-}{\rightleftharpoons}} ArO{-}P(=S)(NHPh){-}\bar{N}Ph \underset{}{\overset{HO^-}{\rightleftharpoons}} ArO{-}P(=S)(^-NPh){-}\bar{N}Ph$$

(**57**)

$$\xrightarrow{k_1} \text{(from monoanion)}, \quad \xrightarrow{k_2} \text{(from dianion)}: \quad ArO^- + PhHN{-}P(=S){=}NPh$$

(**58**)

SCHEME 7.

has a ρ-value of 1.37 and the k_2 pathway, elimination from the dianion, has a ρ^--value of 3.08 indicating greater phenoxide ion character of the leaving group in the k_2 pathway. The use of solvent isotope effects to distinguish between $E1cB$ and $B_{Ac}2$ was questioned.[305]

Si—O bond fission occurs in the hydrolysis of trialkylsilyl esters of phosphorus oxyacids.[306]

In acidic solution the rate of hydrolysis of the phosphonate ester (**59**) is more than 10^5 times greater than that of the corresponding *p*-isomer. This is presumably due to neighbouring-group participation by the amide group.[307]

The monosodium salt of acetonylphosphonic acid (**60**) undergoes thermal decomposition at 150° to acetone and the polymer of metaphosphate. The methyl ester is stable

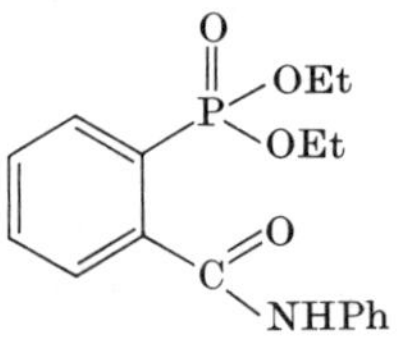

(**59**) 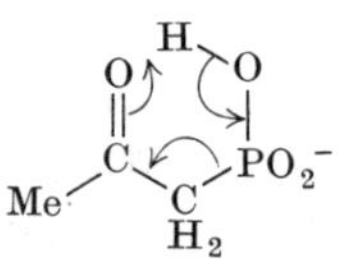 (**60**)

under these conditions and a mechanism analogous to the decarboxylation of β-keto-carboxylic acids was suggested (see **60**).[308]

Hydrolysis and addition of nucleophiles to acyclic acylphosphonates proceeds by attack at the carbonyl–carbon atom.[309]

Alkaline hydrolysis of a series of *O*-alkyl methylphosphothionic acid esters has also been investigated.[310, 311]

The stereochemistry of chlorination of a cyclic phosphorinan has been studied.[312]

There is no phosphinyl-oxygen exchange during the alkaline hydrolysis of phosphinate esters (**61**), and the hydroxide ion attacks only at phosphorus. The rates of hydrolysis of strained cyclic phosphinates show ca. 100-fold acceleration, indicating a pentaco-ordinate intermediate. The Taft parameters for *O*-alkyl substituents yield the large ρ^*-value of 8–11 and a δ-value of ca. 1, suggesting rate-limiting breakdown of the pentaco-ordinate intermediate (**62**).[313]

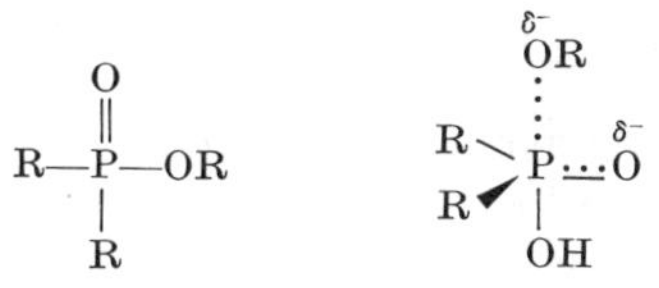

R—P(=O)(R)—OR

(61)

R₂P(OH)(⋯O$^{\delta-}$)(⋯O$^{\delta-}$R)

(62)

The rates of hydrolysis of methyl methylarylphosphinates in perchloric acid solutions show a maximum at ca. 6–7M-acid and occur at the position of substrate protonation, unlike most aryl phosphate esters.[314]

There have been studies of the acid-catalysed[315] and base-catalysed[316] hydrolysis of other phosphinic esters.

A plot of the rates of the acid-catalysed hydrolysis of arylphenylphosphinanilides (**63**) against H_A is linear with slope 1.0. The solvent isotope effect k_{D_2O}/k_{H_2O} is 2.8, the ρ-value with *N*-aryl substituents is –1.7 and the entropy of activation is –25 cal deg^{-1} mol^{-1}. These results are taken to indicate an *A*1-mechanism with dissociation of the protonated species as the rate-limiting step (**64**). For poorer leaving groups it is suggested that the mechanism changes to *A*2.[317]

Ar—P(=O)(Ph)—NHAr′

(63)

Ar—P(=O)(Ph)$^{\delta+}$----$^{\delta+}$NH$_2$Ar

(64)

The acid-catalysed hydrolysis of alkylphenylphosphinanilides is enhanced when the alkyl group is cyclopropyl, in agreement with the transition state (**64**) suggested above.[318]

The rates of the alkaline hydrolysis of various phosphinamides (**65**) are comparable with those for analogous carboxamides. ^{18}O-Studies yielded k(hydrolysis)/k(exchange) = 70, which was taken as evidence for a pentacovalent intermediate. However, the acid-catalysed rates of hydrolysis of phosphinamides are ca. 10^5 times those of carboxamides. The solvent isotope effect k_{D_2O}/k_{H_2O} of 1.3 and an entropy of activation of –35 cal deg^{-1} mol^{-1} are taken to indicate an *A*2 mechanism. The lability of phosphorus amides is attributed to *N*-protonation as opposed to *O*-protonation in carboxamides. A mechanism involving direct displacement rather than a pentaco-ordinate intermediate was favoured.[319]

An unusual order of ligand kinetic axiophilicities has been observed in the alkaline hydrolysis of 1-X-1-alkoxy-2,2,3,4,4-pentamethyl-phosphetanium salts (**66**).[320]

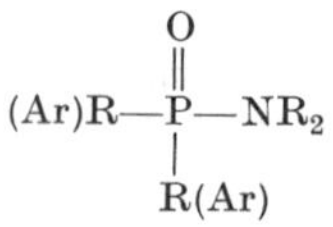

(Ar)R—P(=O)(—NR$_2$)—R(Ar)

(**65**)

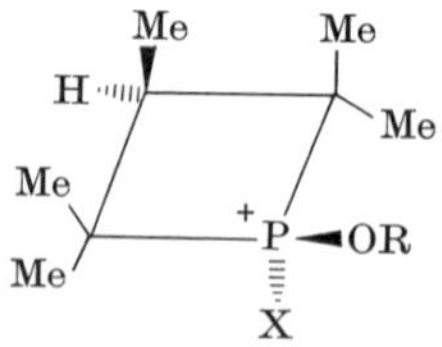

(**66**)

The pH–rate profile for the hydrolysis of 1-methoxy-2,2,3-trimethyl-1-phenylphosphetanium hexafluorophosphate is sigmoid between pH 3 and pH 8. This is attributed to a change in rate-limiting step from hydration at pH ca. 7 to rate-limiting breakdown of a pentacovalent intermediate in acid.[321]

Pentaco-ordinate phosphorus intermediates may be involved only as non-obligatory intermediates in the reaction of nucleophiles with alkoxyphosphonium salts.[322]

Investigations of the stereochemistry of nucleophilic substitution reactions at various phosphorus centres have been reported.[323]

Phosphonate–phosphate rearrangements[324] and the rearrangement of the oxidized adduct of diphenylphosphine and hexafluoroacetone[325] have been studied.

The equilibrium constant for the hydrolysis of ATP to ADP and inorganic phosphate has been calculated as a function of pH and metal-ion concentration and presented in the form of contour diagrams.[326]

The standard free energy of hydrolysis of the terminal pyrophosphate bond of ATP under physiological conditions has been determined.[327]

The heats of hydrolysis of simple aliphatic diesters of phosphoric acid have been determined by using a newly isolated phosphodiesterase[328] as catalyst. The heat of hydrolysis of trimethylene phosphate (1-hydroxyphosphetane 1-oxide) is ca. –3.5 kcal mol^{-1} in sharp contrast to the value of –10 to –14 obtained for 3′,5′-cyclic nucleotides.[329] The value for ethylene phosphate is –6.8, comparable to those of 2′,3′-cyclic nucleotides.[330]

Enzymic Reactions

Values of k_0 for the hydrolysis of substituted phenyl phosphates catalysed by alkaline phosphatase from *E. coli* are constant, but k_0/K_m varies with the substituent to yield a ρ^--value of 0.43. This low value was taken to indicate electrophilic assistance, probably by Zn(II), during phosphorylation or binding steps of the enzyme. The rate-limiting step in phosphorylation is suggested to be breakdown of the pentacovalent intermediate.[331]

The removal of the two molecules of tightly bound inorganic phosphate from *E. coli* alkaline phosphatase transforms the pre-steady-state burst of product release normally seen at pH 5.5 into an instantaneous burst of equal amplitude; an equivalent instant phase is created at pH 8.0 where no burst is normally seen. It is suggested that the enzyme shuttles between two conformations, one of which binds phosphate preferentially and has no catalytic activity.[332]

The alkaline phosphatase-catalysed hydrolysis of *p*-nitrophenyl phosphate is dependent upon groups of p*K* 8.1 and 8.6 in the enzyme which it is suggested are an α-amino-group and ionization of a Zn(II)-co-ordinated water molecule.[333]

Electrostatic interactions between substrate and enzyme are considered important in the potato acid phosphatase-catalysed hydrolysis of phosphate esters.[334]

A phosphoryl-enzyme intermediate has been suggested for the catalysed reactions of prostatic acid phosphatase.[335]

The binding of phosphate groups to the guanidino-groups of Arg-35 and Arg-87 of staphylococcal nuclease is supported by model studies of methylguanidinium phosphates. It is also suggested that the guanidino-groups may facilitate the hydrolytic process by charge-neutralization and/or transition-state stabilization.[336]

The hydrolysis of deoxythymidine 3′-phosphate–5′-*p*-nitrophenyl phosphate catalysed by staphylococcal nuclease exhibits a solvent isotope effect k_{H_2O}/k_{D_2O} of 0.95 for breakdown of the Michaelis complex. This was taken to indicate rate-limiting nucleophilic attack or slow pseudorotation of a pentacovalent intermediate.[337]

Carboxymethylation of bovine pancreatic ribonuclease by reagents with various leaving groups indicates that the nature of the leaving group is not important in the rate-limiting step.[338]

The histidine ring-proton NMR titration curves of chemical shift against pH for ribonuclease S and ribonuclease A are almost identical. The effects of phosphate and sulphate ions are interpreted in terms of carboxyl–histidine interactions in the active site of ribonuclease.[339]

Other studies of ribonuclease A are listed in ref. 340 and of ribonuclease S in ref. 341.

The complete covalent structure of bovine pancreatic deoxyribonuclease A has been determined.[342]

Porcine spleen deoxyribonuclease is completely inactivated by iodoacetate, and the inactive enzyme has one residue of 3-(carboxymethyl)histidine.[343]

The source of oxygen in the C—O—P linkage of the acyl phosphate in transport adenosine triphosphatase has been determined. The acyl phosphate is cleaved by water–oxygen attack on the phosphorus atom, and there is no acyl transfer to an enzyme group and subsequent hydrolysis.[344]

Other studies of adenosine triphosphatases are listed in ref. 345.

There have been several investigations of inorganic pyrophosphatase.[346]

The DNA-joining reaction catalysed by *E. coli* DNA ligase is enhanced by low concentrations of univalent cations, NH_4^+ being most effective. The reaction obeys "ping-pong" kinetics which provides evidence for a covalent intermediate, presumably ligase-adenylate. This is supported by the facts that the rate of the ligase-catalysed diphosphopyridine nucleotide–nicotinamide mononucleotide exchange reaction is unaffected by NH_4^+ and that the exchange reaction is faster than DNA-joining.[347]

Other enzymes studied include cyclic adenosine 3′,5′-monophosphate phosphodiesterase[348] and pancreatic phospholipase.[349] The mode of action of aminoacyl-transfer ribonucleic acid synthetase has been reviewed.[350]

Sulphur-containing Acids

The alkylammonium carboxylate-catalysed hydrolysis of 2,4-dinitrophenyl sulphate in benzene has been rationalized in terms of micellar and general acid–base catalysis. It is suggested that the substrate is held rigidly in the polar cavity of the reversed micelle and that hydrolysis is assisted by "enhanced water activity" and general-acid catalysis.[351]

Product analysis indicates that only alkyl–oxygen fission occurs in the methanolysis of methyl *p*-nitrophenyl sulphate.[352]

Electrophilic impurities appear to be necessary for easy hydrolysis of neutral sulphate ester salts.[353]

Nucleophilic displacement at chlorine has been suggested for reactions of aryl chlorosulphates.[354]

The Brønsted β-value for the pH-independent hydrolysis of aryl *N*-methylamino-sulphates, $MeNHSO_2OAr$, is –1.85. There is no buffer catalysis and the second-order rate constant (kK_a/K_w) is 10^8 times greater than that for hydroxide-ion catalysed hydrolysis of the *N*,*N*-dimethyl derivative. This is consistent with the elimination mechanism shown in equation (4).[355]

$$HO^- + MeNHSO_2OAr \underset{}{\overset{K_a/K_w}{\rightleftharpoons}} Me\bar{N}SO_2OAr \xrightarrow{k} MeNSO_2 + ArO^-$$

$$MeNSO_2 \xrightarrow{H_2O} MeNHSO_3H \qquad (4)$$

The acid-catalysed[356] and base-catalysed[357] hydrolysis of *NN′*-diarylsulphamides have been investigated.

In strongly acidic solutions S—N bond-fission occurs in the hydrolysis of sulphisoxazole, but in weakly acidic solutions C—N bond-fission takes place.[358]

The hydrolysis of 1-tosylimidazole[359] and *N*-tosyl-*p*-toluidine[360] have been studied.

The Brønsted β-value for the reaction of thiophene-2-sulphonyl chloride with substituted anilines is 0.53.[361]

The enthalpies of transfer, from methanol to protic and aprotic solvents, in the transition states for reaction of imidazole with aromatic sulphonyl chlorides have been determined.[362]

The fluoride ion-catalysed hydrolysis of sulphonylmethyl perchlorates, $RSO_2CH_2OClO_3$, has been attributed to general-base catalysis[363] rather than the "water-structure-making" effect of the fluoride ion.[364]

Studies of the reactions of 1,2-diaryl-2-[aryl(methyl)amino]vinyl arenesulfonates,[365] hydroxylamine monosulphonate[366] and sulphoximines,[367] and of the neutral[368] and basic[369] hydrolysis of esters of alkene- and alkane-sulphonic acids have been reported.

The Brønsted β-value for the reactions of substituted phenyl esters of methanesulphinic acid (equation 5) with hydroxide ion is –0.71; but with variation in the structure

$$HO^- + MeSOOAr \rightarrow MeSOOH + ArO^- \qquad (5)$$

of the nucleophile the Brønsted plot is non-linear, the β-value changing from ca. 0.75 for substituted phenoxides to ca. 0 for alkoxides and hydroxide ion. The interpretation of curved Brønsted plots is discussed and attributed to bond-order changes in the transition state or simply to different groups of nucleophiles falling on different regression lines.[370]

The acid- and nucleophile-catalysed oxygen-18 exchange of phenyl benzenethiosulphinate, $PhS(^{18}O)SPh$, has been studied. Comparison of the results with those for other reactions indicates that benzenesulphenic acid, PhSOH, is a very powerful nucleophile towards sulphenyl-sulphur.[371]

The disproportionation of alkyl thiosulphinate *S*-esters has been studied.[372]

Values of pK_{BH^+} have been determined for several sulphinates, sulphoxides and phosphinates.[373]

The reaction of *p*-nitrophenyl triphenylmethanesulphenate with *n*-butylamine is of the second-order in nucleophile, but with benzamidine it is of the first-order. This difference, which is also observed in other reactions,[374] is rationalized without invoking bifunctional catalysis.[375]

Other reactions of sulphenates[376] and sulphenamides[377] have also been reported.

There is a linear relationship between the logarithms of the rate constants for nucleophilic attack upon sulphur in diphenyl α-disulphone and the corresponding values for attack upon 2,4-dinitrophenyl acetate.[378, 37]

Oxidation of dimethyl sulphoxide by bromine has been investigated,[379] and other reactions of sulphoxides have also been reported.[380]

Other Acids

The base-induced decomposition of *N*-alkyl-*N*-nitrosoureas is suggested to involve proton abstraction from the urea-nitrogen.[381]

Other reactions studied include: acid-catalysed nitro-ester hydrolysis,[382] hydrolysis of 2-nitratovaleric acid,[383] dissociation of dimeric nitrosyl chloride adducts of glycols,[384] nitrile oxide formation,[385] rate-limiting protonation of diphenyldiazomethane with *p*-toluenesulphinic acid[386] and rates of reaction between 2,2-diarylcyclopropanecarboxylic acids and diphenyldiazomethane.[387]

References

[1a] A. Kilvinen, "Mechanisms of Substitution at the COX Group" in "The Chemistry of Acyl Halides", S. Patai (Ed.), Interscience, London, 1972, p. 177.

[1b] Y. Murakami and J. Sunamoto, *Kagaku No Ryoiki*, **27**, 716, 631 (1973); *Chem. Abs.*, **79**, 104367, 104368 (1973).

[2] G. A. Rogers and T. C. Bruice, *J. Am. Chem. Soc.*, **95**, 4452 (1973).

[3] J. Hine, D. Ricard and R. Perz, *J. Org. Chem.*, **38**, 110 (1973).

[4] J. P. Guthrie, *J. Am. Chem. Soc.*, **95**, 6999 (1973).

[5] P. Deslongchamps, C. Lebreux and R. Taillefer, *Can. J. Chem.*, **51**, 1665 (1973).

[6] R. E. Barnett, *Accounts Chem. Research*, **6**, 41 (1973).

[7] M. I. Page and W. P. Jencks, *J. Am. Chem. Soc.*, **94**, 8828 (1972).

[8] D. Drake, R. L. Schowen and H. Jayaraman, *J. Am. Chem. Soc.*, **95**, 454 (1973).

[9] C. R. Hopper, R. L. Schowen, K. S. Venkatasubban and H. Jayaraman, *J. Am. Chem. Soc.*, **95**, 3280 (1973).

[10a] R. M. Pollack and T. C. Dumsha, *J. Am. Chem. Soc.*, **95**, 4463 (1973).

[10b] V. Meresaar and S. O. Eriksson, *Acta Chem. Scand.*, **26**, 4186 (1973).

[11] E. A. Castro and R. B. Moodie, *Chem. Comm.*, **1973**, 828.

[12] F. M. Menger and A. C. Vitale, *J. Am. Chem. Soc.*, **95**, 4931 (1973).

[13] R. Hershfield and G. L. Schmir, *J. Am. Chem. Soc.*, **95**, 3994 (1973).

[14] T. Okuyama and G. L. Schmir, *J. Am. Chem. Soc.*, **94**, 8805 (1972).

[15] C. J. Belke, S. C. K. Su and J. A. Shafer, *J. Am. Chem. Soc.*, **93**, 4552 (1971).

[16] T. Okuyama, T. C. Pletcher, D. J. Sahn and G. L. Schmir, *J. Am. Chem. Soc.*, **95**, 1253 (1973).

[17] T. Okuyama, D. J. Sahn and G. L. Schmir, *J. Am. Chem. Soc.*, **95**, 2345 (1973).

[18] Cf. *Org. Reaction Mech.*, **1968**, 379.

[19] P. Y. Bruice and H. G. Mautner, *J. Am. Chem. Soc.*, **95**, 1582 (1973).

[20] L. A. Cohen and S. Takahashi, *J. Am. Chem. Soc.*, **95**, 443 (1973).

[21] F. M. Menger and J. A. Donohue, *J. Am. Chem. Soc.*, **95**, 432 (1973).

[22] C. R. Wasmuth and D. A. Copeland, *J. Am. Chem. Soc.*, **95**, 3808 (1973).

[23] F. Secco, M. Venturini and S. Celsi, *J.C.S. Perkin II*, **1973**, 1544.

[24] V. Gani and P. Viout, *Tetrahedron Letters*, **1972**, 5241.
[25] T. J. Broxton and L. W. Deady, *Tetrahedron Letters*, **1973**, 3915.
[26] C. A. Lewis and R. Wolfenden, *J. Am. Chem. Soc.*, **95**, 6685 (1973).
[27] C. R. Smith and K. Yates, *J. Am. Chem. Soc.*, **94**, 8811 (1972).
[28] B. E. Evans and R. V. Wolfenden, *Biochemistry*, **12**, 392 (1973).
[29] R. L. Schowen, "Mechanistic Deductions from Solvent Isotope Effects" in *Progr. Phys. Org. Chem.*, **9**, 275 (1972).
[30] W. P. Jencks, *Chem. Rev.*, **72**, 705 (1972).
[31] J. E. Critchlow, *J.C.S. Faraday Trans. I*, **68**, 1774 (1972).
[32] Y. Pocker and E. Green, *J. Am. Chem. Soc.*, **95**, 113 (1973).
[33] G. M. Blackburn and J. D. Plackett, *J.C.S. Perkin II*, **1973**, 981.
[34] C. Lopez Castellani, *Rev. Fac. Cienc., Univ. Oviedo*, **13**, 19 (1972); *Chem. Abs.*, **78**, 96774 (1973).
[35] C. K. Sauers, C. A. Marikakis and M. A. Lupton, *J. Am. Chem. Soc.*, **95**, 6792 (1973).
[36] A. Kankaanperä, M. Lahti and R. Aaltonen, *Acta Chem. Scand.*, **27**, 1444 (1973).
[37] See *Org. Reaction Mech.*, **1968**, 377; **1970**, 459.
[38] C. D. Ritchie and P. O. I. Virtanen, *J. Am. Chem. Soc.*, **95**, 1882 (1973).
[39] D. S. Kemp and M. L. Casey, *J. Am. Chem. Soc.*, **95**, 6670 (1973).
[40] C. D. Johnson and K. Schofield, *J. Am. Chem. Soc.*, **95**, 270 (1973).
[41] G. W. Koeppl and A. J. Kresge, *Chem. Comm.*, **1973**, 371.
[42] A. Babadjamian, M. Chanon, R. Gallo and J. Metzger, *J. Am. Chem. Soc.*, **95**, 3807 (1973); R. Fellous and R. Luft, *ibid.*, p. 5593; A. J. Hoefnagel, J. C. Monshouwer, E. G. Snom and B. M. Wepster, *ibid.*, p. 5350; A. J. Hoefnagel and B. M. Wepster, *ibid.*, p. 5357; K. Kalfus and M. Večeřa, *Coll. Czech. Chem. Comm.*, **37**, 3607 (1972); W. H. Richardson, R. S. Smith, G. Snyder, B. Anderson and G. L. Kranz, *J. Org. Chem.*, **37**, 3915 (1972); T. Fujita, C. Takayama and M. Nakajima, *ibid.*, p. 1623; L. L. McCoy and E. E. Riecke, *J. Am. Chem. Soc.*, **95**, 7407 (1973).
[43] J. J. Zimmerman and S.-J. Yau, *J. Pharm. Sci.*, **62**, 902 (1973); *Chem. Abs.*, **79**, 65410 (1973).
[44] N. J. Fina and J. O. Edwards, *Internat. J. Chem. Kinetics*, **5**, 1 (1973).
[45] J. D. Aubort, R. F. Hudson and R. C. Woodcock, *Tetrahedron Letters*, **1973**, 2229.
[46] J. F. Liebman and R. M. Pollack, *J. Org. Chem.*, **38**, 3444 (1973).
[47] S. V. Bogatkov, V. G. Zaslavskii and L. M. Litvinenko, *Dokl. Akad. Nauk SSSR*, **210**, 97 (1973); *Chem. Abs.*, **79**, 77713 (1973).
[48] P. G. Gassman and G. D. Hartman, *J. Am. Chem. Soc.*, **95**, 449 (1973).
[48a] C. B. Sawyer and J. F. Kirsch, *J. Am. Chem. Soc.*, **95**, 7375 (1973).
[49] K. Bowden and A. M. Last, *J.C.S. Perkin II*, **1973**, 358.
[50] See p. 2.
[51] See *Org. Reaction Mech.*, **1972**, 427.
[52] A. Williams, *J.C.S. Perkin II*, **1973**, 1244.
[53] A. F. Hegarty and L. N. Frost, *J.C.S. Perkin II*, **1973**, 1719.
[54] S. Takeuchi, *Bull. Chem. Soc. Japan*, **46**, 1496 (1973); L. W. Brown and A. A. Forist, *J. Pharm. Sci.*, **62**, 145 (1973); T. Vontor and M. Vecera, *Coll. Czech. Chem. Comm.*, **38**, 3139 (1973).
[55] D. DeFilippo, P. Deplano, F. Devillanova, E. F. Trogu and G. Verani, *J. Org. Chem.*, **38**, 560 (1973); J. O. Branstad, G. Ekberg and I. Nilsson, *Acta Pharm. Suec.*, **10**, 1 (1973); *Chem. Abs.*, **78**, 135486 (1973).
[56] Y. Furuya, K. Itoha, O. Shibata and K. Ohkubo, *Chem. Letters (Tokyo)*, **1972**, 971.
[57] R. Bacaloglu and C. A. Bunton, *Tetrahedron.* **29**, 2721, 2725 (1973).
[58] K. Koyano and C. R. McArthur, *Can. J. Chem.*, **51**, 333 (1973).
[59] D. C. Berndt and J. K. Sharp, *J. Org. Chem.*, **38**, 396 (1973).
[60] J. W. Barnett and C. J. O'Connor, *J.C.S. Perkin II*, **1973**, 1331.
[61] S. Rysman de Lockerente, P. VanBrandt and A. Bruylants, *Bull. Cl. Sci., Acad. Roy. Belg.*, **58**, 23 (1972); *Chem. Abs.*, **78**, 15328 (1973).
[62] C. J. Hyland and C. J. O'Connor, *J.C.S. Perkin II*, **1973**, 1402.
[63] C. J. O'Connor and J. W. Barnett, *J.C.S. Perkin II*, **1973**, 1457.
[64] J. W. Barnett and C. J. O'Connor, *J.C.S. Perkin II*, **1973**, 685.
[65] J. W. Barnett and C. J. O'Connor, *J.C.S. Perkin II*, **1972**, 2378.
[66] J. W. Barnett and C. J. O'Connor, *J.C.S. Perkin II*, **1972**, 220.
[67] M. I. Vinnik and L. R. Andreeva, *Zh. Org. Khim.*, **9**, 547 (1973); *Chem. Abs.*, **78**, 146994 (1973); S. Rysman de Lockerente, P. VanBrandt, A. Bruylants, M. De Buyl-Gaudissart and D. Hubert, *Bull. Cl. Sci., Acad. Roy. Belg.*, **58**, 180 (1972); *Chem. Abs.*, **78**, 42382 (1973).
[68] M. Tamme, U. Haldna and H. Kuura, *Organic Reactivity (Tartu)*, **8**, 1152 (1971); H. Kuura, M. Tamme and U. Haldna, *ibid.*, p. 1206; M. Tamme, U. Haldna, and H. Kuura, *ibid.*, **9**, 635, 652, 1030 (1972).

[69] V. Stella and T. Higuchi, *J. Pharm. Sci.*, **62**, 968 (1973).

[70] W. I. Congdon and J. T. Edward, *Can. J. Chem.*, **50**, 3767, 3780, 3921 (1972).

[71] V. A. Dadali, Yu. S. Simanenko and L. M. Litvinenko, *Zh. Org. Khim.*, **9**, 102 (1973); *Chem. Abs.*, **78**, 110086 (1973).

[72] A. Arcoria and S. Fisichella, *J. Org. Chem.*, **38**, 3774 (1973).

[73] G. Alberghina, A. Arcoria, S. Fisichella and G. Scarlata, *Gazz. Chim. Ital.*, **103**, 319 (1973); *Chem. Abs.*, **79**, 91260 (1973).

[74] A. Arcoria, S. Fisichella, G. Scarlata and D. Sciotto, *J. Org. Chem.*, **38**, 32 (1973).

[75] L. V. Kuritsyn and V. M. Kuritsyna, *Zh. Org. Khim.*, **8**, 2148 (1972); *Izv. Vyssh. Ucheb. Zaved., Khim. Khim. Tekhnol.*, **16**, 857 (1973); *Chem. Abs.*, **78**, 57332 (1973); **79**, 77687 (1973); P. W. Hickmott, G. J. Miles, G. Sheppard, R. Urbani and C. T. Yoxall, *J.C.S. Perkin I*, **1973**, 1514; N. K. Vorob'ev, E. A. Chizhova and L. I. Smirnova, *Izv. Vyssh. Ucheb. Zaved., Khim. Khim. Tekhnol.*, **15**, 1800 (1972); *Chem. Abs.*, **78**, 123567 (1973).

[76] V. A. Savelova, T. N. Solomoichenko and L. M. Litvinenko, *Zh. Org. Khim.*, **9**, 110 (1973); *Chem. Abs.*, **78**, 110090 (1973); *Organic Reactivity* (*Tartu*), **9**, 678 (1972).

[77] D. P. N. Satchell and V. F. Shvets, *J.C.S. Perkin II*, **1973**, 995.

[77a] V. V. Korshak, V. A. Vasniev, S. V. Bogatkov, A. I. Tarassov and S. V. Vinogradova, *Organic Reactivity* (*Tartu*), **10**, 389 (1973).

[78] T. G. Bonner and K. Hillier, *J.C.S. Perkin II*, **1973**, 1828.

[79] F. E. Condon, *J. Org. Chem.*, **37**, 3608 (1972); K. A. Connors and K. S. Albert, *J. Pharm. Sci.*, **62**, 845 (1973); L. Joly and G. Hoornaert, *Bull. Soc. Chim. Belges*, **82**, 523 (1973); J. B. Nagy, O. B. Nagy and A. Bruylants, *ibid.*, p. 539; L. M. Litvinenko, A. I. Kirichenko, A. S. Savchenko, L. Ya. Galushko and A. E. Shumeiko, *Ukr. Khim. Zh.* (*Russian Ed.*), **38**, 1024 (1972); *Chem. Abs.*, **78**, 42416 (1973); L. M. Litvinenko, A. I. Kirichenko and A. S. Savchenko, *Ukr. Khim. Zh.* (*Russian Ed.*), **38**, 1136 (1972); *Chem. Abs.*, **78**, 83648 (1973); Z. Szponar, *Rocz. Chem.*, **47**, 779 (1973); *Chem. Abs.*, **79**, 91271 (1973); Z. Szponar and T. Jasinski, *Zesz. Nauk. Wydz. Mat., Fiz. Chem. Univ. Gdanski, Chem.*, **1**, 141 (1971); *Chem. Abs.*, **78**, 110155 (1973); Ya. L. Gol'dfarb and L. I. Belen'kii, *Can. J. Chem.*, **51**, 2174 (1973); R. P. Mariella and K. H. Brown, *ibid.*, p. 2177.

[80] M. N. Sorokin, L. M. Litvinenko and N. M. Oleinik, *Organic Reactivity* (*Tartu*), **9**, 118 (1972); L. M. Litvinenko, N. M. Oleinik and M. N. Sorokin, *Vop. Stereokhim.* **1971**, 43; *Chem. Abs.*, **78**, 3491 (1973).

[81] L. M. Litvinenko, N. M. Oleinik, P. L. Kurchenko and V. K. Yushko, *Ukr. Khim. Zh.* (*Russian Ed.*), **38**, 1235 (1972); *Chem. Abs.*, **78**, 96766 (1973).

[82] V. V. Korshak, S. V. Vinogradova, A. I. Tarasov, V. A. Vasnev and L. V. Morgunova, *Tr. Mask. Khim.-Tekhnol. Inst.*, **1972**, 84; *Chem. Abs.*, **78**, 135272 (1973); L. M. Litvinenko, N. M. Oleinik, Yu. S. Sadovskii and A. N. Daragan, *Zh. Org. Khim.*, **9**, 363 (1973); *Chem. Abs.*, **78**, 110136 (1973).

[83] H. Iwamura, M. Iwamura, M. Imanari and M. Takeuchi, *Tetrahedron Letters*, **1973**, 2325.

[83a] V. V. Kosmynin, Y. A. Sharanin, L. M. Litvinenko and V. A. Saviolova, *Organic Reactivity* (*Tartu*), **9**, 997 (1972); L. M. Litvinenko, V. V. Kosmynin, Y. A. Sharanin and L. P. Drizha, *ibid.*, p. 1017.

[84] M. I. Page, *Chem. Soc. Rev.*, **2**, 295 (1973).

[85] C. Delisi and D. M. Crothers, *Biopolymers*, **12**, 1689 (1973).

[86] M. I. Page and W. P. Jencks, *Proc. Nat. Acad. Sci., USA.*, **68**, 1678 (1971).

[87] J. W. Larsen, *Biochem. Biophys. Res. Comm.*, **50**, 839 (1973).

[88] See *Org. Reaction Mech.*, **1972**, 434.

[89] R. M. Moriarty and T. Adams, *J. Am. Chem. Soc.*, **95**, 4070, 4071 (1973).

[90] A. Dafforn and D. E. Koshland, Jr., *Biochem. Biophys. Res. Comm.*, **52**, 779 (1973).

[91] C. R. Patrick, *Internat. J. Chem. Kinetics*, **5**, 769 (1973).

[92] L. N. Frost and A. F. Hegarty, *Chem. Comm.* **1973**, 82; see also p. 32.

[93] S. S. Minor and R. L. Schowen, *J. Am. Chem. Soc.*, **95**, 2279 (1973).

[94] See *Org. Reaction Mech.*, **1967**, 337; **1968**, 383; **1969**, 447.

[95] A. J. Kresge, *J. Am. Chem. Soc.*, **95**, 3065 (1973).

[96] P. P. Nechaev, Yu. V. Moiseev, E. V. Kamzolkina, Z. V. Gerashchenko, Ya. S. Vygodskii and G. E. Zaikov, *Izv. Akad. Nauk SSSR, Ser. Khim.*, **1972**, 2723; *Chem. Abs.*, **78**, 123565 (1973).

[97] T. Fueno, O. Kajimoto, Y. Nishigaki and T. Yoshioka, *J.C.S. Perkin II*, **1973**, 738.

[98] K. Bowden and A. M. Last, *J.C.S. Perkin II*, **1973**, 345.

[99] K. Bowden and A. M. Last, *J.C.S. Perkin II*, **1973**, 351.

[100] See *Org. Reaction Mech.*, **1972**, 433.

[101] G. H. Henderson and G. Dahlgren, *J. Org. Chem.*, **38**, 754 (1973).

[102] B. Capon, S. T. McDowell and W. V. Raftery, *J.C.S. Perkin II*, **1973**, 1118; see *Org. Reaction Mech.*, **1971**, 433.

[103] T. H. Fife and B. M. Benjamin, *J. Am. Chem. Soc.*, **95**, 2059 (1973).

104 J. E. C. Hutchins and T. H. Fife, *J. Am. Chem. Soc.*, **95**, 2282 (1973).
105 J. E. C. Hutchins and T. H. Fife, *J. Am. Chem. Soc.*, **95**, 3786 (1973); cf. p. 31.
106 A. Tsuji, T. Yamana and Y. Mizukami, *Chem. Pharm. Bull.*, **20**, 2528 (1972); *Chem. Abs.*, **78**, 96807 (1973).
107 N. Tomoto, W. J. Boyle, Jr., and J. F. Bunnett, *J. Org. Chem.*, **37**, 4315 (1972).
108 See *Org. Reaction Mech.*, **1970**, 468.
109 R. Hershfield and G. L. Schmir, *J. Am. Chem. Soc.*, **95**, 7359 (1973).
110 S. Milstein and L. A. Cohen, *J. Am. Chem. Soc.*, **94**, 9158 (1972); see *Org. Reaction Mech.*, **1970**, 467.
111 R. T. Borchardt and L. A. Cohen, *J. Am. Chem. Soc.*, **94**, 9166, 9175 (1972).
112 J. M. Karle and I. L. Karle, *J. Am. Chem. Soc.*, **94**, 9182 (1972).
113 M. M. Werber and Y. Shalitin, *Bio-organic Chemistry*, **2**, 221 (1973).
114 W. Korytnyk, H. Ahrens, N. Angelino and G. Kartha, *J. Org. Chem.*, **38**, 3793 (1973).
115 F. M. Menger and G. Saito, *J. Am. Chem. Soc.*, **95**, 6838 (1973); S. Greenberg and J. G. Moffatt, *ibid.*, p. 4016; A. F. Russell, S. Greenberg and J. G. Moffatt, *ibid.*, p. 4025; D. G. Clark and E. H. Cordes, *J. Org. Chem.*, **38**, 270 (1973); S. Senent, J. Casado and F. Mata, *An. Quím.*, **69**, 13 (1973); *Chem. Abs.*, **78**, 96790 (1973); A. O. Fitton, J. R. Frost, M. K. Zakaria and G. Andrew, *Chem. Comm.* **1973**, 889; F. Notheisz, M. Bartok and V. Remport, *Acta Phys. Chem.*, **18**, 197 (1972); *Chem. Abs.*, **78**, 96810 (1973); D. H. Kim, A. A. Santilli and R. A. Fieber, *J. Heterocyclic Chem.*, **9**, 1347 (1973); J. H. Peet and B. W. Rockett, *J.C.S. Perkin I*, **1937**, 106; G. Cooper and W. J. Irwin, *ibid.*, p. 911.
116 M. I. Page and W. P. Jencks, *J. Am. Chem. Soc.*, **94**, 8818 (1972).
117 J. E. Purdie and N. L. Benoiton, *J.C.S. Perkin II*, **1973**, 1845.
118 E. E. Roets, A. J. Vlietinck, G. A. Janssen and H. Vanderhaeghe, *Chem. Comm.*, **1973**, 484.
119 V. M. Fedoseev and V. S. Churilin, *Zh. Org. Khim.*, **8**, 2434 (1972); *Chem. Abs.*, **78**, 57354 (1973).
120 V. Stella and T. Higuchi, *J. Org. Chem.*, **38**, 1527 (1973); D. A. Tomalia and J. N. Paige, *ibid.*, p. 422; J-L Olive, C. Petrus and F. Petrus, *Bull. Soc. Chim. France*, **1973**, 1138; J. L. Wong and D. O. Helton, *Chem. Comm.*, **1973**, 352; J. M. Coxon, M. P. Hartshorn and W. H. Swallow, *ibid.*, p. 261.
121 J. P. Guthrie, *Can. J. Chem.*, **51**, 3494 (1973); see *Org. Reaction Mech.*, **1972**, 438.
122 R. Smith and C. Tanford, *Proc. Nat. Acad. Sci.*, USA, **70**, 289 (1973).
123 J. P. Guthrie and Y. Ueda, *Chem. Comm.*, **1973**, 898.
124 D. G. Oakenfall, *J.C.S. Perkin II*, **1973**, 1006.
125 C. G. Overberger, R. C. Glowaky and P.-H. Vandewyer, *J. Am. Chem. Soc.*, **95**, 6008 (1973).
126 C. G. Overberger and R. C. Glowaky, *J. Am. Chem. Soc.*, **95**, 6014 (1973).
127 K. Nakajima and K. Okawa, *Bull. Chem. Soc. Japan*, **46**, 1811 (1973).
128 H. J. Brass and M. L. Bender, *J. Am. Chem. Soc.*, **95**, 5391 (1973).
129 D. W. Griffiths and M. L. Bender, *J. Am. Chem. Soc.*, **95**, 1679 (1973).
130 T. Okubo and N. Ise, *J. Org. Chem.*, **38**, 3120 (1973).
131 C. A. Bunton and B. Wolfe, *J. Am. Chem. Soc.*, **95**, 3742 (1973).
132 J. W. Larsen, L. J. Magid and V. Payton, *Tetrahedron Letters*, **1973**, 2663.
133 F. M. Menger, J. A. Donohue and R. F. Williams, *J. Am. Chem. Soc.*, **95**, 286 (1973).
134 D. A. Buckingham, J. Dekkers and A. M. Sargeson, *J. Am. Chem. Soc.*, **95**, 4173 (1973).
135 R. Nakon and R. J. Angelici, *J. Am. Chem. Soc.*, **95**, 3170 (1973).
136 R. W. Hay and P. J. Morris, *J.C.S. Dalton*, **1973**, 56.
137 D. A. Buckingham, F. R. Keene and A. M. Sargeson, *J. Am. Chem. Soc.*, **95**, 5649 (1973).
138 M. M. Werber and Y. Shalitin, *Bio-organic Chemistry*, **2**, 202 (1973).
139 M. Wagatsuma, S. Terashima and S. Yamada, *Tetrahedron*, **29**, 1497 (1973).
140 M. Hatano and T. Nozawa, *Asahi Garasu Kogyo Gijutsu Shoreikai Kenkyu Hokoku*, **20**, 213 (1973); *Chem. Abs.*, **78**, 96787 (1973).
141 A. Y. Girgis and J. I. Legg, *J. Am. Chem. Soc.*, **94**, 8420 (1972).
142 C. J. O'Connor, E. J. Fendler and J. H. Fendler, *J. Am. Chem. Soc.*, **95**, 600 (1973).
143 G. R. Dukes and D. W. Margerum, *J. Am. Chem. Soc.*, **94**, 8414 (1973).
144 A. Noels and P. Lefebvre, *Tetrahedron Letters*, **1973**, 3035.
145 L. A. Bakalo, T. E. Lipatova, T. I. Novikova and I. A. Pronina, *Sin. Fiz.-Khim. Polim.*, **1972**, 21; *Chem. Abs.*, **78**, 57311 (1973).
146 T. E. Lipatova and Yu. N. Nizel'skii, *Zh. Fiz. Khim.*, **46**, 2294 (1972); *Chem. Abs.*, **78**, 3459 (1973).
147 H. Alper and C. C. Huang, *J. Org. Chem.*, **38**, 64 (1973).
148 J. Drapier, A. J. Hubert and Ph. Teyssie, *Tetrahedron Letters*, **1973**, 419.
149 R. G. R. Bacon and A. Karim, *J.C.S. Perkin I*, **1973**, 272, 278.
150 Y. Sasaki and C. U. Pittman, *J. Org. Chem.*, **38**, 3723 (1973).
151 K. Takahashi, G. Hata and A. Miyake, *Bull. Chem. Soc. Japan*, **46**, 1012 (1973).
152 L. S. Hegedus, S. M. Lo and D. E. Bloss, *J. Am. Chem. Soc.*, **95**, 3040 (1973).

[153] A. D. McLachlan, "Protein Structure and Function" in *Ann. Rev. Phys. Chem.*, **23**, 165 (1972); *Enzymes: Structure and Function*, Proc. 8th FEBS Meeting, Vol. 29, North-Holland/American Elsevier, New York, 1972; *Analysis and Simulation of Biochemical Systems*, Proc. 8th FEBS Meeting, Vol. 25, North Holland/American Elsevier, New York, 1972.

[154] J. F. Kirsch, "Mechanism of Enzyme Action", *Ann. Rev. Biochem.*, **42**, 205 (1973); P. F. Knowles, "Application of Magnetic Resonance Methods to the Study of Enzyme Structure and Action" in *Essays in Biochemistry*, **8**, 79 (1972); D. E. Koshland, Jr., *Scientific American*, **229**, 52 (1973).

[155] W. P. Jencks and M. I. Page, Proc. 8th FEBS Meeting, Vol. 29, p. 45, North-Holland/American Elsevier, New York, 1972.

[156] Cf. P. R. Andrews, G. D. Smith and I. G. Young, *Biochemistry*, **12**, 3492 (1973); T. Nowak and A. S. Mildvan, *ibid.*, **11**, 2813 (1972).

[157] R. A. Firestone and B. G. Christensen, *Tetrahedron Letters*, **1973**, 389.

[158] M. Renard and A. R. Fersht, *Biochemistry*, **12**, 4713 (1973).

[159] J. Fastrez and A. R. Fersht, *Biochemistry*, **12**, 1067 (1973).

[160] See *Org. Reaction Mech.*, **1968**, 394; J. H. Wang and L. Parker, *Proc. Nat. Acad. Sci.*, **58**, 2451 (1967).

[161] See *Org. Reaction Mech.*, **1972**, 442; **1969**, 458.

[162] E. C. Lucas, M. Caplow and K. J. Bush, *J. Am. Chem. Soc.*, **95**, 2670 (1973).

[163] See *Org. Reaction Mech.*, **1972**, 441.

[164] B. Zeeberg, M. Caswell and M. Caplow, *J. Am. Chem. Soc.*, **95**, 2734 (1973).

[165] J. Fastrez and A. R. Fersht, *Biochemistry*, **12**, 2025 (1973).

[166] A. R. Fersht, D. M. Blow and J. Fastrez, *Biochemistry*, **12**, 2035 (1973).

[167] P. Valenzuela and M. L. Bender, *J. Biol. Chem.*, **248**, 4909 (1973).

[168] E. Pollock, J. L. Hogg and R. L. Schowen, *J. Am. Chem. Soc.*, **95**, 968 (1973).

[169] D. M. Blow, J. J. Birktoft and B. S. Hartley, *Nature* (*London*), **221**, 337 (1969).

[170] B. Zeeberg and M. Caplow, *J. Biol. Chem.*, **248**, 5887 (1973).

[171] J. E. C. Hutchins and T. H. Fife, *J. Am. Chem. Soc.*, **94**, 8848 (1972); cf. *Org. Reaction Mech.*, **1972**, 442.

[172] R. N. Smith and C. Hansch, *Biochemistry*, **12**, 4924 (1973).

[173] T. H. Marshall and V. Chen, *J. Am. Chem. Soc.*, **95**, 5400 (1973).

[174] A. Dupaix, J.-J.Bechet and C. Roucous, *Biochemistry*, **12**, 2559, 2566 (1973).

[175] W. E. Brown and F. Wold, *Biochemistry*, **12**, 828, 835 (1973).

[176] K. Kurachi, J. C. Powers and P. E. Wilcox, *Biochemistry*, **12**, 771 (1973).

[177] A. L. Fink, *Biochemistry*, **12**, 1736 (1973).

[178] R. P. Taylor, J. B. Vatz and R. Lumry, *Biochemistry*, **12**, 2933 (1973); cf. *Org. Reaction Mech.*, **1970**, 478.

[179] A. Tulinksky, R. L. Vandlen, C. N. Morimoto, N. V. Mani and L. H. Wright, *Biochemistry*, **12**, 4185 (1973).

[180] R. L. Vandlen and A. Tulinksky, *Biochemistry*, **12**, 4193 (1973).

[181] A. H. Hunt and B. Jirgensons, *Biochemistry*, **12**, 4435 (1973).

[182] T. A. Horbett and D. C. Teller, *Biochemistry*, **12**, 1349 (1973).

[183] Y. Y. Lin and J. B. Jones, *J. Org. Chem.*, **38**, 3575 (1973).

[184] See *Org. Reaction Mech.*, **1972**, 445.

[185] L. J. Berliner and S. S. Wong, *J. Biol. Chem.*, **248**, 1118 (1973).

[186] G. P. Royer and R. Uy, *J. Biol. Chem.*, **248**, 2627 (1973).

[187] R. W. Sealock and M. Laskowski, Jr., *Biochemistry*, **12**, 3139 (1973); N. C. Robinson, H. Neurath and K. A. Walsh, *ibid.*, p. 420; J. A. Luthy, M. Praissman, W. R. Finkenstadt and M. Laskowski, Jr., *J. Biol. Chem.*, **248**, 1760 (1973); M. Chavko, M. Bartik and E. Kasafirek, *Coll. Czech. Chem. Comm.*, **37**, 3956 (1972); A. B. Cohen, *J. Biol. Chem.*, **248**, 7055 (1973).

[188] M. L. Bender and M. Philipp, *J. Am. Chem. Soc.*, **95**, 1665 (1973).

[189] See *Org. Reaction Mech.*, **1971**, 449.

[190] R. C. Thompson and E. K. Blout, *Biochemistry* **12**, 44, 51, 57, 66 (1973); R. C. Thompson, *ibid.*, p. 47; see also D. Atlas and A. Berger, *ibid.*, p. 2573.

[191] J. E. Mole and H. R. Horton, *Biochemistry*, **12**, 816 (1973).

[192] P. Campbell and E. T. Kaiser, *J. Am. Chem. Soc.*, **95**, 3735 (1973).

[193] R. W. Furlanetto and E. T. Kaiser, *J. Am. Chem. Soc.*, **95**, 6786 (1973).

[194] M. J. Skalski, S. D. Lewis, E. T. Maggio and J. A. Shafer, *Biochemistry*, **12**, 1884 (1973).

[195] A. A. Kortt and T. Y. Liu, *Biochemistry*, **12**, 320, 328, 338 (1973).

[196] K. Mizusawa and F. Yoshida, *J. Biol. Chem.*, **248**, 4417 (1973).

[197] G. Clement, in *Progr. in Bioorganic Chem.*, **2**, 177 (1972).

[198] J. W. Bunting and C. D. Myers, *Can. J. Chem.*, **51**, 2639 (1973).

199 V. R. Naik and H. R. Horton, *J. Biol. Chem.*, **248**, 6709 (1973).

200 F. H. Carpenter, H. Frederick and J. M. Vahl, *J. Biol. Chem.*, **248**, 294 (1973).

201 B. F. Roufogalis and U. M. Wickson, *J. Biol. Chem.*, **248**, 2254 (1973); N. K. Schaffer, H. O. Michel and A. F. Bridges, *Biochemistry*, **12**, 2946 (1973).

202 J. C. Lee and J. A. Harpst, *Biochemistry*, **12**, 1622 (1973).

203 D. Wynne, S. Ginsberg and Y. Shalitin, *Arch. Biochem. Biophys.*, **154**, 204 (1973).

204a H. L. Brockman, J. H. Law and F. J. Kézdy, *J. Biol. Chem.*, **248**, 4965 (1973).

204b J. W. Lagocki, J. H. Law and F. J. Kézdy, *J. Biol. Chem.*, **248**, 580 (1973).

205 J. Olander, S. F. Bosen and E. T. Kaiser, *J. Am. Chem. Soc.*, **95**, 1616 (1973).

206 D. L. Cybulsky, S. I. Kandel, M. Kandel and A. G. Gornell, *J. Biol. Chem.*, **248**, 3411 (1973).

207 Y. Pocker and N. Watamori, *Biochemistry*, **12**, 2475 (1973).

208 Y. Pocker, M. W. Beug and F. A. Beug, *Biochemistry*, **12**, 2483 (1973).

209 K-T. D. Lin and H. F. Deutsch, *J. Biol. Chem.*, **248**, 1885 (1973).

210 P. H. Haffner and J. E. Coleman, *J. Biol. Chem.*, **248**, 6630 (1973).

211 R. Roskoski, Jr., *Biochemistry*, **12**, 3709 (1973).

212 P. A. Whitney and T. Cooper, *J. Biol. Chem.*, **248**, 325 (1973).

213 C. A. Carlson and K.-H. Kim, *J. Biol. Chem.*, **248**, 378 (1973).

214 O. M. Pitts and D. G. Priest, *Biochemistry*, **12**, 1358 (1973).

215 K. Taguchi and F. H. Westheimer, *J. Am. Chem. Soc.*, **95**, 7413 (1973).

216 J. P. Guthrie and F. Jordan, *J. Am. Chem. Soc.*, **94**, 9132, 9136 (1972).

217 L. K.-M. Lam and D. E. Schmidt, Jr., *Can. J. Chem.*, **51**, 1959 (1973).

218 C. A. Bunton, M. J. Minch, J. Hidalgo and L. S. Sepulveda, *J. Am. Chem. Soc.*, **95**, 3262 (1973).

219 T. S. Straub and M. L. Bender, *J. Am. Chem. Soc.*, **94**, 8875, 8881 (1972).

220 R. G. Button and P. J. Taylor, *J.C.S. Perkin II*, **1973**, 557.

221 Y. Pocker and B. C. Davis, *J. Am. Chem. Soc.*, **95**, 6216 (1973).

222 M. S. Newman and M. C. V. Zwan, *J. Org. Chem.*, **38**, 319 (1973); A. P. Krapcho and A. J. Lovey, *Tetrahedron Letters*, **1973**, 957; P. Muller and B. Siegfried, *ibid.*, p. 3565; R. R. Koganty, M. B. Schambhu and G. A. Digenis, *ibid.*, p. 4511; N. J. Daly and F. Ziolkowski, *Austral. J. Chem.*, **25**, 1453 (1972); T. Cohen, C. K. Shaw and J. A. Jenkins, *J. Org. Chem.*, **38**, 3737 (1973); R. H. Wiley and K. S. Kim, *ibid.*, p. 3582; B. Matuszewski, R. S. Givens and C. Neywick, *J. Am. Chem. Soc.*, **95**, 595 (1973); G. L. Buchanan and G. A. R. Young, *J.C.S. Perkin I*, **1973**, 732.

223 H. Wenck and J. Polster, *Helv. Chim. Acta*, **56**, 2036 (1973).

224 T. O. Pussa, V. M. Nummert (Maramae) and V. A. Palm, *Organic Reactivity (Tartu)*, **9**, 727 (1972).

225 S. Hiraoka, S. Suda and I. Yamada, *Kagaku Kogaku*, **37**, 627 (1973); *Chem. Abs.*, **79**, 65436 (1973).

226 T. O. Pussa, V. M. Nummert (Maramae) and V. A. Palm, *Organic Reactivity (Tartu)*, **9**, 888 (1972), S. V. Bogatkov, L. A. Kundryutskova, L. V. Ponomarenko and E. M. Cherkassova, *ibid.*, **8**, 1014 (1971).

227 R. I. Kruglikova, S. V. Bogatkov, L. N. Zhestkova, L. A. Kundryutskova, B. K. Berestevich and B. V. Unkowsky, *Organic Reactivity (Tartu)*, **8**, 1024 (1971).

228 B. I. Istomin, V. A. Palm and V. M. Nummert, *Organic Reactivity (Tartu)*, **10**, 616 (1973).

229 L. M. Kogan, A. I. Ezrielev and N. B. Monastyrskaya, *Zh. Obshch. Khim.*, **42**, 2264 (1972); *Chem. Abs.*, **78**, 83514 (1973).

230 G. V. Kryshtal', L. A. Yanovskaya and V. F. Kucherov, *Izv. Akad. Nauk SSSR, Ser. Khim.*, **1972**, 2364; *Chem. Abs.*, **78**, 57317 (1973).

231a L. Ya Glinskaya and G. L. Kamalov, *Vop. Stereokhim. Mezhvedom. Respub. Sb.*, **1971**, 154; *Chem. Abs.*, **78**, 123531 (1973).

231b N. J. Cleve and E. K. Euranto, *Acta Chem. Scand.*, **27**, 1841 (1973).

232 G. S. Litvinenko, V. V. Sosnova and D. V. Sokolov, *Izv. Akad. Nauk Kaz. SSR, Ser. Khim.*, **23**, 37 (1973); G. S. Litvinenko, V. V. Sosnova and D. V. Sokolov, *ibid.*, p. 47; *Chem. Abs.*, **78**,123529 (1973); **79**, 17851 (1973).

233 F. A. Kundell, D. J. Robinson and W. J. Svirbely, *J. Phys. Chem.*, **77**, 1552 (1973).

234 A. Cambon and R. Jullien, *Bull. Soc. Chim. France*, **1973**, 2003.

235 N. J. Cleve, *Suomen Kem. B*, **46**, 5 (1973).

236 T. O. Pussa and V. A. Palm, *Organic Reactivity (Tartu)*, **9**, 1223 (1973); V. A. Palm, T. O. Pussa, V. M. Nummert and I. V. Talvik, *ibid.*, **10**, 264 (1973); G. Gambaretto and S. Gliozzi, *Atti Ist Veneto Sci. Lett. Arti, Cl. Sci. Mat. Natur.*, **129**, 47 (1970); *Chem. Abs.*, **78**, 110153 (1973).

237 G. Costeanu and O. Landauer, *Stud. Cercet. Chim.*, **20**, 1393 (1973); *Chem. Abs.*, **78**, 123400 (1973); J. Guenzet, A. Rousset and A. Toumi, *Compt. Rend., C*, **277**, 473 (1973); G. Costeanu and O. Landauer, *Rev. Roum. Chim.*, **17**, 1339 (1972); *Chem. Abs.*, **78**, 3497 (1973); C. W. Thomas and L. L. Leveson, *J.C.S. Perkin II*, **1973**, 20; N. J. Cleve, *Suomen Kem. B.*, **45**, 385 (1972); G. Casteanu

and C. Mateescu, *Stud. Cercet. Chim.*, **20**, 1383 (1972); *Chem. Abs.*, **78**, 110122 (1973); *Rev. Roum. Chim.*, **17**, 1353 (1972); *Chem. Abs.*, **78**, 3479 (1973); S. C. Rakshit and M. K. Sarkar, *J. Indian Chem. Soc.*, **50**, 241 (1973).

238 N. Tuchel and St. Velea, *Farmacia (Bucharest)*, **20**, 645 (1972); *Chem. Abs.*, **79**, 17709 (1973).

239a D. Sybistowicz and K. Stoklosa, *Rocz. Chem.*, **46**, 2061 (1972); *Chem. Abs.*, **78**, 147040 (1973); J. N. Gardner, *Can. J. Chem.*, **51**, 1416 (1973); M. I. Vinnik, I. M. Medvetskaya and L. R. Andreeva, *Zh. Fiz. Khim.*, **47**, 1159 (1973); *Chem. Abs.*, **79**, 52481 (1973); Yu. V. Svetkin, A. N. Minlibaeva and N. A. Akmanova, *Tr. Ural. Univ.*, **1**, 234 (1971); *Chem. Abs.*, **78**, 16072 (1973); Z. Eckstein and G. Matolcsy, *Acta Phytopathol.*, **7**, 139 (1972); *Chem. Abs.*, **78**, 57302 (1973); S. S. Hecht and E. S. Rothman, *J. Org. Chem.*, **38**, 3733 (1973); J. A. Franz and J. C. Martin, *J. Am. Chem. Soc.*, **95**, 2017 (1973); I. Shahak and Y. Sasson, *ibid.*, p. 3440.

239b S. O. Eriksson, M. Jakobsson, and N. A. Jonsson, *Acta Pharm. Suec.*, **10**, 63 (1973); *Chem. Abs.*, **79**, 52523 (1973).

239c L. A. Gardner and J. E. Goyan, *J. Pharm. Sci.*, **62**, 1026 (1973).

240 A. Brown and T. C. Bruice, *J. Am. Chem. Soc.*, **95**, 1593 (1973).

241 See p. 3.

242 V. I. Minkin, L. P. Olekhnovich, Yu. A. Zhdanov, V. V. Kiselev, M. A. Voronov and Z. N. Budarina, *Zh. Org. Khim.*, **9**, 1319 (1973); *Chem. Abs.*, **79**, 91359 (1973).

243 H. Rodé-Gowal and H. Dahl, *Helv. Chim. Acta*, **56**, 2070 (1973); T. Ozeki, T. Nakamura and T. Inden, *Nippon Kagaku Kaishi*, **1973**, 1135; *Chem. Abs.*, **79**, 65424 (1973); S. Oae, Y. Uchida K. Fujimori and S. Kozuka, *Bull. Chem. Soc. Japan*, **46**, 1741 (1973).

244 T. P. Visvanathan, *J. Inst. Chem., Calcutta*, **44**, 115 (1973); *Chem. Abs.*, **78**, 71165 (1973); C. Chuaqui, S. Atala, A. Marquez and H. Rodriguez, *Tetrahedron*, **29**, 1197 (1973).

245 E. Kuhle, *Angew. Chem. Internat. Ed.*, **12**, 630 (1973).

246 J. P. Engstrom and J. C. DuBose, *J. Org. Chem.*, **38**, 3817 (1973).

247 V. A. Kremer and L. A. Zatuchnaya, *Kontr. Tekhnol. Protsessov Obogashch. Polez. Iskop.*, **1971**, 116; *Chem. Abs.*, **78**, 70957 (1973); M. Nagano and K. Tomita, *Chem. Pharm. Bull.*, **20**, 2308 (1972); *Chem. Abs.*, **78**, 71012 (1973).

248 Ya. Yu. Rekshinskii and G. M. Strongin, *Tr. Khim. Khim. Tekhnol.*, **1972**, 147; *Chem. Abs.*, **78**, 147029 (1973).

249 M. S. Sytlin, A. I. Morozov and I. A. Makolkin, *Izv. Vyssh. Ucheb. Zaved., Khim. Khim. Tekhnol.*, **16**, 559 (1973); *Chem. Abs.*, **79**, 17705 (1973).

250 M. I. Karayannis, *Anal. Letters*, **6**, 629 (1973); *Chem. Abs.*, **79**, 91228 (1973); T. R. N. Kutty, A. R. V. Murthy, *Indian J. Technol.*, **10**, 309 (1972); *Chem. Abs.*, **78**, 71128 (1973); H. G. Aurick and H.-G. Scharpenberg, *Chem. Ber.*, **106**, 1881 (1973).

251 B. I. Rubinshtein, Ya. A. Leont'ev, L. A. Morozov and B. F. Ustavshchikov, *Neftekhimiya*, **12**, 589 (1972); *Chem. Abs.*, **78**, 3487 (1973); R. Subramanian and P. B. Rao, *Indian J. Technol.*, **10**, 400 (1972); *Chem. Abs.*, **79**, 4580 (1973); M. F. Sorokin, B. N. Grafkin and N. Yu. Stepanova, *Tr. Mosk. Khim.-Tekhnol. Inst.*, **1972**, 91; *Chem. Abs.*, **78**, 135283 (1973); E. S. Zelikman, L. V. Berezova, Yu. N. Yur'ev and E. P. Tarasenkova, *Zh. Org. Khim.*, **9**, 1140 (1973); *Chem. Abs.*, **79**, 65395 (1973).

252 O. N. Sharma, G. D. Nageshwar and P. S. Mene, *Indian J. Technol.*, **10**, 221 (1972); *Chem. Abs.*, **78**, 15189 (1973); N. G. Polyansku and G. V. Gorbunov, *Zh. Prikl. Khim. (Leningrad)*, **45**, 2282 (1972); *Chem. Abs.*, **78**, 42423 (1973).

253 M. S. Sytilin, A. I. Morozov and I. A. Makolkin, *Zh. Fiz. Khim.*, **46**, 2266 (1972); *Chem. Abs.*, **78**, 3508 (1973).

254 M. Lounasmaa, *Acta Chem. Scand.*, **27**, 715 (1973).

255 C. J. Hyland and C. J. O'Connor, *J.C.S. Perkin II*, **1973**, 223.

256 T.-R. Kim and M.-I. Lim, *Daehan Hwahak Hwoejee*, **17**, 129 (1973); *Chem. Abs.*, **79**, 4606 (1973); T. Nishino, M. Kiyokawa, Y. Miichi and T. Tokuyama, *Bull. Soc. Chem. Japan*, **46**, 253 (1973); G. E. Hawkes and J. H. P. Utley, *J. C. S. Perkin II*, **1973**, 128.

257 A. A. Zalinkin, L. P. Nikitenkova and Yu. A. Strepikheev, *Zh. Obshch. Khim.*, **43**, 902 (1973); *Chem. Abs.*, **79**, 52612 (1973).

258 P. B. Talukdar, S. Banerjee and A. Chakraborty, *Indian J. Chem.*, **10**, 610 (1972); *Chem. Abs.*, **78**, 3448 (1973).

259 H. Collet, A. Commeyras and A. Casadevall, *Tetrahedron*, **28**, 5883 (1972).

260 M. K. Hargreaves and M. A. Khan, *J.C.S. Perkin II*, **1973**, 1204; J. Kovacs, R. E. Cover, R. H. Johnson, T. J. Kalas, G. L. Mayers and J. E. Roberts, *J. Org. Chem.*, **38**, 2518 (1973).

261 M. S. Miller and I. M. Klotz, *J. Am. Chem. Soc.*, **95**, 5694 (1973); R. B. Martin and W. C. Hutton, *ibid.*, p. 4752.

[262] A. A. Forist and D. J. Weber, *J. Pharm. Sci.*, **62**, 318 (1973).
[263] R. H. Manzo and M. M. de Bertorello, *J. Pharm. Sci.*, **62**, 152 (1973).
[264] M. M. Kayser and M. S. Gibson, *Can. J. Chem.*, **51**, 3499 (1973).
[265] L. do. Amaral and S. C. Melo, *J. Org. Chem.*, **38**, 800 (1973); V. I. Stanineto and Yu. I. Gevaza, *Ukr. Khim. Zh* (*Russian Ed.*), **39**, 699 (1973); *Chem. Abs.*, **79**, 91263 (1973).
[266] P. G. Gordon, *Austral. J. Chem.*, **26**, 1771 (1973).
[267] G. G. Arbuzova, I. N. Azerbaev, R. D. Yakubov and G. A. Shitov, *Dokl. Vses. Konf. Khim. Atsetilena, 4th*, **3**, 353 (1972); *Chem. Abs.*, **79**, 77691 (1973).
[268] J. E. W. van Melick, J. W. Scheeren and R. J. F. Nivard, *Rec. Trav. chim.*, **92**, 775 (1973).
[269] M. V. Bhatt and M. Ravindranathan, *J.C.S. Perkin II*, **1973**, 1158.
[270] Y. Nagata, T. Dohmaru and J. Tsurugi, *J. Org. Chem.*, **38**, 795 (1973).
[271] T. Ho and C. M. Wong, *Chem. Comm.*, **1973**, 224.
[272] J. Ratusky, *Coll. Czech. Chem. Comm.*, **38**, 74, 87 (1973).
[273] E. Shami, J. Dudzinski, L. Lachman and J. Tingstad, *J. Pharm. Sci.*, **62**, 1283 (1973).
[274] A. Hosono, K. Fujii, T. Tada, H. Tanaka, Y. Ohgo, Y. Ishido and T. Sato, *Bull. Chem. Soc. Japan*, **46**, 2814 (1973).
[275] W. O. George, D. V. Hassid, and W. F. Maddams, *J.C.S. Perkin II*, **1973**, 952; W. O. George and A. J. Parker, *J.S.C. Perkin II*, **1973**, 954.
[276] M. E. Rennekamp and C. A. Kingsbury, *J. Org. Chem.*, **38**, 3959 (1973).
[277] O. Exner, P. Dembech, G. Seconi and P. Vivarelli, *J.C.S. Perkin II*, **1973**, 1870.
[278] H. Ogura, H. Takayanagi, K. Kubo and K. Furuhata, *J. Am. Chem. Soc.*, **95**, 8056 (1973).
[279] P. Gillespie, F. Ramirez, I. Ugi and D. Marquarding, "Displacement Reactions at Phosphorus(v) Compounds and their Pentacoordinate Intermediates", *Angew. Chem., Internat. Ed.*, **12**, 91 (1973).
[280] E. J. Sampson, J. Fedor, P. A. Benkovic and S. J. Benkovic, *J. Org. Chem.*, **38**, 1301 (1973).
[281] Cf. *Org. Reaction Mech.*, **1970**, 495.
[282] J. J. Steffens, E. J. Sampson, I. J. Siewers and S. J. Benkovic, *J. Am. Chem. Soc.*, **95**, 936 (1973).
[283] J. I. G. Cadogan, D. T. Eastlick, J. A. Challis and A. Cooper, *J.C.S. Perkin II*, **1973**, 1798; J. I. G. Cadogan and D. T. Eastlick, *Chem. Comm.*, **1973**, 238.
[284] Y. Murakami, J. Sunamoto and N. Kanamoto, *Bull. Chem. Soc. Japan*, **46**, 871 (1973).
[285] Y. Murakami and J. Sunamoto, *J.C.S. Perkin II*, **1973**, 1235.
[286] M. M. Mhala and S. Prabha, *Indian J. Chem.*, **10**, 1002 (1972); *Chem. Abs.*, **78**, 110129 (1973).
[287] M. M. Mhala and S. B. Saxena, *Indian J. Chem.*, **10**, 703 (1972); *Chem. Abs.*, **78**, 96791 (1973).
[288] M. M. Mhala and S. Prabha, *Indian J. Chem.*, **10**, 1073 (1972); *Chem. Abs.*, **78**, 110111 (1973).
[289] C. Triantaphylides, G. Peiffer and R. Gester, *Bull. Soc. Chim. France*, **1973**, 1756.
[290] E. P. Lyznicki and T. T. Tidwell, *J. Am. Chem. Soc.*, **95**, 4935 (1973).
[291] I. V. Mel'nichenko, N. Ya Kozlova and A. A. Yasnikov, *Ukr. Khim. Zh.* (*Russian Ed.*), **38**, 1152 (1972); *Chem. Abs.*, **78**, 83540 (1973).
[292] I. V. Mel'nichenko, N. Ya. Kozlova and A. A. Yasnikov, *Ukr. Khim. Zh.* (*Russian Ed.*), **38**, 472 (1972); *Chem. Abs.*, **78**, 3496 (1973).
[293] C. A. Bunton and L. G. Ionescu, *J. Am. Chem. Soc.*, **95**, 2912 (1971).
[294] S. Suzuki, T. Higashiyama and A. Nakahara, *Bioorganic Chem.*, **2**, 145 (1973).
[295] R. Fernández-Prini and D. Turyn, *J.C.S. Faraday I*, **1973**, 1326.
[296] R. J. Brooks, C. A. Bunton and J. M. Hellyer, *J. Org. Chem.*, **38**, 2151 (1973).
[297a] W. Tagaki, Y. Asai and T. Eiki, *J. Am. Chem. Soc.*, **95**, 3038 (1973).
[297b] N. Nelson and E. Racker, *Biochemistry*, **12**, 563 (1973).
[298] V. E. Bel'skii, M. V. Efremova and I. A. Aleksandrova, *Izv. Akad. Nauk SSSR, Ser. Khim.*, **1972**. 2794; *Chem. Abs.*, **78**, 110135 (1973).
[299a] V. P. Evdakov, V. P. Beketov and V. I. Svergun, *Zh. Obshch. Khim.*, **43**, 55 (1973); *Chem. Abs.*, **78**, 123523 (1973).
[299b] M. S. Zetter, H. W. Dodgen and J. P. Hunt, *Biochemistry*, **12**, 778 (1973).
[300a] J. K. Chakrabarti and O. M. Friedman, *J. Heterocyclic Chem.*, **10**, 55 (1973).
[300b] G. W. Allen and P. Haake, *J. Am. Chem. Soc.*, **95**, 8080 (1973); see *Org. Reaction Mech.*, **1971**, 460.
[301] H. C. Froede and D. B. Wilson, *J. Am. Chem. Soc.*, **95**, 1987 (1973).
[302] E. C. F. Ko and R. E. Robertson, *Can. J. Chem.*, **51**, 597 (1973).
[303] P. M. Zavlin, *Khim. Primen. Fosfororg. Soedin. Tr. Konf., 4th*, **1969**, 134; *Chem. Abs.*, **78**, 147023 (1973).
[304] W. C. Archie and F. H. Westheimer, *J. Am. Chem. Soc.*, **95**, 5955 (1973).
[305] A. Williams and K. T. Douglas, *J.C.S. Perkin II*, **1973**, 318.

[306] V. M. D'yakov, M. G. Voronkov and N. F. Orlov, *Izv. Akad. Nauk SSSR, Ser. Khim.*, **1972**, 2484; *Chem. Abs.*, **78**, 83559 (1973).
[307] R. Kluger and J. L. W. Chan, *J. Am. Chem. Soc.*, **95**, 2363 (1973).
[308] R. Kluger, *J. Org. Chem.*, **38**, 2721 (1973).
[309] R. M. Laird and M. J. Spence, *J.C.S. Perkin II*, **1973**, 1434.
[310] J. L. Jarv and A. A. Aaviksarar, *Organic Reactivity* (*Tartu*), **8**, 980 (1971).
[311] J. L. Jarv, A. A. Aaviksarar, N. N. Godovikov and N. A. Morozova, *Organic Reactivity* (*Tartu*), **9**, 695, 828 (1972).
[312] C. L. Bodkin and P. Simpson, *J.C.S. Perkin II*, **1973**, 676.
[313] R. D. Cook, C. E. Diebert, W. Schwarz, P. C. Turley and P. Haake, *J. Am. Chem. Soc.*, **95**, 8088 (1973).
[314] J. F. Bunnett, J. O. Edwards, D. V. Wells, H. J. Brass and R. Curci, *J. Org. Chem.*, **38**, 2703 (1973).
[315] R. D. Cook and K. Abbas, *Tetrahedron Letters*, **1973**, 519.
[316] R. D. Cook and H. K. Norrish, *Tetrahedron Letters*, **1973**, 521.
[317] D. A. Tyssee, L. P. Bausher and P. Haake, *J. Am. Chem. Soc.*, **95**, 8066 (1973).
[318] M. J. Harger, *Chem. Comm.*, **1973**, 774.
[319] T. Koizumi and P. Haake, *J. Am. Chem. Soc.*, **95**, 8073 (1973).
[320] K. E. DeBruin, A. G. Padilla and M. T. Campbell, *J. Am. Chem. Soc.*, **95**, 4681 (1973).
[321] D. G. Gorenstein, *J. Am. Chem. Soc.*, **95**, 8060 (1973).
[322] K. E. DeBruin and S. Chandrasekaran, *J. Am. Chem. Soc.*, **95**, 974 (1973).
[323] E. F. Korman and J. McLick, *Bioorganic Chemistry*, **2**, 179 (1973); W. S. Wadsworth, Jr., S. Larsen and H. L. Horten, *J. Org. Chem.*, **38**, 256 (1973); W. S. Wadsworth, Jr., *ibid.*, p. 2921.
[324] A. N. Pudovik, I. V. Konovalova, G. V. Romanov, R. G. Fitseva and N. P. Burmistrova, *Zh. Obshch. Khim.*, **43**, 41 (1973); *Chem. Abstr.*, **78**, 123716 (1973).
[325] A. F. Janzen and O. C. Vaidya, *Can. J. Chem.*, **51**, 1136 (1973).
[326] K. Shikama and K. Nakamura, *Arch. Biochem. Biophys.*, **157**, 457 (1973).
[327] R. W. Guynn and R. L. Veech, *J. Biol. Chem.*, **248**, 6966 (1973).
[328] J. A. Gerlt and F. H. Westheimer, *J. Am. Chem. Soc.*, **95**, 8166 (1973).
[329] See *Org. Reaction Mech.*, **1969**, 472.
[330] J. M. Sturtevant, J. A. Gerlt and F. H. Westheimer, *J. Am. Chem. Soc.*, **95**, 8168 (1973).
[331] A. Williams, R. A. Naylor and S. G. Collyer, *J.C.S. Perkin II*, **1973**, 25.
[332] W. Bloch and M. J. Schesinger, *J. Biol. Chem.*, **248**, 5794 (1973).
[333] I. Hinberg and K. J. Laidler, *Can. J. Biochem.*, **51**, 1096 (1973).
[334] Y. Murakami, J. Sunamoto and C. Gondo, *Chem. Letters* (*Tokyo*), **1973**, 25.
[335] W. Ostrowski and E. A. Barnard, *Biochemistry*, **12**, 3893 (1973).
[336] F. A. Cotton, E. E. Hazen, Jr., V. W. Day, S. Larsen, J. G. Norman, Jr., S. T. K. Wong and K. H. Johnson, *J. Am. Chem. Soc.*, **95**, 2367 (1973).
[337] B. M. Dunn, C. DiBello and C. B. Anfinsen, *J. Biol. Chem.*, **248**, 4769 (1973).
[338] B. V. Plapp, *J. Biol. Chem.*, **248**, 4896 (1973).
[339] J. S. Cohen, J. H. Griffin and A. N. Schecter, *J. Biol. Chem.*, **248**, 4305 (1973).
[340] W. Saenger, *Angew. Chem. Internat. Ed.*, **12**, 591 (1973); P. H. Haffner and J. H. Wang, *Biochemistry*, **12**, 1608 (1973); D. R. Pollard and J. Nagyvary, *ibid.*, p. 1063.
[341] N. M. Allewell, Y. Mitsui and H. W. Wyckoff, *J. Biol. Chem.*, **248**, 5291 (1973).
[342] T.-H. Liao, J. Salnikow, S. Moore and W. H. Stein, *J. Biol. Chem.*, **248**, 1489 (1973).
[343] R. G. Oshima and P. A. Price, *J. Biol. Chem.*, **248**, 7522 (1973).
[344] A. S. Dahms, T. Kanazawa and P. D. Boyer, *J. Biol. Chem.*, **248**, 6592 (1973).
[345] L. C. Cantley, Jr., and G. G. Hammes, *Biochemistry*, **12**, 4900 (1973); D. C. Pang and F. N. Briggs, *ibid.*, p. 4905; R. L. Post and S. Kume, *J. Biol. Chem.*, **248**, 6993 (1973); A. S. Dahms and P. D. Boyer, *ibid.*, p. 3155 (1973); L. B. Hersh, *ibid.*, p. 7295.
[346] J. W. Sperow, O. A. Moe, J. W. Ridlington and L. G. Butler, *J. Biol. Chem.*, **248**, 2062 (1973); B. S. Cooperman and N. Y. Chiu, *Biochemistry*, **12**, 1665, 1670, 1676 (1973); S. J. Kelly, F. Feldman, J. W. Sperow and L. G. Butler, *ibid.*, p. 3338.
[347] P. Modrich and I. R. Lehman, *J. Biol. Chem.*, **248**, 7502 (1973).
[348] T. S. Teo, T. H. Wang and J. H. Wang, *J. Biol. Chem.*, **248**, 588 (1973).
[349] R. Verger, M. C. E. Mieras and G. H. DeHaas, *J. Biol. Chem.*, **248**, 4023 (1973).
[350] P. R. Schimmel, *Accounts Chem. Research*, **6**, 299 (1973).
[351] C. J. O'Connor, E. J. Fendler and J. H. Fendler, *J. Org. Chem.*, **38**, 3371 (1973).
[352] E. Buncel, A. Raoult and J. F. Wiltshire, *J. Am. Chem. Soc.*, **95**, 799 (1973).
[353] M. B. Goren and M. E. Kochansky, *J. Org. Chem.*, **38**, 3510 (1973).
[354] E. Buncel, A. Raoult and L.-A. Lancaster, *J. Am. Chem. Soc.*, **95**, 5964 (1973).

[355] K. T. Druglas and A. Williams, *Chem. Comm.*, **1973**, 356.
[356] W. J. Spillane, J. A. Barry and F. L. Scott, *J.C.S. Perkin II*, **1973**, 481.
[357] H. Yamaguchi and K. Nakano, *Hiroshima Daigaku Kogakubu Kenkyu Hokoku*, **21**, 23 (1972); *Chem. Abs.*, **79**, 65420 (1973).
[358] R. H. Manzo and M. Martinez de Bertorello, *J. Pharm. Sci.*, **62**, 154 (1973); *Chem. Abs.*, **78**, 70978 (1973).
[359] P. Monjoint and M. Laloi-Diard, *Bull. Soc. Chim. France*, **1973**, 2357.
[360] E. Y. Belayev and L. I. Kotlar, *Organic Reactivity* (*Tartu*), **10**, 274 (1973).
[361] A. Arcoria, E. Maccarone, G. Musumara and G. A. Tomaselli, *J. Org. Chem.*, **38**, 2457 (1973).
[362] O. Rogne, *J.C.S. Perkin II*, **1973**, 823, 1760.
[363] J. Pink, *Tetrahedron Letters*, **1973**, 531.
[364] L. Menninga and J. B. F. N. Engberts, *Tetrahedron Letters*, **1972**, 617.
[365] A. Burighel, G. Modenn and V. Tonellato, *J.C.S. Perkin II*, **1973**, 1021.
[366] M. Hlubuckova and O. Podrouzek, *Chem. Prum.*, **22**, 440 (1972); *Chem. Abstr.*, **78**, 83527 (1973).
[367] S. Oae, Y. Tsuchida and N. Furukawa, *Bull. Chem. Soc. Japan*, **46**, 648 (1973).
[368] R. V. Vizgert and Yu. G. Skrypnik, *Organic Reactivity* (*Tartu*), **9**, 53 (1972).
[369] R. V. Vizgert and Yu. G. Skrypnik, *Organic Reactivity* (*Tartu*), **9**, 425 (1972).
[370] L. Senatore, E. Ciuffarin, A. Fava and G. Levita, *J. Am. Chem. Soc.*, **95**, 2918 (1973).
[371] J. L. Kice and J. P. Cleveland, *J. Am. Chem. Soc.*, **95**, 104, 109 (1973).
[372] E. Block and S. W. Weidman, *J. Am. Chem. Soc.*, **95**, 5046 (1973).
[373] O. Bohman and S. Allenmark, *Tetrahedron Letters*, **1973**, 405; R. Curci, A. Levi, V. Lucchini and G. Scorrano, *J.C.S. Perkin II*, **1973**, 531.
[374] See *Org. Reaction Mech.*, **1966**, 343; **1969**, 442; **1972**, 189.
[375] E. Ciuffarin, L. Senatore and L. Sagramora, *J.C.S. Perkin II*, **1973**, 534.
[376] S. Braverman and D. Reisman, *Tetrahedron Letters*, **1973**, 3563.
[377] V. A. Ignatov and R. A. Akchurina, *Organic Reactivity* (*Tartu*), **9**, 64 (1972).
[378] J. L. Kice and E. Legan, *J. Am. Chem. Soc.*, **95**, 3912 (1973).
[379] B. G. Cox and A. Gibson, *J.C.S. Perkin II*, **1973**, 1355.
[380] T. Durst, K.-C. Tin and M. J. V. Marcil, *Can. J. Chem.*, **51**, 1704 (1973); N. Kunieda and S. Oae, *Bull. Chem. Soc. Japan*, **46**, 1745 (1973).
[381] S. M. Hecht and J. W. Kozarich, *Tetrahedron Letters*, **1972**, 5147.
[382] B. S. Svetlov, V. P. Shelaputina and L. B. Gorodkova, *Kinet. Katal.*, **13**, 880 (1972); *Chem. Abs.*, **78**, 71014 (1973).
[383] M. S. Rusakova, V. A. Podgornova, V. N. Kreitsberg and N. Yu. Musabekova, *Uch. Zap., Yaroslav, Tekhnol. Inst.*, **1971**, 29; *Chem. Abs.*, **78**, 83528 (1973).
[384] R. U. Lemieux, T. L. Nagabhushan and K. James, *Can. J. Chem.*, **51**, 1 (1973).
[385] C. Grundmann, R. K. Bansal and P. S. Osmanski, *Ann. Chem.*, **1973**, 898; *Chem. Abs.*, **79**, 104565 (1973).
[386] M. Kobayashi, H. Minato and H. Fukuda, *Bull. Chem. Soc. Japan.*, **46**, 1266 (1973).
[387] R. R. Kostikov, N. P. Bobko, I. A. Dyakonov and I. A. Favorskaya, *Organic Reactivity* (*Tartu*), **8**, 1092 (1971).

CHAPTER 3

Radical Reactions

D. C. NONHEBEL

Department of Pure and Applied Chemistry, University of Strathclyde

and J. C. WALTON

Department of Chemistry, University of St. Andrews

Introduction

This year has seen the publication of the long-awaited treatise on *Free Radicals* edited by J. K. Kochi.[1] The first volume on *Free Radical Reactions* in the biennial *MTP International Reviews of Science* has also appeared.[2] The rôle of free radicals in biological systems was the subject of a symposium, the papers of which have been published.[3] Other books and reviews of specific aspects of radical chemistry are mentioned in the appropriate Sections of this chapter. Included for the first time in this year's report is a section on Radical Combinations, reflecting the growing interest in these reactions.

Structure, Stereochemistry and Stability

Carbon Radicals[4]

Theoretical approaches to the structures of aliphatic free radicals have been reviewed.[5] Interest continues in factors that favour non-planarity in substituted alkyl radicals. It had been inferred that tertiary alkyl radicals were significantly non-planar, but Symons, from a survey of the existing literature, indicates that any deviation from planarity is small.[6] ESR studies using ^{13}C splitting constants likewise confirm the essential planarity of *tert*-butyl radicals and also of 1-methylcyclopentyl and isopropyl radicals;[7] however, another ESR study refutes these conclusions and favours a non-planar structure for the *tert*-butyl radical.[8] There are indications from the matrix-isolation IR spectrum of iodomethyl that this radical also is planar.[9] The trichloromethyl radical is, however, bent to the extent of 8–10° in line with the effect of electronegative substituents attached to a radical centre.[10] Bending also occurs in the radicals $Me_3M\cdot$ and $Cl_3M\cdot$ (M = Si, Ge or Sn).[10] Analysis of the hyperfine splitting constants for the silyl radical $\cdot SiH_3$ indicates that this is less pyramidal than was previously believed, although it is not quasi-planar.[11] Di- and tri-*tert*-butylmethyl radicals have exceptional stability on account of their inability to disproportionate and because of steric hindrance to dimerization.[12] This steric hindrance to dimerization is appreciably diminished in the radicals $(Me_3Si)_2\dot{C}H$ and $(Me_3Si)_3C\cdot$, both of which are much less stable; this arises because of the larger size of silicon which enables the trimethylsilyl groups to avoid each other on dimerization. 1,1-Di-*tert*-butylalkyl radicals containing β-C—H bonds are also exceptionally stable even though, on paper, they could disproportionate. ESR studies indicate that the preferred conformations of the radicals $Bu^t_2\dot{C}CH_2CH_3$ and $Bu^t_2\dot{C}CH_2CH(CH_3)_2$ are such that the dihedral angle between the p-orbital at the radical centre and the β-C—H bonds are 60° and 76°, respectively,[13] i.e. the degree of overlap between the p-orbital and the β-C—H bond is small; this suggests that stereoelectronic as well as steric factors are important in determining whether or not disproportionation occurs.

Analysis of the ^{13}C-ESR spectra of β-substituted 1,1-di-*tert*-butylethyl radicals, $Bu^t_2\dot{C}CH_2MR^1_n$ ($MR^1_n = CF_3$, CCl_3, $SiMe_3$ or $SiBu^n_3$), shows that they exist in the eclipsed conformation (**1**; $R^2 = Bu^t$).[14] In all cases the spin density on C-α is approximately the same, and hence the extent of hyperconjugative interaction between the

C_α-2*p*-orbital and the C_β —M-σ-bond is also the same. Hyperconjugation has been invoked to explain why the simple β-substituted ethyl radicals $\cdot CH_2CH_2MR^1_n$ ($MR^1_n = SiR_3$, GeR_3 at SnR_3) adopt conformation (**1**; $R^2 = H$).[15] If this were the sole factor in stabilizing this conformation then $\cdot CH_2CH_2CR_3$ radicals would also be expected to exist in this conformation and not as is found in the experimentally observed

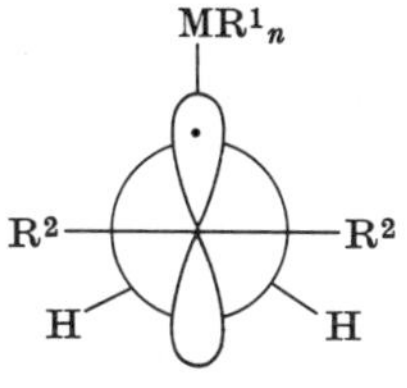

(**1**) 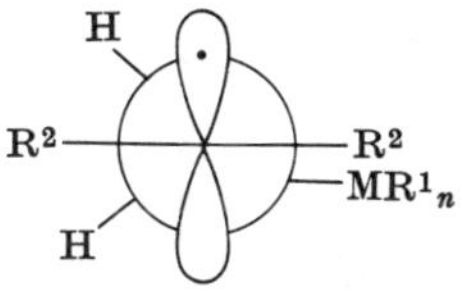 (**2**)

conformation (**2**; $R^2 = H$). This suggests *p–d* homoconjugation is necessary to stabilize the eclipsed conformation. Radicals containing β-phosphorus or arsenic groups show greatly enhanced isotropic coupling to ^{31}P or ^{75}As, as with the corresponding cations, and this is also taken as good evidence for homoconjugation.[16]

Further details of the ESR spectra of 2-chloroalkyl radicals have appeared.[17–19] Unsymmetrical chlorine-bridging has been confirmed by analysis of isotropic *g*-values and β-proton and ^{35}Cl splittings. The chlorine is eclipsed with the *p*-orbital at the radical centre (**3**), and the C—Cl bond is bent towards the *p*-orbital with displacement of the β-protons towards the nodal plane (**4**) in the most stable conformation of the $\cdot CH_2CH_2Cl$ radical. This structure is in agreement with INDO calculations.[20] The symmetry of the

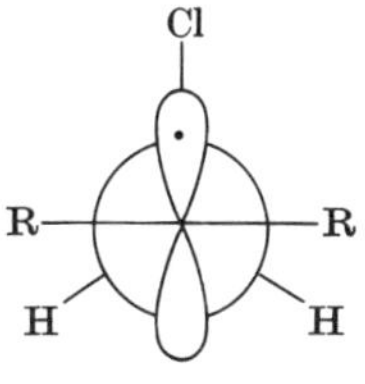

(**3**) 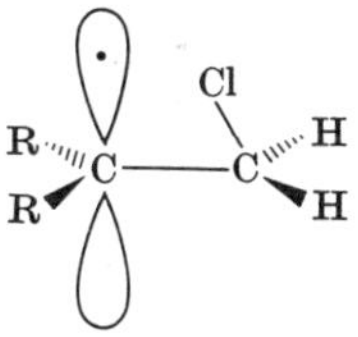 (**4**)

bridged radical is increased by methyl substitution at the α-carbon as indicated by the increase in the ^{35}Cl splitting constant. Further methyl substitution at the β-carbon atom decreases the extent of bridging because of steric interaction between the α- and the β-methyl groups. The ^{35}Cl splitting constant is almost temperature-invariant for $\cdot CMe_2CH_2Cl$, indicating that this radical is in a locked conformation (**3**; R = Me). The degree of bridging is less in 1,2-dichloroalkyl radicals because the radical centre is pyramidal, and also in the $\cdot CH_2SiMe_2Cl$ radical. The ^{35}Cl splitting constant for the latter radical is low and temperature-independent, indicating virtual free rotation about the C—Si bond; the absence of bridging here is consistent with the importance of hyperconjugation in determining conformational preference in radicals; hyperconjugative interaction with the Si—Cl bond would be much less than with the C—Cl bond. CIDNP effects observed during the decomposition of 3-benzoyl-3-bromopropionyl peroxide indicate that the methylene groups in the derived β-bromoethyl radical are non-equivalent, consistently with an unsymmetrical bromine bridge.[21] The preferred conformations of the β-fluoroalkyl radicals $\cdot CH_2CH_2F$, $\cdot CH_2CHF_2$ and $\cdot CH_2CMe_2F$ are

(**5**), (**6**) and (**7**), respectively.[22] Unlike their chloro-analogues there is virtually free rotation in these radicals.

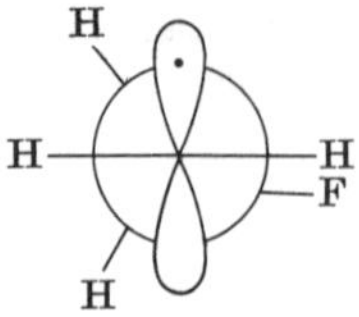

(**5**) 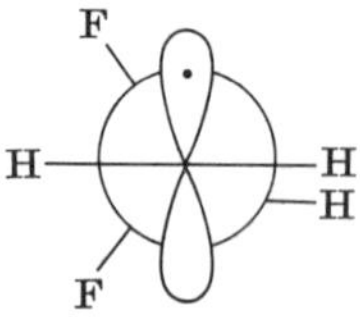 (**6**) 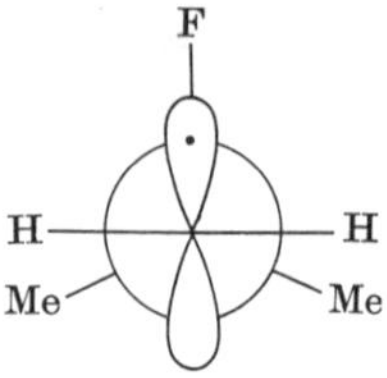 (**7**)

The preferred conformations of the substituted cyclopropylethyl and allylcarbinyl radicals (**8**) and (**9**) are as shown; in both the β-protons become non-equivalent at low temperatures, indicating restricted rotation about the $C_\beta - C_\gamma$ bond.[23,24] Radical (**9**) was formed by hydrogen abstraction from 1,1,2,2-tetramethylcyclopropane and

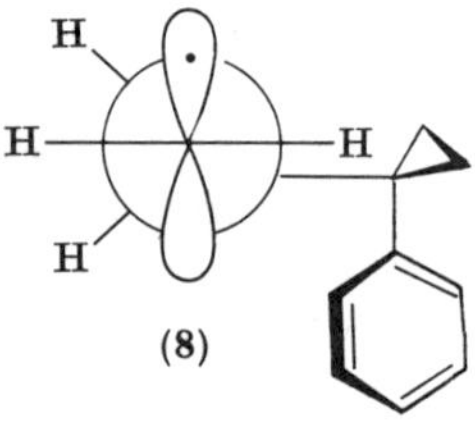

(**8**) 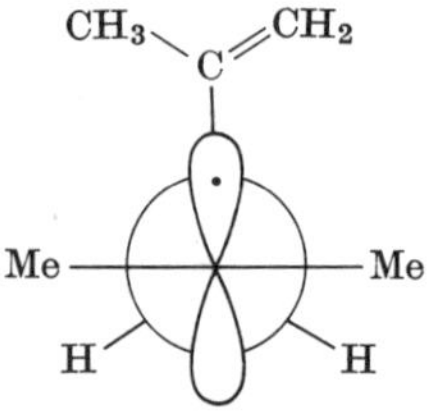 (**9**)

rearrangement of the resulting cyclopropylcarbinyl radical. An earlier report that this reaction gives the 1,1,3,3-tetramethylallyl radical by rearrangement of an initially formed tetramethylcyclopropyl radical was shown to be in error. The conformations of 2-carboxyethyl radicals, $ROCOCH_2CH_2\cdot$ have also been studied.[25]

The adduct of the tributyltin radical to cyclopentadiene has been shown to have the classical structure (**10**) rather than the bridged (**11**) or the π-complex (**12**) structure.[26]

(**10**)  (**11**)  (**12**)

There is significant delocalization on to the halogen atom in the α-haloalkyl radicals $\cdot CHXCOOH$ and $\cdot CHXCONH_2$.[27,29] These radicals and also $\cdot CR(OH)COOH$ radicals are planar.[30] The carboxyl group apparently inhibits the halogen or hydroxyl groups from inducing bending at the radical centre. $\cdot CR^1(OH)COOR^2$ radicals have also been shown to exist in both *cis*- and *trans*-forms (**13** and **14**).[31]

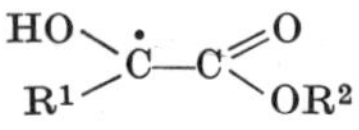

(**13**) 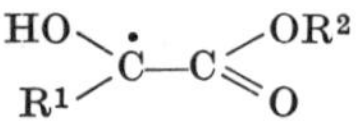 (**14**)

Replacement of fluorine by trifluoromethyl groups has been shown to decrease the deviation from planarity in perfluoroalkyl radicals.[32] Steric effects outweigh electronegativity effects in determining the configuration of these radicals. α,α-Difluorobenzyl radicals have also been shown to be planar and thus they resemble the α,α-difluorobenzyl cation rather than the non-planar anion.[33]

α-Thioalkoxy groups have been shown to be much less effective at inducing bending in alkyl radicals than α-methoxy-groups. This has been inferred from the ratio $a(\beta\text{-Me})/a(\alpha\text{-H})$ which is close to 1.2. This is the typical value for planar radicals and is substantially lower than is found for pyramidal radicals.[34]

Confirmation of a bent structure for the $\cdot CH_2OH$ radical has come from IR studies.[35,36] Bending at the radical centre in oxiranyl radicals has been demonstrated by the observation of the isomeric radicals (**15** and **16**) as distinct species.[37] The pyramidal nature of the *cis*- and *trans*-2,4-dimethyl-1,3-dioxolan-2-yl radicals (**17** and **18**) has been established by their ability to add to alkenes with partial retention of configuration.[38] ESR studies have confirmed the pyramidal character of the 2-methyl-1,3-dioxolan-2-yl radical, and thermodynamic parameters relating to the inversion of this radical have been measured.[39]

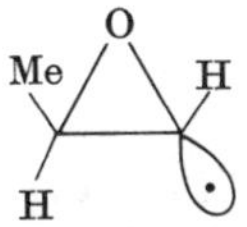

(**15**) 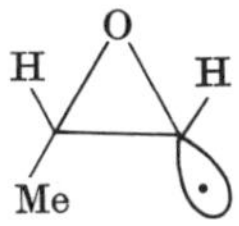 (**16**) 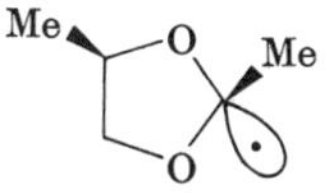 (**17**) 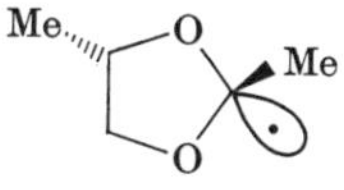 (**18**)

The preferred conformation of disubstituted radicals $\cdot CHYCH_2X$ has been related to whether the radical centre is (*a*) electron-rich, in which case, for a group X that is inductively electron-withdrawing, the eclipsed conformation (**19**) is strongly favoured, or (*b*) electron-deficient, in which case the conformation (**20**) is slightly preferred.[40] These preferences are reversed when X is electron-donating. The results parallel those of

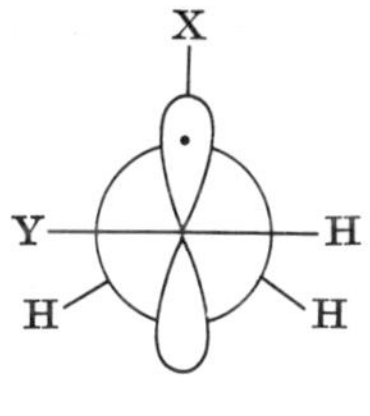

(**19**) 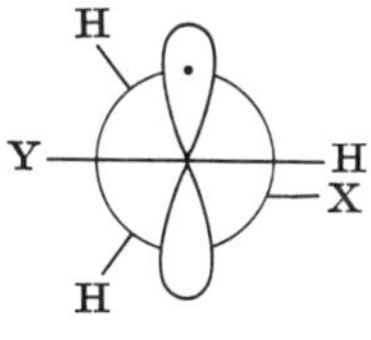 (**20**)

MO calculations for $^{-}CH_2CH_2X$ and $^{+}CH_2CH_2X$, which indicate that, with X electron-withdrawing, the former prefers the eclipsed conformation, and again the reverse holds with change in the electrical character of X.

Studies on the ^{13}C and ^{17}O splittings in the ESR spectra of the 2-oxocyclohexyl radical (**21**) show that there is definitely some delocalization of unpaired spin on to oxygen,[41,42] structure (**21b**) contributing about 15% to the resonance hybrid.

The implication of high configurational stability for styryl radicals in that they can be scavenged before isomerization by bromotrichloromethane but not by carbon tetrachloride has been shown to be misleading in that the β-halostyrenes produced are equilibrated under the reaction conditions.[43] Calculations have been made on the configurational stability of 1-substituted vinyl and cyclopropyl radicals. Substituents

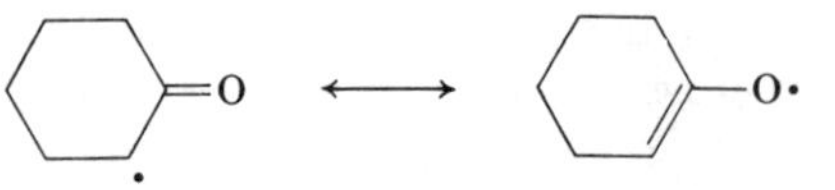

(21a) **(21b)**

possessing unshared electron pairs or C—H bonds increase the configurational stability of the radical.[44a] α-Methoxy-groups confer significantly less configurational stability on cyclopropyl radicals than on vinyl radicals.[44b] In contrast to the majority of vinyl radicals, $H_2C{=}\dot{C}COOH$ radicals are linear;[45] this configuration is presumably favoured by overlap of the *p*-orbital with the π-orbitals of the carboxy-group, yet surprisingly there is little delocalization of spin on to the carboxy-group.

Acyl[7] and iminoyl radicals (•CR=NR′)[46] are both σ-radicals, as indicated by their low *g*-values (less than the free-spin value) and low β-proton coupling constants. The relatively high stability of the iminoyl radical is ascribed to resonance interaction of the unpaired electron with the nitrogen lone pair.

$$R^1\text{—}\dot{C}{=}\ddot{N}\text{—}R^2 \longleftrightarrow R^1\text{—}\ddot{\bar{C}}\text{—}\dot{\overset{+}{N}}\text{—}R^2$$

Analysis of the ^{13}C splittings in the ESR spectra of adamantyl radicals in the solid state confirms that these radicals are pyramidal with approximately 12% of *s*- and 66% of *p*-character at the radical centre.[47,48]

Full details of the ESR spectra of 2-norbornyl radicals such as (**22**; X = CH_2) and their 7-oxa-analogues (**22**; X = O) have been published. These confirm the pyramidal nature of the radical with the C—H bond bent in the *endo*-direction.[49] The radical centre becomes increasingly pyramidal as electronegative groups are introduced at C-3. INDO calculations indicate that the orbital containing the unpaired electron is expanded towards the *exo*-direction, consistently with the preference for attack from this side.[50] The extent of attack from the *exo*-side is reduced in the apobornyl series where the methyl groups at C-7 provide hindrance to attack from that side.[51]

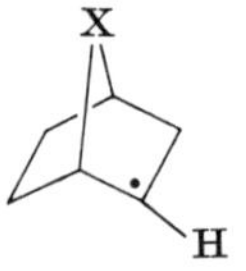

(22)

Decompositions of *tert*-butyl *syn*- and *anti*-norbornen-7-yl peroxyformate afford the same mixture of *syn*- and *anti*-7-*tert*-butoxynorbornene in a ratio of 1:2.7.[52] Deuterium-labelling at the 7-position in the perester indicated that hydrogen-abstraction from the solvent by the norbornenyl radical also occurred preferentially from the *anti*-side of the radical. These results, together with a detailed analysis of the ESR spectrum of the 7-norbornenyl radical, indicate that there is a dynamic equilibrium between the *syn*- and the *anti*-radical with a double-minimum potential well. The potential well is deeper on the *anti*-side, thereby explaining the preference for attack from that side.

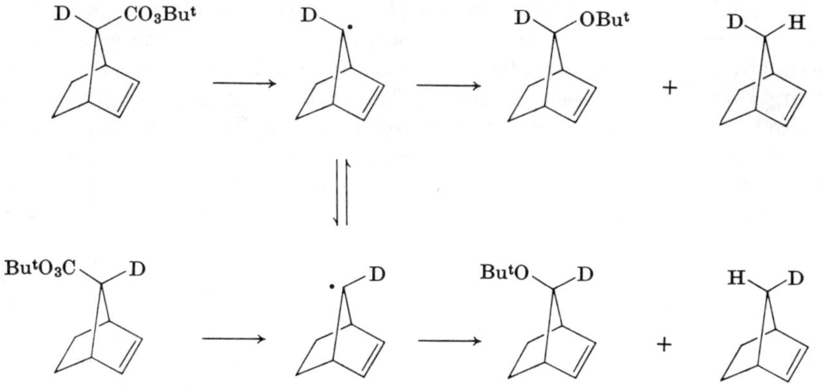

Reviews on triarylmethyl radicals have appeared.[53] It has been pointed out that the published ESR spectra of diphenylmethyl and 9-fluorenyl radicals have smaller proton coupling constants than would be expected and can be ascribed to the radicals (**23**) and (**24**).[54] This has stimulated further interest in this area and resulted in the unambiguous generation and observation of diphenylmethyl radicals,[55] which have significantly

$Ph_2CH\dot{C}Ph_2$

(**23**)

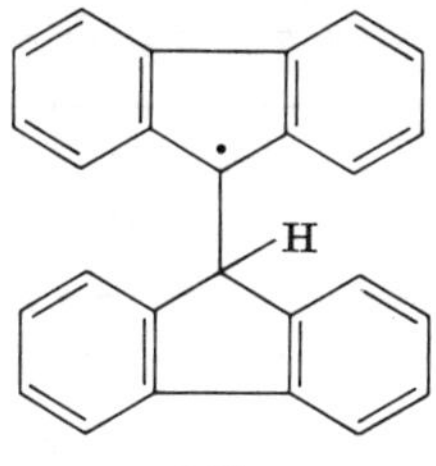

(**24**)

higher coupling constants than (**23**) but slightly less than di-(2-*tert*-butylphenyl)methyl radicals in which there is less delocalization of spin due to twisting of the phenyl groups. The 1,1-diphenylethyl radical has also been examined and the spin density at C-1 shown to be 0.54.[56] The triphenylmethyl radical has been shown to be appreciably twisted in solution,[57] as has the 1-phenoxy-1,1-diphenylmethyl radical.[58]

Stable radicals (**26**) have been generated from 4,4′-diaryl-4,4′-bipyrazolones (**25**). The extent of dissociation is increased by *ortho*- and particularly di-*ortho*-substitution in the 4-aryl groups as a consequence of steric hindrance to dimerization.[59]

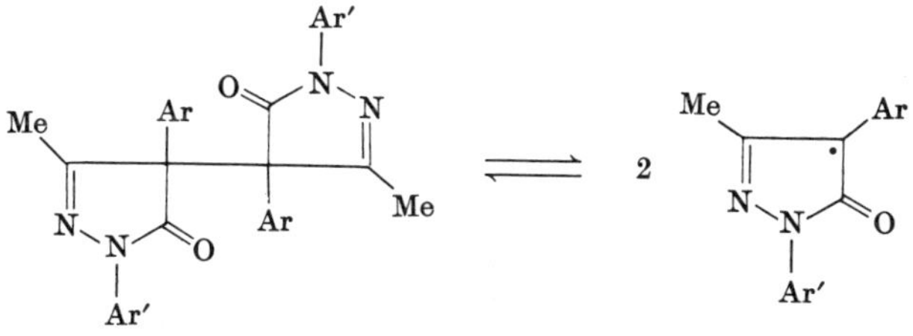

(**25**) ⇌ 2 (**26**)

The stabilization energies of a variety of delocalized radicals have been correlated and discussed.[60] The bond-dissociation energies of propene and propyne have been determined by mass spectrometry, and hence the stabilization energies of allyl and propargyl radicals have been obtained.[61] The stabilization energy of the allyl radical has also been determined by studying the isomerization of hexa-1,3,5-trienes.[62] Methyl substituents stabilize allyl radicals[63] by about 2–3 kcal mol^{-1}, and hydroxyl groups[64] have a somewhat smaller stabilizing influence. Halogens apparently reduce the stabilization energy of allyl radicals, though this could be due more to increased stabilization in the reference compound.[65,66]

α-Cyclopropyl groups have been claimed to exert a significant stabilizing effect on radicals,[67] but this has been disputed by Beckhaus and Rüchardt who showed that there was only a small difference in the activation energies for dissociation of hexaethylethane and hexacyclopropylethane, and that this difference could easily be accounted for by the different steric requirements of the triethylmethyl and tricyclopropylmethyl radicals.[68]

The stability of radicals is affected by the degree of overlap of the *p*-orbital at the radical centre with an adjacent π-system. If this is reduced by steric effects, the stability of the radical is reduced. In accord with this, the relative rates of hydrogen-abstraction from benzocycloalkenes by trichloromethyl radicals are sensitive to ring-size effects

Table 1. Relative rates of hydrogen-abstraction from benzocycloalkenes $C_6H_4(CH_2)_n$ by trichloromethyl radicals

n	4	5	6	7
Rel. rate[a]	0.09	1.65	1.68	0.50

[a]Rate relative to *o*-diethylbenzene = 1.00.

(cf. Table 1).[69] The stability of the radical (**27**; R = Me) is less than that of (**27**; R = H), as judged from the rates of decomposition of the parent azo-compounds; steric interaction between the isopropyl group and the *ortho*-hydrogen atoms prevents the radical attaining a planar conformation.[70]

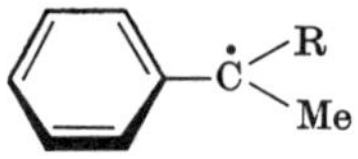

(**27**)

Nitroxides

The evidence for the planarity and for the non-planarity of the NO group in nitroxides has been summarized.[71] Non-planarity has been established in a series of bicyclic nitroxides (**28–32**) by ESR spectroscopy[71,72] and X-ray crystallographic analysis.[73,74]

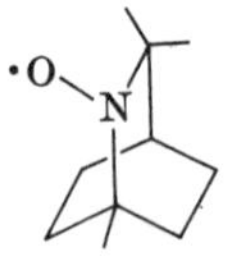

(**28**) 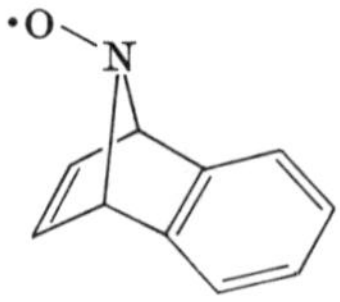 (**29**) 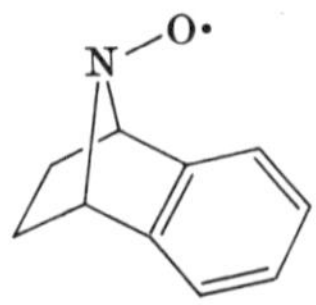 (**30**)

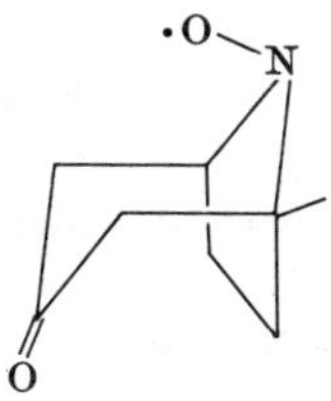

(31) 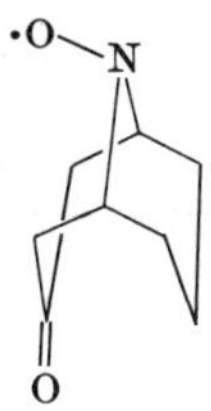 (32)

In (**28**) and probably also in (**29**–**32**) the NO group inverts very rapidly. The extent of bending in (**28**), (**31**) and (**32**) is 35°, 24.9° and 30.5° respectively.

Bicyclic nitroxides are much more stable than their acyclic or monocyclic counterparts, as disproportionation would lead to a nitrone with a bridgehead double bond in contravention of Bredt's rule.[75] Somewhat surprisingly (**34**) is more stable than (**33**): the reverse order of stability is normal in bicyclic compounds. The extra reactivity of (**33**) is attributed, at least in part, to overlap of the backside orbitals of the bridgehead positions with the orbitals of the NO group.

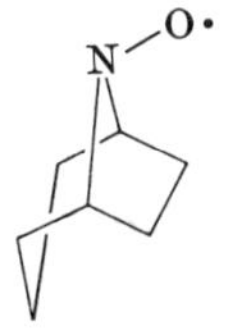

(33) 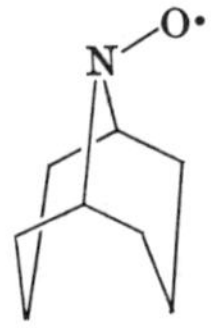 (34)

The stability of pyrrole nitroxides is markedly increased by bulky substituents in the 2- and the 5-position.[76]

Azomethine nitroxides can exist in either of the conformations (**35**) and (**36**). The latter is preferred for steric reasons when $R^2 \neq H$, but when $R^2 = H$ the preferred conformation depends on solvent polarity.[77] Dipolar interactions are weaker in the conformation (**36**), so this is the predominant form in non-polar solvents.[77] When R^2 is large the azomethine and the nitroxide groups are twisted out of coplanarity,[78] which is reflected in a higher value for a_N (NO) and a very small value for $a_N(C=N)$. The spin density on the nitroxide nitrogen is also dependent on the electronic nature of R^2 in (**36**) and of X in (**37**), being decreased by electron-withdrawing substituents.[77] Acyl-*tert*-butylnitroxides can also adopt analogous conformations, that resembling (**36**) being favoured when $R \neq H$.[80]

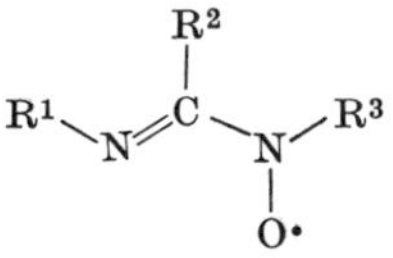

(35) 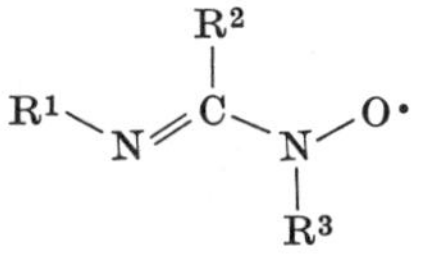 (36) 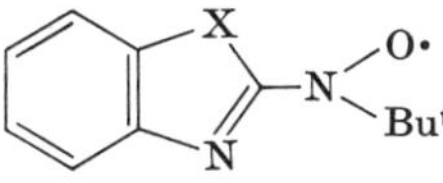 (37)

The ESR spectra of bis(trialkylsilyl)nitroxides (and their germyl counterparts) $(R_3M)_2NO\cdot$ are characterized by very low a_N values, these being consistent with delocalization of unpaired spin into the vacant *d*-orbitals of silicon or germanium.[81] The *g*-values of *N*-alkoxy-*N*-arylnitroxides are anomalously low, suggesting that the nitrogen radical centre is pyramidal.[82]

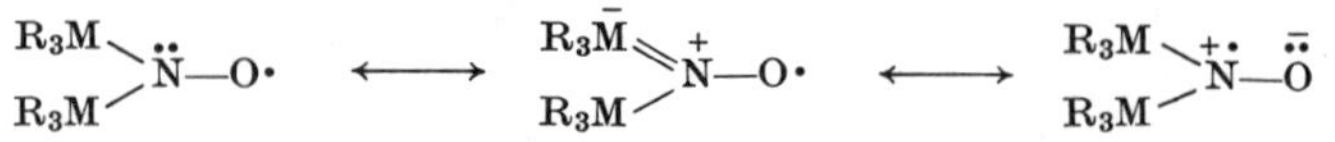

R_3M N—O· ⟷ $R_3\bar{M}$ $\overset{+}{N}$—O· ⟷ R_3M $\overset{+\cdot}{N}$—$\bar{O}$

The nitroxide (**38**) forms stable 1:1 complexes with zinc and other diamagnetic metals.[83] In these complexes there is significantly less spin density on the nitroxide-nitrogen atom and more on the other nitrogen, owing to delocalization of spin on to the metal.

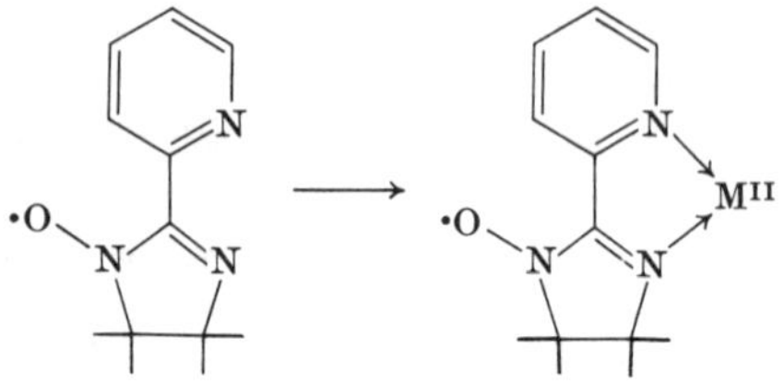

(**38**)

Nitrogen Radicals

Reviews on triarylimidazolyl[84] and hydrazidinyl radicals[85] (including 1,2,4,5-tetra-azapentenyls, verdazyls and tetrazolinyls) have appeared, and the crystal structure of 2,4,6-triphenylverdazyl has been reported.[86] The ESR spectra of 1,2,4,5-tetra-azapentenyls indicates their allylic character.[87]

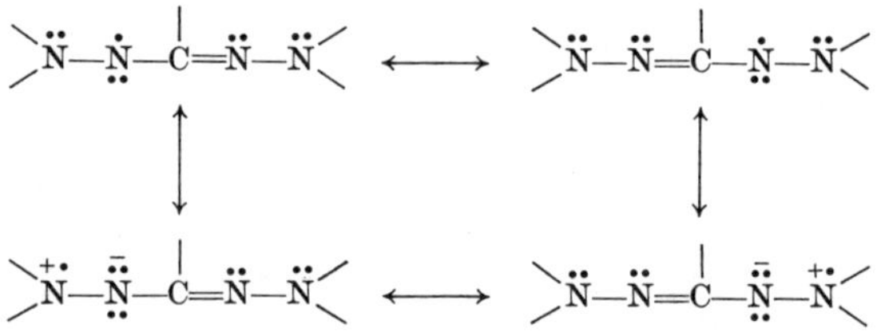

2,2-Dialkylhydrazyl radicals, $R_2N\dot{N}H$, have been shown to be π-radicals.[88,89] The α-hydrogen splitting is less in 2,2-dialkylhydrazyls than in the parent hydrazyl, owing to increased importance of (**39b**) in the resonance hybrid. The alkoxyamino-radicals $RO\dot{N}R$, $RO\dot{N}Ar$ and $RO\dot{N}COOEt$ are also π-radicals.[82] The N-alkoxy-group increases

(**39a**) (**39b**)

delocalization of the unpaired spin. In the N-alkoxy-N-arylamino-radicals both electron-releasing and electron-attracting substituents in the p-position increase the delocalization of the unpaired electron and can accordingly be classified to Walter's S class.

The stability of the bicyclic hydrazyls (**40**–**42**) is in the order (**41**) > (**40**) ≫ (**42**),[90] the much lower stability of (**42**) being due to its ability to disproportionate. There is no evidence for the dimerization of (**40**) or (**41**).

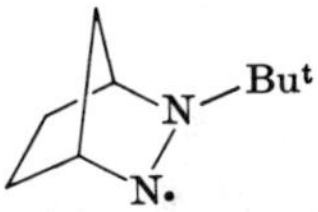

(40) 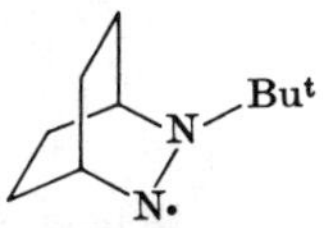 (41) 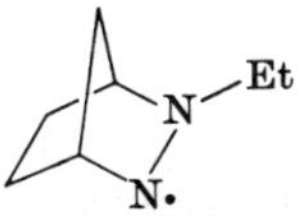 (42)

Conditions for radical stability include steric screening of sites of high electron density, delocalization of unpaired spin, inability to disproportionate and steric hindrance to dimerization, and they are found in the radicals (**43**)–(**49**).[91–96]

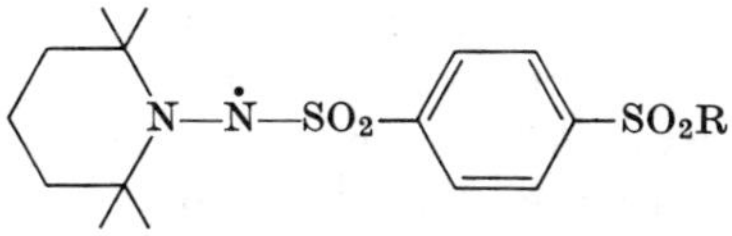

(43)

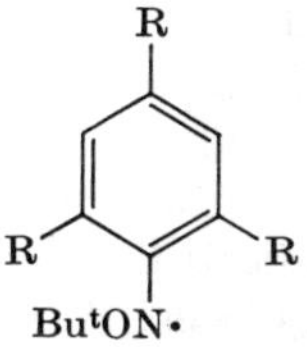

$R = Bu^t$, Ph or OMe

(44)

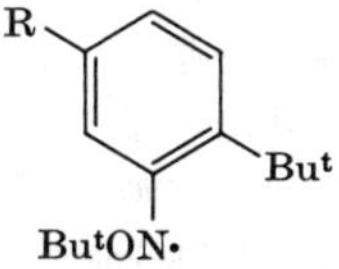

$R = Bu^t$, H

(45)

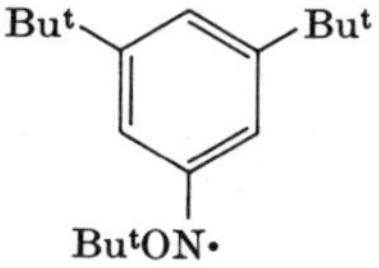

(46)

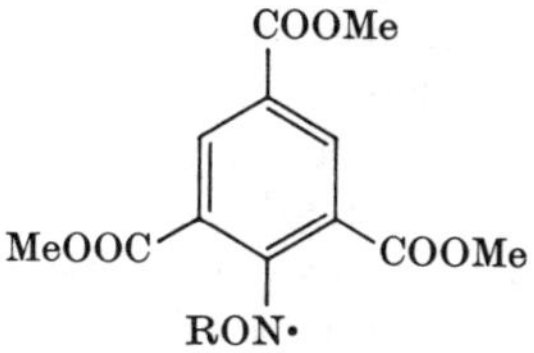

$R = Pr^i$ or Ph_3C

(47)

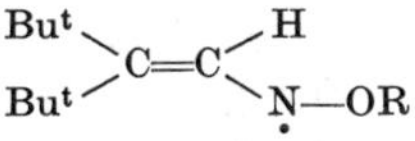

(48)

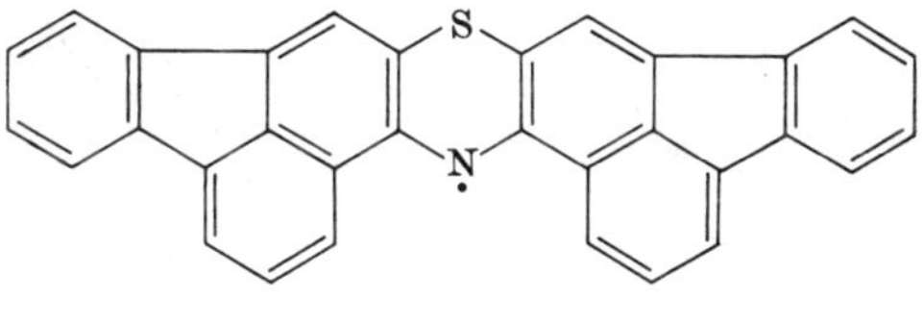

(49)

The unusual stability of the imino-radical (**50**) has been attributed to hyperconjugation;[97] the planar structure and the short C═N bond allow efficient overlap between

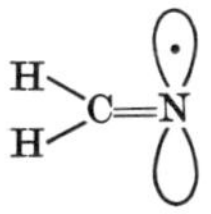

(50)

the *p*-orbital and the σ-C—H bonds, and considerable orbital readjustment would be necessary before dimerization.

Miscellaneous Radicals

The pyramidal nature of trialkylsilyl radicals has been confirmed by demonstration that asymmetric radicals react with retention of configuration at the radical centre.[98]

$$R_3Si^*H + \cdot CCl_3 \rightarrow R_3Si^*\cdot + CHCl_3$$

$$R_3Si^*\cdot + CCl_4 \rightarrow R_3Si^*Cl + \cdot CCl_3$$

This is not the case for $R_3Si\dot{S}iR_2$ radicals, reactions of which proceed with racemization, pointing to the planarity of such radicals.

The tin radical $([Me_3Si]_2CH)_3Sn\cdot$ is unexpectedly stable owing to its reluctance to dimerize or disproportionate for steric reasons and because the Sn—H bond is weak.[99] The ESR spectrum of the $Ph_2\dot{P}{=}S$ radical can be interpreted by simple sp^n hybridization of phosphorus without invoking the participation of the *d*-orbitals of phosphorus.[100] Sulphonyl radicals $RSO_2\cdot$ have been generated by reaction of sulphonyl halides with triethylsilyl radicals;[101] their ESR spectra indicate that they are pyramidal and that there is appreciable hindrance to rotation about the C—S bond. The structure of sulphanyl radicals $(CF_3O)_3S\cdot$ and $(CF_3O)_2\dot{S}F$ have also been examined.[102]

Radical Anions

The spin densities in the radical anions derived from benzocycloalkenes are sensitive to ring-strain effects, increasing at the position α to the aromatic system with increasing ring strain.[103] There is also substantial delocalization of spin on to the alkyl groups or alicyclic ring in 1,8-disubstituted[104] and pericyclic naphthalenes[105] and in pleiadenes,[106] although the degree of spin delocalization is claimed to be insensitive to ring strain. There is disagreement as to whether the radical anion derived from (**51**) is homoaromatic, i.e. can be represented by (**52**).[107, 108]

The ESR spectrum of the radical anion of azulene can be interpreted in terms of a large contribution from structure (**53**) to the resonance hybrid;[109] the calculated spin

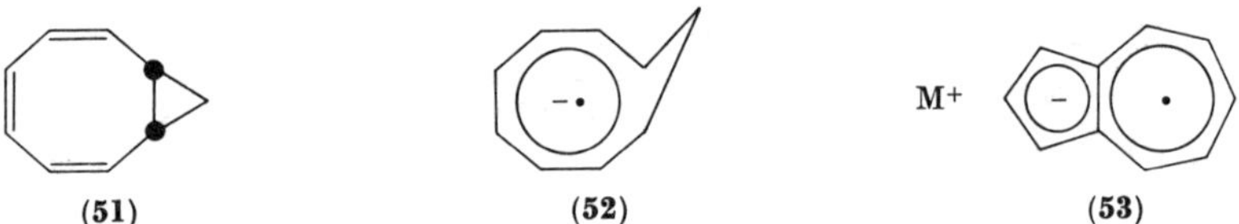

(**51**) (**52**) (**53**)

densities depend somewhat on the counterion, reflecting the importance of ion-pair phenomena. The importance of such effects is even more clearly seen in the radical anion of 5*H*-dibenzo[*a*,*d*]cycloheptene, which can exist in different conformations according to whether it exists as a contact or solvent-separated ion-pair.[110]

ESR studies have established that the unpaired electron is delocalized over the whole system in the radical anion of 2-substituted thieno[3,2-*b*]thiophens (**54**), but in the isomeric 2-substituted thieno[2,3-*b*]thiophens (**55**) the condensed thieno-group is

(**54**) (**55**)

scarcely involved.[111] These results correlate well with the faster phenylation of the first compound and also with the products of phenylation.[112] Conformational preferences for the alkyl groups in 6α-thiathiophenyl radical anions have been determined.[113]

The radical anion of benzocyclo-octatetraene is shown to be present in diphenylmethane radical anions.[115] the spin densities have been calculated for the radical anions of cyclazines,[116] calicenes,[117] diphenylcyclobutene,[118] (π-cyclopentadienyl)-(π-cycloheptatriene)chromium(I)[119] and the trianion radical of heptafulvene.[120] Delocalization of spin on to the metal atom occurs in (**56**).[121]

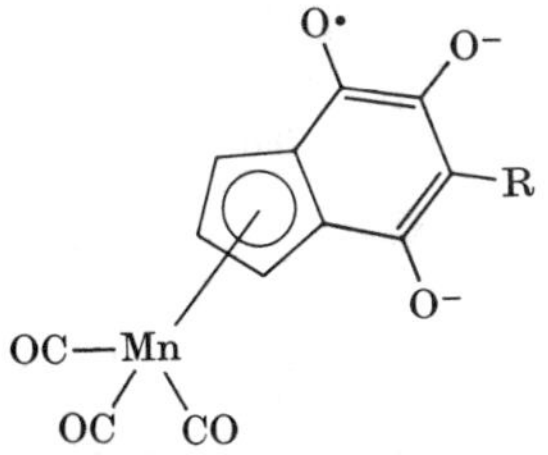

(**56**)

Considerable attention has also been devoted to the radical anions of nitro-compounds. The nitrogen is pyramidal in $MeNO_2^{-}\cdot$ and $EtNO_2^{-}\cdot$ but planar in $PhNO_2^{-}\cdot$.[122] *ortho*-Substituents induce twisting and bending of the nitro-group of substituted nitrobenzene radical anions. Twisting is also observed in the radical anions of 3-nitrotropolones.[123] Studies of the radical anions of *p*-dinitrobenzene[124] nitrobenzobicycloalkanes,[125] and polynitroacenaphthenes[126] have also been reported. The spin density is primarily located in the nitroso-nitrogen atom of the nitrosamine radical anions $R_2N{-}\dot{N}{-}O^-$.[127]

The radical anion of pyrrole (and similarly pyrazole and imidazole) exists in the α-pyrrolenine form (**57**), and that of phenol in the keto-form (**58**).[128] The radical anion of indole takes up mainly the structure (**59**) and thus behaves more like phenol than pyrrole. The observed reversal of the normal tautomeric equilibrium in these cases indicates that α-pyrrolenine and cyclohexadienone are better electron acceptors than pyrrole and phenol.

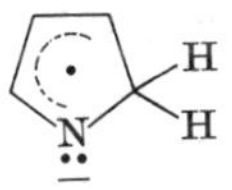

(**57**) 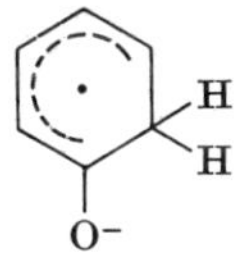 (**58**) 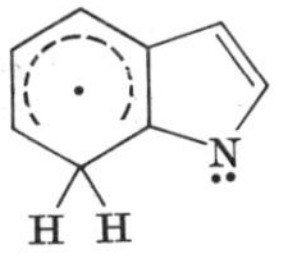 (**59**)

Radical Cations

The extents of delocalization of unpaired spin in the radical cations $PhH_2N^{+}\cdot$, $Ph_2HN^{+}\cdot$ and $Ph_3N^{+}\cdot$ are similar in all cases.[129] Delocalization by the greater number of phenyl groups is counterbalanced by increasing twist of these groups. The nitrogen radical centre is planar in all cases. The stability of *N*-substituted dimethylaniline[130] and 3-(arylamino)indole radical cations[131] is increased by electron-donating substituents. Trimethylsilyl and trimethylgermyl substituents enhance the stability of radical cations by *p*–*p* and *p*–*d* hyperconjugative interactions.[130] In the radical cation of *p*-phenylenediamine the unpaired spin is largely delocalized into the aromatic nucleus.[132]

The radical cations $Ar_3As^{+\cdot}$, unlike their nitrogen analogues, are pyramidal; the unpaired electron is in an essentially 4*p*-orbital but overlap between this and the aromatic rings is so inefficient that there is virtually no delocalization of spin.[133] The stability of radical cations of 2,5-disubstituted *p*-dimethoxybenzene derivatives is enhanced by electron-donating substituents.[134] The preferred conformation of radical cations of 2,5-bis(alkylthio)thiophens is (**60**).[135]

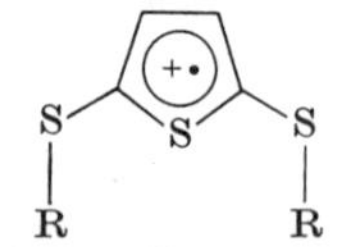

(**60**)

Azo-compounds[136]

Interest continues into the precise mode of decomposition of azoalkanes, particularly on the role of *cis*-azo-compounds in photolysis[137] and on the question of one- or two-bond cleavage.[138] The results of the photolysis of a series of phenylazoalkanes are rationalized in Scheme 1.[139,140] The relative importance of the two routes depends on the stability of the derived radical R·, route b being favoured when R· is very stable ($PhCMe_2$ or $CH_2{=}CHCMe_2$) but not when R = Bu^t or Ph. Isomerization of the *trans*-azo-compound occurs either directly (R = Ph or Bu^t) or by a radical-pair mechanism. The precise energy levels involved in the isomerization process have been calculated.[141]

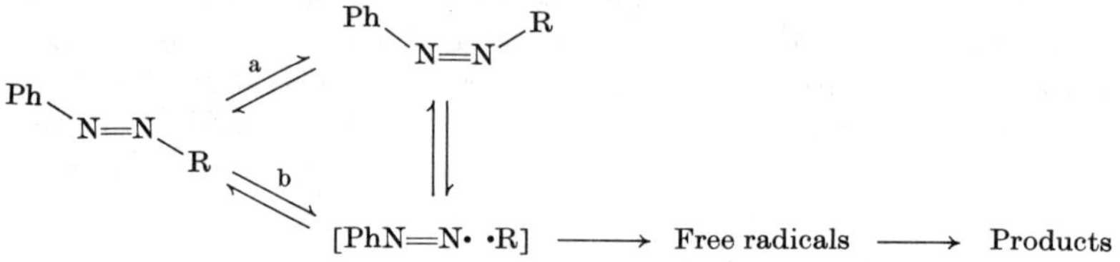

[PhN=N· ·R] ⟶ Free radicals ⟶ Products

SCHEME 1.

Support for essentially one-bond cleavage in the photolysis of phenylazoalkanes when R· is a stable radical comes from the virtual absence of 2,2-diphenylpropane (the phenyl–cumyl radical-pair product).[139] This suggests that scission of the C—N bond in the phenyldiazenyl radical occurs outside the solvent cage; one-bond scission is also implicated by dependence of the rate of decomposition on solvent viscosity.[142] Arylazobisnitriles (**61**) similarly decompose by a one-bond process, evidence coming from the non-stoichiometric evolution of nitrogen and formation of (**62**).[143] Further support for a non-concerted bond cleavage in gas-phase decompositions of azoalkanes has been

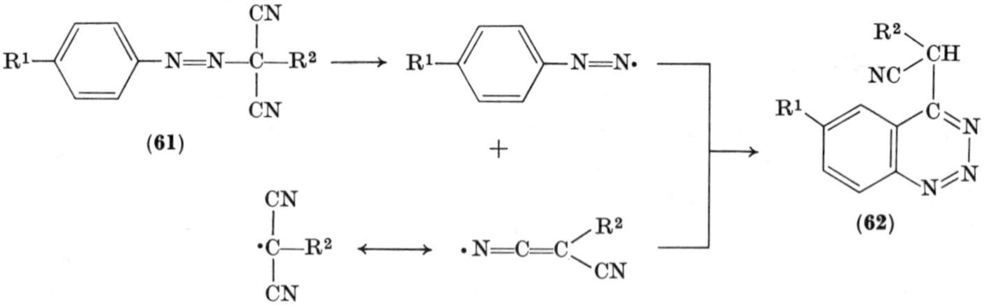

(**61**) (**62**)

obtained from mass-spectroscopic studies.[144] However, for a wide range of unsymmetrical azoalkanes, the results of a comparison of the rates of decomposition of unsymmetrical azoalkanes RN=NR′ with those of the two symmetric analogues RN=NR and R′N=NR′[145] lead to the conclusion that two-bond scission is generally preferred. This is true even when the derived radicals are of very disparate stability, e.g. *tert*-butyl and 1-norbornyl.

The evidence for one-bond cleavage of (**63**) to the diazenyl radical (**64**)[146] has been

$$\underset{(\mathbf{63})}{\text{EtOOC—N=}\dot{\text{N}}\text{—COOEt}} \rightarrow \underset{(\mathbf{64})}{\text{EtOOC—N=N}\cdot} + \cdot\text{COOEt}$$

shown to have been misinterpreted by identification of the spectroscopically observed radical as (**65**).[147] This had also been claimed to have the structure (**66**).[148]

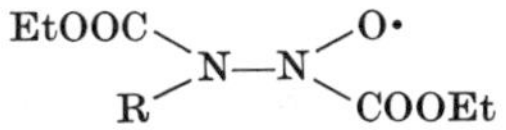

(**65**): (EtOOC)(R)N—N(O•)(COOEt)

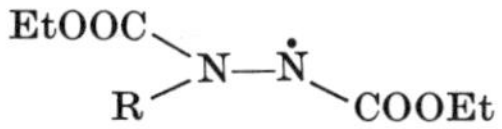

(**66**): (EtOOC)(R)N—Ṅ(COOEt)

Decomposition of acylazonitriles $R^1CON{=}NC(CN)R^2_2$ proceeds by a concerted two-bond process with the acyl C—N bond cleavage less advanced than that of the other C—N bond.[149] Substituents in the benzoyl group of benzoylazonitriles have little influence on the rate of decomposition, indicating that polar effects are unimportant in the transition state. The predominant coupled product (**67**) from these compounds arises from attack by the acyl radical on the more accessible imine-nitrogen atom.

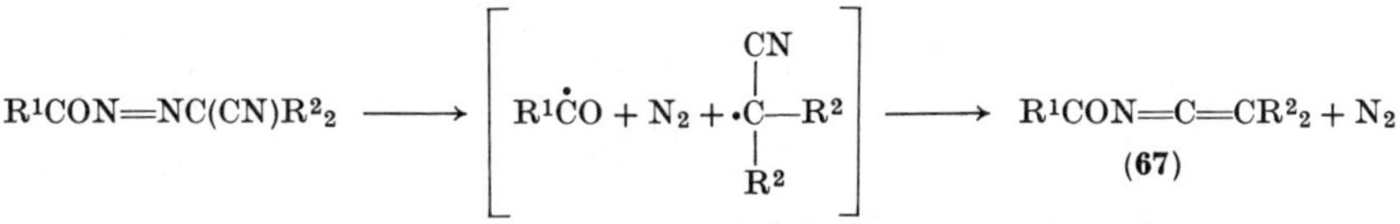

$$R^1CON{=}NC(CN)R^2_2 \longrightarrow \left[R^1\dot{C}O + N_2 + \cdot C(CN)R^2R^2 \right] \longrightarrow \underset{(\mathbf{67})}{R^1CON{=}C{=}CR^2_2} + N_2$$

The rates of decomposition of the azoalkanes (**68**) (see Table 2) and (**69**) have been shown to be sensitive to steric effects.[150,151] Back-strain effects in the parent azo-compound enhance the rate of decomposition. This also seems to be operative in phenyl-

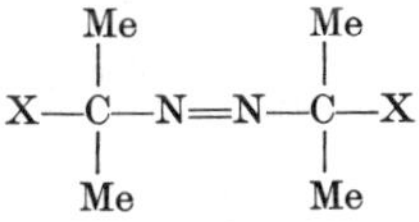

(**68**): X—C(Me)(Me)—N=N—C(Me)(Me)—X

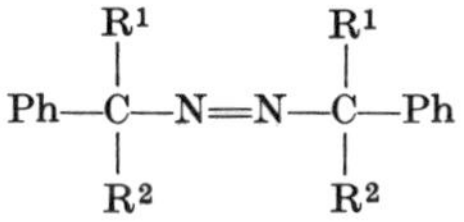

(**69**): Ph—C(R¹)(R²)—N=N—C(R¹)(R²)—Ph

Table 2. Relative rates of decomposition of azoalkanes (**68**) at 100°

X =	H	Me	Bu^t	CH_2Bu^t	OMe	SMe
Relative rate	1	5.6×10^2	5.8×10^3	7.2×10^5	5.8×10^3	1.3×10^7

azonitriles.[152] The rate of decomposition is, however, reduced under certain circumstances when the stability of the radical is reduced by steric interactions that prevent effective delocalization of unpaired spin. The rate of decomposition of azoalkanes is also enhanced by α-thio-groups (X = SR).[150]

The rates of decomposition of 1,1′-disubstituted azocycloalkanes (**70**) are sensitive to ring-strain effects, suggesting that the transition state in these thermolyses bears a closer resemblance to the derived radical than to the starting material.[153,154] The

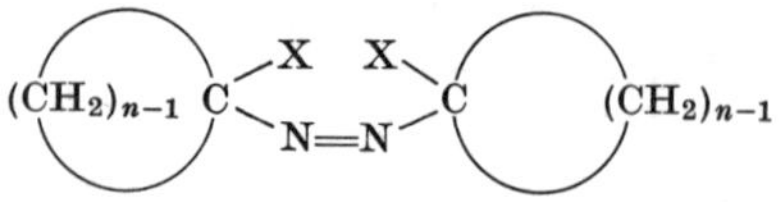

(**70**)

sensitivity of the rate of decomposition to ring size depends on the nature of the α-substituent, being greater the more advanced is bond-breaking in the transition state

Table 3. Relative rates of decomposition (k_5/k_6) (**70**; $n = 5$) and (**70**; $n = 6$)

X =	Ph	CN	Me
k_5/k_6	70	11.5	2.66

(cf. Table 3). An unusual product (**72**) which was isolated from the thermolysis of 1,1′-diphenylazocyclopropane (**71**) results from rearrangement rather than elimination of nitrogen.[154]

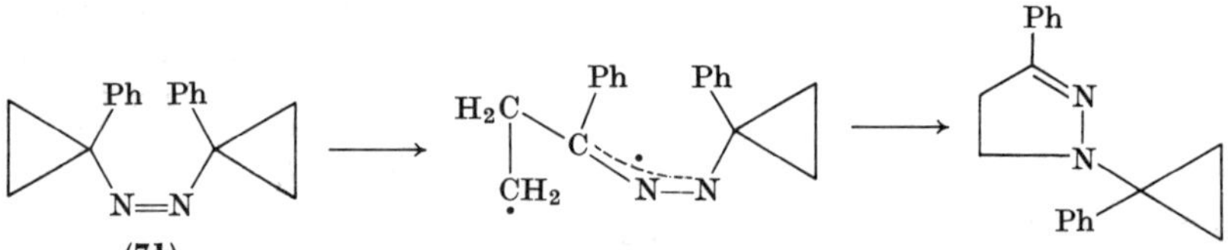

(**71**) (**72**)

The stability of compounds is dependent on the stability of the derived radicals. This accounts for the high thermal stability of divinylazo-compounds (**73**)[155] and for the very low stability of *p*-[(triphenylmethylazo)sulfonyl]toluene (**74**) which cannot even be isolated.[156]

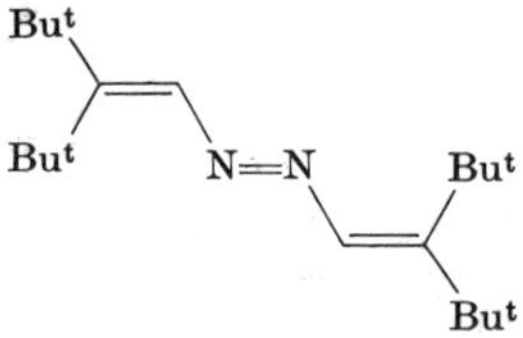

(**73**) 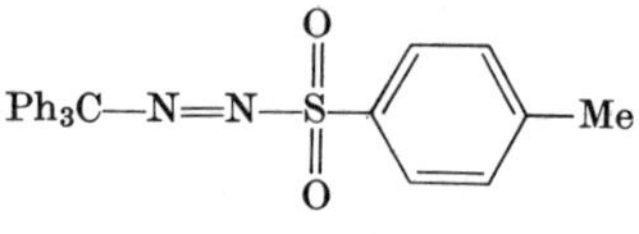 (**74**)

An *endo*-cyclopropane ring as in (**75**) has a dramatic accelerating effect on the decomposition of cyclic azo-compounds. The relative rates of decomposition of (**75**)–(**78**) are

(**75**) (**76**) (**77**) (**78**)

8.8×10^2, 1.1×10^{17}, 1 and 6.7×10^4, respectively. Decomposition in each instance occurs by a biradical mechanism. In (**76**) there is perfect alignment of overlapping orbitals resulting from disrotatory opening of the edge cyclopropane bond, whereas in (**75**) the orbitals are orthogonally orientated with little possibility of overlap.[157]

The rate of rotation of the 2-bromo-9-methylfluorenyl radical in the thermolysis of (**79**)[142] is very much slower ($r < 1$) than that of 1-phenylethyl radicals ($r \cong 15$),[158] which is attributed to the larger moment of inertia of the fluorenyl radical. Rotation is only possible about an axis coplanar with the aromatic system or by translation of the phenyl-diazenyl radical to the opposite face of the fluorenyl radical. The extent of rotation is also reduced in more viscous solvents in which the radical-pair would be more tightly bound.

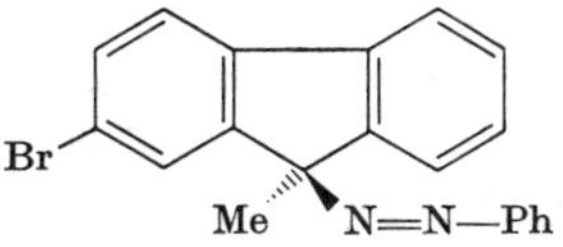

Br

Me N=N—Ph

(**79**)

The nature of the photolysis products of azobisisobutyronitrile (**80**; R = Me) are very dependent on the phase in which photolysis is carried out.[159] In solution there is little disproportionation (<5%) whereas in the crystalline state this was the predominant type of product (>95%). In both cases the dimeric product consisted of both the C—C and the C—N dimer. These results contrast with those obtained for 3,3-azodi-(3-cyanopentane) (**80**; R = Et) for which disproportionation is of little importance either in solution or in the solid phase, the main product being the ketenimine (**81**). This dramatic difference in product composition from the two azo-compounds is attributed

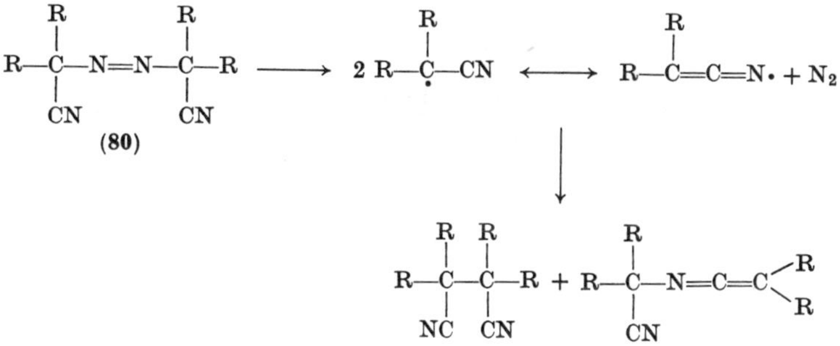

$$\mathrm{R{-}C(R)(CN){-}N{=}N{-}C(R)(CN){-}R} \longrightarrow 2\,\mathrm{R{-}\dot{C}(R){-}CN} \longleftrightarrow \mathrm{R{-}C(R){=}C{=}N\cdot} + \mathrm{N_2}$$

(**80**)

$$\downarrow$$

$$\mathrm{R{-}C(R)(NC){-}C(R)(CN){-}R} + \mathrm{R{-}C(R)(CN){-}N{=}C{=}C(R)_2}$$

(**81**)

to a less compact crystal structure in (**80**; R = Et).[160] Dipolar forces are important in determining product composition and in the less open crystal of azobisisobutyronitrile pivoting of a CN group about an axis perpendicular to the plane of the molecule would be opposed by dipolar forces exerted by CN groups on neighbouring molecules, thus inhibiting dimerization and allowing disproportionation to occur. This would only need rotation about the axis of the CN group to bring the methyl radical into the correct steric environment for disproportionation. These steric constraints are less severe in the more open structure of (**80**; R = Et).

Two types of S_H2 reaction have been encountered in reactions of aryl radicals with phenylazoalkanes involving attack at the azo-group and at the aromatic nucleus, respectively.[161]

$$\text{Ph}\cdot + \text{R—N=N—Ar} \longrightarrow \text{(Ph)(R)N—}\dot{\text{N}}\text{—Ar} \longrightarrow \text{R}\cdot + \text{Ph—N=N—Ar}$$

$$\text{Ph}\cdot + \text{R—N=N—C}_6\text{H}_4\text{—NO}_2 \longrightarrow$$

$$\text{R—N=N—C(Ph)(cyclohexadienyl radical)—NO}_2 \longrightarrow \text{R}\cdot + \text{N}_2 + \text{Ph—C}_6\text{H}_4\text{—NO}_2$$

Peroxides[162]

Further evidence has been obtained which indicates that the C_α—CO bond is little stretched in the transition state for thermolysis of peresters. Thus ring-strain effects have little influence on the rates of decomposition of *tert*-butyl 1-phenylcycloalkane-1-percarboxylates[163a] in contrast to their effect on the decomposition of azocycloalkanes.[163b] α-Cyano-groups have a rate-retarding influence on the decomposition of *tert*-butyl α-cyano-percarboxylates, again indicating that the C_α—CO bond is little stretched in the transition state and that the transition state has appreciable polar character.[164] This result is in very marked contrast to the accelerating effect of α-cyano-groups on the decomposition of azo-compounds. The importance of polar effects in perester decomposition has been further demonstrated from a study of substituent and solvent effects in the thermolysis of *tert*-butyl 2-phenylperisobutyrates.[165] The acyloxy-radical derived from (**82**) apparently has sufficient life-time to abstract a hydrogen atom in an intramolecular process, indicating that the initial step is a one-bond process.[166]

Ph CO$_3$Bu^t (**82**) ⟶ Ph •O O ⟶ Ph H O • O

$$\text{Ar—C(=O)—N(N=O)—OBu}^t \xrightarrow{\;\;\ell\;\;} \text{Ar—C(—O—N=)N—OBu}^t \xrightarrow{-\text{N}_2} [\text{ArCO}_2\cdot\ \cdot\text{OBu}^t] \longrightarrow \text{ArCO}_3\text{Bu}^t \;\text{or}\; \text{ArCOOH} + \text{Bu}^t\text{OH}$$

(**83**)

The decomposition of *N*-nitrosohydroxylamines (**83**) involves an aroyloxy–*tert*-butoxy radical pair,[167] which can either diffuse from the solvent cage or combine to form the perester. The yield of peresters obtained from *N*-nitrosohydroxylamines can be correlated with substituent effects on the aryl group, giving a linear Hammett relationship ($\rho = +0.4$).[168] The sign of ρ is opposite to that encountered in the decomposition of peresters, which is to be expected since the dipolar forces facilitating decomposition should retard combination.

Tracer studies indicate that the extent of oxygen scrambling in di-1-apocamphoryl peroxide is 100 times that in 1-apocamphoryl benzoyl peroxide.[169,170] The extent of scrambling increases with increasing solvent viscosity, which is consistent with scrambling by cage combination of the germinate acyloxy radical-pair. The yields of ester from the two peroxides are 2% and 20%, respectively. These results and others enable three types of peroxide decomposition to be formulated, according to the relative stability of the acyloxy-groups. The first type, in which both acyloxy-radicals are unstable, is exemplified by diapocamphoryl peroxide and diacetyl peroxide; it is characterized by a large amount of oxygen scrambling, as the activation energy for the reverse reaction is

$$R^1{-}\overset{O}{\overset{\|}{C}}{-}O{-}O{-}\overset{O}{\overset{\|}{C}}{-}R^2 \rightleftharpoons [R^1CO_2\cdot\ \cdot O_2CR^2] \longrightarrow [R^1\cdot\ 2\,CO_2\ \ R^2\cdot] \longrightarrow \text{Products}$$

Case 1

very low, and also by a low yield of ester. The other extreme is that in which both acyloxy-radicals are relatively stable, as in dibenzoyl peroxide; the activation energy for recombination will be significant and the radicals will diffuse out of the cage before recom-

$$R^1{-}\overset{O}{\overset{\|}{C}}{-}O{-}O{-}\overset{O}{\overset{\|}{C}}{-}R^2 \rightleftharpoons [R^1CO_2\cdot\ \cdot O_2CR^2] \longrightarrow R^1CO_2\cdot + R^2CO_2\cdot$$

Case 2

bination, leading to a small degree of scrambling and a low yield of ester. In the intermediate case, illustrated by 1-apocamphoryl benzoyl peroxide, the extent of scrambling will be low but the yield of ester will be high, as one acyloxy radical is relatively stable and will react with the radical produced from the less stable radical within the solvent cage. The amount of decomposition outside the solvent cage was significantly less for di-*p*-nitrobenzoyl peroxide than for dibenzoyl peroxide.[171]

$$R^1{-}\overset{O}{\overset{\|}{C}}{-}O{-}O{-}\overset{O}{\overset{\|}{C}}{-}R^2 \rightleftharpoons [R^1CO_2\cdot\ \cdot O_2CR^2] \longrightarrow [R^1CO_2\cdot\ \ CO_2\ \ \cdot R^2] \longrightarrow R^1COOR^2$$

Intermediate case

Dialkylmalonyl peroxides undergo photolysis and thermolysis to give α-lactones.[172,173] A similar process in which one molecule of carbon dioxide is lost from the intermediate diacyloxy-radical occurs with phenylmaleoyl peroxide.[174]

$$\text{(dialkylmalonyl peroxide)} \longrightarrow R_2C(CO_2\cdot)_2 \longrightarrow R_2C\cdot\ \ \cdot O{-}\overset{O}{\overset{\|}{C}} \longrightarrow \text{(α-lactone, } R_2C{-}C(=O){-}O\text{ ring)}$$

The initial alkoxy-radical obtained in the thermolysis of aromatic peracetals (**84**) subsequently reacts in three ways.[175] Geminate reaction gives the perester, fragmentation gives the aldehyde and *tert*-butylperoxy-radicals, while reaction with solvent gives the peroxide.

$$\underset{(\mathbf{84})}{Ar{-}CH(OOBu^t)_2} \longrightarrow \left[Ar{-}CH(O\cdot)(OOBu^t)\ \ \cdot OBu^t\right] \begin{cases} \longrightarrow ArCO_3Bu^t + Bu^tOH \\ \longrightarrow ArCHO + Bu^tO_2\cdot \\ \xrightarrow{SH} ArCH(OH)OOBu^t \end{cases}$$

Cyclic dialkyl peroxides have significantly greater stability than their acyclic analogues. Thus (**85**) is a thousand times as stable as di-*tert*-butyl peroxide.[176] This stabilization is due entirely to entropy effects and indicates that recombination of the initial biradical occurs more easily than its fragmentation. Isotope studies indicate that the preferred mode of decomposition of the initial 1,6-biradical (**86**) involves rotation about the central C—C bond to give the *transoid*-biradical (**87**) which subsequently fragments to ethylene and two molecules of acetone. Fragmentation of the *cisoid*-biradical is a less important competing process.

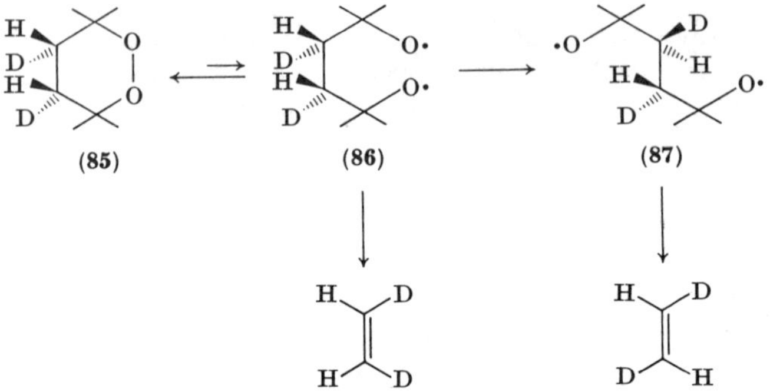

(**85**) (**86**) (**87**)

Kinetic and stereochemical studies on the thermolyses of 1,2-dioxolanes (**88**) indicate that the 1,3-biradical is formed directly and not *via* the 1,5-biradical.[177] Tetramethyl-1,2-

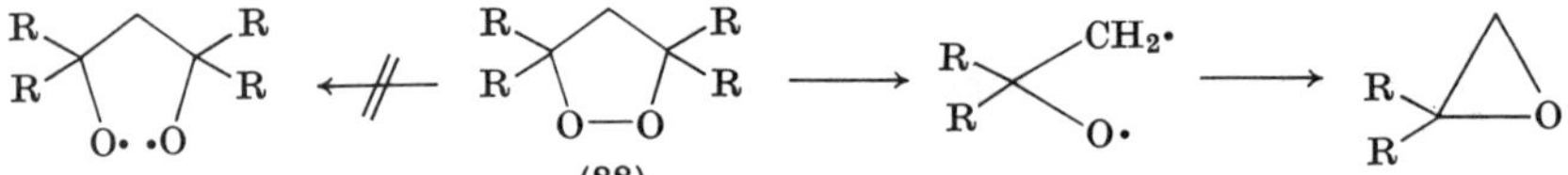

(**88**)

dioxetane (**89**), on either thermolysis or photolysis, gives acetone *via* the triplet biradical (**90**).[178]

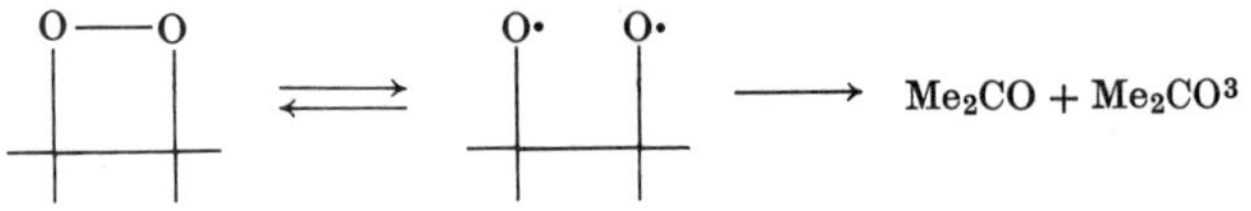

$$\longrightarrow Me_2CO + Me_2CO^3$$

(**89**) (**90**)

Induced decompositions of peresters have been brought about by trialkyltin radicals.[179] Chlorotris(triphenylphosphine)hydridorhodium(I) induces decomposition of di-*tert*-butyl peroxide.[180]

$$HRh(complex) + (Bu^tO)_2 \longrightarrow Bu^tOH + Bu^tO\cdot + Rh(complex)$$

Combination and Disproportionation Reactions[181–183]

The disproportionation–combination ratio for cyclohexyl radicals, generated from cyclohexyl bromide in a rotating cryostat, was found to lie in the range 0.4–1.1, depending on the temperature and the nature of the solid matrix.[184] The lower and upper limits are not very different from the ratios observed in the gas phase and in solution, respectively.

Various allyl radicals were generated by decomposition of di-*tert*-butyl peroxide in the presence of an alkene.[185]

$$\mathrm{Bu^tO\cdot} + (\mathrm{CH_3})\mathrm{RC{=}CR_2} \longrightarrow \mathrm{H_2C{\cdots}C(R){\cdots}CR_2}\ (\text{allyl radical}) + \mathrm{Bu^tOH}$$

$$\mathrm{R = H\ or\ Me}$$

Termination rate constants measured by the ESR technique for allyl, *cis*-1-methylallyl, *trans*-1-methylallyl and 2-methylallyl radicals were found to be in the range $1\text{–}10 \times 10^9\ \mathrm{l\ mol^{-1}\ s^{-1}}$ and the activation energies were in the range 1.2–2.4 kcal $\mathrm{mol^{-1}}$. The experimental results were in good agreement with those of the Noyes diffusion model, except for the rate constant of *cis*-1-methylallyl which was anomalously high at the high-temperature end. Photolysis of ketones or quinones in alcoholic solvents led to ketyl or semiquinone radicals whose termination reactions were followed by an ESR technique.[186] The termination rate constant was inversely proportional to solvent

$$\mathrm{RR'CO} \xrightarrow{h\nu} \mathrm{(RR'CO)^{S,T}}$$

$$\mathrm{(RR'CO)^{T} + DH} \longrightarrow \mathrm{RR'\dot{C}OH + D\cdot}$$

$$\mathrm{2RR'\dot{C}OH} \longrightarrow \text{Products}$$

viscosity except at low viscosities, and termination activation energies ran parallel to the temperature coefficients of viscosity of the solvents. Some evidence was found for rotational restriction. A rough correlation between k_t and radical size was observed for the ketyl radicals, but not for the semiquinone radicals, possibly because they terminate mainly by disproportionation. An improved method of baseline measurement in ESR experiments has been applied to the determination of the combination rate of $\cdot\mathrm{C(CH_3)(OH)COCH_3}$ and $\mathrm{Bu^t\cdot}$ radicals.[187] Diffusion coefficients have been measured by the space intermittency method for cyclohexyl and benzyl radicals.[188] Alkyl radicals (but not benzyl) diffuse much more slowly than the parent hydrocarbons, presumably because of interaction with the solvent.

A kinetic study of the decay of tetra-alkoxyphosphoranyl radicals in cyclopropane

$$\mathrm{Bu^tO\cdot + (RO)_3P} \longrightarrow \mathrm{Bu^tO\dot{P}(OR)_3}$$

$$\mathrm{R = Me,\ Et,\ Pr^i\ or\ Bu^t}$$

revealed that at temperatures above 158K β-scission was the predominant mode of reaction, but that at lower temperatures self-reactions were more important.[189] Self-reaction of phosphoranyl radicals probably yields P—P-coupled products and this

$$\mathrm{Bu^tO\dot{P}(OR)_3} \longrightarrow \mathrm{Bu^t\cdot + OP(OR)_3}$$

$$\mathrm{2\,Bu^tO\dot{P}(OR)_3} \longrightarrow \mathrm{Bu^tO(RO)_3P{-}P(OR)_3OBu^t}$$

$$\mathrm{Bu^t\cdot + Bu^tO\dot{P}(OR)_3} \longrightarrow \mathrm{Bu^t(Bu^tO)P(OR)_3}$$

$$\mathrm{Bu^t\cdot + Bu^tO\dot{P}(OR)_3} \longrightarrow \mathrm{Bu^tH + Me_2C{=}CH_2 + OP(OR)_3}$$

process showed noticeable steric hindrance. Phosphoranyl radicals may react with *tert*-butyl radicals by combination, since stable alkoxyalkylphosphoranes are known, but disproportionation is also a possibility. The rate of combination of $Bu^t\cdot$ with phosphoranyl radicals was found to be about the mean of the rates of self-reaction of $Bu^t\cdot$ and phosphoranyl radicals.

Silyl radical self-reactions have been studied by a flash photolysis–ESR method.[190] The decay of $Me_3Si\cdot$ and $H_3Si\cdot$ radicals was shown to be of the second order. The main product from the self-reaction of $Me_3Si\cdot$ radicals was hexamethyldisilane indicating that combination was the main mode of reaction:

$$Bu^tO\cdot + Me_3SiH \rightarrow Bu^tOH + Me_3Si\cdot$$
$$2Me_3Si\cdot \rightarrow Me_3SiSiMe_3$$

Simple dimerization was also suggested as the predominant reaction of $H_3Si\cdot$ radicals. Coupling of polysilanyl radicals, generated by attack of *tert*-butoxy-radicals on branched-chain polysilanes in benzene solution, is a useful method for the synthesis of branched-chain methylpolysilanes.[191] Only highly branched radicals undergo coupling, other radicals substituting the aromatic solvent.

$$(Me_3Si)_3SiH \xrightarrow[C_6H_6]{(Bu^tO)_2} (Me_3Si)_3SiSi(SiMe_3)_3 \qquad (92\%)$$

Dialkylamino-radicals generated from tetra-alkyltetrazenes decay in solution by second-order processes.[192] The rate constants of the self-reactions of dimethylamino- and

$$R_2N{-}N{=}N{-}NR_2 \xrightarrow[\text{or } h\nu]{\Delta} 2R_2N\cdot + N_2$$
$$2R_2N\cdot \longrightarrow \text{Products}$$
$$R = Me, Et \text{ or } Pr^i$$

diethylamino-radicals are of the order of 10^9–10^{10} l mol^{-1} s^{-1}, suggesting that the process is diffusion-controlled. 2,2,6,6-Tetramethylpiperidyl radicals on the other hand, generated from the corresponding tetrazene, react mainly by hydrogen abstraction from the solvent.

This year has seen more than usual interest in gas-phase combination reactions, mainly because of the current controversy over the combination rate constants for secondary and tertiary alkyl radicals. The combination rate constant for methyl radicals has been determined by monitoring the product ethane by time-resolved mass-spectrometry.[193] The high-pressure value of the rate constant (2.41×10^{10} l mol^{-1} s^{-1}) was in excellent agreement with previous values and there can be no doubt that this rate constant is now known with a high degree of certainty. Recent evidence has indicated combination rate constants for ethyl and isopropyl radicals considerably lower than their collision rates. The combination rate constant for *tert*-butyl has now been determined by Benson's "buffer" method using ethyl and *tert*-butyl iodide.[194] The very low value of $10^{5.4 \pm 1.5}$ l mol^{-1} s^{-1} was obtained, leading to the tentative conclusion that combination rates of alkyl radicals are in the order Me > Pr^i > sec. > tert. Mercury-photosensitized decomposition of various methyl- and dimethyl-substituted butenes leads to the formation of a methyl radical and a substituted allyl radical.[195] The reaction centre in the allyl radical with the greater number of methyl substituents was less reactive in combining with the methyl radical. By making certain assumptions it was possible to derive combination rate constants for isopropyl and *tert*-butyl radicals of $10^{7.5}$ and $10^{6.3}$ l mol^{-1} s^{-1}, respectively, in fair agreement with other recent "low" values. Flash photolysis of 2-methylbut-2-ene produced methyl and 2-methylallyl radicals.[196] The UV absorption spectra of the two radicals were detected, and the combination rate of methyl was found to be 2.4×10^{10} l mol^{-1} s^{-1} and that of 2-methylallyl was 2.6×10^{10} l mol^{-1} s^{-1}. Cross-combination of the two radicals was apparently very inefficient. Methyl and *tert*-butyl radicals were produced in the photolysis of 3,3-dimethylbutan-2-one.[197] The main reactions in the system were combination and disproportionation reactions of methyl, *tert*-butyl and acetyl

$$CH_3COBu^t \xrightarrow{h\nu} Bu^t\cdot + CH_3\dot{C}O \longrightarrow CH_3\cdot + CO$$
$$\searrow \; CH_3\cdot + Bu^t\dot{C}O \longrightarrow Bu^t\cdot + CO$$

$$2\,CH_3\cdot \longrightarrow C_2H_6$$

$$CH_3\cdot + Bu^t\cdot \longrightarrow neo\text{-}C_5H_{12}$$
$$\searrow \; CH_4 + iso\text{-}C_4H_8$$

$$2\,Bu^t\cdot \longrightarrow C_8H_{18}$$
$$\searrow \; iso\text{-}C_4H_{10} + iso\text{-}C_4H_8$$

$$2\,CH_3\dot{C}O \longrightarrow CH_3COCOCH_3$$

$$CH_3\dot{C}O + CH_3\cdot \longrightarrow CH_3COCH_3$$

$$CH_3\dot{C}O + Bu^t\cdot \longrightarrow CH_3CHO + iso\text{-}C_4H_8$$
$$\searrow \; CH_3CO\,Bu^t$$

radicals. Disproportionation was particularly favoured in reactions involving the *tert*-butyl radical. The combination rate constant of $CF_3CCl_2\cdot$ radicals has been determined by chlorination of 2,2-dichloro-1,1,1-trifluoroethane in intermittent light.[198] A study of the pyrolysis of hexachloroethane led to a value for the combination of $CCl_3\cdot$ radicals.[199] In a theoretical study of the combination of $C_2Cl_5\cdot$ radicals the rate constant was calculated by absolute-rate theory and the activation energy by a semiempirical MO method.[200]

Vinyl and vinoxyl radicals were formed in the flash photolysis of divinyl ether.[201] The vinyl radicals were observed to disproportionate, and the vinoxyl radicals fragmented into methyl radicals and CO:

$$\cdot C_2H_3 + \cdot C_2H_3 \rightarrow C_2H_4 + C_2H_2$$
$$C_2H_3O\cdot \rightleftharpoons CH_3\dot{C}O \rightarrow CH_3\cdot + CO$$

Photolysis of methyl[202] and ethyl sulphides[203,204] gave rise to alkyl and thiyl radicals, enabling the combination and disproportionation process of these species to be studied. The primary step in the mercury-photosensitized decomposition of disilane produced $\cdot Si_2H_5$ radicals;[205] the disproportionation–combination ratio of this radical was found to be pressure-dependent. Disproportionation and combination were also observed for carbamoyl radicals, formed in the high-temperature photolysis of formamide.[206]

Atom-abstraction Reactions[207–209]

Hydrogen-atom Abstraction

The reaction of methyl radicals with molecular hydrogen in reflected shock waves shows non-Arrhenius behaviour,[210] as does the reaction of methyl radicals with neopentane.[211] Methyl radicals abstract hydrogen more readily from Si—H than from C—H bonds in methylsilanes, but the expected decrease in activation energy with increasing number of methyl substituents at silicon was not observed.[212] γ-Irradiation of methyl iodide in

3-methylpentane glass produced methyl radicals whose main reaction was to abstract hydrogen from the matrix, giving 3-methylpentyl radicals.[213] The relative rates of hydrogen abstraction from substituted toluenes by *tert*-butyl radicals show a positive ρ value of 0.99 from the Hammett correlation;[214] this is a good indication that *tert*-butyl radicals are rather more nucleophilic than methyl radicals for which a ρ value of about zero has been determined. A study of the reaction of ethyl, *n*-propyl and isopropyl radicals with silanes showed that the *A*-factors for hydrogen abstraction were virtually constant;[215] if a small activation energy was assumed for the radical combination reactions, the activation energies of the hydrogen-abstraction steps could be adequately predicted from BEBO calculations. Kinetic studies of the reactions of trifluoromethyl radicals with chlorosilane[216] and alkanes[217] have appeared; the rates of hydrogen abstraction from specific sites in alkanes were determined by analysis of the product iodides from CF_3I/alkane mixtures.[217] (Trichloromethyl)vinyl radicals have been generated by addition of trichloromethyl radicals to acetylene;[218, 219] and the relative rates of hydrogen abstraction from a variety of substrates by this radical have been correlated by

$$CCl_3\cdot + CH{\equiv}CH \rightarrow CCl_3CH{=}CH\cdot$$
$$CCl_3CH{=}CH\cdot + RH \rightarrow CCl_3CH{=}CH_2 + R\cdot$$

means of a two-parameter Taft equation allowing for both polar and resonance effects.

A synthetic procedure has been devised for the deamination of aromatic amines by pentyl nitrite.[220] The key step involves generation of aryl radicals from the amine which subsequently abstract hydrogen from the solvent:

$$ArNH_2 + RONO \rightarrow H_2O + ArN_2^+\ RO^- \rightarrow Ar\cdot$$
$$Ar\cdot + R'H \rightarrow ArH + R'\cdot$$

Hydrogen abstraction by phenyl radicals, generated from (phenylazo)triphenylmethane, from a series of pentyl phenyl sulphides has been examined.[221] The results suggest a simple two stage mechanism, and relative rates measured by analysis for the product

$$Ph\cdot + PhSCH_2CH_2R \rightarrow PhH + PhSCH_2\dot{C}HR \rightarrow PhS\cdot + CH_2{=}CHR$$

alkenes gave no indication of rate enhancement at positions β to the sulphur; contrary to previous reports, the reactivities at all positions were about the same as in alkanes, so that sulphur lends no detectable anchimeric assistance (cf. p. 87). Decomposition of tri- or tetra-nitromethane in tetrahydrofuran or tetrahydropyran gives a low yield of 2-(trinitromethyl) derivatives of the cyclic ethers by the intermediacy of trinitromethyl radicals.[222] A useful method for reducing polyhalomethyl groups involves photolysing the polyhalomethyl compound in tetrahydrofuran; good yields of the reduced compound were obtained from a variety of polyhaloalkanes and cycloalkanes:[223]

$$RCX_3 + h\nu \rightarrow X\cdot + RCX_2\cdot \xrightarrow{R'H} RCHX_2 + R'\cdot$$

Hydrogen abstraction by trimethylsilyl radicals generated from hexamethyldisilane has also been investigated.[224]

Reports have appeared of the reaction of fluorine atoms with molecular hydrogen[225] and with substituted methanes.[226–228] Relative rates of hydrogen abstraction by ^{18}F from alkanes indicate that the main controlling factor may be the C—H bond dissociation energy;[229] fluorine substituents β to the attacked site had a strong retarding influence. Gas-phase chlorination of halomethanes has been studied by the discharge-flow technique;[230] the activation energy for hydrogen abstraction decreases monotonically as the number of chlorine atoms in the methane increases; reasonable agreement with previous competitive work was obtained, but the absolute values of the Arrhenius parameters indicated that the chlorination of molecular hydrogen (on which previous absolute rate data were based) might need re-examination. Kinetic data for the chlorination of chloroethanes in the gas phase was compared with previous liquid-phase values,[231] and it was found that reactivities of primary and secondary, but not tertiary, hydrogen atoms were enhanced in the liquid phase. Good Hammett correlations have been observed in the chlorination of nuclear-substituted toluenes, diphenylmethanes and ethylbenzenes by sulphuryl chloride and *tert*-butyl hypochlorite,[232] and in brominations with *N*-bromosuccinimide and bromotrichloromethane.[233]

The main products in the reaction of sulphur monochloride with alkanes were alkyl chlorides and di- and poly-sulphides;[234] the relative rates of abstraction from primary, secondary and tertiary sites were virtually the same as in chlorination with molecular chlorine, indicating that the chlorine atom is the main abstracting species:

$$RH + Cl\cdot \rightarrow R\cdot + HCl$$

$$R\cdot + S_2Cl_2 \rightarrow RCl + \cdot S_2Cl$$

The di- and poly-sulphides were probably formed by displacement reactions at sulphur:

$$R\cdot + Y\text{—}(S)_n\text{—}R \rightarrow R\text{—}(S)_n\text{—}R + Y\cdot \qquad (Y = Cl,\ RS\ \text{or}\ SCl)$$

Halogenation of methane and ethane with an equimolar mixture of chlorine and bromine gives good yields of the monobromoalkanes,[235] the process probably involving BrCl as the halogenating agent. Bromination of methane has been used as a method of obtaining $D(CH_3\text{—}Br)$:[236]

$$Br_2 + CH_4 \rightleftarrows HBr + CH_3Br$$

the reaction was carried out in a large excess of hydrogen bromide to prevent polybromination.

In the bromination of 1-bromobutane by bromotrichloromethane the predominance of 1,2-dibromobutane in the products was attributed to a favoured conformation (**91**) in the transfer step.[237] The minor product, 2-bromobutane, was probably formed by a

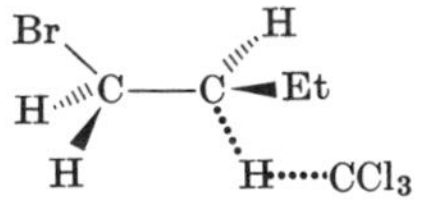

Br, H, H, H, Et, H·····CCl₃

(**91**)

1,2-bromine shift in the intermediate radical. Kinetic studies of the bromination of 1-bromobutane have shown that in the presence of hydrogen bromide a catalysed reaction

$$BrCH_2\dot{C}HCH_2CH_3 \xrightarrow{\ell} \cdot CH_2CHBrCH_2CH_3 \xrightarrow{XH} CH_3CHBrCH_2CH_3$$

occurs that increases the yield of the 1,2-dibromo-product.[238] A number of studies, which compliment the structural ESR work, have shown that *p–p* homoconjugation has chemical significance in the reactions of *β*-haloalkyl radicals. Bromination of (+)-1-bromo-2-methylbutane (**92**) in the presence of DBr gave 1,2-dibromo-2-methylbutane (**93**) and (+)-1-bromo-2-deuterio-2-methylbutane (**94**);[239] the high degree of retention was explained in terms of partial bridging by the bromine atom in the radical intermediate. Photobromination of (+)-(2*S*)-bromobutane (**95**) gave (±)-2,3-dibromobutane and *meso*-2,3-dibromobutane;[240] to explain this result, two bridged forms of the 2,3-bromobutyl radical were postulated, one (**97**) optically active, the other (**96**) not. To test

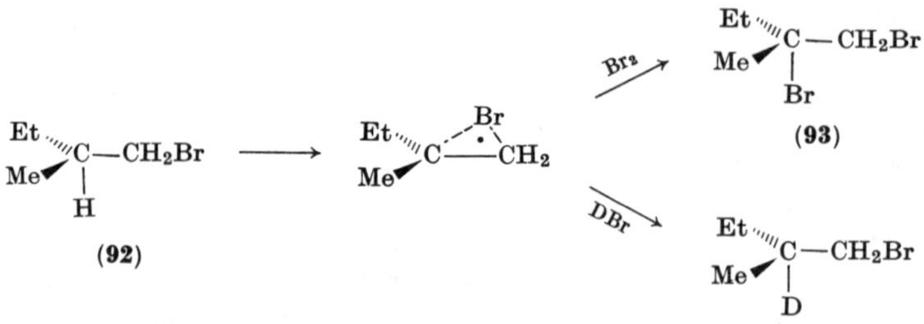

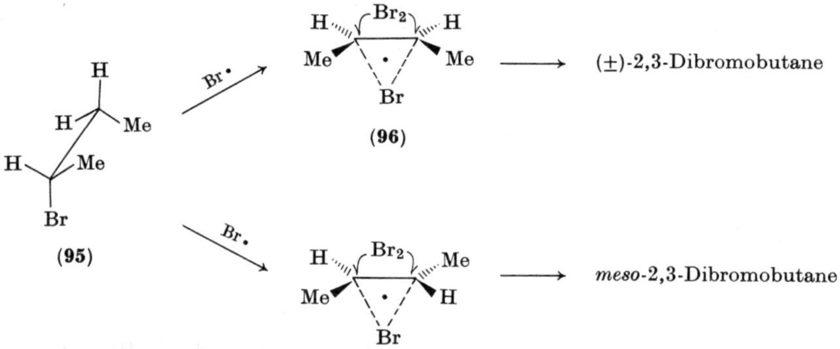

(±)-2,3-Dibromobutane

meso-2,3-Dibromobutane

the possibility of obtaining an optically active product from the chiral bridged intermediate, chlorination of (**95**) was carried out with *tert*-butyl hypochlorite: the chiral (+)-(2*S*)-bromo-(3*R*)-chorobutane (**99**) was obtained together with the (±)-*threo*-2-bromo-3-chloro-isomer (**98**). Attack at position-1 gave a rearranged product, 2-bromo-1-

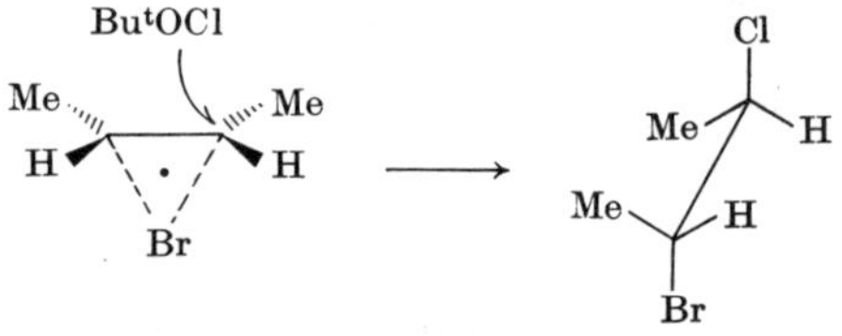

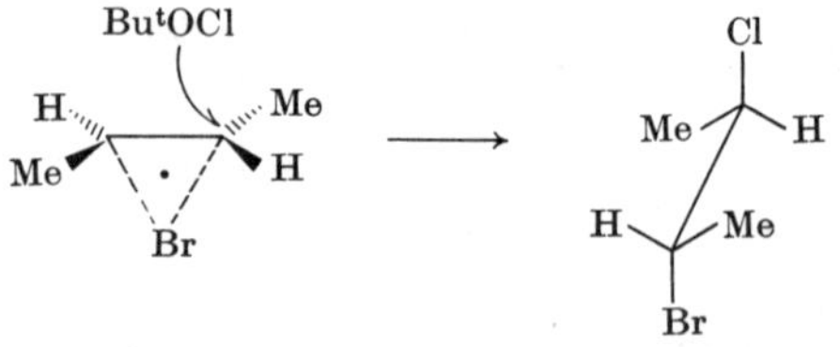

chlorobutane, of high optical purity, which can also be accounted for in terms of an unsymmetrically bridged intermediate (**100**). The main product from the photobromination of (+)-(2*S*)-chlorobutane was (±)-2-bromo-2-chlorobutane, but this was accompanied by small amounts of *erythro*- and *threo*-2-bromo-3-chlorobutanes, which appeared

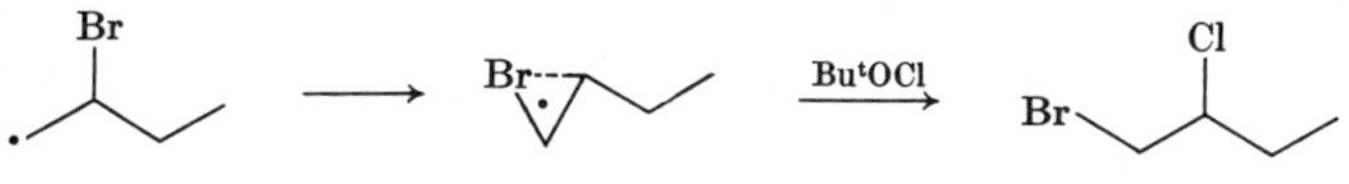

(**100**)

to be optically pure.[240] Partial bridging by the chlorine atom was invoked to account for the latter products.

Gas-phase bromination of halocyclohexanes gave cyclohexene, bromocyclohexane and *trans*-1,2-dibromocyclohexane, in addition to the expected bromohalocyclohexane isomers,[241] decomposition of the starting material to cyclohexene was catalysed by the product HBr, and the other products were formed from the cycloalkene. Similar reactions

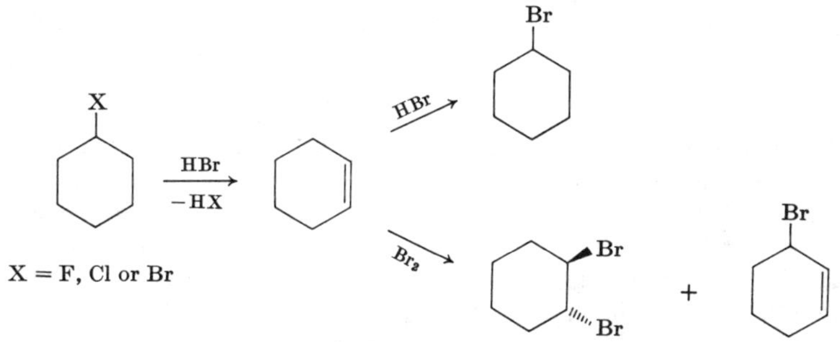

X = F, Cl or Br

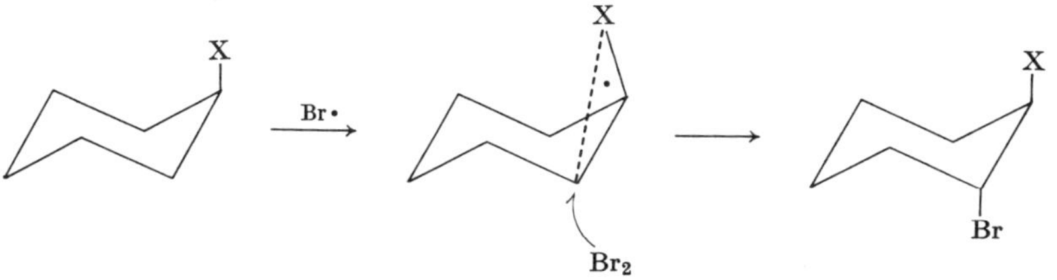

were observed with halocyclopentanes. The relative rates of formation of the products were seriously distorted by the HBr-catalysed decomposition of the bromohalocycloalkanes as well as of the starting material. When the reactions were carried out in a flow reactor, which minimized the contact between HBr and the halocycloalkanes, no catalysed decomposition was observed.[242] Under these conditions there was no increase in the rate of attack at the 2-position, which ruled out anchimeric assistance by the halogen substituents. The 1-bromo-2-halocycloalkane products were formed predominantly with the *trans*-configuration, and *trans*-products were also favoured, although to a lesser extent, as a result of attack at the 3- and the 4-position. Thus, although "bridging" halogen substituents do not lend anchimeric assistance in the gas phase, *p*-*p*-homoconjugation may be responsible for stereochemical control. Polar effects could also be an important factor in controlling the stereochemistry. Chlorination of cyanocyclobutane also gave 1-cyano-2-chlorobutane with a very high *trans*/*cis* ratio, either because of polar effects and/or because of p–π homoconjugation between the unpaired electron of the intermediate radical and the cyano-substituent.[243] Bromination of cyanocyclobutane gave almost exclusively 1-bromo-1-cyanobutane as the product of abstraction; both the starting material and the product underwent a catalysed decomposition in the presence of HBr giving, initially, cyclobutenes which then reacted with bromine or hydrogen

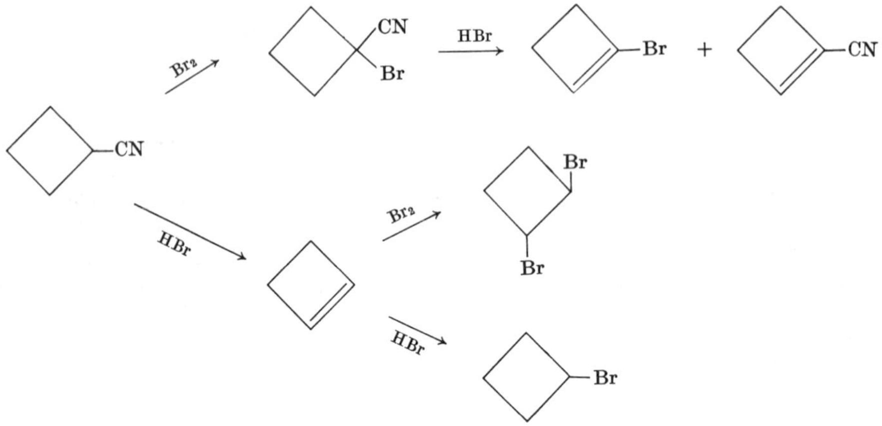

bromide. Treating a number of halogen-substituted alkanes and cycloalkanes in solution with molecular bromine showed that the substituted compounds reacted faster than the parent alkanes;[244] this was attributed to reversibility of the reaction in the presence of HBr, Br· + RH ⇄ R· + HBr, and when the reverse reaction was suppressed by using *N*-bromosuccinimide or very high molecular bromine concentrations the substituted alkanes were less reactive than the parent alkanes.

Iodination of methane,[245] ethane[245] and isopentane[246] has been investigated. Bond dissociation energies for C—H in acrolein[247] and Si—H in trimethylsilane[248] have been determined from the relevant iodination reactions.

Hydroxyl radicals, produced by titration of nitrogen dioxide with atomic hydrogen, abstract hydrogen from hydrogen halides,[249] a steady decrease in the activation energy from HCl to HI was observed. An ESR study of the reaction of hydroxyl radicals with glycerol phosphates in aqueous solution showed that hydrogen was abstracted from all available positions;[250] the intermediate α-phospho-radicals predominantly eliminated phosphate, β-phospho-radicals (**101**) eliminated phosphate or water, and γ-phospho-radicals eliminated mainly water. Ethanolamine and serine phosphates are attacked by

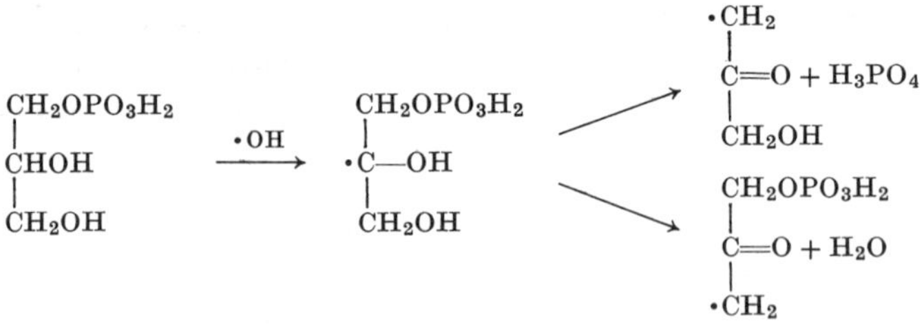

(**101**)

hydroxy-radicals in neutral solution mainly at the position α to the phosphate, the other sites being deactivated by the neighbouring H_3N^+ group.[250] The main product from the reaction of *tert*-butoxy-radicals with cyclopropyltin alkoxides is the cyclopropyl ketone, but a substantial amount of the ring-opened ketone is also formed.[251]

$$c\text{-}C_3H_5\text{-}CH(OSnBu_3)\text{-}R + Bu^tO\cdot \longrightarrow c\text{-}C_3H_5\text{-}\dot{C}(OSnBu_3)\text{-}R \longrightarrow c\text{-}C_3H_5\text{-}C(=O)\text{-}R$$

$$c\text{-}C_3H_5\text{-}\dot{C}(OSnBu_3)\text{-}R \longrightarrow \cdot CH_2CH_2CH{=}C(OSnBu_3)R \longrightarrow CH_3CH_2CH_2C(=O)R$$

tert-Butoxy-radicals abstract hydrogen from polystyrene or polypropene rather less readily than from the monomer itself.[252] A study of hydrogen abstraction by *tert*-butylperoxy-radicals from an extensive series of *m*- and *p*-substituted toluenes gave reasonable correlation with σ or σ^+ constants;[253] the ρ-values were less negative than those determined from free-radical brominations, suggesting that polar effects are less important with peroxy-radicals. Competitive reactions of benzoyloxy-radicals with penta-arylethanes gave rise to a positive ρ^*-value from the correlation with σ^* constants,[254] showing that hydrogen abstraction is favoured by electron-withdrawing substituents. Benzoyloxy-radicals abstract hydrogen from the amino-group of esters of *o*-amino-*p*-*tert*-butylphenol,[255] the products being formed by combination of the resulting amino-radicals with benzoyloxy-radicals or with themselves.

An ingenious method has been developed for studying the kinetics of the reversible hydrogen-abstraction reactions of thiyl radicals.[256] A tritium-labelled thiol is used as the solvent, Q· radicals from the substrate abstract hydrogen from the thiol and the level of radioactivity in QH* is used to evaluate the relative rate of hydrogen abstraction, which is independently corrected for tritium isotope effects. The relative rate constants

$$RS\cdot + QH \rightarrow RSH + Q\cdot$$
$$Q\cdot + RSH^* \rightarrow QH^* + RS\cdot$$

for hydrogen abstraction from substituted methyl- and ethyl-benzenes by cyclohexanethiyl radicals, measured in this way, show an excellent linear correlation with σ^+ constants. Pentachlorobenzenethiyl radicals are believed to be the main chain-carrying species in the reaction of pentachlorobenzenesulphenyl chloride with an alkane;[257] the relative selectivity for hydrogen abstraction from primary, secondary and tertiary sites was virtually the same with this reagent as with pentachlorobenzenethiyl radicals generated from pentachlorobenzene disulphide.

$$C_6Cl_5S\cdot + RH \rightarrow C_6Cl_5SH + R\cdot$$
$$R\cdot + C_6Cl_5SCl \rightarrow RCl + C_6Cl_5S\cdot$$

A pulse radiolysis study of reaction of cyanogen radicals with alkanes indicated that the activation energy for abstraction by this radical is intermediate between that of chlorine and fluorine atoms.[258] An ESR study of the mercury-photosensitized oxidation of amines $R^1R^2CHNHR^3$ indicated that hydrogen abstraction takes place mainly at the secondary carbon atom.[259]

Halogen-atom Abstraction

Photolysis of alkyl iodides at 147 nm produces iodine atoms together with excited alkyl radicals which rapidly decompose into alkenes and hydrogen atoms.[260] Hydrogen atoms, produced in this way, abstract halogen from alkyl halides. The most important factor governing the rate of reaction appears to be the strength of the carbon–halogen bond. The rate of iodine abstraction from *o*-substituted benzenes and from iodoferrocene by phenyl radicals has been investigated:[261] the scatter of points on a Taft plot using *o*-substituent constants σ^*_o indicated steric interactions to be important; however donor substituents reduce the rate, and electron-withdrawing substituents increase it, suggesting transition states such as (**102**) or (**103**); anchimeric assistance from neighbouring bromo- or iodo-substituents was not observed.

$$[\overset{\delta-}{\mathrm{R}}\text{---}\mathrm{I}\text{---}\overset{\delta+}{\mathrm{Ph}}]^{\ddagger} \qquad [\mathrm{R}\text{---}\overset{\delta-}{\mathrm{I}}\text{---}\overset{\delta+}{\mathrm{Ph}}]^{\ddagger}$$

(**102**) (**103**)

Rate constants for the abstraction of chlorine, bromine or iodine from the corresponding *N*-halosuccinimide by benzyl or propyl radicals showed the expected increase from chlorine to bromine to iodine.[262] A study of the rate of chlorine abstraction from

$$\mathrm{R\cdot} + \text{(succinimide)NX} \longrightarrow \mathrm{RX} + \text{(succinimide)N\cdot}$$

alkyl chlorides by trimethylsilyl radicals in the gas phase showed that the activation energies obeyed an approximate Polanyi equation with $\alpha = 0.2$; the *A*-factors showed very little variation.[263]

Dehalogenation of vicinal dihalides by hexa-*n*-butylditin showed some degree of stereoselectivity with a proclivity towards *anti*-elimination on increased ditin concentration.[264] Relative rates of halogen abstraction from a variety of halogen compounds by tri-*n*-butyltin radicals have been measured by an ESR method;[265] trends were interpreted in terms of the strength of the carbon—halogen bond and polar effects, and the reactivity of trimethyltin radicals in the halogen-abstraction reaction can be interpreted in terms of the same two factors.[266]

Theoretical studies include examinations of the Hammett equation[267] and the Polanyi equation,[268] an empirical method for determining activation energies of metathesis reactions based on the group-additivity principle[269] and a simplified method for BEBO calculations.[270]

Intramolecular Hydrogen-transfer

Conformational factors determine whether or not C—O hydrogen-transfer occurs in the lead tetra-acetate oxidation of substituted cyclohexanols; transfer occurs only when the hydrogen on the δ-carbon and the oxygen radical centre of the derived alkoxy-radical are in close proximity. Thus intramolecular hydrogen-transfer occurs readily in the case of the radical derived from *trans*-3,3,5-trimethylcyclohexanol which exists in the conformation (**104**; $R^1 = Me$, $R^2 = H$), but not in the case of the isomeric *cis*-alcohol, for which the preferred conformation is (**105**; $R^1 = H$, $R^2 = Me$);[271] transfer is also not observed when it would require the cyclohexane ring to adopt a boat conformation. Hydrogen transfer occurs stereoselectively in 1,5–C—O migrations in radicals of the type $R^2CH_2CH_2CH(R^1)O\cdot$ as a result of the differing steric requirements of the

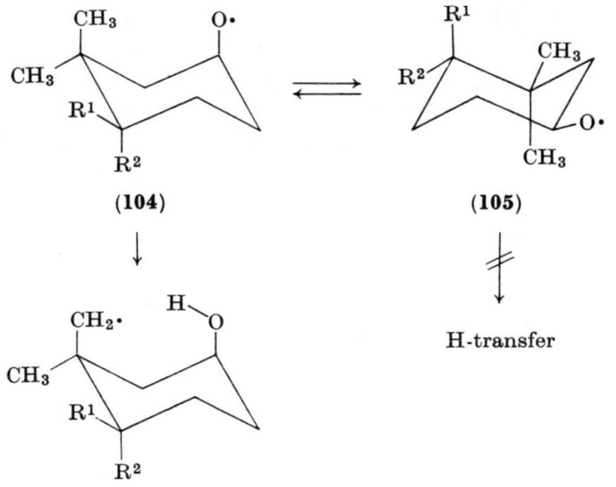

(104) (105)

H-transfer

two possible transition states for migration of the two hydrogen atoms.[272] The nature of the product arising from the photolysis of the nitrite (**106**) depends on whether the reaction is carried out under nitrogen or in oxygen;[273] Scheme 2 indicates the mechanism of formation of the two products.

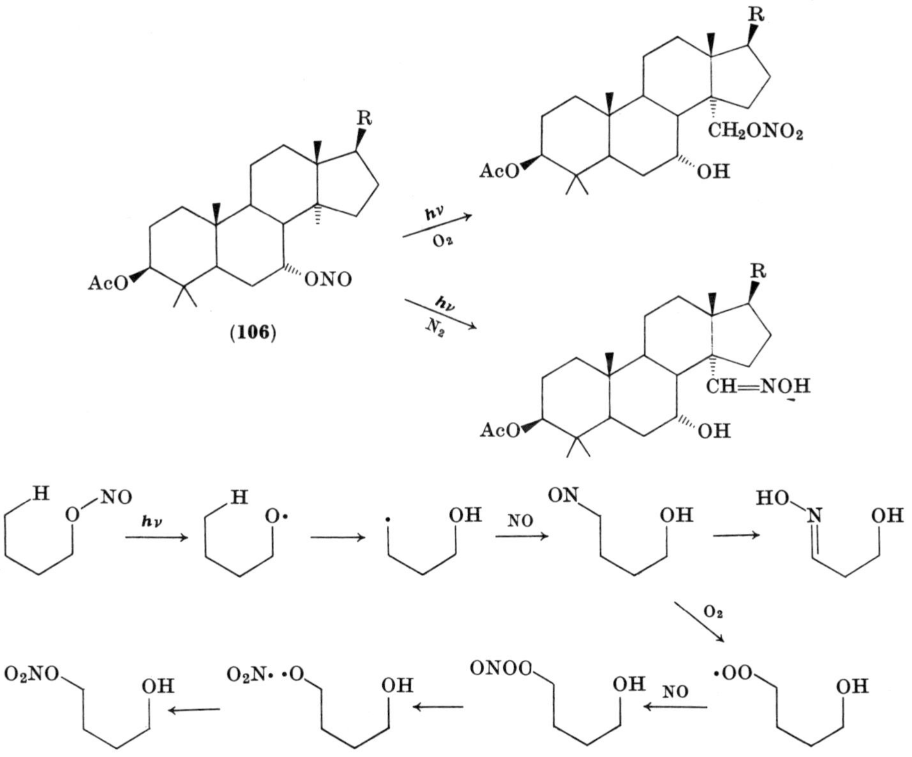

(106)

SCHEME 2.

The extent and site of intramolecular C–C hydrogen-transfer is likewise controlled by steric factors. 1,5-Hydrogen-transfer occurs three times as readily as 1,6-hydrogen-transfer with primary alkyl radicals;[274] increasing length of the alkyl radical also reduces the total extent of transfer. A significant amount of 1,10- and 1,11-transfer occurs for C-12 and C-13 radicals, respectively; models indicate that these radicals can adopt conformations to allow this to occur fairly readily. 1,5-Hydrogen-transfer has also been shown to be an important process in the oxidation of 5,5-dimethyl-2-heptanol (**107**)[275] and in the reduction of the oxaziridine (**108**).[276]

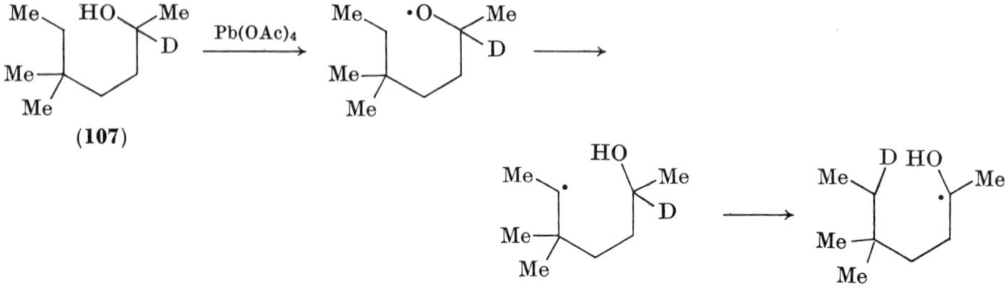

Me HO Me D $\xrightarrow{Pb(OAc)_4}$ Me •O Me D $\longrightarrow$ Me Me

(**107**)

Me • HO Me D Me Me $\longrightarrow$ Me D HO • Me Me Me

NR O $\xrightarrow{Fe^{II}}$ $\dot{N}R$ OH $\longrightarrow$ CONHR $CH_2\cdot$ $\longrightarrow$ • CONHR CH_3

(**108**)

Photochemical Atom-abstractions

CIDNP studies of the photoreduction of quinones in presence of hydrogen-donors confirm that hydrogen-transfer occurs to the triplet state of the carbonyl group.[277,278] CIDNP experiments have also established that photolysis of acetaldehyde (and of acetoin) give the acetyl-α-hydroxyethyl radical-pair which disproportionates to acetaldehyde and vinyl alcohol.[279] The acetophenone enol has been detected in the photolysis of a mixture of acetophenone and phenol;[280] this had previously been implicated in the formation of 1,2-dibenzoylethane in this reaction.[281]

$$2\,CH_3CHO \xrightarrow{h\nu} [CH_3\dot{C}HOH\;\dot{C}OCH_3] \longrightarrow CH_2{=}CHOH + CH_3CHO$$

$$PhCOCH_3 + PhOH \xrightarrow{h\nu} [HO\dot{C}PhCH_3\;\dot{O}Ph] \longrightarrow HOCPh{=}CH_2 + PhOH$$

$$\downarrow PhCOCH_3^{*}$$

$$PhCOCH_2CH_2COPh \longleftarrow PhCO\dot{C}H_2$$

Similar studies have shown that photolysis of benzaldehyde in most solvents gives the benzoyl-α-hydroxybenzyl radical pair, although in propan-2-ol the $Ph\dot{C}HOH$ and $Me_2\dot{C}OH$ radicals are produced.[282] Semiquinone radicals and radical anions have been detected in the photolysis of benzoquinone in ethanol.[283]

Cage effects are observable in the photoreduction of (*S*)-(+)-2-phenylpropiophenone, which is partially racemized after photolysis for a limited time.[284] Cage recombination is also inferred from the quantum inefficiency of the photolysis in presence of a thiol.

The value of k_{rot}/k_{comb} ($\simeq 2.3$) is much less than that encountered in the photolysis of azo-compounds or peresters, which break down by a two-bond process; this may be due to the absence of an inert molecule separating the radical pair.

The precise nature of the hydrogen-transfer step in the photoreduction of benzophenone in benzene has been further studied.[285] Tetrafluorohydrazine has been used to trap the radicals generated in the photoreduction of benzophenone by hydrocarbons:[286] $Ph_2CO^* + RH \rightarrow (HO)Ph_2C\cdot + R\cdot$; with $NF_2\cdot$ or N_2F_4 the former radical gives $Ph_2C(OH)NF_2$, and $R\cdot$ gives RNF_2. In this study the high selectivity of the triplet benzophenone towards hydrogen abstraction was noted; thus with 3-methylpentane more than 95% of the reaction product resulted from attack at the tertiary hydrogen atom.

Cyclopropanols can effect photoreduction of aryl ketones such as 2-acetonaphthone and fluorenone which are unaffected by propan-2-ol.[287] Isotope effects indicate that

$$\text{(cyclopropanol: } R^2, R^2, OH, R^1) + Ar_2CO \xrightarrow{h\nu} R^2\dot{C}(R^2)CH_2C(=O)R^1 + Ar_2\dot{C}OH \longrightarrow \text{Products}$$

the hydroxyl-hydrogen is abstracted and that abstraction is accompanied by ring opening. The photoreduction of ferrocenyl ketones occurs only in the presence of very good hydrogen-donors such as trialkyltin hydrides and not by propan-2-ol.[288]

In the photoreduction of unsymmetrical benzils in propan-2-ol, the more electron-deficient carbonyl group is reduced preferentially;[289] this is attributed in part to the

$$ArCOCOPh \xrightarrow[Me_2CHOH]{h\nu} ArCOCH(OH)Ph + \begin{matrix} ArCOCHPh \\ | \\ ArCOCHPh \end{matrix}$$

relative stabilities of the two ketyl radicals (**109**) and (**110**), but principally to their relative reactivities in coupling; isomerization of the two radicals is assumed to be very fast, and products are derived from the more reactive ketyl radical substituted by the more electron-withdrawing group.

$$\underset{(\mathbf{109})}{Ar\dot{C}(OH)COPh} \rightleftharpoons \underset{(\mathbf{110})}{ArCO\dot{C}(OH)Ph} \longrightarrow \text{Products}$$

Adamantanes can be efficiently photoacetylated at a bridgehead position by means of biacetyl;[290] hydrogen-abstraction is effected by triplet biacetyl; the reaction is much more susceptible to electronic influences ($\rho = -0.71$) than is hydrogen-abstraction by other radical species, which indicates that triplet biacetyl is considerably electrophilic, more so than triplet benzophenone.

Photoreduction of trifluoroacetophenone with alkylbenzenes occurs both by a charge-transfer process followed by proton transfer and by direct hydrogen-abstraction, whereas with acetophenone the latter process predominates.[291] This difference in behaviour is attributed to the electron-withdrawing influence of the α-fluorine atoms which lower the activation energy for charge-transfer more than that for hydrogen-transfer.

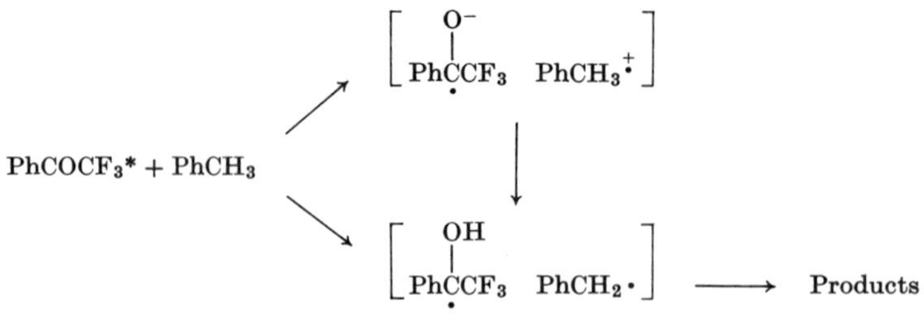

$$PhCOCF_3^* + PhCH_3 \rightarrow \left[Ph\dot{C}(O^-)CF_3 \quad PhCH_3^{+\cdot} \right] \rightarrow \left[Ph\dot{C}(OH)CF_3 \quad PhCH_2\cdot \right] \rightarrow \text{Products}$$

Intramolecular hydrogen-transfer has been observed by CIDNP in the photolysis of cycloalkanones.[292] The reactivity of excimers of alkyl phenyl ketones in intramolecular

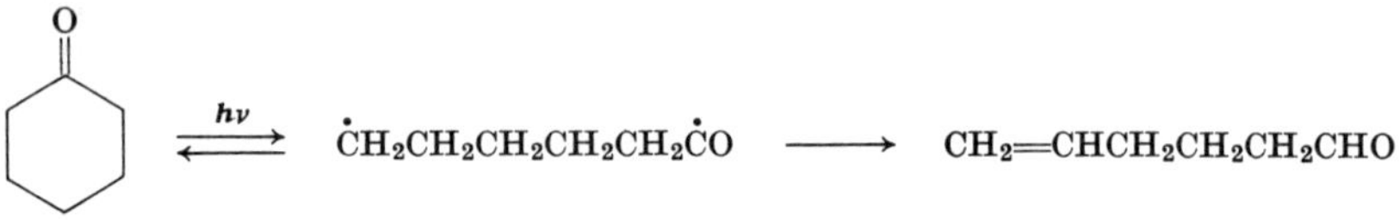

$$\text{cyclohexanone} \underset{}{\overset{h\nu}{\rightleftharpoons}} \dot{C}H_2CH_2CH_2CH_2CH_2\dot{C}O \longrightarrow CH_2{=}CHCH_2CH_2CH_2CHO$$

hydrogen-transfer depends on the preferred conformation of the molecule and on the relative rates of reaction of the alternative conformations.[293] Thus the relative rates of γ-hydrogen-abstraction in the photolysis of (**111**) and (**112**) are 1.2:70 respectively,[294] this difference being attributable solely to differences in the entropies of activation for the two reactions.

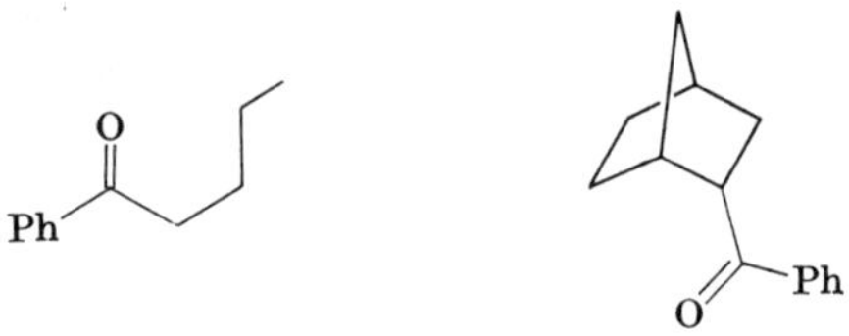

(**111**) (**112**)

Triplet lifetimes are much greater for ketones that cannot undergo ready intramolecular hydrogen-transfer.[295] Quantum yields are also lower for ketones with short triplet lifetimes because of the ease of biradical reversion. Further details have appeared of the application of photoreduction to functionalization of steroids.[296,297]

Intramolecular γ-hydrogen-transfer is also encountered in the photolysis of α-methylene ketones. Abstraction occurs exclusively at the γ- and not the γ'-carbon as the former site is sterically more accessible.[298] β-Abstraction is an alternative process

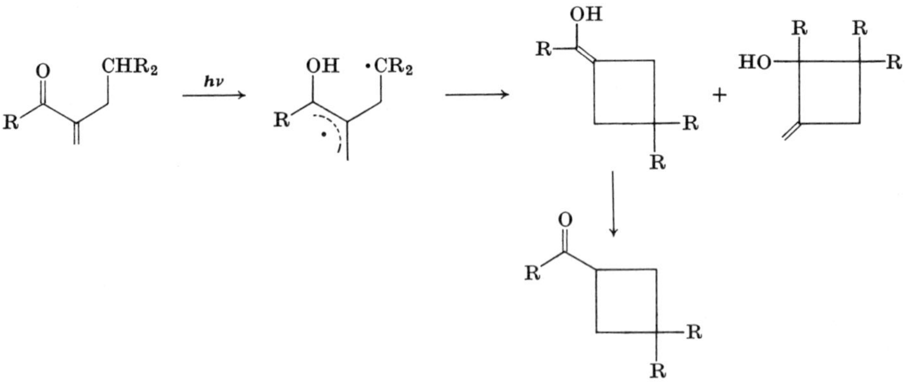

R, O, CHR_2, $h\nu$, OH, $\cdot CR_2$, R, OH, R, R, +, HO, R, R, R, R, O, R, R, R

when the β-hydrogen is tertiary. Transfer of β-hydrogen from C-5 to C-1 of a pent-1-en-3-one system has been observed with photoexcited cyclopent-1-enyl ketones.[299]

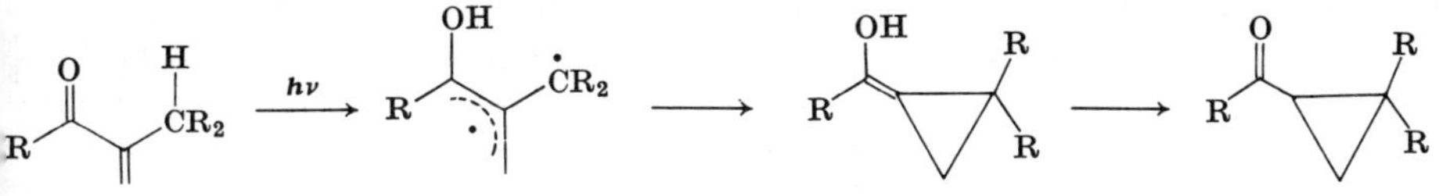

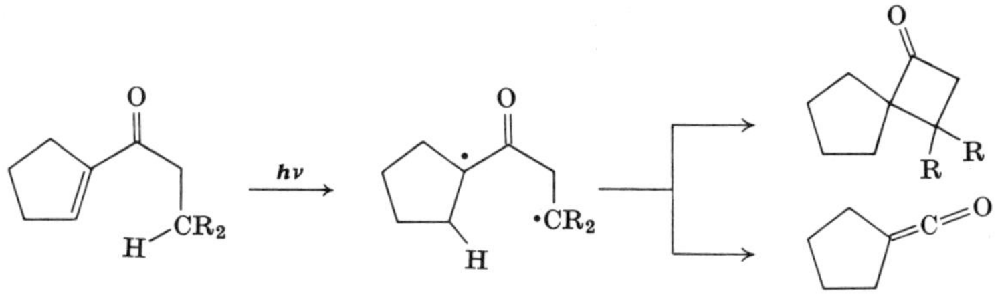

Photoreduction of duryl phenyl ketones in propan-2-ol leads to varying amounts of the cyclobutenol (**113**) in addition to the expected benzhydrol.[300] From a study of the effect of substituents on the yields of these products, it was concluded that cyclobutenol

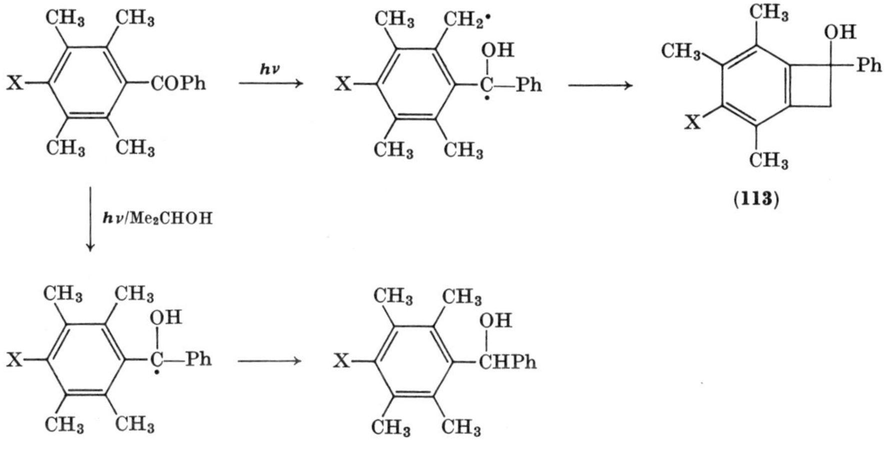

(**113**)

formation involved the π,π^* excited state of the ketone while the n,π^* state was implicated in the production of the benzhydrol. Intramolecular hydrogen transfer to the π,π^* excited state also occurs in the photolysis of certain α,β-unsaturated ketones.[301]

Hydrogen-transfer has been reported for the photolyses of diaryl thioketones,[302] nitro-compounds,[303–305] azo-compounds[306] and 2-phenylquinoxaline.[307] In the last instance reduction in alcohols proceeds *via* an excited singlet state whereas in amines a charge-transfer mechanism operates.

Addition Reactions[309–311]

Intermolecular Additions

The rates of decomposition of chemically activated n-alkyl radicals, produced by addition of hydrogen atoms to the appropriate terminal alkene, have been measured by an internal comparison method and agree well with values calculated by the Marcus–Rice theory.[312] The rate of addition of hydrogen atoms to fluoroalkenes decreases as the number of fluorine substituents increases;[313] a similar effect was noted in the addition of fluorine to fluoroalkenes,[314] the rate decrease being attributed to steric effects. Kinetic studies of the addition of chlorine to perfluoroalkenes indicate a minimum in the rate for the C_5-compound.[315] Chlorination of C_5-, C_6- and C_7-cycloalkenes, indene, acenaphthene and phenanthrene with $PhICl_2$ gave preferential *trans*-addition, but *cis*-addition was observed with *cis*-cyclo-octene.[316] The main products formed in the liquid-phase reaction of atomic nitrogen with propene are HCN and acetonitrile;[317] the imino-radicals (**114**) and (**115**), formed by addition of the nitrogen to one or other end of the propene double bond and followed by a 1,2-hydrogen shift, were postulated as intermediates. Imino-radicals, formed by addition of a fluorine atom to the cyano-carbon of trifluoroacetonitrile, were also postulated in the photolysis of tetrafluorohydrazine with this nitrile.[318]

$$CH_3CH{=}CH_2 \xrightarrow{N} H_2C(\dot{N}\cdot)\text{—}\dot{C}HCH_3 \longrightarrow \underset{(\mathbf{114})}{H\text{—}\dot{N}{=}CCH_2CH_3} \longrightarrow HCN + CH_3CH_2\cdot$$

$$CH_3CH{=}CH_2 \xrightarrow{N} H_2\dot{C}\text{—}C(\dot{N}\cdot)\text{—}CH_3 \longrightarrow \underset{(\mathbf{115})}{CH_3C(=\dot{N})CH_3} \longrightarrow CH_3\cdot + CH_3CN$$

Perfluoroisopropyl radicals add predominantly to the least substituted end of vinyl fluoride and trifluoroethylene, but reaction is very slow with perfluoropropene.[319] In the gase phase, *n*-perfluoroalkyl radicals (up to C_8) showed a modest increase in selectivity with chain length in adding to vinyl fluoride, 1,1-difluoroethylene and trifluoroethylene.[320] Perfluoroalkyl radicals branching at the radical centre showed a steep increase in selectivity, these trends being interpreted in terms of steric and polar factors. $CFBr_2\cdot$ radicals were shown to be more selective than $CF_2Br\cdot$ or $CF_3\cdot$ radicals in addition to fluoroalkenes,[321] this being attributed mainly to steric factors. Comparison of the calculated and the experimental secondary deuterium isotope effects in addition of methyl and trifluoromethyl radicals to ethylene has led to a clearer understanding of the transition state.[322] Cyclohexyl radicals, generated by radiolysis of cyclohexane, add to chloroalkenes, and the resulting radical can abstract a hydrogen atom from cyclohexane or eliminate a chlorine atom;[323] the extent of elimination increases with the number of chlorine atoms in the alkene.

$$\begin{aligned} R\cdot + ECl &\rightarrow RECl\cdot \\ RECl\cdot + RH &\rightarrow REClH + R\cdot \\ RECl\cdot &\rightarrow RE + Cl\cdot \end{aligned}$$

In the telomerization of chlorotrifluoroethylene by carbon tetrachloride the transfer constants were shown to increase with radical-chain length;[324] addition of chloroethanes to styrene gave rise to 1:1 adducts and telomers.

$$\begin{aligned} CX_3CCl_2\cdot + PhCH{=}CH_2 &\rightarrow CX_3CCl_2CH_2\dot{C}HPh \\ CX_3CCl_2CH_2\dot{C}HPh + CX_3CCl_3 &\rightarrow CX_3CCl_2CH_2CHClPh + CX_3CCl_2\cdot \\ X = Cl \text{ or } H & \end{aligned}$$

The yield of the 1:1 adduct decreased as the number of β-chlorine atoms in the chloroethyl radicals decreased,[325] and a correlation between log (yield) and the combined σ^* of the β-substituents was observed. In the copper(II)-catalysed addition of carbon tetrachloride to methyl acrylate and methyl trichloroacetate, 2:1 adducts are the main products,[326] but with chloroform no 2:1 adduct could be detected.

In the presence of zinc, phenacyl bromide and its derivatives give rise to free radicals which can be detected by ESR spectroscopy;[327] the radicals couple to give 1,4-diketones, or in the presence of terminal alkenes add to the terminal site. Trends in the rates of addition of aryl radicals (generated from arylazotriphenylmethanes) to allyl compounds have been interpreted in terms of polar effects.[328] The kinetics of the addition of trichlorosilyl radicals to ethylene in the gas phase has been investigated.[329]

N-Haloacetamides (**116**; X = Cl or Br) react with alkenes either by addition or by allylic abstraction; the yield of adduct was much increased at lower temperatures. Amides with an alkyl substituent on nitrogen, such as (**117**), were much more reluctant

$$\mathrm{RCONHX} + \rangle\mathrm{C{=}C}\langle \xrightarrow{h\nu} \mathrm{RCONH\overset{|}{\underset{|}{C}}{-}\overset{|}{\underset{|}{C}}{-}X} \qquad \mathrm{ClCH_2CON(CH_3)Cl}$$

(**116**) X = Cl or Br (**117**)

to undergo addition, even at 203K.[330] *N*-Bromobistrifluoromethylamine (**118**) gave about 10% of non-terminal addition to allyl chloride in the gas phase;[331] in the liquid phase an ionic reaction was observed.

$$\mathrm{(CF_3)_2NBr + CH_2{=}CHCH_2Cl} \xrightarrow{h\nu} \mathrm{(CF_3)_2NCH_2CHBrCH_2Cl} + \mathrm{(CF_3)_2NCH}\langle^{\mathrm{CH_2Cl}}_{\mathrm{CH_2Br}}$$

(**118**) 90% 10%

Photolysis of perfluoromorpholine (**119**) in the presence of perfluorocyclobutene gave a good yield of the adduct (**120**) together with some of the dimer (**121**) of the perfluoromorpholino-radical; perfluorobicyclobutyl (**122**) was also formed, presumably by attack of fluorine atoms on perfluorocyclobutene, the resulting radical adding to a further cyclobutene unit before abstracting fluorine from (**119**).[332] The adduct $Me_2PCF_2CF_2H$

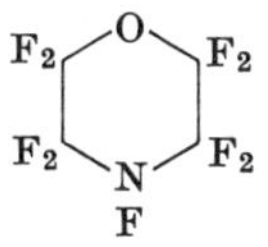

(**119**) 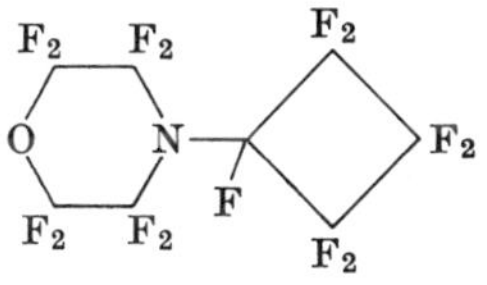 (**120**) 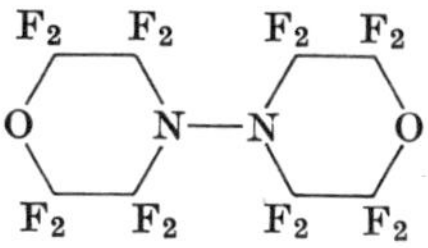 (**121**)

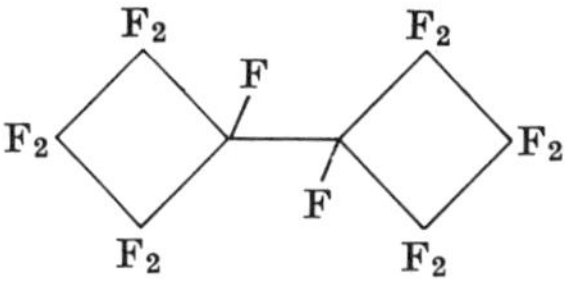

(**122**)

was formed in good yield from thermolysis of dimethylphosphine and tetrafluoroethylene, an unusual initiation step, $Me_2PH + C_2F_4 \rightarrow Me_2P\cdot + \cdot CF_2CF_2H$, being proposed.[333] Tetramethyldiphosphine decomposes thermally by a unimolecular process, and also gives good yields of the adduct $Me_2PCF_2CF_2PMe_2$ with tetrafluoroethylene; the kinetics of the reaction led to an approximate value of 38 kcal mol^{-1} for $D(Me_2P{-}PMe_2)$.[333]

Hydroxy-radicals add to unsaturated alcohols in aqueous solution, but this is accompanied by allylic abstraction, particularly when two conjugated double bonds are present.[334] Bis(trifluoromethyl) trioxide (**123**) readily breaks down at 70° to give trifluoromethoxy-radicals and trifluoromethylperoxy-radicals; when simple species such as SO_2, SF_4 or CO are present, both radicals add to these molecules and the products are formed either by S_H2 reactions at oxygen or by radical combination:[335]

$$CF_3OOOCF_3\ (\mathbf{123}) \rightarrow CF_3OO\cdot + \cdot OCF_3$$
$$CF_3OO\cdot + X \rightarrow CF_3OOX\cdot$$
$$CF_3O\cdot + X \rightarrow CF_3OX\cdot$$
$$CF_3OOX\cdot + CF_3OOOCF_3 \rightarrow CF_3OOXOCF_3 + CF_3OO\cdot$$
$$CF_3OX\cdot + CF_3OOOCF_3 \rightarrow CF_3OXOCF_3 + CF_3OO\cdot$$

When the trioxide was decomposed in the presence of chlorotrifluoroethylene the main products were the alkene epoxide and bis(trifluoromethyl) peroxide, but low yields of the adducts (**124**) and (**125**) were also obtained:[336]

$CF_3OOCF_2CFClOCF_3$ (**124**) $CF_3OCF_2CFClOOCF_3$ (**125**)

Photolysis of benzoic acid in the presence of tetramethylethylene gives a fair yield of the adduct (**126**) together with substantial amounts of isobutyrophenone (**127**) and two other products (**128**) and (**129**) believed to be formed as shown.[337]

PhCOOH + Me₂C=CMe₂ → PhCOO–CMe₂–CMe₂–H (**126**)

→ [intermediate: O, Ph, OH] → PhCO–CHMe₂ (**127**)

→ [Ph–Ċ(OH)₂ + allylic radical] → (**128**) + (**129**)

Both "normal" and "reverse" adducts have been detected in the additions of diols and diacetates to oct-1-ene, but the yields were poor.[338] Photolysis of *tert*-butyl trimethylsilyl peroxide (**130**) gives *tert*-butoxy-radicals and trimethylsilyloxy-radicals; the latter radicals were much more prone than the former to add to alkenes, and the adduct radicals (**131**) were detected by ESR spectroscopy.[339]

$$\underset{(\mathbf{130})}{Me_3SiOOBu^t} \xrightarrow{h\nu} Me_3SiO\cdot + \cdot OBu^t$$

$$Me_3SiO\cdot + RR'C{=}CH_2 \longrightarrow \underset{(\mathbf{131})}{Me_3SiOCH_2\dot{C}RR'}$$

Addition of propylthiyl radicals to cyclohexene[340] and of butylthiyl radicals to alkenes and conjugated dienes[341] has been investigated; the dienes were more reactive than terminal alkenes, which were more reactive than alkenes with internal double bonds. Attack by phenyl radicals on an episulphide produces the radical (**132**) which can fragment to give either of the two possible alkenes (**133**) and (**134**).[342] Photoinitiated bisulphite addition to alkenes was found to proceed by the annexed mechanism. The

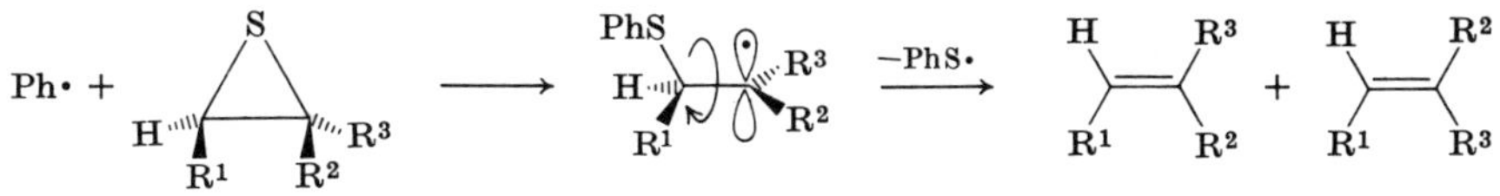

(**132**) (**133**) (**134**)

$$HSO_3^- + In \longrightarrow InH + SO_3^{\bar{\cdot}} \xrightarrow{RCH=CH_2} R\dot{C}HCH_2SO_3^- \xrightarrow{HSO_3^-} RCH_2CH_2SO_3^-$$

$SO_3^{\bar{\cdot}}$ radicals were destroyed by combination to $S_2O_6^{2-}$ or by oxidation to sulphate.[343]

The norbornane skeleton has continued as favourite workhorse amongst stereochemists this year. Several studies have shown that norbornenes, particularly those with substituents in the 5- or 6-*endo*-position, can give *cis*- as well as *trans*-adducts, unlike norbornene itself which gives exclusively *trans*-adducts. Chlorination of 5,5,6,6-tetrafluoronorborn-2-ene (**135**) gave 76% of the *cis*-adduct (**136**) and 24% of the *trans*-adduct (**137**);[344] the preference for the *cis*-product was attributed to coulombic repulsion

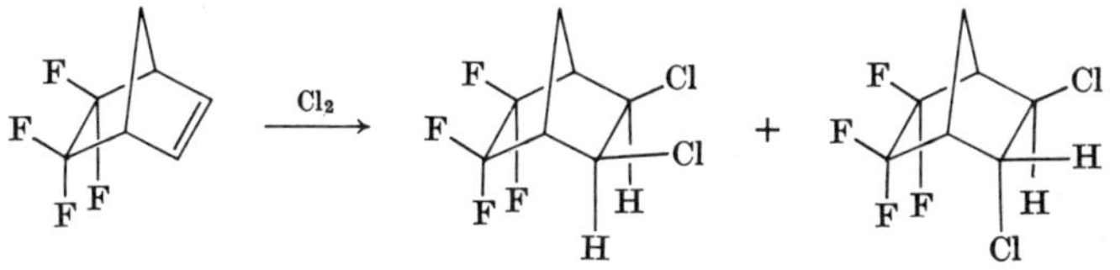

(**135**) (**136**) (**137**)

between the *endo*-fluorine substituents and the incoming chlorine molecule. Bromination of a series of 5,6-substituted norbornenes (**139**), gave exclusively the *exo-cis*-product (**140**) when the *endo*-substituent Y was F, Cl, CF_3 or CF_2Cl. A mixture of (**140**) and the *trans*-dibromide was obtained when Y = H.[345]

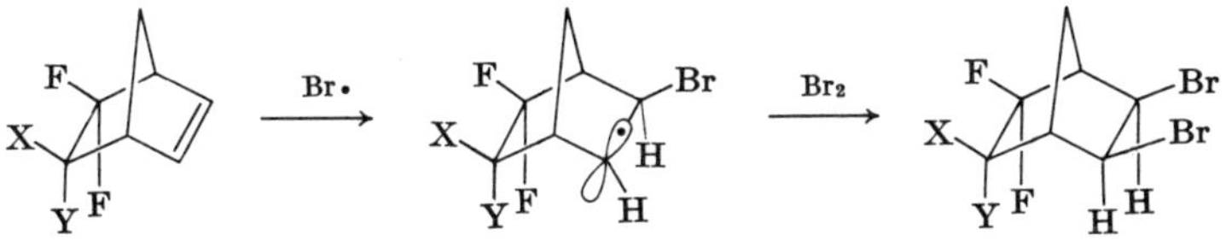

(**138**) (**139**) (**140**)

Exclusive formation of the *exo-cis*-dibromide was also observed from 5,5,6,6-tetracyanonorbornene and *endo-cis*-5,6-dichloronorbornene;[345] attack by bromine atoms came first from the less hindered *exo*-side (**139**), but the *endo*-5,6-substituents shielded the radical (**139**) from *endo*-approach by coulombic repulsion of the incoming bromine molecule. *exo-cis*-Bromination was more exclusive than chlorination because of greater coulombic interaction of *endo*-substituents with bromine. Bromination of *exo-cis*-5,6-dichloronorbornene, which has no *endo*-substituent, gave predominantly, though not exclusively, the *trans*-dibromide, whereas bromination of *trans*-5,6-dichloronorbornene, with one *endo*-chlorine atom, gave a mixture of the *exo-cis*-dibromide and *trans*-dibromide.[346] Addition of $CCl_3\cdot$ or n-$C_3F_7\cdot$ radicals to 5,5,6,6-tetrafluoronorbornene also gave a mixture of *cis*- and *trans*-adducts;[347] the fact that some *cis*-product was formed can also be attributed to coulombic repulsion between the *endo*-fluorine substituents and the incoming radical.

Addition of thiols to norbornenes with one or more substituents at the 5-position led preferentially to sulphides formed by attack at the 2-position;[348] the intermediate radical from addition at the 2-position was considered to be more stable than that from addition at the 3-position. Benzenethiol reacts with norborn-2-en-5-one (**141**) to give the 2-*exo*- and 3-*exo*-sulphides together with the 7-*anti*-sulphide (**142**) which may have

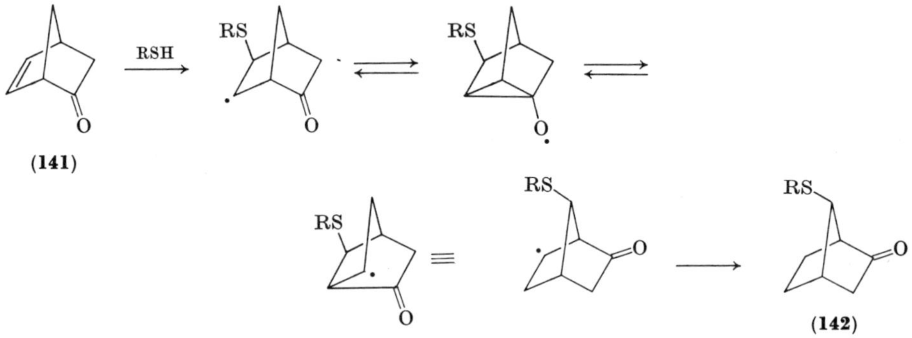

RSH RS RS O O O RS RS RS O O O

(**141**) (**142**)

been formed in a radical homoenolization as shown.[348] 7-Methyl substituents in norbornene exert practically no influence on the direction of attack of thiyl radicals;[349] alkanethiols showed a preference for *exo*-attack on 1,4,7,7-tetramethylnorbornene which increased with the size of the approaching thiyl radical, i.e. in the order $Bu^tS\cdot > Pr^iS\cdot > MeS\cdot$. Arenethiols on the other hand showed an increasing amount of *endo*-product in the order p-$ClC_6H_4S\cdot > p$-$CH_3C_6H_4S\cdot > PhS\cdot$; the trends were interpreted in terms of steric and electrostatic effects.[349] A small increase in the amount of the *endo*-sulphide was observed in the addition of benzenethiol to 7,7-dimethylnorbornene relative to the addition of norbornene itself.[350]

Addition of *tert*-butyl hypochlorite to *endo*-tricyclo[3.2.1.$0^{2.4}$]octene (**143**) gave *trans*- (**144**) and *cis*-adducts (**145**) rather than dichlorides.[351] Adducts were also obtained with *exo*-tricyclo[3.2.1.$0^{2.4}$]octene (**146**) and deltacyclene (**147**); steric effects were

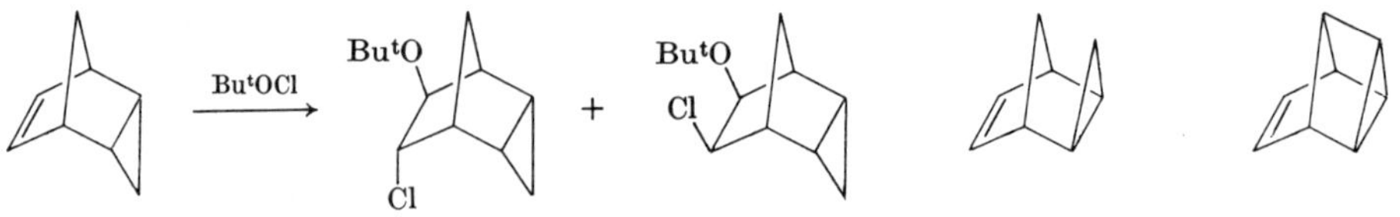

Bu^tOCl Bu^tO Cl + Bu^tO Cl

(**143**) (**144**) (**145**) (**146**) (**147**)

considered to be the main factor controlling the proportions of *cis*- and *trans*-adducts. Photoaddition of the nitrosamine (**148**) to norbornene gave *syn*- and *anti-exo*-3-dimethylaminonorbornanone oxime (**149**) together with *cis*-1,3-diformylcyclopentane monoxime (**150**), probably formed as shown.[352]

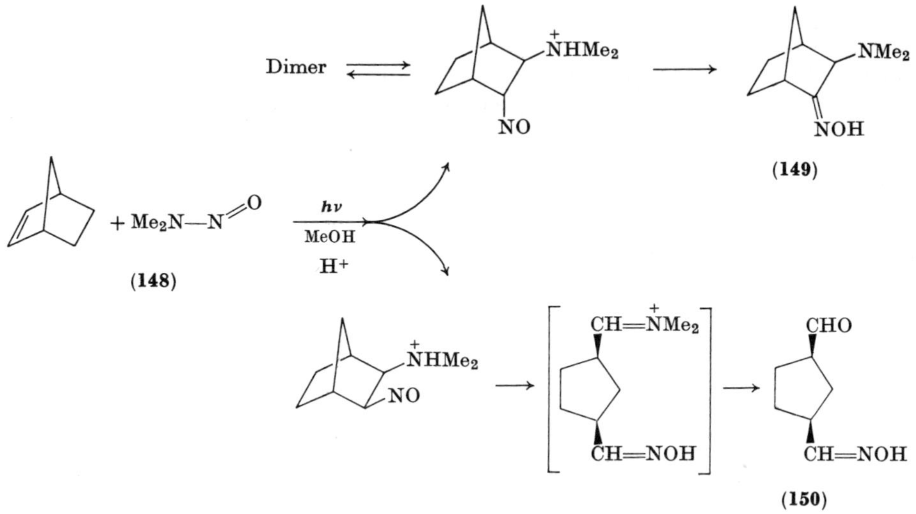

Dimer ⇌ [bicyclic $\overset{+}{N}HMe_2$ / NO] ⟶ [bicyclic NMe_2 / NOH]

(149)

[norbornene] + $Me_2N—N{=}O$ $\xrightarrow[\text{MeOH, } H^+]{h\nu}$

(148)

[bicyclic $\overset{+}{N}HMe_2$ / NO] ⟶ [$CH{=}\overset{+}{N}Me_2$ / $CH{=}NOH$] ⟶ [CHO / $CH{=}NOH$]

(150)

Photolysis of 3,3,3-trifluoropropyne produces trifluoromethyl radicals which subsequently add to the unsubstituted end of the alkyne.[353] A useful synthesis of perfluoro-2,3-dimethylbut-2-ene involves photoaddition of trifluoromethyl iodide to hexafluorobut-2-yne.[354] Hydroxyl radicals add as readily to alkynes as to alkenes.[355] Mono- and di-adducts have been obtained on addition of thiols to alkynes, but the mechanism involved competing radical and ionic pathways.[355]

In the addition of CF_3I and CH_3I to penta-2,3-diene the products were formed by terminal attack of the methyl radical, but 41–49% of central attack was also observed in additions with CCl_3Br.[356] Addition of toluene-*p*-sulphonyl iodide to a series of allenes gave, in most cases, an allylic iodide by central attack by the $ArSO_2\cdot$ radical. Both the recovered diene and the mono-adducts were racemized in the addition of toluene-*p*-sulphonyl iodide to (*S*)-(−)-cyclonona-1,2-diene (**151**);[356] this suggested that the addition step is reversible and that the adduct radicals undergo rapid conformational inter-

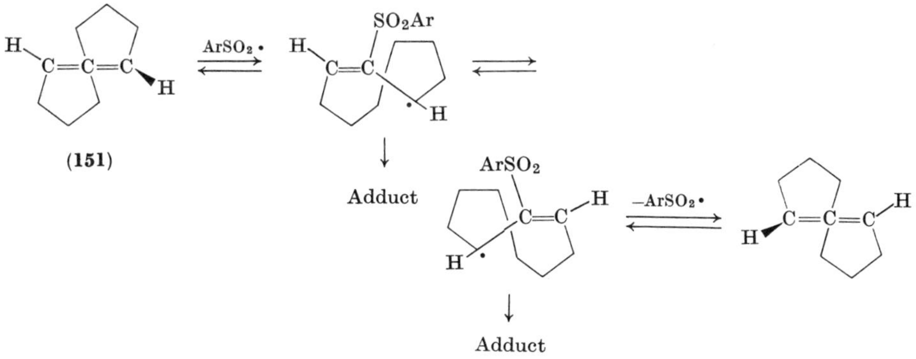

(151) $\underset{\longleftarrow}{\overset{ArSO_2\cdot}{\longrightarrow}}$ [SO_2Ar radical] ⇌ [$ArSO_2$ radical] $\underset{\longleftarrow}{\overset{-ArSO_2\cdot}{\longrightarrow}}$ [enantiomeric allene]

↓ Adduct   ↓ Adduct

conversion. The reversibility in this addition was attributed to conformational prohibition of coplanarity (and hence of allylic stabilization) in 2-cyclononenyl radicals. Addition of *N*-bromobis(trifluoromethyl)amine to allenes has been used as a method for preparing *N*,*N*-bis(trifluoromethyl)amino-substituted allenes.[357] Addition of trimethylhydridotin to conjugated enynes gave 1,2-, 4,3- and 1,4-adducts, together with telomers,[358] the relative yields depending on the concentration ratio of the reactants, indicating reversible addition. The kinetics of the gas-phase addition of acetyl radicals, generated from 3-methylbutan-2-one, to buta-1,3-diene have also been investigated.[359]

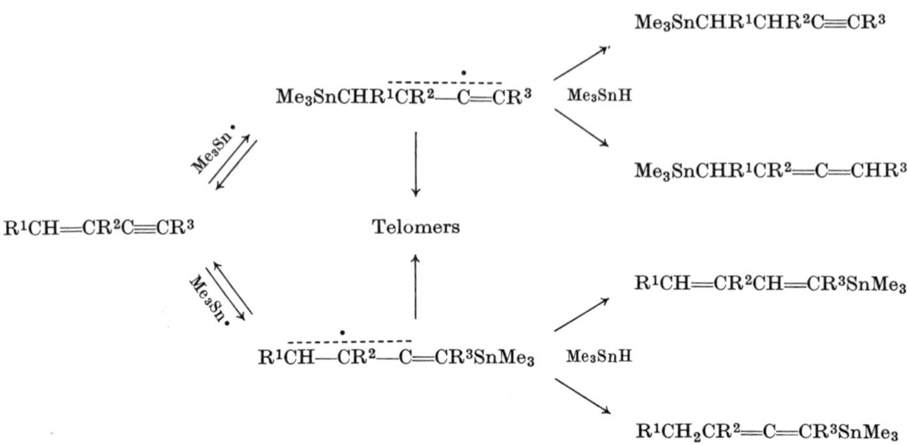

An ESR study of the addition of Group IV metal radicals $R_3M\cdot$ to the carbonyl-oxygen of ketones showed that the ease of addition was in the order $R_3Si\cdot > R_3Ge\cdot > R_3Sn\cdot > R_3Pb\cdot$;[360] for a given Group IV radical the order of reactivity was: diketones > oxalates > ketones > trifluoroacetates > formates > acetates. Triphenylsilyl radicals, generated by photolysis of acetyltriphenylsilane in cyclohexane solution, add to the carbonyl-oxygen of the reactant;[361] the resulting radical can abstract from the solvent to give (**152**) or lose a hydrogen to give the alkene (**153**) which undergoes a further addition sequence, finally giving (**154**). Similar series of reactions were observed with other acylsilanes.

$Ph_3Si\cdot + Ph_3SiCOMe \longrightarrow (Ph_3SiO)(Ph_3Si)\dot{C}{-}CH_3$

$\xrightarrow{RH} (Ph_3SiO)(Ph_3Si)C(Me)H$ (**152**)

$\longrightarrow (Ph_3SiO)(Ph_3Si)C{=}CH_2$ (**153**) $\xrightarrow{Ph_3Si\cdot} (Ph_3SiO)(Ph_3Si)\dot{C}{-}CH_2SiPh_3 \longrightarrow (Ph_3SiO)(Ph_3Si)C{=}C(SiPh_3)H$ (**154**)

Trichlorosilyl radicals add to the carbonyl-oxygen of methyl acetate, and the resulting radicals abstract hydrogen from the parent trichlorosilane.[362a] Tri-*n*-butyltin radicals add to the oxygen of cyclopropyl ketones whose ring subsequently opens;[362b] from the relative proportions of the two ring-opened ketones it was deduced that the cyclopropyl ketones adopt a transoid conformation.

Recent developments in the spin-trapping technique include the first reports of measurement of the rate constants for radical addition to the nitrone $PhCH{=}\overset{+}{N}(O^-)Bu^t$ [363a] and to the nitroso-compound $Bu^tN{=}O$.[363b] The technique used in each case involved competition between bimolecular spin-trapping and a unimolecular fragmentation of known rate. For the nitrone, the bimolecular rate constant for trapping of $PhCO_2\cdot$ (in competition with fragmentation $\rightarrow$ $Ph\cdot + CO_2$) was estimated to be between 10^5 and 10^6 l mol^{-1} sec^{-1} at 40°, and for $Bu^tN{=}O$ the rate constant for scavenging $Bu^tO\dot{C}{=}O$ is 1.1×10^6 l mol^{-1} sec^{-1} at 40°. A rate constant of ca. 6×10^7 l mol^{-1} sec^{-1} has been reported for scavenging of methoxy-radicals by the nitrone at 25° (see ref. 627).

Intramolecular Additions[364]

The addition of CCl_4 to 3,3,4,4-tetrafluorohexa-1,5-diene (**155**) has been initiated with a variety of redox-transfer agents;[365] the straight-chain monoadduct was isolated, together with C_4 (**156**) and C_5 cycloadducts (**157**). An adduct containing C_6 rings (**158**) was formed from the 2-to-1 telomer radical, and C_7 cycloadduct was obtained by the

$CH_2{=}CH(CF_2)_2CH{=}CH_2 \xrightarrow{CCl_4} CCl_3CH_2CHCl(CF_2)_2CH{=}CH_2$ +

(**155**) (**156**) (**157**) (**158**)

action of a copper(I)–n-butylamine complex on the straight-chain adduct. An investigation of the addition of isobutyric acid to α,ω-dienes $CH_2{=}CH(CH_2)_nCH{=}CH_2$ showed that for $n = 1$ or 2 good yields of mono- and di-adduct could be obtained, thus providing a relatively simple synthesis for $\alpha,\alpha\ \alpha',\alpha'$-tetramethylalkanedioic acids;[366] for $n = 3$ or 4, cycloadducts predominated, mainly C_5 from hepta-1,6-diene but including C_6 (**159**) and C_7 cycloadducts (**160**) from octa-1,7-diene. Addition of fully fluorinated iodoalkanes to terminal alkadienes leads to significant cyclization only for hepta-1,6-diene, and this

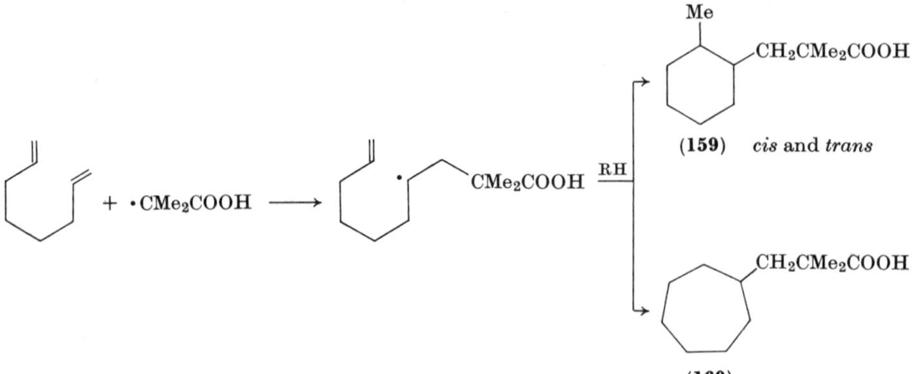

Me

CH_2CMe_2COOH

(**159**) *cis* and *trans*

$+ \cdot CMe_2COOH \longrightarrow$ CMe_2COOH $\xrightarrow{RH}$

CH_2CMe_2COOH

(**160**)

diene is the only one to react faster than alkenes such as hex-1-ene on an equivalent-double-bond basis;[367] the decisive factor appears to be a favourable conformation of the intermediate hepta-1,6-diene adduct-radical.

Intramolecular cyclization of unsaturated thiols (**161**) goes exclusively by path *a* for $n = 3$ or 4, in contrast to the cyclization of the unsubstituted thiols $CH_2{=}CH{-}(CH_2)_nSH$.[368]

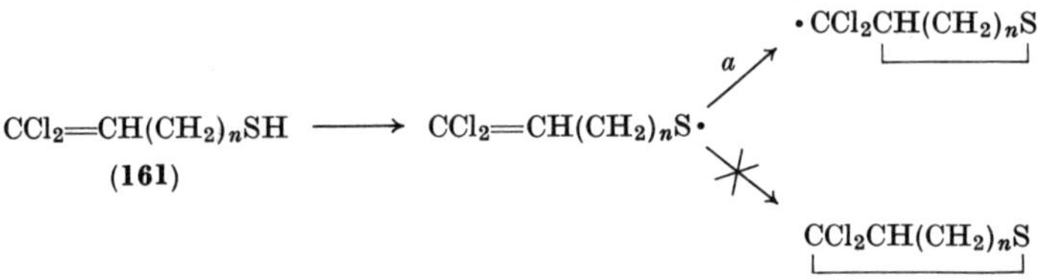

$$CCl_2{=}CH(CH_2)_nSH \longrightarrow CCl_2{=}CH(CH_2)_nS\cdot$$

(**161**)

a → $\cdot CCl_2\overline{CH(CH_2)_nS}$

✗ → $\overline{CCl_2CH(CH_2)_nS}$

Both *trans*- (**162**) and *cis*-deca-5,9-dienyl radicals (**163**), derived by treatment of the appropriate bromodeca-5,9-diene with tri-*n*-butylhydridotin, give predominantly non-cyclized products; that cyclization which does occur gives exclusively the cyclopentane (**164**).[369] 2-(But-3-enyl)cyclohexyl radicals (**165**) also cyclize almost exclusively to the five-membered ring product.[369]

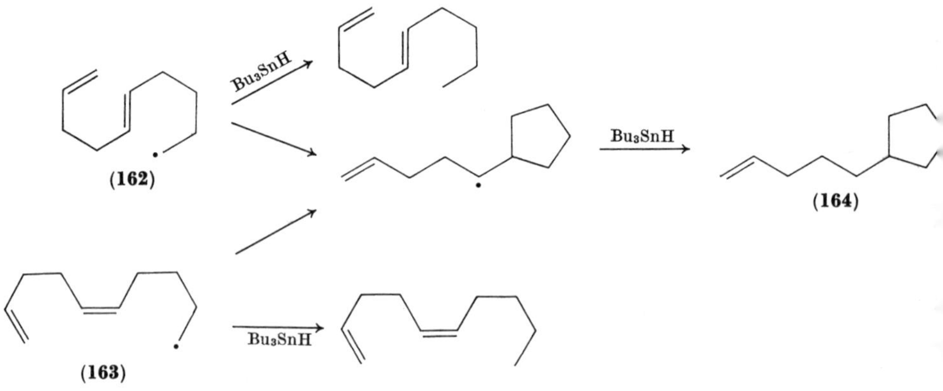

(**162**) $\xrightarrow{Bu_3SnH}$

$\xrightarrow{Bu_3SnH}$ (**164**)

(**163**) $\xrightarrow{Bu_3SnH}$

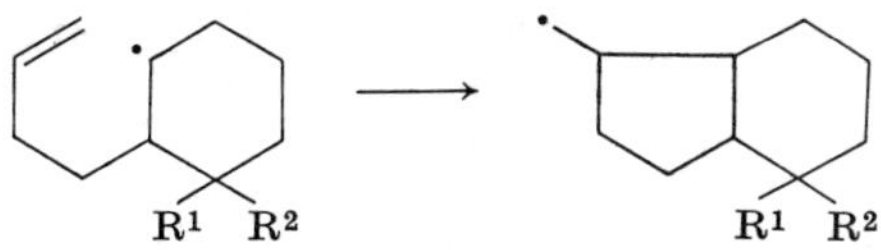

R¹ R² R¹ R²

(165)

Addition of bromoform to *cis,trans*-1,5-cyclodecadiene (**166**) gave a 45% yield of the cyclized *cis*-decalin (**167**) in which the ring-bromine atom was equatorial and *trans* to the ring juncture.[370] The main product from the addition of bromotrichloromethane

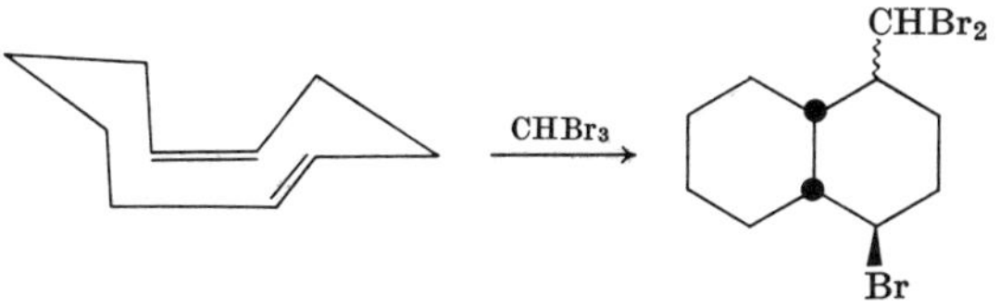

$CHBr_2$

$CHBr_3$

Br

(166) **(167)**

to (**166**) has similar stereochemistry, indicating that radicals preferentially attack the *cis*-double bond of the cyclodecadiene, unlike cationic reagents which attack the *trans*-double bond. Cyclization of a mixture of isomers of the trimethylphenyltridecatrienes (**168**) with benzoyl peroxide (in the absence of transition-metal catalysts) provides a route to tetracyclic members of the triterpene series (**169**).[371] The preference for free-radical intermediates to cyclize to five-membered rings has been used in elucidating the mechanism of the reaction of anions with *p*-nitrobenzyl halides.[372]

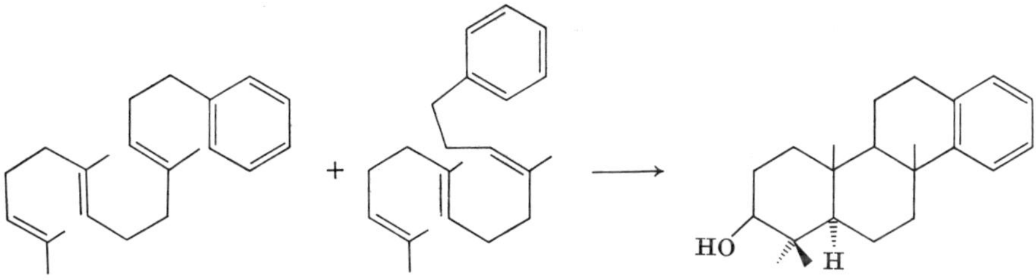

+

HO H

(168) **(169)**

The photochemical addition of hydrogen sulphide to α,ω-dienes has been suggested as a route to medium-sized heterocycles. In particular, treatment of diallyldimethylsilane (**170**) with H_2S gave 5,5-dimethyl-5-thia-1-silacyclo-octane (**172**).[373] The reaction probably proceeds through the monoadduct (**171**) since photolysis of the monoadduct, synthesized independently, also gave (**172**).

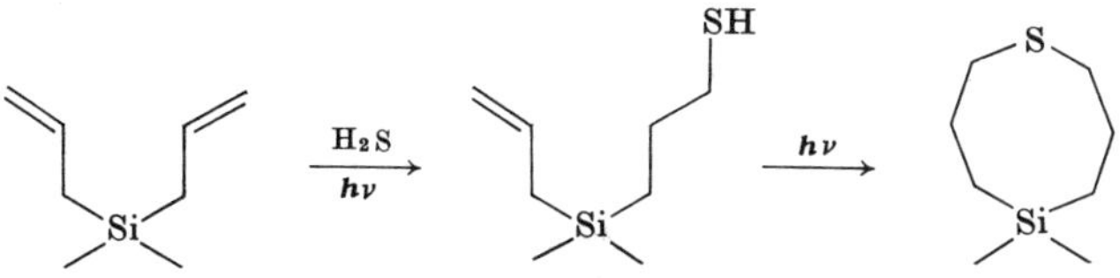

SH S

H_2S $h\nu$ $h\nu$

Si Si Si

(170) **(171)** **(172)**

Photolysis of nitrites of γ-ethylenic alcohols (**173**) gave γ-ethylenic alcohols (or ketones when $R^5 = H$), together with five-membered cyclic products (**174**, **175** and **176**);[374] only five-membered ring compounds were isolated irrespective of the substitu-

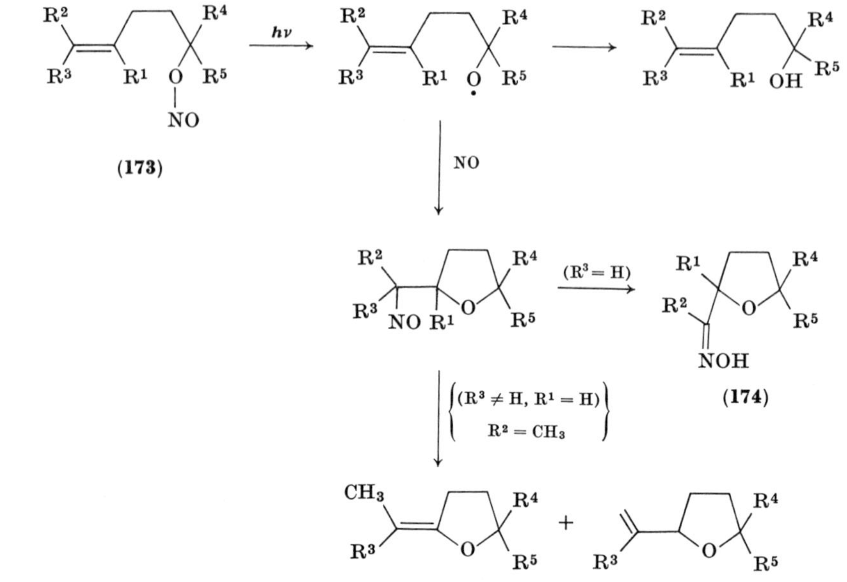

(**173**) (**174**) (**175**) (**176**)

ents. Nitrites of 4-(hydroxymethyl)cyclohex-1-enes (**177**) gave bicyclic ketones (**178**) or olefins (**179**) on photolysis, together with the products of disproportionation, oxidation,

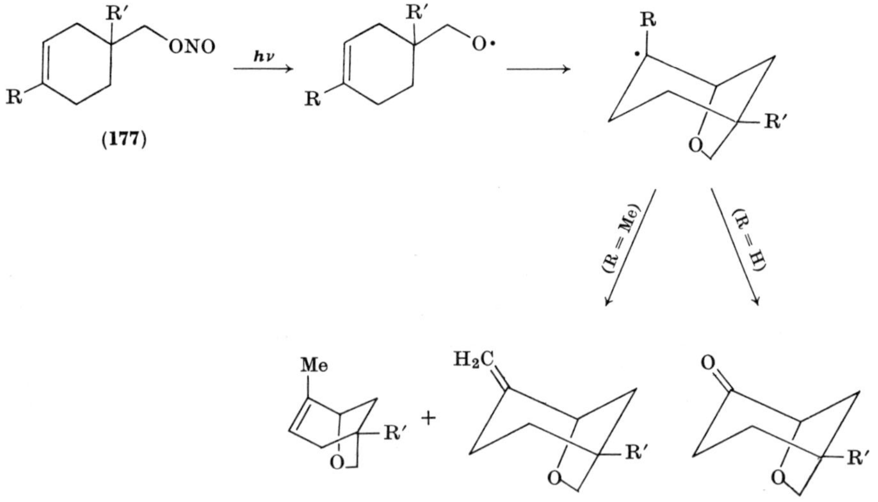

(**177**) (**179**) (**178**)

and hydrogen-transfer of the intermediate cyclohexenylalkoxy-radicals.[375] The substituted hex-5-en-1-yl radical (**184**) was produced by decarboxylating the half-ester of heptenoic acid with lead-tetra-acetate or with copper(II) acetate in pyridine.[376] The main products from the lead tetra-acetate reaction were the dimethylcyclopentane derivatives (**180**) together with some of the cyclohexane product (**182**). In the copper(II) acetate reaction, the elimination product (**181**) predominated together with the bicyclic lactone (**185**) and traces of six-membered elimination products (**183**).

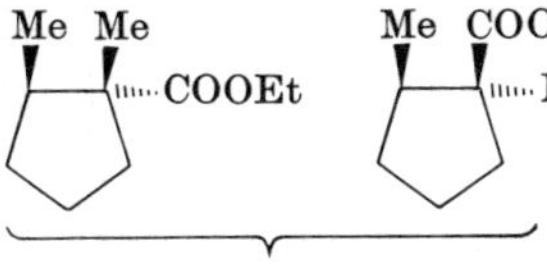

(**180**) 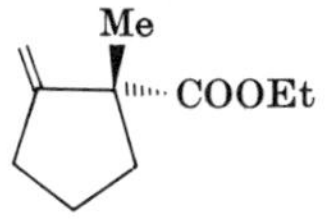 (**181**) 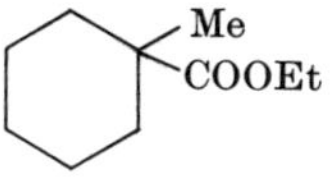 (**182**) 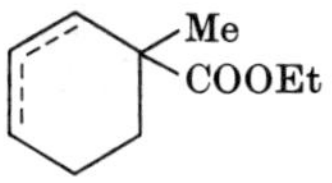 (**183**)

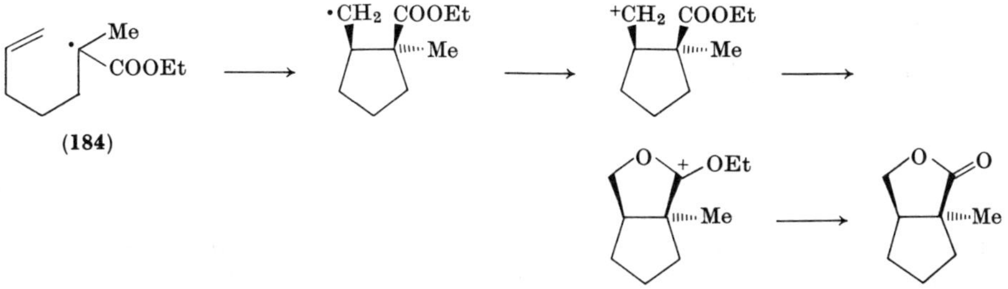

(**184**) (**185**)

Fragmentations

Reaction of phenyl radicals with allylic sulphides (**186**; X = SR) gives good yields of allylbenzene as a result of fragmentation of the intermediate radical:[376a]

$$\mathrm{Ph{\cdot}} + \underset{\textbf{(186)}}{\mathrm{CH_2{=}CHCH_2X}} \rightarrow \mathrm{PhCH_2\dot{C}HCH_2X} \rightarrow \mathrm{PhCH_2CH{=}CH_2} + \mathrm{X{\cdot}}$$

The extent of fragmentation is largely determined by the C—X bond strength, being greater the weaker the bond.

Photolysis of 3β-acetoxycholest-5-en-19-yl nitrite (**187**) gives both (**190**) and (**191**),[377] which arise as a result of fragmentation of the initial alkoxy-radical (**188**) to (**189**) and

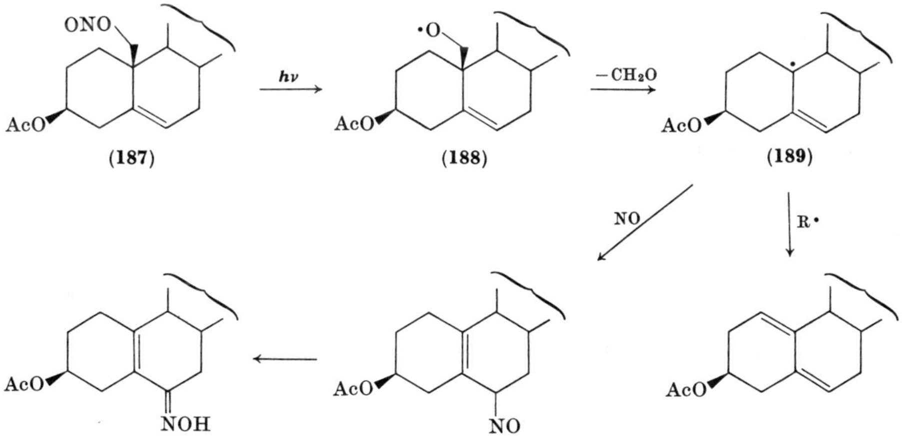

(**187**) (**188**) (**189**)

(**190**) (**191**)

formaldehyde, CIDNP effects have been observed in the fragmentation of (**192**) which gives both (**193**) and (**194**).[378] The cyclopropyloxy-radical (**195**) undergoes fragmentation to the tertiary radical (**196**) and the primary radical (**197**);[379] the radicals were

$$(192) \rightarrow CH_3COOCH_2CH_2CH_2\cdot\ (193) \longrightarrow CH_3COOCH_2CH_2CH_3$$

$$(192) \rightarrow CH_3COCH_2CH_2CH_2O\cdot\ (194) \longrightarrow CH_3COCH_2CH_2CH_2OH$$

$$\xrightarrow{Bu^tO\cdot} (195) \rightarrow \cdot CMe_2CH_2COOMe\ (196);\quad (195) \rightarrow \cdot CH_2CMe_2COOMe\ (197)$$

trapped by nitrosobutane and their relative concentrations were estimated to be 10:1.

The 1,2-elimination of water from the radicals $(HO)_2\dot{C}CH_2OH$ and $HC(OH)_2\dot{C}HOH$ occurs very readily under acid catalysis,[380] e.g. $(HO)_2\dot{C}CH_2OH \rightarrow HOOCCH_2\cdot + H_2O$; decarbonylation of the radical $HOCH_2\dot{C}O$ is a much more efficient process than decarbonylation of the acetyl radical. Loss of nitrous acid occurs very readily from the radicals derived by addition of hydroxy-radicals to 5-nitrofuroic acid and 5-nitrouracil.[381,382] Radical cations derived by oxidation of β-hydroxy-sulphides undergo three different types of fragmentation,[383] as illustrated.

$$H{-}O{-}CH_2{-}CH_2{-}\overset{+\cdot}{S}R \longrightarrow H^+ + CH_2O + \dot{C}H_2SR$$

$$HOCH_2{-}CH(H){-}\overset{+\cdot}{S}R \longrightarrow H^+ + HOCH_2\dot{C}HSR$$

$$R'{-}C(H)(OH){-}CH_2{-}\overset{+\cdot}{S}R \longrightarrow H^+ + R'C(OH){=}CH_2 + \dot{S}R$$

There are several reports of carbon–halogen bond fission from aromatic radical anions.[384–386] The ease of fragmentation in the radical anions of halogenated quinolines, quinoxalines and phenazines is determined by the strength of the C—X bond and by the redox potential of the substrate–radical anion couple;[385] exceptions are the radical anions of 4-(3-chlorostyryl)pyridine and 3-chlorobenzophenone, both of which have very low electron densities on the carbon carrying the chlorine atom. α-Sulphonylnitroalkane radical anions have been found to undergo exceedingly rapid fragmentation;[387] the radical anion (**198**) is so unstable that it cannot be detected by ESR spectroscopy.

$$TosCMe_2NO_2 \xrightarrow{Me_2CNO_2^-} TosCMe_2NO_2^{\overline{\cdot}}\ (198) \longrightarrow Tos^- + Me_2\dot{C}NO_2$$

Homolytic Aromatic Substitution[388]

Aromatic alkylation[389] has been reviewed, as have various anodic aromatic substitutions.[390]

The use of the Hammett $\rho\sigma$ relationship in homolytic aromatic substitutions has been surveyed for all known examples and extended to tritiation.[391] In phenylation of monosubstituted benzenes, the value of log F_p/F_m is positive irrespective of the nature of the substituent, whereas in tritiations it is positive for OH, NH_2 and NO_2 groups but negative for Me, CN, Br and Ac groups. This figure is a measure of the interaction of unpaired spin with the substituent. Orientation in homolytic aromatic substitution can be predicted by using the hyperfine splitting constants of related radical anions and cations.[392]

Factors that favour generation of aryl radicals rather than aryl cations from diazonium salts have been discussed.[393] One new method of generating aryl radicals involves reduction of diazonium salts with copper(I) perchlorate–amine complexes.[394] Aryl radicals have also been generated by reaction of 1-phenylazo-2-naphthol with *p*-chlorobenzoyl nitrite[395] and of N^1-aroyl-N^2-arylhydrazines with sodium hydride and air.[396]

The nature of the substituent X in the 5-position in 2-thiazolyl radicals (**199**) can enhance the weak electrophilicity of the parent radical when X = Br or NO_2 or have the reverse effect when X = Me.[397] 3-Methyl groups also reduce the electrophilic character of 4- and 5-isothiazolyl radicals (**200** and **201**).[398]

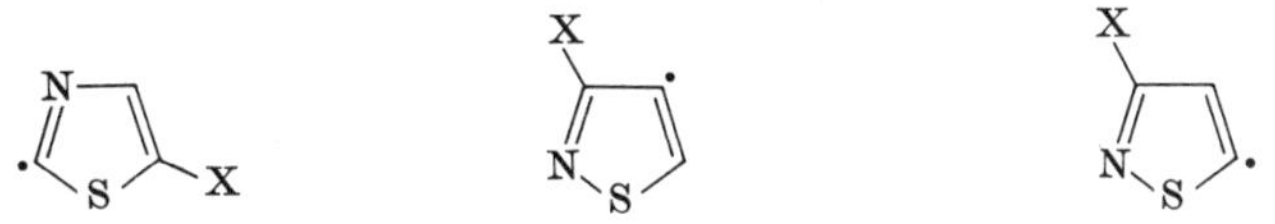

(**199**) (**200**) (**201**)

The isomer ratios in phenylations with (phenylazo)triphenylmethane are somewhat solvent-sensitive.[399] This is claimed to indicate that polar forms make some contribution to the transition state (cf. **202**).

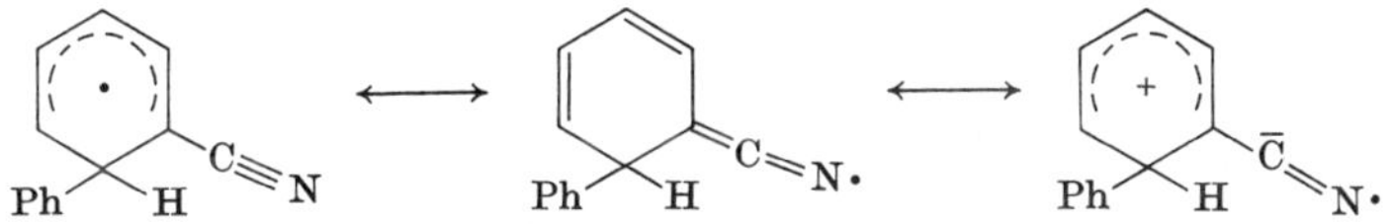

(**202**)

Phenyl radicals display significant nucleophilic behaviour in the phenylation of 4-substituted pyridines,[400] which contrasts with the phenylation of non-protonated pyridines where phenyl radicals appear to have little polar character. This difference in behaviour is attributed to the transition state for protonated and non-protonated bases resembling π- and σ-complexes, respectively. In the π-complex, polar forms make a greater contribution to the transition state (**203**). The relative rates of benzylation of

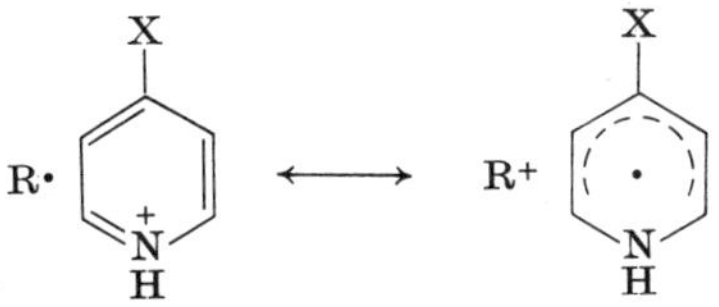

(**203**)

protonated 4-substituted pyridines indicate that the benzyl radical has appreciable nucleophilic character;[401] it is more nucleophilic than the *tert*-butyl radical and much more nucleophilic than the methyl radical. Carbamoyl ($\cdot CONR_2$), α-*N*-amidoalkyl [$RCON(R)CH_2\cdot$] and carboxyl ($\cdot COOH$) radicals display nucleophilic character and give good yields of 2- and 4-substituted products in their reactions with protonated pyridine and quinoline.[402,403]

The amount of attack at the *o*-position of substituted benzenes is dependent on polar effects. Thus there is less *o*-substitution in the reaction of toluene with 2-pyridyl radicals than with 3-pyridyl radicals because of repulsion between the nitrogen lone-pair and the methyl group.[404] The overall reactivity of 2-pyridyl radicals with toluene is also less than that of 3-pyridyl radicals, whereas with nitrobenzene the reverse order of reactivity occurs, consistently with an attractive effect between the lone-pair on nitrogen and the nitro-group. Dichlorophenyl radicals react with *o*-dichlorobenzene preferentially in the 4-position, possibly owing to polar and steric effects.[405]

Tritium atoms display weak electrophilic character in homolytic aromatic substitution.[391] All substituents except cyano and trifluoromethyl enhance the reactivity of substituted benzenes with respect to benzene.[406] Both toluene and anisole undergo preferential attack (ca. 70%) at the *o*-position.[407] Halobenzenes, in addition to undergoing substitution, also undergo halogen replacement.[408] The extent of this can, in the case of bromobenzene, be diminished by addition of copper(II) ions which oxidize the intermediate cyclohexadienyl radical to the cation. Replacement of halogen also occurs in the reactions of halobenzenes with arylthio and arylsulphonyl radicals.[409]

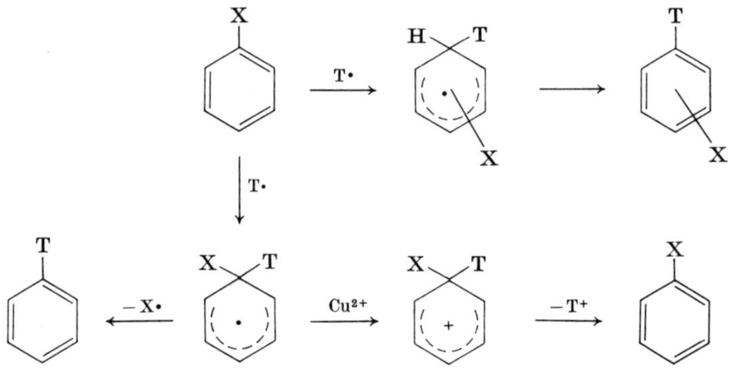

The partial rate factors for the phenylation of polycyclic aromatic hydrocarbons have been shown to be the same whether *N*-nitrosoacetanilide and the Meerwein reaction are the radical sources, except that in the case of phenanthrene more 4-phenylphenanthrene was obtained from the Meerwein reaction.[410] This is attributed to the greater sensitivity of (**204**) to oxidation by copper(II) than by other species. The other less sterically hindered

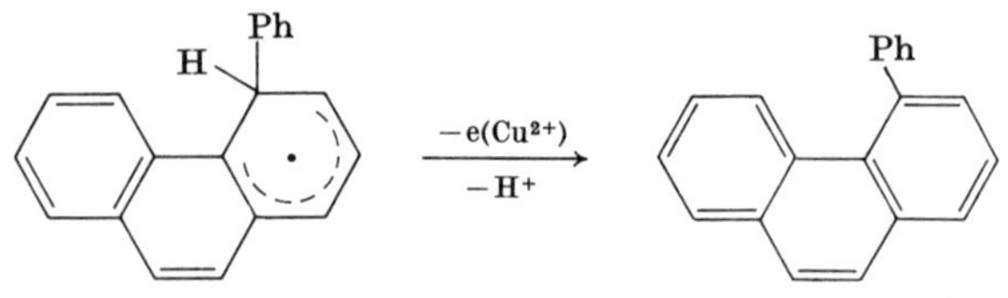

(**204**)

radicals are oxidized with equal ease by all oxidants. Good correlations of reactivity with Hückel localization numbers were obtained. Phenylation of anthracene was concluded to be an irreversible process on account of the absence of isotope effects.

The methylation of methylfurans, methylthiophens[411] and dimethylnaphthalenes[412] have been examined; the points of attack correlate well with HMO calculations. A new method of determining methyl affinities of substituted benzenes has been devised based on the photolysis of methylmercury(II) iodide:[413]

$$\begin{aligned} \mathrm{MeHgI} &\rightarrow \mathrm{Me}\cdot + \mathrm{HgI}\cdot \\ \mathrm{HgI}\cdot &\rightarrow \mathrm{Hg} + \mathrm{I}\cdot \\ \mathrm{Me}\cdot + \mathrm{ArH} &\rightarrow [\mathrm{MeAr}\cdot\mathrm{H}]\cdot \\ [\mathrm{MeAr}\cdot\mathrm{H}]\cdot + \mathrm{I}\cdot &\rightarrow \mathrm{MeAr} + \mathrm{HI} \\ \mathrm{HI} + \mathrm{MeHgI} &\rightarrow \mathrm{MeI} + \mathrm{HgI_2} \end{aligned}$$

From Table 4, which lists some results, it can be seen that there is a wider spread of reactivity than with phenylation, and this is attributable to the smaller reactivity and hence greater selectivity of methyl radicals.

Table 4. Methyl affinities of aromatic substrates

Compound	PhH	PhF	PhCl	PhBr	PhMe	$PhBu^t$	PhOPh	Ph_2
Methyl affinity	1.00	1.51	2.11	3.87	1.46	0.68	4.89	7.47

Phenylation of hexafluorobenzene with dibenzoyl peroxide affords 2,3,4,5,6-pentafluorobiphenyl (**206**) as a result of defluorination of the intermediate cyclohexadienyl radical (**205**).[414] The yield of (**206**) is increased by acids. Smaller quantities of 2,2′,3,4,5,6-hexafluorobiphenyl (**207**) are also formed in the reaction as a result of rearrangement of the radical (**205**) involving fluorine migration.

$C_6F_6 + Ph\cdot \longrightarrow$ (**205**) $\longrightarrow$ [radical with H, F and C_6F_5]

(**205**) $\xrightarrow{(PhCOO)_2 \text{ or } ArCOOH}$ (**206**)

[radical] $\xrightarrow{(PhCOO)_2}$ (**207**)

Radical chlorinations of chlorobenzene and benzonitrile occur via aryl radicals formed by hydrogen abstraction rather than through cyclohexadienyl radicals.[415] Fluorination of benzene can also be accomplished anodically.[416]

$$PhX + Cl\cdot \longrightarrow \cdot C_6H_4X \xrightarrow{Cl_2} ClC_6H_4X$$

Hydroxylation of aromatic compounds gives products with isomer distributions consistent with the electrophilic nature of the hydroxy-radical.[417] The relative rates of hydroxylation of furan, benzene and pyridine are also in agreement with the hydroxy-radical having electrophilic character.[418] The intermediate hydroxycyclohexadienyl radicals formed in the hydroxylation of benzoic acids have been detected;[419] the position of attack correlates with localization energy calculations.

Homolytic aromatic cyanation has been achieved by using cyano-radicals generated by diazotization of cyanamide and by the photolysis of halogen cyanides;[420] isomer ratios are those expected for a radical process. Electrochemical cyanation has also been discussed.[421] The trichloromethylation of ferrocene has been brought about photochemically, the reaction proceeding through a charge-transfer complex.[422] Acyloxylation rather than arylation of arenes takes place in the decomposition of diacyl peroxides in presence of copper(II) chloride.[423,424] In a similar way cerium(IV) salts promote the formation of carbonates in the decomposition of peroxydicarbonates in arenes.[425]

S_H2 Reactions[426, 427]

Dialkylmagnesium compounds and Grignard reagents react with *tert*-butoxy-radicals to give both alkyl and 1-magnesioalkyl radicals, and the ESR spectrum of the solution shows a superposition of the spectra of the two radicals.[428] The alkyl radical is formed

$$R_2CHMgX + Bu^tO\cdot \longrightarrow \begin{cases} R_2\dot{C}H + Bu^tOMgX \\ R_2\dot{C}MgX + Bu^tOH \end{cases}$$

by a homolytic displacement reaction at magnesium, but no ESR evidence was obtained for an intermediate radical having the unpaired electron centred on magnesium. A homolytic mechanism has also been suggested for the displacement of halogen from α-haloesters by PhCdX reagents.[429]

In the reaction of diborane with *tert*-butoxy-radicals there was no evidence for displacement at boron,[430] unlike the analogous reaction with other organoboranes.[431] The only species detected by ESR were *tert*-butyl radicals:

$$Bu^tO\cdot + B_2H_6 \rightarrow Bu^t\cdot + [HOBH_2BH_3 \text{ or } HOBH_2 + BH_3]$$

Tri-*n*-propylaluminium reacts with α,β-unsaturated ketones to give good yields of adducts;[432] the key step in the mechanism is believed to involve radical attack at aluminium, with displacement of a propyl radical, as illustrated:

$$R\cdot + \rangle C{=}\overset{|}{C}{-}\overset{|}{C}{=}O \longrightarrow R{-}\overset{|}{\underset{|}{C}}{-}\overset{|}{\dot{C}}{-}\overset{|}{C}{=}O \longleftrightarrow R{-}\overset{|}{\underset{|}{C}}{-}\overset{|}{C}{=}\overset{|}{C}{-}O\cdot$$

$$R{-}\overset{|}{\underset{|}{C}}{-}\overset{|}{C}{=}\overset{|}{C}{-}O\cdot + R_3Al \longrightarrow R\overset{|}{\underset{|}{C}}{-}\overset{|}{C}{=}\overset{|}{C}{-}OAlR_2 + R\cdot$$

There has been considerable interest in displacement reactions at carbon atoms. Photolysis of hexafluoroacetone in the presence of neopentane at 573K gave CF_3CH_3 amongst other products.[433] The yield of this product was too great to be accounted for solely by combination of methyl and trifluoromethyl radicals and it was suggested that it is also formed in an S_H2 reaction at a fully saturated carbon atom:

$$CF_3\cdot + (CH_3)_4C \rightarrow CF_3CH_3 + (CH_3)_3C\cdot$$

Bromination of *cis*-1,2,3-trimethylcyclopropane (**208**) gave equal proportions of the *S-meso-* (**209**) and (±)- (**210**) ring-opened products, but none of the *R-meso*-product.[434] This implies inversion of configuration in at least one of the two intermediate steps, as illustrated.

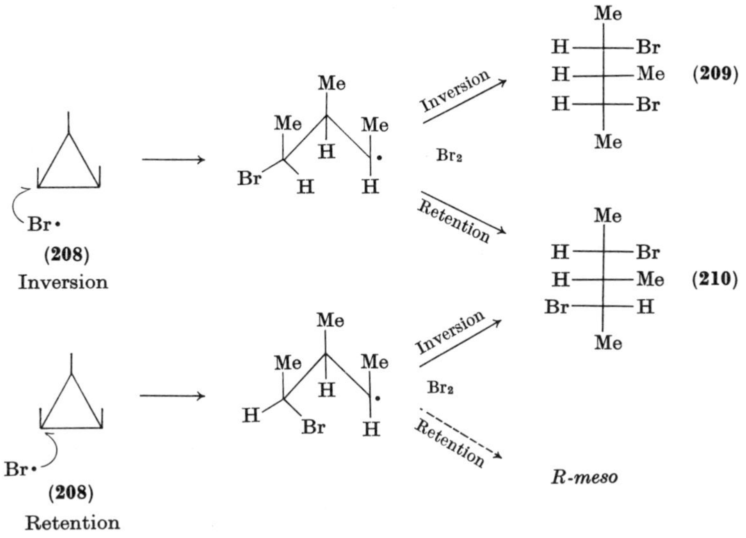

Photobromination of alkylcyclopropanes at 195K was observed to proceed entirely by a radical mechanism, giving exclusively 1,3-dibromoalkanes and none of the bromocyclopropanes;[435] cleavage of the ring took place most readily between the least and the most substituted ring-atoms, *e.g.*:

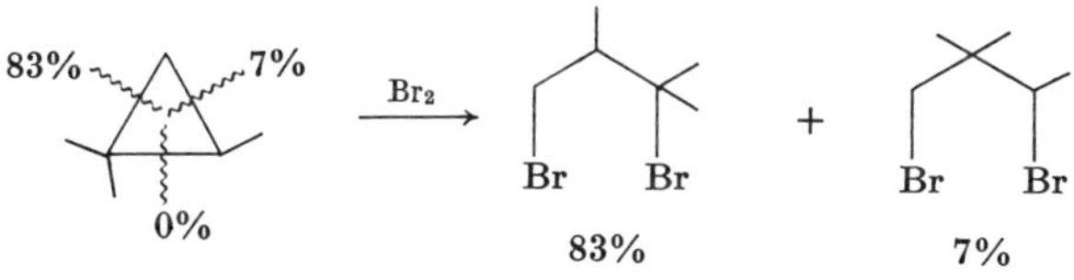

The same rule was not obeyed by alkylarylcyclopropanes which showed preferential cleavage between the carbon atom bearing the aryl substituent and the alkyl-substituted carbon atom. Further evidence that the homolytic substitution at a cyclopropane carbon atom involves inversion of configuration was provided by a study of the bromination of "2,4-dehydroadamantane" (**211**).[435] The first bromine atom enters an equatorial position, and the intermediate bromoadamantyl radical reacts from either side with molecular bromine to give *e,e*- (**212**) and *a,e*-2,4-dibromoadamantane (**213**); none of the diaxial dibromoadamantane was obtained, indicating that the initial attack of the bromine atom occurs with inversion. Cyclization of halopropyl radicals to cyclopropane also occurs in an S_H2 reaction. When 3-iodopropyl radicals are generated in carbon

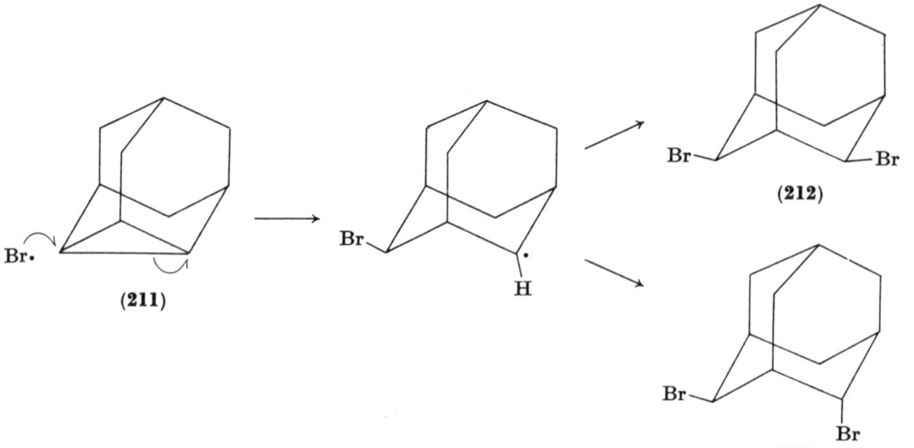

Br· (211) → [Br, H, radical] → Br, Br (212); Br, Br (213)

tetrachloride solution the rate of cyclization was greater than the rate of chlorine abstraction from the solvent:[436]

$$ICH_2CH_2CH_2\cdot \xrightarrow{CCl_4} ICH_2CH_2CH_2Cl + \cdot CCl_3$$

$$ICH_2CH_2CH_2\cdot \longrightarrow \text{cyclo-}C_3H_6 + I\cdot$$

A substantial amount of cyclization was also observed when substituted iodopropyl radicals were generated in the presence of tributyltin hydride,[437] as shown for (**215**); the

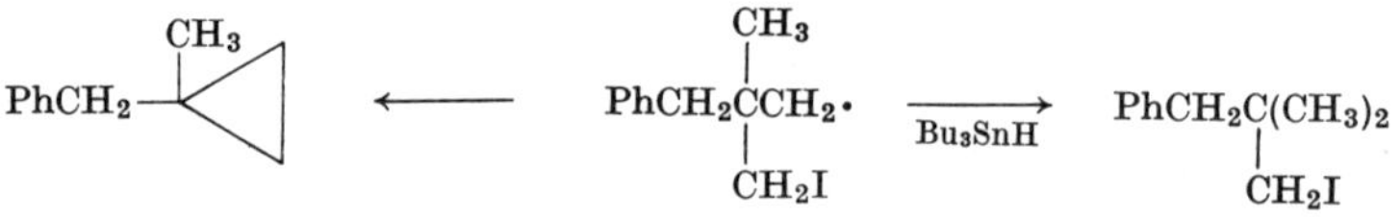

$$PhCH_2\text{-}C(CH_3)(\text{cyclopropane}) \longleftarrow PhCH_2C(CH_3)(CH_2I)CH_2\cdot \ (\mathbf{215}) \xrightarrow{Bu_3SnH} PhCH_2C(CH_3)_2CH_2I$$

proportion of cyclized product depends strongly on the solvent. Allyl sulphides and halides react with phenyl radicals, generated from (phenylazo)triphenylmethane, by a displacement reaction:[438]

$$Ph\cdot + CH_2{=}CHCH_2X \rightarrow PhCH_2CH{=}CH_2 + X\cdot$$
$$(X = SR \text{ or Hal})$$

Best product yields were obtained from allyl compounds with the weakest C—X bonds. Haloalkyl radicals also displaced trimethyltin radicals from allyltrimethyltin in a similar way:[439]

$$R\cdot + CH_2{=}CHCH_2SnMe_3 \rightarrow RCH_2CH{=}CH_2 + \cdot SnMe_3$$
$$RCl + \cdot SnMe_3 \rightarrow R\cdot + Me_3SnCl$$

Several homolytic substitution reactions at tin have been confirmed. Thus the ESR spectra of the displaced alkyl radicals have been observed when *tert*-butoxy-radicals attack a variety of trialkyltin halides,[440,441] e.g.

$$Bu^tO\cdot + R_3SnX \rightarrow Bu^tOSnR_2X + R\cdot$$

Photolysis of hexamethyldilead in carbon tetrachloride solution gave rise to ESR signals from both methyl and trichloromethyl radicals, suggesting an S_H2 reaction at the lead atom:

$$CCl_3\cdot + Me_3PbPbMe_3 \rightarrow Me_3PbPbMe_2CCl_3 + Me\cdot$$

tert-Butoxy-radicals react with tetra-alkyltin compounds to abstract a hydrogen atom, and it is only when the tin atom bears an electronegative substituent, such as halogen or carboxylate, that the *tert*-butoxy-radicals attack the tin centre displacing an alkyl radical.[441] The kinetics of the S_H2 reactions between succinimido-radicals and tetra-alkyltin compounds have now been studied;[442] trends in the rates are interpreted in terms of steric hindrance towards formation of the five-co-ordinate transition state and the stability of the displaced radical.

Decomposition of tetrazenes (**214**) by trialkyltin hydrides is reported to occur by attack of trialkyltin radicals at N-1 or N-4, giving the corresponding tinamine, nitrogen and an amino-radical:[443]

$$\underset{(\mathbf{214})}{RR'N{-}N{=}N{-}NR'R} + R_3Sn\cdot \rightarrow R_3SnNR'R + N_2 + \cdot NR'R$$

the latter radical is scavenged by the trialkyltin hydride or combines with a trialkyltin radical at low hydride concentrations.

Interest has continued this year in phosphoranyl radical chemistry. The temperature-dependence of the ESR spectrum[444] of the radical $\cdot PH_3OBu^t$ shows that the hydrogen atoms are mobile, the energy barrier to exchange being 5.3 kcal mol^{-1}. CNDO calculations gave best agreement with the experimental coupling constants for a trigonal-bypyramidal structure with the alkoxy-group axial and the unpaired electron equatorial.[444] A correlation has been shown between the ^{31}P splitting in the ESR spectra of radicals such as $\cdot P(O)X_2$, $\cdot P(S)X_2$ and $\cdot PX_4$ and J_{P-H} in the NMR spectra of the corresponding phosphoranes.[445] The radical structure is similar to that of the parent phosphorane, the unpaired electron taking the place of the hydrogen atom. The ESR spectra of series of radicals

$$Bu^tO(R)_n\dot{P}Cl_{3-n} \text{ and } Bu^t(RO)_n\dot{P}Cl_{3-n}$$

show that their structures are similar to those of the corresponding phosphoranes,[446] the most electronegative groups occupying the axial positions as shown in (**215**). Spirophosphoranyl radicals (**216**) also have structures close to those of the parent phosphoranes, with the unpaired electron equatorial.[447] They undergo α- or β-scission very slowly, even at high temperatures, but add readily to 2-methyl-2-nitrosopropane and to terminal alkenes.

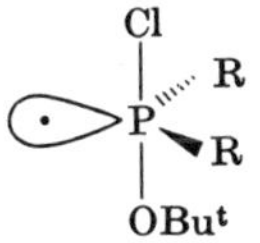

Cl, P, R, R, OBu^t

(**215**)

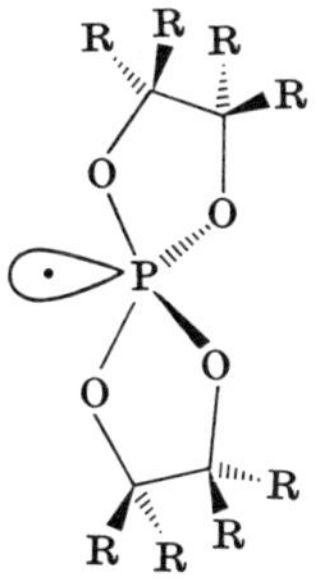

R, R, R, R, O, O, P, O, O, R, R, R, R

(**216**)

From an analysis of the products of alkoxy and thiyl radical reactions with a series of phosphites $XP(OEt)_2$ (X = Cl, Alkyl, Aryl, AcO, or $Bu^n{}_2N$) the relative importance of α- and β-scission has been determined;[448] the relative strengths of the R—A and P—X

$$RA\cdot + XP(OEt)_2 \longrightarrow RA{-}\dot{P}(X)(OEt)(OEt) \begin{cases} \xrightarrow{\alpha\text{-Scission}} RAP(OEt)_2 + X\cdot \\ \xrightarrow[\beta\text{-Scission}]{} R\cdot + A{=}PX(OEt)_2 \end{cases}$$

bonds in the phosphoranyl intermediate were the controlling factors. Rate differences were negligible in the reaction of *tert*-butoxy-radicals with phosphites $XC_6H_4P(OMEe)_2$ X = *p*-MeO, H, *p*-Cl or *m*-CF_3);[449] it was suggested that the enthalpy of formation of the phosphoranyl intermediate (calculated by an empirical method) was the dominant influence. The highly successful ESR method has been used for studying the reaction of *tert*-butoxy-radicals with phosphites $P(OR)_3$;[450] the rate of β-scission increases along the series of R = primary < secondary < tertiary alkyl; bulky alkyl groups were retained preferentially, but it seems probable that equatorial substituents undergo more rapid β-scission. *tert*-Butylperoxy-radicals react with all the arylphosphines $(XC_6H_4)_3P$ (X = *p*-MeO, *p*-Cl, *p*-F, H or *p*-Me) by β-scission at essentially the same rate.[451] For the series Ph_2PCl, Ph_2POCH_3 ,Ph_2PH, Ph_3P, Ph_2PCH_3 the reactivity increased as the electron-donating power of the α-substituent increased; the importance of polar effects in this process was spotlighted by a correlation between the logarithms of the relative rate constants and the Taft σ^* constants of the α-substituents. CIDNP effects have been observed in the reaction of triaryl phosphites with *tert*-butoxy-radicals.[452] The ESR spectra of arsanyl radicals formed from *tert*-butoxy-radicals and tri- or di-phenylalkoxy-arsines have been observed;[453] the arsanyl radicals decompose mainly by α-scission, losing a phenyl radical:

$$Bu^tO\cdot + Ph_2AsOR \rightarrow Ph_2\dot{A}s(OR)OBu^t \rightarrow PhAs(OR)OBu^t + Ph\cdot$$
$$R = Me, Et \text{ or } Bu^t$$

The order of stability of the radicals is:

$$Ph_2\dot{A}s(OMe)OBu^t > Ph_2\dot{A}s(OEt)OBu^t > Ph_2\dot{A}s(OBu^t)_2$$

Perhaps the most interesting development this year has been the discovery of homolytic displacement reactions with platinum(II) complexes. ESR spectra have been recorded for the alkyl radicals (and on the radicals spin-trapped with Bu^tNO) displaced from complexes *cis*-PtR_2X_2 (R = Me, Et, Me_3SiCH_2 or $PhCH_2$; X = $PAlk_3$ or PMe_2Ph where Alk = Et, Pr^n or Bu^n) by $Bu^tO\cdot$ or $PhS\cdot$ radicals:[454]

$$cis\text{-}PtR_2X_2 + Bu^tO\cdot \rightarrow PtR(OBu^t)X_2 + R\cdot$$

Alkyl radicals are also displaced from dialkylplatinum(II) complexes in their reaction with alkenes as illustrated; the alkyl radical added to unused alkene, and the resulting

$$R_2PtL_2 + EtOOC{-}CH{=}CH{-}COOEt \longrightarrow L_2R_2Pt{-}(CHCOOEt{=}CHCOOEt) \longrightarrow L_2Pt{-}(CHCOOEt{=}CHCOOEt) + 2R\cdot$$

R = Me, CD_3 or Et; L = 2,2′-bipyridyl, 1,10-phenanthroline or cyclo-octa-1,5-diene.

adduct radical was detected by the spin-trap technique with $Bu^tNO\cdot$. We can infer from this that homolytic reactions of transition-metal complexes may be more common than has heretofore been realized.

S_H2 processes have also been reported for the reaction of alkyl radicals with peracids[456] and of tin radicals with the esters of unsaturated percarboxylic acids.[457]

Rearrangements

Rearrangement of the neophyl radical has been followed by ESR spectroscopy;[458] the activation energy for the process $PhCMe_2CH_2\cdot \rightarrow \cdot CMe_2CH_2Ph$ has been evaluated as 10 ± 2 kcal mol^{-1} and the rate constant at 290K as 10^3 s^{-1}. *Ab initio* calculations indicate that ring-opening of the cyclopropyl radical is formally forbidden in both the conrotatory and the disrotatory mode, but if there is distortion of the radical a near-disrotatory reaction would be preferred.[459]

However, the calculated activation barrier for this pathway is ca. 40 kcal mol^{-1} greater than the experimental value. *Ab initio* calculations have also been used to calculate the potential energy for the interconversion of cyclopropylmethyl and homoallyl radicals.[460] The calculations predict that the homoallyl radical is more stable by 8.6 kcal mol^{-1} and that the activation energy for its formation is 17.3 kcal mol^{-1}.

ω-(1-Phenylcyclopropyl)-ethyl and -butyl (but not -propyl) radicals (**217**) undergo ring-expansion, albeit to a very limited extent.[461] These rearrangements are the first of their type recorded; they can be considered to involve intramolecular attack on the cyclopropane ring. All three radicals also undergo intramolecular cyclization on to the aromatic ring, the extent of this being greatest for the 3-(1-phenylcyclopropyl)propyl radical. A striking feature of the last reaction is the absence of any product of rearrangement of a cyclized radical (**218**), which is a cyclopropylcarbinyl radical that might be

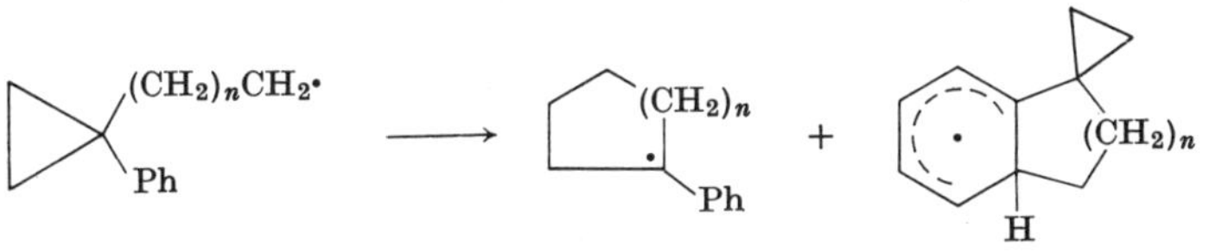

$(CH_2)_nCH_2\cdot$ Ph ⟶ $(CH_2)_n$ Ph + $(CH_2)_n$ H

(**217**) (**218**)

expected to rearrange readily; that it does not is attributed to adoption of the bisected conformation whereas rearrangement would be favoured only from the perpendicular conformation. This result thus throws possible light on the stereoelectronic requirements for the cyclopropylmethyl–homoallyl rearrangement. Stereoelectronic effects are also invoked to explain the isomerization of the *trans*-cyclopropylcarbinyl radical (**219**) to the *cis*-radical (**221**) [*via* ring-opening to (**220**) followed by ring closure] in the photo-

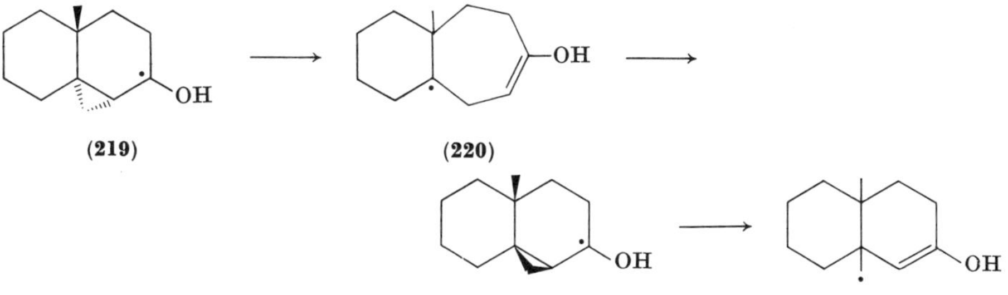

OH ⟶ OH ⟶

(**219**) (**220**)

OH ⟶ OH

(**221**)

reduction of 10-methyl-1,9-methano-*trans*-decal-2-one to 9,10-dimethyl-*cis*-decal-2-one, whereas the *cis*-isomer gave only the *cis*-photoreduced product.[462]

Photolysis of the α-bromo-sulphone (**222**) provides a further example of a norbornenyl–nortricyclyl radical rearrangement.[463,464] The 1-cyclopropyl-1-nitroethyl radical fails to undergo a cyclopropylmethyl–homoallyl rearrangement, probably owing to the

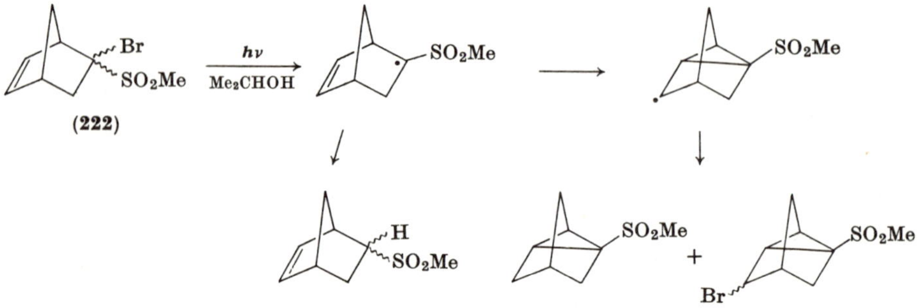

Br
SO₂Me
(**222**)
$h\nu$
Me_2CHOH
SO₂Me
SO₂Me
H
SO₂Me
SO₂Me
+
SO₂Me
Br

low spin density on carbon because of the spin-withdrawing effect of the α-nitro-group.[465] Both bent and linear 2-arylvinyl radicals, $\cdot CH{=}CPh_2$ and $\cdot CAr{=}CPh_2$, failed to undergo rearrangement even though the rearrangement of vinyl cations is well known.[466] A further example of rearrangement occurring in the radical additions to α- and β-pinenes has been reported.[467]

The mechanism of rearrangement of β-acyloxyalkyl radicals has been finally elucidated by using ^{18}O-labelled esters and by kinetic studies.[468] The results indicate that 1,2-acyloxy-migration involves a concerted mechanism involving a five-membered transition state (**223**) and not through the pyramidal dioxolanyl radical (**224**) or the three-membered transition state (**225**) as was previously envisaged.

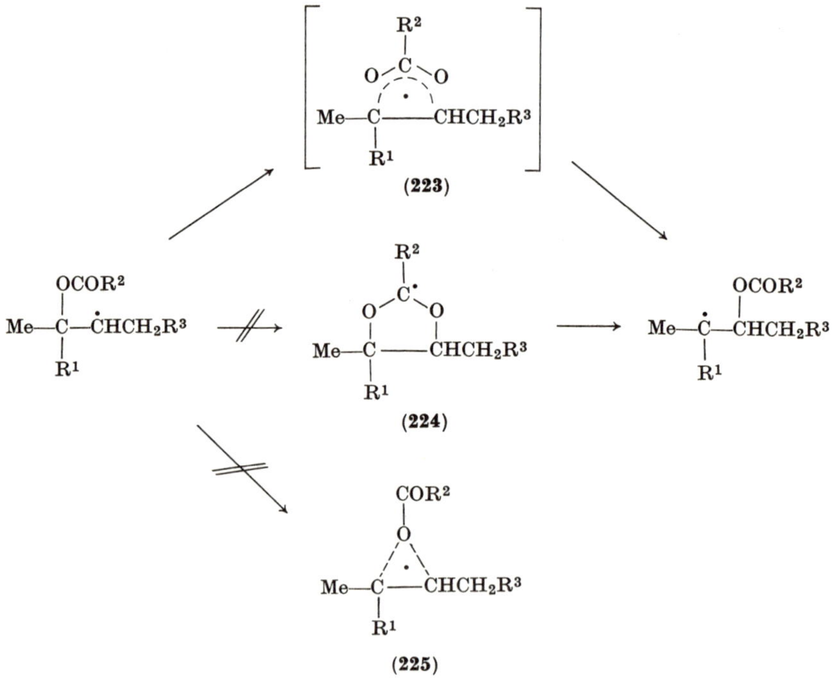

R²
O C O
Me—C——CHCH₂R³
R¹
(**223**)

OCOR²
Me—C—ĊHCH₂R³
R¹

R²
C
O O
Me—C——CHCH₂R³
R¹
(**224**)

OCOR²
Me—Ċ—CHCH₂R³
R¹

COR²
O
Me—C——CHCH₂R³
R¹
(**225**)

1,2-Chlorine migrations occur with extreme ease in $Me_2CClCH_2\cdot$ and $Me_3CCl\dot{C}HMe$ radicals, which rearranged so rapidly that they could not be observed even at $-120°$;[469,470] similarly the $CCl_3CH_2\cdot$ radical cannot be observed. It has been shown that rearrangement of the 2-bromobutyl radical to the 1-bromo-2-butyl radical occurs with retention of configuration.[471] 1,2-Chlorine migration occurs in the addition of bromine to 3,3,3-trichloropropene even at 77K:

$$Br\cdot + CCl_3CH{=}CH_2 \rightarrow CCl_3\dot{C}HCH_2Br \xrightarrow{\ell} \cdot CCl_2CHClCH_2Br$$

these results are consistent with a bridged structure for β-haloalkyl radicals.[472] Bromine migration has been observed in the bromination of bromomenthanes.[473] Much more unusual is the migration of fluorine observed in the pyrolysis of 2,2-bis(difluoroamino)-propane.[474]

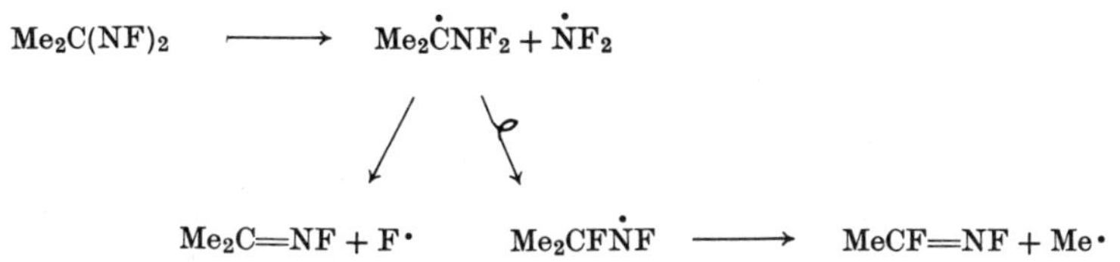

$$Me_2C(NF)_2 \longrightarrow Me_2\dot{C}NF_2 + \dot{N}F_2$$

$$Me_2C{=}NF + F\cdot \qquad Me_2CF\dot{N}F \longrightarrow MeCF{=}NF + Me\cdot$$

Trialkylsilyloxyamino-radicals and their germanium counterparts undergo very ready rearrangement to the corresponding nitroxides.[475]

$$R_3Si\dot{N}OSiR_3 \xrightarrow{\ell} (R_3Si)_2NO\cdot$$

Migration of a nitro-group takes place with the biradical (**226**),[476] and an unusual rearrangement has been encountered in the study of semidione radicals.[477]

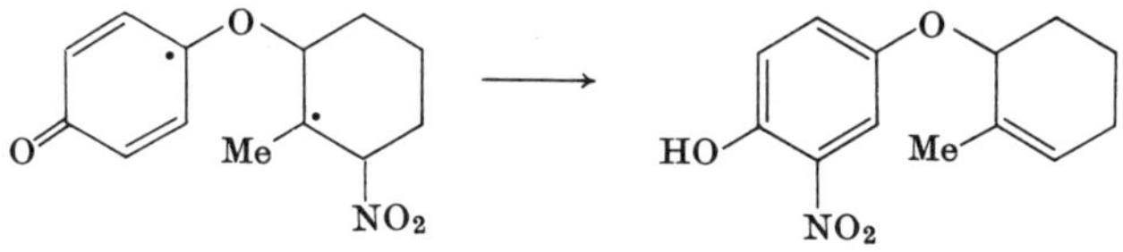

O, Me, NO_2 → O, HO, Me, NO_2

(**226**)

The photochemical rearrangement of *o*-phenoxybenzoic acid to phenyl salicylate has been reported;[478] it involves formation of a biradical followed by migration of a phenyl group.

A radical-pair mechanism has been formulated for the photo-Fries rearrangement of *p*-tolyl *p*-chlorobenzoate;[479] CIDNP studies indicate that an aryloxyaroyl radical pair (**227**) is formed from an excited singlet precursor. A similar mechanism has been proposed

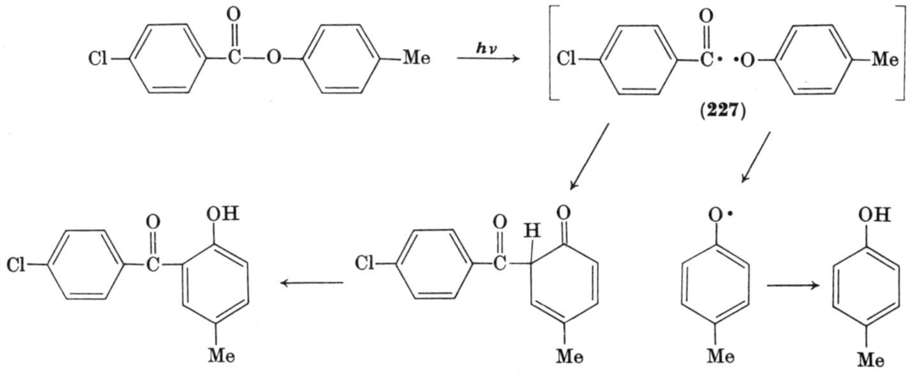

$$Cl{-}C_6H_4{-}C(=O){-}O{-}C_6H_4{-}Me \xrightarrow{h\nu} [Cl{-}C_6H_4{-}\dot{C}(=O)\ \ \cdot O{-}C_6H_4{-}Me]$$

(**227**)

Cl, O, OH, Me ← Cl, O, H, O, Me   O·, Me → OH, Me

for the rearrangement of *N*-arylimides.[480] The photo-Claisen rearrangement of *p*-tolyl 2-methylallyl ether affords the *meta*- as well as the more usual *ortho*- and *para*-rearranged products;[481] CIDNP results indicate that the reaction proceeds through both singlet and triplet radical-pairs, the former affording the *o*- and *p*-products and the latter *m*-products. 2-Phenoxybenzimidazole also undergoes a photo-Fries rearrangement,[482] but aromatic thiol esters do not;[483,484] this is ascribed to localization of the spin density in thiophenoxy-radicals mainly on sulphur.

The extent of the radical nature of the Meisenheimer, Wittig, Stevens and related rearrangements has been questioned, as it is often impossible to observe CIDNP effects in the product. ^{13}C-CIDNP effects have, however, been observed in the rearrangement product (**228**) of the oxime thionocarbamate, establishing that this is formed by a

$$Ar_2C{=}N{-}O{-}C({=}S){-}NMe_2 \longrightarrow Ar_2C{=}N{-}S{-}C({=}O){-}NMe_2$$

(**228**)

radical-pair combination within the solvent cage.[485,486] Radical-pair combination occurs both within and outside the solvent cage in the Wittig rearrangement of benzhydryl hex-5-enyl ether (**229**; R = C_6H_{11}) and in the reaction of lithium benzophenone ketyl (**230**) with hex-5-enyl iodide which gives rise to products resulting from both hexenylation and cyclopentylmethylation;[487] the cyclopentylmethylated products must represent combination outside the solvent cage since the rate of cyclization of hex-5-enyl

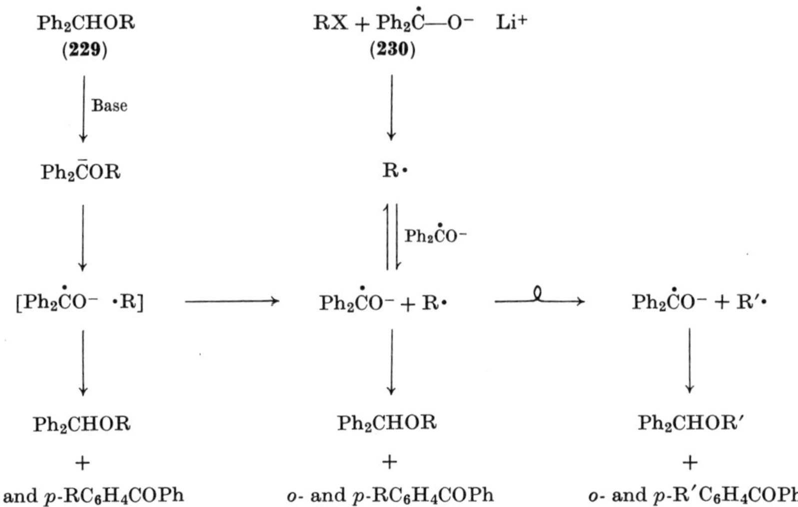

R = hex-5-enyl; R′ = cyclopentylmethyl

radicals is slow compared to the rate of cage combinations. The Meisenheimer rearrangement likewise involves a radical-pair mechanism with combination both within and outside the solvent cage.[488] In the rearrangement of 2-oxyanilinium ylides there is evidence for both a homolytic radical-pair combination, detectable by CIDNP, and a major concerted process.[489] Radical-pair combinations are also involved in the radical rearrangement of arylhydrazonates to N^2,N^2-diarylhydrazides,[490] the rearrangement of (phenylazo)triphenylmethane[491] and the rearrangement of nitrones (**231**) to *O*-alkyl oximes (**234**) and (**235**).[492] In the last reaction the rearrangement is accompanied by isomerization due to the rapid interconversion of the intermediate isomeric iminoxy-radicals (**232** and **233**).

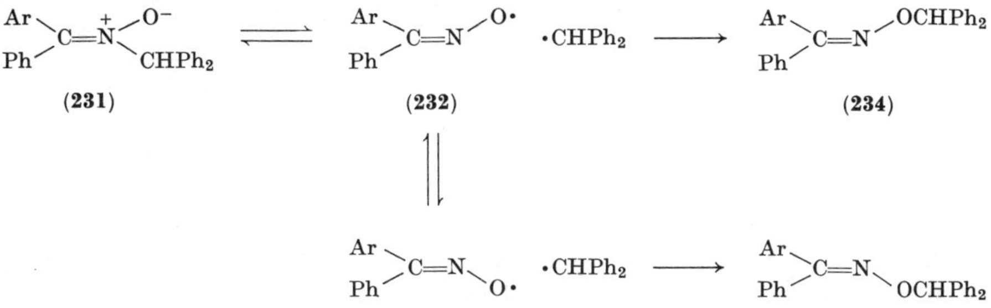

(**231**) (**232**) (**234**)

163 (**233**) (**235**)

Biradicals

The nature of biradicals has been discussed.[493] It has been calculated that 1,3- and 1,4-biradicals are destabilized relative to classical non-interacting systems by 10 and 6 kcal mol^{-1}, respectively,[494] but the origin of this barrier to ring closure remains obscure.

Photolysis of *cis*- and *trans*-α,γ-diphenylbutyrolactone gives the same mixture of *cis*- and *trans*-1,2-diphenylcyclopropanes, indicative of the participation of the triplet biradical (**236**).[495] The biradicals (**237** and **238**) derived from photolysis of *cis*- and

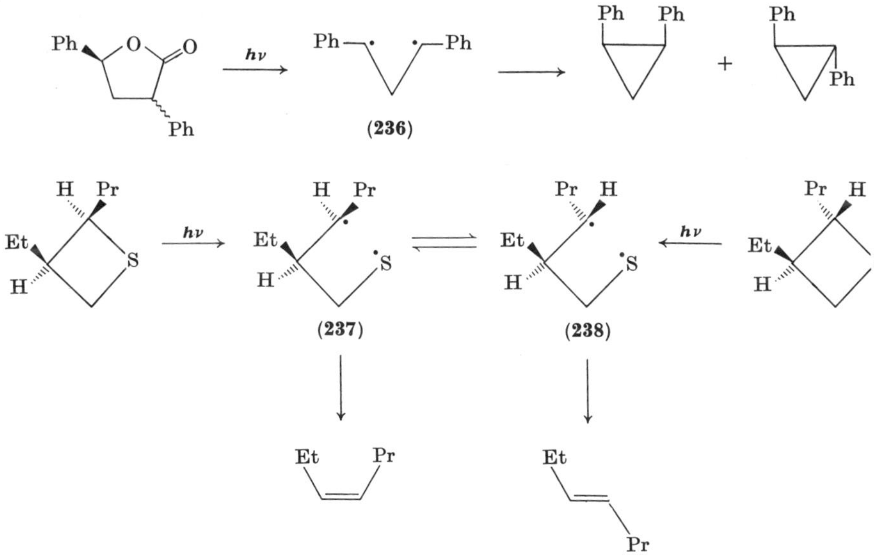

(**236**)

(**237**) (**238**)

trans-3-ethyl-2-propylthietans have much shorter lifetimes, as shown by the fact that the stereochemistry of the product resembles that of the starting material.[496] 1,3-Biradicals are implicated in the epimerizations of substituted cyclopropanes[497] and spiropentanes,[498] the isomerization of *endo*- and *exo*-5-(trimethylsilyl)[2.1.0]pentanes,[499] and the rearrangements of vinylcyclopropanes,[500] acetylcyclopropanes[501] and 1,2-divinylcyclobutane.[502] Photolysis of 1,2,3-thiadiazoles also yields 1,3-biradicals.[503]

An α-(methoxycarbonyl) group confers a stabilization energy of about 4.0 kcal mol^{-1} on a biradical, which is similar to the stabilizing influence of an α-carbonyl group on a simple radical.[504] The 1,4-biradical (**239**) undergoes fragmentation to (**240**) rather than intramolecular cycloaddition such as occurs for 1,3-biradicals.[505]

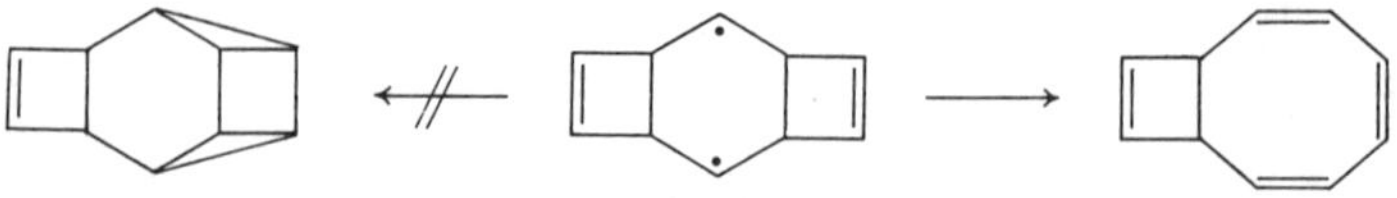

(**239**) (**240**)

1,4-Biradicals are also involved in the pyrolysis of (**241**), which affords acenaphthene.[506] That the radical (**242**) is formed in preference to the 1,4-biradical (**243**), which would be expected to be more stable as both radical centres are benzylic, is attributed to the route to the latter being stereoelectronically unfavourable. 1,4-Cleavage gives benzylic stabilization at one centre at an early stage along the reaction co-ordinate.

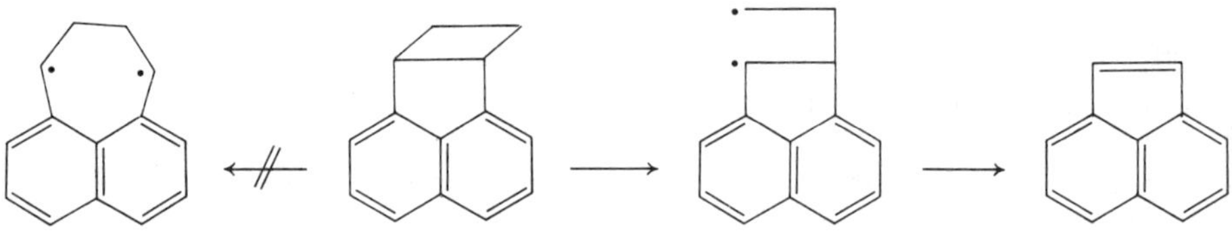

(**243**) (**241**) (**242**)

Crossed cycloaddition products (**245** and **246**) are formed in the acetone-sensitized photochemical reaction of 1,5-diene diesters, whereas the straight cycloaddition products (**247** and **248**) are formed from 1,6-diene diesters.[507] The reactions involve the initial formation of a triplet 1,4-biradical (**244**) which then undergoes addition to the second

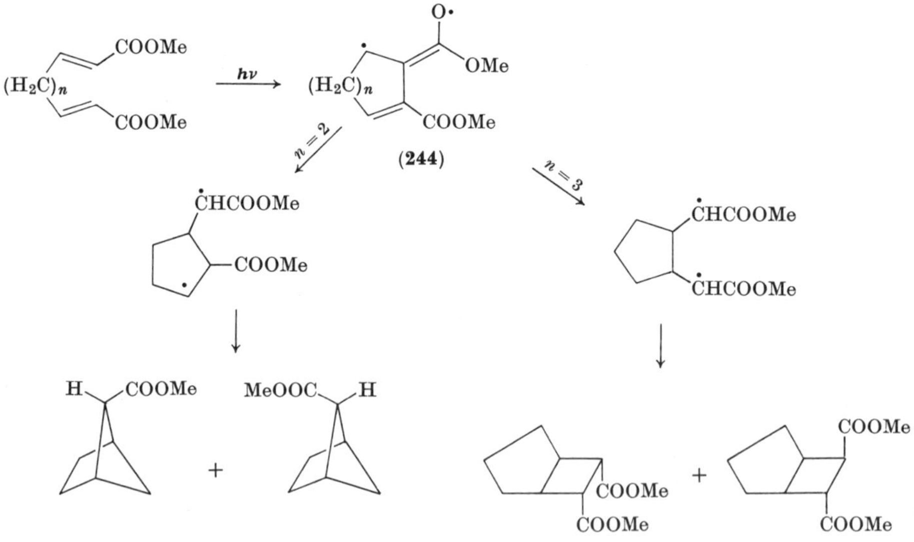

COOMe $(H_2C)_n$ COOMe $h\nu$ O· OMe $(H_2C)_n$ COOMe

(**244**)

$n = 2$ $n = 3$

ĊHCOOMe COOMe ĊHCOOMe ĊHCOOMe

H COOMe MeOOC H COOMe COOMe COOMe COOMe

(**245**) (**246**) (**247**) (**248**)

double bond, followed by cyclization. The direction of addition is controlled by the same factors as in simple intramolecular additions of pent-4-enyl and hex-5-enyl radicals.

On photolysis, cyclobutanones undergo ring expansion to tetrahydrofurans in a reaction involving a vibrationally excited singlet state;[508] cyclopropanes are also formed, the yield of these increasing in the presence of a triplet photosensitizer. Formation of 7-ketonorbornanes on photolysis of bicyclo[3,2,0]hept-3-en-2-ones proceeds with loss of stereochemistry at the migrating centre, which is consistent with a biradical mechanism rather than a concerted suprafacial shift.[509]

Photolysis of thiobenzoate *O*-esters (**249**) results in the generation of 1,4-biradicals *via* the triplet state. The biradical then undergoes intramolecular hydrogen-abstraction to give the biradical (**250**), which either fragments or cyclizes, analogously to the behaviour of ketones on photolysis.[510] Fragmentation is favoured with thioesters because of the significant driving force towards the formation of a carbonyl group.

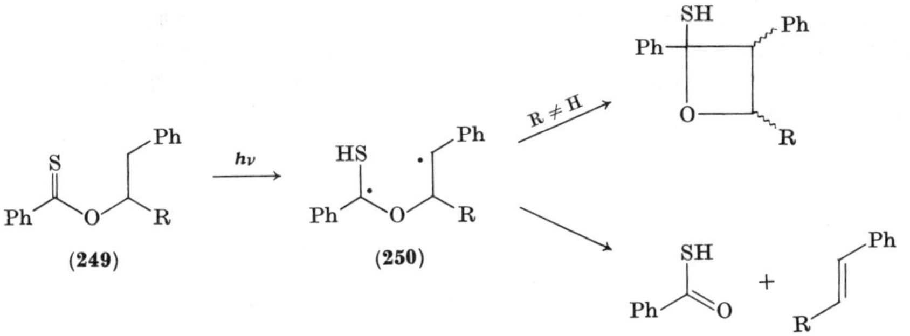

(**249**) $\xrightarrow{h\nu}$ (**250**) $\xrightarrow{R \neq H}$ oxetane; (**250**) → PhC(O)SH + RCH=CHPh

The 2,3-dimethylcyclohexane-1,4-diyl radical (**256**) is proposed as an intermediate in the decomposition of (**251**) or (**252**), which give rise to monomeric (**257**) and (**258**) and dimeric products (**259**) and (**260**). The former pair predominates in the gas phase and

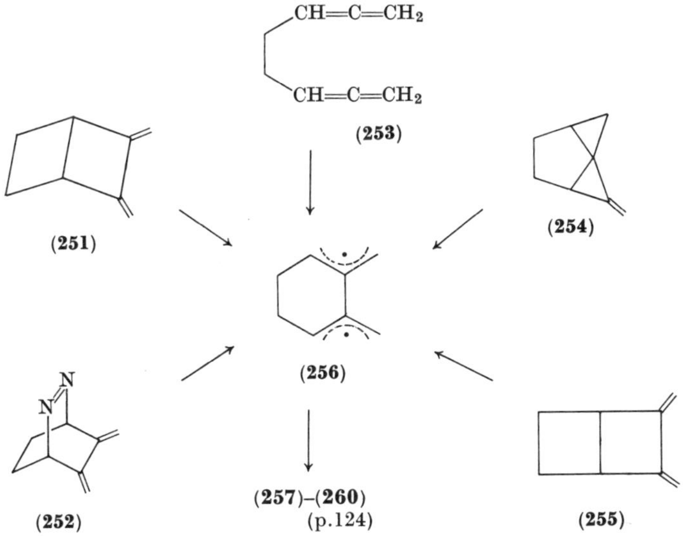

CH=C=CH$_2$ / CH=C=CH$_2$

(**253**)

(**251**)

(**254**)

(**256**)

(**252**)

(**257**)–(**260**)
(p.124)

(**255**)

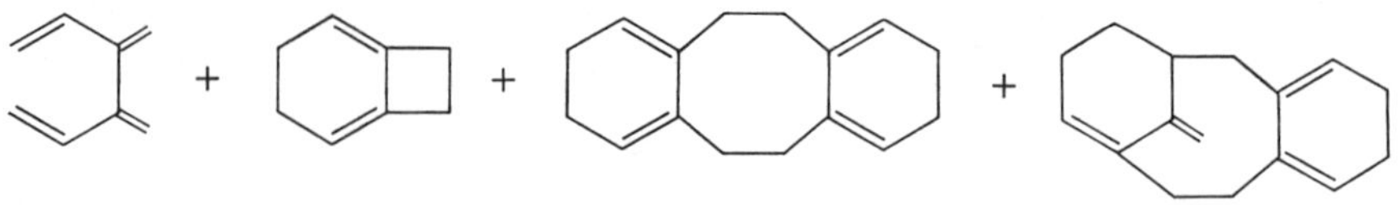

(257) **(258)** **(259)** **(260)**

the latter in solution.[511] CINDP effects confirm that the dimers arise *via* the triplet biradical.[512] In the gas phase, intersystem crossing to the triplet state is slow and the monomers are favoured. The ratio of the two monomers formed is pressure-dependent, with the thermodynamically more stable (**257**) favoured at low pressure. The increasing proportions of (**258**) formed at higher pressures reveal that the initial biradical is formed in a vibrationally excited state. The ratio of the two monomeric products (**257**)/(**258**) also depends on the source of the biradical,[513] the ratio being 1.05:1.00 from (**253**) and (**254**) but 0.45:1.00 from (**255**).[514] These results are interpreted as indicating that the biradical can exist in two extreme conformations, the twisted chair (**261**) and the planar conformations (**262**); the former is derived from (**253**) and (**254**) and gives rise preferentially to (**257**), whereas the latter is derived from (**255**) which affords (**258**) more readily. Reaction occurs before conformational equilibrium is attained.

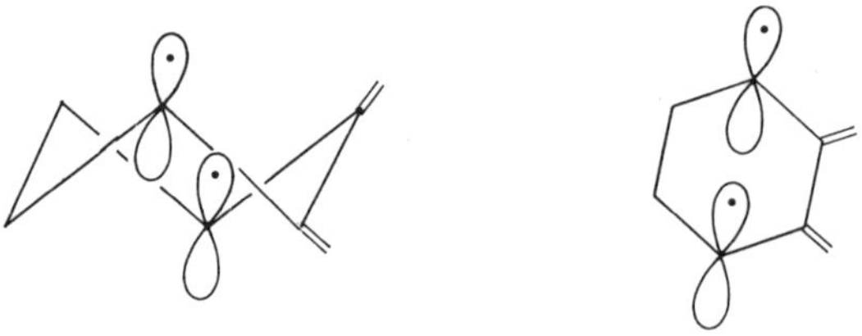

(261) **(262)**

Photolysis of 1*H*-benzotriazoles (**263**) in aromatic solvents gives products typical of homolytic aromatic substitution by the intermediate biradical (**264**).[515] The isomer distribution varies with the relative concentrations of triazole and aromatic compound; this is attributed to different radical multiplicities which lead to different isomer distributions.

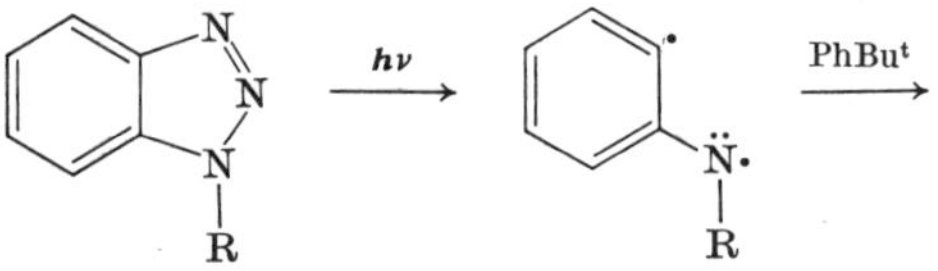

(263) **(264)**

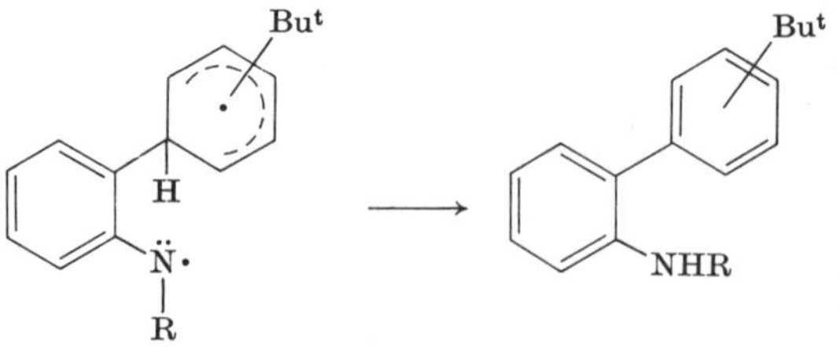

Nitroxides and Nitrogen Radicals

The bicyclic nitroxide (**265**) exists as a dimer in the solid state and in solution it dimerizes reversibly. This reversible dimerization is accompanied by a slow irreversible formation of the dimer (**268**), which arises from disproportionation of the radical to the hydroxylamine (**266**) and the nitrone (**267**), which latter subsequently reacts very rapidly with the nitroxide or the hydroxylamine to give the dimer.[516]

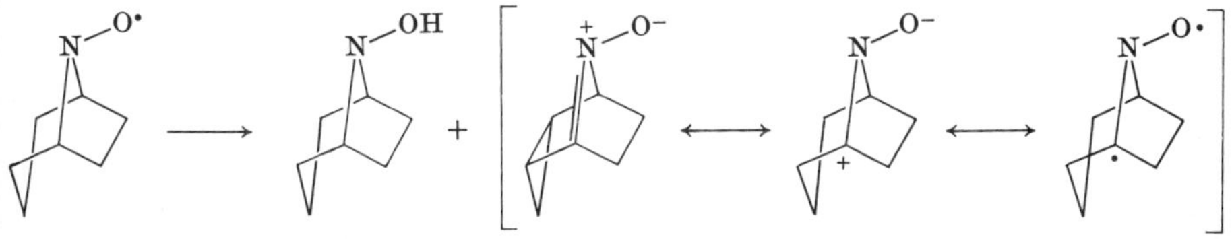

(**265**) (**266**) (**267**)

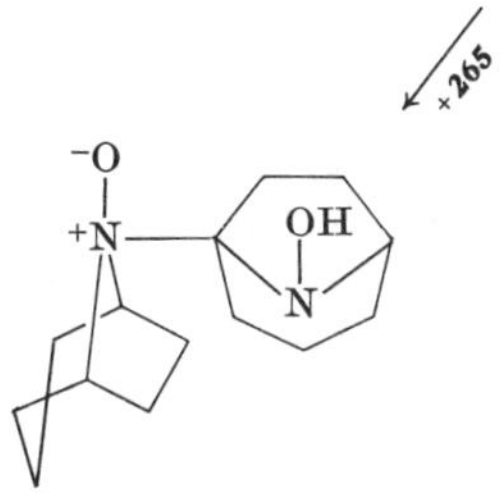

(**268**)

tert-Butyl *p*-halophenyl nitroxides (**269**) decompose to give *p*-benzoquinone *tert*-butylimine *N*-oxide (**270**), the corresponding amine and hydrogen halide or halogen

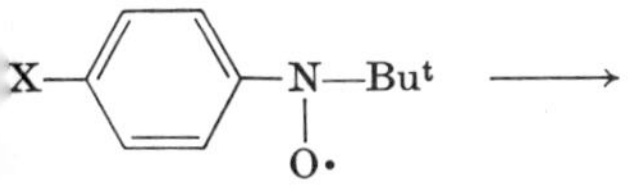

(**269**)

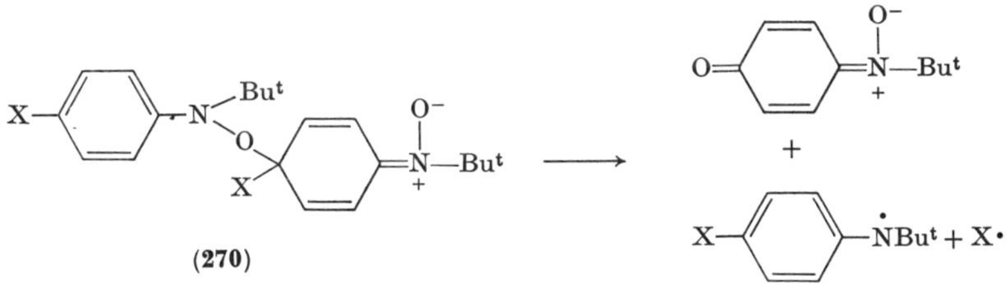

(**270**)

according to the scheme indicated.[517] The relative rates of decomposition are in the order: *p*-I < *p*-F < *p*-Cl < *p*-Br, which is interpreted as indicating that the rate of decomposition is controlled by steric factors, which affect the ease of attack at the *p*-position, and by the strength of the C—X bond. The *m*-substituted nitroxides behave similarly although *O*-to-*o*-C coupling also occurs. *o*-Halophenyl nitroxides are more stable, because the halogen group twists the nitroxide group out of conjugation with the aromatic ring. Decomposition of *N*-*tert*-butyl-*N*-(1,1-dimethyl-prop-2-ynyl)aminoxyl gives an unusual dimeric product containing an extra oxygen atom.[518]

Bistrifluoromethyl nitroxide abstracts hydrogen readily from aldehydes. The resultant acyl radicals are very efficiently scavenged.[519]

Hydrogen-abstraction also occurs readily from alkylbenzenes.[520] Hydrogen abstrac-

$$\mathrm{RCHO} + (\mathrm{CF_3})_2\mathrm{NO}\cdot \rightarrow \mathrm{R\dot{C}O} + (\mathrm{CF_3})_2\mathrm{NOH}$$
$$\mathrm{R\dot{C}O} + (\mathrm{CF_3})_2\mathrm{NO}\cdot \rightarrow \mathrm{RCOON(CF_3)_2}$$

tion from the initial product gives the radical (**271**) which either combines with another nitroxide radical or undergoes fragmentation.

$$\mathrm{ArCH_3} + (\mathrm{CF_3})_2\mathrm{NO}\cdot \longrightarrow \mathrm{ArCH_2}\cdot + (\mathrm{CF_3})_2\mathrm{NOH}$$
$$\mathrm{ArCH_2}\cdot + (\mathrm{CF_3})_2\mathrm{NO}\cdot \longrightarrow \mathrm{ArCH_2ON(CF_3)_2}$$
$$\mathrm{ArCH_2ON(CF_3)_2} \xrightarrow{(\mathrm{CF_3})_2\mathrm{NO}\cdot} \mathrm{Ar\dot{C}HON(CF_3)_2}\ (\mathbf{271}) \rightarrow \mathrm{ArCHO} + \mathrm{\dot{N}(CF_3)_2}\ \text{or}\ \mathrm{ArCH[ON(CF_3)_2]_2}$$

Oxidation of 1-hydroxybenzotriazole yields the nitroxide (**272**), which in aromatic solvents loses nitrogen and gives nitroxides of the type (**273**) by the mechanism outlined.[521] Oxidation of (**272**) gives the nitronyl nitroxide (**274**). The photochemical

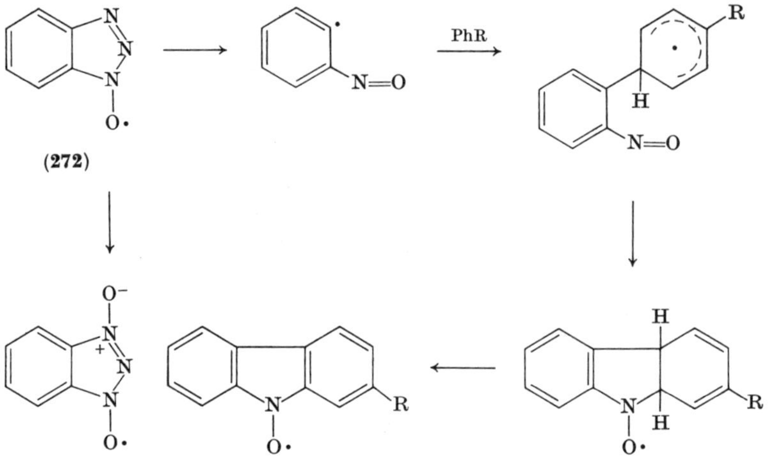

(**272**) (**274**) (**273**)

behaviour of the nitronyl nitroxide (**275**) has been examined;[522] the photoexcited nitroxide initially undergoes intramolecular hydrogen-abstraction.

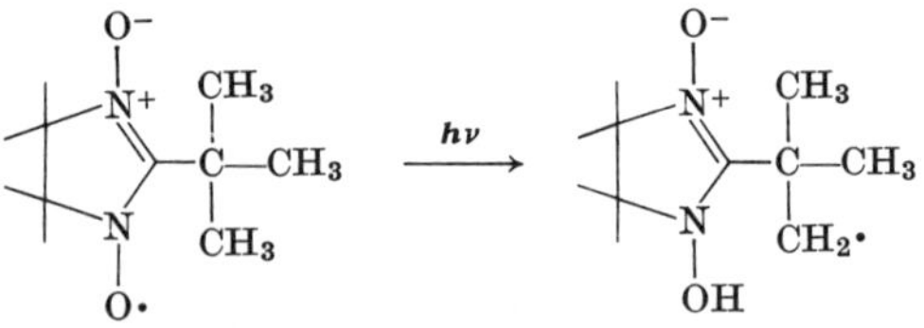

(**275**)

Stable dialkyl nitroxides have been shown to be quite stable also in strong acid. ESR spectroscopy indicates that the protonated nitroxides have structures similar to those of the parent radicals.[523]

Interest in spin traps continues. New spin traps include substituted nitrosobenzenes, nitrosodurene and its deuterated analogue (**276**),[524] 2,4,6-trisubstituted nitrosobenzenes

(277)[524,525] and perfluoronitrosobenzene **(278)**.[526] Of these, **(276)** seems to show most promise, being stable to photolysis and capable of trapping a wide range of radicals

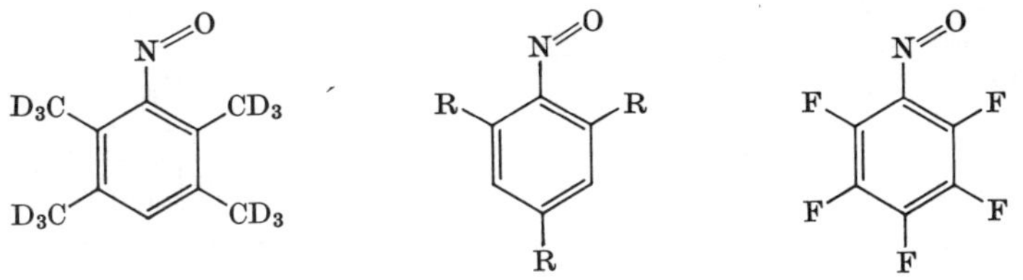

$R = Bu^t$, OMe, COOMe or Cl

(276) **(277)** **(278)**

efficiently; the spectra of the derived nitroxides are simple and are sensitive to the nature of the radical trapped. 2,4,6-Tri-*tert*-butylnitrosobenzene is attacked at nitrogen by primary alkyl radicals but at oxygen by tertiary radicals as a consequence of steric hindrance to attack at nitrogen; reaction occurs at both nitrogen and oxygen with secondary radicals. Mention has also been made of nitrosomethane acting as a spin trap.[527] Phenyl *N*-*tert*-butyl nitrone has been used as a spin trap for radicals and atoms produced in microwave discharges.[528] Measurements of the rate constants for spin-trapping reactions are mentioned on p. 103.

Decomposition of tribenzenesulphenamide in inert solvents gives a quantitative yield of diphenyl disulphide and an almost quantitative yield of nitrogen.[529] The $\cdot N(SPh)_2$

$$2(PhS)_3N \rightleftharpoons 2(PhS)_2N\cdot + 2PhS\cdot$$

$$2(PhS)_2N\cdot \rightleftharpoons (PhS)_2N{-}N(SPh)_2$$

$$(PhS)_2N{-}N(SPh)_2 \longrightarrow 2\,PhS\cdot + PhS{-}N{=}N{-}SPh$$

$$PhS{-}N{=}N{-}SPh \longrightarrow 2\,PhS\cdot + N_2$$

$$2PhS\cdot \longrightarrow PhSSPh$$

radicals thus produced effectively abstract hydrogen from phenols. The resultant phenoxy-radicals couple with $\cdot N(SPh)_2$ radicals to afford cyclohexadienone derivatives **(279)**. These radicals also react initially with arylhydrazines by hydrogen-abstraction

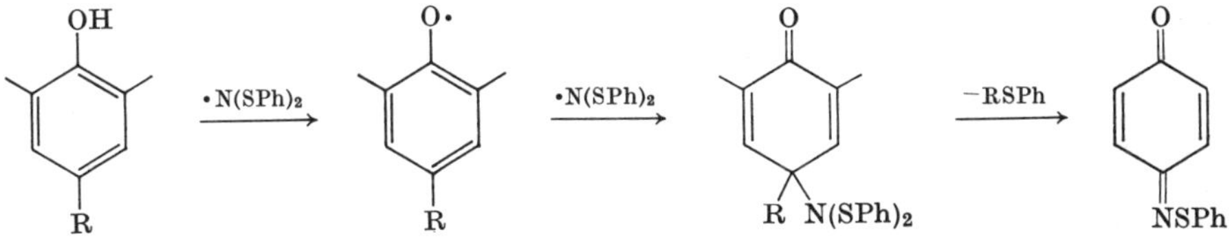

$\xrightarrow{\cdot N(SPh)_2}$ $\xrightarrow{\cdot N(SPh)_2}$ R N(SPh)$_2$ $\xrightarrow{-RSPh}$ NSPh

(279)

and with tetraphenylpyrrole by addition. The latter reaction gives an excellent yield of 2,3,4,5-tetraphenylpyrimidine.[530]

Halogen-abstraction by the 1-ethyl-4-(methoxycarbonyl)pyridinyl radical proceeds by an atom transfer in the case of benzoyl chloride but by electron transfer with *p*-nitrobenzoyl chloride.[531,532] The disappearance of these radicals in aqueous solution is also essentially an electron-transfer process.[533]

Autoxidation[534–536]

Kinetic chain-length in methane autoxidation has been investigated,[537] as have the reactions of oxides of nitrogen with methylperoxy-radicals, generated by photolysis of azomethane in oxygen.[538] Hydrocarbon oxidation in the gas phase has been studied by freezing out the mixture and detecting the radicals trapped in the matrix.[539] Oxygen-containing products obtained in the oxidation of butane were attributed to the reaction of butoxy-radicals formed by self-reaction of two butylperoxy-radicals.[540] Comparison of the gas- and the liquid-phase oxidation of butane indicated that the effect of the phase change was small at 398K.[541] Trimethylborane is a rapid initiator of isobutane oxidation, even under conditions where oxidation does not normally occur.[542] Initiation probably occurs by the sequence:

$$Me_3B + O_2 \longrightarrow Me_2BOO\cdot + Me\cdot \xrightarrow{O_2} MeOO\cdot$$

Alkyldichloroboranes are also very rapidly oxidized by molecular oxygen, giving alkylperoxy-radicals which then react with more alkyldichloroborane. Hydrolysis of the (alkylperoxy)dichloroboranes then gives the alkyl hydroperoxides, and this procedure can be useful for the synthesis of hydroperoxides.[543]

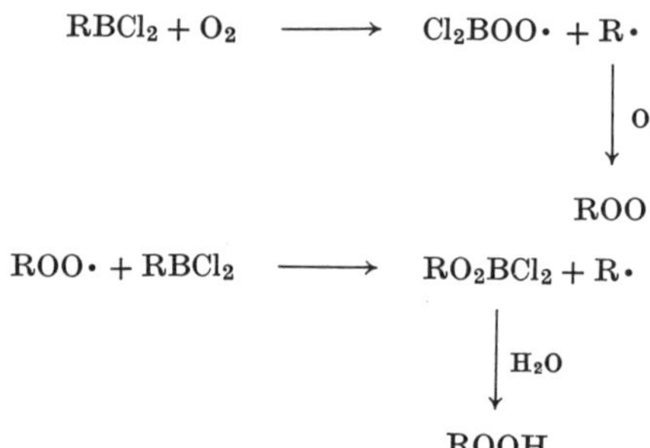

In the autoxidation of trialkylboranes, R_3B, both R• and RO• radicals have been detected by the spin-trap method.[544] The main products in the liquid-phase oxidation of 2,4-dimethylpentane are 2,4-(dihydroperoxy)-2,4-dimethylpentane and 2-hydroperoxy-2,4-dimethylpentane; at 373K the oxidation is about a quarter as fast as the oxidation of isobutane.[545,546] Termination occurs mainly between 4-hydroperoxy-2,4-dimethyl-2-pentylperoxy-radicals. Oxidation of cyclohexane by ozone–oxygen mixtures gave peroxides, cyclohexanone and adipic acid; termination was mainly by disproportionation of cyclohexylperoxy-radicals.[547] The rate constants for abstraction of hydrogen from a variety of organic substrates by tertiary and secondary peroxy-radicals have been correlated with the strengths of the C—H bonds.[548] The propagation rates in the autoxidation of vinyl compounds depend on the structure of the peroxy-radical and that of the vinyl compound; the propensity of peroxy-radicals for addition was increased by electron-withdrawing α-substituents.[549] Zinc dialkyl dithiophosphates inhibit the autoxidation of hydrocarbons by trapping peroxy-radicals, and two mechanisms have been suggested for this inhibition,[550] namely, as in (A) or by direct homolytic substitution at zinc: $ROO\cdot + ZnX_2 \rightarrow ROOZnX + X\cdot$, or as shown below:

$$ROO\cdot + [(RO)_2PS_2]_2Zn \longrightarrow ROO^- + (RO)_2PS_2Zn^+ + (RO)_2P\langle_{S}^{S}\cdot$$

Oxidation of long-chain *n*-alkanes floating on water is photosensitized by addition of 1-naphthol, the products being long-chain alcohols;[551] naphthoxy-radicals (**280**) formed from photoexcited 1-naphthol initiate the process illustrated.

$$\underset{(\mathbf{280})}{C_{10}H_7O\cdot} + RH \longrightarrow R\cdot + C_{10}H_7OH$$

$$R\cdot \xrightarrow{O_2} ROO\cdot \xrightarrow{RH} R\cdot + ROOH \xrightarrow{RH} 2\,ROH$$

The ratio of *cis*- to *trans*-epoxide formed in the co-oxidation of isobutyraldehyde and oct-2-ene showed that they were formed by a radical pathway; the main termination process involved self-reaction of two isopropylperoxy-radicals.[552] The suggestion that carbonyl compounds (which are the main products in the autoxidation of alkenes) are formed by consecutive addition of hydroxy-radicals and oxygen to the alkene, followed by decomposition of the hydroxy-substituted peroxy-radical (**281**) to give two molecules of the carbonyl compound and the chain carrier, is supported by the results of a study of the autoxidation of 2,3-dimethylbutan-2-ol and its ^{18}O-analogue.[553]

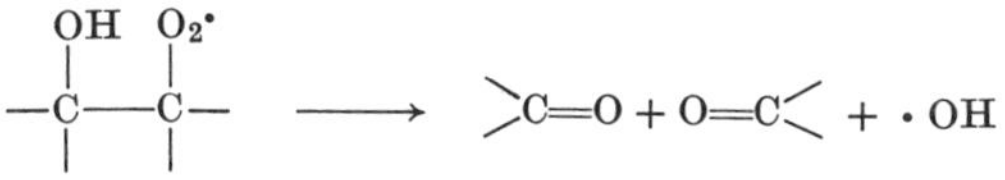

$$\underset{(\mathbf{281})}{-\overset{OH}{\underset{|}{\overset{|}{C}}}-\overset{O_2\cdot}{\underset{|}{\overset{|}{C}}}-} \longrightarrow \rangle C{=}O + O{=}C\langle + \cdot OH$$

Autoxidation of cyclohexene catalysed by cobalt salts involved an initiation step as shown,[554] the first reaction being fast:

$$Co^{2+} + ROOH \rightarrow Co^{3+} + RO\cdot + HO^-$$
$$Co^{3+} + RH \rightarrow Co^{2+} + R\cdot + H^+$$

The rates of reaction of diphenylethylperoxy-radicals with alkyl-substituted benzenes have been determined.[555] Relative rate constants, determined from the autoxidation of di-isopropylbenzene and derivatives, fit a Hammett plot with $\rho = -0.50$.[556]

The amount of $^{16}O^{18}O$ evolved from the decomposition of a mixture of alkylperoxy-radicals $R^{16}O^{16}O\cdot + R^{18}O^{18}O\cdot$ (R = primary, secondary or tertiary alkyl, acetyl or 2-pyridyl), determined by an ESR technique, indicated that head-to-head reaction of the peroxy-radicals is the preferred mode of termination.[557] From a study of the products of the reactions of *sec*-alkylperoxy-radicals generated from 1-ethoxyethyl hydroperoxide and 1,2-diphenylethyl hydroperoxide, and when the corresponding peroxy-radicals were generated in pairs, it was concluded that the tetroxide does not decompose exclusively by the cyclic path (*a*), but that a second decomposition process (*b*) can also occur.[558] The kinetics of the reaction of 2,4,6-tri-*tert*-butylphenoxy-radicals with cumylperoxy-radicals has been investigated.[559] Thermal oxidation of *exo,exo*-3,4,6-triphenylbicyclo[3.1.0]hex-1-ene (**282**) probably proceeds by way of the diradical (**283**) and peroxy-radical (**284**), giving the epoxide (**285**) in good yield.[560] Epoxidation was also observed in the liquid-phase oxidation of *exo*- and *endo*-2-cyanonorborn-5-ene;[561] the peroxy-radical formed by addition of oxygen and a carrier radical to the double bond reacted with more starting material to give two molecules of the epoxide and regenerate the carrier radical.

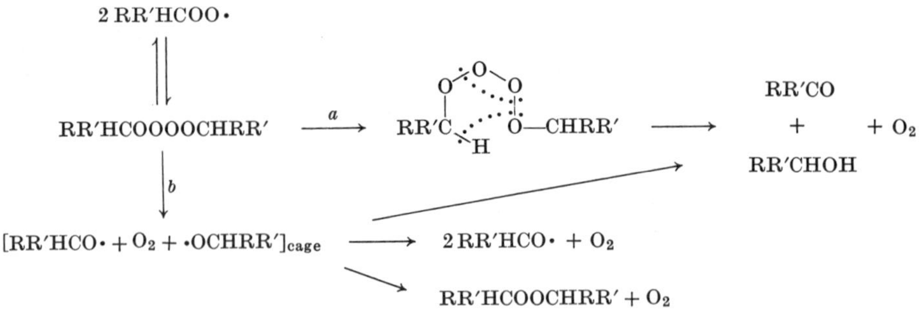

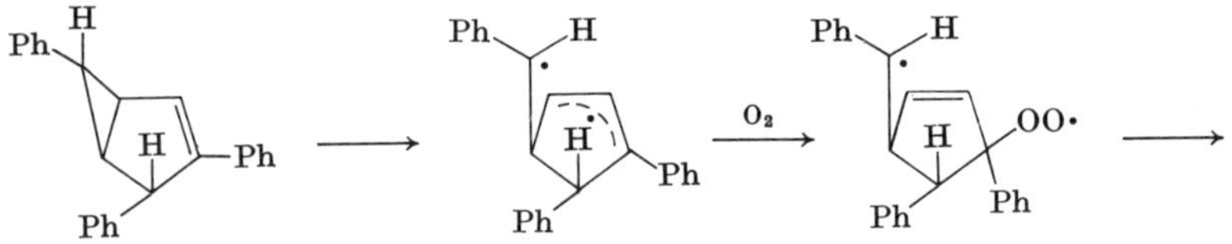

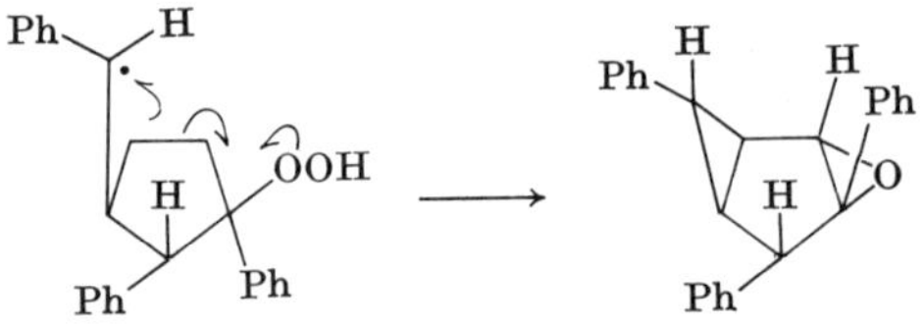

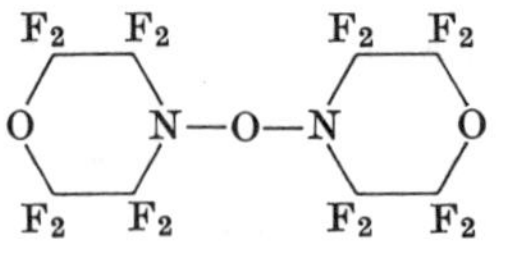

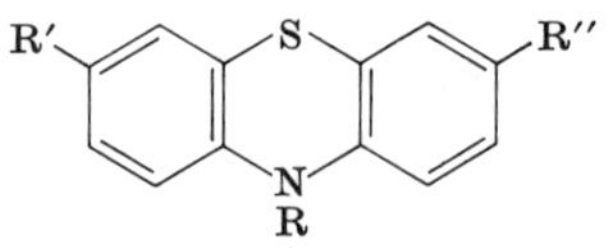

Photo-oxidation of perfluoromorpholine (**119**) gives perfluoro-4,4′-oxydimorpholine (**286**), which breaks up at 433K to give perfluoromorpholin-4-yloxy- and perfluoromorpholin-4-yl radicals.[332] The latter radicals give a variety of products by elimination of fluorine, combination and *β*-scission. Cycloalkylperoxy-radicals react with phenothiazines (**287**; R = H) by abstracting the hydrogen attached to nitrogen. For *N*-methyl-substituted phenothiazines (**287**; R = Me), an electron-transfer process occurs, and for R = CH_2Ph the electron-transfer step is followed by C—N bond cleavage.[562] Autoxidation of blocked tetrahydropteridines such as (**288**) in ^{18}O-labelled water has provided further evidence that the mechanism of ring contraction to the spirohydantoin (**289**) is as shown, the incorporated oxygen coming from the aqueous medium.[563] A kinetic study

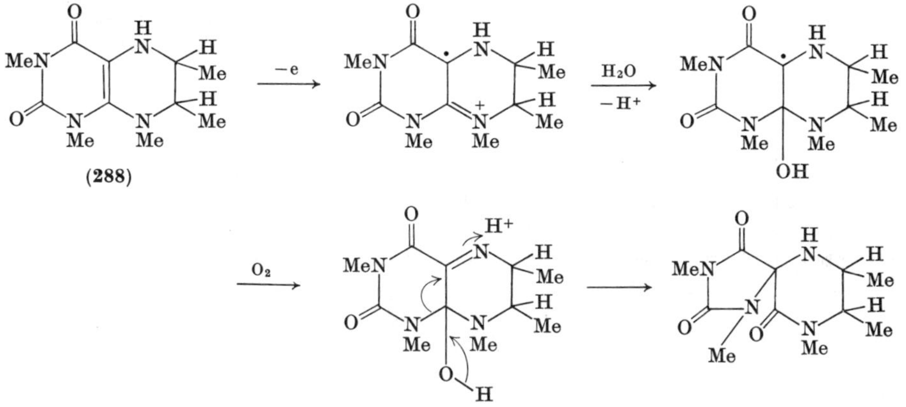

(**288**)

(**289**)

of the autoxidation of tetrahydrobiopterin (**290**) indicated that the hydroperoxide (**291**) was an intermediate in the formation of 7,8-dihydrobiopterin (**292**) which was itself oxidized to biopterin (**293**).[564]

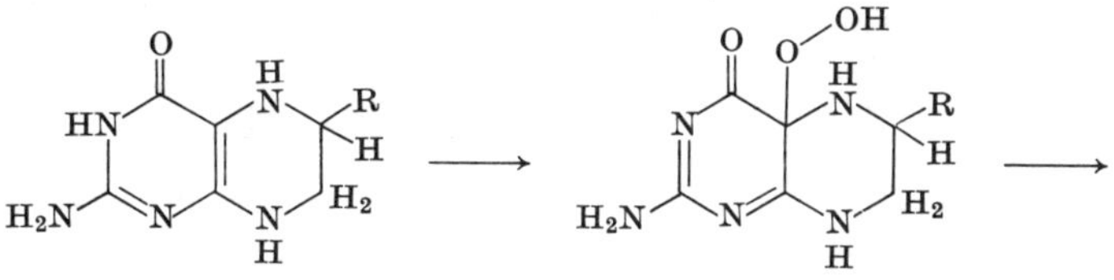

(**290**) (**291**)

R = CH(OH)CH(OH)CH_3

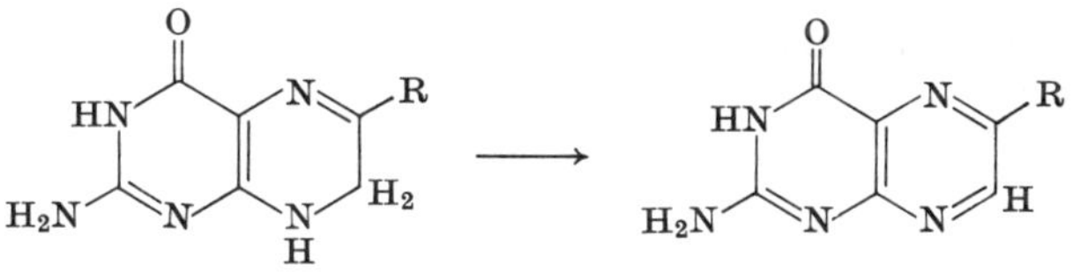

(**292**) (**293**)

A small yield of compound (**295**) was obtained in the autoxidation of 2-(*o*-aminophenyl)-3-methylindole (**294**), and its formation was shown to be dependent on the

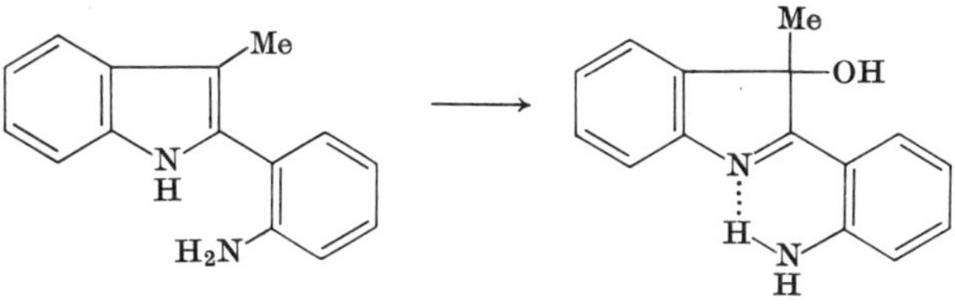

(**294**) (**295**)

presence of the *o*-amino-group in the phenyl ring.[565] The main product from the autoxidation of 2-*tert*-butyl-2,3-diazanorbornane (**296**) was the 2,3-diazanortricyclene derivative (**298**), believed to be formed from the diazenium cation (**297**) which could be isolated as the fluoroborate.[566]

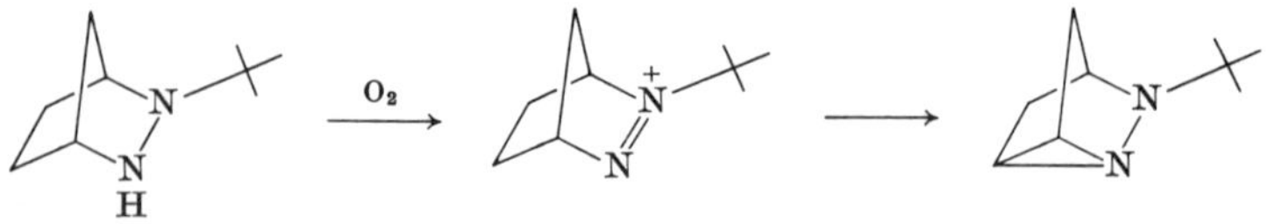

(**296**) (**297**) (**298**)

From a study of the autoxidation of torulosal (**299**), 4-hydroperoxides were shown to be the main products formed by homolytic decarbonylation. These hydroperoxides were postulated as intermediates in the formation of 4-hydroxynorditerpenes.[567]

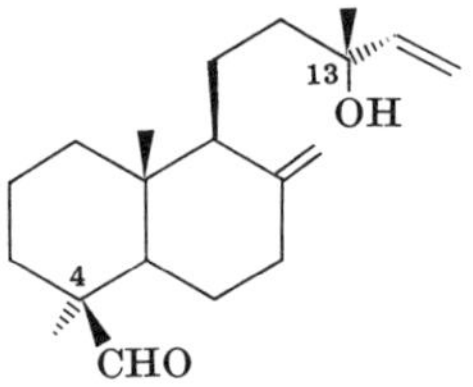

(**299**)

Autoxidation of 4-amino-3-methyl-1-phenyl-5-pyrazoline has been investigated.[568] The kinetics of the catalysis of autoxidation of Ph_3P, Ph_3As and cumene by Fe(III)– and Fe(IV)–dithiolate complexes has also been studied.[569] The products of the oxidation of 2,4,6-triphenyl-λ^3-phosphorin[570] and trimethyl phosphite[571] have been determined. Other autoxidations investigated include those of ethylbenzene in the presence of nitrosobutane,[572] phenyl-substituted phenols,[573] quinones and hydroquinones in alkaline solution,[574] acetoin in presence of ferrocene,[575] neophyl-lithium[576] and ethyl-(triethylsilyl)mercury.[577]

Homolytic Oxidation and Reduction*

General reviews of radical oxidation and reduction have appeared.[578,579] Reviews on specialist aspects include oxidations by cerium(IV)[580] and copper(II) salts,[581] electrochemical oxidations with specific reference to the Kolbe reaction,[582] phenol oxidations[583,584] and one-electron reductions of aromatic cations.[585]

Oxidation of Hydrocarbons

Kinetic studies indicate that oxidation of alkylbenzenes by cobalt(III) or manganese(III) salts proceeds by an electron-transfer mechanism with the initial formation of a radical cation.[586–590] The radical cation (**300**) generated in the oxidation of arenes by cobalt(III) trifluoroacetate undergoes nucleophilic attack by the solvent to give trifluoroacetates, the isomer distribution of which is typical of that encountered in electrophilic aromatic substitution.[591] The radical cation may also react with another molecule of arene to give a dimeric radical cation (**301**). Methylbenzenes also react by loss of a proton from

*See also Chapter 4.

the radical cation to give benzyl radicals which give rise to either benzyl trifluoroacetates or diarylmethanes.

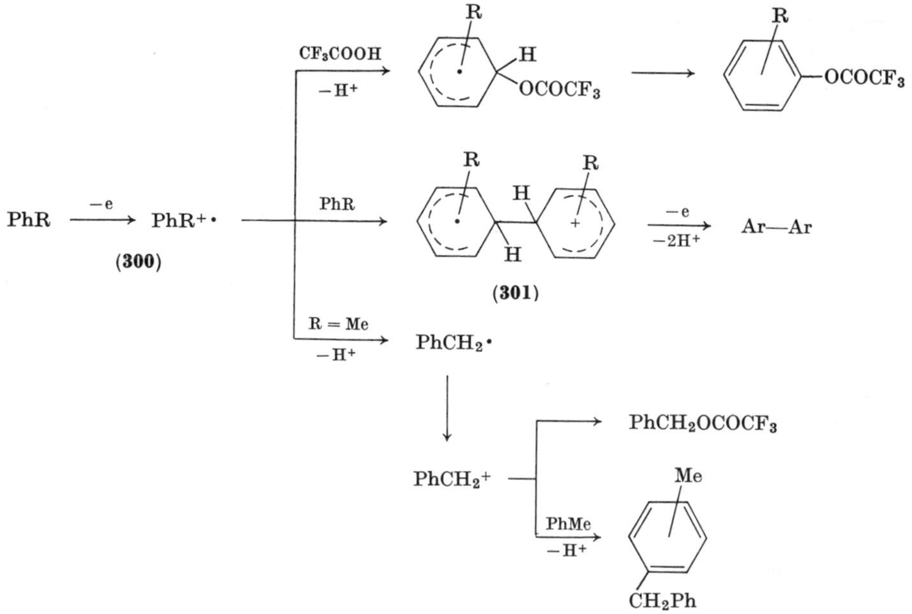

One-electron oxidation of toluene is distinguished from two-electron oxidation by lead(IV) trifluoroacetate by the absence of tolyl esters. The dimeric radical cations (**301**) have been detected in several instances by ESR spectroscopy. These radical cations are also detected in oxidations by thallium trifluoroacetate, implying that this may act as a one-electron oxidant.[592] Lead(IV) acetate in trifluoroacetic acid acts as a one-electron oxidant in its reactions with methylarenes and anisole, giving biaryls and diarylmethanes.[593] The products from these reactions bear close resemblance to those from anodic oxidation.[593a] From the relative ease of oxidation of arenes it is concluded that in the arylation step the radical cation behaves more as a radical than as a cation; thus the intermediate in the oxidation of a mixture of anisole and durene is (**302**) and not (**303**).

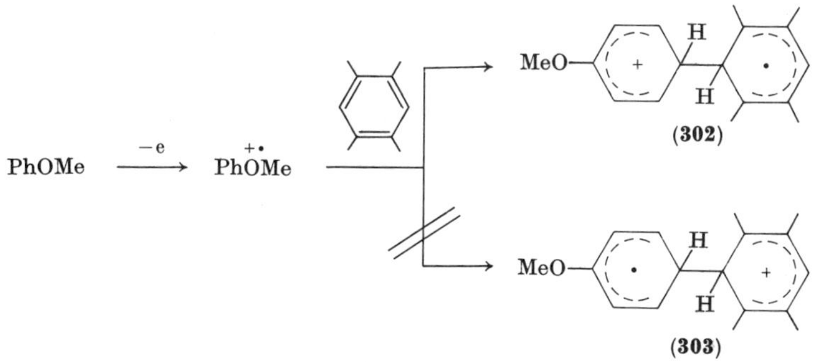

The formation of mixed biaryls in the anodic oxidation of naphthalene and alkylbenzene mixtures arises through attack by the napthalene radical cation on the alkylbenzene, the rate being roughly proportional to the basicity of the alkylbenzene.[594] Anodic coupling of arenes in the presence of strong acids has also been investigated.[595]

The precise mechanism of formation of biaryls in anodic oxidations of arenes has been postulated to involve either attack by the radical cation on the parent arene or dimerization of the radical cation. The anodic oxidation of 3,3′,4-trimethoxybibenzyl at low potentials at which only the monocation radical (**304**) is formed furnishes the dimeric product (**306**).[596] At higher potentials, at which the dication radical (**305**) is generated, this product is accompanied by 2,3,7-tetramethoxyphenanthrene, which is formed rapidly from the dihydrophenanthrene (**307**). It can thus be concluded that the favoured mechanism for intramolecular anodic coupling of arenes involves radical-cation dimerization, though this is not necessarily true for intermolecular oxidative coupling of arenes brought about by transition-metal oxidants (see above). Cerium(IV) oxidations of bibenzyls proceed by a radical-cation mechanism followed by cleavage of the central C—C bond.[597]

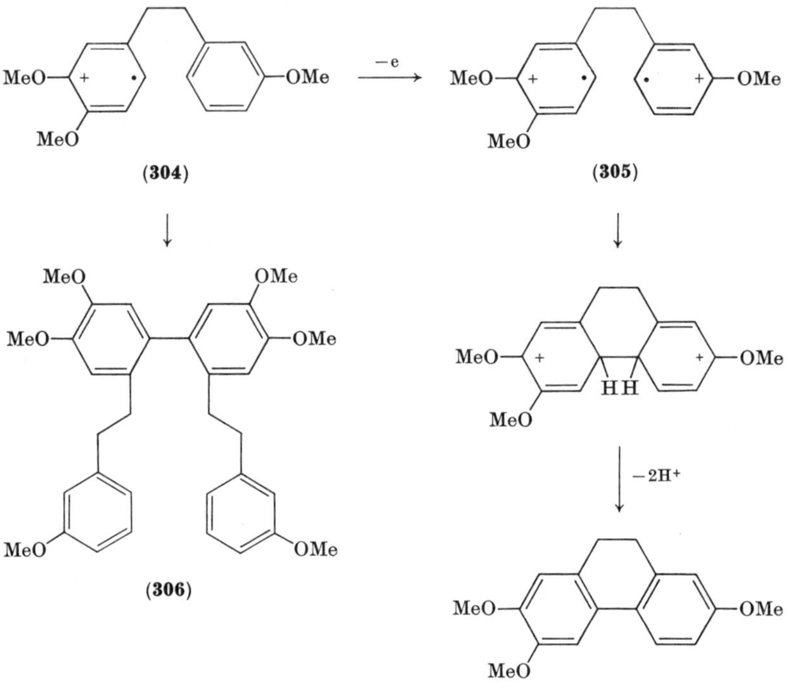

MeO MeO OMe (**304**) —(−e)→ MeO MeO OMe (**305**)

(**304**) ↓ (**306**): MeO MeO OMe OMe MeO OMe

(**305**) ↓ MeO MeO HH OMe —(−2H⁺)→ MeO MeO OMe (**307**)

$$\mathrm{ArCH_2CH_2Ar'} \xrightarrow{\mathrm{Ce^{IV}}} [\mathrm{ArCH_2CH_2Ar'}]^{+\bullet} \longrightarrow \mathrm{ArCH_2\bullet} + \mathrm{ArCH_2^+} \longrightarrow \text{Products}$$

Coupling and halogenation of 1-methoxynaphthalene by copper(II) halides is another example of an electron-transfer oxidation involving the generation of a radical cation.[598] This reaction does not involve the usual role of copper(II) halides which effect the halogenation of aromatic compounds by a ligand-transfer mechanism, as in reactions of pyrene derivatives.[599] Crowded aromatic compounds have also been shown to undergo electron-transfer oxidation by copper(II) chloride,[600] and anodic methoxylation of methoxybenzenes has been reported.[601] Anodic substitution has been shown to proceed by an ECE and not an EEC mechanism, by use of the rotating ring-disc electrode.[602]

Cyclohexane has been oxidized by cobalt(III) salts, showing that such oxidations are not confined to aromatic systems.[603] The mechanism differs from that for the oxidation of arenes in that the rate of reaction is governed by the rate of formation of the radical cation (k_1) rather than by the rate of loss of the proton from the radical cation (k_2).

$$\mathrm{RH + Co^{III}} \underset{k_{-1}}{\overset{k_1}{\rightleftharpoons}} \mathrm{RH^{+\cdot} + Co^{II}}$$

$$\mathrm{RH^{+\cdot}} \xrightarrow{k_2} \mathrm{R\cdot + H^+}$$

Oxidation of Carboxylic Acids

The nature of the anode surface in the Kolbe reaction has considerable influence on the relative amounts of products derived by radical and ionic pathways.[604,605] Surfaces on which adsorption can occur favour the radical pathway. The almost complete racemization encountered in the coupling of radicals derived from optically active acids indicates that there is no need to infer that coupling occurs on the anode surface, it may be between free radicals.[606] The distribution of stereoisomeric dimers and ethers and the dimer:alkene ratio in the anodic oxidation products of 4-substituted cyclohexanecarboxylic acids also suggests that adsorption is unimportant in the coupling process.[607] It is, however, of some significance in the case of unsaturated acids and can influence the product distribution.

Anodic oxidation of benzoic acid in acetonitrile gave acetylanthranilic acid.[608] Cerium(IV) oxidation of malic acid has been reported.[609] Lead(IV) acetate oxidation of cinnamic acids give rise to β-methylstyrenes as a result of coupling of styryl and methyl radicals generated in close proximity.[610] The reaction of p-methoxycinnamic acid takes

$$\mathrm{PhCH{=}CHCOOH + Pb(OAc)_4} \longrightarrow \mathrm{[PhCH{=}\dot{C}H \ldots Pb^{III}{-}OCOCH_3] + CO_2}$$

$$\mathrm{[PhCH{=}\dot{C}H \ldots Pb^{III}{-}OCOCH_3]} \longrightarrow \mathrm{[PhCH{=}\dot{C}H\ \dot{C}H_3] + Pb^{II} + CO_2}$$

$$\mathrm{[PhCH{=}\dot{C}H\ \dot{C}H_3]} \longrightarrow \mathrm{PhCH{=}CHCH_3}$$

a different course, giving the β-acetoxystyrene as the main product;[611a] the intermediate p-methoxystyryl radical is oxidized to the carbonium ion which is then captured by solvent.

$$\mathrm{ArCH{=}CH\cdot} \xrightarrow{\mathrm{Pb^{IV}}} \mathrm{ArCH{=}CH^+} \xrightarrow[-\mathrm{H^+}]{\mathrm{AcOH}} \mathrm{ArCH{=}CHOAc}$$

Oxidation of Phenols

The phenoxy radicals (**309**) derived from coniferyl alcohol (**308**; $R^1 = OMe$, $R^2 = H$, $R^3 = CH_2OH$) and isoeugenol (**308**; $R^1 = OMe$, $R^2 = H$, $R^3 = CH_3$) are trapped by nitrosobutane exclusively at the β-position to give the nitroxides (**310**).[611b] Oxidative coupling of these and related phenols gives C_β—C_β, C_β—C-5 and C_β—O dimers but no

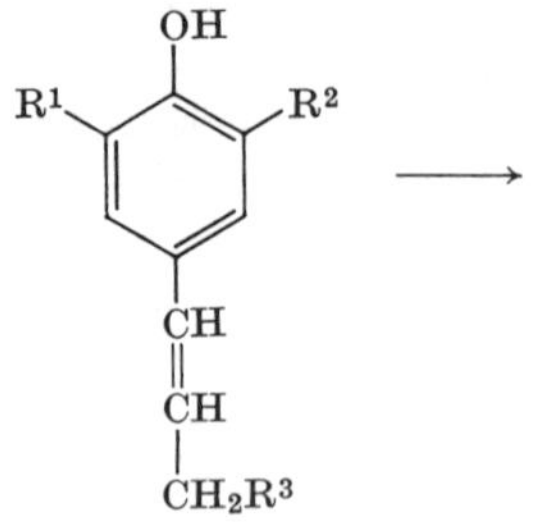

(308)

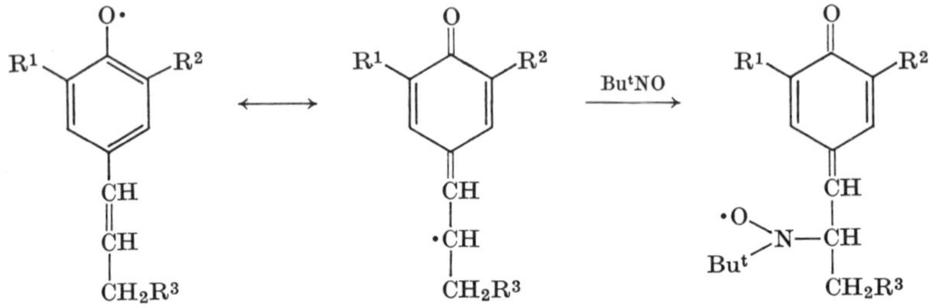

(309)

(310)

C-5—C-5 or C-5—O products.[612–616] The C_β—C_β coupling occurs stereospecifically in the oxidations of *E*-isoeugenol and *E*-2,6-dimethoxy-4-propenylphenol (**308**; $R^1 = R^2 =$ OMe, $R^3 = CH_3$) to give the *threo*-product, whereas the isomeric *Z*-phenols give mixtures of *threo*- and *erythro*-products.[612] This difference in behaviour is attributed to the different steric requirements of the intermediate tail-to-tail charge-transfer complex: for the *E*-phenol, the conformation of the complex leading to the *threo*-dimer is less subject to steric hindrance than that leading to the *erythro*-product. The difference is much less marked for the oxidation of the *Z*-phenols and also for the oxidation of methyl *E*-3,5-di-*tert*-butyl-4-hydroxycinnamate (**310**; $R^1 = R^2 = Bu^t$, $R^3 = COOMe$), which also leads to the both *threo*- and *erythro*-C_β—C_β dimers.[613] This is due to the steric interaction between the *tert*-butyl and aryl groups in the conformation of the complex leading to the *threo*-isomer. The major products from the oxidations of *E*- and *Z*-isoeugenols are the C_β—C-5 and C_β—O dimers, respectively; these arise through a head-to-tail complex; the preference for C_β—C-5 coupling from *E*-isoeugenol is because of steric hindrance between the methyl group and the oxygen for C_β—O coupling, which is much less for *Z*-isoeugenol, thus rendering C_β—O coupling more likely. 4-Arylazo-2,6-di-*tert*-butylphenols also give products arising from coupling in the side chain.[617]

The influence of the nature of the electrode on the products from the anodic oxidation of methylphenols has been investigated;[618] less of the coupled products and more of products derived from phenoxonium ions (4-hydroxycyclohexane-2,5-diones or *p*-benzoquinones) were obtained with a polished lead electrode than with a platinum electrode. In this reaction, the intermediate phenoxy-radicals would be more strongly adsorbed on the lead than on the platinum, facilitating further oxidation to the phenoxonium ion. Anodic oxidation of 2,4,6-tri-*tert*-butylphenol in acetonitrile or acetonitrile–pyridine mixtures gives solvent-derived products.[619] Oxidation of methyl *p*-hydroxycinnamate gives a "Pummerer's ketone-type" product.[620] Oxidation of halophenols gives coupled products in which a halogen has been replaced.[621]

New oxidants examined in phenol oxidation include chromyl chloride,[622] which forms a complex (**311**) that dissociates into radicals. This reagent oxidizes phenols to benzoquinones.

$$ArOH + CrO_2Cl_2 \longrightarrow ArO{-}Cr(OH)(Cl)(=O)Cl \longrightarrow ArO\cdot + CrO(OH)Cl$$

(**311**)

Oxidative coupling with thallium(III) trifluoroacetate has been claimed to give products derived from phenoxonium ions derived by two-electron oxidation.[623] It seems possible, however, that this oxidant acts as a one-electron oxidant.[592]

Reactions of phenols with cyclohexyloxy-radicals,[624] stable phenoxy-radicals[625] and cumyl hydroperoxide[626] have been studied. Reactions of phenols and phenoxides which proceed by an $S_{RN}1$ mechanism have been reported.[626a]

Oxidation of Alcohols

Alkoxy-radicals are produced by oxidation of alcohols with peroxydisulphates or silver nitrate or in photo-oxidations with peroxydisulphate, "Paraquat" or lead(IV) acetate in electron-transfer reactions.[627] The alkoxy-radicals rapidly abstract hydrogen from the alcohol to give α-hydroxyalkyl radicals, as formulated, but these have been trapped with benzylidene nitrone. Oxidation with di-*tert*-butyl peroxide, or photosensitized

$$RCH_2OH \xrightarrow{-e} RCH_2\overset{+\bullet}{O}H \xrightarrow{-H^+} RCH_2O\cdot$$

$$RCH_2OH + Bu^tO\cdot \longrightarrow R\dot{C}HOH + Bu^tOH$$

oxidation with benzophenone, leads directly to the α-hydroxyalkyl radical by hydrogen abstraction. The electron-transfer photosensitized oxidation of phenethyl alcohol with peroxydisulphate or "Paraquat" gives bibenzyl;[628] phenethyloxy-radicals are first obtained and lose formaldehyde to give benzyl radicals which dimerize; benzylidene nitrone but not nitrosobutane can trap the phenethyloxy-radical.

$$PhCH_2CH_2OH \rightarrow PhCH_2CH_2O\cdot$$
$$PhCH_2CH_2O\cdot \rightarrow PhCH_2\cdot + CH_2O$$
$$2PhCH_2\cdot \rightarrow PhCH_2CH_2Ph$$

Oxidations of tertiary alcohols (**312**) by cerium(IV) involve the intermediates (**313**), which break down directly in two ways;[629a] the relative yields of the two ketones (**314**) and (**315**) enabled the relative rates of formation of allyl, benzyl and *tert*-butyl radicals to be estimated as 1:4.4:19.9–62.9. Alkoxy-radicals are not formed.

$$\underset{(\mathbf{312})}{PhCR^1R^2OH} \xrightarrow{Ce^{IV}} \underset{(\mathbf{313})}{PhCR^1R^2OCe^{IV}} \longrightarrow \begin{cases} R^2\cdot + PhCOR^1\ (\mathbf{314}) + Ce^{IV} \\ R^1\cdot + PhCOR^2\ (\mathbf{315}) + Ce^{IV} \end{cases}$$

The alkoxy-radicals derived from lead(IV) acetate oxidation of secondary alcohols undergo intramolecular hydrogen abstraction and thence yield cyclized products.[629b] The same ratio of stereoisomeric products is obtained on oxidation with silver oxide–bromine as with lead(IV) acetate, indicating that the former reaction also proceeds by a radical mechanism.[630] Ring-opened products are obtained on oxidation of cyclobutanols with one-electron oxidants (Mn^{III}, V^{V} or Cr^{IV}), whereas two-electron oxidants (Cr^{VI}) afford cyclobutanone from cyclobutanol.[631,632]

$$\text{1-R-cyclobutanol (OH, R)} \xrightarrow{Mn^{III}} \cdot CH_2CH_2CH_2COR \xrightarrow[H_2O]{Mn^{III}} HOCH_2CH_2CH_2COR$$

Oxidation of a range of simple alcohols with cobalt(III) hydroxide has been examined.[633] Carbon—carbon bond cleavage also occurs in cerium(IV) oxidation of glycols (**316**).[634,635]

$$\underset{(\mathbf{316})}{ArCH(OH)CH(OH)Ar'} \xrightarrow{Ce^{IV}} ArCH(O\cdot)\text{—}CH(OH)Ar \longrightarrow ArCHO + Ar\dot{C}HOH$$

$$Ar\dot{C}HOH \xrightarrow{Ce^{IV}} ArCHO$$

Ring-opening occurs in the iron(III) and copper(II) oxidation of 1-methoxycyclopropanol.[636] The resultant β-propionate radical (**317**) adds to alkenes to give the adduct radical (**318**), reduction of which gives the carbanion (**319**) which subsequently picks up a proton from the solvent.

$$\text{1-methoxycyclopropanol (OH, OMe)} \xrightarrow{Cu^{II}} \dot{C}H_2CH_2COOMe \quad (\mathbf{317})$$

$$\dot{C}H_2CH_2COOMe + MeCOCH{=}CH_2 \longrightarrow \underset{(\mathbf{318})}{MeCO\dot{C}HCH_2CH_2CH_2COOMe} \xrightarrow{Cu^{I}}$$

$$\underset{(\mathbf{319})}{MeCO\bar{C}HCH_2CH_2CH_2COOMe} \xrightarrow{\text{Solvent}} MeCOCH_2CH_2CH_2CH_2COOMe$$

Copper(II) oxidation of α-ketols proceeds through a chelated intermediate.[637] Anodic oxidation of 2-methoxyethanol occurs at the ether rather than the alcohol group.[638]

Oxidation of Amines

Kinetic studies have confirmed that the rate-determining step in the ferricyanide oxidation of tertiary amines involves electron-transfer.[639] The rates of oxidation of cyclic tertiary amines indicate that the transition state is late on the reaction co-ordinate and that the configuration of the nitrogen in the transition state resembles that of the intermediate radical cation. The influence of deuteriation on product distribution indicates that the second step is product-determining. The same radical cation (**320**) is generated by the iron(II) reduction of tertiary-amine oxides;[640] products derived by reduction of the radical cation or from the dication are obtained.

$$R_2\overset{+}{N}(O^-)Bu \xrightarrow{Fe^{II}} \underset{(\mathbf{320})}{R_2\overset{+\bullet}{N}Bu} \longrightarrow R_2\overset{+}{N}H\text{—}CH_2CH_2\dot{C}H_2 \xrightarrow{Fe^{III}} R_2\overset{+}{N}H\text{—}CH_2CH_2\overset{+}{C}H_2 \longrightarrow \text{Products}$$

$$R_2\overset{+\bullet}{N}Bu \ (\mathbf{320}) \xrightarrow{Fe^{II}} R_2NBu$$

Amino-radicals, generated by silver(I) oxidation of chloramines and *N*-chlorosuccinimide, have been trapped by nitrosobutane;[641] they give rise to aldimines. 2,3,5,6-Tetrachloroquinone di-imine is obtained on oxidation of 2,3,5,6-tetrachloro-*p*-phenylenediamine.[642] Anodic oxidation of aniline gives benzidine.[643] Anodic cyanation of diphenylamines[644] and anodic oxidation of 4-dimethylaminopurine,[645] benzaldehyde diphenylhydrazone[646] and Schiff bases[647] have been reported.

Oxidation of amides can occur (*a*) by an electron-transfer mechanism, giving rise to alkylamido-radicals with peroxydisulphate, or (*b*) by hydrogen abstraction with *tert*-butoxy-radicals to give carbamoyl radicals.[648]

$$\mathrm{HCONMe_2 + SO_4^{\bar{\bullet}} \longrightarrow HCO\overset{+\bullet}{N}Me_2 + SO_4^{2-}}$$

$$\mathrm{HCO\overset{+\bullet}{N}Me_2 \longrightarrow {\bullet}CH_2N(CH_3)CHO}$$

$$\mathrm{HCONMe_2 + Bu^tO{\bullet} \longrightarrow {\bullet}CONMe_2 + Bu^tOH}$$

Miscellaneous Oxidations

The reactions of alkyl aryl ketones with iron(III) chloride give mixtures of α-haloketones and coupled products.[649]

$$\mathrm{ArCOCH_2R \xrightarrow{FeCl_3} ArCO\dot{C}HR \longrightarrow ArCOCHRCHRCOAr}$$

$$\mathrm{ArCO\dot{C}HR \xrightarrow{FeCl_3} ArCOCHRCl}$$

Diethyl alkylmalonates react with alkenes and manganese(III) or cobalt(III) acetate in the presence of copper(II) acetate to give products derived from addition of the $\cdot\mathrm{CR(COOEt)_2}$ radical to the alkene.[650]

$$\mathrm{RHC(COOEt)_2 + Mn^{III} \longrightarrow R\dot{C}(COOEt)_2 + Mn^{II} + H^+}$$

$$\mathrm{R\dot{C}(COOEt)_2 + Me_2C{=}CH_2 \longrightarrow Me_2\dot{C}CH_2CR(COOEt)_2}$$

$$\mathrm{Me_2\dot{C}CH_2CR(COOEt)_2 \xrightarrow{Cu(OAc)_2} CH_2{=}CMeCH_2CR(COOEt)_2}$$

The products derived from the reaction of alkenes with manganese(III) acetate in acetic or propionic acid can be attributed to initial formation of the radical $\cdot\mathrm{CHRCOOH}$ which undergoes addition to the alkene.[651] The reaction of manganese(III) acetate in acetic acid with α- and β-methylstyrene in the presence of potassium takes a different course, giving products derived by abstraction of allylic hydrogen.[652]

$$\mathrm{PhCH{=}CHCH_3 \xrightarrow[AcOH]{Mn(OAc)_2\,+\,KBr} PhCH{=}CHCH_2{\bullet}}$$

$$\mathrm{PhCH{=}CHCH_2{\bullet} \longrightarrow PhCH{=}CHCH_2OAc + PhCH(OAc)CH{=}CH_2}$$

Oxidation of acetone with manganese(III) acetate in benzene gives phenylacetone;[653] acetonyl radicals are produced and effect the substitution of benzene.

Oxidation of phenylphosphonous acid with vanadium(V),[654] catalytic oxidation of thiophenol with copper(I) chloride–amine complexes[655] and anodic oxidation of alkyl phenyl sulphides[656] have been reported. Anodic oxidation of trialkylboranes generates alkyl radicals which give rise to dimers.[657]

Metal and Metal-ion Reductions

Reductive coupling of arylalkyl halides has been brought about by vanadium(II).[658] The mechanism resembles that for reductions with chromium(II) except that the first step is an outer-sphere rather than an inner-sphere process:[659]

$$\begin{aligned} RX + V^{II} &\rightarrow R\cdot + V^{III} + X^- \\ R\cdot + V^{II} &\rightarrow RV^{III} \\ RV^{III} + RX &\rightarrow R{-}R + V^{III} + X^- \end{aligned}$$

Reduction of alkyl halides has also been carried out with iron(II) porphyrins,[660] iron(II) haemoglobin and myoglobin;[661] iron(II) cyotochrome is, however, inert. This difference suggests that reaction occurs by an axial inner-sphere process. Chloromethylbenzenes also undergo reductive coupling with iron(II) complexes.[662] In the reductive coupling of phenacyl bromide by zinc, free phenacyl radicals are produced and can be trapped with nitrosobutane.[663] Copper(I) chloride catalyses the addition of benzenesulphonyl chloride and carbon tetrachloride to alkenes:[664]

$$\begin{aligned} RCl + CuCl &\rightarrow R\cdot + CuCl_2 \\ R\cdot + PhCH{=}CH_2 &\rightarrow \cdot CHPhCH_2R \\ \cdot CHPhCH_2R + CuCl_2 &\rightarrow PhCHClCH_2R + CuCl \end{aligned}$$

One-electron reduction of substituted tropylium ions with chromium(II),[665,666] and reductive dimerization of 1,3-cyclohexadiene by sodium in liquid ammonia,[667] have been reported.

Cathodic Reduction

Coupling of the radical anions derived by cathodic reduction of *trans*-cinnamates occurs stereospecifically to give the (±)-dianions such as (**321**).[668] The radical anions approach each other so that the phenyl group of one moiety lies over the hydrogen of the other, thus giving the (±)-dianion.

$$PhCH{=}CHCOOEt \longrightarrow [PhCH{=}CHCOOEt]^{\overset{-}{\cdot}} \longrightarrow \left[\begin{matrix} PhCHCHCOOEt \\ | \\ PhCHCHCOOEt \end{matrix}\right]^{2-} \longrightarrow$$

(**321**)

$$\left[\begin{matrix} PhCHCHCOOEt \\ | \\ PhCHCH_2COOEt \end{matrix}\right]^{-} \longrightarrow$$

O, CO_2Et, H, H, Ph, Ph

Cathodic reduction of carvone leads to three isomeric diketo-dihydro-dimers.[669] Cathodic reduction of acetophenone[670] and of *trans*-3-benzylideneindol-2(3*H*)-one has been reported.[671] Reduction of the *E*-isomer (**322**) occurs more readily than that of the *Z*-isomer (**323**), as the former can approach the electrode more easily.[672]

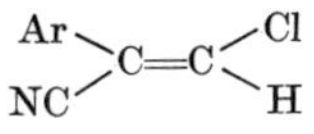

(322) 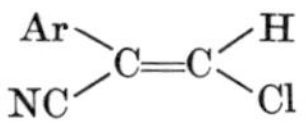 (323)

Benzylic C—F and C—O bonds are cleaved in reductions of the esters (**324**; X = OMe, OAc or F);[673] an electron-withdrawing group is necessary in the aromatic nucleus to stabilize the intermediate radical anion (**325**). Cathodic reduction of 1,2-dithiolylium

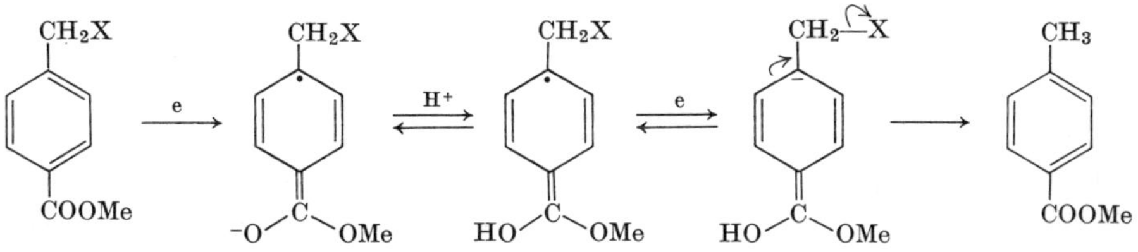

(324) (325)

cations gives 1,2-dithiolyl radicals which are in equilibrium with dimers (**326**);[674] the equilibrium constant depends on the size of the 3,5-substituents R. Electroreductions of sulphonium salts,[675] pyrimidine,[676] purine and imidazole[677] have been reported.

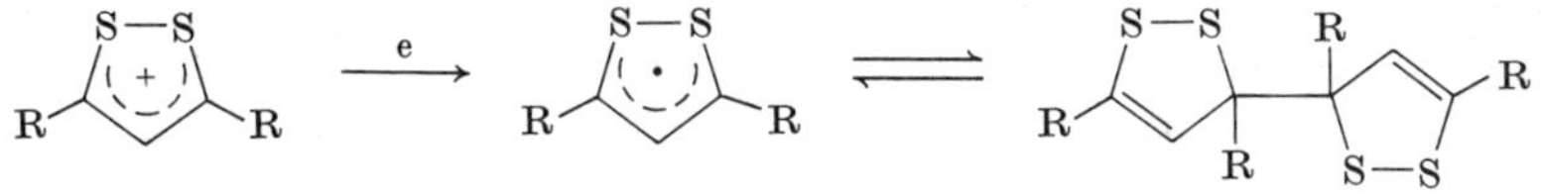

(326)

Reduction of Diazonium Salts

2-Benzoylphenyl radicals have been generated by reduction of the diazonium salt of 2-aminobenzophenone with a copper(I)–amine complex.[678] These radicals couple, or in the presence of ethanol give benzophenone or are oxidized by copper(II) to *o*-hydroxybenzophenone. The Pschorr cyclization reaction has been carried out under a wide range of conditions in protic and aprotic solvents, both with and without catalysts, and also electrochemically;[679] two mechanisms can be put forward to explain the results: (i) a classical homolytic pathway, and (ii) an intramolecular reduction proceeding through an intramolecular charge-transfer complex (**327**) in which electron-transfer proceeds with simultaneous release of nitrogen within the solvent cage.

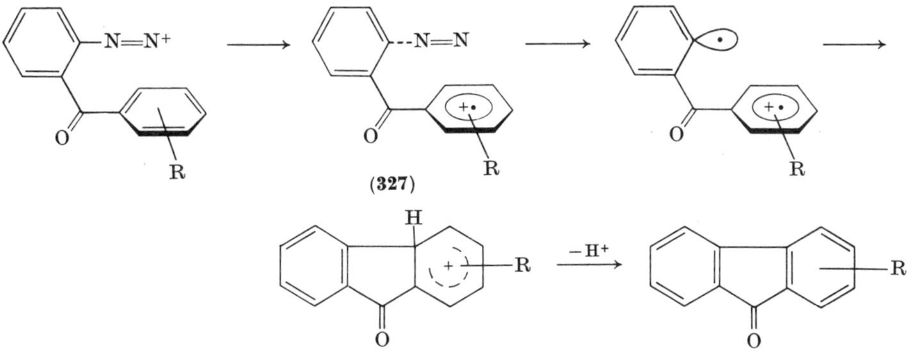

(327)

Fenton's Reaction

Three different modes of reaction are available to organic radicals produced in Fenton's reaction, namely oxidation, dimerization and reduction. Reduction normally occurs with acetonyl radicals, but this may be avoided by inverse addition which affords dimers.[680]

$$CH_3COCH_3 + HO\cdot \longrightarrow CH_3CO\dot{C}H_2 + H_2O$$

$$CH_3CO\dot{C}H_2 \xrightarrow{Fe^{II}} CH_3CO\bar{C}H_2 \xrightarrow{H^+} CH_3COCH_3$$

$$2\,CH_3CO\dot{C}H_2 \longrightarrow CH_3COCH_2CH_2COCH_3$$

Hydroxy-radicals produced in Fenton's reaction add to unsaturated acids at a rate consistent with their having electrophilic character; the resultant radicals undergo reduction:

$$HOOCCH{=}CHCOOH + HO\cdot \longrightarrow HOOCCH(OH)\dot{C}HCOOH$$

$$HOOCCH(OH)\dot{C}HCOOH \xrightarrow[Fe^{II}]{} HOOCCH(OH)\bar{C}HCOOH \longrightarrow HOOCCH(OH)CH_2COOH$$

The α-hydroxyvinyl radicals (**328**) produced by addition to acetylenes undergo reduction by iron(II), or oxidation in the presence of added copper(II); the latter reaction only occurs with terminal alkynes.[681] The vinyl radicals generated from disubstituted

$$R^1C{\equiv}CR^2 + HO\cdot \longrightarrow HOCR^1{=}\dot{C}R^2\ (\mathbf{328})$$

$$HOCR^1{=}\dot{C}R^2 \xrightarrow{Fe^{II}} HOCR^1{=}CHR^2 \longrightarrow R^1COCH_2R^2$$

$$HOCR^1{=}\dot{C}R^2 \xrightarrow[R^1 = H]{Cu^{II}} HOCH{=}CR^2OH \longrightarrow HOCH_2COR^2$$

alkynes are too hindered to allow oxidation to occur, supporting the idea of an intermediate organocopper species in radical oxidations by copper(II). The radical (**329**), generated from *tert*-butyl alcohol, dimerizes, undergoes oxidation with copper(II) to

$$Me_3COH + HO\cdot \longrightarrow \cdot CH_2CMe_2OH\ (\mathbf{329}) \longrightarrow HOCMe_2CH_2CH_2CMe_2OH$$

$$\cdot CH_2CMe_2OH\ (\mathbf{329}) \xrightarrow{Cu^{II}} HOCH_2CMe_2OH \qquad \cdot CH_2CMe_2OH\ (\mathbf{329}) \xrightarrow{O_2} Me_2CO$$

2-methylpropane-1,2-diol, or in presence of oxygen gives acetone.[682] The reaction takes a different course if a substrate that can complex with the iron(II) is added to the system;[683] the yield of dimeric product then becomes very low in many instances, indicating the absence of the radical (**329**).

The role of perhydroxyl radicals in Fenton's reaction has also been examined.[684–686]

Hydroxylation of aromatic ethers has been effected by using iron(II)/EDTA/oxygen or copper(I)/ascorbic acid/oxygen systems, attack occurring preferentially at the *o*-position with (**330**; X = CH_2 or CO, Y = OMe or NMe_2) owing to a chelating effect.[687,688]

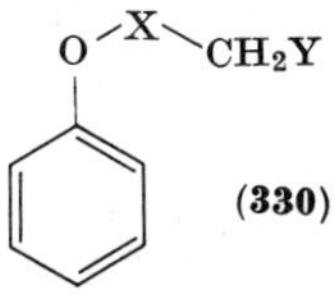

(330)

Radical Ions and Electron-transfer Processes

Bromobenzene undergoes a very rapid reaction with the enolate anion of acetone on irradiation.[689] The reaction proceeds by an $S_{RN}1$ mechanism[689a] and constitutes the first example of a new mode of initiation of such reactions:

$$\begin{aligned} \text{Electron source} + \text{PhBr}\ (h\nu) &\rightarrow \text{PhBr}^{\overline{\cdot}} + \text{Residue} \\ \text{PhBr}^{\overline{\cdot}} &\rightarrow \text{Ph}\cdot + \text{Br}^- \\ \text{Ph}\cdot + \text{CH}_3\text{COCH}_2^- &\rightarrow [\text{PhCH}_2\text{COCH}_3]^{\overline{\cdot}} \\ [\text{PhCH}_2\text{COCH}_3]^{\overline{\cdot}} + \text{PhBr} &\rightarrow \text{PhCH}_2\text{COCH}_3 + \text{PhBr}^{\overline{\cdot}} \end{aligned}$$

Similar reactions with other anions, including those of fluorene, anethole and anilene, occur with either photochemical or alkali-metal initiation;[690] the anilide anion undergoes both *C*- and *N*-arylation. Methoxide catalyses the $S_{RN}1$ reaction between 4-bromoisoquinoline and thiophenoxide;[691] the formation of 4-(phenythio)isoquinoline is accompanied by the formation of isoquinoline, the latter arising by reaction of the 4-isoquinolyl radical with methoxide:

$$\begin{aligned} \text{Ar}\cdot + \text{CH}_3\text{O}^- &\rightarrow \text{ArH} + \text{CH}_2\text{O}^{\overline{\cdot}} \\ \text{ArBr} + \text{CH}_2\text{O}^{\overline{\cdot}} &\rightarrow \text{ArBr}^{\overline{\cdot}} + \text{CH}_2\text{O} \end{aligned}$$

The photoreduction of haloaromatic compounds with borohydride occurs in near-quantitative yields, an $S_{RN}1$ mechanism being formulated:[692]

$$\begin{aligned} \text{PhX}\ (h\nu) \rightarrow [\text{PhX}]^* &\rightarrow \text{Ph}\cdot + \text{X}\cdot \\ \text{X}\cdot + \text{BH}_4^- &\rightarrow \text{BH}_3^{\overline{\cdot}} + \text{X}^- + \text{H}^+ \\ \text{Ph}\cdot + \text{BH}_4^- &\rightarrow \text{PhH} + \text{BH}_3^{\overline{\cdot}} \\ \text{PhX} + \text{BH}_3^{\overline{\cdot}} &\rightarrow \text{PhX}^{\overline{\cdot}} + \text{BH}_3 \\ \text{PhX}^{\overline{\cdot}} &\rightarrow \text{Ph}\cdot + \text{X}^- \end{aligned}$$

Cyclohex-3-enyl radicals (**331**) can be efficiently scavenged by nucleophiles such as ethoxide or azide so as effectively to prevent their dimerization.[693] A similar mechanism

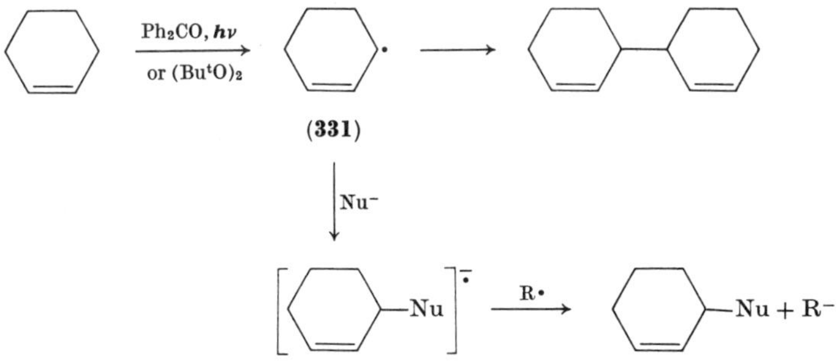

$Ph_2CO, h\nu$ or $(Bu^tO)_2$

(331)

Nu^-

Nu, R·, Nu + R^-

seems to be involved in the reactions of anthraquinone-1-sulphonates with hydroxide,[694] 2-chloro- and 2-methyl-anthraquinones with methoxide,[695] *p*-chloronitrobenzene and alkoxynitrobenzenes with alkoxides,[696,697] and 4-substituted 2,4,6-tri-*tert*-butylcyclohexa-2,5-dienones with phenoxide,[698,699] diethylamine[700] or tetramethylphenylenediamine.[701] In the last instance 2,4,6-tri-*tert*-butylphenoxy-radicals were detected by their ESR spectra. The reaction of 2,4,6-trisubstituted phenoxy-radicals with diethylamine or triethylamine proceeds by an electron-transfer process.[702]

Naphthalene radical anions[703,704] react with phenylacetonitrile by either proton- or electron-transfer;[705] the electron-transfer is promoted by polar solvents that favour loosening of the ion-pair by solvation of the cation.

$$PhCH_2CN + C_{10}H_8^{\overline{\cdot}}, Na^+ \longrightarrow \begin{cases} Ph\bar{C}HCN \\ CN^- + PhCH_2\cdot \longrightarrow PhCH_3 \end{cases}$$

Alkyl radicals are generated in the reaction of sodium naphthalenide with benzoates and benzenesulphonates:[706]

$$PhCOOR + C_{10}H_8^{\overline{\cdot}}, Na^+ \longrightarrow \cdot CPh(OR)ONa \longrightarrow PhCOONa + R\cdot$$

$$PhSO_2R + C_{10}H_8^{\overline{\cdot}}, Na^+ \longrightarrow Ph\text{—}\dot{S}(ONa)(=O)\text{—}R \longrightarrow PhSO_2Na + R\cdot$$

Hex-5-enyl radicals generated in this way give rise to both hex-1-ene and methylcyclopentane, formed by cyclization of the hex-5-enyl radicals. In the reductive dehalogenation of aryl halides by sodium naphthalenide or sodium, an aryl radical is produced by fragmentation of the initial radical anion:[707]

$$ArCl \rightarrow ArCl^{\overline{\cdot}} \rightarrow Ar\cdot + Cl^-$$

Formation of the reduced compound involves either direct hydrogen-abstraction or reduction to the anion followed by proton-abstraction. Experiments with *o*-chlorobenzylideneaniline in isopropyl methyl ether indicate that the former route, ($Ar\cdot + HS \rightarrow ArH + S\cdot$), is preferred from the relative ease of hydrogen-abstraction from the primary and tertiary sites in the ether.[708]

Reduction of di-*tert*-butyldiaziridinone (**332**) with sodium naphthalenide or butyllithium, or cathodically, affords *N,N'*-di-*tert*-butylurea;[709] reaction proceeds through the radical anion (**333**) which is formed in a concerted process. The rate of reaction of potassium naphthalenide with hydrocarbons is proportional to the acidity of the hydrocarbon.[710] The 1-(dimethylamino)naphthalene radical anion is an alternative radical-anion reducing agent, which should prove useful as products derived from it can be readily extracted.[711] Reaction of trimethylsilyl sodium with hydrocarbons provides a

$$\underset{\mathbf{(332)}}{Bu^tN\text{—}C(=O)\text{—}NBu^t} \xrightarrow{e} \underset{\mathbf{(333)}}{Bu^t\dot{N}CO\bar{N}Bu^t} \xrightarrow[2H^+]{e} Bu^tNHCONHBu^t$$

simple way of generating radical anions:[712]

$$Me_3Si^- + ArH \rightarrow Me_3Si\cdot + ArH^{\overline{\cdot}}$$

Nitrogen dioxide has been used to effect the single-electron oxidation of organometallic compounds.[713] Charge-transfer complexes have been observed between galvinoxyl and electron-acceptors such as dichlorodicyano-1,4-benzoquinone (DDQ).[714] One-electron transfer occurs in the reaction of the $(CF_3)_3C^-$ anion with triarylmethyl halides:[715]

$$(CF_3)_3C^- + Ph_3CCl \longrightarrow [(CF_3)_3C\cdot \ Ph_3C\cdot] + Cl^- \xrightarrow{SH} (CF_3)_3CH$$
$$[(CF_3)_3C\cdot \ Ph_3C\cdot] \longrightarrow p\text{-}(CF_3)_3CC_6H_4CHPh_2$$

Carbon tetrachloride can effect the dimerization of the 2-nitropropane anion;[716] the reaction proceeds by an electron-transfer mechanism with the generation of the 2-nitro-

$$Me_2\bar{C}NO_2 + CCl_4 \longrightarrow Me_2\dot{C}NO_2 + CCl_4^{\overline{\cdot}}$$

(334)

$$Me_2\dot{C}NO_2 \xrightarrow{Me_2CNO_2^-} [Me_2C(NO_2)\text{—}CMe_2(NO_2)]^{\overline{\cdot}} \xrightarrow{CCl_4} Me_2C(NO_2)\text{—}CMe_2(NO_2)$$
$$Me_2\dot{C}NO_2 \xrightarrow{Me_2\dot{C}NO_2} Me_2C(NO_2)\text{—}CMe_2(NO_2)$$

isopropyl radical (**334**). Radicals are also generated by an electron-transfer process in the reaction between α-alkyl-*p*-nitrobenzyl chlorides with carbanions:[717]

$$ArCHClR \xrightarrow{Me_2CNO_2^-} [ArCHClR]^{\overline{\cdot}} \longrightarrow \cdot CHArR + Cl^-$$

The radical polymerization of styrene has been initiated by a bis-[(–)-ephedrine]copper[II] chelate–carbon tetrachloride system which likewise involves initial electron-transfer to carbon tetrachloride.[718] One electron-transfer occurs between trimethyl phosphite and quinones, the rate of reaction depending on the redox potential of the quinone:[719]

$$Q + P(OMe)_3 \rightarrow Q^{\overline{\cdot}} + P(OMe)_3^{+\cdot}$$

The behaviour of the three isomeric fluorobenzonitrile radical anions depends on the substitution pattern[720] in contrast to the radical anions of chloro- and bromo-benzonitrile which all give benzonitrile.[721] Both the *ortho*- and *para*-species initially yield the dimeric dianion and thence benzonitrile and 4,4′-dicyanobiphenyl, respectively; the *meta*-radical anion behaves differently, losing cyanide to give fluorobenzene. Steric hindrance prevents simple dimerization of the radical anion of 1-*tert*-butyl-2-phenylacetylene which affords four dimers.[722] The radical anion (**335**) is remarkably easily cleaved,[723]

(335)

the extra electron in an anti-bonding orbital raising the energy. Oxidation of aromatic nitro-compounds by α- and β-hydroxy-alkyl radicals[724] and by alkali metals[725] proceeds by way of nitro-aromatic radical anions. The radical anions generated by reaction of fluorenylidenephenylphosphorane, *N*-phenyltriphenylphosphine imine or diphenyl sulphone with sodium all undergo loss of a phenyl group to give stable radicals.[726]

The first step in the formation of Grignard reagents involves electron-transfer from the magnesium to the alkyl halide.[727] The resultant radical–ion pair may collapse to the Grignard reagent with complete retention of configuration in the case of an optically active halide; alternatively, halide transfer gives a loose radical-pair which either combines to give the racemic Grignard reagent or the organic radical escapes to give radical products such as those of disproportionation. Evidence for the loose radical pair

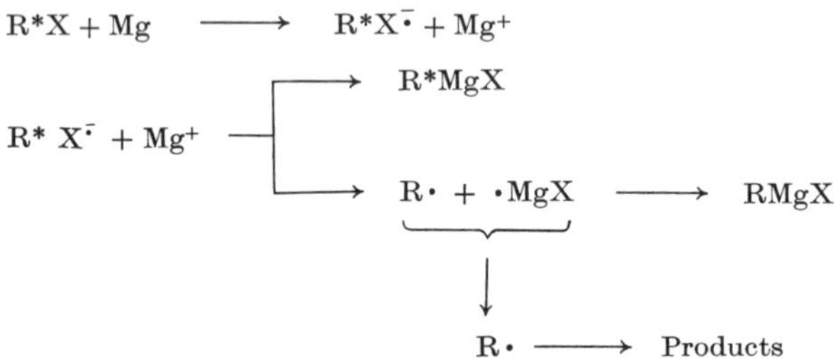

is provided by the observation of CIDNP;[728] the extent of direct formation of the Grignard reagent from the radical–ion pair is increased in more basic ethers such as tetrahydrofuran which can solvate the radical anion. CIDNP techniques have established that metal salts in an oxidized state[729] and halogens,[730] which may oxidize trace-metal impurities in the magnesium, promote the reaction of Grignard reagents with alkyl halides by a radical mechanism. Ethyl but not phenyl Grignard reagents react with carbon tetrachloride and chloroform in the presence of cyclohexene by a radical pathway.[731] Radical processes are also important in the reaction of Grignard reagents with unsaturated ketones[732] and with (2-bromoalkoxy)silanes.[733] Triphenylmethyl-lithium reacts with α,β-unsaturated ketones in an electron-transfer process.[734] Radicals are also involved in the reaction of *tert*-butyl-lithium with acid chlorides[735] and of methyl-lithium with 7-chloro-norbornenes and -norbornadienes.[736] The coupling of cyclopropyl-(1-halocyclopropyl)acetylene with organolithium compounds is likewise a radical process.[737] Organomercury compounds undergo single-electron oxidation with nitronium fluoroborate[738] [$\rightleftarrows RHgX^{+\bullet}(NO_2)BF_4^-$]. Reaction of a tetra-alkyl-lead with a copper(II) halide involves an inner-sphere alkyl-transfer with the generation of a metastable species RCuX which undergoes homolysis to alkyl radicals.[738a]

$$R_4Pb + CuX_2 \rightarrow R_3PbX + RCuX$$
$$RCuX \rightarrow R\cdot + CuX$$

The adducts (**336**) formed between Grignard reagents and thioketones undergo homolysis to free radicals which have been detected by ESR spectroscopy:[739,740]

$$R^1_2C{=}S + R^2MgX \rightarrow \underset{\mathbf{(336)}}{R^1_2C(MgX)SR^2} \rightleftharpoons \cdot CR^1_2(SR^2) + MgX$$

In the reaction of the dithione (**337**) with Grignard reagents, the bicyclobutane (**338**) and the rearranged compound (**339**) are formed in addition to the expected products.[741]

Radical chlorinations proceeding through nitrogen radical cations (Hoffman–Löffler and Minisci reactions) have been reviewed.[742]

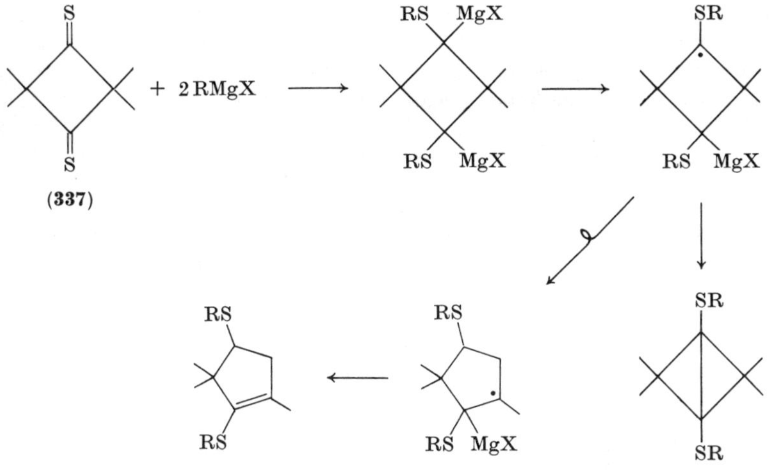

+ 2 RMgX ⟶ ⟶

(337)

(339) **(338)**

The redox behaviour of quaternary salts derived from bipyridyls, bipyridylethylenes,[743] 2,7-diazapyrene and diazoniapentaphens,[744] and of bipyrylium and bithiopyrylium salts have been examined.[745]

The photochemically induced decarboxylation of carboxylate anions is effected by the bipyridylium salts, "Paraquat" and "Diquat" at a wavelength longer than that absorbed by either reactant.[746] This suggests the intermediacy of a charge-transfer complex.

$$PQ^{2+} + RCOO^- \rightleftharpoons [PQ^{2+}, {}^-OOCR \longleftrightarrow PQ^{+\bullet}, \bullet OOCR] \xrightarrow{h\nu}$$

$$PQ^{+\bullet} + RCOO\bullet \longrightarrow R\bullet + CO_2$$

(PQ = 'Paraquat')

The tris-(*p*-bromophenyl)aminium ion oxidizes 3,3′,4,4′-tetramethoxybibenzyl to 3,3′,4,4′-tetramethoxyphenanthrene in an electron-transfer process.[747] The bimolecular self-reactions of aminium radicals $R_2HN^{+\bullet}$ have been examined by ESR methods.[748] The rates of coupling of *p*-substituted triphenylaminium ions $ArPh_2N^{+\bullet}$ give a good correlation with σ^+ constants for all except strongly electron-releasing groups;[749] these substituents enhance the stability of the radical cations by delocalizing the unpaired spin preferentially around the substituted benzene ring. The dimerization of the phenothiazine[750] and phenoxselenin radical cations[751] has been examined.

The indene radical cation, which is generated by charge-transfer excitation of complexes with anhydrides, gives *anti*-head-to-head cyclobutane dimers.[752] Photosensitized conversion of 1,1-diphenylethylene into 1,2,3,4-tetrahydro-1,1,4-triphenylnaphthalene involves generation of the $[Ph_2C{=}CH_2]^{+\bullet}$ radical cation, which is then thought to attack a second molecule of diphenylethylene. The resulting radical cation adduct then undergoes intramoledular cyclization.[753]

Aromatic amines can induce the photodecomposition of alkyl halides;[754] electron-transfer occurs between the excited singlet state of the amine and the alkyl halide, giving alkyl radicals which either abstract hydrogen from the solvent or react with the aminium ion. The degree of charge-transfer in the transition state is low as the reaction is relatively solvent-insensitive. *N*-Arylglycines undergo photosensitized decarboxylation with aromatic hydrocarbons, reaction proceeding by an electron-transfer mechanism followed by proton abstraction.[755] Products from the reaction of phenylglycine with anthracene include methylaniline and dihydroanthracene:

$$ArH + PhNHCH_2COOH \rightarrow ArH^{\cdot-} + Ph\overset{+\cdot}{N}HCH_2COOH \rightarrow ArH_2 + CO_2 + PhNHMe$$

Acridine and other heterocyclic compounds induce photodecarboxylation of phenylacetic acid and benzilic acid by a similar mechanism.[756] Ferricyanide has also been found to induce the photochemical decarboxylation of glycine and alanine peptides in neutral or weakly acid media:[757]

$$RCOO^- + [Fe(CN)_6]^{3-} \xrightarrow{h\nu} RCOO\cdot + [Fe(CN)_6]^{4-}$$
$$RCOO\cdot \longrightarrow R\cdot + CO_2$$

In stronger acid, hydrogen-abstraction by hydroxyl radicals, produced by electron-transfer from a water ligand to the photoexcited ferricyanide, occurs preferentially:

$$HO\cdot + RCH_2COOH \rightarrow \cdot CHRCOOH + H_2O$$

Oxidation of dialkyl sulphides with hydroxyl radicals involves electron-transfer and not hydrogen-abstraction as usually occurs.[758] The resultant radical cation reacts with another molecule of dialkyl sulphide to give a dimeric radical cation, the ease of formation of which is determined by steric factors:

$$R_2S + HO\cdot \rightarrow R_2S^{+\cdot} + OH^-$$
$$R_2S^{+\cdot} + R_2S \rightarrow [R_2SSR_2]^{+\cdot}$$

Pyrolysis

Gas-phase pyrolysis of a number of hydrocarbons, including ethane[759] (and fluoro-[760] and chloro-derivatives[761] thereof), propene,[762,763] propane,[764] but-1-ene,[765] pent-1-ene,[766] pent-2-ene,[767] benzene,[768] and chlorobenzene[769] has been investigated. Radical mechanisms have been proposed for the thermal decompositions of trifluoroacetaldehyde,[770] biacetyl,[771] allyl oxalates,[772] and benzyl triphenylacetate.[773] Two contrasting reaction pathways have been identified in the thermolysis of pinanes (**340**);[774] substituents that stabilize the radical at C-1 favour pathway *a* (e.g. nopinone, $R^1 = O$,

R1 R2 (**340**) *a* *b* R1 R2 R1 R2 R1 R2 R1 R2

$R^2 = H,H$); substituents that stabilize the radical at C-4 favour path *b* (e.g. *cis*-verbanone, $R^1 = Me,H$, $R^2 = O$). Two main primary processes have been identified in the thermolysis of ethyltrichlorosilane and ethyltrimethylsilane, one producing ethylene and trichloro- or trimethyl-silane, and the other producing ethyl radicals.[775]

$$C_2H_4 + HSiX_3 \xleftarrow{\Delta} C_2H_5SiX_3 \xrightarrow{\Delta} C_2H_5\cdot + \cdot SiX_3$$
$$X = Cl \text{ or } Me$$

A carbene was proposed as the reactive intermediate in the decomposition of trifluoro-1,1,2,2-tetra(fluoroethyl)silane.[776] Bis(trimethylsilyl)mercury decomposes mainly by a

$$CHF_2CF_2SiF_3 \xrightarrow{\Delta} \left[CHF_2CF \overset{F}{\underset{SiF_3}{\vdots}} \right]^* \longrightarrow CHF_2CF\text{:} + SiF_4$$

second-order process in solution, giving trimethylsilyl radicals, hexamethyldisilane and mercury;[777] a lower limit on $D(Me_3Si—HgSiMe_3)$ of 48 kcal mol^{-1} was set from the small amount of first-order decomposition.

The mechanism of the thermal decomposition of azomethane was investigated,[778] and low-pressure pyrolysis studies on azoethane, azoisopropane and 2,2′-azoisobutane have been published.[779] The reactive intermediate in the gas-phase pyrolysis of cyclopropylamine is probably a propylideneamine,[780] and not a diradical as previously proposed. 1-Acyl-2-alkyldi-imides (**341**) decompose in toluene to free radicals which then combine producing *N*-acetylketene imines as the main products.[781]

$$RCON{=}NCR'_2CN \quad \textbf{(341)}$$
$$R = Ph, Me \text{ or } PhCH_2;\ R'_2 = Me_2, (CH_2)_4, (CH_2)_5 \text{ or } (CH_2)_6$$

1,2-Bis(difluoroamino)propane decomposes thermally by a mechanism that is essentially the reverse of that for addition of tetrafluorohydrazine to propene.[782] The first step in the pyrolysis of 2,2-bis(difluoroamino)propane is loss of a difluoroamino-radical; the resulting 2-(difluoroamino)isopropyl radical can lose fluorine to give (**342**) as the main product, or, after fluorine migration, lose methyl radicals to give (**343**).

$$CH_3C(NF_2)_2CH_3 \xrightarrow{-NF_2\cdot} CH_3\dot{C}(NF_2)CH_3 \xrightarrow{-F\cdot} (CH_3)_2C{=}NF \quad \textbf{(342)}$$
$$\downarrow$$
$$(CH_3)_2CF{—}\dot{N}F \longrightarrow CH_3CF{=}NF + CH_3\cdot \quad \textbf{(343)}$$

The decompositions of 2-benzoxazolinone (**344**; X = O), 2-benzimidazolinone (**344**; X = NH) and 2-benzothiazolinone (**344**; X = S) on pyrolysis show certain similarities to the decomposition pathways on electron-impact.[783] Products from the thermal

H
N
=O (**344**)
X

decomposition of (**345**) were accounted for in terms of radicals (**346**) and (**347**) as intermediates.[784]

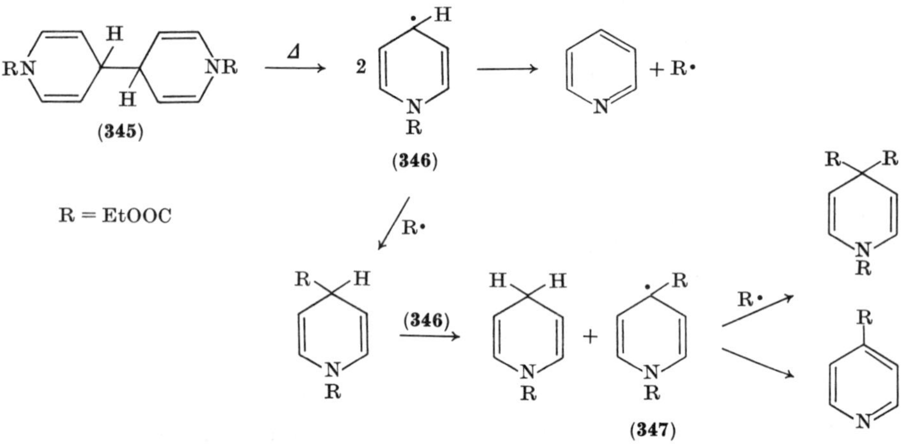

(**345**) (**346**) (**347**)

R = EtOOC

Thermal decomposition of dimethyl sulphoxide gives mainly methyl radicals and sulphur monoxide;[785] the methyl radicals abstract hydrogen from the sulphoxide, and the minor products are accounted for by radical–radical reactions. Pyrolysis of trimethylthallium leads to stepwise loss of the methyl groups;[786] $D(Me_2Tl—CH_3) = 36$ kcal mol^{-1} was determined by the toluene-carrier technique. Tricyclopentadienylalkyluranium compounds Cp_3Ur decompose by a homolytic mechanism to give RH, by abstraction of hydrogen from the cyclopentadienyl rings, as the main product.[787] The main product from the pyrolysis of *N*-(2-hydroxyethyl)aniline borate (**348**) was indole, which was

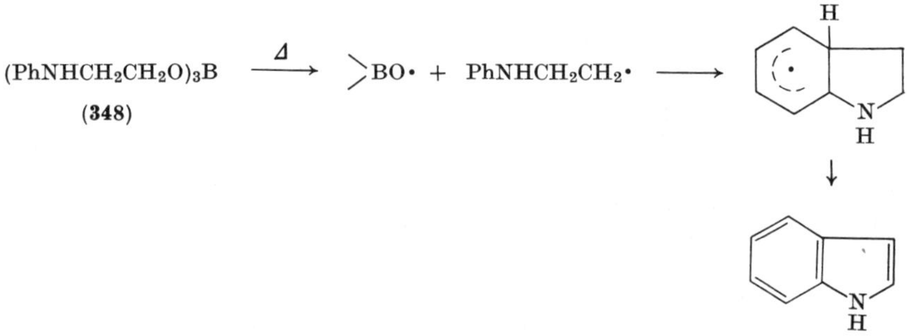

$(PhNHCH_2CH_2O)_3B \xrightarrow{\Delta} >BO\cdot + PhNHCH_2CH_2\cdot \longrightarrow$

(**348**)

probably formed as shown;[788] the process is similar to that for the decomposition of the corresponding phosphate, acetate, formate, benzoate and benzenesulphonate esters. Arylthio-methanes and -ethanes (**349**) decompose to give initially tri(arylthio)methyl

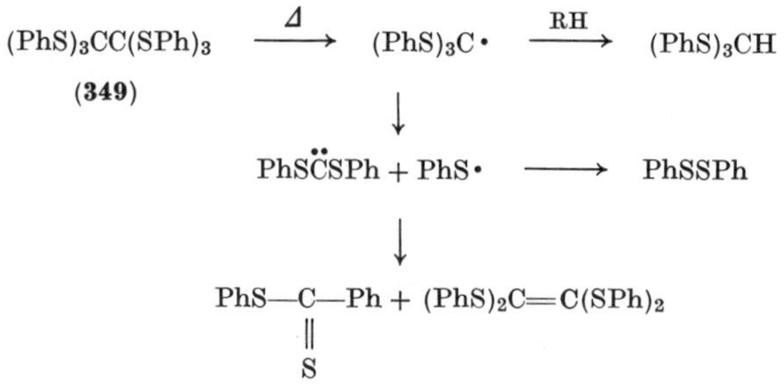

$$(PhS)_3CC(SPh)_3 \xrightarrow{\Delta} (PhS)_3C\cdot \xrightarrow{RH} (PhS)_3CH$$

(**349**)

$$PhS\ddot{C}SPh + PhS\cdot \longrightarrow PhSSPh$$

$$PhS—C(=S)—Ph + (PhS)_2C{=}C(SPh)_2$$

radicals; these may react with solvent or decompose further by loss of another arylthiyl radical.[789]

Radiolysis

Hydroxy-radicals abstract hydrogen mainly from the alkyl chain in aliphatic alcohols, but there is some attack at the OH group, particularly in *tert*-butyl alcohol.[790] The rate of hydroxy-radical reaction with aromatic carboxylate anions is nearly at the diffusion limit, but $O^{\cdot-}$ reacts less rapidly.[791] It was proposed that the site of hydrogen-atom addition to unsaturated acids and amides is controlled by the stability of the resulting radical.[792] A variety of radicals derived from γ-radiolysis of glycine have been studied by ESR spectroscopy.[793] Hydroxy-radicals and hydrogen atoms add to the furan ring which then opens in basic solution.[794] The structures of the radicals formed by addition of hydrogen atoms to imidazole and 1,2,4-triazole have been examined by ENDOR spectroscopy.[795] Hydroxy-radicals add to ring-carbon atoms in pyrrole, imidazole and related compounds,[796] and also in pyrimidine bases,[797] rather than abstracting hydrogen. The radical anions formed by γ-radiolysis of 5-halouracils may lose halide ion, giving uracil after reaction with the solvent; alternatively, they may pick up a proton, particularly when the halogen is chlorine.[798,799]

Carbon—carbon bond scission, rather than C—S scission, predominates in the γ-radiolysis of polypropylene sulphide.[800] Study has been made of radicals formed in the radiolysis of amines,[801] nitriles[802] and *tert*-butyl alcohol.[803]

Miscellaneous

Photolysis of 2-*exo*-bromonorbornane (**350**; X = Br) in ether affords norbornane whilst the 2-iodo-compound (**350**; X = I) furnishes a mixture of 2-norbornene and nortricyclene.[804] In both cases reaction proceeds by homolysis of the C—X bond and the subsequent difference in behaviour is attributable to the fact that the bromine atom is "hot" and abstracts hydrogen from the solvent cage, whereas with iodine the radical-pair has sufficient lifetime to enable electron-transfer to occur. The norbornene and nortricyclene arise by way of the 2-norbornyl cation (**352**).

(350) $\xrightarrow{h\nu}$ (351) $\cdot$X $\xrightarrow{X = I}$ (352) + I^-

(351) $\xrightarrow[(X = Br)]{RH}$ norbornane

(352) $\rightarrow$ 2-norbornene + nortricyclene

Photodehydrodimerization of benzimidazole gives only the unsymmetrical dimers.[805] The photodecomposition of carboxylic acids gives rise to $\cdot$COOH radicals;[806] the residual radicals can undergo cage disproportionation with the $\cdot$COOH radicals or they can dimerize or disproportionate.

$$\mathrm{PhCH(OH)COOH} \longrightarrow \mathrm{Ph\dot{C}HOH} + \cdot\mathrm{COOH}$$
$$\mathrm{Ph\dot{C}HOH} + \cdot\mathrm{COOH} \longrightarrow \mathrm{PhCH_2OH} + \mathrm{CO_2}$$
$$2\,\mathrm{Ph\dot{C}HOH} \longrightarrow \mathrm{PhCH_2OH} + \mathrm{PhCHO}$$
$$2\,\mathrm{Ph\dot{C}HOH} \longrightarrow \mathrm{PhCH(OH)CH(OH)Ph}$$

Photolysis of benzoin ethers affords high yields of glycol ethers:[807]

$$\mathrm{PhCOCHPh(OR)} \xrightarrow{h\nu} \mathrm{Ph\dot{C}O} + \mathrm{\dot{C}HPh(OR)}$$
$$\mathrm{\dot{C}HPh(OR)} \longrightarrow \mathrm{PhCH(OR)CHPh(OR)}$$

Photolyses of nitronitrosoalkanes effects homolysis of the C—NO bond.[808,809] The behaviour of *tert*-butyl methyl ether and di-*tert*-butyl ether on photolysis has been reported.[810] The kinetics of the iodide-catalysed dimerization of pseudo-acid chlorides of *o*-benzoylbenzoic acids have been described.[811]

ESR Spectroscopy

Use of double mixing in the study of alcohol-derived radicals by the hydrogen peroxide–titanium(III) system has been shown to be superfluous.[812] The use of ESR spectroscopy in determination of atom and radical concentrations has been reviewed.[813] Details of a computer programme for interpreting ESR spectral data for organic radicals have appeared.[814] An improved model for calculation of coupling constants for aromatic radicals is based on a slight distortion of the aromatic nucleus from a regular hexagonal shape.[815] CNDO calculations have been developed for the calculation of g-values of radicals.[816] The method gives good results for both σ- and π-radicals. The methylene protons in $\mathrm{XYZCCH_2N(Bu^t)O\cdot}$ radicals are magnetically non-equivalent and appear as a 1:1:1:1 quartet as a consequence of the chiral nature of the CXYZ group.[817] Spin labels have been trapped in organic inclusion crystals to examine anisotropic molecular motion;[818] this technique has potential use in spin-labelling studies of biological membranes.[819] Spin-exchange studies have been carried out on binitroxide[820] and bihydrazyl radical.[821]

CIDNP[822, 823] and CIDEP Methods

The CIDNP method has again proved capable of providing answers to a wide range of chemical problems. Low concentrations of lanthanide reagents $\mathrm{Eu(fod)_3}$ and $\mathrm{Pr(fod)_3}$ can be used to shift selectively the resonance lines of products with atoms containing lone-pair electrons. In the thermal decomposition of benzoyl propionyl peroxide the CIDNP resonance lines due to the methylene group in the product $\mathrm{PhCO_2CH_2CH_3}$ were shifted downfield by $\mathrm{Eu(fod)_3}$ and upfield by $\mathrm{Pr(fod)_3}$.[824] Two radical-pairs were shown to be important in the photochemistry of propionaldehyde from a study of CIDNP evidence;[825] radical-pair (**353**) predominated in deuterioacetonitrile solution, but radical pair (**354**) was more important in perfluoromethylcyclohexane. In the

$$\mathrm{CH_3CH_2CHO} \xrightarrow{h\nu} \mathrm{CH_3CH_2CHO^T} \longrightarrow [\mathrm{CH_3CH_2\cdot\ \cdot CHO}]^{\mathrm{T}}\ (\mathbf{353}) \longrightarrow \text{Products}$$
$$\mathrm{CH_3CH_2CHO^T} \xrightarrow{\mathrm{CH_3CH_2CHO}} [\mathrm{CH_3CH_2\dot{C}HOH\ \ \dot{C}OCH_2CH_3}]^{\mathrm{T}}\ (\mathbf{354}) \longrightarrow \text{Products}$$

photolysis of pivalaldehyde, radical pairs $[Me_3C\cdot \ \cdot CHO]^{T,F}$ and $[Me_3C\cdot \ \cdot CMe_3]^F$ account for the spin polarization observed in the products.[826] CIDNP evidence has also been useful in elucidating the mechanism of glycollaldehyde photolysis[827] (see also the section on photochemical atom abstractions). The polarization of isobutene formed on photolysis of di-*tert*-butyl disulphide indicated that it had been formed by disproportionation of a caged triplet pair of *tert*-butyl and *tert*-butylperthiyl radicals. *tert*-Butyl radicals escaping from the cage also gave a polarized product.[828] Trichloromethyl radicals, produced by decomposition of bis(trichloroacetyl) peroxide, reacted with tetramethylethylene to give chloroform and 4,4,4-trichloro-2,3,3-trimethylbut-1-ene (**355**), both of which showed CIDNP effects;[829] at high alkene concentration hydrogen abstraction from the methyl groups was observed. CIDNP effects observed in 1-arylpentenes and 1,2-diarylethanes, obtained from the reaction of BuLi with $ArCH_2Cl$, confirmed that they were formed from butyl and arylmethyl radicals.[830] CIDNP studies have also indicated that

•CCl3 CCl3 •CCl3 HCCl3 + CCl3 (**355**)

•CCl3 •CCl3 + HCCl3

radicals are involved in some substitution reactions of diazonium salts,[831] and ^{13}C- and ^{15}N-CIDNP studies of the thermolysis of diazoaminobenzene have been reported.[832]

From the ^{19}F-CIDNP study of the products of the decomposition of bis(pentafluorobenzoyl) peroxide it was deduced that the *o*- and *m*-fluorine coupling constants of the pentafluorophenyl radical were positive and that the *para*-coupling constant was negative. By a similar technique, a_F-*ortho* and a_F-*para* in fluorophenyl radicals were shown to be positive.[833] This technique has been applied to determining the signs of ^{19}F-couplings in a variety of fluoroaromatic radicals.[834] The ^{19}F-CIDNP spectrum indicated that α-fluorobenzyl radicals dimerize by way of a pair of diastereoisomeric fluoromethylenecyclohexa-2,4-dienes (**356**) which subsequently rearrange to the difluorobibenzyl (**357**).[835]

H F H CHFPh and F H H CHFPh ⟶ PhCHFCHFPh

(**356**) (**357**)

Treatment of pentafluorobenzyl chloride with butyl-lithium gave 1-(pentafluorophenyl)pentane and 1,2-bis(pentafluorophenyl)ethane;[836,837] ^1H- and ^{19}F-CIDNP spectra showed that the products came from singlet pairs of pentafluorobenzyl and butyl radicals; when the reaction was done in the presence of a high magnetic field the proportion of 1-(pentafluorophenyl)pentane increased; the strong magnetic field retarded singlet triplet transitions, thus increasing the probability of recombination.

^{31}P-CIDNP spectra have been used as a probe into the mechanism of the reactions of (phenylazo)triphenylmethane[838] and various peroxides[838,839] with dialkyl, trialkyl and triaryl phosphites.

In CIDEP spectra the extent of polarization depends on the rate of separation of the initial radical-pair and on the strength of the interaction causing the intersystem mixing. The polarization may be enhanced if the pairs are constrained to remain together, as in exciplexes. Photochemical reduction of benzophenone by tertiary amines was thought to proceed by an exiplex mechanism; strong emission was observed, confirming this hypothesis.[840] CIDEP emission observed from the durosemiquinone–phenoxy radical pair was found to be in conflict with the current radical-pair theory.[841] CIDEP has also been the subject of a number of theoretical studies.[842–844]

References

1 J. K. Kochi (Ed.), *Free Radicals*, Vols. I and II, Wiley-Interscience, New York, 1973.
2 W. A. Waters (Ed.), *Free Radical Reactions, MTP (Med. Tech. Publ. Co.) Int. Rev. Sci.: Org. Chem., Ser. One, Vol. 10.*
3 *Symposium on Free Radical Pathology, Fed. Proc.*, **32**, 1859–1908 (1973).
4 Cf. Reference 1, Volume II, Chaps. 18, 19, 20.
5 P. Hedvig, *Radiat. Chem. Macromol.*, **1**, 55 (1972); *Chem. Abs.*, **78**, 109989 (1973).
6 M. C. R. Symons, *Tetrahedron Letters*, **1973**, 207.
7 H. Paul and H. Fischer, *Helv. Chim. Acta*, **56**, 1575 (1973).
8 D. E. Wood and R. F. Sprecher, *Mol. Phys.*, **26**, 1311 (1973).
9 D. W. Smith and L. Andrews, *J. Chem. Phys.*, **58**, 5222 (1973).
10 R. V. Lloyd and M. T. Rogers, *J. Am. Chem. Soc.*, **95**, 2459. (1973); L. D. Kispert and M. T. Rogers, *J. Chem. Phys.*, **58**, 2065 (1973).
11 I. Biddles and A. Hudson, *Mol. Phys.*, **25**, 707 (1973).
12 G. D. Mendenhall and K. U. Ingold, *J. Am. Chem. Soc.*, **95**, 3422 (1973).
13 W. Schreiner and A. Berndt, *Tetrahedron Letters*, **1973**, 3411.
14 D. Griller and K. U. Ingold, *J. Am. Chem. Soc.*, **95**, 6459 (1973).
15 See *Org. Reaction Mech.*, **1972**, 280–281.
16 M. C. R. Symons, *Chem. Phys. Letters*, **19**, 61 (1973).
17 See *Org. Reaction Mech.*, **1973**, 281.
18 J. Cooper, A. Hudson and R. A. Jackson, *Tetrahedron Letters*, **1973**, 831.
19 K. S. Chen, I. H. Elson and J. K. Kochi, *J. Am. Chem. Soc.*, **95**, 5341 (1973).
20 I. Biddles and A. Hudson, *Chem. Phys. Letters*, **18**, 45 (1973).
21 J. H. Hargis and P. B. Shevlin, *Chem. Comm.*, **1973**, 179 (1973).
22 I. Biddles, J. Cooper, A. Hudson, R. A. Jackson and J. T. Wiffen, *Mol. Phys.*, **25**, 225 (1973).
23 M. P. Doyle, P. W. Raynolds, R. A. Barents, T. R. Bade, W. C. Danen and C. T. West, *J. Am. Chem. Soc.*, **95**, 5988 (1973).
24 H. S. Chen, D. J. Edge and J. K. Kochi, *J. Am. Chem. Soc.*, **95**, 7036 (1973).
25 D. J. Edge and J. K. Kochi, *J. Am. Chem. Soc.*, **95**, 2635 (1973).
26 T. Kawamura and J. K. Kochi, *J. Organometal. Chem.*, **47**, 79 (1973).
27 S. P. Mishra, G. W. Neilson and M. C. R. Symons, *J.C.S. Faraday II*, **69**, 1425 (1973).
28 S. P. Mishra, G. W. Neilson and M. C. R. Symons, *J. Am. Chem. Soc.*, **95**, 605 (1973).
29 G. W. Neilson and M. C. R. Symons, *Chem. Comm.*, **1973**, 711.
30 L. Bonazzola, C. Hesse-Bezot and J. Roncin, *Chem. Phys. Letters*, **20**, 479 (1973).
31 A. Samuni, D. Behar and R. W. Fessenden, *J. Phys. Chem.*, **77**, 777 (1973).
32 R. V. Lloyd and M. T. Rogers, *J. Am. Chem. Soc.*, **95**, 1512 (1973).
33 L. D. Kispert, H. Liu and C. U. Pittmann, *J. Am. Chem. Soc.*, **95**, 1657 (1973).
34 B. C. Gilbert, J. P. Larkin and R. O. C. Norman, *J.C.S. Perkin II*, **1973**, 272
35 M. E. Jacox and D. E. Milligan, *J. Mol. Spectroscopy*, **47**, 148 (1973).
36 *Org. Reaction Mechanisms*, **1971**, 280.
37 G. Behrens and D. Schutte-Frohlinde, *Angew. Chem. Internat. Ed.*, **12**, 932 (1973).
38 S. O. Kobayashi and O. Simamura, *Chem. Letters (Tokyo)*, **1973**, 695.
39 S. O. Kobayashi and O. Simamura, *Chem. Letters (Tokyo)*, **1973**, 699.
40 A. J. Dobbs, B. C. Gilbert and R. O. C. Norman, *J. Mag. Resonance*, **11**, 100 (1973).
41 D. M. Camaioni and D. W. Pratt, *J. Am. Chem. Soc.*, **94**, 9258 (1972).
42 D. M. Camaioni, H. F. Walter and D. W. Pratt, *J. Am. Chem. Soc.*, **95**, 4057 (1973).
43 L. A. Singer and S. S. Kim, *Tetrahedron Letters*, **1973**, 1705.
44a R. C. Bingham and M. J. S. Dewar, *J. Am. Chem. Soc.*, **95**, 7180, 7182 (1973); C. U. Pittmann, L. D. Kispert and T. B. Patterson, *J. Phys. Chem.*, **77**, 494 (1973).
44b T. Ando, A. Yamishata, M. Matsumoto and T. Ishihara, *Chem. Letters (Tokyo)*, **1973**, 1133.
45 G. W. Neilson and M. C. R. Symons, *J.C.S. Perkin II*, **1973**, 1405.

[46] W. C. Danen and C. T. West, *J. Am. Chem. Soc.*, **95**, 6872 (1973).
[47] S. P. Mishra and M. C. R. Symons, *Tetrahedron Letters*, **1973**, 2267.
[48] Cf. *Org. Reaction Mech.*, **1972**, 282.
[49] T. Kawamura, T. Koyama and T. Yonezawa, *J. Am. Chem. Soc.*, **95**, 3226 (1973).
[50] See *Org. Reaction Mech.*, **1970**, 315.
[51] M. Gruselle and D. Lefort, *Tetrahedron*, **29**, 3035 (1973).
[52] J. K. Kochi, P. Bakusis and P. J. Krusic, *J. Am. Chem. Soc.*, **95**, 1516 (1973).
[53] D. V. Ioffe and L. S. Efros, *Reakts. Sposobast. Org. Soedin.*, **9**, 1165 (1972); *Chem. Abs.*, **79**, 77565 (1973).
[54] F. A. Neugebauer and W. R. Groh, *Tetrahedron Letters*, **1973**, 1005.
[55] A. R. Bassindale, A. J. Bowles, A. Hudson, R. A. Jackson, K. Schreiner and A. Berndt, *Tetrahedron Letters*, **1973**, 3185.
[56] K. Schreiner and A. Berndt, *Tetrahedron Letters*, **1973**, 4083.
[57] K. Schreiner, A. Berndt and F. Bauer, *Mol. Phys.*, **26**, 929 (1973).
[58] J. Tino and F. Szöcs, *Coll. Czech. Chem. Comm.*, **38**, 1343 (1973).
[59] R. Hittel, M. Rosner and D. Wagner, *Chem. Ber.*, **106**, 2767 (1973).
[60] K. W. Egger and A. T. Cocks, *Helv. Chim. Acta*, **56**, 1516, 1537 (1973).
[61] D. K. Sen Sharma and J. L. Franklin, *J. Am. Chem. Soc.*, **95**, 6562 (1973).
[62] W. von E. Doering and G. H. Beasley, *Tetrahedron*, **29**, 2231 (1973).
[63] A. S. Rodgers and M. C. R. Wu, *J. Am. Chem. Soc.*, **95**, 6913 (1973).
[64] Z. B. Alfassi and D. M. Golden, *Internat. J. Chem. Kinetics* **5**, 295 (1973); A. B. Trenwith, *J.C.S. Faraday II*, **69**, 1737 (1973).
[65] Z. B. Alfassi, D. M. Golden and S. W. Benson, *Internat. J. Chem. Kinetics*, **5**, 155 (1973).
[66] A. B. Trenwith, *Internat. J. Chem. Kinetics*, **5**, 67 (1973).
[67] *Org. Reaction Mech.*, **1970**, 320.
[68] H. D. Beckhaus and C. Rüchardt, *Tetrahedron Letters*, **1973**, 1971.
[69] R. H.-W. Wong and G. J. Gleicher, *J. Org. Chem.*, **38**, 1957 (1973).
[70] W. Duismann and C. Rüchardt, *Chem. Ber.*, **106**, 1083 (1973).
[71] I. Morishima, K. Yoshikawa, K. Bekki, M. Kohno and K. Arita, *J. Am. Chem. Soc.*, **95**, 5815 (1973).
[72] I. Morishima, K. Yoshikawa, K. Bekki, M. Kohno and K. Arita, *J. Am. Chem. Soc.*, **95**, 5815 (1973).
[73] A. Capiomont, *Acta Cryst.*, **B29**, 1720 (1973).
[74] A. Capiomont, B. Chiou and J. Lajzerowicz, *Acta Cryst.*, **B29**, 322 (1973).
[75] G. D. Mendenhall and K. U. Ingold, *J. Am. Chem. Soc.*, **95**, 6395 (1973).
[76] G. Cauquis, P.-J. Grossi, A. Rassat and D. Serve, *Tetrahedron Letters*, **1973**, 1863.
[77] H. G. Aurich, W. Deutsch and H. Forster, *Chem. Ber.*, **106**, 2854 (1973).
[78] H. G. Aurich, W. Deutsch, A. Lotz and H. Forster, *Chem. Ber.*, **106**, 2832 (1973).
[79] H. G. Aurich, A. Lotz and W. Weiss, *Chem. Ber.*, **106**, 2845 (1973).
[80] H. G. Aurich and J. Trösken, *Chem. Ber.*, **106**, 3483 (1973).
[81] R. West and P. Boukjouk, *J. Am. Chem. Soc.*, **95**, 3983 (1973).
[82] W. C. Danen, C. T. West and T. T. Kensler, *J. Am. Chem. Soc.*, **95**, 5716 (1973).
[83] B. E. Wagner, J. N. Helbert, R. D. Bates and E. H. Poindexter, *Chem. Comm.*, **1973**, 748.
[84] B. S. Tanaseichuk, *Khim. Geterotsikl. Soedin.*, **1972**, 1299; *Chem., Abst.*, **78**, 42293 (1973).
[85] F. A. Neugebauer, *Angew. Chem. Internat. Ed.*, **12**, 455 (1973).
[86] D. W. Williams, *Acta Cryst.*, **B29**, 96 (1973).
[87] F. A. Neugebauer, *Chem. Ber.*, **106**, 1716 (1973).
[88] V. Malatesta and K. U. Ingold, *J. Am. Chem. Soc.*, **95**, 6110 (1973).
[89] D. E. Wood, C. E. Wood and W. A. Latham, *J. Am. Chem. Soc.*, **94**, 9278 (1972).
[90] S. F. Nelson and R. T. Landis, *J. Am. Chem. Soc.*, **95**, 6454 (1973).
[91] A. T. Balaban and R. Istrăţoiu, *Tetrahedron Letters*, **1973**, 1879.
[92] S. Terabe and R. Konaka, *J.C.S. Perkin II*, **1973**, 369.
[93] W. Ahrens and A. Berndt, *Tetrahedron Letters*, **1973**, 4281.
[94] N. Negoiţa, R. Baican and A. T. Balaban, *Tetrahedron Letters*, **1973**, 1877.
[95] W. Ahrens, K. Weiser and A. Berndt, *Tetrahedron Letters*, **1973**, 3141.
[96] J. Brandt, G. Fauth, W. H. Franke and M. Zander, *Chem. Ber.*, **106**, 1711 (1973).
[97] M. C. R. Symons, *Tetrahedron*, **29**, 615 (1973).
[98] L. H. Sommer and L. A. Ulland, *J. Org. Chem.*, **37**, 3878 (1972).
[99] P. J. Davidson, A. Hudson, M. F. Lappert and P. W. Lednor, *Chem. Comm.*, **1973**, 829.
[100] M. Geoffroy, *Helv. Chim. Acta*, **56**, 1552 (1973).
[101] A. G. Davies, B. P. Roberts and B. R. Sanderson, *J.C.S. Perkin II*, **1973**, 626.
[102] J. R. Morton and K. F. Preston, *J. Phys. Chem.*, **77**, 2645 (1973).

103 S. E. Bates and R. D. Rieke, *J. Org. Chem.*, **37**, 3866 (1972).
104 R. F. C. Claridge and B. M. Peake, *Austral. J. Chem.*, **25**, 2353 (1972).
105 S. F. Nelsen and J. P. Gillespie, *J. Am. Chem. Soc.*, **95**, 2940 (1973).
106 S. F. Nelsen and J. P. Gillespie, *J. Am. Chem. Soc.*, **95**, 1874 (1973).
107 S. Winstein, G. Moshuk, R. Rieke and M. Ogliaruso, *J. Am. Chem. Soc.*, **95**, 2624 (1973).
108 L. B. Anderson, M. J. Broadhurst and L. A. Paquette, *J. Am. Chem. Soc.*, **95**, 2198 (1973).
109 H. Sakurai, H. Umino and A. Okad, *Chem. Letters (Tokyo)*, **1973**, 671.
110 B. T. Tabner and J. H. Zdysiewicz, *J.C.S. Perkin II*, **1973**, 811.
111 G. F. Pedulli, M. Tiecco, A. Alberti and G. Martelli, *J.C.S. Perkin II*, **1973**, 1816.
112 *Org. Reaction Mech.*, **1972**, 314.
113 F. Gerson, J. Heinzer and M. Stavaux, *Helv. Chim. Acta*, **56**, 1845 (1973).
114 J. R. Dodd, *Tetrahedron Letters*, **1973**, 3943.
115 J. D. Young, G. R. Stevenson and N. L. Bauld, *J. Am. Chem. Soc.*, **94**, 8790 (1972).
116 F. Gerson, J. Jachimowicz, B. Kowert and D. Leaver, *Helv. Chim. Acta*, **56**, 258 (1973); F. Gerson, J. Jachimowicz and D. Leaver, *J. Am. Chem. Soc.*, **95**, 6702 (1973).
117 S. Konishi, S. Niizuma, H. Kokuburi and M. Koizumi, *Bull. Chem. Soc. Japan*, **46**, 2279 (1973).
118 I. B. Goldberg, *Org. Mag. Res.*, **5**, 29 (1973).
119 C. Eischenbroich, F. Gerson and F. Stohler, *J. Am. Chem. Soc.*, **95**, 6956, (1973).
120 N. L. Bauld, C. S. Chang and J. H. Eilert, *Tetrahedron Letters*, **1973**, 153.
121 J. J. McDonnell and D. J. Pochopieu, *J. Org. Chem.*, **37**, 4064, (1972).
122 B. C. Gilbert and M. Trenwith, *J.C.S. Perkin II*, **1973**, 2010.
123 Y. Ikegami and S. Seto, *Chem. Letters (Tokyo)*, **1973**, 155.
124 R. G. Parrish, G. S. Hall and W. M. Gulick, *Mol. Phys.*, **26**, 1121 (1973).
125 S. Terabe and R. Konaka, *J. Am. Chem. Soc.*, **95**, 4976 (1973).
126 B. C. Webb, C. H. J. Wells and J. A. Wilson, *J.C.S. Perkin II*, **1973**, 681.
127 G. R. Stevenson, J. G. Concepcion and J. Castillo, *J. Phys. Chem.*, **77**, 611 (1973).
128 P. Kasai and D. McLeod, *J. Am. Chem. Soc.*, **94**, 6872 (1972); **95**, 27 (1973).
129 F. A. Neugebauer, S. Bamberger and W. R. Groh, *Tetrahedron Letters*, **1973**, 2247.
130 M. J. Drews, P. S. Wong and P. R. Jones, *J. Am. Chem. Soc.*, **94**, 9122 (1972).
131 P. Bruni, M. Colonna and L. Greci, *Tetrahedron*, **29**, 185 (1973).
132 E. Meyer, R. B. Girling and R. E. Hester, *Chem. Comm.*, **1973**, 192.
133 A. R. Lyons and M. C. R. Symons, *J. Am. Chem. Soc.*, **95**, 3483 (1973).
134 A. Nishinga, H. Hayashi and T. Matsuura, *Chem. Letters (Tokyo)*, **1973**, 141.
135 C. M. Camaggi, L. Lumazzi and G. Placucci, *J.C.S. Perkin I*, **1973**, 1491.
136 Ref. 1, Vol. I, Chapter 3.
137 P. S. Engel and C. Steel, *Accounts Chem. Res.*, **6**, 275 (1973).
138 *Org. Reaction Mech.*, **1971**, 287; **1972**, 291.
139 N. A. Porter and L. J. Marnett, *J. Am. Chem. Soc.*, **95**, 4361 (1973).
140 N. A. Porter and M. O. Funk, *Chem. Comm.*, **1973**, 263.
141 N. C. Baird and J. R. Swenson, *Can. J. Chem.*, **51**, 3097 (1973).
142 R. A. Jackson and S. Seltzer, *J. Am. Chem. Soc.*, **95**, 938 (1973).
143 R. Kerber and O. Nuyken, *Makromol. Chem.*, **164**, 183 (1973).
144 S. Yamashita and T. Hayakawa, *Bull. Chem. Soc. Japan.* **46**, 2290 (1973).
145 J. Hinz, A. Oberlinner and C. Rüchardt, *Tetrahedron Letters*, **1973**, 1975.
146 See *Org. Reaction Mech.*, **1972**, 292.
147 V. Malatesta and K. U. Ingold, *Tetrahedron Letters*, **1973**, 3311.
148 L. J. Marnett, P. Smith and N. A. Porter, *Tetrahedron Letters*, **1973**, 1081.
149 T. R. Lynch, F. N. MacLachlan and J. L. Suschitzky, *Can. J. Chem.*, **51**, 1378 (1973).
150 J. W. Timberlake, A. W. Garner and M. L. Hodges, *Tetrahedron Letters*, **1973**, 309.
151 W. Duisman and C. Rüchardt, *Chem. Ber.*, **106**, 1083 (1973).
152 Cf. ref. 143.
153 J. Hinz and C. Rüchardt, *Annalen*, **765**, 94 (1974).
154 J. Bonnekessel and C. Rüchardt, *Chem. Ber.*, **106**, 2890 (1973).
155 W. Ahrens and A. Berndt, *Angew. Chem. Internat. Ed.*, **12**, 655 (1973).
156 M. Kojima, N. Kamigata, H. Minato and M. Kobayashi, *Bull. Chem. Soc. Japan*, **46**, 2501 (1973).
157 E. L. Allred and K. J. Voorhees, *J. Am. Chem. Soc.*, **95**, 620 (1973).
158 *Org. Reaction Mech.*, **1970**, 321, 322.
159 A. B. Jaffe, K. J. Skinner and J. M. McBride, *J. Am. Chem. Soc.*, **94**, 8510 (1972).
160 A. B. Jaffe, D. S. Malament, E. P. Slisz and J. M. McBride, *J. Am. Chem. Soc.*, **94**, 8515 (1972).
161 H. Lui and J. Warkentin, *Can. J. Chem.*, **51**, 1148 (1973).
162 E. G. E. Hawkins in ref. 2, p. 95; E. G. E. Hawkins, *Angew. Chem. Internat. Edn.*, **12**, 783 (1973).

[163a] J. Bonnekessel and C. Rüchardt, *Chem. Ber.*, **106**, 2890 (1973).
[163b] Cf. p. **76**.
[164] C. Rüchardt and R. Pantke, *Chem. Ber.*, **106**, 2542 (1973).
[165] J. P. Eversum and J. C. DuBose, *J. Org. Chem.*, **38**, 3817 (1973).
[166] M. P. Doyle, P. W. Raynolds, R. A. Barents, T. R. Bade, W. C. Danen and C. T. West, *J. Am. Chem. Soc.*, **95**, 5988 (1973).
[167] See *Org. Reaction Mech.*, **1969**, 307.
[168] T. Koenig, *Tetrahedron Letters*, **1973**, 3487.
[169] K. Fujimori and S. Oae, *Tetrahedron*, **29**, 65 (1973).
[170] See *Org. Reaction Mech.*, **1972**, 288.
[171] I. S. Voloshanovskii, Yu. N. Anisimov and S. S. Ivanchev, *Zh. Obsch. Khim.*, **43**, 354 (1973); *Chem. Abs.*, **78**, 147078 (1973).
[172] W. Adam and R. Rucktäschel *J. Org. Chem.*, **37**, 4128 (1972).
[173] M. M. Martin, F. T. Hammer and E. Zador, *J. Org. Chem.*, **38**, 3422 (1973).
[174] M. M. Martin and J. M. King, *J. Org. Chem.*, **38**, 1588 (1973).
[175] Y. L. Fan and R. G. Shaw, *J.C.S. Perkin II*, **1973**, 1805.
[176] W. Adam and J. Sanabia, *Angew. Chem. Internat. Ed.*, **12**, 843 (1973).
[177] W. Adam and N. Duran, *J. Org. Chem.*, **38**, 1434 (1973).
[178] N. J. Turro and P. Lechtken, *Pure Appl. Chem.*, **33**, 363 (1973).
[179] Cf. p. **117**.
[180] L. Kim and K. C. Dewhirst, *J. Org. Chem.*, **38**, 2722 (1973).
[181] M. J. Gibian and R. C. Corkey, *Chem. Rev.*, **73**, 441 (1973).
[182] T. Koenig and H. Fischer in ref. 1, Vol. I, p. 157.
[183] A. D. Stepukjovich and V. I. Babayan, *Usp. Khim.*, 41, 1610 (1972); *Chem. Abst.* **78**, 3354 (1973).
[184] J. E. Bennet, L. H. Gale, E. J. Howard and B. Mile, *J.C.S.Faraday I*, **69**, 1655 (1973).
[185] H. J. Hefter, C-H. S. Wu and G. S. Hammond, *J. Am. Chem. Soc.*, **95**, 851 (1973).
[186] P. B. Ayscough and R. C. Sealy, *J.C.S. Perkin II*, **1973**, 543.
[187] E. J. Hamilton and H. Fischer, *J. Phys. Chem.*, **77**, 723 (1973).
[188] R. D. Burkhart and R. J. Wang, *J. Am. Chem. Soc.*, **95**, 7023 (1973).
[189] G. B. Watts, D. Griller and K. U. Ingold, *J. Am. Chem. Soc.*, **94**, 8784 (1972).
[190] P. P. Gaspar, A. D. Haizlip and K. Y. Choo, *J. Am. Chem. Soc.*, **94**, 9032 (1972).
[191] M. Ishikawa, A. Nakamura and M. Kumada, *J. Organometal. Chem.*, **59**, C11 (1973).
[192] J. R. Roberts and K. U. Ingold, *J. Am. Chem. Soc.*, **95**, 3228 (1973).
[193] F. K. Truby and J. K. Rice, *Internat. J. Chem. Kinetics*, **5**, 721 (1973).
[194] R. Hiatt and S. W. Benson, *Internat. J. Chem. Kinetics*, **5**, 385 (1973).
[195] D. C. Montague, *Internat. J. Chem. Kinetics*, **5**, 513 (1973).
[196] F. Bayrakceken, J. H. Brophy, R. D. Fink and J. E. Nicolas, *J.C.S. Faraday I*, **69**, 228 (1973).
[197] H. M. Frey and I. C. Vinall *Internat. J. Chem. Kinetics*, **5**, 523 (1973).
[198] R. F. Cullison, R. C. Pogue and M. L. White, *Internat. J. Chem. Kinetics*, **5**, 415 (1973).
[199] M. L. White and R. R. Kuntz, *Internat. J. Chem. Kinetics*, **5**, 187 (1973).
[200] A. Ts. Sarkisyan, *Arm. Khim. Zh.* **25**, 639 (1972); *Chem. Abs.*, **78**, 123659 (1973).
[201] K. O. MacFadden and C. L. Currie, *J. Chem. Phys.*, **58**, 1213 (1973).
[202] D. R. Tycholiz and A. R. Knight, *J. Am. Chem. Soc.*, **95**, 1726 (1973).
[203] C. S. Smith and A. R. Knight, *Can. J. Chem.*, **51**, 780 (1973).
[204] D. D. Carlson and A. R. Knight, *Can. J. Chem.*, **51**, 1410 (1973).
[205] T. L. Pollock, H. S. Sandhu, A. Jodham and O. P. Strausz, *J. Am. Chem. Soc.*, **95**, 1017 (1973).
[206] T. Yokota and R. A. Back, *Internat. J. Chem. Kinetics*, **5**, 37 (1973).
[207] G. A. Russell, in ref. 1, p. 275.
[208] M. V. Basilevskii and E. A. Trasman, *Russian Chem. Rev.* **1972**, 1.
[209] N. C. Deno, *Methods Free-Radical Chem.*, **3**, 135 (1972).
[210] T. C. Clark and J. E. Dove, *Can. J. Chem.*, **51**, 2155 (1973).
[211] P. D. Pacey, *Can. J. Chem.*, **51**, 2415 (1973).
[212] R. E. Berkley, I. Safarik, H. E. Gunning and O. P. Strausz, *J. Phys. Chem.*, **77**, 1734 (1973).
[213] E. D. Sprague, *J. Phys. Chem.*, **77**, 2066 (1973).
[214] W. A. Pryor, W. H. Davis and J. P. Stanley, *J. Am. Chem. Soc.*, **95**, 4754 (1973).
[215] R. E. Berkley, I. Safarik, O. P. Strausz and H. E. Gunning, *J. Phys. Chem.*, **77**, 1741 (1973).
[216] N. L. Arthur and B. R. Harman, *Austral. J. Chem.*, **26**, 1269 (1973).
[217] M. H. Arican, E. Potter and D. A. Whytock, *J.C.S. Faraday I*, **69**, 1811 (1973).
[218] I. B. Afanas'ev, N. G. Baranova and G. I. Samokhvalov, *Zh. Org. Khim.*, **8**, 2449 (1972); *Chem. Abs.*, **78**, 83575 (1973).
[219] I. B. Afanas'ev, N. G. Baranova, and G. I. Samokhvalov, *Internat. J. Chem. Kinetics*, **5**, 477 (1973).

220 J. I. G. Cadogan and G. A. Molina, *J.C.S. Perkin I*, **1973**, 541.
221 J. T. Hepinstall and J. A. Kampmeier, *J. Am. Chem. Soc.*, **95**, 1904 (1973).
222 I. A. Leenson, G. B. Sergeev, O. P. Shitov, S. L. Ioffe and V. A. Tartakovskii, *Izv. Akad. Nauk SSSR, Ser. Khim.*, **1973**, 1149; *Chem. Abs.*, **79**, 65447 (1973).
223 N. Mitsuo, T. Kunieda and T. Takizawa, *J. Org. Chem.*, **38**, 2255 (1973).
224 I. M. T. Davidson and A. B. Howard, *Chem. Comm.*, **1973**, 323.
225 G. C. Schatz, J. M. Bowman and A. Kupperman, *J. Chem. Phys.*, **58**, 4023 (1973).
226 R. L. Johnson, K. C. Kim and D. W. Setser, *J. Phys. Chem.*, **77**, 2499 (1973).
227 J. W. Bozzelli and M. Kaufman, *J. Phys. Chem.*, **77**, 1748 (1973).
228 T. L. Pollock and W. E. Jones, *Can. J. Chem.*, **51**, 2041 (1973).
229 R. L. Williams and F. S. Rowland, *J. Phys. Chem.*, **77**, 301 (1973).
230 M. A. A. Clyne and R. F. Walker, *J.C.S. Faraday I*, **69**, 1547 (1973).
231 C. J. Martens, J. A. Franklin, M. Godfroid, M. Delvaux and J. Verbeyst, *Internat. J. Chem. Kinetics*, **5**, 539 (1973).
232 K. H. Lee and T. O. Teo, *J.C.S. Perkin II*, **1973**, 689, 1617.
233 K. H. Lee, *J.C.S. Perkin II*, **1973**, 693.
234 D. D. Tanner and B. G. Brownlee, *Can. J. Chem.*, **51**, 3366 (1973).
235 J. D. van Drumpt, *Rec. Trav. Chim.*, **92**, 161 (1973).
236 K. C. Ferguson, E. N. Okafo and E. Whittle, *J.C.S. Faraday I*, **69**, 295 (1973).
237 J. H. Hargis, *J. Org. Chem.*, **38**, 346 (1973).
238 J. Ph. Soumillion, P. Dejaifre and A. Bruylants, *Tetrahedron Letters*, **1973**, 3615.
239 K. J. Shea and P. S. Skell, *J. Am. Chem. Soc.* **95**, 283 (1973).
240 P. S. Skell, R. R. Pavlis, D. C. Lewis and K. J. Shea, *J. Am. Chem. Soc.*, **95**, 6735 (1973).
241 D. S. Ashton, J. M. Tedder, J. C. Walton, A. Nechvatal and I. K. Stoddart, *J.C.S. Perkin I*, **1973**, 846.
242 D. S. Ashton, J. M. Tedder, M. D. Walker and J. C. Walton, *J.C.S. Perkin II*, **1973**, 1346.
243 D. S. Ashton, H. Singh, J. M. Tedder, J. C. Walton and E. A. Watt, *J.C.S. Perkin II*, **1973**, 125.
244 D. D. Tanner, J. E. Rowe, T. Pace and Y. Kosugi, *J. Am. Chem. Soc.*, **95**, 4705 (1973).
245 M. D. Loberg, K. A. Krohn and M. J. Walsh, *J. Am. Chem. Soc.*, **95**, 5496 (1973).
246 S. V. Adel'son, *Zh. Fiz. Khim.*, **46**, 1996 (1972); *Chem. Abs.*, **78**, 15207 (1973).
247 Z. B. Alfassi and D. M. Golden, *J. Am. Chem. Soc.*, **95**, 319 (1973).
248 R. Walsh and J. M. Wells, *Chem. Comm.*, **1973**, 513.
249 G. A. Takacs and G. P. Glass, *J. Phys. Chem.*, **77**, 1948 (1973).
250 A. Samuni and P. Neta, *J. Phys. Chem.*, **77**, 2425 (1973).
251 J.-Y. Godet, M. Pereyre, J.-C. Pommier and D. Chevolleau, *J. Organometal. Chem.*, **55**, C15 (1973).
252 E. Niki and Y. Kamiga, *J. Org. Chem.*, **38**, 1403 (1973).
253 J. A. Howard and J. H. B. Chenier, *J. Am. Chem. Soc.*, **95**, 3054 (1973).
254 J. C. Bevington, P. Hyde and M. Johnson, *Tetrahedron*, **29**, 399 (1973).
255 F. N. Mazitova and V. V. Abushaeva, *Izv. Akad. Nauk SSSR*, **1972**, 2449.
256 W. A. Pryor, G. Gojon and J. P. Stanley, *J. Am. Chem. Soc.*, **95**, 945 (1973).
257 D. D. Tanner, N. Wada and B. G. Brownlee, *Can. J. Chem.*, **51**, 1870 (1973).
258 G. E. Bullock and R. Cooper, *J.C.S. Faraday I*, **68**, 2185 (1973).
259 A. A. Baum, L. A. Karnischky, D. McLeod and S. H. Kasai, *J. Am. Chem. Soc.*, **95**, 617 (1973).
260 R. E. Rebbert, S. G. Lias and P. Ausloos, *Internat. J. Chem. Kinetics*, **5**, 893 (1973).
261 W. C. Danen, D. G. Saunders and K. A. Rose, *J. Am. Chem. Soc.*, **95**, 1612 (1973).
262 A. G. Davies, B. P. Roberts and J. M. Smith, *J.C.S. Perkin II*, **1972**, 2221.
263 P. Cadman, G. M. Tilsley and A. F. Trotman-Dickenson, *J.C.S. Faraday I*, **69**, 914 (1973).
264 H. G. Kuivila and C. H-C. Pian, *Tetrahedron Letters*, **1973**, 2561.
265 J. Cooper, A. Hudson and R. A. Jackson, *J.C.S. Perkin II*, **1973**, 1056.
266 D. A. Coates and J. M. Tedder, *J.C.S. Perkin II*, **1973**, 1570.
267 W. A. Pryor, T. H. Lin, J. P. Stanley and R. W. Henderson, *J. Am. Chem. Soc.*, **95**, 6993 (1973).
268 R. R. Baldwin and R. W. Walker, *J.C.S. Perkin II*, **1973**, 361.
269 Z. B. Alfassi and S. W. Benson, *Internat. J. Chem. Kinetics*, **5**, 879 (1973).
270 T. C. Clark and J. E. Dove, *Can. J. Chem.*, **51**, 2147 (1973).
271 J. Bosnjak, V. Andrejević, Ž. Čeković and M. Lj. Mihailović, *Tetrahedron*, **28**, 6031 (1972).
272 M. Lj. Mihailović, S. Gojković and S. Konstantinović, *Tetrahedron*, **29**, 3675 (1973).
273 J. Allen, R. B. Boar, J. F. McGhie and D. H. R. Barton, *J.C.S. Perkin I*, **1973**, 2402.
274 J. Y. Nedelec and D. Lefort, *Tetrahedron Letters*, **1973**, 5073.
275 S. Milosavljević, D. Jeremić and M. Lj. Mihailović, *Tetrahedron*, **29**, 3547 (1973).
276 E. G. E. Hawkins, *J.C.S. Perkin I*, **1973**, 2155.
277 K. Marujama and G. Takahashi, *Chem. Letters (Tokyo)*, **1973**, 295.

[278] K. Marujama, T. Otsuki, A. Takuwa and S. Arakawa, *Bull. Chem. Soc. Japan*, **46**, 2470 (1973).
[279] B. Black and H. Fischer, *Helv. Chim. Acta*, **56**, 506 (1973).
[280] S. M. Rosenfeld, R. G. Lawler and H. R. Ward, *J. Am. Chem. Soc.*, **95**, 946 (1973).
[281] H.-D. Becker, *J. Org. Chem.*, **32**, 2140 (1967).
[282] P. W. Atkins, J. M. Frimston, P. G. Frith, R. G. Gurd and K. A. McLauchlan, *J.C.S. Faraday II*, **69**, 1542 (1973).
[283] H. Yoshida, K. Hayashi and T. Warashina, *Bull. Chem. Soc. Japan*, **45**, 3515 (1972).
[284] F. D. Lewis and J. G. Magyar, *J. Am. Chem. Soc.*, **95**, 5973 (1973).
[285] D. I. Schuster and T. M. Weil, *J. Am. Chem. Soc.*, **95**, 4091 (1973).
[286] C. Gotzmer, K. F. Mueller and M. J. Cziesla, *J. Org. Chem.*, **38**, 2964 (1973).
[287] C. H. De Puy, H. L. Jones and W. M. Moore, *J. Am. Chem. Soc.*, **95**, 477 (1973).
[288] J. D. Coyle and G. Marr, *J. Organometal. Chem.*, **60**, 153 (1973).
[289] Y. Ogata, K. Takagi and Y. Fuji, *J. Org. Chem.*, **37**, 4026 (1972).
[290] I. Tabushi, S. Kojo and Z. Yoshida, *Tetrahedron Letters*, **1973**, 2329.
[291] P. J. Wagner and R. A. Leavitt, *J. Am. Chem. Soc.*, **95**, 3669 (1973).
[292] G. L. Closs and C. E. Doubleday, *J. Am. Chem. Soc.*, **94**, 9142 (1972).
[293] F. D. Lewis and R. W. Johnson, *J. Am. Chem. Soc.*, **94**, 8914 (1972).
[294] F. D. Lewis, R. W. Johnson and D. R. Kory, *J. Am. Chem. Soc.*, **95**, 6470 (1973).
[295] F. D. Lewis and R. W. Johnson, *Tetrahedron Letters*, **1973**, 2557.
[296] R. Breslow, S. Baldwin, T. Flechter, P. Kalicky, S. Liu and W. Washburn, *J. Am. Chem. Soc.*, **95**, 3251 (1973).
[297] See *Org. Reaction Mech.*, **1971**, 477.
[298] R. A. Cormier, W. H. Schreiber and W. C. Agosta, *J. Am. Chem. Soc.*, **95**, 4873 (1973).
[299] A. B. Smith and W. C. Agosta, *J. Am. Chem. Soc.*, **95**, 1961 (1973).
[300] F. A. Lamb and B. M. Vittimberga, *J. Org. Chem.*, **38**, 3520 (1973).
[301] J. Gloor, G. Bernardelli, R. Gerdil and K. Schaffner, *Helv. Chim. Acta*, **56**, 2520 (1973).
[302] N. Kito and A. Ohno, *Bull. Chem. Soc. Japan*, **46**, 2487 (1973).
[303] D. Dopp and E. Brugger, *Chem. Ber.*, **106**, 2166 (1973).
[304] A. Cu and A. C. Testa, *J. Phys. Chem.*, **77**, 1487 (1973).
[305] G. G. Wubbels, J. W. Jordan and N. S. Mills, *J. Am. Chem. Soc.*, **95**, 1281 (1973).
[306] V. Řehak, F. Novák and I. Čepčiansky, *Coll. Czech, Chem. Comm.*, **38**, 697 (1973).
[307] N. Baumann, *Helv. Chim. Acta*, **56**, 2227 (1973).
[309] J.-M. Surzur and M.-P. Bertrand, *Bull. Soc. Chim. France*, **1973**, 1861.
[310] V. A. Azovskaya and E. N. Prilezhaeva, *Russian Chem. Rev.*, **1972**, 516.
[311] Y. L. Chow, *Accounts Chem. Research*, **6**, 354 (1973).
[312] E. A. Hardwidge, B. S. Rabinovitch and R. C. Ireton, *J. Chem. Phys.*, **58**, 340 (1973).
[313] L. Teng and W. E. Jones, *J.C.S. Faraday I*, **69**, 189 (1973).
[314] R. Foon and G. P. Reid, *J. Phys. Chem.* **77**, 1193 (1973).
[315] Z. R. Alberto, J. J. Cosa and E. H. Staricco, *Internat. J. Chem. Kinetics*, **5**, 321 (1973).
[316] M.-C. Lasne, S. Masson and A. Thuillier, *Bull. Soc. Chim. France*, **1972**, 4592.
[317] J. J. Havel and P. S. Skell, *J. Org. Chem.*, **37**, 3744 (1972).
[318] J. C. Wazny and C. L. Bumgardner, *Tetrahedron Letters*, **1973**, 3763.
[319] G. L. Fleming, R. N. Haszeldine and A. E. Tipping, *J.C.S. Perkin I*, **1973**, 574.
[320] D. S. Ashton, A. F. Mackay, J. M. Tedder, D. C. Tipney and J. C. Walton, *Chem. Comm.*, **1973**, 496.
[321] J. P. Sloan, J. M. Tedder and J. C. Walton, *J.C.S. Faraday I*, **69**, 1143 (1973).
[322] I. Safarik and O. P. Strausz, *J. Phys. Chem.*, **76**, 3613 (1972).
[323] A. Horowitz and L. A. Rajbenbach, *J. Am. Chem. Soc.*, **95**, 6308 (1973).
[324] B. Boutevin and Y. Pietrasanta, *Tetrahedron Letters*, **1973**, 887.
[325] J.-P. Rabat and J.-L. Vernet, *Compt. Rendu*, **276**, 1699 (1973).
[326] Y. Mori and J. Tsuji, *Tetrahedron*, **29**, 827 (1973).
[327] E. Ghera, D. H. Perry and S. Shoua, *Chem. Comm.*, **1973**, 858.
[328] K. Takayama, M. Kosugi, and T. Migita, *Chem. Letters* (*Tokyo*), **1973**, 193.
[329] T. Dohmaru, Y. Nagata and J. Tsurugi, *Chem. Letters* (*Tokyo*), **1973**, 1031.
[330] D. Touchard and J. Lessard, *Tetrahedron Letters*, **1973**, 3827.
[331] D. H. Coy, R. N. Haszeldine, M. J. Newlands and A. E. Tipping, *J.C.S. Perkin II*, **1973**, 1062.
[332] R. E. Banks, J. A. Parker, M. J. Sharp and G. F. Smith, *J.C.S. Perkin I*, **1973**, 5.
[333] R. Brandon, R. N. Haszeldine and P. J. Robinson, *J.C.S. Perkin II*, **1973**, 1295, 1301.
[334] M. Simic, P. Neta and E. Hayon, *J. Phys. Chem.*, **77**, 2662 (1973).
[335] F. A. Hohorst, D. D. DesMarteau, L. R. Anderson, D. E. Gould and W. B. Fox, *J. Am. Chem. Soc.*, **95**, 3866 (1973).
[336] F. A. Hohorst and D. D. DesMarteau, *Chem. Comm.*, **1973**, 386.

337 T. S. Cantrell, *J. Am. Chem. Soc.*, **95**, 2714 (1973).
338 B. Maillard, M. Cazaux and R. Lalande, *Compt. Rend.*, **277**, 711 (1973).
339 D. J. Edge and J. K. Kochi, *J.C.S. Perkin II*, **1973**, 182.
340 J. Esser and J. A. Stone, *Can. J. Chem.*, **51**, 192 (1973).
341 D. W. Grattan, J. M. Locke and S. R. Wallis, *J.C.S. Perkin I*, **1973**, 2264.
342 J. K. Weseman, R. Williamson, J. L. Greene and P. B. Shevlin, *Chem. Comm.*, **1973**, 901.
343 B. N. Tyutyunnikov, Z. I. Bukhshtab, E. P. Yushchenko and L. V. Gasyuk, *Visn. Kharkiv. Politekh. Inst.*, **60**, 67 (1971); *Chem. Abs.*, **78**, 28836 (1973).
344 B. E. Smart, *J. Org. Chem.*, **38**, 2039 (1973).
345 B. E. Smart, *J. Org. Chem.*, **38**, 2027 (1973).
346 B. E. Smart, *J. Org. Chem.*, **38**, 2366 (1973).
347 B. E. Smart, *J. Org. Chem.*, **38**, 2035 (1973).
348 D. I. Davies, D. J. A. Pearce and E. C. Dart, *J.C.S. Perkin I*, **1973**, 433.
349 M. J. Parrott and D. I. Davies, *J.C.S. Perkin I*, **1973**, 2205.
350 H. C. Brown, J. H. Kawakami, and K.-T. Liu, *J. Am. Chem. Soc.*, **95**, 2209 (1973).
351 P. K. Freeman and R. S. Raghavan, *J. Org. Chem.*, **37**, 3670 (1973).
352 Y. L. Chow, S. C. Chen, and T. Majelsky, *Chem. Comm.*, **1973**, 827.
353 D. F. Howarth and A. G. Sherwood, *Can. J. Chem.*, **51**, 1655 (1973).
354 H. H. Evans, R. Fields, R. N. Haszeldine and M. Illingworth, *J.C.S. Perkin I*, **1973**, 649.
355 R. Mantione and H. Normant, *Bull. Soc. Chim. France*, **1973**, 2261.
356 L. R. Byrd and M. J. Caserio, *J. Org. Chem.*, **37**, 3881 (1972).
357 D. H. Coy, R. N. Haszeldine, M. J. Newlands and A. E. Tipping, *J.C.S. Perkin I*, **1973**, 1066.
358 M. L. Poutsma and P. A. Ibarbia, *J. Am. Chem. Soc.*, **95**, 6000 (1973).
359 M. V. Encina and E. A. Lissi, *J.C.S. Faraday I*, **69**, 1505 (1973).
360 J. Cooper, A. Hudson and R. A. Jackson, *J.C.S. Perkin II*, **1973**, 1933.
361 A. G. Brook and J. M. Duff, *Can. J. Chem.*, **51**, 352 (1973).
362a Y. Nagata, T. Dohmaru and J. Tsurugi, *J. Org. Chem.*, **38**, 795 (1973).
362b J.-Y. Godet and M. Pereyre, *Compt. Rend.* **277**, 211 (1973).
363a E. G. Janzen, C. A. Evans and Y. Nishi, *J. Am. Chem. Soc.*, **94**, 8236, 1972; see *Org. Reaction Mech.*, **1972**, 348.
363b M. J. Perkins and B. P. Roberts, *Chem. Comm.*, **1973**, 173.
364 J.-M. Surzur and M.-P. Bertrand, *Bull. Soc. Chim. France*, **1973**, 1861.
365 P. Piccardi, P. Massardo, M. Modena and E. Santoro, *J.C.S. Perkin I*, **1973**, 982.
366 M. A. M. Bradney, A. D. Forbes and J. Wood, *J.C.S. Perkin II*, **1973**, 1655.
367 N. O. Brace, *J. Org. Chem.*, **38**, 3167 (1973).
368 R. G. Petrova, V. V. Stepushkin and R. Kh. Freidlina, *Dokl. Akad. Nauk SSSR*, **202**, 125 (1972).
369 A. L. J. Beckwith and G. Phillipou, *Chem. Comm.*, **1973**, 280.
370 J. G. Traynham and H. H. Hsieh, *J. Org. Chem.*, **38**, 868 (1973).
371 J. V. Lallemand, M. Julia and D. Mansuy, *Tetrahedron Letters*, **1973**, 4461.
372 M. Barreau and M. Julia, *Tetrahedron Letters*, **1973**, 1537.
373 K. E. Koenig and Wm. P. Weber, *Tetrahedron Letters*, **1973**, 3151.
374 M.-P. Bertrand and J.-M. Surzur, *Bull. Soc. Chim. France*, **1973**, 2393.
375 R. Nouguier and J.-M. Surzur, *Bull. Soc. Chim. France*, **1973**, 2399.
376 M. Julia, J.-M. Salard and J.-C. Chottard, *Bull. Soc. Chim. France*, **1973**, 2479.
376a T. Migita, M. Kosugi, K. Takayama and Y. Nakagawa, *Tetrahedron*, **29**, 51 (1973).
377 Y. Watanabe and Y. Mizuhara, *Chem. Comm.*, **1973**, 752.
378 A. U. Ignatenko, A. U. Kessenikh, V. G. Glukhovtsev and M. A. Nadtochii, *Org. Mag. Res.*, **5**, 219 (1973).
379 B. H. Bakker, G. J. A. Schilder, T. Reints Bok, H. Steinberg and Th. J. de Boer, *Tetrahedron*, **29**, 93 (1973).
380 D. Steenken and D. Schulte-Frohlinde, *Tetrahedron Letters*, **1973**, 653.
381 P. Neta and C. L. Greenstock, *Chem. Comm.*, **1973**, 309.
382 C. L. Greenstock, I. Dunlop and P. Neta, *J. Phys. Chem.*, **77**, 1187 (1973).
383 B. C. Gilbert, J. P. Larkin and R. O. C. Norman, *J.C.S. Perkin II*, **1973**, 272.
384 K. Alwair and J. Grimshaw, *J.C.S. Perkin II*, **1973**, 1150.
385 K. Alwair and J. Grimshaw, *J.C.S. Perkin II*, **1973**, 1811.
386 J. G. Smith and J. Ho, *J. Org. Chem.*, **37**, 4260 (1972).
387 J. J. Zeilstra and J. B. F. N. Roberts, *Rec. Trav. chim.*, **92**, 954 (1973).
388 M. J. Perkins in ref. 1, Vol. II, Chapter 16.
389 R. O. C. Norman, *Chem. Ind.* (*London*), **1973**, 874.

[390] L. Eberson and K. Nyberg, *Accounts Chem. Research*, **6**, 106 (1973); M. Ya. Fioshin, L. A. Mirkind and M. Zh. Zhurinov, *Russian Chem. Rev.*, **42**, 293 (1973).
[391] W. A. Pryor, T. H. Lin, J. P. Stanley and R. W. Henderson, *J. Am. Chem. Soc.*, **95**, 6993 (1973).
[392] N. D. Epiotis, *J. Am. Chem. Soc.*, **95**, 3188 (1973).
[393] H. Zollinger, *Accounts Chem. Research.*, **6**, 335 (1973).
[394] A. H. Lewin and R. J. Michl, *J. Org. Chem.*, **38**, 1126 (1973).
[395] J. I. G. Cadogan, C. D. Murray and J. T. Sharp, *Chem. Comm.*, **1973**, 572.
[396] T. W. Bentley, J. A. John, R. A. W. Johnstone, P. J. Russell and L. H. Sutcliffe, *J.C.S. Perkin II*, **1973**, 1039.
[397] G. Vernin, H. J. M. Dou and J. Metzger, *J.C.S. Perkin II*, **1973**, 1093.
[398] G. Vernin, J. C. Poite, C. Riou, H. J. M. Dou and J. Metzger, *Bull. Soc. Chim. France*, **1973**, 1822.
[399] M. Kobayashi, H. Minato and N. Watanabe, *Bull. Chem. Soc. Japan*, **46**, 546 (1973).
[400] A. Clerici, F. Minisci and A. Porta, *Gazzetta*, **103**, 171 (1973).
[401] A. Clerici, F. Minisci and O. Porta, *Tetrahedron*, **29**, 2775 (1973).
[402] A. Arnone, M. Cecere, R. Galli, F. Minisci, M. Perchinunno, A. Porta and G. Gardini, *Gazzetta*, **103**, 13 (1973).
[403] R. Bernardi, T. Caronna, R. Galli, F. Minisci and M. Perchinunno, *Tetrahedron Letters*, **1973**, 643.
[404] G. Filippi, H. J. M. Dou, G. Vernin and J. Metzger, *J. Heterocyclic Chem.*, **10**, 259 (1973).
[405] G. Sundstrom, *Acta Chem. Scand.*, **27**, 600 (1973).
[406] C. L. Brett and V. Gold, *J.C.S. Perkin II*, **1973**, 1437.
[407] C. L. Brett, V. Gold and G. Perez, *J.C.S. Perkin II*, **1973**, 1450.
[408] C. L. Brett and V. Gold, *J.C.S. Perkin II*, **1973**, 1453.
[409] L. Benatti, C. M. Camaggi and G. Zanardi, *J.C.S. Perkin I*, **1972**, 2817.
[410] S. C. Dickermann, W. M. Feigenbaum, M. Fryd, N. Milstein, G. O. Vermont, I. Zimmerman and J. F. McOmie, *J. Am. Chem. Soc.*, **95**, 4624 (1973).
[411] M. Janda, J. Stogl, I. Stibor, M. Nemec and P. Vopatrna, *Tetrahedron Letters*, **1973**, 637.
[412] H. Eustathapoulos, J. Rinaudo and J. M. Bonnier, *Bull. Soc. Chim. France*, **1973**, 2377.
[413] S. J. Hammond and G. H. Williams, *J.C.S. Perkin II*, **1973**, 484.
[414] R. Bolton and J. P. B. Sandall, *Chem. Comm.*, **1973**, 286.
[415] R. Louw, G. in't Veld and W. Dorrepaal, *J.C.S. Perkin II*, **1973**, 650.
[416] I. N. Rozhkov, A. V. Bukhtiarov and I. L. Knunyants, *Bull. Acad. Sci. USSR*, **1972**, 1082.
[417] M. K. Eberhardt and M. Yoshida, *J. Phys. Chem.*, **77**, 589 (1973).
[418] O. S. Savel'eva, L. G. Shevchuk and N. A. Vysotsraya, *Zh. Org. Khim.*, **9**, 737 (1973); *Chem. Abs.*, **79**, 4645 (1973).
[419] T. Chiga, T. Kishimoto and E. Tomita, *J. Phys. Chem.*, **77**, 330 (1973).
[420] S. Nilsson, *Acta Chem. Scand.*, **27**, 329 (1973).
[421] L. Eberson, S. Nilsson and B. Rietz, *Acta Chem. Scand.*, **26**, 3870 (1973).
[422] T. Akiyama, A. Sugimori and H. Hermann, *Bull. Chem. Soc. Japan*, **46**, 1855 (1973).
[423] S. Hashimoto, W. Koike and Y. Ohakata, *Nippon Kagaku Kaishi*, **1973**, 986, 1139; *Chem. Abs.*, **79**, 31221, 65450 (1973).
[424] S. Hashimoto, T. Kimoto, T. Hayashi and H. Watanabe, *Nippon Kagaku Kaishi*, **1973**, 1143; *Chem. Abs.*, **79**, 77745 (1973).
[425] M. E. Kurz, E. S. Steele and R. L. Vecchio, *Chem. Comm.*, **1973**, 109.
[426] A. G. Davies and B. P. Roberts in ref. 1, Vol. I, p. 547.
[427] S. H. Ong in ref. 2, Vol. 10, p. 71.
[428] K. S. Chen, J-P. Battioni and J. K. Kochi, *J. Am. Chem. Soc.*, **95**, 4439 (1972).
[429] P. R. Jones and S. J. Costanzo, *J. Org. Chem.*, **38**, 3189 (1973).
[430] K. U. Ingold, *J.C.S. Perkin II*, **1973**, 420.
[431] See *Org. Reaction Mech.*, **1972**, 326.
[432] G. W. Kabalka and R. F. Daley, *J. Am. Chem. Soc.*, **95**, 4428 (1973).
[433] R. A. Jackson and M. Townson, *Tetrahedron Letters*, **1973**, 193.
[434] G. G. Maynes and D. E. Applequist, *J. Am. Chem. Soc.*, **95**, 856, (1973).
[435] K. J. Shea and P. S. Skell, *J. Am. Chem. Soc.*, **95**, 6728 (1973).
[436] R. F. Drury and L. Kaplan, *J. Am. Chem. Soc.*, **95**, 2217 (1973).
[437] M. S. Newman, G. S. Cohen, R. F. Cunico and L. W. Dauernheim, *J. Org. Chem.*, **38**, 2760 (1973).
[438] T. Migita, M. Kosugi, K. Takayama and Y. Nakagawa, *Tetrahedron* **29**, 51 (1973).
[439] M. Kosugi, K. Kurino, K. Takayama and T. Migita, *J. Organometal. Chem.*, **56**, C11 (1973).
[440] J. Cooper, A. Hudson and R. A. Jackson, *J.C.S. Perkin II*, **1973**, 1056.
[441] A. G. Davies and J. C. Scaiano, *J.C.S. Perkin II*, **1973**, 1777.
[442] A. G. Davies, B. P. Roberts and J. M. Smith, *J.C.S. Perkin II*, **1972**, 2221.

443 J. Hollaender, W. P. Neumann and H. Lind, *Chem. Ber.*, **106**, 2395 (1973).
444 P. J. Krusic and P. Meakin, *Chem. Phys. Letters*, **18**, 347 (1973).
445 A. G. Davies, R. W. Dennis, D. Griller, K. U. Ingold and B. P. Roberts, *Mol. Phys.*, **26**, 989 (1973).
446 D. Griller and B. P. Roberts, *J.C.S. Perkin II*, **1973**, 1339.
447 D. Griller and B. P. Roberts, *J.C.S. Perkin II*, **1973**, 1416.
448 W. G. Bentrude, E. R. Hansen, W. A. Khan, T. B. Min and P. E. Rogers, *J. Am. Chem. Soc.*, **95**, 2286 (1973).
449 W. G. Bentrude, J. L. Fu and P. E. Rogers, *J. Am. Chem. Soc.*, **95**, 3625 (1973).
450 A. G. Davies, D. Griller and B. P. Roberts, *J.C.S. Perkin II*, **1973**, 2224.
451 E. Furimsky and J. A. Howard, *J. Am. Chem. Soc.*, **95**, 369 (1973).
452 Ya. A. Levin, A. V. Il'yasov, E. I. Gol'dfarb and E. I. Vorkanova, *Izvest. Akad. Nauk SSSR* (**1972**), 1624.
453 E. Furimsky, J. A. Howard and J. R. Morton, *J. Am. Chem. Soc.* **95**, 6574 (1973).
454 D. J. Cardin, M. F. Lappert and P. W. Lednor, *Chem. Comm.*, **1973**, 350.
455 N. G. Hargreaves, R. J. Puddephatt, L. H. Sutcliffe and P. J. Thompson, *Chem. Comm.*, **1973**, 861.
456 J. Fossey, *Tetrahedron Letters*, **1973**, 1127.
457 U. Christen and W. P. Neumann, *Chem. Ber.*, **106**, 421 (1973).
458 E. J. Hamilton and H. Fischer, *Helv. Chim. Acta*, **56**, 795 (1973).
459 L. Farnell and W. G. Richards, *Chem. Comm.*, **1973**, 334.
460 W. J. Hehre, *J. Am. Chem. Soc.*, **95**, 2643 (1973).
461 M. P. Doyle, P. W. Raynolds, R. A. Barents, T. R. Bade, W. C. Danen and C. T. West, *J. Am. Chem. Soc.*, **95**, 5988 (1973).
462 G. W. Schaffer, *J. Org. Chem.*, **38**, 2842 (1973).
463 M. Oku and J. C. Philips, *J. Am. Chem. Soc.*, **95**, 6495 (1973).
464 See *Org. Reaction Mech.*, **1969**, 340, 341.
465 T. A. B. M. Bolsman and T. J. de Boer, *Tetrahedron*, **29**, 3579 (1973).
466 J. K. Haynes and J. A. Kampmeier, *J. Org. Chem.*, **37**, 4167 (1972).
467 M. Julia and D. Mansuy, *Compt. Rend.(C)*, **276**, 1049 (1973).
468 A. L. J. Beckwith and C. B. Thomas, *J. C. S. Perkin II*, **1973**, 861.
469 J. Cooper, A. Hudson and R. A. Jackson, *Tetrahedron Letters*, **1973**, 831.
470 K. S. Chen, I. H. Elson and J. K. Kochi, *J. Am. Chem. Soc.*, **95**, 5341 (1973).
471 P. S. Skell, R. R. Pavlis, D. C. Lewis and K. J. Shea, *J. Am. Chem. Soc.*, **95**, 6735 (1973).
472 Cf. pp. **63** and **86**.
473 R. M. Carman and B. N. Venzke, *Austral. J. Chem.*, **26**, 571 (1973).
474 D. S. Ross, J. Mill and M. B. Hill, *J. Am. Chem. Soc.*, **94**, 8776 (1972).
475 R. West and P. Boukjouk, *J. Am. Chem. Soc.*, **95**, 3983 (1973).
476 A. Hassner and D. J. Blythin, *J. Org. Chem.*, **37**, 4209 (1972).
477 G. A. Russell and K. Schmitt, *J. Am. Chem. Soc.*, **94**, 8918 (1973).
478 N. C. Young, P. Kumler and S. S. Yang, *J. Org. Chem.*, **37**, 4022 (1972).
479 W. Adam, J. A. de Sanabia and H. Fischer, *J. Org. Chem.*, **38**, 2571 (1973).
480 Y. Katsuhara, H. Maruyama, Y. Shigemitsu and Y. Odaira, *Tetrahedron Letters*, **1973**, 1323.
481 W. Adam, H. Fischer, H.-J. Hansen, H. Heimgarten, H. Schmid and H.-R. Waespe, *Angew. Chem. Internat. Ed.*, **12**, 662 (1973).
482 P. D. Hobbs and P. D. Magnus, *J.C.S. Perkin I*, **1973**, 469.
483 J. R. Grunwell, N. A. Marron and S. L. Hankan, *J. Org. Chem.*, **38**, 1559 (1973).
484 K. Praefke, *Tetrahedron Letters*, **1973**, 973.
485 C. Brown, R. F. Hudson and A. J. Lawson, *J. Am. Chem. Soc.*, **95**, 6501 (1973).
486 See *Org. Reaction Mech.*, **1972**, 325.
487 J. F. Garst and C. D. Smith, *J. Am. Chem. Soc.*, **95**, 6807 (1973).
488 J. P. Lorand, R. W. Grant, P. A. Samuel, E. M. O'Connell, J. Zaro, J. Pilotte and R. W. Wallace, *J. Org. Chem.*, **38**, 1813 (1973).
489 W. D. Ollis, I. O. Sutherland and Y. Thebtaranonth, *Chem. Comm.*, **1973**, 654.
490 A. F. Hegarty, J. A. Kearney and F. L. Scott, *J.C.S. Perkin II*, **1973**, 1422.
491 D. B. Pendergrass, D. Y. Curtin and I. C. Paul, *J. Am. Chem. Soc.*, **94**, 8722 (1972).
492 T. S. Dobashi and E. J. Grubbs, *J. Am. Chem. Soc.*, **95**, 5070 (1973).
493 L. Salem, *Pure Appl. Chem.*, **33**, 317 (1973).
494 L. M. Stephenson, T. A. Gibson and J. J. Brauman, *J. Am. Chem. Soc.*, **95**, 2849 (1973).
495 R. S. Givens and W. T. Oettle, *J. Org. Chem.*, **37**, 4325 (1972).
496 D. R. Dice and R. P. Steer, *Chem. Comm.*, **1973**, 106.
497 A. B. Chmurny and D. J. Cram, *J. Am. Chem. Soc.*, **95**, 4237 (1973).

[498] J. J. Gajewski and L. T. Burka, *J. Am. Chem. Soc.*, **94**, 8865 (1972).
[499] A. J. Ashe, *J. Am. Chem. Soc.*, **95**, 818 (1973).
[500] J. M. Simpson and H. G. Richey, *Tetrahedron Letters*, **1973**, 2545.
[501] A. T. Cocks and K. W. Egger, *J.C.S. Perkin II*, **1973**, 199.
[502] J. A. Berson and P. B. Dervan, *J. Am. Chem. Soc.*, **95**, 267 (1973).
[503] K.-P. Zeller, H. Meier and E. Müller, *Annalen*, **766**, 32 (1972).
[504] E. N. Cain and R. K. Jolly, *J. Am. Chem. Soc.*, **95**, 4791 (1973).
[505] H.-D. Martin and M. Hekmann, *Angew. Chem. Internat. Ed.*, **12**, 572 (1973).
[506] S. F. Nelsen and J. P. Gillespie, *J. Am. Chem. Soc.*, **95**, 1874 (1973).
[507] J. R. Scheffer and R. A. Wostradowski, *J. Org. Chem.*, **37**, 4317 (1972).
[508] D. R. Morton and N. J. Turro, *J. Am. Chem. Soc.* **95**, 3946 (1973).
[509] R. L. Cargill, A. B. Sears, J. Boehm and M. R. Willcott, *J. Am. Chem. Soc.*, **95**, 4346 (1973).
[510] J. Wirz, *J.C.S. Perkin II*, **1973**, 1307.
[511] W. R. Roth and G. Erker, *Angew. Chem. Internat. Ed.*, **12**, 503 (1973).
[512] W. R. Roth, M. Heiber and G. Erker, *Angew. Chem. Internat. Ed.*, **12**, 504 (1973).
[513] W. R. Roth and G. Erker, *Angew. Chem. Internat. Ed.*, **12**, 505 (1973).
[514] W. Grimme and H.-J. Rother, *Angew. Chem. Internat. Ed.*, **12**, 505 (1973).
[515] P. Claus, T. Doppler, N. Gakis, M. Georgarakis, H. Geizendanner, P. Gilgen, H. Heimgastner, B. Jackson, M. Marky, N. S. Narasimhan, H. J. Rosenkranz, A. Wunderli, H.-J. Hansen and H. Schmid, *Pure Appl. Chem.*, **33**, 339 (1973).
[516] G. D. Mendenhall and K. U. Ingold, *J. Am. Chem. Soc.*, **95**, 6390, 6395 (1973).
[517] A. Calder, A. R. Forrester and S. P. Hepburn, *J.C.S. Perkin I*, **1973**, 456.
[518] R. L. Craig, J. Murray-Rust, P. Murray-Rust and J. S. Roberts, *Chem. Comm.*, **1973**, 751.
[519] R. E. Banks, D. R. Choudhury and R. N. Haszeldine, *J.C.S. Perkin I*, **1973**, 80.
[520] R. E. Banks, D. R. Choudhury and R. N. Haszeldine, *J.C.S. Perkin I*, **1973**, 1092.
[521] H. G. Aurich and W. Weiss, *Chem. Ber.*, **106**, 2408 (1973).
[522] L. Call and E. F. Ullman, *Tetrahedron Letters*. **1973**, **961**.
[523] V. Malatesta and K. U. Ingold, *J. Am. Chem. Soc.*, **95**, 6404 (1973).
[524] S. Terabe, K. Kuruma and R. Konaka, *J.C.S. Perkin II*, **1973**, 1252.
[525] S. Terabe and S. Konaka, *J.C.S. Perkin II*, **1973**, 369.
[526] R. A. Abramovitch and S. R. Challard, *J. Heterocyclic Chem.*, **10**, 683 (1973).
[527] G. A. Akakumov, V. K. Cherkasov and I. D. Petrovskaya, *Izv. Akad. Nauk SSSR, Ser. Khim.*, **1973**, 1341; *Chem. Abs.*, **79**, 104531 (1973).
[528] E. G. Janzen, T. Kasai and K. Kuwata, *Bull. Chem. Soc. Japan*, **46**, 2061 (1973).
[529] D. H. R. Barton, I. A. Blair, P. D. Magnus and R. K. Norris, *J.C.S. Perkin I*, **1973**, 1031.
[530] D. H. R. Barton, I. A. Blair, P. D. Magnus, and R. K. Norris, *J.C.S. Perkin I*, **1973**, 1031.
[531] M. Mohammed, *Austral. J. Chem.*, **26**, 229 (1973).
[532] See *Org. Reaction Mech.*, **1971**, 297.
[533] E. M. Kosower, A. Teuerstein and A. J. Swallow, *J. Am. Chem. Soc.*, **95**, 6127 (1973).
[534] Yu. A. Alexandrov, *J. Organometal. Chem.*, **55**, 1 (1973).
[535] O. N. Karpukhin and E. M. Slobodeslkaya, *Russian Chem. Rev.*, **1973**, 173.
[536] E. T. Denisov, *Russian Chem. Rev.*, **1973**, 157.
[537] A. I. Poroikova, *Vses. Konf. Kinet. Mekh. Gazofazn. Reakts., 2nd*, **1971**, 13; *Chem. Abs.*, **78**, 15255 (1973).
[538] C. H. Spicer, A. Villa, H. A. Wiebe and J. Heicklen, *J. Am. Chem. Soc.*, **95**, 13 (1973).
[539] M. A. Beibutyan, T. A. Garibyan, R. R. Grigoryan, A. A. Mantashyan and A. S. Saakyan, *Vses. Konf. Kinet. Mekh. Gazofaz. Reakts., 2nd*, **1971**, 5; *Chem. Abs.*, **78**, 15219 (1973).
[540] R. R. Baldwin and R. W. Walker, *Combust. Flame*, **21**, 55 (1973); *Chem. Abs.*, **79**, 104554 (1973).
[541] D. L. Allara, T. Mill, F. R. Mayo, H. Richardson and K. Irwin, *Am. Chem. Soc., Div. Petrol., Chem. Prepr.*, **16**, B31 (1971); *Chem. Abs.*, **78**, 15279 (1973).
[542] C. F. Cullis and S. C. W. Hook, *J.C.S. Perkin II*, **1973**, 803.
[543] M. M. Midland and H. C. Brown, *J. Am. Chem. Soc.*, **95**, 4069 (1973).
[544] T. Sato, K. Hibino, N. Fukumaru, and T. Otsu, *Chem. Ind.* (*London*), **1973**, 745.
[545] T. Mill and G. Montorsi, *Am. Chem. Soc., Div. Petrol., Chem. Prepr.*, **16**, B23 (1971); *Chem. Abs.*, **78**, 15265 (1973).
[546] T. Mill and G. Montorsi, *Internat. J. Chem. Kinetics*, **5**, 119 (1973).
[547] L. G. Galimova, V. D. Komissarov, and E. T. Danisov, *Izv. Akad. Nauk. SSSR, Ser. Khim.*, **1973**, 307; *Chem. Abs.*, **79**, 4689 (1973).
[548] S. Korcek, J. H. B. Chenier, J. A. Howard and K. U. Ingold, *Can. J. Chem.*, **50**, 2285 (1972).
[549] J. A. Howard, *Can. J. Chem.*, **50**, 2298 (1972).

550 J. A. Howard, Y. Ohkatsu, J. H. B. Chenier and K. U. Ingold, *Can. J. Chem.*, **51**, 1543 (1973).
551 A. E. Klein and N. Pilpel, *J.C.S. Faraday I*, **69**, 1729 (1973).
552 G. Montorsi, G. Caprara, G. Pregaglia and G. Messina, *Internat. J. Chem. Kinetics*, **5**, 777 (1973).
553 D. J. M. Ray, A. Redfern and D. J. Waddington, *J.C.S. Perkin II*, **1973**, 540.
554 E. Bordier, *Bull. Soc. Chim. France*, **1973**, 2621.
555 I. A. Opeida, R. V. Kucher and N. P. Mytsyk, *Neftekhimiya*, **13**, 235 (1973); *Chem. Abs.*, **79**, 41671 (1973).
556 Y. Ogata and M. Haba, *J. Org. Chem.*, **38**, 2779 (1973).
557 J. E. Bennett and J. A. Howard, *J. Am. Chem. Soc.*, **95**, 4008 (1973).
558 D. Lindsay, J. A. Howard, E. C. Horswill, L. Iton, K. U. Ingold, "T. Cobbley and A. Ll," *Can. J. Chem.*, **51**, 870 (1973).
559 A. R. Griva and E. T. Denisov, *Internat. J. Chem. Kinetics*, **5**, 869 (1973).
560 A. Padwa and L. Brodsky, *Tetrahedron Letters*, **1973**, 1045.
561 T. V. Finippova, E. A. Blyumberg, L. I. Kas'yan, Ya. L. Letuchii and L. A. Sil'chenko, *Dokl. Akad. Nauk SSSR*, **210**, 644 (1973); *Chem. Abs.*, **79**, 52570 (1973).
562 H. D. Burrows, T. J. Kemp and M. J. Welbourn, *J.C.S. Perkin II*, **1973**, 969.
563 J. A. Blair and A. J. Pearson, *Tetrahedron Letters*, **1973**, 1681.
564 J. A. Blair and A. J. Pearson, *Tetrahedron Letters*, **1973**, 203.
565 B. Robinson and M. V. Zubair, *Tetrahedron*, **29**, 1429 (1973).
566 S. F. Nelson and R. T. Landis, *J. Am. Chem. Soc.*, **95**, 2719 (1973).
567 R. Caputo, L. Mangoni, L. Previtera and R. Iaccario, *Tetrahedron*, **29**, 2047 (1973).
568 N. F. Usacheva, Yu. G. Oranskii and M. S. Khaikin, *Zh. Nauch. Prikl. Fotogr. Kinematogr.*, **17**, 416 (1972); *Chem. Abs.*, **78**, 71036 (1973).
569 N. Sutin and J. K. Yandell, *J. Am. Chem. Soc.*, **95**, 4847 (1973).
570 A. Hettche and K. Dimroth, *Chem. Ber.*, **106**, 1001 (1973).
571 Y. Ogata and M. Yamashita, *Bull. Chem. Soc. Japan*, **46**, 2208 (1973).
572 J. A. Massen and Th. J. de Boer, *Rec. Trav. chim.*, **92**, 185 (1973).
573 C. R. H. I. de Jonge, H. J. Hageman, W. G. B. Huysmens and W. J. Mÿs, *J.C.S. Perkin II*, **1973**, 1276.
574 J. A. Pedersen, *J.C.S. Perkin II*, **1973**, 424.
575 J. Lubach and W. Drenth, *Rec. Trav. chim.*, **92**, 403 (1973).
576 E. J. Panek and F. M. Whitesides, *J. Am. Chem. Soc.*, **94**, 8768 (1972).
577 V. N. Glushakova, Yu. A. Alexandrov, and G. A. Razuvaev, *J. Organometal. Chem.*, **40**, 43 (1972).
578 J. K. Kochi and ref. 1, Vol. I, Chapter 11.
579 J. S. Littler in ref. 2, p. 237.
580 T.-L. Ho, *Synthesis*, **1973**, 347.
581 W. G. Nigh in *Oxidation in Organic Chemistry, Part B* (ed. W. S. Trahanovsky), Academic, New York, 1973, p. 1.
582 H.-G. Gilde in *Methods in Organic Chemistry*, Vol. III (ed. E. S. Huyser), Dekker, New York, 1973, p. 1.
583 F. R. Hewgill in ref. 2, p. 167.
584 P. D. McDonald and G. A. Hamilton in ref. 581, p. 97.
585 M. Siskin in ref. 582, p. 83.
586 J. Hanotier, M. Hanotier-Bridoux and P. de Radzitsky *J.C.S. Perkin II* **1973**, 381.
587 J. Hanotier and M. Hanotier-Bridoux, *J.C.S. Perkin II*, **1973**, 1035.
588 Y. Kamiya and M. Kashima, *Bull. Chem. Soc. Japan*, **46**, 905 (1973).
589 M. Hronec and V. Vesely, *Coll. Czech. Chem. Comm.*, **38**, 1095, 1227 (1973); *Chem. Zvesti*, **27**, 94 (1973).
590 D. I. Dimitrov, A. D. Stetanova and L. A. Philipov, *Monatsh.*, **104**, 579 (1973).
591 J. K. Kochi, R. T. Tang and T. Bernath, *J. Am. Chem. Soc.*, **95**, 7114 (1973).
592 I. H. Elson and J. K. Kochi, *J. Am. Chem. Soc.*, **95**, 5060 (1973).
593 R. O. C. Norman, C. B. Thomas and J. S. Willson, *J.C.S. Perkin I*, **1973**, 325.
593a Cf. K. Nyberg, *Acta Chem. Scand.*, **24**, 1609 (1970).
594 K. Nyberg, *Acta Chem. Scand.*, **27**, 503 (1973).
595 L. Eberson, K. Nyberg and H. Sternerup, *Acta Chem. Scand.*, **27**, 1679 (1973).
596 A. Ronlán, O. Hammerich and V. D. Parker, *J. Am. Chem. Soc.*, **95**, 7132 (1973).
597 W. S. Trahanovsky and D. W. Brixius, *J. Am. Chem. Soc.*, **95**, 6778 (1973).
598 S. R. Bansal, D. C. Nonhebel and J. M. Mancilla, *Tetrahedron*, **29**, 993 (1973).
599 A. D. Mosnaim, M. E. Wolf, I. Saavedra, A. M. Amaro, G. Cordano and D. C. Nonhebel, *Tetrahedron Letters*, **1973**, 1491.

[600] D. L. Fields, T. H. Regan and D. P. Maier, *J. Org. Chem.*, **38**, 407 (1973).
[601] N. L. Weinberg and B. Belleau, *Tetrahedron*, **29**, 279 (1973).
[602] K. Sasaki and H. Ohishi, *Electrochim. Acta*, **17**, 2391 (1972).
[603] A. Onopchenko and J. G. D. Schulz, *J. Org. Chem.* **38**, 3729 (1973).
[604] M. P. J. Brennan and R. Brettle, *J.C.S. Chem. Comm.*, **1973**, 257.
[605] F. Brunox and J.-E. Dubois, *Bull. Soc. Chim. France*, **1973**, 2270.
[606] L. Eberson and G. Ryde-Petterson, *Acta Chem. Scand.*, **27**, 1159 (1973).
[607] G. E. Hawkes, J. H. P. Uttley and G. B. Yates, *Chem. Comm.*, **1973**, 305.
[608] Y. Matsuda, K. Kimura, C. Iwakura and H. Tamura, *Bull. Chem. Soc. Japan*, **46**, 430 (1973).
[609] R. Dayal and G. U. Bakore, *Indian J. Chem.*, **10**, 1165 (1972).
[610] B. Danieli, P. Manitto, F. Ronchetti and G. Russo, *Chem. Ind.* (*London*), **1973**, 430.
[611a] B. Danieli, F. Ronchetti and G. Russo, *Chem. Ind.* (*London*), **1973**, 1067.
[611b] I. J. Miller and G. J. Smith, *Tetrahedron Letters*, **1973**, 2277.
[612] K. V. Sarkanen and A. F. A. Wallis, *J.C.S. Perkin I*, **1973**, 1869.
[613] K. V. Sarkanen and A. F. A. Wallis, *J.C.S. Perkin I*, **1973**, 1878.
[614] A. F. A. Wallis, *Austral. J. Chem.*, **26**, 1571 (1973).
[615] A. F. A. Wallis, *Austral. J. Chem.*, **26**, 585 (1973).
[616] I. J. Miller, *Tetrahedron Letters*, **1972**, 4955.
[617] E. Mauda, *Bull. Chem. Soc. Japan*, **46**, 2160 (1973).
[618] A. Nilsson, A. Ronlán and V. D. Parker, *J.C.S. Perkin I*, **1973**, 2337.
[619] G. Popp and N. C. Reitz, *J. Org. Chem.*, **37**, 3646 (1972).
[620] H.-P. Husson, C. Poupat, B. Rodriguez and P. Potier, *Tetrahedron*, **29**, 1405 (1973).
[621] D. A. Bolon, *J. Org. Chem.*, **38**, 1471 (1973).
[622] M. J. Leigh and J. A. Strickson, *J.C.S. Perkin II*, **1973**, 1476.
[623] M. A. Schwartz, B. F. Rose and B. Vishnuvajjala, *J. Am. Chem. Soc.*, **95**, 612 (1973).
[624] K. I. Ivanov, T. N. Kulikovskaya, V. K. Savinova, E. S. Panfilova, V. P. Zhakovskay and M. G. Semenova, *Proc. Acad. Sci.*(*USSR*), **205**, 626 (1972).
[625] S. G. Kukes, N. N. Bubnov, A. I. Prokof'ev, S. P. Solodovnikov, A. A. Volod'kin, G. A. Nikiforov and V. V. Ershov, *Izv. Akad Nauk SSSR, Ser. Khim.*, **1973**, 684; *Chem. Abs.*, **79**, 4632 (1973).
[626] V. S. Martem'yanov and E. T. Denisov, *Izv. Akad. Nauk SSSR, Ser. Khim.*, **1972**, 2191; *Chem. Abs.*, **78**, 57420 (1973).
[626a] See p. **143**.
[627] A. Ledwith, P. J. Russell and L. H. Sutcliffe, *Proc. Roy. Soc., A.*, **332**, 151 (1973).
[628] A. Ledwith, P. J. Russell and L. H. Sutcliffe, *J.C.S. Perkin II*, **1973**, 630.
[629a] W. S. Trahanovsky and D. B. Macaulay, *J. Org. Chem.*, **38**, 1497 (1973).
[629b] See p. **90**.
[630] M. M. Green, J. M. Moldowan and J. G. Mcgrew, *Chem. Comm.*, **1973**, 451.
[631] J. Roček and A. E. Radkowsky, *J. Org. Chem.*, **38**, 89 (1973).
[632] J. Roček and A. E. Radkowsky, *J. Am. Chem. Soc.*, **95**, 7123 (1973).
[633] L. Syper, *Rocz. Chem.*, **47**, 43 (1973).
[634] W. S. Trahanovsky, J. R. Gilmore and P. C. Heaton, *J. Org. Chem.*, **38**, 760 (1973).
[635] P. G. Sant, *Vijnana Parishad Anusandhan Patrika*, **15**, 151 (1972); *Chem. Abs.*, **78**, 135352 (1973).
[636] S. E. Schaafsma, R. Jorritsma, H. Steinberg and T. J. de Boer, *Tetrahedron Letters*, **1973**, 827.
[637] H. A. Connon, B. G. Sheldon, K. E. Harding, L. E. Letterman, D. C. Fulton and W. G. Nigh, *J. Org. Chem.*, **38**, 2020 (1973).
[638] S. D. Ross, J. E. Barry, M. Finkelstein and E. J. Rudd, *J. Am. Chem. Soc.*, **95**, 2193 (1973).
[639] J. R. Lindsay Smith and L. A. V. Mead, *J.C.S. Perkin II*, **1973**, 206.
[640] J. R. Lindsay Smith, R. O. C. Norman and A. G. Rowley, *J.C.S. Perkin I*, **1973**, 566.
[641] O. E. Edwards, D. H. Paskovich and A. H. Reddoch, *Can. J. Chem.*, **51**, 978 (1973).
[642] M. Hedayatullah, J.-C. Raoult and L. Denivelle, *Bull. Chem. Soc. France*, **1973**, 2702.
[643] L. Dunsch, E. Keller and G. Henze, *Z. Chem.*, **13**, 222 (1973).
[644] K. Yoshida and T. Fueno, *J. Org. Chem.*, **37**, 4145 (1972).
[625] H. Sata and M. Masui, *J.C.S. Perkin II*, **1973**, 1640.
[646] G. Barbey, M. Genies, M. Libert and C. Caullet, *Bull. Soc. Chim. France*, **1973**, 1942.
[647] M. Masui and H. Ohmori, *J.C.S. Perkin II*, **1973**, 1112.
[648] A. Arnone, M. Cecere, R. Galli, F. Minisci, M. Perchimunno, O. Porta and G. Gardini, *Gazz. Chim. Ital.*, **103**, 13 (1973).
[649] H. Inoue, M. Sakata and E. Inoto, *Bull. Chem. Soc. Japan*, **46**, 2211 (1973).
[650] G. I. Nikishin, M. G. Vinogradov and T. M. Fedorova, *Chem. Comm.*, **1973**, 693.
[651] M. Okano, *Chem. Letters* (*Tokyo*), **1973**, 165.

[652] A. Kasahara, R. Saito and T. Izumi, *Bull. Chem. Soc. Japan*, **46**, 2610 (1973).
[653] M. G. Vinogradov, S. P. Verenchikov and G. I. Nikishin, *Bull. Acad. Sci. USSR*, **1972**, 1626.
[654] K. K. S. Gupta, J. K. Chakladar, B. B. Pal and D. C. Mukherjee, *J.C.S. Perkin II*, **1973**, 926.
[655] B. W. Brooks and R. M. Smith, *Chem. Ind.* (*London*), **1973**, 326.
[656] S. Torii, Y. Matsuyama, K. Kamasaki and K. Uneyama, *Bull. Chem. Soc. Japan*, **46**, 2912 (1973).
[657] T. Taguchi, M. Itoh and A. Suzuki, *Chem. Letters* (*Tokyo*), **1973**, 719.
[658] T. A. Cooper, *J. Am. Chem. Soc.*, **95**, 4158 (1973).
[659] See *Org. Reaction Mech.*, **1969**, 347.
[660] R. S. Wade and C. E. Castro, *J. Am. Chem. Soc.*, **95**, 226 (1973).
[661] R. S. Wade and C. E. Castro, *J. Am. Chem. Soc.*, **95**, 231 (1973).
[662] K. Onuma, J. Yamashita and H. Hashimoto, *Bull. Chem. Soc. Japan*, **46**, 333 (1973).
[663] E. Ghera, D. H. Perry and S. Shousa, *Chem. Comm.*, **1973**, 858.
[664] A. Orochov, M. Assche and D. Vofsi, *J.C.S. Perkin II*, **1973**, 1000.
[665] K. Okamoto, K. Komatsu, S. Tsukada and O. Murai, *Bull. Chem. Soc. Japan*, **46**, 1780 (1973).
[666] K. Okamoto, K. Komatsu, O. Murai and O. Sakaguchi, *Tetrahedron Letters*, **1973**, 4989.
[667] D. Y. Myers, R. R. Grabbe and P. D. Gardner, *Tetrahedron Letters*, **1973**, 533.
[668] L. H. Klem and D. R. Olson, *J. Org. Chem.*, **38**, 3390 (1973).
[669] J. Grimshaw and J. T.-Grimshaw, *J.C.S. Perkin I*, **1973**, 2584.
[670] A. Bewick and H. P. Cleghorn, *J.C.S. Perkin II*, **1973**, 1410.
[671] L. G. Chatten, R. W. Daisley and C. J. Olliff, *J.C.S. Perkin II*, **1973**, 469.
[672] G. Le Guillanton and A. Daver, *Bull. Soc. Chim. France*, **1973**, 709.
[673] J. P. Coleman, Nasur-ud-din, H. G. Gilde, J. H. P. Uttley, B. C. L. Weedon and L. Eberson, *J.C.S. Perkin II*, **1973**, 1903.
[674] K. Bachgaard, V. D. Parker and C. T. Pederson, *J. Am. Chem. Soc.*, **95**, 4373 (1973).
[675] T. Shono, T. Akazawa and M. Mitani, *Tetrahedron*, **29**, 817 (1973).
[676] E. van't Land and D. Van der Meer, *Rec. Trav. chim.*, **92**, 409 (1973).
[677] P. J. Elving, S. J. Pace and J. E. O'Reilly, *J. Am. Chem. Soc.*, **95**, 647 (1973).
[678] A. H. Lewin and R. J. Michl, *J. Org. Chem.*, **38**, 1126 (1973).
[679] F. F. Gadallah, A. A. Cantu and R. M. Elofson, *J. Org. Chem.*, **38**, 2386 (1973).
[680] C. Walling and G. M. El-Taliawi, *J. Am. Chem. Soc.*, **95**, 844 (1973).
[681] C. Walling and G. M. El-Taliawi, *J. Am. Chem. Soc.*, **95**, 848 (1973).
[682] D. L. Ingles, *Austral. J. Chem.*, **26**, 1015 (1973).
[683] D. L. Ingles, *Austral. J. Chem.*, **26**, 1021 (1973).
[684] C. Walling and A. Goosen, *J. Am. Chem. Soc.*, **95**, 2987 (1973).
[685] G. G. Jayson, B. J. Parsons and A. J. Swallow, *J.C.S. Faraday I*, **1973**, 236.
[686] D. Meisel, G. Czapski and A. Samuri, *J. Am. Chem. Soc.*, **95**, 4148 (1973).
[687] M. Julia and P. E. Bost, *Compt. Rend.*, *C*, **276**, 1195 (1973).
[688] M. Julia, P. E. Bost and F. Rickalaus, *Compt. Rend.*, *C*, **275**, 577 (1972).
[689] R. A. Rossi and J. F. Bunnett, *J. Org. Chem.*, **38**, 1407 (1973).
[689a] See *Org. Reaction Mech.*, **1971**, 337.
[690] R. A. Rossi and J. F. Bunnett, *J. Org. Chem.*, **38**, 3020 (1973).
[691] J. A. Zoltewicz and T. M. Oestreich, *J. Am. Chem. Soc.*, **95**, 6863 (1973).
[692] J. A. Barltrop and D. Bradbury, *J. Am. Chem. Soc.*, **95**, 5085 (1973).
[693] D. Y. Myers, G. G. Stroebel, B. R. Ortiz de Montellano and P. D. Gardner, *J. Am. Chem. Soc.*, **95**, 5832 (1973).
[694] J. O. Morley, *J.C.S. Perkin II*, **1973**, 1626.
[695] M. V. Shternskis, L. V. Bryukovetskaya, N. M. Katkova, I. I. Vil'kis and S. M. Shein, *Zhur. Org. Khim*, **9**, 1242 (1973); *Chem. Abs.*, **79**, 65459 (1973).
[696] S. M. Shein, L. V. Bryukhovetskaya and T. M. Ivanova, *Izvest. Akad. Nauk SSSR, Ser. Khim.*, **1973**, 1594; *Chem. Abs.*, **79**, 104537 (1973).
[697] L. V. Bryukhovestkaya, L. V. Mironova and S. M. Shein, *Izvest. Akad. Nauk SSSR, Ser. Khim.*, **1973**, 1601; *Chem. Abs.*, **79**, 104539 (1973).
[698] S. G. Kukes, N. N. Bubnov, A. I. Prokof'ev, S. P. Solodnikov, A. A. Volod'kin, G. A. Nikifororov and V. V. Ershov, *Izv. Akad. Nauk SSSR, Ser. Khim.*, **1973**, 184; *Chem. Abs.*, **78**, 135207 (1973).
[699] G. A. Razuvaev, G. A. Abakumov and E. S. Klimov, *Dokl. Akad. Nauk SSSR*, **206**, 788 (1972); *Chem. Abs.*, **78**, 42482 (1972).
[700] N. N. Bubnov, S. G. Kukes, A. I. Prokof'ev, S. P. Solodovnikov, A. A. Volod'kin and V. V. Ershov, *Bull. Acad. Sci. USSR*, **1972**, 2032.
[701] N. N. Bubnov, S. G. Kukes, A. I. Prokof'ev, S. P. Solodovnikov, D. Kh. Rasuleva, A. A. Volod'kin and V. V. Ershov, *Prov. Acad. Sci.* (*USSR*), **204**, 463 (1972).

702 S. G. Kukes, N. N. Bubnov, A. I. Prokof'ev, S. P. Solodovnikov, I. B. Khristman, A. A. Volod'kin, G. A. Nikiforov and U. V. Ershov, *Izv. Akad. Nauk SSSR, Ser. Khim.*, **1973**, 686; *Chem. Abs.*, **79**, 4626 (1973).
703 See *Org. Reaction Mech.*, **1971**, 338.
704 Ref. 1, Vol. 1, Chapter 9.
705 S. Bank and S. P. Thomas, *Tetrahedron Letters*, **1973**, 305.
706 H. Stetter and K.-A. Lehmann, *Annalen*, **1973**, 499.
707 Cf. p. **143**.
708 J. G. Smith and I. Ho., *J. Org. Chem.*, **38**, 3601 (1973).
709 A. J. Fry, W. E. Britton, R. Wilson, F. D. Greene and J. G. Pacifici, *J. Org. Chem.*, **38**, 2620 (1973).
710 R. N. Nasirov and S. P. Solodovnikov, *Bull. Acad. Sci. USSR.*, **1972**, 2047.
711 S. Bank and M. Platz, *Tetrahedron Letters*, **1973**, 2097.
712 H. Sakurai, A. Okada, H. Umino and M. Kiro, *J. Am. Chem. Soc.*, **95**, 955 (1973).
713 R. H. Reimann and E. Singleton, *J. Organometal. Chem.*, **57**, C75 (1973).
714 G. R. Stevenson and R. Concepcion, *Tetrahedron Letters*, **1973**, 2105.
715 N.I. Delyagina, B. L. Dyatkin, I. L. Knunyants, N. N. Bubnov and B. Ya. Medvedev, *Chem. Comm.*, **1973**, 456.
716 S. Limatibul and J. W. Watson, *J. Org. Chem.*, **37**, 4491 (1973).
717 M. Barreau and M. Julia, *Tetrahedron Letters*, **1973**, 1537.
718 J. Bartoň, P. Werner and J. Vlčková, *Makromol. Chem.*, **172**, 77 (1973).
719 Y. Ogata and M. Yamashta, *J. Org. Chem.*, **38**, 3423 (1973).
720 K. J. Houser, D. E. Bartak and M. D. Hawley, *J. Am. Chem. Soc.*, **95**, 6033 (1973).
721 See *Org. Reaction Mech.*, **1972**, 345.
722 Z. Csürös, P. Caluwe and M. Szwarc, *J. Am. Chem. Soc.*, **95**, 6171 (1973).
723 S. F. Nelsen and J. P. Gillespie, *J. Org. Chem.*, **38**, 3592 (1973).
724 C. L. Greenstock and I. Dunlop, *J. Am. Chem. Soc.*, **95**, 6917 (1973).
725 V. Kalynaraman and M. V. George, *J. Org. Chem.*, **38**, 507 (1973).
726 T. Kaufmann, G. Ruckelhaus and D. Glindemann, *Chem. Ber.*, **106**, 1618 (1973).
727 H. M. Walborsky and M. S. Aronoff, *J. Organometal. Chem.*, **51**, 31 (1973).
728 H. W. H. J. Bodewitz, C. Blomberg and F. Bickelhapt, *Tetrahedron*, **29**, 719 (1973).
729 R. B. Allen, R. G. Lawler and H. R. Ward, *J. Am. Chem. Soc.*, **95**, 1692 (1973).
730 R. B. Allen, R. G. Lawler and H. R. Ward, *Tetrahedron Letters*, **1973**, 3303.
731 M. Davis, L. W. Deady, A. J. Finch and J. F. Smith, *Tetrahedron*, **29**, 349 (1973).
732 R. Gelin, S. Gelin and A. Dehondt, *Bull. Soc. Chim. France*, **1973**, 2692.
733 P. Fostein, B. Delmond and J.-C. Pommier, *J. Organometal. Chem.*, **61**, C11 (1973).
734 R. A. Lee and W. Reusch, *Tetrahedron Letters*, **1973**, 968.
735 G. J. Abruscato and T. T. Tidwell, *J. Org. Chem.*, **37**, 4151 (1972).
736 R. Alexander and D. I. Davies, *J.C.S. Perkin I*, **1973**, 83.
737 D. Merkel and G. Kobrich, *Chem. Ber.*, **106**, 2025, 2040 (1973).
738 V. I. Stanko, A. N. Kashin and I. P. Beletskaya, *J. Organometal. Chem.* **56**, 111 (1973).
738a N. A. Clinton and J. K. Kochi, *J. Organometal. Chem.*, **56**, 243 (1973).
739 M. Daggoneau, *Compt. Rend.*, *C*, **276**, 1683 (1973).
740 M. Daggoneau and J. Vialle, *Tetrahedron Letters*, **1973**, 3017.
741 M. Daggoneau and J. Vialle, *Tetrahedron Letters*, **1973**, 3675.
742 N. C. Deno in ref. 582, p. 135.
743 S. Hünig, J. Gross and W. Schenk, *Annalen*, **1973**, 324.
744 S. Hünig, J. Gross, E. F. Lier and H. Quast, *Annalen*, **1973**, 339.
745 S. Hünig, B. J. Garner, G. Ruider and W. Scheuk, *Annalen*, **1973**, 1036.
746 J. R. Barnett, A. S. Hopkins and A. Ledwith, *J.C.S. Perkin II*, **1973**, 80.
747 L. J. Kricka and A. Ledwith, *J.C.S. Perkin I*, **1973**, 294.
748 V. Malatesta and K. U. Ingold, *J. Am. Chem. Soc.*, **95**, 6400 (1973).
749 S. C. Creason, J. Wheeler and R. F. Nelson, *J. Org. Chem.*, **37**, 4440 (1972).
750 P. Hanson and R. O. C. Norman, *J.C.S. Perkin II*, **1973**, 264.
751 G. Cauquis and M. Maurey-Mey, *Bull. Soc. Chim. France*, **1973**, 291.
752 S. Farid and S. E. Shealer, *Chem. Comm.*, **1973**, 677.
753 R. A. Neuntenfel and D. R. Arnold, *J. Am. Chem. Soc.*, **95**, 4080 (1973).
754 L. L. Miller, R. S. Narang and G. D. Nordblom, *J. Org. Chem.*, **38**, 340 (1973).
755 D. R. G. Brimage and R. S. Davidson, *J.C.S. Perkin I*, **1973**, 496.
756 D. R. G. Brimage, R. S. Davidson and P. R. Steiner, *J.C.S. Perkin I*, **1973**, 526.
757 I. Rosenthal, R. Poupko and D. Elad, *J. Phys. Chem.*, **77**, 1944 (1973).

758 B. C. Gilbert, D. K. C. Hodgeman and R. O. C. Norman, *J.C.S. Perkin II*, **1973**, 1748.
759 A. Burcat, G. B. Skinner, R. W. Crossley and K. Scheller, *Internat. J. Chem. Kinetics*, **5**, 345 (1973).
760 W. Tsang, *Internat. J. Chem. Kinetics*, **5**, 643 (1973).
761 M. U. C. Sekhar, G. E. Millward and E. Tschuikow-Roux, *Internat. J. Chem. Kinetics*, **5**, 363 (1973).
762 M. Simon and M. H. Back, *Can. J. Chem.*, **51**, 2934 (1973).
763 M. Taniuchi, *Bull. Inst. Chem. Res., Kyoto Univ.*, **50**, 660 (1972); *Chem. Abs.*, **79**, 31270 (1973).
764 L. Zalotai, T. Berces, and F. Marta, *Magyar Kem. Folyoirat*, **78**, 549, 556 (1973); *Chem. Abs.*, **78**, 71124, 71125 (1973).
765 H. Shibatani and H. Kinoshita, *Nippon Kagaku Kaishi*, **1973**, 1005; *Chem. Abs.*, **79**, 31271 (1973).
766 H. Shibatani and H. Kinoshita, *Nippon Kagaku Kaishi*, **1973**, 336; *Chem. Abs.*, **78**, 123731 (1973).
767 H. Shibatani and A. Amana, *Nippon Kagaku Kaishi*, **1972**, 2119; *Chem. Abs.*, **78**, 57469 (1973).
768 R. Louw and H. J. Lucas, *Rec. Trav. chim.*, **92**, 55 (1973).
769 R. Louw, J. W. Rothiuzen and R. C. C. Wegman, *J.C.S. Perkin II*, **1973**, 1635.
770 M. T. H. Liu, L. F. Louchs and R. C. Michaelson, *Can. J. Chem.*, **51**, 2292 (1973).
771 H. Knoll, K. Scherzer, and G. Geiseler, *Internat. J. Chem. Kinetics*, **5**, 271 (1973).
772 R. Louw, M. van den Brink and H. P. W. Vermeeren, *J.C.S. Perkin II*, **1973**, 1327.
773 W. S. Trahanovsky, D. E. Zabel and M. L-S. Louie, *J. Org. Chem.*, **38**, **1973**, 757.
774 J. M. Coxon, R. P. Garland and M. P. Hartshorn, *Austral. J. Chem.*, **25**, 2409 (1972).
775 C. Eaborn, J. M. Simmie and I. M. T. Davidson, *J. Organometal. Chem.* **47**, 45 (1973).
776 R. N. Haszeldine, P. J. Robinson and W. S. Williams, *J.C.S. Perkin II*, **1973**, 1013, 1018.
777 C. Eaborn, R. A. Jackson and R. W. Walsingham, *J.C.S. Perkin II*, **1973**, 366.
778 Y. Paquin and W. Forst, *Internat. J. Chem. Kinetics*, **5**, 691 (1973).
779 M. J. Perona, P. C. Beadle and D. M. Golden, *Internat. J. Chem. Kinetics*, **5**, 495 (1973).
780 K. A. W. Parry and P. J. Robinson, *Internat. J. Chem. Kinetics*, **5**, 27 (1973).
781 T. R. Lynch, F. N. MacLachlan and J. L. Suchitsky, *Can. J. Chem.*, **51**, 1378 (1973).
782 D. S. Ross, T. Mill and M. E. Hill, *J. Am. Chem. Soc.*, **94**, 8776 (1972).
783 M. L. Thomson and D. C. DeJongh, *Can. J. Chem.*, **51**, 3313 (1973).
784 P. Atlani, J. F. Biellmann, R. Briere and A. Rassat, *Tetrahedron*, **28**, 5805 (1972).
785 F. C. Thyrion and G. Debecker, *Internat. J. Chem. Kinetics*, **5**, 583 (1973).
786 S. J. Price, J. P. Richard and R. C. Rumfeldt, *Can. J. Chem.*, **51**, 1397 (1973).
787 T. J. Marks, A. M. Seyam and J. R. Kolb, *J. Am. Chem. Soc.*, **95**, 5529 (1973).
788 M. Petinaux and J. P. Aure, *Bull. Soc. Chim. France*, **1973**, 2490.
789 D. Seebach and A. K. Beck, *Chem. Ber.*, **105**, 3892 (1972).
790 K.-D. Asmus, H. Möckel and A. Henglein, *J. Phys. Chem.*, **77**, 1218 (1973).
791 M. Simic, M. Z. Hoffman and M. Ebert, *J. Phys. Chem.*, **77**, 1117 (1973).
792 G. W. Neilson and M. C. R. Symons, *J.C.S. Perkin II*, **1973**, 397.
793 M. V. Serdobov and M. C. R. Symons, *J.C.S. Perkin II*, **1973**, 1808.
794 R. H. Schuler, G. P. Laroff and R. W. Fessenden, *J. Phys. Chem.*, **77**, 456 (1973).
795 P. Gloux and B. Lamotte, *Mol. Phys.*, **25**, 161 (1973); B. Lamotte and P. Gloux, *J. Chem. Phys.*, **59**, 3365 (1973).
796 A. Samuri and P. Neta, *J. Phys. Chem.*, **77**, 1629 (1973).
797 E. Haydon and M. Simic, *J. Am. Chem. Soc.*, **95**, 1029 (1973).
798 K. Bhatia and R. H. Schuler, *J. Phys. Chem.*, **77**, 1888 (1973).
799 See *Org. Reaction Mech.*, **1972**, 363.
800 A. V. Raven, P. Hesse and H. Heusinger, *Makromol. Chem.*, **163**, 215 (1973).
801 M. C. R. Symons, *J.C.S. Perkin II*, **1973**, 797.
802 S. P. Mishra and M. C. R. Symons, *J.C.S. Perkin II*, **1973**, 395.
803 L. A. Tikhomirov, *Khim. Vys. Energ.*, **7**, 246 (1973).
804 P. J. Krupp, T. H. Jones and G. S. Poindexter, *J. Am. Chem. Soc.*, **95**, 5420 (1973).
805 E. H. Cole, G. Crank and A.-S. Sheikh, *Tetrahedron Letters*, **1973**, 2659, 2687.
806 H. C. A. Van Beek, P. M. Heertjes and K. Schaafsma, *Rec. Trav. chim.*, **92**, 1189 (1973).
807 S. P. Pappas and A. Chattopadhyay, *J. Am. Chem. Soc.*, **95**, 6484 (1973).
808 T. A. B. M. Bolsman and T. J. de Boer, *Tetrahedron*, **29**, 3579 (1973).
809 B. G. Gowenlock, G. Kresze and J. Pfab, *Tetrahedron*, **29**, 3587 (1973).
810 H.-P. Schuchmann and C. von Sonntag, *Tetrahedron*, **29**, 1811, 3351 (1973).
811 M. V. Bhatt and M. Ravindranathan, *J.C.S. Perkin II*, **1973**, 1158.
812 D. Meisel, G. Czapski and A. Samuni, *J.C.S. Perkin II*, **1973**, 1702.
813 A. A. Westenberg, *Progr. Reaction Kinetics*, **7**, 23 (1973).
814 C. Pomponui and A. T. Balaban, *Rev. Roum. Chem.*, **18**, 1173 (1973).
815 D. Greatorex, H. D. B. Jenkins and T. J. Kemp, *J.C.S. Perkin II*, **1973**, 1762.

[816] N. D. Chuvylkin and G. M. Zhidomirov, *Mol. Phys.*, **27**, 1233 (1973).
[817] B. C. Gilbert and M. C. Trenwith, *J.C.S. Perkin II*, **1973**, 1834.
[818] G. B. Birrell, S. P. Van and O. H. Griffith, *J. Am. Chem. Soc.*, **95**, 2451 (1973).
[819] V. N. Parmon, A. I. Kokorin, G. M. Zhidomirov and K. I. Zamaraev, *Mol. Phys.*, **26**, 1565 (1973).
[820] A. B. Shapiro, M. G. Goldfield and E. G. Rozantzev, *Tetrahedron Letters*, **1973**, 2183.
[821] V. I. Koryakov, V. A. Gubznov, A. K. Chirkov and R. O. Matevosyan, *Dokl. Akad. Nauk SSSR*, **206**, 649 (1972); *Chem. Abs.*, **78**, 15115 (1973).
[822] A. P. Lepley and G. L. Closs, *Chemically Induced Magnetic Polarization*, Wiley, London, 1973.
[823] R. G. Lawler and H. R. Ward in *Determination of Organic Structures by Physical Methods*, Vol. 5 (ed. F. C. Nachod and J. J. Zuckerman), Academic Press, New York, 1973.
[824] J. Bargon *J. Am. Chem. Soc.*, **95**, 941 (1973).
[825] H. E. Chen, S. P. Vaish and M. Cocivera, *Chem. Phys. Letters*, **22**, 576 (1973).
[826] H. E. Chen, A. Groen and M. Cocivera, *Can. J. Chem*, **.51**, 3032 (1973).
[827] K.-G. Seifert and J. Bargon, *Angew. Chem. Internat. Ed.*, **12**, 763 (1973).
[828] S. M. Rosenfeld, R. G. Lawler and H. R. Ward, *J. Am. Chem. Soc.*, **94**, 9255 (1972).
[829] H. Y. Loken, R. G. Lawler and H. R. Ward, *J. Org. Chem.*, **38**, 106 (1973).
[830] T. V. Leshina, R. Z. Sagdeev, M. A. Kamkha, S. M. Shein and Yu. N. Molin, *Zh. Org. Khim.*, **8**, 2366 (1972); *Chem. Abs.*, **78**, 71035 (1973).
[831] N. N. Bubnov, B. Ya. Medvedev, L. A. Polyakova, K. V. Bilevich and O. Yu. Okhlobystin, *Org. Mag. Resonance*, **5**, 437 (1973); I. P. Grageerov, A. F. Levit, L. A. Kiprianova and A. I. Buchachenko, *ibid.*, **5**, p. 445; V. I. Savin, I. D. Temyachev and Yu. P. Kitaev, *ibid.*, p. 449.
[832] E. Lippmaa, T. Saluvere, T. Pehkz and A. Olivson, *Org. Mag. Resonance*, **5**, 429 (1973).
[833] H. D. Roth and M. L. Kaplan, *J. Am. Chem. Soc.*, **95**, 262 (1973).
[834] M. R. Brinkman, D. Bethell and J. Hayes, *J. Chem. Phys.*, **59**, 3431 (1973).
[835] D. Bethell, M. R. Brinkman and J. Hayes, *Chem. Comm.*, **1972**, 1324.
[836] R. Z. Sagdeev, T.V. Leshina, M. A. Kamkha, S. M. Shein and Yu. N. Molin, *Izv. Akad. Nauk SSSR, Ser. Khim.*, **1972**, 2128; *Chem. Abs.*, **78**, 15203 (1973).
[837] R. Z. Sagdeev, K. M. Salikov, T. V. Leshina, M. A. Kamkha, S. M. Shein and Yu. N. Molin, *Pis'ma Zh. Eksp. Teor. Viz.*, **16**, 599 (1972); *Chem. Abs.*, **78**, 57366 (1973).
[838] Ya. A. Levin, A. V. Il'yasov, E. I. Goldfarb and E. I. Vorkunova, *Org. Magnetic Resonance*, **5**, 487, 497 (1973).
[839] D. G. Pobedimsky, P. A. Kirpichnikov, Yu. Yu. Samitov and E. I. Goldfarb, *Org. Magnetic Resonance*, **5**, 503 (1973).
[840] P. W. Atkins, K. A. McLauchlan and P. W. Percival, *Chem. Comm.*, **1973**, 121.
[841] S. K. Wong, D. A. Hutchinson and J. K. S. Wan, *J. Am. Chem. Soc.*, **95**, 623 (1973); *J. Chem. Phys.*, **58**, 985 (1973).
[842] P. W. Atkins, A. J. Dobbs and K. A. McLauchlan, *Chem. Phys. Letters*, **22**, 209 (1973).
[843] P. W. Atkins and E. A. Moore, *Mol. Phys.*, **25**, 825 (1973).
[844] J. B. Pedersen and J. H. Freed, *J. Chem. Phys.*, **58**, 2746 (1973).

CHAPTER 4

Oxidation and Reduction*

T. W. BENTLEY

Department of Chemistry, University College of Swansea

Oxidation by Metallic Ions[1]

The second volume of the well-known monograph[2] on oxidation in organic chemistry includes sections on oxidation by Cu(II), mechanisms of phenol coupling, oxidation by Tl(III), and oxidation by ruthenium tetroxide; there is also an advance undergraduate text.[2a]

Chromium[3]

The long-standing question of the nature of the transition state of chromic acid oxidation of alcohols to ketones has been reopened. (The background to this problem is discussed in ref. 4.) A study of the temperature-dependence of pseudo-first-order rates of oxidation of protio- and deuterio-cyclohexanol and di-*tert*-butylcarbinol gave the results shown in Table 1, which indicate that the unhindered cyclohexanol behaves as expected for a

* One-electron oxidations and reductions are also discussed in Chapter 3.

Table 1. Activation parameters for pseudo-first-order oxidation of alcohols[5]

	Cyclohexanol (**1**)		Di-*tert*-butylcarbinol (**2**)	
	protio	deuterio	protio	deuterio
E_A(kcal/mole)	12.6 ± 0.1	13.6 ± 0.1	14.1 ± 0.1	17.1 ± 0.1
$-\Delta S^{\ddagger}$ (e.u.)	27.5 ± 0.5	28.0 ± 0.7	20.3 ± 0.8	14.9 ± 0.9
A	3.27×10^7	2.44×10^7	1.22×10^9	18.4×10^9

symmetrical transition state (**1**), with the difference in activation energies (ΔE_A) for the protio- and deuterio-compounds equal to the difference in zero-point energy for the respective C—H and C—D bonds. In contrast, the results for the hindered di-*tert*-butylcarbinol are consistent with quantum-mechanical tunnelling in the hydrogen-transfer transition state (**2**), in which the hydroxylic solvent is required to complete the hydrogen-transfer. The proposal of two extreme transition states (**1** and **2**) for electron-transfer in the decomposition of chromate esters casts doubt on structure–reactivity correlations based on the assumption of identical mechanisms for the Cr(VI) oxidation of all alcohols.

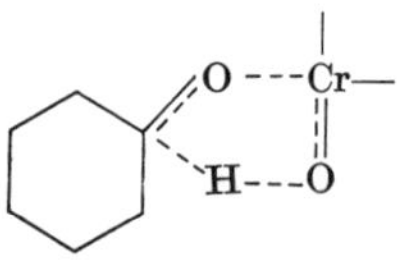

(**1**) 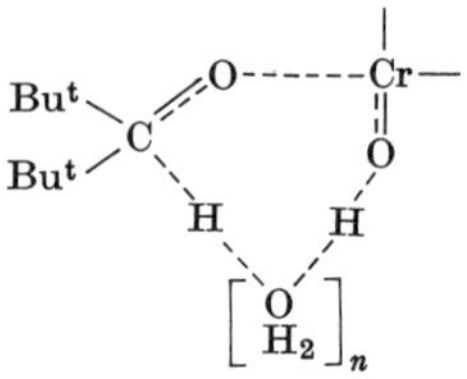 (**2**)

Full details have appeared of the marked accelerations, first observed last year,[6] in the chromic acid co-oxidation of propan-2-ol and oxalic acid.[7] The phenomenon is called "three-electron oxidation" because the mechanism is thought to involve formation of a termolecular Cr(VI) complex, which decomposes to Cr(III) in a single step. As might be expected, the ability to accelerate oxidation of propan-2-ol is not restricted to oxalic acid but is shared by a large number of other substrates, particularly those able to form relatively stable bidentate complexes with chromic acid, e.g. dicarboxylic acids, hydroxy-acids [e.g. equation (1)], and aldehydo- or keto-acids.[8]

The products obtained in the chromic acid oxidation of propan-2-ol and 2-hydroxy-2-methylbutyric acid are explained by reactions (1)–(3).[9] Decomposition of the termolecular complex (**3**) yields the radical anion of carbon dioxide (**4**), which is oxidized by Cr(VI) to yield carbon dioxide and Cr(V) [see reaction (2)]. Cr(V) reacts preferentially with

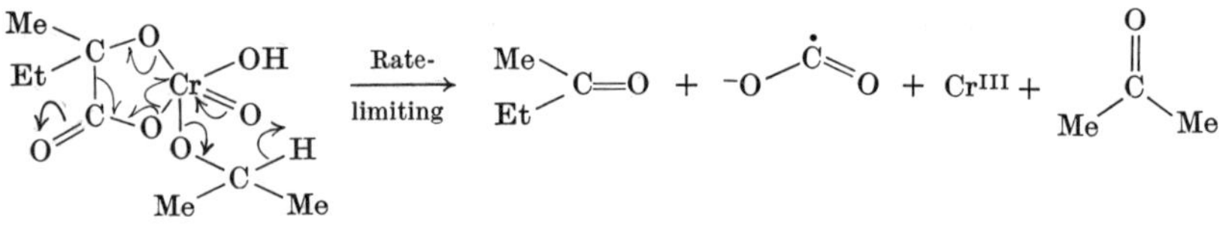

$$(\mathbf{3}) \xrightarrow[\text{limiting}]{\text{Rate-}} \text{MeEtC=O} + {}^{-}\text{O}\!-\!\dot{\text{C}}\!=\!\text{O}\ (\mathbf{4}) + \text{Cr}^{\text{III}} + \text{Me}_2\text{C=O} \quad (1)$$

$$^{-}O-\dot{C}=O + Cr^{VI} \longrightarrow CO_2 + Cr^{V} \quad (2)$$

$$Me_2CHOH + Cr^{V} \longrightarrow Me_2C{=}O + Cr^{III} \quad (3)$$

another molecule of propan-2-ol [reaction (3)]. Slightly more acetone is formed than the predicted ratio of 2:1:1 for acetone:ethyl methyl ketone:carbon dioxide. In the presence of the free-radical scavenger acrylonitrile, no carbon dioxide is produced and acetone and ethyl methyl ketone are produced in approximately equimolar amounts, indicating that the free-radical intermediate (**4**) is trapped and only reaction (1) yields ketone products. Careful study of the stoichiometry of the co-oxidation of propan-2-ol and oxalic acid is also consistent with the proposed mechanism.[6,10]

There has been a surprising development in work on the Cr(VI) oxidation of strained secondary alcohols. Whereas cyclobutanol and norbornan-7-ol react normally by rate-limiting cleavage of the α-C—H bond, cyclopropanol (**5**; R = H) reacts rapidly to give 3-hydroxypropionaldehyde.[11] The unusually rapid rates, substituent effects and products of Cr(VI) oxidation of cyclopropanols (**5**) are explained by reactions (4)–(6).[11]

$$\text{cyclo-}C_3H_4(OH)R + H_2CrO_4 \rightleftharpoons \text{cyclo-}C_3H_4(OCrO_3H)R \quad (4)$$

(**5**)

$$\text{cyclo-}C_3H_4(OCrO_3H)R \xrightarrow[\text{limiting}]{\text{Rate-}} {}^{+}CH_2CH_2C(=O)R + Cr(IV) \quad (5)$$

$$ {}^{+}CH_2CH_2C(=O)R \longrightarrow HOCH_2CH_2C(=O)R \quad (6)$$

There is now detailed evidence for the idea[12] that aldehydes are oxidized as the hydrate rather than the free carbonyl compound.[13] Cr(VI) oxidations of bridgehead alcohols[14] and phenylphosphonous[15] acid have been examined. The rates of Cr(VI) oxidation of substituted benzhydrols in acetic acid are linearly proportional (slope 0.14) to the $-pK_{R^+}$-values of the alcohols.[16]

Thorough investigation of the chromic acid oxidation of cyclobutanol shows that the reaction leading to the opening of the cyclobutane ring is due to Cr(IV);[17,18] although Cr(V) reacts to give cyclobutanone rather than a ring-cleavage product, the reaction is slow and enough Cr(V) may accumulate to make its bimolecular disproportionation to Cr(IV) and Cr(VI) important. The first stable Cr(IV) alkoxide of a secondary alcohol was prepared by reaction of $Cr(OBu^t)_4$ with 3,3-dimethylbutan-2-ol;[19] tetrakis (1,2,2-trimethylpropoxy)chromium(IV) is very sensitive to both air and moisture, but remarkably stable in their absence; the bulky *tert*-butyl groups appear to prevent formation of the conformation required for hydrogen-transfer in the oxidation step.

In the oxidation of norbornene by chromyl chloride,[20] only 5% of the isolated products correspond to those predicted by Freeman's mechanism;[21] the major product

(>74%) is the *exo-cis*-chlorohydrin. Chromyl chloride oxidizes phenols to quinones via solid non-stoichiometric intermediates.[22]

Lead

Oxidation of alcohols by lead tetra-acetate has been reviewed by Mihailovic and Partch, and that of olefins by Moriarty;[23] Sheldon and Kochi reviewed oxidative decarboxylation by the same reagent.[24]

In a novel $Pb(OAc)_4$ oxidation, the semicarbazone of *p*-(dimethylamino)benzaldehyde (**6**) yields *p*-(dimethylamino)benzoyl cyanide (**8**) via the oxadiazoline (**7**).[25]

$$Me_2NC_6H_4CH{=}NNHCONH_2 \xrightarrow{Pb(OAc)_4} \text{(7)} \xrightarrow{Pb(OAc)_4} Me_2NC_6H_4COCN$$

(7): ring N=N–C(H)(C₆H₄NMe₂)–O–C(=NH)–

(**6**) (**7**) (**8**)

Although *C*-phenyl-*N*-benzylnitrone (**10**) is the initial product of oxidation of *N*,*N*-dibenzylhydroxylamine (**9**) with $Pb(OAc)_4$, an interesting rearrangement takes place during subsequent oxidation; the ^{14}C-labelled nitrone (**10**) reacts with two moles of $Pb(OAc)_4$ to give (**11**) with the label almost equally distributed between the two α-phenyl-carbon atoms.[26]

$$PhCH_2N(OH)CH_2Ph \xrightarrow{Pb(OAc)_4} Ph\overset{*}{C}H{=}N(\rightarrow O)CH_2Ph \xrightarrow{Pb(OAc)_4} Ph\overset{*}{C}(OAc)ON{-}\overset{*}{C}H(OAc)Ph$$

(**9**) (**10**) (**11**)

When methyl-substituted benzenes are oxidized at low temperatures by $Pb(OAc)_4$ in trifluoroacetic acid/dichloromethane solution to biaryls and diarylmethanes,[27] the first step is thought to be formation of an aromatic radical cation, paralleling the electrochemical ECE mechanism. Oxidation of trityloxyamine (Ph_3CONH_2) with $Pb(OAc)_4$ at −78° gives triphenylmethyl peroxide ($Ph_3COOCPh_3$) in 85% yield, indicating that the triphenylmethoxy-radical ($Ph_3CO\cdot$) is an intermediate in the reaction.[28]

Steric effects in the $Pb(OAc)_4$ oxidation of a series of alkyl-substituted cyclohexanols have been examined by careful analysis of the products.[29] Similar research, including ^2H-labelling, has been carried out on aliphatic alcohols.[30,31] Other work with $Pb(OAc)_4$ includes oxidation of hydroxy-acids,[32,33] enamines,[34] *O*-alkylhydroxylamines,[35] and aromatic anils.[36]

Manganese

Although hypomanganate esters have often been considered as possible intermediates in permanganate oxidations,[37] they have only recently been detected spectroscopically in the oxidation of α,β-unsaturated carboxylic acids.[38,39] The mechanism [equation (7)] of oxidation of cinnamic acid involves a second-order reaction (k_2), forming the Mn(v) intermediate (**12**), which decomposes in a subsequent slower first-order reaction (k_1) to to Mn(III) and benzaldehyde. Both α- and β-deuterio-*trans*-cinnamic acid show inverse kinetic isotope effects (0.76) on k_2, suggesting a hybridization change from sp^2 to sp^3. (The corresponding isotope effects on k_1 are 1.09).[38] Mn(v) intermediates, e.g. (**12**), are oxidized by permanganate to Mn(VI) species.[39]

$$\text{C}_6\text{H}_5\text{CH=CHCOOH} + \text{MnO}_4^- \xrightarrow{k_2} \text{C}_6\text{H}_5\text{CH}-\text{CHCOOH (cyclic O–Mn(=O)(O}^-\text{)–O ester)} \; \textbf{(12)} \xrightarrow{k_1}$$

$$\text{C}_6\text{H}_5\text{CHO} + \text{HCOCOOH} + \text{MnO}_2^- \quad (7)$$

Detailed kinetic studies of the permanganate oxidation of acetylenedicarboxylic acid show the different reactivities of the undissociated acid, the monoanion and the dianion.[40] The stoichiometry of the reaction of one mole of permanganate with maleic and fumaric acids in aqueous acid is:

$$\text{HOOCCH=CHCOOH} + \text{MnO}_4^- + 4\text{H}^+ \rightarrow \text{OHCCH(OH)COOH} + \text{Mn(III)} + \text{CO}_2 + 2\text{H}_2\text{O}$$

The rate-limiting step is *cis*-attack on the double bond, resulting in a short-lived cyclic Mn(v) intermediate, similar to (**12**). Whilst it is not a large effect, steric hindrance to *cis*-attack could account for the 14-fold difference in reactivity between maleic and fumaric acid.[41] The kinetics and mechanism of oxidation of maleic and fumaric acid to carbon dioxide and formic acid,[42] and of substituted propynes to diketones,[43] have also been investigated.

Linear free-energy relationships and kinetic isotope effects are explained by hydride-transfer from the organic substrate to the oxidant during oxidation of benzyl alcohols,[44] mandelic acids,[45] aliphatic aldehydes (oxidized via the aldehyde hydrate)[46] and 1,3-diols.[47] Further studies of the kinetics of oxidation of formic acid by Mn(VII) have also been published.[48]

Thallium[49,50]

Because the more important oxidation states of thallium are Tl(I) and Tl(III), oxidations with Tl(III) are commonly considered to involve two-electron transfer mechanisms.[49,50] However, there is now good evidence from ESR spectroscopy that Tl(III) can act as a one-electron oxidant. Observation of cation radicals during the thallation of arenes, equation (8), with thallium tris(trifluoroacetate) may be explained by reactions (9)–(11):[51]

$$\text{Tl(OOCCF}_3)_3 + \text{ArH} \rightarrow \text{ArTl(OOCCF}_3)_2 + \text{CF}_3\text{COOH} \quad (8)$$

$$\text{Tl(III)} + \text{ArH} \rightleftharpoons [\text{ArHTl(III)}] \quad (9)$$

$$[\text{ArHTl(III)}] \rightleftharpoons [\text{ArH}^{\bullet+}\text{Tl(II)}] \quad (10)$$

$$[\text{ArH}^{\bullet+}\text{Tl(II)}] \rightarrow \begin{cases} \text{Tl(II)} + \text{ArH}^{\bullet+}, \text{ etc.} \\ {}^+\text{Ar}\langle^{\text{H}}_{\text{Tl}} \xrightarrow{-\text{H}^+} \text{ArTl(III)} \end{cases} \quad (11)$$

There is still no *direct* evidence that Tl(II) species are intermediates in these thallation reactions, owing to the instability of Tl(II) species which readily disproportionate to

Tl(III) and Tl(I). But one-electron oxidation by Tl(III) is now established as a viable alternative to two-electron oxidation and may, for example, explain the recently reported phenol-coupling by Tl(III).[52]

The rates of oxythallation of double bonds [k_1, equation (12)] parallel those for hydroxymercuration, and the intermediate (**13**) has been detected spectroscopically.[53] Detailed studies[53,54] support the currently accepted mechanism of oxidation of olefins by Tl(III), as in equations (12) and (13). With more highly substituted olefins,[55] or with phenylethenes,[56] oxidation by Tl(III) is accompanied by rearrangement. The products of oxidation of diarylacetylenes in several solvents have been examined.[57]

$$\text{Tl(III)} + CH_2{=}CHR + H_2O \xrightarrow{k_1} \underset{\textbf{(13)}}{[TlCH_2CHROH]^{2+}} + H^+ \qquad (12)$$

$$\textbf{(13)} \xrightarrow{k_2} \begin{cases} \text{Tl(I)} + CH_3COR + H^+ \\ \xrightarrow{H_2O} \text{Tl(I)} + HOCH_2CHROH + H^+ \end{cases} \qquad (13)$$

New aspects of the oxidation of acetophenones with Tl(III) include ^{14}C-labelling studies of the oxidative rearrangement to methyl arylacetates,[58] and a Hammett correlation ($\rho^+ = -0.70$) of the rates.[59] Details of the oxidation of chalcones to benzils,[60] and the oxidative cleavage of glycols[61] have been reported.

Other Metals

In the Ce(IV) nitrate oxidation of 1,2-diarylethanes, rate-limiting formation of cation radical intermediates, followed by cleavage of the central carbon–carbon bond to produce a benzyl radical and a benzyl cation, explains the formation of the three major products (ArCHO, $ArCH_2OH$ and $ArCH_2ONO_2$).[62] The relative rates of formation of radicals by oxidative cleavage of tertiary alcohols with Ce(IV) nitrate were calculated as: allyl (1): benzyl (4.4); *tert*-butyl (19.9–62.9).[63] Electronic and steric effects on the oxidative cleavage of 1,2-glycols[64] and the oxidation of malic acid[65] have also been studied.

An unusual reagent, VOF_3 in trifluoroacetic acid, has been used to convert (±)-laudanosine (**14**) into (±)-glaucine (**15**);[66] such oxidative coupling of non-phenolic substrates has previously only been achieved by electrochemical methods.

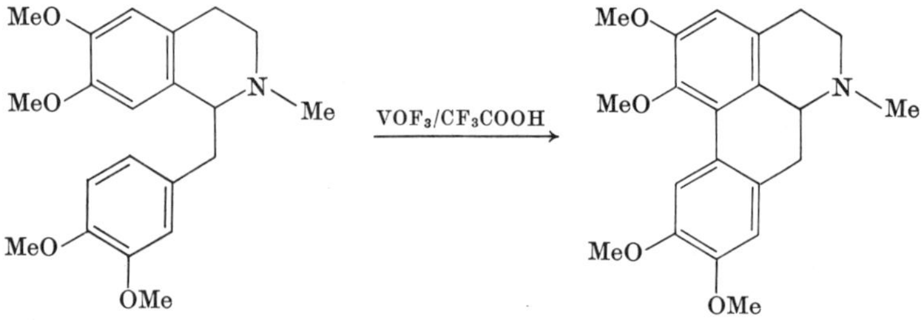

MeO, MeO, MeO, OMe, N–Me — VOF_3/CF_3COOH → MeO, MeO, MeO, OMe, N–Me

(**14**) (**15**)

Like Cr(IV) and Mn(III), V(V) is a one-electron oxidizing agent, which reacts with cyclobutanol (**16**) to give the hydroxy-aldehyde (**17**).[67] The mechanism of oxidation of (**16**) by V(V) or Mn(III) is similar to that of oxidation by Ce(IV), described last year.[68]

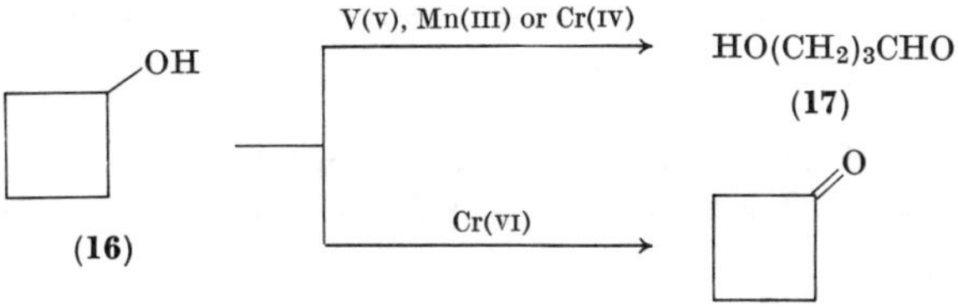

$$\text{(16)} \xrightarrow{\text{V(v), Mn(III) or Cr(IV)}} \text{HO(CH}_2)_3\text{CHO (17)}; \quad \text{(16)} \xrightarrow{\text{Cr(VI)}} \text{cyclobutanone} \qquad (14)$$

Oxidation of phenylphosphonous acid by V(v) occurs according to the stoichiometry of equation (15):[69]

$$PhPO_2H_2 + 2V(v) + 3H_2O = PhPO_3H_2 + 2V(IV) + 2H_3O^+ \qquad (15)$$

The reaction is catalysed by H^+ and it appears that V(v) acts as a two-electron oxidizing agent in perchloric acid (compare ref. 15); radicals were detected on polymerization of acrylamide in sulphuric acid but not in perchloric acid. Other V(v) oxidations have been reported.[70,71]

Oxidation of the benzophenone (**18**) with $Fe(CN)_6^{3-}$ in alkaline solution yields mainly the xanthone (**19**) and less than 5% of the isomeric xanthone (**20**).[72] Above pH 10, oxidation of the trianion of (**18**) yields a resonance-stabilized dianion radical, which cyclizes and loses a proton en route to the product (**19**). Reaction of the dianion is significant below pH 10. As expected for a reaction in which two highly charged negative ions must come together to bring about the initial oxidation, there is a large increase in the observed pseudo-first-order rate constant as the ionic strength of the solution increases. Interestingly, the reaction is subject to specific cation catalysis; the order $Cs^+ > K^+ > Na^+$ indicates that Cs^+ facilitates electron-transfer.

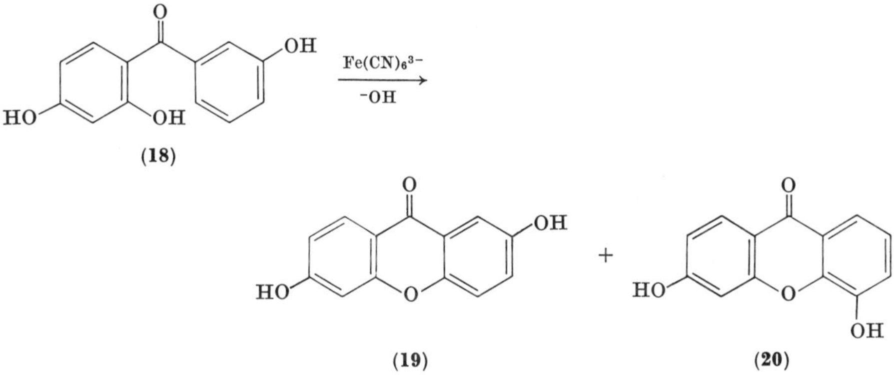

$$\text{(18)} \xrightarrow[\text{-OH}]{Fe(CN)_6^{3-}} \text{(19)} + \text{(20)}$$

On the basis of products isolated and ESR spectra of stable free radical intermediates, a mechanism of ferricyanide oxidation of the azo-compounds (**21**) has been proposed.[73]

As expected, the rate-limiting step in the oxidation of *N*-methyl- and *N*-(deuteriomethyl)-di-*n*-butylamine is electron-transfer from the nitrogen to $Fe(CN)_6^{3-}$, with

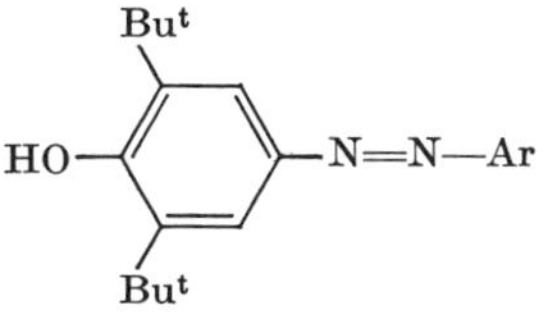

$$\text{3,5-Bu}^t_2\text{-4-HO-C}_6\text{H}_2\text{—N=N—Ar} \qquad (21)$$

$k_H/k_D = 1.04$.[74] Ferricyanide oxidation has also been used to prepare a series of reactive intermediates, aryldiazenes, by oxidation of arylhydrazines.[75] Similar oxidations of dihydropyridines[76] and pyrrole derivatives[77] have been discussed.

Equation (16) was deduced, from kinetic data at pH 4–6, for the initial rate of Cu(II) oxidation of the monohydroascorbate anion HA^- [where $H_2A \equiv$ (**22**)]:[78]

$$-(d[HA^-]/dt)_{t=0} = k_1[Cu(II)]^2_T[HA^-] + k_2[Cu(II)]^2_T[A^{2-}] \tag{16}$$

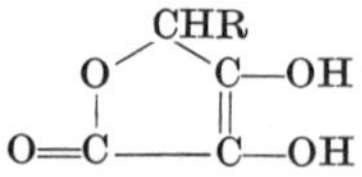

(**22**) [structure: five-membered lactone ring — CHR, O, O=C, C–OH, C–OH]

The kinetics of Cu(II) oxidation of a series of α-ketols in buffered aqueous pyridine have been interpreted in terms of rate-limiting enolization of an intermediate Cu(II)–ketol complex (**23**); the final product is C_6H_5COCHO.[79] Earlier work on the Cu(II)-

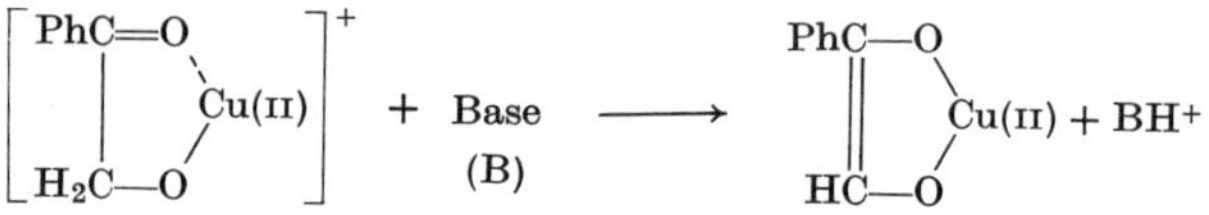

$$\left[\begin{matrix} PhC{=}O \\ | \quad \searrow Cu(II) \\ H_2C{-}O \end{matrix}\right]^+ (\mathbf{23}) + \text{Base (B)} \longrightarrow \begin{matrix} PhC{-}O \\ \| \quad \searrow Cu(II) \\ HC{-}O \end{matrix} + BH^+ \tag{17}$$

promoted cyclization of heteroaromatic compounds has been extended to more crowded systems.[80]

A detailed study of the kinetics and mechanism of the oxidative substitution of arenes by Co(III) trifluoroacetate shown in equations (18) and (19) has been reported.[81] Co(III)

$$C_6H_6 + 2Co(III)(OOCCF_3)_3 \rightarrow C_6H_5OOCCF_3 + 2Co(II)(OOCCF_3)_2 \tag{18}$$

$$-d[Co(III)]/dt = k[C_6H_6][Co(III)] \tag{19}$$

is a powerful one-electron oxidant and the slow step of the reaction is thought to be formation of a cation radical [equation (20)]; this is supported by the kinetics [equation (19)], the observation of cation radical intermediates and the rate-retarding effect of Co(II) trifluoroacetate. Despite the weakly nucleophilic properties of trifluoroacetic acid, competitive attack of another benzene molecule on the cation radical (**24**) is important only when an excess of benzene is present, and the products are formed in accord with equations (21) and (22). Strong support for this mechanism (20–22) comes from a comparison of the oxidations of toluene with Pb(IV) and with Co(III). Oxidation of toluene

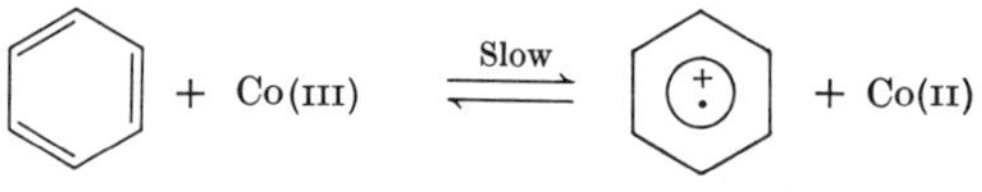

$$C_6H_6 + Co(III) \underset{}{\overset{\text{Slow}}{\rightleftharpoons}} C_6H_6^{+\cdot}\ (\mathbf{24}) + Co(II) \tag{20}$$

$$C_6H_6^{+\cdot}\ (\mathbf{24}) + CF_3COOH \longrightarrow \text{[cyclohexadienyl radical, H / } OOCCF_3\text{]} + H^+ \tag{21}$$

$$\text{[cyclohexadienyl radical, H / } OOCCF_3\text{]} + Co(III) \longrightarrow C_6H_5OOCCF_3 + H^+ + Co(II) \tag{22}$$

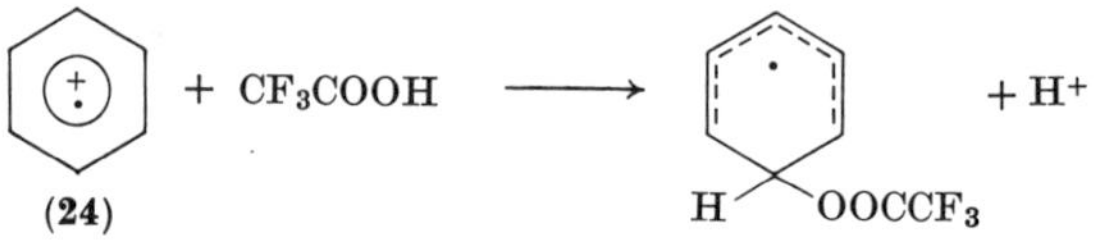

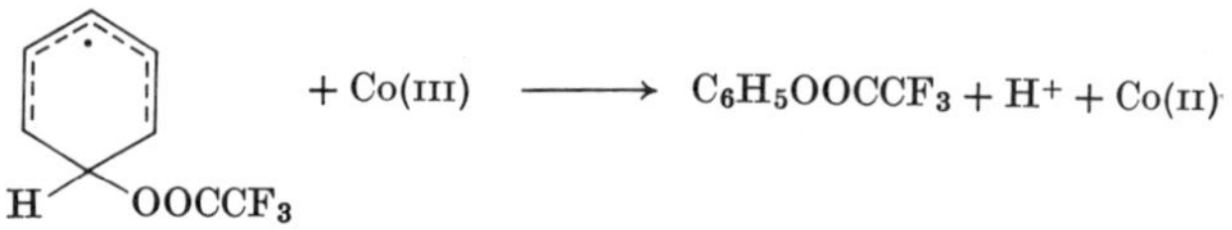

with Co(III) yields the benzyl ester ($C_6H_5CH_2OOCCF_3$), dimers $(C_7H_7)_2$ and trimers, etc., whereas Pb(IV) yields mainly nuclear-substituted tolyl esters derived from tolyl-lead intermediates.

Kinetic isotope effects in the oxidation of alcohols by silver carbonate have been determined by a double-labelling technique, which may be utilized in similar heterogeneous systems; k_H/k_D was calculated to be 3.0 by partial oxidation of a mixture of ^{14}C-labelled and ^{2}H-labelled alcohols.[82]

Tetrachloroauric(III) acid ($HAuCl_4$) stereospecifically oxidizes methionine to methionine sulphoxide, probably by way of a sulphide complex of Au(III) which is slowly reduced to Au(I).[83] A radical-chain mechanism has been proposed for oxidative addition of α-bromo-esters to Ir(I) complexes,[84] and the products of oxidation of benzoylhydrazones with nickel peroxide have been examined.[85]

Various oxidants, e.g. Cr(VI), promote the oxidation of ethylene by Pd(II) acetate in acetic acid.[86] A study of isotope effects confirms that hydroxymetallation, as in equation (12), is the rate-limiting step in the oxidation of ethylene with either Pd(II) or Tl(III).[87]

Oxidation by Molecular Oxygen[87a]

Hydrocarbons[88]

Of particular importance to "unambiguous" syntheses using ^{13}C- and ^{14}C-labels is the observation that up to 32% scrambling of the 1-carbon atom can occur during catalytic dehydrogenation of 1-methylcyclohexene.[89] Oxidative dehydrogenation of isopentenes on phosphoric acid catalysts has also been studied.[90]

Considerable interest continues in the cobalt ion-catalysed oxidation of hydrocarbons.[91] In the case of Co(III), oxidation in the absence of oxygen can yield different products;[81, 92] low temperature oxidation of *n*-alkanes by Co(III) acetate is activated by strong acids,[92] and similar results are obtained for Mn(III) and Co(III) oxidations of alkylbenzenes.[93, 94] Whilst Co(III) is known to form cation radicals with arenes as in equation (20), the major pathway for oxidation of both alkanes and alkylbenzenes is suggested to involve electron transfer from the C—H σ-bond to the Co(III) ion.[95]

In addition to acetic acid (84% yield), several other products are observed in the low-temperature, Co(III)-promoted, oxidation of *n*-butane. Unlike high-temperature (155°) oxidations, the reaction is not affected by Mn(III).[96] Bromide ion facilitates oxidations by Co(II),[91] exemplified by oxidations of alkylbenzenes[97] and other substrates.[98]

Addition of acetylacetone or EDTA increases the yield of coupling product (RR) in Pd(II)-catalysed oxidation of hydrocarbons [equations (23)–(26)].[99]

$$2RH + Pd(II) \rightarrow RPdR \tag{23}$$

$$RPdR \rightarrow RR + Pd \tag{24}$$

$$RH + Pd + O_2 \rightarrow RPdOOH \tag{25}$$

$$RPdOOH + RH \rightarrow RPdR + HOOH \tag{26}$$

The function of the EDTA is to remove contaminating metal ions, such as iron, nickel and chromium, which inhibit the reaction.

Several papers report aspects of heterogeneous catalytic oxidation,[100] including catalysis by molecular sieves containing transition-metal ions,[101] and autoxidation of alkanes in the presence of boric oxide.[102]

An unusual term, "triangular pathway", describes formation of CO_2 directly from ethylene and also via the intermediacy of ethylene oxide, during Ag(I)-catalysed oxidation.[103] Trimethylborane initiates the oxidation of isobutane;[104] and autoxidations of propane,[105] propene,[106] cyclohexane,[107] norbornene[108] and benzene derivatives[109] have also been studied.

As autoxidation is often an undesired reaction, possible inhibitors have received attention. Antioxidative properties of 4-alkoxy-2,6-diphenylphenols (AH), in the inhibition of autoxidation of polypropene at 180°, are enhanced by the sulphoxide (**25**) because the phenoxyl radical (A·) is regenerated by donation of a hydrogen atom by the sulphenic acid (**26**):[110]

$$\underset{(\mathbf{25})}{C_6H_5CH_2CH_2SOCH_2CH_2C_6H_5} \rightarrow \underset{(\mathbf{26})}{C_6H_5CH_2CH_2SOH} + C_6H_5CH{=}CH_2$$

The rate constants for reaction of ethylbenzeneperoxy-radicals with phenols have been determined by a chemiluminescene technique.[111]

Other Autoxidations

Correlation of product distributions with the composition of the aggregates of neophyl-lithium indicates that the unsolvated tetrameric aggregates of neophyl-lithium autoxidize in major part by a path involving free neophyl radicals, while the solvated dimeric aggregates do not.[112] The proposed mechanism of oxidation of the anion (**27**), shown in equation (27), is consistent with spin-trapping studies,[113] and a mechanism for base-

$$ArNHNHCOC_6H_5 \xrightarrow{NaH} \underset{(\mathbf{27})}{ArNH\bar{N}COC_6H_5} \xrightarrow{O_2} ArNHN(O{-}O^-)COC_6H_5 \rightarrow ArNHNO + C_6H_5COO^- \rightarrow ArN{=}NOH \rightarrow Ar\cdot \quad (27)$$

catalysed oxidative decarbonylation of the flavone (**28**) to (**29**) has been suggested.[114]

$$(\mathbf{28}) \xrightarrow[DMF-O_2]{Bu^tO^-K^+} (\mathbf{29}) + CO$$

Further attention has been given to the mechanism of oxidation of tetrahydropteridines,[115] dihydroalloxazines (using $^{18}OH_2$),[116] and *p*-phenylenediamines.[117]

The rate equation for oxidation of ferrocene in acidic medium is:[118]

$$\nu = k[Fe(C_5H_5)_2]^2[O_2][H^+]^2$$

Autoxidations of triphenylphosphine and triphenylarsine, catalysed by $Fe(mnt)_2^-$ and $Fe(mnt)_3^{2-}$ where mnt^{2-} is *cis*-1,2-dicyanoethylene-1,2-dithiolate, have been studied.[119] Acenaphthenequinone is thought to initiate the autoxidation of trimethyl phosphite by one-electron transfer.[120]

A refreshing mechanistic problem is the formation of chill haze in beer, which involves complex anion radical intermediates.[121]

Among the research on oxidations of alcohols,[122] there is an investigation of α-methylbenzyl alcohol by ^{18}O-labelling and kinetics.[123] A similar study on adipic diesters using ^{14}C-labelling showed that the CO and CO_2 arose from the acid residue.[124] The kinetics of oxidation of benzaldehyde in benzene,[125] and of co-oxidation of acetaldehyde and styrene,[126] have been measured.

The unusual rates of oxygen consumption by the succinate–Methylene Blue reductase is not a specific property of the enzyme system and is probably due to the Methylene Blue.[127] In the oxidative modification of certain proteins by Fe(II) and oxygen, the products depend markedly on ionic constituents of the reaction medium.[128] As part of a study of the metabolism of alcohols in mice, the kinetic isotope effects for oxidation of ethanol and propan-2-ol by liver alcohol dehydrogenases have been measured.[129] Recent aspects of oxidations by horseradish peroxidase include studies of 3-indoleacetic acid,[130] of alkyl group migrations in ring-substituted anilines,[131] and of a three-substrate "ping-pong" mechanism for oxidation of phenol.[132]

Evidence has been presented against participation of singlet oxygen in several enzymic oxygenations; a free-radical mechanism was preferred.[133]

Ozonation and Ozonolysis

There is now an impressive consensus of opinion and experimental evidence to show that the mechanism of ozonolysis, recently proposed by Story and his co-workers,[134] is incorrect. The claim[135] that the dioxetane (**31**) is an intermediate in the ozonolysis of ethylidenecyclohexane (**30**) in pinacolone is disputed, both because the dioxetane (**31**) has been independently synthesized and shown to be capable of surviving the reaction conditions, and because it was not detected in the reaction mixture.[136,137] Careful reinvestigation of the ozonolysis of (**30**) in pentane and cyclohexanone shows that the yield of ozonide (**32**) is reduced from 86.6% to 2.9% (not 0%, as previously reported[134]).

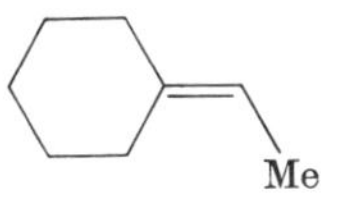

Me O—O CH Me O—O O Me

(**30**) 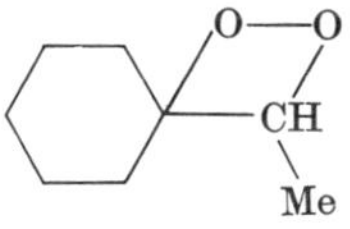 (**31**) 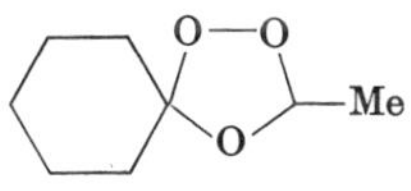 (**32**)

Thus in energetic terms, the effect of change in solvent is only a few kcal mol^{-1}, which may divert the commonly accepted zwitterion intermediate ($>C{=}O^{+}{-}O^{-}$) to "cyclohexanone peroxides".[138] Also the mechanism of Story and his co-workers[134] is not consistent with recent ^{14}C-labelling experiments.[138]

More attention has been given to the intriguing ability of olefin stereochemistry to influence the amounts and stereochemistry of the ozonide products. Ozonation in the presence of ^{18}O-labelled aldehyde leads to some ^{18}O-label in the peroxide bridge of the ozonide, which indicates that there is an additional pathway to the ozonide.[139] This view is supported by an independent synthesis of the zwitterion ($>C{=}O^{+}{-}O^{-}$),[140] which reacts with ^{18}O-labelled aldehyde to give the ozonide, *without* significant labelling of the peroxide bridge of the ozonide.[141] As microwave-spectroscopic studies show that ozonides of propene and *trans*-but-2-ene have the same half-chair conformation as that of ethylene,[142] modifications of the Bauld–Bailey mechanism[143] are being considered.

An unstable ozonide intermediate, probably a 1,2,3-trioxolane, has been observed by low-temperature IR measurements during the liquid-phase ozonation of *trans*-di-isopropylethylene.[144] Similar research on alkynes indicates the presence of an unstable precursor (possibly **33**) of the anyhydride (**34**).[145]

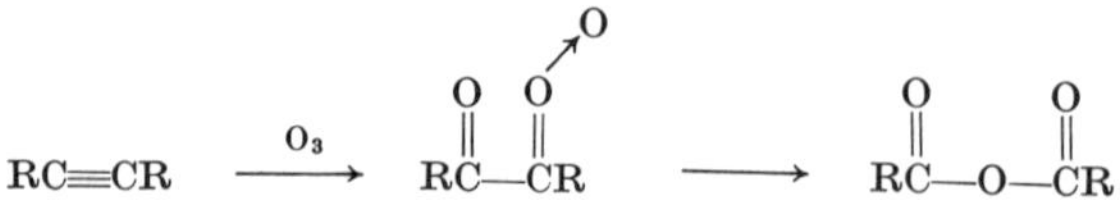

$$RC{\equiv}CR \xrightarrow{O_3} \underset{(\mathbf{33})}{RC(=O){-}C(=O{\rightarrow}O)R} \longrightarrow \underset{(\mathbf{34})}{RC(=O){-}O{-}C(=O)R}$$

Kinetic investigations of ozonation have dealt with substituent effects in a series of phenylethylenes[146] and competition experiments with olefin pairs.[147] Gas-phase ozonolysis of $Cl_2C{=}CH_2$ has been studied,[148] and photoionization mass spectrometry has been applied to low-pressure gas-phase reactions of simple olefins with ozone.[149] Ozonolysis of benzene yields some phenol,[150a] and reactions of hindered allenes with ozone to form epoxides have been noted.[150b]

Other Oxidations

Peracids and Peroxides

For a review of α-peroxy-amines see ref. 151.

Electrophilic oxidation of octamethylnaphthalene (**35**) yields the methyl-shifted product (**36**) (70%) and the aryl-shifted product (**37**) (16%).[152] The nature of the peracid affects the relative migrations of alkyl groups in the Baeyer–Villiger reaction of a series

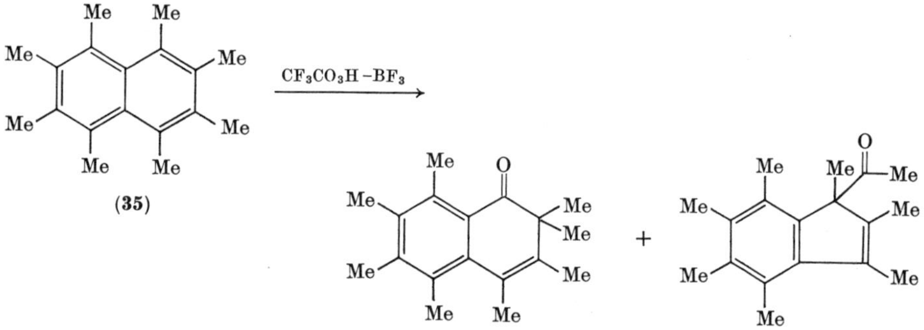

$$(\mathbf{35}) \xrightarrow{CF_3CO_3H-BF_3} (\mathbf{36}) + (\mathbf{37})$$

of primary-alkyl methyl ketones. In rearrangements induced by CF_3CO_3H, neopentyl migrates 13 times more readily than ethyl, but only 4 times faster when peroxymaleic acid is used.[153]

Complex oxidative rearrangements of heterocyclic compounds have been observed: by ^{14}C-labelling, it was shown that the pteridine (**38**) rearranges to the purine (**39**),[154] and the conversion of the pyrimidine (**40**) to the *s*-triazine (**41**) has been rationalized by an eight-step mechanism.[155]

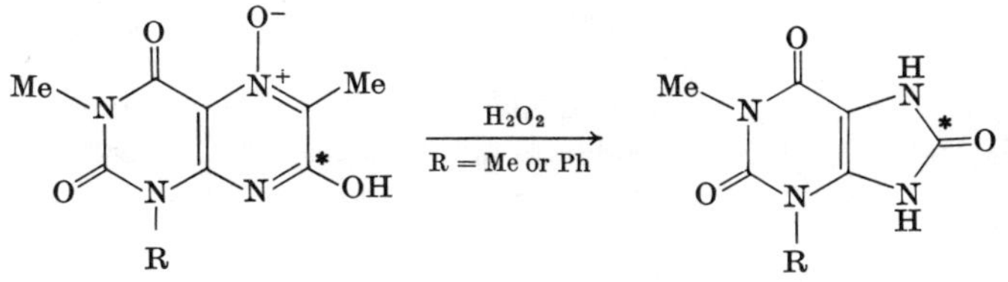

$$\text{(38)} \xrightarrow[\text{R = Me or Ph}]{H_2O_2} \text{(39)}$$

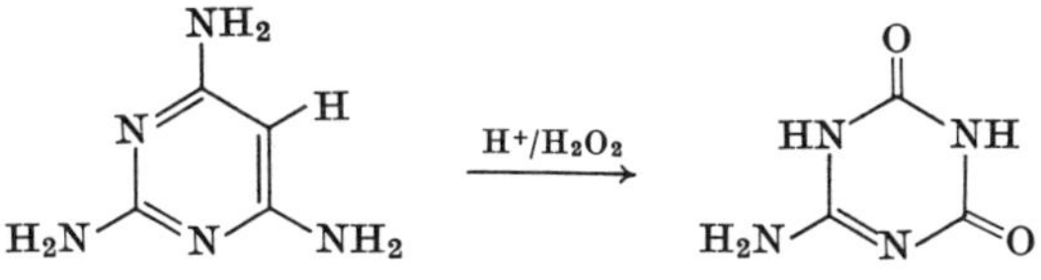

$$\text{(40)} \xrightarrow{H^+/H_2O_2} \text{(41)}$$

Detailed research on the peracid oxidation of imines has been published [equation (28)].[156] Under most conditions, k_1 is rate-limiting and direct formation of the oxazirane (**42**) does not occur. Nitrones are minor products formed by nucleophilic attack of the nitrogen lone pair of the C=N on the peracid-oxygen atom.[156,157]

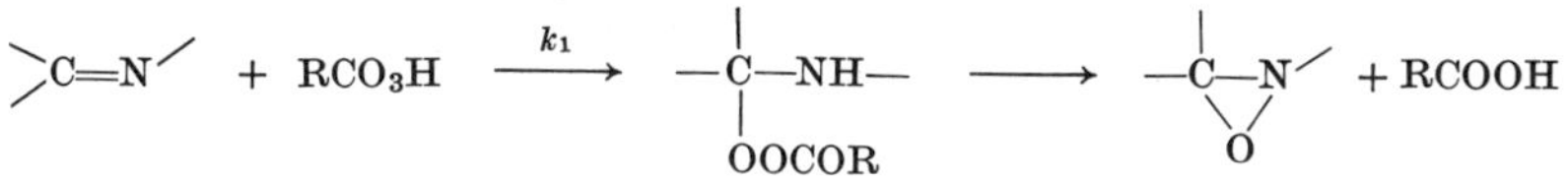

$$>C{=}N{-} + RCO_3H \xrightarrow{k_1} -\underset{\text{OOCOR}}{\overset{|}{C}}-NH- \longrightarrow \text{(42)} + RCOOH \quad (28)$$

By approaching the oxirene ⇌ oxocarbene equilibrium [equation (29)] from two directions, the intermediacy of oxirenes (**43**) in the oxidation of acetylenes has been deduced.[158,159] In support of this conclusion, the relative rates of perbenzoic acid

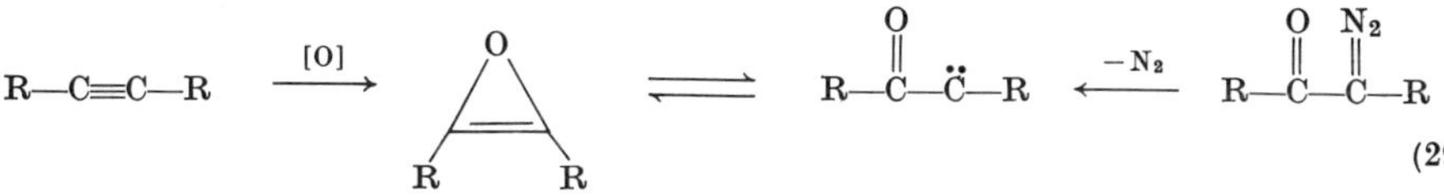

$$R{-}C{\equiv}C{-}R \xrightarrow{[O]} \text{(43)} \rightleftharpoons R{-}\overset{O}{\overset{\|}{C}}{-}\ddot{C}{-}R \xleftarrow{-N_2} R{-}\overset{O}{\overset{\|}{C}}{-}\overset{N_2}{\overset{\|}{C}}{-}R \quad (29)$$

oxidation of substituted phenylacetylenes correlate with σ^+ (slope = −1.40); the negative ρ-value suggests a mechanism involving electrophilic attack of the peroxidic oxygen to form the oxirene ring (**43**).[160] More work has been reported on the peracid oxidation of simple allenes, which proceeds through a reactive intermediate allene oxide.[161]

The rate of oxidation of the stilbene (**44**) by peracetic acid to give benzil (**45**) is given by equation (30). As the reaction is accelerated by acid, retarded by NaOAc and stopped

$$\nu = (k_2 + k'_2 h_0)[(\mathbf{44})][CH_3CO_3H] \quad (30)$$

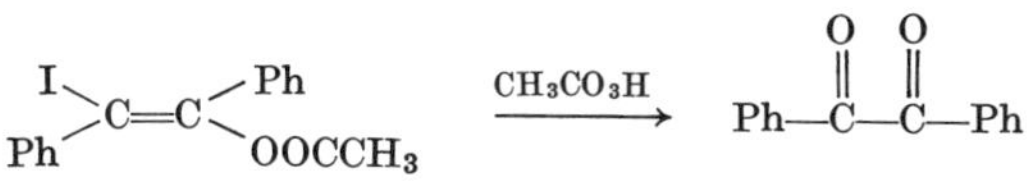

$$\text{(44)} \xrightarrow{CH_3CO_3H} Ph{-}\overset{O}{\overset{\|}{C}}{-}\overset{O}{\overset{\|}{C}}{-}Ph \ \text{(45)}$$

by strong base, attack by peracid on iodine to give a vinyl cation intermediate has been proposed.[162]

Various aspects of oxidations with alkaline hydrogen peroxide have been reported including the Dakin reaction,[163] oxidation of aldohexoses,[164] ketoses[165] and propionaldehyde,[166] and co-oxidation of *N*-benzyl-1,4-dihydronicotinamide and phenylglyoxal.[167]

A free-radical mechanism is probably involved in the reaction of peroxyacetyl nitrate with aldehydes.[168] Further studies have appeared on the kinetics and mechanism of epoxidation of aliphatic olefins by *tert*-butyl hydroperoxide in the presence of hexacarbonylmolybdenum as catalyst.[169] The kinetics of base-catalysed oxidation of sulphoxides with 1-methyl-1-phenylethyl hydroperoxide,[170] and kinetic isotope effects on the oxidation of 2,3-dimethylbut-2-ene with various oxidants have been reported.[171]

Halogens

Earlier work on the mechanism of oxidation of sulphides by bromine has been extended to sulphoxides; see equations (31)–(33). The first step [equation (31)] is retarded by bromide ion and the second step [equation (32)] is base (e.g. acetate)-catalysed.[172] Bromine

$$Br_2 + Me_2SO \underset{\text{Fast}}{\overset{K}{\rightleftharpoons}} Me_2Br\overset{+}{S}{=}O + Br^- \tag{31}$$

$$Me_2Br\overset{+}{S}{=}O + H_2O + B \underset{\text{Slow}}{\overset{k_1}{\longrightarrow}} Me_2S({=}\overset{+}{O}H)O + BH^+ + Br^- \tag{32}$$

$$Me_2S({=}\overset{+}{O}H)O \xrightarrow{\text{Fast}} Me_2SO_2 + H^+ \tag{33}$$

can achieve the overall conversion: $>C{=}S \rightarrow >C{=}O$, in the oxidation of thioanilides.[173]

Correlation of the rates of bromine oxidation of eleven primary alcohols in aqueous acetic acid with σ^* gives $\rho^* = -2.84 \pm 0.06$ at 25°, indicating a slow hydride-transfer from alcohol to bromine.[174] Two groups have investigated the oxidation of secondary alcohols to the corresponding tetrahydrofuran derivative by bromine–silver oxide.[175,176] For the oxidation of a series of dimethylbenzylamines to aldehydes by means of bromine [equation (34)], a mechanism involving base-catalysed loss of the benzylic proton and

$$PhCH_2N(CH_3)_2 + Br_2 + H_2O \rightarrow PhCHO + (CH_3)_2NH + 2HBr \tag{34}$$

two-electron transfer to bromine to give $PhCH{=}\overset{+}{N}Me_2$ has been proposed.[177]

A mechanism suggested for oxidation of mandelic acid and related compounds by hypobromous acid is shown in equations (35) and (36).[178]

$$PhCH(OH)COOH + HOBr \rightleftharpoons PhCH(OBr)COOH + H_2O \tag{35}$$

$$Ph{-}\underset{\substack{|\\ O-Br}}{CH}{-}CO{-}O{-}H \xrightarrow{\text{Slow}} PhCHO + CO_2 + H^+ + Br^- \tag{36}$$

Oxidative decarboxylation of several α-amino-acids by hypochlorous acid yields a mixture of the corresponding nitrile (major) and aldehyde (minor product);[179] and the oxidation of dimethylamine by sodium chlorite has been studied.[180]

Cleavage of hydroxamic acids (RCONHOH) by periodate yields the reactive intermediate *C*-nitrosocarbonyl compounds (RCONO), which can be trapped by dienes.[181]

Iodination occurs during periodate oxidation of 3-aminophenols, e.g. (**46**)→(**47**).[182] There are differences in the relative proportions of oxidation and cleavage of simple glycols, depending on the oxidant [Cr(VI) or bromate].[183]

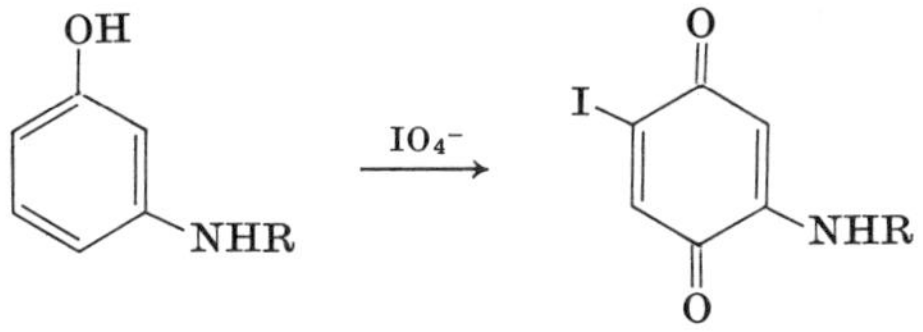

(**46**) (**47**)

Electrochemical Oxidation[184]

Many anodic oxidations have been reported and only a small selection will be noted here. Full details of the oxidation of *n*-alkanes in fluorosulphonic acid have been published.[185] In the presence of acetic acid and SbF_5, oxidation of both C_2H_6 and Me_3CH gives the α,β-unsaturated ketone $Me_2C{=}CHCOMe$, probably via the *tert*-butyl cation and isobutene. There is much interest in the oxidation of aromatic hydrocarbons,[186,187] carboxylic acids,[188] and their derivatives,[189] and phenols.[190]

In the anodic oxidation of methoxybibenzyls, intramolecular cyclization is thought to occur via dication diradicals.[191] Aspects of oxidations of 2-methoxyethanol,[192] diphenylamine[193] and aromatic Schiff's bases[194] have been discussed. Electrochemical oxidation of benzophenone hydrazone yields mainly the azine ($Ph_2C{=}NN{=}CPh_2$), whereas chemical oxidation yields some Ph_2CN_2 in addition to azine.[195]

Miscellaneous Oxidations

During nitric acid oxidation of hexadecanoic acid at low conversions (<6%), 77% of the diacids produced are C_{10}–C_{15}, showing a high selectivity for attack at positions remote from the carboxyl group.[196] Other aspects of nitric acid oxidations includes cyclohexanol[197] and aromatic oximes,[198] and various organic compounds have been oxidized in the liquid phase by NO_2.[199]

Following the recently suggested mechanism for selenium dioxide oxidation of olefins [equation (37)] and the evidence for the [2,3]-sigmatropic shift (step c),[200] support for the

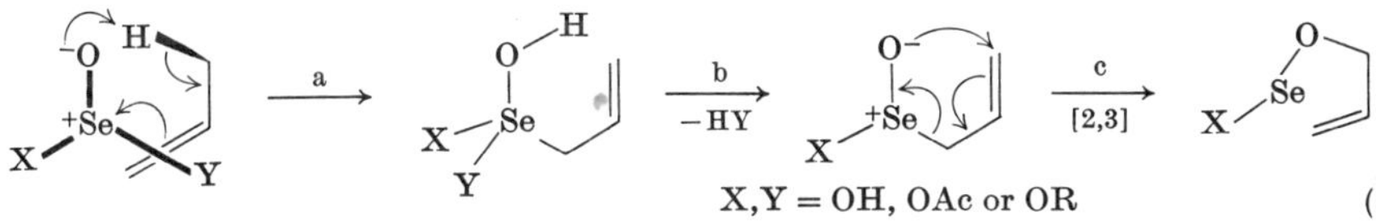

X,Y = OH, OAc or OR (37)

initial ene reaction (step a) has been obtained by trapping experiments. On reaction with 0.5 equiv. of SeO_2 in Bu^tOH, 1-isopulegol (**48**; R = H) yields the seleninolactone (**49**) (39% yield), presumably by intramolecular trapping of the half-ester (**48**; R = SeO_2H).[201a] As there are strong orientational effects in the ene reactions of other enophiles, the unusual orientational and stereochemical features of SeO_2 oxidations are probably associated with step a [equation (37)]. A careful study of the products of oxidation of (+)-liminonene by SeO_2/H_2O_2 has been reported.[201b]

The [2,3]sigmatropic rearrangement in equation (37) is reminiscent of the recently

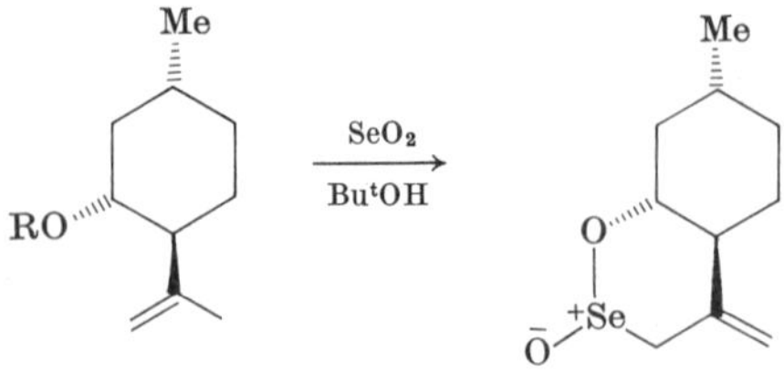

Me Me

$\xrightarrow[\text{Bu}^t\text{OH}]{\text{SeO}_2}$

RO O

$\bar{O}$ $^+$Se

(48) **(49)**

exploited fragmentation of selenoxides to form olefins.[202] These reactions are typified by the convenient synthesis of α,β-unsaturated carbonyl compounds from their saturated analogues by the sequence:

$$\text{PhCOCH}_2\text{CH}_3 \xrightarrow[\text{(ii)PhSeBr}]{\text{(i) LiNPr}^i_2} \text{PhCOCH(SePh)CH}_3 \xrightarrow{\text{NaIO}_4} \text{PhCOCH}(\overset{+}{\text{Se}}\overset{-}{\text{O}}\text{Ph})\text{CH}_3 \xrightarrow[\text{1h}]{20^\circ} \underset{85\%}{\text{PhCOCH=CH}_2} + [\text{PhSeOH}]$$

Solvent effects on the rates of oxidation of aromatic amines by potassium peroxydisulphate indicate charge development in the transition state.[203] Ag(I)-catalysed oxidations of *o*-cresol[204] and of aniline[205] are inhibited by allyl alcohol, indicating a free-radical pathway.[206]

An unusual cyclization (**50** → **51**) has been observed,[207] and further studies have appeared of the oxidative cyclization of 2-substituted anilines with (diacetoxyiodo)benzene in toluene.[208]

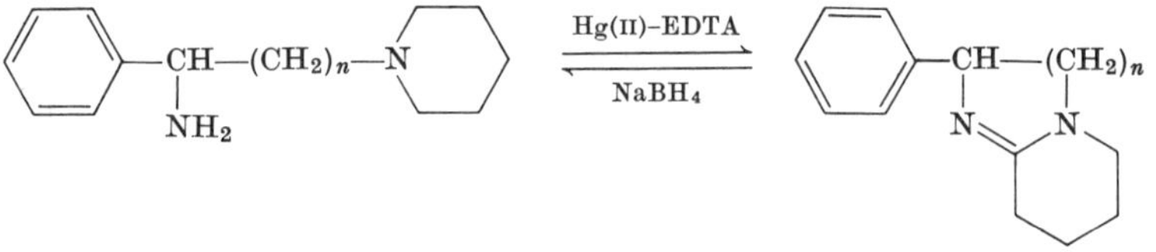

CH—$(CH_2)_n$—N NH$_2$

$\underset{\text{NaBH}_4}{\overset{\text{Hg(II)-EDTA}}{\rightleftharpoons}}$

CH—$(CH_2)_n$ N N

(50) **(51)**

The primary products from reaction of benzylamines with nitrosobenzene are azoxybenzene and imines from oxidation of the starting amines.[209] Other reports include a short review of the oxidation of monocyclic pyrroles,[210] and a study of the stereochemistry of oxidation of 1-phenylethyl *p*-tolyl sulphide.[211] Oxidations of the following substrates were also noted: ethanol and *tert*-butyl alcohol by Fenton's reagent,[212] aldoses by chloramine T,[213] π-allylpalladium chloride by *p*-benzoquinone,[214] substituted benzyl alcohols by *N*-bromosuccinimide ($\rho = -1.77$),[215] hindered phenols by Lewis acids,[216] *N*-alkylhydroxylamines by 1,1-diphenyl-2-picrylhydrazyl,[217] and 2,3-dihydrobenzofurans by activated carbon.[218]

Mechanistic information about biologically interesting systems can be obtained from studies of inhibitors. For example, 1-amino-2-alkynes are capable of inactivating plasma amine oxidase, and it is thought that the amine is first bound to the enzyme as a Schiff's base, followed by rearrangement of the triple bond to an allene, which adds a nucleophile from the enzyme to form a covalent adduct.[219] It has also been shown that 3-bromoallylamine is an irreversible inhibitor of mitochondrial monoamine oxidase.[220] Other studies include the action of some cholinomimetic agents on the oxidation of benzidine and *o*-dianisidine,[221] and the oxidation of *p*-cresol by lactoperoxidase compound II.[222]

Mercury ^3P-photosensitized decomposition of primary and secondary carbinamines (–CHNHR) gives imines in high yield, by a hydrogen-abstraction mechanism,[223] which similarly accounts for the oxidation of cyclopropanols by photoexcited benzophenone.[224] Detailed coverage of photo-oxidation, photoreduction, and the reactions of singlet oxygen is given elsewhere.[225]

Reductions[226]

Metal Hydride Reduction

It is known that sodium borohydride reductions of alkyl-substituted cyclohexanones are second-order within the limits of experimental detection, with the first hydride transfer, k_1 [equation (38)] rate-limiting. The change in the *cis/trans* product ratio during the course of the reaction can be explained if the fourth hydride-transfer, k_4 [equation (39)], is the second slowest.[227] Bulky intermediate borohydride species build up during

$$NaBH_4 + \text{Ketone} \xrightarrow{k_1} NaBH_3OR \tag{38}$$

$$NaBH(OR)_3 + \text{Ketone} \xrightarrow{k_4} NaB(OR)_4 \tag{39}$$

the reaction and compete with $NaBH_4$ for the remaining ketone. As alkoxy-substitution is known to increase (electronically) the hydride-donor ability of boron, there is a balance of steric and electronic effects.

The stereochemistry of reduction of ketones by complex metal hydrides has continued as an active research area.[228] There are indications that complex formation to the carbonyl group by a metal cation occurs and that the stereochemical outcome is affected by such complex formation.[229] Another two detailed* interpretations of stereoselectivity for alkyl aryl ketones[230] and cyclohexanones[231] are noted.

Although $LiAlH_4$ is thought to displace or replace halide from alkyl halides by either an S_N2 or an S_N1 process, depending on the substrate, it has now been shown that "unreactive" vinyl and bridgehead halides do react.[232] Also cyclopropyl halides react without rupture of the ring. Possible mechanistic explanations of the above results include: (i) direct attack by hydride on the halogen atom to yield the appropriate carbanion, which can then capture environmental hydrogen, or (ii) reaction of $LiAlH_4$ (behaving as a Lewis acid) with the alkyl halide to form an incipient carbenium ion, which is rapidly attacked by hydride in a front-side manner.

Reduction of dibenzyl ketoxime (**52**) with $LiAlH_4$ in tetrahydrofuran yields the aziridine (**53**) (92%) and the amine (**54**) (8%); this amine catalyses the formation of the

$PhCH_2C(=N{-}OH)CH_2Ph$ (**52**) $\xrightarrow{\text{Fast}}$ [Intermediate] → aziridine (Ph, H on one C; CH_2Ph, H on the other C; NH) (**53**) + $PhCH_2CH(NH_2)CH_2Ph$ (**54**)

* Although this work is extensive the magnitude of the "effects" discussed both here and in related papers published over the last ten years is usually of the order of a few kcal $mole^{-1}$ or even less. Such small effects may well be altered or even reversed by variations of solvent and temperature, or by other small changes not included in the "theory".

aziridine.[233] A mechanism has been proposed for the $LiAlD_4$ reduction of the oxime (**55**) to the deuterium-labelled aziridine (**56**).[234] Other substrates subjected to $LiAlH_4$ reduction include cyclohexene oxides with β-hydroxy- and methoxy-substituents,[235]

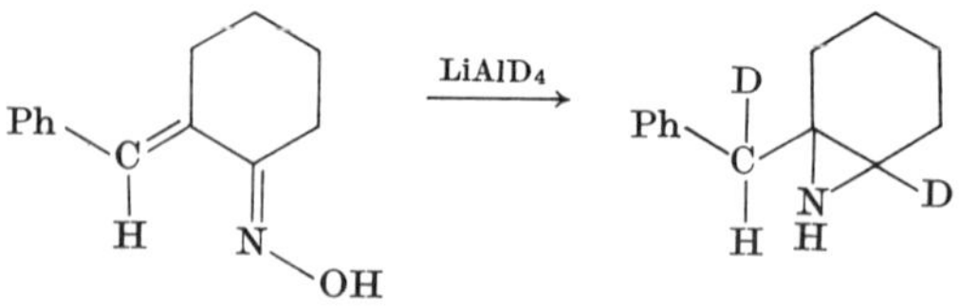

$LiAlD_4$

(**55**) (**56**)

oxadiazole derivatives,[236] and substituted *N*-phenylbenzanilides.[237] The stereoselectivity of seven aluminium hydride reagents in the reduction of α-ketols have been compared,[238] and from competitive reductions with $LiAlH(OBu^t)_3$, the order of reactivity of ketones was found to be: conjugated ketone < non-conjugated enone < cyclic saturated ketone.[239]

A dark-brown mixture ("reagent I"), prepared by addition of two molar equivalents of $LiAlH(OCH_3)_3$ to a suspension of CuI in tetrahydrofuran, is a useful reducing agent for various bromides and methanesulphonates and is probably a hydride analogue of the well-known lithium dialkylcuprates. Investigation of the stereochemistry of the reaction, with $LiAlD(OCH_3)_3$ used for preparation of reagent I, showed that reduction of both *exo*- and *endo*-bromonorbornane proceeds with complete retention, whereas the corresponding methanesulphonates were completely inverted (Scheme 1).[240]

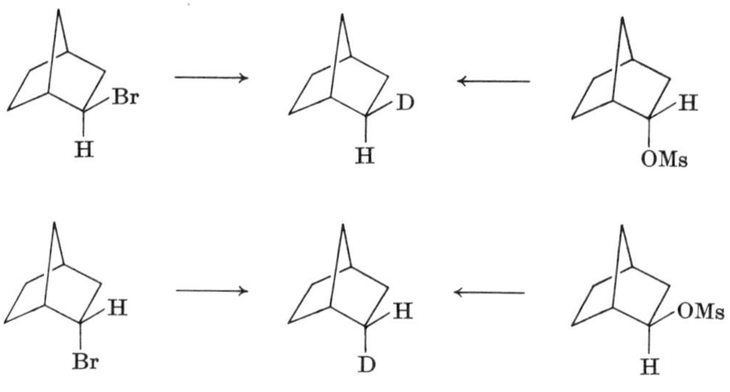

SCHEME 1.

Another new reducing agent, tetrabutylammonium cyanoborohydride in hexamethylphosphoramide has equally interesting properties, because it selectively reduces primary alkyl iodides to the corresponding hydrocarbons; for example, 1-iododecane is reduced much more rapidly than the corresponding bromide, chloride or toluene-*p*-sulphonate. Remarkably, the addition of acid (e.g. 0.1M) to the reaction medium permits selective reduction of aldehydes to be performed in the presence of most other functional groups including ketones and iodides.[241]

In the $NaBH_4$ reductions of bicyclo[3.2.1]octan-2- and -3-one, the kinetics are of the second order, and the product distributions do not change during the course of the reaction.[242] $KBH(OPr^i)_3$ is a highly selective reducing agent for ketones,[243] and lithium di-*n*-butylcuprate reduces ketones to secondary alcohols, whereas lithium dimethylcuprate yields the corresponding tertiary alcohol.[244] The effect of a crown ether on the stereochemistry of $NaBH_4$ reduction of a few cyclic ketones has also been studied.[245] During the $NaBH_4$ reduction of *N*-benzylnicotinamide chloride a charge-transfer compound is formed which loses hydrogen.[246]

A mechanism suggested for the borohydride reduction of organomercurials is shown in equation (40); by using $NaBD_4$, it has been shown that the reaction proceeds with retention of configuration.[247]

$$H_3B^-\text{—}H + RHgY \rightarrow [H_3BY]^- + RHgH \rightarrow RH + Hg(0) \qquad (40)$$

Triphenyltin hydride and acetyl chloride in benzene reduces the carbonyl group in acylferrocenes to methylene by way of an acetate intermediate.[248] The products formed by reduction of various *gem*-dihalocyclopropane derivatives with Bu_3SnH (via a radical), sodium in dimethyl sulphoxide (via a carbanion) and $LiAlH_4$, have been compared,[249] and a similar study of reductive cleavage of phenylhydrazones of α-keto-acids to amino-acids has been reported.[250]

Both $NaBH_4$ and B_2H_6 reduce α-D-hexopyranosulosides to the products of *gluco*-configuration, whereas catalytic hydrogenation yields a mixture.[251] Other work on selective reduction includes a study of the reduction of amides to amines with B_2H_6.[252]

Although it is generally accepted that the stereochemistry of hydroboration of an olefin is *cis*, this has only recently been established for an acyclic olefin (*cis*-but-2-ene, by use of B_2D_6 in tetrahydrofuran). After oxidation by H_2O_2 less than 1% of the product of *trans*-hydroboration was present[253] (i.e. within the limit of NMR analysis, the reaction is stereospecifically *cis*).

A detailed kinetic study of the reduction of substituted acetophenones and benzaldehydes with morpholineborane shows that for acetophenones, $k_H/k_D = 1.23$ and $\rho = +1.64$, and for benzaldehydes $k_H/k_D = 1.47$ and $\rho = +1.95$. As the rates are insensitive to the composition of the aqueous alcohol solvent, all the kinetic data are accommodated by a mechanism involving a four-centre transition state.[254]

Dissolving Metal Reduction

Reductive dehalogenation of aryl halides by sodium or sodium naphthalenide appears to involve abstraction of hydrogen atoms from the solvent by the intermediate aryl radical, because when isopropyl methyl ether is the solvent abstraction occurs preferentially on the α-hydrogen of the isopropyl group.[255] Organometallic complexes of definite stoichiometries may be obtained by reduction of dialkylmagnesium compounds in hydrocarbons by alkali metals.[256]

The various products of reductions of nitrobenzene and nitrosobenzene with lithium in tetrahydrofuran are similar, whereas azoxybenzene gave exclusively azobenzene.[257] Further studies of the reductive dimerization of Schiff's bases by sodium in tetrahydrofuran,[258] and the reduction of acetophenone by metal/ammonia in the presence of *tert*-butyl alcohol,[259] have been reported.

A comparison has been made of relative migratory aptitudes of substituents on C-2 in the Clemmensen reduction of 2-phenylindan-1-ones (**57**);[260] the isotope effect k_H/k_D (**57**; R = H or D) is 1.57, and the order of migratory aptitudes is: H > phenyl > methyl, which is explained by the mechanism shown in Scheme 2 where Zn* denotes the metal surface.

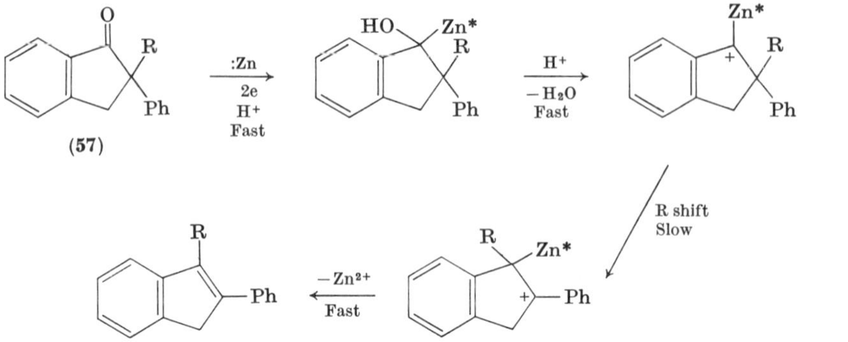

(57) —[:Zn, 2e, H^+, Fast]→ HO, Zn*, R, Ph —[H^+, $-H_2O$, Fast]→ Zn*, R, +, Ph —[R shift, Slow]→ R, Zn*, +, Ph —[$-Zn^{2+}$, Fast]→ R, Ph

SCHEME 2.

Electrochemical and Other Reductions

Reduction of di-*tert*-butyldiaziridinone (**58**) to the urea (**60**), both electrochemically and by sodium naphthalenide, probably proceeds through the anion radical (**59**) and the dianion, with further reduction of (**59**) faster than the initial reduction.[261] Similarly, electroreduction of benzophenone in anhydrous liquid ammonia occurs in two reversible

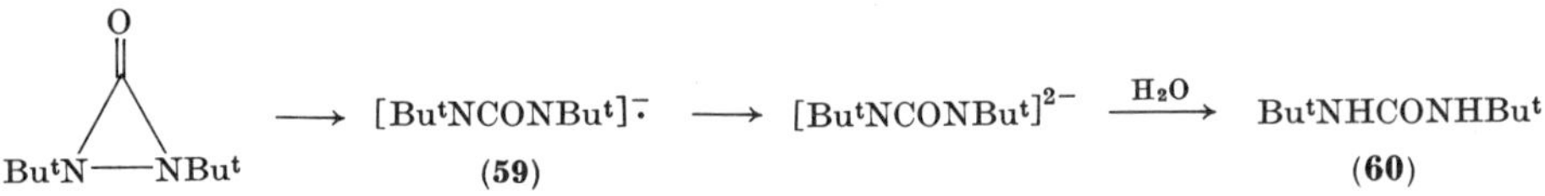

$$\underset{(58)}{Bu^tN{-}NBu^t\ (C{=}O)} \longrightarrow \underset{(59)}{[Bu^tNCONBu^t]^{\overset{\cdot}{-}}} \longrightarrow [Bu^tNCONBu^t]^{2-} \xrightarrow{H_2O} \underset{(60)}{Bu^tNHCONHBu^t}$$

one-electron steps, producing the anion radical and the dianion. In the presence of alcohol, the first step of the reduction is unaffected and the second becomes irreversible, suggesting a rapid reaction of the dianion with the proton-donor.[262]

The simple empirical rules (Table 2), which provide estimates of the reduction potentials of α,β-unsaturated aldehydes, ketones and esters within ± 0.1 V of the measured values, will be useful in estimating the feasibility of selective reductions and conjugate additions.[263]

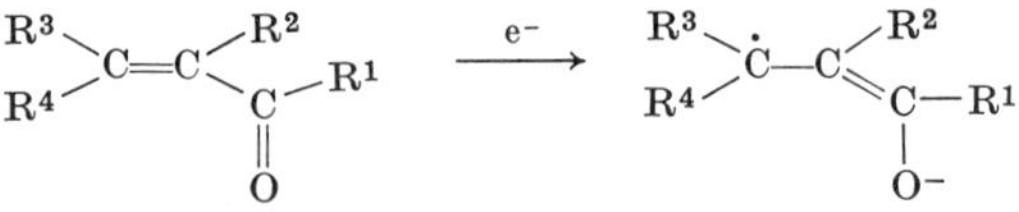

$$R^3R^4C{=}CR^2{-}C({=}O){-}R^1 \xrightarrow{e^-} R^3R^4\dot{C}{-}CR^2{=}C(O^-){-}R^1 \qquad (41)$$

The effects of geometry and substituents on the ease of reduction of dibenzoylethylenes and dibenzoylcyclopropanes has also received attention,[264] and the reduction potentials of a series of substituted oxindoles have been correlated with σ-values.[265] Polarographic reduction of the triene (**61**) and related compounds is not a good criterion of homoaromaticity in the anion radicals but may help to evaluate ground-state conformations.[266]

Table 2. Empirical rules for estimating reduction potentials against the standard calomel electrode in aprotic media [equation (41), base value ($R^1 = R^2 = R^3 = R^4 = H$), −1.9 V]

Substituent	Increment for reduction potential, V R^1	R^2	R^3 or R^4
Alkyl group	−0.1	−0.1	−0.1
First alkoxy-group	−0.3	0	−0.3
First phenyl-group	+0.4	+0.1	+0.4

Further studies of the reductive cleavage of carbon—halogen bonds have been carried out. Electrochemical reduction of 1,2-dibromobenzocyclobutene (**62**) generates benzocyclobutadiene (**63**), which dimerizes faster than it is further reduced to the benzocyclobutadiene anion radical.[267] 2,2-Dichloronorbornane and the epimeric bromochloro-

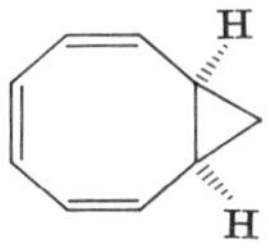

H H

(**61**)

Br Br

(**62**)

(**63**)

compounds (**66**) are electrochemically reduced to the *endo*-chloride (**65**) and nortricyclene (**67**). From the radio-labelled dichloride (**64**), the *endo*-chloride (**65**) is formed with only 7(+1)% retention of label. This was taken to indicate a preference for *exo*-approach to the electrode surface. A possible reaction intermediate is the carbanion of the *endo*-chloride (**65**), which could abstract a proton to give (**65**) or eject chloride ion to give a carbene, followed by transannular insertion to give (**67**).[268]

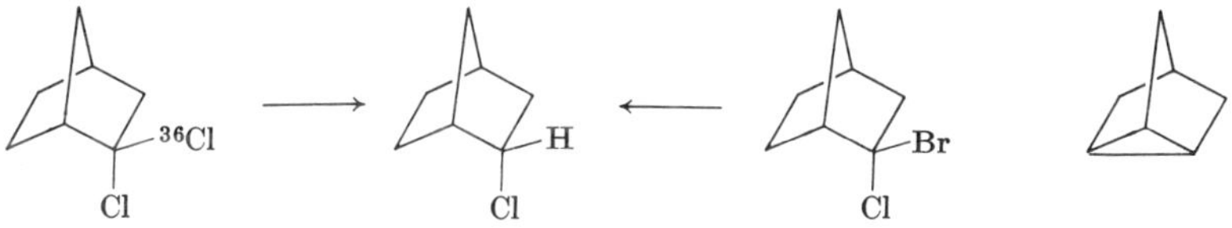

^{36}Cl Cl

(**64**)

H Cl

(**65**)

Br Cl

(**66**)

(**67**)

Details of the protonation, reduction and dimerization of species related to purine, pyrimidine and imidazole,[269] of cytosine derivatives,[270] and of nicotinamide derivatives[271] have been investigated.

Other electrochemical studies include the formation of the pinacol derivative of acetophenone,[272] the reduction of optically active acetylphenylcarbinol [PhCH(OH)-$COCH_3$],[273] the substituent effect on reduction of some perbenzoic acids,[274] the relation between reduction potential and structure for aromatic sulphones and homologous alkyl toluene-*p*-sulphonates,[275] and the mechanisms of polarographic reduction of *S,S*-diphenylsulfoximine.[276]

Analysis of the rates and products of Meerwein–Ponndorf–Verley reduction of eight mono- and bi-cyclic ketones permits a systematic comparison of substituent effects and stereochemistry in the vicinity of the carbonyl group.[277] Surprisingly, the concentration of reactants (ketone and aluminium isopropoxide) also influence the proportion of epimeric alcohol in the products, and the stereospecificity of reduction was more pronounced in dilute than in concentrated solutions. This may be caused by concentration-dependent association of $Al(OPr^i)_3$ with Pr^iOH. Cryoscopic and NMR techniques have been used in a further study of the interconversion reactions of dimeric, trimeric and tetrameric forms of $Al(OPr^i)_3$.[278]

There is some evidence for an electron-transfer component in the reaction of *sec*-butyl *p*-nitrobenzenesulphonate with thiophenoxide, although the reaction proceeds with complete inversion of configuration; the butyl spin adduct nitroxide of *tert*-butylphenylnitrone was detected by ESR and the phenylthiyl (•SPh) radical was detected by styrene polymerization.[279] It appears that trityl carbanion and trityl chloride react to give two trityl radicals in a single-electron transfer process.[280]

Reductive coupling of aryl-alkyl halides by the V(II) complex, $VCl_2(py)_4$, proceeds in essentially quantitative yields for substituted benzyl chlorides and bromides, and the general stoichiometry is summarized in equation (42). For benzotrichloride ($PhCCl_3$) in

$$2V(II) + 2RX \rightarrow R{-}R + 2V(III) + 2X^- \tag{42}$$

tetrahydrofuran the reaction rate is given by equation (43), with $\Delta H^{\ddagger} = 17.3$ kcal mol^{-1} and $\Delta S^{\ddagger} = -3.5$ e.u.:

$$-d[VCl_2(py)_4]/dt = k_2[VCl_2(py)_4][PhCCl_3] \tag{43}$$

It was argued that the products could not be formed by coupling of free radicals because products of hydrogen-abstraction (RH) are not observed. Possibly, an (arylalkyl)-vanadium(III) intermediate is formed, which is rapidly oxidized by a second molecule of the organic halide.[281]

Reduction of *meso*-stilbene dibromide with hydrazine yields bibenzyl as well as *trans*-stilbene and α-bromostilbene,[282] whereas most reagents dehalogenate vicinal dihalides to alkenes. As reductions carried out in the presence of cyclohexene yield cyclohexane at the expense of bibenzyl, di-imide (N_2H_2) is thought to be a reaction intermediate which hydrogenates olefins to alkanes. More complicated reactions, in the presence of added nucleophiles, were also interpreted mechanistically.[282]

Copper salts solubilized in alkylpolyamine solvents promote the homogeneous reduction of nitroalkanes to oximes by carbon monoxide.[283] The reaction is thought to proceed by attack of nitroalkane anion on a Cu(I) carbonyl complex.

Organomercury compounds are reduced by NO_2BF_4 in tetrahydrothiophen dioxide to the corresponding hydrocarbons and nitro-derivatives (amongst other products).[284] A detailed study of kinetics and substituent effects in the reduction of *N*-benzylideneaniline by formic acid has been reported,[285] as have other aspects of formic acid reductions.[286] The kinetics of reduction of some perbenzoic acids by iodide ions have been measured.[287]

A mechanism has been proposed for the conversion of α,β-unsaturated α-nitrocarboxylates (**68**) with triethyl phosphite to the β-diethoxy-phosphinyl compound (**69**).[288] Reductive cyclization of (**70**) to (**72**) can occur by photolysis of the triethyl phosphite derivative (**71**).[289] Treatment of the *N*-oxide (**73**) with base yields "reduced" derivative (**74**) with simultaneous "oxidation" of the 4-alcohol group.[290]

The reduction of *N*-methylacridinium ion by nicotinamide adenine dinucleotide and its derivatives has been examined as a model enzymic reaction.[291] Cytochrome c, a small

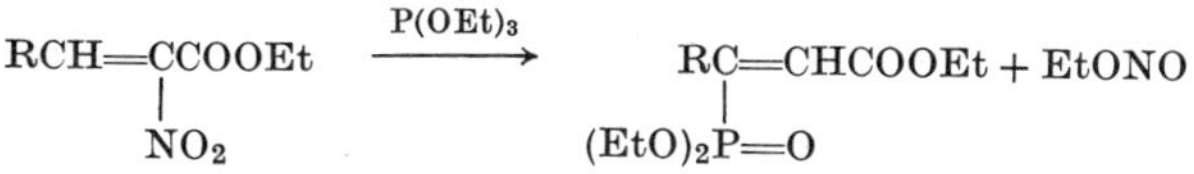

$$\underset{\textbf{(68)}}{RCH{=}C(NO_2)COOEt} \xrightarrow{P(OEt)_3} \underset{\textbf{(69)}}{RC(P(O)(OEt)_2){=}CHCOOEt} + EtONO$$

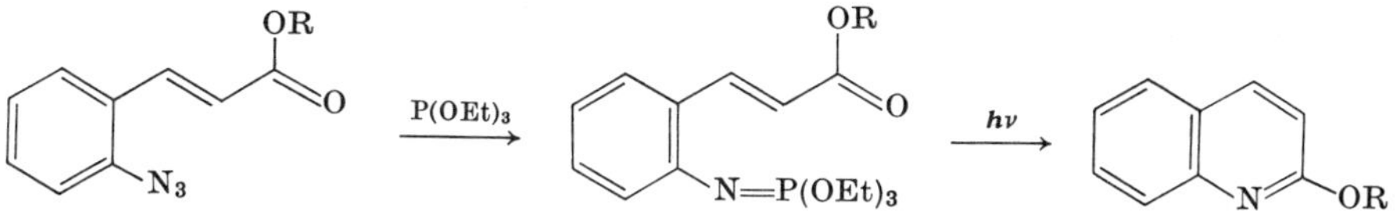

(70) → P(OEt)₃ → **(71)** → hν → **(72)**

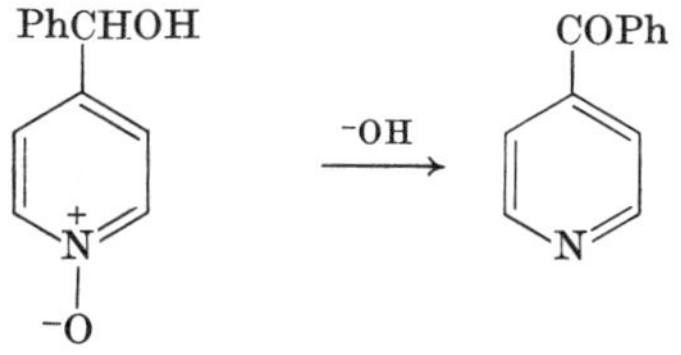

(73) → ⁻OH → **(74)**

(molecular weight ~12,400), relatively stable heme protein, is reduced by Cr(II) in a reaction that has been followed kinetically by a flow technique. As the reduction is catalysed by iodide, azide or thiocyanate ions a remote electron-transfer pathway is proposed.[292] [Heme proteins, e.g. iron(II) human hemoglobin and sperm-whale myoglobin, may be oxidized by alkyl halides to iron(III) proteins without protein denaturation.][293]

Hydrogenation and Hydrogenolysis

The complex $(Ph_3P)_2(CO)_2RuCl_2$, formed directly from $RuCl_3$, PPh_3 and CO, catalyses the selective homogeneous hydrogenation of cyclodeca-1,5,9-triene to cyclodecene.[294] In the presence of added PPh_3, the rates of hydrogenations of alkenes and alkadienes catalysed by this complex are decreased in the order, conjugated dienes > non-conjugated dienes > terminal alkenes > internal alkenes,[295] and these results were rationalized by using an extension of accepted mechanisms.

The effect of pressure of hydrogen on the stereochemistry of homogeneous hydrogenation of *tert*-butylmethylenecyclohexane can aid in distinguishing between various possible reaction schemes.[296] When $RhCl(PPh_3)_3$ is used as catalyst, the *cis/trans*-ratio of products was independent of pressure, but with $HRhCO(PPh_3)_3$ the *cis/trans*-ratio varied from 0.25 at 0.1 atm. to the "limiting value" of about 2 at 30 atm.[297] The mechanistic interpretation did not, however, take account of the most recent evidence for the mechanisms of these hydrogenations.[298]

With $RhCl(PPh_3)_3$ as catalyst, the saturated cyclic ether, 1,4-dioxan, donates hydrogen to an olefin (e.g. cyclopentene) to give stoichiometric amounts of cyclopentane and dioxen [equation (44)].[299] Isotopic scrambling and H/D exchange during homogeneous

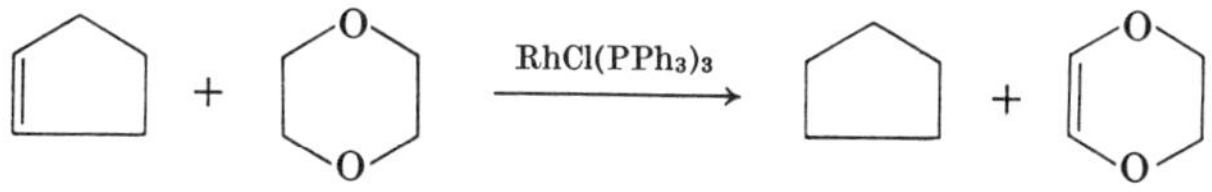

$$\text{cyclopentene} + \text{1,4-dioxan} \xrightarrow{RhCl(PPh_3)_3} \text{cyclopentane} + \text{dioxen} \qquad \textbf{(44)}$$

deuteriation of cycloalkenes has been observed.[300] Potassium hydroxide promotes the reduction of cyclohexanone by hydrogen-transfer from pentan-2-ol in the presence of $(PPh_3)_3RuCl_2$.[301] Homogeneous catalytic hydrogenation of methacrylic acid in the presence of pentacyanocobaltate(II) has also been studied.[302]

There has been an interesting semantic development in the field of heterogeneous hydrogenations. The term "haptophilicity", based on a non-existent Greek word, is intended to express the tendency of a functional group to cling to a catalyst surface and cause hydrogenation of a nearby double bond to be *cis* with respect to that functional group,[303] (e.g. the CH_2OH group strongly directs *cis*-hydrogenation[304]).

Heterogeneous catalyses of cyclopentene and di-*tert*-butylacetylene on various catalysts have been compared.[305] Other studies of heterogeneous catalytic hydrogenation include the stereochemistry of reduction of substituted methylenecyclohexanes,[306] substituted cyclopentenes,[307] substituted cyclohexanones,[308] and carbohydrates.[309] Aspects of the reductions of cinnamaldehyde,[310] methyl esters of C_{10-18} fatty acids,[311] dienones,[312] acetone[313] and aromatic nitro-compounds,[314] have also been discussed.

On using optically active benzylamine derivatives, it was found that hydrogenolysis occurred with inversion of configuration.[315]

The Pt/C-catalysed rearrangement of trimethylcyclopentane to dimethylcyclohexane has been observed.[316]

References

[1] J. K. Kochi, *Oxidation–Reduction Reactions of Free Radicals and Metal Complexes* in *Free Radicals* (J. K. Kochi, Ed.), Vol. 1, Wiley, New York, 1973, Chapter II, pp. 591–683.

[2] *Oxidation in Organic Chemistry*, Part B (W. S. Trahanovsky, Ed.), Academic Press, New York, 1973.

[2a] K. L. Rinehart, *Oxidation and Reduction of Organic Compounds*, Prentice Hall, Englewood Cliffs, N.J., 1972.

[3] Y. Ogata, *Kagaku No Ryoiki*, **27**, 317 (1973); *Chem. Abs.*, **78**, 146900 (1973).

[4] R. Stewart, *Oxidation Mechanisms*, W. A. Benjamin, New York, 1964, pp. 37–46.

[5] H. Kwart and J. H. Nickle, *J. Am. Chem. Soc.*, **95**, 3394 (1973).

[6] F. Hasan and J. Roček, *J. Am. Chem. Soc.*, **94**, 3181 (1972); *Org. Reaction Mech.*, **1972**, 535.

[7] F. Hasan and J. Roček, *J. Am. Chem. Soc.*, **94**, 9073 (1972).

[8] F. Hasan and J. Roček, *J. Org. Chem.*, **38**, 3812 (1973).

[9] F. Hasan and J. Roček, *J. Am. Chem. Soc.*, **95**, 5421 (1973).

[10] F. Hasan and J. Roček, *J. Am. Chem. Soc.*, **94**, 8946 (1972).

[11] J. Roček, A. M. Martinez and G. E. Cushmac, *J. Am. Chem. Soc.*, **95**, 5425 (1973).

[12] J. Roček, *Tetrahedron Letters*, **5**, 1 (1959).

[13] J. Roček and C.-S. Ng, *J. Org. Chem.*, **38**, 3348 (1973).

[14] J. J. Cawley and V. T. Spaziano, *Tetrahedron Letters*, **1973**, 4719.

[15] K. K. Sen Gupta and J. K. Chakladar, *J.C.S. Perkin II*, **1973**, 929.

[16] D. G. Lee and M. Raptis, *Tetrahedron*, **29**, 1481 (1973).

[17] J. Roček and A. E. Radkowsky, *J. Am. Chem. Soc.*, **90**, 2986 (1968).

[18] J. Roček and A. E. Radkowsky, *J. Am. Chem. Soc.*, **95**, 7123 (1973).

[19] G. Dyrkacz and J. Roček, *J. Am. Chem. Soc.*, **95**, 4756 (1973).

[20] F. W. Bachelor and V. O. Cheriyan, *Tetrahedron Letters*, **1973**, 3291; see also F. W. Bachelor and V. O. Cheriyan, *Chem. Comm.*, **1973**, 195.

[21] F. Freeman and N. J. Yamachika, *J. Am. Chem. Soc.*, **94**, 1214 (1972); *Org. Reaction Mech.*, **1972**, 536.

[22] M. J. Leigh and J. A. Strickson, *J.C.S. Perkin II*, **1973**, 1476.

[23] M. Lj. Mihailović and R. E. Partch, *Selective Organic Transformations* (B. S. Thyagarajan, Ed.), Vol. 2, Interscience, New York, 1972, pp. 97–182; R. M. Moriarty, *ibid.*, pp. 183–237.

[24] R. A. Sheldon and J. K. Kochi, *Organic Reactions*, **19**, 279 (1972).

[25] P. Knittel and J. Warkentin, *Can. J. Chem.*, **50**, 4066 (1972).

[26] L. A. Neiman, S. V. Zhukova and V. A. Tyurikov, *Tetrahedron Letters*, **1973**, 1889.

[27] R. O. C. Norman, C. B. Thomas and J. S. Willson, *J.C.S. Perkin I*, **1973**, 325.

[28] A. J. Sisti and S. Milstein, *J. Org. Chem.*, **38**, 2408 (1973).
[29] J. Bošnjak, V. Andrejević, Z. Čeković and M. Lj. Mihailović, *Tetrahedron*, **28**, 6031 (1972).
[30] S. Milosavljević, D. Jeramić and M. Lj. Mihailović, *Tetrahedron*, **29**, 3547 (1973).
[31] M. Lj. Mihailović, S. Gojković, and S. Konstantinović, *Tetrahedron*, **29**, 3675 (1973).
[32] Y. Pocker and B. C. Davis, *J. Am. Chem. Soc.*, **95**, 6216 (1973).
[33] B. Shanker, S. K. Banerjee, and O. P. Sachdeo, *Z. Naturforsch.*, **28B**, 375 (1973).
[34] F. Corbani, B. Rindone and C. Scolastico, *Tetrahedron*, **29**, 3253 (1973).
[35] F. A. Carey and L. J. Hayes, *J. Org. Chem.*, **38**, 3107 (1973).
[36] A. Catto, F. Corbani, B. Rindone and C. Scolastico, *Tetrahedron Letters*, **1973**, 2723.
[37] K. B. Wiberg and K. A. Saegebarth, *J. Am. Chem. Soc.*, **79**, 2822 (1957); K. B. Wiberg and R. D. Geer, *ibid.*, **88**, 5827 (1966); *Org. Reaction Mech.*, **1967**, 424.
[38] D. G. Lee and J. R. Brownridge, *J. Am. Chem. Soc.*, **95**, 3033 (1973).
[39] K. B. Wiberg, C. J. Deutsch and J. Roček, *J. Am. Chem. Soc.*, **95**, 3034 (1973).
[40] L. Simándi and M. Jáky, *J.C.S. Perkin II*, **1972**, 2326.
[41] L. I. Simándi and M. Jáky, *J.C.S. Perkin II*, **1973**, 1856.
[42] M. Jáky, L. I. Simándi, L. Maros and I. Molnar-Perl, *J.C.S. Perkin II*, **1973**, 1565.
[43] L. I. Simándi and M. Jáky, *J.C.S. Perkin II*, **1973**, 1861.
[44] K. K. Banerji, *J.C.S. Perkin II*, **1973**, 435.
[45] K. K. Banerji, *Tetrahedron*, **29**, 1401 (1973).
[46] K. K. Banerji, *Indian J. Chem.*, **11**, 242 (1973); *Chem. Abs.*, **79**, 41673 (1973).
[47] P. Nath, K. K. Banerji and G. V. Bakore, *J. Indian Chem. Soc.*, **50**, 30 (1973); *Chem. Abs.*, **79**, 52568 (1973).
[48] P. V. Subba Rao, *Z. Phys. Chem. (Leipzig)*, **252**, 276 (1973).
[49] A. McKillop and E. C. Taylor, *Advan. Organometal. Chem.*, **11**, 147 (1973).
[50] R. J. Ouellette in ref. 2a (Rinehart), p. 135.
[51] I. H. Elson and J. K. Kochi, *J. Am. Chem. Soc.*, **95**, 5060 (1973).
[52] M. A. Schwartz, B. F. Rose and B. Vishnuvajjala, *J. Am. Chem. Soc.*, **95**, 612 (1973).
[53] J. E. Byrd and J. Halpern, *J. Am. Chem. Soc.*, **95**, 2586 (1973).
[54] P. Abley, J. E. Byrd and J. Halpern, *J. Am. Chem. Soc.*, **95**, 2591 (1973).
[55] A. McKillop, J. D. Hunt, F. Kienzle, E. Bigham and E. C. Taylor, *J. Am. Chem. Soc.*, **95**, 3635 (1973).
[56] L. Nadon, M. Tardat, M. Zador and S. Fliszár, *Can. J. Chem.*, **51**, 2366 (1973).
[57] A. McKillop, O. H. Oldenziel, B. P. Swann, E. C. Taylor and R. L. Robey, *J. Am. Chem. Soc.*, **95**, 1296 (1973).
[58] A. McKillop, B. P. Swann and E. C. Taylor, *J. Am. Chem. Soc.*, **95**, 3340 (1973).
[59] N. C. Khandual, K. K. Satpathy and P. L. Nayak, *Proc. Indian Acad. Sci., Sect. A*, **77**, 163 (1973); *Chem. Abs.*, **79**, 77774 (1973).
[60] A. McKillop, B. P. Swann, M. E. Ford and E. C. Taylor, *J. Am. Chem. Soc.*, **95**, 3641 (1973).
[61] A. McKillop, R. A. Raphael and E. C. Taylor, *J. Org. Chem.*, **37**, 4204 (1972).
[62] W. S. Trahanovsky and D. W. Brixius, *J. Am. Chem. Soc.*, **95**, 6778 (1973).
[63] W. S. Trahanovsky and D. B. Macaulay, *J. Org. Chem.*, **38**, 1497 (1973).
[64] W. S. Trahanovsky, J. R. Gilmore and P. C. Heaton, *J. Org. Chem.*, **38**, 760 (1973).
[65] R. Dayal and G. V. Bakore, *Indian J. Chem.*, **10**, 1165 (1972); *Chem. Abs.*, **78**, 123687 (1973).
[66] S. M. Kupchan, A. J. Liepa, V. Kameswaran and R. F. Bryan, *J. Am. Chem. Soc.*, **95**, 6861 (1973).
[67] J. Roček and A. E. Radkowsky, *J. Org. Chem.*, **38**, 89 (1973).
[68] K. Meyer and J. Roček, *J. Am. Chem. Soc.*, **94**, 1209 (1972); *Org. Reaction Mech.*, **1972**, 538.
[69] K. K. Sen Gupta, J. K. Chakladar, B. B. Pal and D. C. Mukherjee, *J. C. S. Perkin II*, **1973**, 926.
[70] L. Kalvoda, *Coll. Czech. Chem. Comm.*, **37**, 4046 (1972).
[71] K. B. Yatsimirskii and B. G. Zhelyazkova, *Teor. Eksp. Khim.*, **8**, 641 (1972); *Chem. Abs.*, **78**, 42506 (1973).
[72] P. D. McDonald and G. A. Hamilton, *J. Am. Chem. Soc.*, **95**, 7752 (1973).
[73] E. Manda, *Bull. Soc. Chem. Japan*, **46**, 2160 (1973).
[74] J. R. L. Smith and L. A. V. Mead, *J.C.S. Perkin II*, **1973**, 206.
[75] S. Mannen and H. A. Itano, *Tetrahedron*, **29**, 3497 (1973).
[76] O. M. Grishin and A. A. Yasnikov, *Zh. Obshch. Khim.*, **43**, 1342 (1973); *Chem. Abs.*, **79**, 77779 (1973).
[77] K. Tomita and N. Yoshida, *Bull. Chem. Soc. Japan.*, **45**, 3160 (1972); *Chem. Abs.*, **78**, 15230 (1973).
[78] K. Hayakawa, S. Minami and S. Nakamura, *Bull. Chem. Soc. Japan*, **46**, 2788 (1973).
[79] H. A. Connon, B. G. Sheldon, K. E. Harding, L. E. Letterman, D. C. Futton and W. G. Nigh, *J. Org. Chem.*, **38**, 2020 (1973).

[80] D. L. Fields, T. H. Regan and D. P. Maier, *J. Org. Chem.*, **38**, 407 (1973).
[81] J. K. Kochi, R. T. Tang and T. Bernath, *J. Am. Chem. Soc.*, **95**, 7114 (1973).
[82] F. J. Kakis, *J. Org. Chem.*, **38**, 2536 (1973).
[83] E. Bordignon, L. Cattalini, G. Natile and A. Scatturin, *Chem. Comm.*, **1973**, 878.
[84] J. A. Labinger, A. V. Kramer and J. A. Osborn, *J. Am. Chem. Soc.*, **95**, 7908 (1973).
[85] K. S. Balachandran and M. V. George, *Tetrahedron*, **29**, 2119 (1973).
[86] P. M. Henry, *J. Org. Chem.*, **38**, 1681 (1973).
[87] P. M. Henry, *J. Org. Chem.*, **38**, 2415 (1973).
[87a] J. A. Howard, *Homogeneous Liquid Phase Autoxidations*, in ref. 1, Vol. 2, Chapter 12, pp. 3–62.
[88] C. F. Cullis, *Nat. Bur. Stand. (U.S.), Spec. Pub.*, **357**, 111 (1972); *Chem. Abs.*, **78**, 135105 (1973).
[89] J. L. Marshall, D. E. Müller and A. M. Ihrig, *Tetrahedron Letters*, **1973**, 3491.
[90] Yu. N. Usov, E. V. Skvortsova, T. G. Vaismub and I. K. Kuchkaeva, *Neftekhimiya*, **12**, 481 (1972); *Chem. Abs.*, **78**, 15271 (1973).
[91] *Org. Reaction Mech.*, **1972**, 543.
[92] J. Hanotier, Ph. Camerman, M. Hanotier-Bridoux and P. de Radzitzky, *J.C.S. Perkin II*, **1972**, 2247; see also R. T. Savel'yanova, M. N. Manakov, A. E. Gref and N. N. Lebedev, *Tr. Mosk. Khim.-Tekhnol. Inst.*, **1972**, 39; *Chem. Abs.*, **78**, 135356 (1973).
[93] J. Hanotier, M. Hanotier-Bridoux and P. de Radzitzky, *J.C.S. Perkin II*, **1973**, 381; J. Hanotier and M. Hanotier-Bridoux, *ibid.*, p. 1035.
[94] V. N. Aleksandrov, S. S. Gitis, V. I. Ovchinnikov, I. M. Sosonkin and T. A. Simonova, *Zh. Obshch. Khim.*, **43**, 123 (1973); *Chem. Abs.*, **78**, 123669 (1973).
[95] A. Onopchenko and J. G. D. Schulz, *J. Org. Chem.*, **38**, 3729 (1973).
[96] A. Onopchenko and J. G. D. Schulz, *J. Org. Chem.*, **38**, 909 (1973).
[97] A. M. Ivanov and V. I. Tanashchuk, *Zh. Fiz. Khim.*, **46**, 2290 (1972); *Chem. Abs.*, **78**, 15288 (1973); S. R. Trusov and O. Neilands, *Neftekhimiya*, **13**, 69 (1973); *Chem. Abs.*, **78**, 135372 (1973); V. N. Aleksandrov, S. S. Gluzman, S. S. Gitis, V. Kh. El'man, G. S. Golubev, V. V. Khomin, A. A. Sidorova and L. S. Kovalev, *Neftekhimiya*, **12**, 558 (1972); *Chem. Abs.*, **78**, 15284 (1973); M. Hronec and V. Vesely, *Coll. Czech. Chem. Comm.*, **38**, 1226 (1973).
[98] V. K. Kondratov, and N. D. Rus'yanova, *Neftekhimiya*, **13**, 76 (1973); *Chem. Abs.*, **78**, 135363 (1973); M. S. Brodskii, M. Ya. Gervits, V. I. Trofimov and V. A. Nefedov, *Neftekhimiya*, **12**, 862 (1972); *Chem. Abs.*, **78**, 110233 (1973); M. Hronec and V. Vesely, *Coll. Czech. Chem. Comm.*, **38**, 1095 (1973).
[99] H. Iataaki and H. Yoshimoto, *J. Org. Chem.*, **38**, 76 (1973).
[100] J. Haber, *Z. Chem.*, **13**, 241 (1973); B. S. Khromykh, V. P. Shatalov, V. B. Grigor'ev and V. I. Eremeev, *Khim. Prom. (Moscow)*, **48**, 665 (1972); *Chem. Abs.*, **78**, 15302 (1973); I. G. Tishchenko and L. S. Novikov, *Vestn. Beloruss, Univ.*, **2**, 25 (1972); *Chem. Abs.*, **78**, 123685 (1973).
[101] R. Rudham and M. K. Sanders, *J. Catal.*, **27**, 287 (1972); *Chem. Abs.*, **78**, 42535 (1973).
[102] L. Lena and J. Metzger, *Kinet. Katal.*, **14**, 136 (1973); *Chem. Abs.*, **78**, 123677 (1973).
[103] G. Greco, F. Gioia and F. Alfani, *Chim. Ind. (Milan)*, **54**, 990 (1972); *Chem. Abs.*, **78**, 57406 (1973).
[104] C. F. Cullis and S. C. W. Hook, *J.C.S. Perkin II*, **1973**, 803.
[105] M. M. Aleksishvili, S. S. Polyak and V. Ya Shtern, *Vses. Konf. Kinet. Mekh. Gazofaz. Reakts., 2nd.*, **1971**, 8; *Chem. Abs.*, **78**, 15275 (1973).
[106] E. S. Artsis, E. I. Evzerikhim, S. S. Polyak and V. Ya. Shtern, *Kinet. Katal.*, **13**, 1119 (1972); *Chem. Abs.*, **78**, 42527 (1973).
[107] N. Takamitsu and T. Hamamoto, *Nippon Kagaku Kaishi*, **1972**, 1587; *Chem. Abs.*, **78**, 28915 (1973); P. I. Shaboldo, V. A. Proskuryakov and V. M. Potekhin, *Zh. Prikl. Khim. (Leningrad)*, **45**, 1852 (1972); *Chem. Abs.*, **78**, 42533 (1973).
[108] V. V. Voronenkov and L. F. Titova, *Vop. Stereokhim.*, **1972**, 100; *Chem. Abs.*, **79**, 4675 (1973).
[109] Y. Ogata and M. Haba, *J. Org. Chem.*, **38**, 2779 (1973); O. M. Niazyan, E. A. Poladyan, A. A. Mantashyan and A. B. Nalbandyan, *Vses. Konf. Kinet. Mekh. Gazafazn. Reakts., 2nd.*, **1971**, 24; *Chem. Abs.*, **78**, 15256 (1973).
[110] C. R. H. I. de Jonge, H. J. Hageman, W. G. B. Huysmans and W. J. Mijs, *J.C.S. Perkin II*, **1973**, 1276; M. Prusiková, L. Jiračková and J. Pospišil, *Coll. Czech. Chem. Comm.*, **37**, 3788 (1972).
[111] N. A. Azatyan, G. V. Karpukhina, I. S. Belostotskaya and N. L. Komissarova, *Neftekhimiya*, **13**, 435 (1973); *Chem. Abs.*, **79**, 77750 (1973); T. V. Lomteva, G. V. Karpukhina, and Z. K. Maizus, *Izv. Akad. Nauk SSSR, Ser. Khim.*, **1973**, 930; *Chem. Abs.*, **79**, 17786 (1973).
[112] E. J. Panek and G. M. Whitesides, *J. Am. Chem. Soc.*, **94**, 8768 (1972).
[113] T. W. Bentley, J. A. John, R. A. W. Johnstone, P. J. Russell and L. H. Sutcliffe, *J.C.S. Perkin II*, **1973**, 1039; cf. *Org. Reaction Mech.*, **1971**, 349.

114 A. Nishinaga and T. Matsuura, *Chem. Comm.*, **1973**, 9.
115 J. A. Blair and A. J. Pearson, *Tetrahedron Letters*, **1973**, 203, 1681.
116 H. I. X. Mager and W. Berends, *Tetrahedron Letters*, **1973**, 4051; *Org. Reaction Mech.*, **1972**, 542.
117 E. Pelizzetti and G. Saini, *J.C.S. Perkin II*, **1973**, 1766; J. F. Corbett, *J. Soc. Cosmet. Chem.*, **23**, 683 (1972); *Chem. Abs.*, **78**, 83602 (1973).
118 J. Lubach and W. Drenth, *Rec. Trav. chim.*, **92**, 586 (1973).
119 N. Sutin and J. K. Yandell, *J. Am. Chem. Soc.*, **95**, 4847 (1973).
120 Y. Ogata and M. Yamashita, *Bull. Chem. Soc. Japan*, **46**, 2208 (1973).
121 M. Mocek and P. J. Richardson, *J. Inst. Brew., London*, **78**, 459 (1972); *Chem. Abs.*, **78**, 71095 (1973).
122 T. Kowalska, *Pr. Nauk. Univ. Slask. Katowicach*, **1972**, 87; *Chem. Abs.*, **78**, 123680 (1973); L. N. Kurina, G. I. Sterligova and N. V. Savel'eva, *Zh. Fiz. Khim.*, **46**, 2020 (1972); *Chem. Abs.*, **78**, 15416 (1973); S. Matsuda, A. Uchida and T. Yamazi *Nippon Kagaku Kaishi*, **1973**, 296; *Chem. Abs.*, **78**, 123660 (1973).
123 T. Vidoczy, E. Danoczy and D. Gal, *Magyar Kem. Folyoira*, **79**, 258 (1973); *Chem. Abs.*, **79**, 52596 (1973).
124 M. N. Fedorishcheva, M. M. Konoplyannik, V. E. Agabekov and N. I. Mitskevich, *Vesti Akad. Nauk Belarus SSR, Ser. Khim. Nauk*, **1972**, 29; *Chem. Abs.*, **78**, 15301 (1973).
125 T. Takeuchi and T. Osa, *Nippon Kagaku Kaishi*, 2374 (1972); *Chem. Abs.*, **78**, 71068 (1973).
126 S. A. Maslov, S. A. Vasil'chenko, E. A. Blyumberg and V. G. Dryuk, *Izv. Akad. Nauk SSSR, Ser. Khim.*, **1972**, 2467; *Chem. Abs.*, **78**, 96881 (1973).
127 P. Zbořil and V. Dadák, *Coll. Czech. Chem. Comm.*, **38**, 2453 (1973).
128 G. Taborsky, *Biochemistry*, **12**, 1341 (1973).
129 H. Gershman and R. H. Abeles, *Arch. Biochem. Biophys.*, **154**, 659 (1973).
130 H. Yamazaki and I. Yamazaki, *Arch. Biochem. Biophys.*, **154**, 147 (1973).
131 P. B. Baker, V. R. Holland and B. C. Saunders, *Tetrahedron*, **29**, 85 (1973).
132 D. J. Danner, P. J. Brignac, D. Arceneux and V. Patel, *Arch. Biochem. Biophys.*, **156**, 759 (1973).
133 J. I. Teng and L. L. Smith, *J. Am. Chem. Soc.*, **95**, 4060 (1973).
134 P. R. Story, J. A. Alford, J. R. Burgess and W. C. Ray, *J. Am. Chem. Soc.*, **93**, 3042, 3044 (1971); *Org. Reaction Mech.*, **1971**, 527–528.
135 P. R. Story, E. A. Whited and J. A. Alford, *J. Am. Chem. Soc.*, **94**, 2143 (1972); *Org. Reaction Mech.*, **1972**, 543.
136 K. R. Kopecky, P. A. Lockwood, J. E. Filby and R. W. Reid, *Can. J. Chem.*, **51**, 468 (1973).
137 P. S. Bailey, unpublished results cited in ref. 136.
138 D. R. Kerur and D. G. M. Diaper, *Can. J. Chem.*, **51**, 3110 (1973).
139 R. W. Murray and R. Hagen, *J. Org. Chem.*, **36**, 1103 (1971); *Org. Reaction Mech.*, **1971**, 528.
140 R. W. Murray and A. Suzui, *J. Am. Chem. Soc.*, **95**, 3343 (1973); see also R. Nahavandi, F. Razmara and M. P. Stevens, *Tetrahedron Letters*, **1973**, 301.
141 R. W. Murray and D. P. Higley, *J. Am. Chem. Soc.*, **95**, 7886 (1973).
142 R. P. Lattimer, C. W. Gillies and R. L. Kuczkowski, *J. Am. Chem. Soc.*, **95**, 1348 (1973); cf. CNDO/2 and EHT calculations: F. A. Rouse, *ibid.*, p. 3460.
143 N. L. Bauld, J. A. Thompson, C. E. Hudson and P. S. Bailey, *J. Am. Chem. Soc.*, **90**, 1822 (1968).
144 W. G. Alcock and B. Mile, *Chem. Comm.*, **1973**, 575; cf. *Org. Reaction Mech.*, **1972**, 534.
145 W. B. DeMore and C.-L. Lin, *J. Org. Chem.*, **38**, 985 (1973).
146 H. Henry, M. Zador and S. Fliszár, *Can. J. Chem.*, **51**, 3398 (1973).
147 E. R. Altwicker and J. Basila, *Tetrahedron*, **29**, 1969 (1973).
148 L. A. Hull, I. C. Hisatsune and J. Heicklen, *Can. J. Chem.*, **51**, 1504 (1973).
149 R. Atkinson, B. J. Finlayson and J. N. Pitts, Jr., *J. Am. Chem. Soc.*, **95**, 7592 (1973).
150a V. D. Komissarov and I. N. Komissarova, *Izv. Akad. Nauk SSSR, Ser. Khim.*, **1973**, 677; *Chem. Abs.*, **79**, 4670 (1973).
150b J. K. Crandall and W. W. Conover, *Chem. Comm.*, **1973**, 340.
151 E. G. E. Hawkins, *Angew. Chem. Internat. Ed.*, **12**, 783 (1973).
152 H. Hart and A. Oku, *J. Org. Chem.*, **37**, 4274 (1972).
153 M. A. Winnik and V. Stoute, *Can. J. Chem.*, **51**, 2788 (1973).
154 W. Hutzenlaub, H. Yamamoto, G. B. Barlin and W. Pfleiderer, *Chem. Ber.*, **106**, 3203 (1973).
155 H. Yamamoto and W. Pfleiderer, *Chem. Ber.*, **106**, 3194 (1973).
156 Y. Ogata and Y. Sawakik *J. Am. Chem. Soc.*, **95**, 4687 (1973).
157 Y. Ogata and Y. Sawaki, *J. Am. Chem. Soc.*, **95**, 4692 (1973).
158 P. W. Concannon and J. Ciabattoni, *J. Am. Chem. Soc.*, **95**, 3284 (1973).
159 K. M. Ibne-Rasa, R. H. Pater, J. Ciabattoni and J. O. Edwards, *J. Am. Chem. Soc.*, **95**, 7894 (1973).

160 Y. Ogata, Y. Sawaki and H. Inoue, *J. Org. Chem.*, **38**, 1044 (1973).
161 J. K. Crandall, W. H. Machleder and S. A. Sojka, *J. Org. Chem.*, **38**, 1149 (1973).
162 Y. Ogata and I. Urasaki, *J. Org. Chem.*, **38**, 100 (1973).
163 M. B. Hocking, *Can. J. Chem.*, **41**, 2384 (1973).
164 H. S. Isbell, H. L. Frush and E. T. Martin, *Carbohydrate Res.*, **26**, 287 (1973).
165 H. S. Isbell and H. L. Frush, *Carbohydrate Res.*, **28**, 295 (1973).
166 S. E. Kharin, P. Ya. Bachurin, A. A. Kniga and G. S. Sorokina, *Ferment. Spirit. Prom.*, **1973**, 41; *Chem. Abs.*, **79**, 91335 (1973).
167 S. P. Ponomarenko and A. A. Yasnikov, *Dopov. Akad. Nauk Ukr. RSR, Ser. B*, **34**, 936 (1972); *Chem. Abs.*, **78**, 28927 (1973).
168 P. H. Wendschuh, C. T. Pate and J. N. Pitts, Jr., *Tetrahedron Letters*, **1973**, 2931.
169 T. N. Baker, G. J. Mains, M. N. Sheng and J. G. Zajacek, *J. Org. Chem.*, **38**, 1145 (1973).
170 Y. Ogata and S. Suyama, *J.C.S. Perkin II*, **1973**, 755.
171 K. R. Kopecky and J. H. van de Sande, *Can. J. Chem.*, **50**, 4034 (1972).
172 B. G. Cox and A. Gibson, *J.C.S. Perkin II*, **1973**, 1355.
173 H. Kunzek and G. Barnikow, *Z. Chem.*, **13**, 175 (1973).
174 K. K. Banerji, *Indian J. Chem.*, **11**, 244 (1973); *Chem. Abs.*, **79**, 41672 (1973).
175 N. M. Roscher and E. J. Jedziniak, *Tetrahedron Letters*, **1973**, 1049.
176 A. Deluzarche, P. Rimmelin and J.-M. Sommer, *Bull. Soc. Chim. France*, **1973**, 1810.
177 D. G. Lee and R. Srinivasan, *Can. J. Chem.*, **51**, 2546 (1973).
178 P. Aukett and I. R. L. Barker, *J.C.S. Perkin II*, **1973**, 965.
179 W. E. Pereira, Y. Hoyano, R. Summons, V. A. Bacon and A. M. Duffield, *Biochem. Biophys. Acta*, **313**, 170 (1973).
180 M. Albert, A. Maghea and C. Panait, *An. Univ. Bucuresti Chim.*, **20**, 153 (1971); *Chem. Abs.*, **79**, 52569 (1973).
181 G. W. Kirby and J. G. Sweeny, *Chem. Comm.*, **1973**, 704.
182 H. Wunderer, *Arch. Pharm.*, **306**, 380 (1973).
183 A. C. Chatterji, *Indian J. Chem.*, **10**, 831 (1972); *Chem. Abs.*, **78**, 96875 (1973).
184 M. Fleischmann and D. Pletcher, *Adv. Phys. Org. Chem.*, **10**, 155 (1973); Electrochemistry, Specialist Periodical Reports, Vol. 3.
185 J. Bertram, J. P. Coleman, M. Fleischmann and D. Pletcher, *J.C.S. Perkin II*, **1973**, 374; see also D. B. Clark, M. Fleischmann and D. Pletcher, *J.C.S. Perkin II*, **1973**, 1578.
186 O. Hammerich and V. D. Parker, *Electrochim. Acta*, **18**, 537 (1973); *Chem. Abs.*, **79**, 104556 (1973).
187 For example see K. Nyberg, *Acta Chem. Scand.*, **27**, 503 (1973).
188 R. N. Renaud and D. E. Sullivan, *Can. J. Chem.*, **51**, 772 (1973).
189 M. Janda, J. Šrogl, M. Němec and H. Janoušová, *Coll. Czech. Chem. Comm.*, **38**, 1221 (1973).
190 I. G. Arzamanova, R. M. Logvinenko, Ya. A. Gurvish and A. E. Grinberg, *Zh. Obshch. Khim.*, **42**, 2137 (1972); *Chem. Abs.*, **78**, 83597 (1973).
191 A. Ronlán, O. Hammerich and V. D. Parker, *J. Am. Chem. Soc.*, **95**, 7132 (1973).
192 S. D. Ross, J. E. Barry, M. Finkelstein and E. J. Rudd, *J. Am. Chem. Soc.*, **95**, 2193 (1973).
193 K. Yoshida and T. Fueno, *J. Org. Chem.*, **37**, 4145 (1972).
194 M. Masui and H. Ohmori, *J.C.S. Perkin II*, **1973**, 1112.
195 G. Barbey, J. Huguet and C. Caullet, *Compt. Rend., Ser. C*, **275**, 435 (1972); *Chem. Abs.*, **78**, 15290 (1973).
196 N. C. Deno, C. Pierson and S. Seyhan, *J. Am. Chem. Soc.*, **95**, 6857 (1973).
197 A. V. Dzhaparidze, I. Ya Lyubyanitskii and N. G. Bekauri, *Tr. Gruz. Politekh. Inst.*, **8**, 29 (1971); *Chem. Abs.*, **78**, 15291 (1973); A. V. Dzhaparidze, I. Ya. Lyubyanitskii and N. G. Bekauri, *Tr. Gruz. Politekh. Inst.*, **1972**, 93; *Chem. Abs.*, **78**, 123683 (1973).
198 B. Unterhalt and V. Pindur, *Chimia*, **27**, 210 (1973).
199 B. S. Svetlov, B. A. Lur'e and G. E. Kornilova, *Gorenie Vzryv, Mater. Vses. Simp., 3rd*, **1971**, 780; *Chem. Abs.*, **78**, 123676 (1973).
200 K. B. Sharpless and R. F. Lauer, *J. Am. Chem. Soc.*, **94**, 7154 (1972); *Org. Reaction Mech.*, **1972**, 549.
201a D. Arigoni, A. Vasella, K. B. Sharpless and H. P. Jensen, *J. Am. Chem. Soc.*, **95**, 7917 (1973).
201b C. W. Wilson and P. E. Shaw, *J. Org. Chem.*, **38**, 1684 (1973).
202 K. B. Sharpless and R. F. Lauer, *J. Am. Chem. Soc.*, **95**, 1697 (1973); K. B. Sharpless, M. W. Young and R. F. Lauer, *Tetrahedron Letters*, **1973**, 1979; K. B. Sharpless, R. F. Lauer and A. Y. Teranishi, *J. Am. Chem. Soc.*, **95**, 6137 (1973); H. J. Reich, I. L. Reich and J. M. Renga, *ibid.*, p. 5813; D. L. J. Clive, *Chem. Comm.*, **1973**, 695.

[203] A. Sabesan and N. Venkatasubramanian, *Indian J. Chem.*, **10**, 1092 (1972); *Chem. Abs.*, **78**, 110246 (1973).
[204] L. D. Sharma and S. P. Srivastava, *Indian J. Chem.*, **11**, 17 (1973); *Chem. Abs.*, **79**, 4682 (1973).
[205] R. C. Gupta and S. P. Srivastava, *Indian J. Chem.*, **10**, 706 (1972); *Chem. Abs.*, **78**, 96892 (1973).
[206] See also *Org. Reaction Mech.*, **1972**, 547.
[207] H. Möhrle and S. Mayer, *Arch. Pharm.*, **306**, 209 (1973).
[208] L. K. Dyall and J. E. Kemp, *Austral. J. Chem.*, **26**, 1969 (1973).
[209] D. W. Lamson, R. Sciarro, D. Hryb and R. O. Hutchins, *J. Org. Chem.*, **38**, 1952 (1973).
[210] G. P. Gardini, *Adv. Heterocyclic Chemistry*, **15**, 67 (1973).
[211] K. Nishihata and M. Nishio, *J.C.S. Perkin II*, **1973**, 758.
[212] D. L. Ingles, *Austral. J. Chem.*, **26**, 1015, 1021 (1973).
[213] M. C. Agrawal and S. P. Mushram, *J.C.S. Perkin II*, **1973**, 762.
[214] N. G. Satsko, A. P. Belov and I. I. Moiseev, *Katal. Reakts. Zhidk. Faze, Tr. Vses. Konf., 3rd.*, **1971**, 441; *Chem. Abs.*, **79**, 41675 (1973); N. G. Satsko, A. P. Belov and I. I. Moiseev, *Kinet. Katal.*, **13**, 892 (1972); *Chem. Abs.*, **78**, 83584 (1973).
[215] N. S. Srinivasan and N. Venkatasubramanian, *Indian J. Chem.*, **10**, 1014 (1972); *Chem. Abs.*, **78**, 96890 (1973).
[216] V. D. Pokhodenko, V. A. Khizhnyi, V. G. Koshechko and O. I. Shkrebtii, *Dokl. Akad. Nauk SSSR*, **210**, 640 (1973); *Chem. Abs.*, **79**, 52595 (1973).
[217] H. Sayo, S. Ozaki and M. Masui, *Chem. Pharm. Bull.*, **21**, 415 (1973); *Chem. Abs.*, **78**, 135233 (1973).
[218] E. A. Karakhanov, M. V. Vagavob and E. A. Viktorova, *Dokl. Akad. Nauk SSSR*, **206**, 118 (1972); *Chem. Abs.*, **78**, 15294 (1973).
[219] R. C. Hevey, J. Babson, A. L. Maycock and R. H. Abeles, *J. Am. Chem. Soc.*, **95**, 6125 (1973).
[220] R. R. Rando, *J. Am. Chem. Soc.*, **95**, 4438 (1973).
[221] E. Kunec-Vajic and K. Weber, *Acta Pharm. Jugoslav.*, **23**, 79 (1973); *Chem. Abs.*, **79**, 65499 (1973).
[222] R. J. Maguire and H. B. Dunford, *Can. J. Chem.*, **51**, 1721 (1973).
[223] A. A. Baum, L. A. Karnischky, D. McLeod, Jr., and P. H. Kasai, *J. Am. Chem. Soc.*, **95**, 617 (1973).
[224] C. H. DePuy, H. L. Jones and W. M. Moore, *J. Am. Chem. Soc.*, **95**, 477 (1973).
[225] A. A. Gorman, *Photochemistry*, Specialist Periodical Reports, Chemical Society, London, Vol. **4**, p. 708.
[226] H. O. House, *Modern Synthetic Reactions*, 2nd edn., Benjamin, Menlo Park, California, 1972.
[227] M. T. Wuesthoff, *Tetrahedron*, **29**, 791 (1973).
[228] G. J. Karabatsos, C. Zioudrou and I. Moustakali, *Tetrahedron Letters*, **1972**, 5289; C. Zioudrou, P. Chrysochou, G. J. Karabatsos, D. Herlem, and R. N. Nipe, *ibid.*, p. 5293; L. Gruber, I. Tömösközi and L. Ötvös, *Tetrahedron Letters*, **1973**, 811; C. Bernard, M. T. Maurette and A. Lattes, *ibid.*, p. 2305; O. Červinka and O. Kříž, *Coll. Czech. Chem. Comm.*, **38**, 294 (1973); see also O. Kříž, J. Macháček and O. Štrouf, *ibid.*, p. 2072; O. Kříž, B. Cásenský and O. Štrouf, *ibid.*, p. 2076.
[229] E. C. Ashby, J. R. Boone and J. P. Oliver, *J. Am. Chem. Soc.*, **95**, 5427 (1973).
[230] C. Alvarez-Ibarra, F. Fernández-González, A. Garcia-Martinez, R. Pérez-Ossorio and M. L. Quiroga, *Tetrahedron Letters*, **1973**, 2715.
[231] J. Klein, *Tetrahedron Letters*, **1973**, 4307.
[232] C. W. Jefford, D. Kirkpatrick and F. Delay, *J. Am. Chem. Soc.*, **94**, 8905 (1972).
[233] H. Tanida, T. Okada and K. Kotera, *Bull. Chem. Soc. Japan*, **46**, 934 (1973).
[234] J. R. Dimmock, W. A. Turner, P. J. Smith and R. G. Sutherland, *Can. J. Chem.*, **51**, 427 (1973).
[235] B. C. Hartman and B. Rickborn, *J. Org. Chem.*, **37**, 4246 (1972).
[236] Y. Boyer, M. Sélim and P. Rumpf, *Bull. Soc. Chim. France*, **1973**, 1060.
[237] B. L. Fox and R. J. Doll, *J. Org. Chem.*, **38**, 1136 (1973).
[238] J. A. Katzenellenbogen and S. B. Bowlus, *J. Org. Chem.*, **38**, 627 (1973).
[239] H. Haubenstock and P. Quezada, *J. Org. Chem.*, **37**, 4067 (1972).
[240] S. Masamune, P. A. Rossy and G. S. Bates, *J. Am. Soc.*, **95**, 6452 (1973).
[241] R. O. Hutchins and D. Kandasamy, *J. Am. Chem. Soc.*, **95**, 6131 (1973).
[242] E. Volpi, G. Biggi and F. Pietra, *J.C.S. Perkin II*, **1973**, 571.
[243] C. A. Brown, S. Krishnamurthy and S. C. Kim, *Chem. Comm.*, **1973**, 391.
[244] L. T. Scott and W. D. Cotton, *Chem. Comm.*, **1973**, 320.
[245] T. Matsuda and K. Koida, *Bull. Chem. Soc. Japan*, **46**, 2259 (1973).
[246] A. B. Uzienko and Yasnikov, *Ukr. Khim. Zh.* (*Russian Ed.*), **38**, 1289 (1972); *Chem. Abs.*, **78**, 96871 (1973).
[247] F. R. Jensen, J. J. Miller, S. J. Cristol and R. S. Beckley, *J. Org. Chem.*, **37**, 4341 (1972).
[248] H. Patin and R. Dabard, *Bull. Soc. Chim. France*, **1973**, 2756, 2760, 2764.

[249] J. Hatem and B. Waegell, *Tetrahedron Letters*, **1973**, 2019, 2023.
[250] N. H. Khan and H. R. Kidwai, *J. Org. Chem.*, **38**, 822 (1973).
[251] R. U. Lemieux, K. James, and T. L. Nagabhushan, *Can. J. Chem.*, **51**, 27 (1973).
[252] H. C. Brown and P. Heim, *J. Org. Chem.*, **38**, 912 (1973); see also K. Ishizumi, S. Inaba and H. Yamamoto, *J. Org. Chem.*, **37**, 4111 (1972).
[253] G. W. Kabalka and N. S. Bowman, *J. Org. Chem.*, **38**, 1607 (1973).
[254] T. C. Wolfe and H. C. Kelly, *J.C.S. Perkin II*, **1973**, 1948.
[255] J. G. Smith and I. Ho, *J. Org. Chem.*, **38**, 3601 (1973).
[256] D. B. Malpass and J. F. Eastham, *J. Org. Chem.*, **38**, 3718 (1973).
[257] V. Kalyanaraman and M. V. George, *J. Org. Chem.*, **38**, 507 (1973).
[258] J. G. Smith and I. Ho, *J. Org. Chem.*, **38**, 2776 (1973).
[259] M. Narisada and F. Watanabe, *J. Org. Chem.*, **38**, 3887 (1973).
[260] S. A. Galton and R. Abbas, *J. Org. Chem.*, **38**, 2008 (1973).
[261] A. J. Fry, W. E. Britton, R. Wilson, F. D. Greene and J. G. Pacifici, *J. Org. Chem.*, **38**, 2620 (1973).
[262] A. Demortier and A. J. Bard, *J. Am. Chem. Soc.*, **95**, 3495 (1973).
[263] H. O. House, L. E. Huber and M. J. Umen, *J. Am. Chem. Soc.*, **94**, 8471 (1972).
[264] W. F. Winecoff, F. L. O'Brien and D. W. Boykin, Jr., *J. Org. Chem.*, **38**, 1474 (1973).
[265] L. G. Chatten, R. W. Daisley and C. J. Olliff, *J.C.S. Perkin II*, **1973**, 469.
[266] L. B. Anderson, M. J. Broadhurst and L. A. Paquette, *J. Am. Chem. Soc.*, **95**, 2198 (1973).
[267] R. D. Rieke and P. M. Hudnall, *J. Am. Chem. Soc.*, **95**, 2646 (1973).
[268] A. J. Fry and R. G. Reed, *J. Am. Chem. Soc.*, **94**, 8475 (1972).
[269] P. J. Elving, S. J. Pace and J. E. O'Reilly, *J. Am. Chem. Soc.*, **95**, 647 (1973).
[270] J. W. Webb, B. Janik and P. J. Elving, *J. Am. Chem. Soc.*, **95**, 991 (1973).
[271] K. S. V. Santhanam and P. J. Elving, *J. Am. Chem. Soc.*, **95**, 5482 (1973).
[272] M. P. J. Brennan and O. R. Brown, *J.C.S. Faraday I*, **1973**, 132; A. Bewick and H. P. Cleghorn, *J.C.S. Perkin II*, **1973**, 1410.
[273] M. Fedoroňko, *Coll. Czech. Chem. Comm.*, **37**, 3897 (1972).
[274] M. Venturini and F. Secco, *J. C.S. Perkin II*, **1973**, 491.
[275] R. Gerdil, *Helv. Chim. Acta*, **56**, 196 (1973).
[276] H. Hoffmann and A. Gescher, *Arch. Pharm.*, **306**, 492 (1973).
[277] V. Hach, *J. Org. Chem.*, **38**, 293 (1973).
[278] D. C. Kleinschmidt, V. J. Shiner, Jr., and D. Whittaker, *J. Org. Chem.*, **38**, 3334 (1973).
[279] S. Bank and D. A. Noyd, *J. Am. Chem. Soc.*, **95**, 8203 (1973).
[280] H. E. Zieger, I. Angres and L. Maresca, *J. Am. Chem. Soc.*, **95**, 8201 (1973).
[281] T. A. Cooper, *J. Am. Chem. Soc.*, **95**, 4158 (1973).
[282] J. E. Gordon and V. S. K. Chang, *J. Org. Chem.*, **38**, 3062 (1973).
[283] J. F. Knifton, *J. Org. Chem.*, **38**, 3296 (1973).
[284] V. I. Stanko, A. N. Kashin and I. P. Beletskaya, *J. Organometal. Chem.*, **56**, 111 (1973).
[285] M. Sekiya, K. Mori, K. Ito and K. Suzuki, *Tetrahedron*, **29**, 57 (1973).
[286] K. Ito, H. Oba and M. Sekiya, *Chem. Pharm. Bull.*, **20**, 2112 (1972); *Chem. Abs.*, **78**, 16087 (1973).
[287] F. Secco and M. Venturini, *J.C.S. Perkin II*, **1972**, 2305.
[288] C. Shin, Y. Yonezawa, K. Katayama and J. Yoshimura, *Bull. Chem. Soc. Japan*, **46**, 1727 (1973).
[289] S. A. Foster, L. J. Leyshon and D. G. Saunders, *Chem. Comm.*, **1973**, 29.
[290] C. W. Muth, J. C. Patton, B. Battacharya, D. L. Giberson, C. A. Ferguson, *J. Heterocyclic Chem.*, **9**, 1299 (1972).
[291] D. J. Creighton, J. Hajdu, G. Mooser and D. S. Sigman, *J. Am. Chem. Soc.*, **95**, 6855 (1973).
[292] J. K. Yandell, D. P. Fay and N. Sutin, *J. Am. Chem. Soc.*, **95**, 1131 (1973).
[293] R. S. Wade and C. E. Castro, *J. Am. Chem. Soc.*, **95**, 226, 231 (1973).
[294] D. R. Fahey, *J. Org. Chem.*, **38**, 80 (1973).
[295] D. R. Fahey, *J. Org. Chem.*, **38**, 3343 (1973).
[296] S. Siegel, M. Dunkel, G. V. Smith, W. Halpern and J. Cozort, *J. Org. Chem.*, **31**, 2802 (1966).
[297] S. Siegel and D. W. Ohrt, *Tetrahedron Letters*, **1972**, 5155.
[298] *Org. Reaction Mech.*, **1972**, 563.
[299] T. Nishiguchi, K. Tachi and K. Fukuzumi, *J. Am. Chem. Soc.*, **94**, 8916 (1972).
[300] B. Bayerl, M. Wahren and J. Graefe, *Tetrahedron*, **29**, 1837 (1973).
[301] V. Z. Sharf, L. Kh. Freidlin, V. N. Krutii and T. V. Lysyak, *Izv. Akad. Nauk SSSR, Ser. Khim.*, **1972**, 2195; *Chem. Abs.*, **78**, 71087 (1973).
[302] A. Zakjariev, V. Ivanova and D. Shopov, *Izv. Otd. Khim. Nauki, Bulg. Akad. Nauk*, **5**, 483 (1972); *Chem. Abs.*, **78**, 110253 (1973).

[303] H. W. Thompson and R. E. Naipawer, *J. Am. Chem. Soc.*, **95**, 6379 (1973).
[304] H. W. Thompson, *J. Org. Chem.*, **36**, 2577 (1971); *Org. Reaction Mech.*, **1971**, 556.
[305] R. L. Burwell, Jr., D. Barry and H. H. Kung, *J. Am. Chem. Soc.*, **95**, 4466 (1973).
[306] S. Mitsui, K. Gohke, H. Saito, A. Nanbu and Y. Senda, *Tetrahedron*, **29**, 1523 (1973).
[307] S. Mitsui, Y. Senda, H. Susuki, S. Sekiguchi and Y. Kumagai, *Tetrahedron*, **29**, 3341 (1973).
[308] S. Mitsui, H. Saito, Y. Yamashita, M. Kaminaga and Y. Senda, *Tetrahedron*, **29**, 1531 (1973).
[309] R. U. Lemieux, K. James, T. L. Nagabhushan and Y. Ito, *Can. J. Chem.*, **51**, 33 (1973).
[310] T. Sakai and N. Ohi, *Nippon Kagaku Kaishi*, **1972**, 2259; *Chem. Abs.*, **78**, 83600 (1973).
[311] F. V. Linchevskii, Z. A. Min'kova and Yu. V. Ganin, *Khim. Prom.* (*Moscow*), **48**, 651 (1972); *Chem. Abs.*, **78**, 15299 (1973).
[312] L. Kh. Freidlin, N. V. Borunova and L. I. Gvinter, *Dokl. Akad. Nauk SSSR*, **207**, 1141 (1972); *Chem. Abs.*, **78**, 123686 (1973).
[313] O. B. Aliev, O. A. Narimanbekov and Z. I. Sheikhova, *Azerb. Khim. Zh.*, **1972**, 25; *Chem. Abs.*, **78**, 42532 (1973).
[314] K. Glinka and K. Slon, *Przem. Chem.*, **52**, 283 (1973); *Chem. Abs.*, **79**, 41668 (1973); K. Taya and Y. Takagi, *Tokyo Gakugei Daigaku Kiyo, Dai-4-Bu*, **24**, 142 (1972); *Chem. Abs.*, **78**, 15296 (1973); V. P. Shmonina and K. S. Kulazhanov, *Izv. Akad. Nauk Kaz. SSR, Ser. Khim.*, **23**, 63 (1973); *Chem. Abs.*, **79**, 65507 (1973).
[315] Y. Sugi and S. Mitsui, *Tetrahedron*, **29**, 2041 (1973).
[316] O. V. Bragin, T. G. Olfer'eva, A. V. Kazanskaya-Koperina and A. L. Liberman, *Kinet. Katal.*, **13**, 922 (1972); *Chem. Abs.*, **78**, 83563 (1973).

CHAPTER 5

Carbenes and Nitrenes

T. L. Gilchrist

Department of Organic Chemistry, University of Liverpool

The first volume of a new series containing critical reviews of topics in carbene chemistry has been published.[1] Carbene chemistry is also reviewed in a book on reactive intermediates.[2]

Structure

Molecular-orbital calculations on the structures of several carbenes and nitrenes have been presented. The carbenes studied include methylene,[3,4] fluorocarbene,[4] difluorocarbene,[4,5] methylcarbene,[6,7] cyclohexylidene,[6] cyclopropylidene,[8] and formylcarbene.[6,9–11] Formylcarbene (**1**) is a particularly intriguing species because of the possible types of rearrangement it can undergo, and also because of its structural relationship to a 1,3-dipole (**2**). Structures (**1**) and (**2**) can be regarded merely as resonance forms, or as valence tautomers, depending upon whether or not their geometry is different.

$$H\ddot{C}CHO \qquad H\overset{+}{C}{=}CH\overset{-}{O}$$

(**1**) (**2**)

The calculations indicate that the 1,3-dipolar structure (**2**) is not a good representation of the ground state of formylcarbene.[9,10] On the other hand, some cycloaddition reactions ascribed to acylcarbenes are more characteristic of 1,3-dipoles.[12,12a] The opposite is true of the formal 1,3-dipole benzonitrile oxide where a cycloaddition has been described that is more in keeping with its representation as phenylnitrosocarbene.[13]

Calculations predict a triplet ground state for acylnitrenes, but alkoxycarbonylnitrenes may have a singlet ground state.[14] Substituted aminonitrenes are also predicted to have triplet ground states at optimum geometry, although the chemistry of these species seems more typical of the singlet state; it is suggested that the reactions reflect a singlet state that undergoes intersystem crossing only very slowly.[15] A septet ground state is suggested for the triscarbene (**3**) on the basis of its ESR spectrum.[16]

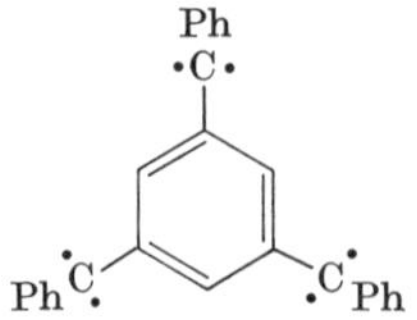

Ph
•C•

Ph C: .C Ph

(**3**)

Methods of Generation

Carbenes

A wide-ranging review of the generation and reactions of carbenes from diazo-compounds[17] and a survey of the reactions of diazomalonic esters[18] have appeared. The generation of carbenes is discussed in several other reviews: one concerns their production from cyclic precursors,[19] and others deal with the formation of dihalocarbenes from haloforms and epoxides[20,21] or by phase-transfer catalysis.[21] The last method has now been used as an efficient route to dibromocarbene[22] and to fluoroiodocarbene.[23] Another useful reagent for α-elimination is lithium 2,2,6,6-tetramethylpiperidide, which reacts with benzyl chloride in the presence of olefins to give phenylcyclopropanes in good yield.[24]

A number of papers include evidence for the thermal generation of carbenes from organometallic precursors; these are listed in Table 1.

Table 1. Methods of generating carbenes.

Carbene	Precursor	Ref.
CF_2	$(CF_3)_2Hg$	25
	$(CF_3)_3PF_2$	26
	$PhHgCF_3$	27
$CFCl$	$PhHgCCl_2F$	28
$CFBr$	$PhHgCBr_2F$	29
CCl_2	$C_6H_{11}HgCCl_3$	30
	Cl_3GeCCl_3	31
	F_3SiCCl_3	26
	Cl_3SiCCl_3	32
	$(EtO)_3SiCCl_3$	32
$CBrCl$	$C_6H_{11}HgCBr_2Cl$	30
$CFCHF_2$	$F_3SiCF_2CHF_2$	33
	$Me_3SiCF_2CHF_2$	34
$CFCF_3$	$PhHgCFBrCF_3$	35
$CFCO_2R$	$PhHgCFClCO_2R$	36
CPh_2	Cl_3GeCPh_2Cl	31

An efficient new method of generating difluorocarbene involves the reaction of phosphonium salts $[R_3\overset{+}{P}CF_2Br]Br^-$ with fluoride ions.[37] Difluorocarbene is also available from the pyrolysis of 1,1-difluorocyclopropanes,[26] and diphenylcarbene from the photolysis of 1,1-diphenylcyclopropanes. Photolysis of *cis*- and *trans*-2,3-diacetoxy-1,1-diphenylcyclopropane gave the olefins stereospecifically; the reaction was also suppressed by triplet sensitization, indicating that the extrusion of diphenylcarbene is a concerted reaction of a species in the singlet excited state.[38]

Vinylcarbenes (**4**) have been generated by thermolysis,[39] photolysis,[40,41] and metal-catalysed[42] opening of cyclopropenes; in some cases there is evidence that the ring opening is reversible. Details of the analogous thermal ring opening of 2,3-diphenylazirine to a vinylnitrene (**5**) are reported.[43]

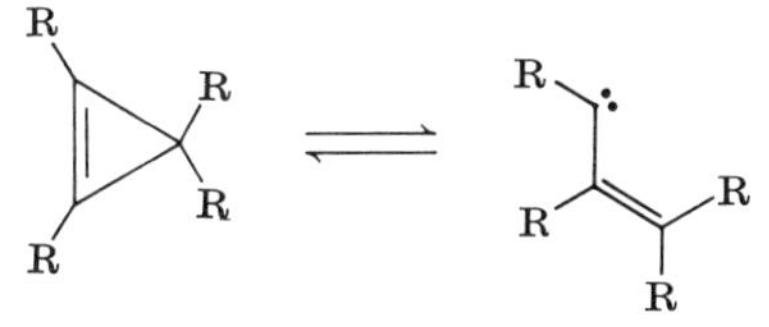

(**4**)

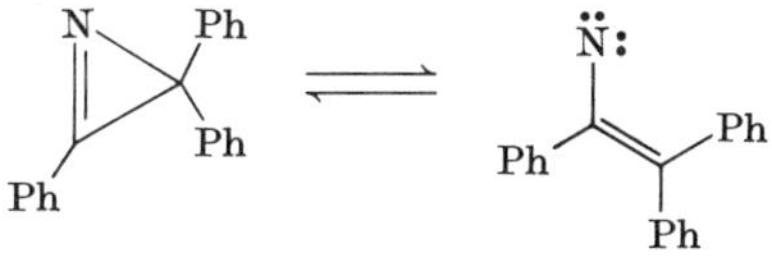

(**5**)

An account of the generation of C_1, C_2 and C_3 species from a carbon arc and their interaction with organic substrates has appeared,[44] and there is another example of the reaction of "chemically" generated monoatomic carbon formed by the thermolysis of tetrazole-5-diazonium chloride.[45] New methods of generating vinylidenecarbenes are illustrated by the routes shown to the carbenes (**6**),[46] (**7**)[47] and (**8**).[48]

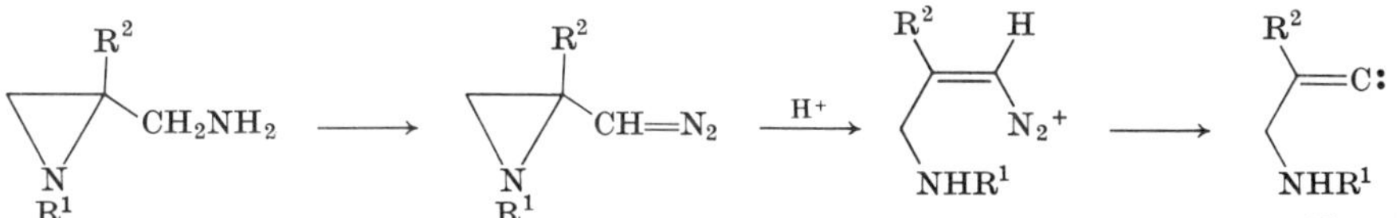

(**6**)

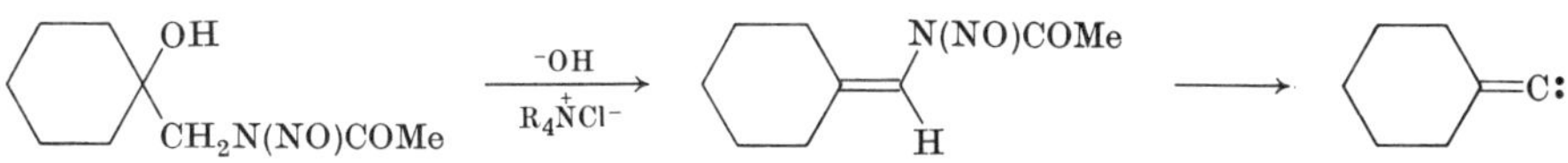

(**7**)

$$MeCHXCH{=}CHC{\equiv}CH \xrightarrow{BuO^-} MeCH{=}CHCH{=}C{=}C\colon$$

(**8**)

Details of the generation of dithiolium carbenes from carbon disulphide and alkynes have been published.[49] A new route to ketocarbenes is claimed in the reaction of αα-dibromo-ketones with zinc;[12,50] with olefins the products isolated were dihydrofurans rather than cyclopropanes.[12]

Other reactions in which carbene intermediates have been proposed include the reaction of trinitromethylsilver with chlorotrimethylsilane (in which dinitrocarbene is invoked),[51] the hydrolysis of aminomalonitriles (involving aminocyanocarbenes),[52] and the photolysis of benzoylmethylenetriphenylphosphorane.[53] Open-chain diazo-intermediates, which subsequently decompose to carbenes and undergo cyclization reactions, are proposed in the photolysis of 3,6-diphenylpyridazine 1-oxide (**9**)[54] and the pyrolysis of the benzodiazepine (**10**).[55]

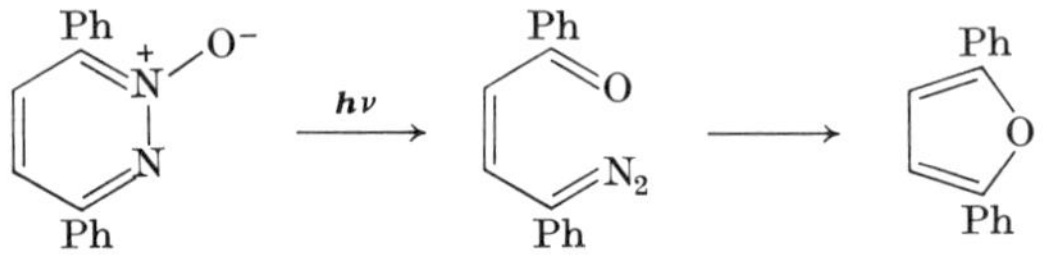

(**9**)

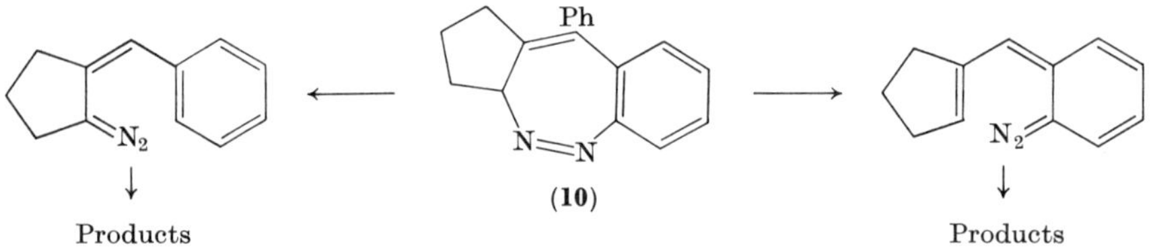

Products

(**10**)

Products

Calculations and discussion of the least-motion paths for the formation of methylene from ketene have been published.[56]

Nitrenes

The most important contributions here have been negative ones, in the sense that careful work has cast doubt on accepted nitrene mechanisms of some established reactions. One involves the production of carbazole from 2-azidobiphenyl and 2-nitrosobiphenyl. It has now been shown that the thermolysis of the azide and the deoxygenation of the nitroso-compound do not involve a common (nitrene) intermediate. The reactions of the azide (**11**) and the nitroso-compound (**12**) were carried out together, and 2- and 4-methylcarbazole were isolated. A common intermediate would require the same ratio of dideuteriated to undeuteriated material in the two carbazoles, but the ratios were different.[57]

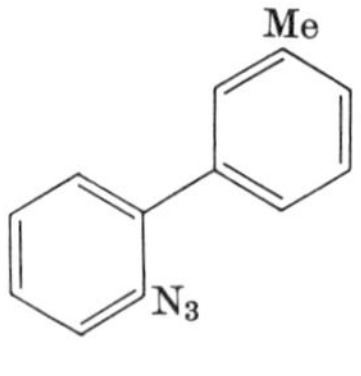

(**11**) 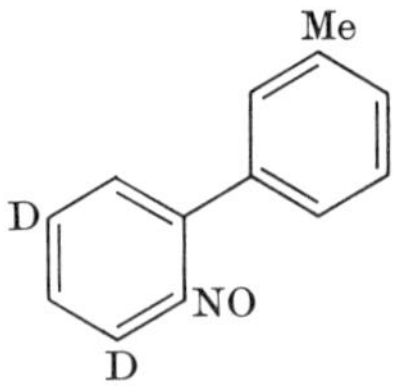 (**12**)

Another nitrene mechanism to be rejected is that proposed[58] for the formation of sulphoximides by the oxidation of sulphonamides in dimethyl sulphoxide. In the absence of dimethyl sulphoxide, the sulphonamide is unaffected by lead tetra-acetate, so it must be the sulphoxide that is attacked by the oxidant.[59] The intermediacy of *O*-nitrenes in methoxylamine oxidations has also been questioned.[60] Clearly, nitrene mechanisms proposed for deoxygenation and oxidation reactions must be examined critically.

N-Sulphonyliminopyridinium ylides have been investigated as possible precursors of sulphonylnitrenes; sulphonamides, possibly formed from triplet nitrenes, were isolated.[61] Wavelength can be an important variable in determining product ratios in the photolysis of aryl azides.[62]

Cycloaddition

A comprehensive review concerning selectivity in carbene addition,[63] and brief surveys of alkoxycarbonylnitrenes[64] and of synthetic applications of carbene reactions[65] have been published.

Theoretical treatment of the carbene addition process[3,66,67] includes an explanation for the preferred *syn*-selectivity of many such addition reactions.[67] The addition of chloro(methylthio)carbene to unsymmetrical olefins leads to the formation of both possible cyclopropanes.[68] Chlorocarbene reacts with vinyl ethers to give mainly the *syn*-adducts.[69]

Carbenes that are formed by photolysis of diazo-compounds (**13**) are rapidly deactivated to their ground states, probably because of an internal "heavy atom" effect.[70] This modifies their chemistry, and by the stereospecificity of their addition to olefins provides a useful probe for determining the spin multiplicity of the ground state.

$XCY{=}N_2$ (**13**)

$X = RHg, Cl, Br, I$

$Y = CN, CO_2R$

$X, Y = CBr{=}CBr{-}CBr{=}CBr$

The addition of cycloheptatrienylidene (**14**) to penta-1,3-diene and to styrene is rationalized as involving nucleophilic attack by the carbene.[71] In its reaction with 1,2-dicyanocyclo-octatetraene, however, a more complex mechanism, involving electron-transfer from the carbene, is suggested.[72] A similar mechanism is proposed for the reaction of the dibenzo-derivative of (**14**) with tetracyanoethylene;[73] in this case, the radical anion of tetracyanoethylene can be detected. Other cycloaddition reactions of carbene (**14**)[74] and of its 10π-electron analogue (**15**)[75] have been described.

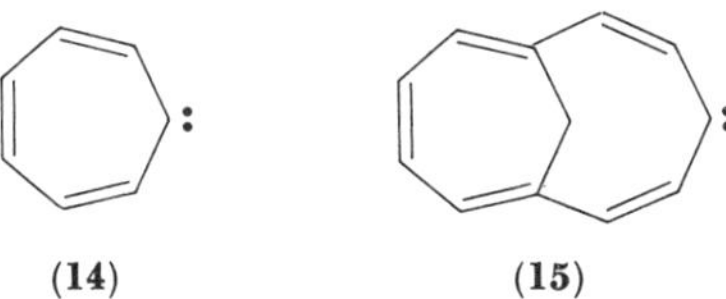

(**14**) (**15**)

The relative rates of addition of dichlorocarbene to methylenecyclohexanes with and without a methylenedioxy-group in the 4-position failed to show any evidence for prior ylide formation with the oxygen function.[76] Full reports of the reaction of diphenylcarbene with alkenes and alkynes,[77] and of the oxidation of *N*-aminophthalimide in the presence of alkynes,[78] have appeared. In a study of the addition of alkenes to phthalimidonitrene, their relative reactivity is related to the extent of the frontier orbital interactions.[79] With activated aromatic compounds, phthalimidonitrene gives mainly either CH insertion products or azepines, depending upon the source of the nitrene.[80] Addition reactions of benzoylnitrene[81] and of ethoxalylnitrene, EtOCO·CON,[82] have also been reported. Carbonyl azides act as formal 1,3-dipoles in their addition to isocyanates.[83]

In an important contribution, Salomon and Kochi have investigated the mechanism of copper-catalysed cyclopropanation of olefins.[84] Copper(I) trifluoromethanesulphonate appears to catalyse the reaction by prior co-ordination of the olefin; the complex then reacts directly with the diazo-compound. With other copper(I) salts for which olefin co-ordination is likely to be less important, an alternative mechanism, involving reaction of the olefin with a copper–carbenoid complex, is proposed. Apparent catalysis by copper(II) is shown to be due to copper(I) formed *in situ*. Similar mechanistic conclusions have been reached by another group after studying copper-catalysed addition of ethoxycarbonylcarbene to dicyclopentadiene.[85] With hexa-2,4-dienes, the selectivity of addition of this carbenoid is mainly steric in origin.[86]

A free-radical chain mechanism is proposed for the oxygen-accelerated cyclopropanation of olefins by the diethylzinc–dihalomethane system.[87] A similar carbenoid, from tri-iodomethane and diethylzinc, has been used for the ring expansion of alkylbenzenes.[88] Copper-catalysed decomposition of aryldiazocarbonyl compounds leads to intramolecular addition to the aromatic ring;[89,90] for example, the diazo-ketone (**16**) is a precursor of azulene.[90] Details of the cyclopropanation of olefins using haloacetic esters and copper(I) isocyanide complexes have appeared.[91]

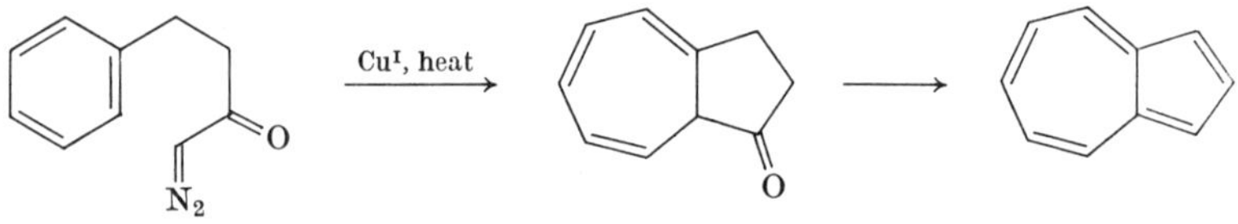

N_2 O Cu^I, heat O

(**16**)

Other cycloadditions include the reactions of cyclopentadienylidenes with alkenes, alkynes and benzene,[92] and those of methoxycarbonylnitrene with cyclobutenes, cyclopropenes and methylenecyclopropane.[93] An attempt to produce tetrahedrane by pyrolysis of the *trans*-butenedial bistosylhydrazone salt (**17**) was unsuccessful, although the product (acetylene) could have been formed from tetrahedrane.[94] Tetramethylbicyclobutane, formed by intramolecular addition, was one of the products of pyrolysis of the tosylhydrazone salt (**18**).[95]

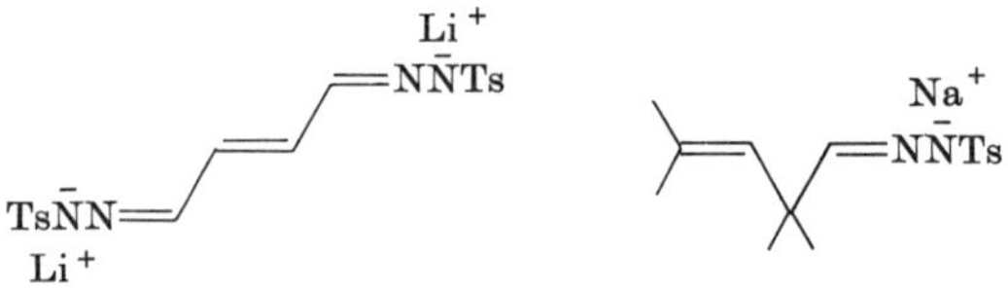

Li^+ =$N\bar{N}Ts$ Ts$\bar{N}$N= Li^+ Na^+ =$N\bar{N}Ts$

(**17**) (**18**)

Addition reactions of carbene analogues include the formation of silacyclopent-3-ene from silylene and butadiene,[96] and the reaction of molybdenocene and tungstenocene with alkenes and alkynes to form three-membered ring derivatives.[97]

Insertion and Abstraction

The mechanism of insertion of singlet methylene into hydrogen has been subjected to a very detailed analysis using trajectory calculations, which give a dynamic picture of the colliding molecules.[98] Although the results are in accord with earlier theoretical predictions, they indicate that insertion can take place over a wider range of initial conditions and by a much more complex pathway than static "minimum-energy-path" analysis would suggest. The reaction of methylene with hydrogen has also been the subject of other calculations.[99]

Calculations by the MINDO/2 method on the reaction of singlet and triplet methylene with methane predict two initial approach pathways for each species.[3] The singlet can approach along the axis of a CH bond of methane, either in an eclipsed (**19**) or in a staggered conformation (**20**). The triplet can also approach along the CH axis, or along an axis midway between two CH bonds of methane; the latter approach is predicted to lead to insertion by the triplet. Insertion of methylene has been analysed by Zimmerman, using his "MO following" concept,[66] and the mechanism of insertion of nitrenes has also been discussed.[14]

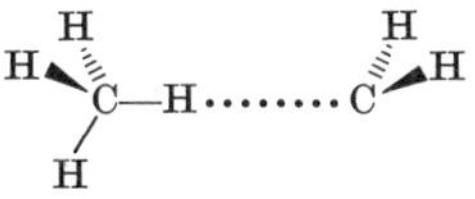

(**19**) 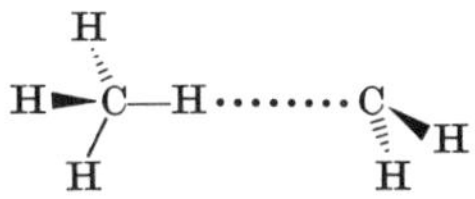 (**20**)

Further experimental work on the selectivity of the insertion of methylene into CH bonds of hydrocarbons has been presented.[100] When ketene is irradiated in cyclopropane, the excess energy of the methylcyclopropane so formed is related to the wavelength of the light used.[101]

A useful application of intramolecular insertion is to the selective functionalization of a tertiary CH bond; this is illustrated by the conversion of the diazo-ester (**21**) into the diacid (**22**).[102] Other intramolecular insertions have also been shown to be selective, leading to the formation of three-[103] and five-membered rings.[104]

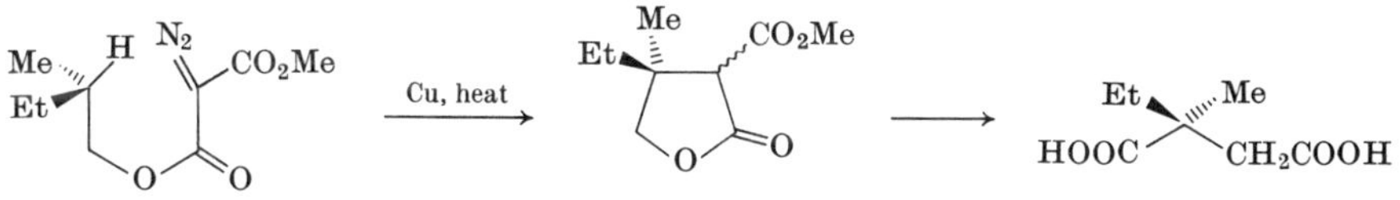

Cu, heat

(**21**) (**22**)

The insertion of dichlorocarbene into the benzylic CH bond of (+)-2-phenylbutane is a concerted process, during which a small positive charge develops at the benzylic carbon atom. The kinetic deuterium isotope for the insertion is 2.5.[105] Dichlorocarbene insertion reactions into SiC and β-CH bonds of silanes have also been studied.[106] The synthesis of organometallic compounds by carbene insertion has been reviewed.[107]

The deuterium-labelled cyclopentylcarbene (**23**) reacts by 1,3-insertion to give the deuteriocyclopropanes (**24**) and (**25**), the ratio of which is highly dependent on the carbene precursor.[108] The *exo*-isomer (**24**) is the major isomer formed in tosylhydrazone decompositions, as required if insertion takes place preferentially through a transition state (**26**). With organometallic precursors, the *endo*-isomer (**25**) predominates, indicating a different mechanism, such as a double displacement reaction. The carbenoids

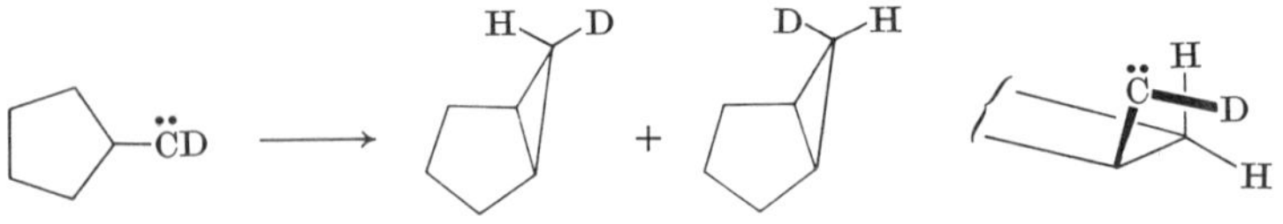

(23) (24) (25) (26)

(**27**) and (**28**) also give different insertion products, showing that the free carbenes are not formed.[109]

Abstraction competes with addition in the reaction of diphenylcarbene with alkenes.[77] Aryldiazirines have been used as sources of arylcarbenes for photolabelling of biological receptor sites by abstraction–recombination reactions.[110]

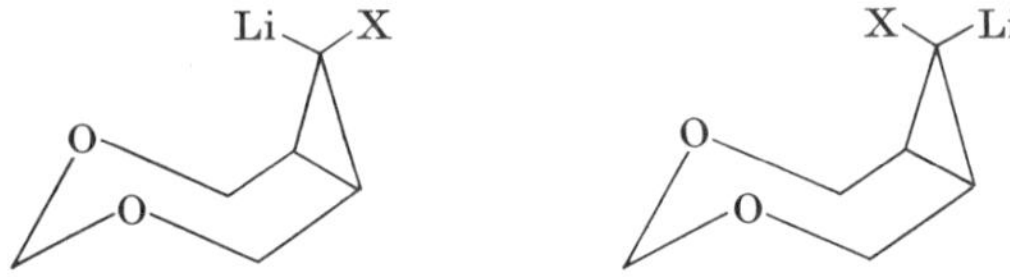

(27) (28)

The CIDNP technique has been used to investigate the reaction of triplet ethoxycarbonylcarbene with decalin.[111] Insertion reactions of ethoxalylnitrene[82] and of methanesulphonylnitrene[112] into CH and OH bonds, and of silylene into SiH and GeH bonds,[113] have also been reported.

Aromatic Substitution

Reactions of this type involving carbenes and nitrenes are now surveyed in an annual Specialist Periodical Report.[114]

Dimethyl diazomalonate gives ring-substitution products when it decomposes in alkylbenzenes, but its reaction with benzene is less favourable than insertion into cyclohexane.[115] Addition reactions followed by rearrangement are suggested to account for the substitution of phthalimodonitrene into alkoxybenzene derivatives[80] and for formation of the quinoline derivative (**30**) as a product of the pyrolysis of the azide (**29**).[116] *NN*-Dimethylaniline undergoes substitution at the 2- and the 4-position by pentafluorophenylnitrene.[117]

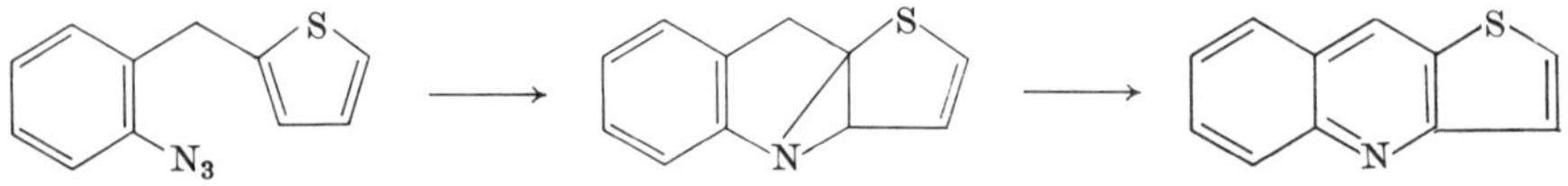

(29) (30)

Decomposition of the diazonium betaine (**31**) in benzene leads to the formation of the substitution product (**33**); a carbene (**32**) is suggested as an intermediate.[117a]

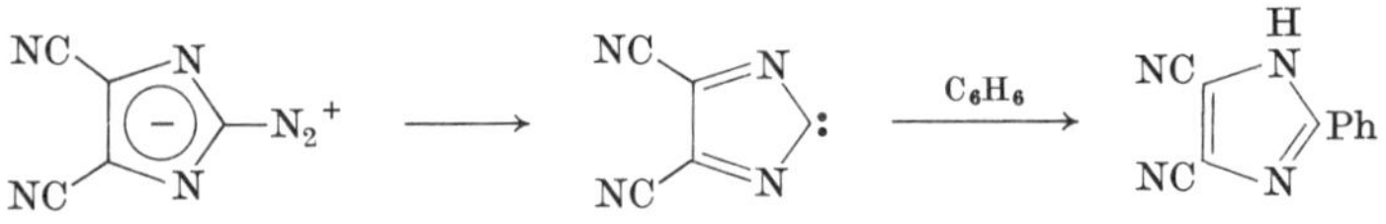

(31) (32) (33)

Reaction with Nucleophiles

The reactions of carbenes with bivalent sulphur compounds have been reviewed.[118] Reactions of this type involving diazocyclopentadiene,[119] diazo-ketones,[120] and diazo-esters[121,122] have been described; they proceed through ylides. With vinyl sulphides, the intermediate ylide (**34**) reacts by 1,5-hydrogen transfer and extrusion of an olefin.[122]

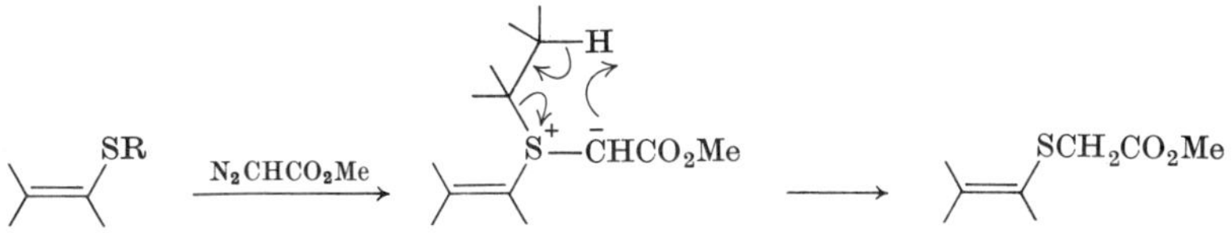

N_2CHCO_2Me ; $\overset{+}{S}—\bar{C}HCO_2Me$; SCH_2CO_2Me

(**34**)

Sulphoxides similarly give sulphoxonium ylides as primary products with diazo-ketones and copper catalysts;[120,123] sulphides are also formed. Products consistent with the formation of cyclic ylides (**36**) are reported from the decomposition of the tosyl-hydrazone salts (**35**);[124] other attempts to observe intramolecular reactions of this

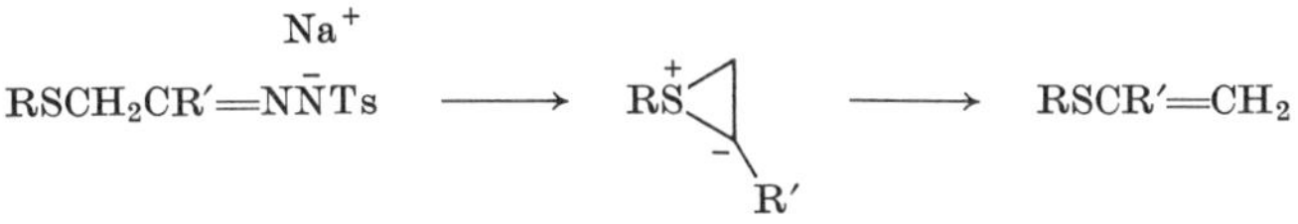

$$RSCH_2CR'{=}N\bar{N}Ts\ (Na^+) \longrightarrow R\overset{+}{S}\text{(ring, }\bar{C}R') \longrightarrow RSCR'{=}CH_2$$

(**35**) (**36**)

type were unsuccessful, however.[125] Sulphonium ylides may also be intermediates in the reaction of singlet monoatomic carbon with cyclic thioethers.[126]

There are several examples of carbene insertion into the OH bond of alcohols, which probably involve intermediate oxygen ylides.[127–129] Rhodium catalysis of diazo-ester decomposition in hydroxylic solvents makes the insertion particularly efficient.[129] The reaction of dichlorocarbene with 1,2-glycols gives olefins as primary products; a possible mechanism involves a cyclic dialkoxycarbene intermediate (**37**), generated from the

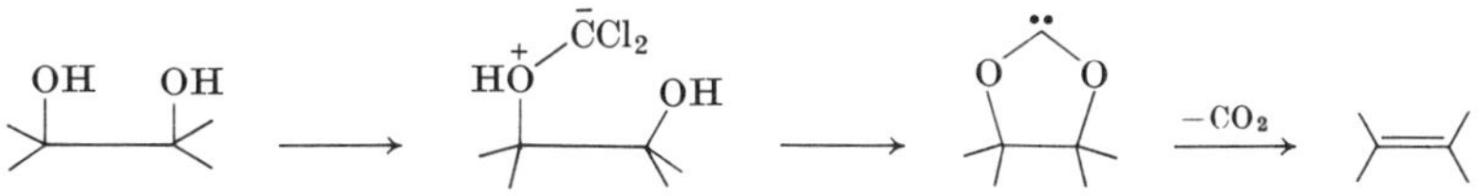

OH OH ⟶ $H\overset{+}{O}$–$\bar{C}Cl_2$ OH ⟶ (O, O ring carbene) $\xrightarrow{-CO_2}$ olefin

(**37**)

oxygen ylide.[130] Another useful synthetic procedure which probably involves ylide intermediates is the conversion of amides, thioamides and oximes into nitriles by the reaction with dichlorocarbene.[131] Further examples of the cleavage of aziridines with dichlorocarbene are reported;[132] azirines also react by ring opening to give dichloro-imines (**38**).[133]

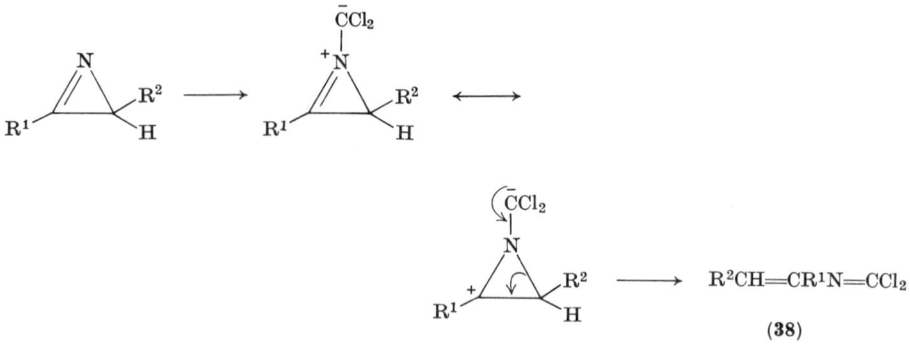

$$R^2CH{=}CR^1N{=}CCl_2$$

(38)

Other carbene reactions that probably involve nucleophilic attack include the insertion of methylene and dichlorocarbene into SiCl bonds,[134] and the reaction of 4-diazo-2,6-di-*tert*-butylcyclohexadienone (**39**) with halogenated solvents (for which a CIDNP study is reported[135]) and with thiophen, which gives the unsaturated ketones (**40**).[136] Examples of cycloaddition of carbenes to carbonyl compounds[137] may involve intermediate oxygen ylides.

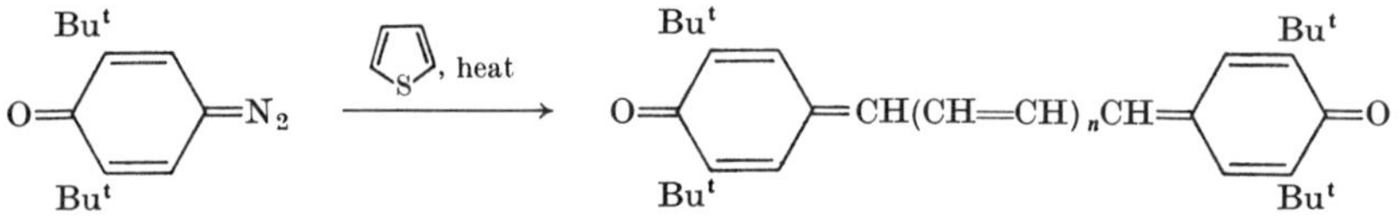

(**39**) (**40**) ($n = 1$ or 3)

Ethoxycarbonylnitrene acts as an electrophile in its reaction with ethers,[138] with 1-azidohexane[139] and with triphenylarsine;[59] the last reaction is also observed with other nitrenes. Pentafluorophenylnitrene, generated thermally from the azide, reacts at the nitrogen atom of aniline to give products derived from the ylide (**41**).[140] Similar reactions

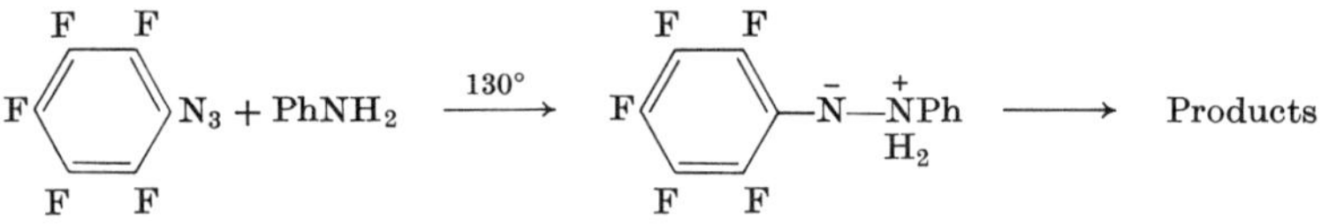

$$C_6F_5N_3 + PhNH_2 \xrightarrow{130^\circ} C_6F_5\bar{N}{-}\overset{+}{N}H_2Ph \longrightarrow \text{Products}$$

(**41**)

are reported for other aryl azides bearing electron-withdrawing substituents and electron-rich amines.[62,141] Phenylnitrene is proposed as an intermediate in the palladium chloride-catalysed reaction of nitrobenzene with carbon monoxide, which gives phenyl isocyanate and diphenylurea.[142]

Rearrangement

Full accounts of the work by W. M. Jones' group[143] and by Crow and Paddon-Row[144] on the rearrangement of phenylcarbene and related intermediates have appeared. The latter authors conclude that the rearrangements occur from a pool of rapidly interconverting intermediates. These degenerate interconversions occur more rapidly than hydrogen shifts, which in turn are more rapid than ring contraction. Carbene–carbene rearrangements (**42**)–(**43**)[145] and (**44**)–(**45**),[146] related to the ring expansion of phenylcarbene, have been reported. The formation of benzocyclobutene and styrene in the high-temperature pyrolysis of methyl *p*-tolylacetate is cited as evidence for the generation and

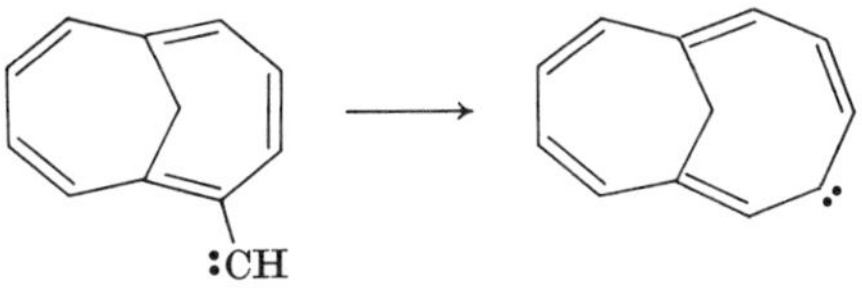

:CH

(**42**) (**43**)

CH

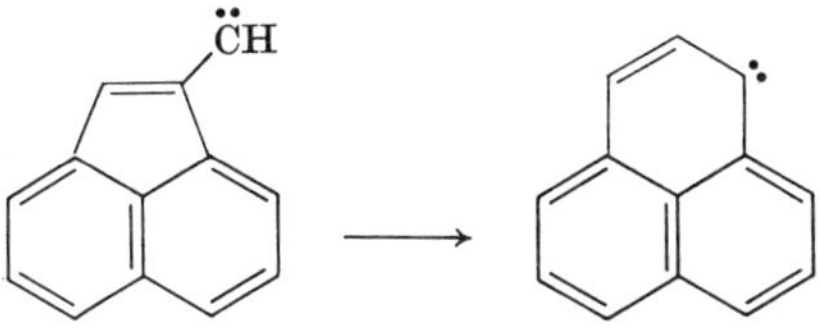

(**44**) (**45**)

rearrangement of *p*-tolylcarbene,[147] and the gas-phase rearrangement of quinolylnitrenes into isoquinolylnitrenes is now also ascribed to a ring expansion–contraction sequence rather than to a fragmentation and re-combination.[148]

When 2-azidopyridine 1-oxide (**46**) is heated, 2-cyano-1-hydroxypyrrole is formed; this is ascribed to a ring opening followed by re-cyclization; but because of the low

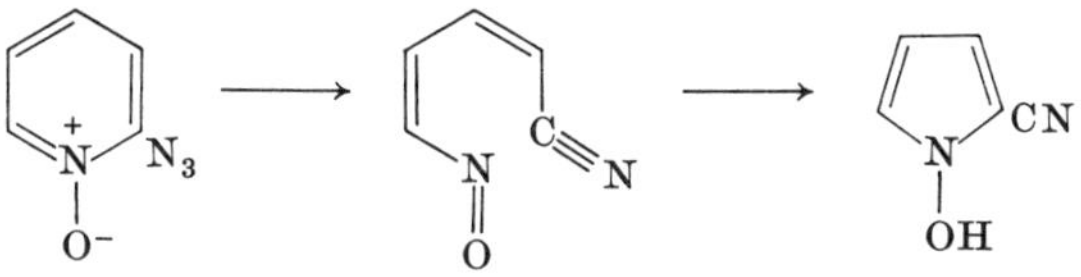

N^+ N_3 O^- → N=O, C≡N → N–OH, CN

(**46**)

pyrolysis temperature (90°) it is not certain that nitrenes are involved in these reactions.[149] There are further examples of the ring expansion of arylnitrenes in the presence of amines to give azepines.[62,150]

The status of oxirenes (**47**) as reaction intermediates in diazo-ketone decompositions is the subject of several theoretical studies. On the basis of extended Hückel calculations, oxirenes are predicted to be intermediates only in photochemical decompositions.[11] This prediction is not borne out by results obtained last year on gas-phase

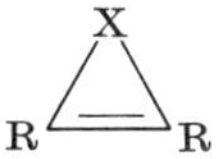

X R R

(**47**), X = O; (**48**), X = NR′

pyrolyses of diazo-ketones, in which oxirenes were implicated,[151] or by other calculations[9,10] which suggest that oxirenes are energetically accessible intermediates. All the calculations predict that the Wolff rearrangement of formylcarbene to ketene is a process of little or no activation energy, and this is confirmed by a study of the matrix photolysis of diazoacetaldehyde:[152] even in an ethylene matrix, ketene is the sole organic product, indicating that formylcarbene is hardly a discrete intermediate. With ethyl diazoacetate, however, Wolff rearrangement to ethoxyketene is substantially reduced in hydrocarbon matrices. Full papers have appeared with details of the work of two groups on the diazo-ketone–oxirene problem,[153] and of a third group on the related 1*H*-azirines (**48**) (NR′ = phthalimido).[78,154] The work on 1*H*-azirines involving the gas-phase pyrolysis of 1,2,3-triazoles has been extended to 1-alkyltriazoles (**49**); imino-carbenes are suggested as intermediates; these undergo Wolff rearrangement and also 1,4-hydrogen transfer to give isoquinolines (**50**). 1*H*-Azirines are also implicated as intermediates in these reactions.[155]

Ph, Ph, N, N, N, CH_2R (**49**) → Ph, Ph, NCH_2R → $Ph_2C{=}C{=}NCH_2R$ → Ph_2CCN (CH_2R); → H, Ph, Ph, N, CHR → Ph, N, R (**50**)

Other studies of the Wolff rearrangement[40,156–158] include examples of migration of methoxyl groups.[114,156] The rearrangement is suppressed in diazo-ketones containing a "heavy atom" substituent.[70]

Mechanistic studies are reported of the photochemical ring expansion of cyclic ketones to oxycarbenes [e.g., (**51**) → (**52**)][159] and the thermal reverse process.[160] There is a loss of stereochemistry in the ring contraction, so the reaction is partly or completely a stepwise one. Calculations on the rearrangement of cyclopropylidene to allene,[8] as well as further examples of the reaction[161,162] have been described. Side reactions are apparent when the cyclopropane ring is substituted with *tert*-butyl groups; cyclopropenes and bicyclobutanes are also formed.[161] A related reaction is the ring contraction of cyclobutylidenes to methylenecyclopropane derivatives.[163]

The 1,2-hydrogen shift in methylcarbene was subjected to detailed analysis.[6,7,66] In the conformation (**53**), the hydrogen atoms (arrowed) adjacent to the vacant *p*-orbital

O; $h\nu$, heat; O; H, H, H, H

(**51**) (**52**) (**53**)

can migrate with little or no activation, but movement of the third hydrogen is much less favourable; in cases where rotation about the carbon–carbon bond is inhibited, hydrogen migration could therefore be stereospecific; in cyclohexylidene, for example, the axial hydrogen atoms adjacent to the carbene centre should migrate preferentially. The 1,2-hydrogen migration has been shown to be a very efficient intramolecular trapping process for phenylmethylcarbene.[164] The competition between hydrogen and phenyl migration in the photolysis or thermolysis of the diazo-compound (**54**) is dependent on the mode of decomposition; it is proposed that migration of the phenyl group takes place in the triplet carbene, whereas hydrogen migration occurs exclusively in the singlet.[165] The diazo-compounds (**54**) and (**55**) give the same ratio of stereoisomers of 1,2-diphenylpropene; the authors invoke an excited singlet state (**56**) of the olefin as a discrete intermediate in the reactions.

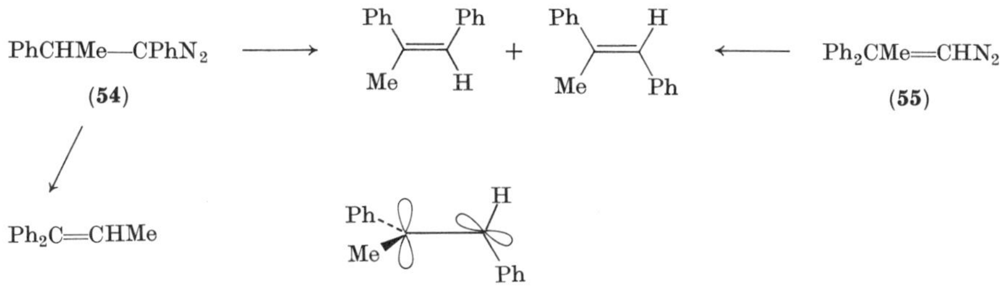

PhCHMe—CPhN_2 (**54**) ⟶ (Ph)(Me)C=C(Ph)(H) + (Ph)(Me)C=C(H)(Ph) ⟵ Ph_2CMe=CHN_2 (**55**)

(**54**) ⟶ Ph_2C=CHMe

(**56**)

The 1,2-migration of alkyl groups in carbenes has been used to generate highly unstable olefins as transient intermediates. The carbene (**57**), for example, rearranges in the gas phase to the olefin (**58**) which then undergoes a retro Diels–Alder reaction (by cleavage at the arrowed bonds), or dimerizes.[166] A similarly strained olefin is generated from 1-adamantylcarbene.[167] With the α-silylcarbene (**59**), one reaction involves methyl migration to give the ester (**60**) containing an unstable silicon–carbon double bond and this is then attacked by the solvent.[127]

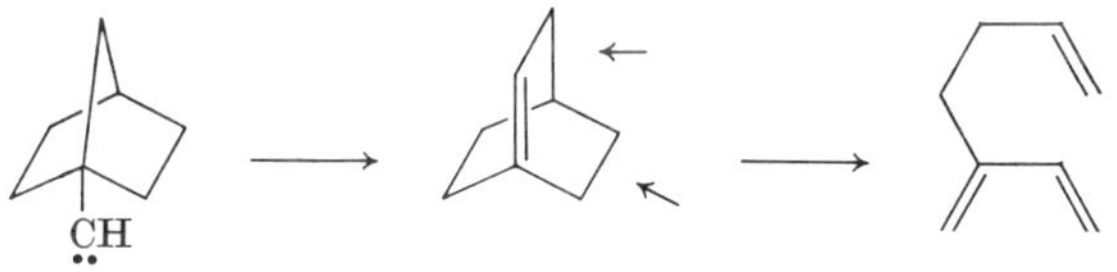

(**57**) ⟶ (**58**) ⟶

$Me_3Si\ddot{C}CO_2Et$ (**59**) ⟶ $Me_2Si{=}CMeCO_2Et$ (**60**) $\xrightarrow{\text{ROH}}$ $Me_2Si(OR)CHMeCO_2Et$

The photochemical ring opening of 3,3-dimethyl-1,2-diphenylcyclopropene (**61**) to the vinylcarbene (**62**) is reversible, but the carbene also reacts by 1,4-hydrogen transfer.[41] Other examples of cyclopropene–vinylcarbene equilibria have also been described.[39] 1,4-Migration of alkyl groups also takes place in the cyclic carbenes (**63**), but this has been shown to be a free-radical process rather than a concerted intramolecular reaction;[168] it provides a useful route to [7]paracyclophane.[169]

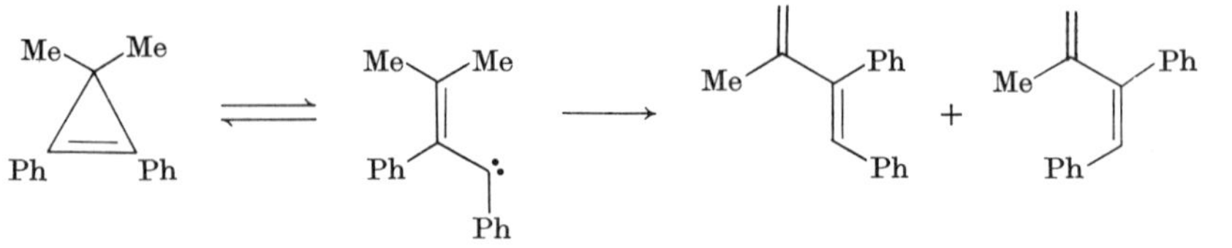

Me Me Ph Ph (61) ⇌ Me Me Ph Ph (62) → Me Ph Ph + Me Ph Ph

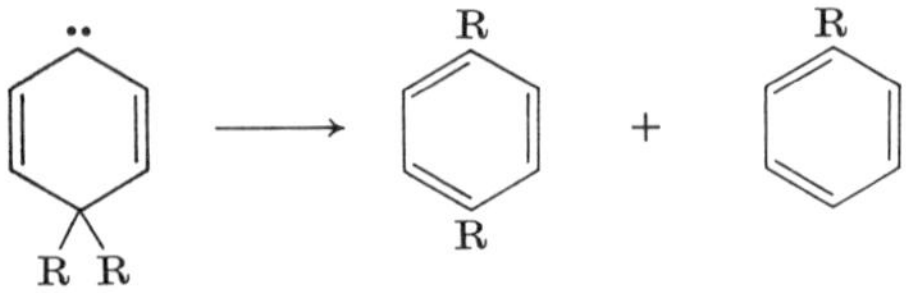

(63) R = Me, Et; R, R = $(CH_2)_7$

The carbenes (**64**) and (**65**), generated by vacuum-pyrolysis, give markedly different rearrangement products; a 1,2-alkyl-shift is the major reaction of the *exo*-isomer (**64**), whereas the products of the *endo*-isomer are consistent with a "foiled methylene" structure in which the bivalent carbon is bent towards the three-membered ring.[170]

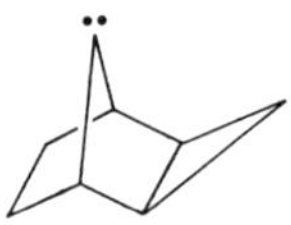

(64) 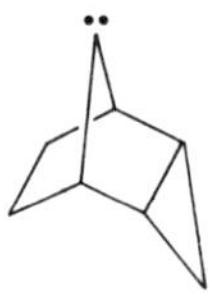 (65)

Other reactions in which 1,2-alkyl shifts are implicated include the reactions of bromomethylenecycloalkenes with base,[171] the isomerization of aminonitrenes to azoalkanes,[172] and the ring expansion of the phenalenide ion (**66**) with dichloro[$^2H_2$]methane and butyl-lithium.[173]

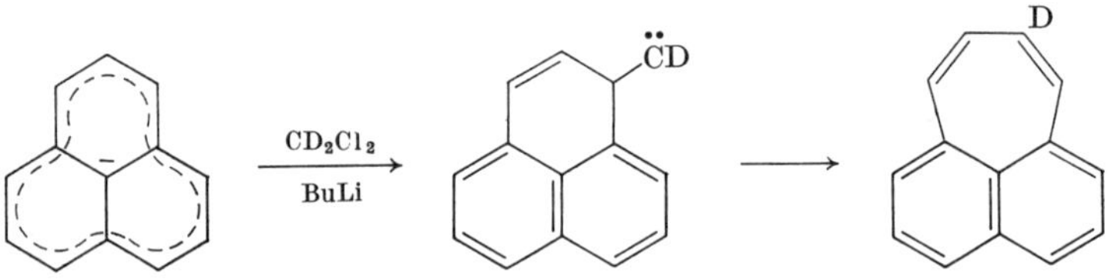

CD_2Cl_2 / BuLi

(66)

Fragmentation

A general olefin synthesis, similar to the Corey–Winter synthesis based on cyclic thionocarbonates, involves the pyrolysis of the corresponding carbonate tosylhydrazone salts [e.g. (**67**) → (**68**).[174] Differences in product distribution between this method and that based on the thionocarbonates indicate that the latter may not involve free carbenes. A variant of the reaction invokes the desulphurization of thionocarbonates with

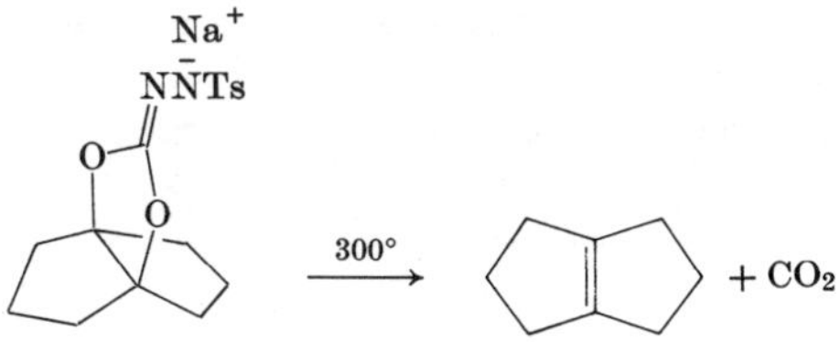

Na$^+$
N$\bar{N}$Ts

300°

(67) **(68)**

$+ CO_2$

bis(cyclo-octa-1,5-diene)nickel.[175] The related fragmentation of cyclic alkoxycarbenes (**69**) is stereospecific and involves the extrusion of a molecule of ketene.[160]

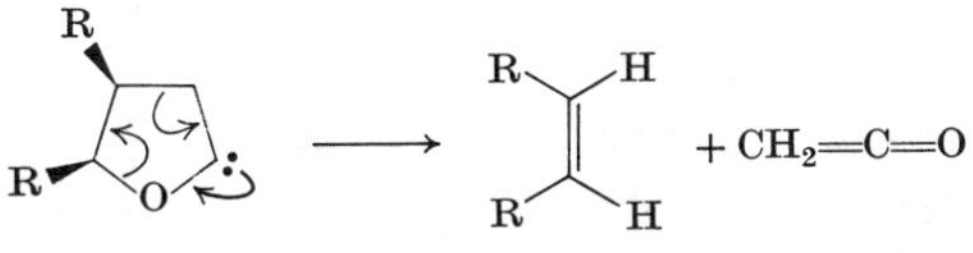

$+ CH_2{=}C{=}O$

(69)

The pyrolytic ring cleavage of azidocyclopropanes (**70**) occurs under mild conditions and involves the stereospecific formation of an olefin, together with a nitrile and nitrogen;[176] ring expansion is a competing reaction. Similarly, chrysanthemylcarbenes (**71**) undergo competing ring expansion and fragmentation. A conformational effect in the carbenes is invoked to explain the greater tendency of these carbenes than of cyclopropylcarbene itself to undergo fragmentation.[177]

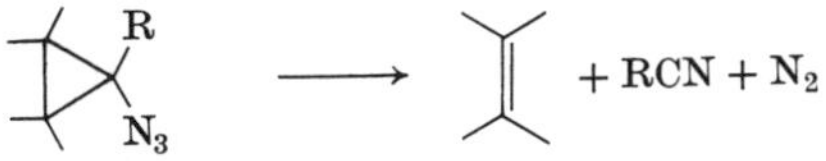

$+ RCN + N_2$

(70)

$+ C_2H_2$

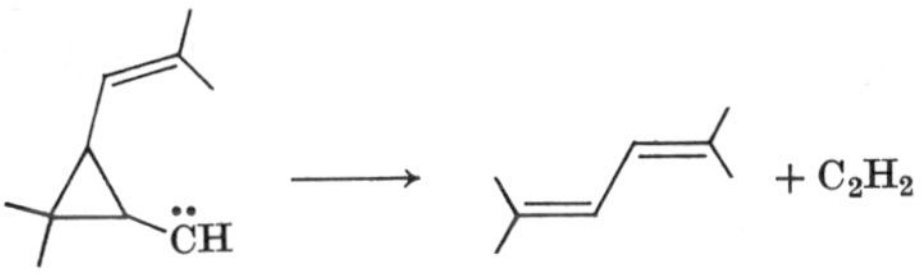

(71) (*cis* and *trans*)

Further examples of the fragmentation of heteroaromatic azides include those of 4-azido-1,5-diphenyl-1,2,3-triazole (**72**) and related 4-azidopyrazoles.[178]

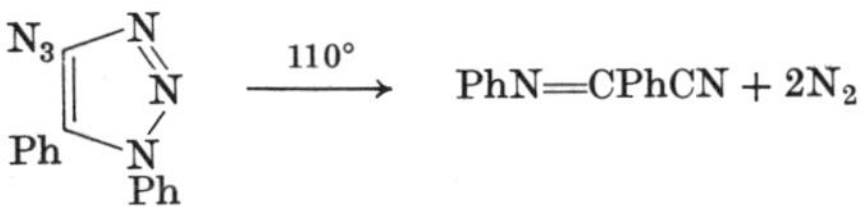

110°

$PhN{=}CPhCN + 2N_2$

(72)

Other Reactions

The thermal[179] and photochemical[180] decomposition of diphenyldiazomethane in the presence of oxygen has been investigated; in both cases, a reaction involving the addition of the triplet carbene to oxygen is suggested. In the presence of aldehydes, ozonides (**73**) are isolated. Reactions of diazo-ketones with sulphur dioxide have also been reported.[181] Dimethoxycarbene acts as a nucleophile in its reaction with isocyanates and isothiocyanates.[182]

$$Ph_2CN_2 \xrightarrow[O_2]{h\nu} Ph_2CO_2\cdot \longrightarrow Ph_2C{=}\overset{+}{O}{-}\overset{-}{O} \xrightarrow{RCHO} Ph_2C\langle O{-}O, O\rangle CHR$$

(**73**)

Transition-metal Complexes*

Transition-metal carbene and nitrene complexes are now covered in a comprehensive annual review.[183] A second review of carbene complexes by Lappert and his co-workers includes, not only isolable complexes, but also those proposed as reaction intermediates.[184] It has been suggested that the chemistry of the ligands is more in keeping with a stabilized carbonium-ion structure than with a carbene structure.[185]

When the carbene complex (**74**) is treated with phenyl-lithium at –78°, the diphenylcarbene complex (**75**) is formed; isolation of this product shows that nucleophilic character is not essential in the carbene ligands of such complexes.[186]

$$(CO)_5W{=}C\begin{matrix}Ph\\OMe\end{matrix} \xrightarrow{PhLi} (CO)_5W{=}CPh_2$$

(**74**) (**75**)

The reaction between 1,2,3-thiadiazoles and selenadiazoles with nonacarbonyldi-iron leads to the formation of carbene complexes.[187] Complexes are also obtained from the cleavage of electron-rich olefins.[188] Of the many carbene complexes formed by nucleophilic addition to carbonyl ligands, the hydroxycarbene complexes (**76**)[189] and the ylide complex (**77**)[190] are noteworthy. Anions generated at the carbon α to the carbenoid

$$(CO)_5M{=}C\begin{matrix}OH\\R\end{matrix} \qquad (CO)_5Cr{=}C\begin{matrix}OMe\\CH{=}PPh_3\end{matrix}$$

(**76**) (**77**)

carbon have been used to produce new complexes of elaborated structure.[191] Carbene complexes have also found a use as amino-protecting groups in peptide synthesis.[192]

The reactions of the complex (**78**) with boron trihalides[193] and with diazabicyclo-[2.2.2]octane[194] gave the new complexes (**79**) and (**80**), respectively; (**79**) contains a formal "carbyne" ligand. In the presence of triphenylphosphine and hydrochloric acid, (**78**) reacts with displacement of the ligand, giving the phosphonium salt $[Ph_3\overset{+}{P}CHPhOMe]Cl^-$.[195] Other examples of ligand displacement in carbene complexes include the formation of imines from aminocarbene complexes,[196] olefin formation by displacement and reaction with diazomethane[197] or with pyridinium ylides,[198] oxidative

* Synthesis of new complexes by standard routes and investigations of their structure by physical methods are not included in this survey.

$$(CO)_5CrC(OMe)Ph \ (\mathbf{78}) \xrightarrow{BX_3} XCr(CO)_4CR + CO \ (\mathbf{79})$$

$$(CO)_5CrC(OMe)Ph \ (\mathbf{78}) \xrightarrow{\text{N(CH}_2\text{CH}_2)_3\text{N}} (CO)_5\overset{-}{Cr}C(OMe)(Ph)—\overset{+}{N}(CH_2CH_2)_3N \ (\mathbf{80})$$

displacement of pyranylidene ligands to give α-pyrones,[198] insertion reactions with silanes,[199] and cyclopropane formation with olefins.[200,201] The mechanisms of the last two reactions have been studied, and in each case there was evidence against dissociation of the complexes to give free carbenes. Thus, Cooke and Fischer found that the cyclopropane formed from the optically active carbene complex (**81**) and diethyl fumarate was also optically active.[200] With the *N*-acylimines (**82**), oxazolines were formed.[202]

$$(CO)_4Cr(=PPhMePr)=C(OMe)Ph \ (\mathbf{81}) + EtO_2CCH=CHCO_2Et \longrightarrow \text{cyclopropane: } C(Ph)(OMe),\ CH(CO_2Et),\ CH(CO_2Et)$$

$$(CO)_5Cr=C(OR)R' + R''CON=C(CF_3)_2 \ (\mathbf{82}) \longrightarrow \text{oxazoline: } (F_3C)_2C\text{–}C(R')(OR)\text{–}O\text{–}C(R'')=N$$

Transient carbene complexes have been suggested as intermediates in the catalysed ring opening of bicyclobutanes[203] and of 1,3,3-trimethylcyclopropene,[204] and in the molybdenum-catalysed oxidation of acetylenes.[205]

References

[1] M. Jones, Jr. and R. A. Moss (Eds.), "Carbenes", Vol. 1, Interscience, New York, 1973.
[2] D. Bethell, in S. P. McManus (Ed.) "Organic Reactive Intermediates", Academic Press, New York, 1973, p. 61.
[3] N. Bodor, M. J. S. Dewar and J. S. Wasson, *J. Am. Chem. Soc.*, **94**, 9095 (1972).
[4] V. Menendez and J. M. Figuera, *An. Quím.*, **69**, 1 (1973); *Chem. Abs.*, **78**, 110316 (1973).
[5] S. Rothenberg and H. F. Schaefer, *J. Am. Chem. Soc.*, **95**, 2095 (1973).
[6] N. Bodor and M. J. S. Dewar, *J. Am. Chem. Soc.*, **94**, 9103 (1972).
[7] V. Menendez and J. M. Figuera, *Chem. Phys. Letters*, **18**, 426 (1973).
[8] N. Bodor, M. J. S. Dewar and Z. B. Maksic, *J. Am. Chem. Soc.*, **95**, 5245 (1973).
[9] A. C. Hopkinson, *J.C.S. Perkin II*, **1973**, 794.
[10] M. J. S. Dewar and C. A. Ramsden, *Chem. Comm.*, **1973**, 688.
[11] I. G. Csizmadia, H. E. Gunning, R. K. Gosavi and O. P. Strausz, *J. Am. Chem. Soc.*, **95**, 133 (1973).
[12] L. T. Scott and W. D. Cotton, *J. Am. Chem. Soc.*, **95**, 5416 (1973).
[12a] B. M. Trost and P. L. Kinson, *Tetrahedron Letters*, **1973**, 2675.
[13] G. Lo Vecchio, G. Grassi, F. Risitano and F. Foti, *Tetrahedron Letters*, **1973**, 3777.
[14] P. F. Alewood, P. M. Kazmaier and A. Rauk, *J. Am. Chem. Soc.*, **95**, 5466 (1973).
[15] N. C. Baird and R. F. Barr, *Can. J. Chem.*, **51**, 3303 (1973).

[16] T. Takui and K. Itoh, *Chem. Phys. Letters*, **19**, 120 (1973).
[17] W. J. Baron, M. R. DeCamp, M. E. Hendrick, M. Jones, R. H. Levin and M. B. Sohn, in ref. 1, p. 1.
[18] B. W. Peace and D. S. Wulfman, *Synthesis*, **1973**, 137.
[19] G. W. Griffin and N. R. Bertoniere, in ref. 1, p. 305.
[20] J. Buddrus, *Angew. Chem. Internat. Edn.*, **11**, 1041 (1972).
[21] J. Dockx, *Synthesis*, **1973**, 441.
[22] L. Skattebøl, G. A. Abskharoun and T. Greibrokk, *Tetrahedron Letters*, **1973**, 1367; M. Makosza and M. Fedorynski, *Syn. Comm.*, **3**, 305 (1973).
[23] P. Weyerstahl, R. Mathias and G. Blume, *Tetrahedron Letters*, **1973**, 611.
[24] R. A. Olofson and C. M. Dougherty, *J. Am. Chem. Soc.*, **95**, 581 (1973).
[25] I. L. Knunyants, Y. F. Komissarov, B. L. Dyatkin and L. T. Lantseva, *Izv. Akad. Nauk SSSR*, **1973**, 943.
[26] J. M. Birchall, R. N. Haszeldine and D. W. Roberts, *J.C.S. Perkin I*, **1973**, 1071.
[27] D. Seyferth and S. P. Hopper, *J. Org. Chem.*, **37**, 4070 (1972).
[28] D. Seyferth and G. J. Murphy, *J. Organometal. Chem.*, **49**, 117 (1973).
[29] D. Seyferth and S. P. Hopper, *J. Organometal. Chem.*, **51**, 77 (1973).
[30] D. Seyferth and C. K. Haas, *J. Organometal. Chem.*, **46**, C33 (1973).
[31] S. P. Kolesnikov, B. L. Perlmutter and O. M. Nefedov, *Izv. Akad. Nauk SSSR*, **1973**, 1418.
[32] E. Lee and D. W. Roberts, *J.C.S. Perkin II*, **1973**, 437.
[33] R. N. Haszeldine, P. J. Robinson and W. J. Williams, *J.C.S. Perkin II*, **1973**, 1013.
[34] R. N. Haszeldine, C. Parkinson and P. J. Robinson, *J.C.S. Perkin I*, **1973**, 1018.
[35] D. Seyferth and G. J. Murphy, *J. Organometal. Chem.*, **52**, Cl (1973).
[36] D. Seyferth and R. A. Woodruff, *J. Org. Chem.*, **38**, 4031 (1973).
[37] D. J. Burton and D. G. Naae, *J. Am. Chem. Soc.*, **95**, 8467 (1973).
[38] S. S. Hixson, *J. Am. Chem. Soc.*, **95**, 6144 (1973).
[39] R. D. Streeper and P. D. Gardner, *Tetrahedron Letters*, **1973**, 767; C. Wentrup and P. Müller, *Tetrahedron Letters*, **1973**, 2915; E. J. York, W. Dittmar, J. R. Stevenson and R. G. Bergman, *J. Am. Chem. Soc.*, **95**, 5680 (1973).
[40] L. Schrader and W. Hartmann, *Tetrahedron Letters*, **1973**, 3995.
[41] J. A. Pincock, R. Morchat and D. R. Arnold, *J. Am. Chem. Soc.*, **95**, 7536 (1973).
[42] R. Weiss and S. Andrae, *Angew. Chem. Internat. Edn.*, **12**, 150 (1973).
[43] J. H. Bowie and B. Nussey, *J.C.S. Perkin I*, **1973**, 1693.
[44] P. S. Skell, J. J. Havel and M. J. McGlinchey, *Accounts Chem. Res.*, **6**, 97 (1973).
[45] S. Kammula and P. B. Shevlin, *J. Am. Chem. Soc.*, **95**, 4441 (1973).
[46] G. Szeimies, *Chem. Ber.*, **106**, 3695 (1973).
[47] M. S. Newman and Z. u. Din, *J. Org. Chem.*, **38**, 547 (1973).
[48] J. Gore and A. Doutheau, *Tetrahedron Letters*, **1973**, 253.
[49] H. D. Hartzler, *J. Am. Chem. Soc.*, **95**, 4379 (1973).
[50] L. T. Scott and W. D. Cotton, *J. Am. Chem. Soc.*, **95**, 2708 (1973).
[51] S. L. Ioffe, L. M. Makarenkova, M. V. Kashutina, V. A. Tartakovskii, N. N. Rozhdestvenskaya, L. I. Kovalenko and V. G. Isagulyants, *Zh. Org. Khim.*, **9**, 905 (1973).
[52] L. de Vries, *J. Org. Chem.*, **38**, 2604 (1973).
[53] R. R. da Silva, V. G. Toscano and R. G. Weiss, *Chem. Comm.*, **1973**, 567.
[54] K. B. Tomer, N. Harrit, I. Rosenthal, O. Buchardt, P. L. Kumler and D. Creed, *J. Am. Chem. Soc.*, **95**, 7402 (1973).
[55] R. McEwan and J. T. Sharp, *Chem. Comm.*, **1973**, 85.
[56] H. Basch, *Theor. Chim. Acta.*, **28**, 151 (1973).
[57] P. K. Brooke, R. B. Herbert and F. G. Holliman, *Tetrahedron Letters*, **1973**, 761.
[58] T. Ohashi, K. Matsunaga, M. Okahara and S. Komori, *Synthesis*, **1971**, 96.
[59] J. I. G. Cadogan and I. Gosney, *Chem. Comm.*, **1973**, 586.
[60] F. A. Carey and L. J. Hayes, *J. Org. Chem.*, **38**, 3107 (1973).
[61] R. A. Abramovitch and T. Takaya, *J. Org. Chem.*, **38**, 3311 (1973).
[62] R. A. Odum and G. Wolf, *Chem. Comm.*, **1973**, 360.
[63] R. A. Moss, in ref. 1, p. 153.
[64] H. S. M. Abdul, *Pak. J. Sci. Ind. Res.*, **15**, 258 (1973); *Chem. Abs.*, **79**, 65679 (1973).
[65] T. L. Gilchrist, *Chem. Ind.* (*London*), **1973**, 881.
[66] H. E. Zimmerman, *Accounts Chem. Res.*, **5**, 393 (1972).
[67] R. Hoffmann, C. C. Levin and R. A. Moss, *J. Am. Chem. Soc.*, **95**, 629 (1973).
[68] R. A. Moss and F. G. Pilkiewicz, *Synthesis*, **1973**, 209.

69 S. M. Shostakovskii, L. N. Aksenova, V. I. Erofeev and V. S. Aksenov, *Dokl. Vses. Konf. Khim. Atsetilena, 4th*, **2**, 354 (1972); *Chem. Abs.*, **79**, 78165 (1973).

70 P. S. Skell, S. J. Valenty and P. W. Humer, *J. Am. Chem. Soc.*, **95**, 5041 (1973); P. S. Skell and S. J. Valenty, *ibid.*, p. 5042; S. J. Valenty and P. S. Skell, *J. Org. Chem.*, **38**, 3937 (1973); E. T. McBee and K. J. Sienkowski, *ibid.*, p. 1340; M. Reetz, U. Schöllkopf and B. Bánhidai, *Ann. Chem.*, **1973**, 599.

71 E. E. Waali and W. M. Jones, *J. Am. Chem. Soc.*, **95**, 8114 (1973).

72 K. Saito and T. Mukai, *Tetrahedron Letters*, **1973**, 4885.

73 Y. Yamamoto, S.-I. Murahashi and I. Moritani, *Tetrahedron Letters*, **1973**, 589.

74 E. E. Waali and W. M. Jones, *J. Org. Chem.*, **38**, 2573 (1973).

75 R. A. LaBar and W. M. Jones, *J. Am. Chem. Soc.*, **95**, 2359 (1973).

76 R. A. Moss and C. B. Mallon, *Tetrahedron Letters*, **1973**, 4481.

77 W. J. Baron, M. E. Hendrick and M. Jones, *J. Am. Chem. Soc.*, **95**, 6286 (1973).

78 D. J. Anderson, T. L. Gilchrist, G. E. Gymer and C. W. Rees, *J.C.S. Perkin I*, **1973**, 550.

79 H. Person, F. Tonnard, A. Foucaud and C. Fayat, *Tetrahedron Letters*, **1973**, 2495.

80 D. W. Jones, *Chem. Comm.*, **1973**, 67.

81 Y. Hayashi and D. Swern, *J. Am. Chem. Soc.*, **95**, 5205 (1973).

82 T. Shingaki, M. Inagaki, M. Takebayashi and W. Lwowski, *Bull. Chem. Soc. Japan*, **45**, 3567 (1972).

83 S. M. A. Hai and W. Lwowski, *J. Org. Chem.*, **38**, 2442 (1973).

84 R. G. Salomon and J. K. Kochi, *J. Am. Chem. Soc.*, **95**, 3300 (1973).

85 T. Sato, T. Mori and J. Shinoda, *Bull. Chem. Soc. Japan*, **46**, 1833 (1973).

86 P. H. Mazzocchi and H. J. Tamburin, *J. Org. Chem.*, **38**, 2221 (1973).

87 S. Miyano and H. Hashimoto, *Bull. Chem. Soc. Japan*, **46**, 892 (1973).

88 S. Miyano and H. Hashimoto, *Chem. Comm.*, **1973**, 216; *Bull. Chem. Soc. Japan*, **46**, 3257 (1973).

89 H. Ledon, G. Linstrumelle and S. Julia, *Tetrahedron*, **29**, 3609 (1973).

90 L. T. Scott, *Chem. Comm.*, **1973**, 882.

91 T. Saegusa, K. Yonezawa, I. Murase, T. Konoike, S. Tomita and Y. Ito, *J. Org. Chem.*, **38**, 2319 (1973).

92 H. Dürr and W. Bujnoch, *Tetrahedron Letters*, **1973**, 1433; H. Dürr, B. Ruge and T. Erhardt, *Ann. Chem.*, **1973**, 214; H. Dürr and H. Kober, *Chem. Ber.*, **106**, 1565 (1973).

93 D. H. Aue, H. Iwahashi and D. F. Shellhamer, *Tetrahedron Letters*, **1973**, 3719; D. H. Aue, R. B. Lorens and G. S. Helwig, *ibid.*, p. 4795.

94 L. B. Rodewald and H. Lee, *J. Am. Chem. Soc.*, **95**, 623 (1973).

95 D. P. G. Hamon and H. P. Hopton, *Tetrahedron Letters*, **1973**, 1675.

96 G. P. Gennaro, Y.-Y. Su, O. F. Zeck, S. H. Daniel and Y.-N. Tang, *Chem. Comm.*, **1973**, 637.

97 J. L. Thomas, *J. Am. Chem. Soc.*, **95**, 1838 (1973).

98 I. S. Y. Wang and M. Karplus, *J. Am. Chem. Soc.*, **95**, 8160 (1973).

99 H. Kollmar, *Tetrahedron*, **28**, 5893 (1972); J. N. Murrell, J. B. Pedley and S. Durmaz, *J.C.S. Faraday II*, **69**, 1370 (1973).

100 M. L. Halberstadt and J. Crump, *J. Photochem.*, **1**, 295 (1973); *Chem. Abs.*, **79**, 4608 (1973).

101 G. B. Kistiakowsky and B. B. Saunders, *J. Phys. Chem.*, **77**, 427 (1973).

102 H. Ledon, G. Linstrumelle and S. Julia, *Tetrahedron Letters*, **1973**, 25.

103 P. K. Freeman, T. D. Ziebarth and V. N. M. Rao, *J. Org. Chem.*, **38**, 3823 (1973); R. B. Reinarz and G. J. Fonken, *Tetrahedron Letters*, **1973**, 4013.

104 I. T. Jacobson, *Chem. Scripta*, **2**, 127 (1972); *Chem. Abs.*, **78**, 3780 (1973).

105 D. Seyferth and Y. M. Cheng, *J. An. Chem. Soc.*, **95**, 6763 (1973).

106 D. Seyferth, Y. M. Cheng and D. D. Traficante, *J. Organometal. Chem.*, **46**, 9 (1973); D. Seyferth, H.-M. Shih, J. Dubac. P. Mazerolles and B. Serres, *ibid.*, **50**, 39 (1973).

107 M. Weidenbruch, *Chem.-Ztg.*, **97**, 355 (1973).

108 E. A. Hill, *J. Org. Chem.*, **37**, 4008 (1972).

109 K. G. Taylor and J. Chaney, *J. Am. Chem. Soc.*, **94**, 8924 (1972).

110 R. A. G. Smith and J. R. Knowles, *J. Am. Chem. Soc.*, **95**, 5072 (1973).

111 M. R. Brinkman, D. Bethell and J. Hayes, *Tetrahedron Letters*, **1973**, 989.

112 T. Shingaki, M. Inagaki, N. Torimoto and M. Takebayashi, *Chem. Letters*, **1972**, 1181.

113 M. D. Sefcik and M. A. Ring, *J. Am. Chem. Soc.*, **95**, 5168 (1973).

114 S. R. Challand, *Aromat. Heteroaromatic Chem.*, **1**, 260 (1973).

115 H. Ledon, G. Linstrumelle and S. Julia, *Bull. Soc. Chim. France*, **1973**, 2065.

116 G. R. Cliff, G. Jones and J. M. Woollard, *Tetrahedron Letters*, **1973**, 2401.

117 R. A. Abramovitch and S. R. Challand, *J. Heterocyclic Chem.*, **10**, 683 (1973).

117a W. A. Sheppard and O. W. Webster, *J. Am. Chem. Soc.*, **95**, 2695 (1973).

[118] W. Ando, *Int. J. Sulphur Chem., B*, **7**, 189 (1972).
[119] W. Ando, J. Suzuki, Y. Saiki and T. Migita, *Chem. Comm.*, **1973**, 365; W. Ando, Y. Saiki and T. Migita, *Tetrahedron*, **29**, 3511 (1973).
[120] J. Quintana, M. Torres and F. Serratosa, *Tetrahedron*, **29**, 2065 (1973).
[121] T. Migita and W. Ando, *Asahi Garasu Kogyo Gijutsu Shoreikai Kenkyu Hokoku*, **20**, 317 (1972); *Chem. Abs.*, **78**, 123954 (1973).
[122] W. Ando, H. Fujii, T. Takeuchi, H. Higuchi, Y. Saiki and T. Migita, *Tetrahedron Letters*, **1973**, 2117.
[123] M. Takebayashi, T. Kashiwada, M. Hamaguchi and T. Ibata, *Chem. Letters*, **1973**, 809.
[124] I. Ojima and K. Kondo, *Bull. Chem. Soc. Japan*, **46**, 1539 (1973).
[125] P. Y. Johnson, E. Koza and R. E. Kohrman, *J. Org. Chem.*, **38**, 2967 (1973).
[126] P. S. Skell, K. J. Klabunde, J. H. Plonka, J. S. Roberts and D. L. Williams-Smith, *J. Am. Chem. Soc.*, **95**, 1547 (1973).
[127] W. Ando, T. Hagiwara and T. Migita, *J. Am. Chem. Soc.*, **95**, 7518 (1973).
[128] V. E. Platonov, N. G. Malyuta and G. G. Yakobson, *Izv. Akad. Nauk SSSR*, **1972**, 2819; P. Scheiner, E. Stockel, D. Cruset and R. Noto, *J. Org. Chem.*, **37**, 4207 (1972); J. M. Duff and A. G. Brook, *Can. J. Chem.*, **51**, 2869 (1973).
[129] R. Paulissen, H. Reimlinger, E. Hayez, A. J. Hubert and P. Teyssié, *Tetrahedron Letters*, **1973**, 2233.
[130] P. Stromquist, M. Radcliffe and W. P. Weber, *Tetrahedron Letters*, **1973**, 4523.
[131] T. Saraie, T. Ishiguro, K. Kawashima and K. Morita, *Tetrahedron Letters*, **1973**, 2121.
[132] V. I. Markov and A. E. Polyakov, *Zh. Org. Khim.*, **9**, 1759 (1973); A. G. Giumanini, *Boll. Chim. Farm.*, **112**, 6 (1973); *Chem. Abs.*, **79**, 66107 (1973).
[133] A. Hassner, J. O. Currie, A. S. Steinfeld and R. F. Atkinson, *J. Am. Chem. Soc.*, **95**, 2982 (1973).
[134] M. Weidenbruch and C. Pierrard, *Angew. Chem. Internat. Ed.*, **12**, 500 (1973).
[135] G. A. Nikiforov, S. A. Markaryan, L. G. Plekhanova, B. D. Sviridov, S. V. Rykov, V. V. Ershov, A. L. Buchachenko, T. Pehk, T. Saluvere and E. Lippmaa, *Org. Mag. Res.*, **5**, 339 (1973).
[136] L. G. Plekhanova, G. A. Nikiforov, V. V. Ershov and E. P. Zakharov, *Izv. Akad. Nauk. SSSR*, **1973**, 846.
[137] Y.-N. Kuo and M. J. Nye, *Can. J. Chem.*, **51**, 1995 (1973); G. Ege and G. Jooss, *Chem. Ber.*, **106**, 1678 (1973).
[138] T. Shingaki, M. Inagaki, N. Torimoto and M. Takebayashi, *Chem. Letters*, **1972**, 859.
[139] H. H. Gibson, C. H. Bundy and H. R. Gaddy, *Tetrahedron Letters*, **1973**, 3801.
[140] R. E. Banks and A. Prakash, *Tetrahedron Letters*, **1973**, 99.
[141] E. F. V. Scriven, H. Suschitzky and G. V. Garner, *Tetrahedron Letters*, **1973**, 103.
[142] F. J. Weigert, *J. Org. Chem.*, **38**, 1316 (1973).
[143] W. M. Jones, R. C. Joines, J. A. Myers, T. Mitsuhashi, K. E. Krajca, E. E. Waali, T. L. Davis and A. B. Turner, *J. Am. Chem. Soc.*, **95**, 826 (1973).
[144] W. D. Crow and M. N. Paddon-Row, *Austral. J. Chem.*, **26**, 1705 (1973).
[145] P. H. Gebert, R. W. King, R. A. LaBar and W. M. Jones, *J. Am. Chem. Soc.*, **95**, 2357 (1973).
[146] T. T. Coburn and W. M. Jones, *Tetrahedron Letters*, **1973**, 3903.
[147] W. J. Baron and M. R. DeCamp, *Tetrahedron Letters*, **1973**, 4225.
[148] R. F. C. Brown, F. Irvine and R. J. Smith, *Austral. J. Chem.*, **26**, 2213 (1973).
[149] R. A. Abramovitch and B. W. Cue, *J. Org. Chem.*, **38**, 173 (1973).
[150] F. R. Atherton and R. W. Lambert, *J.C.S. Perkin I*, **1973**, 1079.
[151] S. A. Matlin and P. G. Sammes, *J.C.S. Perkin I*, **1972**, 2623; see *Org. Reaction Mech.*, **1972**, 382.
[152] A. Krantz, *Chem. Comm.*, **1973**, 670.
[153] J. Fenwick, G. Frater, K. Ogi and O. P. Strausz, *J. Am. Chem. Soc.*, **95**, 124 (1973); P. W. Concannon and J. Ciabattoni, *J. Am. Chem. Soc.*, **95**, 3284 (1973).
[154] T. L. Gilchrist, G. E. Gymer and C. W. Rees, *J.C.S. Perkin I*, **1973**, 555.
[155] T. L. Gilchrist, G. E. Gymer and C. W. Rees, *Chem. Comm.*, **1973**, 835.
[156] K. P. Zeller, *Chem.-Ztg.*, **97**, 37 (1973).
[157] K. P. Zeller, H. Meier and E. Müller, *Tetrahedron*, **28**, 5831 (1972); Z. Arnold and J. Šauliová, *Coll. Czech. Chem. Comm.*, **38**, 2641 (1973).
[158] A. C. Day, A. N. McDonald, B. F. Anderson, T. J. Bartczak and O. J. R. Hodder, *Chem. Comm.*, **1973**, 247.
[159] D. R. Morton and N. J. Turro, *J. Am. Chem. Soc.*, **95**, 3947 (1973); see also J. M. Denis and J. M. Conia, *Tetrahedron Letters*, **1973**, 461.
[160] A. M. Foster and W. C. Agosta, *J. Am. Chem. Soc.*, **95**, 608 (1973).
[161] D. W. Brown, M. E. Hendrick and M. Jones, *Tetrahedron Letters*, **1973**, 3951; F. D. Greene, R. L. Camp, V. P. Abegg and G. O. Pierson, *ibid.*, p. 4091.

[162] R. B. Reinarz and G. J. Fonken, *Tetrahedron Letters*, **1973**, 4591.
[163] R. D. Miller and D. L. Dolce, *Tetrahedron Letters*, **1973**, 4403.
[164] W. D. Crow and M. N. Paddon-Row, *Tetrahedron Letters*, **1973**, 2217.
[165] M. Pomerantz and T. H. Witherup, *J. Am. Chem. Soc.*, **95**, 5977 (1973).
[166] A. D. Wolf and M. Jones, *J. Am. Chem. Soc.*, **95**, 8209 (1973).
[167] M. Fărcasiu, D. Fărcasiu, R. T. Conlin, M. Jones and P. von R. Schleyer, *J. Am. Chem. Soc.*, **95**, 8207 (1973).
[168] T. E. Berdick, R. H. Levin, A. D. Wolf and M. Jones, *J. Am. Chem. Soc.*, **95**, 5087 (1973).
[169] A. D. Wolf, V. V. Kane, R. H. Levin and M. Jones, *J. Am. Chem. Soc.*, **95**, 1680 (1973).
[170] S.-I. Murahashi, K. Okumura, T. Kubota and I. Moritani, *Tetrahedron Letters*, **1973**, 4197.
[171] K. L. Erickson, *J. Org. Chem.*, **38**, 1463 (1973); see also K. L. Erickson and S. B. Gordon, *ibid.*, p. 1431.
[172] B. V. Ioffe and L. A. Kartsova, *Tetrahedron Letters*, **1973**, 623.
[173] I. Murata and K. Nakasuji, *Tetrahedron Letters*, **1973**, 47.
[174] W. T. Borden, P. W. Concannon and D. I. Phillips, *Tetrahedron Letters*, **1973**, 3161.
[175] M. F. Semmelhack and R. D. Stauffer, *Tetrahedron Letters*, **1973**, 2667.
[176] G. Szeimies, U. Siefken and R. Rinck, *Angew. Chem. Internat. Ed.*, **12**, 161 (1973); G. Szeimies and J. Harnisch, *Chem. Comm.*, **1973**, 739.
[177] T. Sasaki, S. Eguchi, M. Ohno and T. Umemura, *J. Org. Chem.*, **38**, 4095 (1973).
[178] P. A. S. Smith and H. Dounchis, *J. Org. Chem.*, **38**, 2958 (1973).
[179] M. Contineanu, R. N. Schindler and T. Onescu, *Z. Phys. Chem.* (*Frankfurt*), **82**, 217 (1973).
[180] R. W. Murray and A. Suzui, *J. Am. Chem. Soc.*, **95**, 3343 (1973).
[181] M. Tanaka, T. Nagai and N. Tokura, *J. Org. Chem.*, **37**, 4106 (1972); **38**, 1602 (1973).
[182] R. W. Hoffmann, K. Steinbach and B. Dittrich, *Chem. Ber.*, **106**, 2174 (1973).
[183] J. A. Connor, *Organometal. Chem.*, **1**, 254 (1972).
[184] D. J. Cardin, B. Cetinkaya, M. J. Doyle and M. F. Lappert, *Chem. Soc. Rev.*, **2**, 99 (1972).
[185] A. Davidson and D. L. Reger, *J. Am. Chem. Soc.*, **94**, 9237 (1972).
[186] C. P. Casey and T. J. Burkhardt, *J. Am. Chem. Soc.*, **95**, 5833 (1973).
[187] G. N. Schrauzer and H. Kisch, *J. Am. Chem. Soc.*, **95**, 2501 (1973); see *Org. Reaction Mech.*, **1972**, 386.
[188] B. Cetinkaya, P. Dixneuf and M. F. Lappert, *Chem. Comm.*, **1973**, 206; D. J. Cardin, B. Cetinkaya and M. F. Lappert, *J.C.S. Dalton*, **1973**, 514.
[189] E. O. Fischer, G. Kreis and F. R. Kreissl, *J. Organometal. Chem.*, **56**, C37 (1973).
[190] D. K. Mitchell and W. C. Kaska, *J. Organometal. Chem.*, **49**, C73 (1973).
[191] C. O. Casey, R. A. Boggs and R. L. Anderson, *J. Am. Chem. Soc.*, **94**, 8947 (1972).
[192] K. Weiss and E. O. Fischer, *Chem. Ber.*, **106**, 1277 (1973).
[193] E. O. Fischer, G. Kreis, C. G. Kreiter, J. Müller, G. Huttner and H. Lorenz, *Angew. Chem. Internat. Ed.*, **12**, 564 (1973).
[194] F. R. Kreissl, E. O. Fischer, C. G. Kreiter and K. Weiss, *Angew. Chem. Internat. Ed.*, **12**, 563 (1973).
[195] U. Schubert and E. O. Fischer, *J. Organometal. Chem.*, **51**, C11 (1973).
[196] J. A. McCleverty and M. M. M. da Mota, *J.C.S. Dalton*, **1973**, 2571; J. A. Connor and P. D. Rose, *J. Organometal. Chem.*, **46**, 329 (1972).
[197] C. P. Casey, S. H. Bertz and T. J. Burkhardt, *Tetrahedron Letters*, **1973**, 1421.
[198] T. L. Gilchrist, R. Livingston, C. W. Rees and E. von Angerer, *J.C.S. Perkin I*, **1973**, 2535.
[199] J. A. Connor, P. D. Rose and R. M. Turner, *J. Organometal. Chem.*, **55**, 111 (1973); J. A. Connor, J. P. Day and R. M. Turner, *Chem. Comm.*, **1973**, 578.
[200] M. D. Cooke and E. O. Fischer, *J. Organometal. Chem.*, **56**, 279 (1973).
[201] E. O. Fischer and K. H. Dötz, *Chem. Ber.*, **105**, 3966 (1972).
[202] E. O. Fischer, K. Weiss and K. Burger, *Chem. Ber.*, **106**, 1581 (1973).
[203] W. G. Dauben, A. J. Kielbania and K. N. Raymond, *J. Am. Chem. Soc.*, **95**, 7166 (1973); R. Noyori, *Tetrahedron Letters*, **1973**, 1691; P. G. Gassman and R. R. Reitz, *J. Organometal. Chem.*, **52**, C51 (1973).
[204] T. Shirafuji and H. Nozaki, *Tetrahedron*, **29**, 77 (1973).
[205] S. A. Matlin and P. G. Sammes, *J.C.S. Perkin I*, **1973**, 2851.

CHAPTER 6

Nucleophilic Aromatic Substitution

M. R. CRAMPTON

Department of Chemistry, Durham University

Mechanisms and reactivity in aromatic nucleophilic substitution have been excellently summarized by Bernasconi;[1] he deals mainly with the widely applicable addition–elimination (S_NAr) mechanism and particularly with reactions involving amines as nucleophiles. The nucleophilic dediazoniation of arenediazonium ions is generally considered to occur by an S_N1 mechanism; however, Zollinger[2] concludes from evidence obtained in the arylation of aromatic compounds that the existence of aryl cations is questionable; a bimolecular rate-limiting step, consistent with S_NAr or S_N2-like mechanisms, is indicated in this case.

Aryl halides are common reactants in nucleophilic substitution reactions. Bunnett[3] has summarized the interesting and varied possibilities for their reaction with basic media; these include substitution by the radical-chain $S_{RN}1$ mechanism, and isomerization (base-catalysed halogen dance). The phenylation of a number of carbanions, R^-, by treatment with bromobenzene and an alkali metal in liquid ammonia is believed to occur by the $S_{RN}1$ mechanism.[4] The initial electron transfer may be stimulated

$$\text{Electron donor} + \text{ArX} \rightarrow [\text{ArX}]^{\overline{\cdot}} + \text{Residue}$$
$$[\text{ArX}]^{\overline{\cdot}} \rightarrow \text{Ar}\cdot + \text{X}^-$$
$$\text{Ar}\cdot + \text{R}^- \rightarrow [\text{ArR}]^{\overline{\cdot}}$$
$$[\text{ArR}]^{\overline{\cdot}} + \text{ArX} \rightarrow \text{ArR} + [\text{ArX}]^{\overline{\cdot}}$$

photochemically: thus under the influence of UV radiation, halobenzenes react with acetone to give phenylacetone.[5] Application of these newer methods of aromatic substitution leads to an improved synthesis of cephalotoxinone by internal cyclization of an iodo-ketone precursor.[6] Shein and his co-workers have shown that radical anions are formed during the reactions with base of a number of aromatic compounds, including *p*-substituted chlorobenzenes,[7] 2-substituted anthraquinones[8] and alkoxy-substituted nitrobenzenes.[9] Free-radical initiators have been found to promote substitution in 1-chloro-2-nitrobenzene.[10] Also reported[11] is the fact that, although 4-bromoisoquinoline reacts with sodium thiophenoxide—produced from thiophenol and sodium methoxide—by a normal S_NAr mechanism, in the presence of an excess of methoxide a radical-chain pathway is favoured; this accounts for the formation of isoquinoline in addition to the expected 4-(phenylthio)isoquinoline.

A previously offered explanation of the α-effect (enhanced reactivity of nucleophiles with an unshared electron pair adjacent to the nucleophilic centre) has been applied to reactions of basic oximes;[12] it is proposed that lone pair–lone pair repulsion leads to energy-level splitting so that when electron-attracting substituents are present the oximate ion behaves as an α-nucleophile. In an alternative theoretical explanation[13] it is proposed that the observed increased reactivity of α-nucleophiles is derived from the formation of aromatic-like transition states.

The S_NAr Mechanism

The reactions of primary and secondary amines with activated aromatic molecules containing a replaceable group X proceed through intermediates such as (**1**). It is now generally accepted that the base-catalysed path to products proceeds by a fast acid-base equilibrium to give the conjugate base of (**1**) followed by rate-limiting expulsion of X

$+ R^1R^2NH \underset{k_{-1}}{\overset{k_1}{\rightleftharpoons}}$ X, $^+NR^1R^2$(H), NO_2, NO_2 $\xrightarrow[k_3(B)]{k_2}$ NR^1R^2, NO_2, NO_2

(**1**)

with general acid catalysis by BH. Less attention has been paid to the uncatalysed breakdown of the intermediate. A mechanism analogous to that of the base-catalysed path appears possible, with solvent playing the role of B and the lyonium ion the role of BH. However, Bernasconi[1] has argued that a more likely pathway involves intramolecular acid-catalysis of leaving-group expulsion; the transition state would have structure (**2**), or probably (**3**) in hydroxylic solvents. The evidence favouring this

H, X, NR^1R^2, NO_2, NO_2 (**2**)

R, O, H, H, X, NR^1R^2, NO_2, NO_2 (**3**)

O, NMe, NO_2, NO_2 (**4**)

mechanism is derived in part from rate measurements with the spiro-adduct (**4**) which serves as a model for the anionic intermediate in nucleophilic substitution by amines.[14]

A comparison has been made of the kinetics of reaction of 2,4-dinitrofluoro- and 2,4-dinitrochloro-benzene with substituted anilines[15] and *N*-methylanilines[16] in methanol and acetonitrile as solvent; initial base attack (k_1) is usually rate-determining although in acetonitrile loss of fluoride ion can become kinetically significant. Breakdown of the intermediate is the slow step in the reaction of 4-substituted 2-nitrofluorobenzenes with piperidine in aprotic media; here intramolecular hydrogen-bonding in the intermediate (**5**) is postulated.[17] The reaction between *p*-anisidine and 1-X-2,4-dinitrobenzenes (X = Cl, Br, $MeSO_2$ or $PhSO_2$) is catalysed by DABCO (1,4-diaza[2.2.2]bicyclooctane). For this reaction where bond-breaking is rate-limiting a large "element effect" (effect of varying the leaving group) might be expected; however, the rate of the DABCO-catalysed reaction was found to vary only within a factor of nine for the leaving groups examined.[18]

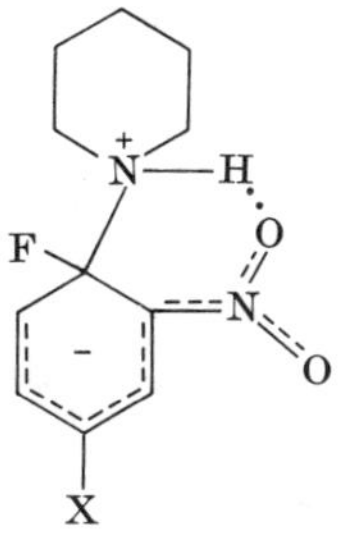

(**5**) 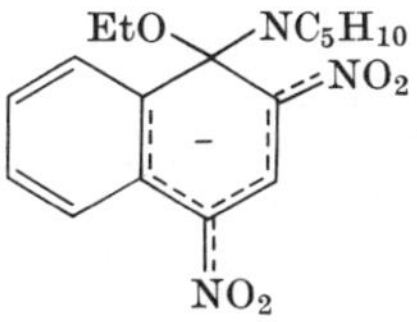 (**6**) 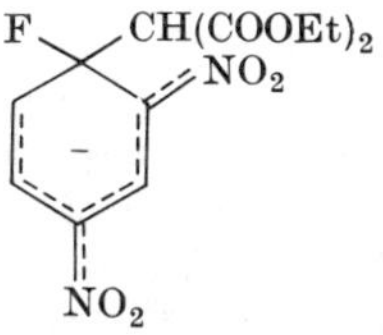 (**7**)

The reactions of substituted anilines, *p*-$RC_6H_4QC_6H_4NH_2$-*p*, with picryl chloride have been studied. When Q = CO_2 or NHCO, Hammett data indicate that transmission of the electronic effects of R to the reaction centre is small.[19] However, when Q = S, NH or SO_2, larger substituent effects are observed.[20]

In favourable cases the intermediates in substitution reactions may be observed spectroscopically. Thus the ^{1}H-NMR spectrum of (**6**) has been reported[21] in the reaction of 2,4-dinitro-1-naphthyl ethyl ether with piperidine in dimethyl sulphoxide (DMSO). The visible spectra of 1-fluoro-2,4-dinitrobenzene in reaction with the sodium salt of diethyl malonate in DMSO indicate initial fast formation of an intermediate, thought to be (**7**), and slower formation of products;[22] a kinetic analysis has been made. It has also been claimed[23] that, in media rich in DMSO, adducts such as (**8**) are detectable by visible spectroscopy in the reactions of 1-chloro-2,4-dinitrobenzene and 2,4-dinitroanisole with hydroxide ion. However, it seems possible that the spectra obtained with λ_{max} at *ca.* 520 nm result from hydroxide addition at C-5 rather than the chloro-substituted position.

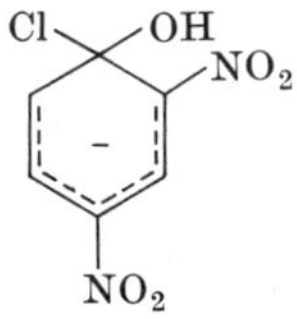

(**8**) 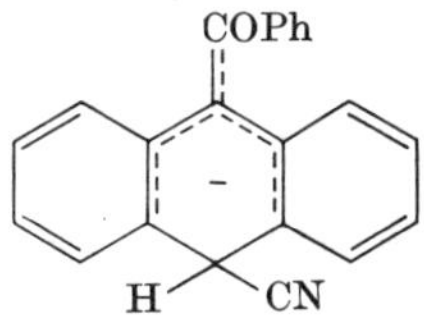 (**9**)

An intermediate formed by addition at the 4-position has been detected during amination of 1-nitronaphthalene by hydroxylamine.[24] The reaction of cyanide ion with dimethylformamide solutions of 9-benzoylanthracene yields intensely blue solutions from which a variety of products including 9-cyano- and 9,10-dicyano-anthracene may be isolated;[25] it is thought that reaction proceeds by initial formation of the adduct (**9**).

There have been two reports of substitutions by ambident nucleophiles. In the case of the nitrite ion[26] the relative importance of the two modes of displacement, O attack and N attack, depends on the halogen displaced, oxygen attack increasing with increasing electronegativity of the halogen. Similarly, in reactions of thiocyanate ion[27] the relative rates of reaction by SCN^- and NCS^- depend on the "hardness" of the leaving group. Abnormal transition states, involving interaction between entering and leaving groups, are likely in the reactions of thiocyanate with 1-fluoro-2,4-dinitrobenzene and with 2,4-dinitrophenyl tosylate.

The high *o*/*p* reactivity ratio found in the reactions of fluoronitrobenzenes with potassium *tert*-butoxide in *tert*-butyl alcohol is attributed[28] to specific stabilization of the transition state for the reaction of the *o*-nitro-compound by potassium ion bridging between the nucleophile and the nitro-group. Significantly the *o*/*p* ratio is close to unity in the presence of crown ether which forms complexes with the potassium ions. In contrast, the variation in *o*/*p* ratios found in the reactions of mono- and di-halonitrobenzenes with a series of alkoxide ions has been attributed to steric exclusion of solvation in the transition states of the reactions.[29]

The kinetics of the decomposition of *p*-dinitrobenzene in ethylene glycol containing base show[30] the rate-limiting step to be base addition, and the results were used to define a kinetic J_- acidity function. The effects of solvent change on the Hammett ρ value for reaction of 4-substituted 1-chloro-2-nitrobenzene derivatives with toluene-*p*-thiolate have been examined.[31] Aromatic nucleophilic substitutions are often speeded up enormously on transfer from protic to dipolar aprotic solvent; hexamethylphosphoramide is an excellent solvent. These protic–dipolar-aprotic solvent effects usually reflect changes in the enthalpy rather than the entropy of activation;[32] nevertheless, entropy effects may sometimes be important. Thus the main reason for the increase in rate of halogen exchange between potassium iodide and 1-iodo-2,4-dinitrobenzene on transfer from methanol to dipolar aprotic solvents is an initial-state entropy effect.[33] A kinetic study[34] of ^{36}Cl exchange between lithium chloride and di- or tri-nitro-derivatives of dichlorobenzenes indicates that both polar and steric effects are important, twisting of nitro-groups from ring-coplanarity resulting in rate reductions. Kinetic studies have also been reported for exchange of ^{131}I with *o*-iodohippuric acid[35] and *o*-iodobenzoic acid;[36] the latter reaction is susceptible to heterogeneous catalysis at a glass surface.[37]

Treatment of polysubstituted aromatic nitro-compounds with sodium borohydride results in nucleophilic displacement of a nitro-group by hydride;[38] the reaction requires the presence of electron-withdrawing groups to activate the ring towards attack and the presence of bulky substituents adjacent to the nitro-group to prevent its conjugation with the ring; in the absence of these substituents, reduction of the nitro-group occurs. *o*-Dinitrobenzene gives nitrobenzene on treatment with borohydride.[39]

Cuprous ions catalyse the ammonolysis of iodobenzoic acids,[40] whereas palladium(II) salts catalyse the conversion of aryl halides into the nitriles in a convenient synthetic route.[41] The rate of reaction of phenyl phenylmercury sulphide with picryl halides to give phenyl 2,4,6-trinitrophenyl sulphide is only slightly affected by solvent change, and a four-centred concerted mechanism involving mercury-assisted loss of chloride is suggested.[42] In the reaction of 2-bromobenzophenone with Grignard reagents substantial replacement of bromine by the organic group of the Grignard occurs[43] and reaction is thought to occur by intramolecular substitution in a ketone–Grignard complex.

Reports have also appeared on the reactions of 2-chloro-5-nitrobenzenesulphonate with phenoxide ions,[44] of pentafluorobenzene with piperidine,[45] of anthraquinones with ammonia[46] and of anthraquinone-1,8-disulphonate with aqueous sodium hydroxide.[47] The latter reaction yields eight main products and may involve a radical mechanism.

Heterocyclic Systems

Nucleophilic substitutions in heterocyclic systems have been reviewed.[48–50] There have been a number of studies of substitution by the addition–elimination mechanism. A comparison of fluorine *v.* chlorine displacement by aromatic amines in halogeno-1,3,5-triazines indicates that decomposition of the intermediate may be rate-determining in the displacement of fluorine.[51] The relative reactivities of six alkoxides have been compared by measuring their rates of reaction with 2-chloro-4,6-dipiperidino-1,3,5-triazine;[52] the effects of ion-pairing in the metal alkoxides are indicated here as they are also in reaction of alkoxides with 1-methyl-3-nitro-5-halogenotriazoles.[53] Kinetic studies have been made of the substitution of halogen by aliphatic amines in 1,2,4-triazoles;[54] and solvent effects on rates of reaction of 2-halopyrimidines with piperidine have been reported.[55] Other kinetic studies have been reported for the hydrolysis of the triazine herbicide, cyanazine,[56] and for the reaction of substituted pyrroles.[57]

Liquid ammonia has proved to be a popular solvent for studying the reactions of heterocyclic compounds. Reactions more complicated than simple substitution may occur. Thus in ammonia in the presence of metal amides isomerization of 2-halothianaphthenes occurs to give the 3-halo-isomers, which are stable under the reaction conditions.[58] Bromine migrations in several five-membered hetarenes have also been reported.[59] These are further examples of a previously noted transhalogenation mechanism (base-catalysed halogen dance) in which proton-abstraction from the substrate gives a carbanion which then removes positive halogen from a second substrate molecule. Incidentally a re-investigation[60] of the reaction of 3-bromothianaphthene with piperidine has shown that the major product is the expected 3-piperidinothianaphthene rather than the cine-substitution product; the small amount of cine-substitution occurring may take place by a transbromination mechanism rather than via 2,3-thianaphthyne. The initial reaction during amination of several dibromoquinolines is also believed to be bromine migration though in this case aminated products are formed via 2-bromo-3,4-quinolyne.[61]

The aminations of phenyl-substituted halogenopyrimidines by potassium amide in ammonia have been studied by van der Plas and his co-workers. By use of ^{15}N-labelled compounds they have shown that reaction may occur by an ANRORC (Addition Nucleophile Ring Opening Ring Closure) mechanism in competition with the more usual addition–elimination and elimination–addition mechanisms.[62] In the case of 4-bromo-6-phenylpyrimidine (**10**) the ANRORC mechanism involves attack by base at the 2-position to yield (**11**), and the nitrogen of the added amide ion is eventually incorporated into the ring of the substitution product (**12**). This mechanism is related to the Dimroth

(**10**) $\xrightarrow{NH_2^-}$ (**11**) $\longrightarrow$ (**12**) + Br^-

rearrangement, where a ring-nitrogen exchanges place with an imino-group in the α-position; this mechanism is operative in the conversion of 1-methyladenosine to N^6-methyladenosine.[63] The presence of a phenyl group at the 2-position in 4-halo-2,6-diphenylpyrimidines inhibits amide attack at this ring-position and thus reduces the importance of the ANRORC mechanism.[64] The reaction of 4-bromo-6-phenylpyrimidine with lithium isopropylamide to give 4-isopropylamino-6-phenylpyrimidine is not a simple replacement, but proceeds by an ANRORC mechanism combined with a Dimroth rearrangement.[65] In related work[66] it has been shown that the ring contraction of 2-chloropyrazine into 2-cyanoimidazole involves initial attack of amide ion at the unsubstituted 3-position. Also the reaction of 4-chloroquinazoline with base gives an open-chain compound resulting from nucleophilic attack at an unsubstituted position, together with substitution products.[67]

Hydrogen–deuterium exchange at the 2- and the 6-position of the *N*-methylpyridinium ion is catalysed by amines in methanol. However, it has been shown,[68] at least in the cases of morpholine and DABCO, that the function of amine is to produce methoxide ions ($Am + CH_3OD \rightleftharpoons AmD^+ + CH_3O^-$), which are responsible for deprotonation of the substrate. [3-D]Pyridine is formed by treatment of the product of the Ullman reaction of 3-bromopyridine with acidified deuterium oxide.[69] Quantitative incorporation of dueterium at C-6 in all common pyrimidine nucleosides has been achieved[70] by treatment with deuteriated base in [D_6-]DMSO; the reaction is thought to involve direct proton abstraction from the substrate. Measurements with 5-substituted pyrimidine *N*-oxides show that the rates of H–D exchange at the 6-position depend on the inductive effect of the substituent.[71] NMR has been used to examine the rates and orientations of H–D exchange on the ring of hydroxyquinolines[72] and also to study the influence of electron pairs at *peri*-positions on exchange in quinoline and 1-5-naphthyridines.[73] The isotopic exchange of *N*-methyl-2-isoxazolinium salts, prepared by the action of dimethyl sulphate on 2-isoxazolines, may involve solvent addition and ring-opening rather than direct proton loss.[74] Other examples of base-catalysed proton exchange in aromatic systems will be found in Chapter 7, and the important implications on the validity of selectivity–reactivity relations resulting from a study of the base-catalysed rearrangement of benzisoxazoles are discussed in Chapter 2.

The fluoride ion-induced reactions of pentafluoropyridine and tetrafluoropyridazine with several fluorinated olefins have been studied.[75] Pentafluoropyridine and tetrafluoroethylene give perfluoro-4-ethyl-, -2,4-diethyl- and -2,4,5-triethylpyridine; rearrangement of the kinetically controlled trisubstituted product could not be achieved, in contrast to the previously reported behaviour of the trisperfluoroisopropyl analogue. However, reaction with perfluorocyclohexene and perfluoroisobutene gave the thermodynamically stable 2,4,6-trisubstituted products. From reaction with chlorotrifluoroethylene, both chlorotetrafluorethyl and pentafluoroethyl derivatives were obtained; the latter was suggested to result from displacement of chloride ion in a pseudo-S_N2' process involving attack of fluoride ion on the aromatic ring.

Meisenheimer and Related Complexes

Interest in Meisenheimer complexes continues to expand and diversify. The values for the stoichiometric equilibrium constants for formation of 1,1-complexes (**13**) from several substitituted anisoles in dilute methanolic methoxide solutions depend on the particular cation present, an effect which has been attributed to stabilization of the complexes by ion association.[76] Larger effects are produced in the presence of the bivalent calcium or barium ions.[77] It is often found that stabilization of σ-complexes is produced by DMSO. A calorimetric study[78] of the reaction of 1,3,5-trinitrobenzene with sodium thiophenoxide in methanol-DMSO mixtures shows that enthalpy rather than entropy effects are dominant; thus the heat of reaction becomes more exothermic as the DMSO concentration increases while there is little change in reaction entropy. Thio-analogues of Meisenheimer complexes have been identified by NMR spectroscopy;[79] the reaction of ethyl thiopicrate with sodium ethanethiolate yields a mixture of the isomeric adducts (**14**) and (**15**). Both are transient species because of specific rapid replacement of the *p*-nitro-group by ethanethiolate. In marked contrast to the corresponding oxygen compounds, the 1,3-adduct (**15**) is thermodynamically more stable than the 1,1-adduct (**14**).

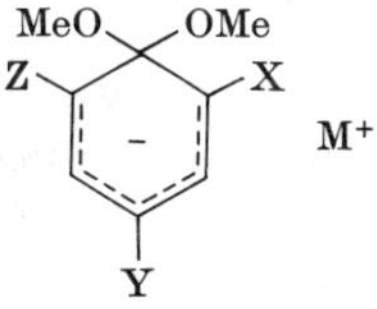

MeO OMe Z X M+ Y

(**13**)

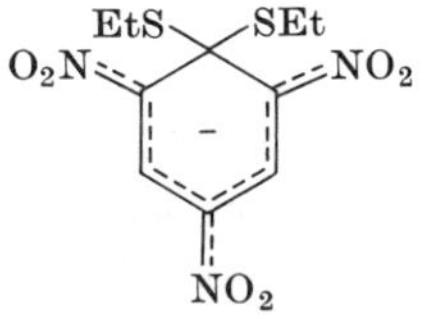

EtS SEt O_2N NO_2 NO_2

(**14**)

SEt O_2N NO_2 H SEt NO_2

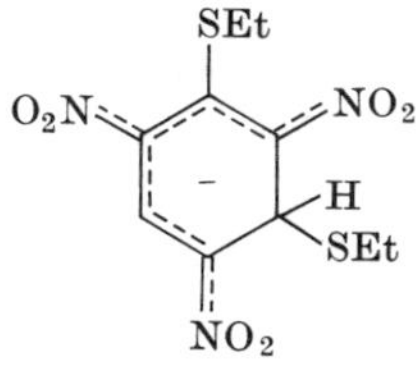

(**15**)

Development[80] of a flow-technique to measure the NMR spectra of transient species promises to be extremely useful in the studies of short-lived Meisenheimer complexes. The method has already been used to examine methoxide attack on 1-cyano-3,5-dinitrobenzene where isomeric addition at C-2 or C-4 is possible; initially a mixture of the two products in similar concentration is produced, but at equilibrium the C-2 isomer is favoured; in general the addition of methoxide to 1-X-3,5-dinitrobenzenes yields thermodynamically more stable adducts by addition at C-2. The equilibrium constants for complex formation have been determined by means of a J_M acidity function in methanol–DMSO mixtures;[81] the visible spectra obtained initially on mixing were originally interpreted[81] as indicating exclusive formation of the C-4 isomer, but in the light of the NMR results of Fyfe *et al.* it seems likely that they correspond to a mixture of C-2 and C-4 adducts; it appears that methoxide ions show little kinetic preference for any one of the three unsubstituted ring positions. A comparison has been made by use of a stopped-flow kinetic technique of methoxide addition to 1-X-3,5-dinitrobenzenes and 2-X-4,6-dinitroanisoles;[82] addition at unsubstituted carbon atoms is much faster than at the methoxyl-substituted position. In a related study[83] the adducts formed from 3,5-dinitropyridine with base have been investigated; here too isomeric addition is possible, with the C-2 adduct having greater thermodynamic stability.

The complex produced from potassium methoxide and 2-methoxytropone in methanol–DMSO has been characterized[84] by NMR as (**16**). 2-Methoxy-5-nitrotropone gives a related complex, addition again occurring at the methoxyl-substituted position.[85] Methoxide addition to 1,3-dinitroazulene gives (**17**); however, 1-nitro-3-(trifluoroacetyl)-azulene gives a hemiacetal by methoxide addition at the carbonyl group.[86] Complexes from 2,4-disubstituted thiophens and selenophens have been prepared (substituent

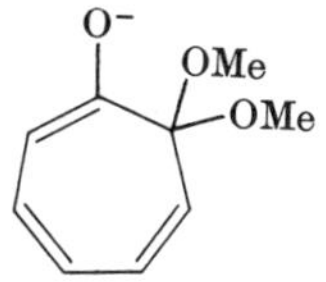

O^- OMe OMe

(**16**)

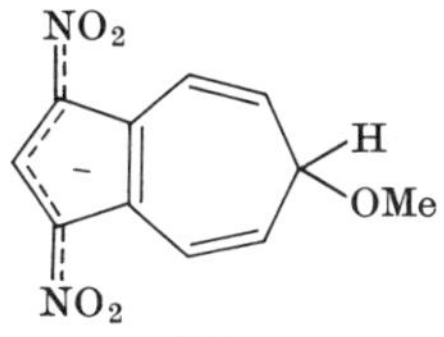

NO_2 H OMe NO_2

(**17**)

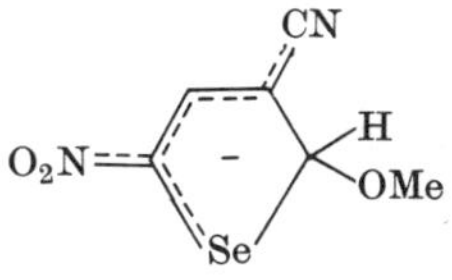

CN O_2N H OMe Se

(**18**)

NO_2 or CN).[87] Generally, in media rich in DMSO, addition occurs at C-5 to give for example (**18**); by contrast, in methanol, methoxide adds to the cyano-group of 2-cyano-4-nitrothiophene to give a carboxyimidate;[88] thus changing the solvent may in some cases change the mode of interaction. It was also found that under basic conditions in MeOD, ring-hydrogen at C-5 is rapidly exchanged for deuterium. Similar exchange at C-3 is slow.[89] Because of this, the structure of σ-complexes cannot be deduced from measurements with compounds deuteriated at C-5.

In basic media, *N*-2-(hydroxyethyl)-*N*-methyl-2,4-dinitroaniline is converted into a spiro-complex (**4**). Rate and equilibrium constants associated with complex formation have been determined in water–DMSO mixtures.[14] Similar spiro-complexes have been prepared from *N*-(2-hydroxyethyl)-*N*-methylpicramide and *N*-(2-hydroxyethyl)-*N*-methyl-2,4-dinitronaphthylamine by the action of methanolic methoxide.[90] Acidification of solutions of these complexes under controlled conditions gives 2-(methylamino)-ethyl nitroaryl ethers which are trapped as their ammonium salts;[91] these aryl ethers cannot be prepared by other methods and undergo rapid conversion by a Smiles rearrangement into the arylamines.

Two kinetic studies by stopped-flow spectrophotometry of the interaction of 1,3,5-trinitrobenzene with sulphite ions have appeared,[92,93] In a careful piece of work Bernasconi and Bergstrom[92] show that three different σ-complexes are formed: a 1:1 complex and two isomeric 1:2 complexes. Confirmation that the latter are the expected *cis*- and *trans*-isomers has been obtained by NMR measurements in water[94] and in deuterium oxide.[95] Several mechanistic studies of sulphite or bisulphite addition to heterocyclic systems of biochemical importance have been reported. Addition to isoalloxazines[96] may occur at nitrogen to give (**19**) or at carbon to give (**20**). The bisulphite-catalysed deamination of cytidine 5′-phosphate involves an initial fast equilibrium to give the adduct (**21**) which is then converted into the uridine adduct (**22**); the rate-dependence

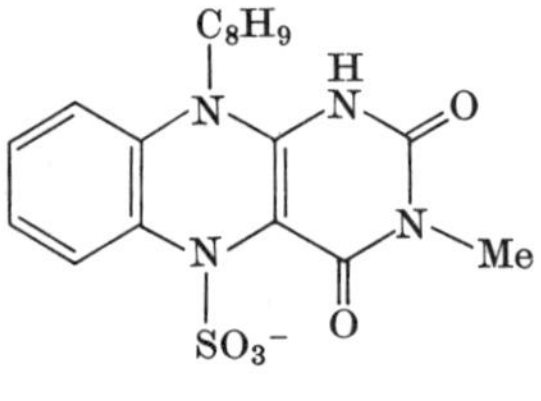

(**19**) 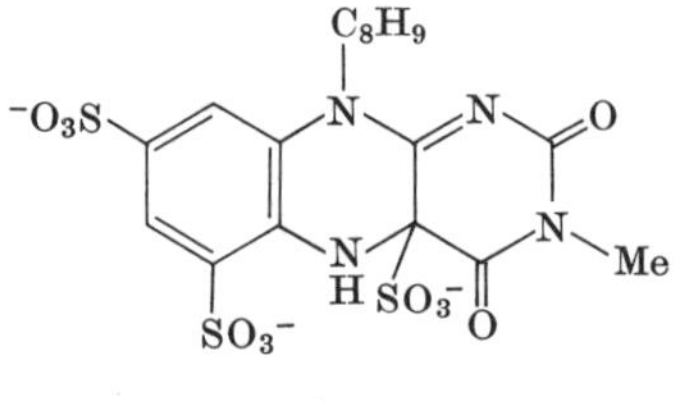 (**20**)

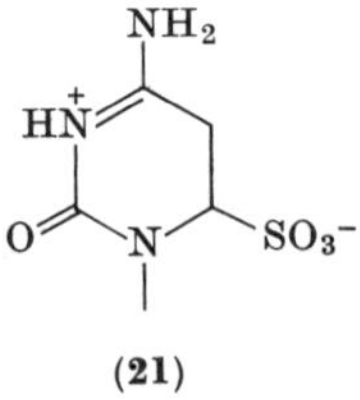

(**21**) 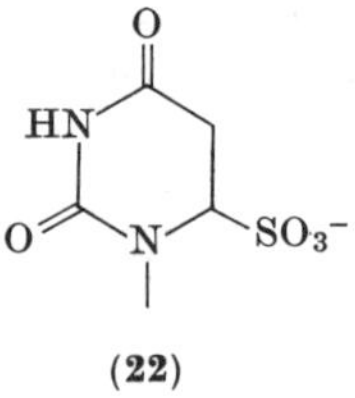 (**22**)

on bisulphite concentration indicates that bisulphite participates in the hydrolytic step as well as the adduct-forming step.[97] In a related study[98] it was found that derivatives of 5-bromouridine are debrominated by aqueous bisulphite via a bisulphite adduct. The reactions in deuterium oxide provide a convenient method for preparing 5-deuteriated compounds.

In aqueous solution a number of *N*-methyl heterocyclic cations yield neutral pseudo-bases, e.g. (**23**) from 2-methyl-4-nitroisoquinolinium ion; the pH–rate profiles of these reactions show that kinetically pseudo-base formation may involve attack of either a water molecule or hydroxide ion on the heterocyclic cation, depending on the acidity of the solution; rate constants and activation parameters have been determined.[99] In liquid ammonia, analogous adducts are produced by the covalent amination of quaternized heterocyclics.[100] The extent of amination of a given compound greatly exceeds its hydration. Addition of potassium amide to ammoniacal solutions of quinoline or isoquinoline yields anionic σ-complexes; thus (**24**) is formed from isoquinoline, while in the case of quinoline the C-2 adduct is kinetically and the C-4 adduct thermodynamically favoured.[101] 2-Chloro-3,6-diphenylpyrazine gives the adduct (**25**), and replacement of

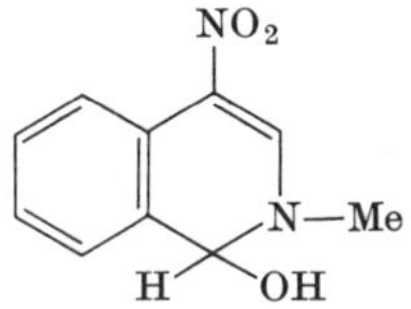

(**23**) 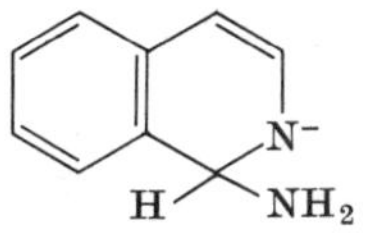 (**24**) 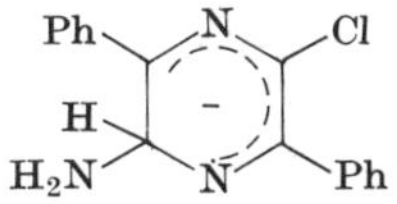 (**25**)

chlorine to give the expected 2-amino-3,6-diphenylpyrazine proceeds slowly.[102] Similarly 4-chloropyrimidine initially yields the 6-amino-adduct.[103] The last two reactions are analogous to those of activated chlorobenzenes with nucleophiles where addition occurs first at an unsubstituted ring position and replacement of chlorine is relatively slow.

The reactions of methylpteridines with several nucleophiles yield addition products in some cases. However, treatment of 6,7-dimethyl- or 7-methyl-pteridine (**26**) with aqueous alkali or with aliphatic amines gave dimeric products,[104] the latter being formed by attack of a carbanion (produced by proton loss from the methyl group of one molecule) at the 4-position of a second molecule. A full report of the catalysis by adenosine deaminase of the covalent hydration of pteridine has appeared.[105]

There have been several studies of complexes formed from electron-deficient aromatic compounds and carbanions. The methyl groups in di- and tri-nitrotoluenes are deprotonated by base to yield carbanions which attack 1,3,5-trinitrobenzene to give σ-complexes.[106] The acetonate ion, generated from acetone and base, gives adducts such as (**27**) (Janovsky reaction).[107] Isomeric adducts are in some cases produced from substituted dinitrobenzenes.[108] The condensation of aldehydes with 1,3-dinitrobenzene has

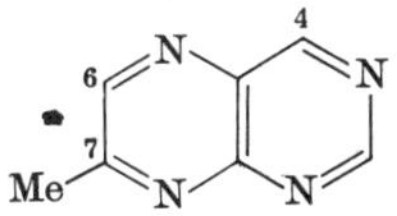

(**26**) 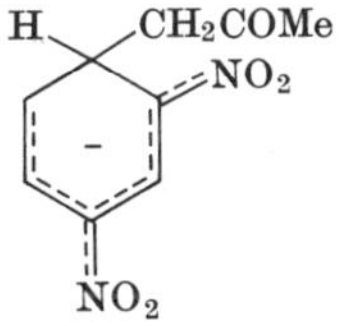 (**27**)

also been reported.[109] Kinetic studies of the decomposition of (**27**), including investigation of solvent effects and substituent effects, have been reported.[110] In a separate kinetic study the rate-determining step in the Janovsky reaction of acetone[111] or cyclohexanone[112] with 1,3,5-trinitrobenzene is shown to be addition of carbanion to the aromatic ring rather than deprotonation of ketone by base. It has been known for some time that the initial condensation of carbanions with activated arenes may be followed by internal cyclization to yield stable bicyclic nitronates such as (**29**). Work particularly by Strauss and his co-workers continues in this field.[113] The reactions of 1,3,5-trinitrobenzene with cycloalkanones and with phloroglucinol yield structurally similar tricyclic

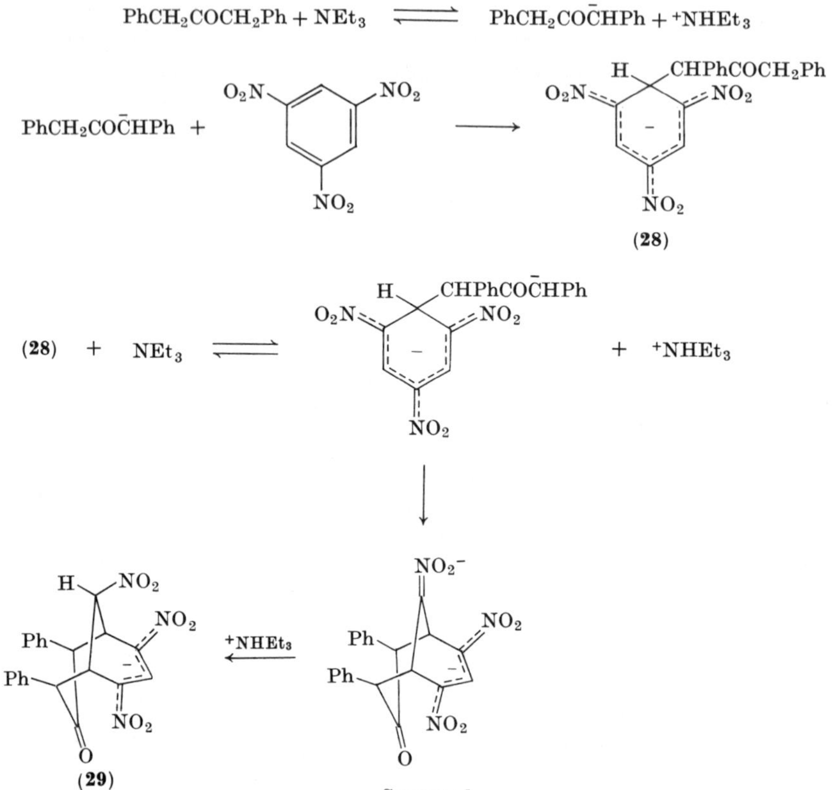

$$PhCH_2COCH_2Ph + NEt_3 \rightleftharpoons PhCH_2CO\bar{C}HPh + {}^+NHEt_3$$

(**28**)

(**29**)

SCHEME 1.

nitropropene nitronates which may be protonated to yield nitronic acids.[114] Reaction of 2-substituted 1,3,5-trinitrobenzene derivatives with dimethyl acetone-1,3-dicarboxylate yields mainly the 2-substituted 3-nitropropene nitronates although some of the isomeric bridgehead-substituted product is obtained.[115] In the naphthalene series benzobicyclic nitronates are produced.[116] A detailed kinetic study of a typical reaction, that of 1,3,5-trinitrobenzene with dibenzyl ketone in the presence of triethylamine in DMSO, has been reported;[117] this reaction occurs in two stages (Scheme 1), a fast condensation to give (**28**), and slow cyclization to (**29**). In contrast, the reaction of 3,5-dinitroacetophenone with acetone and diethylamine yields *N*,*N*-diethyl-4-methyl-6,8-dinitro-2-naphthylamine rather than a bicyclic nitronate;[118] initial addition of acetonate ion to the ring is followed by reaction with the carbonyl group of the dinitroacetophenone to yield a ring-closed product. A similar reaction occurs with 3,5-dinitrobenzaldehyde.

Surfactants may have marked effects on complex formation. Thus a large increase in the rate of decomposition of 1,1-dimethoxy-2,4,6-trinitrocyclohexadienate in benzene is brought about by reversed micelles;[119] this is attributed to solubilization of the polar substrate in hydrophilic cavities in the micelles where hydrogen-bonding, proton-transfer, and enhanced water activity aid product formation. Cyanide-ion addition to *N*-substituted 3-carbamoylpyridinium ions yields neutral products; the effects of polyelectrolytes and surfactants on rate and equilibrium constants depend critically on the nature of the substituent, and hydrophobic interactions are important.[120,121] The rate of base-catalysed decomposition of the hydride-ion adduct of 1,3,5-trinitrobenzene to give tetranitroazoxybenzene is greatly increased by bovine serum albumin.[122]

The rate of aromatic proton exchange of 1,3,5-trinitrobenzene (TNB) with base in dimethylformamide (DMF)–deuterium oxide decreases with increasing proportion of DMF; this unusual behaviour results from competition between loss of proton and addition of hydroxide ion to the ring.[123] The nucleophilicity of a series of solvents has been compared by competitive nucleophilic exchange in σ-complexes of TNB.[124] Complex formation between TNB and imidazole probably involves σ-complex formation,[125] while that between 2,6-dibromo-4-nitrophenol and aromatic amines involves only hydrogen-bonding.[126]

Benzyne and Related Intermediates

The production and chemistry of *p*-benzyne, which has a diradical structure, and of the 1,4-dehydrocycloheptatrienyl system have been summarized.[127] The infrared spectrum of *o*-benzyne isolated in an argon matrix at 8 K has been measured.[128]

The relative reactivities of amide ion and ammonia towards the two reactive positions of 4-chlorobenzyne have been determined by an *ad eosdem* (to the same products) competition method.[129] In the case of amide ion, rates for attack on the two positions are equal, but ammonia reacts more rapidly at carbon *para* to chlorine than at carbon *meta* to chlorine by a factor of 4.92. Overall, amide ion is considerably more reactive than ammonia.

A full report on the reactions of benzyne and tetrachlorobenzyne (**30**) with $\alpha\beta$-unsaturated aldehydes has appeared.[130] Cinnamaldehyde yields the flav-3-ene (**32**).

Cl Cl Cl Cl Ph * O → Cl Cl Cl Cl * Ph O → Cl Cl Cl Cl * O Ph

(**30**) (**31**) (**32**)

Using ^{14}C, ^{2}H and alkyl groups as labels it has been shown that the key step is cycloaddition to give a benzoxete derivative (**31**) which then undergoes ring opening and electrocyclic closure to the observed product. Benzyne can react with substituted vinylcyclopropanes by three main modes: "ene reaction", when readily accessible allylic hydrogens are present; [2 + 2]-cycloaddition; and, if the only allylic hydrogen atoms present are attached to a cyclopropane ring, by a novel "ene reaction" involving a cyclopropane-methine hydrogen.[131] Reaction of ethinylcyclopropane (**33**) by the last of these processes yields[132] α-cyclopropylidenestyrene (**34**). The reaction of benzyne with *cis*-cyclononene gives exclusively the "ene" addition product although a similar reaction with *trans*-cyclononene gives the [2 + 2]-addition product.[133]

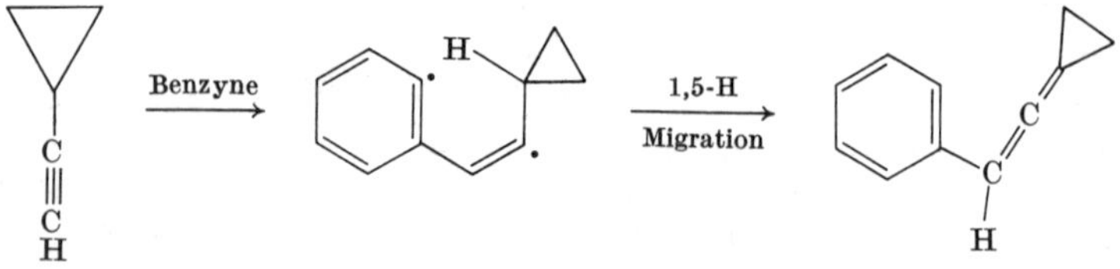

(33) —Benzyne→ —1,5-H Migration→ (34)

A convenient synthesis of phenanthridine is based on the reaction of haloanils with potassium amide in ammonia;[134] the mechanism involves addition of amide across the azomethine linkage followed by cyclization via a benzyne intermediate; the synthesis is widely applicable for the preparation of substituted phenanthridines.[135] Also reported[136] are synthetic applications of the condensation of benzyne with the enolates of a variety of aliphatic and aromatic ketones. Diverse products are formed, including alkyl-substituted aromatic ketones, anthrones and anthranols and substituted benzyl alcohols. The ready cleavage of ethers by benzyne in chlorocarbon solvents has been reported[137] (diethyl ether yields phenetole), while the reaction of benzyne with anions derived from 1-alkynyl sulphides yields products resulting from attack by sulphur.[138]

The decomposition of arenediazonium ions may in certain cases lead to arynes. A mechanistic study[139] has shown the involvement of the betaine (**35**) in the formation

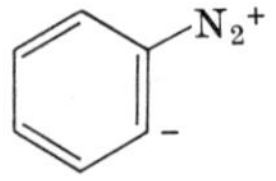

(**35**)

of benzyne from benzenediazonium ions produced from 1-phenyl-3,3-dimethyltriazenes.* In contrast, the decomposition of 2,3,4,5-tetrahalogenophenyltriazenes does not lead to arynes. The reactions of (*o*-halophenyl)trimethylsilanes with potassium *tert*-butoxide or tetramethylammonium fluoride in aprotic solvents yield benzyne.[140] Benzyne intermediates are implicated in the reaction of cyclopropyl-lithium with halobenzenes in ether[141] and in the pyrolysis of phthalic acid and its methyl derivatives.[142]

The formation of 3,4-pyridyne in the gas phase by flash photolysis of 3-diazoniopyridine-4-carboxylate is reported.[143] The rate of dimerization, to give a diazabiphenylene, is similar to that of benzyne. The preparation and reactions of 3,4-pyridyne and 2-fluoro-3,4-pyridyne in ether solutions have been described.[144] Aminations of several methyl-substituted halopyridines involve substituted 3,4-pyridynes as intermediates,[145] and hetarynes are also involved in the aminations of 2-substituted derivatives of 3-bromo-1,5-naphthyridine and 3-bromoquinoline.[146]

Other Reactions

Aromatic substitutions brought about by electrochemical techniques have been summarized.[147] Anodic oxidation of methoxybenzenes in alkaline methanol affords diketals by a direct discharge mechanism;[148] 1,4-dimethoxybenzene yields (**36**), the diketal of *p*-benzoquinone. Electrochemical oxidation of 2,4,6-tri-*tert*-butylphenol in the presence of pyridine gives substituted products from which the betaine (**37**) can be obtained.[149] The isomer distributions of aromatic nitriles formed by five separate methods, including anodic cyanation, have been determined.[150]

* In a very recent Note this mechanism has been revised with the agreement of the original authors (J. G. Cadogan, C. D. Murray and J. T. Sharp, *Chem. Comm.*, **1974**, 133).

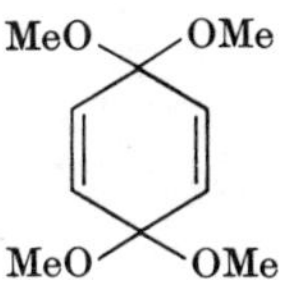

(36) 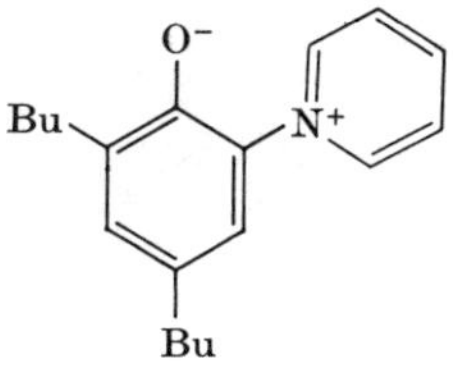 (37)

A comprehensive review of the properties and reactions of cycloheptatrienones and related compounds has appeared.[151] Reactions of these compounds with nucleophiles are complicated because of the possibility of attack at ring positions other than that bearing a replaceable group.[152] Thus 2-Y-cycloheptatrienones (**38**) may yield substitution products resulting from nucleophilic addition at C-2 (**39**) or at C-7 (**40**), or ring-contracted products such as (**41**). In general, electron-attracting Y groups which are

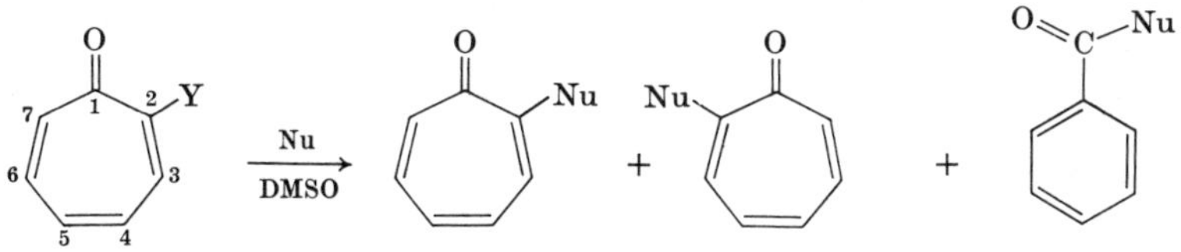

(38) (39) (40) (41)

capable of stabilizing an anionic intermediate (**42**) favour attack at C-7. With electron-releasing Y groups, nucleophilic attack normally occurs at C-2. Several pathways leading to ring-contracted products are available.[153] In the case of 2-methoxytropone a further possibility for nucleophilic reaction is demethylation.[154]

Large rate-enhancements in the formation of chromans (**43**) from a series of 3-(*o*-hydroxyphenyl)propan-1-ol monoesters are brought about by appropriate alkyl-substitution in the aromatic ring or in the side chain. These rate increases result from the interlocking of alkyl groups, which increases the population of the conformer most resembling the transition state.[155]

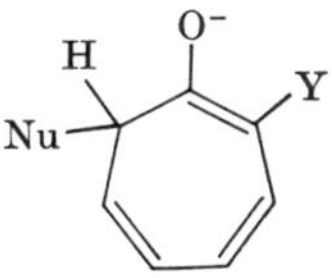

(42) 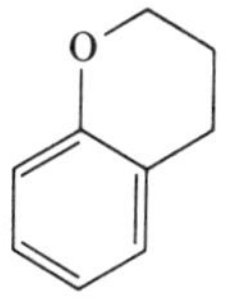 (43)

Base-catalysed cyclization has been used in synthesis of pyrrolo[1,2-*d*]-*as*-triazine (**44**)[156] and *N*-hydroxybenzimidazolones (**45**).[157] Cyclization of *N*-(*p*-nitrobenzyl)-*N*-(*p*-toluenesulphonyl)-*o*-nitroaniline by sodium methoxide gives a mixture of 1-hydroxy- and 1-methoxy-2-(*p*-nitrophenyl)benzimidazole (**46**); the latter compound results from methylation of its hydroxy-analogue by methyl toluene-*p*-sulphonate formed during the reaction.[158] Study of the intermediate steps in the synthesis of benzazepine shows it to involve trimerization of 5-bromoindole, condensation of the trimer with dimedone and

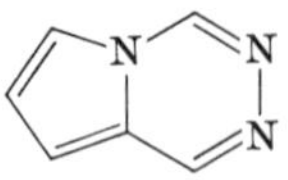

(**44**) 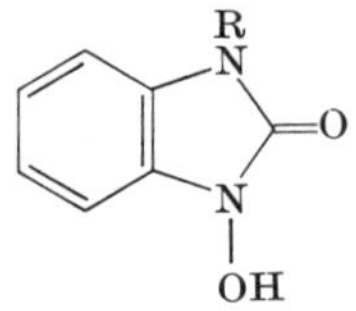 (**45**) 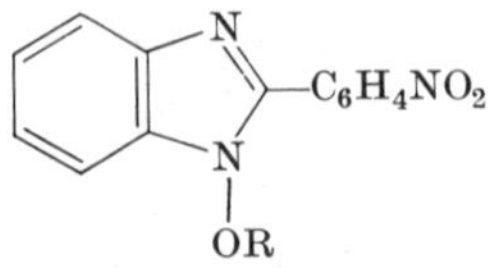 (**46**) R = H or Me

cyclization.[159] Syntheses of the fully conjugate azulene-like heterocycle 4-methyl-4*H*-cyclopenta[*b*]quinoline [160] and of 1-benzothienoselenopyrylium perchlorate[161] have been reported. 2-Arylthiazolo[3,2-*a*]pyridinium 3-oxides (**47**) have been synthesised;[162] derivatives containing a nitro-group in the pyridine ring undergo hydroxide-ion addition in alkaline media which facilitates their hydrolysis.

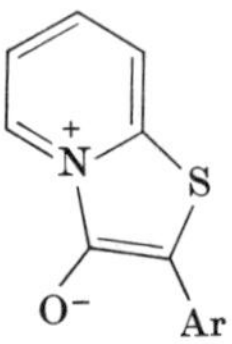

(**47**) 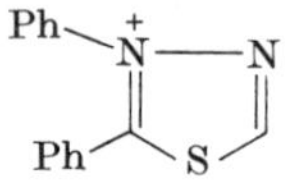 (**48**)

2,3-Diphenyl-1,3,4-thiadiazol-3-ium (**48**) undergoes specific hydroxide ion-catalysed hydrolysis in the presence of amines.[163] Reaction of thiopyrylium cations with amines under mild conditions gives ring-opened products[164] probably resulting from nucleophilic attack at C-2. The conversion of halopyridinium salts into pyridones in good yield occurs by reaction with DMSO; the reaction may involve attack by DMSO, followed by an oxidation–reduction mechanism.[165] Reaction of 4-pyridylpyridinium chloride with sodium cyanide gives a dimerized product from which 4,4′-bipyridine may be obtained.[166]

Anionic substitution in the benzene ring of *N*-acetoxyquinoxalinium perchlorates proceeds by a novel mechanism involving prior attack on the heterocyclic ring.[167]

The mechanisms of hydrolysis of chlorobenzenes in the presence of heterogeneous catalyst have received attention.[168]

The base-catalysed H–D exchange rates for the aromatic and benzylic centres of several acenaphthenes and related hydrocarbons have been measured,[169] the results serving as a probe of carbanion stability and reactivity. Other studies have involved the reductive dehalogenation of *o*-chlorobenzylideneaniline[170] and the alkylation of thiophosphinic amides.[171]

References

1 C. F. Bernasconi, *MTP Internat. Rev. Sci.: Org. Chem., Ser.* 1, **3**, 33 (1973).

2 H. Zollinger, *Accounts Chem. Research*, **6**, 335 (1973).

3 J. F. Bunnett, *Accounts Chem. Research*, **5**, 139 (1972).

4 R. A. Rossi and J. F. Bunnett, *J. Org. Chem.*, **38**, 3020 (1973).

5 R. A. Rossi and J. F. Bunnett, *J. Org. Chem.*, **38**, 1407 (1973).

6 M. F. Semmelhack, R. D. Stauffer and T. D. Rogerson, *Tetrahedron Letters*, **1973**, 4519.

7 S. M. Shein, L. V. Bryukhovetskaya and T. M. Ivanova, *Izv. Akad. Nauk SSSR, Ser. Khim.*, **1973**, 1594; *Chem. Abs.*, **79**, 104537 (1973); I. I. Bil'kis and S. M. Shein, *Izv. Akad. Nauk SSSR, Ser.* Khim., **1973**, 961; *Chem. Abs.*, **79**, 52546 (1973).

[8] M. V. Shtemshis, L. V. Bryukhovetskaya, N. M. Katkova, I. Bil'kis and S. M. Shein, *Zh. Org. Khim.*, **9**, 1242 (1973); *Chem. Abstr.*, **79**, 65459 (1973).
[9] L. V. Bryukhovetskaya, L. V. Mironova and S. M. Shein, *Izv. Akad. Nauk SSSR, Ser. Khim.*, **1973**, 1601; *Chem. Abs.*, **79**, 104539 (1973).
[10] A. A. Ponomarenko and N. A. Tsybina, *Zh. Org. Khim.*, **9**, 1016 (1973); *Chem. Abs.*, **79**, 52482 (1973).
[11] J. A. Zoltewicz and T. M. Oestreich, *J. Am. Chem. Soc.*, **95**, 6863 (1973).
[12] J. D. Aubort, R. F. Hudson and R. C. Woodcock, *Tetrahedron Letters*, **1973**, 2229.
[13] J. F. Liebman and R. M. Pollack, *J. Org. Chem.*, **38**, 3444 (1973).
[14] C. F. Bernasconi and R. H. de Rossi, *J. Org. Chem.*, **38**, 500 (1973).
[15] J. Kaválek, J. Haasová and V. Štěrba, *Coll. Czech. Chem. Comm.*, **37**, 3333 (1972).
[16] J. Kaválek, J. Kubias and V. Štěrba, *Coll. Czech. Chem. Comm.*, **37**, 4041 (1972).
[17] J. Kaválek and V. Štěrba, *Coll. Czech. Chem. Comm.*, **38**, 884 (1973).
[18] B. Lamm and J. Lammert, *Acta Chem. Scand.*, **27**, 191 (1973).
[19] N.I. Titarenko and L. S. Fedorova, *Visn. Khar'kiv. Univ.*, **1971**, 95; *Chem. Abs.*, **78**, 57329 (1973); L. S. Fedorova, Mai Thi Thao and M. I. Titarenko, *Visn. Khar'kiv. Univ.*, 92 (1971); *Chem. Abs.*, **78**, 57342 (1973); N. I. Titarenko and L. S. Fedorova, *Visn. Khar'kiv. Univ.*, **1972**, 76; *Chem. Abs.*, **78**, 123775 (1973).
[20] R. S. Cheshko, L. M. Litvinenko, V. A. Dadali and T. T. Nguyen, *Ukr. Khim. Zh.* (*Russ. Ed.*), **39**, 184 (1973); *Chem. Abstr.*, **78**, 146986 (1973).
[21] S. Sekiguchi, T. Itagaki, T. Hirose, K. Matsui and K. Sekine, *Tetrahedron*, **29**, 3527 (1973).
[22] K. T. Leffek and P. H. Tremaine, *Can. J. Chem.*, **51**, 1659 (1973).
[23] Y. Hasegana and T. Abe, *Chem. Letters* (*Tokyo*), **1972**, 985.
[24] N. A. Pankova, S. S. Gitis, A. Ya. Kaminskii and E. G. Kaminskaya, *Sin. Anal. Strukt. Org. Soedin.*, **1969**, 19; *Chem. Abs.*, **78**, 42440 (1973).
[25] N. A. Goeckner and H. R. Snyder, *J. Org. Chem.*, **38**, 481 (1973).
[26] D. H. Rosenblatt, W. H. Dennis and R. D. Goodwin, *J. Am. Chem. Soc.*, **95**, 2133 (1973).
[27] D. E. Giles and A. J. Parker, *Austral. J. Chem.*, **26**, 273 (1973).
[28] F. Del Cima, G. Biggi and F. Pietra, *J.C.S., Perkin II*, **1973**, 55.
[29] T. O. Bamkole, J. Hirst and E. I. Udoessien, *J.C.S., Perkin II*, **1973**, 110.
[30] L. Aiyar and K. K. Kundu, *J.C.S., Perkin II*, **1973**, 146.
[31] P. Carniti, P. Beltrame and S. Cabiddu, *J.C.S., Perkin II*, **1973**, 1430.
[32] B. G. Cox and A. J. Parker, *J. Am. Chem. Soc.*, **95**, 402, 408 (1973).
[33] B. Dacre, D. F. Maundrell and S. E. Rees, *J.C.S., Perkin II*, **1973**, 1060.
[34] P. H. Gore, S. D. Hammond and D. F. C. Morris, *J.C.S., Perkin II*, **1973**, 1883.
[35] S.-J. Yeh, C.-T. Chen and K.-Y. Chiou, *Ho Tzu K'o Hsueh*, **8**, 12 (1971); *Chem. Abs.*, **79**, 77667 (1973).
[36] C.-T. Chen, M.-C. Wu and S.-J. Yeh, *Ho Tzu K'o Hsueh*, **8**, 17 (1971); *Chem. Abs.*, **79**, 77666 (1973).
[37] V. Spěvaček, *Tetrahedron*, **29**, 2285 (1973).
[38] D. W. Lamson, P. Ulrich and R. O. Hutchins, *J. Org. Chem.*, **38**, 2928 (1973).
[39] K. J. Bird, I. D. Rae and A. M. White, *Austral. J. Chem.*, **26**, 1683 (1973).
[40] E. I. Tomilenko and F. M. Vainshtein, *Ukr. Khim. Zh.* (*Russ. Ed.*), **39**, 52 (1973); *Chem. Abs.*, **78**, 123572 (1973).
[41] K. Takagi, T. Okamoto, Y. Sakakibara and S. Oka, *Chem. Letters* (*Tokyo*), **1973**, 471.
[42] C. Dell'Erba, G. Guanti, M. Novi and G. Leandri, *J.C.S., Perkin II*, **1973**, 1879.
[43] C. Jongsma and F. Bickelhaupt, *Rec. Trav. chim.*, **92**, 1143 (1973).
[44] I. V. Kudryashov and I. A. Balakirev, *Zh. Vses. Khim. Obshchest.*, **17**, 586 (1972); *Chem. Abs.*, **78**, 28781 (1973).
[45] S. M. Shein and P. P. Rodionov, *Organic Reactivity* (*Tartu*), **8**, 964 (1971).
[46] J. Griffiths and C. Hawkins, *Chem. Comm.*, **1973**, 111.
[47] J. O. Morley, *J.C.S., Perkin II*, **1973**, 1626.
[48] G. B. Barlin, *Aromat. Heteroaromat. Chem.*, **1**, 230 (1973); *Chem. Abs.*, **79**, 104361 (1973).
[49] I. C. Lee, *Hwahak Kwa Kongop Ui Chinbo*, **12**, 310 (1973); *Chem. Abs.*, **79**, 4445 (1973).
[50] H. Arai, *Sekiyu Gakkai Shi*, **16**, 507 (1973); *Chem. Abs.*, **79**, 104375 (1973).
[51] T. M. Chinh, J. Kaválek and M. Večeřa, *Coll. Czech. Chem. Comm.*, **37**, 3328 (1972); J. Kaválek, T. M. Chinh, V. Mikan, V. Štěrba and M. Večeřa, *ibid.*, **38**, 1935 (1973).
[52] H. Clauss and H. Suhr, *Chem. Ber.*, **106**, 3398 (1973).
[53] N. N. Mel'nikova, M. S. Pevzner and L. I. Bagal, *Zh. Org. Khim.*, **9**, 799 (1973); *Chem. Abs.*, **79**, 4595 (1973).

[54] N. N. Mel'nikova, M. S. Pevzner and L. I. Bagal, *Organic Reactivity* (*Tartu*), **9**, 561, 574 (1972).
[55] S. M. Shein, O. A. Zaguloyeva, A. I. Shvets and V. P. Mamayev, *Organic Reactivity* (*Tartu*), **9**, 895, 918 (1972).
[56] N. P. H. Brown, C. G. L. Furmidge and B. T. Grayson, *Pestic. Sci.*, **3**, 669 (1972); *Chem. Abs.*, **78**, 83513 (1973).
[57] L. V. Kazanskaya, T. A. Melent'eva and V. M. Berezovskii, *Zh. Obshch. Khim.*, **42**, 2523 (1972); *Chem. Abs.*, **78**, 110110 (1973).
[58] M. G. Reinecke and T. A. Hollingworth, *J. Org. Chem.*, **37**, 4257 (1972).
[59] D. A. de Bie, H. C. van der Plas, G. Geurtsen and K. Nijdam, *Rec. Trav. chim.*, **92**, 245 (1973).
[60] M. G. Reinecke, W. B. Mohr, H. W. Adickes, D. H. de Bie, H. C. van der Plas and K. Nijdam, *J. Org. Chem.*, **38**, 1365 (1973).
[61] H. J. den Hertog and D. J. Buurman, *Rec. Trav. chim.*, **92**, 304 (1973).
[62] J. de Valk and H. C. van der Plas, *Rec. Trav. chim.*, **91**, 1414 (1972); **92**, 471 (1973); A. P. Kroon and H. C. van der Plas, *ibid.*, **92**, 1020 (1973).
[63] M. H. Wilson and J. A. McCloskey, *J. Org. Chem.*, **38**, 2247 (1973).
[64] J. de Valk and H. C. van der Plas, *Rec. Trav. chim.*, **92**, 145 (1973); J. de Valk, H. C. van der Plas and J. W. A. de Bode, *ibid.*, **92**, 442 (1973).
[65] H. C. van der Plas and A. Koudijs, *Rec. Trav. chim.*, **92**, 711 (1973).
[66] P. J. Lont and H. C. van der Plas, *Rec. Trav. chim.*, **92**, 311, 449 (1973).
[67] J. de Valk, H. C. van der Plas, F. Jansen and A. Koudijs, *Rec. Trav. chim.*, **92**, 460 (1973).
[68] J. A. Zoltewicz and V. W. Cantwell, *J. Org. Chem.*, **38**, 829 (1973).
[69] M. Goshaev, O. S. Ostroshchenko, A. S. Sadykov and F. G. Kamaev, *Izv. Akad. Nauk Turkm. SSSR, Ser. Fiz.-Tekh., Khim. Geol. Nauk*, **1973**, 116; *Chem. Abs.*, **78**, 158571 (1973).
[70] J. A. Rabi and J. J. Fox, *J. Am. Chem. Soc.*, **95**, 1628 (1973).
[71] S. A. Krueger and W. W. Paudler, *J. Org. Chem.*, **37**, 4188 (1972).
[72] V. P. Lezina, A. U. Stepanyants, L. D. Smirnov and N. A. Andronova, *Izv. Akad. Nauk SSSR, Ser. Khim.*, **1972**, 2267; *Chem. Abs.*, **78**, 71180 (1973).
[73] J. A. Zoltewicz and A. A. Sale, *J. Am. Chem. Soc.*, **95**, 3928 (1973).
[74] A. Belly, C. Petrus and F. Petrus, *Bull. Soc. Chim. France*, **1973**, 1390.
[75] R. D. Chambers and M. Y. Gribble, *J.C.S. Perkin I*, **1973**, 1405, 1411; R. D. Chambers, M. Y. Gribble and E. Marper, *J.C.S. Perkin I*, **1973**, 1710, 1716.
[76] M. R. Crampton and H. A. Khan, *J.C.S. Perkin II*, **1972**, 2286.
[77] M. R. Crampton and H. A. Khan, *J.C.S. Perkin II*, **1973**, 1103.
[78] J. W. Larsen, K. Amin, S. Ewing and L. L. Magid, *J. Org. Chem.*, **37**, 3857 (1972).
[79] G. Biggi and F. Pietra, *Chem. Comm.*, **1973**, 229; *J.C.S. Perkin I*, **1973**, 1980.
[80] C. A. Fyfe, M. Cocivera and S. N. H. Damji, *Chem. Comm.*, **1973**, 743.
[81] M. R. Crampton and H. A. Khan, *J.C.S. Perkin II*, **1973**, 710.
[82] F. Millot and F. Terrier, *Compt. Rend., C*, **276**, 327 (1973).
[83] A.-P. Chatrousse and F. Terrier, *Bull. Soc. Chim. France*, **1972**, 4549.
[84] G. Biggi, C. A. Veracini and F. Pietra, *Chem. Comm.*, **1973**, 523.
[85] T. Abe and T. Asao, *Tetrahedron Letters*, **1973**, 1327.
[86] C. M. Lok, M. E. Den Boer and J. Cornelisse, *Rec. Trav. chim.*, **92**, 340 (1973).
[87] C. Paulmier, M.-P. Simonnin, A.-P. Chatrousse and F. Terrier, *Tetrahedron Letters*, **1973**, 1123.
[88] G. Doddi, G. Illuminati and F. Stegel, *Tetrahedron Letters*, **1973**, 3221.
[89] M.-P. Simonnin, F. Terrier and C. Paulmier, *Tetrahedron Letters*, **1973**, 2803.
[90] S. Sekiguchi and T. Shiojima, *Bull. Chem. Soc. Japan*, **46**, 693 (1973).
[91] C. F. Bernasconi, R. H. de Rossi and C. L. Gehriger, *J. Org. Chem.*, **38**, 2838 (1973).
[92] C. F. Bernasconi and R. G. Bergstrom, *J. Am. Chem. Soc.*, **95**, 3603 (1973).
[93] M. Sasaki, *Chem. Letters* (*Tokyo*), **1973**, 205.
[94] M. R. Crampton and M. J. Willison, *Chem. Comm.*, **1973**, 215.
[95] M. J. Strauss and S. P. B. Taylor, *J. Am. Chem. Soc.*, **95**, 3813 (1973).
[96] L. Hevesi and T. C. Bruice, *Biochemistry*, **12**, 290 (1973); T. C. Bruice, L. Hevesi and S. Shinkai, *ibid.*, **12**, 2083 (1973).
[97] M. Sono, Y. Wataya and H. Hayatsu, *J. Am. Chem. Soc.*, **95**, 4745 (1973).
[98] J.-L. Fourrey, *Bull. Soc. Chim. France*, **1972**, 4580.
[99] J. W. Bunting and W. G. Meathrel, *Can. J. Chem.*, **51**, 1965 (1973).
[100] J. A. Zoltewicz, T. M. Oestreich, J. K. O'Halloran and L. S. Helmick, *J. Org. Chem.*, **38**, 1949 (1973).
[101] J. A. Zoltewicz, L. S. Helmick, T. M. Oestreich, R. W. King and P. E. Kandetzki, *J. Org. Chem.*, **38**, 1947 (1973).

[102] P. J. Lont, H. C. van der Plas and A. van Veldhuizen, *Rec. Trav. chim.*, **92**, 708 (1973).
[103] J. P. Geerts, H. C. van der Plas and A. van Veldhuizen, *Rec. Trav. chim.*, **92**, 1232 (1973).
[104] A. Albert and H. Mizuno, *J.C.S. Perkin I*, **1973**, 1974.
[105] B. E. Evans and R. V. Wolfenden, *Biochemistry*, **12**, 392 (1973).
[106] S. S. Gitis, A. Ya. Kaminskii, Yu. D. Grudtsyn and E. E. Gol'teuzen, *Dokl. Akad. Nauk SSSR*, **206**, 102 (1972); *Chem. Abs.*, **78**, 3477 (1973); E. E. Gol'teuzen, Yu. D. Grudtsyn and A. Ya. Kaminskii, *Zh. Org. Khim.*, **8**, 1916 (1972); *Chem. Abs.*, **78**, 42411 (1973).
[107] K. Kohashi and Y. Ohkura, *Kagaku No Ryoiki*, **27**, 303 (1973); *Chem. Abs.*, **78**, 146901 (1973); E. E. Gol'teuzen, A. Ya Kaminskii, Z. V. Todres, S. S. Gitis and D. N. Kursanov. *Zh. Vses. Khim. Obshchest.*, **17**, 458 (1972); *Chem. Abs.*, **78**, 28790 (1973); S. S. Gitis, I. G. L'vovich and A. Ya Kaminskii, *Sin. Anal. Strukt. Org. Soedin.*, **2**, 25 (1969); *Chem. Abs.*, **78**, 15696 (1973).
[108] E. E. Gol'teuzen, S. S. Gitis and A. Ya. Kaminskii, *Sin. Anal. Strukt. Org. Soedin*, **1971**, 121; *Chem. Abs.*, **78**, 28745 (1973).
[109] S. S. Gitis and A. Ya. Kaminskii, *Sin. Anal. Strukt. Org. Soedin*, **1969**, 3; *Chem. Abs.*, **78**, 3885 (1973).
[110] L. N. Savinova, S. S. Gitis, A. Ya. Kaminskii, L. V. Illarionova and E. A. Bronstein, *Organic Reactivity* (*Tartu*), **9**, 529, 616, 1139 (1972).
[111] N. Obi and M. Kimura, *Chem. Pharm. Bull.*, **20**, 2295 (1972); *Chem. Abs.*, **78**, 70976 (1973).
[112] N. Obi, H. Kakizaki and M. Kimura, *Chem. Pharm. Bull.*, **21**, 235 (1973); *Chem. Abs.*, **78**, 147025 (1973).
[113] K. Kohashi, Y. Ohkura and T. Momose, *Chem. Pharm. Bull.*, **21**, 118 (1973).
[114] M. J. Strauss, S. P. B. Taylor and H. Shindo, *J. Org. Chem.*, **37**, 3658 (1972).
[115] M. J. Strauss and S. P. B. Taylor, *J. Org. Chem.*, **38**, 1330 (1973).
[116] M. J. Strauss and S. P. B. Taylor, *J. Org. Chem.*, **38**, 856 (1973).
[117] M. J. Strauss, H. F. Schran and R. R. Bard, *J. Org. Chem.*, **38**, 3394 (1973).
[118] S. R. Alpha, *J. Org. Chem.*, **38**, 3136 (1973).
[119] J. H. Fendler and S. A. Chang, *J. Am. Chem. Soc.*, **95**, 3273 (1973).
[120] J. Baumrucker, M. Calzadilla, M. Centeno, G. Lehrmann, M. Urdaneta, P. Lindquist, D. Dunham, E. J. Fendler, M. Price, B. Sears and E. H. Cordes, *J. Am. Chem. Soc.*, **94**, 8164 (1972).
[121] T. Okubo and N. Ise, *J. Am. Chem. Soc.*, **95**, 4031 (1973).
[122] R. P. Taylor and J. B. Vatz, *J. Am. Chem. Soc.*, **95**, 5819 (1973).
[123] E. Buncel and E. A. Symons, *J. Org. Chem.*, **38**, 1201 (1973).
[124] E. A. Bronshtein, S. S. Gitis, A. Ya. Kaminskii, Yu. D. Grudtsyn and E. E. Gol'teuzen, *Sin. Anal. Strukt. Org. Soedin*, **1971**, 110; *Chem. Abs.*, **78**, 158740 (1973).
[125] F. Takahashi, M. Okamoto and H. Maeda, *Nippon Kagaku Kaishi*, **1972**, 2017; *Chem. Abs.*, **78**, 57200 (1973).
[126] S. A. Savintseva and E. A. Krulikooskaya, *Tr. Konf. Molodykh Uch.-Khim. Goroda Tomska 1st*, **1967**, 242; *Chem. Abs.*, **78**, 42600 (1973).
[127] R. G. Bergman, *Accounts Chem. Res.*, **6**, 25 (1973).
[128] O. L. Chapman, K. Mattes, C. L. McIntosh, J. Pacansky, G. V. Calder and G. Orr, *J. Am. Chem. Soc.*, **95**, 6134 (1973).
[129] J. F. Bunnett and J. K. Kim., *J. Amer. Chem. Soc.*, **95**, 2254 (1973).
[130] H. Heaney, J. M. Jablonski and C. T. McCarty, *J.C.S., Perkin I*, **1972**, 2903.
[131] V. Usieli and S. Sarel, *J. Org. Chem.*, **38**, 1703 (1973).
[132] V. Usieli and S. Sarel, *Tetrahedron Letters*, **1973**, 1349.
[133] G. Mehta, *Indian J. Chem.*, **10**, 592 (1972); *Chem. Abs.*, **78**, 3772 (1973).
[134] S. V. Kessar, R. Gopal and M. Singh, *Tetrahedron*, **29**, 167 (1973).
[135] S. V. Kessar, D. Pal and M. Singh, *Tetrahedron*, **29**, 177 (1973).
[136] P. Caubere and G. Guillaumet, *Bull. Soc. Chim. France*, **1972**, 4643, 4649.
[137] G. D. Richmond and W. Spendel, *Tetrahedron Letters*, **1973**, 4557.
[138] L. Brandsma, S. Hoff and H. D. Verkruijsse, *Rec. Trav. chim.*, **92**, 272 (1973).
[139] P. C. Buxton and H. Heaney, *Chem. Comm.*, **1973**, 545.
[140] R. F. Cunico and E. M. Dexheimer, *J. Organometal. Chem.*, **59**, 153 (1973).
[141] W. Kurtz and F. Effenberger, *Chem. Ber.*, **106**, 560 (1973).
[142] J. M. Patterson, C. Y. Shiue and W. T. Smith, *J. Org. Chem.*, **38**, 387 (1973).
[143] J. Kramer and R. S. Berry, *J. Am. Chem. Soc.*, **94**, 8336 (1972).
[144] M. Mallet, F. Marsais, G. Queguiner and P. Pastour, *Compt. Rend., Ser. C*, **275**, 1439 (1972).
[145] L. Van der Does and H. J. den Hertog, *Rec. Trav. chim.*, **91**, 1403 (1972).
[146] J. Pomorski, H. J. den Hertog, D. J. Buurman and H. N. Dekker, *Rec. Trav. chim.*, **92**, 970 (1973).
[147] L. Eberson and K. Nyberg, *Accounts Chem. Res.*, **6**, 106 (1973).

[148] N. L. Weinberg and B. Belleau, *Tetrahedron*, **29**, 279 (1973).
[149] G. Popp and N. C. Reitz, *J. Org. Chem.*, **37**, 3646 (1972).
[150] S. Nilsson, *Acta Chem. Scand.*, **27**, 329 (1973).
[151] F. Pietra, *Chem. Rev.*, **73**, 293 (1973).
[152] G. Biggi, F. Del Cima and F. Pietra, *J. Am. Chem. Soc.*, **95**, 7101 (1973).
[153] G. Biggi, A. J. de Hoog, F. Del Cima and F. Pietra, *J. Am. Chem. Soc.*, **95**, 7108 (1973).
[154] G. Biggi, F. Del Cima and F. Pietra, *Tetrahedron Letters*, **1973**, 183.
[155] R. T. Borchardt and L. A. Cohen, *J. Am. Chem. Soc.*, **94**, 9166 (1972).
[156] J. P. Cress and D. M. Forkey, *Chem. Comm.*, **1973**, 35.
[157] D. B. Livingstone and G. Tennant, *Chem. Comm.*, **1973**, 96.
[158] H. McNab and D. M. Smith, *J.C.S. Perkin I*, **1973**, 1310.
[159] E. Gonzalez and R. Sarlin, *Combt. Rend., Ser. C.*, **277**, 117 (1973).
[160] J. J. Eisch, F. J. Gadek and G. Gupta, *J. Org. Chem.*, **38**, 431 (1973).
[161] A. Tadino, L. Christiaens, M. Renson and P. Cagniant, *Bull. Soc. Chim. Belges*, **81**, 595 (1972).
[162] P. E. Fjeldstad and K. Undheim, *Acta Chem. Scand.*, **27**, 1963 (1973).
[163] P. B. Talukdar, S. Banerjee and A. Chakraborty, *Indian J. Chem.*, **10**, 929 (1972); *Chem. Abs.*, **78**, 83509 (1973).
[164] Z. Yoshida, H. Sugimoto, T. Sugimoto and S. Yoneda, *J. Org. Chem.*, **38**, 3990, (1973).
[165] R. E. Lyle and M. J. Kane, *J. Org. Chem.*, **38**, 3740 (1973).
[166] R. H. Reuss and L. J. Winters, *J. Org. Chem.*, **38**, 3993 (1973).
[167] D. B. Livingstone and G. Tennant, *Chem. Ind.* (*London*), **1973**, 848.
[168] A. Krause, *Sci. Pharm.*, **40**, 131, (1972); *Chem. Abs.*, **78**, 42459 (1973); A. Malik, B. Preising, J. Wojcik and B. Zaborowski, *Przem. Chem.*, **52**, 99 (1973); *Chem. Abs.*, **79**, 17700 (1973).
[169] D. H. Hunter and J. B. Stothers, *Can. J. Chem.*, **51**, 2884 (1973).
[170] J. G. Smith and I. Ho, *J. Org. Chem.*, **37**, 4260 (1972).
[171] J. Boedeker, *J. Organometal. Chem.*, **56**, 255 (1973).

CHAPTER 7

Electrophilic Aromatic Substitution

A. R. BUTLER

Department of Chemistry, St. Salvator's College, University of St. Andrews

Introduction

There have been two reviews of recent work on electrophilic aromatic substitution by Taylor[1] and several studies concerning five-membered rings.[2] Electrophilic attack on π-complexes such as ferrocene has been surveyed.[3]

Semiempirical models have been formulated that predict accurately the effect of substituents on the susceptibility of aromatic compounds to electrophilic attack,[4] but MINDO calculations of certain indices of reactivity (electron densities, localization energies) show no better correlation with electrophilic reactivity than do the same quantities calculated by the simpler Hückel method.[5] A CNDO/2 calculation of electrophilic attack on the furan ring, using electrostatic potentials, shows that the carbon atom undergoing attack becomes almost tetrahedral.[6] In the solvolysis of 1-(2-benzofuryl)ethanol derivatives there is more resonance interaction with substituents at the 5-position than is predicted by the use of $\sigma_m{}^+$ constants; the Dewar–Grisdale equation[7] has been applied to this reaction.[8] CNDO/2 calculations indicate that substituents from the 5- and the 6-position of 1-(3-benzofuryl)ethanol should exert similar effects on the sidechain, and this has been confirmed by a study of the solvolyses of the *p*-nitrobenzoates.[9] Values of σ^+ constants for the Me_3Si group at the *o*-, *m*- and *p*-positions have been determined by the pyrolysis of 1-arylethyl acetates; the values differ from those in other reactions, reflecting differing resonance demands and steric factors.[10] It has been found unnecessary to apply the Yukawa–Tsuno equation to a number of electrophilic substitution reactions as the Stock–Brown extended-selectivity treatment gives unique σ^+ values; however, curved plots are obtained for the reactions of benzothiophen and naphthalene, indicating variable σ^+ values.

The generally accepted view of the mechanism of electrophilic aromatic substitution has been challenged: the reactivities of different positions in substituted benzenes follow the order of hyperfine coupling constants of the ESR spectra of the radical cations.[12] Does this prove that the reaction involves radical cations (Scheme 1)? It would be interesting to see this tested on a substrate that gives unexpected substitution products.

Electrophilic attack by the ion $[C_6H_7Fe(CO)_3]^+$ on reactive substrates such as indole leads to, e.g. [5-(indol-3-yl)cyclohexa-1,3-diene]tricarbonyliron.[13]

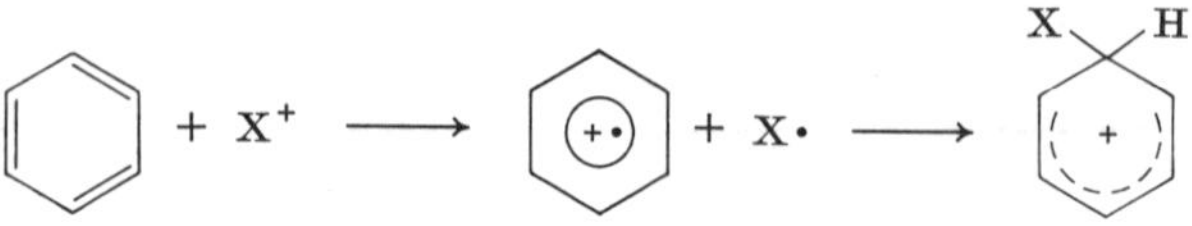

SCHEME 1.

The original designation[14] of the reactive sites in methoxyferrocene as C-2 and C-5 has been confirmed.[15] Attack on 8-substituted 5,6,12-triazadibenzo[*a,e*]pentalene (**1**) occurs at the 8-position, in agreement with HMO and SCF-MO calculations.[16] The most reactive positions in 4-hydroxybenzo[*b*]thiophen and its 3-methyl derivative are C-5 and C-7,[17] while for the 3-methyl derivative of 4-methoxybenzo[*b*]thiophen it is the 2-position.[18] The effect of the cyclopropyl substituent on aromatic reactivity depends on the nature of the transition state; when it is late (e.g. bromination), reactivity depends upon whether a bisected conformation can be adopted to give maximum stabilization of the Wheland intermediate; thus (**2**) reacts faster than (**3**) and attack is at the *p*-position. Nitration, with an early transition state, does not depend upon the conformation of the cyclopropyl group, and (**2**) and (**3**) react at similar rates to give the *o*-substituted product.[19] Most annelated rings activate the β-position towards electrophilic attack but, when $n = 6$ in (**4**), attack is at the α-position; if the annelated ring is seven-membered,

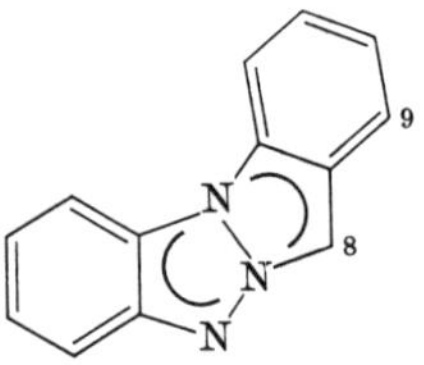

(**1**) 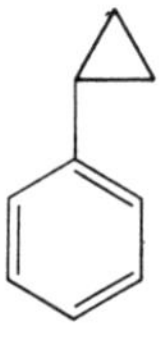 (**2**) 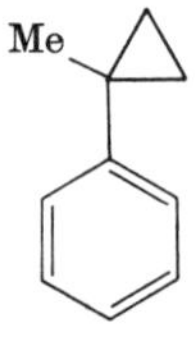 (**3**) 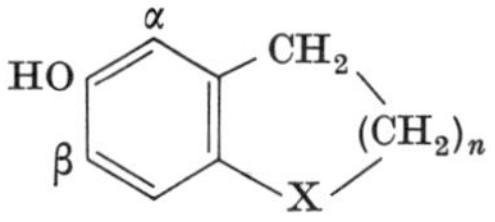 (**4**)

then the β-position is more activated in hydrogen exchange.[20] Electrophilic attack of λ^5-phosphorins results in loss of CH_2OH^+ (Scheme 2).[21]

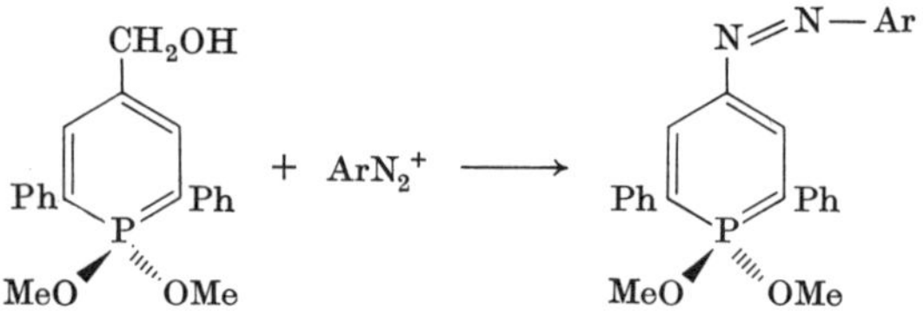

SCHEME 2.

Sulphonation

In the sulphonation of toluene by H_2SO_4 in nitrobenzene, the attacking species is thought to be $H_2S_2O_7$.[22] The isomer ratios observed in the sulphonation of benzenesulphonic acid and some polymethyl derivatives vary with the acid composition, probably indicating a changing mechanism.[23] On the other hand, such changes with other substrates have been ascribed to solvation effects.[24] In very concentrated sulphuric acid, sulphonation of benzenesulphonic acid occurs *via* the protonated acid;[23] with 3-phenylpropane-1-sulphonic[25] and 2-phenylethane-1-sulphonic acid[26] it is both the un-ionized acids and the conjugate bases that undergo reaction; partial rate factors for the $CH_2CH_2SO_3H$ group are reported for the latter case. Several substituted benzenesulphonic acids isomerize when heated with sulphuric acid.[27] The sulphonation of neopentylbenzene at the *o*-position is, not surprisingly, subject to steric hindrance.[28] Studies of the sulphonation of biphenyl[29] and fluorobenzene[30] have been reported. Reaction of indole occurs predominantly at the 3-position if the 2-position is blocked, as would be expected; 2,3-dimethylindole does not react.[31]

Sulphonation (and concurrent hydrogen exchange) of *N,N*-dimethylanisidine (both *para* and *meta*) occurs *via* the cation; surprisingly, positional discrimination in hydrogen exchange and sulphonation is about the same, although the former reaction is 100 times faster; the group NMe_3^+ does not exert any special influence on the *o*-position.[32] The intermediate (**5**) is formed in the sulphonation of 1*H*-pyrazolo[3,4-*b*] quinoxaline by SO_3 in pyridine.[33]

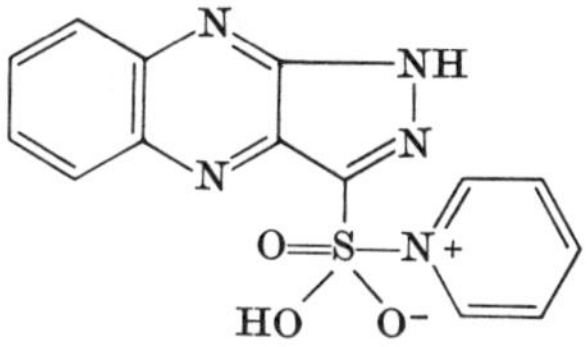

(**5**)

Methanesulphonic acid will sulphonylate polymethylbenzenes but the reaction is subject to steric hindrance and is accompanied by isomerization, methylation and the all too familiar tar formation.[34a] Varying substrate and positional selectivity in aromatic sulphonylation have been interpreted in terms of the position of the transition state along the reaction profile.[34b]

Nitration

There have been two reviews of aromatic nitration.[35] Some *ab initio* calculations for the interaction of NO_2^+ and ethylene have shown that the bridge structure (**6**) has some stability.[36] The reaction of methyl nitrate and benzene in the gas phase gives a nitronium

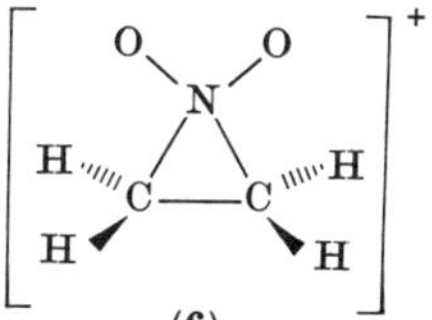

(**6**)

ion–benzene complex[37] and it has been suggested that this has a related bridged structure (**7**), which may lose CH_2O to give nitrobenzene. The relative rates of nitration of toluene

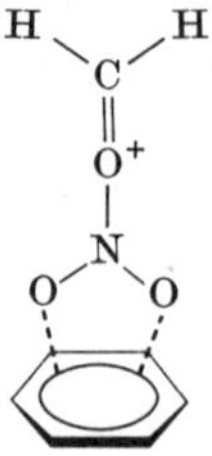

(**7**)

and benzene by NO_2BF_4 depend on the solvent; the low value reported by Olah was found for solvents such as *n*-hexane but with more polar solvents a much higher value is obtained;[38] the crucial role of solvation in solution chemistry is illustrated by this simple study.

A small hydrogen kinetic isotope effect has been found in the nitration of a number of sterically hindered alkylbenzenes,[39] and it has been shown that nitration of dinitroxylene is subject to steric hindrance.[40] The absence of any 5-substituted product in the nitration of 2-alkyl-1-methyl and 2-alkyl-1,4-dimethyl derivatives of tetrahydroquinolines is consistent with an explanation of the Mills–Nixon effect in terms of strain in bonds common to the fused rings,[41] and the explanation of the Mills–Nixon effect promoted elsewhere[42] cannot apply here. Reactions of organomercury compounds with some nitrating agents have been shown to be radical reactions.[43] The rates of nitration of a number of substituted benzene compounds have been used to define a new set of σ^+ constants.[44]

The groups CH_2PO_3H and $CH_2CH_2PO_3H$ are both mainly *p*-directing but some *o*-product is obtained; the attenuating effect of a methylene group on substituent effects is discussed.[45] In compounds of formula $Ph(CH_2)_nPh$, nitration in one phenyl ring has little effect on the other when $n > 2$;[46] this fact was used earlier by Ridd[47] in a study of the nitration of dibenzyl by nitronium salts. Nitration of 1-phenyl-1*H*-pyrazolo-[3,4-*b*]quinoxalines by NHO_3/Ac_2O occurs in the phenyl rather than in the heterocyclic ring.[48] Reaction between cold NHO_3 and (methoxymethyl)naphthalene results in oxidation as well as nitration, e.g. 2-(methoxymethyl)naphthalene gives 8-nitro-2-naphthaldehyde.[49] Although the thiophen ring is very susceptible to nitration, reaction of nitric acid and *trans*-2-styrylthiophen gives mainly 2-(β-nitrostyryl)thiophen with only a small amount of attack on the thiophen ring.[50] The unprotonated form of 1-chloroanthraquinone participates in nitration.[51]

Nitration may be effected by metal nitrates in Ac_2O: the rate of reaction depends on the metal ion.[52] Titanium nitrate appears to be electrophilic as a nitrating agent. It is unlikely that free NO_2^+ is formed and the mechanism is thought to involve attack of benzene on the nitrogen of the nitrate (Scheme 3).[53a] The use of $NaNO_3$–KNO_3 in a

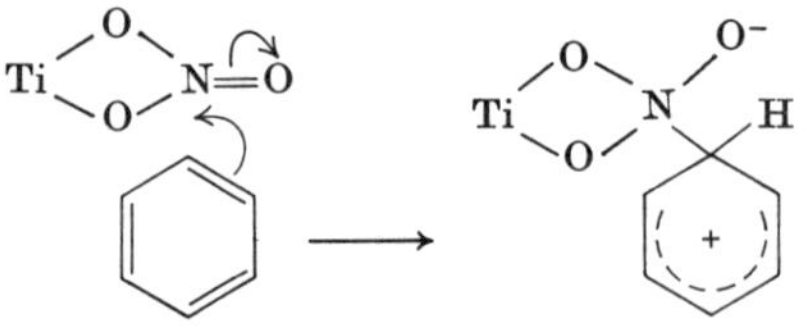

SCHEME 3.

pyrosulphate–sulphate melt at 250° as a nitrating agent has been investigated in some detail; a kinetic study, in which benzenesulphonate was used as substrate indicated a mechanism involving an intermediate formed from NO_2^+ and the benzenesulphonate ion, which decomposed to give nitrobenzene and SO_3.[53b] Dinitrogen tetraoxide is an effective, if inconvenient, nitrating agent; measurement of relative reactivities did not permit the effective species to be identified with certainty, but it is clear that nitrosation plays a significant role with highly activated substrates.[54]

More addition compounds formed during aromatic nitration have been reported. At room temperature nitration (HNO_3/Ac_2O) of 1,4-dimethylnaphthalene results in sidechain attack (**8**) but at –40° an adduct (**9**) is obtained.[55] With toluene, *cis*- and *trans*-forms of (**10**) are obtained and decompose to give *p*-tolyl acetate.[56] Analogous

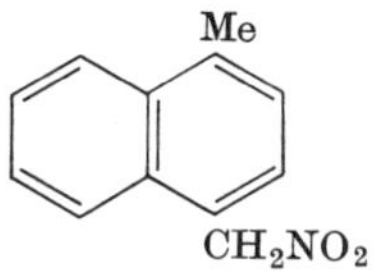

(**8**) 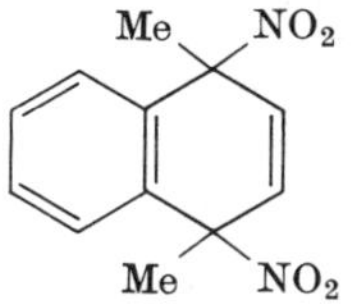 (**9**) 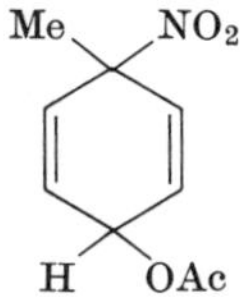 (**10**)

adducts, (**11**) and (**12**), are obtained with 2,3- and 3,4-dimethylbenzonitrile,[57] while 3,4-dimethylanisole gives (**13**).[58] Attack at the carbon bearing a methyl group (*ipso*-attack) is slight unless activated by *o*- and *p*-methyl groups or their equivalent. Sidechain nitration occurs only on a methyl group which is *para* to another methyl group activated

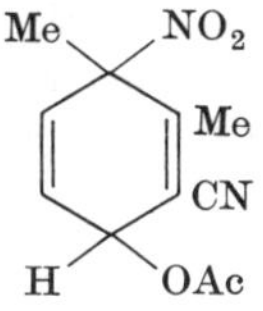

(**11**) 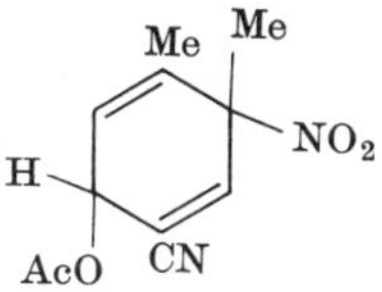 (**12**) 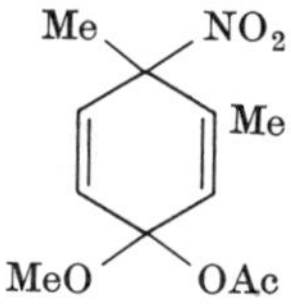 (**13**)

towards *ipso*-attack, e.g. nitration of pseudocumene gives (**14**).[59] The nitro-group in nitration adducts is mobile, as has been shown by an analysis of the products resulting from the action of acetic acid on the compound (**15**).[60] Intermediates have also been detected in the reaction of indones with nitrogen oxides.[61]

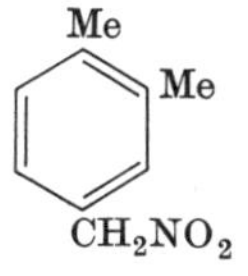

(**14**) 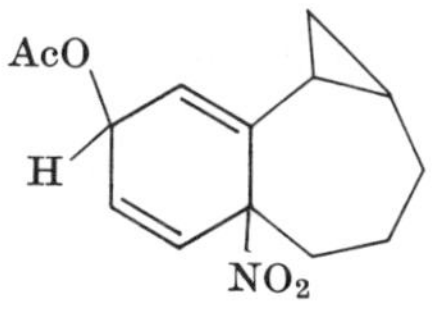 (**15**)

Nitration of *p*-iodoanisole is known to lead to nitrosodeiodination,[62] but with polyalkyliodobenzenes it appears that nitration is accompanied by nitrodeiodination.[63] Nitration of cyclotriveratrylene (**16**) by an excess of acetyl nitrate gives a mixture of (**17**) and (**18**), the initial product being (**19**) which undergoes further cleavage; the unsubstituted compound, 10,15,-dihydro-5*H*-tribenzo[*a*,*d*,*g*]cyclononene, does not undergo cleavage, so that the ready reaction of (**16**) must be due to stabilization of the veratryl cation by the methoxy-groups.[64]

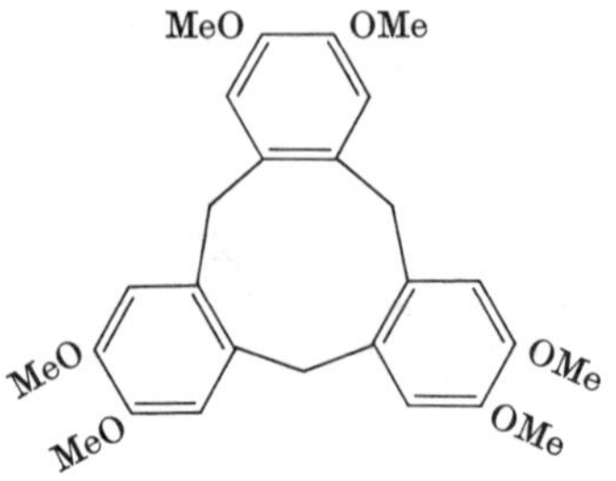

MeO OMe MeO MeO OMe OMe

(**16**)

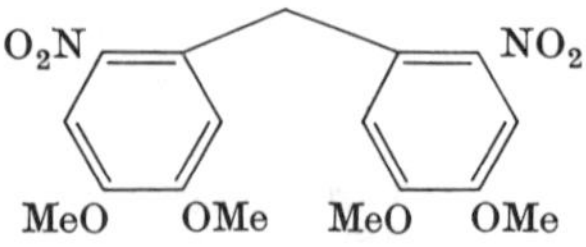

O_2N NO_2 MeO OMe MeO OMe

(**17**)

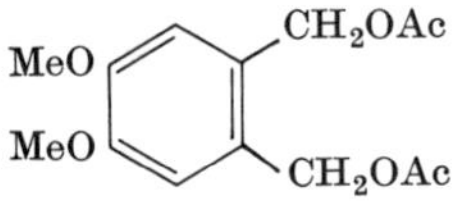

MeO MeO CH_2OAc CH_2OAc

(**18**)

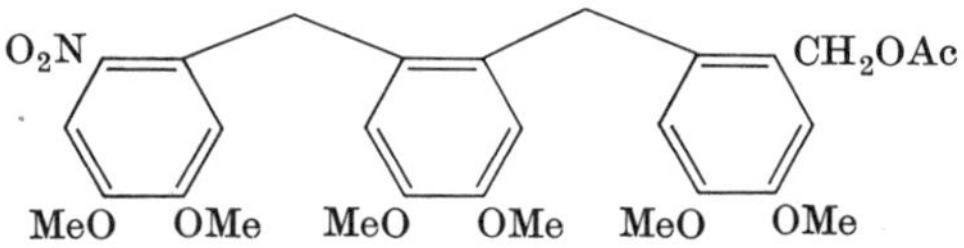

O_2N CH_2OAc MeO OMe MeO OMe MeO OMe

(**19**)

Nitrosation

Nitrosation of *p*-substituted phenols gives the 2-nitro-compound, and the rate-determining step is loss of a proton from dienone intermediate (**20**) and subsequent oxidation. For this reaction there is a low ρ value, which is characteristic of rate-determining proton

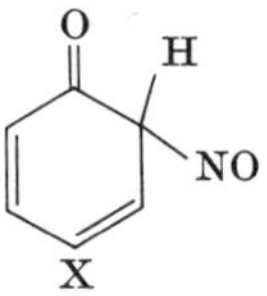

O H NO X

(**20**)

loss;[65] at high pH dienone formation becomes the slow step. Nitrosation is general-base catalysed and, with halide ions present, nitrosyl halides are formed and can effect nitrosation.[66] With resorcinol and its *O*-methyl derivative the product appears to be an oxime and either proton loss or reaction of undissociated substrate and $H_2NO_2^+$ is rate-determining, depending on the pH. For 1,3-dimethoxybenzene, the slow step is decomposition of the dienone intermediate to give dimethoxynitrosobenzene; reaction of $H_2NO_2^+$ and substrate is almost diffusion-controlled for this substrate, and similarly for all substrates.[67] Nitrosation of various indoles in dilute acid is brought about by a variety of species, depending on the conditions. The rate-determining step depends on the basicity of the substrate: for less basic ones it is proton loss, but with others it may be either formation of reagent or diffusion of reactants.[68] The transfer of a nitroso-group from *N*-nitrosodiphenylamine to *N*-methylaniline and other nucleophiles is acid-catalysed; in some cases HNO_2 is an intermediate but protonation is not the slow step.[69]

Azo Coupling

Reviews of the reactivity of diazonium ions[70] and of diazotisation[71] have appeared. In the reaction of pyrroles[72] and dimethylamine[73] with benzenediazonium salts both bases react in the unprotonated form; however, at high pH pyrrole can react as the conjugate base. The rate-determining step in the reaction between a diazonium ion and OH^- is desolvation of the hydroxide ion and so substituents in the diazonium ion have very little effect.[74] The reverse reactions, splitting at the N—O bond in benzenediazonium hydroxides to give diazonium ions, is subject to general-acid catalysis and substituents do have a substantial effect here.[75] Coupling with 1-naphthol occurs at either the 2- or the 4-position: the latter is subject to general-base catalysis, but not the former, possibly because of some intramolecular effect of the proximate OH group. In attack at the 2-position there is a change of rate-determining step from formation of a Wheland intermediate to its deprotonation as the substituent becomes less electron-withdrawing.[76] It has been suggested that the rate of coupling with 1-naphthol-4-sulphonic acid is governed by the presence or absence of water molecules in the transition state, as indicated by the activation parameters;[77] however, this criterion must be treated with some caution: coupling of *o*-substituted diazonium salts with 2-naphthol-3,6-disulphonic acid may be subject to steric hindrance.[78] The initial reaction between 2-methoxy-4-nitrobenzenediazonium ions and skatole is azo coupling, but this is followed by ring opening and deformylation to give (**21**).[79]

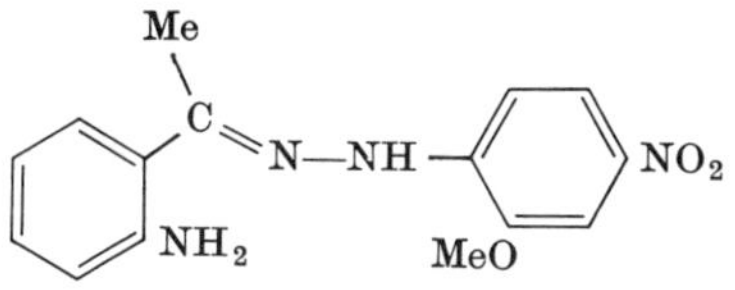

Me
C=N—NH NO$_2$
NH$_2$ MeO

(**21**)

Decomposition of benzenediazonium tetrafluoroborate in 2,2,2-trifluoroethanol gives fluorobenzene and 2,2,2-trifluoroethyl ether. In the presence of benzene some biphenyl is formed but the mechanism is not free-radical; it is more likely to be attack by benzene, acting as a nucleophile, on the diazonium ion at the carbon carrying the N_2^+ group and loss of molecular nitrogen.[80] Electrophilic substitution proceeds normally with 6a-thiathiophthens but reaction with benzenediazonium tetrafluoroborates results in rearrangement to give hypervalent heterocyclic systems.[81]

Halogenation

Formation of a 1:1 complex is the first step in the chlorination of phenol by molecular chlorine: breakdown of this intermediate follows a number of parallel paths.[82] From a study of the chlorinated products it is clear that hypochlorite in nitromethane does not form molecular chlorine.[83] *N*-Chloropolymaleimide will chlorinate aromatic but not aliphatic compounds.[84]

Positive bromine (Br^+) is not responsible for bromination by hypobromous acid in acid solution. The electrophile is thought to be H_2OBr^+ and this is formed, not directly from hypobromous acid, but by protonation of a substrate–HOBr complex formed in a pre-equilibrium step.[85] Bromination of naphthalene by molecular bromine was found to exhibit a kinetic hydrogen isotope effect which varied with the bromide ion concentration, as required by a two-step mechanism.[86]

Ring bromination of $PhSiMeCl_2$ and related compounds is catalysed by iodine but, in general, bromodesilylation is faster; iododesilylation may also compete and complex formation is an important step in the mechanism.[87] Similar results are obtained with germyl and methylgermyl substituents.[88] All these substituents, as well as silylmethyl groups, are *o*/*p*-directing in ring bromination.[89]

Bromination of aromatic hydroxy-compounds catalysed by $TiCl_4$ occurs at the *o*-position owing to formation of a complex.[90]

Mesitylene may be brominated by α,α,α-tribromoacetophenone and $AlBr_3$: a number of routes have been suggested.[91] Bromination of (**22**) occurs at the 9-position,[92] as predicted by MO calculations.[93]

At low iodide concentration the slow step in the acid-catalysed iodination of phenol is attack of I_2 on the substrate to give (**23**), but at higher concentrations (**23**) is formed in a reversible process and the slow step is deprotonation.[94] Solvents affect the mechanism

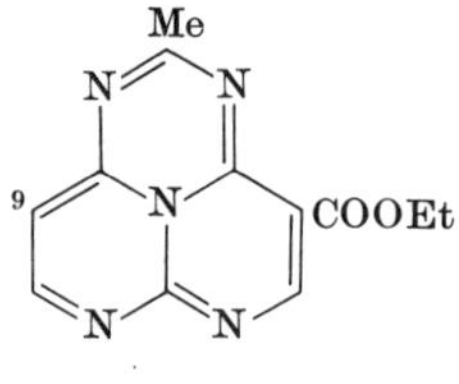

(**22**) 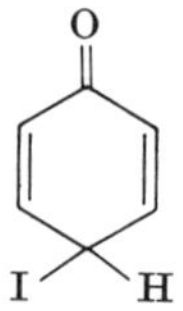 (**23**)

of iodination of 1-phenyl-2,3-dimethyl-5-pyrazolone.[95] The Jacobsen reaction of iodo-pseudocumenes has been examined: the same mixture of di-iodo-compounds results from both the 3- and the 5-iodo-compound as they are interconverted before reacting.[96] The effect of substituents on iodination by iodine in H_2SO_4, where the electrophile is I_3^+, has been examined ($\rho = -6.4$); proton loss is a slow step in this reaction.[97] It is claimed that iodination by iodine and pyridine in nitromethane involves slow formation of an iodine–pyridine complex.[98]

A cage mechanism has been suggested for the fluorination of phenol by perfluoropiperidine,[99] and the electrochemical fluorination of benzenes and pyridines has been reported.[100]

Studies of the halogenation of methoxy-aromatic ketones,[101] 4-substituted phenols,[102] aminopyridines,[103] phenyloxazoles [e.g. (**24**)],[104] and 8-methoxyquinoline[105] have been reported.

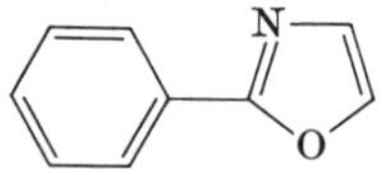

(**24**)

Iodine exchange between tetrachlorotetraiodofluorescein and ICl is thought to be electrophilic.[106] Cyanogen chloride reacts with phenyltrimethylstannane in the presence of $AlCl_3$ to give benzonitrile, but there is no reaction with phenyltrimethylsilane; the product of reaction with cyanogen bromide is, in contrast, bromobenzene, and this differing behaviour of the two cyanogen compounds has been noted elsewhere.[107] Studies of electrochemical cyanation[108] and thiocyanation[109] have been reported.

Rearrangement of (**25**), a possible intermediate in the bromination of 2,6-di-*tert*-butylphenol, to 4-bromo-2,6-di-*tert*-butylphenol exhibits a large hydrogen-isotope effect and the mechanism is thought to be attack by sodium acetate on an unprotonated form of the dienone.[110]

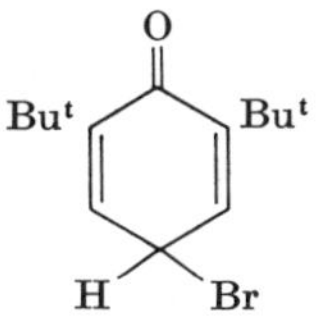

(**25**)

Metal Cleavage

In the cleavage of trimethylsilyl derivatives of 1,2-dihydrodrobenzocyclobutene (**26**) by OH^- in aqueous DMSO, the 4-isomer is less reactive than the 3-isomer, owing to the

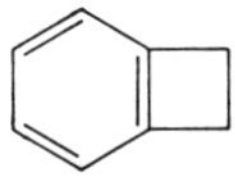

(**26**)

effect of ring strain on hybridization of the orbitals of the atoms of the four-membered ring.[111]

Metallation

The kinetics of the very fast metallation of 9-phenylfluorene at low temperatures has been studied.[112] Mercuric trifluoroacetate and BF_3–etherate effect aromatic mercuration,[113] and lead tetrakis(trifluoroacetate) alone is a reagent for direct aromatic plumbylation,[114] which occurs *para* to a halogen substituent. Lithiation of quinoline proceeds by attack at C-2 to give a 1,2-dihydro-1-lithio-intermediate.[115] Metallation often occurs *ortho* to an *N*- or *O*-containing substituent owing to co-ordination preceding aromatic attack.[116] With 2-*tert*-butylanisole metallation is greatly reduced owing to steric interference in the co-ordinated complex.

Decarboxylation*

From the variation of rate with solvent composition it has been concluded that decarboxylation of 2- and 4-pyridylacetic acid occurs *via* the zwitterionic form.[117]

Friedel–Crafts and Related Reactions

A mixture of methyl chloroformate and silver hexafluoroantimonate is a methylating agent: with anisole the main pathway is formation of the dimethylphenyloxonium ion (**27**) which reacts with a further molecule of anisole.[118] Activated aromatic compounds

* See also Chapter 2.

are readily alkylated by *tert*-butyl trifluoroacetate in trifluoroacetic acid under catalysis by strong acids; under these conditions *o*-substituted anisoles rearrange to the *p*-compound. Tri-*tert*-butylanisole undergoes debutylation in neat trifluoroacetic acid, the intermediate (**28**) having been proposed.[119] Benzo[*b*]furan (**29**) may be arylated at

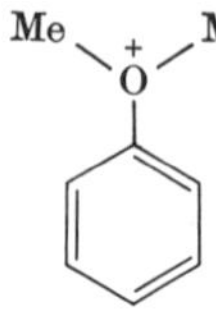

(**27**) 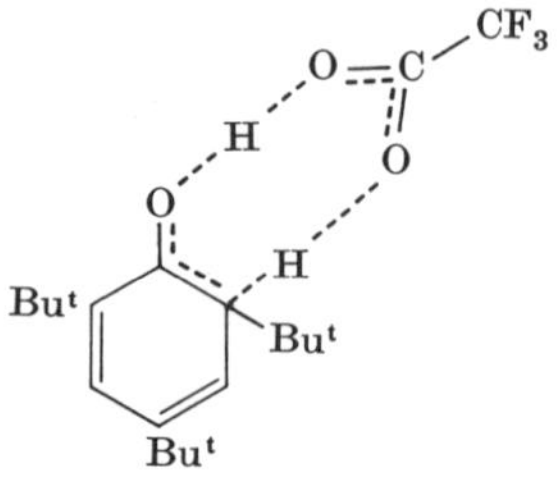 (**28**) 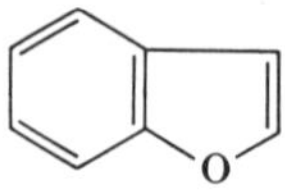 (**29**)

the 2-position by arylpalladium chloride and there appears to be no hydride shift from the 2- to the 3-position.[120] Mechanistic studies of the transaralkylation of α-chloro-3,4-dimethylcumene and toluene have been reported,[121] and it has been shown that trifluoromethanesulphonic acid catalyses the Friedel–Crafts reactions of benzyl acetate.[122] In the alkylation of benzene by 8-methylnon-1-ene the amount of isomerization was found to depend upon the catalyst used, indicating, presumably, varying degrees of separation of the carbocation.[123] Low substrate and positional selectivities were observed in the alkylation by phosphorus esters and $AlCl_3$.[124] A number of studies of the alkylation of phenols have been described.[125] Exclusive *o*-alkylation is obtained in the reaction of aromatic amines with *tert*-butyl hypochlorite and dimethyl sulphide, the proposed mechanism being shown in Scheme 4.[126]

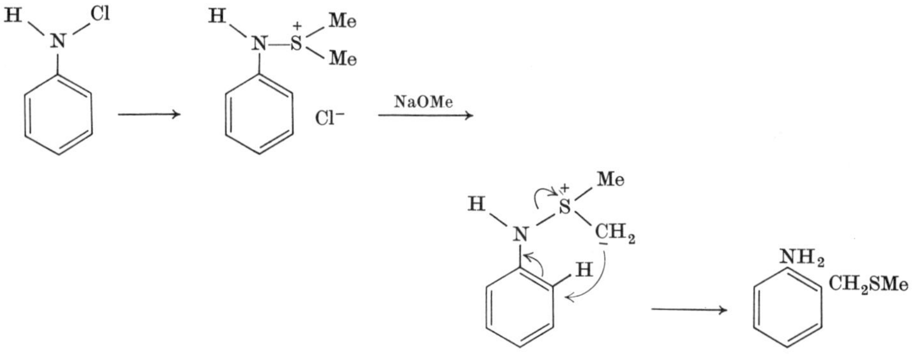

Scheme 4.

One of the most important developments in organic chemistry in the future will be the use of electrochemical methods and this technique has been used to effect acylation by the anodic oxidation of 2,3,5,6-tetramethoxyquinol diacetate; an acylium ion is formed.[127] Similar ions have been detected in a solution of Ac_2O in concentrated acid.[128] However, the kinetics of acylation by this reagent are difficult to interpret owing to partial orders. No kinetic hydrogen isotope effect was detected, and the identity of various by-products obtained led to the suggestion that protonated Ac_2O was involved.[129] Acetylation at the β-position of furan correlates well with σ^+ constants when the Yukawa–Tsuno equation is used.[130] A variety of factors appear to control orientation in Friedel–Crafts acylations.[131] The slow step in the benzoylation of toluene is attack of the complex PhCOCl–$FeCl_3$ on the substrate.[132] Evidence has also been presented to suggest that Friedel–Crafts acylations are reversible.[133]

Factors affecting the ease of cyclization under Friedel–Crafts conditions have been examined.[134] In an attempt to establish the absolute configuration of (**30**), the cyclization of 2-phenylcyclopropanecarboxylic acid to (**31**) was shown to be accompanied by racemization; this is thought to occur at the homoketenyl cation stage (**32**), as (**31**) does not racemize under the reaction conditions.[135] Cinnamic acid and thionyl chloride react

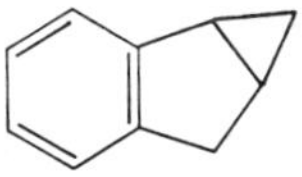

(**30**) 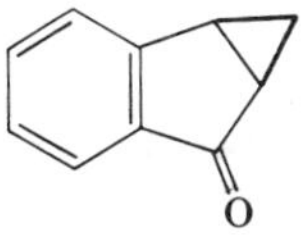 (**31**) 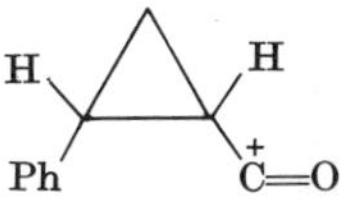 (**32**)

to give a substituted benzo[*b*]thiophen and the episulphide (**33**) is thought to be an intermediate.[136] The cyclization of cyclohept-4-ene-1-carbonyl to 2-*endo*-chlorobicyclo-[3.2.1]oct-1-ene is stereospecific, suggesting formation of the non-classical ion (**34**) as an intermediate.[137a] Cyclization of *o*-(arylthio)phenyl-substituted carbinols involves

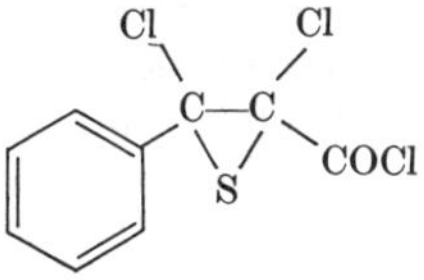

(**33**) 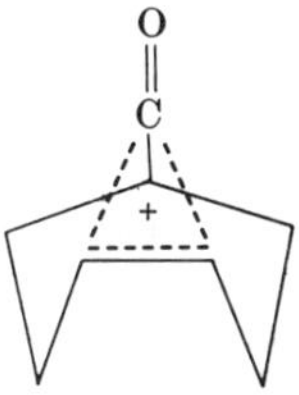 (**34**)

formation of a carbenium ion and there is a 1,2-sulphur shift following formation of the spiro-intermediate (**35**).[137b] Cyclization of anilides of di-(*p*-tolyl)glycollic acid is an electrophilic process.[138]

There have been further studies of the reactions of indoles. Cyclization of (**36**) is known to occur by formation of the spiroindolenine (**37**), and the tosylate has been found to react in the same way.[139] However, although the main route in the cyclization of 4-(6-methoxyindol-3-yl)butanol to tetrahydro-7-methoxycarbazole is the same, there is a

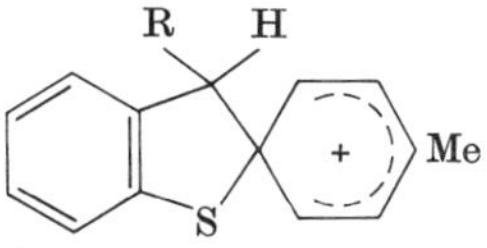

(**35**) 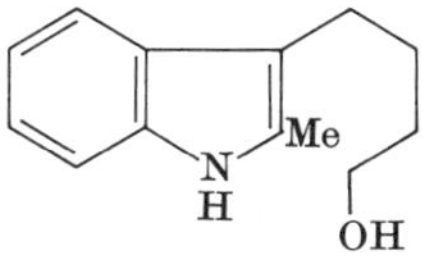 (**36**) 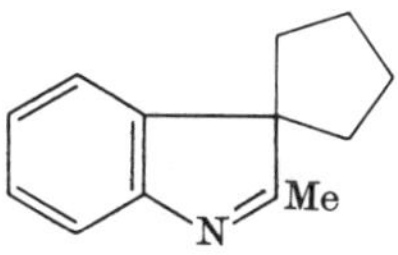 (**37**)

minor route involving direct attack at the 2-position, owing to activation by the OMe group;[140] from a study of the rearrangement of isolated indolenines it is probable that this minor route is also operative in the simple alkylation of indoles.[141]

Further study has shown that ionic Friedel–Crafts reactions in the gas phase are more complex than had been hoped. Such reactions are normally characterized by high positional selectivity but the isopropyl group appears to be an exception.[142] Positional selectivity was found to increase at higher pressures, and the large amount of the *meta*-isomer observed previously appears to be due to isomerization of the Wheland intermediate.[143] The *tert*-butyl ion generated by radiolysis in the gas phase reacts with toluene to give *p*- and *m*-*tert*-butyltoluene, the relative amounts depending on the pressure and on the concentration of added base (ethanol) needed to remove a proton from the Wheland intermediate; at high pressures the products are kinetically controlled and similar to those obtained in the liquid-phase reaction; in contrast to previous studies, this reaction showed a high substrate selectivity, probably because a much weaker electrophile is involved.[144]

The isomer distribution is aromatic acetoxylation is the opposite of that in normal electrophilic aromatic substitution and a π-complex is thought to intervene.[145] Partial rate factors for nitrophenylsulphonoxylation have been reported.[146] Other Friedel–Crafts reactions have been studied.[147]

The role of carbocations in electrophilic reactions has been reviewed.[148] The triphenylmethyl cation is a very selective electrophile[149] and reacts *via* a charge-transfer transition state.[150] Phenyl cations are responsible for phenylation through decomposition of benzenediazonium trifluoroactate.[151] Benzenium ions, obtained from polymethylbenzenes and $AlCl_3$, have been examined by NMR spectroscopy.[152]

It is difficult to obtain monosubstituted products of [2.2]metacyclophane because of the ease of oxidation, but electrophilic formylation at the 4-position proceeds cleanly.[153] Other studies of formylation have also appeared.[154] Formylation of 2-carbonyl-substituted phenols by dichloromethyl methyl ether and $TiCl_4$ occurs at the position *ortho* to the hydroxy-group; this has been suggested to be due to initial formation of a six-membered ring (**38**)[90] in which partial bond-fixation similar to that in naphthalene has a pronounced directing effect.

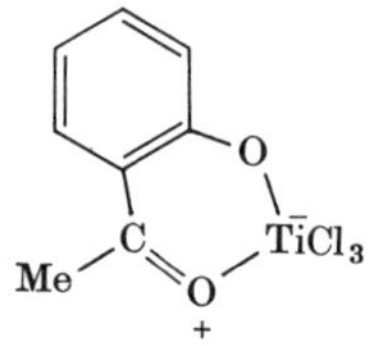

O
$\bar{T}iCl_3$
Me C O
+

(**38**)

Hydrogen Exchange

This reaction, perhaps because of its simplicity, continues to attract attention. In a superacid (trifluoromethanesulphonic acid) the reaction is 10^{11} times faster than in trifluoroacetic acid. This enormous increase is definitely due to enhanced reactivity of the electrophile and is not a medium effect.[155] All available data for aromatic hydrogen exchange have been collected and standardized to 100° and pH 0; this permits an exact comparison to be made of the susceptibility of many aromatic compounds to electrophilic attack but in some cases the computation has involved considerable extrapolation and there is uncertainty in these values.[156] Data for azaindoles[157] and methylpyrazoles[158] have been determined for inclusion. Using only substituted benzenes there is an excellent correlation between the standardized rate constants and σ_p^+ values ($\rho = -7.11$); this modifies slightly a previously quoted value ($\rho = -8.2$).[159] If all the data are included the correlation is less good.[160]

Exchange at both the 1- and the 2-position in naphthalene has been examined as a function of H_2SO_4 concentration. In both cases log k is a linear function of H_0 but the slope of the plot for the 2-position is greater than that for the 1-position, while $\Delta S^\ddagger$ is less negative for the former. These differences are explained in terms of varying hydration of the transition state.[161] In the excited state the reactivities are reversed.[162]

Exchange on purine and 9-substituted purines (**39**) at position 8 involves protonation at N-7 and attack by OH^-; at high pH there is attack on the neutral species.[163] With hypoxanthines (**40**) most of the exchange occurs at C-8 but methylation modifies this behaviour; both the free base and the anion react.[164] An ylide mechanism has been proposed for exchange at the 2-position of benzimidazoles.[165] The usual A–S_E2 mechanism operates with furan, 2-methylfuran[166] and phenylenediamines.[167] Addition of cysteine anion at the 4-position occurs with cysteine-catalysed hydrogen exchange at the 5-position of various uridine derivatives (**41**).[168]

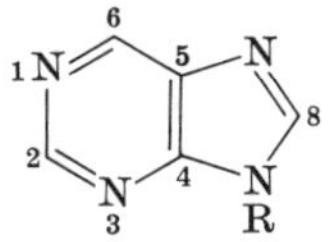

(**39**) 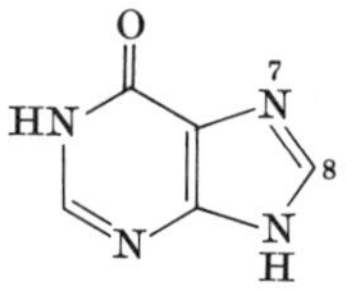 (**40**) 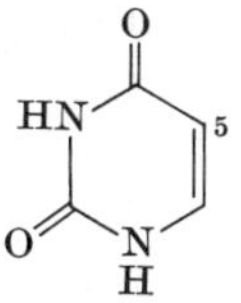 (**41**)

In hydrogen exchange in polymethylbenzenes the additivity principle breaks down owing to buttressing, which prevents hyperconjugative contribution to Wheland-intermediate stabilization.[169] The deactivating effect of the pentafluorophenyl group is greater at the *m*- than at the *p*-position and the effect of five fluorine atoms is not additive.[170] Values of σ^+ for both positions in thiophen, furan, selenophen and pyrrole in aqueous methanol have been measured.[171]

Hydrogen exchange on co-ordinated ligands has been studied[172] as well as in chlorotoluene,[173] quinolines,[174] and imidazole and histidine.[175] Further studies of exchange on thiophen in highly basic media have been reported.[176] In the presence of hexapyridine-magnesium di-iodide (a Lewis acid) porphyrins undergo exchange at the *meso*-positions with no demetallation.[177] A number of procedures for preparing deuteriated compounds have been reported.[178]

Protonation of naphthalene in superacid gives an equilibrating mixture of 1- and 2-naphthalenium ions,[179] and intramolecular 1,2-hydrogen shifts occur in difluoro- and dimethyl-benzenium ions.[180] In weak superacid media mono- and di-hydroxybenzenes are *O*-protonated but at higher temperatures *C*-protonation occurs.[181] Calculation of electron densities in substituted benzenium ions predicts precisely the ratio $\log f_o / \log f_p$ for hydrogen exchange with electron-donating substituents.[182]

Miscellaneous Reactions

There is nucleophilic assistance by ethanol, but not by Cl^-, in the protolysis of (**42**),[183] and a mechanism for the protolysis of substituted 3,3-dimethyl-1-phenyltriazene derivatives has been proposed.[184] Amination of toluene by monochloramine and $AlCl_3$ goes by an addition–elimination mechanism (Scheme 5). However, adamantyl cations are

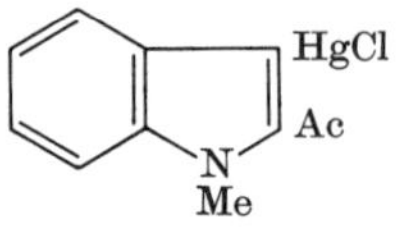

(**42**)

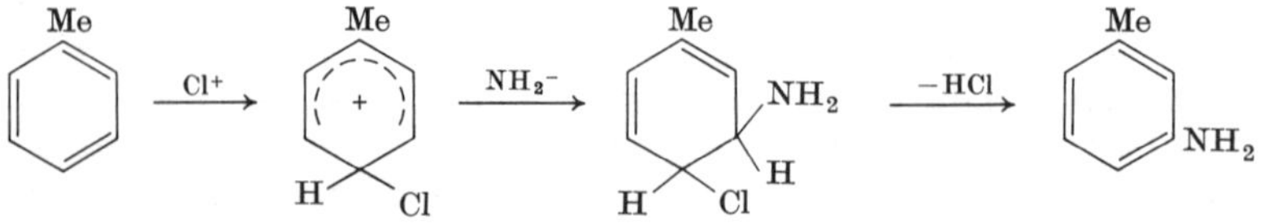

SCHEME 5.

involved in the amination of adamantane by trichloramine.[185] The acid-catalysed rearrangement of (**43**) to (**44**) is an *A*-1 reaction and solvation factors are important in understanding the effect of acidity on rate.[186] Reaction of *p*-nitrosoanisole with aniline to give *p*-nitrosodiphenylamine involves attack on the protonated aniline to give (**45**); as the solvent (methanol) appears to be important the transition state leading to (**45**) is thought to have the structure (**46**).[187]

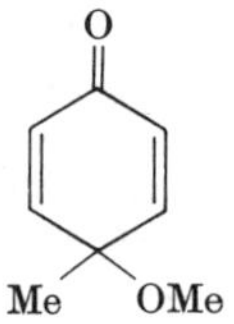

(**43**) 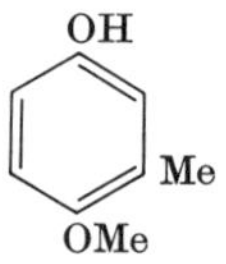 (**44**) 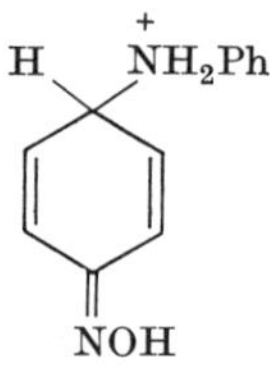 (**45**)

References

1 R. Taylor, *MTP (Med. Tech. Publ. Co.) Internat. Rev. Sci.: Org. Chem., Ser.* 1, **3**, 1 (1973); *Aromatic, Heteroaromatic Chem.*, **1**, 176 (1973).
2 G. Marino, *Chim. Ind. (Milan)*, **55**, 349 (1973); *Chem. Abs.*, **79**, 65266 (1973); G. Marino, *Pr. Nauk. Inst. Chem. Technol. Naft Wegla Politech Wroclaw*, **1973**, 89; *Chem. Abs.*, **79**, 77564 (1973); G. Marino, *Khim. Geterotsikl. Soedin.*, **1973**, 579; *Chem. Abs.*, **79**, 52418 (1973).
3 M. I. Bruce, *Organometal. Rev. B*, **9**, 99 (1972).
4 D. A. Forsyth, *J. Am. Chem. Soc.*, **95**, 3594 (1973).
5 J. N. Murrell, W. Schmidt and R. Taylor, *J.C.S. Perkin II*, **1973**, 179.
6 P. Politzer, R. A. Donnelly and K. C. Daiker, *Chem. Comm.*, **1973**, 617.
7 M. J. S. Dewar and P. J. Grisdale, *J. Am. Chem. Soc.*, **84**, 3539 (1962).
8 D. S. Noyce and R. W. Nichols, *J. Org. Chem.*, **37**, 4306 (1972).
9 D. S. Noyce and R. W. Nichols, *J. Org. Chem.*, **37**, 4311 (1972).
10 E. Glyde and R. Taylor, *J.C.S. Perkin II*, **1973**, 1632.
11 S. Clementi, P. Linda, and C. D. Johnson, *J.C.S. Perkin II*, **1973**, 1250.
12 E. B. Pedersen, T. E. Petersen, K. Torssell and S.-O. Lawesson, *Tetrahedron*, **29**, 579 (1973).
13 L. A. P. Kane-Maguire and C. A. Mansfield, *Chem. Comm.*, **1973**, 540.
14 P. L. Pauson, G. R. Knox and I. G. Morrison, *J.C.S.*, (*C*), **1967**, 1847.
15 D. W. Slocum and C. R. Ernst, *Tetrahedron Letters*, **1972**, 5217.
16 O. Tsuge and H. Samura, *Chem. Letters (Tokyo)*, **1973**, 175.
17 K. Clarke, R. M. Scrowston and T. M. Sutton, *J.C.S. Perkin I*, **1973**, 1196.
18 K. Clarke, R. M. Scrowston and T. M. Sutton, *J.C.S. Perkin I*, **1973**, 623.

[19] W. Kurtz, P. Fischer and F. Effenberger, *Chem. Ber.*, **106**, 525 (1973); P. Fischer, W. Kurtz and F. Effenberger, *ibid.*, p. 549.
[20] K. G. Svensson, H. Selander, M. Karlsson and J. L. G. Nilsson, *Tetrahedron*, **29**, 1115 (1973); H. Selander and J. L. G. Nilsson, *Acta Chem. Scand.*, **26**, 3377 (1973).
[21] W. Schäfer and K. Dimroth, *Angew. Chem. Internat. Ed.*, **12**, 753 (1973).
[22] O. I. Kachurin and A. P. Zaraiskii, *Ukr. Khim. Zh.* (*Russian Ed.*), **38**, 1079 (1972); *Chem. Abs.*, **78**, 83552 (1973).
[23] A. Koeberg-Telder and H. Cerfontain, *J.C.S. Perkin II*, **1973**, 633.
[24] O. I. Kachurin and A. P. Zaraiskii, *Ukr. Khim. Zh.* (*Russian Ed.*), **39**, 12 (1973); *Chem. Abs.*, **78**, 110368 (1973); A. P. Zaraisky and O. I. Kachurin, *Organic Reactivity* (*Tartu*), **10**, 72 (1973).
[25] H. Cerfontain and Z. R. H. Schaasberg-Nienhuis, *J.C.S. Perkin II*, **1973**, 1413.
[26] A. Koeberg-Telder, Z. R. H. Nienhuis and H. Cerfontain, *Can. J. Chem.*, **51**, 462 (1973).
[27] V. P. Leshchev, V. V. Kharitonov and A. A. Spryskov, *Izv. Vyssh. Ucheb. Zaved., Khim. Khim. Tekhnol.*, **16**, 253 (1973); *Chem. Abs.*, **79**, 4714 (1973); B. I. Karavaev, A. A. Spryskov and A. G. Zakharov, *Izv. Vyssh. Ucheb. Zaved., Khim. Khim. Tekhnol.*, **15**, 1356 (1972); *Chem. Abs.*, **78**, 57451 (1973).
[28] C. Ris, Z. R. H. Schaasberg-Nienhuis and H. Cerfontain, *Tetrahedron*, **29**, 3165 (1973).
[29] A. P. Zaraiskii and O. I. Kachurin, *Zh. Org. Khim.*, **9**, 991 (1973); *Chem. Abs.*, **79**, 41588 (1973).
[30] A. P. Zaraiskii and O. I. Kachurin, *Ukr. Khim. Zh.* (*Russian Ed.*), **39**, 117 (1973); *Chem. Abs.*, **78**, 147064 (1973).
[31] G. F. Smith and D. A. Taylor, *Tetrahedron*, **29**, 669 (1973).
[32] J. R. Blackborow, *J.C.S. Perkin II*, **1972**, 2387.
[33] B. Kohlstock and G. Henseke, *Z. Chem.*, **13**, 100 (1973).
[34a] E. G. Willard and H. Cerfontain, *Rec. Trav. Chim.*, **92**, 793 (1973).
[34b] G. A. Olah, S. Kobayashi and J. Nishimura, *J. Am. Chem. Soc.*, **95**, 564 (1973).
[35] S. R. Hartshorn and K. Schofield, *Progr. Org. Chem.*, **8**, 278 (1973); T. Yashida, *Kogyo Kayaku*, **33**, 52 (1972); *Chem. Abs.*, **78**, 3351 (1973).
[36] F. Bernardi and W. J. Hehre, *J. Am. Chem. Soc.*, **95**, 3078 (1973).
[37] S. A. Benezra, M. K. Hoffmann and M. M. Bursey, *J. Am. Chem. Soc.*, **92**, 7501 (1970).
[38] S. Sekiguchi, A. Hirose, S. Kato and K. Matsui, *Bull. Chem. Soc. Japan*, **46**, 646 (1973).
[39] K. Olsson, *Acta Chem. Scand.*, **26**, 3555 (1972).
[40] Ya. I. Leitman, E. F. Zaslavets and M. A. Ilyushin, *Zh. Prikl. Khim.* (*Leningrad*), **46**, 366 (1973); *Chem. Abs.*, **78**, 135212 (1973).
[41] J. H. P. Utley and T. A. Vaughan, *J.C.S. Perkin II*, **1972**, 2343.
[42] A. Streitwieser, G. R. Ziegler, P. C. Mowery, A. Lewis and R. G. Lawler, *J. Am. Chem. Soc.*, **90**, 1357 (1968).
[43] A. N. Kashin, I. P. Beletskaya and V. I. Stanko, *Zh. Org. Khim.*, **9**, 207 (1973); V. I. Stanko, A. N. Kashin, and I. P. Beletskaya, *ibid.*, p. 208; *Chem. Abs.*, **78**, 123541, 110132 (1973).
[44] G. P. Sharnin and I. F. Falyakhov, *Zh. Org. Khim.*, **9**, 730 (1973); *Chem. Abs.*, **79**, 4759 (1973).
[45] T. A. Modro and A. Piekos, *Tetrahedron*, **29**, 2561 (1973).
[46] D. N. Butovetskii, I. F. Falyakhov and G. P. Sharnin, *Tr. Kazan. Khim.-Tekhnol. Inst.*, **1972**, 87; *Chem. Abs.*, **79**, 17697 (1973).
[47] See *Org. Reaction Mech.*, **1970**, 227.
[48] B. Kohlstock, G. Henseke and R. Starke, *Z. Chem.*, **13**, 11 (1973).
[49] D. G. Markees, *Helv. Chim. Acta*, **56**, 1382 (1973).
[50] A. Arcoria, E. Maccarone and G. A. Tomaselli, *J. Heterocyclic Chem.*, **10**, 191 (1973).
[51] M. Remes and M. Adamek, *Chem. Prum.*, **22**, 499 (1972); *Chem. Abs.*, **78**, 28751 (1973).
[52] K. Fukunaga and M. Kimura, *Nippon Kagaku Kaishi*, **1973**, 1306; *Chem. Abs.*, **79**, 77700 (1973).
[53a] D. W. Amos, D. A. Baines and G. W. Flewett, *Tetrahedron Letters*, **1973**, 3191.
[53b] S. M. Schlegel and J. J. Robinson, *J. Am. Chem. Soc.*, **95**, 665 (1973).
[54] G. R. Underwood, R. S. Silverman and A. Vanderwalde, *J.C.S. Perkin II*, **1973**, 1177.
[55] A. Fischer and A. L. Wilkinson, *Can. J. Chem.*, **50**, 3988 (1972).
[56] A. Fischer and J. N. Ramsay, *J.C.S. Perkin II*, **1973**, 237.
[57] A. Fischer and C. C. Grieg, *Chem. Comm.*, **1973**, 396.
[58] A. Fischer and D. R. A. Leonard, *Chem. Comm.*, **1973**, 300.
[59] D. J. Blackstock, A. Fischer, K. E. Richards and G. J. Wright, *Austral. J. Chem.*, **26**, 775 (1973).
[60] R. C. Hahn and M. B. Groen, *J. Am. Chem. Soc.*, **95**, 6128 (1973).
[61] B. Aleksiev and P. Nishanyan, *God. Vissh. Khimikotekhnol. Inst., Sofia*, **15**, 267 (1968); *Chem. Abs.*, **78**, 3991 (1973).
[62] See *Org. Reaction Mech.*, **1972**, 210.

[63] K. Olsson and P. Martinson, *Acta Chem. Scand.*, **26**, 3549 (1972).
[64] T. Sato, T. Akima and K. Uno, *J.C.S. Perkin I*, **1973**, 891.
[65] B. C. Challis and R. J. Higgins, *J.C.S. Perkin II*, **1972**, 2365.
[66] B. C. Challis and R. J. Higgins, *J.C.S. Perkin II*, **1973**, 1597.
[67] J. Jahelka, V. Štěrba and K. Valter, *Coll. Czech. Chem. Comm.*, **38**, 877 (1973).
[68] B. C. Challis and A. J. Lawson, *J.C.S. Perkin II*, **1973**, 918.
[69] B. C. Challis and M. R. Osborne, *J.C.S. Perkin II*, **1973**, 1526.
[70] H. Zollinger, *Accounts Chem. Research*, **6**, 335 (1973).
[71] S. Radu, *Stud. Cercet. Chim.*, **20**, 1287 (1972); *Chem. Abs.*, **78**, 109975 (1973).
[72] K. Mitsumura, Y. Hashida, S. Sekiguchi and K. Matsui, *Bull. Chem. Soc. Japan*, **46**, 1770 (1973).
[73] M. Remeš, J. Diviš, V. Zvěřina and M. Matrka, *Coll. Czech. Chem. Comm.*, **38**, 1049 (1973).
[74] V. Beránek, V. Štěrba and K. Valter, *Coll. Czech. Chem. Comm.*, **38**, 257 (1973).
[75] J. Jahelka, O. Macháčková and V. Štěrba, *Coll. Czech. Chem. Comm.*, **38**, 706 (1973).
[76] S. Kishimoto, O. Manabe, H. Hachiro and N. Hirao, *Nippon Kagaku Kaishi*, **1972**, 2132; *Chem. Abs.*, **78**, 70984 (1973).
[77] B. Demian, *Bull. Soc. Chim. France*, **1973**, 769.
[78] Z. V. Belitskaya, M. V. Plakidina and T. L. Bagal, *Organic Reactivity* (*Tartu*), **8**, 1046 (1971).
[79] T. F. Spande and G. G. Glenner, *J. Am. Chem. Soc.*, **95**, 3400 (1973).
[80] P. Burri and H. Zollinger, *Helv. Chim. Acta*, **56**, 2204 (1973).
[81] R. M. Christie, A. S. Ingram, D. H. Reid and R. G. Webster, *Chem. Comm.*, **1973**, 92.
[82] E. Grimley and G. Gordon, *J. Phys. Chem.*, **77**, 973 (1973).
[83] V. L. Heasley, G. E. Heasley, D. M. Ingle, P. D. Davis and T. L. Rold, *J. Org. Chem.*, **38**, 2549 (1973).
[84] C. Yaroslavsky and E. Katchalski, *Tetrahedron Letters*, **1972**, 5173.
[85] H. M. Gillow and J. H. Ridd, *J.C.S. Perkin II*, **1973**, 1321.
[86] A. Ehrlich and E. Berliner, *J. Org. Chem.*, **37**, 4186 (1972).
[87] J. Včelák and V. Chvalovský, *Coll. Czech. Chem. Comm.*, **38**, 1055 (1973).
[88] J. Včelák and V. Chvalovský, *Coll. Czech. Chem. Comm.*, **37**, 3615 (1973).
[89] J. Včelák and V. Chvalovský, *Coll. Czech. Chem. Comm.*, **38**, 3623 (1973).
[90] T. M. Cresp, M. V. Sargent, J. A. Elix and D. P. H. Murphy, *J.C.S. Perkin I*, **1973**, 340.
[91] V. A. Smrchek, V. F. Traven and B. I. Stepanov, *Zh. Org. Khim.*, **9**, 358 (1973); *Chem. Abs.*, **78**, 110128 (1973).
[92] O. Ceder and K. Rosén, *Acta Chem. Scand.*, **27**, 359 (1973).
[93] O. Ceder and J. F. Witte, *Acta Chem. Scand.*, **26**, 635 (1972).
[94] E. Grovenstein, N. S. Aprahamian, C. J. Bryan, N. S. Gnanapragasam, D. C. Kilby, J. M. McKelvey and R. J. Sullivan, *J. Am. Chem. Soc.*, **95**, 4261 (1973).
[95] M. Orban, E. Koros and K. Krudy-Sandor, *Magyar Kem. Foly.*, **78**, 618 (1972); *Chem. Abs.*, **78**, 83517 (1973).
[96] H. Suzuki and T. Sugiyama, *Bull. Chem. Soc. Japan*. **46**, 586 (1973); H. Suzuki and Y. Haruta, *ibid.*, p. 589.
[97] J. Arotsky, A. C. Darby and J. B. A. Hamilton, *J.C.S. Perkin II*, **1973**, 595.
[98] C. R. Das and A. N. Bose, *J. Indian Chem. Soc.*, **49**, 463 (1972).
[99] V. R. Polishchuk and L. S. German, *Tetrahedron Letters*, **1972**, 5169.
[100] Y. Inoue, S. Nagase, K. Kodaira, H. Baba and T. Abe, *Bull. Chem. Soc. Japan*, **46**, 2204 (1973).
[101] J.-J. Aaron, J.-E. Dubois, F. Krausz and R. Martin, *J. Org. Chem.*, **38**, 300 (1973).
[102] J. Kulič and M. Večeřa, *Coll. Czech. Chem. Comm.*, **37**, 3610 (1972).
[103] G. J. Fox, J. D. Hepworth and G. Hallas, *J.C.S. Perkin I*, **1973**, 68.
[104] I. Simiti, E. Chindris and I. Schwartz, *Rev. Chim.* (*Bucharest*), **23**, 460 (1972); *Chem. Abs.*, **78**, 28787 (1973).
[105] H. Gershon, M. W. McNeil and S. G. Shulman, *J. Org. Chem.*, **37**, 4078 (1972).
[106] V. I. Stanko and N. G. Iroshnikova, *Isotopenpraxis*, **8**, 252 (1972); *Chem. Abs.*, **78**, 3453 (1973).
[107] E. H. Bartlett, C. Eaborn and D. R. M. Walton, *J. Organometal. Chem.*, **46**, 267 (1972).
[108] K. Yoshida and T. Fueno, *J. Org. Chem.*, **38**, 1045 (1973).
[109] V. R. Kartashov, E. V. Skorobogatova and I. V. Bodrikov, *Zh. Org. Khim.*, **9**, 214 (1973); *Chem. Abs.*, **78**, 110264 (1973).
[110] P. B. D. de la Mare and A. Singh, *J.C.S. Perkin II*, **1973**, 59.
[111] C. Eaborn, A. A. Najam and D. R. M. Walton, *J. Organometal. Chem.*, **46**, 255 (1972).
[112] M. A. Hamid, *Can. J. Chem.*, **50**, 3761 (1972).
[113] B. G. Gnedin and G. P. Chudinova, *Izv. Vyssh. Ucheb. Zaved., Khim. Khim. Tekhnol.*, **16**, 730 (1973); *Chem. Abs.*, **79**, 41623 (1973).
[114] D. de Vos, J. Wolters and A. van der Gen, *Rec. Trav. chim.*, **92**, 701 (1973); D. de Vos, J. Spierenburg, and J. Wolters, *ibid.*, **91**, 1465 (1972).

[115] D. I. C. Scopes and J. A. Joule, *J.C.S. Perkin I*, **1972**, 2810.
[116] D. W. Slocum and F. E. Stonemark, *J. Org. Chem.*, **38**, 1677 (1973); D. W. Slocum and B. P. Koonsviteky, *ibid.*, p. 1675.
[117] R. G. Button and P. J. Taylor, *J.C.S. Perkin II*, **1973**, 557.
[118] P. Beak, J. T. Adams, P. D. Klein, P. A. Szczepanik, D. A. Simpson, and S. G. Smith, *J. Am. Chem. Soc.*, **95**, 6027 (1973).
[119] U. Svanholm and V. D. Parker, *J.C.S. Perkin I*, **1973**, 562.
[120] A. Kasahara, T. Izumi, M. Yodono, R. Saito, T. Takeda and T. Sugawara, *Bull. Chem. Soc. Japan*, **46**, 1220 (1973).
[121] R. J. Larrain, R. S. Ramirez and F. R. Diaz, *Rev. Latinoamer. Quim.*, **3**, 135 (1972); *Chem. Abs.*, **78**, 70999 (1973).
[122] K. Nyberg, *Chem. Scripta*, **4**, 143 (1973).
[123] H. R. Alul and G. J. McEwan, *J. Org. Chem.*, **37**, 4157 (1972).
[124] G. Sosnovsky and M. W. Shende, *Z. Naturforsch.*, **27**, 1339 (1972).
[125] S. T. Kumok, Ya. A. Gurvich, E. L. Styskin and A. A. Grinberg, *Zh. Vses. Khim. Obshchest.*, **17**, 460 (1972); *Chem. Abs.*, **78**, 28810 (1973); L. Baiocchi, M. Giannangeli and G. Palazzo, *Gazz. Chim. Ital.*, **103**, 71 (1973); *Chem. Abs.*, **79**, 52526 (1973); N. Yoneda, A. Naka, H. Ohtsuka and K. Aomura, *Nippon Kagaku Kaishi*, **1972**, 2385; *Chem. Abs.*, **78**, 71003 (1973); G. P. Yakunina, A. P. Lysenko, V. G. Plyusnin and M. I. Zelentsova, *Zh. Fiz. Khim.*, **47**, 1452 (1973); *Chem. Abs.*, **79**, 91283 (1973).
[126] P. G. Gassman and G. Gruetzmacher, *J. Am. Chem. Soc.*, **95**, 588 (1973); P. G. Gassman and C. T. Huang, *ibid.*, p. 4453.
[127] J. H. P. Utley and G. B. Yates, *Chem. Comm.*, **1973**, 473.
[128] A. Germain and A. Commeyras, *Bull. Soc. Chim. France*, **1973**, 2532.
[129] A. Germain, A. Commeyras and A. Casadevall, *Bull. Soc. Chim. France*, **1973**, 2527, 2537.
[130] J. P. Girault, P. Scribe and G. Dana, *Bull. Soc. Chim. France*, **1973**, 1760.
[131] J. P. Girault, P. Scribe and G. Dana, *Tetrahedron*, **29**, 413 (1973); L. J. Kricka and A. Ledwith, *J.C.S. Perkin I*, **1973**, 859; G. Valkanas, *Chem. Chron.*, **1**, 255 (1972); *Chem. Abs.*, **79**, 65435 (1973).
[132] V. A. Ustinov, G. S. Mironov and M. I. Farberov, *Zh. Org. Khim.*, **9**, 733 (1973); *Chem. Abs.*, **79**, 4558 (1973).
[133] I. Agranat and D. Avnir, *Chem. Comm.*, **1973**, 362.
[134] A. A. Khalof and R. M. Roberts, *J. Org. Chem.*, **37**, 4227 (1972).
[135] G. R. Elling, R. C. Hahn and G. Schwab, *J. Am. Chem. Soc.*, **95**, 5659 (1973).
[136] A. J. Krubsack and T. Higa, *Tetrahedron Letters*, **1973**, 125.
[137a] G. N. Fickers and K. C. Kemp, *Chem. Comm.*, **1973**, 84.
[137b] G. Capozzi, G. Melloni and G. Modena, *J.C.S. Perkin I*, **1973**, 2250.
[138] V. S. Shklayev and A. V. Milyutia, *Organic Reactivity* (*Tartu*), **9**, 602 (1972).
[139] A. H. Jackson and B. Naidoo, *J.C.S. Perkin II*, **1973**, 548.
[140] R. Iyer, A. H. Jackson, P. V. R. Shannon and B. Naidoo, *J.C.S. Perkin II*, **1973**, 872; R. Iyer, A. H. Jackson and P. V. R. Shannon, *ibid.*, p. 878.
[141] G. Casnati, A. Dossena and A. Pochini, *Tetrahedron Letters*, **1972**, 5277.
[142] S. Takamuku, K. Iseda and H. Sakurai, *J. Am. Chem. Soc.*, **95**, 2420 (1973).
[143] F. Cacace and E. Possagno, *J. Am. Chem. Soc.*, **95**, 3397 (1973).
[144] F. Cacace and P. Giacomello, *J. Am. Chem. Soc.*, **95**, 5851 (1973).
[145] L. Eberson and L. Gomez-Gonzalez, *Acta Chem. Scand.*, **27**, 1162, 1249, 1255 (1973).
[146] R. L. Dannley, J. E. Gagen and K. Zak, *J. Org. Chem.*, **38**, 1 (1973); R. L. Dannley and W. R. Knipple, *ibid.*, p. 6.
[147] M. I. Farberov, G. S. Mironov, Yu. A. Moskvichev, G. N. Timoshenko and V. A. Ustinov, *Dokl. Akad. Nauk SSSR*, **210**, 128 (1973); *Chem. Abs.*, **79**, 52530 (1973); A. Sammour and M. Elhashash *J. Prakt. Chem.*, **314**, 906 (1972); G. S. Mironov, M. I. Farberov, I. V. Budnii and V. D. Shein, *Uch. Zap., Yaroslav. Teckhnol. Inst.*, **1971**, 21; *Chem. Abs.*, **78**, 123557 (1973).
[148] G. A. Olah, *Angew. Chem. Internat. Ed.*, **12**, 173 (1973).
[149] M. K. Eberhardt and G. Chuchani, *J. Org. Chem.*, **37**, 3649 (1972).
[150] M. K. Eberhardt and G. Chuchani, *J. Org. Chem.*, **37**, 3654 (1972).
[151] N. Kamigata, R. Hisada, H. Minato and M. Kobayashi, *Bull. Chem. Soc. Japan*, **46**, 1016 (1973).
[152] Y. Okami, N. Otani, D. Katoh and S. Hamanaka, *Bull. Chem. Soc. Japan*, **46**, 1860 (1973).
[153] A. Maquestiau, Y. van Haverbeke, R. Flammang, N. Flammang-Barbieux and N. Clerbois, *Tetrahedron Letters*, **1973**, 3259.
[154] M. Iwata and S. Emoto, *Chem. Letters* (*Tokyo*), **1973**, 477; C. Beguin, C. Coulombeau and S. Hamman, *Compt. Rend., C.*, **275**, 1351 (1972).
[155] R. Taylor and T. J. Tewson, *Chem. Comm.*, **1973**, 836.

[156] A. El-Anani, J. Banger, G. Bianchi, S. Clementi, C. D. Johnson and A. R. Katritzky, *J.C.S. Perkin II*, **1973**, 1065.
[157] A. El-Anani, S. Clementi, A. R. Katritzky and L. Yakhontov, *J.C.S. Perkin II*, **1973**, 1072.
[158] S. Clementi, P. P. Forsythe, C. D. Johnson and A. R. Katritzky, *J.C.S. Perkin II*, **1973**, 1675.
[159] L. M. Stock and H. C. Brown, *Adv. Phys. Org. Chem.*, **1963**, 96.
[160] S. Clementi and A. R. Katritzky, *J.C.S. Perkin II*, **1973**, 1077.
[161] C. G. Stevens and S. J. Strickler, *J. Am. Chem. Soc.*, **95**, 3918 (1973).
[162] C. G. Stevens and S. J. Strickler, *J. Am. Chem. Soc.*, **95**, 3922 (1973).
[163] J. A. Elvidge, J. R. Jones, C. O'Brien, E. A. Evans and H. C. Sheppard, *J.C.S. Perkin II*, **1973**, 1889.
[164] D. Lichtenberg and F. Bergmann, *J.C.S. Perkin I*, **1973**, 789.
[165] J. A. Elvidge, J. R. Jones, C. O'Brien, E. A. Evans and J. C. Turner, *J.C.S. Perkin II*, **1973**, 432.
[166] P. Salomaa, A. Kaukaanperä, E. Nikander, K. Kaipainery and R. Aaltonen, *Acta Chem. Scand.*, **27**, 153 (1973).
[167] J. C. Grivas, *J. Org. Chem.*, **38**, 1204 (1973).
[168] Y. Wataya, H. Hayatsu and Y. Kawazoe, *J. Am. Chem. Soc.*, **94**, 8927 (1972).
[169] K. E. Richards, A. L. Wilkinson and G. J. Wright, *Austral. J. Chem.*, **25**, 2369 (1972).
[170] R. Taylor, *J.C.S. Perkin II*, **1973**, 253.
[171] K. Schwellick and K. Unverferth, *J. Prakt. Chem.*, **314**, 603 (1972).
[172] D. N. Kursanov, V. N. Setkina, A. I. Khatami and M. N. Nefedova, *Dokl. Akad. Nauk SSSR*, **207**, 111 (1972); *Chem. Abs.*, **78**, 71183 (1973); D. N. Kursanov, V. N. Setkina, N. K. Baranetskaya, V. I. Zdanovich, A. I. Yurtanov and K. N. Anisimov, *Izv. Akad. Nauk SSSR, Ser. Khim.*, **1973**, 1359; *Chem. Abs.*, **79**, 91277 (1973).
[173] P. P. Alukhanov, I. S. Temnova, V. R. Kalinachenko and L. M. Yakimenko, *Zh. Obshch. Khim.*, **43**, 896 (1973); *Chem. Abs.*, **79**, 52531 (1973).
[174] V. P. Lezina, A. U. Stepanyants, L. D. Smirnov, N. A. Andronova and K. M. Dyumaev, *Akad. Nauk SSSR, Ser. Khim.*, **1972**, 2092; *Chem. Abs.*, **78**, 3461 (1973).
[175] J. H. Bradbury, B. E. Chapman and F. A. Pellegrino, *J. Am. Chem. Soc.*, **95**, 6139 (1973).
[176] Yu. I. Ranneva and A. I. Shatenshtein, *Zh. Obshch. Khim.*, **42**, 1812, 2250 (1972); *Chem. Abs.*, **78**, 3553, 83659 (1973).
[177] G. W. Kenner, K. M. Smith and M. J. Sutton, *Tetrahedron Letters*, **1973**, 1303.
[178] N. H. Werstiuk and T. Kadai, *Can. J. Chem.*, **51**, 1485 (1973); G. E. Calf and J. L. Garnett, *Tetrahedron Letters*, **1973**, 511; U. Klabunde and G. W. Parshall, *J. Am. Chem. Soc.*, **94**, 9081 (1972).
[179] G. A. Olah, G. D. Matrescu and Y. K. Mo, *J. Am. Chem. Soc.*, **95**, 1865 (1973).
[180] G. A. Olah and Y. K. Mo, *J. Am. Chem. Soc.*, **94**, 9241 (1972).
[181] G. A. Olah and Y. K. Mo, *J. Org. Chem.*, **38**, 353 (1973).
[182] H. V. Ansell, J. Le Guen and R. Taylor, *Tetrahedron Letters*, **1973**, 13.
[183] L. V. Pepekin, M. I. Vinnik, A. M. Boganov, A. I. Pavlyuchenko, L. G. Yudin and A. N. Kost, *Zh. Org. Khim.*, **8**, 2582 (1972); *Chem. Abs.*, **78**, 83536 (1973).
[184] V. Zverina, J. Divis, M. Remes and M. Matrka, *Chem. Prum.*, **22**, 454 (1972); *Chem. Abs.*, **78**, 151576 (1973).
[185] J. W. Strand and P. Kovacic, *J. Am. Chem. Soc.*, **95**, 2977 (1973).
[186] V. P. Vitullo and E. A. Logue, *J. Org. Chem.*, **38**, 2265 (1973).
[187] Y. Furuya and K. Teraoka, *Bull. Chem. Soc. Japan*, **46**, 679 (1973).

CHAPTER 8

Carbonium Ions[1]

R. Baker

Chemistry Department, The University, Southampton

Bicyclic and Polycyclic Systems

Derivatives of Norbornane and Related Compounds[2]

The effects of increasing electron demand on the rates of solvolysis of 2-aryl-3-methyl-2-butyl and 1-aryl-1-cyclopropyl-1-ethyl *p*-nitrobenzoates in 80% aqueous acetone have been examined: participation by the carbon–carbon bonds of the cyclopropane ring was found to increase dramatically with increasing electron demand at the cationic centre; correlation against σ^+ constants ($\rho - 4.65$ and -2.78, respectively) indicated that stabilization by the cyclopropyl group is a linear function of the electron demand of the incipient carbonium ion. Application of a similar treatment to the rates of solvolysis of the 2-aryl-2-*exo*-norbornyl *p*-nitrobenzoates indicated absence of participation by the 1,6 carbon–carbon bonds in the tertiary aryl norbornyl derivatives.[3] Similarly, the π-electrons do not participate in the solvolysis of 2-aryl-2-norbornenyl *p*-nitrobenzoates in 80% aqueous acetone.[4] However, the amount of rearranged product, 1-aryl-3-nortricyclanol, increases as the stability of the carbonium-ion centre decreases, indicating involvement of the π-electrons of the double bond in the product-determining step.

A computer programme has been applied to the rearrangement of ^{14}C-labelled camphor to 3,4-dimethylacetophenone in sulphuric acid;[5] a mechanism excluding 3,2-*endo* hydroxyl shifts[6] is completely consistent with the isotopic data. The involvement of a 3,2-*endo* methyl shift in the acid-catalysed racemization of [8-^{13}C]camphene[7] has also been questioned.[8] Full details have been reported of the nitrous acid deamination of substituted secondary norbornyl derivatives in which the cation has been formed by Wagner–Meerwein rearrangement;[9] although these cations appear to have a special property no evidence was obtained that they are of higher energy than if they had been formed on solvolysis.

The reversible interconversions of substituted cyclohexenyl and bicyclo[2.2.1]oct-2-yl cations have been studied.[10] The cations (**1**) and (**2**) are interconverted by a Wagner–Meerwein shift even faster than the 3,2-hydride shift, and a minimum of 6.9 kcal/mol difference in the transition-state free energies for *exo*- and *endo*-hydride shifts was observed. In the interconversion of (**3**) and (**4**), involving two 1,2-hydride shifts and a ring closure or opening by the 1–7 π-route, the ring closure was suggested as being the slow step.

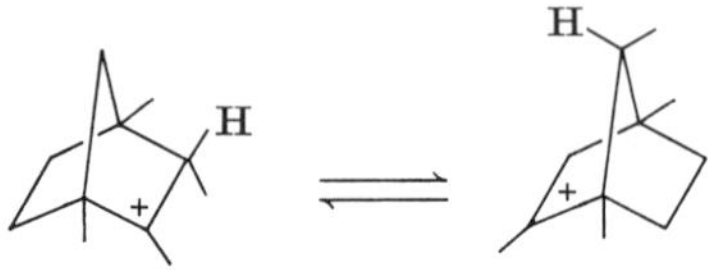

H H + +

(**1**) (**2**)

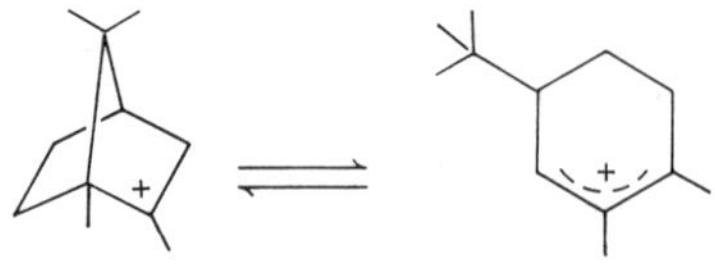

+ +

(**3**) (**4**)

Since the γ-deuterium isotope effects for solvolysis of the 7-halonorbornyl *p*-bromobenzenesulphonates are slightly larger than those observed for *exo*-norbornyl *p*-bromobenzenesulphonate, it is suggested that these isotope effects do not originate from C-1—C-6 bond participation.[11] A number of ionic intermediates have been implicated in the addition of acetic [^{2}H]acid (CH_3COOD) to norbornadiene catalysed by sulphuric acid.[12]

Calculations have been made on the non-classical stabilization in the 2,3-diaza-7-norbornyl cation.[13] The steric influence of the 7,7-dimethyl substituents on norbornene in addition reactions has been fully reported.[14]

Cyclohex-3-ene-1-carboxylic acid is formed on reaction of diazonorcamphor with dilute acetic acid in ether.[15] A full report of the solvolytic fragmentation studies of 7-substituted *exo*- and *endo*-5,6-(*o*-phenylene)-2-norbornyl toluene-*p*-sulphonates[16] and a description of the solvolytic behaviour of 7,7-dimethoxybicyclo[2.2.1]hept-2-yl toluene-*p*-sulphonates[17] have appeared. The extensive fragmentation observed in the former reactions was attributed to the favourable alignment of the participating bonds. In the latter, MeO-4 and C-1–C-6 participation were suggested to determine the course of reaction for the *endo*- and *exo*-derivatives, respectively.

Almost complete rearrangement (98%) of (**5**) to (**7**) occurs in the acetolysis of (**5**) and the major product is (**8**) (64.8%) consistent with involvement of the intermediate (**6**).[18] Formation of (**9**) (27.9%) and (**10**) (6.5%) indicates the importance of an anchimerically unassisted route from (**7**), and traces of other products, (**11**) (0.4%), (**12**) (0.34%) and (**13**) (0.07%), were suggested to arise from cation (**14**). Together with other products, (**17**) (22.5%) is formed on acetolysis of (**15**); this may arise via (**16**).

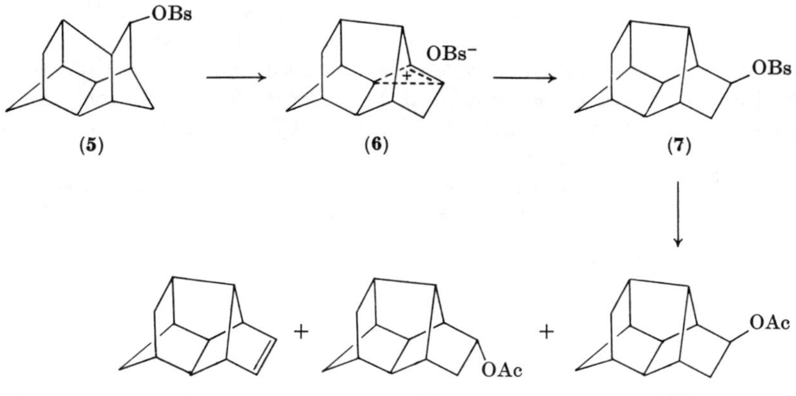

(5) (6) (7)

(10) (9) (8)

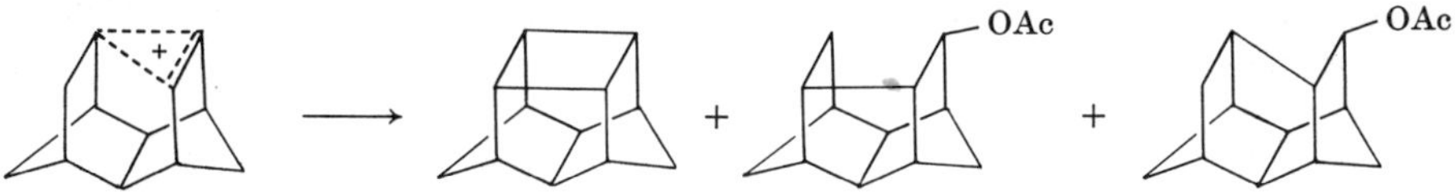

(14) (11) (12) (13)

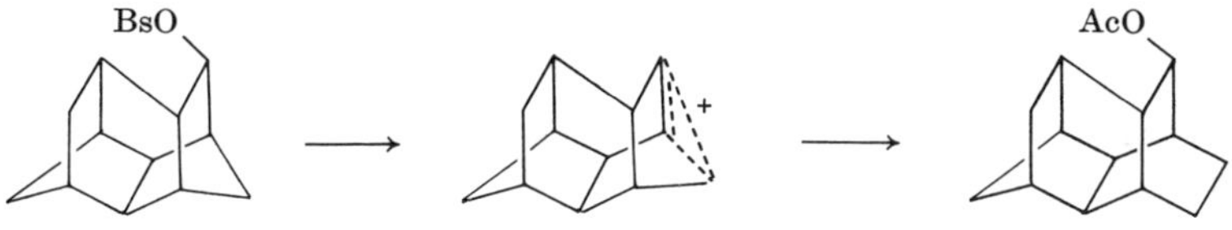

(15) (16) (17)

A cationic intermediate has been implicated in the photolysis of 2-*exo*-iodonorbornane whereas the photoproducts from the corresponding bromide are those expected from radical intermediates.[19] Rates of, and product distributions from, phenolysis of *exo*- and *endo*-2-norbornyl *p*-toluenesulphonate have been determined.[20]

Other studies include the decomposition of stereoisomeric norbornanediol mesylates,[21] rearrangement of 2,3-dichloro-2-norbornene oxide,[22] pyrolysis of cyclic sulphites from 2,3-*endo*- and *exo*-bornanediols[23] and the rearrangement of 10-isobornyl and 4-methyl-10-isobornyl sultone.[24]

Other Bicyclic Systems

endo- and *exo*-2-Bicyclo[3.1.0]hexyl toluene-*p*-sulphonate (**18**) undergo reaction in acetic acid at similar rates and give, within experimental error, an identical product mixture of (**19**) (16%), (**20**) (37%) and (**21**) (38%).[25,26] Reaction of the corresponding

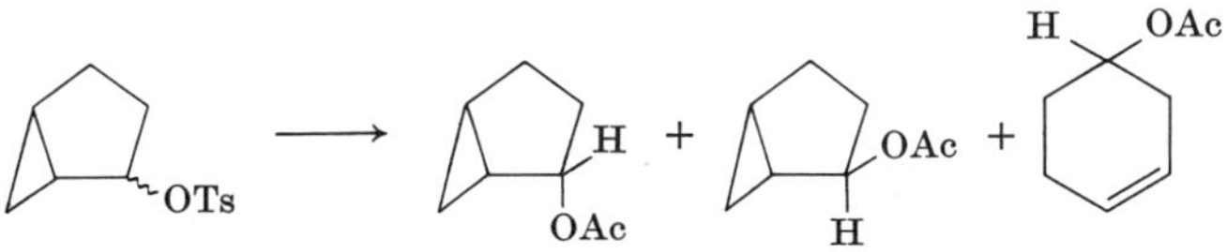

(18) (19) (20) (21)

3,5-dinitrobenzoates in 80% aqueous acetone gave similar results. Although experimental difficulties were encountered, essentially no deuterium scrambling was observed on solvolysis of the C-2-labelled derivatives, and introduction of a 5-methyl substituent produced a similar rate acceleration of about 20 for both isomers. The results were interpreted as indicating the formation of a single bisected homoallylic cyclopropylcarbinyl cation intermediate in the hydrolysis of both derivatives.

Bicyclo[2.1.0]pentane has been found to have a low reactivity towards acetolysis in contrast to other reactions of this compound, and the results were compared with the acetolysis of other bicyclo[*n*,1,0]-compounds.[27]

Reaction of 2-chlorobicyclo[2.1.1]hexane with SbF_5–SO_2ClF at –125° gave a cation in solution with three single peaks at δ 8.32 (2H), 3.70 (6H) and 2.95 (1H).[28] This indicates that the six methylene protons are equivalent, and the activation enthalpy for this equilibration was estimated as >7 kcal/mol. The minimum activation enthalpy for hydrogen shift was estimated as 13 kcal/mol and the results were suggested to be consistent with equilibration of the bridged ions (**22**). At –97° a slow rearrangement to the cyclohexenyl cation was observed with a further rearrangement to the methylcyclopentenyl cation at a higher temperature.

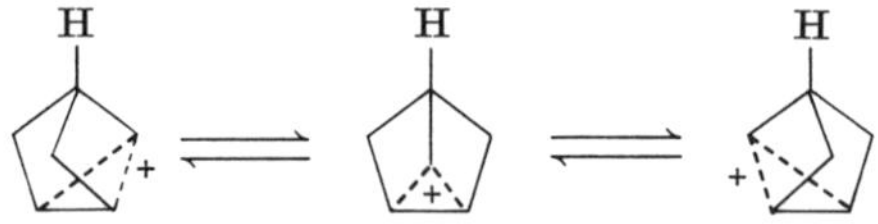

(**22**)

Treatment of the ions (**23**) with triethylamine gave 1,2,3,4,6-pentamethyl-5-methylenetricyclo[2.2.0.0^{2,6}]hexane (**24**).[29] Reaction with hydroxide is, however, faster and attack at the bridge-carbon atom takes place to yield (**25**). At –90° in CH_2Cl_2, bromine attacks hexamethyl-Dewar-benzene exclusively from the *endo*-direction, producing

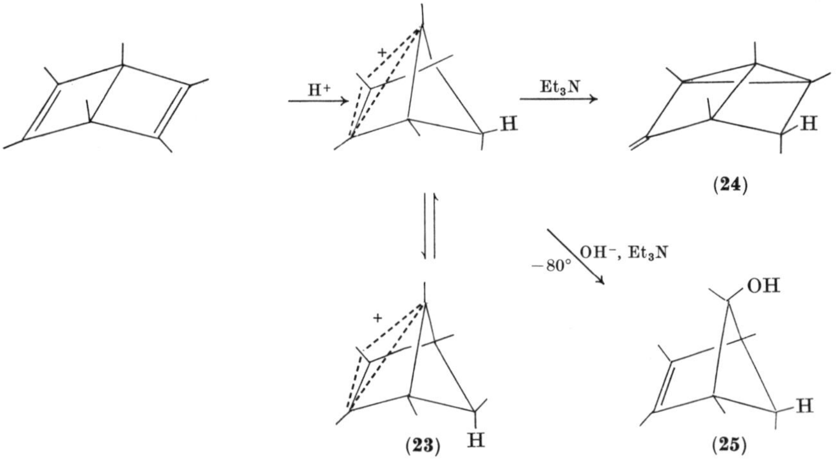

(**24**) (**23**) (**25**)

the *endo*-5-bromo-1,2,3,4,5,6-hexamethylbicyclo[2.1.1]hexenyl cation (**26**).[30] Reaction of (**26**) with an excess of sodium methoxide in methanol gives (**27**), but (**28**) is produced with KOH–EtOH at –80°; (**29**) can be obtained by reaction of (**26**) with lithium aluminium hydride at –70° in ether.[31] It was also demonstrated that the *exo*-5-hydroxy-ion analogous to (**26**) could be generated from hexamethyl-Dewar-benzene epoxide, and reaction of the former with KOH–EtOH at –80° gives (**28**). These nucleophilic reactions at C-2 and C-3 were suggested to be kinetically controlled at the site of lowest electron density.

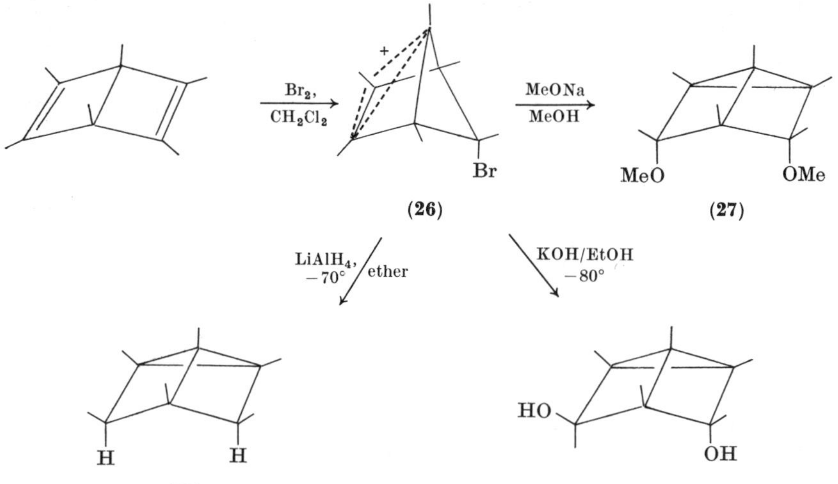

Br_2, CH_2Cl_2 — MeONa, MeOH — MeO, OMe — Br

(**26**) (**27**)

$LiAlH_4$, –70° ether — KOH/EtOH –80°

H H — HO OH

(**29**) (**28**)

The 9-deuteriobicyclo[4.2.1]nona-2,4,7-trien-9-yl cation (**30**) has been shown to be converted into 2-deuterioindene (**33**) at 74° in dimethyl sulphoxide, and the proposed mechanism involved (**31**) and (**32**).[32]

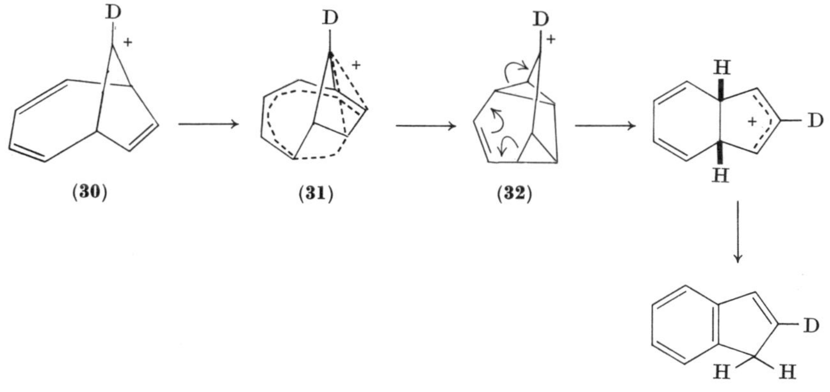

(**30**) (**31**) (**32**)

(**33**)

Rate studies for (**34–37**) in 80% aqueous acetone have been interpreted without invoking differences in the equilibria to the corresponding cyclopropyl analogues.[33] By extrapolation from these results it was suggested that the influence of conformation upon the rates of solvolysis of cycloheptatrien-7-ylmethanol derivatives is negligible.

The rate of perchloric acid-catalysed hydrolysis of (**38**) is 10^5 times slower than that of 1-ethoxycyclohexene and this has been attributed to inhibition of resonance of the bridging oxygen atom with the positive bridgehead carbon in (**39**).[34]

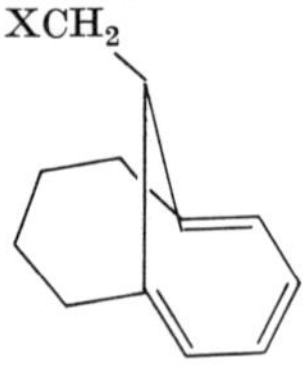

XCH_2 CH_2X CH_2X CH_2X

(34) 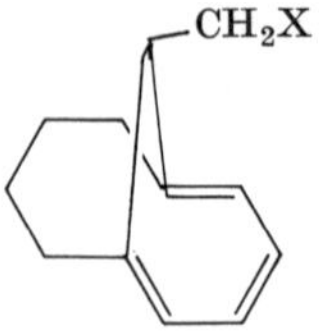 **(35)** 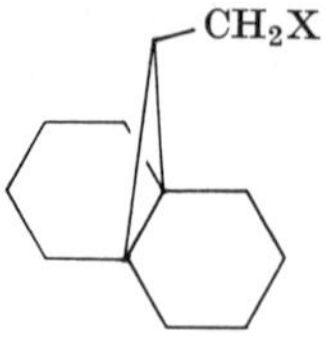 **(36)** 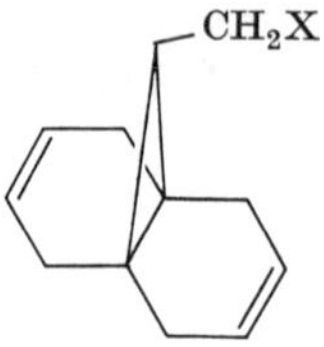 **(37)**

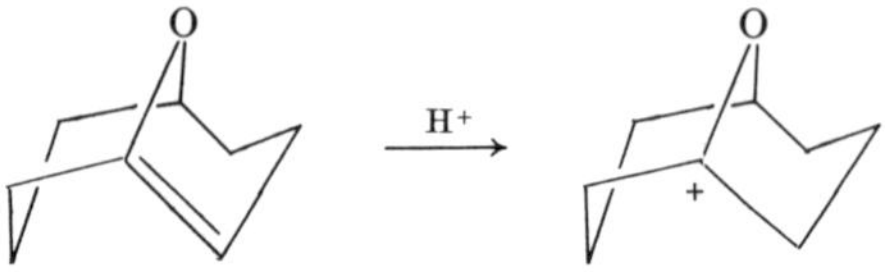

O O H^+

(38) **(39)**

Carbon–hydrogen participation has been observed in the solvolysis of 6-substituted bicyclo[3.3.1]non-3-yl toluene-*p*-sulphonates and it has been demonstrated that this can be enhanced by steric effects.[35]

The Ag(I)-promoted reactions of bicylobutanes in methanol have been shown to deviate significantly from normal second-order kinetic behaviour and follow an autocatalytic pathway accompanied by acid production.[36] Thus, in reaction of either (**40**) or (**41**) similar products were obtained if either anhydrous $AgClO_4$ or 70% $HClO_4$ in 1:1 methanol–benzene was used. A decrease in concentration of Ag^+ or bicyclobutane increased the time required for onset of the exceedingly rapid reaction. It was concluded that methanol provides conditions in which bicyclobutanes react with silver salts to form acid.

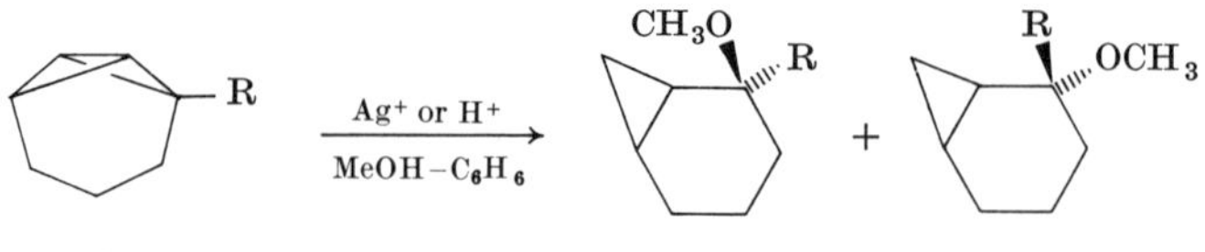

R $\xrightarrow[\text{MeOH}-C_6H_6]{Ag^+ \text{ or } H^+}$ CH_3O R + R OCH_3

(40)
(R = H or CH_3)

$\xrightarrow[\text{MeOH}]{Ag^+ \text{ or } H^+}$ CH_3O H

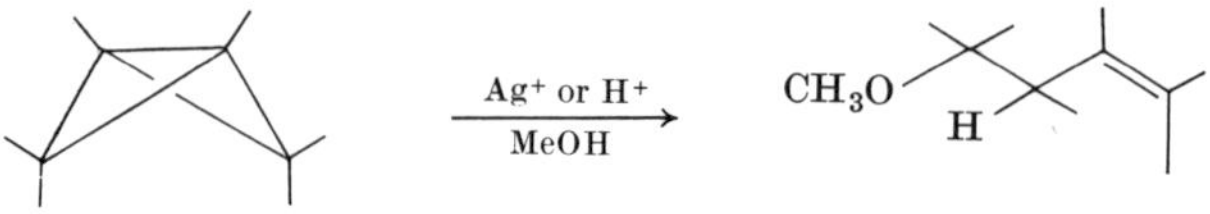

(41)

Essentially similar results have been reported on reaction of tricyclo[4.1.0.0^{2,7}]heptanes with dicarbonylchlorido rhodium dimer in methanol.[37] The mechanism of acid production has been examined and it was found that, with sodium hydrogen carbonate present, the pH of the solution dropped to 5.0 when the catalyst was added to the bicyclobutane in solution and then rapidly increased. In the absence of sodium hydrogen carbonate the pH dropped from 8.0 to 2.5 when the Rh(I) catalyst was added. The nature of this acidic species has been established as (**42**) by IR evidence.[38] Initially the highly strained bicyclobutane was suggested to co-ordinate with the transition metal, after which methanol attacked the electrophilic carbonyl group to yield (**42**) since the bicyclobutane portion acts as a strong accepting ligand. It was consistently found that larger amounts

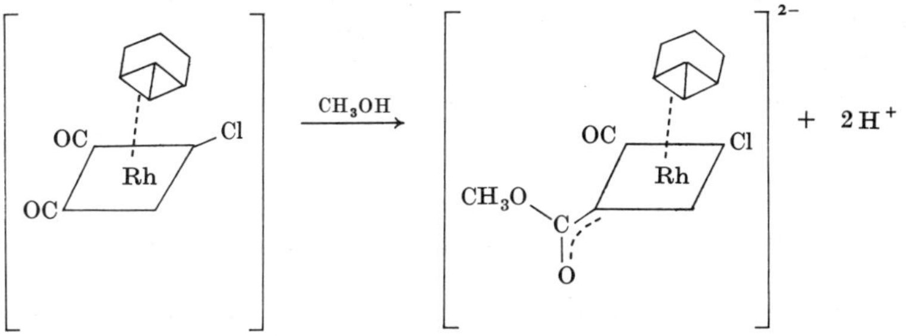

(**42**)

of dienes were formed in the presence of sodium hydrogen carbonate. For reactions catalysed by $PdCl_2(PhCN)_2$ it was suggested that formation of methyl ethers is the result of an acid-catalysed ring opening where the acid is derived from reaction of π-allylpalladium complexes with methanol.[38]

The high specificity of the 1-alkyltricycloheptane to bicyclohept-6-ene rearrangement has been noted.[39] Thus when (**43**) and (**46**) are treated with $AgClO_4$ the major products are (**44**) and (**45**) from the former and (**47**) and (**48**) from the latter, together with a series of other products in both cases. Similar results were obtained when a methoxy-group replaced methyl at C-3, although this substitution had a small rate-retarding effect. Faster rates were also found when the 3-substituent was stereoproximal to the angular methyl group. This was interpreted as due to an easier approach to the opposite side of the molecule by the Ag^+ ion.

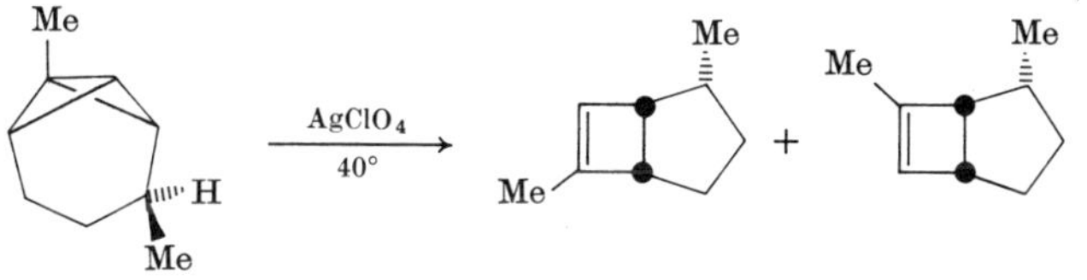

(**43**) (**44**) (46%) (**45**) (27%)

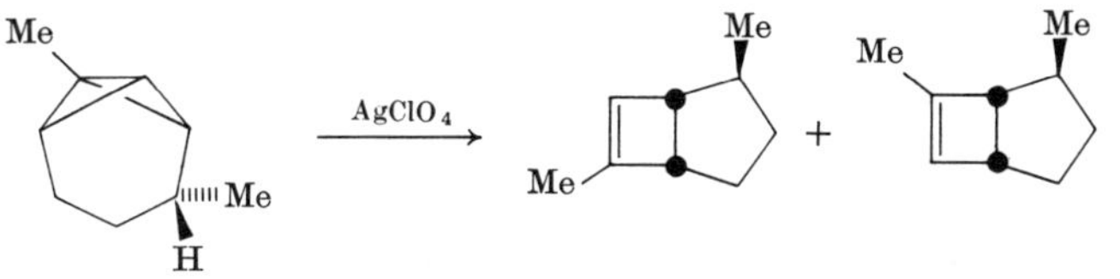

(**46**) (**47**) (38%) (**48**) (26%)

Silver-catalysed rearrangement of dimethyl *anti*-tricyclo[3.1.0.0^{2,4}]hexane-1,2-dicarboxylate has been reported.[40]

Reactions of bicyclo[1.1.0]butanes have been discussed in terms of the nature of the resulting allylcarbene–metal complexes formed.[41]

Details of the rearrangement of 1-(*N*,*N*-dichloroamino)apocamphene by aluminium chloride have appeared.[42] Evidence has been obtained for the formation of gaseous bicycloalkylium ions from reaction of helium tritiide ions with gaseous bicyclo[*n*.1.0]alkenes.[43] The photolysis of 1-alkoxybicyclo[*n*.1.0]alkane-2-diazonium ions has also been studied.[44]

A dipolar intermediate having the structural features of a *trans*-1,3-bishomotropylium ion is suggested to be important in the addition of tetracyanoethylene to *cis*-bicyclo[6.1.0]nonatriene;[45] reaction with chlorosulphonyl isocyanate has also been investigated.[46] Uniparticulate electrophilic addition has been suggested as a probe for bicycloaromatic and antibicycloaromatic carbonium-ion character.[47]

Polycyclic Systems

Following the stability–selectivity relationship for the solvolysis of *p*-substituted 2-adamantyl arenesulphonates a relationship has been demonstrated for alkyl chlorides in a binary nucleophilic solvent.[48] Rates of ethanolysis of 1- and 2-adamantyl arenesulphonates have been shown to be similar for *meta*- and *para*-substituents, indicating that steric rather than electronic effects are of major importance.[49] In the same study it was established that the Hammett σ values for *m*- and *p*-nitro-substituents of arenesulphonates are solvent-dependent.

Alkyl substituents in the 1-position of 2-adamantyl toluene-*p*-sulphonates have been shown to have a strong influence on the reaction rate and on the proportion of rearrangement on solvolysis in 60% aqueous acetone.[50] The results were consistent with the importance of the electronic effect of the alkyl group on the bridged ion (**49**). No significant effect was observed by substitution of an aryl group at the 1-position.[51]

The solvolyses of the 2-adamantyl derivatives (**50**) and (**51**) have been studied and the rates discussed in terms of the effects of 1,3-diaxial interactions.[52]

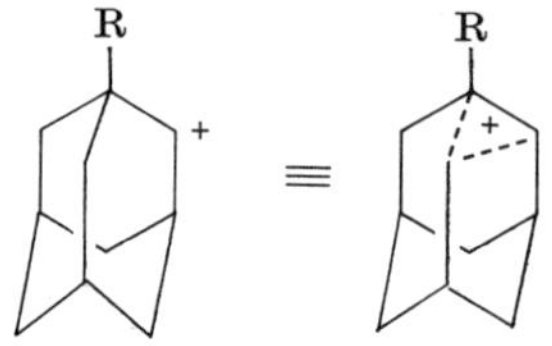

R R + ≡

(**49**)

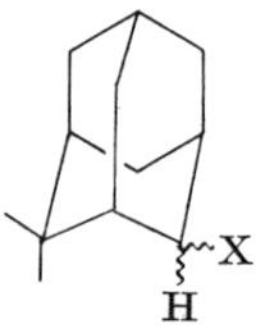

X H X Me X = OTs or OPNB

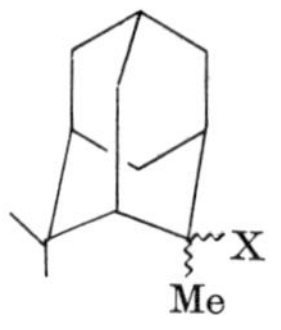

(**50**) (**51**)

A comparison of the solvolysis rates of 2-adamantyl and 3,3-dimethyl-2-butyl (pinacolyl) *p*-bromobenzenesulphonate has been made to test their relative response to solvent change.[53] With trifluoroacetic acid as the reference solvent the expression:

$$\frac{(k_{\text{solvent}}/k_{\text{TFA}})_{\text{pinacolyl}}}{(k_{\text{solvent}}/k_{\text{TFA}})_{\text{2-AdOTs}}}$$

has been calculated in order to test the degree of nucleophilic unassisted solvolysis in both systems. If unassisted solvolysis (k_c) is complete for both systems the expression should be unity and be independent of solvent. The results ranged upwards to 16 for 80%

aqueous ethanol but it was suggested that, compared with those in other systems, the values were small. Variations in the ion-pair chemistry with solvent change were discussed as constituting a possible origin of the results for the two systems.

Based on calculated heats of formation of tricyclodecanes and the transition state energies for the predicted carbonium ion intermediates, predictions have been made for the Lewis acid-catalysed rearrangements of a series of tricyclodecanes to adamantane.[54]

A new type of diamondoid hydrocarbon, (**53**) called [2]diadamantane has been prepared by reaction of 2,2′-binoradamantane (**52**) with aluminium bromide.[55]

Strong evidence has been obtained for an intramolecular 1,3-hydride shift in an acid-catalysed rearrangement in the adamantane nucleus.[56] The ketone (**55**) was obtained by generation of the cation (**54**) from the glycol, epoxide or a spiroketone in 50% sulphuric acid and steps 1 and 5 were suggested (on the basis of deuterium-labelling results) to be intramolecular.

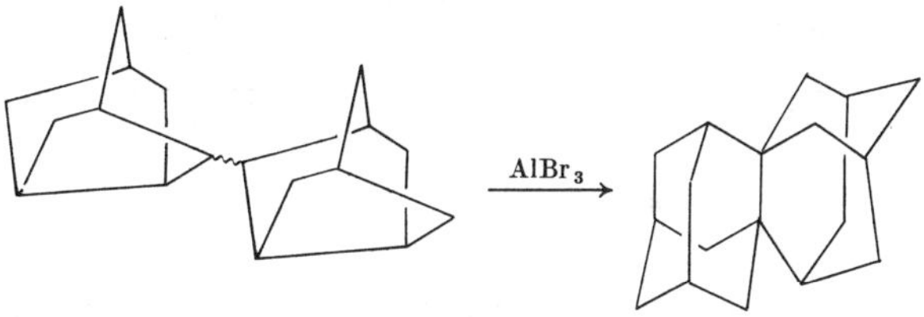

$AlBr_3$

(**52**) (**53**)

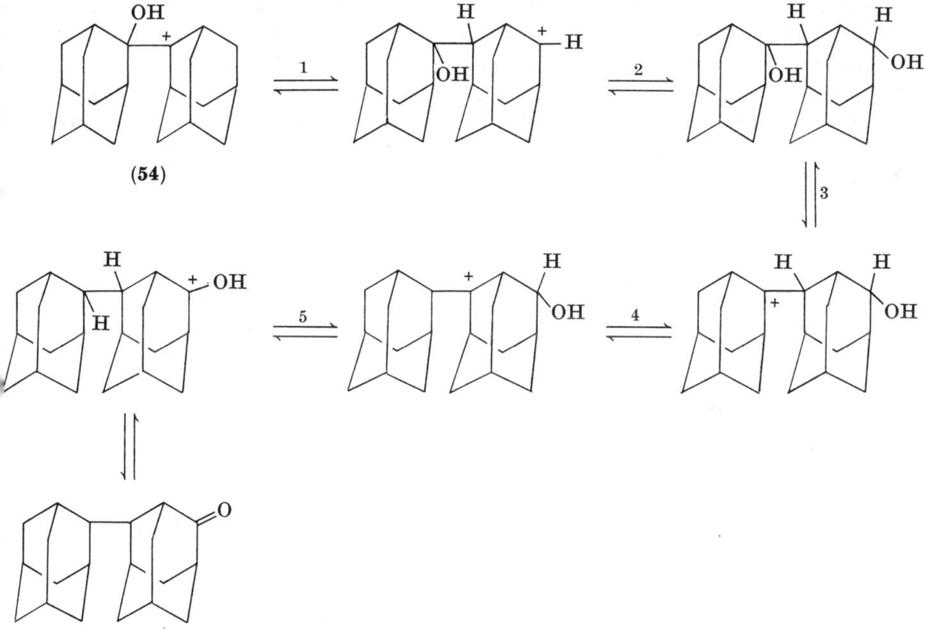

OH + H + H 1 OH 2 H H OH OH

(**54**)

3

H + OH H 5 + H OH 4 H + H OH

O

(**55**)

Anodic acetamidation of adamantane and 1-haloadamantanes,[57] the anodic fragmentation of adamantane derivatives,[58] amination of adamantane with monochloramine–aluminium chloride,[59] and halogen exchange of bridgehead adamantyl halides in the presence of aluminium halides in halogenated solvents[60] have been reported. Some reactions of 5-substituted 1,3-dehydroadamantanes have been studied and thermochemical data obtained.[61]

Treatment of bicyclo[3.3.1]nonane-2,6-diol with concentrated sulphuric acid gives 2-oxa-adamantane,[62] and Koch–Haaf carboxylation of 1-(1-adamantyl)ethanol afforded a mixture of 3-ethyladamantanecarboxylic acid and 2-(1-adamantyl)propanoic acid.[63]

Silver-ion assisted acetolysis of 2-*exo*-bromo-5-protoadamantanone (**56**) has been found to give 2-*exo*-acetoxy-7-isotwistanone (**57**).[64] It was suggested that *endo*-attack of nucleophile on the 2-protoadamantyl cation is not competitive with the 1,2-carbon shift of the C-1–C-10 σ-bond.

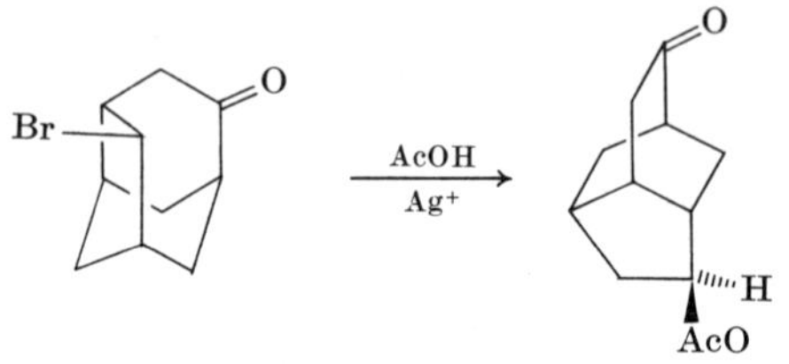

Br, O, AcOH, Ag^+, O, H, AcO

(**56**) (**57**)

Either 1- or 2-noradamantanol, or noradamantane, was formed in the reaction of tetracyclo[4.3.0.0^{2,4}0^{3,7}]nonane with sulphuric acid depending on the conditions.[65]

A carbonium-ion intermediate (**59**) has been implicated in solvolysis of the *anti*-tricyclo[5.2.0.0^{2,5}]nona-3,8-dien-6-yl derivative (**58**), which yields mainly (**60**) on reaction in acetic acid.[66] The reactivity of (**58**) is less than that of *anti*-7-norbornenyl toluene-*p*-sulphonate, possibly owing to the extensive structural changes required to reach (**59**); it was suggested that a different carbonium ion could be formed initially with subsequent rearrangement. A degenerate rearrangement (**61a**) to (**61b**) involving (**62**) was observed at 10° in fluorosulphonic acid.

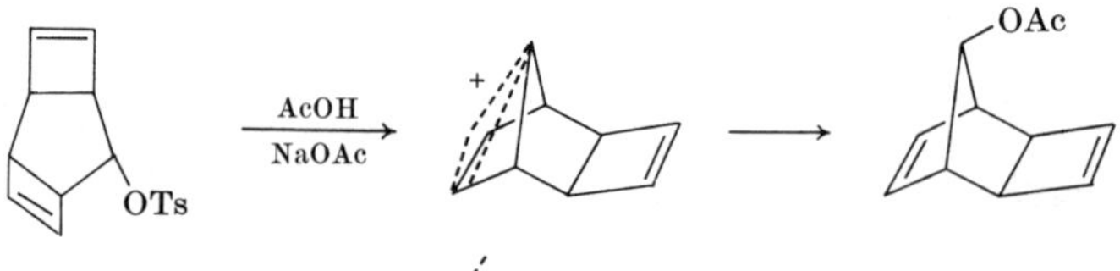

OTs, AcOH, NaOAc, +, OAc

(**58**) (**59**) (**60**)

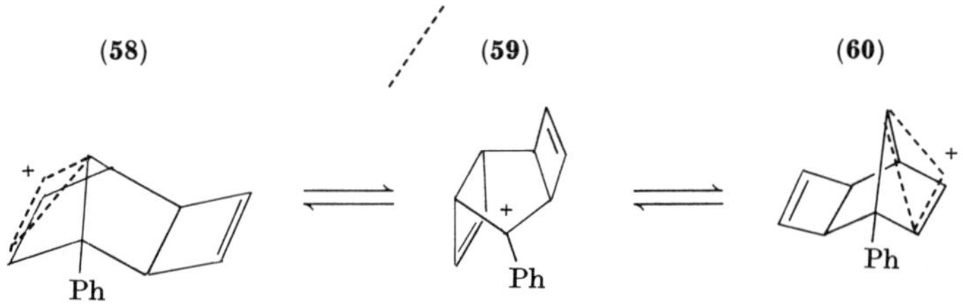

+, Ph, +, Ph, +, Ph

(**61a**) (**62**) (**61b**)

The cation (**64**) obtained from (**63**) in FSO_3H has been shown to have a five-fold degeneracy attained by a double circumambulatory process with an extremely low barrier to rotation.[67] The question of (**64c**) being the structure of the cation rather than a transition state has been discussed and it has been termed the [3.5.3]-armilenium cation.

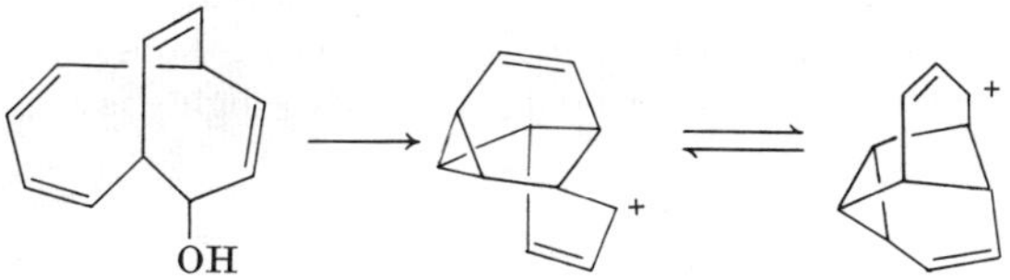

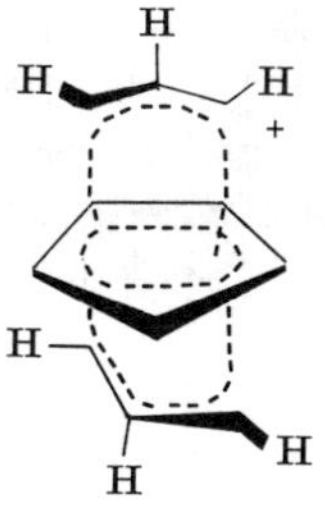

(**63**) (**64a**) (**64b**) (**64c**)

Evidence against (**66**) having bishomoantiaromatic character has been obtained.[68] Treatment of (**65**) with 60:40% diglyme–water containing sodium borohydride gave a mixture of (**69**) (11%), (**70**) (13%), (**71**) (35%) and cycloheptatriene (13%); the cations (**66**), (**67**) and (**68**) must be the precursors of these products. Reaction of (**72**), however, gave only (**71**) and cycloheptatriene under similar conditions, so that ions are being trapped at a different stage of equilibration, compared to that for (**65**). The intermediacy of a single bishomoantiaromatic ion (**73**) was, therefore, excluded. Additional evidence was obtained from rate studies.

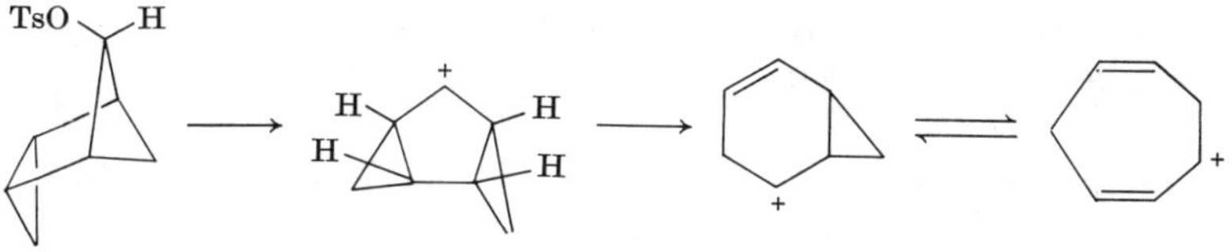

(**65**) (**66**) (**67**) (**68**)

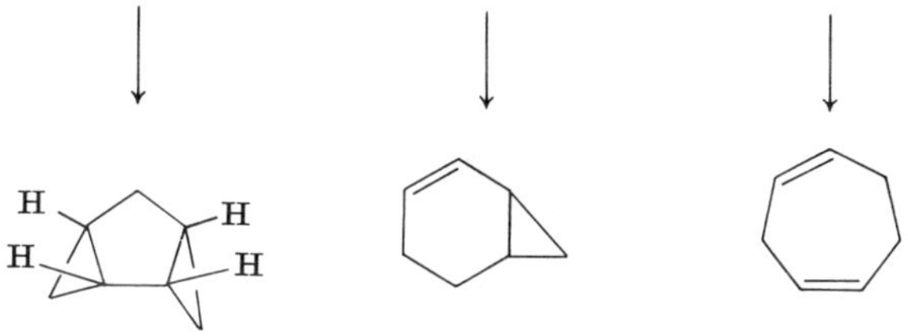

(**69**) (**70**) (**71**)

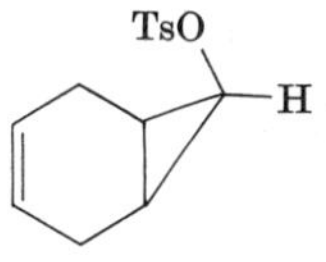

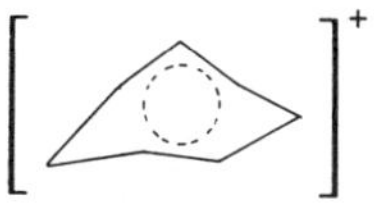

(**72**) (**73**)

Further studies on 3,3-diaryltricyclo[3.2.1.0^{2,4}]octane derivatives have been reported,[69] and the effect of 1,5-bridging on the solvolytic behaviour of bicyclo[3.1.0]hex-2-yl cations has been examined.[70]

The two homoallylic cations derived from the four possible tricyclo[4.2.0.0^{2,4}]octan-5-yl 3,5-dinitrobenzoates (**74**) and (**75**) have been shown to be conformationally distinguishable.[71] In aqueous acetone, (**74**) gave identical mixtures of (**76a**) and the internally returned ester (**76b**), whilst (**75**) gave a mixture of (**77**) (30–34%), (**78**) (24%), (**79**) (33–41%), (**75b**-OH) (4–5%) and (**75a**-OH) (2.5%). It was suggested that the different products from the two systems arise from the different alignment of the homoallylic *p*-orbital in the cation with the internal cyclobutane ring. In (**75**), this is highly favourable for a 1,2-carbon shift.

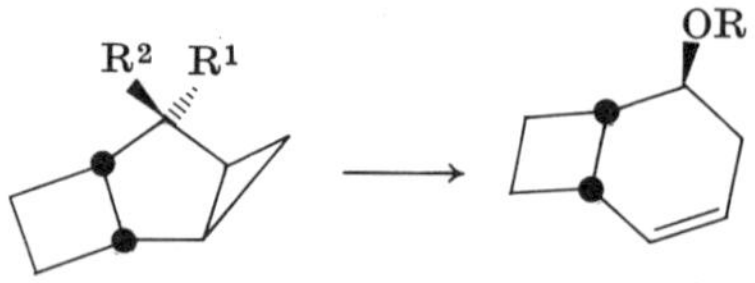

(**74**) a, R^2 = ODNB, R^1 = H; b, R^2 = H, R^1 = ODNB

(**76**) a, R = H; b, R = DNB

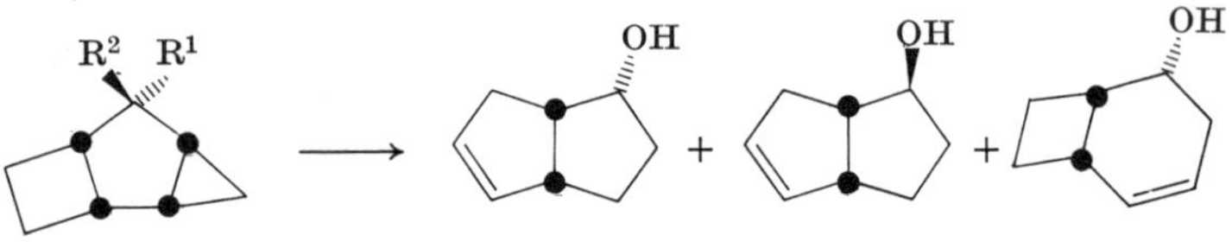

(**75**) a, R^2 = ODNB, R^1 = H; b, R^2 = H, R^1 = ODNB

(**77**) (**78**) (**79**)

Cyclopropane participation observed in solvolysis of *exo*- and *endo*-deltacyclyl *p*-bromobenzenesulphonates has been found not to occur in solvolysis of the C-5-substituted analogues (**80**) and (**81**); substantial amounts of *endo*-product were found.[72]

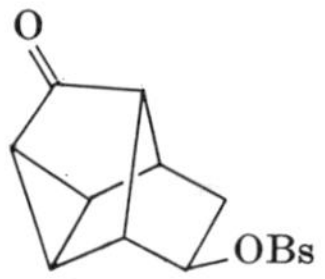

(**80**) 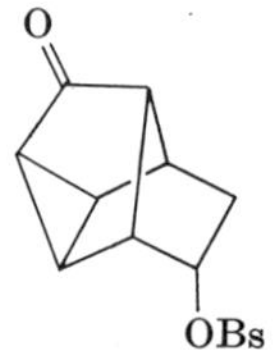 (**81**)

Silver acetate-promoted solvolysis of 1-bromomethyl-3,6-dibenzotricyclo[3.3.0.0^{2,8}]-octadiene,[73] acid-catalysed rearrangement of 3,3-dimethoxytricyclo[3.2.0.0^{2,7}]heptane[74] and the pinacol rearrangement of tricyclo[3.3.1.0^{2,6}]nonane-2,6-diol[75] have been investigated.

Other studies relate to acid-catalysed rearrangements of tricyclo[4.3.2.0]undecanones[76] and caryolan-1-ol,[77] and the rearrangement of cations derived from monodechloroisodrin and monodechloroaldrin.[78]

Participation by Aryl Groups

Trifluoroacetolysis of *m*-bromophenethyl and *p*-nitrophenethyl *p*-nitrobenzenesulphonate have been shown to proceed mainly by a k_Δ pathway.[79] Carbon-14 and deuterium isotope effects for solvolysis of neophyl *p*-bromobenzenesulphonate are consistent with a k_Δ process.[80,81] The factors responsible for the change in rate ratio of 0.24 to 1770 for phenethyl toluene-*p*-sulphonate/ethyl toluene-*p*-sulphonate from ethanol to trifluoroacetic acid have been studied.[82] Kinetic studies have shown that k_s is decreased for both systems relative to Fk_Δ as solvent nucleophilicity is decreased. Enhancement of k_Δ is also due to a change in response of the two pathways with the ionizing power of the solvent.

Trifluoroethanol has been shown to enhance phenyl participation to an equivalent degree to that in acetic acid in solvolysis of phenethyl toluene-*p*-sulphonate.[83] Essentially complete scrambling has been observed on reaction of [1-^{14}C]phenethyl iodide with silver trifluoromethanesulphonate.[84] Carbon-14 isotope effects have been measured in the solvolysis[85] and silver-catalysed ionization[86] of 2-bromo-1,1,2-triphenylethanol.

An NMR study to estimate the migratory aptitudes of substituted phenyl groups in 9-aryl-9,10-dimethylphenanthrenonium ions has been reported.[87]

Bridged ions have been found to be less important in solvolysis of *erythro*-derivatives of 1,2-diphenylpropyl bromide compared to that of the *threo*-isomers in 80% aqueous ethanol.[88] Ar_3-5 and Ar_3-6 participation has been suggested as occurring in the buffered acetolysis of 2-(4- and 2-(6-azulenyl)ethyl arenesulphonates and 3-(4-azulenyl)propyl *p*-nitrobenzenesulphonate,[89] and participation by heterocyclic components in the solvolytic rearrangement of 2-arylethyl toluene-*p*-sulphonates has been studied.[90]

Rearrangements and cyclizations to tetralin derivatives have been observed in reactions of ω-arylalkenes with trifluoroacetic acid.[91] Kinetic measurements indicate anchimeric assistance in reaction of 5-phenylpent-1-ene.

An *exo*/*endo* ratio of 5,700 was found in the solvolysis of (**82**) and (**83**) in buffered acetic acid, which compares with 15,000 for the analogous benzobicyclo[2.2.1]heptene derivatives.[92] The data were inconsistent with σ-bond participation although p-π participation could be in operation.

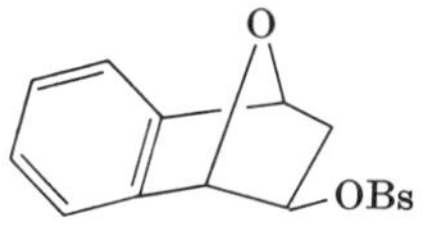

O OBs O OBs

(**82**) 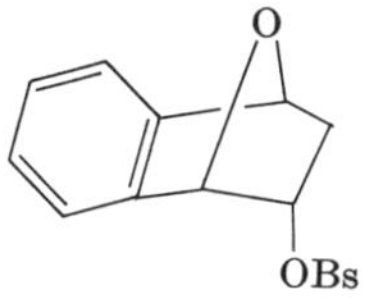 (**83**)

Further details of the effect of complexation of a chromium tricarbonyl unit on solvolysis of phenylalkyl systems has been reported.[93] Other studies include the interaction of phenols with the tri-*p*-methoxyphenylmethyl cation[94] and Friedel–Crafts cycloalkylation of mono- and di-phenyl-substituted alcohols.[95]

Participation by Double and Triple Bonds

Double-bond Participation

The first example has been reported of an olefin cyclization in which the complete steroid nucleus is produced as the primary product and which also results in the stereospecific formation of an optically active tetracyclic substance easily convertible into a natural steroid. Thus, (**84**) gave (**85**) in 65% yield in the presence of trifluoroacetic acid and ethylene carbonate in difluoroethane; (**85**) has been converted into progesterone.[96]

Stereospecific total syntheses of (±)-taondiol methyl ether (**87**) from (**86**)[97] and of estrone have appeared.[98] Cyclization of (**88**) in the presence of stannic chloride gave a mixture of (**89**) and (**90**) in 59% and 12% yield, respectively; (**89**) was converted into

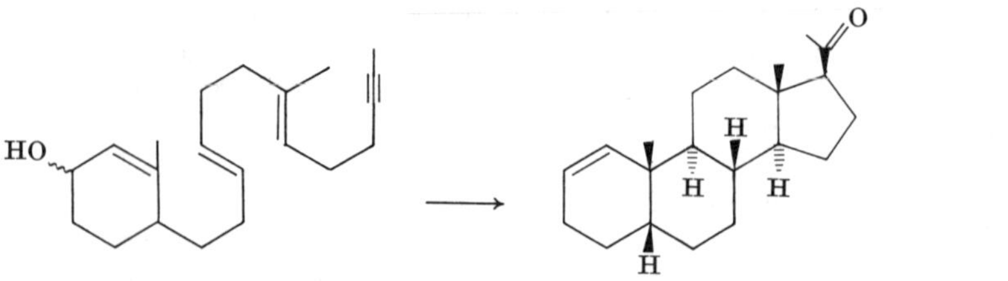

HO (84) → (85)

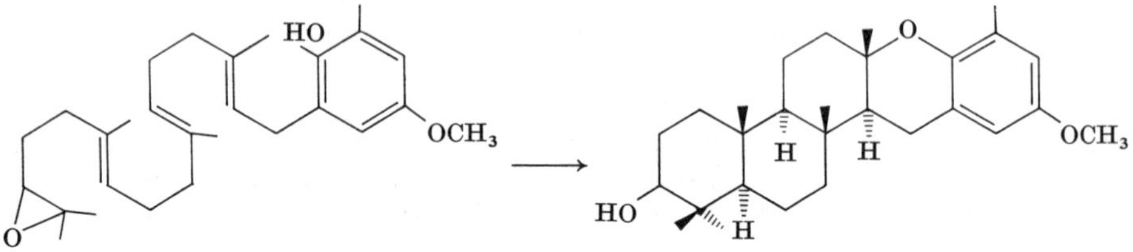

(86) HO OCH$_3$ → (87) HO OCH$_3$

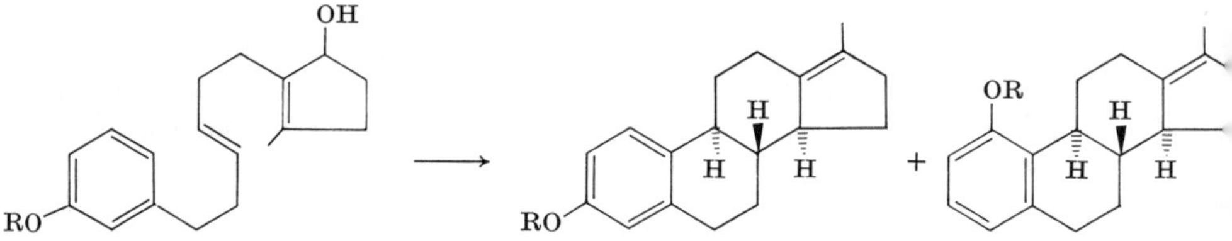

(88) OH RO → (89) RO + (90) OR

estrone and the overall yield from the precursor to (**88**) was 22%. Further studies have provided evidence that the cyclization of (**88**) is concerted, in that the *o*/*p* ratio in the cyclization of (**91**) to (**92**) and (**93**) depends on the leaving group;[99] the conclusion was that this would require some degree of bond formation between the aromatic ring and

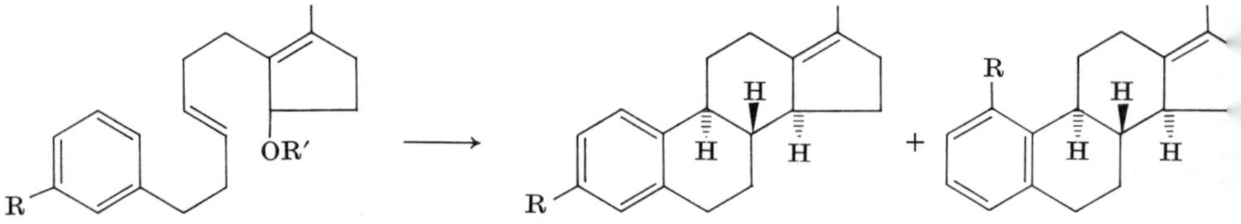

(91) R OR′ → (92) R + (93) R

C-9 (and also C-8 and C-14) before complete separation of the leaving group and cation; further evidence was obtained from a study of the effect of different substituents on the cyclization process in which the relative rates increased with increasing nucleophilicity of the aromatic nucleus.

The stereoselective formation of 10-*epi*-β-vetivone from (**94**) has been reported.[100] This spiro-alkylation was suggested to result from approach of the allylic cation from the face of the molecule opposite the pseudo-axial C-10 methyl group.

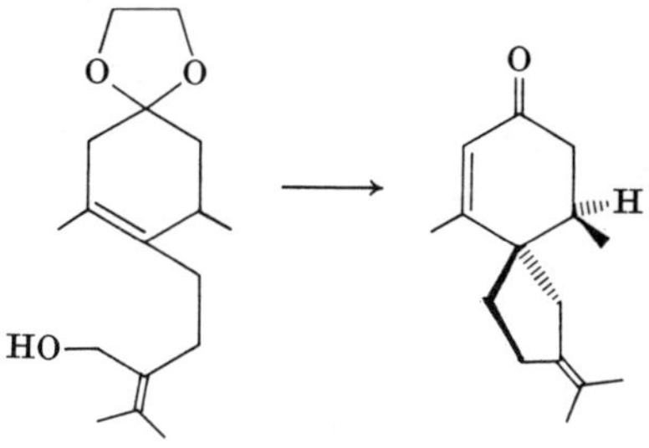

(94)

Full details of the stereospecific cyclization of diene acetals has appeared.[101] A paper describing the competitive participation of a double bond and a cyclopropane ring in solvolysis of *cis*-bicyclo[5.1.0]oct-5-en-3-yl toluene-*p*-sulphonate has been published; 80% and 20% of the reaction occurs by double-bond and cyclopropyl participation, respectively.[102]

In the solvolysis of 7-norbornyl- and 7-norbornenyl-derivatives and of the corresponding 7-methyl derivatives, the methyl/hydrogen rate ratio was found to vary from 1.3×10^8 for the saturated norbornyl system to 44 for the 7-norbornenyl *anti-p*-nitrobenzoate.[103]

1,2,4,5,6-Pentamethyl-3-methylenetricyclo[$2.2.0.0^{2,6}$]hexane reacts at the double bond with FSO_3H and at the cyclopropane ring with HCl/CH_2Cl_2.[104]

Greater than 90% cyclization occurs in solvolysis of (**95**) in 2,2,2-trifluoroethanol to yield (**96a**) (45%) and (**96b**) (45%) together with 5% of uncyclized material and unidentified material (5%) in 65% total yield.[105]

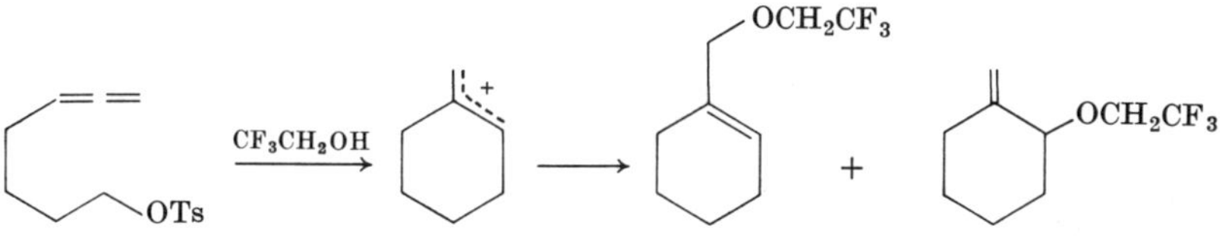

CF_3CH_2OH ; OCH_2CF_3 ; OTs

(95) **(96a)** **(96b)**

Solvolyses of the allenic toluene-*p*-sulphonates (**97**) ($m = 0$–4) have been studied and shown to yield varying amounts of the products (**98**)–(**101**).[106]

The influence of various factors on the solvolysis of an number of exocyclic primary and tertiary allylic 3,5-dinitrobenzoates have been fully discussed;[107] constriction of the endocyclic C—C—C angle diminishes the rates of solvolysis, and both σ–π and π–π delocalization are important, depending upon suitable orientations.

It has been demonstrated that five and six-membered cycloalkanecarboxylic acids (**103**) can be synthesized by intramolecular chloro-olefin annelation from (**102**).[108] Other interesting synthetic applications which have appeared are the acid-catalysed cyclization of germacrone in 80% aqueous acetic acid, to give a cadinane structure,[109] a synthesis of (+)-cedrene and (±)-cedrol using a synchronous double-annulation process[110] and a synthesis of a prostanoid precursor by cyclization of (**104**).[111]

The axial epimer of 6-oxo-$\Delta^{4a(5)}$-*trans*-2-decalyl toluene-*p*-sulphonate solvolyses more slowly than the equatorial epimer in formic or acetic acid; this was attributed to participation of the $\Delta^{5(10)}$-double bond in solvolysis of the latter derivative.[112] Reports

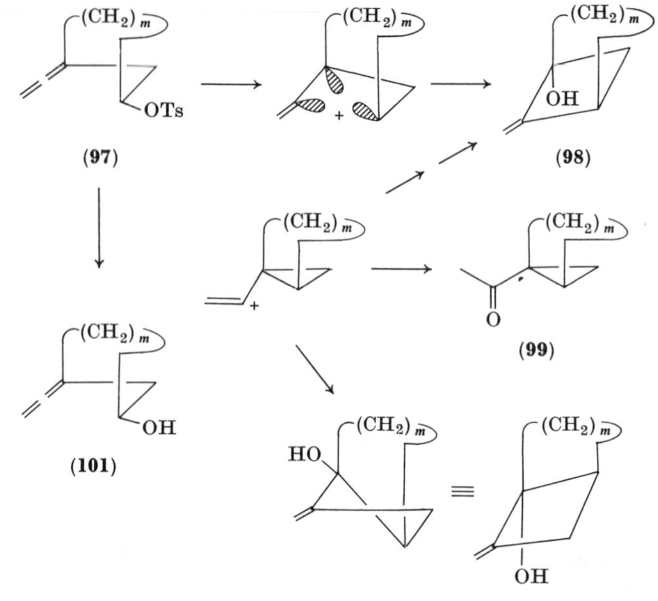

(97) (98) (99) (100) (101)

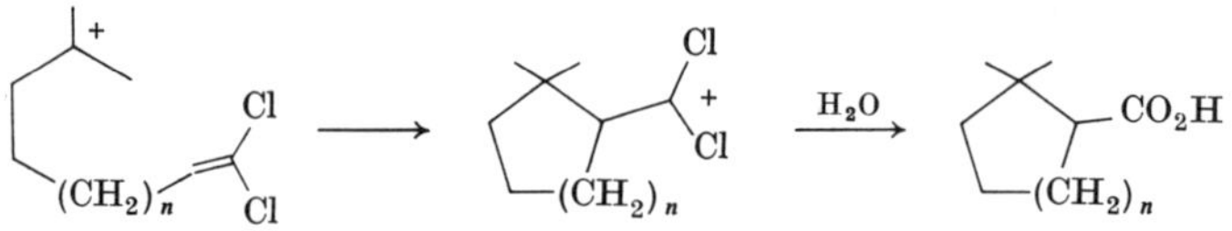

(102) (103)

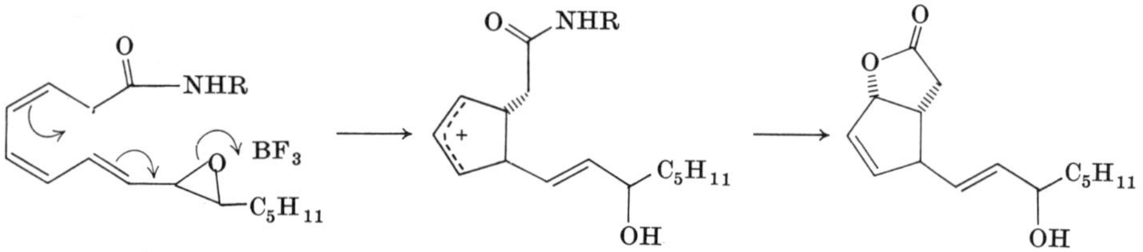

(104)

have appeared of the acetolysis of 2-(5*H*-dibenzo[*a*,*d*]cyclohepten-5-yl)ethyl toluene-*p*-sulphonate,[113] acid-catalysed cyclization of 2-(but-3-enyl)-1-phenylcyclohexanols,[114] oxidative cyclization of 1,5-dienes induced by mercuric salts[115] and acid-catalysed cyclization of 2-(but-3-enyl)-1-vinylcyclopentanol.[116]

The homocyclopropenium cation (**105**) has been implicated in the solvolysis of cyclo-octatetraenylmethyl chloride, which would account for formation of *o*-tolyl-acetaldehyde after conrotatory opening of the cyclobutene ring in (**106**); a substantial amount of direct displacement also occurs.[117] Solvolysis of (**107**) in acetic acid at 75° gave (**111**) and, at low buffer concentrations, 1,2-dihydronaphthalene (**112**), naphthalene and tetralin;[118] the rate of solvolysis was five times that of the fully saturated analogue and a homoallyl–cyclopropylmethyl interconversion (**108**) was suggested to be in operation in solvolysis of (**107**), followed by formation of (**109**) and (**110**); product resulting from direct displacement was also formed, and products from methyl-substituted (**107**) were consistent with the proposed mechanism.

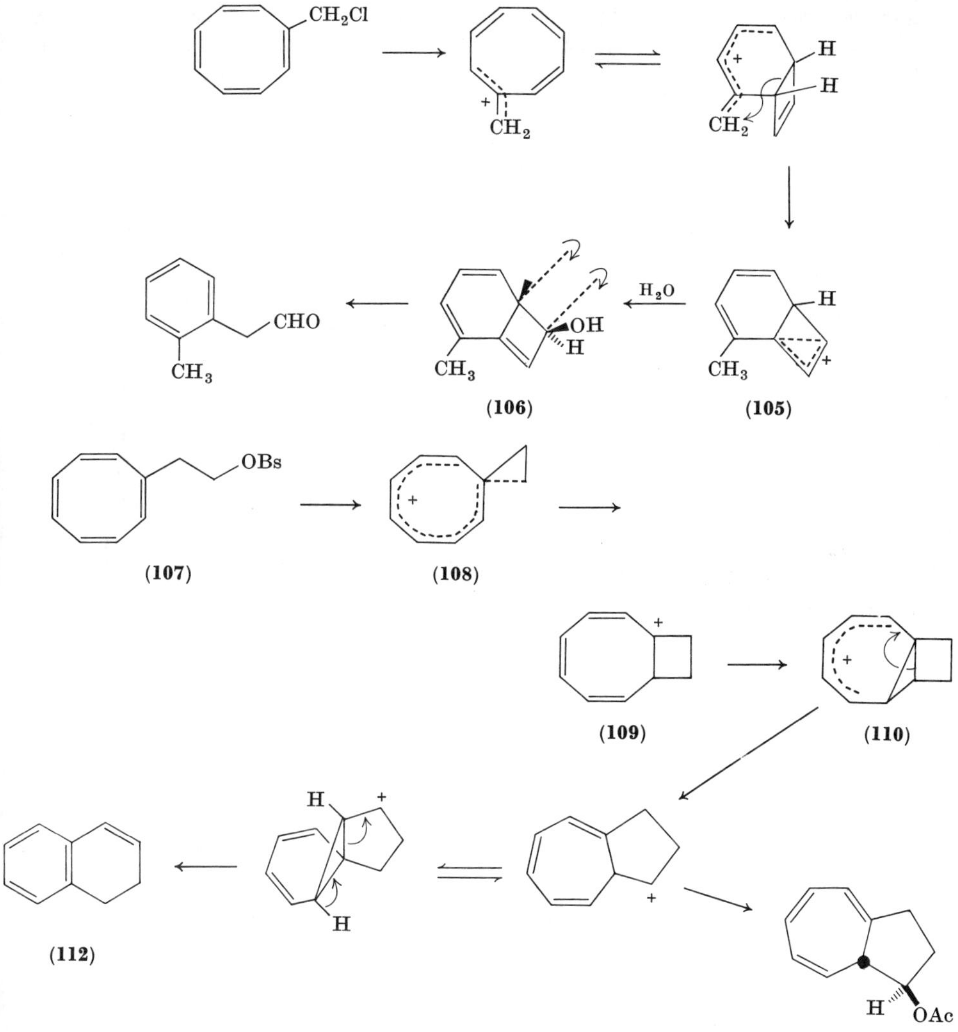

(**106**) (**105**)

(**107**) (**108**)

(**109**) (**110**)

(**112**) (**111**)

Acetolysis of *exo*-3-deuterio-*exo*-2-norbornenyl *p*-bromobenzenesulphonate has shown that unsymmetrical intermediates are not required for this reaction.[119] Double-bond participation has been observed to facilitate the hydrolysis of ketols.[120,121]

The intermediate 7-(triphenylphosphonio)norbornen-7-ylium dication has been detected in the reaction of 7-methoxy-7-(triphenylphosphonio)norbornene in HSO_3F–SbF_5.[122]

An alternative molecular-orbital-based model has been used to account for instability or stability of antihomoaromatic and homoaromatic molecules.[123]

Triple-bond Participation

In the absence of a good nucleophile the vinyl cation (**114**), formed by triple bond participation in reaction of (**113**) in methylene chloride containing trifluoroacetic acid, undergoes rearrangement to (**115**);[124] abstraction of chloride from the solvent then occurs with formation of (**116**). The relief of torsional strain in the conversion of a 6/5 to a 6/6 *trans*-fused ring system was suggested as a factor in the rearrangement.

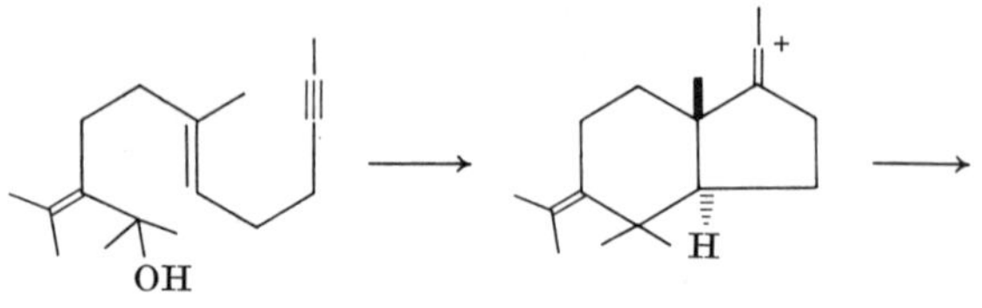

OH

(**113**) (**114**)

Cl

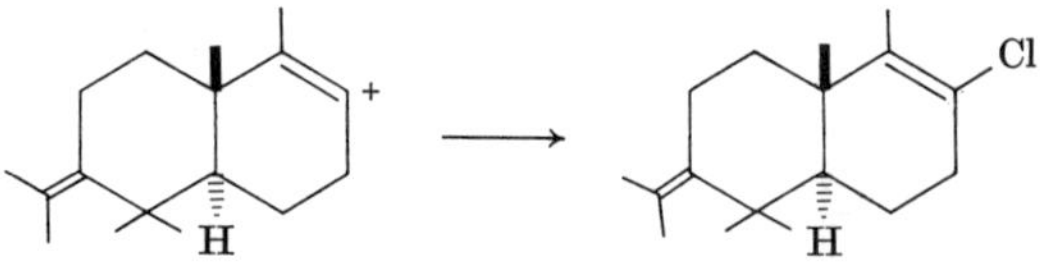

(**115**) (**116**)

It has also been demonstrated that vinyl cations can be trapped by nitroalkanes to form oxime ethers (**117**)[125] and these have been used as intermediates in a stereospecific synthesis of (±)-testosterone benzoate.[126]

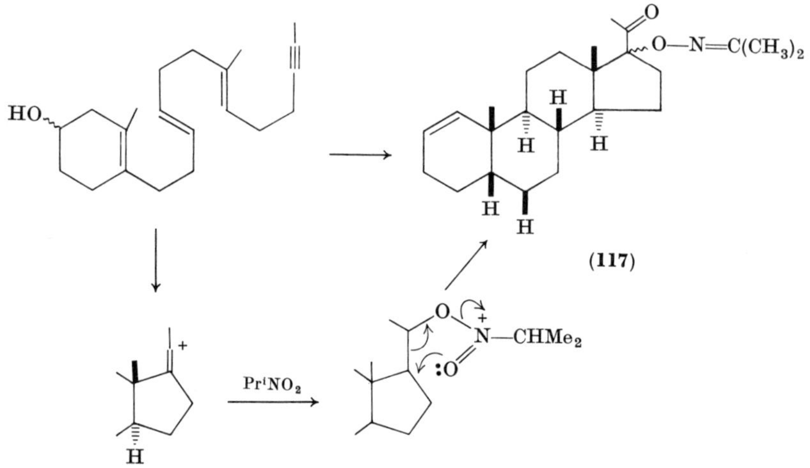

HO

O—N=C(CH$_3$)$_2$

(**117**)

Pr^{i}NO$_2$

N—CHMe$_2$

Reactions of Small-ring Compounds

Cyclopropylmethyl Derivatives

Two communications have appeared expressing different opinions regarding charge delocalization and participation by phenyl and cyclopropyl groups in carbonium ions. Whilst ^{13}C studies indicate more delocalization into the former group,[127] solvolysis studies confirm the larger rate enhancement of the latter group in similar systems.[128] Olah and Westerman[127] have suggested that no conflict exists here but that the cyclopropyl group may be more effective in participation whilst a phenyl group may provide greater delocalization.

An NMR study has clearly distinguished between the stabilization provided by cyclopropyl compared to other cycloalkyl and alkyl groups.[129] Evidence has been provided that a bicyclo[3.1.0]hex-2-yl cation is the primary rearrangement product in the conversion of cyclopropyl-substituted allylic cations into cyclohexenyl and dienylic cations;[130] the reaction has been described as an allowed $\pi 2a + \sigma 2a$ or $\pi 2s + \sigma 2a$ cycloaddition.

A significant reactivity difference has been found in the solvolysis of (**118**) and (**119**) in 70% aqueous acetone. The enhanced reactivity of (**118**) (a factor of 160) was attributed to conjugative factors.[131] A much lower energy difference was estimated for the cycloheptatriene–norcaradiene equilibrium than previously (see page 265).

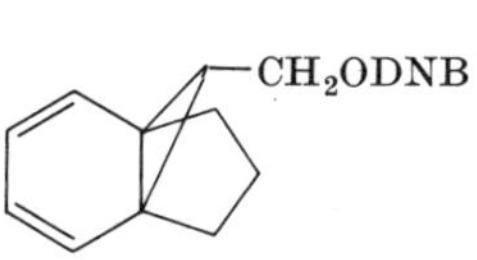

CH_2ODNB DNBOCH_2

(**118**) 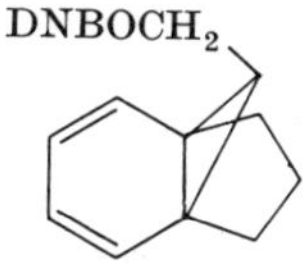 (**119**)

A common intermediate is indicated for solvolyses of (**120**), (**121**) and (**122**) which leads to formation of the same product mixture consisting of (**120**-OH) (78%) and (**122**-OH) (22%).[132]

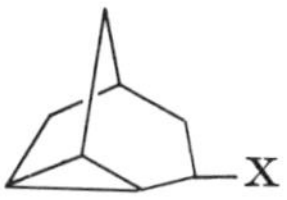

X X X

(**120**) 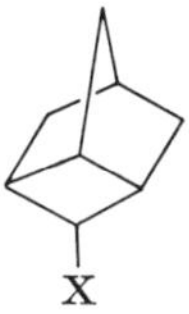 (**121**)  (**122**)

Although the rates of solvolysis of (**123**) and (**124**) were similar, the different products indicated that ring expansion of a cyclobutylmethyl- to a cyclopentyl-type cation is extremely selective.[133] It was suggested that the cations (**125**) and (**127**) were formed by rearrangement and that stabilization by the cyclopropyl ring could occur for (**125**) ($\equiv$**126**).

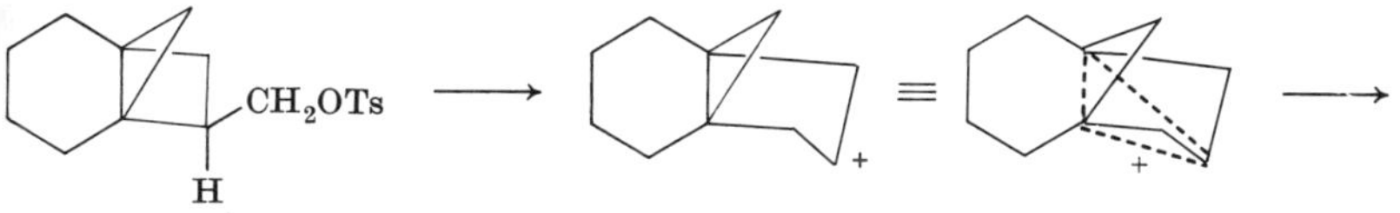

CH_2OTs H ⟶ ≡ ⟶

(**123**) (**125**) (**126**)

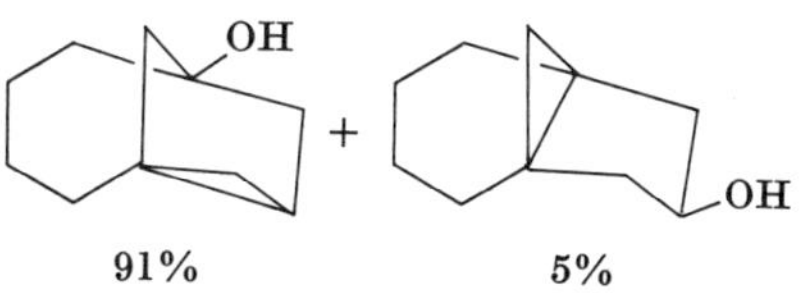

OH + OH

91% 5%

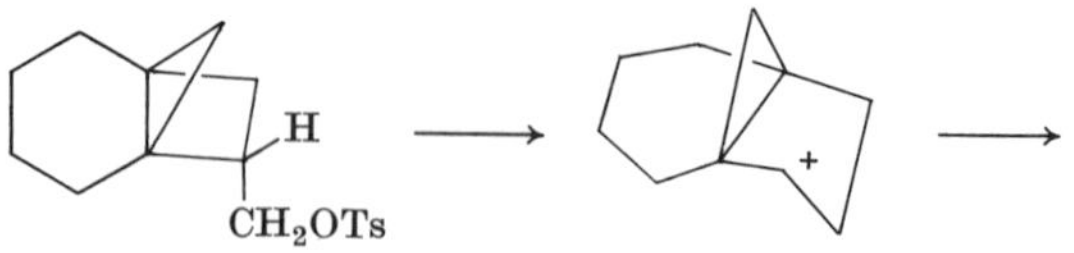

(124) (127)

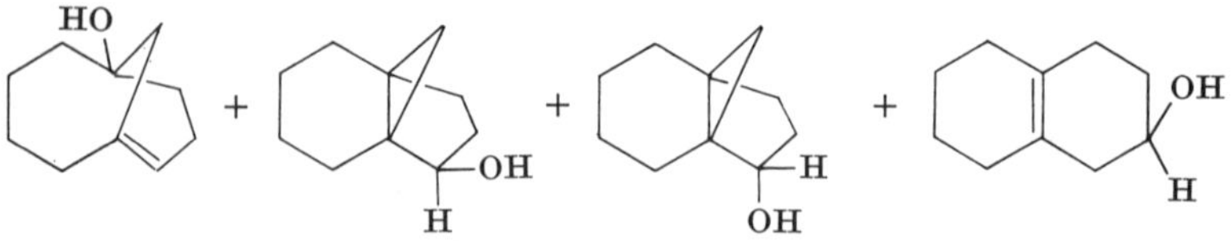

56% 12% 12% 11%

Secondary deuterium isotope effects have been reported for the solvolysis of cyclobutyl and cyclopropylmethyl methanesulphonates.[134]

Participation by More Remote Cyclopropane Rings

Rate and product studies have indicated cyclopropyl participation in the solvolysis of (**128**), and (**129**) was suggested as the intermediate.[135] Rearrangement and participation was also observed in the solvolysis of (**130**) where the four-membered ring is involved.

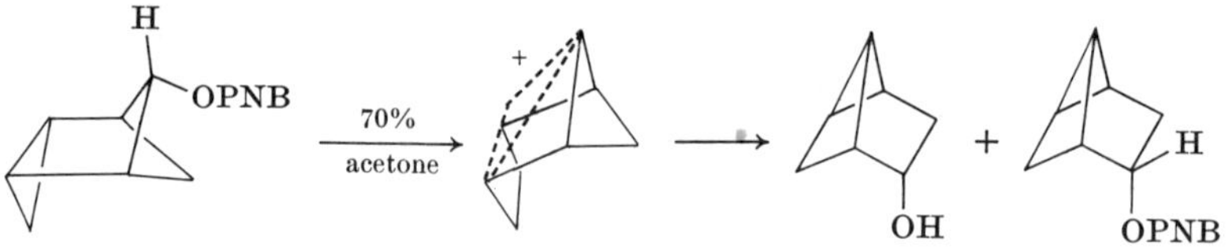

(128) (129)

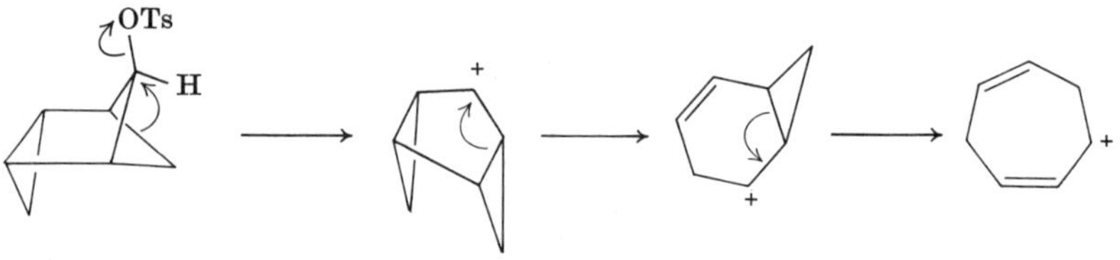

(130)

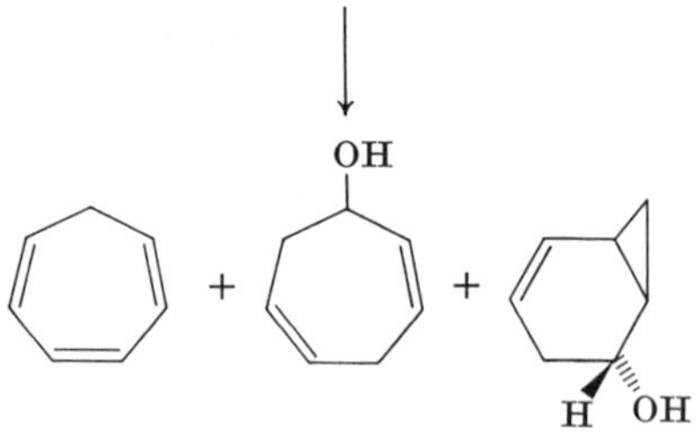

Rearrangement was observed in the hydrolysis of the ketal of *endo-exo*-tetracyclo-[3.3.1.0^{2,4}0^{6,8}]nonan-9-one[136] in contrast to earlier studies.[121] UV and CD evidence has been reported for the transmission of electronic effects through the cyclopropane ring in some arylcyclopropanes.[137]

Reactions of Cyclopropyl and Cyclobutyl derivatives

Reactions of alkylcyclopropanes with bromine in the presence of *N*-bromosuccinimide (a HBr scavenger) are reported.[138] Ring opening and rearrangement of the intermediate cations occurs; a mixture of 1,4-, 1,5- and 1,6-dibromoalkanes is produced from *n*-butylcyclopropane.

A detailed investigation has been made into silver-assisted solvolysis of the dibromides (**131**) and (**132**);[139] the participation of an intermediate with a bridgehead double bond has been established.

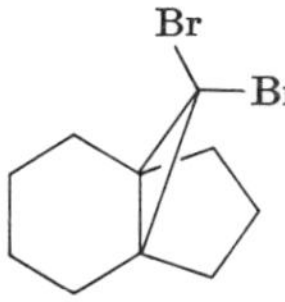

(**131**) 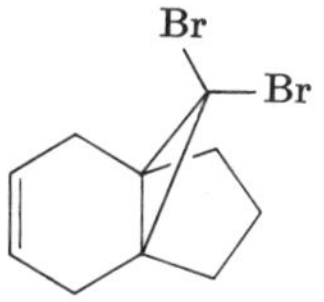 (**132**)

(**132**) $\xrightarrow{Ag^+}$ $\xrightarrow{H^+}$ $\longrightarrow$ CHBr + HO Br H OH

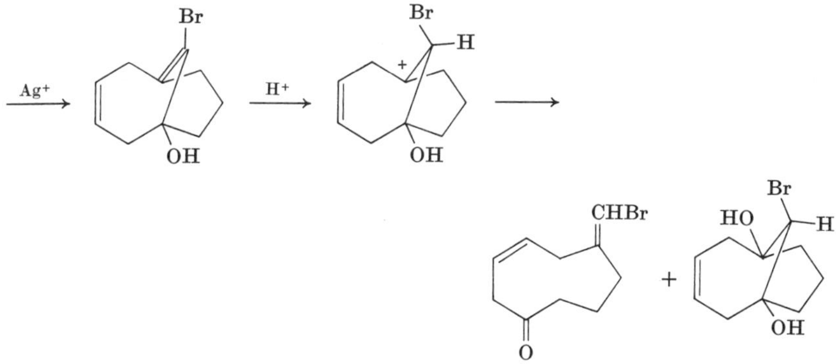

The solvolysis of 2,3-diphenylcyclopropyl chlorides,[140] the *cis-trans*-isomerization of 1,2-diphenylcyclopropane in the presence of acid[141] and the dimerization of dimethylcyclopropanes[142] have also been studied.

The geometries and energies of various $C_3H_5^+$ cations have been determined by *ab initio* molecular-orbital methods.[143] Solvolysis and racemization of cyclopropanes have been discussed in terms of carbonium ion–carbanion intermediates.[144] Solvolytic rearrangement of 1,2,3-tri-*tert*-butyl-3-(dichloromethyl)cyclopropane and 1,2,3-tri-*tert*-butyl-*cis*-3,4-dichlorocyclobutane have been fully reported.[145]

Protonated Cyclopropane Intermediates[146]

More scrambling to C-3 than to C-4 has been found in the trifluoroacetolysis of [1-^{14}C]-propylmercuric perchlorate, indicating intervention of equilibrating edge-protonated cyclopropanes.[147,148] Therefore a discrepancy still exists between different sets of calculations for the respective stability of edge- and corner-protonated cyclopropanes.[149]

A protonated cyclopropane intermediate has been suggested for the formation of *trans*-1,2-dimethylcyclopropane on thermal decomposition of 2-methylbutyl chloroformate.[150]

In reactions of a number of cyclopropanes with mercuric acetate and trifluoroacetate in methanol it has been shown that: (1) the stereochemistry of attack by an electrophile is controlled by its attack on the least substituted bond of the ring; (2) the nucleophile reacts with inversion, and (3) inversion predominates slightly in electrophilic attack in symmetrical systems.[151] The results are consistent with edge attack on the ring, giving a corner-protonated cyclopropane.

Extended Hückel calculations have been made for electrophilic ring-opening of substituted cyclopropanes.[152] *exo*-Norbornyl formate is the sole main product on addition of formic acid to nortricyclene.[153]

Metallocenylmethyl Cations and Other Derivatives

The structure of the α-ferrocenyl carbonium ions has been intensively examined by a number of physical methods. On the basis of ^{1}H-NMR studies, the Cais model, which involves no movement of the Fe atom but a displacement of the cationic centre by a molecular distortion across C-2–C-5, has been favoured.[154,155] Studies by ^{13}C-NMR[156] and ^{1}H-NMR and Mössbauer spectroscopy[157] are reported, the pK_{R^+} of the cations being determined.[158] Diastereotopism of ^{1}H and ^{13}C nuclei was observed in the 1-ferrocenyl-2-methylpropyl cation due to its planar chirality.[159] Enantiomeric forms of the 1-ferrocenylmethyl cation have been generated in acidic solvents, and the free-energy barrier to rotation around the Fc–$\overset{+}{\mathrm{C}}$HMe bond has been determined.[160]

The rearrangement of (**133**) to (**135**) has been suggested to involve the short-lived *spiro*-diene intermediate (**134**) rather than sequential methyl and hydrogen shifts; double-shift rearrangements of other secondary ferrocenylcarbonium ions have also been observed.[161]

Fc—C$^+$(H)(CMe$_3$) (**133**) ⟶ (**134**) [Fe$^+$, H] ⟶ Fc—$\overset{+}{\mathrm{C}}$(Me)—CHMe$_2$ (**135**)

Other studies include the intramolecular cyclization of 1,1′-bis(α-hydroxyisopropyl)-ferrocene,[162] solvolysis of cumyl chlorides co-ordinated with the chromium tricarbonyl group,[163] and an investigation into the stability of α-(alkynyl)dicobalt hexacarbonyl carbonium ions.[164]

By a study of acid-catalysed equilibration of a series of 1,3-diarylallyl alcohols and 3-aryl-1-ferrocenylallyl ethers it has been shown that greater thermodynamic stability was obtained by conjugation of a carbon–carbon double bond with a ferrocenyl group rather than with an aryl group.[165]

Stable Carbonium Ions and their Reactions

The tetracyclic alcohol (**138**) has been shown to ionize in FSO_3H–SO_2ClF at temperatures below −50° to the pyramidal cation (**139**),[166] which is an analogue of $(CH)_5^+$ and of (**137**) which it had earlier been suggested might be formed by loss of X^- from (**136**).[167] In connection with this, the crystal structure of 2,3,6,7,7,8-hexamethyl-1,5-diphenyl-tetracyclo[3.3.0.0^{2,8}0^{3,6}]octan-4-one has been determined.[168]

Interest in $(CH)_5^+$ has been stimulated from an experimental and theoretical point of view. Thus, CNDO calculations have indicated that (**137**) should be the most stable

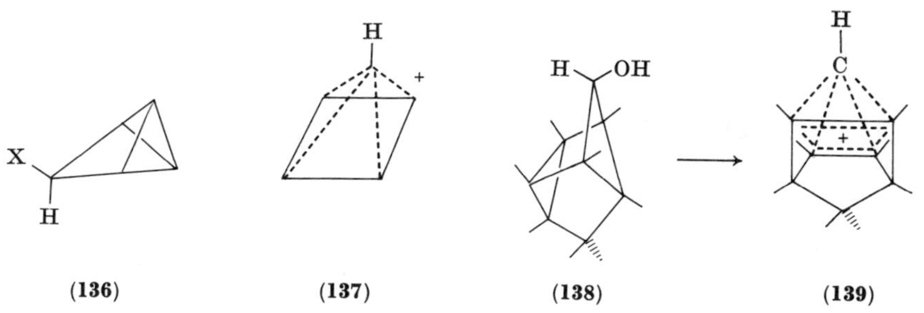

(136) **(137)** **(138)** **(139)**

structure on the $(CH)_5^+$ energy hyperplane, but that there is also a second energy minimum which could be that for (**140**), (**141**) or (**142**); (**143**) and (**144**) were predicted not to be stable species.[169] In contrast, a MINDO/3 study indicated a distorted form of (**141**) to be more stable than (**137**).[170]

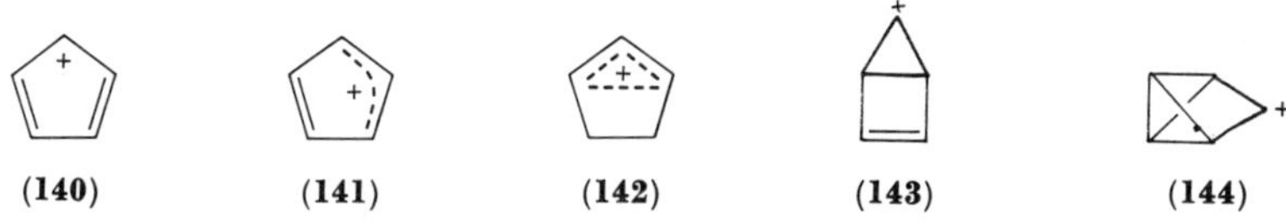

(140) **(141)** **(142)** **(143)** **(144)**

Experimental results have so far proved inconclusive regarding the intervention of an intermediate analogous to (**137**). In aprotic solvents (**145**) rearranges rapidly to (**146**) and in MeOH further reaction occurs to yield (**147**) and (**149**) (the latter being shown to be formed from **148**);[171] a mechanism similar to that for the cyclopropylmethyl system was suggested as sufficient to explain the observations. Similarly, the NMR spectra of the series of alcohols analogous to (**145**), (**146**) and (**148**) in SO_2ClF–FSO_3H at −78° could be explained either as a solution of an intermediate equivalent to (**137**) or as a series of equilibrating structures (**150a–d**).[172]

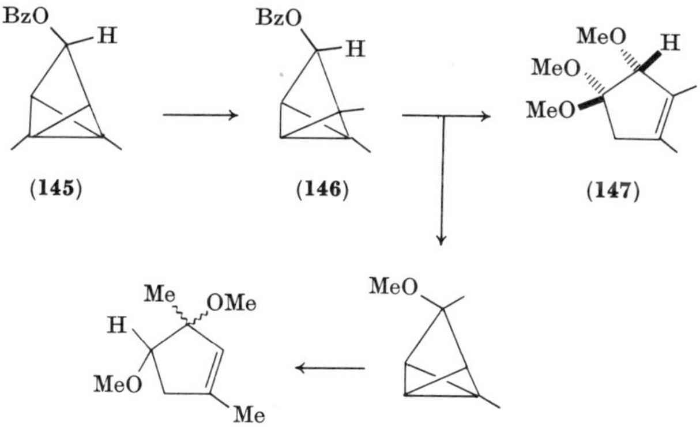

BzO H BzO H MeO MeO MeO H

(145) **(146)** **(147)**

Me OMe H MeO Me MeO

(149) **(148)**

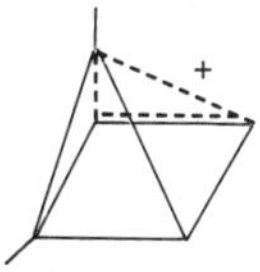

(150a) 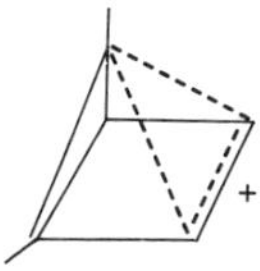 **(150b)** 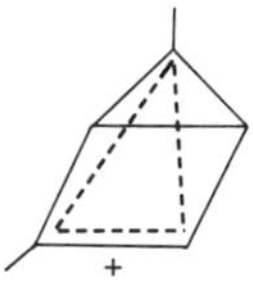 **(150c)** 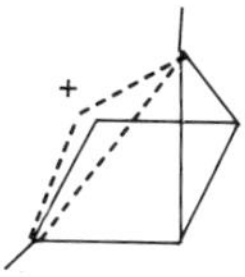 **(150d)**

Deuterium scrambling in the solvolysis of (**151**) has indicated formation of (**152**) and the ^{13}C-NMR and ^{1}H-NMR of the cation generated in super-acid is consistent with (**152**) or the equilibrating pair (**153**).[173]

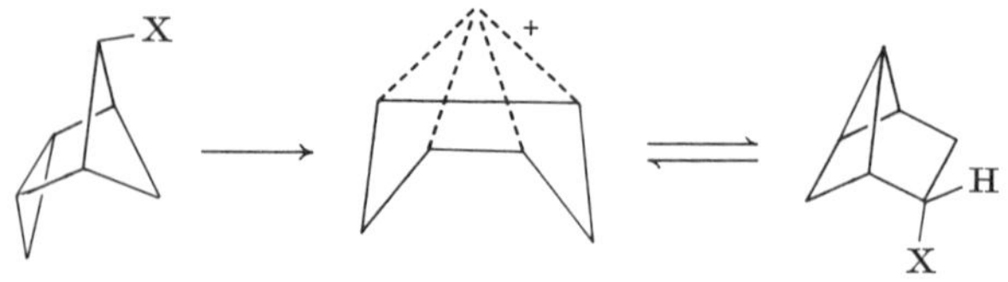

(**151**) (**152**)

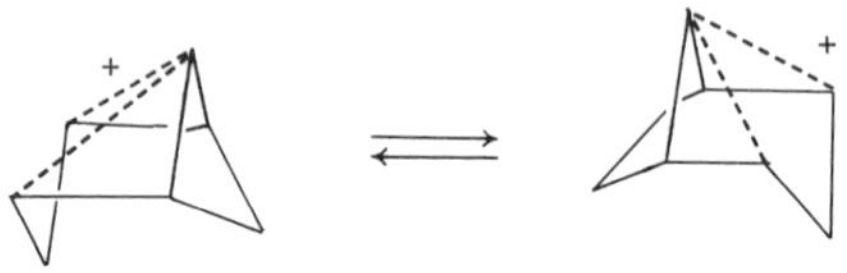

(**153a**) (**153b**)

The cation obtained by interaction of 5-bromocyclopentadiene with SbF_5 at 78°K has been shown to be a triplet.[174]

At temperatures above –50°, the cation (**139**) has been shown to rearrange to (**154**).[175] With (**155**) in FSO_3H–SO_2ClF at –90°, rearrangement to (**156**) takes place, followed by further rearrangement to the analogue of (**139**).[176] In the latter case the initial ionization must be involved with, or followed by, the 1,2-shift of a cyclobutane bond; ionization of (**138**) would involve, in contrast, participation of a cyclopropane bond. At temperatures between –50° and –100°, (**156**) undergoes a reversible exchange which equilibrates

(**139**) ⟶ 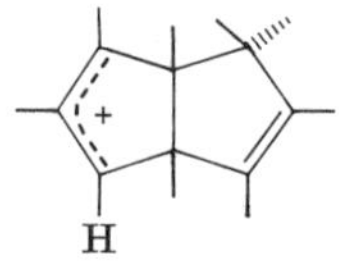 (**154**) (**155**) (**156**)

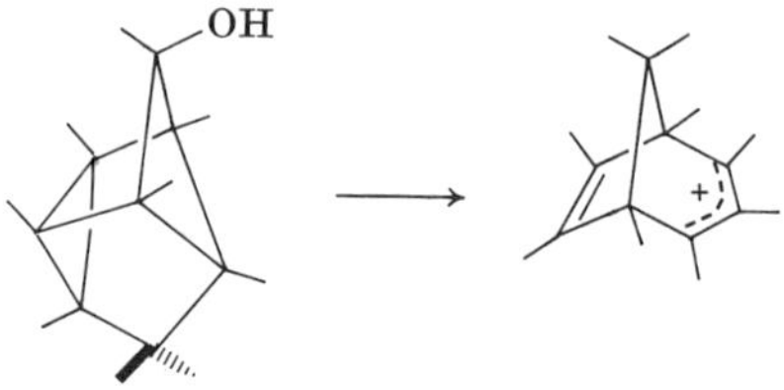

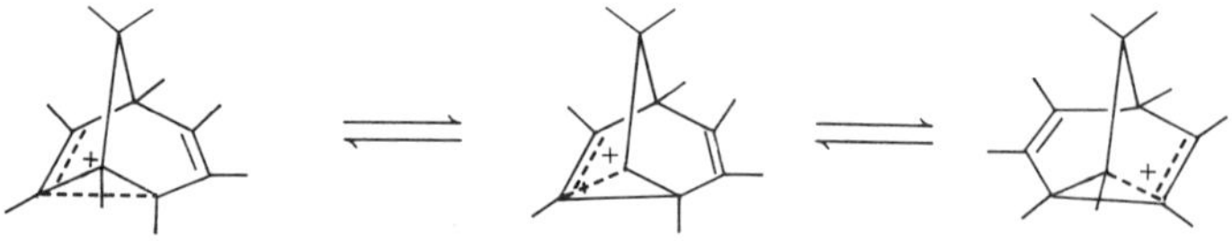

(**157a**) (**157b**) (**157c**)

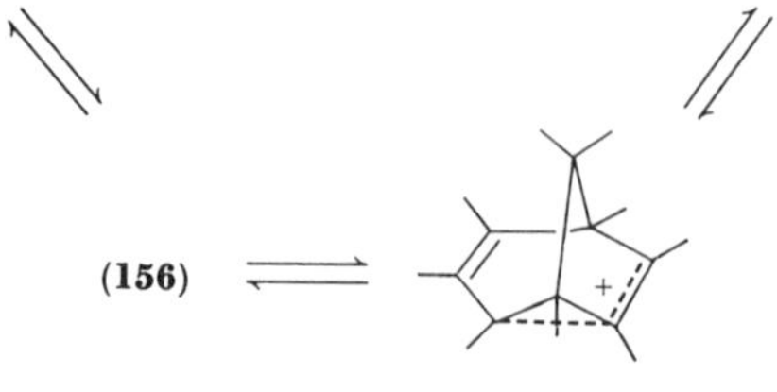

(**156**)

methyl groups at positions 2, 3, 4, 6 and 7 but does not involve the bridgehead methyl groups (1 and 5); a circumambulation process was suggested involving (**157**). The cation (**156**) was also shown to undergo a 1,2-methano-bridge shift in strong acid at a rate that is slow relative to circumambulation[177] but faster than a further rearrangement to a bicyclo[3.3.0]octadien-2-yl cation (analogous to **154**) which occurs irreversibly.[178]

Evidence has been presented that in strong acid solutions ions of the type (**158**) are converted into a stable dication (**159**) with approximate C_{5v} symmetry.[179] *Ab initio* molecular-orbital calculations have been made on this $(C\text{-}CH_3)_6{}^{2+}$ species.[180] PMR and CMR data have been obtained for a number of 5-substituted 1,2,3,4,5,6-hexamethylbicyclo[2.1.1]hexenyl cations, and the conclusion that the positive charge resides on C-2 and C-3 is in agreement with reaction of nucleophiles at these positions.[181] Reaction of (**159**) with triethylamine has been shown to yield 1,2,5,6-tetramethyl-3,4-dimethylenetricyclo[3.1.0.0^{2,6}]hexane.[182]

OH 2+

(**158**) (**159**)

The existence of barbaryl cations undergoing fast degenerate rearrangements has been proved by NMR studies.[183] The cations (**160**) and (**161**) have been generated from their corresponding chlorides in SbF_5–SO_2ClF at –78°.[184] Reaction with carbon monoxide followed by quenching with sodium methoxide in methanol gave a mixture of the corresponding carboxylic acid and methoxy derivatives. Evidence for formation of (**162**) from the dihalide was obtained but the spectroscopic and chemical observations were also consistent with the presence of a 5-chlorotrishomobarrelyl monocation with fast intermolecular exchange of halogen.

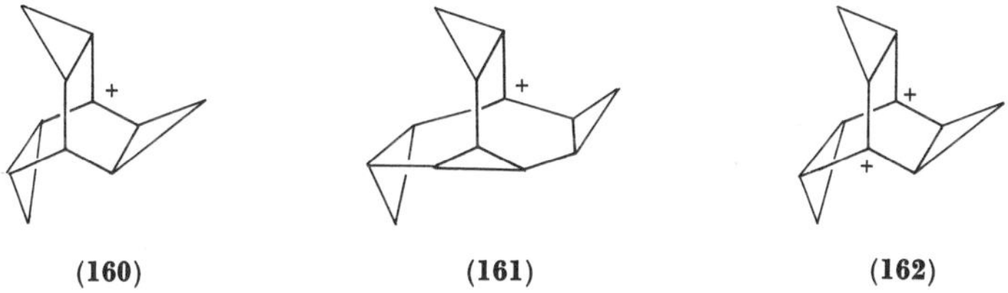

(**160**) (**161**) (**162**)

The 3-homoadamantyl cation has been generated in FSO_3H–SbF_5–SO_2ClF from 1-adamantylmethyl chloride and shown to be classical in nature and more stable than the 1-homoadamantyl cation.[185] Significant charge delocalization has been observed by CMR and PMR in 3-nortricyclyl cations even though bridging interactions must be extremely limited in these rigid systems;[186] it was considered that these cations may be limiting examples of charge delocalization with little nuclear movement, and the continuous nature of charge delocalization with or without nuclear reorganization was stressed. The absence of 1,2-hydrogen or methyl shifts has been reported for the 1-acenaphthenium ions, probably owing to the geometry of the system which prevents formation of the correct alignment without considerable strain.[187]

Evidence has been provided that protonation of *cis*-bicyclo[6.1.0]nonatriene involves initial formation of a *trans*-cation (**163**) which undergoes conformational inversion to the *cis*-analogue (**164**) under thermodynamic control;[188] stereoselectivity was not apparent in the initial protonation.

Ring inversion has been shown not to occur in the [4-^{2}H]homotropylium cation, and the minimum value $\Delta F^\ddagger$ for the "allowed" circumambulatory rearrangement was estimated as >27 kcal/mol.[189] *Ab initio* molecular-orbital theory has been applied to the modes of degenerate rearrangements in homotropylium and bicyclo[3.1.0]hex-3-en-2-yl cations.[190]

(**163**) (**164**)

Protonation of tricarbonyl-*cis*-cyclononatetraeneiron in FSO_3H–SO_2ClF occurs at C-6, the β-position of the double bond adjacent to the bound diene. This is analogous to the protonation of other cyclic triene and tetraene complexes.[191]

Isomeric 1,2,4,5-tetramethylpentadienyl cations have been observed in FSO_3H–SO_2ClF; these cyclize to cyclopentenyl cations from the U geometry consistent with Woodward–Hoffmann rules.[192] The *cis*-1,3,4,5-tetramethylcyclopentenyl cation rearranges to the *trans*-isomer and the 1,2,3,4-tetramethylcyclopentenyl cation in aqueous sulphuric acid by two different base-catalysed processes.[193]

Stereoselective formation of indanyl cations has been observed from acyclic phenyl-substituted allylic cations.[194] An NMR study of 2,4-dimethyl-2-pentyl, 2,5-dimethyl-2-hexyl and 2,6-dimethyl-2-heptyl cations in SO_2ClF has demonstrated the occurrence of direct 1,3-, and 1,4-, and 1,5-hydride shifts.[195] The magnitude of the energy barriers was 1,4 > 1,3 > 1,5 which was suggested to be a reflection of steric and conformational factors in the systems. A number of other hydride-transfer reactions have also been studied.[196–198]

The 1,4-bicyclo[2.2.2]octyl dication has been formed from 1,4-dichlorobicyclo[2.2.2]-octane in SbF_5–SO_2ClF.[199] Despite the two positive charges the bridgehead carbons are closer than in the parent hydrocarbon or monocation. This was attributed to hyperconjugative transfer of electron density to the positive centres which then participate in symmetry-allowed 1,4-bonding.

Formation of a [16]annulenediyl dication,[200] a fluorinated long-lived allyl cation,[201] a methylketenylacylium cation,[202] di- and tri-halonium ions[203] and the 8-cycloheptatrienylheptafulvenyl cation[204] have been reported.

A molecular-beam method for the preparation of carbonium-ion solutions has been developed.[205] Stable mercurinium ions[206] and halogenated phenyldifluorocarbenium ions[207] have been prepared; and other studies have dealt with the charge distribution in arylcyclobutenyl cations.[208] carbonylation in $SbCl_5$–SO_2 systems,[209] hydride-transfer reactions involving carbon–, boron–, and aluminium–hydrogen bonds to triphenylcarbonium ions,[210] and oxidation of alkanes to stable carbonium ions in SbF_5.[211]

Other Reactions

The reactivity variations in the solvolysis of aryldineopentylcarbinyl, aryl-*tert*-butylneopentylcarbinyl and aryldi-*tert*-butylcarbinyl *p*-nitrobenzoate have been discussed in terms of steric hindrance to resonance stabilization and relief of strain.[212] Steric retardation in acid-catalysed dehydration of alkyldi-*tert*-butylcarbinols in anhydrous acetic acid has been attributed to steric hindrance to solvation.[213] The rate enhancements observed in solvolysis of the corresponding *p*-nitrobenzoates were attributed to relief of front (*F*)-strain.

Remote inductive effects have been observed in solvolysis of 6-substituted *trans*-2*ax*-decalyl toluene-*p*-sulphonates.[214] Dipole–dipole interactions operating through space were suggested to be responsible for the different ρ^* values for the 6-equatorial and 6-axial derivatives. In the solvolysis of (**165**),[215] intervention of a classical ion rather than a bridged ion is indicated by the equal distribution of the trideuteriomethyl group in the diastereotopic positions in the product (**166**).

A small amount of a product (**168**) with inversion of configuration at C-13 has been observed in solvolysis of D-homo-5α-androstan-17$\alpha\beta$-yl toluene-*p*-sulphonate (**167**);[216] a similar result was found in solvolysis of (**169**) and a mechanism was proposed involving an initial 1,2-migration to give a C-homo-structure with a planar configuration at C-13.[217]

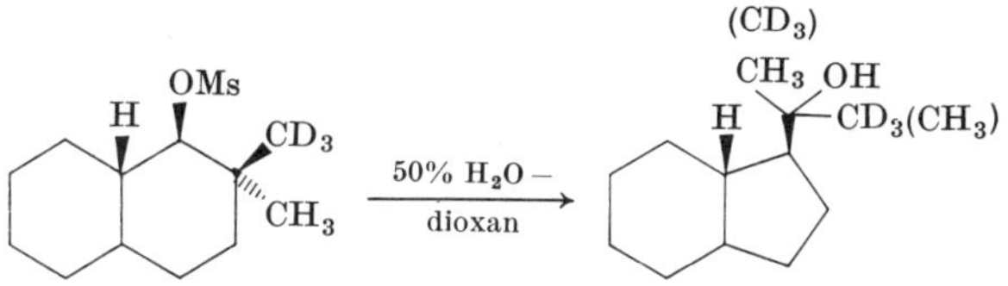

OMs H CD_3 CH_3 —50% H_2O–dioxan→ (CD_3) CH_3 OH H $CD_3(CH_3)$

(**165**) (**166**)

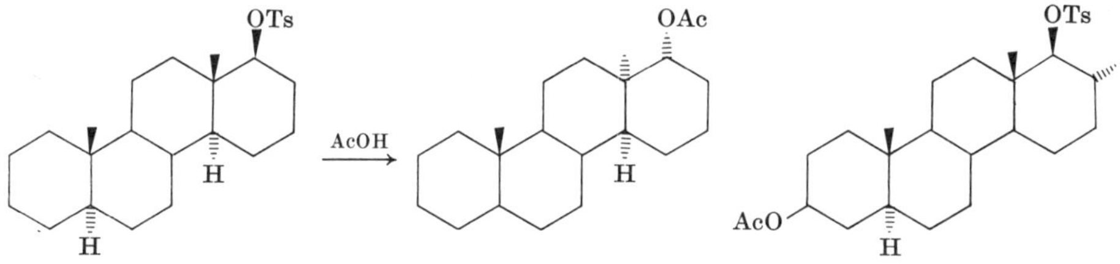

OTs H H —AcOH→ OAc H

(**167**) (**168**) (**169**)

AcO H OTs

The formolysis of a 20α-tosyloxy-steroid[218] and solvolysis of methyl 13-isopropyl-15β-tosyloxy-17-noratis-13-en-18-oate[219] have been investigated. The transmission of substituent effects has been examined in solvolysis of a number of heterocyclic derivatives.[220–226]

In a stereochemical study of product formation from 4-*tert*-butylcyclohexyl cations generated in a number of ways,[227] axial bond formation was found to be preferred in reactions in which the stereochemistry of the leaving group is unimportant. Similar products were obtained from acetolysis of *cis*- and *trans*-4-*tert*-butyl-1-cyclohexyl trifluoromethanesulphonate.[228]

The migratory aptitudes of methyl and ethyl groups in rearrangements of cyclohexadienones have been compared.[229] 3,5-Dinitrobenzoate is reported to be the most useful leaving group for the generation of nitrenium ions.[230]

Reactions of amines, sulphite ions and hypochlorite ions with triarylmethyl, tropylium and aryldiazonium ions in aqueous solution have been shown to be correlated by log $k_n = N_+ + \log k_0$.[231]

Theoretical calculations are reported on a number of alkyl carbonium ions,[232] the ethyl and hydroxymethyl carbonium ion,[233] arylmethyl cations [234] and a series of 2-substituted 1,3-dioxolan-2-ylium cations.[235]

The electrochemical determination of pk_R^+ for cyclopentadienyl cations is consistent with the antiaromaticity of $C_5H_5^+$.[236] Other reports describe a determination of pK_a's by reduction-potential measurements,[237] the ionization of triphenylmethanols in sulphuric acid,[238] the fluorine magnetic resonance spectra of some *p*-fluorophenyl carbonium ions,[239] the effect of fluorine on the stability of carbonium ions,[240] spectral characteristics of the odd polyenylcarbonium ions,[241] the effect of substituents on the stability of the methylene carbonium ion[242] and measurements of chlorine kinetic isotope effects.[243]

The synthesis of five- and seven-membered rings from allyl cations has been reviewed.[244] Formation of the 2-methoxyallyl cation from the corresponding halides is slow owing to the electron-withdrawing effect of the methoxy-group, but the cycloaddition of 2-methoxyallyl cation to furan and cyclopentadiene has been achieved under controlled experimental conditions.[245] Cycloaddition of cyclic oxyallyl ions, generated from dibromides in the presence of a zinc–copper couple, to furan and cyclopentadiene has also been reported.[246]

$\alpha\beta$-Unsaturated ketones are formed by anodic oxidation of alkanes in fluorosulphonic acid containing carboxylic acids.[247] The anodic oxidation of alkanes in acetonitrile and trifluoroacetic acid,[248] the anodic oxidation of 2-methoxyethanol,[249] the photolysis of triphenylmethyl and related cations in sulphuric acid,[250] and one-electron reduction of substituted tropylium ions with Cr(II) in 10% hydrochloric acid have been studied.[251] In the last case the rates correlate with the stability of the cations (pK_R^+), which suggests that the readiness to accept one electron is parallel to the reactivity of a cation towards the electron pair of a nucleophile.

Carbonium ion stabilities have been determined by cyclotron resonance spectroscopy.[252] Photodissociation of gaseous $C_7H_8^+$ cations[253] and toluene parent cations,[254] rearrangements and unimolecular decompositions of benzenoid $C_9H_{11}^+$ ions,[255] ion molecule reactions[256] of CH_5^+ and $C_2H_5^+$ and the reaction of CH_3^+ with D_2 have been investigated.[257] $C_7H_7^+$ has been shown to be formed by two different reaction pathways in an electron-impact study of benzyl phenylacetate.[258]

In the presence of trityl salts, alkyl trityl ethers disproportionate to aldehydes and ketones,[259] and alcohols can be converted into amides with chlorodiphenylmethylium hexachloroantimonate in nitrile solvents.[260]

Solvent participation has been studied in the acetolysis of ethers;[261] predominant retention observed in the silver ion-catalysed solvolysis of an alkyl bromide in acetonitrile.[262] Acetolysis of 3-equatorial-, 3-axial- or 1-axial-sulphonated 2-ketones of tetracyclic diterpenoids results in attack at the 1-axial position, and a common delocalized enol cation intermediate was suggested.[263]

Alkoxycarbonium ions have been observed in the acid-catalysed decomposition of organic hydroperoxides;[264] a transient intermediate detected during the fast polymerization of styrene by perchloric acid was interpreted as the polystyryl carbonium ion.[265]

Other reports discuss the rearrangement of benzobicyclo[3.2.2]nona-2,6-diene by aluminium chloride,[266] micellar catalysis for carbonium ion reactions,[267] radiochemical studies of reactions of free methyl cations with some methyl ethers,[268] hydroxymercuration of alkenes,[269] reactions of trifluoromethanesulphonates,[270] reactions of cyclic alcohols with boron trifluoride,[271] solvolysis of *tert*-butyl chloride in phenol,[272] Koch–Haaf carboxylations,[273] the effect of added anions on the rate of reduction of the tropylium ion with Cr(II)[274] use of silicon hydrides in carbonium ion reactions,[275] the mechanism and stereochemistry of 1,4-diol ring closure to tetrahydrofuran[276] and reactions of nucleophiles with π-allyliron cations.[277]

References

[1] "Carbonium Ions," G. A. Olah (Ed.), John Wiley & Sons Ltd., New York, 1973; G. A. Olah, "Carbocations and Electrophilic Reactions", *Angew. Chem. Internat. Ed*, **12**, 173 (1973); "Carbocation and Onium Ion Reagents", *Aldrichchimica Acta*, **6**, 7 (1973); *Am. Chem. Soc., Div. Petrol Chem., Prepr.*, **16**, C11 (1971); *Chem. Abs.*, **78**, 28633 (1973); N. D. Epiotis, "Sigmatropic Reactions and Ionic Rearrangements", *J. Am. Chem. Soc.*, **95**, 1206 (1973).

[2] H. C. Brown, *Accounts Chem. Research*, **6**, 377 (1973).

[3] E. N. Peters and H. C. Brown, *J. Am. Chem. Soc.*, **95**, 2397 (1973).

[4] E. N. Peters and H. C. Brown, *J. Am. Chem. Soc.*, **95**, 2399 (1973).

[5] C. J. Collins and C. K. Johnson, *J. Am. Chem. Soc.*, **95**, 4766 (1973).

[6] O. R. Rodig and R. J. Sysko, *J. Am. Chem. Soc.*, **94**, 6475 (1972).

[7] C. W. David, B. W. Everling, R. J. Kilian, J. B. Stothers and W. R. Vaughan, *J. Am. Chem. Soc.*, **95**, 1256 (1973).

[8] C. J. Collins and M. H. Lietzke, *J. Am. Chem. Soc.*, **95**, 6842 (1973).

[9] C. J. Collins and B. M. Benjamin, *J. Org. Chem.*, **37**, 4358 (1972); see *Org. Reaction Mech.*, **1970**, 5.

[10] L. Huang, K. Ranganayakakulu and T. S. Sorensen, *J. Am. Chem. Soc.*, **95**, 1936 (1973).

[11] N. H. Werstiuk, G. Timmins and F. P. Capelli, *Can. J. Chem.*, **51**, 3473 (1973).

[12] T. C. Morrill and B. E. Greenwald, *J. Org. Chem.*, **38**, 616 (1973).

[13] W. W. Schoeller and M. E. Hendrick, *Tetrahedron Letters*, **1973**, 2035.

[14] H. C. Brown, J. H. Kawakami and K. T. Liu, *J. Am. Chem. Soc.*, **95**, 2209 (1973); see *Org. Reaction Mech.*, **1970**, 9.

[15] M. Hanack and J. Dolde, *Ann. Chem.*, **1973**, 1557; see *Org. Reaction Mech.*, **1972**, 5.

[16] R. Baker and J. C. Salter, *J.C.S. Perkin II*, **1973**, 150; see *Org. Reaction Mech.*, **1970**, 9

[17] P. G. Gassman, J. L. Marshall and J. G. Macmillan, *J. Am. Chem. Soc.*, **95**, 6319 (1973).

[18] R. K. Howe and S. Winstein, *J. Org. Chem.*, **38**, 2797 (1973).

[19] P. J. Kropp, T. H. Jones and G. S. Poindexter, *J. Am. Chem. Soc.*, **95**, 5420 (1973).

[20] K. Okamoto, T. Kinoshita and Y. Ito, *Bull. Chem. Soc. Japan*, **46**, 2905 (1973).

[21] E. W. Robb and E. K. Onsager, *J. Org. Chem.*, **37**, 4013 (1972).

[22] G. Nagendrappa and K. Griesbaum, *Chem. Ind.* (*London*), **1973**, 902.

[23] R. F. C. Cole, J. M. Coxon and M. P. Hartshorn, *Austral. J .Chem.*, **26**, 1595 (1973).

[24] D. R. Dimmel and W. Y. Fu, *J. Org. Chem.*, **38**, 3778 (1973).

[25] E. C. Friedrich, M. A. Saleh and S. Winstein, *J. Org. Chem.*, **38**, 860 (1973).

[26] E. C. Friedrich and M. A. Saleh, *J. Am. Chem. Soc.*, **95**, 2617 (1973).

[27] K. B. Wiberg, K. C. Bishop and R. B. Davidson, *Tetrahedron Letters*, **1973**, 3170.

[28] G. Seybold, D. Vogel, M. Saunders and K. B. Wiberg, *J. Am. Chem. Soc.*, **95**, 2045 (1973).

[29] H. Hogeveen and P. W. Kwant, *Tetrahedron Letters*, **1972**, 5357.

[30] H. Hogeveen and P. W. Kwant, *Tetrahedron Letters*, **1973**, 423.

[31] H. Hogeveen and P. W. Kwant, *Tetrahedron Letters*, **1973**, 1351.

[32] D. C. Sanders and H. Schechter, *J. Am. Chem. Soc.*, **95**, 6858 (1973).

[33] L. A. Paquette and G. L. Thompson, *J. Am. Chem. Soc.*, **95**, 2364 (1973).

[34] C. B. Quinn and J. R. Wiseman, *J. Am. Chem. Soc.*, **95**, 1342 (1973).

[35] L. Stéhelin, L. Kanellias and G. Ourisson, *J. Org. Chem.*, **38**, 847, 851 (1973).

[36] L. A. Paquette, S. E. Wilson, G. Zon and J. A. Schwartz, *J. Am. Chem. Soc.*, **95**, 9222 (1972).

[37] P. G. Gassman and R. K. Reitz, *J. Am. Chem. Soc.*, **95**, 3057 (1973).

[38] W. G. Dauben, A. J. Kielbania and K. N. Raymond, *J. Am. Chem. Soc.*, **95**, 7166 (1973).

[39] G. Zon and L. A. Paquette, *J. Am. Chem. Soc.*, **95**, 4456 (1973).

[40] A. Wissner and J. Meinwald, *J. Org. Chem.*, **38**, 1697 (1973).

[41] R. Noyori, *Tetrahedron Letters*, **1973**, 1691.

[42] R. D. Fischer, T. D. Bogard and P. Kovacic, *J. Am. Chem. Soc.*, **95**, 3646 (1973).

[43] F. Cacace, A. Guarino and M. Speranza, *J. C.S. Perkin II*, **1973**, 66.

[44] W. Kirmse and J. Alberti, *Chem. Ber.*, **106**, 236 (1973).

[45] L. A. Paquette, M. J. Broadhurst, L. K. Read and J. Clardy, *J. Am. Chem. Soc.*, **95**, 4639 (1973).

[46] L. A. Paquette, M. J. Broadhurst, C. Lee and J. Clardy, *J. Am. Chem. Soc.*, **95**, 4647 (1973).

[47] L. A. Paquette and M. J. Broadhurst, *J. Org. Chem.*, **38**, 1893 (1973).

[48] J. M. Harris, A. Becker and D. C. Clark, *Tetrahedron Letters*, **1973**, 3813.

[49] D. N. Kevill, K. C. Kolwyck, D. M. Shold and C. B. Kim, *J. Am. Chem. Soc.*, **95**, 6022 (1973).

[50] D. Lenoir, *Chem. Ber.*, **106**, 2366 (1973).

[51] D. Lenoir, *Chem. Ber.*, **106**, 78 (1973).

[52] D. Faulkner, M. A. McKervey, D. Lenoir, C. A. Senkler and P. von R. Schleyer, *Tetrahedron Letters*, **1973**, 705.

[53] J. E. Nordlander, R. R. Gruetzmacher and F. Miller, *Tetrahedron Letters*, **1973**, 927.

[54] E. M. Engler, M. Farcasiu, A. Sevin, J. M. Cense and P. von R. Schleyer, *J. Am. Chem. Soc.*, **95**, 5769 (1973).

[55] W. D. Graham, P. von R. Schleyer, E. W. Hagaman and E. Wenkert, *J. Am. Chem. Soc.*, **95**, 5785 (1973).

[56] E. Boelema, J. H. Wieringa, H. Wynberg and J. Strating, *Tetrahedron Letters*, **1973**, 2377.

[57] V. R. Koch and L. L. Miller, *Tetrahedron Letters*, **1973**, 693.

[58] L. L. Miller, V. R. Koch, T. Koenig and M. Tuttle, *J. Am. Chem. Soc.*, **95**, 5075 (1973).

[59] J. W. Strand and P. Kovacic, *J. Am. Chem. Soc.*, **95**, 2977 (1973).

[60] J. W. McKinley, R. E. Pincock and W. B. Scott, *J. Am. Chem. Soc.*, **95**, 2030 (1973).

[61] W. B. Scott and R. E. Pincock, *J. Am. Chem. Soc.*, **95**, 2040 (1973).

[62] N. V. Averina and N. S. Zefirov, *Chem. Comm.*, **1973**, 197.

[63] J. A. Peters and H. Van Bekkum, *Rec. Trav. chim.*, **63**, 379 (1973).

[64] R. K. Murray and T. K. Morgan, *Tetrahedron Letters*, **1973**, 3299.

[65] J. S. Wishnok, P. von R. Schleyer, E. Funke, G. D. Pandit, R. O. Williams and A. Nickon, *J. Org. Chem.*, **38**, 539 (1973).

[66] R. M. Coates and K. Yano, *J. Am. Chem. Soc.*, **95**, 2203 (1973).

[67] M. J. Goldstein and S. A. Kline, *J. Am. Chem. Soc.*, **95**, 935 (1973).

[68] P. G. Gassman and X. Creary, *J. Am. Chem. Soc.*, **95**, 6852 (1973).

[69] J. W. Wilt and T. P. Malloy, *J. Org. Chem.*, **38**, 277 (1973).

[70] P. G. Gassman, R. N. Steppel and E. A. Armour, *Tetrahedron Letters*, **1973**, 3287.

[71] L. A. Paquette, O. Cox, M. Oku, R. P. Henzel and J. A. Schwartz, *Tetrahedron Letters*, **1973**, 3295.

[72] P. K. Freeman and B. K. Stevenson, *J. Am. Chem. Soc.*, **95**, 2890 (1973); see *Org. Reaction Mech.*, **1970**, 8.

[73] S. J. Cristol, G. C. Schloemer, D. R. James and L. A. Paquette, *J. Org. Chem.*, **37**, 3852 (1972).

[74] A. S. Kende and J. Y. C. Chu, *J. Org. Chem.*, **38**, 2252 (1973).

[75] R. Bishop and W. Parker, *Tetrahedron Letters*, **1973**, 2375.

[76] N. P. Peet, R. L. Cargill and D. F. Bushey, *J. Org. Chem.*, **38**, 1218 (1973).

[77] D. Baines, C. Eck and W. Parker, *Tetrahedron Letters*, **1973**, 3933.

[78] C. H. M. Adams, D. J. Cawley and K. Mackenzie, *J.C.S. Perkin II*, **1973**, 909.

[79] T. Ando, N. Shimizu, S. G. Kim, Y. Tsuno and Y. Yukawa, *Tetrahedron Letters*, **1973**, 117.

[80] Y. Yukawa, S. G. Kim and H. Yamataka, *Tetrahedron Letters*, **1973**, 373.

[81] H. Yamataka, S. G. Kim, T. Ando and Y. Yukawa, *Tetrahedron Letters*, **1973**, 4767.

[82] F. L. Schadt and P. von R. Schleyer, *J. Am. Chem. Soc.*, **95**, 7860 (1973).

[83] D. S. Noyce, R. L. Castenson and D. A. Meyers, *J. Org. Chem.*, **37**, 4222 (1972).

[84] C. C. Lee and D. Unger, *Canad. J. Chem.*, **51**, 1494 (1973).

[85] J. Laureillard, A. Laurent and E. Laurent, *Bull. Soc. Chim. France*, **1973**, 232.

[86] J. Laureillard, A. Laurent and E. Laurent, *Bull. Soc. Chim. France*, **1973**, 242.

[87] V. G. Shubin, D. V. Korchagina, G. I. Borodkin, B. G. Derendyaev and V. Koptyug, *Zh. Org. Khim.*, **9**, 1031 (1973); *Chem. Abs.*, **79**, 41567 (1973); *Tetrahedron Letters*, **1973**, 539.

[88] C. A. Kingsbury and D. C. Best, *Bull. Chem. Soc. Japan*, **45**, 3440 (1972).

[89] R. N. McDonald, N. L. Wolfe and H. E. Petty, *J. Org. Chem.*, **38**, 1106 (1973).

[90] D. S. Noyce and R. L. Castenson, *J. Am. Chem. Soc.*, **95**, 1247 (1973).

[91] T. J. Mason and R. O. C. Norman, *J.C.S. Perkin II*, **1973**, 1840.

[92] L. A. Paquette and I. R. Dunkin, *J. Am. Chem. Soc.*, **95**, 3067 (1973).

[93] R. S. Bly, R. A. Mateer, K. K. Tse and R. L. Veazey, *J. Org. Chem.*, **38**, 1518 (1973).

[94] C. A. Bunton and S. K. Huang, *J. Am. Chem. Soc.*, **95**, 2701 (1973).

[95] A. A. Khalaf and R. M. Roberts, *J. Org. Chem.*, **37**, 4227 (1972).

[96] B. E. McCarry, R. L. Markezich and W. S. Johnson, *J. Am. Chem. Soc.*, **95**, 4416 (1973).

[97] A. G. Gonzalez, J. D. Martin and M. L. Rodriquez, *Tetrahedron Letters*, **1973**, 3657.

[98] P. A. Bartlett and W. S. Johnson, *J. Am. Chem. Soc.*, **95**, 7501 (1973).

[99] P. A. Bartlett, J. I. Brauman, W. S. Johnson and R. A. Volkmann, *J. Am. Chem. Soc.*, **95**, 7502 (1973).

[100] P. M. McCurry and R. K. Singh, *Tetrahedron Letters* **1973**, 1155.

[101] A van der Gen, K. Wiedhaup, J. J. Swoboda, H. C. Dunathan and W. S. Johnson, *J. Am. Chem. Soc..*, **95**, 2656 (1973).

[102] J. B. Lambert, A. P. Jovanovich, J. W. Hamersma, F. R. Koeng and S. S. Oliver, *J. Am. Chem. Soc.*, **95**, 1570 (1973); see also *Org. Reaction Mech.*, **1970**, 26.

[103] P. G. Gassman and J. M. Pascone, *J. Am. Chem. Soc.*, **95**, 7801 (1973).
[104] H. Hogeveen and P. W. Kwant, *Tetrahedron Letters*, **1972**, 5361.
[105] M. H. Sekera, B. A. Weissman and R. G. Bergman, *Chem. Comm.*, **1973**, 679.
[106] M. Bertrand and C. Santelli-Rouvier, *Bull. Soc. Chim. France*, **1973**, 1800.
[107] T. J. Mason, M. J. Harrison, J. A. Hall and G. D. Sargent, *J. Am. Chem. Soc.*, **95**, 1849 (1973).
[108] P. T. Lansbury and R. C. Stewart, *Tetrahedron Letters*, **1973**, 1569.
[109] M. Iguchi, M. Niwa and S. Yamamura, *Tetrahedron Letters*, **1973**, 1687.
[110] E. J. Corey and R. D. Balanson, *Tetrahedron Letters*, **1973**, 3153.
[111] E. J. Corey, G. W. J. Fleet and M. Kato, *Tetrahedron Letters*, **1973**, 3963.
[112] M. Tanida, S. Yamamoto and K. Takada, *J. Org. Chem.*, **38**, 2792 (1973).
[113] M. Voicu, F. Badea and A. Voicu, *Rev. Roum. Chim.*, **18**, 131 (1973).
[114] K. E. Harding, R. C. Ligon, C. Y. Tseng and T. C. Wu, *J. Org. Chem.*, **38**, 3478 (1973).
[115] M. Julia and E. C. Gazquez, *Bull. Soc. Chim. France*, **1973**, 1796.
[116] K. E. Harding and W. D. Nash, *Tetrahedron Letters*, **1972**, 4973.
[117] L. A. Paquette and K. A. Henzel, *J. Am. Chem. Soc.*, **95**, 2724 (1973).
[118] L. A. Paquette and K. A. Henzel, *J. Am. Chem. Soc.*, **95**, 2726 (1973).
[119] S. J. Cristol and D. A. Beimborn, *J. Am. Chem. Soc.*, **95**, 3651 (1973).
[120] D. Wege and S. P. Wilkinson, *Austral. J. Chem.*, **26**, 1751 (1973).
[121] G. Lamaty, A. Malaval, J. D. Roque and P. Geneste, *Bull. Soc. Chim. France*, **1972**, 4563; see also *ibid.*, **1972**, 4567.
[122] P. Schipper, W. A. M. Castenmiller, J. W. de Haan and H. M. Buck, *Chem. Comm.*, **1973**, 574.
[123] W. J. Hehre, *J. Am. Chem. Soc.*, **95**, 5807 (1973).
[124] W. S. Johnson, M. B. Gravestock, R. J. Parry and D. A. Okorie, *J. Am. Chem. Soc.*, **94**, 8604 (1972).
[125] D. R. Morton, M. B. Gravestock, R. J. Parry and W. S. Johnson, *J. Am. Chem. Soc.*, **95**, 4417 (1973).
[126] D. R. Morton and W. S. Johnson, *J. Am. Chem. Soc.*, **95**, 4419 (1973).
[127] G. A. Olah and P. W. Westerman, *J. Am. Chem. Soc.*, **95**, 7530 (1973).
[128] H. C. Brown and E. N. Peters, *J. Am. Chem. Soc.*, **95**, 2400 (1973).
[129] T. S. Sorensen, I. J. Miller and K. Ranganayakulu, *Austral. J. Chem.*, **26**, 311 (1973).
[130] K. Rajeswari and T. S. Sorensen, *J. Am. Chem. Soc.*, **95**, 1239 (1973).
[131] P. Warner and S. L. Lu, *J. Am. Chem. Soc.*, **95**, 5099 (1973).
[132] M. Geisel, C. A. Grob, W. Santi and W. Tschudi, *Helv. Chim. Acta*, **56**, 1055 (1973).
[133] P. G. Gassman and E. A. Armour, *J. Am. Chem. Soc.*, **95**, 6129 (1973).
[134] B. Goricnik, Z. Majerski, S. Borcic and D. E. Sunko, *J. Org. Chem.*, **38**, 1881 (1973).
[135] P. G. Gassman and X. Creary, *J. Am. Chem. Soc.*, **95**, 2729 (1973).
[136] R. Bicker, H. Kessler and A. Steigel, *Tetrahedron Letters*, **1973**, 2371.
[137] G. Montaudo and C. G. Overberger, *J. Org. Chem.*, **38**, 804 (1973).
[138] J. C. Day, K. J. Shea and P. S. Skell, *J. Am. Chem. Soc.*, **95**, 5089 (1973).
[139] P. Warner, J. Fayos and J. Clardy, *Tetrahedron Letters*, **1973**, 4473; cf. *Org. Reaction Mech.*, **1972**, 35.
[140] J. W. Hausser and J. T. Uchic, *J. Org. Chem.*, **37**, 4087 (1972).
[141] I. P. Stepanova, N. M. Loim, Z. N. Parnes and Y. S. Shabarov, *Zh. Org. Khim.*, **9**, 521 (1973): *Chem. Abs.*, **78**, 158691 (1973).
[142] J. Tanaka, T. Katagiri, K. Takabe and T. Sakamoto, *Bull. Chem. Soc. Japan*, **45**, 3147 (1972).
[143] L. Radom, P. C. Hariharan, J. A. Pople and P. von R. Schleyer, *J. Am. Chem. Soc.*, **95**, 6531 (1973).
[144] E. W. Yankee, F. D. Badea, N. E. Howe and D. J. Cram, *J. Am. Chem. Soc.*, **95**, 4210 (1973).
[145] A. E. Feiring and J. Ciabattoni, *J. Am. Chem. Soc.*, **95**, 5266 (1973).
[146] "*Evidence for Protonated Cyclopropane Intermediates from Studies of Stable Solutions of Carbonium Ions*," M. Saunders, P. Vogel, E. L. Hagan and J. Rosenfeld, *Accounts Chem. Research*, **6**, 53 (1973).
[147] C. C. Lee, S. Vassie and E. C. F. Ko, *J. Am. Chem. Soc.*, **94**, 8931 (1972).
[148] C. C. Lee, A. J. Cessna, E. C. F. Ko and S. Vassie, *J. Am. Chem. Soc.*, **95**, 5688 (1973).
[149] See *Org. Reaction Mech.*, **1972**, 37.
[150] W. E. Dupuy, E. A. Goldsmith and H. R. Hudson, *J.C.S. Perkin II*, **1973**, 74.
[151] C. H. DePuy and R. H. McGirk, *J. Am. Chem. Soc.*, **95**, 2366 (1973).
[152] A. H. Andrist, *J. Am. Chem. Soc.*, **95**, 7531 (1973).
[153] J. Paasivirta, *Acta Chem. Scand.*, **27**, 374 (1973).
[154] R. G. Sutherland, J. R. Sutton and W. M. Horspool, *Tetrahedron Letters*, **1973**, 3283.
[155] T. D. Turbitt and W. E. Watts, *J. Organometal. Chem.*, **49**, C30 (1973).
[156] G. H. Williams, D. D. Traficante and D. Seyferth, *J. Organometal. Chem.*, **60**, C53 (1973).
[157] J. J. Dannenberg, M. K. Levenberg and J. H. Richards, *Tetrahedron*, **29**, 1575 (1973).
[158] J. Tirouflet, E. Laviron, C. Moise and Y. Mugnier, *J. Organometal. Chem.*, **50**, 241 (1973).

[159] V. I. Sokolov, P. V. Petrovskii and O. Reutov, *J. Organometal. Chem.*, **59**, C27 (1973).
[160] T. D. Turbitt and W. E. Watts, *Chem. Comm.*, **1973**, 182; N. M. D. Brown, T. D. Turbitt and W. E. Watts, *J. Organometal. Chem.*, **46**, C19 (1972).
[161] T. D. Turbitt and W. E. Watts, *J. Organometal. Chem.*, **57**, C78 (1973).
[162] M. Hisatome and K. Yamakawa, *Chem. Comm.*, **1973**, 199.
[163] A. Z. Kreindlin, *Izv. Akad. Nauk SSSR, Ser. Khim.*, **1973**, 952; *Chem. Abs.*, **79**, 41585 (1973).
[164] K. M. Nicholas and R. Pettit, *J. Organometal. Chem.*, **44**, C21 (1972).
[165] A. M. Easton, M. J. A. Habib, J. Park and W. Watts, *J.C.S. Perkin II*, **1972**, 2290.
[166] H. Hart and M. Kuzuya, *J. Am. Chem. Soc.*, **94**, 8958 (1972).
[167] W. D. Stohrer and R. Hoffmann, *J. Am. Chem. Soc.*, **94**, 1661 (1972).
[168] C. G. Biefeld, H. A. Eick and H. Hart, *Tetrahedron Letters*, **1973**, 4507.
[169] H. Kollmar, H. O. Smith and P. von R. Schleyer, *J. Am. Chem. Soc.*, **95**, 5835 (1973).
[170] M. J. S. Dewar and R. C. Haddon, *J. Am. Chem. Soc.*, **95**, 5836 (1973).
[171] S. Masamune, M. Sakai and H. Ona, *J. Am. Chem. Soc.*, **94**, 8955 (1972).
[172] S. Masamune, M. Sakai, H. Ona and A. J. Jones, *J. Am. Chem. Soc.*, **94**, 8956 (1972).
[173] S. Masamune, M. Sakai, A. V. Kemp-Jones, H. Ona, A. Vernot and T. Nakashima, *Angew. Chem. Internat. Ed.*, **12**, 769 (1973).
[174] M. Saunders, R. Berger, A. Jaffe, J. M. McBride, J. O'Neill, R. Breslow, J. M. Hoffman, C. Perchonork, E. Wasserman, R. S. Hutton and V. J. Kuck, *J. Am. Chem. Soc.*, **95**, 3017 (1973).
[175] H. Hart and M. Kuzuya, *Tetrahedron Letters*, **1973**, 4123.
[176] H. Hart and M. Kuzuya, *J. Am. Chem. Soc.*, **95**, 4096 (1973).
[177] M. Kuzuya and H. Hart, *Tetrahedron Letters*, **1973**, 3887.
[178] M. Kuzuya and H. Hart, *Tetrahedron Letters*, **1973**, 3891.
[179] H. Hogeveen and P. W. Kwant, *Tetrahedron Letters*, **1973**, 1665.
[180] H. T. Jonkman and W. C. Nieuwport, *Tetrahedron Letters*, **1973**, 1671.
[181] H. Hogeveen and P. W. Kwant, *J. Am. Chem. Soc.*, **95**, 7315 (1973).
[182] H. Hogeveen and P. W. Kwant, *Tetrahedron Letters*, **1973**, 3747.
[183] P. Ahlberg, *Chem. Ser.*, **2**, 231 (1972); *Chem. Abs.*, **78**, 96901 (1973).
[184] A. de Meijere and O. Schallner, *Angew. Chem. Internat. Ed.*, **12**, 399 (1973).
[185] G. A. Olah and G. Liang, *J. Am. Chem. Soc.*, **95**, 194 (1973).
[186] G. A. Olah and G. Liang, *J. Am. Chem. Soc.*, **95**, 3792 (1973).
[187] G. A. Olah, G. Liang and P. Westerman, *J. Am. Chem. Soc.*, **95**, 3698 (1973).
[188] L. A. Paquette, M. J. Broadhurst, P. Warner, G. A. Olah and G. Liang, *J. Am. Chem. Soc.*, **95**, 3386 (1973).
[189] J. A. Berson and J. A. Jenkins, *J. Am. Chem. Soc.*, **94**, 8907 (1972).
[190] W. J. Hehre, *J. Am. Chem. Soc.*, **94**, 8908 (1972).
[191] E. J. Reardon and M. Brookhart, *J. Am. Chem. Soc.*, **95**, 4311 (1973); cf. *Org. Reaction Mech.*, **1971**, 39.
[192] N. W. K. Chiu and T. S. Sorensen, *Can. J. Chem.*, **51**, 2776 (1973).
[193] N. W. K. Chiu and T. S. Sorensen, *Can. J. Chem.*, **51**, 2783 (1973).
[194] C. U. Pittman and W. C. Miller, *Can. J. Chem.*, **95**, 2947 (1973).
[195] M. Saunders and J. J. Stofko, *J. Am. Chem. Soc.*, **95**, 252 (1973).
[196] D. M. Brouwer and A. A. Kiffen, *Rec. Trav. chim.*, **92**, 809 (1973).
[197] P. van Pelt and H. M. Buck, *Rec. Trav. chim.*, **92**, 1057 (1973).
[198] D. M. Brouwer and A. A. Kiffen, *Rec. Trav. chim.*, **92**, 689 (1973).
[199] G. A. Olah, G. Liang, P. von R. Schleyer, E. M. Engler, M. J. S. Dewar and R. C. Bingham, *J. Am. Chem. Soc.*, **95**, 6829 (1973).
[200] J. F. M. Oth, D. M. Smith, U. Prange and G. Schröder, *Angew. Chem. Internat. Ed.*, **12**, 327 (1973).
[201] R. D. Chambers, R. S. Matthews and A. Parkin, *Chem. Comm.*, **1973**, 509.
[202] K. Conrow and D. L. Morris, *Chem. Comm.*, **1973**, 5.
[203] G. A. Olah, Y. K. Mo, E. G. Melby and H. C. Lin, *J. Org. Chem.*, **38**, 367 (1973).
[204] I. Fleming, *J.C.S. Perkin I*, **1973**, 1019.
[205] M. Saunders, D. Cox and W. Ohlmstead, *J. Am. Chem. Soc.*, **95**, 3018 (1973).
[206] G. A. Olah and P. R. Clifford, *J. Am. Chem. Soc.*, **95**, 6067 (1973).
[207] G. A. Olah and Y. K. Mo, *J. Org. Chem.*, **38**, 2686 (1973).
[208] A. E. van der Hout-Lodder, J. W. de Haan, L. J. M. van de Ven and H. M. Buck, *Rec. Trav. chim.*, **92**, 1040 (1973).
[209] M. Yoshimura, M. Nojima and N. Tokwa, *Bull. Chem. Soc. Japan*, **46**, 2164 (1973).
[210] G. A. Olah and J. J. Svoboda, *J. Am. Chem. Soc.*, **95**, 3794 (1973).
[211] J. Lukas, P. A. Kramer and A. P. Kouwenhoven, *Rec. Trav. chim.*, **92**, 44 (1973).
[212] H. Tanida and H. Matsumura, *J. Am. Chem. Soc.*, **95**, 1586 (1973).

[213] J. E. Dubois and J. S. Lomas, *Tetrahedron Letters*, **1973**, 1791.
[214] H. Tanida, S. Yamamoto and K. Takeda, *J. Org. Chem.*, **38**, 2077 (1973).
[215] R. M. Coates and S. K. Chung, *J. Org. Chem.*, **38**, 3677 (1973).
[216] I. Khattak, D. N. Kirk, C. M. Peach and M. A. Wilson, *Chem. Comm.*, **1973**, 341.
[217] F. B. Hirschmann and H. Hirschmann, *J. Org. Chem.*, **38**, 1270 (1973).
[218] S. S. Deshmane and H. Hirschmann, *J. Org. Chem.*, **38**, 748 (1973).
[219] W. A. Ayer and P. D. Deshpande, *Can. J. Chem.*, **51**, 77 (1973).
[220] D. S. Noyce and S. A. Fike, *J. Org. Chem.*, **38**, 3318 (1973).
[221] D. S. Noyce and G. T. Stowe, *J. Org. Chem.*, **38**, 3762 (1973).
[222] D. S. Noyce and S. A. Fike, *J. Org. Chem.*, **38**, 2433 (1973).
[223] D. S. Noyce, J. A. Virgilo and B. Bartman, *J. Org. Chem.*, **38**, 2657 (1973).
[224] D. S. Noyce and J. A. Virgilo, *J. Org. Chem.*, **38**, 2660 (1973).
[225] D. S. Noyce and R. W. Nichols, *J. Org. Chem.*, **37**, 4306 (1972).
[226] D. S. Noyce and R. W. Nichols, *J. Org. Chem.*, **37**, 4311 (1972).
[227] S. D. Elakovich and T. G. Traynham, *J. Org. Chem.*, **38**, 873 (1973).
[228] T. G. Traynham and S. D. Elakovich, *Tetrahedron Letters*, **1973**, 155.
[229] J. W. Pilkington and A. J. Waring, *Tetrahedron Letters*, **1973**, 4345.
[230] P. G. Gassman and G. D. Hartman, *J. Am. Chem. Soc.*, **95**, 449 (1973).
[231] C. D. Ritchie and P. O. I. Virtanen, *J. Am. Chem. Soc.*, **95**, 1882 (1973); cf. *Org. Reaction Mech.*, **1972**, 47.
[232] I. J. Miller, *Austral. J. Chem.*, **26**, 301 (1973).
[233] L. M. Tel, S. Wolfe and I. G. Csizmadia, *Internat. J. Quantum Chem.*, **7**, 475 (1973); *Chem. Abs.*, **79**, 41771 (1973).
[234] N. G. Raman and R. G. Jesaitis, *J.C.S. Perkin II*, **1973**, 1063.
[235] C. U. Pittman, Jr., J. B. Patterson, Jr., and L. D. Kispart, *J. Org. Chem.*, **38**, 471 (1973).
[236] R. Breslow and S. Mazur, *J. Am. Chem. Soc.*, **95**, 584 (1973).
[237] R. Breslow and W. Chu, *J. Am. Chem. Soc.*, **95**, 411 (1973).
[238] J. N. Ride and P. A. H. Wyatt, *J.C.S. Perkin II*, **1973**, 746.
[239] D. G. Farnum and D. S. Patton, *J. Am. Chem. Soc.*, **95**, 7728 (1973).
[240] S. V. Kulkarni, R. Schure and R. Filler, *J. Am. Chem. Soc.*, **95**, 1859 (1973).
[241] H. V. Navangul and P. E. Blatz, *J. Am. Chem. Soc.*, **95**, 1508 (1973).
[242] P. A. Kollman, W. F. Trager, S. Rothenberg and J. E. Williams, *J. Am. Chem. Soc.*, **95**, 458 (1973).
[243] C. R. Turnquist, J. W. Taylor, E. P. Gimsrud and R. C. Williams, *J. Am. Chem. Soc.*, **95**, 4133 (1973).
[244] H. M. R. Hoffman, *Angew. Chem. Internat. Ed.*, **12**, 819 (1973).
[245] A. E. Hill, G. Greenwood and H. M. R. Hoffman, *J. Am. Chem. Soc.*, **95**, 1338 (1973).
[246] S. Ito, H. Ohtani and S. Amiya, *Tetrahedron Letters*, **1973**, 1737.
[247] J. Bertram, J. P. Coleman, M. Fleischmann and D. Fletcher, *J.C.S. Perkin II*, **1973**, 374.
[248] D. B. Clark, M. Fleischmann and D. Pletcher, *J.C.S. Perkin II*, **1973**, 1578.
[249] S. D. Ross, J. E. Barry, M. Finkelstein and E. J. Rudd, *J. Am. Chem. Soc.*, **95**, 2193 (1973).
[250] E. D. Owen and D. M. Allen, *J.S.C. Perkin II*, **1973**, 95.
[251] K. Okamoto, K. Komatsu, O. Murai and O. Sakaguchi, *Tetrahedron Letters*, **1972**, 4989.
[252] T. B. McMahon, R. J. Blint, D. P. Ridge and J. L. Beauchamp, *J. Am. Chem. Soc.*, **94** 8934 (1972).
[253] R. C. Dunbar and E. W. Fu, *J. Am. Chem. Soc.*, **95**, 2716 (1973).
[254] R. C. Dunbar, *J. Am. Chem. Soc.*, **95**, 472 (1973).
[255] N. A. Uccela and D. H. Williams, *J. Am. Chem. Soc.*, **94**, 8778 (1972).
[256] A. S. Blair and A. G. Harrison, *Can. J. Chem.*, **51**, 1645 (1973).
[257] A. G. Harrison and B. G. Keyes, *Can. J. Chem.*, **51**, 1265 (1973).
[258] M. K. Hoffman and J. C. Wallace, *J. Am. Chem. Soc.*, **95**, 5064 (1973).
[259] M. P. Doyle, D. J. DeBruyn and D. J. Scholten, *J. Org. Chem.*, **38**, 625 (1973).
[260] D. H. R. Barton, P. D. Magnus and R. N. Young, *Chem. Comm.*, **1973**, 331.
[261] S. Coffi-Nketsia and A. Kergomard, *Bull. Soc. Chim. France*, **1973**, 2115.
[262] T. Cohen and J. Solash, *Tetrahedron Letters*, **1973**, 2513.
[263] K. H. Pegal, L. P. L. Piacenzi and C. P. Gorst-Allman, *Tetrahedron Letters*, **1973**, 4053.
[264] R. A. Sheldon and J. A. van Doorn, *Tetrahedron Letters*, **1973**, 1021.
[265] M. de Sorgo, D. C. Pepper and M. Szwarc, *Chem. Comm.*, **1973**, 419.
[266] C. E. Low and R. M. Roberts, *J. Org. Chem.*, **38**, 1909 (1973).
[267] J. Baumrucker, M. Calzadilla and E. H. Cordes, *React. Kinet. Micelles, Proc. Am. Chem. Soc. Symp.*, **1972**, 25; *Chem. Abs.*, **78**, 158444 (1973).
[268] V. D. Nefedov, E. N. Sinotova, G. P. Akulov and M. V. Korsakov, *Zh. Org. Khim.* **9**, 629 (1973); *Chem. Abs.*, **78**, 147019 (1973).
[269] P. Abley, J. E. Byrd and J. Halpern, *J. Am. Chem. Soc.*, **95**, 2591 (1973).

[270] C. D. Beard, K. Baum and V. Grakauskas, *J. Org. Chem.*, **38**, 3673 (1973).
[271] H. R. Hudson and P. A. Karam, *J.C.S. Perkin II*, **1973**, 1141.
[272] M. Bataille and J. Landais, *Compt. Rend.*, (*C*), **276**, 1307 (1973).
[273] J. A. Peters and H. van Bekkum, *Rec. Trav. chim.*, **92**, 379 (1973).
[274] K. Okamoto, K. Komatsu, S. Tsukada and O. Murai, *Bull. Chem. Soc. Japan*, **46**, 1785 (1973).
[275] V. A. Tsyryapkin, N. M. Loim, Z. N. Parnes and D. N. Kursanov, *Zh. Org. Khim.*, **8**, 2342 (1972); *Chem. Abs.*, **78**, 70987 (1973).
[276] J. Jacobus, *J. Org. Chem.*, **38**, 402 (1973).
[277] T. H. Whitesides, R. W. Arhart and R. W. Slaven, *J. Am. Chem. Soc.*, **95**, 5792 (1973).

CHAPTER 9

Nucleophilic Aliphatic Substitution

I. D. R. STEVENS

Chemistry Department, University of Southampton

Ion-pair Phenomena and Borderline Mechanisms

Sneen[1] has reviewed his work on the general ion-pair hypothesis[2] and has produced evidence that it operates also in the hydrolysis of (4-methoxybenzyl)dimethylsulphonium salts. The effect of added azide ion is normal in that it gives a linear increase in azide-to-alcohol ratio with azide ion concentration; but the effect of added iodide ion is to produce an increase in rate, the extent of which is not linear in iodide ion concentration, an effect that can be explained satisfactorily only by using the ion-pair hypothesis.[3]

Thermodynamic parameters ($\Delta G°$, $\Delta H°$ and $\Delta S°$) have been calculated for the ionization of methyl, ethyl, isopropyl, and *tert*-butyl halides in water and also for their dissociation. The values for the four methyl halides are all much larger than those previously calculated[4] and are also larger than the observed $\Delta G^{\ddagger}$ for the hydrolyses. The ethyl halides also have $\Delta G°$ for ionization larger than $\Delta G^{\ddagger}$ for hydrolysis, which rules out a mechanism involving rate-determining attack on an ion-pair for these compounds. For the isopropyl halides the difference is in the same direction, but its magnitude (1–3 kcal mol^{-1}) is sufficiently small to leave the possibility open. The *tert*-butyl halides have $\Delta G°$(ionization) < $\Delta G^{\ddagger}$(hydrolysis) and therefore do hydrolyse by ionization. The discrepancies from the results discussed in ref. 4 are enormous (45–60 kcal mol^{-1}), and there is no very obvious reason why this should be so; it is suggested that it may arise from the assumption made earlier that the reactions of trityl carbocation with water and hydroxide ion make good models.[5]

From the difference in the ROEt/ROH ratio observed between the solvolysis of benzhydryl *p*-nitrobenzoate in 50% acetone/(25 + X)% ethanol/(25 – X)% water and the reaction of diphenyldiazomethane with *p*-nitrobenzoic acid in the same mixed solvent, it has been argued that the two processes do not have the same product-determining step. Since the second reaction has been shown to involve both intimate and solvent-separated ion-pairs,[6] it is suggested that this requires a direct S_N2 component for benzhydryl *p*-nitrobenzoate solvolysis as well as the S_N1 one.[7]

Grob *et al.*[8] have warned of the dangers of accepting the Sneen hypothesis just because of its simplicity and has produced evidence in favour of an S_N2-type reaction for the tertiary system (**1**). For (**1**; R = H) the rate of solvolysis is nearly identical with that of the carbon analogue (**1**; NMe_2 = $CHMe_2$), showing that there is no anchimeric assistance in the rate-determining step, and 16% of the cyclic salt is produced. With change of R to $-(CH_2)_5-$ the rate increases 70-fold with respect to the carbon analogue and 100% of cyclic salt is produced. While this could be taken as indicating rate-determining attack on an ion-pair, taken with the ρ-value of –1.91 for the system (**3**) (*m*-substituted aryl groups), and the Winstein–Grunwald *m*-value of 0.236, the authors' conclusion of a loose S_N2 or true borderline transition state seems justified, particularly since the carbon analogue of (**3**) shows a ρ (versus σ^+) of –4.33 more consonant with a LIM process. Similarly, the difference in ρ-values for the k_Δ and k_s processes [–3.8 (σ^+) and –0.18 (σ), respectively] observed in solvolysis of (**4**) in dimethylformamide (DMF) has been used to justify the absence of reaction by a Sneen-type scheme[9] (see also ref. 41).

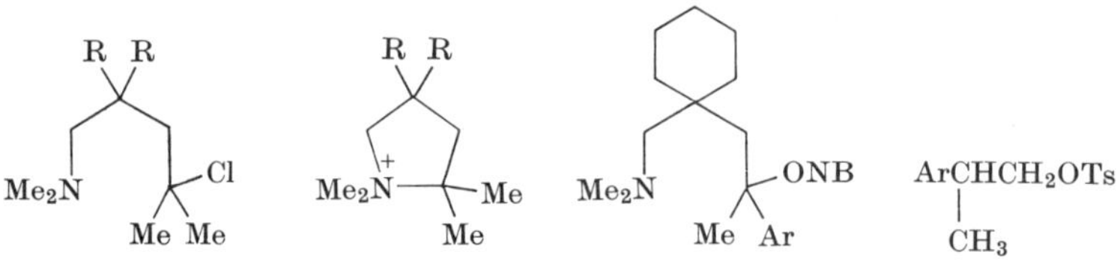

(**1**) (**2**) (**3**) (**4**) $ArCH(CH_3)CH_2OTs$

ONB = *p*-nitrobenzoate

Pinacolyl toluene-*p*-sulphonate has been advocated as a model system for investigation of ion-pair return. The ratio of the rate of solvolysis for pinacolyl toluene-*p*-sulphonate in a given solvent to that in trifluoroacetic acid should be identical with that for other compounds where only a k_c process is rate-determining. A comparison with 2-adamantyl toluene-*p*-sulphonate shows that $(k_{solvent}/k_{TFA})$Pin OTs/$(k_{solvent}/k_{TFA})$ 2-AdOTs varies from 16 (80% EtOH) to 2.5 (CF_3CH_2OH). It is suggested that the difference shows how the partitioning of the 2-adamantyl–toluene-*p*-sulphonate ion-pair varies with solvent. The authors[10] note that return increases as solvent Y decreases, and therefore that product is formed by nucleophilic attack at the solvent-separated ion-pair stage (see also p. 299 and ref. 41).

By comparing the polarimetric and conductometric rates for solvolysis of 1-phenylethyl chloride in 97% aqueous CF_3CH_2OH, and 60% and 80% aqueous ethanol, and evaluating the ratio of k_2/k_{-1} from the reaction of styrene with HCl in trifluoroethanol, a complete dissection of the fate of the intimate ion-pairs formed in the solvolysis has been performed.[11] The kinetics of racemization of 1-phenylethyl chloride catalysed by metal halides in ether have been measured: the rate equation contains terms in catalyst concentration from the first to the fourth power, and this has been explained in terms of ion-pairs requiring solvation either by ether or by catalyst.[12] Benzyl and substituted benzyl chlorides and bromides give non-linear mY plots in ethanol/trifluoroethanol mixtures; the bromide/chloride rate ratio varies with solvent composition and it is adduced that this is due to differences in the ratio of return to solvolysis of the intimate ion-pair.[13] The effect of complexing a tricarbonylchromium group on to the aryl group of substituted benzyl chlorides is nearly to suppress the S_N1 component.[14] Halide exchange of substituted benzyl chlorides in acetonitrile gives non-linear Hammett plots.[15] The kinetics of the solvolysis of benzyl and benzhydryl bromide in alcoholic solvents have been interpreted in terms of mixed S_N1 and S_N2 mechanisms.[16] The polar exchange of benzhydryl iodides with iodine in hexane solution is of the second order in iodine concentration and has a ρ (versus σ^+) of -5.2; a transition state with one iodine molecule solvating the cation, and the other solvating the iodide ion, has been adduced.[17] The large negative $\Delta S^\ddagger$ observed in the ethanolysis of the 4-, 4,4′- and 4,4′,4″-nitro-substituted trityl chlorides has led to the suggestion that they react by an S_N2-type process. However, the reactions are faster in ethanol (the more polar solvent) than in 30% dioxan–70% ethanol, indicating a considerable ionic character.[18]

Further studies of racemization, ^{18}O-exchange and conversion into thioethers of 4-chloro- and 4-methyl-benzhydryl thiocarbonate in acetonitrile have been reported. The percent return observed decreases as the carbocation stability increases.[19]

It has been argued that the insensitivity of the percent elimination to change in solvent for the reaction of cyclohexyl bromide with $Pr^n{}_4N^+$ PhS^- shows that the two reactions follow very similar paths; for an overall rate change of nearly 500, the percent elimination changes from 54 to 57; product-determining attack on an ion-pair is postulated.[20] The solvolyses of *cis*- and *trans*-4-*tert*-butylcyclohexyl toluene-*p*-sulphonates in slightly (<1%) aqueous *N*-methylacetamide gives more retained alcohol from the *trans*- than from the *cis*-starting material. Recovered toluene-*p*-sulphonate shows 5.9% of isomerization from *trans* to *cis*, but only 0.3% from *cis* to *trans*; it is suggested that solvent attack on the intimate ion-pair is slower for *trans*- than for *cis*-substrate, allowing OTs^- to compete effectively. As the water concentration increases from 0.06% to 0.94%, the amount of retained alcohol from *trans*-substrate decreases from 28.7% to 2%, this change being due, it is proposed, to the effect of water in depolymerizing the solvent, making it more nucleophilic.[21] Both *cis*- and *trans*-4-*tert*-butylcyclohexyl trifluoromethanesulphonates acetolyse 10^4 times faster than the corresponding toluene-*p*-sulphonates; despite the difference in rate, the product proportions are nearly identical for the two leaving groups, pointing strongly to the intervention of similar ion-pairs in both cases.[22]

Further studies on the transmission of substituent effects in heteroaryl systems have been reported.[23–25] Whereas in most cases rates correlate with σ^+, substituents at the 6-position of the pyridine nucleus in 1-methyl-1-(2′- or 3′-pyridyl)ethyl chloride do not do so[23] and, although for 1-(2X-5-thiazolyl)ethyl chlorides there is a correlation with $\sigma_p{}^+$, solvolysis of the corresponding 2X-4-thiazolyl compounds gives no correlation with either σ_m or $\sigma_m{}^+$.[24] Methanolyses and hydrolyses of 1-methyl-1-(2′-, 3′- and 4′-pyridyl)-ethyl chloride have been used to evaluate σ^+-values for the nitrogen in pyridine.[26]

It has been shown that neglect of ion-pair association for an ionic nucleophile can lead to mixed kinetics even when the reaction is entirely of the second order.[27]

Solvent and Medium Effects[28]

A kinetic investigation of phase-transfer catalysis has shown that for *n*-octyl chloride and bromide the reaction is of the first order in halide and also in the concentration of phase-transfer catalyst in the organic phase.[28a]

The volume of activation for the quaternization of pyridine with benzyl bromide has been measured in twelve solvents. A plot of log k is linear against $\Delta V^{\ddagger}$ and also against the $\Delta G^{\ddagger}$ for other quaternizations. $\Delta V^{\ddagger}$ is also linear with respect to its pressure derivative in the various solvents, and an equation has been derived that permits an estimate of the solvation number of the transition state.[29] A study of the effect of water on the exchange of $Li^{82}Br$ with butyl bromide in acetone, at low water concentrations, has shown that salt–ion pairs are inactive and that the overall rate is depressed because strong hydrogen-bonding of bromide ion to water lowers the ion activity.[30]

Further work on the effect of cryptates for complexing metal ions has been published. Crystalline complexes have been isolated that have been shown by X-ray methods to have the metal ion within the cavity of the cryptate. The relative stabilities for Ba^{2+} and K^+ can be varied from 100:1 to 1:200 by change of cavity size and shape.[31]

Micellar catalysis of organic reactions continues to attract attention. For 2-phenylpropyl bromide with aqueous hydroxide, the effect is to reduce the S_N1 and increase the S_N2 and $E2$ components. The effect of the micelle is to lower the activity of the water, increase that of OH^- and lower the effective dielectric constant in its vicinity. Micellar salts containing a hydroxy-group β to the ammonium centre are more effective than those without it.[32] Water-insoluble organic compounds increase the viscosity of cetyltrimethylammonium bromide solutions in the presence of bromide ions, whereas water-soluble compounds do not.[33] The hydrolysis of 2-*tert*-butyl-3-phenyloxaziridine is accelerated by negatively, but not by positively, charged micelles; a mechanism involving protonation followed by absorbtion on the micelle is invoked.[34]

The effect of solvent and metal-ion on the reaction of isopropoxide ion with *n*-hexyl iodide has been studied.[35] The free energies of transfer of reactants and transition states have been evaluated for the reaction of *n*-hexyl toluene-*p*-sulphonate with five tetrabutylammonium salts, for transfer from methanol to dimethyl sulphoxide (DMSO). The faster rates for azide, bromide and chloride are due to desolvation of the anions; there is almost no effect on the thiocyanate reaction; and for reaction with iodide the transition state is better solvated in DMSO, leading to faster reaction.[36]

The thermodynamic parameters for transfer of single ions from propylene carbonate to seven other solvents have been examined to test the assumptions involved. It is recommended that the assumption that $\Delta G_{tr}(Ph_4As^+) = \Delta G_{tr}(Ph_4B^-)$ be used.[37] The NMR chemical shifts of $^{35}Cl^-$, $^{79}Br^-$ and $^{127}I^-$ depend strongly on the solvent but very little on the counter-ion; a correlation was found between the shift and the ultraviolet absorbtion energy of the charge-transfer to solvent.[38]

Isotope Effects[39a]

Calculations of α-deuterium isotope effects and their temperature-dependence have revealed an additivity relation for the quantum-mechanical temperature-dependent term and its associated *A*-factor; this relation gives a basis for fitting force-fields to the observed isotope effects.[39b] Fractionation factors for H/D exchange and carbon isotope exchange have been calculated by methods developed previously; it was found that the H/D fractionation factor is affected in a regular way by substituents on the α-carbon, but very little by groups on the β-carbon atom; similar results are found for carbon isotope effects, and the use of these fractionation factors for equilibrium and kinetic isotope effects is illustrated; for the most part the calculated and experimental values are in good agreement, in particular the variation in the α-deuterium isotope effect with leaving group (F, Cl, Br or I) is well accounted for.[40]

Kinetic isotope effects for both α- and β-deuterium have been measured for the aqueous trifluoroethanolysis and ethanolysis of cyclopentyl *p*-bromobenzenesulphonate. The results are interpreted on the basis of the Winstein scheme (Scheme 1). The α-k_H/k_D (1.24 and 1.15) shows that k_2 is rate-determining in trifluoroethanol and k_s and

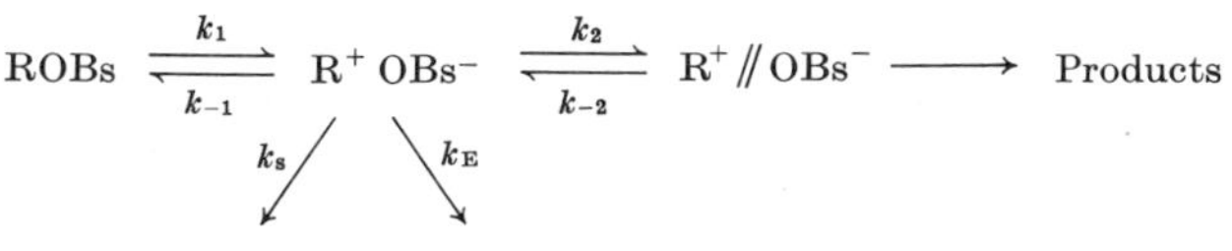

$$\mathrm{ROBs} \underset{k_{-1}}{\overset{k_1}{\rightleftharpoons}} \mathrm{R^+\,OBs^-} \underset{k_{-2}}{\overset{k_2}{\rightleftharpoons}} \mathrm{R^+ /\!/ OBs^-} \longrightarrow \text{Products}$$

$$\mathrm{R^+\,OBs^-} \xrightarrow{k_s} \qquad \mathrm{R^+\,OBs^-} \xrightarrow{k_E}$$

(Bs = p-$BrC_6H_4SO_2$-)

SCHEME 1.

k_E rate-determining in ethanol. Assuming that there will be no isotope effect on the solvent separation step k_2, the authors dissect the overall effects into those on k_1, k_s and k_E for both α-D and β-D. For the ionization step the effects of α-D, *cis*-β-D, and *trans*-β-D are nearly identical (1.24, 1.23 and 1.21, respectively). Further the overall α-k_H/k_D for nucleophilic attack on the intimate ion-pair is 1.14, indicating that another mechanism is necessary for those cases where this value is less than unity.[41] Secondary deuterium isotope effects have been measured for chlorine isotope exchange in acetonitrile.[42]

The temperature-dependence of the volume of activation for hydrolysis of benzyl chloride, isopropyl bromide, and methanesulphonyl chloride is larger in deuterium oxide than in water. It is concluded that this shows that heavy water is more structured than light water.[43] The $\Delta C_p^{\ddagger}$ for the hydrolyses of 4-chloro-, 4-methyl- and 4-nitro-benzyl chloride have been measured. That for the 4-methyl compound is almost identical with that for *tert*-butyl chloride, indicating a similarity of mechanism, while those for the other two compounds are nearly the same as that for benzyl chloride and methyl chloride. However, the solvent isotope effect is virtually identical for the 4-nitro- and the 4-methyl-benzyl chloride. The authors conclude that solvent isotope effects are not helpful for S_N reactions.[44]

Kinetic isotope effects for ^{14}C and ^2H have been measured for acetolysis and trifluoroacetolysis of neophyl *p*-bromobenzenesulphonate. The $^{12}C/^{14}C$ effect at C-1 is identical with the maximum reported value for S_N2 reactions, indicating, it is suggested, important phenyl participation, and yet the α-k_H/k_D value (1.247 in CF_3CO_2H) shows a LIM S_N1 process; no explanation is offered.[45] Exchange with $H_2{}^{18}O$, and with ^{14}C isotope effects, have also been used to dissect the product-forming step and evaluate the degree of participation in the hydrolysis of 2-bromo-1,1,2-triphenylethanol in 80% dioxan and in its silver nitrate-catalysed reaction. Incorporation of ^{18}O occurs solely at the tertiary position in the glycol product, showing that this comes only from O-3 participation, and this shows a k^{14}/k^{12} at C-1 of 0.978. Ketone is formed with $k^{14}/k^{12} = 1.002$ at C-1, and 0.98 at C-2, and it is suggested that this arises solely by phenyl participation with a well-developed bond to C-1 and that to C-2 nearly broken in the transition state;

maximum effects are observed in the dehydration of 3-methyl-3-phenylbutan-2-ol, which occurs with exclusive phenyl migration; a symmetrical phenonium ion-like transition state is postulated.[46]

The Bigeleisen–Mayer heavy-atom approximation has been used to calculate the $^{35}Cl/^{37}Cl$ isotope effect for the reaction of butyl chloride with a thiophenoxide anion. The observed and calculated values for $[\Delta H^{\ddagger}(^{35}Cl) - \Delta H^{\ddagger}(^{37}Cl)]$ are in good agreement, and it is calculated that $k(^{35}Cl)/k(^{37}Cl)$ should be larger for S_N1 than for S_N2 reactions (1.0106 versus 1.0089).[47]

Neighbouring-group Participation*

Participation by Ether and Hydroxyl Groups

Lithium dichloromethylide reacts with ketones to give the lithium salts (**5**) which are stable up to 80°, after which they react to give epoxides (**6**). These rearrange thermally to give the aldehydes (**7**) with exclusively chlorine migration.[48] However, when salts (**5**) are treated with lithium piperidide, the products are ketones (**8**), which probably arise by a carbenoid path involving O-3 participation of the lithium alkoxide.[49] The

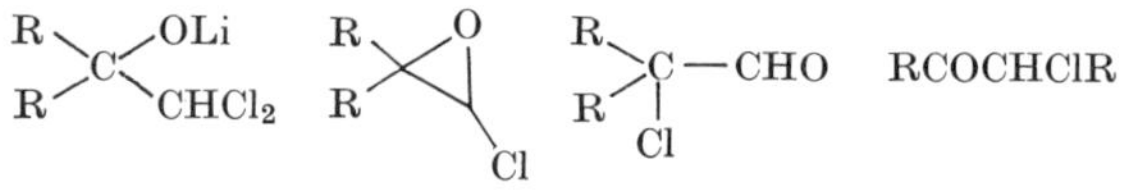

RCOCHClR

(**5**) (**6**) (**7**) (**8**)

effects of α- and β-aryl-substitution on the rates of ring-closure of epoxides have been studied, by using 1-aryl- and 2-aryl-2-haloethanols as substrates; in all these cases the rates correlate against σ with ρ-values of 0.34 and 1.61 for ring-closure, which for the β-aryl compounds indicate a substantial development of charge before attack by oxygen.[50]

All four isomers of 2-methylnorbornane-2,3-diol monomethanesulphonate have been treated with base; both isomers with an *exo*-2-hydroxyl group give *exo*-epoxide; of the two isomers with an *endo*-2-hydroxyl group, the isomer (**9**) gives the epoxide (**10**) with potassium *tert*-butoxide in *tert*-butyl alcohol but gives 3-methylnorban-2-one with sodium hydride in benzene; however, its epimer at C-3 gives the ketone under both sets of conditions. It is suggested that, in benzene, ionization of (**9**) to the classical ion is followed by methyl migration.[51] The ylide formed from chloromethyl di-isopropylphosphonate and butyl-lithium reacts with aldehydes to give principally the *trans*-epoxides (**11**); the intermediate chlorohydrins, which may be isolated by quenching

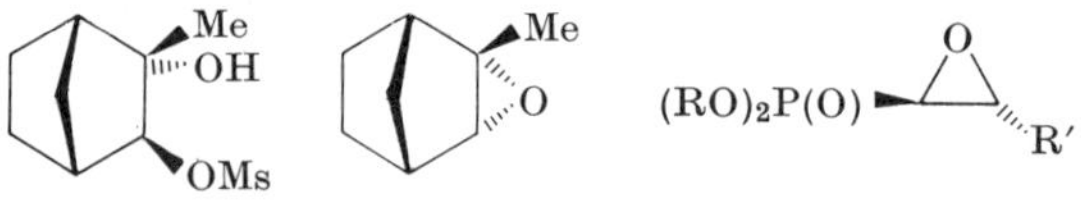

(**9**) (**10**) (**11**)

at –80°, consist of a mixture of isomers, showing that the addition is reversible with ring-closure both rate- and product-determining.[52] A similar glycidic ester-type condensation occurs when *p*-tolyl chloromethyl sulphide reacts with non-enolizable aldehydes

* See also **Chapter 8**.

in the presence of base (Bu^tOK or Dabco), affording α-thioepoxides.[53] Participation of O-3 is also observed when β-hydroxyalkylmercuric chlorides are treated with base,[54] and participation has been proposed to account for the alkaline cleavage of nucleosides in anhydrous dioxan;[55] it is also involved in the studies reported in refs. 56, 57 and 83.

It is suggested that the formation of (**13**) from the diazo-ketone (**12**) occurs by participation of O-4.[58] 1,3-Diols react with phosphorous hexamethyltriamide in carbon tetrachloride to give intermediate phosphonium salts (**14**), which are converted into oxetans on treatment with alkoxides.[59]

$$ArOCMe_2COCHN_2 \xrightarrow[THF]{H^+} ArOCH_2COCMe{=}CH_2$$

R, OH, R′, R″, Cl^-, $\overset{+}{O}P(NMe_2)_3$

(**12**) (**13**) (**14**)

The mechanism originally proposed for the conversion of 1-*O*-acetyl-4-*O*-alkanesulphonyl-α-D-glucopyranoses into 1,4-anhydro-β-D-galactopyranoses by sodium azide in aprotic solvents, involving O-3 participation, ring-contraction and attack by the C-1-oxygen at C-5, has been questioned. The α-L-talopyranose (**15**) also undergoes the reaction and gives (**16**). In (**15**) the leaving group at C-4 is incorrectly oriented for ring

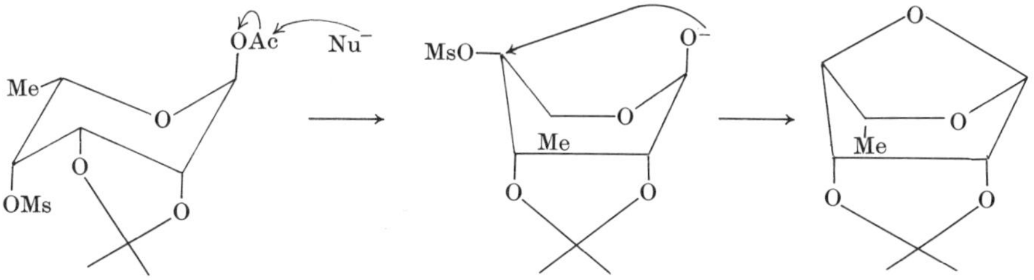

OAc, Nu^-, Me, O, OMs, O, O; MsO, O^-, O, Me, O, O; O, O, Me, O, O

(**15**) (**16**)

contraction (or leaves C-5 *trans* with respect to the C-1-oxygen), and so a route involving initial 1-deacetylation and direct displacement of the 4-methanesulphonyl group is proposed, not only for (**15**) but for the other cases as well.[60]

The diketone ditoluene-*p*-sulphonate (**17**) is reduced by sodium borohydride in ethanol to give (**18**) and some (**19**), Although (**18**) is converted into (**19**) under the conditions shown, it is unaffected by sodium methoxide in methanol.[61] Favourable conformational

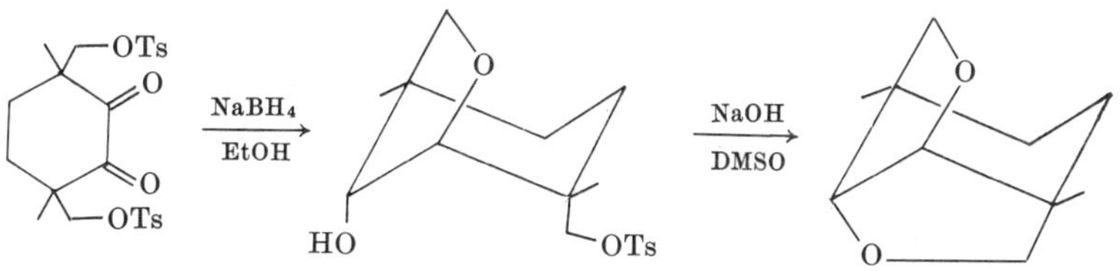

OTs, O, O, OTs $\xrightarrow[EtOH]{NaBH_4}$ O, HO, OTs $\xrightarrow[DMSO]{NaOH}$ O, O

(**17**) (**18**) (**19**)

factors can facilitate the participation by phenolic oxygen in the displacement of X from the 4-(*o*-hydroxyphenyl)butyl-X system by up to 2×10^5; the effects of leaving group and basicity have been studied in relation to the effect of methylation of the aromatic ring, and the conformation has been determined by X-ray crystallography.[62]

Participation by Thioether and Thiol Groups

Ion-cyclotron resonance spectroscopy has been used to study the effects of neighbouring groups in the 2-X-ethanol and 2-X-propanol systems. The ease of formation of the ion $[MH - H_2O]^+$ parallels that found for solvolysis, with a "neighbouring group" order of Br > SH > SMe > OMe $\gg$ Cl, F. However, only in the case of the sulphur groups is there any evidence for a thi-iranium ion. For methoxyl and bromine the evidence supports migration of hydrogen to give the $CH_3CH{=}X^+$ ion. For the 2-X-cyclopentanol system, the *trans*-isomers give the $[MH - H_2O]^+$ ion more readily than do the *cis*-isomers.[63]

Rates of solvolysis and activation parameters have been evaluated for the four stereoisomers of (**20**) in 95% ethanol. The isomer (**20**) reacts 2×10^3 more rapidly than the isomer (**21**); but there is no correlation between the preferred conformations of the four isomers and their relative reactivity.[64]

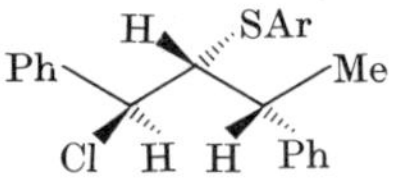

(**20**) 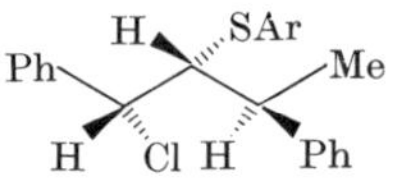 (**21**)

In the presence of sodium carbonate, the pyridine derivative (**22**) is converted into a 3:2 mixture of (**23**) and (**24**), and it is suggested that both are formed by the intermediacy of the thi-iranium ion (**25**), which would involve a front-side conversion of (**25**) into (**23**).[65]

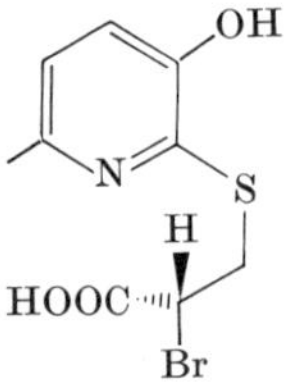

(**22**) 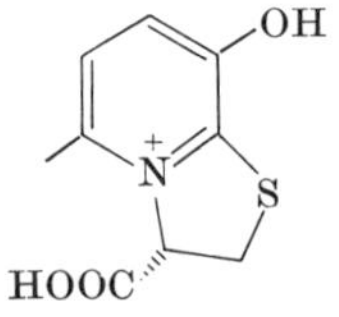 (**23**) 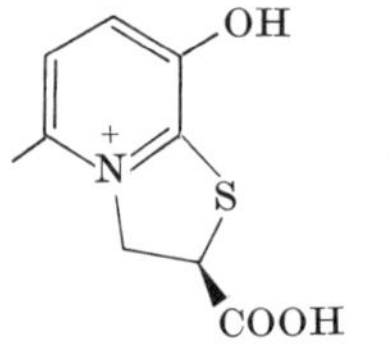 (**24**) 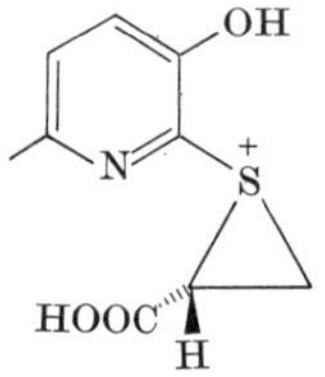 (**25**)

Evidence has been presented to substantiate the claim that, just as for S-3 participation in saturated compounds, so also for vinyl cations the thi-irenium intermediate can be diverted by chloride ion to give sulfenyl chloride. Thus (**26**) [X = 2,4,6-$(NO_2)_3$-$C_6H_2SO_3$] is converted into (**26**; X = Cl) by HCl in benzene; but in the presence of *p*-TolC≡C-Tol-*p*, (**27**) is formed.[66] 6-Chloro-9-thiabicyclo[3.3.1]non-2-ene is converted into 9-thiabicyclo[4.2.1]nona-2,4-diene *via* a thi-iranium ion.[67] S-3 participation has also been studied in the system (**28**; R = Me or Ph).[68]

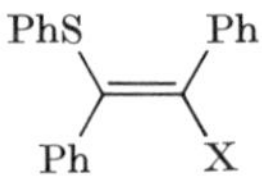

PhNHCOCH(SR)CR′MeCl

(**26**) 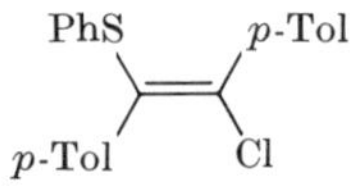 (**27**)  (**28**)

A kinetic study of ω-haloalkyl *p*-tolyl sulphides and ω-halo-ω-phenylalkyl *p*-tolyl sulphides has shown that in both series the rate of S-3 closure is greater than that of S-5 closure and that this is due, not to an entropy, but an enthalpy difference. The effect of the phenyl group adjacent to the leaving group is to increase the rate by an entropy change.[69]

The methyl groups in (**29**; $R^1 = R^2 =$ Me) are equivalent at 25° from the NMR point of view, although those in the compound without the arylthio-group are non-equivalent up to above 200°. At low temperatures the methyls in (**29**; $R^1 = R^2 =$Me) become non-equivalent and a similar non-equivalence is shown by "aromatic" methyl groups when Ar = *p*-tolyl or *p*-methoxyphenyl. However, the rate of exchange is different for the two types of methyl group, showing that a symmetrical ion is not involved. This is also borne out by the small rate difference between (**29**; $R^1 = R^2 =$ H) and (**29**; $R^1 = R^2 =$ Me), viz. 2:1. The "bell-clapper" process (**29a**) ⇌ (**29b**) is invoked to account for this and an S_N2-like transition state is postulated. Attack on C-10 of the anthracene nucleus occurs readily and leads, for example, to (**30**); it is suggested that this may be due to

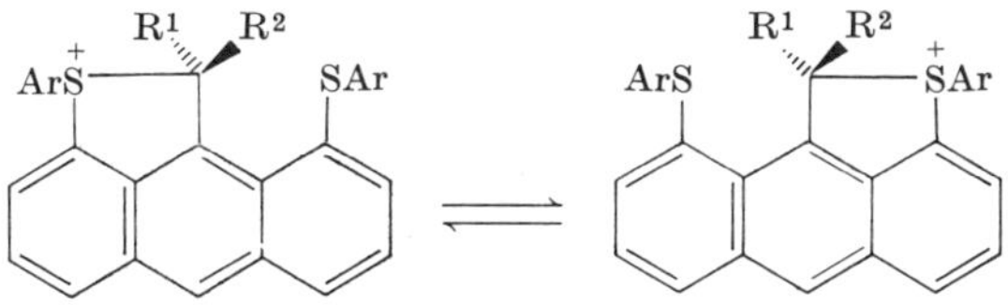

(**29a**) (**29b**) 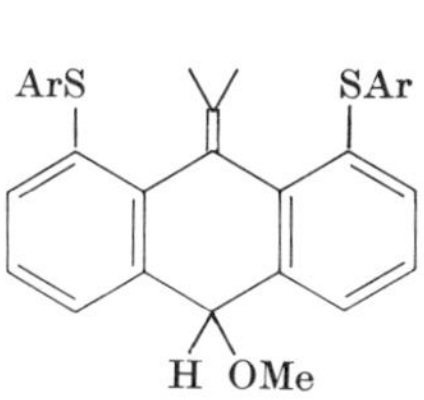 (**30**)

trapping of the symmetrical pentaco-ordinate species of an S_N2 transition state, although the authors found no other evidence for such a symmetrical ion.[70]

Participation by Halogen

Evidence for hyperconjugation by carbon—halogen bonds has come from an analysis of the photoelectron spectra of the benzyl halides. It is found that the change in excitation energy from the HOMO of the benzyl halides compared to that of the e_{1g} band of benzene is proportional to the reciprocal of the difference between the π_1 energy and the C—X binding energy in the methyl halide.[71]

Participation by chlorine has been invoked to account for the ease of conversion of (**31**; Ar = *o*-ClC_6H_4) into (**32**) in the presence of acid.[72] 1-Chloro-2-fluoroethane ionizes in $SO_2ClF–SbF_5$ to give a mixture of (**33**) and (**34**); participation by chlorine is more important than hydride migration.[73] The 1,2-diaxial to 1,2-diequatorial rearrangement

$ArCH(OH)CMeCl_2$ (**31**) $ArCHClCOMe$ (**32**) (**33**) $CH_3\overset{+}{C}HCl$ (**34**)

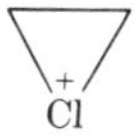

of *vic*-dihalides has been examined by MINDO/2 calculations as a possible candidate for inclusion in the orbital-symmetry-allowed class of dyotropic rearrangements. It is concluded that the reaction is more probably ionic, as previously postulated, than dyotropic.[74]

Two further examples of Br-4 participation have been demonstrated. The compounds

(**35**; X = Br or F) afford the corresponding brometanium ions (**36**; X = Br or F) in SbF_5–SO_2ClF. When X = Br, only a single NMR line is observed, whose width varies with temperature, indicating that all three bromines contribute in turn to the stabilization of (**36**; X = Br).[75] The temperature-dependence of the ^{13}C-NMR shift for the chlorolanium ion (**37**) shows that it is in equilibrium with the tertiary cation (**38**).

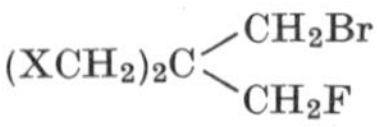

$(XCH_2)_2C(CH_2Br)(CH_2F)$ (**35**) X = Br or F

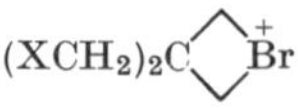

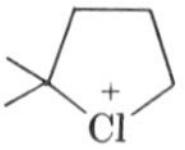

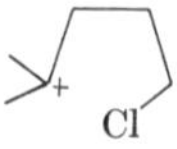

$(XCH_2)_2C$ brometanium ring (**36**)

(**37**) (**38**)

As might be expected, the monomethyl analogue shows almost no temperature-dependent shift. When 1,6-dichlorohexane is ionized in SO_2–SbF_5, the first detectable ion is the 2-ethylchlorolanium ion, which slowly isomerizes to (**37**), demonstrating once more the lack of Cl-7 and Cl-6 participation.[76] However, it is possible to make the iodonium and bromonium ions by treatment of the 1,5-dihalides with MeF–SbF_5, to give the monomethylhalonium salts, $MeX^+(CH_2)_5X$, followed by dissolution in SO_2–SbF_5. The six membered-ring ions slowly isomerize to the corresponding 2-methyl five-membered ones.[77]

Participation by Carbonyl Groups

The effect of a substituent Z on the solvolysis of the ester (**39**) suggests participation by the unsolvated carbonyl group in trifluoroacetolysis *via* an ion (**40**), but *via* the solvated carbonyl group (**41**) in formic acid. The overall rate is about the same as the k_Δ-term in neopentyl solvolysis, indicating an acceleration of ca. 10^7 due to participation.[78]

p-$ZC_6H_4COCMe_2CH_2OTs$ (**39**)

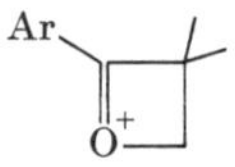

(**40**) 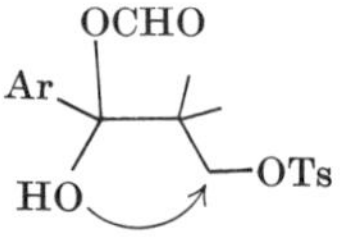 (**41**)

The conversion of (**42**) into (**43**) and (**44**) discussed in 1971[79] has been re-investigated by two groups in the light of the finding that the ketene (**45**) gives (**44**) by a $\pi 2s + \pi 2a$ process.[80] One group has shown that the ketene (**45**) is efficiently trapped by 5,6-dihydropyran, and that when (**42**) reacts in the presence of 5,6-dihydropyran the reaction

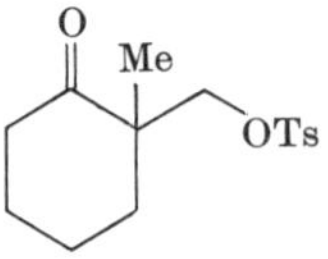

(**42**) 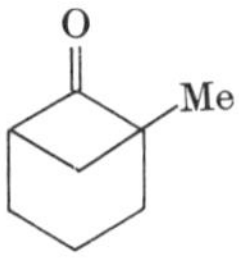 (**43**) 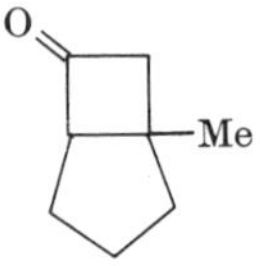 (**44**) 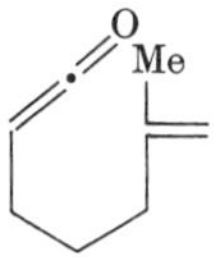 (**45**)

proceeds normally and no trapping product is formed.[81] The other group have utilized the labelled ketone (**46**), which would afford the symmetrical ketene (**47**) and result in scrambling of the labelled methyl group. When (**46**) was treated with base, products analogous to (**43**) and (**44**) were formed without any label-scrambling.[82] The two groups concur in concluding that the mechanism is as previously proposed.

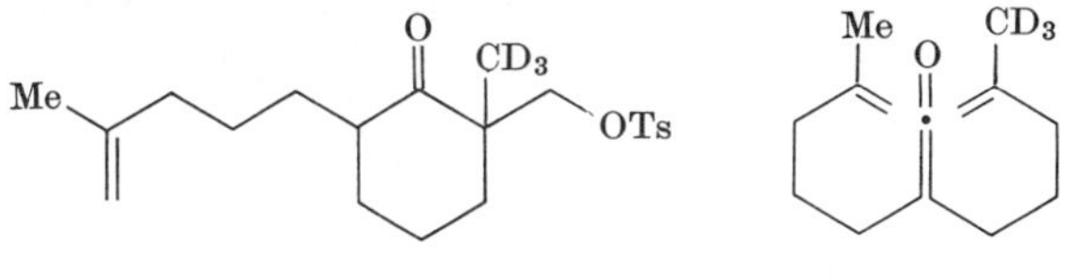

(**46**) (**47**)

5,5,5-Trichloropent-3-en-2-one reacts with ethyl sodioacetoacetate by Michael addition, followed by O-5 participation of the enolate (of the acetoacetate) to give dihydrofurans and thence, by elimination, furans.[83]

Participation by Ester, Carboxyl and Amide Groups

The rates of hydrolysis of *o*- and *p*-dichloromethylbenzoic acid in alkali have been measured in water, 50% dioxan and 60% dioxan. The ratio of the rates (*ortho*/*para*) varies from 0.83 to 110, and it is therefore suggested that the process is S_N1-like, with the *o*-carboxylate exerting only an electrostatic interaction.[84]

The X-ray crystal structure of 1,4,4,5,5-pentamethyl-1,3-dioxolanium perchlorate shows that the ring is planar with the *gem*-dimethyl groups eclipsed with one another.[85] Neighbouring acetoxyl participation in (**48**) occurs exclusively with loss of the *endo*-acetoxy-group in CF_3SO_3H. The ring-contraction to give (**49**) occurs with retention of configuration at C-5, indicating that O-3 participation is not involved.[86]

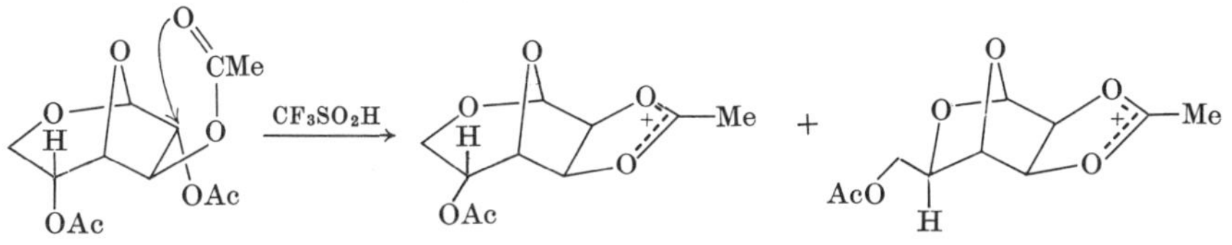

(**48**) (**49**)

The relative stabilities of the 2-substituted 1,3-dioxolanium ions derived from 1,5-anhydro-D-arabinitol by treatment with HF in acetonitrile have been estimated by use of NMR spectra. For aromatic 2-substituents, stability increases with electron-donating ability and the methyl group falls between phenyl and *p*-nitrophenyl.[87] Solvent-separated dioxolanium–halide ion-pairs are involved in the equilibrium of 1,3-dihalo-2-propyl acetate and 2,3-dihalo-1-propyl acetate.[88] Neighbouring acetoxyl participation in the addition of hypochlorous acid to 3β-acetoxycholest-4-ene gives 4β-acetoxy-5α-chlorocholestan-3β-ol and 3β-acetoxy-4α-chlorocholestan-5β-ol by acetoxyl-5 and -6 participation, respectively. The ratio of AcO-5 to AcO-6 is about 2:1 and is nearly the same for benzoyloxy-participation. However, the 3β-ethoxycarbonyl group participates overwhelmingly by way of a six-membered ring.[89] Acetoxyl-6 participation is also involved in the BF_3-catalysed rearrangements of the 1-acetoxy-3,4-epoxy-pentanes and -hexanes.[90]

6-Benzamido-6-deoxy-1,2-*O*-isopropylidene-3,5-di-*O*-methanesulphonyl-α-D-glucofuranose reacts in DMF with participation of the 5-benzamido-group to give the phenyloxazoline.[91]

Participation by Neighbouring Amino, Silyl and Phosphorus Groups

A CNDO/2 investigation of the rotational barriers in 2-(trimethylsilyl)ethyl and 2-(trimethylgermyl)ethyl cations has shown that the conformers with the C—Si or C—Ge bond eclipsed with the vacant *p*-orbital are, respectively 40 or 47.7 kcal mol^{-1} more stable than those where they are orthogonal. The stabilization is almost entirely due to hyperconjugative interaction, $p\pi$–$d\pi$ overlap accounting for less than 2 kcal mol^{-1} of it.[92]

3-Chloro-1-ethylpiperidine and 2-(chloromethyl)-1-ethylpyrrolidine react with alkoxides in alcohol (or with aqueous base) to give the same mixture of products; the intermediacy of the ion (**50**) is used to account for this.[93] *N*-(2-Chloroethyl)-*p*-nitrobenzenesulphonamide reacts with base to give *N*-(2-hydroxyethyl)aniline by way of an intermediate aziridine (which may be isolated) and two Smiles' rearrangements.[94] *cis*- and *trans*-1-*tert*-Butyl-2-methylazetidin-3-yl toluene-*p*-sulphonate with KCN in methanol each give the corresponding nitrile with retention of configuration, and solvolyse in 60% acetone also with retention. The *cis*-2-methyl compound reacts 350 times faster than the *trans*-isomer, and also 29 times faster than the 2-H compound although the activation enthalpy is lower for the last-mentioned compound; it is suggested that there is greater anchimeric assistance in the *cis*-2-methyl compound, resulting in less solvation and hence a more favourable $\Delta S^{\ddagger}$, and also that in the bridged ion-pair (**51**), the *endo*-methyl group prevents solvent approach. The *trans*-2-methyl compound reacts more slowly owing, it is suggested, to eclipsing interaction between methyl and *tert*-butyl groups in the ion-pair (**52**).[95]

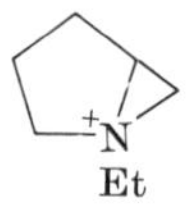

(**50**) 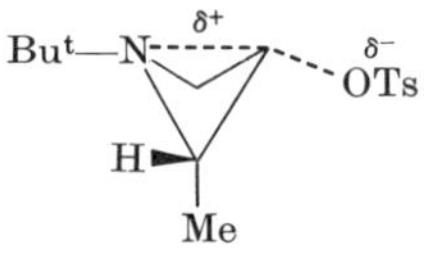 (**51**) 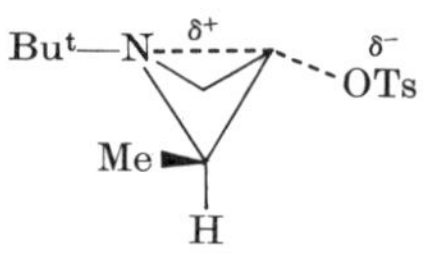 (**52**)

The rates of ring closure and activation parameters have been measured for the series $ArNH(CH_2)_nX$ and *p*-$C_6N_4MeSO_2NH(CH_2)_nX$ (X = Cl or Br), and also for the ω-haloalkylamines; the chloride/bromide rate ratios and the small effect of *N*-phenyl and *N*-(2,6-dimethylphenyl) substitution show that for all cases there is very little C—N bond formation in the transition state; in all the series the rate of pyrrolidine formation is greater than that of aziridine formation, the difference being due entirely to the difference in $\Delta H^{\ddagger}$. Aziridine formation is faster than azetidine formation and this is also almost entirely due to an enthalpy difference, and not to an entropy one as has been the classical argument.[96] Pyridine-2-thiones react with *cis*- and *trans*-2-bromocrotonic acids to give salts (**53**) by Michael addition of sulphur, followed by N-5 participation (cf. ref. 65).[97] The salt (**54**) reacts with base to give 3,5-ethanoquinuclidine,[98]

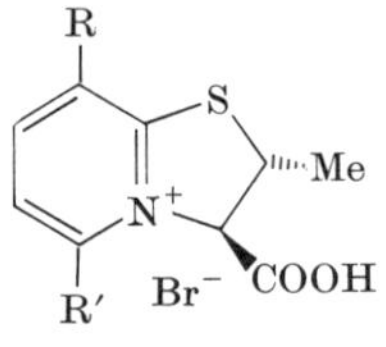

(**53**) 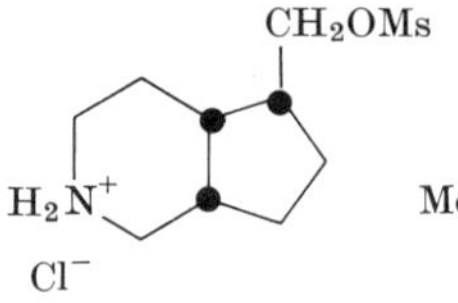 (**54**) 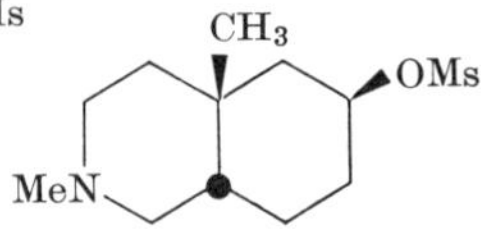 (**55**)

and (**55**) affords the 1,8-dimethyl-1-azatwistanium ion as its methanesulphonate,[99] both by N-6 participation.

Participation by the phosphoric group may occur in the conversion of (**56**; R = PO_3H_2, X = Cl) into (**57**; R = PO_3H_2) by aqueous alkali,[100] and also in that of (**56**; R = Ph, X = NH_2) into (**57**; R = Ph) by nitrous acid.[101] Further proof has appeared of the migration by the diphenylphosphinyl group in the acetolysis of 2-(diphenylphosphinyl)-1,2-dimethylpropyl methanesulphonate.[102] The balance of factors between methyl and diphenylphosphinyl migration appears to depend in particular on the stability of the carbocation left behind. Thus (**58**) with acid gives the tertiary alcohol formed by methyl migration from the *tert*-butyl group, (**59**) gives the alkene derived by diphenylphosphinyl migration, and (**60**) gives a mixture of 25% of methyl- and 75% of diphenylphosphinyl-migrated products.[103]

$H_2O_3P-CR(OH)CH_2X$ (**56**) $RCOCH_2PO_3H_2$ (**57**) $Bu^tCH(OH)CMe_2—P(O)Ph_2$ (**58**)

$[Ph_2P(O)—CMe_2]_2CHOH$ (**59**) $Ph_2P(O)CMe_2CH(OH)CMe_2CH_2—P(O)Ph_2$ (**60**)

The solvolysis kinetics of *o*-nitrobenzohydrazonyl bromide in 50% dioxan show that participation by the nitro-group occurs.[104]

Participation by Neighbouring Carbanion

Potassiohexamethyldisilazane reacts with α-bromoketals of the type (**61**) in benzene to give five- and six-membered rings as in (**62**; n = 2 or 3), and the bisketal (**63**; X = Br) gives (**64**).[105] Lithiohexamethyldisilazane reacts with (**63**; X = Br) in benzene to give the *trans*-isomer of (**64**), but with (**63**; X = OTs) to give (**64**), and (**64**) is also formed when (**63**; X = Br) reacts in THF. Because (**65**) cyclizes to give only the *trans-syn-trans*-perhydroanthracene skeleton regardless of the metal ion, cyclization to give

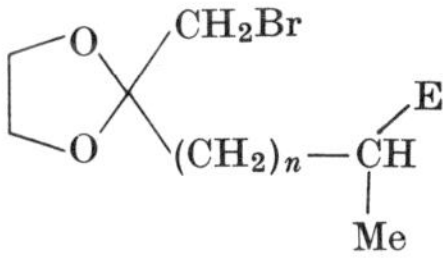

(**61**) 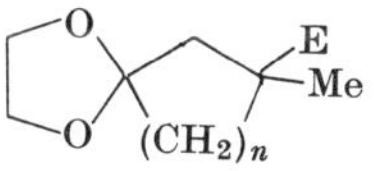 (**62**) 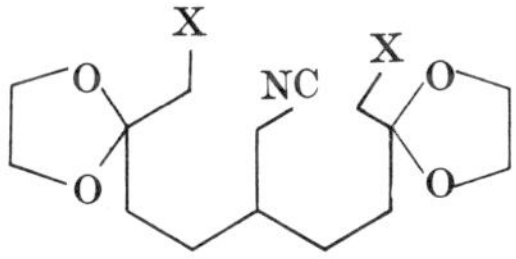 (**63**)

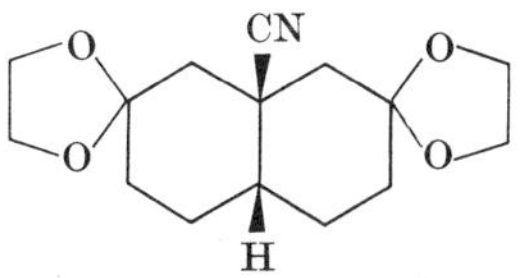

(**64**) 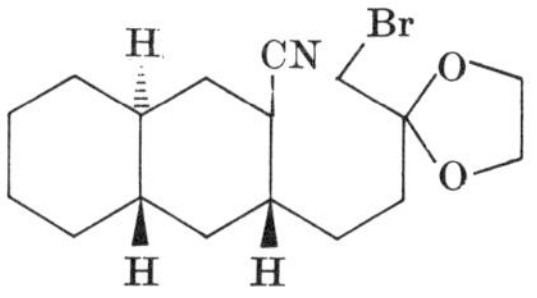 (**65**)

E = COOEt or CN

cis-ring junctions must arise from the conformation (**66**) and the rate of closure to *cis*-must be greater than that to *trans*-product. Lithium changes the stereochemical outcome owing to its greater requirement for solvation, which means that the liberated counter-ion must be formed in close proximity to it; this is possible for (**67**), but not for (**66**), except when X = OTs, a bifunctional anion. Solvation of Li^+ by THF restores the order found

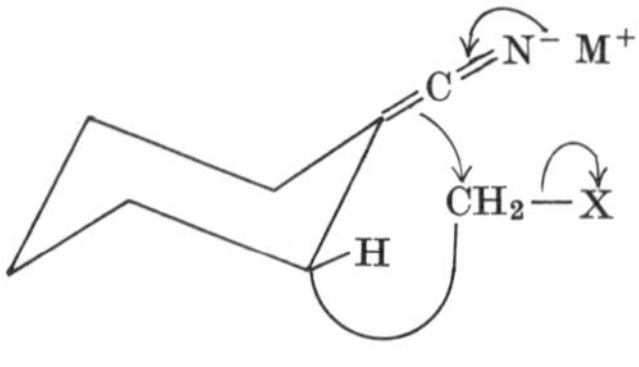

(**66**) 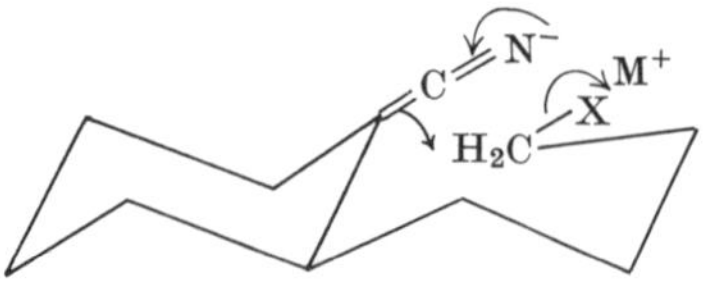 (**67**)

with potassium. Perhydroindane systems are formed with greater than 80% *cis*-stereochemistry regardless of the metal ion.[106] The allylic dibromide (**68**) reacts with potassium *tert*-butoxide in *tert*-butyl alcohol to give the semibullvalene (**69**).[107]

Sulphones with two α-methylene groups (**70**) react with CCl_4–KOH in *tert*-butyl

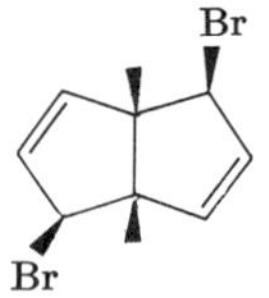

(**68**) 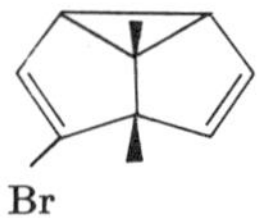 (**69**)

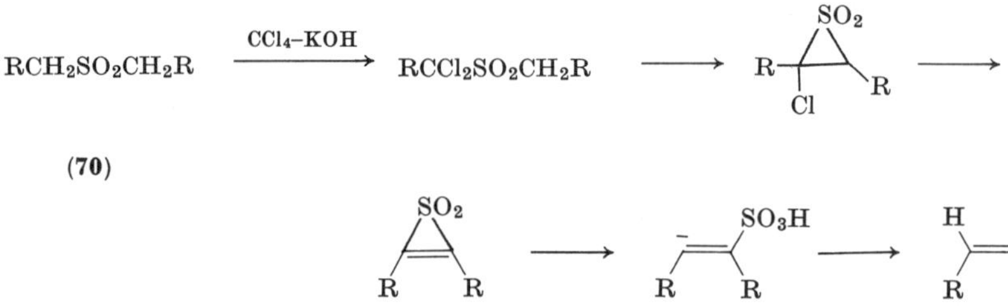

$$RCH_2SO_2CH_2R \xrightarrow{CCl_4\text{–}KOH} RCCl_2SO_2CH_2R \longrightarrow \longrightarrow$$

(**70**)

alcohol to give vinylsulphonic acids by the route shown.[108] The transformation of α,α′-dibromodibenzyl sulphone (**71**) to stilbene induced by triphenylphosphine is stereospecific, the (±)-isomer giving *trans*- and the *meso*-isomer giving *cis*-stilbene, and yet this is a two-step reaction, for the intermediate carbanion can be trapped with methanol to give α-bromodibenzyl sulphone. The authors argue that the majority of 1,3-eliminations go by two steps, viz. formation of a carbanion, initially with retention of configuration, followed by ring-closure with inversion. The sulphone (**71**) reacts with a large number of reagents to give stilbenes, with greater or lesser degrees of stereospecificity, which Bordwell and Jarvis discuss at length.[109]

Neighbouring Carbon and Hydrogen

Hydrogen scrambling in ethyl cations generated in the gas phase has been studied by using CD_3CH_2I and CH_3CD_2I and quenching the cations with isobutane or perdeuterioisobutane. The deuterium distribution in the product is found to be statistical up to pressures of 1000 torrs, from which a $\Delta G^{\ddagger}$ for hydrogen shift of less than 5 kcal mol^{-1} is calculated.[110]

Migratory aptitudes have been evaluated for cyclopropyl groups in the silver nitrate-catalysed rearrangement of (**72**). For R = Me, the cyclopropyl migrates 48 times more readily. 2-Methyl substitution in the cyclopropyl group causes it to migrate three times faster than the unsubstituted analogue, and 2,3-dimethyl substitution causes a ten-fold increase. Because the effect of methyl substitution is multiplicative, the authors favour

(**72**) (**73**)

a transition state such as (**73**).[111] The labelled 3-hydroxyoxetanol (**74**) affords the hydroxyketone (**75**), and therefore by the route shown. The preferential migration of hydroxymethyl over methyl is surprising.[112] Reaction of the cyclopropanol (**76**) with manganese dioxide gives the azetidinone (**77**) with retention of configuration.[113] The full paper on the reaction of diazomethane with, and the Tiffeneau–Demjanov ring

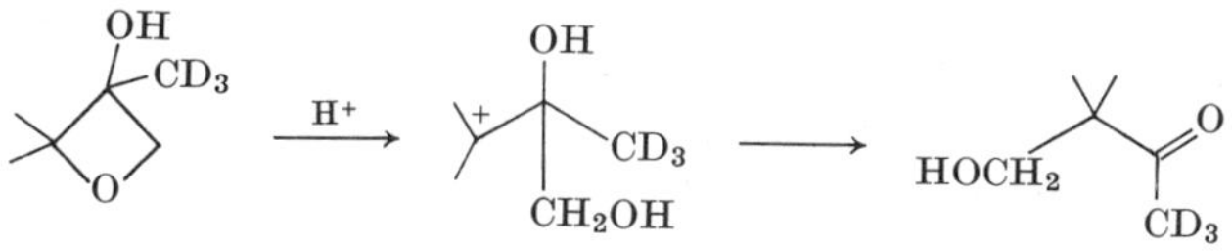

(**74**) (**75**)

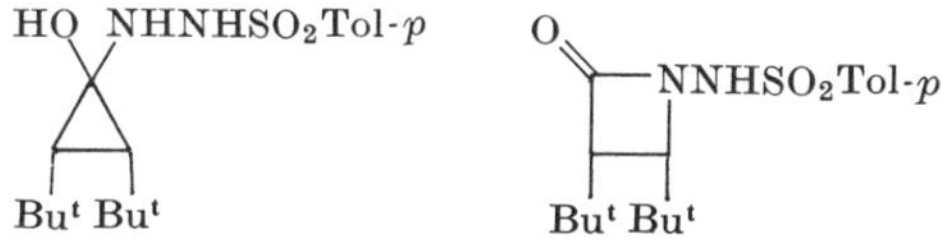

(**76**) (**77**)

expansion of, 17β-hydroxy-5α-androstan-3-one and the corresponding 5α-cholestan-3-one has been published.[114] The use of diazonium betaines, formed by pyrolysis of the adducts of arenesulphonyl azides with olefins, in ring-contraction and ring-expansion has been exemplified; the method compares favourably with others that have been used.[115] Perhydroindane-*cis*- and -*trans*-8,9-diol both undergo pinacol rearrangement with contraction exclusively of the six-membered ring.[116] Ring contraction of the decalol derivative (**78**) in a number of solvents occurs with stereochemical scrambling of the labelled methyl group,[117] which contrasts with the stereospecificity found for the 4,4-dimethylcholestan-3β-yl derivatives reported last year.[118] The formolysis of 3β-acetoxy-17α-[^{2}H]-20α-(toluene-*p*-sulphonyl)-5α-pregnane (**79**) affords mainly the Δ^{13}-olefin formed by hydrogen and methyl shift. However, about 15% of retained 20α-formate and 8% of (**80**) are formed. It is postulated that the bridged ion (**81**) is formed and partitions between retained formate and conversion into ion (**82**), which affords (**80**).[119] The epimer of (**79**), the 20β-compound, undergoes stereospecific ring expansion

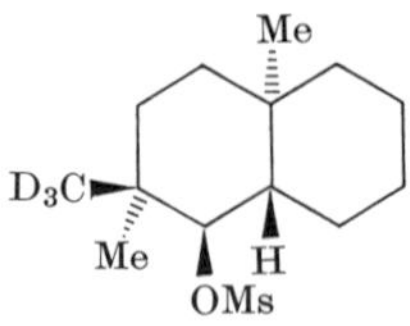

(**78**) 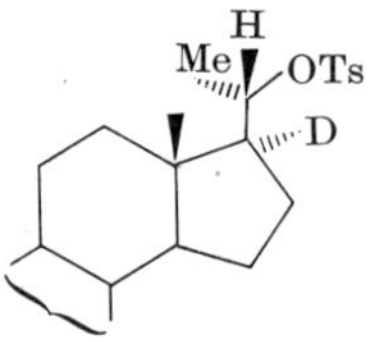 (**79**) 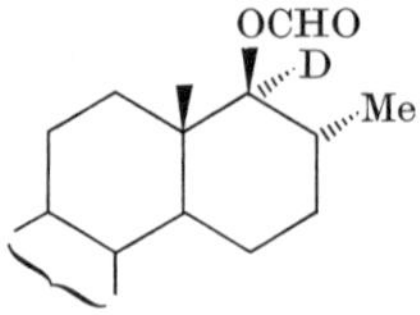 (**80**)

Formate ← (**79**) (**81**) (**82**)

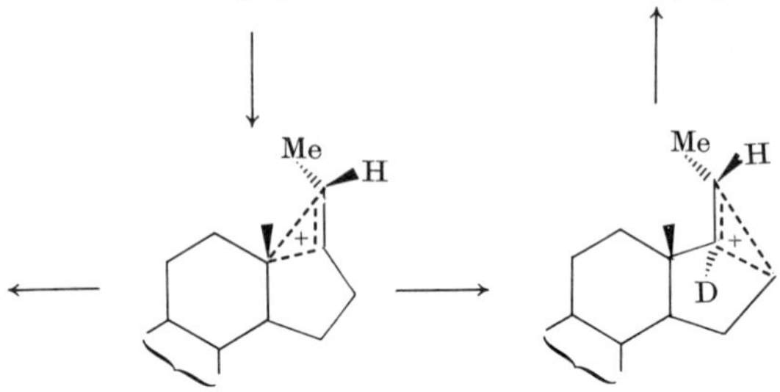

to (**83**), which formolyses to give retained formate (**83**; OCHO in place of OTs) and (**84**) with inversion at both position 13 and position 17. A process involving participation of the C-13—C-14 bond followed by sequential ring contractions to regenerate the pregnane skeleton and finally re-expansion of ring D is proposed.[120] It is not clear to the Reviewer why this should occur stereoselectively in the observed manner.

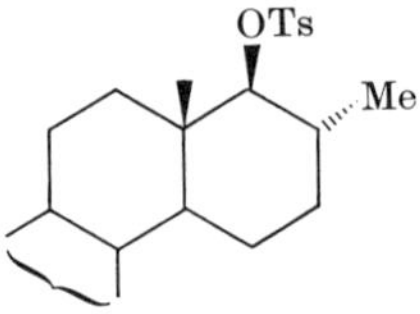

(**83**) 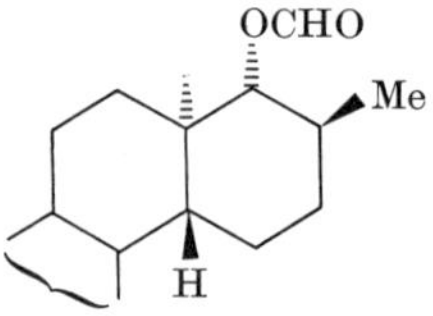 (**84**)

The fact that there is no change in the steric course of the backbone rearrangement between A-noreuphenone and the D-homo-analogue has been used to argue for a thermodynamic control of the process.[121] The retro-Westphalen rearrangement of 5β-methyl-10α,14β-cholest-8-en-2α-ol takes place with epimerization at C-14, to give the *trans-anti-trans*-BCD geometry in the product.[122] Backbone rearrangement of methylparavallarine takes place to give the $\Delta^{8,9}$-14β-compound if the 18-methyl group is not involved.[123]

By using 1-[^{2}H]-cyclohexyl toluene-*p*-sulphonate and *trans*-2,2,6-[$^2H_3$]-cyclohexyl toluene-*p*-sulphonate, Lambert and Putz[124] have dissected the routes by which unsubstituted cyclohexyl derivatives are solvolysed. From the isotope effect (α $k_H/k_D \pm 1.22$) they infer a rate-determining attack on the ion-pair, in each of the three solvents studied. In acetic acid, cyclohexyl acetate is formed 20% from a hydrogen-shifted ion and 80% from the unrearranged ion-pair, the unrearranged product being completely inverted. Cyclohexene, the major product (80%), is formed 95% from the unrearranged ion-pair. The partitioning of unrearranged and rearranged ion-pairs between S_N and E, 16:76 and

4:4, respectively, is almost identical with the values found previously for 4-*tert*-butylcyclohexyl toluene-*p*-sulphonate in acetic acid, 19:74 and 2:6, respectively. The authors therefore suggest that in both cases reaction takes place with the leaving group equatorial. As would be expected, more rearrangement is observed in formic acid, 40% for S_N and 10–15% for elimination, and still more (50%) in trifluoroacetic acid. Surprisingly, however, the overall S_N/E ratio is higher in the latter than in the former solvent, and higher even than in acetic acid.

Deamination and Related Reactions

Moss *et al.*[125] have reviewed the work of his group on micellar effects, and the full paper on these has been published.[126]

The reaction of the potassium diazotates from optically active α-methylbenzylamine and 2-octylamine with ammonium tetrafluoroborate in ammonia and ammonia–HMPT affords amines of mainly inverted configuration and alcohols of mainly retained configuration; the ion-pair scheme discussed in 1970[127] is used to explain the results. A summary of all the work in this field shows that the nature of the anion is not very important, but that the stability of the cation plays a rôle; thus for products arising by internal return, overall retention is observed and the stereospecificity increases with cation stability, the α-methylbenzyl cation giving ca. 82% of retention and 2-octyl cation ca. 68%; the reverse obtains with the products of solvolysis, where overall inversion is observed, ca. 70% for the α-methylbenzyl and ca. 82% for the 2-octyl cation.[128]

Deamination of nerylamine and geranylamine in a number of solvents gives much less cyclized product than does solvolysis of the corresponding chlorides, phosphates and pyrophosphates. Substantial amounts of alcohols are formed even in glacial acetic acid. Partition of an intermediate diazonium hydroxide to diazonium ion–hydroxide ion and diazonium ion–acetate ion pairs is postulated (as ref. 127) with cyclized product arising only from the carbocation–acetate ion pairs which become solvent-separated.[129]

Deamination of 3*S*-(**85**) in aqueous perchloric acid leads to the formation of 4*S*-(**86**) as 33% of the products, in 86–88% optical purity. When carried out in deuterium oxide, the (**86**) formed contains 4 atom-% of deuterium, consistent with the intermediacy of the protonated cyclopropane (**87**) and the related cyclopropanol. However, to explain the optical result, 7% of (**86**) must be formed via (**87**), and so direct interconversion of the edge-protonated cyclopropanes (**87**) and (**88**) is postulated;[130] it should be noted that

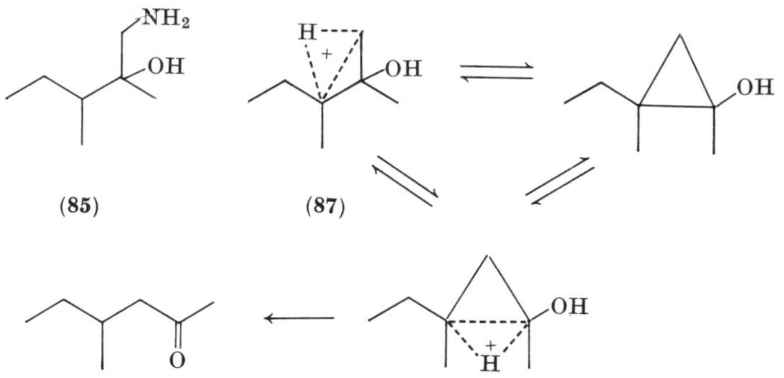

NH2 OH (**85**)

H + OH (**87**)

OH

OH H + (**88**)

O (**86**)

cyclopropane protonated at a single corner would suffice. Similar deamination of 3*S*-2,3-dimethylpentylamine (**89**; R = H) gives 5.5% each of 3*S*-(**90**; R = H) and (**91**); the (**90**; R = H) is formed with 87% retention of configuration and the amount of (**91**) is too large to be explained by two successive 1,2 hydride shifts since 1,1,2-trimethylbutylamine gives only 3.5% of (**91**) on deamination. The protonated cyclopropane (**92**) is used to

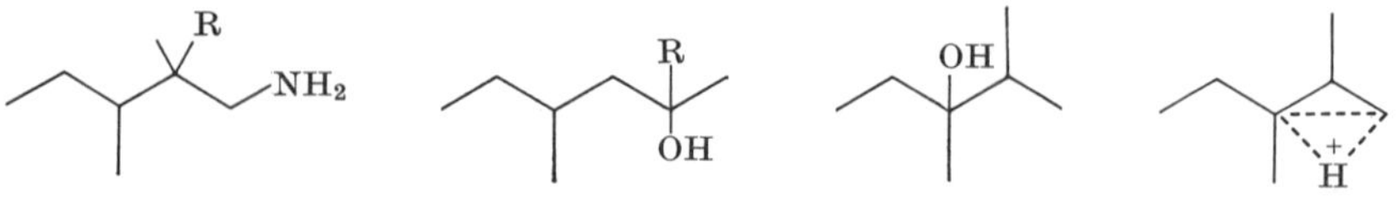

(**89**) (**90**) (**91**) (**92**)

explain both the racemization and the formation of (**91**). When the carbocation generated by migration of the *sec*-butyl group is tertiary, however, migration occurs with nearly complete retention of configuration, S-(**89**; R = Me) giving *S*-(**90**; R = Me) of 98% purity.[131] The alkyl chain in dihydro-2-methylcitronellylamine migrates with complete retention of configuration when deaminated in aqueous solution.[132] The deamination of D-glucosaminitol has also been studied.[133]

Formation of 1-phenylprop-3-yne-1-diazotate (**93**) (from the nitrosourethane by treatment with sodium carbonate in methanol) leads exclusively to (**94**), whereas the corresponding 3-phenylprop-3-yne-1-diazotate affords a mixture of (**94**) and (**95**). The equilibration of diazonium ions and diazo-compounds shown in Scheme 2 is used to explain the results. However, similar treatment of (**93**) with 0.5M-sodium methoxide gives exclusively methyl 3-phenylpropionate by a reaction involving intramolecular addition of the diazotate group to the acetylene.[134] Prop-3-yne-1-diazotate and its 3-alkyl-substituted derivatives react analogously.[135]

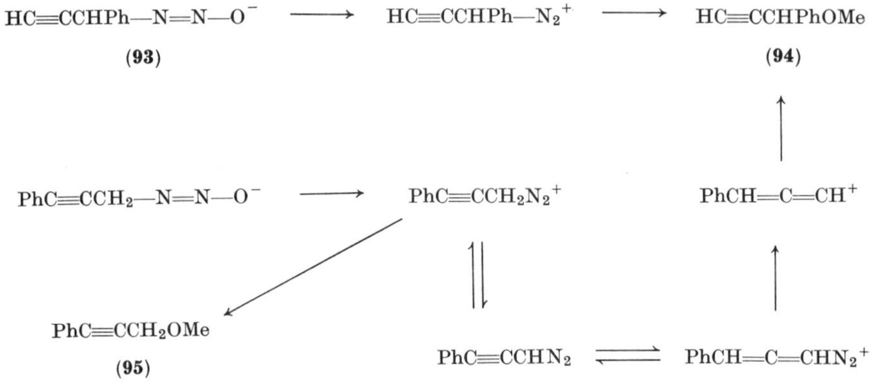

HC≡CCHPh—N=N—O⁻ (**93**) → HC≡CCHPh—N_2^+ → HC≡CCHPhOMe (**94**)

PhC≡CH_2—N=N—O⁻ → PhC≡$CCH_2N_2^+$ → PhC≡CCH_2OMe (**95**)

PhC≡$CCH_2N_2^+$ ⇌ PhC≡$CCHN_2$ ⇌ PhCH=C=CHN_2^+ → PhCH=C=CH^+ → (**94**)

SCHEME 2.

In the presence of acid, reaction of oximes with ^{15}N-labelled nitrous acid gives N_2O and N_2, each with one labelled nitrogen atom, and small amounts of NO. In the absence of an excess of acid, N_2O and NO are the major gaseous products. The proposed route, which takes into account the formation of $N_2{}^{18}O$ when ^{18}O-labelled nitrous acid is employed, is shown in Scheme 3.[136]

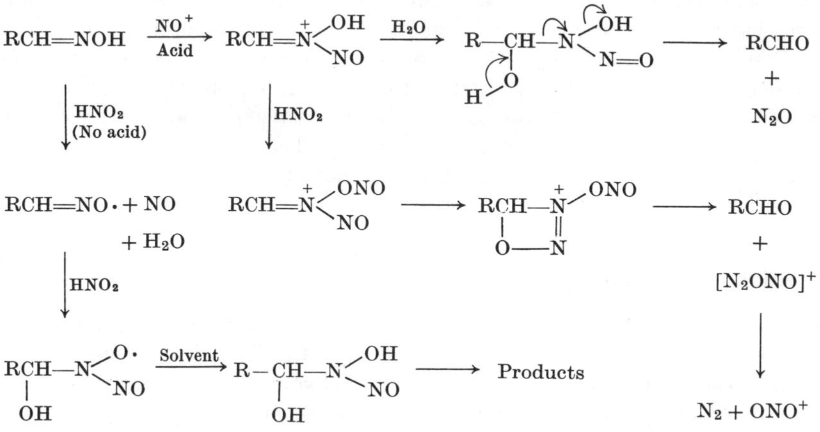

$$RCH{=}NOH \xrightarrow[\text{Acid}]{NO^+} RCH{=}\overset{+}{N}(OH)(NO) \xrightarrow{H_2O} R{-}CH(OH){-}N(OH)(N{=}O) \longrightarrow RCHO + N_2O$$

$$RCH{=}NOH \xrightarrow[\text{(No acid)}]{HNO_2} RCH{=}NO\cdot + NO + H_2O \xrightarrow{HNO_2} RCH(OH){-}N(O\cdot)(NO) \xrightarrow{\text{Solvent}} R{-}CH(OH){-}N(OH)(NO) \longrightarrow \text{Products}$$

$$RCH{=}\overset{+}{N}(OH)(NO) \xrightarrow{HNO_2} RCH{=}\overset{+}{N}(ONO)(NO) \longrightarrow RCH{-}\overset{+}{N}(ONO){=}N{-}O\ (\text{ring}) \longrightarrow RCHO + [N_2ONO]^+ \longrightarrow N_2 + ONO^+$$

SCHEME 3.

Reactions of Aliphatic Diazo-compounds

Diazonorcamphor reacts with aq. acetic acid in ether to give tricyclanone, 3-hydroxynorbornanone (9:1 *endo*:*exo*) and 2-hydroxybicyclo[2.2.1]heptan-7-one (2:1 *endo*:*exo*), together with small amounts of cyclohex-3-enecarboxylic acid. Similar results are obtained with anhydrous hydrochloric acid in ether and with anhydrous hydrofluoric acid in dichloromethane (halides formed instead of alcohols), except for the 7-keto-products which are entirely *exo*-substituted. Protonation from the *exo*-side is postulated, followed by formation of an open α-keto-carbocation which subsequently affords the products.[137] No explanation of the high *endo*/*exo* ratio for the 3-X-2-ketone is advanced, and the possibility of *endo*-protonation,[138] followed by a competition between direct displacement and participation of the 4,5-bond, is not considered. In any case the product ratios are strikingly different from those reported last year.[138a]

The kinetics of the reaction of diphenyldiazomethane with toluene-*p*-sulphinic acid in dichloromethane, benzene, acetonitrile, ethanol, dioxan, and dimethyl sulphoxide

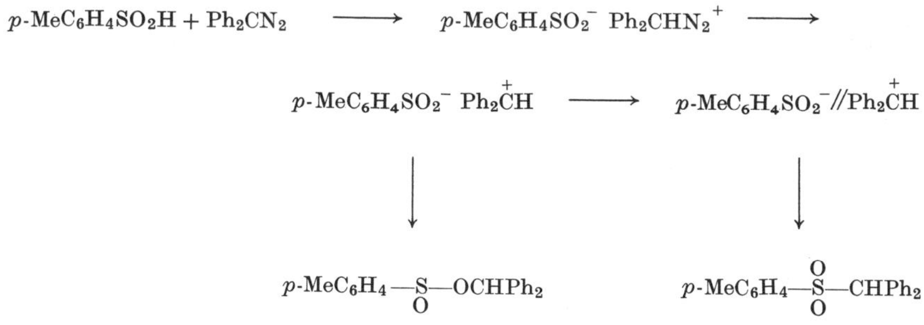

$$p\text{-}MeC_6H_4SO_2H + Ph_2CN_2 \longrightarrow p\text{-}MeC_6H_4SO_2^-\ Ph_2CHN_2^+ \longrightarrow p\text{-}MeC_6H_4SO_2^-\ Ph_2\overset{+}{C}H \longrightarrow p\text{-}MeC_6H_4SO_2^-//Ph_2\overset{+}{C}H$$

$$p\text{-}MeC_6H_4SO_2^-\ Ph_2\overset{+}{C}H \longrightarrow p\text{-}MeC_6H_4{-}S(O){-}OCHPh_2\ \textbf{(96)}$$

$$p\text{-}MeC_6H_4SO_2^-//Ph_2\overset{+}{C}H \longrightarrow p\text{-}MeC_6H_4{-}S(O)_2{-}CHPh_2\ \textbf{(97)}$$

SCHEME 4.

(DMSO) have been measured, and activation parameters have been evaluated for the last two. The rates decrease by a factor of 3×10^4 on change from dichloromethane to dimethyl sulphoxide. Reaction with [*O*-^{2}H]toluene-*p*-sulphinic acid gives a k_H/k_D of 3.0, consistent with a rate-determining proton-transfer. The authors suggest that the solvent effect is caused by the DMSO solvatively de-dimerizing the acid and not solvating the anion, thus lowering its proton-donating ability. The product changes from 100% sulphinate (**96**) in dichloromethane to 100% sulphone (**97**) in DMSO, and the ion-pair Scheme 4 is used to account for the results.[139]

Fragmentation Reactions

Little of importance has been reported this year. The fragmentation of 2-(trialkoxylsilyl)-ethyl chlorides and bromides is dependent on pH. In alkali, hydrolysis of the first alkoxy-group is rate-determining, while in acid this hydrolysis is postulated to be rapid with rate-determining attack of water on the (trihydroxysilyl)ethyl halide.[140]

N—OMgBr MgBr Ph D D → ≡N + D D → CN D

(**98**)

Camphor oxime reacts with two molecules of phenylmagnesium bromide, to give the nitrile (**98**) by the route shown.[141] Fragmentation reactions are also observed in the reports of refs. 142 and 143.

Displacement Reactions at Elements Other than Carbon

Silicon, Germanium, Tin and Lead

Unlike open chain R_3Si^*Cl compounds which undergo displacement with inversion of configuration, 1-chloro-1-phenyl-1-sila-acenaphthene reacts with $LiAlH_4$, aqueous hydroxide ion or powdered potassium hydroxide in benzene, in all cases with retention of configuration. It is suggested that this is due to angle strain at the silicon, where the α-NpSiCH$_2$ angle is only 93.4°, as shown by X-ray methods.[144] 1-Chloro-3,5,7-trimethyl-1,3,5,7-tetrasila-adamantane reacts rapidly with nucleophiles, chlorine being displaced; an S_N1-Si reaction is suggested.[145] The kinetics of the reduction of optically active silanes with di-isobutylalane, whose stereochemistry was reported last year, have been measured; for reductions involving retention (R_3SiF in hexane and R_3SiOPr^i in ether), the rate is first-order in silane and in alane, whereas for reductions proceeding with inversion (R_3SiF in ether and R_3SiCl in ether), the rate is both second- and third-order in the alane, while remaining first-order in the silane; the authors[146] find this consistent with their previously expressed view that retention occurs by a four-centre S_Ni-Si process, whereas inversion requires electrophilic catalysis and hence more than one equivalent of reagent. On the other hand, because (**99**; X = OMe or F) reacts with variety of alkylmagnesium bromides and with allyl-lithium with retention of configuration and with a leaving-group rate-ratio, k_F/k_{OMe}, that is nearly constant, and because similar results are found for (**100**; Y = OMe, R, H or D), a mechanism involving rate-determining formation of a five-co-ordinate intermediate, followed by loss of the leaving group is proposed, and the S_Ni-Si mechanism of ref. 146 is ruled out.[147] It should be noted that there are some serious discrepancies in the argument, viz. k_{OMe}/k_H for (**100**) with butyl-lithium in hexane is 40 and k_F/k_{OMe} with butyl-lithium in ether is 50. Possible support for a five-co-ordinate species comes from a study of the solvolyses of p-$XC_6H_4SiMe_2OPh$; in acid, a ρ-value of −0.57 is consistent with a five-co-ordinate intermediate, but the ρ-value of +0.19 in alkaline solution is not, and an S_N2-Si mechanism is proposed.[148]

Further work on displacement by allylic anions has been reported. (**101**; Z = H) reacts with allylmagnesium bromide with inversion in ether, but with retention of

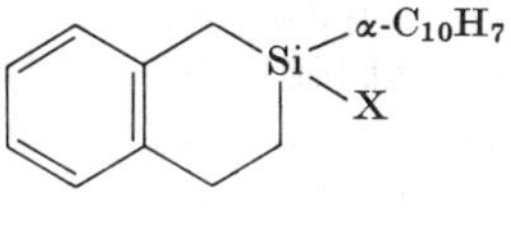

(Si bearing α-$C_{10}H_7$ and X in a ring fused to a benzene ring)

(**99**)

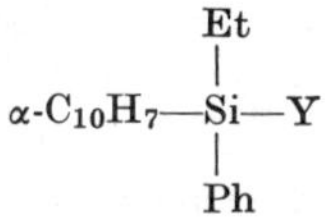

α-$C_{10}H_7$—Si(Et)(Ph)—Y

(**100**)

α-$C_{10}H_7$—Si(Ph)(OMenthyl)—Z

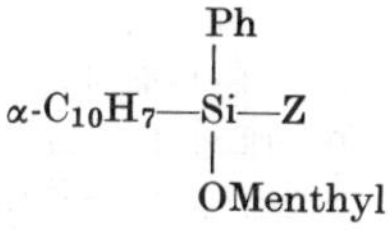

(**101**)

configuration in THF, as does allyl-lithium in ether. (**101**; Z = OMe) reacts with allylmagnesium bromide in ether with inversion of configuration and loss of menthyl oxide, and in THF with retention and displacement of methoxide. The hardness or softness of the nucleophile controls the stereochemistry, it is suggested: hard groups react with retention and soft ones with inversion; thus in the more basic solvent THF, the hardness of allylmagnesium bromide is increased, leading to retention, whereas, if the "acid" $MgBr_2$ is also added, it becomes soft again and inversion is again observed.[149]

Two new syntheses of optically active silanes have been reported. The first of these is the asymmetric reduction of methoxysilanes with lithium aluminium hydride–alkaloid complexes; thus racemic (ethyl)(methoxy)(methyl)phenylsilane with $LiAlH_4$ in the presence of (+)-cinchonine gives (+)-(ethyl)(methyl)phenylsilane (41% optically pure) and the (−)-enantiomer of the starting methoxysilane; the method is particularly useful for alkyl- and benzyl-silanes.[150] The other involves use of tris(triphenylphosphine)rhodium chloride and tris(triphenylphosphine)ruthenium dichloride as catalysts for the conversion of diarylsilanes into (alkoxy)diarylsilanes with optically active alcohols; asymmetric inductions of up to 48% are reported.[151]

Optically active silanes R_3Si^*Z are converted into the R_3Si^*H compounds by Grignard reagents in the presence of bis(triphenylphosphine)nickel dichloride. For Z = OMe or F, reaction takes place with retention of configuration, but when Z = Cl the reaction proceeds with inversion. The suggested mechanism is shown in Scheme 5.[152]

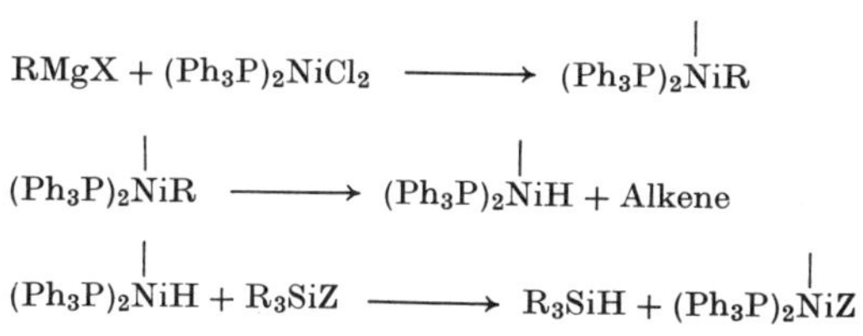

$$RMgX + (Ph_3P)_2NiCl_2 \longrightarrow (Ph_3P)_2\overset{|}{Ni}R$$

$$(Ph_3P)_2\overset{|}{Ni}R \longrightarrow (Ph_3P)_2\overset{|}{Ni}H + \text{Alkene}$$

$$(Ph_3P)_2\overset{|}{Ni}H + R_3SiZ \longrightarrow R_3SiH + (Ph_3P)_2\overset{|}{Ni}Z$$

SCHEME 5.

Phosphorus, Arsenic and Antimony[153]

It has been suggested that the permutational isomerization of five-co-ordinate phosphorus compounds may take place by a bimolecular mechanism, and Scheme 6 illustrates this for RPF_4. This mechanism is held to be preferable to others because it utilizes the

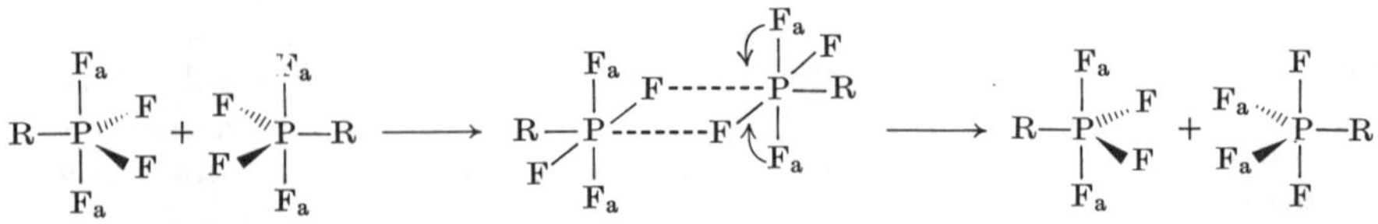

(**102**)

SCHEME 6.

known octahedral geometry rather than the postulated tetragonal pyramid [but see discussion of (**103**) below] and also because it can readily account for the range of activation energies found for the process, since rates of dimerization must depend on the molecules involved. Further, if the intermediate (**102**) is formed with sufficient energy, it can become symmetrical, which would explain the observed intermolecular exchange of halide.[154]

An X-ray structure determination of (**103**) has shown it to have the tetragonal pyramid structure hitherto only postulated as the transition state for the pseudorotation process.[155]

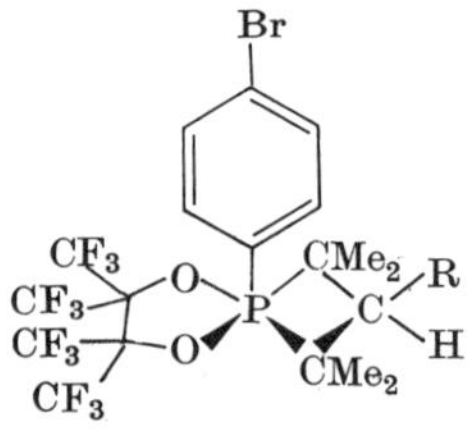

(**103**)
(R = H or Me)

Optically active trisubstituted phosphine oxides are converted into the corresponding phosphine sulphides by boron trisulphide with 68–97% retention of configuration.[156]

An example of neighbouring-group participation at phosphorus has been reported. The mixed hydroxamic phosphonic anhydride (**104**) is hydrolysed extremely rapidly by base to give the *O*-ethyl phosphonate and arylamine. The reaction is not affected by steric crowding at the aryl group, and the proposed mechanism is as shown.[157]

The cleavage reactions of tetra-alkylphosphonium salts by hydroxide ion in DMSO show that, although methyl groups are cleaved in preference to ethyl groups, there is no preference for cleavage of ethyl rather than higher *n*-alkyl groups.[158] The same study has shown that cleavage of 1,1-diethylphospholanium and 1,1-diethylphosphetanium salts occurs almost exclusively with ring-cleavage, although the six-membered rings behave normally. The rates of cleavage are virtually invariant with alkyl substitution except for the four- and five-membered-ring compounds which react four and eight times faster than the tetramethylphosphonium salt. Both rate and product effects are explained by the ring strain which makes formation of the five-co-ordinate intermediate easier and

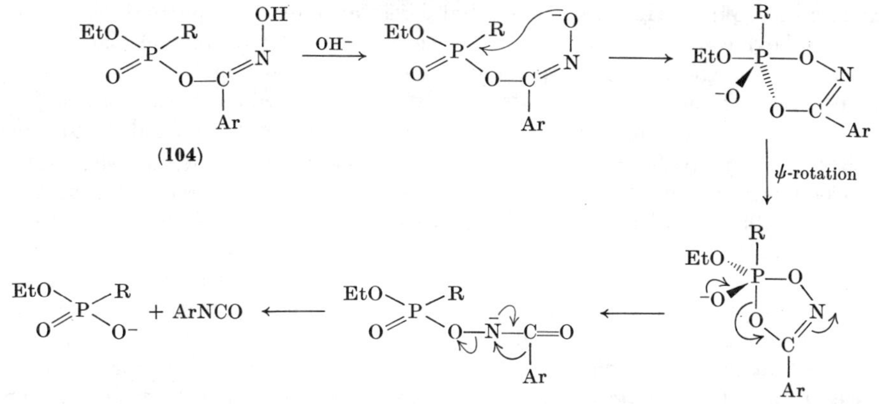

(104)

requires the ring to be apical-equatorial, i.e. with one bond always in the correct orientation for elimination.[158] By making the assumption that both attack of hydroxide and loss of leaving group occur from apical positions,[159] the alkaline hydrolysis of *cis*- and *trans*-1-X-1-alkoxy-2,2,3,4,4-pentamethylphosphetanium salts (**105**) has been used to derive a set of kinetic apicophilicities. Cl > SMe > *O*-Alkyl > NMe_2. With X = Cl or SMe, the products are almost entirely (**106**) and (**107**) in a ratio of 36 : 62, indicating that

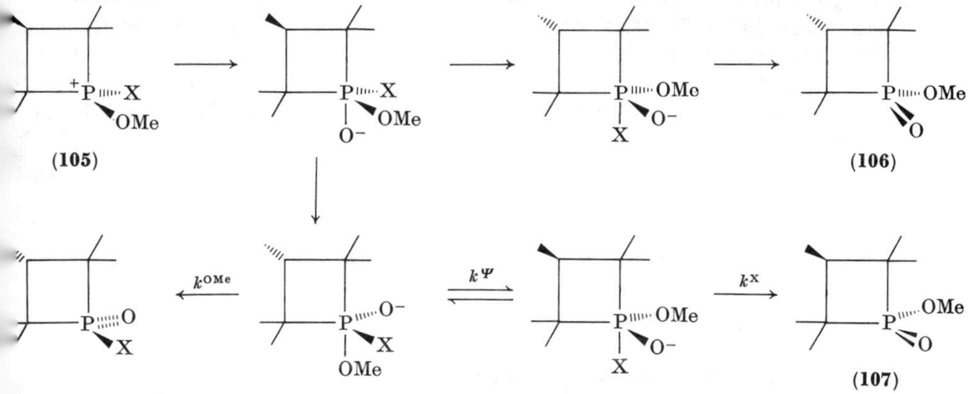

(105) (106) (107)

the rate constant $k^{\psi}.k^{X}$ must be much greater than k^{OMe}. It is suggested that this is largely due to the k^{ψ} term and the π-donor ability of the group X in comparison with that of OMe. Good π-donors are known to overlap with phosphorus better when equatorial than apical, and OMe is a better π-donor than either Cl or SMe.[160] The salt (**108**) is hydrolysed under alkaline conditions to mixtures of (**109**) with retention of configuration and (**110**) with inversion. The proportion of (**109**) to (**110**) depends on the size of the group R but is independent of that of R′; a mechanism involving attack opposite OR, followed by competitive loss of OR^- and pseudorotation is proposed; the rate of the latter process is known to be insensitive to the size of alkoxyl groups, whereas the former would increase with the size of R, giving the observed increase in the formation of (**110**) at the

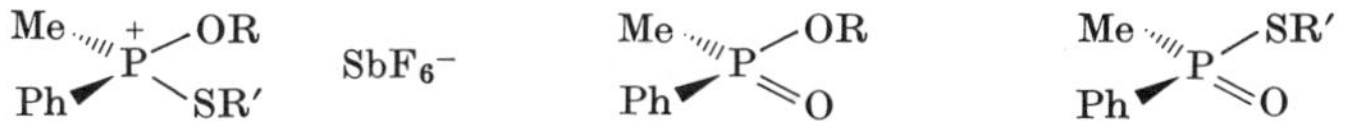

(108) (109) (110)

expense of (**109**).[161] However, *O*-menthyl *S*-methyl phenylphosphonothiolate reacts with methoxide in methanol to give *O*-menthyl *O*-methyl phenylphosphonate with inversion of configuration, corresponding to methoxide attack opposite the SMe group. In order to see if any attack opposite the OR group occurred (perhaps reversibly), the racemization and methanolysis of the *O*-methyl *S*-methyl compound, and the exchange of the *O*-[$^2H_3$]methyl *S*-methyl compound were studied; the relative rates showed that attack opposite SMe is preferred by 5:1 over that opposite OMe. No reason for the difference with (**108**) could be adduced.[162] Because of the large steric effects, the ρ value of 2–2.4 (per aryl group), the solvent isotope effect ($k_{H_2O}/k_{D_2O} = 3$), and low m (Y) value of 0.3, a six-co-ordinate intermediate transition or state [$(ArO)_5POH]^-$ has been proposed for the hydrolysis of penta-aryloxyphosphoranes.[163] Rate constants have been measured for the alkaline hydrolysis of trialkoxyphosphine oxides, sulphides and selenides.[164] The fact that the most acidic group is lost on alkaline hydrolysis of phosphonium salts has been used to evaluate the acidities of the 2-furyl, 2-thienyl, 2-furylmethyl and 2-thenyl groups relative to the benzyl group.[165]

The stereochemistry of displacement at phosphorus in 2-chloro-*cis*-5-(chloromethyl)-5-methyl-2-oxo-1,3,2-dioxaphosphorinane has been studied. Piperidine and thiophenoxide react with inversion, while substituted phenoxides give mixtures of *cis*- and *trans*-isomers; the amount of inversion observed is inversely proportional to the basicity of the phenol used.[166] The ratio of inversion to retention also varies with added salts and with solvent, the proportion of inversion increasing with tetramethylammonium salts and decreasing with lithium salts. A mechanism is proposed that involves co-ordination of metal to phosphoryl-oxygen and attack opposite the leaving group to give inversion, in competition with one of attack opposite the oxygen, pseudorotation and hence retention;[167] however this would require inversion to increase with co-ordinating ability of the metal ion, the opposite of what is observed.

Stopped-flow techniques have enabled the methanolysis of bromodiphenylphosphine, Ph_2PBr, to be studied. The rate is both second- and third-order in methanol and is catalysed by pyridine, being first-order in pyridine. A mechanism involving addition of methanol, followed by rate-determining loss of HBr, catalysed by methanol is postulated.[168]

Sulphur

For nucleophilic substitution at bivalent sulphur, an extensive investigation of the cleavage of diaryl disulphides by cyanide ion in aqueous *tert*-butyl alcohol has been carried out (equation 1). Hammett plots for Ar^1 (Ar^2 constant) gave ρ-values of 1.70

$$Ar^1SSAr^2 + CN^- \longrightarrow Ar^1SCN + Ar^2S^- \quad (1)$$

and 1.89 with the points for p-$MeOC_6H_4$ and p-$NH_2C_6H_4$ lying above, and those for p-$MeCOC_6H_4$ and p-$NO_2C_6H_4$ lying below the lines. For variation in the leaving group Ar^2 (Ar^1 constant) a ρ-value of 1.97 was found. The investigators were unable to decide between an S_N2-S mechanism with substantial charge on the leaving group, and one involving reversible formation of an intermediate followed by rate-determining cleavage.[169] However, from a study of the nucleophilic cleavage of p-nitrophenyl triphenylmethanesulphenate with butylamine and benzamidine, the kinetic Scheme 7 has been

$$RNH_2 + ArOSCPh_3 \underset{k_{-1}}{\overset{k_1}{\rightleftharpoons}} \text{Intermediate} \xrightarrow[RNH_2]{k_B} \text{Product}$$

SCHEME 7.

proposed. The reaction is second-order in butylamine concentration but only first-order in benzamidine concentration. Benzamidine cannot operate by a "push-pull" mechanism for an S_N2-S process, and it is suggested that, for butylamine, proton transfer in (**111**) is slow, so that $k_B[RNH]_2$ is less than k_{-1} and rate-determining, while for benzamidine proton transfer in (**112**) is much faster owing to the accessibility and therefore k_1 is rate-determining with $k_B[RNH_2]$ much greater than k_{-1}.[170]

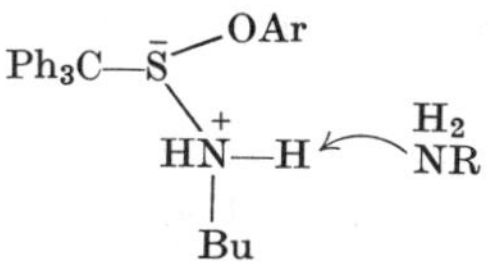

(**111**)

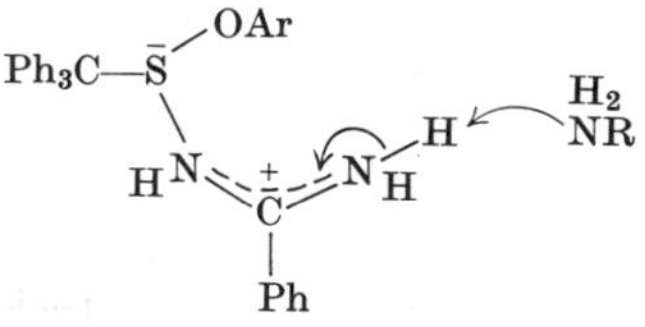

(**112**)

The full paper on the acid- and nucleophile-catalysed ^{18}O-exchange and racemization of PhSS(O)Ph has been published,[171] and a study of its disproportionation to diphenyl disulphide and $PhSS(O)_2Ph$ has been reported.[172] The hydrolysis of 1,2-dithia-acenaphthene 1-oxide is 10^3 times slower than that of PhSS(O)Ph; this has been ascribed to the rigidity of the former.[173] Methyl arenethiosulphonate reacts with morpholine to give the sulphenamide 10^5 times faster than does the *tert*-butyl ester, but the equilibrium constants are almost identical.[174] Other reactions involving nucleophilic displacement at bivalent sulphur are reported in refs. 175 and 176.

Further work of significance has been published on the mechanism of the conversions of sulphoxides into sulphilimines in which *N*,*N'*-bis(toluene-*p*-sulphonyl)sulphur di-imide was used, as reported in 1971.[177] The cyclic sulphoxide (**113**) reacts in pyridine in a third-order reaction, second-order in (TosNSNTos), with inversion of configuration, and in benzene in a second-order reaction with retention of configuration. This oxide (**113**) reacts in each case about 10^2 times faster than the open-chain analogue (**114**),

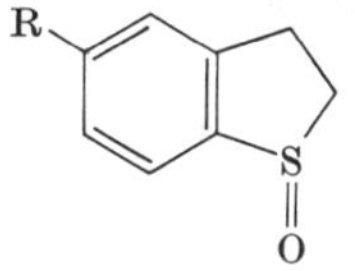

(**113**) R = H or Br

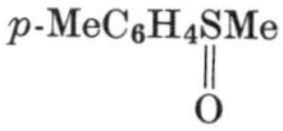

(**114**)

and this is consistent with rate-determining formation of the six- and four-membered transition states previously postulated. In dichloromethane, the conversion of sulphoxide into sulphilimine results in racemization of both at a measurable rate, the sulphoxide more rapidly than the sulphilimine. The mechanism outlined in Scheme 8 is proposed.[178] The same conversion, but with toluene-*p*-sulphonyl isocyanate as reagent, proceeds mainly with inversion, and a mechanism similar to Scheme 8 is proposed. The effect of added *tert*-butyl methyl sulphide is to increase the optical purity of the product, and it is proposed that it acts by trapping the 1,4-dipole analogous to (**115**), ensuring its reaction with a second molecule of p-$TolSO_2NCO$, instead of cyclizing and giving product of retained configuration.[179]

Diaryl sulphoxides react with triethyloxonium fluoroborate to give *O*-ethyl salts which are hydrolysed by hydroxide ion to sulphoxides of 93–99% inverted configuration.[180]

Neighbouring carboxyl participation is observed in the hydrolysis of *N*-(toluene-*p*-sulphonylmethyl)-2-carboxyphenylsulphilimine in aqueous acid. Sulphoxide is formed

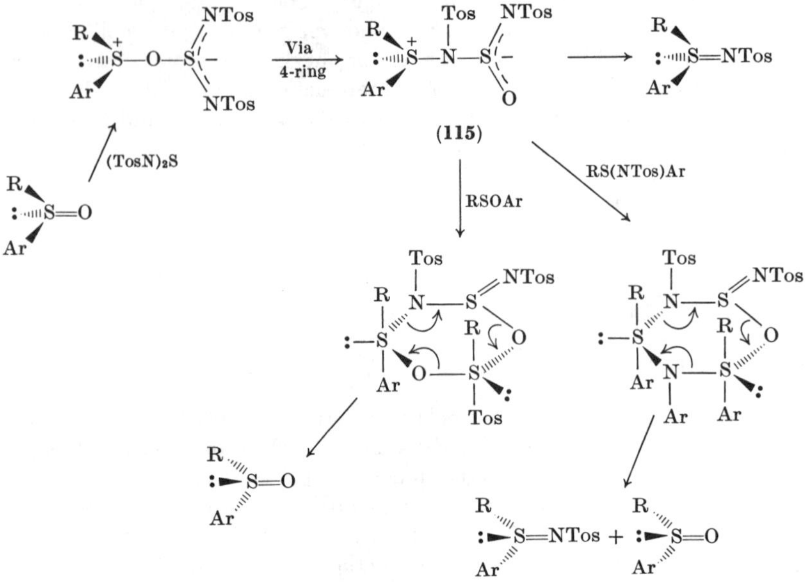

SCHEME 8.

with retention of configuration, whereas the corresponding ester reacts with inversion of configuration.[181]

Sulphoxides of the type $R^1S(O)CHR^2R^3$, where $R^1 = Bu^t$ or PhCHMe, react with *N*-halosuccinimides in chloroform–ethanol to give the halide of R^1 and the ethylsulphinate of the other radical. Reaction takes place with overall retention at R^1, although with extensive racemization. A mechanism analogous to that discussed last year for the racemization and exchange of sulphoxides and chlorination of sulphides is proposed.[182]

Further evidence in favour of the mechanism involving diprotonation followed by rate-determining loss of water for the exchange and racemization of sulphoxides in concentrated sulphuric acid has been reported.[183]

The nucleophilic displacement reactions of aryl methanesulphinates show a Brønsted β for the leaving group of –0.71. That for the nucleophiles is non-linear, being about 0.75 for phenoxides and virtually zero for hydroxide. The ρ-value for similar displacement of benzyl arenesulphinate is +1.56, which indicates substantial bonding of nucleophile to sulphur. The investigators conclude that Brønsted β-values are of little mechanistic significance owing to the combination of effects that can operate on the bond being formed and on that being broken, as well as the extent to which each is affected by the various other factors.[184]

Arenesulphonyl chlorides react with imidazole in methanol or acetonitrile with ρ-values of 1.41 and 1.43, respectively. The enthalpies of transfer from methanol to acetonitrile of ground state and transition state have been evaluated; for benzenesulphonyl chloride this has been done for nine solvents in all; it is found that the reactants are desolvated in the aprotic solvents, owing to lack of hydrogen bonding, while the transition states are better solvated. An S_N2-S mechanism is therefore proposed.[185] Other studies of arenesulphonyl chloride displacement have been reported.[186,187] *p*-Tolylmethylsulphoximine is reduced by diphenyl disulphide, or by elemental sulphur, to the sulphoxide with retention of configuration.[188]

The hydrolysis of *N*,*N*′-diarylsulphamides in aqueous acetone is almost independent of hydrogen ion concentration in the pH range 1.05–3.05 and has a Hammett ρ-value of 1.03; a nucleophilic attack by water on the neutral sulphamide is proposed as mechanism.[189]

Other Elements

Reduction of α-halo-sulphones, $ArCHXSO_2Ph$, with triphenylphosphine in aqueous DMF occurs by nucleophilic attack at halogen. The rates correlate against σ with ρ-values of 2.23(Cl), 5.97(Br) and 6.29(I), indicating substantial release of charge on to the benzylic carbon in the transition state. The rate constants for the halogens fall in the order $k^{Br} > k^{I} > k^{Cl}$, and it is suggested that the lower rate for the iodides is due to the decreased P—I bond strength in the product outweighing the reduced C—I bond strength in the starting material.[190] Similar attack at halogen probably accounts for the ready reduction of α-halo-esters by sodium alkanethioate.[191]

Reduction of 1-chloro-1-fluorocyclopropanes in the bicyclo[4.1.0]heptane and tricyclo[4.2.1.0^{2,4}]nonane series by lithium aluminium hydride occurs by attack at chlorine and with predominant retention of configuration in the bicyclic compounds but gives the more stable of the two epimeric fluorides in the tricyclic series. Use of lithium aluminium deuteride showed that the intermediate carbanion is protonated by the hydrochloric acid formed in the first step, faster than the latter can escape.[192]

A nitrenium ion has been postulated as the intermediate in the cyclization of *N*-(2-biphenylyl)hydroxylamine to carbazole.[193]

The displacement of tropylidene from tricarbonyltropylium-chromium, -molybdenum and -tungsten by alkyl cyanides occurs by nucleophilic attack at the metal.[194]

Cysteine and cysteamine are reported to react with hydrogen peroxide by nucleophilic attack of sulphur on oxygen, with hydrogen bonding of the other oxygen atom to the ammonium group.[195]

Ambident Nucleophiles

The effect of 18-crown-6-polyethers on the alkylation of the enolate of ethyl acetoacetate has been studied as a function of counter-ion and solvent, and rate constants for each component have been evaluated. The effect is greatest in non-polar solvents and is such as to reduce the amount of *O*-alkylation. The degree of dissociation of the enolate induced by the crown ether is greatest for the potassium and least for the sodium salt. It is suggested that the C/O ratio varies with the solvation of the free enolate ion, and that therefore the electrophilicity of the solvents should parallel this ratio. This leads to the series MeCN > DMSO > DMF > sulpholane > 1-methylpyrrolidone.[196]

The effect of added tetraglyme and $Bu_4N^+ Br^-$ on the alkylation of potassium acetylacetonate with *n*-propyl and isopropyl iodide in THF has been investigated. The results have been interpreted in terms of intimate, glymated, and solvent-separated ion-pairs, in addition to free ions.[197] However, since kinetic arguments are based only on yields of products, the results should not be too strongly relied on.

The alkylation of the dilithium salts of $\alpha\beta$-unsaturated acids occurs at the γ-position to the extent of 40–60% if this position is unsubstituted, but almost exclusively at the α-position if the γ-carbon bears an alkyl group.[198]

Specific control of the direction of enolization and alkylation of the monoenol ethers of 1,3-diketones can be achieved by using lithium di-isopropylamide in THF.[199] *O*-Methylation of phenols with methyl iodide–sodium hydride in THF has been investigated.[200]

Substitution at Vinylic Carbon

This year again the majority of the work published concerns investigations of vinyl carbocations, although work of significance to the addition–elimination and elimination–addition mechanisms has been carried out.

Further evidence for the non-classical nature of the ion derived by solvolysis of cyclobut-1-enyl sulphonate has been presented. 2-Methylcyclobutenyl nonafluorobutanesulphonate is solvolysed in aqueous ethanol with an *m* of 0.67 at a rate 140 times that of the unsubstituted compound; only 1% of cyclopropyl methyl ketone is formed, the rest of the product being 2-methylcyclobutanone and its enol ether.[201] The additional stability afforded by a phenyl group results in solvolysis of 2-phenylcyclobutenyl bromide (**116**) giving >80% of cyclopropyl phenyl ketone. (**117**) is solvolysed only to cyclopropyl phenyl ketone and 1-phenylbut-3-en-1-yne, showing that while the cyclobutenyl cation isomerizes to the cyclopropylidene cation (**118**), the reverse is not true.

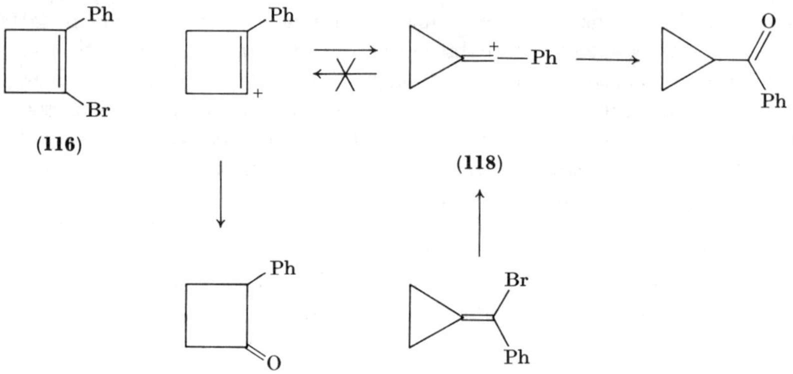

(**116**) (**118**) (**117**)

Both (**116**) and (**117**) show significant rate acceleration relative to 1-bromo-2-methyl-1-phenylpropene (20 and 10^4 times, respectively).[202] The rate-enhancing effect of a cyclopropyl group (4.2×10^4) is even greater than that of a phenyl group, as determined by the solvolysis of (**119**).[203] The silver acetate-assisted ionization of (**120**) and (**121**) (R = cyclopropyl or 1-ethylbutyl) give ratios of cyclopropyl-containing to cyclobutyl-containing products that depend on the nature of R but not on the original geometry, even though the latter affects the rate of reaction; this suggests that two discrete steps are involved, rate-determining and product-forming, and from the way in which the ratio of cyclobutyl to cyclopropyl products changes it is concluded that the latter is controlled by electronic effects and that the ion (**122**) undergoes a concerted twist and shift (arrows) to give the allylic cyclobutenyl ion directly;[204] this would not accord with the evidence of Santelli and Bertrand reported in 1971.[205]

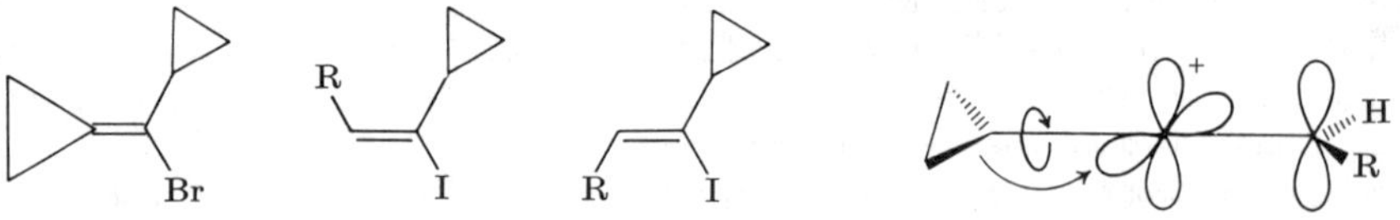

(**119**) (**120**) (**121**) (**122**)

R = 3-hexyl or cyclo-C_3H_5

Further investigations of isotope effects in vinyl cation solvolysis have shown that the orientation of the deuterium relative to the leaving group modifies the effect considerably. Thus the acetolysis of (**123**; $R^1 = D$, $R^2 = H$) has a k_H/k_D of 1.10, whereas a deuterium atom *trans* to the leaving group (**123**; $R^1 = H$, $R^2 = D$) has a much larger effect $k_H/k_D = 1.57$; this difference argues quite forcefully for a bent transition state for the solvolysis.[206] Surprisingly, no isotope effect was found for the solvolysis of (**124**; $R^1 = H$, $R^2 = D$)

R^1 OSO_2F / R^2 Ph

(123)

R^1 Br / R^2 C_6H_4OMe-*p*

(124)

and (**124**; $R^1 = D$, $R^2 = H$) in aqueous ethanol.[207] While some diminution due to the change in leaving group would be expected, the total lack of an effect is astounding. Deuterium substitution on a sp^3 carbon atom, but still β to the leaving group, as in (**125**), (**126**) and (**127**) ($R^1 = CD_3$, $R^2 = CH_3$), also has a substantial effect ($k_{CH_3}/k_{CD_3} = 1.46$, 1.47, and 1.16, respectively), while the effect of a γ-deuterium atom (**126**; $R^1 = CH_3$, $R^2 = CD_3$) is inverse ($k_{CH_3}/k_{CD_3} = 0.90$). The low β-isotope effect in (**127**) is due to

Me OSO_2CF_3 / Me R^1

(125)

Ph OSO_2CF_3 / R^2 R^1

(126)

Ph R^1 / R^2 OSO_2CF_3

(127)

participation by the phenyl group, and this is supported by the γ-isotope effect, k_{CD_3}/k_{CD_3} for (**127**; $R^1 = CH_3$, $R^2 = CD_3$) which is 1.04[208] (see also ref. 212). In the allenyl system (**128**), both α- and δ-deuterium isotope effects have been measured: for $R^1 = D$, $R^2 = CH_3$, an isotope effect of 1.22 is observed, which is substantially larger than that found in saturated compounds (k_H/k_D for Lim solvolysis of alkyl bromides is 1.125). Use has been made of fractionation factors to estimate a maximum value for solvolysis of vinyl bromide, and a value of 1.32 is predicted. For (**128**; $R^1 = H$ or Bu^t, $R^2 = CD_3$), isotope effects of 1.23 (k_{CH_3}/k_{CD_3}) were found, a reasonable value for a transition state containing a substantial contribution from the form (**129**).[209]

$Bu^t(R^2)C{=}C{=}C(Br)R^1$

(128)

$Bu^t(R^2)C{=}C{=}\overset{+}{C}R^1 \longleftrightarrow Bu^t(R^2)\overset{+}{C}{-}C{\equiv}CR^1$

(129)

The effect of a β-bromide in the solvolysis of 2-bromo-1,2-di-(*p*-methoxyphenyl)vinyl bromide has been found to be rate-retarding, but much more so when *cis* than when *trans* to the leaving group. Common-ion depression is observed, supporting the vinyl cation mechanism proposed.[210] The effect of a 2-substituent (R) in solvolysis of 1-(*p*-methoxyphenyl)-2-R-vinyl bromide is to increase the rate for the *cis*- relative to that for the *trans*-isomer by a steric effect, for a *tert*-butyl group has an effect 30 times greater than a *p*-methoxyphenyl group and 200 times greater than a methyl group. It is suggested that this is due to twisting of the 1-(*p*-methoxyphenyl) group out of conjugation with the double bond and into conjugation with the developing *p*-orbital.[207] Further work on the triarylvinyl bromides suggests that they give open carbocations on solvolysis in aqueous ethanol and that the rate of rearrangement relative to that of return varies with the substituents.[211]

Phenyl migration in (**126**) and (**127**) ($R^1 = R^2 = CH_3$) has been shown by deuterium labelling to occur to the extent of 95% and 98% for the *trans*-isomer (**127**) and 28% and 68% for (**126**) depending on the solvent; (**127**) reacts 20 and 62 times faster than (**126**) in 60% ethanol and absolute trifluoroethanol, respectively. Re-isolation of unchanged (**127**) shows that scrambling of the label occurs in it, but without any *cis–trans*-isomerization. Recovered (**126**) shows no evidence of scrambling, and the mechanism of Scheme 9 is proposed.[212] Anchimeric assistance has also been proposed to account for

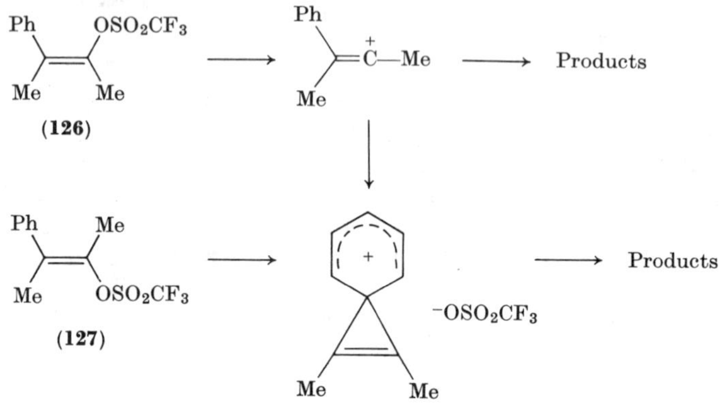

Ph OSO$_2$CF$_3$ / Me Me (**126**) ⟶ Ph, Me, $\overset{+}{C}$—Me ⟶ Products

Ph Me / Me OSO$_2$CF$_3$ (**127**) ⟶ [spiro cation] $^-$OSO$_2$CF$_3$, Me Me ⟶ Products

SCHEME 9.

the acceleration of 2,2-di-(*p*-methoxyphenyl)-1-phenylvinyl bromide over 1,2,2-triphenylvinyl bromide in the silver ion-assisted acetolysis. *cis–trans*-Isomerization observed in the acetolysis of 2-(*p*-methoxyphenyl)-1,2-diphenylvinyl bromide catalysed by silver acetate suggests that the intermediate vinyl cation is capable of removing bromide from silver bromide.[213]

Overlap of the conjugated double bond with the developing charge is easier in cycloocta-1,3-dien-2-yl than in cyclohepta-1,3-dien-2-yl trifluoromethanesulphonate.[214] A similar enhancement of overlap is responsible for the 6800-fold rate-acceleration observed in the solvolysis of (**130**) as compared with (**131**), the fused ring being held in the correct orientation for overlap.[215]

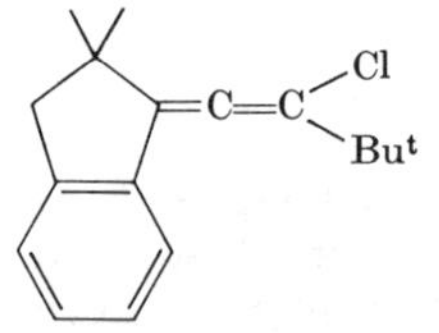

=C=C(Cl)(Bu^t) (**130**)

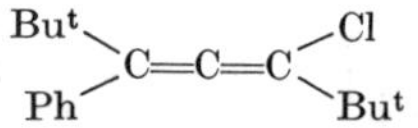

Bu^t(Ph)C=C=C(Cl)Bu^t (**131**)

The attempted formation of $CH_2{=}CH^+$ by solvolysis of vinyl nonafluorobutanesulphonate led only to acetylene by an $E2$ mechanism.[216]

Whereas, in 80% ethanol, 4-methoxystyryl bromide and chloride react by a vinyl cation process, with $k_{Br}/k_{Cl} = 54$, in acetic acid the chloride reacts by an electrophilic addition–elimination route to the extent of at least 35%, as evidenced by the k_{Br}/k_{Cl} ratio of about 0.5.[217]

Further work on amine-catalysed displacement in the 2-cyano-3-(*p*-dimethylaminophenyl)-3-X-acrylonitrile system has been described. A nucleophilic addition–elimination mechanism requiring amine catalysis of the second step in acetonitrile is proposed.[218] The stereochemistry of displacement by the Ad_N-*E* route from the ethyl 3-halo-*cis*- and -*trans*-crotonates and -crotononitriles depends on the halogen being displaced. Chloride, bromide and iodide are displaced with retention of stereochemistry, but with fluoride some loss of stereospecificity is found, which is ascribed to the increased lifetime of the intermediate carbanion.[219] Other examples of Ad_N-*E* processes have been reported.[220,221]

An Ad_E-*E* process has been proposed to account for the conversion of vinylboranes into iodides with retention of stereochemistry and to bromides with inversion of stereochemistry. The suggested mechanisms are outlined in the reactions shown.[222]

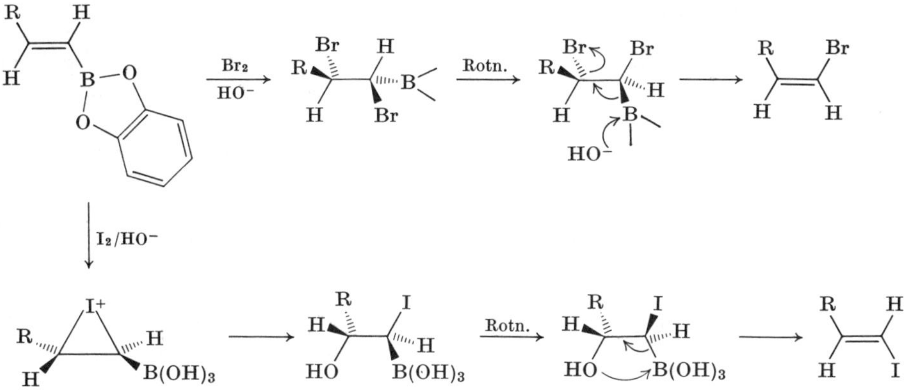

Br_2 / HO^- Rotn. I_2/HO^- Rotn.

Hydrolysis of $Ar^1CCl{=}NN{=}CHAr^2$ in aqueous dioxan goes by way of a vinyl cation. Both aryl groups affect the rate, Ar^1 with a ρ-value of -2.3 and Ar^2 with one of -1.2. At high pH, those compounds with $-M$ groups give evidence of a competitive S_N2-like process, perhaps Ad_N-*E*.[223] In DMSO, the *O*-methyloximes (**132**; Ar = Ph or p-$NO_2C_6H_4$, X = Br or Cl) react with sodium methoxide to give the hydroxamic esters (**133**) with predominant inversion of configuration, starting either from the *E*- (shown) or the *Z*-isomer. An Ad_N-*E* mechanism is suggested, with the additional constraint that the rate of elimination must be faster than pyramidal inversion of nitrogen.[224]

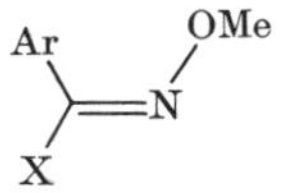

(**132**) X = Br or Cl

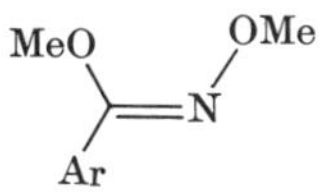

(**133**)

The elimination–addition route of displacement continues to be investigated. Use of [1-^{14}C]cycloheptenyl halides has shown that elimination occurs to give both the 1,2-diene and the cycloalkyne, by competitive processes.[225] Investigations with 2- and 3-halobicyclo[3.2.1]oct-2-enes show that here the cycloalkyne path is favoured and that attack occurs preferentially at the 3-position (10:1). Use of a stronger base (sodium pyrrolidide instead of potassium *tert*-butoxide) favours the formation of an allene.[226] Bromomethylenecyclopropane reacts in molten potassium *tert*-butoxide to give *tert*-butoxymethylenecyclopropane and 1-*tert*-butoxycyclobutene; it is suggested that the latter arises via cyclopropylidenecarbene, whose isomerization to cyclobutyne is followed by addition.[227] Unexpectedly, under the same conditions, bromomethylenecyclobutane gives mainly 1-bromocyclopentene.[228]

Isomerization of terminal to internal acetylenes takes place by a stepwise protonation–elimination process with intermediate vinyl cations.[229] The basicity of acetylenes is higher in the excited (electronic) state than in the ground state.[230] The displacement of halogen from 1-aryl 2-haloethynes takes place by an Ad_N-E process.[231]

cis- and *trans*-β-Bromostyrene react with cuprous cyanide and chloride to furnish *cis*- and *trans*-cinnamonitrile and *cis*- and *trans*-β-chlorostyrene, respectively, each with complete retention of configuration.[232]

Reactions of α-Halocarbonyl Compounds

The silver hexafluoroantimonate-catalysed formation of α-keto-carbocations has been further investigated. Alcohols react with PhCHBrCOPh with nearly quantitative replacement of bromide; reaction does not occur via the hemiketal, for the ethylene ketal of PhCHBrCOPh reacts with exclusive migration of phenyl.[233] 1-Acyl-1-bromocyclohexanes react with $AgSbF_6$ to give cations of either the 2-oxabicyclo[2.2.2]- or the 6-oxabicyclo[3.2.1]-octane series depending on whether the initial cyclohexane has a 4- or a 3-alkyl substituent, respectively.[234]

The Perkow reaction of α-chloro-ketones with trialkyl phosphites has $\Delta S^\ddagger$ in the range –30 to –60 e.u., and the mechanism of reactions (2) is proposed, with the decomposition of (**134**) rate-determining.[235] The Perkow reaction of trimethyl phosphite with 2-chloro-*N*-methylacetoacetamide in benzene or acetonitrile also shows a very large $\Delta S^\ddagger$ (–47 and –45 e.u., respectively).[236]

$$(RO)_3P + R'C(=O)CHClR'' \longrightarrow (RO)_3\overset{+}{P}{-}C(O^-)(R')CHClR'' \rightarrow \rightarrow \text{Products} \qquad (2)$$

(**134**)

The reaction of 2,2-dihaloaldehydes with methoxide in methanol leads to approximately 1:1 mixtures of the 2,2-dimethoxy-aldehyde and the corresponding 1,1-dimethoxy-2-ketone. The proposed mechanism is shown in Scheme 10. Treatment of (**135**), made by an independent route, under the reaction conditions gives a 1:1, *cis–trans*-mixture of epoxides (**136**), which open to give the products, *trans*-(**136**) giving aldehyde and *cis*-(**136**) giving ketone.[237]

Phenacyl fluoride reacts with sodium hydride in benzene to give 1,2,3-tribenzoylcyclopropane and 3-fluoro-2,4-diphenylfuran. The furan is formed by aldol condensation, followed by O-5 participation of the carbonyl group to displace fluoride ion; with sodium methoxide, the intermediate aldol cyclizes to an epoxide.[238] The corresponding keto-epoxide derived from *o*- or *p*-nitrophenacyl bromide reacts with sodium methoxide to give a bisepoxide by successive 1,3 displacements.[239]

Ketones $RCH_2COCH{=}CHCCl_3$ undergo a 1,3-shift of chlorine catalysed by aluminium

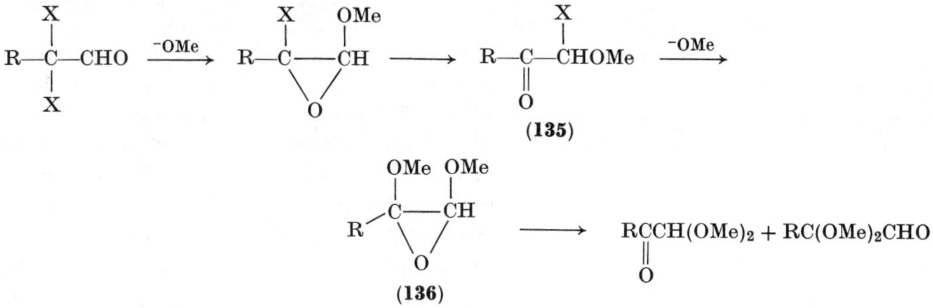

$$\mathrm{R{-}CX_2{-}CHO} \xrightarrow{^{-}\mathrm{OMe}} \mathrm{R{-}CX(O)CH{-}OMe\ (epoxide)} \longrightarrow \mathrm{R{-}C(=O){-}CHX{-}OMe}\ \mathbf{(135)} \xrightarrow{^{-}\mathrm{OMe}}$$

$$\mathrm{R{-}C(OMe)(O)CH(OMe)}\ \mathbf{(136)} \longrightarrow \mathrm{RC(=O)CH(OMe)_2 + RC(OMe)_2CHO}$$

SCHEME 10.

chloride and a thermal 1,5-shift of chlorine, by way of the enolate, to give $RCHClCOCH_2$-$CH{=}CCl_2$.[240]

Other work on α-halo-ketones has included studies of the reaction of the oxime of phenacyl bromide with morpholine,[241] and the kinetics of the reaction of phenacyl bromide with phenylthioureas[242] and with the anions of aromatic carboxylic acids.[243] The rates of the reactions of amines with substituted phenacyl bromides are correlated by the Hammett equation.[244]

S_N2 Processes and Other Reactions

A virial partitioning method has been used to assign the calculated activation energy for the $F^- + CH_3F$ and $CN^- + CH_3F$ reaction into the contributions from the various fragments involved. It is calculated that the nucleophiles and the carbon undergoing substitution are stabilized at the transition state (relative to the ground state), and the leaving group is strongly destabilized. Thus for the $(CNCH_3F)^-$ transition state, the δ^- fluoride contributes 163 kcal mol^{-1} to the total E_a of 22.1 kcal mol^{-1}. The hydrogen atoms are found to be raised in energy owing to compression of charge density, and this results in transfer of charge from hydrogen to carbon, which is calculated at +0.463 for the $(FCH_3F)^-$ and +0.316 for $(NCCH_3F)^-$ reactions. The geometry of the transition state for the latter is close to that of the products.[245]

A Hammett–Taft treatment has been applied to the solvolysis rates of (**137**) in ethanol,

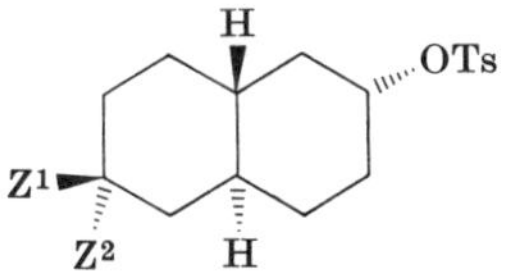

(**137**)

acetic acid and trifluoroacetic acid, for $Z^2 = H$, $Z^1 =$ OMe, Ph, Cl, CN or COOMe, for $Z^1 = H$, $Z^2 =$ OMe, Ph, Cl or CN, and for $Z^1, Z^2 = O$. A correlation of the type $\rho^*\sigma^*(a^l + b^m + c^n + \ldots)$ was used, summing σ^* over all the paths with the usual attenuation factors. The correlation coefficients were not very good if both axial and equatorial substituents were taken together (0.8–0.9), but plots of each separately gave better straight lines, with significant differences in ρ^* for the two sets of substituents except in trifluoroacetic acid. This was ascribed to a through-space, dipole–dipole interaction of the axial substituents. For the equatorial ones, the ρ^*-values are −0.74, −1.98 and −4.59 in ethanol, acetic acid and trifluoroacetic acid, respectively, and a similar trend is shown by the axial substituents: –1.30, –2.63 and –4.25. Decreasing nucleophilic assistance by the solvent is invoked to account for this.[246] The effect of neighbouring fluorine on the displacement of bromide by chloride and iodide has been studied: reaction occurs with inversion of stereochemistry and rates are slowed 50 times in open-chain and cyclohexyl compounds, but 2×10^3 times for cyclopentyl compounds.[247]

The ^{1}H-NMR chemical shift of the methylene protons in substituted benzyl chlorides and aryl chloromethyl sulphides varies with the concentration of added tetraethylammonium chloride in acetonitrile; it is suggested that this is due to complex formation ($ArCH_2Cl–Cl$), and the equilibrium for this complex formation is found to correlate with Hammett's σ-value. Because the ρ-values for complex formation (0.59 and 0.63 for $ArSCH_2Cl$ and $ArCH_2Cl$, respectively) are comparable to that for exchange of the benzyl chlorides (0.57), it is proposed that the complex lies on the route to the S_N2 transition state.[248] The change in α-deuterium isotope effect for the 80% aqueous ethanolysis of substituted benzyl chlorides, from 1.04 for the 4-methyl compound to 1.16 for the 2,4,6-tri-*tert*-butyl one, has been interpreted as showing loss of solvent participation at the transition state.[249] Activation parameters have been determined for the displacements of bromide from α-methylbenzyl and phenethyl bromide by chloride, bromide or iodide in acetone.[250] The solvolysis, racemization and exchange of 2-chlorobutane with chloride ion in DMF have been studied.[251] Enthalpies of activation for the hydrolysis of ethyl iodide in aqueous acetone have been measured.[251a] Non-linear Hammett plots are found for the ^{36}Cl exchange of *p*-substituted cinnamyl chlorides and the rates are faster than those of the corresponding phenethyl chlorides by ca. 10^2. The *m*-substituted isomers give rates which correlate with σ_m ($\rho = -0.19$), and a correlation with the same ρ-value is found for both *m*- and *p*-substituted 1-aryl-3-chloropropynes.[252] Activation energies have been evaluated for the hydrolysis of allyl chloride in the presence of a CaO/Cr_2O_3 catalyst.[253] The ratios of hydrolysis and chloride exchange for 1,2-*trans*-acetylglycosyl halides have been measured; it is suggested that exchange occurs by direct S_N2 reaction and that hydrolysis occurs by way of ion-pair formation and an S_N1 process.[254] The displacement kinetics have been measured for the reaction of aliphatic alcohols with dimeric tri-*O*-acetyl-2-deoxy-2-nitroso-α-D-glucopyranosyl chloride in DMF.[255] Toluene-*p*-sulphonate to bromide rate ratios have been determined for the displacement reactions of $PhCOCMe_2CH_2X$ by potassium aryl oxide in DMSO.[256]

Kinetic equations have been derived for the parallel substitution reactions of *p*-nitrobenzenediazotate with methyl iodide. Mixed kinetics are found and interpreted as simultaneous competitive S_N1 and S_N2 reactions.[257]

The variation in the toluene-*p*-sulphonate to iodide rate ratios with nucleophile in acetonitrile has been ascribed to an HSAB effect, the soft nucleophile selenocyanate having a ratio of 4.7×10^{-3}, while the hard one, acetate, has a ratio of 0.54. These effects are reduced by a factor of 10^2 by changing the solvent to methanol.[258] Anhydrous magnesium iodide in ether has been found to be an excellent agent for nucleophilic displacement: 3β-cholestanyl toluene-*p*-sulphonate is converted into the 3α-iodide in 95% yield, and a similar yield is found for the 2-adamantyl system.[259] The gas-phase nucleophilicities of H^-, F^-, OH^- and NH_2^- have been evaluated for displacement on methyl chloride, by a flowing afterglow technique. Nucleophilicities are found to fall in the order $H^- > NH_2^- \geqslant OH^- \approx F^-$, whereas the gas-phase basicities are in the order $NH_2^- > H^- > OH^- > F^-$.[260] The nucleophilicities of 15 amines and 3 anions have been correlated by the equation $\log k_n = N_+ + \log k_0$ for the cation–anion recombination reactions with substituted trityl cations, cycloheptatrienyl cation, and aryldiazonium ions.[261] Lithium triethylborohydride is a powerful nucleophilic hydride donor, reacting 40 times faster than lithium aluminium hydride and 20 times faster than thiophenoxide towards *n*-octyl chloride. *exo*-2-Norbornyl bromide gives [2-*endo*-^{2}H]norbornane with the deuteriated reagent.[262] The nucleophilicities of the imidazole (in histidine), the thiol (in cysteine), the aryloxide (in tyrosine) and the sulphide (in methionine) are nearly identical with those of simple models.[263] Negligible salt effects are found in the methylation of sodium methoxide, phenoxide and *p*-nitrophenoxide with methyl iodide.[264] The free-ion rates for chloride and bromide displacement on ethyl toluene-*p*-sulphonate in hexamethylphosphoric triamide are independent of the nature of the counter-ion.[264a]

Chlorodiphenylcarbinyl hexachloroantimonate converts alcohols in acetonitrile into the corresponding acetamides. That cholesterol, borneol and menthol all give products of retained configuration, while cholestan-3β-ol gives a mixture and isoborneol reacts with inversion, speaks for a Ritter-type mechanism.[265] On the other hand, the axial *trans*-2-decalyl bromide reacts with silver perchlorate or tetrafluoroborate in acetonitrile to give the acetamide with at least 98% retention of configuration. Complexing of acetonitrile to the silver ion and its low nucleophilicity are suggested as reasons for the high specificity.[266] The iron–carbene complex $[(h^5\text{-}C_5H_5)Fe(CO)(PPh_3)\{C(OEt)CH_3\}]^+ BF_4^-$ reacts with sodium iodide to give ethyl iodide and the acetyl complex.[267] The reaction of benzylamines and Mannich bases with triethyl phosphite occurs by an S_N2 process.[268] Diethyl phosphite is dealkylated by metal and alkylammonium chlorides, bromides and iodides by attack on the ethyl group.[269] A similar S_N2 mechanism has been proposed for the alkaline hydrolysis of *O,O'*-dialkyl *S*-methyl thiophosphates and *O*-alkyl *S*-ethylmethylthiophosphonates.[270]

The reaction of trimethylstannyl-lithium with 2-chlorobutane occurs with complete inversion of configuration.[271] Similar complete inversion has been reported for the reaction of (*S*)-*sec*-butyl toluene-*p*-sulphonate with (diphenylcopper)lithium, and for (dimethylcopper)lithium with *cis*- and *trans*-4-*tert*-butylcyclohexyl toluene-*p*-sulphonate. *exo*-Norbornyl halides and toluene-*p*-sulphonate give mainly *exo*- but some *endo*-products, together with nortricyclene. Because the products are S_N2-like and the kinetics of the second order, a mechanism involving attack by copper to give a square-planar R_3Cu(III) species, followed by decomposition with radical coupling, is proposed.[272] (Dialkylcopper)-lithiums react with α-acetylenic epoxides to give allenyl alcohols.[273]

The dimer (**138**) of tetrachloroallene undergoes sequential S_N2' reaction and elimination with ammonia, as shown in Scheme 11.[274] Other S_N2' reactions are reported in refs. 275 and 276.

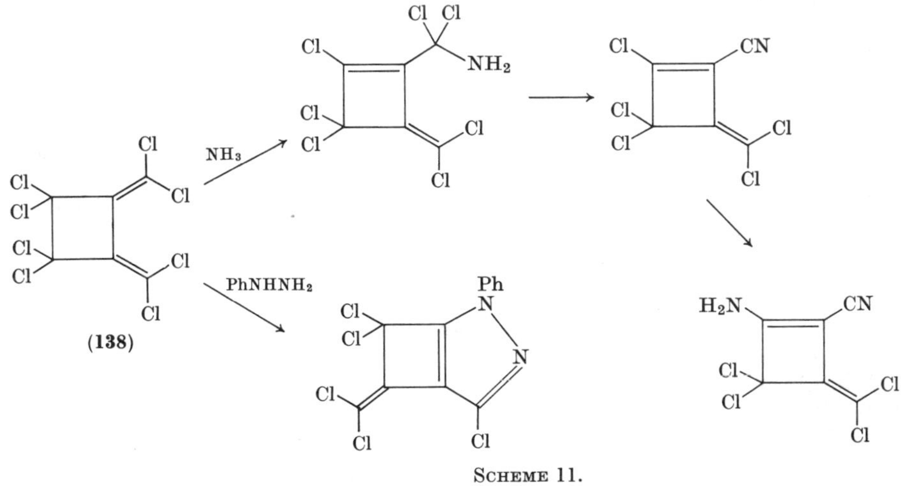

SCHEME 11.

Cyclic ethers, up to tetrahydrofuryl, and allyl ethers are cleaved by 1,1-diphenylhexyllithium. Butyl vinyl ether reacts at the vinyl carbon.[277] *O*-Methyltropolone is demethylated by quinuclidine, toluene-*p*-thiolate and Dabco in hexamethylphosphoric triamide.[278]

Dialkyldimethylammonium iodides are demethylated by acid anhydrides at 200°, by a reverse Menschutkin reaction followed by acylation.[279] The activation energies for the dealkylation of 1,5-bis(trialkylammonio)pent-2-yne dibromides have been determined.[280] Methiodides are dequaternized by refluxing in DMF.[281]

Quaternization continues to attract attention. The effect of solvent on the methylation rate of 3,6-dialkylpyridines has been investigated.[282] The rate of methylation of azoles and their benzologues correlates with basicity.[283] Reaction of scopolamine with optically active halides is only efficient in the presence of ethylene oxide.[284] Methylation of thiazoles has been examined in nitrobenzene and alcohols,[285] and the alkylation of 2,1-benzoisothiazoles has been studied.[286] The methylation of 1,3,4-thiadiazoles in acetonitrile has been investigated[287] and that of 5,6-dihydro-2-thiouracils in acetone has been found to correlate with Hammett's σ-constants.[288]

Reactions of haloacetic acids, esters and amides with amines, anions and bases have received further examination.[289]

Halogen exchange at the bridgehead positions of norbornyl, bicyclo[2.2.2]octyl, and adamantyl halide is strongly catalysed by aluminium bromide formed from aluminium powder and bromine.[290] 7-Chloro-7-methyl- and 7-chloro-7-phenylbicyclo[3.3.1]nonane undergo ready replacement of chloride by nucleophiles; a four-centre mechanism is proposed.[291] The acetolysis of ethers catalysed by perchloric acid and acetyl perchlorate has been studied.[292] Further work on the metal ion-catalysed reactions of alkyl halides has been reported.[293] Activation parameters have been measured for the hydrolysis of (trialkylsilyl)alkyl vinyl ethers and thioethers.[294]

2-Phenyl- and 2-(methoxycarbonyl)-2-azaspiropentane react with nucleophiles by an S_N2 route, in contrast to the reactions of the 2,3-diphenyl compound.[295] *N*-(Ethoxycarbonyl)-7-azabicyclo[4.1.0]heptane undergoes stereospecific ring-opening by BF_3 in acetonitrile.[296] Nucleophilic ring-opening of oxetanes by amines has been studied.[297]

A great deal of work on the ring-opening of epoxides has again been reported. Their reactions with halide ions have been reviewed.[298] Nucleophilic ring-opening by anilines correlates with σ^-;[229] attack by phenylselenoxide gives allylic alcohols.[299a] Other nucleophilic reactions have also been studied.[300] Acid-catalysed openings have attracted the most attention.[301]

A perturbational molecular-orbital treatment of the α-effect in oximate anions has shown why these show the effect only when an electron-withdrawing group is attached to the nitrogen,[302] and an explanation of the α-effect in terms of an aromatic transition state has been advanced.[303] The *N*-chlorobenzenesulphonamide anion shows an α-effect of about 7×10^3 in the S_N2 reaction of methyl methanesulphonate.[304]

The triphenylphosphine–carbon tetrachloride reagent converts 1,3-di-*O*-acylglycerols into the corresponding 2-chloro-compounds with only a trace of the 3-chloro-isomer, in contrast to the action of lithium chloride on the corresponding 1,3-di-*O*-acyl-2-glycerol toluene-*p*-sulphonate which occurs almost exclusively with neighbouring acyl participation to give the di-*O*-acyl-3-chloropropane-1,2-diol. The optically active ether-ester (**139**) reacts with at least 96% inversion of configuration. Because there is no neighbouring-group participation, an ion-pair mechanism is ruled out and it is therefore proposed that the intermediate previously postulated, $Ph_3P(Cl)OR$, reacts with simultaneous cleavage of the P—Cl and C—O bonds and concerted formation of the new

C—Cl bond.[305] The Reviewer finds it extremely difficult to visualize such a process which must involve a four-membered di-apical ring at trigonal-bipyramidal carbon, for it has already been shown that external nucleophiles do not compete.[306] The intermediate salt (**140**) can be isolated from the reaction of hexamethylphosphorous triamide with carbon tetrachloride and alcohols; it undergoes ready nucleophilic displacement by

$CH_2OC_{18}H_{37}$
|
CHOH
|
CH_2OCOPh

(**139**)

Me, Me — C — $CH_2O\overset{+}{P}(NMe_2)_3\ Cl^-$, CH_2OH

(**140**)

anions;[307] the reaction has been applied to replace selectively the 6-hydroxyl group of methyl hexosides.[308] Triphenylphosphine or triphenyl phosphite reacts with alcohols in the presence of *N*-halosuccinimides to give the corresponding halides with inversion of configuration; a mechanism analogous to that of the Ph_3P–CCl_4 reaction is proposed, but with nucleophilic attack of halide on the phosphonium salt Ph_3P^+R or $(PhO)_3P^+R$.[309]

Dimethyl sulphide and bromine convert alcohols into the corresponding bromides also with inversion of configuration. The proposed mechanism involves formation of the dimethylsulphoxonium salt followed by nucleophilic attack of bromide ion.[310]

The full papers on the concurrent racemization and solvolysis of methyl 1-cyano-2,2-diphenylcyclopropanecarboxylate in methanol, and on the epimerization and nucleophilic ring-opening of methyl *Z*- and *E*-1-cyano-2-phenylcyclopropanecarboxylate also in methanol. Catalysis by bromide ion and other nucleophiles, together with solvent effects, confirm the mechanism of zwitterion formation previously proposed.[311] Nucleophilic ring-opening of cyclopropane is also proposed in the formation of 2-acetyl-4-methoxytricyclene form 2-acetylquadricyclene with alkaline methanol,[312] and also for conversion of 1-(2,2-dimethoxy-3-methyl- and -3,3-dimethyl-cyclopropyl) phenyl sulphone into 1-(3,3,3-trimethoxy-2-methyl- and -2,2-dimethyl-propyl) phenyl sulphone.[313]

References

1 R. A. Sneen, *Accounts Chem. Research*, **6**, 46 (1973).
2 *Org. Reaction Mech.*, **1972**, 53; **1971**, 50; **1970**, 63; **1969**, 72; **1966**, 44.
3 R. A. Sneen, G. R. Felt and W. C. Dickson, *J. Am. Chem. Soc.*, **95**, 638 (1973).
4 *Org. Reaction Mech.*, **1971**, 50, ref. 6.
5 M. H. Abraham, *J.C.S. Perkin II*, **1973**, 1893; *Chem. Comm.*, **1973**, 51.
6 S. Winstein and A. Diaz, *J. Am. Chem. Soc.*, **88**, 1318 (1966).
7 D. J. McLennan and P. L. Martin, *Tetrahedron Letters*, **1973**, 4215.
8 C. A. Grob, K. Seckinger, S. W. Tam and R. Traber, *Tetrahedron Letters*, **1973**, 3051.
9 H. Loupy and J. Seyden-Penne, *Tetrahedron*, **29**, 1015 (1973).
10 J. E. Nordlander, R. R. Gruetzmacher and F. Miller, *Tetrahedron Letters*, **1973**, 927.
11 V. J. Shiner, S. R. Hartshorn and P. C. Vogel, *J. Org. Chem.*, **38**, 3604 (1973).
12 R. M. Evans and R. S. Satchell, *J.C.S. Perkin II*, **1973**, 643.
13 D. A. da Roza, L. J. Andrews and R. M. Keefer, *J. Am. Chem. Soc.*, **95**, 7003 (1973).
14 A. Ceccon and S. Sartori, *J. Organometal. Chem.*, **50**, 161 (1973).
15 J. Hayami, N. Tanaka and N. Hihara, *Bull. Inst. Chem. Res. Kyoto Univ.*, **50**, 354 (1972); *Chem. Abs.*, **78**, 96996 (1973).
16 I. G. Murgulescu and I. Demetrescu, *Rev. Roum. Chim.*, **18**, 355, 523 (1973).
17 J. A. Hemmingson and R. M. Noyes, *J. Am. Chem. Soc.*, **94**, 9148 (1972).
18 E. L. Kulin and K. T. Leffek, *Can. J. Chem.*, **51**, 687 (1973).
19 J. L. Kice and G. C. Hanson, *J. Org. Chem.*, **38**, 1410 (1973); cf. *Org. Reaction Mech.*, **1970**, 62.

[20] W. T. Ford and R. J. Hauri, *J. Am. Chem. Soc.*, **95**, 7381 (1973).
[21] S. Saito, T. Yabuki, T. Moriwake and K. Okamoto, *Bull. Chem. Soc. Japan*, **46**, 1795 (1973).
[22] J. G. Traynham and S. D. Elakovich, *Tetrahedron Letters*, **1973**, 155.
[23] D. S. Noyce and J. A. Virgilio, *J. Org. Chem.*, **38**, 2661 (1973).
[24] D. S. Noyce and S. A. Fike, *J. Org. Chem.*, **38**, 2433, 3316, 3318, 3321 (1973).
[25] D. S. Noyce, and R. W. Nichols, *J. Org. Chem.*, **37**, 4306, 4311 (1972).
[26] T. J. Broxton, G. L. Butt, L. W. Deady, S. H. Toh, R. D. Topsom, A. Fischer and M. W. Morgan, *Can. J. Chem.*, **51**, 1620 (1973).
[27] P. Beronius, A.-M. Nilsson and A. Holmgren, *Acta Chem. Scand.*, **26**, 3173 (1972).
[28] J. Dockx, *Synthesis*, **1973**, 441.
[28a] C. M. Starks and R. M. Owens, *J. Am. Chem. Soc.*, **95**, 3613 (1973).
[29] Y. Kondo, M. Ohnishi and N. Tokura, *Bull. Chem. Soc. Japan*, **45**, 3579 (1973).
[30] A. Holmgren and P. Beronius, *Acta Chem. Scand.*, **26**, 3881 (1972).
[31] B. Dietrich, J. M. Lehn and J. P. Sauvage, *Chem. Comm.*, **1973**, 15; *Tetrahedron*, **29**, 1647 (1973); B. Dietrich, J. M. Lehn, J. P. Sauvage and J. Blanzat, *ibid.*, p. 1629.
[32] C. Lapinte and P. Viout, *Tetrahedron Letters*, **1973**, 1113; V. Gani, C. Lapinte and P. Viout, *ibid.*, p. 4435; V. Gani and C. Lapinte, *ibid.*, p. 2775.
[33] J. W. Larsen, L. J. Magid and V. Payton, *Tetrahedron Letters*, **1973**, 2663.
[34] C. J. O'Connor, E. J. Fendler and J. H. Fendler, *J.C.S. Perkin II*, **1973**, 1900.
[35] M. Chastrette and H. Gauthier-Countani, *Bull. Soc. Chim. France*, **1973**, 363.
[36] R. Fuchs and L. L. Cole, *J. Am. Chem. Soc.*, **95**, 3194 (1973).
[37] B. G. Cox and A. J. Parker, *J. Am. Chem. Soc.*, **95**, 402 (1973).
[38] T. R. Stengle, Y.-C. E. Pan and C. H. Langford, *J. Am. Chem. Soc.*, **94**, 9037 (1973).
[39] (*a*) R. L. Schowen, *Progr. Phys. Org. Chem.*, **9**, 275 (1972); (*b*) M. E. Schneider and M. J. Stern, *J. Am. Chem. Soc.*, **95**, 1355 (1973).
[40] S. R. Hartshorn and V. J. Shiner, *J. Am. Chem. Soc.*, **94**, 9002 (1972).
[41] K. Humski, V. Sendijarević and V. J. Shiner, *J. Am. Chem. Soc.*, **95**, 7722 (1973).
[42] N. Tanaka, A. Kaji and J. Hayami, *Chem. Letters (Tokyo)*, **1972**, 1223.
[43] C. S. Davis and J. B. Hyne, *Can. J. Chem.*, **51**, 1687 (1973).
[44] K. M. Kosky, R. E. Robertson and W. M. J. Strachan, *Can. J. Chem.*, **51**, 2958 (1973).
[45] H. Yamataka, S.-G. Kim, T. Ando and Y. Yukawa, *Tetrahedron Letters*, **1973**, 4767.
[46] J. Laureillard, A. Laurent and E. Laurent, *Bull. Soc. Chim. France*, **1973**, 232, 242, 249.
[47] C. R. Turnquist, J. W. Taylor, E. P. Grimsrud and R. C. Williams, *J. Am. Chem. Soc.*, **95**, 4133 (1973).
[48] G. Köbrich, J. Grosser and W. Werner, *Chem. Ber.*, **106**, 2610, 2620 (1973).
[49] G. Köbrich and J. Grosser, *Chem. Ber.*, **106**, 2626 (1973).
[50] A. C. Knipe, *J.C.S. Perkin II*, **1973**, 589.
[51] E. W. Robb and E. K. Onsager, *J. Org. Chem.*, **37**, 4013 (1972); cf *Org. Reaction Mech.*, **1971**, 61; **1972**, 65.
[52] G. Lavielle, M. Carpentier and P. Savignac, *Tetrahedron Letters*, **1973**, 173.
[53] D. F. Tavares and R. E. Estep, *Tetrahedron Letters*, **1973**, 1229.
[54] R. A. Kretchmer, R. A. Conrad and E. D. Mihelich, *J. Org. Chem.*, **38**, 1251 (1973).
[55] H. Follmann, *Tetrahedron Letters*, **1973**, 397.
[56] L. V. Aleksanyan and V. F. Shvets, *Tr. Mosk. Khim.-Tekhnol. Inst.*, **1972**, 51; *Chem. Abs.*, **78**, 135275 (1973).
[57] A. S. Gudkova, E. I. Troyanskii and O. A. Reutov, *Dokl. Akad. Nauk SSSR*, **210**, 855 (1973); *Chem. Abs.*, **79**, 77715.
[58] N. A. Nelson, *J. Org. Chem.*, **38**, 3798 (1973).
[59] B. Castro and C. Selve, *Tetrahedron Letters*, **1973**, 4459.
[60] J. S. Brimacombe, J. Minshall and L. C. N. Tucker, *Chem. Comm.*, **1973**, 142.
[61] H. R. Kruger, H. Marschall, P. Weyerstahl and F. Nerdel, *Chem. Ber.*, **106**, 2255 (1973).
[62] R. T. Borchardt and L. A. Cohen, *J. Am. Chem. Soc.*, **94**, 9166 (1972).
[63] J. K. Kim, M. C. Findlay, W. G. Henderson and M. C. Caserio, *J. Am. Chem. Soc.*, **95**, 2184 (1973).
[64] G. A. Underwood and C. A. Kingsbury, *J.C.S. Perkin II*, **1973**, 947.
[65] K. Undheim and G. A. Ulsaker, *Acta Chem. Scand.*, **27**, 1390 (1973).
[66] G. Modena, G. Scorrano and U. Tonellato, *J.C.S. Perkin II*, **1973**, 493; cf *Org. Reaction Mech.*, **1971**, 63, 100.
[67] P. H. McCabe and C. M. Livingstone, *Tetrahedron Letters*, **1973**, 3029.
[68] L. Rasteikiene, L. Gurvichene and Y. Kulis, *Zh. Vses. Khim. Obshchest.*, **17**, 701 (1972); *Chem. Abs.*, **78**, 83560 (1972).

[69] R. Bird and C. J. M. Stirling, *J.C.S. Perkin II*, **1973**, 1221.
[70] J. C. Martin and R. J. Basalay, *J. Am. Chem. Soc.*, **95**, 2572 (1973).
[71] H. Schmidt and A. Schweig, *Tetrahedron Letters*, **1973**, 981.
[73] B. L. Jensen and R. E. Counsell, *J. Org. Chem.*, **38**, 835 (1973).
[73a] G. A. Olah, D. A. Beal and P. W. Westerman, *J. Am. Chem. Soc.*, **95**, 3387 (1973).
[74] M. T. Reetz, *Tetrahedron*, **29**, 2189 (1973).
[75] J. H. Exner, L. D. Kershner and T. E. Evans, *Chem. Comm.*, **1973**, 361; cf *Org. Reaction Mech.*, **1970**, 77; **1971**, 64.
[76] P. M. Henrichs and P. E. Peterson, *J. Am. Chem. Soc.*, **95**, 7449 (1973); B. R. Bonazza and P. E. Peterson, *J. Org. Chem.*, **38**, 1010 (1973).
[77] P. E. Peterson, B. R. Bonazza and P. M. Henrichs, *J. Am. Chem. Soc.*, **95**, 2222 (1973).
[78] P. Hodgson and S. Warren, *Chem. Comm.*, **1973**, 756.
[79] *Org. Reaction Mech.*, **1971**, 66.
[80] *Org. Reaction Mech.*, **1972**, 182.
[81] R. H. Bisceglia and C. J. Cheer, *Chem. Comm.*, **1973**, 165.
[82] S. Wolff and W. C. Agosta, *Chem. Comm.*, **1973**, 771.
[83] A. Takeda, S. Tsuboi, T. Moriwaae and E. Hirata, *Bull. Chem. Soc. Japan*, **45**, 3685 (1972).
[84] V. P. Vitullo and N. R. Grossman, *J. Org. Chem.*, **38**, 179 (1973).
[85] H. Paulsen and R. Dammeyer, *Chem. Ber.*, **106**, 2324 (1973).
[86] P. L. Durette, P. Köll, H. Meyborg and H. Paulsen, *Chem. Ber.* **106**, 2333 (1973).
[87] S. Jacobsen, I. Lundt and C. Pedersen, *Acta Chem. Scand.*, **27**, 453 (1973).
[88] R. G. Pews and R. A. Davis, *Chem. Comm.*, **1973**, 269.
[89] R. Lorne and S. Julia, *Bull. Soc. Chim. France*, **1973**, 1357.
[90] J. M. Coxon, M. P. Hartshorn and W. H. Swallow, *Chem. Comm.*, **1973**, 261.
[91] M. Miljković, T. Satoh, M. Konopka, E. A. Davidson and D. Miljković, *J. Org. Chem.*, **38**, 716 (1973).
[92] H. L. Hase and A. Schweig, *Tetrahedron*, **29**, 1759 (1973).
[93] C. F. Hammer, M. McC. Ali and J. D. Weber, *Tetrahedron*, **29**, 1767 (1973).
[94] A. C. Knipe, *Tetrahedron Letters*, **1973**, 3031.
[95] R. H. Higgins and N. H. Cromwell, *J. Am. Chem. Soc.*, **95**, 120 (1973).
[96] R. Bird, A. C. Knipe and C. J. M. Stirling, *J.C.S. Perkin II*, **1973**, 1215.
[97] K. Undheim and R. Lie, *Acta Chem. Scand.*, **27**, 1749 (1973).
[98] H. J. Trede, E. F. Jenny and K. Heusler, *Tetrahedron Letters*, **1973**, 3425.
[99] S. Sicsic and N.-T. Luong-Thi, *Tetrahedron Letters*, **1973**, 169.
[100] O. T. Quimby, W. A. Cilley, J. B. Prentice and D. A. Nicholson, *J. Org. Chem.*, **38; 1867** (1973).
[101] G. Richtarski and P. Mastalerz, *Tetrahedron Letters*, **1973**, 4069.
[102] D. Howells and S. Warren, *J.C.S. Perkin II*, **1973**, 1472; cf. *Org. Reaction Mech.*, **1971**, 72.
[103] D. Howells and S. Warren, *J.C.S. Perkin II*, **1973**, 1645.
[104] J. B. Aylward, *Malaysian J. Sci.*, **1**, 145 (1972); *Chem. Abs.*, **79**, 77645 (1973).
[105] G. Stork, J. O. Gardner, R. K. Boeckman and K. A. Parker, *J. Am. Chem. Soc.*, **95**, 2014 (1973).
[106] G. Stork and R. K. Boeckman, *J. Am. Chem. Soc.*, **95**, 2016 (1973).
[107] R. Askani and H. Sönmez, *Tetrahedron Letters*, **1973**, 1751.
[108] C. Y. Meyers, L. L. Ho, G. J. McCollum and J. Branca, *Tetrahedron Letters*, **1973**, 1843.
[109] F. G. Bordwell and B. B. Jarvis, *J. Am. Chem. Soc.*, **95**, 3585 (1973); cf. *Org. Reaction Mech.*, **1972**, 73.
[110] P. Ausloos, R. E. Rebbert, L. W. Sieck and T. O. Tiernan, *J. Am. Chem. Soc.*, **94**, 8939 (1972).
[111] T. Shono, K. Fujita and S. Kumai, *Tetrahedron Letters*, **1973**, 3393.
[112] J. Kagan and J. T. Przybytek, *Tetrahedron*, **29**, 1163 (1973).
[113] F. D. Green, R. L. Camp, V. P. Abegg and G. O. Pierson, *Tetrahedron Letters*, **1973**, 4091.
[114] J. B. Jones and P. Price, *Tetrahedron*, **29**, 1941 (1973); *Chem. Comm.*, **1969**, 1478.
[115] R. A. Wohl, *Tetrahedron Letters*, **1973**, 3111; *J. Org. Chem.*, **38**, 3862 (1973).
[116] B. P. Mundy and R. D. Otzenberger, *J. Org. Chem.*, **38**, 2109 (1973).
[117] R. M. Coates and S. K. Chung, *J. Org. Chem.*, **38**, 3677 (1973).
[118] *Org. Reaction Mech.*, **1972**, 76.
[119] S. S. Deshmane and H. Hirschmann, *J. Org. Chem.*, **38**, 748 (1973); cf. *Org. Reaction Mech.*, **1971**, 77.
[120] F. B. Hirschmann and H. Hirschmann, *J. Org. Chem.*, **38**, 1270 (1973).
[121] J. L. Zundel, G. Wolff and G. Ourisson, *Bull. Soc. chim. France*, **1973**, 3206.
[122] A. Ambles, J. C. Jacquesy and R. Jacquesy, *Bull. Soc. chim. France*, **1973**, 2865.
[123] J. Thierry, F. Frappier, M. Païs, A. Montagnac, F. X. Jarreau and R. Goutarel, *Bull. Soc. chim. France*, **1972**, 4753.
[124] J. B. Lambert and G. J. Putz, *J. Am. Chem. Soc.*, **95**, 6313 (1973).

[125] R. A. Moss, C. J. Talkowski, D. W. Reger and W. L. Sunshine, *React. Kinet. Micelles, Proc. Amer. Chem. Soc. Symp.* **1972**, 99; *Chem. Abs.*, **78**, 158450 (1973).
[126] R. A. Moss, C. J. Talkowski, D. W. Reger and C. E. Powell, *J. Am. Chem. Soc.*, **95**, 5215 (1973); cf. *Org. Reaction Mech.*, **1970**, 88; **1971**, 79; **1972**, 78.
[127] *Org. Reaction Mech.*, **1970**, 88.
[128] R. A. Moss, P. E. Schueler and T. B. K. Lee, *Tetrahedron Letters*, **1973**, 2509.
[129] C. A. Bunton, D. L. Hachey and J. P. Leresche, *J. Org. Chem.* **37**, 4036 (1972).
[130] W. Kirmse and W. Gruber, *Chem. Ber.* **106**, 1365 (1973).
[131] W. Kirmse, W. Gruber and J. Knist, *Chem. Ber.*, **106**, 1376 (1973).
[132] T. Shono, K. Fujita and S. Kumai *Tetrahedron Letters*, **1973**, 3123.
[133] T. Bando and Y. Matsushima, *Bull. Chem. Soc. Japan*, **46**, 593 (1973).
[134] W. Kirmse and A. Engelmann, *Chem. Ber.*, **106**, 3086 (1973); cf. *Org. Reaction Mech.*, **1971**, 80.
[135] W. Kirmse, A. Engelmann and J. Heese, *Chem. Ber.*, **106**, 3073 (1973).
[136] J. M. Kliegman and R. K. Barnes, *J. Org. Chem.*, **37**, 4223 (1972).
[137] M. Hanack and J. Dolde, *Liebig's Annalen*, **1973**, **1557**.
[138] *Org. Reaction Mech.*, **1972**, 80.
[138a] *Org. Reaction Mech.*, **1972**, 5.
[139] M. Kobayashi, H. Minato and H. Fukuda, *Bull. Chem. Soc. Japan*, **46**, 1266 (1973).
[140] M. J. Gregory, *J.C.S. Perkin II*, **1973**, 1699.
[141] P. R. Chaabouni and A. Laurent, *Tetrahedron Letters*, **1973**, 1061.
[142] S. Ranganathan and H. Raman, *Tetrahedron Letters*, **1973**, 411.
[143] D. H. Aue, M. J. Meshishnek and D. F. Shellhamer, *Tetrahedron Letters*, **1973**, 4799.
[144] D. N. Roark and L. H. Sommer, *J. Am. Chem. Soc.*, **95**, 969 (1973).
[145] G. D. Homer and L. H. Sommer, *J. Organometal. Chem.*, **49**, C13 (1973).
[146] L. H. Sommer, C. M. Golino, D. N. Roark and R. D. Bush, *J. Organometal. Chem.*, **49**, C3 (1973); cf. *Org. Reaction Mech.*, **1972**, 84.
[147] R. J. P. Corriu and B. J. L. Henner, *Chem. Comm.*, **1973**, 116.
[148] B. Bøe, *J. Organometal. Chem.*, **57**, 255 (1973).
[149] R. J. P. Corriu and G. Lanneau, *Bull. Soc. chim. France*, **1973**, 3102.
[150] A. Holt, A. W. P. Jarvie and G. J. Jervis, *J.C.S. Perkin II*, **1973**, 114.
[151] R. J. P. Corriu and J. J. E. Moreau, *Tetrahedron Letters*, **1973**, 4469; *Chem. Comm.*, **1973**, 38.
[152] R. J. P. Corriu and B. Meunier, *J. Organometal. Chem.*, **60**, 31 (1973).
[153] P. Gillespie, F. Ramirez, I. Ugi and D. Marquarding, *Angew. Chem. Internat. Ed.*, **12**, 91 (1973).
[154] J. I. Musher, *Tetrahedron Letters*, **1973**, 1093.
[155] J. A. Howard, D. R. Russell and S. Trippett, *Chem. Comm.*, **1973**, 856.
[156] B. E. Maryanoff, R. Tang and K. Mislow, *Chem. Comm.*, **1973**, 273.
[157] J. I. G. Cadogan, D. T. Eastlick, J. A. Challis and A. Cooper, *J.C.S. Perkin II*, **1973**, 1798.
[158] K. L. Marsi and J. E. Oberlander, *J. Am. Chem. Soc.*, **95**, 200 (1973).
[159] *Org. Reaction Mech.*, **1970**, 96.
[160] K. E. De Bruin, A. G. Padilla and M.-T. Campbell, *J. Am. Chem. Soc.*, **95**, 4681 (1973).
[161] K. E. De Bruin and D. M. Johnson, *J. Am. Chem. Soc.*, **95**, 4675 (1973); cf. *Org. Reaction Mech.*, **1971**, 88.
[162] K. E. De Bruin and D. M. Johnson, *J. Am. Chem. Soc.*, **95**, 7921 (1973).
[163] W. C. Archie and F. H. Westheimer, *J. Am. Chem. Soc.*, **95**, 5955 (1973).
[164] V. E. Bel'skii, N. N. Bezzubova, M. V. Efremova and I. A. Nuretdinov, *Zh. Obshch. Khim.*, **43**, 1255 (1973); *Chem. Abs.*, **79**, 77692 (1973).
[165] D. W. Allen, S. J. Grayson, I. Harness, B. G. Hutley and I. W. Mowat, *J.C.S. Perkin II*, **1973**, 1912; cf. *Org. Reaction Mech.*, **1972**, 87.
[166] W. S. Wadsworth, S. Larsen and H. L. Horten, *J. Org. Chem.*, **38**, 256 (1973).
[167] W. S. Wadsworth, *J. Org. Chem.*, **38**, 2921 (1973).
[168] L. J. Stangeland, *Acta Chem. Scand.*, **27**, 1503 (1973).
[169] D. A. R. Happer, J. W. Mitchell and G. J. Wright, *Austral. J. Chem.*, **26**, 121 (1973).
[170] E. Ciuffarin, L. Senatore and L. Sagramora, *J.C.S. Perkin II*, **1973**, 534.
[171] J. L. Kice and J. P. Cleveland, *J. Am. Chem. Soc.*, **96**, 104 (1973); see *Org. Reaction Mech.*, **1970**, 103.
[172] J. L. Kice and J. P. Cleveland, *J. Am. Chem. Soc.*, **95**, 109 (1973).
[173] S. Tamagaki, H. Hirota and S. Oae, *Bull. Chem. Soc. Japan*, **46**, 1247 (1973).
[174] E. Ciuffarin, L. Senatore and G. Giovannini, *J.C.S. Perkin II*, **1972**, 2314.
[175] S. Braverman and D. Reisman, *Tetrahedron Letters*, **1973**, 3563.
[176] C. Christophersen and P. Carlsen, *Tetrahedron Letters*, **1973**, 211.
[177] *Org. Reaction Mech.*, **1971**, 91.

[178] F. G. Yamagishi, D. R. Rayner, E. T. Zwicker and D. J. Cram, *J. Am. Chem. Soc.*, **95**, 1916 (1973).
[179] D. C. Garwood, M. R. Jones and D. J. Cram, *J. Am. Chem. Soc.*, **95**, 1925 (1973).
[180] R. Annunziata, M. Cinquini and S. Colonna, *J.C.S. Perkin I*, **1973**, 1231.
[181] O. Bohman and S. Allenmark, *Tetrahedron Letters*, **1973**, 405; cf. *Org. Reaction Mech.*, **1970**, 103; **1972**, 90, ref. 293.
[182] F. Jung and T. Durst, *Chem. Comm.*, **1973**, 4; cf. *Org. Reaction Mech.*, **1972**, 89.
[183] N. Kunieda and S. Oae, *Bull. Chem. Soc. Japan*, **46**, 1745 (1973); cf. *Org. Reaction Mech.*, **1972**, 90.
[184] L. Senatore, E. Ciuffarin, A. Fava and G. Levita, *J. Am. Chem. Soc.*, **95**, 2918 (1973).
[185] O. Rogne, *J.C.S. Perkin II*, **1973**, 823, 1760.
[186] A. Arcoria, E. Maccarone, G. Musumarra and G. A. Tomaselli, *J. Org. Chem.*, **38**, 2457 (1973).
[187] W. K. Kim and I. Lee, *Daehan Hwahak Hwoejee*, **17**, 163 (1973); *Chem. Abs.*, **79**, 41582 (1973).
[188] S. Oae, Y. Y. Tsuchida and N. Furukawa, *Bull. Chem. Soc. Japan*, **46**, 648 (1973).
[189] W. J. Spillane, J. A. Barry and F. L. Scott, *J.C.S. Perkin II*, **1973**, 481.
[190] B. B. Jarvis and J. C. Saukaitis, *J. Am. Chem. Soc.*, **95**, 7708 (1973); *Tetrahedron Letters*, **1973**, 709.
[191] Y. Aufavre, M. Verny and R. Vessière, *Bull. Soc. chim. France*, **1973**, 1373.
[192] C. W. Jefford, U. Burger, M. H. Laffer and nT. Kabengele, *Tetrahedron Letters*, **1973**, 2483; cf. *Org. Reaction Mech.*, **1971**, 104.
[193] T. B. Patrick and J. A. Schield, *Tetrahedron Letters*, **1973**, 445.
[194] K. M. Al-Kathumi and L. A. P. Kane-Maguire, *J.C.S. Dalton*, **1973**, 1683.
[195] J. P. Barton, J. E. Packer and R. J. Sims, *J.C.S. Perkin II*, **1973**, 1547.
[196] A. L. Kurts, P. I. Dem'yanov, I. P. Beletskaya and O. A. Reutov, *Zh. Org. Khim.*, **9**, 1313 (1973); *Chem. Abs.*, **79**, 91398 (1973); A. L. Kurts, S. M. Sakembaeva, I. P. Beletskaya and O. A. Reutov, *Zh. Org. Khim.*, **9**, 1553 (1973).
[197] F. Chastrette, M. Chastrette and G. Santana-Tavares, *Bull. Soc. chim. France*, **1973**, 368.
[198] P. E. Pfeffer, L. S. Silbert and E. Kinsel, *Tetrahedron Letters*, **1973**, 1163.
[199] G. Stork and R. L. Danheiser, *J. Org. Chim.*, **38**, 1775 (1973).
[200] B. A. Stoochnoff and N. L. Benoiton, *Tetrahedron Letters*, **1973**, 21.
[201] K. Subramanian and M. Hanack, *Tetrahedron Letters*, **1973**, 3365; cf. *Org. Reaction Mech.*, **1972**, 93.
[202] J. L. Derocque, F. B. Sundermann, N. Youssif and M. Hanack, *Liebig's Annalen*, **1973**, 419.
[203] W. E. Heyd and M. Hanack, *Angew. Chem. Internat. Ed.*, **12**, 318 (1973).
[204] D. R. Kelsey and R. G. Bergman, *Chem. Comm.*, **1973**, 589.
[205] *Org. Reaction Mech.*, **1971**, 23.
[206] D. D. Maness and L. D. Turrentine, *Tetrahedron Letters*, **1973**, 755.
[207] Z. Rappoport, A. Pross and Y. Apeloig, *Tetrahedron Letters*, **1973**, 2015.
[208] P. J. Stang and T. E. Dueber, *J. Am. Chem. Soc.*, **95**, 2686 (1973).
[209] M. D. Schiavelli and D. E. Ellis, *J. Am. Chem. Soc.*, **95**, 7916 (1973).
[210] Z. Rappoport and M. Atidia, *J.C.S. Perkin II*, **1973**, 2316; cf. *Org. Reaction Mech.*, **1970**, 106.
[211] Z. Rappoport and Y. Houminer, *J.C.S. Perkin II*, **1973**, 1506.
[212] P. J. Stang and T. E. Dueber, *J. Am. Chem. Soc.*, **95**, 2683 (1973).
[213] Z. Rappoport, A. Gal and Y. Houminer, *Tetrahedron Letters*, **1973**, 641.
[214] E. Lamparter and M. Hanack, *Chem. Ber.*, **106**, 3216 (1973).
[215] M. D. Schiavelli, P. L. Timpanaro and R. Brewer, *J. Org. Chem.*, **38**, 3054 (1973).
[216] L. Eckes, L. R. Subramanian and M. Hanack, *Tetrahedron Letters*, **1973**, 1967.
[217] Z. Rappoport and A. Gal, *J.C.S. Perkin II*, **1973**, 301.
[218] Z. Rappoport and P. Peled, *J.C.S. Perkin II*, **1973**, 616; cf. *Org. Reaction Mech.*, **1971**, 100; **1972**, 94.
[219] J.-C. Chalchat, F. Théron and F. Vessière, *Bull. Soc. chim. France*, **1973**, 2500; cf. *Org. Reaction Mech.*, **1970**, 109.
[220] J.-F. Normant, R. Sauvêtre and J. Villieras, *Compt. Rend., C*, **277**, 469 (1973).
[221] P. C. Billot, R. G. Barker, K. Mackenzie and P. R. Young, *Tetrahedron Letters*, **1973**, 3059.
[222] H. C. Brown, T. Hamaoka and N. Ravindran, *J. Am. Chem. Soc.*, **95**, 6456 (1973).
[223] J. Cronin, A. F. Hegarty, P. A. Cashell and F. L. Scott, *J.C.S. Perkin II*, **1973**, 1708.
[224] J. E. Johnson, E. A. Nalley and C. Weidig, *J. Am. Chem. Soc.*, **95**, 2051 (1973).
[225] A. T. Bottini, K. A. Frost, B. R. Anderson and V. Dev, *Tetrahedron*, **29**, 1975 (1973).
[226] A. T. Bottini and B. Anderson, *Tetrahedron Letters*, **1973**, 3321; P. K. Freeman and T. A. Hardy, *Tetrahedron Letters*, **1973**, 3317.
[227] K. L. Erickson and S. B. Gordon, *J. Org. Chem.*, **38**, 1431 (1973).
[228] K. L. Erickson, *J. Org. Chem.*, **38**, 1463 (1973).
[229] B. J. Barry, W. J. Beale, M. D. Carr, S. Hei and I. Reid, *Chem. Comm.*, **1973**, 177.
[230] T. Wooldridge and T. D. Roberts, *Tetrahedron Letters*, **1973**, 4007.

[231] P. Beltrame, P. L. Beltrame, M. G. Cattania and M. Simonetta, *J.C.S. Perkin II*, **1973**, 63.
[232] R. Lapouyade, M. Daney, M. Lapenue and H. Bouas-Laurent, *Bull. Soc. chim. France*, **1973**, 720.
[233] D. Baudry and M. Charpentier-Morize, *Tetrahedron Letters*, **1973**, 3013.
[234] J. P. Bégué and D. Bonnet, *Compt. Rend.*, *C*, **276**, 1695 (1973); cf. *Org. Reaction Mech.*, **1971**, 102.
[235] E. M. Gaydou, G. Buono and R. Freze, *Bull. Soc. chim. France*, **1973**, 2284; cf. *Org. Reaction Mech.*, **1972**, 97.
[236] J. Konecny, R. Dousse and J. Rosalies, *Helv. Chim. Acta*, **55**, 3048 (1972).
[237] A. Fougerousse and J. J. Riehl, *Tetrahedron Letters*, **1973**, 3593.
[238] E. Elkik and H. Assadi-Far, *Bull. Soc. chim. France*, **1973**, 202.
[239] E. Campaigne and J. H. Hutchinson, *J. Heterocyclic Chem.* **10**, 229 (1973).
[240] F. de Champs and S. Léger, *Ann. Chim.* (*Paris*), **7**, 411 (1972).
[241] J. H. Smith, J. H. Heidema, E. T. Kaiser, J. B. Wetherington and J. W. Moncrief, *J. Am. Chem. Soc.*, **94**, 9274 (1972); J. H. Smith, J. H. Heidema and E. T. Kaiser, *ibid.*, p. 9276.
[242] G. B. Behera, R. C. Acharya and M. K. Rout, *Indian J. Chem.*, **11**, 81 (1973); *Chem. Abs.*, **79**, 4586 (1973).
[243] P. Mishra, P. L. Nayak and M. K. Rout, *Indian J. Chem.*, **11**, 452 (1973); *Chem. Abs.*, **79**, 77716 (1973).
[244] L. M. Litvinenko, A. F. Popov, Z. P. Ghelbina and E. V. Kirillov, *Organic Reactivity* (*Tartu*), **10**, 185 (1973).
[245] R. F. W. Bader, A. J. Duke and R. R. Messer, *J. Am. Chem. Soc.*, **95**, 7715 (1973); cf. *Org. Reaction Mech.*, **1972**, 97.
[246] H. Tanida, S. Yamamoto and K. Takeda, *J. Org. Chem.*, **38**, 2077 (1973).
[247] Y. Gounelle and D. Solgadi, *Bull. Soc. chim. France*, **1973**, 3019.
[248] J. Hayami, N. Tanaka, N. Hihara and A. Kaji, *Tetrahedron Letters*, **1973**, 385.
[249] L. R. C. Barclay, J. R. Mercer and J. C. Hudson, *Can. J. Chem.*, **50** 3965 (1972).
[250] W. S. Jeong, *J. Korean Nucl. Soc.*, **5**, 26 (1973); *Chem. Abs.*, **79**, 52533 (1973).
[251] M. Cavazza, G. Guerra and A. Fava, *Gazzetta*, **103**, 361 (1973).
[251a] V. I. Kokochashvili and L. M. Sepiashvili, *Tr. Tbilis. Univ.*, **1971**, 21; *Chem. Abs.*, **78**, 57340 (1973).
[252] J. Hayami, N. Tanaka and A. Kaji, *Bull. Chem. Soc. Japan*, **46**, 954 (1973).
[253] M. K. Mardoyan, R. M. Flid, Yu. A. Treger and S. M. Gabrielyan, *Uch. Zap., Erevan. Univ., Estestv. Nauki*, **1972**, 63; *Chem. Abs.*, **78**, 158587 (1973).
[254] G. Pass, G. O. Phillips and A. Samce, *J.C.S. Perkin II*, **1973**, 932.
[255] R. U. Lemieux, Y. Ito, K. James and T. L. Nagabhushan, *Can. J. Chem.*, **51**, 7 (1973).
[256] V. S. Karavan, N. A. Venediktova, A. G. Ivanenko and T. I. Temnikova, *Zh. Org. Khim.*, **9**, 544 (1973); *Chem. Abs.*, **78**, 147000 (1973).
[257] A. P. Voropaeva and Ya. Bryuske, *Tr. Tambov. Inst. Khim. Mashinostr.*, **1971**, 143; Ya. E. Bryuske and A. P. Voropaeva, *Tr. Tambovskogo Inst. Khim. Mashinostr.*, **1970**, 124; *Chem. Abs.*, **78**, 3462, 42449 (1973).
[258] L. B. Engemyr and J. Songstad, *Acta Chem. Scand.*, **26**, 4179 (1972).
[259] J. Gore, P. Place and M. L. Roumestant, *Chem. Comm.* **1973**, 821.
[260] L. B. Young, E. Lee-Ruff and D. K. Bohme, *Chem. Comm.*, **1973**, 35.
[261] C. D. Ritchie and P. O. I. Virtanen, *J. Am. Chem. Soc.*, **95**, 1882 (1973); cf. *Org. Reaction Mech.*, **1972**, 59.
[262] H. C. Brown and S. Krishnamurthy, *J. Am. Chem. Soc.*, **95**, 1669 (1973).
[263] C. C. Price, H. Akimoto and R. Ho, *J. Org. Chem.*, **38**, 1538 (1973).
[264] H. Ginsburg, R. Guibé and G. Née, *Bull Soc. chim. France*, **1972**, 4590.
[264a] P. Müller and B. Siegfried, *Helv. Chim. Acta*, **55**, 2965 (1972).
[265] D. H. R. Barton, P. D. Magnus and R. N. Young, *Chem. Comm.*, **1973**, 331.
[266] T. Cohen and J. Solash, *Tetrahedron Letters*, **1973**, 2513.
[267] A. Davison and D. L. Reger, *J. Am. Chem. Soc.*, **94**, 9237 (1972).
[268] B. E. Ivanov, V. F. Zheltukhin and L. A. Valitova, *Khim. Primen. Fosfororg. Soedin., Tr. Konf., 4th*, **1969**, 204; *Chem. Abs.*, **78**, 135260 (1973).
[269] V. V. Orlovskii, B. A. Vovsi and A. E. Mishkevich, *Zh. Obshch. Khim.*, **42**, 1930 (1972); *Chem. Abs.*, **78**, 42458 (1973).
[270] N. N. Bezzubova, V. E. Bel'skii and V. N. Eliseenkov, *Sb. Nekot. Probl. Org. Khim., Mater. Nauch. Sess., Inst. Org. Fiz. Khim., Akad. Nauk SSSR*, **1972**, 53; *Chem. Abs.*, **78**, 28804, (1973).
[271] A. Rahm and M. Pereyre, *Tetrahedron Letters*, **1973**, 1333; cf. *Org. Reaction Mech.*, **1972**, 91.
[272] C. R. Johnson and G. A. Durat, *J. Am. Chem. Soc.*, **95**, 7783 (1973).
[273] P. R. O. de Montellano, *Chem. Comm.*, **1973**, 709.
[274] D. Kristol and R. Shapiro, *J. Org. Chem.*, **38**, 1470 (1973).

275 D. Plouin, C. Coeur and R. Glénat, *Bull. Soc. chim. France*, **1973**, 1043.
276 M. V. Mavrov, A. P. Radionov and V. F. Kucherov, *Tetrahedron Letters*, **1973**, 759.
277 G. Köbrich and A. Baumann, *Angew. Chem. Internat. Ed.*, **12**, 856 (1973).
278 G. Biggi, F. Del Cima and F. Pietra, *Tetrahedron Letters*, **1973**, 183.
279 S. Gerszberg, R. T. Gaona, H. Lopez and J. Comin, *Tetrahedron Letters*, **1973**, 1269.
280 A. T. Babayan, E. S. Ananyan and G. T. Babayan, *Doklady Akad. Nauk Arm. SSSR*, **56**, 27 (1973); *Chem. Abs.*, **79**, 77644 (1973).
281 D. Aumann and L. W. Deady, *Chem. Comm.*, **1973**, 32.
282 H. Lund and V. Lund, *Acta Chem. Scand.*, **27**, 383 (1973).
283 L. W. Deady, *Austral. J. Chem.*, **26**, 1949 (1973).
284 A. Donetti and E. Bellora, *Tetrahedron Letters*, **1973**, 3573.
285 G. B. Behera, J. N. Kar, R. C. Acharya and M. K. Rout, *J. Org. Chem.*, **38**, 2164 (1973).
286 M. Davis, E. Homfeld and K. S. L. Srivastava, *J.C.S. Perkin I*, **1973**, 1863.
287 H. Lund, *Acta Chem. Scand.*, **27**, 391 (1973).
288 L. A. Ignatova, G. L. Ovetchkina and B. V. Unkovsky, *Organic Reactivity* (*Tartu*), **10**, 83 (1973).
289 R. C. Duty and C. F. Milewski, *J. Org. Chem.*, **38**, 2453 (1973); L. Jasinskas, A. Urbonas and J. Dienyte, *Liet. TSR Aukst. Mokyklu Mokslo Darb., Chem. Chem. Technol.*, **13**, 93 (1971); *Chem. Abs.*, **78**, 3417 (1973); L. A. Perelman, A. F. Popov and L. M. Litvinenko, *Organic Reactivity* (*Tartu*), **9**, 964 (1972); L. A. Perelman, A. F. Popov, L. M. Litvinenko and I. I. Zaslavskaya, *ibid.*, p. 936.
290 J. W. McKinley, R. E. Pincock and W. B. Scott, *J. Am. Chem. Soc.*, **95**, 2030 (1973).
291 L. Baiocchi, M. Giannangeli and G. Palazzo, *Gazzetta*, **103**, 61 (1973); A. M. Bianchi, V. Rosnati, A. Saba, F. Soccolini and G. Lecchi, *ibid.*, p. 79.
292 S. Coffi-Nketsia and A. Kergomard, *Bull. Soc. chim. France*, **1973**, 2115.
293 E. S. Rudakov, V. V. Zamashchikov and V. D. Belayev, *Organic Reactivity* (*Tartu*), **9**, 43 (1972); E. S. Rudakov and I. V. Kozhevnikov, *ibid.*, p. 184; V. V. Zamashchikov, and E. S. Rudakov, *ibid.*, pp. 286, 292, 313.
294 I. I. Tsykhanaskaya, F. P. L'vova, N. N. Vlasova and M. G. Voronkov, *Doklady Vses. Konf. Khim. Atsetilena, 4th*, **2**, 444 (1972); *Chem. Abs.*, **79**, 77650 (1973).
295 D. H. Aue, R. B. Lorens and G. S. Helwig, *Tetrahedron Letters*, **1973**, 4795.
296 T. Hiyama, H. Koide, S. Fujita and H. Nozaki, *Tetrahedron*, **29**, 3137 (1973).
297 H. Rieff, D. Dieterich, R. Braden and H. Ziemann, *Liebigs Annalen*, **1973**, 365.
298 J. Buddrus, *Angew. Chem. Internat. Ed.*, **11**, 1041 (1972).
299 I. Dobáš and J. Eichler, *Coll. Czech. Chem. Comm.*, **38**, 2602 (1973).
299a K. B. Sharpless and R. F. Lauer, *J. Am. Chem. Soc.*, **95**, 2697 (1973).
300 J. Halbych, T. Trnka and M. Černý, *Coll. Czech. Chem. Comm.*, **38**, 2151 (1973); B. L. Vorobyov, A. L. Shapiro and T. E. Zhesko, *Organic Reactivity* (*Tartu*), **10**, 281 (1973); P. Bouchet and P. Coquelet, *Bull. Soc. chim. France*, **1973**, 3153, 3159; D. H. R. Barton and Y. Houminer, *Chem. Comm.*, **1973**, 839; T. E. Zhesko, G. N. Mishenkova, V. S. Moshenko, G. S. Lubovsky and A. L. Shapiro, *Organic Reactivity* (*Tartu*), **10**, 306 (1973); B. L. Vorobyov and A. L. Shapiro, *ibid.*, pp. 290, 298; L. N. Finyakin, V. V. Kafarov, M. F. Sorokin, L. G. Shoda and G. V. Onosov, *Tr. Mosk. Khim.-Tekhnol. Inst.*, **1972**, 80; *Chem. Abs.*, **78**, 135217 (1973); J. L. Coke and R. S. Shue, *J. Org. Chem.*, **38**, 2210 (1973); B. C. Hartman and B. Rickborn, *ibid.*, **37**, 4246 (1972); R. P. Thummel and B. Rickborn, *ibid.*, pp. 3919, 4250; K. Jankowski and R. Harvey, *Can. J. Chem.*, **50**, 3930 (1972).
301 H. Kakiuchi and T. Iijima, *Bull. Chem. Soc. Japan*, **46**, 1568 (1973); H. Ohtaka, M. Morisaki and N. Ikekawa, *J. Org. Chem.*, **38**, 1688 (1973); R. A. Wohl and J. Cannie, *ibid.*, p. 1787; J. G. Pritchard and I. A. Siddiqui, *J.C.S. Perkin II*, **1973**, 452; M. D. Carr and C. D. Stevenson, *ibid.*, p. 518; D. L. Whalen, *J. Am. Chem. Soc.*, **95**, 3432 (1973); I. Al-Wahib, V. F. Shvets and T. V. Kozenkova, *Tr. Mosk. Khim.-Tekhnol. Inst.*, **1972**, 54; *Chem. Abs.*, **78**, 135235 (1973); M. F. Sorokin and V. G. Kolosov, *Izv. Vyssh. Ucheb. Zaved., Khim. Khim. Tekhnol.*, **16**, 886 (1973); *Chem. Abs.*, **79**, 77710 (1973); D. G. Smith and D. J. H. Smith, *Tetrahedron Letters*, **1973**, 1249; P. Crotti, B. Macchia and F. Macchia, *Tetrahedron*, **29**, 155 (1973); A. Balsamo, P. Crotti, B. Macchia and F. Macchia, *ibid.*, p. 199; A. Balsamo, P. Crotti, B. Macchia and F. Macchia, *ibid.*, p. 2183; G. Belluchi, G. Berti, B. Macchia and F. Macchia, *Gazzetta*, **103**, 345 (1973); G. Belluchi, G. Berti, G. Ingrosso and E. Mastorilli, *Tetrahedron Letters*, **1973**, 3911; S. Kumazawa, T. Sakakibara, R. Sudoh and T. Nakagawa, *Angew. Chem. Internat. Ed.*, **12**, 921 (1973); S. Ducher and J. Peyronnet, *Ann. Chim.* (*Paris*), **5**, 415 (1970); A. V. Willi, *Helv. Chim. Acta*, **56**, 2094 (1973); L. Bang and G. Ourisson, *Tetrahedron*, **29**, 2097 (1973); M. Weissenberg, D. Lavie and E. Glotter, *ibid.*, p. 353; J. P. Nagarkatti and K. R. Ashley, *Tetrahedron Letters*, **1973**, 4599; C. Berse and P. Bessette, *Can. J. Chem.*, **50**, 4061 (1972).
302 J. D. Aubort, R. F. Hudson and R. C. Woodcock, *Tetrahedron Letters*, **1973**, 2229.

[303] J. F. Liebman and R. M. Pollack, *J. Org. Chem.*, **38**, 3444 (1973).
[304] J. H. Beale, *J. Org. Chem.*, **37**, 3871 (1972).
[305] R. Aneja, A. P. Davies and J. A. Knaggs, *Chem. Comm.*, **1973**, 110; cf. *Org. Reaction Mech.*, **1972**, 103; **1971**, 109.
[306] *Org. Reaction Mech.*, **1971**, 109.
[307] B. Castro, M. Ly, and C. Selve, *Tetrahedron Letters*, **1973**, 4455.
[308] B. Castro, Y. Chapleur and B. Gross, *Bull. Soc. chim. France*, **1973**, 3034.
[309] A. K. Bose and B. Lal, *Tetrahedron Letters*, **1973**, 3937.
[310] N. Furukawa, T. Inoue, T. Aida and S. Oae, *Chem. Comm.*, **1973**, 212.
[311] E. W. Yankee, F. D. Badea, N. E. Howe and D. J. Cram, *J. Am. Chem. Soc.*, **95**, 4210 (1973); E. W. Yankee, B. Spencer, N. E. Howe and D. J. Cram, *J. Am. Chem. Soc.*, **95**, 4220 (1973); cf. *Org. Reaction Mech.*, **1970**, 117.
[312] G. F. Koser and S.-M. Yu, *J. Org. Chem.*, **38**, 1755 (1973).
[313] W. E. Parham, W. D. McKown, V. Nelson, S. Katigaeshi and N. Ishikawa, *J. Org. Chem.*, **38**, 1361.

CHAPTER 10

Carbanions and Electrophilic Aliphatic Substitution

R. B. Boar

Department of Chemistry, Chelsea College, University of London

Carbanion Structure and Stability

^{7}Li- and ^{1}H-NMR spectra of 9-(1-methylpentyl)fluorenyl-lithium revealed a regular increase in cation–anion interaction with decreasing solvent polarity.[1] The ratio of products from protonation of anions (**1**) correlates with the charge distribution as measured by PMR spectroscopy according to the equation $\log([\mathbf{2}]/[\mathbf{3}]) = 1.75(\tau_{H_a} - \tau_{H_b})$.[2]

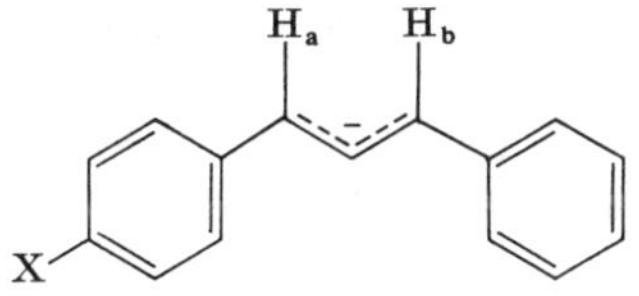

H_a H_b

X

(**1**)

$\longrightarrow$

p-$XC_6H_4CH_2CH{=}CHPh$ (**2**)

+

p-$XC_6H_4CH{=}CHCH_2Ph$ (**3**)

CMR shifts have been measured for various pentadienyl anions, but correlation with known nucleophilicities is not decisive.[3] The PMR spectra of lithiated methylpyridines,[4] and the optical and magnetic properties of hexahelicene anions[5] have been discussed. Observation of 3-alkylallylic Grignard reagents by PMR spectroscopy indicated a rapidly equilibrating mixture of *Z* and *E* primary stereoisomers only, the percentage of *Z*-isomer decreasing as the size of the alkyl substituent increased.[6] Analogous studies of the spectra of neopentylallyl-lithium[7] and bis(neopentylallyl)magnesium[8] have appeared. The effect of varying R on the rate of rearrangement of cyclohex-3-enyl Grignard reagents (**4**) to the more stable primary cyclopentenylmethyl system (**5**) suggests that concerted four-centre addition may be involved.[9,10] By suitable alkylation the equilibrium composition of 3-butenyl Grignard reagents can be changed from ~99.9% of (**7**) when R = Me, R′ = H, to ~99.9% of (**6**) when R = R′ = Me.[11]

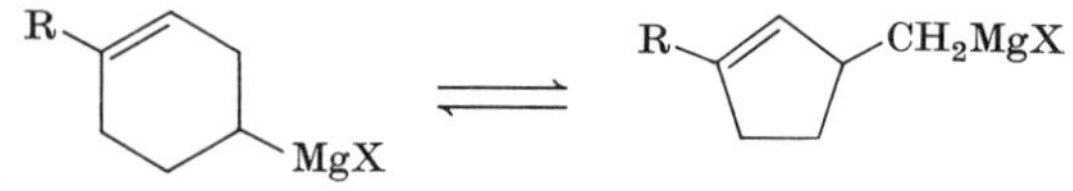

(4) ⇌ (5)

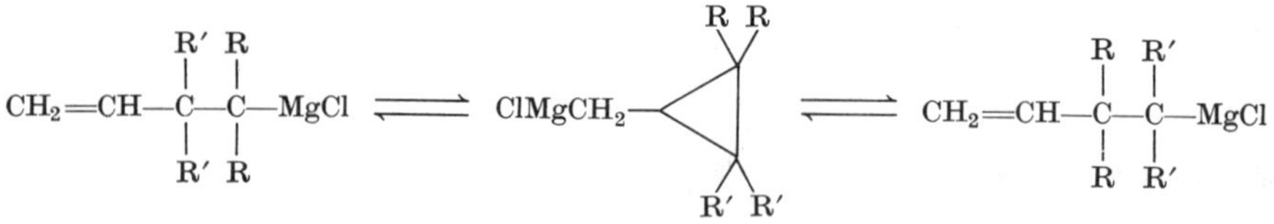

(6) ⇌ (7)

IR spectroscopy suggested the structure $Li_2C{=}C{=}C{=}C{=}CLi_2$ for the perlithiated species from penta-1,3-diyne and *n*-butyl-lithium/*N*,*N*,*N'*,*N'*-tetramethylethylenediamine (TMEDA),[12] and allenic (non-polar solvents) or sesquiacetylenic (polar solvents) structures for prop-2-ynyl dianions.[13] The sites of metallation of diacetylenes have been investigated.[14] UV spectroscopy indicated two preferred conformers for carbanions $RNHCO\bar{C}(NO_2)_2$, the position of the equilibrium depending on solvent and the nature of R.[15]

Aromaticity

Reaction of isobutene with *n*-butyl-lithium/TMEDA produces the methylallyl anion and then, more rapidly, the 6π-electron trimethylenemethane dianion.[16] The pK_a's of further[17] 2*H*-thiopyran 1,1-dioxides have been measured, but definite evidence for the role of aromaticity in the 6π-electron anions is still lacking.[18] Polarography and kinetic acidity measurements[19] revealed no significant stabilization (spiroaromaticity) in the spiro[2.3]hexadienyl anion (**8**) compared with a series of spiro[2.3]hex-4-en-6-yl anions.[20]

The aromatic (by NMR) [17]annulenyl anion has notable thermal stability, being unchanged after one hour at 100°.[21] The conformation and the π-bond structure of [18]annulene dianion have been assessed by NMR spectroscopy.[22] The stability of carbanions (**9a**) and (**9b**) were compared by measurement (NMR spectroscopy) of the

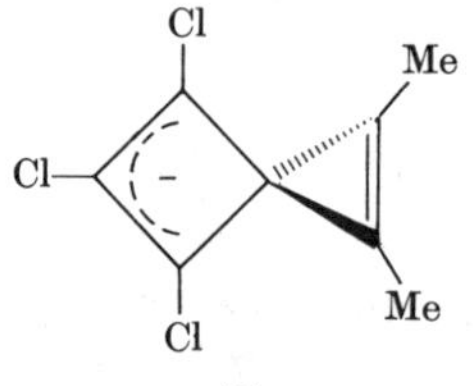

(**8**)

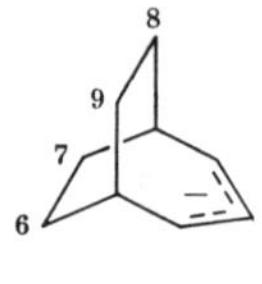

(**9**) **a** 6-ene
b 6,8-diene

rate of their pseudo-first-order deprotonation of an ether solvent,[23] and by $KOBu^t$/DMSO-catalysed detritiation of the corresponding tritiated hydrocarbons.[24] In the two studies (**9b**) was estimated to be the more stable by 2.5 and 4.0 kcal mol^{-1}, respectively. Since the difference in stability of (**9a**) and (**9b**) is >5.6 kcal mol^{-1},[24] previous predictions[25] are upheld to the effect that laticyclic/longicyclic stabilization is of less practical significance than its constant companion homoaromatic stabilization.

NMR spectroscopy indicated bishomoaromaticity in the dianion of 2,4-diphenylbicyclo[3.2.1]oct-6-en-3-one, although deuteriation yielded only the 2,4-$^2H_2$ derivative.[26]

Photolysis of either (**10**) or (**11**) led to a 1:3 mixture of the two.[27] NMR investigation indicated that the preferred conformation of 8,8-dimethylcyclo-octatrienyl anion (**10**) is a folded one,[27] similar to that for 8,8-dimethylcyclononatrienyl anion.[28] The related $4n\pi$-systems, benzocycloheptatrienyl anion[29] and bicyclo[5.4.1]dodecapentaenyl anion[30] each maintain paramagnetic ring currents, although in the latter case the effect is reduced by adoption of a twisted conformation. The anion (**12**) from *cis*-bicyclo[6.1.0]-nona-2,4,6-triene had no substantial diamagnetic or paramagnetic ring current, in keeping with its being considered a "$4n+1$" π-system.[31] The NMR spectrum of the

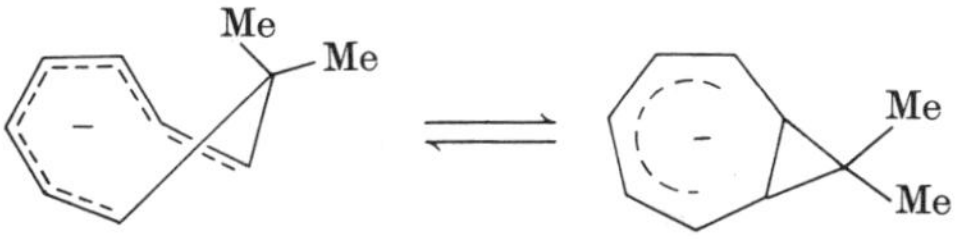

(**10**) (**11**) 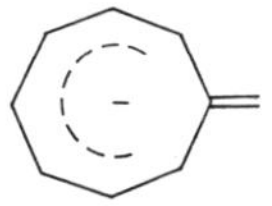 (**12**) 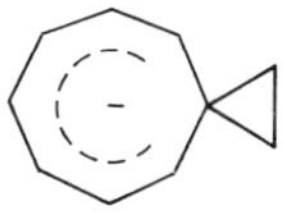 (**13**)

related spiro[2.7]decatrienyl anion (**13**) was, however, best explained by invoking a diagmagnetic ring current, suggesting spirocyclopropyl conjugation.[32] Carbanion (**13**) is stable at –30°, rearranging to bicyclo[6.2.0]deca-2,4,6-trienyl dianion at 0°, but even at –65° the only observable anion from spiro[2,5]octadiene was the phenethyl carbanion.[33]

Reactions of Carbanions

Attention has been drawn to the potential generality of dipole-stabilized carbanions $\bar{C}-\ddot{X}-Y{=}Z \leftrightarrow \bar{C}-\overset{+}{X}{=}Y-\bar{Z}:$.[34,35] Formation of α-carbanions of *N*-nitrosamines has been demonstrated[34,36] and applied to the synthesis of α-substituted secondary amines and, by reaction with non-enolizable nitriles, of substituted triazoles.[37] Metallation of RNMe(NO) occurs selectively at the methyl group for R = Et or Pr[i] but is not regio-specific for R = $PhCH_2$.[36] Although *N*-benzoyldibenzylamine readily formed an anion with lithium di-isopropylamide, this was found not to be a general reaction of *N*-benzoylamines.[34] Azoxy-compounds readily formed α-carbanions (as **14**) with lithium

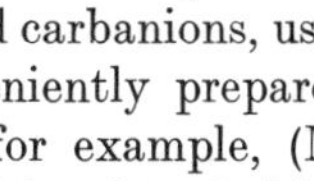

$$RN{=}\overset{\overset{O^-}{|}}{N^+}-\bar{C}HR$$

(**14**)

bistrimethylsilylamide, an observation that should considerably extend the chemistry of this functional group.[38] Vinylboranes may be metallated by lithium 2,2,6,6-tetra-methylpiperidide.[39] Silyl-substituted carbanions, useful in the synthesis of alkenes from carbonyl compounds,[40] were conveniently prepared by cleavage of a second C—Si bond by methoxide in HMPA,[41] for example, $(Me_3Si)_2CH_2 + {}^-OMe \rightarrow Me_3Si\bar{C}H_2 + Me_3SiOMe$). Formation of the Si—O bond probably provides the driving force for this reaction, since methyl-lithium does not behave similarly.

The mechanism of the formation of 1-haloalkynes from acetylenes and sodium hypohalites has been discussed in detail.[42] Other base-catalysed reactions to have received attention are: the hydrolysis of (methylthio)purines;[43] the condensation of 2- and 3-methoxybutyrates with ureas;[44] the reactions of primary and secondary alcohols with the active methylene groups of certain polycyclic hydrocarbons;[45] the reaction of 2-methyl-1,4-naphthoquinone with active-methylene compounds (Craven reaction);[46]

the rearrangement of 1,2,3,4,5-pentaphenylcyclopenta-2,4-dien-1-ol to 2,3,4,5,5-pentaphenylcyclopent-2-en-1-one;[47] and the Favorskii rearrangement of 3-(α-haloacyl)-indoles.[48] The rate-determining step in the rearrangement of 5-(carboxymethylene)-hydantoin (**15**) to orotic acid (**16**) is addition of OH^- to C-4.[49] Proton-removal from N-3 slows the rearrangement by a factor of 3,000.

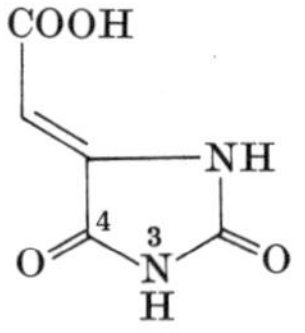

(**15**) 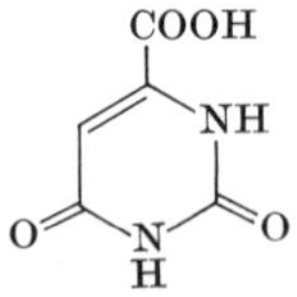 (**16**)

The effect of alkyl substituents R (*o*-, *m*-, or *p*-Pr^i or Bu^t) on the rate of Smiles rearrangement of the sulphones (**17**) has been discussed.[50]

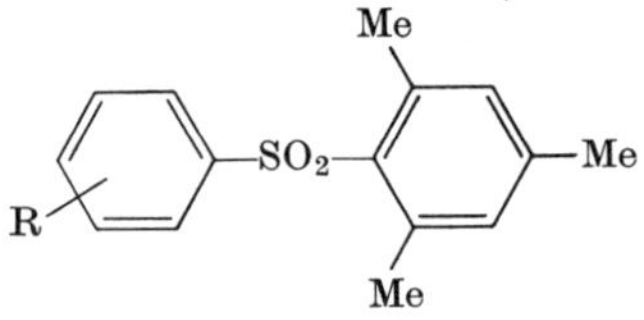

(**17**)

Relative rates of isomerization of hex-1-, -2-, and -3-yne and hexa-1,2- and -2,3-diene by $KOBu^t/Bu^tOH$ were found to be 1,2- > 1- ≫ 2,3- ≫ 3- > 2-. Product analyses during the initial stages of each isomerization support a stepwise alkyne–allene rearrangement (1- ⇌ 1,2- ⇌ 2-, etc.).[51] Similar results were obtained with sodamide in liquid ammonia except that in this more basic system the alk-1-yne is removed from the isomerization by conversion into acetylide ion. Formation of 4% of *trans*-oct-3-ene in the reduction of oct-4-yne by sodium in liquid ammonia was accounted for by isomerization of starting material by sodamide formed during the reaction;[52] the problem was avoided by performing the reaction in the presence of ethanol.

Rates of ring opening of the anions derived from the 1-cyanocyclopropyl derivatives (**18a**), (**19a**) and (**20**) (**18a** > **19a** ≫ **20**) were determined by observing the visible spectra of the allylic anions produced.[53] These kinetic data suggest that the opening of cyclopropyl anions by a conrotatory mode (a course sterically forbidden to **20**) is strongly

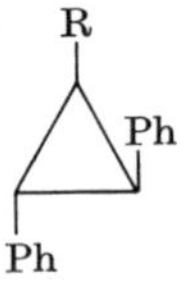

(**18**) 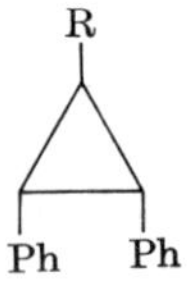 (**19**) 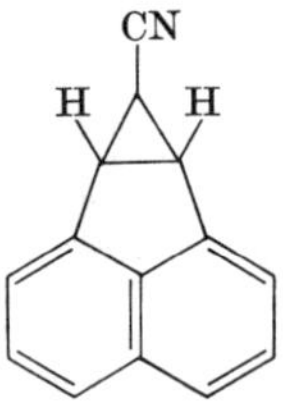 (**20**) 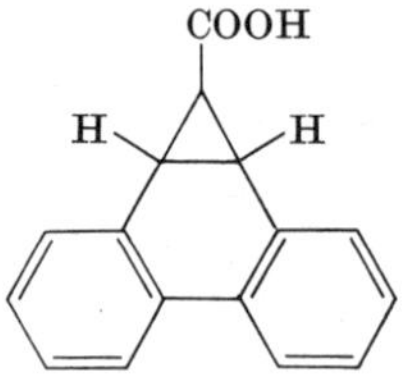 (**21**)

a, R = CN; **b**, R = COOH

favoured over other mechanisms. Similar data for ring opening of the anions of the cyclopropane-1-carboxylic acid derivatives (**18b**), (**19b**) and (**21**) (**18b** $\gg$ **19b** $\gg$ **21**) were complicated by secondary factors affecting the formation of dianions.[54]

*Enolates**

Traditional disadvantages of the aldol condensation are avoided when the aldehyde is added to the preformed lithium enolate of the ketone in diethyl ether or dimethoxyethane containing zinc chloride;[55] the true aldol product is then intercepted as the metal chelate (**22**). Where diastereoisomers of the aldol product are possible, the observed preferential formation of the *threo*-isomer corresponds to adoption by the chelate (**22**) of the conformation of minimum energy. Predominant axial alkylation of cyclohexanone enolates (**23**; R > H) can be understood if the enolate is twisted so as to relieve interaction between R and OM, and the twisting occurs in the direction that avoids R and OM becoming

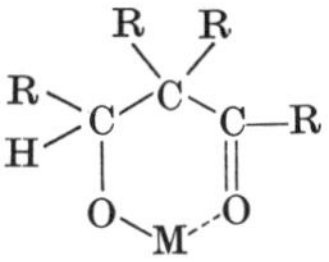

(**22**)

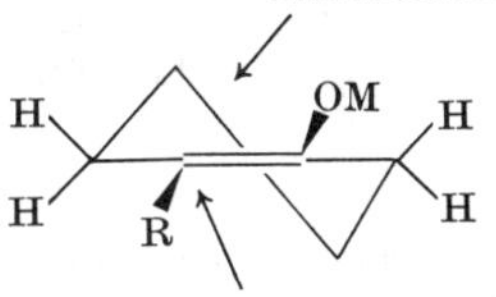

Axial attack

Equatorial attack

(**23**)

eclipsed with the adjacent pseudo-equatorial hydrogen atoms.[56] Factors affecting the mode of reaction of magnesium enolates with α,β-unsaturated ketones have been discussed.[57] The effects of varying the metal (M) or the solvent on the relative contributions of structures (**24**), (**25a**) and (**25b**) to metal enolates have been assessed from a spectroscopic examination of a large number of enolates.[58] In kinetically controlled reactions, *O*-acylation is the preferred reaction of species (**24**) and (**25a**); *C*-acylation is increased for enolates that exist in form (**25b**). The effect of crown polyethers on the ratio of *C*- to

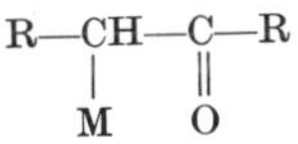

(**24**)

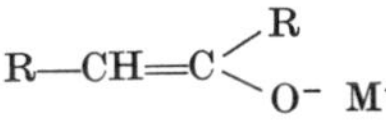

(**25**)

a, Solvent-separated ions

b, Contact ion pairs

O-alkylation of ethyl acetoacetate by ethyl tosylate has been discussed.[59] When R is phenyl or 2-naphthyl, the anion of the 1,3-indandione (**26**) was alkylated by haloacetic acids exclusively on carbon, but when R was 1-naphthyl both *C*- and *O*-alkylation were observed.[60] The virtues of ^{13}C-NMR for monitoring anionic-exchange processes have been expounded and applied to the study of aryl–hydrogen exchange in acenaphthene, and of homoenolization in fenchone (**27**) (sequential deuteriation at C-6, then C-8; other positions, note particularly C-9, were not involved).[61] The presence of two enolizable hydrogen atoms in (**28a**) and (**28b**) did not preclude homoenolization involving C-6, since

* See also Chapter 1.

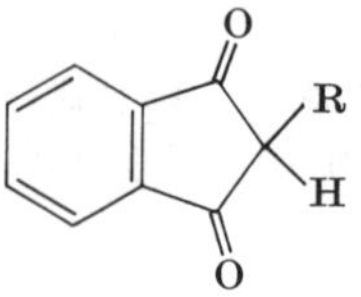

(26) 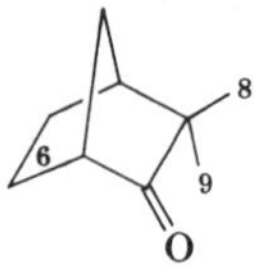 **(27)**

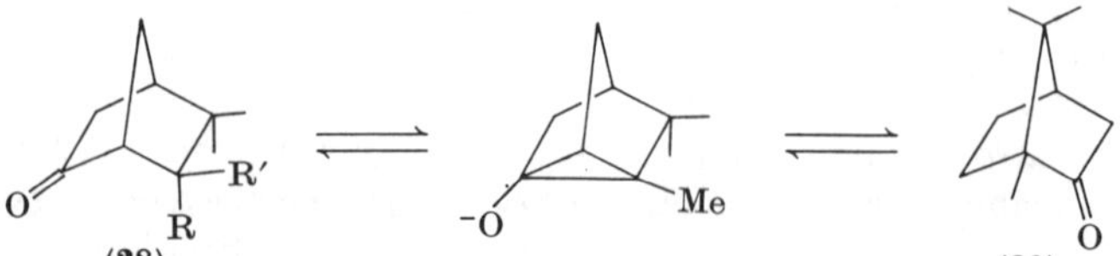

(28)

a, R = H, R′ = Me
b, R = Me, R′ = H

(29)

treatment of either compound with KOBu^t/Bu^{t}OH at 250° achieved both epimerization at C-6 and rearrangement to camphor (**29**).[61]

Anions α to Sulphur

α-Sulphonyl carbanions have been the subject of two reviews.[62] Unsaturated derivatives (7-ene; 2,4- or 2,7-diene: 2,4,7-triene) of the bridged bicyclic sulphone (**30**) all formed coloured dianions extremely readily. Extended Hückel calculations showed a good correlation with the observed spectra of these dianions.[63] Very large solvent effects (up to 10^3) have been observed for the relative rates of exchange of the same two protons in some conformationally rigid sulphoxides.[64] In (**31**) the orientations of H_1 and H_2 with

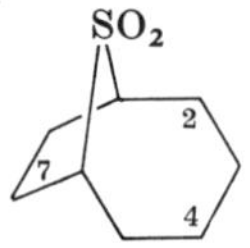

(30) 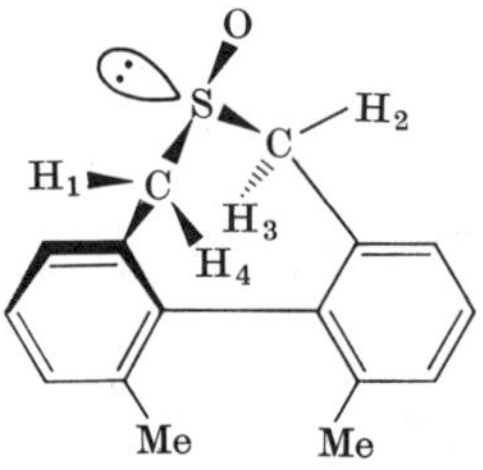 **(31)**

respect to the S—O bond differ by only 40°, but their rates of exchange varied by a factor of 200 (KOBu^t/Bu^{t}OD) and 1,000 (CD_3ONa/CD_3OD). Partial exchange with KOBu^t/[$^2H_6$]DMSO at room temperature of the sulphoxides (**32a**) and (**32b**) yielded the following

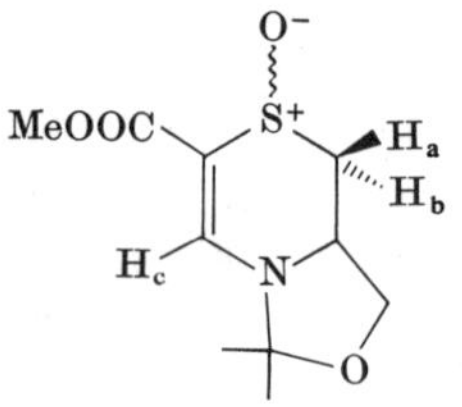

(32)

a, α-Sulphoxide
b, β-Sulphoxide

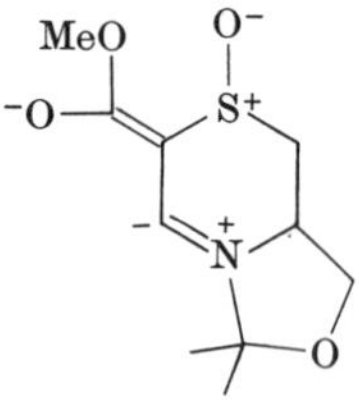

(33)

data: (**32a**) H_a 16, H_b 54, and H_c 69% exchanged; (**32b**) H_c 50% exchanged, H_a and H_b unchanged;[65] the ylide structure (**33**) was proposed as a contributing factor to the enhanced acidity of H_c. For hydroxide-ion catalysed removal of a proton from methyl 1-phenethyl sulphoxide in D_2O, $k_{exchange}/k_{epimerization} = 1.30$, a result consistent with the intervention of either two rapidly inverting pyramidal anions or a single planar anion.[66]

The optically active *p*-tolylsulphinyl carbanion (**34**) has been found to add to imines with extreme stereospecificity.[67] Desulphurization with Raney nickel completes a novel route to optically active amines. Anions of *N*-(*p*-tolylsulphonyl)sulphoximines (**35**) react with electrophilic double bonds with alkylidene-transfer;[68] thus, ketones give epoxides, imines give aziridines, and α,β-unsaturated ketones gave α,β-cyclopropyl ketones.

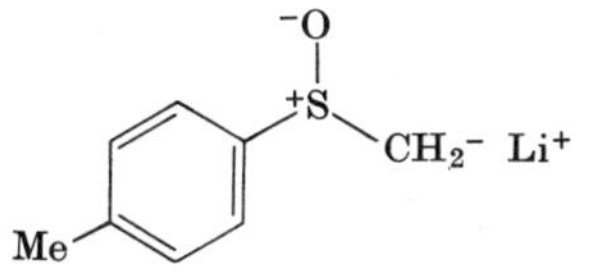

(**34**) 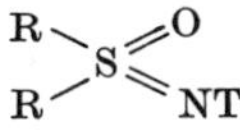 (**35**) 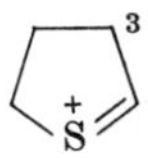 (**36**)

There is general agreement that the stereochemistry of chlorination[69] and bromination[70] of thiane 1-oxides is explained by a mechanism involving initial halogenation at sulphur, *trans*-diaxial elimination of hydrogen halide, and then axial attack of halide at C-2 (Scheme 1). In an analogous mechanism for the halogenation of tetrahydrothiophen, the ratio of 2-halo- to 2,3-dihalo-product has been found to depend on the relative tendencies of the intermediate (**36**) to add halide or to lose a proton from C-3.[71]

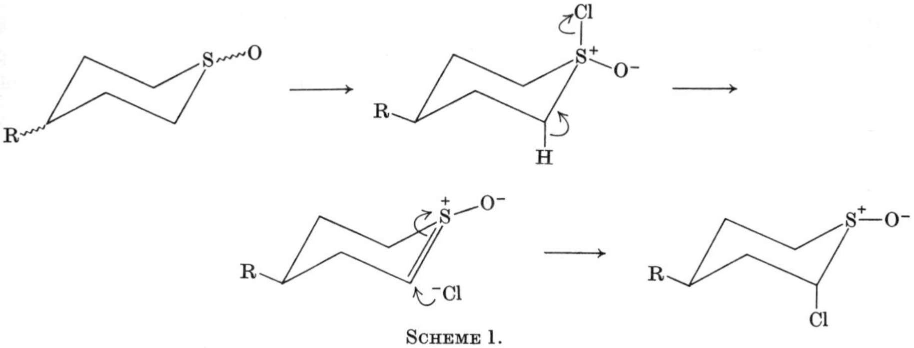

SCHEME 1.

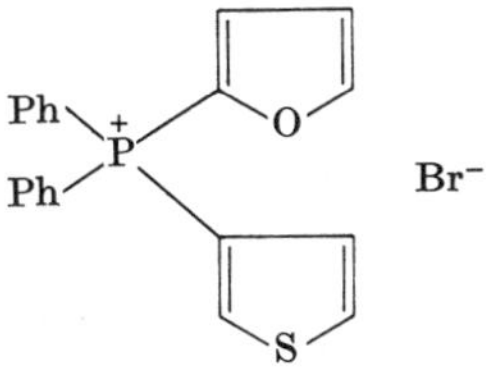

(**37**) 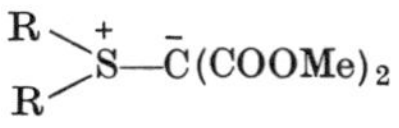 (**38**) 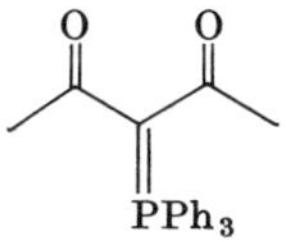 (**39**)

Evidence has been presented that the nature of anion–cation interaction in the carbanion of dibenzyl sulphide could affect the relative tendencies for rearrangements of Stevens–Sommelet type to occur.[72] The formation of furan and thiophen in a ratio of 1:3 on hydrolysis of the phosphonium salt (**37**) was taken to indicate greater stabilization of the developing 2-thienyl carbanion as a consequence of sulphur 3*d*-orbital stabilization.[73] Disulphide-catalysed transylidation reactions of the stable ylide (**38**) have been studied.[74]

Dianions

Like β-ketophosphonium salts,[75] β-ketophosphonates form dianions which undergo alkylation exclusively at the γ-position.[76] Monoalkylation of the ylide (**39**) is conveniently achieved by formation of the reactive dianion, reaction with electrophile at $-78°$, and quenching of the resultant monoanion with water.[77] The undefined species that results from treatment of α,α-dilithiobenzyl phenyl sulphone with magnesium iodide reacts with aldehydes and ketones to give α,β-unsaturated phenyl sulphones in good yield.[78] Regiospecificity during the reactions of the dianion of crotonic acid with electrophiles has been reassessed.[79]

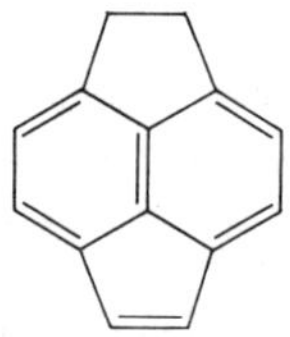

HOOC

COOH

(**40**) 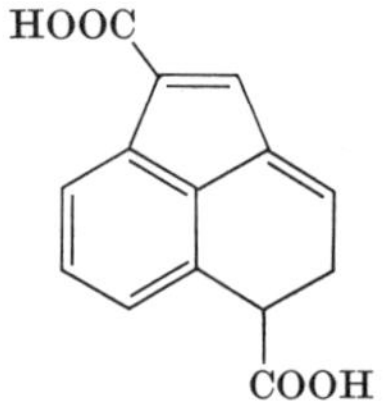 (**41**)

The dianion formed by deprotonation of dihydropyracylene (**40**) reacts with electrophiles exclusively in the five-membered ring, in accord with M.O. predictions for this peripheral 14π-electron species.[80] The fulvene (**41**) is the surprising[81] major product of carboxylation of acenaphthylenyl dianion.[82] The dihydroanthracene (**42**) is converted

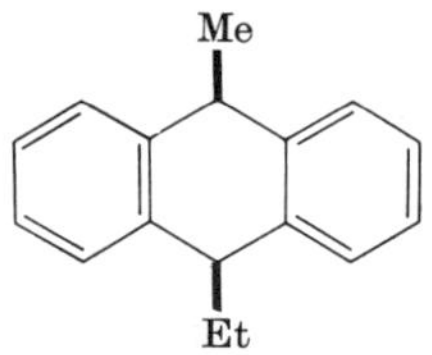

Me

Et

(**42**)

by *n*-butyl-lithium/TMEDA into a dianion which with cadmium chloride (electron acceptor) yields the corresponding anthracene quantitatively.[83] 9,10-Dihydrophenanthrene, however, reacts with two moles of *n*-butyl-lithium/TMEDA to give a coloured intermediate (postulated structure **43**) which is unaffected by metal salts but reacts with water or deuterium oxide to give unlabelled phenanthrene in quantitative yield. The formation of minor amounts of 1-ethyldihydronaphthalenes during reactions of sodium or lithium naphthalene in THF results from the reaction of naphthalene dianion with ethylene, itself formed, together with acetaldehyde enolate,[84] by base cleavage of THF.[85] Rates of metallation by dianions of aromatic hydrocarbons and by monoanions of their

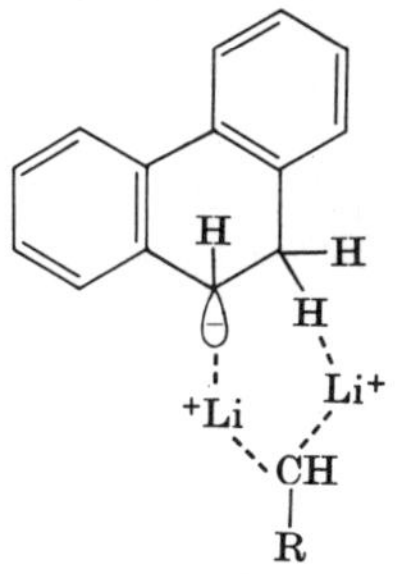

(**43**)

dihydro-derivatives have been measured and compared with analogous data for radical anions.[86] Dimethylarenes can be converted quantitatively into α,α'-dianions by *n*-pentylsodium/TMEDA.[87]

Proton Transfer and Hydrogen Isotope Effects[88]

Rate and equilibrium constants for proton transfer from arylethylmalonate monoanions to hydroxide ion gave a linear correlation with σ°,[89] indicating the importance of electronic rather than simply steric factors in this process. Consistent results were obtained when proton transfer from methyl 4-nitrovalerate to pyridine or 2,4,6-trimethylpyridine was measured by iodination and racemization techniques, and the very large isotope effects observed with the latter base were substantiated.[90] Variations in equilibrium and kinetic parameters for proton-transfer from di-(*p*-nitrophenyl)methane to ethoxide ion in acetonitrile–ethanol mixtures have been interpreted in terms of the Miller–Parker theory of solvation effects.[91] The kinetic isotope effect for proton-transfer reactions between (*p*-nitrophenyl)nitromethane and aliphatic amines is extremely dependent on solvent;[92] for reaction with tetramethylguanidine, k_H/k_D varied from 45 ± 2 (toluene) to 11.7 ± 0.2 (dichloromethane); for the former solvent the ratio of the Arrhenius *A*-factors ($A_D/A_H = 31.8$) was also unusually high, suggesting that tunnelling is important. Evidence has been presented that substantiates previous predictions concerning kinetic solvent isotope effects ($k_{ROH}/k_{ROD} < 1$ for alkoxide-catalysed reactions, but >1 for nitrogen-base catalysis).[93]

Chloroform is the first carbon acid for which proton-transfer is fully normal in the Eigen[94] sense, even for reactions with hydroxide ion.[95] For the reaction of chloroform with hydroxide ion, k_H/k_D was measured by comparing the rates of uptake of tritium by $CHCl_3$ and $CDCl_3$ from an aqueous solution during the initial zero-order phase of the reaction; at 25°, a value of 1.42 was found. Phenylacetylene[96] and disulphonyl-activated carbon acids[97] are two further[98] species for which proton-transfer is normal except when occurring to hydroxide ion. The existence of the "hydroxide ion anomaly" has been ascribed to the fact that formation of an encounter complex before proton-transfer will require a greater expenditure of energy for hydroxide ion, which is strongly solvated by hydrogen bonding, than for most other bases;[96] as the ability of the proton donor to form a hydrogen bond with hydroxide ion increases, so this energy difference decreases until in the limit (apparently reached with chloroform) hydroxide ion is seen to behave normally.

Isomerization (suprafacial, conducted-tour mechanism), racemization, and isoinversion (antarafacial, intramolecular proton-transfer via achiral contact ion-pair) have been demonstrated for ion pairs from 3-*tert*-butyl-1-methylindene (**44**) and tertiary

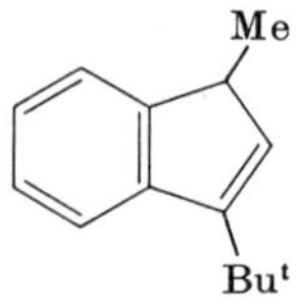

(44) 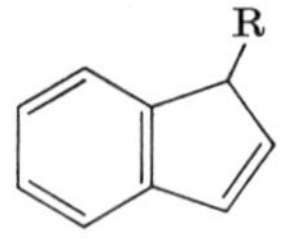 (45)

amines.[99] The relative importance of these processes depend significantly on whether the amine is capable of charge delocalization (e.g. pentamethylguanidine) or not (e.g. triethylenediamine). Isomerization of (**44**) by amidines is stereospecific in benzene or dioxan, but the extreme lack of stereospecificity in DMSO requires the intermediate ion-pair to have a collapse ratio of >31.[100] Relative rates of isomerization of the substituted 1-methylindenes (**45**; R = Et, Me, CH_2OMe, CH_2OAc or CH_2SAc) by triethylamine in benzene are 0.34, 1, 29.7, 143 or 121, respectively.[101] Treatment of [**45**; R = $Me_2C(OAc)$] with triethylamine in methanol containing triethylammonium acetate gave concurrent 1,2- and 1,4-elimination and 1,3-proton-transfer; [102] isotope effects suggest that all three processes occur via a common ion-pair intermediate.

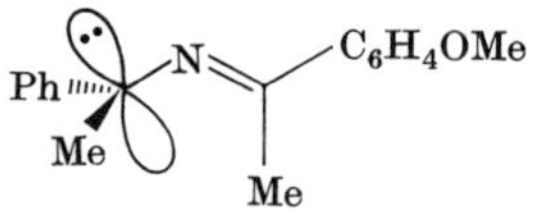

(46) 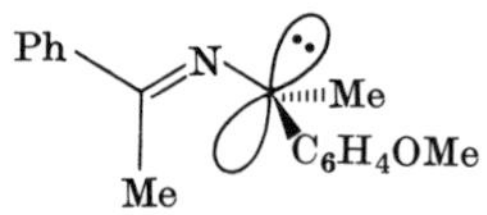 (47)

The intermediacy of two equilibrating carbanions (**46**) and (**47**) has been proposed to explain the uncharacteristically low stereospecificity in the base-catalysed interconversion of *N*-(α-methyl-4-methoxybenzylidene)-1-phenylethylamine and *N*-(α-methylbenzylidene)-1-(*p*-methoxyphenyl)ethylamine.[103] When Bu^t replaces C_6H_4OMe,[104] the second carbanion (**47**) is no longer available and stereospecificity is again observed. A preliminary account has appeared of the complex kinetics that result when the carbanion intermediate in the hydrogen–deuterium exchange of *N*-neopentylidenebenzylamine is intercepted by nitrobenzene.[105]

Comparison of the base-catalysed hydrogen–deuterium exchange reactions of (**48**) and (**49**) (NR_2 has previously been shown to be equivalent to OMe) showed that the pyridyl-nitrogen atom introduces an isoinversion component (Scheme 2) into the racemization of (**48**).[106]

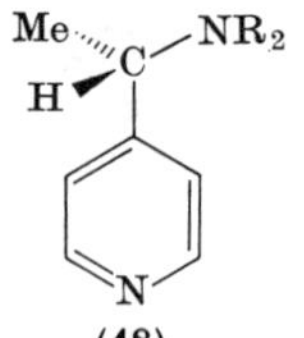

(48) 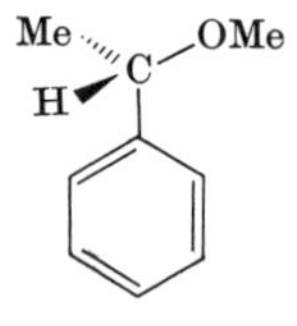 (49)

Exchange reactions of isobutane and butane with deuterium on alumina proceed at similar rates. In the former case one, and in the latter case four, hydrogen atoms are exchanged much less rapidly than the others, an observation consistent with carbanionic intermediates.[107] Hydrogen–deuterium exchange of cyclohexane with a nickel–alumina catalyst has been studied.[108] σ-Bonded alkenyls are proposed intermediates in the highl

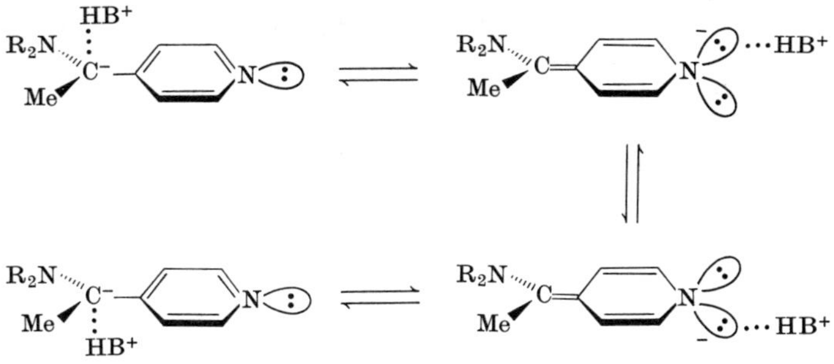

SCHEME 2.

selective exchange of the vinylic and allylic hydrogen atoms of alkenes under catalysis by gallium oxide.[109] The Ge—H bond of triphenylgermane has been shown to undergo rapid exchange with alcohols in the presence of base;[110] and the kinetics of base-catalysed hydrogen exchange in acetylenic compounds have been studied.[111]

Gas-phase proton-transfer reactions in which both reactant and product carbanions are delocalized typically proceed unusually slowly. It has now been shown that methanol can act as a homogeneous catalyst for the reaction of allyl anion and toluene.[112]

Absolute and Relative Acidities

Equilibrium ion-pair acidities of arylmethanes[113] and of certain heterocyclic aromatic compounds[114] toward caesium cyclohexylamide in cyclohexylamine have been determined. The acidity *decrease* of 0.4 pK unit produced by phenyl substitution at C-5 of 5H-dibenzo[a,d]cycloheptadiene (**50**) was attributed to an overall decrease in delocalization resulting from conformational changes caused by interaction of the new substituent

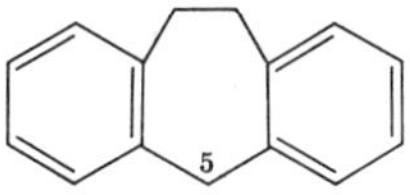

(**50**)

with the neighbouring *peri*-hydrogen atoms.[113] A further decrease of 2.0 pK units on going from the 5-phenyl to the 5-methyl derivative supports this idea. Deductions based on the interrelationship of k_D/k_T and k_H/k_T indicated that in proton exchange catalysed by lithium cyclohexylamide internal return is greater for triphenylmethane than toluene.[115] This probably reflects different steric constraints encountered during the respective diffusion steps. Kinetic acidities of arylmethanes toward lithium cyclohexylamide were combined with known pK's to give a linear Brønsted correlation with $\alpha = 0.31$;[116] the isotope effects, being high ($k_H/k_D \approx 11$),[115] indicated a pyramidal transition state in which delocalization of the carbanion is only partly developed.[116]

The method of determining pK_a through an electrochemical cycle[117] has been reported in detail and extended to trialkylcyclopropenes.[118] The cycle may also be applied in reverse when pK_{R^+} rather than pK_a is the less accessible quantity.[119] Two conspicuous results, pK's of 62–65 for trialkylcyclopropenes and pK_{R^+} for the cyclopentadienyl cation of −40, could each signify antiaromatic destabilization.

Various aspects of the acidity of nitro-[120–123] and other[124] alkanes have been investigated. The β-keto-sulphones (**51**) have pK_a's that follow a linear correlation with σ for variation of both X (ρ + 2.01) and Y (ρ + 2.35).[125]

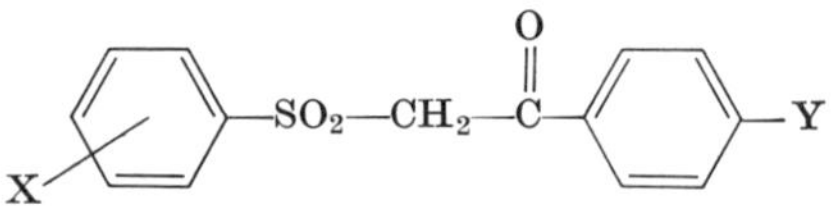

(**51**)

Line-shape analysis of the NMR peak of solvent ammonia during the reaction of 4-methylpyridine with amide ion required the former to have p$K_a \sim 29$.[126] Competition experiments then gave the pK_a of 2-methylpyridine as ~31. Compounds containing the 2,4-dimethylpyridine unit are metallated specifically at the less acidic 2-methyl group by *n*-butyl-lithium in ether–hexane, but at the expected 4-methyl group by amide ion–ammonia or lithium di-isopropylamide.[127] Only under the former conditions is complex formation (as **52**) likely.

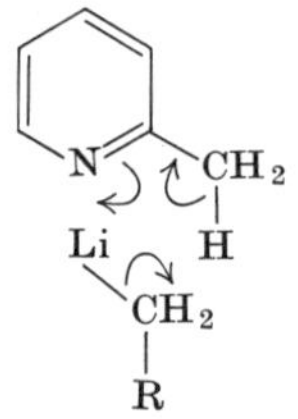

(**52**)

Hydrogen–deuterium exchange rates of the methyl-hydrogen atoms of the following compounds have been reported: 1-methylpyridinium iodides;[128] methyl-naphthalenes and quinolines and their *N*-oxides;[129] 6-methyl-*s*-triazine-2(1*H*),4(3*H*)-diones;[130] and various methylated aromatic nitrogen heterocycles.[131]

Electrophilic Reactions of Hydrocarbons

Reviews have appeared on: Friedel–Crafts acylation of alkenes;[132] electrophilic substitution at alkanes and in alkylcarbonium ions;[133] and carbocations and electrophilic reactions.[134] Olah and his co-workers have described further reactions that they consider to involve attack by an electrophile on the front lobe of a saturated C—H or C—C bond via a triangular three-centre bonded carbonium-ion transition state.[135] Thus, temperature-dependent PMR spectra for reactions between alkylcarbenium ions and their parent alkanes under stable ion conditions (low temperature, superacids) indicated that, in all but the most hindered ($Me_3CCHMe_2 + Me_3C\overset{+}{C}Me_2$) of such systems, fast intermolecular hydrogen-exchange was occurring.[136] Alkylation of alkane by alkylcarbenium ion was a concurrent though much slower reaction; deuterium-labelling indicated that intermediate alkene formation was not involved. Comparison of observed and computed spectra for the process $Me_3\overset{*}{C}H + Me_3C^+ \rightleftharpoons Me_3\overset{*}{C}^+ + Me_3CH$ indicated an activation energy of 4.3 ± 0.6 kcal mol^{-1}. The reactivity of methyl and ethyl fluoride–antimony pentafluoride complexes towards C—C and C—H bonds in alkanes is comparable to that of stable alkylcarbenium ions.[137] With these bulky electrophiles the order of σ-bond reactivity observed in the protolysis of alkanes in superacids (tertiary C—H > C—C > secondary C—H > primary C—H)[138] is not followed. Thus, isobutane and the methyl fluoride–antimony pentafluoride complex form isopentane and neopentane in a ratio of 20:1, indicating preferential attack at the primary C—H bond. The reaction between methyl cation and methane has been investigated by use of a crossed-beam electron-volt accelerator.[139] Labelling studies within the system pentamethylbenzyl cation–isobutane–strong acid showed that hydride-transfer from alkane to benzyl cation and proton-exchange between alkane and acid were occurring with comparable velocities.[140] Formation of HD from H_2 or D_2 in superacids at room temperature requires the intermediacy of $(H,D)_3^+$ ions.[141]

The protolytic behaviour of substituted alkylbenzenes in superacid media has been examined as a measure of π and n versus σ electron-donor ability.[142] The latter is most favoured when electron-withdrawing groups or steric crowding decreases the otherwise dominant π-aromatic donor ability. Scheme 3 contains typical illustrations with the system HF–SbF_5–SO_2ClF.

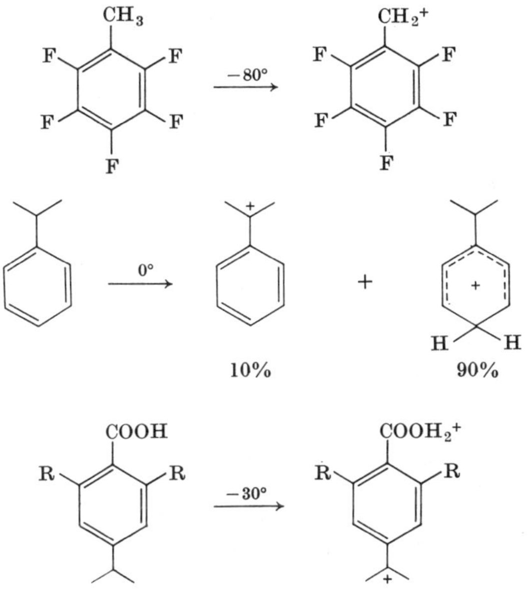

CH$_3$ F F F F F —$-80°$→ CH$_2^+$ F F F F F

0° → (+) 10% + (+) H H 90%

COOH R R —$-30°$→ COOH$_2^+$ R R (+)

R = H or Pr^i

SCHEME 3.

Anhydrous $AgSbF_6$ in dichloromethane solution catalyses electrophilic bromination of tertiary and reactive secondary C—H bonds in alkanes;[143] bromolysis of C—C bonds was not observed. Electrophilic chlorination of alkanes [144] has been extended to include catalysis by $AlCl_3$ and $AgSbF_6$;[145] these form bulky, less electrophilic, species that react with reactive C—H but not C—C bonds. Alkanes are readily chlorinated by sulphuryl chloride in sulpholane.[146] The product from adamantane contained > 97.5% of the 1-chloro-isomer, indicating an ionic mechanism. Dialkylaryloxonium ions, stable products of the reaction of alkoxybenzenes with alkyl fluoroantimonates in SO_2ClF at −70°, are powerful alkylating agents for both π- and n-donor bases.[147]

A comprehensive study of the reaction of simple cyclopropanes with mercuric acetate and trifluoroacetate led to the following conclusions: (1) the stereochemistry of the reaction of the electrophile is generally determined by its attack on the least substituted bond of the ring; (2) the nucleophile reacts almost exclusively with inversion; and (3) in a symmetrical system where all ring bonds are identical, inversion is slightly favoured in the electrophilic attack.[148] Theoretical studies on the protonation of cyclopropanes and other strained hydrocarbons have appeared.[149] The low reactivity of bicyclo[2.1.0]pentane toward acid-catalysed acetolysis has been discussed in terms of the relative pathways available for protonation of cyclopropanes and cyclobutanes.[150] A thorough study of the bromination of *n*-butylcyclopropane under ionic conditions has appeared.[151] In addition to tribromides formed via bromoalkenes, 1,2-, 1,3-, 1,4-, 1,5- and 1,6-dibromoheptane were formed in relative yields of 7, 20, 32, 24, and 17% respectively.

"1,3-Dehydroadamantanes" (**53**) are extremely reactive towards radical or electrophilic attack;[152] with bromine in diethyl ether at –70° the bromo etherate tribromide (**54**) was formed, which with carbonate, iodide or cyanide ion gave 1-bromo-3-ethoxyadamantane.

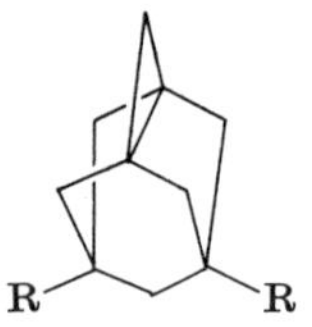

R R

(**53**)

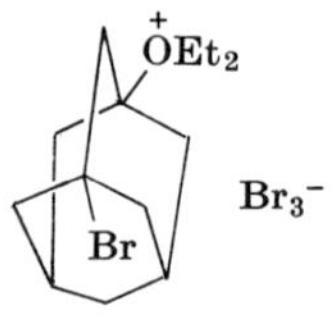

$\overset{+}{O}Et_2$ Br_3^- Br

(**54**)

Organometallics

The effect of aggregate composition on the competitive metallation of indene by *tert*-butyl-lithium and isopropyl-lithium has been examined.[153] *tert*-Butyl-lithium tetramer is ten times less reactive than *tert*-butyl-lithium in mixed aggregates with isopropyl-lithium. For the maximum yield of *addition* product from alkyl-lithiums and aldehydes or ketones, an ether or hexane solution of the carbonyl compound should be added to the lithium reagent at –78°.[154] The synthesis of 1,5-dienes involving generation of allyl-lithium reagents *in situ* from allylic mesitoates and lithium has been detailed[155] and extended to the synthesis of secondary and tertiary alcohols from aldehydes and ketones (Scheme 4), but yields were generally inferior to those obtainable by Reformatsky-type procedures.

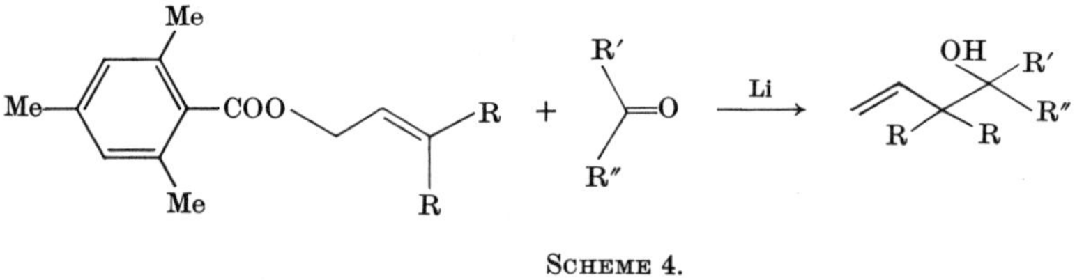

Me Me COO R R + R′ R″ O Li OH R′ R″ R R Me

SCHEME 4.

Reactions of ketones and nitriles with a large excess of methylmagnesium bromide give pseudo-first-order rate constants that are independent of the initial ketone concentration if the reagent is prepared from *pure* magnesium and an excess of methyl bromide;[156] formation of non-addition products is eliminated concurrently. A series of papers on factors affecting the kinetics of Grignard reactions has appeared,[157–161] and the Zerewitinoff reaction has also been studied.[162]

Participation by a vacant *p*-orbital on the non-displaced boron atom in the ion (**55**) has been invoked to explain the 100-fold rate enhancement of mercurideboronation of $XCH_2B(OMe)_2$ by mercuric chloride in methanol when X is changed from H to $B(OMe)_2$.[163]

The change to predominant axial attack when 4-*tert*-butylcyclohexanone, for example, reacts with *two or more* equivalents of trialkylaluminium has been explained by postulating that interaction between the *complexed* carbonyl group and the 2,6-equatorial substituents is more unfavourable than compression between the incoming reagent and the 3,5-diaxial substituents.[164] The reactions of acetaldehyde with AlR_3, AlR_3·organic donor, and $[R_2AlXM][AlR_3]$ (X = O, NPh or S; M = Li or Na) have been compared.[165]

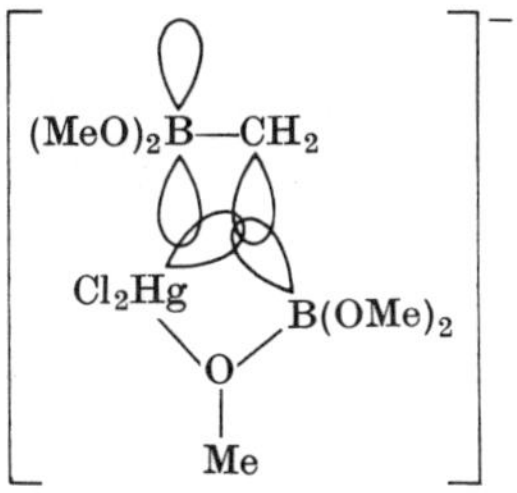

(**55**) 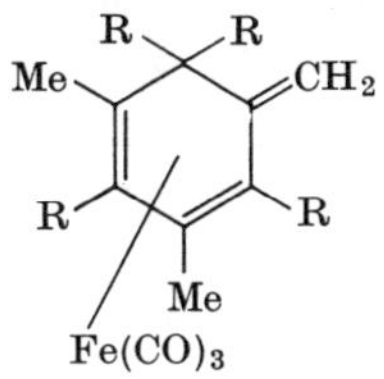 (**56**)

The neutral complex (**56**) has been proposed to account for the fact that the corresponding tricarbonylcyclohexadienyliron complexes (R = H or Me) undergo hydrogen–deuterium exchange exclusively at the 2- and 6-methyl groups.[166]

The aromatic substitution of alkenes using Pd(II) salts has been extended to include furan, thiophen, selenole,[167] and benzo[*b*]furan.[168] For the last-mentioned compound reaction with [β,β-$^2H_2$]styrene gives a monodeuteriated product, ruling out the occurrence of a hydride shift. In contrast, reaction of ferrocenylpalladium chloride complex with [β,β-$^2H_2$]styrene gives a mixture of mono- and di-deuteriated product, indicating that hydride shifts are occurring.[169] The mixtures of products obtained from cycloalkenes and benzene–palladium acetate may be accounted for by formation of the expected 1-phenylcycloalkene and subsequent isomerization or disproportionation;[170] thus, cyclo-octene gives 1-, 3-, 4- and 5-phenylcyclo-octene (6, 8, 12 and 2%, respectively), whereas cyclohexene gives biphenyl and phenylcyclohexane. Phenylation of compounds containing the Ph—C=C—NO_2 unit occurs preferentially on the carbon not bearing the nitro-group.[171] Reaction of styrene with Ph_3M–$Pd(OAc)_2$ gives *trans*-stilbene for M = P, As, Sb or Bi, but (**57**) for M = N.[172]

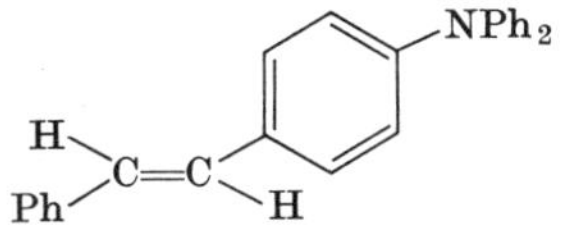

(**57**)

Rate constants for the reaction of mercuric carboxylates, $Hg(OCOR)_2$, with tetra-alkyltins in methanol increase in the order $R = Bu^t < Et < Me < Ph < ClCH_2CH_2 < MeOCH_2 < ClCH_2$, suggesting an S_E2(open) rather than S_E2(cyclic) mechanism.[173] Further, for reaction between mercuric acetate and R_4Sn in methanol, the observed rate constants are $R = Me > Et > Pr^n > Bu^n > Bu^i >$ neopentyl $> Pr^i$;[174] it was suggested that this sequence, which is different from that for the S_E2(open) mechanism with inversion of configuration,[175] is typical of S_E2(open) reactions with retention of configuration at the site of substitution. By using this difference in steric sequences of reactivity, stereochemical assignments were made for a number of S_E2(open) reactions. It has been shown that for the reaction of tetraethyltin with mercuric salts the increase in $\Delta G^\ddagger$ which occurs on change of solvent from methanol to *tert*-butyl alcohol–methanol mixtures reflects an increase in the free energy of the transition state.[176] The rate of electrophilic cleavage of a given lead–alkyl bond in a tetra-alkyl-lead has been shown to be significantly dependent on the substitution pattern of the trialkyl-lead leaving group.[177]

Rate constants for the insertion of sulphur dioxide into the trimethylaryltins (**58**) were conveniently measured by NMR spectroscopy.[178] A good correlation with σ^+ was obtained. With the trimethylbenzyltins (**59**), sulphur dioxide was inserted exclusively into the Sn—benzyl bond;[179] notably, the reaction was little affected by the nature of X.

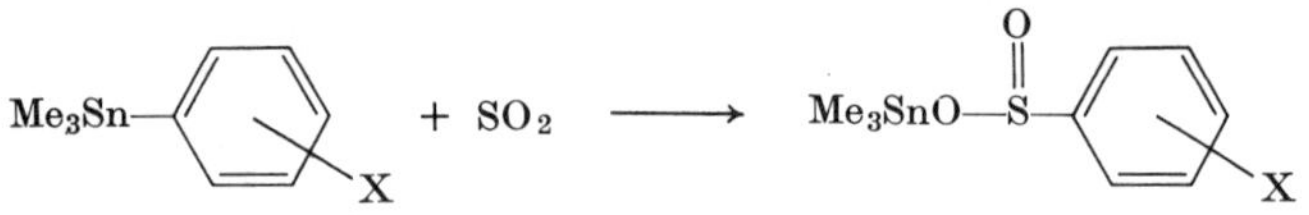

$$Me_3Sn{-}C_6H_4X \; (\mathbf{58}) + SO_2 \longrightarrow Me_3SnO{-}S(=O){-}C_6H_4X$$

$$Me_3SnCH_2{-}C_6H_4X \quad (\mathbf{59})$$

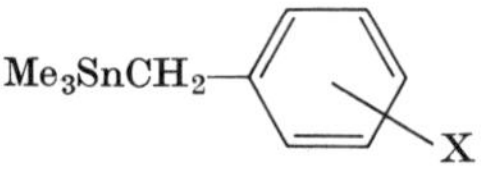

Other Reactions

Further investigations[180] related to the bromopicrin reaction have shown that bromopicrin is formed in high yield from 2-nitroethanol and sodium hypobromite;[181] proton-removal from 2-nitroethanol is the rate-determining step, and 2,2-dibromo-2-nitroethanol is an intermediate; the process (**60**) is indicated, since the corresponding methoxy-compound is stable under the same conditions[181] and dibromo(nitro)phenylmethane, $PhCBr_2(NO_2)$, and its *p*-nitro-derivative are stable to hydroxide ion.[182] Correlation of the rates of ionization of *p*-substituted α-nitrotoluenes with σ rather than σ^- has been observed for a further range of substituents.[182]

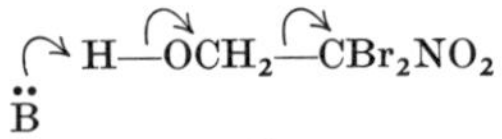

$$\ddot{B}\curvearrowright H{-}OCH_2{-}CBr_2NO_2 \quad (\mathbf{60})$$

Isolation of triazenes (ArN=N—NHR) after reaction of an arenediazonium salt with a primary aliphatic amine requires the presence of an electron-withdrawing group in Ar.[183] Otherwise, penta-azadiene [ArN=NN(R)—N=NAr] formation is favoured. Alkynediazoates (**61**) undergo intramolecular addition to give esters (**62**) by a mechanism reminiscent of that of the Favorskii rearrangement;[184] when phenyl substitution is present, protonation to the diazonium ion (**63**; R = H, R′ = Ph or *vice versa*) is a competing reaction.[185] The Schmidt reaction has been the subject of a further study.[186]

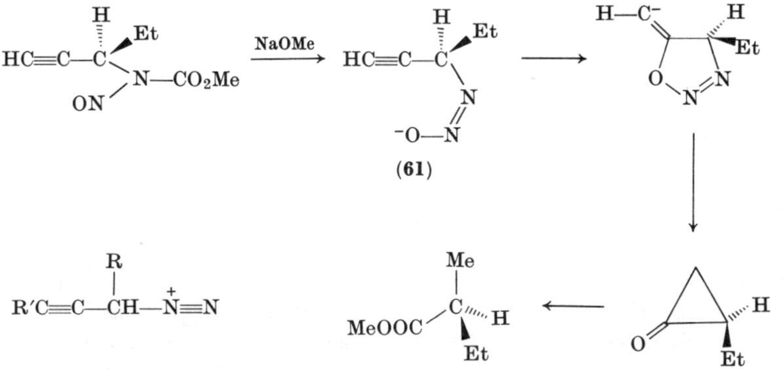

(**61**)

(**63**)

(**62**)
88% inversion

Aldehydic 4-pyridylhydrazones are another[187] class of compound for which isomerization to the *Z*-isomer is the rate-determining step in bromination.[188] For pyridine-2-carbaldehyde 4-pyridylhydrazone the *Z*-isomer (**64**) could be isolated (stabilization by hydrogen bonding) and was shown to undergo rapid bromination, of the first order in both hydrazone and bromine. The surprising observation has been made that heating 4-pyridylhydrazones of aromatic aldehydes at 200–250° with sodium alkoxides results in N—N bond-cleavage and formation of 4-alkylaminopyridines and aldoximes.[189]

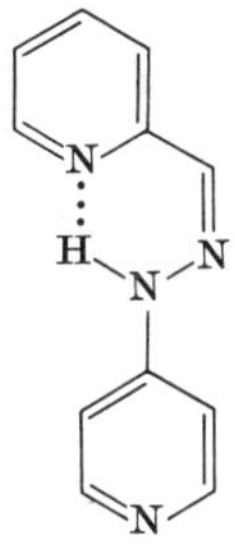

(**64**)

The ratio of monomer to dimer produced in the reduction of benzaldehyde anils by sodium in THF varied with the substitution of the benzaldehyde component and with the duration of the reaction;[190] a mechanism consistent with these observations is shown in Scheme 5.

Bis(chloromethyl) sulphide has been established as the key intermediate in the formation of chloromethanesulphonyl chloride by chlorination of 1,3,5-trithian in water–acetic acid;[191] recognition of the nature of the subsequent steps of this reaction led to useful syntheses of α-halo sulphoxides, sulphinyl chlorides and sulphonyl chlorides.[192] Chlorination of carbon disulphide in an aqueous medium has been studied.[193]

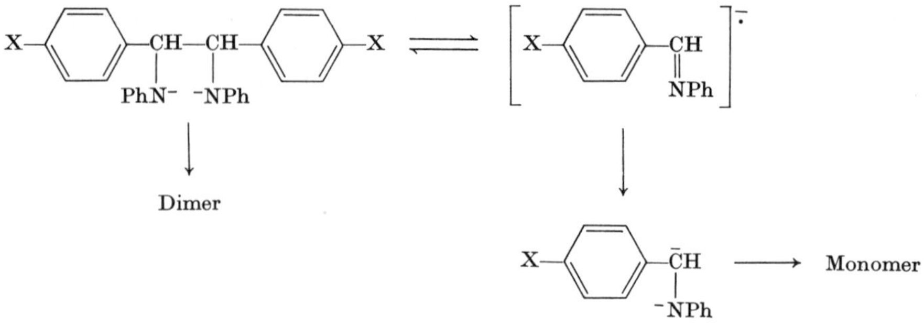

SCHEME 5.

2,3-Dimethyl-2,3-dinitropropane is the major product of the exothermic reaction between carbon tetrachloride and lithium or potassium 2-nitropropane in DMSO;[194] a mechanism involving electron-transfer to CCl_4 has been proposed. Evidence has been obtained for the role of carbanionic intermediates in the oxidation of nitroethane by D-amino-acid oxidase.[195] Other reactions that have received attention are: the nitration of decalins,[196] stilbenes,[197] and *trans*-2-styrylthiophen;[198] electrochemical reduction of fluorene;[199] and base cleavage of 1,2,3-selenadiazoles[200] and 1,4-thiazin-3-ones.[201]

References

1 M. M. Exner, R. Waack and E. C. Steiner, *J. Am. Chem. Soc.*, **95**, 7009 (1973).
2 R. J. Bushby and G. J. Ferber, *Chem. Comm.*, **1973**, 407.
3 R. B. Bates, S. Brenner, C. M. Cole, E. W. Davidson, G. D. Forsythe, D. A. McCombs and A. S. Roth, *J. Am. Chem. Soc.*, **95**, 926 (1973).
4 K. Takahashi, K. Konishi, M. Ushio, M. Takaki and R. Asami, *J. Organometal. Chem.*, **50**, 1 (1973).
5 S. I. Weissman and R. Chang, *J. Am. Chem. Soc.*, **94**, 8683 (1972).
6 D. A. Hutchison, K. R. Beck, R. A. Benkeser and J. B. Grutzner, *J. Am. Chem. Soc.*, **95**, 7075 (1973).
7 W. H. Glaze, J. E. Hanicak, J. Chaudhuri, M. L. Moore and D. P. Duncan, *J. Organometal. Chem.*, **51**, 13 (1973).
8 W. H. Glaze and C. R. McDaniel, *J. Organometal. Chem.*, **51**, 23 (1973).
9 E. A. Hill and G. E.-M. Shih, *J. Am. Chem. Soc.*, **95**, 7764 (1973).
10 A. Maercker and R. Geuß, *Chem. Ber.*, **106**, 773 (1973).
11 A. Maercker, P. Guthlein and H. Wittmayr, *Angew. Chem. Internat. Ed.*, **12**, 774 (1973).
12 T. L. Chwang and R. West, *J. Am. Chem. Soc.*, **95**, 3324 (1973).
13 J. Klein and J. Y. Becker, *Chem. Comm.*, **1973**, 576.
14 J. Klein and J. Y. Becker, *J.C.S. Perkin II*, **1973**, 599.
15 V. K. Krylov and I. V. Tselinsky, *Organic Reactivity* (*Tartu*), **9**, 743 (1972).
16 J. Klein and A. Medlik, *Chem. Comm.*, **1973**, 275.
17 *Org. Reaction Mech.*, **1970**, 131; **1971**, 113.
18 G. Gaviraghi and G. Pagani, *J.C.S. Perkin II*, **1973**, 50.
19 *Org. Reaction Mech.*, **1972**, 107.
20 M. F. Semmelhack, R. J. DeFranco, Z. Margolin and J. Stock, *J. Am. Chem. Soc.*, **95**, 426 (1973).
21 G. Schroder, G. Plinke, D. M. Smith and J. F. M. Oth, *Angew. Chem. Internat. Ed.*, **12**, 325 (1973).
22 J. F. M. Oth, E. P. Woo and F. Sondheimer, *J. Am. Chem. Soc.*, **95**, 7337 (1973).
23 M. V. Moncur and J. B. Grutzner, *J. Am. Chem. Soc.*, **95**, 6449 (1973).
24 M. J. Goldstein and S. Natowsky, *J. Am. Chem. Soc.*, **95**, 6451 (1973).
25 M. J. Goldstein and R. Hoffmann, *J. Am. Chem. Soc.*, **93**, 6193 (1971).
26 G. B. Trimitsis, E. W. Crowe, G. Slomp and T. L. Helle, *J. Am. Chem. Soc.*, **95**, 4333 (1973).
27 S. W. Staley and N. J. Pearl, *J. Am. Chem. Soc.*, **95**, 2731 (1973).
28 S. W. Staley and N. J. Pearl, *J. Am. Chem. Soc.*, **95**, 3437 (1973).
29 S. W. Staley and A. W. Orvedal, *J. Am. Chem. Soc.*, **95**, 3382 (1973).

[30] S. W. Staley and A. W. Orvedal, *J. Am. Chem. Soc.*, **95**, 3384 (1973).
[31] S. W. Staley and G. M. Cramer, *J. Am. Chem. Soc.*, **95**, 5051 (1973).
[32] S. W. Staley and W. G. Kingsley, *J. Am. Chem. Soc.*, **95**, 5804 (1973).
[33] S. W. Staley, G. M. Cramer and W. G. Kingsley, *J. Am. Chem. Soc.*, **95**, 5052 (1973).
[34] R. R. Fraser, G. Boussard, I. D. Postescu, J. J. Whiting and Y. Y. Wigfield, *Can. J. Chem.*, **51**, 1109 (1973).
[35] P. Beak and R. Farney, *J. Am. Chem. Soc.*, **95**, 4771 (1973).
[36] D. Seebach and D. Enders, *Angew. Chem. Internat. Ed.*, **11**, 1101 (1972).
[37] D. Seebach and D. Enders, *Angew. Chem. Internat. Ed.*, **11**, 1102 (1972).
[38] R. A. Moss and G. M. Love, *Tetrahedron Letters*, **1973**, 4701.
[39] R. Kow and M. W. Rathke, *J. Am. Chem. Soc.*, **95**, 2715 (1973); *Org. Reaction Mech.*, **1972**, 133.
[40] D. J. Peterson, *J. Org. Chem.*, **33**, 780 (1968).
[41] H. Sakurai, K. Nishiwaki and M. Kira, *Tetrahedron Letters*, **1973**, 4193.
[42] R.-R. Lii and S. I. Miller, *J. Am. Chem. Soc.*, **95**, 1602 (1973); *Org. Reaction Mech.*, **1970**, 126.
[43] U. Reichman, F. Bergman, D. Lichtenberg and Z. Neiman, *J. Org. Chem.*, **38**, 2066 (1973).
[44] J. D. Fissekis and F. Sweet, *J. Org. Chem.*, **38**, 1963 (1973).
[45] J. Douris and A. Mathieu, *Bull. Soc. Chim. France*, **1973**, 709.
[46] H. J. Kallmayer, *Arch. Pharm.*, **306**, 257 (1973).
[47] A. K. Youssef and M. A. Ogliaruso, *J. Org. Chem.*, **38**, 2023 (1973).
[48] J. Bergman and J.-E. Backvall, *Tetrahedron Letters*, **1973**, 2899.
[49] B. A. Ivin, G. V. Rutkovskii and E. G. Sochilin, *Zh. Org. Khim.*, **8**, 1951 (1972); *Chem. Abs.*, **78**, 28980 (1973).
[50] V. N. Drozd, O. I. Trifonova and V. V. Sergeichuk, *Zh. Org. Khim.*, **9**, 156 (1973); *Chem. Abs.*, **78**, 110285 (1973).
[51] M. D. Carr, L. H. Gan and I. Reid, *J.C.S. Perkin II*, **1973**, 668.
[52] M. D. Carr, L. H. Gan and I. Reid, *J.C.S. Perkin II*, **1973**, 672.
[53] M. Newcomb and W. T. Ford, *J. Am. Chem. Soc.*, **95**, 7186 (1973).
[54] W. T. Ford and M. Newcomb, *J. Am. Chem. Soc.*, **95**, 6277 (1973).
[55] H. O. House, D. S. Crumrine, A. Y. Teranishi and H. D. Olmstead, *J. Am. Chem. Soc.*, **95**, 3310 (1973).
[56] H. O. House and M. J. Umen, *J. Org. Chem.*, **38**, 1000 (1973).
[57] J. Bertrand, N. Cabrol, L. Gorrichon-Guigon and Y. Maroni-Barnaud, *Tetrahedron Letters*, **1973**, 4683.
[58] H. O. House, R. A. Auerbach, M. Gall and N. P. Peet, *J. Org. Chem.*, **38**, 514 (1973).
[59] A. L. Kurts, S. M. Sakembaeva, I. P. Beletskaya and O. A. Reutov, *Dokl. Akad. Nauk. SSSR*, 210, 144 (1973); *Chem. Abs.*, **79**, 52628 (1973); A. L. Kurts, P. I. Dem'yanov, I. P. Beletskaya and O. A. Reutov, *Zh. Org. Khim.*, **9**, 1313 (1973); *Chem Abs.*, **79**, 91398 (1973).
[60] P. Hrnciar and M. Melichercik, *Coll. Czech. Chem. Comm.*, **38**, 1200 (1973).
[61] D. H. Hunter, A. L. Johnson, J. B. Stothers, A. Nickon, J. L. Lambert and D. F. Covey, *J. Am. Chem. Soc.*, **94**, 8582 (1972).
[62] B. S. Thyagarajan, *Mech. React. Sulphur Compounds*, **4**, 115 (1969); *Chem. Abs.*, **78**, 3345 (1973); M. Gresser, *Mech. React. Sulphur Compounds*, **4**, 29 (1969); *Chem. Abs.*, **78**, 3346 (1973).
[63] L. A. Paquette, R. H. Meisinger and R. Gleiter, *J. Am. Chem. Soc.*, **95**, 5414 (1973).
[64] R. R. Fraser, F. J. Schuber and Y. Y. Wigfield, *J. Am. Chem. Soc.*, **94**, 8795 (1972).
[65] A. J. Anderson, J. Kitchin and R. J. Stoodley, *Tetrahedron Letters*, **1973**, 3379.
[66] M. B. D'Amore and J. I. Brauman, *Chem. Comm.*, **1973**, 398; see also *Org. Reaction Mech.*, **1972**, 118, and R. Viau and T. Durst, *J. Am. Chem. Soc.*, **95**, 1346 (1973).
[67] G. Tsuchihashi, S. Iriuchijima and K. Maniwa, *Tetrahedron Letters*, **1973**, 3389.
[68] C. R. Johnson, R. A. Kirchhoff, R. J. Reischer and G. F. Katekar, *J. Am. Chem. Soc.*, **95**, 4287 (1973).
[69] S. Iriuchijima, M. Ishibashi and G. Tsuchihashi, *Bull. Chem. Soc. Japan*, **46**, 921 (1973); J. Klein and H. Stollar, *J. Am. Chem. Soc.*, **95**, 7437 (1973); S. Bory, R. Lett, B. Moreau and A. Marquet, *Compt. Rend., C.*, **276**, 1323 (1973).
[70] S. Iriuchijima and G. Tsuchihashi, *Bull. Chem. Soc. Japan*, **46**, 929 (1973).
[71] G. E. Wilson and R. Albert, *J. Org. Chem.*, **38**, 2160 (1973).
[72] J. F. Biellman and J. L. Schmitt, *Tetrahedron Letters*, **1973**, 4615.
[73] D. W. Allen, S. J. Grayson, I. Harness, B. G. Hutley and I. W. Mowat, *J.C.S. Perkin II*, **1973**, 1912.
[74] H. Matsuyama, H. Minato and M. Kobayashi, *Bull. Chem. Soc. Japan*, **46**, 2845 (1973).

[75] *Org. Reaction Mech.*, **1972**, 116.
[76] P. A. Grieco and C. S. Pogonowski, *J. Am. Chem. Soc.*, **95**, 3071 (1973); P. A. Grieco and R. S. Finkelhar, *J. Org. Chem.*, **38**, 2909 (1973).
[77] M. P. Cooke and R. Goswami, *J. Am. Chem. Soc.*, **95**, 7891 (1973).
[78] V. Pascali, N. Tangari and A. Umani-Ronchi, *J.C.S. Perkin I*, **1973**, 1166.
[79] P. E. Pfeffer, L. S. Silbert and E. Kinsel, *Tetrahedron Letters*, **1973**, 1163.
[80] B. M. Trost, D. Buhner and G. M. Bright, *Tetrahedron Letters*, **1973**, 2787.
[81] *Org. Reaction Mech.*, **1972**, 109.
[82] T. S. Cantrell, *Tetrahedron Letters*, **1973**, 1803.
[83] R. G. Harvey, L. Nazareno and H. Cho, *J. Am. Chem. Soc.*, **95**, 2376 (1973).
[84] See, P. Tomboulian, D. Amick, S. Beare, K. Dumke, D. Hart, R. Hites, A. Metzger and R. Nowak, *J. Org. Chem.*, **38**, 322 (1973).
[85] J. C. Carnahan and W. D. Closson, *J. Org. Chem.*, **37**, 4469 (1972).
[86] M. I. Terekhova, E. S. Petrov and A. I. Shatenshtein, *Zh. Org. Khim.*, **9**, 857 (1973); *Chem. Abs.*, **79**, 41586 (1973).
[87] G. B. Trimitsis, A. Tuncay, R. D. Beyer and K. J. Ketterman, *J. Org. Chem.*, **38**, 1491 (1973).
[88] "Proton Transfer Reactions in Highly Basic Media", J. R. Jones, *Prog. Phys. Org. Chem.*, **9**, 241 (1972); R. P. Bell, "*The Proton in Chemistry*", 2nd edn., Chapman and Hall, London, 1973.
[89] T. Fueno, O. Kajimoto, Y. Nishigaki and T. Yoshioka, *J.C.S. Perkin II*, **1973**, 738.
[90] H. Wilson, J. D. Caldwell and E. S. Lewis, *J. Org. Chem.*, **38**, 564 (1973).
[91] J.-H. Kim and K. T. Leffek, *Can. J. Chem.*, **51**, 2805 (1973).
[92] E. F. Caldin and S. Mateo, *Chem. Comm.*, **1973**, 854.
[93] J.-O. Levin, *Chem. Scripta*, **4**, 85 (1973).
[94] M. Eigen, *Angew. Chem. Internat. Ed.*, **3**, 1 (1964).
[95] Z. Margolin and F. A. Long, *J. Am. Chem. Soc.*, **95**, 2757 (1973).
[96] A. J. Kresge and A. C. Lin, *Chem. Comm.*, **1973**, 761.
[97] F. Hibbert, *J.C.S. Perkin II*, **1973**, 1289.
[98] *Org. Reaction Mech.*, **1972**, 121.
[99] J. Almy, D. H. Hoffman, K. C. Chu and D. J. Cram, *J. Am. Chem. Soc.*, **95**, 1185 (1973).
[100] P. Ahlberg and F. Ladhar, *Chem. Scripta*, **3**, 31 (1973).
[101] P. Ahlberg, *Chem. Scripta*, **3**, 183 (1973).
[102] P. Ahlberg, *Chem. Scripta*, **4**, 33 (1973).
[103] R. D. Guthrie and J. L. Hedrick, *J. Am. Chem. Soc.*, **95**, 2971 (1973); *Org. Reaction Mech.*, **1971**, 123.
[104] D. A. Jaeger and D. J. Cram, *J. Am. Chem. Soc.*, **93**, 5153 (1971).
[105] R. D. Guthrie, L. G. Burdon and F. L. Lovell, *J. Org. Chem.*, **38**, 3114 (1973).
[106] D. A. Jaeger, M. D. Broadhurst and D. J. Cram, *J. Am. Chem. Soc.*, **95**, 7525 (1973).
[107] P. J. Robertson, M. S. Scurrell and C. Kemball, *Chem. Comm.*, **1973**, 799.
[108] N. V. Nekrasov and B. S. Gudkov, *Izv. Akad. Nauk SSSR, Ser. Khim.*, **1973**, 25; *Chem. Abs.*, **78**, 147037 (1973).
[109] F. B. Carleton, H. A. Quinn and J. L. Rooney, *Chem. Comm.*, **1973**, 231.
[110] C. Eaborn and I. D. Jenkins, *Chem. Comm.*, **1973**, 780.
[111] L. V. Timokhina, A. S. Nakhmanovich and A. I. Borisova, *Izv. Sib. Otd. Akad. Nauk SSSR, Ser. Khim. Nauk*, **1973**, 109; *Chem. Abs.*, **79**, 41591 (1973).
[112] J. I. Brauman, C. A. Lieder and M. J. White, *J. Am. Chem. Soc.*, **95**, 927 (1973).
[113] A. Streitwieser, J. R. Murdoch, G. Hafelinger and C. J. Chang, *J. Am. Chem. Soc.*, **95**, 4248 (1973).
[114] A. Streitwieser and P. J. Scannon, *J. Am. Chem. Soc.*, **95**, 6273 (1973).
[115] A. Streitwieser, P. H. Owens, G. Sonnichsen, W. K. Smith, G. R. Ziegler, H. M. Niemeyer and T. L. Kruger, *J. Am. Chem. Soc.*, **95**, 4254 (1973).
[116] A. Streitwieser, M. R. Granger, F. Mares and R. A. Wolf, *J. Am. Chem. Soc.*, **95**, 4257 (1973).
[117] *Org. Reaction Mech.*, **1969**, 129; **1970**, 129.
[118] R. Breslow and W. Chu, *J. Am. Chem. Soc.*, **95**, 411 (1973).
[119] R. Breslow and S. Mazur, *J. Am. Chem. Soc.*, **95**, 584 (1973).
[120] A. Pihl and A. Talvik, *Organic Reactivity* (*Tartu*). **9**, 355 (1972).
[121] H. Timotheus, V. Timotheus and E. Loodmaa, *Organic Reactivity* (*Tartu*), **9**, 1164 (1972).
[122] A. I. Morozov, M. S. Sytilin and I. A. Makolkin, *Zh. Fiz. Khim.*, **46**, 2537 (1972); *Chem. Abs.*, **78**, 42422 (1973).
[123] A. Talvik and A. Pihl, *Organic Reactivity* (*Tartu*), **10**, 495 (1973).
[124] A. Talvik, *Organic Reactivity* (*Tartu*), **9**, 269 (1972).
[125] R. T. Amel and P. J. Marek, *J. Org. Chem.*, **38**, 3513 (1973).

[126] J. A. Zoltewicz and L. S. Helmick, *J. Org. Chem.*, **38**, 658 (1973).
[127] E. M. Kaiser, G. J. Bartling, W. R. Thomas, S. B. Nichols and D. R. Nash, *J. Org. Chem.*, **38**, 71 (1973).
[128] I. F. Tupitsyn, N. N. Zatsepina and A. V. Kirova, *Organic Reactivity* (*Tartu*), **9**, 232 (1972).
[129] N. N. Zatzepina, J. F. Tupitsyn, V. P. Dushina, Y. M. Kapustin and Y. L. Kaminsky, *Organic Reactivity* (*Tartu*), **9**, 762 (1972).
[130] M. Safta, F. Chiraleu, A. T. Balaban and G. Ostrogovich, *Rev. Roum. Chim.*, **17**, 2055 (1972); *Chem. Abs.*, **78**, 71011 (1973).
[131] C. Weiss, F. Hoppner, S. Becker and W. Blaschke, *Tetrahedron*, **29**, 3071 (1973).
[132] J. K. Groves, *Chem. Soc. Periodical Report*, **1**, 73 (1972).
[133] D. M. Brouwer and H. Hogeveen, *Prog. Phys. Org. Chem.*, **9**, 179 (1972).
[134] G. A. Olah, *Angew. Chem. Internat. Ed.*, **12**, 173 (1973).
[135] *Org. Reaction Mech.*, **1971**, 124.
[136] G. A. Olah, Y. K. Mo and J. A. Olah, *J. Am. Chem. Soc.*, **95**, 4939 (1973).
[137] G. A. Olah, J. R. DeMember and J. Shen, *J. Am. Chem. Soc.*, **95**, 4952 (1973).
[138] G. A. Olah, Y. Halpern, J. Shen and Y. K. Mo, *J. Am. Chem.*, **95**, 4960 (1973).
[139] J. Weiner, G. P. K. Smith, M. Saunders and R. J. Cross, *J. Am. Chem. Soc.*, **95**, 4115 (1973).
[140] P. van Pelt and H. M. Buck, *Rec. Trav. chim.*, **92**, 1057 (1973).
[141] G. A. Olah, J. Shen and R. H. Schlosberg, *J. Am. Chem. Soc.*, **95**, 4957 (1973).
[142] G. A. Olah and Y. K. Mo, *J. Am. Chem. Soc.*, **95**, 6827 (1973).
[143] G. A. Olah and P. Schilling, *J. Am. Chem. Soc.*, **95**, 7680 (1973).
[144] *Org. Reaction Mech.*, **1972**, 125.
[145] G. A. Olah, R. Renner, P. Schilling and Y. K. Mo, *J. Am. Chem. Soc.*, **95**, 7686 (1973).
[146] I. Tabushi, Z. Yoshida and Y. Tamaru, *Tetrahedron*, **29**, 81 (1973).
[147] G. A. Olah and E. G. Melby, *J. Am. Chem. Soc.*, **95**, 4971 (1973).
[148] C. H. DePuy and R. H. McGirk, *J. Am. Chem. Soc.*, **95**, 2366 (1973); see also *Org. Reaction Mech.*, **1970**, 144.
[149] J.-M. Lehn and G. Wipff, *Chem. Comm.*, **1973**, 747; A. H. Andrist, *J. Am. Chem. Soc.*, **95**, 7531 (1973).
[150] K. B. Wiberg, K. C. Bishop and R. B. Davidson, *Tetrahedron Letters*, **1973**, 3169.
[151] J. C. Day, K. J. Shea and P. S. Skell, *J. Am. Chem. Soc.*, **95**, 5089 (1973).
[152] R. E. Pincock, J. Schmidt, W. B. Scott and E. J. Torupka, *Can. J. Chem.*, **50**, 3958 (1972).
[153] W. Peascoe and D. E. Applequist, *J. Org. Chem.*, **38**, 1510 (1973).
[154] J. D. Buhler, *J. Org. Chem.*, **38**, 904 (1973).
[155] J. A. Katzenellenbogen and R. S. Lenox, *J. Org. Chem.*, **38**, 326 (1973); *Org. Reaction Mech.*, **1972**, 128.
[156] E. C. Ashby, H. M. Neumann, F. W. Walker, J. Laemmle and L.-C. Chao, *J. Am. Chem. Soc.*, **95**, 3330 (1973).
[157] J. Koppel, J. Loit, M. Luuk and A. Tuulmets, *Organic Reactivity* (*Tartu*), **8**, 1163 (1971).
[158] J. Koppel and A. Tuulmets, *Organic Reactivity* (*Tartu*), **9**, 410 (1972).
[159] S. Truuvalja and A. Tuulmets, *Organic Reactivity* (*Tartu*), **10**, 205 (1973).
[160] M. Luuk and A. Tuulmets, *Organic Reactivity* (*Tartu*), **10**, 525 (1973).
[161] M. Luuk and A. Tuulmets, *Organic Reactivity* (*Tartu*), **10**, 535 (1973).
[162] C. Tuzun and E. Erdik, *Commun. Fac. Sci. Univ. Ankara, Ser. B*, **19**, 35 (1972); *Chem. Abs.*, **79**, 77675 (1973).
[163] D. S. Matteson and P. G. Allies, *J. Organometal. Chem.*, **54**, 35 (1973).
[164] J. Laemmle, E. C. Ashby and P. V. Roling, *J. Org. Chem.*, **38**, 2526 (1973).
[165] T. Araki, K. Hayakawa, T. Aoyagi, Y. Nakano and H. Tani, *J. Org. Chem.*, **38**, 1130 (1973).
[166] V. G. Shubin, R. N. Berezina and V. N. Piottukh-Peletski, *J. Organometal. Chem.*, **54**, 239 (1973).
[167] R. Asano, I. Moritani, Y. Fujiwara and S. Teranishi, *Bull. Chem. Soc. Japan*, **46**, 663 (1973).
[168] A. Kasahara, T. Izumi, M. Yodono, R. Saito, T. Takeda and T. Sugawara, *Bull. Chem. Soc. Japan*, **46**, 1220 (1973).
[169] A. Kasahara and T. Izumi, *Bull. Chem. Soc. Japan*, **46**, 665 (1973).
[170] M. Yamamura, I. Moritani, A. Sonoda, S. Teranishi and Y. Fujiwara, *J.C.S. Perkin I*, **1973**, 203.
[171] K. Yamamura, S. Watarai and T. Kinugasa, *Chem. Letters* (*Tokyo*), **1973**, 91.
[172] R. Asano, I. Moritani, Y. Fujiwara and S. Teranishi, *Bull. Chem. Soc. Japan*, **46**, 2910 (1973).
[173] M. H. Abraham, D. F. Dadjour and C. J. Holloway, *J. Organometal. Chem.*, **52**, C27 (1973); see also M. H. Abraham, "Electrophilic Substitution at a Saturated Carbon Atom,", in C. H. Bamford and C. F. H. Tipper (Eds.), "*Comprehensive Chemical Kinetics*", Elsevier, Amsterdam, Vol. 12, 1973.
[174] M. H. Abraham and P. L. Grellier, *J.C.S. Perkin II*, **1973**, 1132.

175 F. R. Jensen and D. D. Davis, *J. Am. Chem. Soc.*, **93**, 4048 (1971).
176 M. H. Abraham and F. J. Dorrell, *J.C.S. Perkin II*, **1973**, 444.
177 N. A. Clinton, H. C. Gardner and J. K. Kochi, *J. Organometal. Chem.*, **56**, 227 (1973).
178 C. W. Fong and W. Kitching, *J. Organometal. Chem.*, **59**, 213 (1973).
179 C. J. Moore and W. Kitching, *J. Organometal. Chem.*, **59**, 225 (1973).
180 cf. *Org. Reaction Mech.*, **1970**, 222.
181 R. I. Aylott, A. R. Butler, D. S. B. Grace and H. McNab, *J.C.S. Perkin II*, **1973**, 1107.
182 S. P. Avery and A. R. Butler, *J.C.S. Perkin II*, **1973**, 1110.
183 T. P. Ahern and K. Vaughan, *Chem. Comm.*, **1973**, 701.
184 W. Kirmse, A. Engelmann and J. Heese, *Chem. Ber.*, **106**, 3073 (1973).
185 W. Kirmse and A. Engelmann, *Chem. Ber.*, **106**, 3086 (1973).
186 A. S. Enin, G. I. Koldobskii, V. A. Ostrovskii and L. I. Bagal, *Zh. Org. Khim.*, **8**, 1895 (1972); *Chem. Abs.*, **78**, 42663 (1973).
187 Cf. *Org. Reaction Mech.*, **1971**, 133.
188 A. F. Hegarty, P. J. Moroney and F. L. Scott, *J.C.S. Perkin II*, **1973**, 1466.
189 L. N. Yakhontov and M. F. Marshalkin, *Tetrahedron Letters*, **1973**, 2807.
190 J. G. Smith and I. Ho, *J. Org. Chem.*, **38**, 2776 (1973).
191 J. S. Grossert and R. F. Langler, *Chem. Comm.*, **1973**, 49.
192 J. S. Grossert, W. R. Hardstaff and R. F. Langler, *Chem., Comm.*, **1973**, 50.
193 A. D. Raskina, V. I. Zetkin, E. V. Zakharov, V. I. Kosorotov, I. M. Kolesnikov and R. V. Dzhagatspanyan, *Khim. Prom.* (*Moscow*), **48**, 812 (1972); *Chem. Abs.*, **78**, 42424 (1973).
194 S. Limatibul and J. W. Watson, *J. Org. Chem.*, **37**, 4491 (1972).
195 D. J. T. Porter, J. G. Voet and H. J. Bright, *J. Biol. Chem.*, **248**, 4400 (1973).
196 J. Cressely, C. Tanielian and A. Ulrich, *Bull. Soc. Chim. France*, **1973**, 1087, 1090, 1095.
197 J.-C. Dore and C. Viel, *Compt. Rend., C.*, **276**, 1675 (1973).
198 A. Arcoria, E. Maccarone and G. A. Tomaselli, *J. Heterocyclic Chem.*, **10**, 191 (1973).
199 J. R. Jezorek, A. Lagu, T. M. Seigel and H. B. Mark, *J. Org. Chem.*, **38**, 788 (1973).
200 I. Lalezari, A. Shafiee and M. Yalpani, *J. Org. Chem.*, **38**, 338 (1973).
201 S. Hoff, A. P. Blok and E. Zwanenburg, *Rec. Trav. Chim.*, **92**, 879 (1973).

CHAPTER 11

Elimination Reactions

A. C. KNIPE

Chemistry Department, The New University of Ulster

Stereochemistry and Orientation in *E*2 Reactions

The effect of base association, by ion–ion and ion–dipole interactions, on the stereochemistry of *E*2 reactions has received particular attention.[1–6]

It was previously found that variation of base concentration can exert a substantial effect on the steric course of the bimolecular elimination reaction of (**1**) with Bu^tOK and this was attributed to ion-pairing of the base.[7] This view is supported by the observation[1] that for reaction in Bu^tO^-–Bu^tOH there is a considerable decrease in the ratio of *trans*- to *cis*-olefin (**2**) as the cation is changed from Na^+ to heavier alkali-metal ions. The decrease is even more pronounced for Me_4N^+, and similar, but less marked, trends obtain for (**3**). Since *trans*-(**2**) and -(**3**) are formed[7] mainly by *syn*-elimination and the corresponding *cis*-olefins by *anti*-elimination the results represent the differentiating effect of the counterion on the two mechanisms.

(**1**) → *cis*- and *trans*-(**2**) + *cis*- and *trans*-(**3**)

The divergent stereoselectivity (*syn* → *trans*-; *anti* → *cis*-olefin) found for elimination, $R^1CHXCH_2R^2 + RO^-M^+ \rightarrow R^1CH{:}CHR^2$, has been reviewed.[2] It has already been demonstrated[8] that this dichotomy can occur in the absence of steric effects to which the stereochemistry has previously been attributed. For leaving groups with unshared electron pairs (X = OTs or halide; $R^1 = R^2 =$ Bu), contact ion-pairs of an alkoxide base promote the *syn*-mechanism and preferential *cis*-olefin formation in the *anti*-component, while free alkoxide ion promotes almost exclusively an *anti*- → *trans*-mechanism. In contrast, where $X = R_3N^+$, free alkoxide ions are mainly responsible for the above dichotomy. A general explanation of these phenomena has now been proposed in terms of ion–dipole and ion–ion interactions between the leaving group and the participating base.[2] Preferred interactions (**4**) and (**6**) may reasonably account for conditions conducive to *syn* → *trans* stereospecificity for electronegative and cationic leaving groups, respectively. In order to account for the stereochemistry of *anti*-eliminations, non-linear approach of the base has been proposed. For the *syn*-pathway the unfavourable eclipsing of alkyl groups [(**5**) and (**7**)] ensures the promotion of *trans*-olefin formation [via (**4**) and (**6**)], there being little steric interaction with the approaching base. For the proposed *anti*-mechanism, however, the latter effect may be sufficient to favour *anti* → *cis*-configurations (**9**) and (**11**).

$s \rightarrow t$ (**4**) $s \rightarrow c$ (**5**) $s \rightarrow t$ (**6**) $s \rightarrow c$ (**7**)

$a \rightarrow t$ (**8**) $a \rightarrow c$ (**9**) $a \rightarrow t$ (**10**) $a \rightarrow c$ (**11**)

Also in keeping with the promotion of *anti*-elimination of 'onium salts by free rather than ion-paired alkoxide ion are the relative olefin proportions[4] from *anti*-elimination reactions of *sec*-butyltrimethylammonium toluene-*p*-sulphonate with Bu^tO^-–Bu^tOH. In marked contrast with earlier results[9] for *sec*-butyl bromide and tosylate, the percentage of but-1-ene and the *trans*-/*cis*-but-2-ene ratios are insensitive to changes in base concentration or to the addition of dicyclohexyl-18-crown-6.

The *trans*-/*cis*-but-2-ene ratios obtained upon reaction of 2-iodobutane with a series of phenoxide, carboxylate and alkoxide bases in DMSO are high (3.0–3.8) and insensitive to variation in base strength.[10] Positional orientation is, however, base-dependent, the proportion of but-1-ene increasing from 5.8 to 20% as the pK_a of the conjugate acid increases from 8.9 to 29.2. Plots of free-energy differences between transition states for formation of terminal and internal olefins, $\Delta\Delta F^\ddagger$ (but-1-ene – *trans*-but-2-ene) and $\Delta\Delta F^\ddagger$ (but-1-ene – *cis*-but-2-ene), versus the pK_a values are linear over a range of 20 pK_a units and clearly demonstrate a fundamental control of positional orientation by the base strength of oxyanion bases. For reactions in solvents of low polarity, however, base association exerts a profound influence on both positional and geometric orientation in eliminations from *sec*-butyl halides and toluene-*p*-sulphonates[6] and provides an explanation for (1) the orientation dichotomy for base-promoted eliminations in alcoholic and dipolar aprotic solvents, (2) the high proportions of alk-1-ene and low *trans*-/*cis*-alk-2-ene ratios found in reactions with $R^tO^-M^+$–R^tOH or $R^tO^-M^+$–hydrocarbon and (3) the *trans*-/*cis*-alk-2-ene ratios of less than unity which are observed in reactions of 2-alkylarenesulphonates with KOH and alkoxides in Bu^tOH.

Upon reaction with Bu^{t}OK–DMSO and with lithium dicyclohexylamide in Et_2O–hexane, chlorocyclodecane forms 97% of *cis*- and 96% of *trans*-cyclodecane, respectively.[11] It has now been found[5] that under the former reaction conditions the *trans*-cyclodecene, formed initially, isomerizes to 96% of *cis*-cyclodecene which is not therefore the product of a stereospecific *anti* → *cis*-elimination promoted by dissociated base.

The diastereoisomeric sulphones (**12**) and (**13**) undergo predominant *syn*-elimination to (**14**) and (**15**), respectively, upon reaction with PhOK–dioxan.[3] A proposed pseudocyclic transition state featuring reaction of the ion-paired base is supported by the dramatic increase in the contribution of *anti*-elimination upon addition of a crown ether. Leaving-group interaction with ion-paired base is not, however, the sole factor responsible for *syn*-elimination, promoted by Bu^{t}ONa–C_6H_6, in this system.

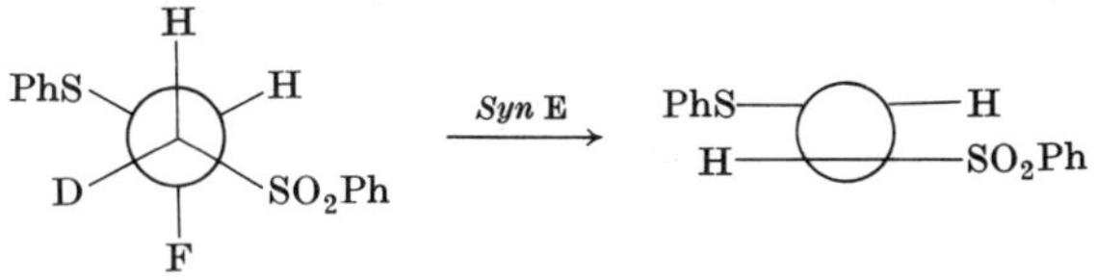

(**12**) —*Syn* E→ (**14**)

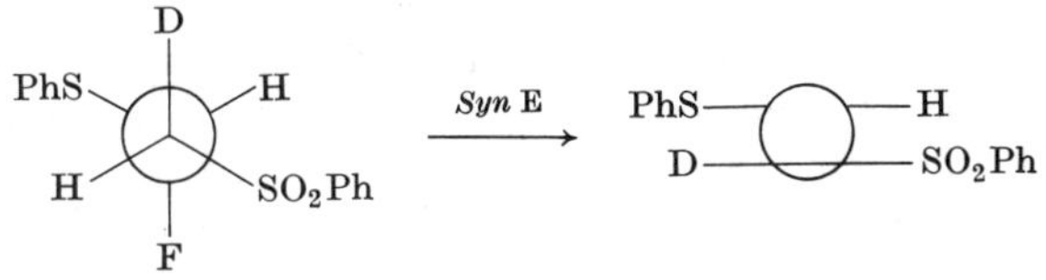

(**13**) —*Syn* E→ (**15**)

β-Deuterium isotope effects have been calculated for *E*2 transition states (**17**–**19**) in which the base, β-H and β-carbon form a right or an acute angle.[12] For a full bond model (in which the total bonding to hydrogen remains constant) of (**16**) the maximum isotope effect (*ca.* 6.1) is typical of those normally found for *anti*-eliminations, while for (**17**) and (**18**) the values (*ca.* 0.7 and 0.45) are inconsistent with nearly all observed isotope effects for *E*2 reactions. Results for the full bond model (**19**), however, agree (k_H/k_D = *ca.* 2.3) with the relatively small isotope effects usually found for *syn*-eliminations.[13] Calculations based on a half-bond model gave larger isotope effects which for (**17**) and (**18**) were still lower than many reported for *anti*-eliminations. An alternative to recently proposed *anti*-elimination transition states [cf. (**8**)–(**11**)] has therefore been suggested. Observed *trans*/*cis*-ratios may be explained in terms of a modified electrostatic model whereby the preferred conformation is that which permits most favourable electrostatic interaction between the base (or cation–base pair) and the leaving group by virtue of the dielectric constant of the intervening space. Thus (**23**) would be preferred over (**20**), (**21**) or (**22**) where RO is negative and X positive, or where the base is associated with a cation that can interact with a neutral X such as halide or tosylate.

In apparent disagreement with the theory of the variable *E*2 transition state, which predicts an increase of carbanion character and consequent decrease of incipient double bond formation as the bond to the leaving group becomes stronger, it has been found[14] that for reactions of α-methylphenethyl halides with EtONa–EtOH the *trans*/*cis*-olefin ratios lie in the order I ⩾ Br ≃ Cl < F. This may indicate that in a phenyl-activated

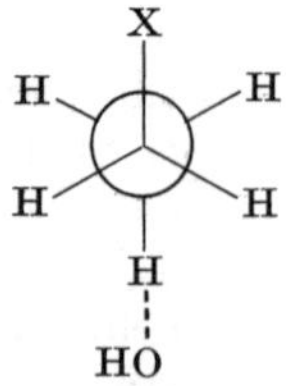

(16) 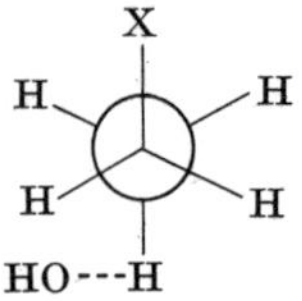 (17) 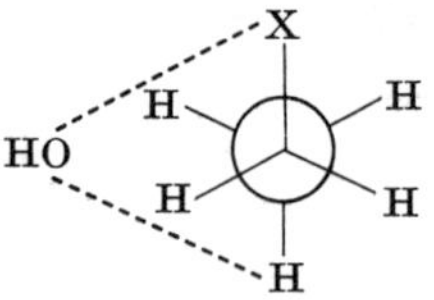 (18) 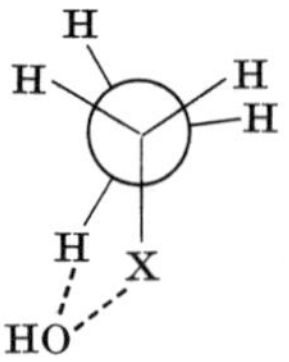 (19)

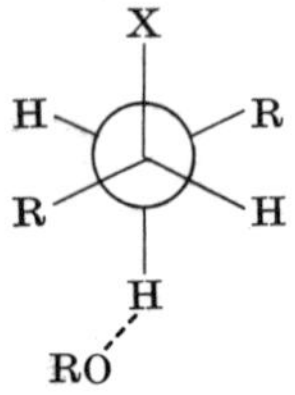

(20) 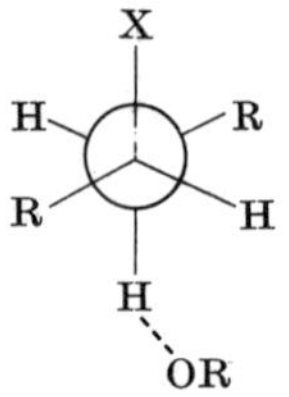 (21) 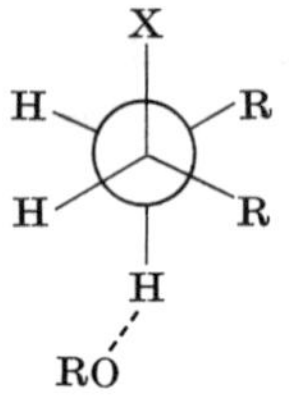 (22) 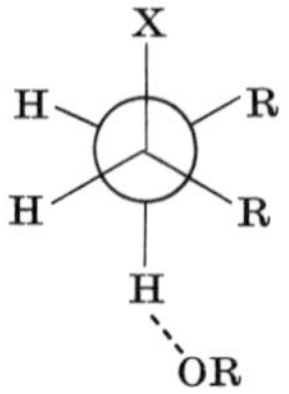 (23)

system a shift towards the *E*1*cB* extreme may be accompanied by an increase in the influence of eclipsing effects.

By analogy with the behaviour of 1-halocyclohexenes, the most important initial reaction of 1-halocycloheptenes with Bu^tOK–DMSO is dehydrohalogenation to 1,2-cycloheptadiene.[15] The kinetics and stereochemistry of elimination reactions between MeO^-–MeOH and the epimeric *trans*-2-decalyl tosylates,[16] and the direction of base-catalysed elimination of di- and tri-mesylated N^3-benzyluridines, have been studied.[17]

The *E*1*cB* Mechanism

A review of anionic β-elimination has especially featured *E*1*cB* reactions.[18]

The effects of structural factors on the rate constants for phenylsulphonyl-activated *E*1*cB* reactions of (**24**–**30**), according to equation (1), have been determined.[19] For (**30**), ionization is rate-determining; however, upon change to the (probably) poorer leaving group Me_3N^+ (**24**), k_1 increases and k_2/k_{-1} decreases, thereby allowing a pre-equilibrium mechanism to manifest itself as the buffer concentration is raised. Comparison of (**24**) and (**25**) reveals that α-Ph increases k_1 which β-Ph (**26**) causes a 35-fold reduction of k_1. Reduction of ionization rate (k_1) also accompanies an increase in size of the leaving group [(**25**) → (**28**)]. These results are in agreement with an earlier suggestion[20] that the meagre accelerative effect of a β-phenyl group, when associated with a bulky activating group,

$$\mathrm{X{-}\overset{\displaystyle H}{\overset{|}{C}}{-}C{-}Z} + \mathrm{B:} \underset{k_{-1}}{\overset{k_1}{\rightleftharpoons}} \mathrm{X{-}\bar{C}{-}C{-}Z} + \mathrm{BH^+} \xrightarrow{k_2} \mathrm{X{-}C{=}C} + \mathrm{Z:} \quad \ldots(1)$$

$PhSO_2CHR^1CHR^2\overset{+}{N}R^3_2R^4$

(**24**) $R^2 = Ph; R^3 = R^4 = Me; R^1 = H$
(**25**) $R^1 = R^2 = H; R^3 = R^4 = Me$
(**26**) $R^1 = Ph; R^3 = R^4 = Me; R^2 = H$
(**27**) $R^1 = Ph; R^3 = Et; R^4 = Me; R^2 = H$
(**28**) $R^1 = R^2 = H; R^3 = Et; R^4 = Me$
(**29**) $R^1 = R^2 = H; R^3 = Me; R^4 = Ph$

(**30**) $PhSO_2CH_2CHPhOAc$
(**31**) $PhSO_2CH_2CH_2X$
(**32**) $PhSO_2CD_2CH_2X$
(**33**) $PhSO_2CD_2CD_2X$
(**a**) X = F
(**b**) X = Cl
(**c**) X = Br

is steric in origin. An interesting observation, which also supports earlier work,[20] is the relative constancy of the ratio k_2/k_{-1} for (**24**), (**26**), (**27**) and (**29**). This implies little double-bond character in the elimination transition state since both reprotonation (k_{-1}) and elimination (k_2) are comparably affected by α- or β-phenyl substitution or by change of leaving group.

As a complement to earlier stereochemical studies the kinetics of reaction of halosulphones (**31–33**) with MeO^-–MeOH and with Et_3N in acetonitrile and benzene have been determined in an attempt to locate the position of the elimination mechanism in the *E*2–*E*1*cB* spectrum.[21] For the bromide and chloride in methanol the pre-equilibrium *E*1*cB* mechanism has been ruled out since no H–D exchange occurs and the β-deuterium isotope effects (5.0 and 3.6, respectively) are higher than expected in the event of internal return. The results are consistent with an *E*2 process in which C—H bond breakage leads to C—X bond breakage which is not well developed (k_{Br}/k_{Cl} = ca. 4). It has been argued that for the latter leaving group there is less accumulation of negative charge at C_β and consequently less C—H breakage, as shown by the larger isotope effect. By an alternative explanation in terms of an irreversible *E*1*cB* mechanism it is difficult to rationalize the differences in isotope effects for the bromide and chloride, which react at comparable rates. In contrast, the low isotope effect (2) for the fluoride is not inconsistent with an internal return–*E*1*cB* mechanism. The case against *E*1*cB* reaction of (**31b**) and (**31c**) under analogous conditions, or in Et_3N–C_6H_6, is unconvincing.

Upon reaction with *n*-BuLi in Et_2O, (**34**) is believed to form the dianion (**35**) which reacts according to Scheme 1. An alternative route to (**36**), whereby the monoanion of (**34**) undergoes irreversible *E*1*cB* elimination, is considered to be less likely.[22]

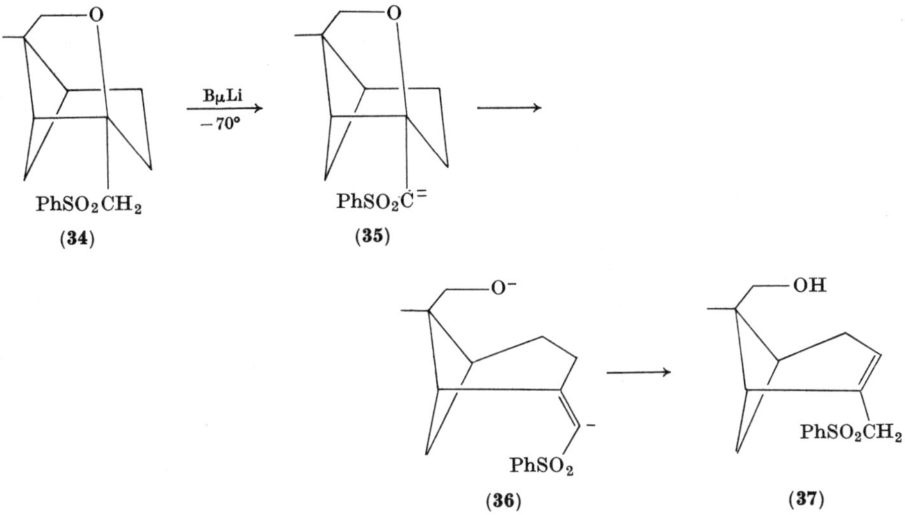

$PhSO_2CH_2$ (**34**) —BuLi, −70°→ $PhSO_2C^=$ (**35**) → $PhSO_2$ (**36**) → $PhSO_2CH_2$ (**37**)

SCHEME 1.

Of particular interest, in view of the recent suggestion that base-initiated 1,2-elimination reactions are rare,[23] are results that suggest that the dehydrochlorination of (**38**) proceeds through a carbanion intermediate even though prior exchange of the benzylic proton does not occur.[24] Observation that (**38**) undergoes dehydrochlorination 68,000-fold faster than (**39**) in EtO^-–EtOH, while in marked contrast (**40**) eliminates HF only

17 times faster than (**41**), prompted a study of isotopic exchange and primary β-isotope effects (k_H/k_D) for this series. The isotope effect (2.8) for (**38**) suggests a highly unsymmetrical transition state or an intermediate carbanion undergoing internal return, while a negligible isotope effect for (**40**) (determined indirectly since H–D exchange occurs

(**38**) $C_6H_5CHClCF_2Cl$ (**40**) $C_6H_5CHClCF_3$ (**42**) $C_6H_5CTClCF_3$

(**39**) $C_6H_5CH_2CH_2Cl$ (**41**) $C_6H_5CH_2CH_2F$ (**43**) $C_6H_5CCl{=}CF_2$

faster than elimination) is consistent with substantial return in the latter mechanism. Hammett ρ-values for elimination of HCl from (**38**) and (**39**), and of HF from (**40**) and (**41**) are 3.94, 2.61, 3.73 and 3.12, respectively, while the value is 3.94 for both detritiation of (**42**) and reaction of (**43**) in EtO^-–EtOH. It has therefore been concluded that reactions of (**38**)–(**41**) proceed via an intermediate carbanion and that the effects of internal return can be substantial.

Pyridine-induced decomposition of β-nitroalkyl nitrates to nitro-olefins also appears to proceed by an *E*1*cB* mechanism when the nitro- and nitrate groups are attached to primary and secondary carbon atoms, respectively, but it may be concerted when the leaving group is attached to a tertiary carbon.[25]

$$ZC(NO_2)_2CH_2CH_2CN \xrightarrow[-HNO_2]{OH^-} ZC(NO_2){=}CHCH_2CN$$

(**44a**) Z = NO_2 (**45**)

(**44b**) Z = F

(**44c**) Z = Cl

The α-fluorine effect[26] has been applied to distinguish between possible concerted and carbanion mechanisms for hydroxide-catalysed conversion of (**44**) into (**45**) by elimination of nitrous acid.[27] A process in which a ground-state carbon atom bearing a fluorine substituent gains *s*-character on going to the transition state should be energetically disfavoured relative to the non-fluorine-substituted carbon; thus for concerted *E*2 elimination of (**44**) activation enthalpies should increase in the order (**44a**) < (**44c**) < (**44b**). The order obtained (**44a**) < (**44b**) < (**44c**) suggests, however, that neither C—NO_2 bond breaking nor double-bond formation occurs to a great extent in the transition state, yet during reaction in a buffered D_2O–dioxan solution neither (**44a**) nor (**45a**) undergoes deuterium exchange. The results are consistent with an irreversible *E*1*cB* mechanism.

For the overall reaction (**46**) → (**47**) the Hammett parameters, $\rho_X = +0.64$ (Y = H) and $\rho_Y = +3.17$ (X = H), favour the E1*cB* mechanism[28] (B) since the $B_{AC}2$ mechanism (A) would involve less acyl—O bond cleavage in the transition state.[29] For each member of a series (X = *p*-NO_2) K_{a_1} has been estimated and for the corresponding k_2 values $\rho_Y = +2.90$. This is consistent with the high degree of acyl—O cleavage already found for related *E*1*cB* reactions. As the leaving group ability is reduced, the mechanism apparently changes from *E*1*cB* to $B_{AC}2$, there being a marked reduction in slope of a plot of log k_2 versus pK_a (of R^2OH) above pK_a = 12.5. The pH–rate profile for conversion of (**51**) into *o*-phenyleneurea, in aqueous dioxan, has been attributed to rate-determining formation of an intermediate isocyanate [from (**51**) or its conjugate acid] which is trapped intramolecularly by the *o*-amino-group.

A change in mechanism from *E*1*cB* to bimolecular substitution, when the pK_a of the leaving alcohol exceeds ca. 17, has also been proposed for alkaline hydrolysis of alkyl

$$\begin{array}{ccccc} R^1NHCO_2R^2 & \underset{\text{Path A}}{\overset{k_1[OH^-]}{\rightleftarrows}} & R^1NHCO_2^- + R^2O^- & \longrightarrow & R^1NH_2 + CO_2 \\ \mathbf{(46)} & & \mathbf{(47)} & & \mathbf{(48)} \\ \downarrow\uparrow & & \uparrow \text{fast}, \ HO^- \text{ or } H_2O & & \\ R^1\bar{N}CO_2R^2 & \underset{\text{Path B}}{\overset{k_2}{\longrightarrow}} & R^1N{=}C{=}O + R^2O^- & & \\ \mathbf{(49)} & & \mathbf{(50)} & & \end{array}$$

$R^1 = XC_6H_4$; $R^2 = YC_6H_4$ **(51)** = (**46**, X = o-NH_2; Y = p-NO_2)

N-phenylcarbamates.[30] Results for both alkyl and aryl *N*-phenylcarbamates fit the same linear free-energy relationship between log k and pK_a (7–16) of the leaving alcohol ($\beta = -1.15$) while for *N,N*-disubstituted carbamates a much lower slope is found ($\beta = -0.25$). Reaction of the latter series is also relatively insensitive to ring substituents on departing phenoxide (in keeping with rate-determining formation of a tetrahedral intermediate) while for the former a dependence on σ^- and high ρ-value (2.86) has previously been attributed to rate-determining C—OAr cleavage[28] following N—H, ionization.

*E*1*cB* mechanisms have also been implicated in the base-hydrolysis of substituted phenyl phosphoro- and phosphorothio-diamidates,[31,32] hydrolysis of aryl *N*-methyl-amino-sulphonates,[33] arylamine-catalysed formation of isocyanates from *N*-aryl-carbamoyl chlorides in acetonitrile[34] and *tert*-amine-catalysed dehydration of bicyclic β-ketols.[35] Intermediate *E*1*cB* reactions occur during coupling between an α-diazo-β-disulphone and the sodium salt of a β-disulphone[36] and during *tert*-amine-catalysed tricyanovinylation of cyclopentadienylidenetriarylphosphoranes with cyclohexyl 1,2,2-tricyanovinyl ether.[37] Addition of *n*-BuLi to 7-chloronorbornadiene is favoured over metallation at the 7-position and dehalogenation reactions of the intermediate anion feature interesting skeletal rearrangements.[38]

The *E*2*C* Mechanism

The *E*2*C* mechanism has again been vigorously challenged[39,40] since there is little evidence of marked steric retardation of base approach to C_α. A statistically corrected retardation of only 6–12-fold has been determined for reactions of sterically hindered (**52**), relative to unhindered (**53**), with tetrabutylammonium chloride in acetone.[39] This

H Me Me OTs

(52)

H H H OTs

(53)

is very much less than the deceleration, by three powers of ten, expected of a reaction in which chloride ion approaches C_α in a fashion substantially similar to an S_N2 displacement, yet explicable in terms of the 1,3-diaxial interaction between a methyl group and the chloride ion attacking H_β according to the conventional $E2$ mechanism. These and related results indicate that any interaction of chloride ion with C_α is geometrically different from that in an S_N2 transition state and that good correlations[40] between log k_S and log k_E cannot be taken as evidence that in the two transition states the interaction with C_α is "much the same".

Hindrance to dehydrobromination of 2-bromo-3,3-dimethylbutane by the inherent neopentylic steric effect has also been determined for reactions of methanolic methoxide and thioethoxide ions, by comparing kinetic results with those for 2-bromopropane which is comparatively unhindered.[41] Thioethoxide/methoxide elimination rate ratios for reaction with the hindered and the unhindered substrate are 1.7 and <0.55, respectively, there being no indication of greater neopentyl steric hindrance to thioethoxide-induced elimination. Upon statistical correction it is found that elimination from the hindered substrate is more rapid with both reagents, while the neopentyl effect retards the thioethoxide and methoxide substitution reactions by 28,000- and at least 1,300-fold, respectively. It has therefore been concluded that the $E2C$ mechanism for thioethoxide-promoted elimination should also be rejected. An alternative ion-pair mechanism[23] is considered improbable since no rearrangement occurs during dehydrobromination of 2-bromo-3,3-dimethylbutane.

Dehydrochlorination of 1,1-diaryl-2,2,2-trichloroethanes (DDT-type compounds) promoted by chloride and thiolate ions has recently been studied[42] since this substrate type is more acidic than those for which $E2C$ reactions have generally been reported and yet sufficiently hindered at C_α to retard competitive S_N2 reaction. For dehydrochlorination by chloride ion in acetone the ρ value $= 1.31$, which is significantly different from that for dehydrobromination of 1-aryl-2-bromopropanes by Br^- in acetone ($\rho = 0.48$) or for the $E1cB$ dehydrochlorination of Ar_2CHCCl_3 by EtO^-–EtOH ($\rho = 2.34$). It has been concluded[43] that the ρ-value is not large enough for reaction to follow the $E1cB$-mechanism but that this is the first example of a halide-promoted $E2H$ reaction.

The steric and electronic effects of base on the rates and product proportions from *tert*-pentyl chloride have been determined by use of thiolate bases which are much weaker than the conjugate base of the solvent (ROH) and are known to react more rapidly than alkoxides with tertiary alkyl halides.[44] Under these conditions results are uncomplicated by change of solvent or by reaction of its lyate ion. The low sensitivity of the relative yields of 2-methylbut-1-ene (ca. 27%) and 2-methylbut-2-ene (ca. 73%) to the nature of the thiolate ion (*n*-BuS, *sec*-BuS, *tert*-BuS, *p*-MeC_6H_4S, C_6H_5S, etc., in EtOH) suggests that the degree of base–substrate interaction is insufficient for differences in base strength to have an appreciable effect. The absence of any steric effect within the series of butyl thiolates is particularly striking and the insensitivity of rates to base structure gives rise to a low Brønsted β value (<0.19). A decrease in proportion of 1-alkene with increasing ability of a halogen leaving group is as previously noted for oxygen bases and reflects increasing double-bond character in the transition state. These results for thiolate-promoted eliminations can be interpreted according to the variable transition state theory or the $E2H$–$E2C$ theory since both can accommodate a loose transition state with high double-bond character. Nucleophilic interaction between C_α and the base may well explain the elimination efficacy of the weakly basic thiolate anions, otherwise there is no compelling evidence on which to accept or reject the $E2C$ mechanism. An ion-pair mechanism[23] is also considered to be improbable for these reactions since it is difficult to reconcile the dependence of orientation on leaving group.

By measurement of olefin proportions formed upon treatment of 4-methyl-2-pentyl, 2-methyl-3-pentyl, 2-pentyl and 3-pentyl bromides with Bu_4NBr in acetone, Et_4NF in acetone and DMF, and with Bu^{t}OK in DMF, it has been shown that halide ion-promoted elimination reactions do not give unusually high *trans/cis*-olefin ratios.[45] These results are in conflict with the popular view that the magnitude of abnormally high *trans/cis*-ratios is diagnostic of the extent of *E2C*-like character.

The kinetics of reactions of cyclohexyl bromide with tetra-*n*-propylammonium thiophenoxide in MeOH, DMF and molten triethyl-*n*-hexylammonium triethyl-*n*-hexylborate have been studied.[46]

Pyrolytic Elimination Reactions

Investigation of the rates of pyrolysis of secondary acetates has been extended to include 1,2-dimethylpropyl acetate[47] and the series[48] Bu^{t}CHR(OAc) where R = Me, Et and Pr^i. The effect of β-unsaturation has been estimated for 1-*tert*-butylbut-3-enyl acetate.[49]

Activation parameters for the pyrolysis of isomeric *S*-butyl thiolacetates suggest that the reaction mechanism is similar to that of the corresponding sulphur-free acetates.[50] Although the thermochemistry of acetate pyrolysis is more favourable than that of thiolacetates by ca. 20 kcal mol^{-1}, the rate constants are only (6–55)-fold greater. A smaller sensitivity of thiol-ester pyrolysis to α-methylation may indicate less incipient charge development at that centre. This should imply greater double-bond formation in the transition state. The higher ratio of but-2-ene to but-1-ene obtained from *sec*-butyl thiolacetate may therefore reflect greater stabilization of the incipient double bond by the alkyl group in this case than in that of the acetate.

Products of decomposition of benzyl triphenylacetate and of benzyl diphenyl-*p*-tolylacetate have been reported,[51] and a preferred *syn*-stereochemistry has been found for photoeliminations from cycloalkyl (cyclopentyl through cyclodecyl) and from *cis*- and *trans*-isomers of 4-*tert*-butylcyclohexyl, 2-methylcyclohexyl and 2-methylcyclo-octyl phenylacetates.[52,53] Intramolecular hydrogen-bonding may play an important role in the decomposition of hydrogen phthalate esters in solution.[54]

Kinetics of thermal decomposition of *tert*-butylperoxy and *tert*-butylperoxymethyl esters of tetrahydrobenzoic acid and of their epoxy-derivatives have been followed.[55] For the first-order thermal decomposition[56] of $RC_6H_4CO_2CH_2CH_2NO_2$ a decrease in basicity of the carbonyl group leads to decrease of $\Delta S^\ddagger$ and increase of $\Delta H^\ddagger$. A study of unsymmetrical carbonates has provided evidence of steric hindrance to pyrolysis of di-*tert*-alkyl carbonates and established that the incipient charge development at C_α is comparable with that for acetate pyrolyses.[57]

Thermochemical estimates have been used, in preference to CNDO/2 calculations, to explain the different gas-phase thermolyses undergone by the isomeric lactones α-coumaranone and phthalide and by *o*-phenylene carbonate and sulphite which extrude CO, CO_2, CO_2 and SO, respectively.[58]

A wide range of alkyl *N*-(methoxycarbonyl)sulphamate esters (**55**) have been prepared[59] by reaction of the corresponding alcohol with (methoxycarbonylsulphamoyl)-triethylammonium hydroxide inner salt (**54**). These readily decompose, over a range of solvent basicities, to yield olefins and (**56**) (E_i mechanism) from secondary and tertiary alcohols, and urethanes (S_N2 mechanism) from primary alcohols. Rate and stereochemical studies are consistent with initial rate-limiting ion-pair formation followed by fast *cis*-β-proton transfer to the departing anion. The Saytzeff product predominates. This promises to be a convenient route for alcohol dehydration since the derivatives can be easily prepared, even from sterically crowded alcohols, and rearrangement of the

$$-\underset{H}{\overset{|}{C}}-\overset{|}{\underset{|}{C}}-OH + MeO_2C\bar{N}SO_2\overset{+}{N}Et_3 \longrightarrow -\underset{H}{\overset{|}{C}}-\overset{|}{\underset{|}{C}}-OSO_2\bar{N}CO_2Me \;\; H\overset{+}{N}Et_3 \xrightarrow{\Delta}$$

(54) **(55)**

$$>C=C< + {}^{-}OSO_2NHCO_2Me + H\overset{+}{N}Et_3$$

(56)

intermediate carbonium ion should be minimized by the good proton nucleophilicity of the departing anion in solvents of low polarity.

N-Chlorosulphonylurethanes of tertiary alcohols, and of other alcohols, ROH, in which R^+ is a moderately stable carbonium ion, decompose by heterolytic dissociation to $ClSO_2NHCO_2^-$ which subsequently decarboxylates before recombining with the counterion R^+ to yield monoalkylsulphamyl chlorides.[60] The reaction sequence provides an alternative to the Ritter reaction for conversion of tertiary alcohols into *tert*-alkylamines (readily obtained from the sulphamyl chloride) and has an advantage in that concentrated acid conditions are not employed.

Gas-phase pyrolysis of *N*-*tert*-butylacetamide occurs by competitive formation of isobutene and ketene upon elimination of acetamide and *tert*-butylamine, respectively.[61] The negligible effect of *N*-substituents on the activation parameters reported for pyrolysis of several ethyl *N*,*N*-dialkylcarbamates supports a six-centered transition state in which cyclization is through carbonyl-oxygen rather than through nitrogen.[62] Rates of decomposition of *N*-monosubstituted dithiocarbamic acids $RNHCS_2H$ (R = Me, Pr, Ph, Et or $PhCH_2$) to isothiocyanate and bisulphide ion are proportional to the mole fraction of the anion form over a wide pH range.[63]

Hammett studies of the decarboxylation of 2- and 5-substituted pyridinecarboxylic acids[64,65] and of *o*- and *p*-substituted benzoic acids[66] have been interpreted; and $^{13}C/^{12}C$ isotope effects for liquid-phase decarboxylation of *all* of the [mono-^{13}C]malonic acids[67] indicate that the concerted reaction involves both carbonyl groups and that bond-breaking between the leaving carbonyl-carbon and the central carbon is almost complete.

Decarboxylations of tetrahydroacridinedionecarboxylic acids,[68] phthalic acid[69] and [2.2.1]bicycloheptaneperoxycarboxylic acids[70] have also been studied and it has been found that Schiff bases of β-keto-acids, where the carbonyl function is present on the bridgehead of a bicyclic system, decarboxylate *ca.* 10^6 times faster than the corresponding keto-acids in non-polar solvents.[71]

A study of specifically labelled *N*,*N*-dimethylcyclo-octylamine oxides has established unequivocally that the Cope elimination is 100% *syn* stereospecific.[72] The rates of Cope elimination in two homologous series of open-chain amine oxides, $RCHXCH_2\cdot C_5H_{11}$ and $RCH_2CHX\cdot C_5H_{11}$ [$X = (CH_3)_2N\rightarrow O$; R = H, Me, Et, Pr^n, Pr^i or Bu^t), and in a complementary series of α-, β- and γ-methyl-substituted *N*,*N*-dimethylheptylamine oxides, range over six powers of ten and increase with increasing alkyl substitution.[73] Relief of ground-state strain is believed to control the reaction rate, which is largely dependent on the energy required for C_α—N bond breaking in the transition state. There is a correlation between σ^* and log k for *trans*-alkene formation. A Hammett plot for a series of *N*-methyl-(1-arylbutyl)amine oxides, however, reveals that the alkyl substituent effect is not of polar origin. The formation of 5-decyl-2-methylisoxazolidine on thermal decomposition of dimethyldodecylamine oxide has been interpreted.[74]

Studies of pyrolytic eliminations on solid surfaces have included the dehydrocyanation of cyclobutane-1,2-dicarbonitrile vapour on granular base,[75] formation of alkenyl ethers on chromotographic elution of aldehyde dimethyl acetals on adsorptive gas–liquid columns,[76] dehydrocyclization of heptane over platinum on non-acidic alumina by direct formation of the six-carbon ring with subsequent aromatization,[77] kinetics of gas-phase dehydration of methanol on a sulphonated ion-exchanger,[78] catalytic dehydration of *cis*-2-methylcyclohexan-1-ol, hexan-2-ol[79] and of diastereoisomers[80] of D,L-[3-$^2H_1$]-butan-2-ol, and elimination of carboxylic acids from *sec*-alkyl carboxylates.[81]

Two mechanisms of decomposition of phosphite ozonides to phosphate esters and singlet excited molecular oxygen have been reported.[82] Pseudorotation occurs during reaction of those ozonides without geometrical constraint, while simple oxygen extrusion occurs without rearrangement from ozonides derived from phosphites with small rings or bicyclic structures.

Results of pyrolytic gas-chromatography suggest that cracking of phenols occurs by initial decarbonylation to cyclopentadienes.[83] The kinetics and mechanism of formation of diphenylketene from azibenzil have been studied with[84] and without[85] solvent. Thermolyses of phenyl glycosides,[86] of phenylalanine and its low-temperature pyrolysis products,[87] of methyl *cis*- and *trans*-3,4-diphenyl-1-pyrazoline-3-carboxylate,[88] of 5-substituted isoxazolines,[89] of bromonitrobornane,[90] of 4,4-dimethyl-2-piperidino-thiazol-5-(4*H*)-one[91] and of 1,2-bis(hydroxydiphenylmethyl) compounds[92] have been investigated. The mechanism of elimination of ethane from [bis(dimethylphenylphosphine)iodotrimethyl]platinum(IV) has been established by deuterium-labelling experiments.[93]

Radical mechanisms have been proposed for deoxygenation and desulphurization of cyclic ethers and sulphides[94] and for pyrolysis and photolysis of diastereomeric 2,3-bis(phenylthio)butanes.[95]

Unimolecular thermolyses of alkyl allyl ethers have previously been studied qualitatively and the clean formation of an olefin and a carbonyl-containing product is well established. Quantitative studies[96] have now shown that activation parameters for thermolysis of $CH_2{=}CR^1CHR^2OCR^3R^4R^5$ are little dependent on variation of R^1–R^4, as expected for a concerted pericyclic retro-ene mechanism. The absence of polar substituent effects ($\rho = 0$) found for thermolysis of *p*-substituted allyl benzyl ethers is in contrast to earlier work and in keeping with the non-polar mechanism. On the basis of a maximum theoretical isotope effect found for reaction of allyl benzyl, benzyl propargyl and allyl isopropyl ethers it has further been argued that C—H bond-making and -breaking is symmetrical;[97] this precludes a completely planar transition state (such as that proposed for related thermolyses of esters[98]) since the required departure from linearity of hydrogen transfer would considerably affect k_H/k_D and its temperature-dependence; the identity of this temperature-dependence and the striking similarity of activation parameters for allylic and propargylic ethers (as well as for β-hydroxyolefins and β-hydroxyacetylenes[99]) suggests that the respective transition states have a common structure.

It has been argued[99] that the 1,5-sigmatropic hydrogen shift, which occurs upon thermolysis of β-hydroxyacetylenes, proceeds through a planar six-membered transition state in close analogy with the corresponding olefinic system. For both systems, the effects on rate of aryl substituents at the C-1 position are identical ($\rho = -0.35$), and previously reported differences in alkyl substituent effects must therefore be attributed to steric effects alone. Surprisingly, the reaction of each acetylene is faster than that reported for the corresponding olefin, despite the additional energy requirement for formation of the six-membered transition state of the more linear system. It has therefore been suggested that the transition state for the acetylene is planar, while that for the olefins is chair-like.

Cycloreversions, including thermal fragmentation of [4.4.2]propella-2,4-dienes and gas-phase eliminations involving $2\pi + 2\sigma$ cycloreversions, have been examined by a molecular-orbital symmetry approach.[100] Variants of the polar transition state predicted for $2\pi + 2\sigma$ cycloreversions have already been proposed. The product pairs, ketene with ethyl isobutenyl ether (65–80%) and 1,1-dimethylketene with ethyl vinyl ether (35–20%), are obtained upon pyrolysis of 3-ethoxy-2,2-dimethylcyclobutan-1-one in the gas phase;[101] the activation parameters for the overall decomposition, when compared with those for cyclobutanone, indicate a stabilizing effect of the 3-alkoxy-group of 11.5 kcal mol^{-1}. These results are also consistent with development in the transition state of an essentially zwitterionic charge for $2s + 2a$ ketene-plus-olefin addition–elimination. Decomposition of 2-chlorobutanone has also been studied.[102]

Intermediate carbene, ylide and ketimine have been proposed to account for thermolyses of phthalide,[103] betaines[104] and 3-hydroxy-1,2,3-benzotriazin-4-one,[105] respectively. It has now been shown that acetic acid catalyses the formation of isopropenylmethyl ether from 2,2-dimethoxypropane in the gas phase.[106] This is the first report of HOAc-catalysed ether decomposition, for which HBr and HCl are respectively 2×10^5 and 4×10^3 times more effective towards this substrate.

Several dehydrohalogenation studies have been reported.[107–111] Equilibrium constants for formation of isocyanates from arylcarbamoyl chlorides in acetonitrile follow a Hammett correlation with σ^- ($\rho = -0.52$) and are increased by the steric effect of *o*-substituents,[107] which retard the back-reaction. The forward rate constants are, however, insensitive to electronic effects, thus excluding a mechanism involving extensive proton-transfer in the transition state. The electronic effects on the back-reaction suggest that C—Cl bond-formation is slightly ahead of C—H bond-formation and a concerted four-centre transition state is favoured. Rates of α,α- and α,β-elimination of HF or DF from vibrationally excited CD_3CHF_2, CH_3CHF_2 and CD_3CH_2F (produced from radical recombination reactions) have been interpreted by RRKM theory,[108] and isotope effects for decomposition of chemically activated CF_3CH_3 and CF_3CD_3 have likewise been predicted.[109] Thermal decomposition of trialkyl-(2,3,3,3-tetrafluoropropyl)ammonium hydroxides[110] and of 3-chlorobut-1-ene and *trans*-1-chlorobut-2-ene[111] have been studied.

An analogy between gas-phase decomposition of alkyl cyanides and of the corresponding alkyl chlorides is supported by observation[112] of dehydrocyanation in the rate order EtCN < Pr^iCN < Bu^tCN. The increased rate of elimination of isothiocyanic acid (with formation of buta-1,3-diene) caused by unsaturation in the carbon skeleton of but-3-enyl and 1-methylprop-2-enyl isothiocyanates has been attributed to allylic weakening of the C—N bond rather than conjugation with the incipient double bond.[113] The effect of vinylation on rate is comparable with that previously reported for allylic chlorides. Vinyl group acceleration has also been estimated for 1-*tert*-butylbut-3-enyl acetate.[49]

The effect of R and R′ substituents on rates of homolytic decomposition of acyl alkyl diimides, RCON:NC(CN)R'_2, in toluene has been determined.[114] Activation parameters for gas-phase decomposition of *N*-allylcyclohexylamine to cyclohexylimine and propene are consistent with a cyclic transition state.[115]

Oxygen-labelling experiments have revealed a new sulphoxide fragmentation reaction of benzoyloxymethyl phenyl sulphoxide.[116]

Other Topics

An excellent review of elimination reactions has been published.[117]

Steric destabilization of the transition state leading to a desolvated carbonium ion is believed to account for retardations of ca. 20-, 110- and 200-fold observed for acid-catalysed dehydration of alkyl di-*tert*-butylcarbinols, in anhydrous acetic acid, where $R = Bu^i$, Bu^t and neopentyl, respectively.[118] The steric effects of bulky alkyl groups (Bu^t and/or neopentyl) on solvolysis of dialkylarylcarbinyl *p*-nitrobenzoates in aqueous acetone have been explained in terms of hindrance to aryl resonance and relief of reactant strain energy.[119] Nitrogen kinetic isotope effects for hydrolysis of substituted benzene-diazonium cations in the presence of added nucleophiles have been interpreted in terms of an S_N1 mechanism.[120a] The *p*-methoxy-derivative is, however, believed to react by a bimolecular route previously proposed.[120b] De-ethanolation of an acetylenic ethoxy-β-oxo-ester[121] and the role of vinyl cations in solvolytic dehydrobromination of *cis*- and *trans*-α-bromo-4,4′-dimethoxystilbene[122] have been studied.

Deuterium kinetic isotope effects and Hammett relationships have been determined in order to elucidate the nature of the transition state for *E*2 elimination of 'onium substrates.[123–125] For reaction[123] of the ions (**57**) in EtO^-–EtOH at 60° the β-deuterium isotope effects (k_H/k_D) decrease from 5.91 ($Y = OCH_3$) to 4.15 ($Y = CF_3$) with increasing electron-withdrawing power of Y. It has already been shown that concerted *E*2 elimination occurs[126] and, in keeping with the small value of ρ (+1.33), this is believed to be the first such example wherein the proton is less than half transferred in the transition state. Previous reports have been of increasing k_H/k_D upon increasing electron-withdrawal and this is the trend even for the closely related (2-arylethyl)trimethylammonium ions[127] for which $\rho = +3.77$. Although H_β of (**57**; Y = H) is appreciably more acidic than that of (**58**; R = H), its rate of elimination is slower; this may be a result of steric hindrance of base approach to H_β.

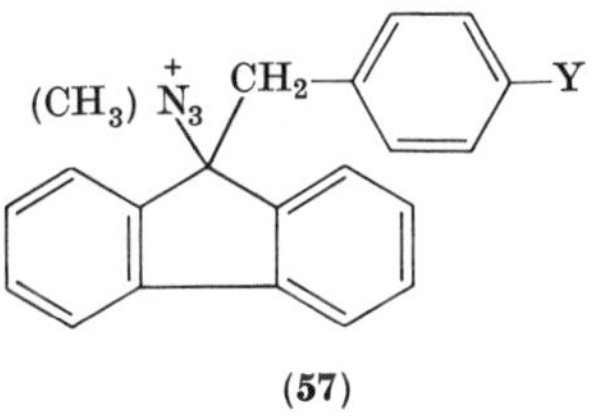

(**57**)

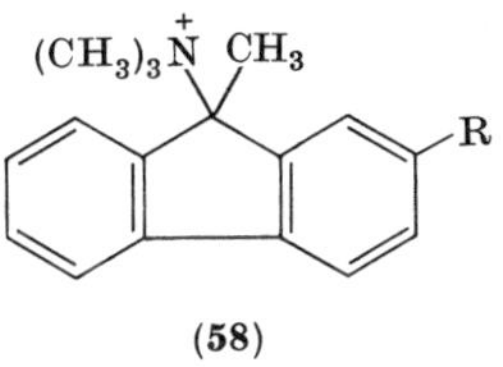

(**58**)

Rate coefficients for elimination reactions of *p*-substituted dimethyl phenethyl-sulphonium and trimethylammonium ions and their 2,2-dideuterio-analogues in Pr^iOH/Pr^iO^- have been compared[125] with those in $EtOH/EtO^-$. Interpretation of results proved to be less ambiguous than for Bu^tO^- Bu^tOH in which pronounced ion-association effects and possible proton tunnelling occur[128] (cf. a recent study[129] of β-deuterium kinetic isotope effects for *E*2 reactions of arylethyl bromides with Bu^tO^-/Bu^tOH). Isotope effects for both 'onium salts increase with the basicity of the medium (EtOH $\rightarrow$ $Pr^iOH \rightarrow Bu^tOH$), and decreasing proton-transfer to base in the transition state is implied. In each medium the isotope effects for the sulphonium salts are larger than for their ammonium analogues and the ρ-value is small. This is consistent with less proton transfer in the transition state for elimination of the better leaving group. The marked increase in ρ for the sulphonium series upon change from EtOH ($\rho = 2.64$) to Pr^iOH ($\rho = 3.47$) is, however, in marked contrast to the relative insensitivity of substituent effects in the ammonium series ($\rho = 3.77$ and 3.87, respectively), which may therefore involve a transition state close to the carbanion extreme. While the ρ-values imply an increase in carbanion character of the sulphonium transition state as the medium is changed from EtOH to Pr^iOH, the variation of k_H/k_D ($5.07 \rightarrow 6.73$) reveals a concomitant decrease in proton transfer. This requires that the accompanying decrease in C—leaving group bond cleavage is even more marked, and this is to be further investigated.

Hammett correlations for reaction of *p*-substituted phenethyl bromides with sodium phenoxide ($\rho = 2.64$) and *p*-nitrophenoxide ($\rho = 1.84$) in DMF indicate that the carbanion character of the transition state (but not necessarily the degree of C—H bond stretching) is more pronounced for the former nucleophile which is more basic than the latter by six powers of ten.[130] It is therefore probable that factors other than base strength account for the decrease of ρ previously reported[131] for reactions of 2-arylethyl bromides and 1-aryl-2-propyl bromides upon change from EtO^-–EtOH to Bu^tO^-–Bu^tOH.

Interpretation of recent results,[132] which were believed to support the view that *E*2 eliminations are rare, has been challenged,[133] as anticipated.[132]

Benzisoxazoles (**59**) undergo rapid base-catalysed isomerizations to salicylonitriles (**60**) by an irreversible *E*2 mechanism. Benzo-substituents can cause considerable

$$B + X\text{-}(\mathbf{59}) \longrightarrow X\text{-}C_6H_3(C{\equiv}N)(O^-)\ (\mathbf{60}) + BH^+$$

(**59**) (**60**)

variation in reactivity without change of the geometry or gross electronic structure of this system, which has therefore been chosen to test the dependence of the Brønsted coefficient on substrate reactivity.[134] The catalytic constants (k) for reactions of 55 combinations of eight benzisoxazoles and twelve tertiary amines in water at 30° vary over a range of 10^5, yet quantitatively fit the equation $\log k = 0.721 \mathrm{p}K_{\text{amine}} + 0.614(14 - \mathrm{p}K_{\text{cyanophenol}}) - 11.9$. Kinetic isotope effects for three 3-deuteriated benzisoxazoles showed no significant variation over a range of catalytic-constant values of 10^9. These results indicate that the proton transfer is characterized by a sensitivity to substituent change which is invariant over appreciable changes in transition state energy and that, in general, the Brønsted coefficient need not necessarily be regarded as a local constant which varies continuously between zero and one over a sufficiently large $\mathrm{p}K_a$ range.[135] This is in marked contrast to the view generally held that selectivity towards reagents decreases with increasing substrate reactivity.

Reactions of 1-chlorobi(cyclopropyl) with $NaNH_2$ in liquid ammonia,[136] cyclopropylmethyl tosylate[137] with Bu^tOK–DMSO, vinylacetylenic halides with amines,[138] and 1,4-benzoquinone dihalides[139] in $BF_3{\cdot}Et_2O$, and the kinetics of dehydrochlorination[140] of 2,3,4- and 2,3,3-trichlorobut-1-ene by methanolic NaOH, have been studied.

The effect of micelles on S_N2 and *E*2 reactions of 1-bromo-2-phenylpropane[141,142] and on decarboxylation of carboxylic acids[143,144] has been determined, and decarboxylation of *trans*-cinnamic acid,[145] 2-hydroxy-carboxylic acids,[146] α-amino-acids,[147] oxaloacetic acid[148] and ammonium trichloroacetate[149] have been investigated.

Methyltriphenoxyphosphoniumiodide (MTPI) in hexamethylphosphoramide (HMPA) is a mild reagent system for selective dehydration of secondary alcohols without concomitant rearrangement.[150] The faster rate of dehydration of *trans*- than of *cis*-4-*tert*-

butylcyclohexanol is in keeping with a mechanism whereby the alcohol forms the corresponding inverted iodide which then undergoes dehydroiodination.

Dehydration of Bu^tOH by dialkoxydiarylsulphuranes is believed to involve dissociative ligand exchange followed by a rate-determining fragmentation with incipient development of the carbonium ion.[151] A mechanism has also been proposed for the dehydration of amides to nitriles promoted by phosphoronitrilic chloride;[152] and the stereoselective dehydration of D,L-*erythro*- and D,L-*threo*-[3-$^2H_1$]butan-2-ol over metal phosphates has been correlated with the molecular dimension of the catalyst.[80]

The pH-dependence of dehydration and rearrangement[153] of prostaglandins E_1 and E_2 and of the elimination of CH_2O and H_2O from *N*-hydroxymethylated 4,6-diamino-2-methoxy-*s*-triazine[154] has also been investigated.

A kinetic study of amine-catalysed β-ketol dehydration has identified the role of intermediate imines and the condition for maximum catalytic effect.[155] Formation of Schiff bases from pyridoxal and amino-acids and the pH-dependence of the stability of their complexes with metal ions have been explored in an attempt to explain the catalytic effects of Cu(II) and oxovanadium(IV) ions on the β-elimination reaction of *O*-phosphothreonine in aqueous pyridoxal solutions.[156] The role of palladium complexes in catalysis of $NaAlH_2(OC_2H_4OCH_3)_2$-promoted dehalogenation of aryl and alkyl halides,[157] and the decomposition of vinyl acetate[158] and of hydration of enol acetates,[159] have been investigated; titanocene has been used to promote dehalogenation of certain aliphatic halides.[160]

The contributions of competitive $\alpha'\beta$-, β- and α-elimination mechanisms of alkyllithium-induced formation of ethylene from diethyl ether,[161] and the 1,1-elimination of HCN from *o*-nitrobenzyl cyanide[162] under electron impact have been reported. Elimination of ketene from acetanilide molecular ion proceeds via a four-membered transition state.[163]

Rate constants for α-proton abstraction, α-elimination, hydrolysis and etherification of *o*-nitrobenzyl halides in 50% (v/v) alkaline-aqueous dioxan have been interpreted[164] and the benzoylphenylcarbene generated by reaction of α,α-dibromodeoxybenzoin with zinc dust has been trapped by cycloaddition to *trans*-stilbene.[165]

Upon reaction with Bu^tOK–DMSO, for short periods, (**61a**), (**61b**) and (**61c**) form (**65a**), (**65b**) and (**65c**), respectively.[166] These bridged ketones undergo irradiation-induced decarbonylation to cyclo-octatetraenes (**64**) and triplet-sensitized photoisomerization to barbaralones. Likewise sulphones (**61d–g**) react with strong base to form (**65d–g**), respectively, which in turn undergo thermal or photochemical extrusion of SO_2. In each case the proposed product of dehydrohalogenation is a zwitterion (**62** or **63**) which undergoes intramolecular cycloaddition to the proximate diene system (Scheme 2, see following page).

Perfluoromethacrylic acid derivatives have been prepared from substituted alkoxyperfluoroisobutenes by elimination of alkyl fluoride upon reaction with Lewis acids.[167] A perepoxide is not a common intermediate of dehydrobromination of isomeric β-bromo-hydroperoxides.[168]

Anchimeric assistance by sulphur in an elimination sequence can account for the exclusive formation of (**67**) from (**66**) during reflux in collidine.[169]

There have been studies of acid-catalysed, thermal and photochemical reactions of *N*-nitroso-5*H*-dibenz[*b,f*]azepine,[170] elimination of HSCN upon reaction of aminoazoles with benzoyl and ethoxycarbonyl isothiocyanates,[171] reaction of triethyl phosphite with Mannich bases,[172] hydrolysis of 4,5-diphenylisosydnone,[173] conversion of *N*-(arylaminomethyl)succinimides into *N*-methylarylamines[174] on treatment with sodium

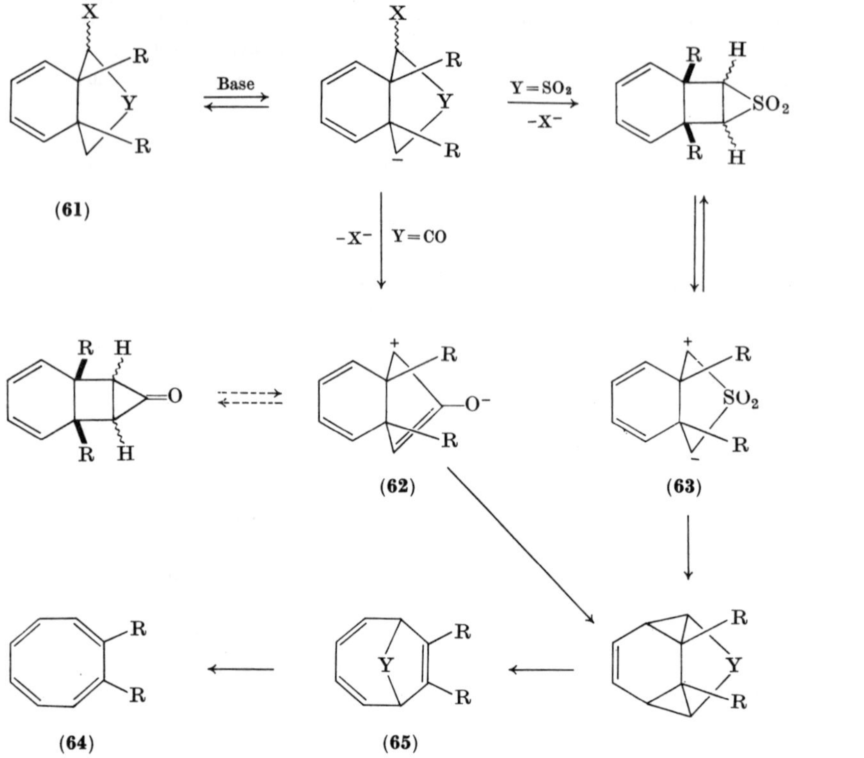

(61) (62) (63) (64) (65)

Y = CO, X = Br: **a**, R = H; **b**, R = Me; **c**, R, R = —$(CH_2)_4$—

Y = SO_2, X = Cl: **d**, R = Me; **e**, R, R = —$(CH_2)_2$—; **f**, R, R = —$(CH_2)_3$—; **g**, R, R = —(CH:CHCH:CH)—

SCHEME 2.

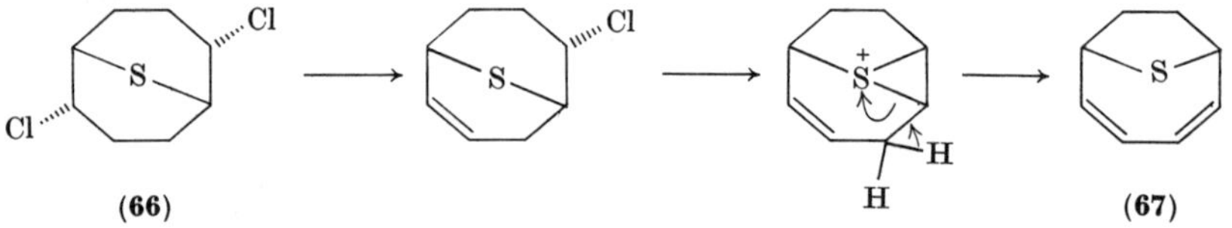

(66) (67)

borohydride in DMSO and of lithium diethylamide-induced rearrangements of diene monoepoxides.[175] The ability of epoxides to convert haloalkanes into alkenes in the presence of chloride ions has been attributed to *E*2 elimination promoted by 2-chloroethoxide ions.[176]

The stereochemistry of reactions of halohydrin derivatives and *vic*-dibromides[177] with BuLi and the kinetics of nucleophilic dehalogenations[178] of 1-chloro-1,2-diphenylethane have been investigated. Reaction of thiosulphate ion with *vic*-dibromides in DMSO has been found to give olefins in high yield by *trans*-stereospecific debromination.[179]

N-Phenylbenzamidine has been used as a mild and selective dehydrobrominating agent,[180] and the kinetics of the Perkin reaction,[181] the intermediacy of sulphenyl chlorides in the conversion of 3-phenylpropanoic acids into benzo[*b*]thiophens[182] and the inductive effect on acid-induced decomposition of dithiocarbamates[183] have been studied. Bu^tCl is not an intermediate of HCl-catalysed decomposition of Bu^tOPr^i to isobutene.[184] Addition–elimination sequences have been proposed[185] to account for the reactions of 6-chloro-and 5-bromo-1,3-dimethyluracil with NaCN in DMF.

Kinetic investigations of the mechanism of γ-elimination of chlorine and phosphate groups upon hydrolysis of 2-chloro-3-pyridyl methyl phosphate[186] and of the complex Hammett relationships for formation of substituted styrene oxides upon reaction of 2- and 1-aryl-2-haloethanols in aqueous alkali[187] have been reported. The stereochemistry of carbanionic 1,3-elimination reactions of *dl*- and *meso*-$PhCHBrSO_2CHBrPh$ induced by Ph_3P in benzene, or on being heated in DMF, has been interpreted.[188] It is concluded that most, if not all, 1,3-eliminations occur by two-stage mechanisms and that either double-inversion or retention–inversion geometry will be the rule for base-initiated reactions.

References

1 M. Svoboda and J. Závada, *Coll. Czech. Chem. Comm.*, **37**, 3902 (1972).
2 J. Závada, M. Pánková, M. Svoboda and M. Schlåsser, *Chem. Comm.*, **1973**, 168.
3 V. Fiandanese, G. Marchese, F. Naso and O. Sciacovelli, *J.C.S. Perkin II*, **1973**, 1336.
4 R. A. Bartsch, *J. Org. Chem.*, **38**, 846 (1973).
5 R. A. Bartsch and T. A. Shelly, *J. Org. Chem.*, **38**, 2911 (1973).
6 R. A. Bartsch, G. M. Pruss, D. M. Cook, R. L. Buswell, B. A. Bushaw and K. E. Wiegers, *J. Am. Chem. Soc.*, **95**, 6745 (1973).
7 See *Org. Reaction Mech.*, **1972**, 139.
8 See *Org. Reaction Mech.*, **1972**, 140.
9 See *Org. Reaction Mech.*, **1972**, 141.
10 R. A. Bartsch, G. M. Pruss, B. A. Bushaw and K. E. Wiegers, *J. Am. Chem. Soc.*, **95**, 3405 (1973).
11 J. G. Traynham, D. B. Stone and J. L. Couvillion, *J. Org. Chem.*, **32**, 510 (1967).
12 W. H. Saunders, *Chem. Comm.*, **1973**, 850.
13 Cf. *Org. Reaction Mech.*, **1970**, 148.
14 S. Alunni and E. Baciocchi, *Tetrahedron Letters*, **1973**, 205.
15 A. T. Bottini, K. A. Frost II, B. R. Anderson, and V. Dev, *Tetrahedron*, **29**, 1975 (1973).
16 D. Brunel, A. Casadevall, E. Casadevall, and C. Largeau, *Bull. Soc. Chim. France*, **1973**, 1325.
17 T. Sasaki, K. Minamoto and H. Suzuki, *J. Org. Chem.*, **38**, 598 (1973).
18 M. Szkoda and L. Prajer-Janczewska, *Wiad. Chem.*, **27**, 167 (1973); *Chem. Abs.*, **78**, 146915 (1973).
19 K. N. Barlow, D. R. Marshall and C. J. M. Stirling, *Chem. Comm.*, **1973**, 175.
20 R. P. Redman and C. J. M. Stirling, *Chem. Comm.*, **1970**, 633.
21 V. Fiandanese, G. Marchese and F. Naso, *J.C.S. Perkin II*, **1973**, 1538.
22 N. Bosworth and P. Magnus, *J.C.S. Perkin I*, **1973**, 2319.
23 F. G. Bordwell, *Accounts Chem. Research*, **5**, 374 (1972).
24 H. F. Koch, D. B. Dahlberg, A. G. Toczko and R. L. Solsky, *J. Am. Chem. Soc.*, **95**, 2029 (1973).
25 W. M. Cummings and K. L. Kreuz, *J. Org. Chem.*, **37**, 3929 (1972).
26 L. A. Kaplan and H. B. Pickard, *Chem. Comm.*, **1969**, 1500; *J. Am. Chem. Soc.*, **93**, 3447 (1971).
27 L. A. Kaplan and N. E. Burlinson, *J. Org. Chem.*, **37**, 3932 (1972).
28 Cf. A. Williams, *J.C.S. Perkin II*, **1972**, 808.
29 A. F. Hegarty and L. N. Frost, *J.C.S. Perkin II*, **1973**, 1719.
30 A. Williams, *J.C.S. Perkin II*, **1973**, 1244.
31 A. Williams and K. T. Douglas, *J.C.S. Perkin II*, **1973**, 318.
32 Cf. A. Williams and K. T. Douglas, *J.C.S. Perkin II*, **1972**, 1454.
33 K. T. Douglas and A. Williams, *Chem. Comm.*, **1973**, 356.
34 R. Bacaloglu and C. A. Bunton, *Tetrahedron*, **29**, 2725 (1973).
35 D. J. Hupe, M. C. R. Kendall, G. T. Sinner and T. A. Spencer, *J. Am. Chem. Soc.*, **95**, 2260 (1973).

[36] G. Heyes and G. Holt, *J.C.S. Perkin I*, **1973**, 189.
[37] M. P. Naan, A. P. Bell, and C. D. Hall, *J.C.S. Perkin II*, **1973**, 1821.
[38] S.-I. Murahashi, K.-I. Hino, Y. Maeda and I. Moritani, *Tetrahedron Letters*, **1973**, 3005.
[39] J. F. Bunnett and D. L. Eck, *J. Am. Chem. Soc.*, **95**, **1897** (1973).
[40] See *Org. Reaction Mech.*, **1968**, 149; **1972**, 146.
[41] J. F. Bunnett and D. L. Eck, *J. Am. Chem. Soc.*, **95**, 1900 (1973).
[42] See *Org. Reaction Mech.*, **1970**, 164; **1972**, 144.
[43] O. R. Jackson, D. J. McLennan, S. A. Short and R. J. Wong, *J.C.S. Perkin II*, **1972**, 2308.
[44] D. S. Bailey and W. H. Saunders, Jr., *J. Org. Chem.*, **38**, 3363 (1973).
[45] I. N. Feit, A. M. Capobianco and T. W. Cooke, *Tetrahedron Letters*, **1973**, 2799.
[46] W. T. Foed and R. J. Hauri, *J. Am. Chem. Soc.*, **95**, 7381 (1973).
[47] G. Chuchani, I. Martin and A. Maccoll, *J.C.S. Perkin II*, **1973**, 663.
[48] G. Chuchani, G. Martin, N. Barroeta and A. Maccoll, *J.C.S. Perkin II*, **1972**, 2239.
[49] G. Chuchani, S. P. de Chang and L. Lombana, *J.C.S. Perkin II*, **1973**, 1961.
[50] D. B. Bigley and R. E. Gabbott, *J.C.S. Perkin II*, **1973**, 1293.
[51] W. S. Trahanovsky, D. E. Zabel and M. L.-S. Louie, *J. Org. Chem.*, **38**, 757 (1973).
[52] M. L. Yarchak, J. C. Dalton and W. H. Saunders, Jr., *J. Am. Chem. Soc.*, **95**, 5224, (1973).
[53] M. L. Yarchak, J. C. Dalton and W. H. Saunders, Jr., *J. Am. Chem. Soc.*, **95**, 5228 (1973).
[54] R. M. Ottenbrite, J. W. Brockington and K. G. Rutherford, *J. Org. Chem.*, **38**, 1186 (1973).
[55] A. M. Ustinova, A. A. Turovskii, R. V. Kucher and A. E. Batog, *Zh. Org. Khim.*, **9**, 78 (1973): *Chem. Abs.*, **78**, 110302 (1973).
[56] R. S. Stepanov, E. S. Buka and V. N. Shanko, *Organic Reactivity* (*Tartu*), **9**, 583 (1972); *Chem. Abs.*, **78**, 135397 (1973).
[57] D. B. Bigley and C. M. Wren, *J.C.S. Perkin II*, **1972**, 2359.
[58] C. Wentrup, *Tetrahedron Letters*, **1973**, 2919.
[59] E. M. Burgess, H. R. Penton and E. A. Taylor, *J. Org. Chem.*, **38**, 26 (1973).
[60] J. B. Hendrickson and I. Joffee, *J. Am. Chem. Soc.*, **95**, 4083 (1973).
[61] A. Maccoll and S. S. Nagra, *J.C.S. Faraday I*, **1973**, 1108.
[62] N. J. Daly, G. M. Heweston and F. Ziolkowski, *Austral. J. Chem.*, **26**, 1259 (1973).
[63] F. Takami, K. Tokuyama, S. Wakahara and T. Maeda, *Chem. Pharm. Bull.*, **21**, 1311 (1973); *Chem. Abs.*, **79**, 77812 (1973).
[64] R. J. Moser and E. V. Brown, *J. Org. Chem.*, **37**, 3938 (1972).
[65] R. J. Moser and E. V. Brown, *J. Org. Chem.*, **37**, 3941 (1972).
[66] D. Sitamanikyam and E. V. Sundaram, *Indian J. Chem.*, **10**, 1011 (1972); *Chem., Abs.*, **78**, 110300 (1973).
[67] A. G. Loudon, A. Maccoll and D. Smith, *J.C.S. Faraday I*, 894 (**1973**).
[68] R. B. Highet and J. F. Biellmann, *J. Org. Chem.*, **37**, 3731 (1972).
[69] K. A. Malyshevskaya and N. M. Dubrovskaya, *Izv. Vyssh. Ucheb. Zaved. Khim. Khim. Tekhnol.*, **15**, 1515 (1972); *Chem. Abs.*, **78**, 42426 (1973).
[70] M. Gruselle and D. Lefort, *Tetrahedron*, **29**, 3035 (1973).
[71] K. Taguchi and F. H. Westheimer, *J. Am. Chem. Soc.*, **95**, 7413 (1973).
[72] R. D. Bach, D. Andrzejewski and L. R. Dusold, *J. Org. Chem.*, **38**, 1742 (1973).
[73] J. Závada, M. Pánková and M. Svoboda, *Coll. Czech. Chem. Comm.*, **38**, 2102 (1973).
[74] R. G. Laughlin, *J. Am. Chem. Soc.*, **95**, 3295 (1973).
[75] D. M. Gale and S. C. Cherkofsky, *J. Org. Chem.*, **38**, 475 (1973).
[76] M. B. Evans and B. Williamson, *Chromatographia*, **5**, 264 (1972); *Chem. Abs.*, **78**, 96918 (1973).
[77] B. H. Davis, *J. Catal.*, **29**, 398 (1973); *Chem. Abs.*, **79**, 41619 (1973).
[78] Le N. Thanh, K. Setínek and L. Beránek, *Coll. Czech. Chem. Comm.*, **37**, 3878 (1972).
[79] P. Canesson and M. Blanchard, *Bull. Soc. Chim. France*, **1973**, 2839.
[80] K. Thomke and H. Noller, *Z. Naturforsch.*, **27**, 1462 (1972).
[81] D. Ali, P. Kripylo and H. Prinzler, *J. Prakt. Chem.*, **315**, 47 (1973).
[82] L. M. Stephenson and D. E. McClure, *J. Am. Chem. Soc.*, **95**, 3074 (1973).
[83] R. Spielmann and C. A. Cramers, *Chromatographia*, **5**, 295 (1972); *Chem. Abs.*, **78**, 57472 (1973).
[84] M. Contineanu, T. Oncescu and R. N. Schindler, *Rev. Roum. Chim.*, **18**, 549 (1973); *Chem. Abs.*, **79**, 31278 (1973).
[85] M. Contineanu, T. Oncescu and R. N. Schindler, *Rev. Roum. Chim.*, **18**, 753 (1973); *Chem. Abs.*, **79**, 65551 (1973).
[86] F. Shafizadeh, M. H. Meshreki and R. A. Susott, *J. Org. Chem.*, **38**, 1190 (1973).
[87] J. M. Patterson, N. F. Haider, E. P. Papadopoulos and W. T. Smith, *J. Org. Chem.*, **38**, 663 (1973).

88 J. P. Deleux, G. Leroy and J. Weiler, *Tetrahedron*, **29**, 1135 (1973).

89 Y. Iwakura, K. Uno, S. Shiraishi, M. Yuyama and Y. Kihara, *Yuki Gosei Kagaku Kyokai Shi*, **30**, 894 (1972); *Chem. Abs.*, **79**, 65548 (1973).

90 S. Ranganathan and H. Raman, *Tetrahedron Letters*, **1973**, 411.

91 M. Kato, A. Kobayashi and S. Umemoto, *Chem. Letters* (*Tokyo*), **1972**, 1129.

92 T. Toda, K. Saito and T. Mukai, *Chem. Letters* (*Tokyo*), **1973**, 1123.

93 M. P. Brown, R. J. Puddephatt and C. E. E. Upton, *J. Organometal. Chem.*, **49**, C61 (1973).

94 P. S. Skell, K. J. Klabunde, J. H. Plonka, J. S. Roberts and D. L. Williams-Smith, *J. Am. Chem. Soc.*, **95**, 1547 (1973).

95 P. B. Shevlin and J. L. Greene, *J. Am. Chem. Soc.*, **94**, 8447 (1972).

96 H. Kwart, J. Slutsky and S. F. Sarner, *J. Am. Chem. Soc.*, **95**, 5234 (1973).

97 K. Kwart, J. Slutsky and S. F. Sarner, *J. Am. Chem. Soc.*, **95**, 5242 (1973).

98 See *Org. Reaction Mech.*, **1972**, 148–149.

99 A. Viola, R. J. Proverb, B. L. Yates and J. Larrahondo, *J. Am. Chem. Soc.*, **95**, 3609 (1973).

100 N. D. Epiotis, *J. Am. Chem. Soc.*, **95**, 1191 (1973).

101 K. W. Egger, *J. Am. Chem. Soc.*, **95**, 1745 (1973).

102 J. Metcalfe and E. K. C. Lee, *J. Am. Chem. Soc.*, **95**, 4316 (1973).

103 U. E. Wiersum and T. Nieuwenhuis, *Tetrahedron Letters*, **1973**, 2581.

104 R. A. Champa and D. L. Fishel, *Can. J. Chem.*, **51**, 2750 (1973).

105 P. Ahern, T. Navratil and K. Vaughan, *Tetrahedron Letters*, **1973**, 4547.

106 D. A. Kairaitis, V. R. Stimson and J. W. Tilley, *Austral. J. Chem.*, **26**, 761 (1973).

107 R. Bacaloglu and C. A. Bunton, *Tetrahedron*, **29**, 2721 (1973).

108 K. C. Kim, D. W. Setser and B. E. Holmes, *J. Phys. Chem.*, **77**, 725 (1973).

109 B. D. Neely and H. Carmichael, *J. Phys. Chem.*, **77**, 307 (1973).

110 A. T. Babayan and L. Kh. Gamburyan, *Dokl. Akad. Nauk Arm.*, **55**, 181 (1972); *Chem. Abs.*, **78**, 158701 (1973).

111 J.-P. Nogier and B. Blouri, *Bull. Soc. Chim. France*, **1973**, 758.

112 P. N. Dastoor and E. V. Emovon, *Can. J. Chem.*, **51**, 366 (1973).

113 N. Barroeta and R. Mazzali, *J.C.S. Perkin II*, **1973**, 839.

114 T. R. Lynch, F. N. MacLachlan and J. L. Suschitzky, *Can. J. Chem.*, **51**, 1378 (1973).

115 K. W. Egger, *J.C.S. Perkin II*, **1973**, 2007.

116 T. J. Maricich, R. A. Jourdenais and C. K. Harrington, *J. Am. Chem. Soc.*, **95**, 2378 (1973).

117 A. F. Cockerill in "Comprehensive Chemical Kinetics", C. H. Bamford and C. F. H. Tipper (Eds.), Elsevier, London, 1973, p. 163.

118 J.-E. Dubois and J. S. Lomas, *Tetrahedron Letters*, **1973**, 1791.

119 H. Tanida and H. Matsumura, *J. Am. Chem. Soc.*, **95**, 1586 (1973).

120a A. G. Loudon, A. Maccoll and D. Smith, *J.C.S. Faraday I*. **1973**, 899.

120b E. S. Lewis, L. D. Hartung and B. M. McKay, *J. Am. Chem. Soc.*, **1969**, **91**, 419.

121 V. N. Sapunov, S. S. Yufit, V. F. Kucherov and I. A. Esikov, *Kinet. Katal.*, **14**, 604 (1973); *Chem. Abs.*, **79**, 77678 (1973).

122 Z. Rappoport and M. Atidia, *J.C.S. Perkin II*, **1972**, 2316.

123 P. J. Smith and S. K. Tsui, *J. Am. Chem. Soc.*, **95**, 4760 (1973).

124 P. J. Smith and S. K. Tsui, *Tetrahedron Letters*, **1973**, 61.

125 A. F. Cockerill and W. J. Kendall, *J.C.S. Perkin II*, **1973**, 1352.

126 See *Org. Reaction Mech.*, **1970**, 161.

127 L. J. Steffa and E. R. Thornton, *J. Am. Chem. Soc.*, **89**, 6149 (1967).

128 W. H. Saunders, D. G. Bushman and A. F. Cockerill, *J. Am. Chem. Soc.*, **90**, 1775 (1968).

129 L. F. Blackwell, P. D. Buckley, K. W. Jolley and A. K. H. MacGibbon, *J.C.S. Perkin II*, **1973**, 169.

130 S. Alunni and E. Baciocchi, *Tetrahedron Letters*, **1973**, 4665.

131 C. H. DePuy, D. L. Storm, J. T. Frey and C. G. Naylor, *J. Org. Chem.*, **35**, 2746 (1970).

132 See *Org. Reaction Mech.*, **1972**, 144–145.

133 W. H. Saunders, *Tetrahedron Letters*, **1972**, 5129.

134 D. S. Kemp and M. L. Casey, *J. Am. Chem. Soc.*, **95**, 6670 (1973).

135 See M. Eigen, *Discuss. Faraday Soc.*, No. 39, 7 (1965).

136 L. Fitjer and J.-M. Conia, *Angew. Chem. Internat. Ed.*, **12**, 332 (1973).

137 W. R. Dolbier and J. H. Alonso, *Chem. Comm.*, **1973**, 394.

138 M. G. Voskanyan, G. G. Khudoyan and Sh. O. Badanyan, *Dokl. Uses. Konf. Khim. Atsetilena, 4th*, **1**, 449 (1972); *Chem. Abs.*, **79**, 52492 (1973).

139 R. K. Norris and S. Sternhill, *Austral. J. Chem.*, **26**, 333 (1973).

140 A. Kanala, J. Vojtko and M. Hrusovsky, *Chem. Zvesti*, **26**, 457 (1972); *Chem. Abs.*, **78**, 3481 (1973).
141 V. Gani, C. Lapinte and P. Viout, *Tetrahedron Letters*, **1973**, 4435.
142 C. Lapinte and P. Viout, *Tetrahedron Letters*, **1973**, 1113.
143 L. K.-M. Lam and D. E. Schmidt, *Can. J. Chem.*, **51**, 1959 (1973).
144 C. A. Bunton, M. J. Minch, J. Hidalgo and L. Sepulveda, *J. Am. Chem. Soc.*, **95**, 3262 (1973).
145 B. Danieli, P. Manitto, F. Ronchetti and G. Russo, *Chem. Ind.* (*London*), **1973**, 430.
146 Y. Pocker and B. C. Davis, *J. Am. Chem. Soc.*, **95**, 6216 (1973).
147 W. E. Pereira, Y. Hoyano, R. E. Summons, V. A. Bacon and A. M. Duffield, *Biochem. Biophys. Acta*, **313**, 170 (1973).
148 H. Ito, H. Kobayashi and K. Nomiya, *J.C.S. Faraday I*, **1973**, 113.
149 Z. Pawlask and T. Jasinski, *Zesz. Nauk Wydz. Mat., Fiz., Chem., Uniw. Gdanski, Chem.*, **1**, 59 (1971); *Chem. Abs.*, **78**, 123740 (1973).
150 R. O. Hutchins, M. G. Hutchins and C. A. Milewski, *J. Org. Chem.*, **37**, 4190 (1972).
151 L. J. Kaplan and J. C. Martin, *J. Am. Chem. Soc.*, **95**, 793 (1973).
152 J. C. Graham and D. H. Marr, *Can. J. Chem.*, **50**, 3857 (1972).
153 D. C. Monkhouse, L. Van Campen and A. J. Aguiar, *J. Pharm. Sci.*, **62**, 576 (1973); *Chem. Abs.* 151597, **78** (1973).
154 T. Tashiro, *Makromol. Chem.*, **167**, 249 (1973); *Chem. Abs.*, **79**, 52525 (1973).
155 D. G. Hupe, M. C. R. Kendall and T. A. Spencer, *J. Am. Chem. Soc.*, **95**, 2271 (1973).
156 Y. Murakami, H. Kondo and A. E. Martell, *J. Am. Chem. Soc.*, **95**, 7138 (1973).
157 I. Simůnek and M. Kraus, *Coll. Czech. Chem. Comm.*, **38**, 1786 (1973).
158 P. M. Henry, *J. Org. Chem.*, **38**, 3596 (1973).
159 P. M. Henry, *J. Org. Chem.*, **38**, 2766 (1973).
160 A. Merijanian, T. Mayer, J. F. Helling and F. Klemick, *J. Org. Chem.*, **37**, 3945 (1972).
161 A. Maercker and W. Demuth, *Angew. Chem. Internat. Ed.*, **12**, 75 (1973).
162 A. Kalir, E. Sali and A. Mandelbaum, *J.C.S. Perkin II*, **1972**, 2262.
163 M. Ohashi, T. Kizaki, K. Tsujimoto, Y. Shida and Y. Yamada, *Shitsuryo Bunseki*, **21**, 85 (1973) *Chem. Abs.*, **79**, 31166 (1973).
164 Y. Riad and A. A. Abdallah, *Egypt. J. Chem.*, **15**, 37 (1972); *Chem. Abs.*, **79**, 77694 (1973).
165 L. T. Scott and W. D. Cotton, *J. Am. Chem. Soc.*, **95**, 5416 (1973).
166 L. M. Paquette, R. H. Meisinger and R. E. Wingard, *J. Am. Chem. Soc.*, **95**, 2230 (1973).
167 I. L. Knunyants, Yo. G. Abduganiev, E. M. Rochlin, P. O. Okulevich and N. I. Karpushina, *Tetrahedron*, **29**, 595 (1973).
168 J. E. Baldwin and O. W. Lever, Jr., *Chem. Comm.*, **1973**, 343.
169 P. H. McCabe and C. M. Livingston, *Tetrahedron Letters*, **1973**, 3029.
170 M. R. Bendall, J. B. Bremner and J. F. W. Fay, *Austral. J. Chem.*, **25**, 2451 (1972).
171 H. J. Schrepfer, L. Capuano and H.-L. Schmidt, *Chem. Ber.*, **106**, 2925 (1973).
172 B. E. Ivanov, V. F. Zheltukhin and L. A. Valitova, *Khim. Primen. Fosfororg. Soedin., Tr. Konf.*, 4th **1969**, 204; *Chem. Abs.*, **78**, 135260 (1973).
173 A. J. Buglass and J. G. Tillett, *J.C.S. Perkin II*, **1973**, 1687.
174 S. B. Kadin, *J. Org. Chem.*, **38**, 1348 (1973).
175 R. P. Thummel and B. Rickborn, *J. Org. Chem.*, **37**, 4250 (1972).
176 J. Buddrus and W. Kimpenhaus, *Chem. Ber.*, **106**, 1648 (1973).
177 T. Sugita, K. Nishimoto and K. Ichikawa, *Chem. Letters* (*Tokyo*), **1973**, 607.
178 E. Baciocchi and C. Lillocci, *J.C.S. Perkin II*, **1973**, 38.
179 K. M. Ibne-Rasa, A. R. Tahir and A. Rahman, *Chem. Ind.* (*London*), **1973**, 232.
180 E. J. Parish and D. H. Miles, *J. Org. Chem.*, **38**, 3800 (1973).
181 K. G. Krachanov, *Dokl. Bolg. Akad. Nauk*, **25**, 1657 (1972); *Chem. Abs.*, **78**, 96786 (1973).
182 A. J. Krubsack and T. Higa, *Tetrahedron Letters*, **1973**, 4515.
183 D. De Filippo, P. Deplano, F. Devillanova, E. F. Trogu and G. Verani, *J. Org. Chem.*, **38**, 560 (1973
184 N. J. Daly, L. P. Steele and V. R. Stimson, *Austral. J. Chem.*, **26**, 767 (1973).
185 S. Senda, K. Hirota and T. Asao, *Tetrahedron Letters*, **1973**, 2647.
186 Y. Murakami, J. Sunamoto and N. Kanamoto, *Bull. Chem. Soc. Japan*, **46**, 1730 (1973).
187 A. C. Knipe, *J.C.S. Perkin II*, **1973**, 589.
188 F. G. Bordwell and B. B. Jarvis, *J. Am. Chem. Soc.*, **95**, 3585 (1973).

CHAPTER 12.I

Addition Reactions. I. Polar Addition

B. V. Smith

Department of Chemistry, Chelsea College, University of London

Electrophilic Addition

In a theoretical discussion, electrophilic addition to olefins has been considered in terms of a $2\pi + 2\sigma$ cycloaddition; however, a stepwise process is not automatically ruled out, and the nature of bonding at the two sites in *cis*-addition has been discussed.[1] A classification of cycloadditions that includes cationic and anionic processes has been made and the reactions have been reviewed.[2]

Calculations of the perturbation of ethylene by a proton indicate that interaction begins at 3.7 Å separation, and for the ethyl cation the contribution of bridged and open structures in the addition of X^- has been discussed.[3] Calculations of the potential energy surfaces for addition to acetylene and fluoroacetylene show that the preferred approach of the electrophile is at right angles to the molecular axis and that the ion $H_2C{=}C^+{-}F$ is favoured in addition to $HC{\equiv}CF$.[4]

A further report cites the use of norbornene and 7,7-dimethylnorbornene as a mechanistic probe for distinguishing between additions with cyclic or non-cyclic transition states. The steric influence of the methyl groups inhibits complex-formation with Ag^+ and addition of NOCl and CCl_2.[5] All known additions are stereospecifically *exo*-oriented.

The extended Hammett equation has been tested for electrophilic and nucleophilic addition to olefins.[6] Bridged and open structures are considered for the former, according to the process in question.

Reviews of electrophilic addition of halogen and hypohalous acid to multiple-bond systems have appeared.[7,8] The comparison between donor co-ordination and addition reactions has been made.[9]

Halogen and Related Addition

The chlorination of cyclo-octatetraene has been examined and the successive steps have been elucidated. The use of $SbCl_5$ (2 equivalents) gives the salt (**1a**) of the *endo*-8-chlorohomotropylium ion which decomposes to give *cis*-dichloride (**2**), and eventually the *trans*-dichloride (**3**); thus the addition of Cl_2 is presumed to give (**1b**) in the primary step.[10] Addition of chlorine to dihydrofurans (**4**) gives *cis*- and *trans*-addition products (*cis:trans* = 1:1 when R = H; but 3:1 when R = Ph), a result that is interpreted by steric control imposed by R.[11] A transannular effect is claimed in the reaction between chlorine and *cis*-cyclo-octene.[12] The metal chloride-catalysed chlorination of isomeric tetrachlorocyclohexenes has been noted.[13]

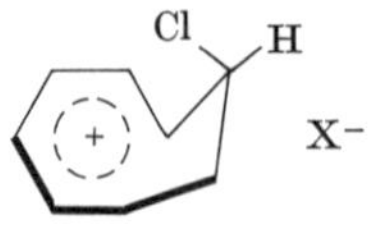

(**1a**) X = $SbCl_6$
(**1b**) X = Cl

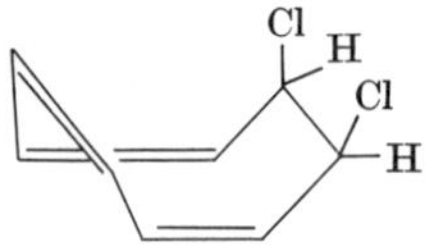

(**2**) 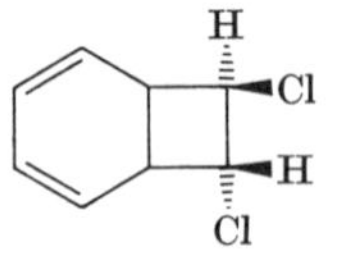 (**3**) 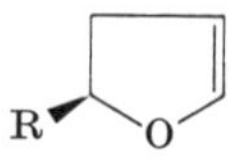 (**4**)

Addition of chlorine acetate to cinnamic acid derivatives, phenanthrene, acenaphthylene and cyclohexene afforded different proportions of acetoxy-chlorides from those formed by chlorination in acetic acid.[14] These differences were attributed to polarity differences in the reaction sequences; ClOAc was considered to give "Cl^+ type sequence" and Cl_2 a "zwitterionic sequence".

Other work reported includes the reaction of acetylene with $CuCl_2$ complexes,[15] a calorimetric study of addition to olefins,[16] formation of halohydrins[17] and the solvent-dependence of halogen additions to cyclopentadiene.[18]

Competitive bromination of alkenes has been studied[19] and the effect of concentration on relative reactivities has been noted.[20] A steric effect in tri-*tert*-butylethylene renders it less reactive than cyclohexene.[21] Addition of bromine to $CH_2{=}CXCH_2Cl$ (where X = H, F, Cl or Br) yields the order H ≫ F > Cl ≈ Br in both trifluoroacetic acid and 50% methanol–H_2O;[22] these results were held to support a bridged ion in the transition state, the relative position of fluorine being interpreted as indicating the contribution by the ion (**5**). The kinetics of addition of bromine to hex-1-ene and 3-(*p*-bromophenyl)-prop-1-ene show second-order dependence on olefin concentration (in CCl_4) and second- and first-order dependence, respectively on bromine concentration.[23] The addition of bromine to $Bu^tC{\equiv}CH$ is complex and rearrangement of the initial bromonium ion is postulated.[24]

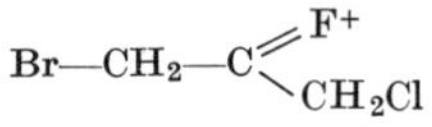

$Br{-}CH_2{-}C({=}F^+){-}CH_2Cl$

(**5**)

Yates and his co-workers[25] have made considerable contributions to the subject of electrophilic additions. The relative ease of formation of carbonium ions and vinyl cations has been compared for addition of halogen and of ArSCl and for acid-catalysed hydration of olefin/acetylene pairs.[25] For the series $R^1CH{=}CHR^2$ and $R^1C{\equiv}CR^2$ where (*a*) R^1 = Ph, R^2 = H, (*b*) R^1 = Ph, R^2 = Me, (*c*) R^1 = *n*-Bu, R^2 = H and (*d*) $R^1 = R^2$ = Et the rate ratios for addition ($k_{olefin}/k_{acetylene}$) were ca. 10^3–10^5 (Br_2), 10^2–10^5 (Cl_2), 2–20 (ArSCl) and 0.6–17 (H_3O^+). The values for $PhCH{=}CH_2/PhC{\equiv}CH$ were 2,590 (Br_2) and 0.7 (H_2O^+). The differences between ρ values for Cl_2, Br_2 and H_3O^+ addition to olefins (–3.2 to –4.8) and for ArSCl addition (–2.4) suggest similar charge development in the first three and lower charge development for ArSCl addition. The rate differences between olefin and acetylene pairs is interpreted in the light of stabilization differences since it is argued that the vinyl carbonium ion should not be strongly solvated. Th contrast between addition and solvolysis reactions leading to a vinyl carbonium ion drew comment.

Bromination of ring-substituted styrenes has been studied and second- and third-order rate constants have been measured.[26] Reaction constants are very similar ($\rho_{k_2} = -4.8$; $\rho_{k_3} = -4.6$), which is taken to mean the similarity of cationic intermediates. The product spread under conditions of "k_2" or "k_3" attack is also similar, and the reaction pathway in Scheme 1 is suggested for the k_3 process.

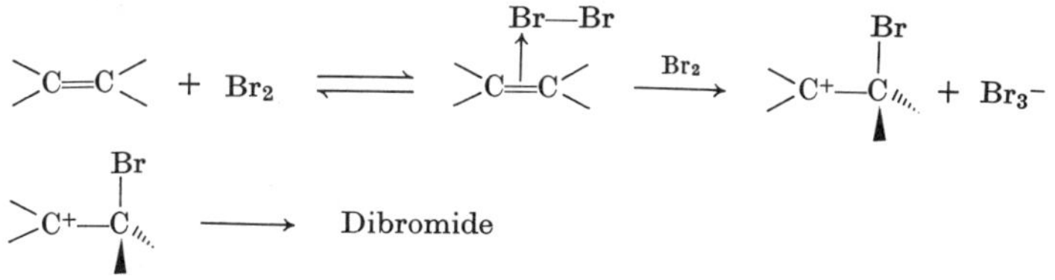

$$>C{=}C< + Br_2 \rightleftharpoons >C{=}C<\cdots Br{-}Br \xrightarrow{Br_2} >C^{+}{-}C(Br) + Br_3^{-}$$

$$>C^{+}{-}C(Br) \longrightarrow \text{Dibromide}$$

Scheme 1.

The intervention of Br_4 ($2Br_2 \rightleftharpoons Br_4$) is rejected on the ground that Br_2---HOAc interaction is assumed to be stronger. The second-order process is enthalpy-controlled but the third-order process is entropy-controlled. In a further study, based on a thermochemical–kinetic approach to transition-state structure in addition reactions, the enthalpies of isomerization and rates of bromination for *cis–trans*-olefin pairs were measured;[27] the addition was stereospecific in some cases but variable in others; nevertheless, it was concluded that a bridged transition state could lead to either a bridged or an open intermediate, dependent upon structure. A neighbouring group effect in the addition of Br_2 to $PhCH{=}CHCH_2NHCOC_6H_4NO_2$-*p* leads to a cyclic product.[28] Halogenation of *cis*- and *trans*-$PhCH{=}CHBu^t$ in CCl_4 gives non-stereospecific addition; an unsymmetrical bridged bromonium ion is discussed but chlorination involves a more open ion;[29] this paper also analyses in detail the populations of *threo*- and *erythro*-products in the light of an NMR study. Bromine addition, in acetic acid, to *p*-$RC_6H_4CMe{=}CH_2$ has a reaction constant $\rho = -5.69$ and an unsymmetrical bridged cation is proposed.[30] Both *cis*- and *trans*-trimethylsilylstyrene add bromine at low temperatures;[31] the dibromides lose the elements of $BrSiMe_3$ at room temperature in such a way that *cis*- and *trans*-β-bromostyrene, respectively, are formed; participation between the silicon atom and the adjacent positive charge is invoked to explain this result (e.g. Scheme 2, p. 384).

The bromination of disubstituted *trans*-stilbenes in CCl_4 or MeOH has been analysed in terms of LFER;[32] additivity of effect in CCl_4 was observed (bromonium-like intermediate) but not always in the case of MeOH (carbonium-like intermediate). The stereochemistry of addition to 4,4′-disubstituted *cis*-stilbenes is a function of the substituent(s) since 4,4′dimethoxystilbene gives only the *meso*-dibromide.[33] *trans*-Addition is observed with electron-withdrawing groups, and these differences are discussed in terms of either a benzylic cation or a bromonium ion's being favoured. Both *cis*- and *trans*-$PhCOCH{=}CHCOPh$ and some of their derivatives add in *trans*-fashion in the reaction between solid olefins and bromine vapour.[34] Addition of HOBr and HOCl to $PhCH{=}CHCOOH$ (and its methyl ester) has been reported, and the difference between Br and Cl in bridging ability has been used to explain the differing stereospecificities.[35]

A considerable volume of work on halogenation of cyclohexene and its derivatives has been reported. Assistance by the medium and the salt as well as steric effects from axial groups at C-4 and C-5 have been noted.[36,37] Solvent and reagent effects for 3- and 4-substituted cyclohexenes (R = 3-Me, 3-Bu^t, or 3-COOMe; 4-Me or 4-Bu^t) in bromination

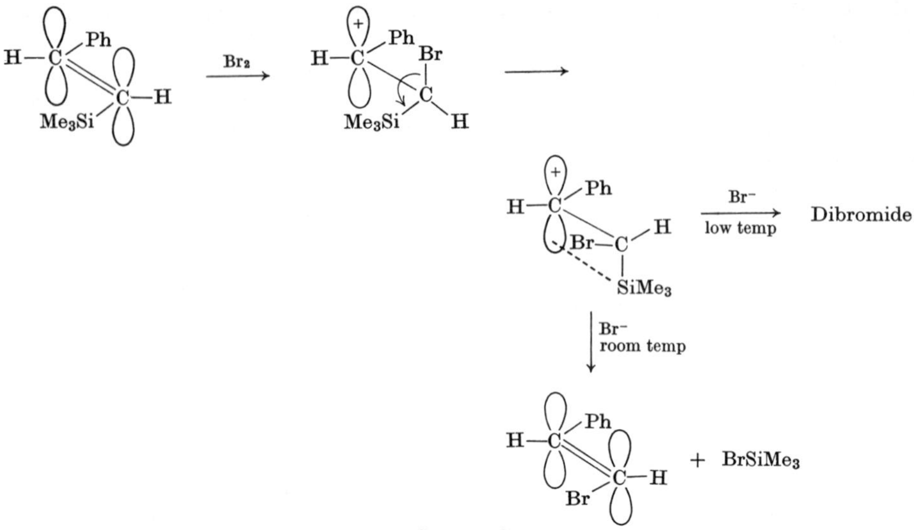

H—C Ph, C—H, Me_3Si $\xrightarrow{Br_2}$ H—C(+) Ph, Br, C, Me_3Si, H ⟶

H—C(+) Ph, Br—C H, $SiMe_3$ $\xrightarrow[\text{low temp}]{Br^-}$ Dibromide

$\downarrow$ Br^- room temp

H—C Ph, C—H, Br + $BrSiMe_3$

SCHEME 2.

have been discussed in terms of opening of the bromonium ion by nucleophilic attack (*a*) antiparallel to the pseudo-axial bond at C-6, forming a diaxial product or (*b*) parallel to the pseudo-axial bond at C-3, leading to a diequatorial product (Scheme 3). With Bu^t at

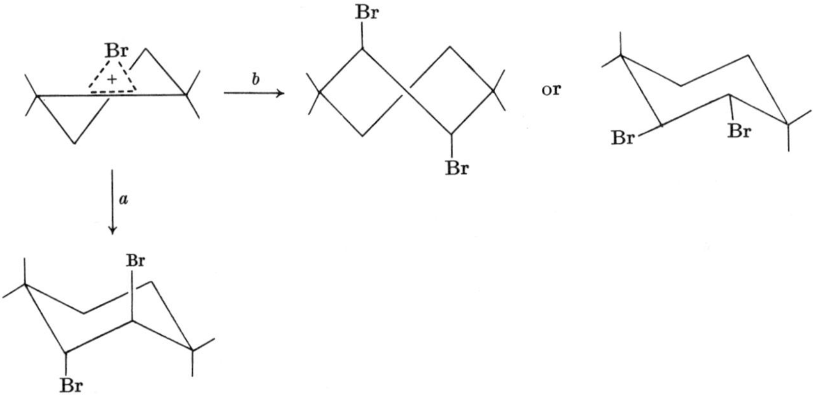

Br, + $\xrightarrow{b}$ Br, Br or Br, Br

$\downarrow a$

Br, Br

SCHEME 3.

position 4, diaxial addition predominates, and vice versa for position 3; however, the product proportions are markedly different in the presence of bases, and this was assumed to indicate the reversibility of the initial step under these conditions.[38] The bromination of 1-phenylcyclohexene is complex.[39] Other work reported includes the bromination of cyclohexene (and epoxide),[40] the generation of "IF" and "BrF" and their *trans*-addition to cyclohexene, methylenecyclohexane, indene and acenaphthylene,[41] the use of $SbCl_5$ or $Et_3O^+SbCl_6^-$ as a halogenating agent[42] (in this case, a variety of chloro-products is formed from cyclohexene) and the bromination of cholesterol derivatives containing dienyl and enone systems.[43,44] Bromination of semibullvalene shows *cis-exo*-addition.[45]

The effect of added ionic bromides and substituents on bromination of unsaturated sulphones has also been studied.[46,47]

A study of ionic addition of bromine to *cis*- and *trans*-1,2-dimethylcyclopropane[48] has been challenged[49] with regard to rate of reaction and product identity; the complexities of the rearrangement reactions which accompany ionic addition were discussed, and it was reported that the initial bromine attack occurs with inversion of configuration.

Thiocyanogen chloride (ClSCN) adds to olefins in acetic acid, forming chloro-thiocyanates and acetoxy-thiocyanates. For ethylene, *cis*: and *trans*-but-2-ene, cyclohexene and *trans*-Δ^2-octalin stereospecific *trans*-addition was observed;[50] it was interpreted as due to a sulphur-bridged ion, e.g. (**6**). For *cis*- and *trans*-stilbene and acenaphthylene *trans*-addition was also observed, but an "open" ion was considered more likely in these cases.[51]

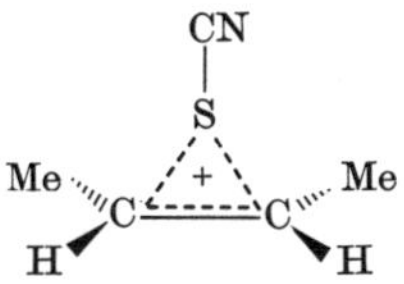

(**6**)

Addition of Hydrogen Halides

The addition of HCl to hexamethyl-Dewar-benzene is stereospecific.[52]

Hydrochlorination of 1-phenylbuta-1,3-diene and the analogous allene gives $PhCH{=}CH{-}CHClCH_3$, via an allylic carbonium ion.[53] Addition to phenylallene (and some of its derivatives) yields only the cinnamyl chloride.[54] At low hydrogen chloride concentration (in acetic acid) the reaction is first-order in HCl, rising to second-order at higher [HCl]. In the high acid concentration region, the reaction constant ρ^+ (–4.20) is consistent with formations of a twisted α-vinylbenzyl cation in the transition state.

Addition of DBr/DCl to $PhCH{=}CHSiMe_3$ gives deuteriostyrenes stereospecifically by way of an unstable intermediate adduct.[55] The participation of silicon with the adjacent positive centre is invoked, as in bromination (see above[31]).

Other work reported is the addition of HCl to biquinones,[56] of HX to alkynes bonded to Pt complexes,[57] of HX to $Me_2NC{\equiv}C{-}COR$ (R = MeO, CH_3, or H),[58] of HCl to piperylene,[59] to $CH_2{=}CHCN$ in the gas phase,[60] to C_2H_2 in presence of Cu salts[61] and of anhydrous HCl to $HC{\equiv}CCH_2R$ (R = Me, Et, Pr or H).[62] In the last of these, cyclic products accompanied chloro-olefins.

Hydration and Related Reactions

The acid-catalysed hydration of sabinene (**7**) and α-thujene (**8**) proceeds through a common ion (**9**) to give almost identical product compositions; (**7**) reacts 23 times faster than (**8**), and this is attributable to the difference in ground-state energies.[63] Addition of alcohols ROH (R = Me[64] or other alkyl[65,66,67]) to various olefins with acid catalysis has been noted. Whereas hydration occurs in aqueous sulphuric acid, in a super-acid (HSO_3F–SbF_5) hydrogen-transfer from alkanes is observed.[68] The addition of CH_3COOD to norbornadiene gives the products shown in Scheme 4.[69] It is suggested that the ion (**10**) is important in determining the product ratios.

Addition of trifluoroacetic acid to ω-arylalk-1-enes gives tetralins and derived trifluoroacetates;[70] in the case of 5-phenylpent-1-ene, anchimeric assistance from the

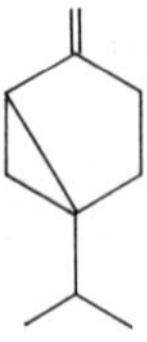

(**7**) 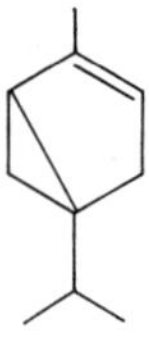 (**8**) 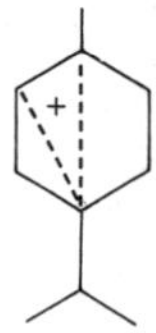 (**9**)

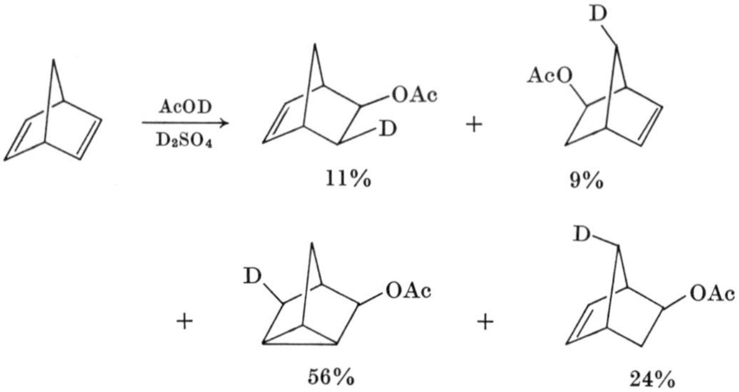

AcOD / D_2SO_4

11% 9% 56% 24%

SCHEME 4.

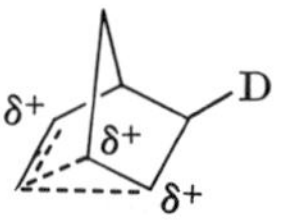

(**10**)

phenyl group is observed. The addition of trifluoroacetic acid to *cis-trans*-cyclododeca-1,5-diene gives, after hydrolysis, pure *cis*-1-*cis*-decalol;[71] H_2SO_4/CH_3CN (the Ritter reaction) and $Hg(N_3)_2$ also give *cis*-oriented products. Greater stereoselectivity was found with cationic than with radical additions. The acid-catalysed addition of acetic acid to a conjugated enyne ester gives rise to furans.[72] Pd(II) complexes catalyse addition of ROH to acetylene.[73]

Acid-catalysed hydrolysis of enol ethers has been reported,[74,75] and incorporation of deuterium has been shown not to result from exchange. Further, distribution of D in the crotonaldehyde formed differs for the *cis/trans* pair (**11**). Acid-catalysed ring closures are reported for a substituted pyrimidine[76] and bicyclo-octenecarboxylic acids.[77]

$$CH_2{=}CH{-}CH{=}CHOEt$$

(**11**)

The isomerization of cyclohepta-3,5-dienone in FSO_3H may occur through a dication; in H_2SO_4 a different product is formed,[78] almost certainly by a different mechanism.

Hydroboration by means of chloroborane is highly regioselective; thus hex-1-ene affords 99.5% of hexan-1-ol. This reagent has been recommended as being more satisfactory than disiamylborane. Relative rates of hydroboration have been measured for oct-1-ene, *cis*- and *trans*-oct-2-ene and cyclohexene.[80] Tri-*tert*-butylethylene reacts very slowly.[21] Racemic allenes react with (+)-tetra-3-pinanyldiborane to form the vinyl- (ca. 70%) and allyl-borane (ca. 30%);[81] the recovered allene showed enrichment in the *R*-enantiomer. Hydroborations of 1,4- and 1,5-dienes give good yields of expected diols.[82] Addition of R_3B and R_2BOR' to quinone and methylene systems has been reported,[83,84] and the stereoselectivity of halogenoboronation of PhC≡CH has been investigated.[85]

The role of steric effects in oxymercuration has been stressed.[86] A mercurinium ion is suggested in this paper[86] and in a report on the reaction with α,β-unsaturated ketones and esters;[87] for the esters, stereospecificity is observed, but not for the ketones. Oxymercuration with a variety of alcohols,[88] with different mercury salts[89] for olefins and dienes,[90] has been recorded. The addition of $Hg(OAc)_2$ and $Hg(OCOCF_3)_2$ to cyclopropanes has been studied[91] with the conclusion that the electrophile attacks the least substituted bond, this being followed by inversion in the attack of the nucleophile.

The kinetics of addition of peracetic acid to olefins have been measured;[92] for acetylenes[93,94] the results are consistent with the formation of an oxirene intermediate (**12**). For allenes, the role of cyclopropanones as intermediates in epoxidation has been discussed.[95]

The catalytic effect of $Mo(CO)_6$ on epoxidation[96] has been interpreted, by analogy with the Michaelis–Menten treatment of enzyme reactions, as due to the presence of MO---HOOR species. Epoxidation of propene,[97] as well as of styrene and cyclohexene,[98] has been reported. The stereo- and regio-selectivities in epoxidations by *tert*-butyl hydroperoxide catalysed by vanadium acetylacetonate or $Mo(CO)_6$ have been determined for geraniol (**13**) and cyclohex-3-enol (**14**); for geraniol the observed product (**15**) is formed in 93% yield.[99] Peracid oxidation of cyclic acetylenes gives bicyclic ketones

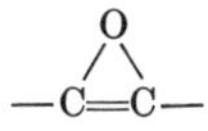

(**12**)

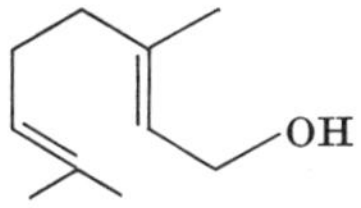

OH

(**13**)

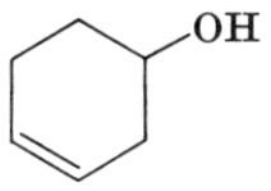

OH

(**14**)

O

OH

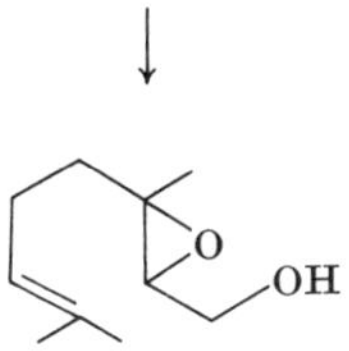

(**15**)

accompanied by some unsaturated and ring-contracted products in proportions depending markedly on ring-size of the starting material.[100] Epoxidation of C=N has been shown to involve a two step process, in which initial addition of peroxyacid is followed by elimination of carboxylic acid; a competing one-step formation of nitrones may also occur.[101]

Miscellaneous Electrophilic Additions

The kinetics of addition of NOCl to nuclear substituted styrenes have been reported.[102] The stereochemistry of addition of ions of the type R^+ or RCO^+ (SbF_6^-, BF_4^-) to olefins has been studied,[103] as has the addition of diarylmethyl halides to diarylethylenes.[104]

The role of cationic intermediates in acid-catalysed cyclizations which have some relations to biogenetic processes has been reported.[105,106] Other work includes the addition of IN_3,[107] of trinitrobenzenesulphonyl hypohalites,[108] and of *p*-chlorobenzenesulphenyl chloride to olefins,[109] of 2,4-dinitrobenzenesulphenyl chloride to phenylallene[110] [affording 95% of PhCH=C(SAr)CH_2Cl], of SO_2 to enol ethers,[111] of $C_6N_5SO_2Cl$ and CCl_4 to substituted styrenes[112] (some evidence for chain reactions being proposed in the latter case; both reactions are catalysed by copper halide), of ClSCl to nitriles,[113] of $(ClCH_2CO)_2O$ to methyl isocyanide.[114] Chlorosulphonyl isocyanate has been used to probe the addition pathways of alkenylidenecyclopropanes;[115] the sensitivity of the two processes (Scheme 5) to ring substituents is marked. The same addend has been used

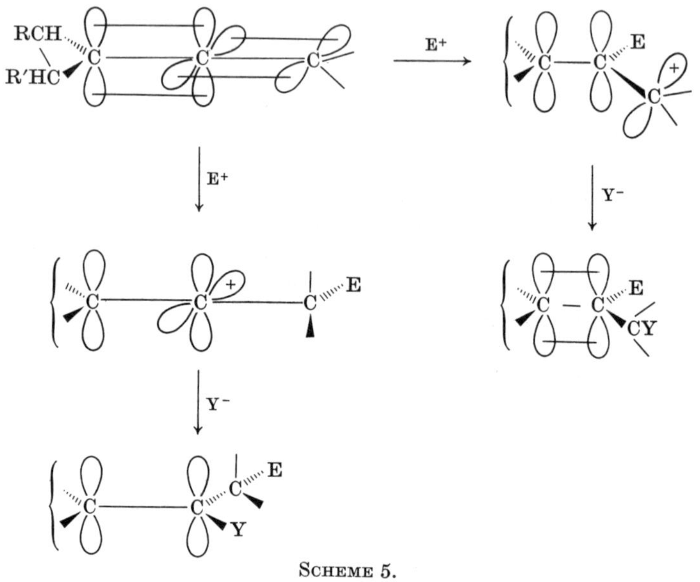

SCHEME 5.

with bicyclo[6,1,0]nonatriene, and a bishomotropylium ion invoked;[116] a similar pathway is proposed for addition of tetracyanoethylene.[117]

The interaction of olefins and acetylenes with organometallic compounds or metallic salts has attracted much interest. The intermediacy of an alkyne–π-complex is suggested for the addition of Et_3Al to PhC≡CH.[118] Aluminium alkyls (AlR_3) have been reacted with PhCN[119] ($R = Bu^i$), Ph_2CO[120] (R = Me) and non-conjugated dienes[121] ($R = Bu^i$). Pd(II)-catalysed exchange, isomerizations and addition processes have been reviewed.[122,123] The arylation of olefins with PhI, catalysed by Pd complexes, is interpreted as an addition–elimination sequence of a metal complex in a low valency state;[124] certain ligands, e.g. Ph_3P, inhibit the initial co-ordination of the olefin. Chloroarylation of conjugated dienes has been interpreted as an intra-complex redox reaction of a Cu(I)–diazonium ion–olefin complex.[125] Silver ion catalyses the addition of methanol to bicyclobutanes;[126] and Ni(0) catalyses the addition of phenol to butadiene.[127] The stability of copper(I) triflate–olefin complexes has been established[128] and their role in the addition of $RCHN_2$ to olefins has been proposed. Acetylenes, e.g. $CH_3C{\equiv}CCH_3$ have been shown to react with chloro(phenyl)palladium.[129]

The stereochemistry of non-catalytic addition of deuterium to cyclopentadiene has been studied.[130] The hydrogenation of cyclopentenes,[131] alkyl-substituted methylenecyclohexanones,[132] alkylcyclohexanones,[133] *cis*-cinnamic acid,[134] and simple alkynes[135] has been reported. Hydrogenation of octynes is said to yield *trans*-addition from oct-3-yne but predominantly *cis*-addition from the 2- and 4-alkynes;[136] catalyst effects are noted.[137,138] The system $Co_2(CO)_8$–PBu_3–H_2 is an effective reducing system,[139] as is $RhCl(PPh_3)_3/H_2$[140] and a cobalt hydrocarbonyl.[141] Tetra-alkylammonium salts of the ions $SnCl_3^-$ and $GeCl_3^-$ are solvents for olefins and platinum-type catalysts and reduction, isomerization, etc., is brought about.[142] Selective reduction of dienes and trienes by Co and Ru complexes is possible.[143,144] Hydroformylation[145–147] and hydrocyanation[148] use zero- or low-valent metal complexes as catalyst.

The addition of silicon compounds, in some cases catalysed by metal complexes, has been studied for $PhSiHCl_2$ (isomeric butenes[149]), siloxanes (allyl halides[150]) Cl_3SiH (styrene[151]) Et_3SiH (alkynes[152]), $Ph_2Si{=}CH_2$ (RCHO or R_2CO[153]), Me_3SiOH (dienes[154]) and $R_3Si(CH_2)_nOH$ (PhNCO and $CH_2{=}CO$[155]).

Other reactions studied include addition of $COCl_2/HCl$ to nitriles,[156] addition to enol ethers[157] and the catalyst-induced polymerization of acetylenes.[158a]

An elegant synthesis of prismane involves, as the first step, addition of 4-phenyltriazolinedione to benzovalene.[158b] The mechanism of adduct formation was deduced by a deuterium-labelling experiment to be that indicated in Scheme 6.

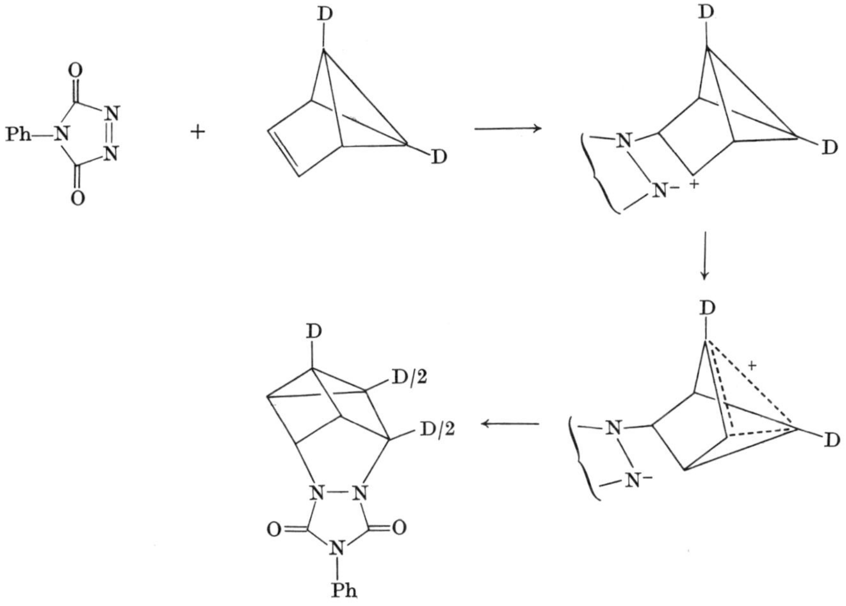

SCHEME 6.

Nucleophilic Addition

Catalysis of the Michael reaction by trialkylphosphines[159] and the use of metal complexes in the aldol condensation[160] have been studied. The formation of lithium derivatives from $R^1SCH{=}CHR^2 + Bu^s\,Li$ has been observed and adducts of these with RCHO and with epoxides have been characterized.[161] RLi adds to allylic alcohols; with cyclopentenol an alkylcyclopentene is formed in which the double bond is located between C-1 and C-2 of the original alcohol.[162] Addition of $LiRC(COOEt)_2$ (R = H or Me) and LiAcCHCOOEt to co-ordinated olefins [h^5-$C_5H_5Fe(CO)_2$(olefins)]$^+$ has been assigned *trans*-stereochemistry.[163] Addition of CN^- to aldehydes[164] and Schiff's bases[165] has been reported. In the latter case the production of optically active amino-acids from asymmetric Schiff's bases in variable yield (9–58%) and optical purity (22–58%) was accompanied by some fractionation in the working up. The dimerization of CH_2=CHCN is catalysed by metal ions and bases.[166]

Stereospecific addition of lithium dialkylamides to butadiene gives 1-dialkylamino-*cis*-but-2-ene;[167] the details of addition are discussed and a role for species R_2N^-:R_2NH (ratio 1:2) is proposed. Similar work on the addition of amines ($LiNH_2$-catalysed) to olefins,[168] to co-ordinated olefins[169] (Pt or Pd complexes) and to α,β-unsaturated ketones[170] has been noted. Miscellaneous reactions include additions of aziridines[171] with ring expansion, ring closure of α,β-unsaturated phenylhydrazones involving stereoselective enamine–imine tantomerism,[172] the kinetics of addition of piperidine to α,β-unsaturated acrylonitriles,[173] addition of aniline to tetracyanoethylene[174] and 2-(dicyanomethylene)indane-1,3-dione (**16**),[175] cyclization of an *N*-chloroamine (**17**),[176] reaction of carboxylic acids with aldimines and nitriles,[177] addition of PhN^-–$^+AsPh_3$ to

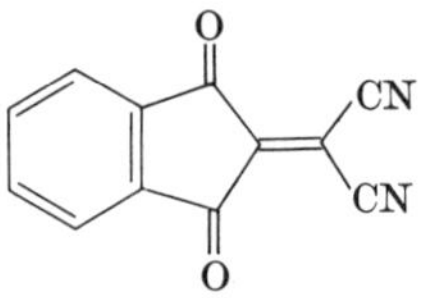

(**16**) 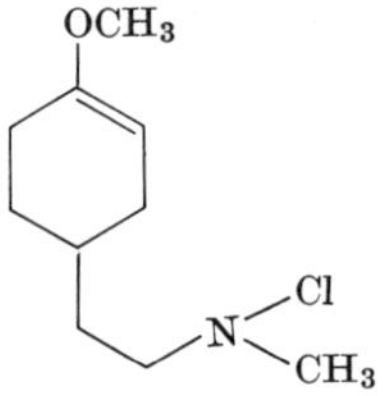 (**17**)

R_2CO and Ph_3C—C≡N^+–O^- to ROOCC≡CCOOR,[178] nucleophilic attack of *m*- and *p*-$XC_6H_4NH_2$ on PhNO[179] (which gives azobenzenes) reaction of enamines with PhCH=$CHNO_2$[180] and 1,4-addition of iminocarbamates to cyclohexa-1,3-diene.[181]

Additions of nitrogen-containing reactants to triple-bonded system include those of primary and secondary amines to PhC≡COX[182] (X = OR or NH_2), isocyanides to activated acetylenic systems,[183] Michael additions to Me_2NC≡CCOR[184] (R = H, Me or OMe), phenylhydrazones[185] and R_3GeNMe_2[186] to ROOCC≡CCOOR and of triazolyl anions to miscellaneous substrates.[187]

Reactions of sulphur nucleophiles include those of mercapto-ester anions with activated cyclohexenes[188] (*trans*-diaxial addition), thiophenol to 4-*tert*-butyl-1-cyanocyclohexene[189] (other additions being also discussed and contrasted), and of substituted (alkyl) benzenethiolates with *N*-ethylmaleimide,[190] base-catalysed (optically active polymeric amines) asymmetric addition of $Me(CH_2)_{11}SH$ to α,β-unsaturated C=O and C≡N-systems,[191] and addition of RSH to R^1C≡CR^2 (R^1 = Ac, R_2 = H, R = Me;[192] R^1 = R^2 = MeOOC, R = XC_6H_4[193]).

Base-catalysed addition of phenols to aryl vinyl sulphones is contrasted with the more rapid reaction of $ArS.^-$[194] Hydroxide ion catalyses the hydration of H_2C=CRCHO;[195] addition of methoxide ion or methanol has been discussed with regard to steric effects in RC≡CH,[196] and addition to activated triple bonds, e.g. $Me_2S^+CH_2C$≡CH.[197] HC≡COOMe with MOAc (M = Li or Na) affords AcOCH=CHCOOMe and (**18**).[198] Addition of water or alcohols to CH_2=CHCN and CH_3CH=CHCN (with Pt–phosphine complexes as catalysts),[199] 2-cyanopyridine (with transition-metal catalysis, not observed for the 3- and 4-cyano-compounds),[200] C_6H_5CN,[201] and aryl isocyanates[202] has also been studied.

Kinetic studies of the addition of MeMgBr[203] and Me_2Mg[204] to PhCN have given evidence for two pathways for the MeMgBr reaction, corresponding to the intervention of Me_2Mg and MeMgBr. Deuteriated cyclopentylmagnesium bromide (**19**) gives k_H/k_D ca. 2 for addition to a substituted acetophenone.[205] The products of reaction of *n*-BuMgBr and CH_3NO_2 arise by both N- and C-addition.[206] Conjugate addition to PhCH=CHCHO and related systems is observed with R_2Cd; organomagnesium reagents favour 1,2 addition.[207] EtZnCl with H_2C=CClCOOMe gives, via intermediate (**20**), $EtCH_2$CHClCOOMe and the cyclized product (**21**).[208] Finally, electronic and steric control of addition of $PhCH_2MgCl$ to $LM(CO)_x$ (L = CO or phosphine; M = Fe, W, Mo or Cr) has been discussed.[209]

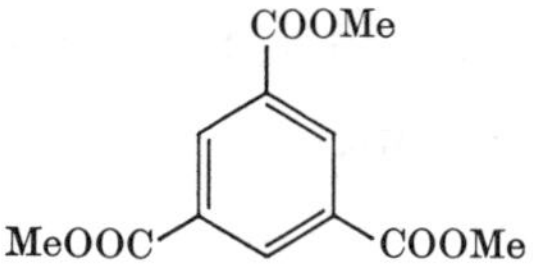

(**18**)

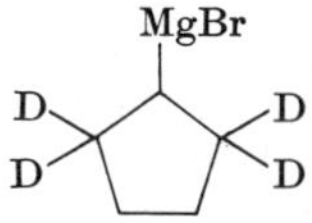

(**19**)

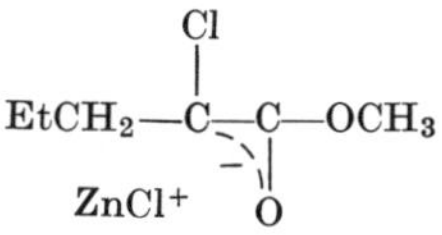

(**20**)

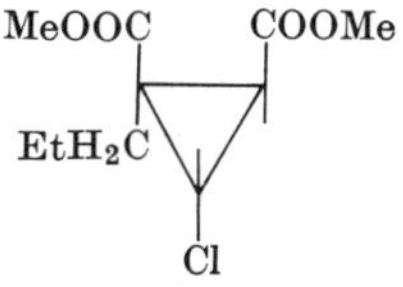

(**21**)

References

[1] N. D. Epiotis, *J. Am. Chem. Soc.*, **95**, 1191 (1973).
[2] R. R. Schmidt, *Angew. Chem. Internat. Ed.*, **12**, 212 (1973).
[3] D. A. Dixon and W. N. Lipscomb, *J. Am. Chem. Soc.*, **95**, 2853 (1973).
[4] D. T. Clark and D. B. Adams, *Tetrahedron*, **29**, 1887 (1973).
[5] H. C. Brown, J. H. Kamakami and K. T. Line, *J. Am. Chem. Soc.*, **95**, 2209 (1973).
[6] M. Charton and B. J. Charton, *J. Org. Chem.*, **38**, 1631 (1973).
[7] L. S. Bognolavskaya, *Usp. Khim.*, **41**, 1591 (1972); *Chem. Abs.*, **78**, 3353 (1973).
[8] V. S. Aksenov, *Sovrem. Probl. Org. Khim.*, **1971**, 3; *Chem. Abs.*, **78**, 3359 (1973).
[9] J. F. Harrod, C. A. Smith and K. A. Than, *J. Am. Chem. Soc.*, **94**, 8321 (1972).
[10] R. Huisgen and J. Gasteiger, *Angew. Chem. Internat. Ed.*, **11**, 1104 (1972).
[11] G. Dana and C. Roos, *Bull. Soc. Chim. France*, **1973**, 371.
[12] T. I. Usmanov, O. A. Subbotin, A. D. Litmanovich and N. M. Sergeev, *Zh. Org. Khim.*, **9**, 428 (1973); *Chem. Abs.*, **78**, 110277 (1973).
[13] K. Shinoda, *Nippon Kagaku Kaishi*, **1973**, 983; *Chem. Abs.*, **79**, 41621 (1973).
[14] P. B. D. de la Mare, M. A. Wilson and M. J. Rosser, *J.C.S. Perkin II*, **1973**, 1480.
[15] A. S. Kostyushin, S. M. Brailovskii, O. N. Temkin, R. M. Flid and G. A. Ermakova, *Uch. Zap. Mosk. Inst. Tonkoi Khim. Tekhnol.*, **1**, 73 (1971); *Chem. Abs.*, **78**, 3443 (1973).
[16] A. P. Shvedchikov, V. I. Gol'danskii, E. V. Lumer, G. B. Sergeev, B. G. Dzantiev, A. M. Kaplan, S. V. Pavlova, V. V. Smirnov and M. A. Vetrova, *Zh. Obshch. Khim.*, **42**, 2753 (1972); *Chem. Abs.*, **78**, 96780 (1973).
[17] V. I. Matrozov, I. Kh. Markish, Z. P. Kazamatkina and G. L. Groshev, *Tr. Khim. Khim. Tekhnol.*, **1972**, 156; *Chem. Abs.*, **78**, 147016 (1973).
[18] G. Henblein and M. Helbig, *Tetrahedron*, **29**, 3247 (1973).
[19] G. Mouvier, D. Grossjean and J.-É. Dubois, *Bull. Soc. Chim. France*, **1973**, 1729.

[20] D. Grossjean, G. Mouvier and J.-É. Dubois, *Bull. Soc. Chim. France*, **1973**, 1735.
[21] G. J. Abruscato and T. T. Tidwell, *J. Org. Chem.*, **37**, 4151 (1972).
[22] S. R. Hooley and D. L. H. Williams, *J.C.S., Perkin II*, **1973**, 1053.
[23] Yu. A. Serguchev, G. B. Sergeev, V. V. Smirnov and E. A. Shilov, *Ukr. Khim. Zh.* (*Russian Ed.*), **38**, 1156 (1972); *Chem. Abs.*, **78**, 83538 (1973).
[24] W. L. Collier and R. S. Macomber, *J. Org. Chem.*, **38**, 1367 (1973).
[25] K. Yates, G. H. Schmid, T. W. Regulski, D. G. Garratt, H. Leung and R. McDonald, *J. Am. Chem. Soc.*, **95**, 160 (1973).
[26] K. Yates, R. S. McDonald and S. A. Shapiro, *J. Org. Chem.*, **38**, 2460 (1973).
[27] K. Yates and R. S. McDonald, *J. Org. Chem.*, **38**, 2465 (1973).
[28] S. P. McManus and R. A. Hames, *Tetrahedron Letters*, **1973**, 4549.
[29] R. J. Abraham and J. R. Monasterios, *J.C.S. Perkin I*, **1973**, 1446.
[30] I. V. Bodrikov, N. G. Bronnikova and I. S. Okrokova, *Zh. Org. Khim.*, **9**, 953 (1973); *Chem. Abs.*, **79**, 52539 (1973).
[31] K. E. Koenig and W. P. Weber, *Tetrahedron Letters*, **1973**, 2533.
[32] J.-É. Dubois and M. F. Ruasse, *J. Org. Chem.*, **38**, 493 (1973).
[33] M. P. Hartshorn, M. C. A. Opie and J. Vaughan, *Austral. J. Chem.*, **26**, 917 (1973).
[34] E. Hajoudis, *Israel J. Chem.*, **11**, 63 (1973).
[35] P. B. D. de la Mare and M. A. Wilson, *J.C.S. Perkin II*, **1973**, 653.
[36] J.-É. Dubois, J. Guillo and X. Q. Huynh, *J. Chim. Phys., Physiocochim. Biol.*, **69**, 1489 (1972); *Chem. Abs.*, **78**, 57359 (1973).
[37] J.-É. Dubois and P. Fresnet, *Tetrahedron*, **29**, 3407 (1973).
[38] P. L. Barili, G. Bellucci, F. Marioni, I. Morelli and V. Scartoni, *J. Org. Chem.*, **37**, 4353 (1972).
[39] P. L. Barili, G. Bellucci, F. Marioni, I. Morelli and V. Scartoni, *J. Org. Chem.*, **38**, 3472 (1973).
[40] M. M. Movsumzade, G. B. Sergeev, A. L. Shabanov and V. V. Smirnov, *Dok. Akad. Nauk. SSSR*, **1972**, 206; *Chem. Abs.*, **78**, 3464 (1973).
[41] L. D. Hall and D. L. Jones, *Can. J. Chem.*, **51**, 2902 (1973).
[42] F. Akiyama, T. Horie and M. Matsuda, *Bull. Chem. Soc. Japan*, **46**, 1888 (1973).
[43] P. B. D. de la Mare and B. N. B. Hannan, *J.C.S. Perkin II*, **1973**, 1086.
[44] P. B. D. de la Mare and B. N. B. Hannan, *J.C.S. Perkin II*, **1973**, 1586.
[45] L. A. Paquette, H. G. Birnberg, J. Clardy and B. Parkinson, *Chem. Comm.*, **1973**, 129.
[46] Sp. Shanmuganathan and V. Subramanian, *Proc. Indian Acad. Sci., Sect. A*, **76**, 72 (1972); *Chem. Abs.*, **78**, 70972 (1973).
[47] Sp. Shanmuganathan and V. Subramanian, *Proc. Indian Acad. Sci., Sect. A*, **76**, 79 (1972); *Chem. Abs.*, **78**, 70970 (1973).
[48] J. B. Lambert and B. A. Iwanetz, *J. Org. Chem.*, **37**, 4082 (1972).
[49] J. C. Day, K. J. Shea and P. S. Skell, *J. Am. Chem. Soc.*, **95**, 5089 (1973).
[50] R. G. Guy and I. Pearson, *J.C.S. Perkin I*, **1973**, 281.
[51] R. G. Guy and I. Pearson, *J.C.S. Perkin II*, **1973**, 1359.
[52] V. Burger and A. Delay, *Helv. Chim. Acta*, **56**, 1345 (1973).
[53] K. Izawa, T. Okuyama, T. Sakagami and T. Fueno, *J. Am. Chem. Soc.*, **95**, 6752 (1973).
[54] T. Okuyama, K. Izawa and T. Fueno, *J. Am. Chem. Soc.*, **95**, 6749 (1973).
[55] K. E. Koenig and W. P. Weber, *J. Am. Chem. Soc.*, **95**, 3418 (1973).
[56] C. Brown, A. R. Forrester and R. H. Thomson, *Tetrahedron*, **29**, 3059 (1973).
[57] M. A. Bennett, G. B. Robertson, P. O. Whimp and T. Yoshida, *J. Am. Chem. Soc.*, **95**, 3028 (1973).
[58] A. Niederhauser and M. Neuenschwander, *Helv. Chim. Acta*, **56**, 1331 (1973).
[59] G. Rajalo, V. Demant'eva, R. Taht, I. B. Kudryavtsev and K. Laats, *Eesti NSV Tead. Akad. Toim., Keem., Geol*, **21**, 310 (1972); *Chem. Abs.*, **78**, 71017 (1973).
[60] I. Sh. Yakobishvili, S. D. Mekntiev and O. A. Narimanbekov, *Dokl. Akad. Nauk Azerb. SSR*, **28**, 34 (1972); *Chem. Abs.*, **78**, 110126 (1973).
[61] A. S. Tarkhanyan, L. A. Gasparyan, T. K. Manukyan, S. S. Kazayan and M. G. Ierysalimskaya, *Khim. Atsetilena, Tr. Vses. Konf., 3rd*, **1968**, 261; *Chem. Abs.*, **79**, 4551 (1973).
[62] K. Griesbaum and M. El-Abed, *Chem. Ber.*, **106**, 2001 (1973).
[63] M. A. Cooper, C. M. Holden, P. Loftus and D. Whittaker, *J.C.S. Perkin II*, **1973**, 665.
[64] J. Guenzet, N. N. Chieu and M. Camps, *Compt. Rend., Ser. C.*, **277**, 391 (1973).
[65] B. A. Trofimov, O. N. Vylegzhanin and M. G. Voronkov, *Dokl. Akad. Nauk SSSR*, **207**, 1137 (1972); *Chem. Abs.*, **78**, 123543 (1973).
[66] M. F. Sorokin and N. E. Kuz'mitskaya, *Tr. Mosk. Khim. Tekhnol. Inst.*, **1972**, 32 (1973); *Chem. Abs.*, **78**, 135215 (1973).
[67] O. N. Vylegzhanin, B. A. Trofimov and N. A. Nedolya, *Dokl. Vses. Konf. Khim, Atsetilena, 4th*, **2**, 426 (1973); *Chem. Abs.*, **79**, 77652 (1973).

[68] G. M. Kramer and R. J. Panciroo, *J. Org. Chem.*, **38**, 349 (1973).
[69] T. C. Morrill and B. E. Greenwald, *J. Org. Chem.*, **38**, 616 (1973).
[70] T. J. Mason and R. O. C. Norman, *J.C.S. Perkin II*, **1973**, 1840.
[71] J. G. Traynham and H. H. Hsieh, *J. Org. Chem.*, **38**, 868 (1973).
[72] I. A. Esikova, S. S. Yufit, V. N. Sapunov, L. M. Suslova and V. F. Kucherov, *Dokl. Vses. Konf. Khim. Atsetilena, 4th*, **2**, 228 (1972); *Chem. Abs.*, **79**, 31177 (1973).
[73] N. G. Mekhryakova, O. L. Kaliya, G. M. Shulyakovskii, L. L. Kirsanova, A. N. Nyrkova, A. A. Khorkin, L. N. Reshetova, O. N. Temkin and R. M. Flid, *Dokl. Vses. Konf. Khim. Atsetilena*, **2**, 257 (1972); *Chem. Abs.*, **79**, 77653 (1973).
[74] T. Okuyama, T. Sakagami and T. Fueno, *Tetrahedron*, **29**, 1503 (1973).
[75] L. R. Fedor, N. C. De and S. K. Gurwara, *J. Am. Chem. Soc.*, **95**, 2905 (1973).
[76] W. Hauswirth, B. S. Hahn and S. Y. Wang, *Biochem. Biophys. Res. Comm.*, **48**, 1614 (1972); *Chem. Abs.*, **78**, 15190 (1973).
[77] T. Adams and R. M. Moriarty, *J. Am. Chem. Soc.*, **95**, 4071 (1973).
[78] K. E. Hine and F. E. Child, *J. Am. Chem. Soc.*, **95**, 3289 (1973).
[79] H. C. Brown and N. Ravindran, *J. Org. Chem.*, **38**, 182 (1973).
[80] R. H. Fish, *J. Org. Chem.*, **38**, 158 (1973).
[81] W. R. Moore, H. W. Anderson and S. D. Clark, *J. Am. Chem. Soc.*, **95**, 835 (1973).
[82] E. Negishi and H. C. Brown, *J. Am. Chem. Soc.*, **95**, 6757 (1973).
[83] G. W. Kabalka, *Tetrahedron*, **29**, 1159 (1973).
[84] J. Hooz and J. N. Bridson, *J. Am. Chem. Soc.*, **95**, 602 (1973).
[85] J. R. Blackborow, *J.C.S. Perkin II*, **1973**, 1989.
[86] R. D. Bach and R. F. Richter, *Tetrahedron Letters*, **1973**, 4099.
[87] M. C. Cabaleiro, A. D. Ayala and M. D. Johnson, *J.C.S. Perkin II*, **1973**, 1207.
[88] A. P. Kreshkov, L. N. Balyatinskaya and S. M. Chesnokova, *Zh. Obshch. Khim.*, **43**, 166 (1973). *Chem. Abs.*, **78**, 123521 (1973).
[89] T. Sugita, K. Honda, O. Itoh and K. Ichikawa, *Bull. Inst. Chem. Res., Kyoto Univ.*, **50**, 216 (1972); *Chem. Abs.*, **78**, 28789 (1973).
[90] G. Casalbore and S. Valcher, *J. Electroanal. Chem. Interfacial Electrochem.*, **42**, 45 (1973); *Chem. Abs.*, **78**, 147159 (1973).
[91] C. H. De Puy and R. H. McGirk, *J. Am. Chem. Soc.*, **95**, 2366 (1973).
[92] M. M. Khalil and W. Pritzkow, *J. Prakt. Chem.*, **315**, 58 (1973).
[93] Y. Ogata, Y. Sawaki and H. Inoue, *J. Org. Chem.*, **38**, 1044 (1973).
[94] K. M. Ibne-Rasa, R. H. Pater, J. Ciabattoni and J. O. Edwards, *J. Am. Chem. Soc.*, **95**, 7894 (1973).
[95] J. K. Crandall, W. H. Marchleder and S. A. Sojka, *J. Org. Chem.*, **38**, 1149 (1973).
[96] T. N. Baker, G. J. Mains, M. N. Sheng and J. G. Zajarcek, *J. Org. Chem.*, **38**, 1145 (1973).
[97] I. E. Pokrovskaya, I. K. Alferova and Kh. E. Khcheyan, *Neftekhimiya*, **13**, 247 (1973); *Chem. Abs.*, **79**, 41667 (1973).
[98] V. G. Dryuk, M. S. Malinovskii, S. P. Shamrovskaya, S. A. Vasil'chenko and S. A. Maslov, *Zh. Org. Khim.*, **9**, 1228 (1973); *Chem. Abs.*, **79**, 65477 (1973).
[99] K. B. Sharpless and R. C. Michaelson, *J. Am. Chem. Soc.*, **95**, 6136 (1973).
[100] P. W. Concannon and J. Ciabattoni, *J. Am. Chem. Soc.*, **95**, 3284 (1973).
[101] Y. Ogata and Y. Sawaki, *J. Am. Chem. Soc.*, **95**, 4687, 4692 (1973).
[102] I. Hahnemann and W. Pritzkow, *J. Prakt. Chem.*, **315**, 345 (1973).
[103] M. T. Mustafaeva, V. A. Kuzovkin, V. A. Smit, A. V. Semenovskii and V. F. Kucherov, *Dokl. Akad. Nauk SSSR*, **208**, 1103 (1973). *Chem. Abs.*, **78**, 147250 (1973).
[104] A. B. Sakla, W. Tadros, S. I. Aziz and A. A. A. Helmy, *J. Prakt. Chem.*, **314**, 793 (1972).
[105] G. Buchi and W. Pickenhagen, *J. Org. Chem.*, **38**, 894 (1973).
[106] P. A. Bartlett and W. S. Johnson, *J. Am. Chem. Soc.*, **95**, 7501, 7502 (1973).
[107] T. Sasaki, K. Kanematsu and Y. Yukimoto, *Chem. Letters* (*Tokyo*), **1972**, 1005.
[108] P. Bassi and V. Tonellato, *J.C.S. Perkin I*, **1973**, 669.
[109] G. H. Schmid and D. G. Garratt, *Can. J. Chem.*, **51**, 2463 (1973).
[110] K. Izawa, T. Okuyama and T. Fueno, *J. Am. Chem. Soc.*, **95**, 4090 (1973).
[111] M. M. Rogić and J. Vitrone, *J. Am. Chem. Soc.*, **94**, 8642 (1972).
[112] A. Orochov, M. Asscher and D. Vofsi, *J.C.S. Perkin II*, **1973**, 1000.
[113] H. Kristinsson, *Tetrahedron Letters*, **1973**, 4489.
[114] P. Krivinka and J. Houzl, *Coll. Czech. Chem. Comm.*, **37**, 4035 (1972).
[115] D. J. Basto, A. F. T. Chen, G. Ciurdaru and L. A. Paquette, *J. Org. Chem.*, **38**, 1015 (1973).
[116] L. A. Paquette, M. J. Broadhurst, C. Lee and J. Clardy, *J. Am. Chem. Soc.*, **95**, 4647 (1973).
[117] L. A. Paquette, M. J. Broadhurst, L. K. Read and J. Clardy, *J. Am. Chem. Soc.*, **95**, 4639 (1973).
[118] P. E. M. Allen and R. M. Lough, *J.C.S. Faraday I*, **1973**, 849.

[119] G. Giacomelli and L. Lardicci, *J.C.S. Perkin II*, **1973**, 1129.
[120] H. M. Neumann, J. Laemmle and E. C. Ashby, *J. Am. Chem. Soc.*, **95**, 2597 (1973).
[121] A. A. Antonov, N. S. Nametkin and V. I. Smetanyuk, *Neftekhimiya*, **12**, 875 (1972); *Chem. Abs.*, **78**, 70997 (1973).
[122] P. M. Henry, *Accounts Chem. Research*, **6**, 16 (1973).
[123] J. Tsuji, *Accounts Chem. Research*, **6**, 8 (1973).
[124] K. Mori, T. Mizoroki and A. Ozaki, *Bull. Chem. Soc. Japan*, **46**, 1505 (1973).
[125] N. I. Ganushchak, V. D. Golik and I. V. Migaichuk, *Zh. Org. Khim.*, **8**, 2356 (1972); *Chem. Abs.*, **78**, 123556 (1973).
[126] L. A. Paquette, S. E. Wilson, G. Zou and J. A. Schwartz, *J. Am. Chem. Soc.*, **94**, 9222 (1972).
[127] F. J. Weigert and W. C. Drinkard, *J. Org. Chem.*, **38**, 335 (1973).
[128] R. G. Salomon and J. K. Kochi, *J. Am. Chem. Soc.*, **95**, 1889 (1973).
[129] T. Hosokawa, C. Calvo, H. B. Lee and P. M. Maitlis, *J. Am. Chem. Soc.*, **95**, 4914 (1973).
[130] F. A. L. Anet and F. Leyendecker, *J. Am. Chem. Soc.*, **95**, 156 (1973).
[131] S. Mitsui, Y. Senda, H. Suzuki, S. Sekiguchi and Y. Kumagai, *Tetrahedron*, **29**, 3341 (1973).
[132] S. Mitsui, K. Gohke, H. Saito, A. Nanbu and Y. Senda, *Tetrahedron*, **29**, 1523 (1973).
[133] S. Mitsui, H. Saito, Y. Yamashita, M. Kaminaga and Y. Senda, *Tetrahedron*, **29**, 1531 (1973).
[134] J. Bastero, C. J. Groenenboom, H. van Bekkum and L. J. van Reijen, *Rec. Trav. chim.*, **92**, 219 (1973).
[135] A. M. Sokol'skaya, E. N. Bosyakova, S. A. Ryabinina and D. V. Sokol'skii, *Zh. Fiz. Khim.*, **46**, 2939 (1972); *Chem. Abs.*, **78**, 57586 (1973).
[136] V. Ružička, V. Zapletal, A. Koppová and J. Soukup, *Coll. Czech. Chem. Comm.*, **38**, 2466 (1973).
[137] A. Koppová, V. Zapletal, V. Ružička and J. Soukup, *Coll. Czech. Chem. Comm.*, **38**, 2472 (1973).
[138] V. Zapletal, A. Koppová, V. Ružička and J. Soukup, *Coll. Czech. Chem. Comm.*, **38**, 2727 (1973).
[139] G. Pregaglia, A. Andreetta, G. Gregorio, G. Ferrari, G. Montrasi and R. Ugo, *Chim. Ind. (Milan)*, **55**, 203 (1973); *Chem. Abs.*, **79**, 31186 (1973).
[140] S. Siegel and D. W. Ohrt, *Tetrahedron Letters*, **1972**, 5155.
[141] P. D. Taylor and M. Orchin, *J. Org. Chem.*, **37**, 3913 (1972).
[142] G. W. Parshall, *J. Am. Chem. Soc.*, **94**, 8716 (1972).
[143] L. I. Gvinter, L. Kh. Freidlin, N. V. Borunova and G. Ya Starodubskaya, *Izv. Akad. Nauk SSSR, Ser. Khim.*, **1972**, 2381; *Chem. Abs.*, **78**, 57419 (1973).
[144] D. R. Fahey, *J. Org. Chem.*, **38**, 80 (1973).
[145] G. Csontos, B. Heil, L. Marko and P. Chini, *Hung. J. Ind. Chem.*, **1**, 53 (1973); *Chem. Abs.*, **79**, 41608 (1973).
[146] V. Yu. Gankin and V. A. Dvinin, *Kinet. Katal.*, **14**, 191 (1973); *Chem. Abs.*, **78**, 135274 (1973).
[147] B. Heil, G. Csontos and L. Marko, *Wiss. Z. Tech. Hochsch. Chem. "Carl Schorlemmer" Leuna-Merseburg*, **72**, 15 (1973); *Chem. Abs.*, **79**, 4692 (1973).
[148] A. A. Khorkin, A. Ermakova, D. N. Temkin, R. M. Flid and V. S. Kozina, *Kinet Katal.*, **14**, 336 (1973); *Chem. Abs.*, **79**, 17706 (1973).
[149] V. N. Tarasenko, G. V. Odabashyan and A. I. Lakhtikov, *Tr. Mosk. Khim.-Tekhnol. Inst.* **1972**, **138**; *Chem. Abs.*, **78**, 135228 (1973).
[150] T. A. Barry, F. A. Davis and P. J. Chiesa, *J. Org. Chem.*, **38**, 838 (1973).
[151] P. Svoboda, P. Sedlmayer and J. Hetflejš, *Coll. Czech. Chem. Comm.*, **38**, 1783 (1973).
[152] V. B. Pukhnarevich, L. I. Kopylova and T. B. Trofimov, *Dokl. Vses. Konf. Khim. Atsetilena, 4th*, **2**, 103 (1972); *Chem. Abs.*, **79**, 77649 (1973); V. B. Pukhnarevich, L. I. Kopylova, B. A. Trofimov and M. G. Voronkov, *Zh. Obshch. Khim.*, **43**, 593 (1973); *Chem. Abs.*, **79**, 31175 (1973); V. B. Pukhnarevich, S. P. Sushchinskaya, V. A. Pestunovich and M. G. Voronkov, *Zh. Obshch. Khim.*, **43**, 1283 (1973); *Chem. Abs.*, **79**, 77689 (1973).
[153] D. N. Roark and L. H. Sommer, *Chem. Comm.*, **1973**, 167.
[154] M. Čapka, P. Svoboda and J. Hetflejš, *Coll. Czech. Chem. Comm.*, **38**, 1242 (1973).
[155] J. Pola, V. Bažant and V. Choalovský, *Coll. Czech. Chem. Comm.*, **37**, 3885 (1972).
[156] M. Ohoka, S. Yanagida, K. Sugahara, T. Fujita, M. Okahara and S. Komori, *Bull. Chem. Soc. Japan*, **46**, 1275 (1973).
[157] B. A. Trofimov, G. A. Kalabin and O. N. Vylegjanin, *Organic Reactivity (Tartu)*, **8**, 956 (1971).
[158a] M. Neuenschivander, G. Bart and A. Niederhauser, *Chimia*, **1973**, 73.
[158b] T. J. Katz and N. Acton, *J. Am. Chem. Soc.*, **95**, 2738 (1973).
[159] D. A. White and M. M. Baizer, *Tetrahedron Letters*, **1973**, 3597.
[160] H. O. House, D. S. Cruvirine, A. Y. Teranishi and H. D. Olmstead, *J. Am. Chem. Soc.*, **95**, 3310 (1973).
[161] K. Oshima, K. Shimoji, H. Takahashi, H. Yamamoto and H. Nozaki, *J. Am. Chem. Soc.*, **95**, 2674 (1973).

162 J. K. Crandall and A. C. Clark, *J. Org. Chem.*, **37**, 4236 (1972).
163 A. Rosan, M. Rosenblum and J. Tancrede, *J. Am. Chem. Soc.*, **95**, 3062 (1973).
164 H. Stetter and M. Schreckenberg, *Angew. Chem. Internat. Ed.*, **12**, 81 (1973).
165 K. Harada and T. Okawara, *J. Org. Chem.*, **38**, 707 (1973).
166 Y. Watanabe and M. Takeda, *Bull. Chem. Soc. Japan*, **46**, 883 (1973).
167 T. Narita, N. Imai and T. Tsuruta, *Bull. Chem. Soc. Japan*, **46**, 1242 (1973).
168 R. J. Schlott, J. C. Falk and K. W. Narducy, *J. Org. Chem.*, **37**, 4243 (1972).
169 D. Hollings, M. Green and D. V. Claridge, *J. Organometal. Chem.*, **54**, 399 (1973); A. De Renzi, G. Paiaro, A. Panunzi and V. Ramano, *Chim. Ind.* (*Milan*), **55**, 248 (1973); *Chem. Abs.*, **79**, 31185 (1973); D. R. Coulson, *J. Org. Chem.*, **38**, 1483 (1973).
170 A. Le Berre and A. Delacroix, *Bull. Soc. Chim. France*, **1973**, 640, 647, 2404.
171 R. A. Wohl and D. F. Headley, *J. Org. Chem.*, **37**, 4401 (1972).
172 H. Ferres, M. S. Hamdam, and W. R. Jackson, *J.C.S. Perkin II*, **1973**, 936.
173 G. Dienys, L. Arkhipova, R. Pupenyte, M. Davidevicíute and L. Kunskaite, *Organic Reactivity* (*Tartu*), **9**, 349 (1972).
174 P. G. Farrell and P.-N. Ngô, *J.C.S. Perkin II*, **1973**, 974.
175 Z. Rapporport and D. Ladkani, *J.C.S. Perkin II*, **1973**, 1045.
176 R. Tadayoni, A. Heumann, R. Furstoss and B. Waegell, *Tetrahedron Letters*, **1973**, 2879.
177 D. Marquarding, *Angew. Chem. Internat. Ed.*, **12**, 79 (1973).
178 P. Frøyen, *Acta Chem. Scand.*, **27**, 141 (1973).
179 R. A. Yunes, A. J. Terenzani, O. D. Andrich and C. A. Scarabino, *J.C.S. Perkin II*, **1973**, 696.
180 F. P. Colonna, E. Valentin, G. Pitacco and A. Risaliti, *Tetrahedron*, **29**, 3011 (1973).
181 G. Krow, R. Rodebaugh, R. Cormosin, W. Figures, H. Panella, G. De Vicoris and M. Grippi, *J. Am. Chem. Soc.*, **95**, 5273 (1973).
182 E. Åkerblom, *Chem. Scripta*, **3**, 232 (1973).
183 T. R. Oakes and D. J. Donovan, *J. Org. Chem.*, **38**, 1319 (1973).
184 A. Niederhauser, A. Frey and M. Neuenschwander, *Helv. Chim. Acta*, **56**, 944 (1973).
185 M. K. Saxena, M. N. Gudi and M. V. George, *Tetrahedron*, **29**, 101 (1973).
186 M. Rivière-Baudet and J. Satgé, *J. Organometal. Chem.*, **56**, 159 (1973).
187 Y. Tanaka and S. I. Miller, *Tetrahedron*, **29**, 3285 (1973).
188 A. van Bruijnsvoort, E. R. de Waard, J. L. van Bruijnsvoort-Meray and H. O. Huisman, *Rec. Trav. Chim.*, **92**, 937 (1973).
189 R. A. Abramovitch, M. M. Rogic, S. S. Singer and N. Venkateswaran, *J. Org. Chem.*, **37**, 3577 (1972).
190 D. S. Garwood, *J. Org. Chem.*, **37**, 3797 (1972).
191 S. Ohashi and S. Inoue, *Makromol. Chem.*, **160**, 69 (1972); *Chem. Abs.*, **78**, 42754 (1973).
192 V. S. Bozdanov, I. L. Mikhelashvili and E. N. Prilezhaeva, *Izv. Akad. Nauk SSSR, Ser, Khim.*, **1972**, 2374; *Chem. Abs.*, **78**, 28756 (1973).
193 K. Undheim and R. Lie, *Acta Chem. Scand.*, **27**, 595 (1973).
194 P. de Maria and A. Fini, *J.C.S. Perkin II*, **1973**, 1773.
195 J.-E. Vik, *Acta Chem. Scand.*, **27**, 251 (1973).
196 V. I. Laba, A. A. Kron, S. P. Sitnikova and E. N. Prilezhaeva, *Izv. Akad. Nauk SSSR, Ser. Khim.*, **1972**, 2129; *Chem. Abs.*, **78**, 3463 (1973).
197 J. W. Batty, P. D. Howes and C. J. M. Stirling, *J.C.S. Perkin I*, **1973**, 59, 65.
198 G. F. Dvorko and N. M. Soboleva, *Kinet. Katal*, **13**, 854 (1972); *Chem. Abs.*, **78**, 71013 (1973).
199 M. Polievka and J. Balko, *Petrochemia*, **12**, 155 (1972). *Chem. Abs.*, **79**, 31174 (1973); M. A. Bennett and T. Yoshida, *J. Am. Chem. Soc.*, **95**, 3030 (1973).
200 K. Watanabe, S. Komiya and S. Suzuki, *Bull. Chem. Soc. Japan*, **46**, 2792 (1973).
201 M. Yasumoto, K. Yanagiya and M. Kurabayashi, *Bull. Chem. Soc. Japan*, **46**, 2798, 2804, 2809 (1973).
202 D. Martin, A. Berger, H.-J. Nidas and R. Bacaloghi, *J. Prakt. Chem.*, **315**, 274 (1973); S. A. Lammiman and R. S. Satchell, *J.C.S. Perkin II*, **1972**, 2300.
203 E. C. Ashby, Li-C. Chao and H. M. Neumann, *J. Am. Chem. Soc.*, **95**, 4896 (1973).
204 E. C. Ashby, Li-C. Chao and H. M. Neumann, *J. Am. Chem. Soc.*, **95**, 5186 (1973).
205 S. E. Rudolph, L. F. Charbonneau and S. G. Smith, *J. Am. Chem. Soc.*, **95**, 7083 (1973).
206 S. Wawzonek and J. V. Kempf, *J. Org. Chem.*, **38**, 2763 (1973).
207 M. Gocmen, G. Soussan and P. Fréon, *Bull. Soc. Chim. France*, **1973**, 1310.
208 T. Tsuruta and Y. Kawakami, *Tetrahedron*, **29**, 1173 (1973).
209 M. Y. Darensbourg, H. L. Conder, D. J. Darensbourg and C. Hasday, *J. Am. Chem. Soc.*, **95**, 5919 (1973).

CHAPTER 12.II

Addition Reactions. II. Cycloaddition

R. C. Storr

Department of Organic Chemistry, University of Liverpool

Epiotis has examined the effect of configuration interaction on the stereoselectivity of cycloadditions, using an orbital symmetry approach,[1] and obtained conclusions similar to those reported last year[2] from perturbation studies. Substituent effects can change the stereoselectivity for 2+2-cycloadditions. Thus, the correlation diagram for the $\pi^2s + \pi^2s$ addition of a pair of monoenes without appreciable electron donor–acceptor interactions (Figure 1) is changed to that shown in Figure 2 for a donor–acceptor pair. For Figure 1 the large separation of ψ^2 and ψ^3 means a high energy barrier for the process, whereas in Figure 2 the proximity of ψ^2 and ψ^3 means a low energy barrier and a relaxation of orbital-symmetry prohibition. The high *cis-cis* stereoselectivity observed in 2+2-cycloadditions of electron-rich to electron-deficient olefins can therefore be explained by concerted $\pi^2s + \pi^2s$ addition. In terms of the aromatic transition-state approach such additions can occur through a resonance-stabilized antiaromatic transition state (cf. push–pull-stabilized cyclobutadienes). The preference for the $\pi^4s + \pi^2s$ mode of addition of dienes to monoenes is not affected by configuration interaction. The implications of configuration interaction for molecular additions ($\sigma + \pi$, electrophilic),[1] electrocyclic reactions[3] and sigmatropic and ionic rearrangements[4] were also discussed in detail; in general, such interaction can reverse the stereoselectivity of $4n$-electron but not that of $(4n+2)$-electron pericyclic processes.[5]

A relatively simple, qualitative picture of reactivity, regiospecificity and periselectivity for cycloadditions in terms of interaction between appropriate frontier molecular orbitals (FMO) has emerged. Perturbation MO (PMO) theory indicates that the principle stabilization in the transition state for a cycloaddition arises from interaction of the HO and LV pair of addend orbitals. The closer these are in energy, the greater the stabilization and the reactivity. PMO theory also indicates that the HO and LV orbitals will combine in the sense dictated by preferential interaction between the larger terminal orbital coefficients. With this basis the regioselectivity of Diels–Alder reactions,[6,7] the enhanced reactivity, regioselectivity and *endo*-specificity of Lewis acid-catalysed Diels–Alder reactions[8,9] and the reactivity, regioselectivity and perispecificity in 1,3-dipolar cycloadditions[10,11] can be rationalized. The contribution by Houk and his co-workers is especially noteworthy and their papers on the Diels–Alder reaction[6] and 1,3-dipolar cycloaddition[10] contain a valuable discussion of, and generalizations for, the

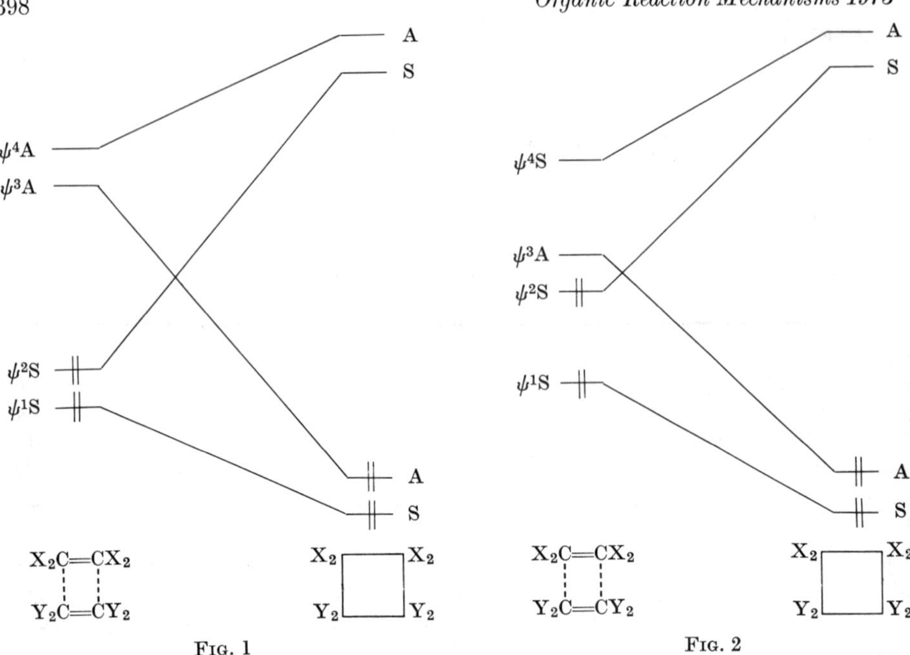

FIG. 1 FIG. 2

FIGS. 1 and 2. Correlation diagrams for 2 + 2-cycloaddition. ψ^1, ψ^2 etc. correspond to transition-state complex molecular orbitals:

orbital energies of a variety of dienes, 1,3-dipoles and dipolarophiles, and the effect of substituents on these orbital energies and coefficients.

Taking the Diels–Alder reaction as a specific example: the approximate energies and coefficients for typical alkenes and 1-substituted dienes are shown in Figures 3 and 4.[6]

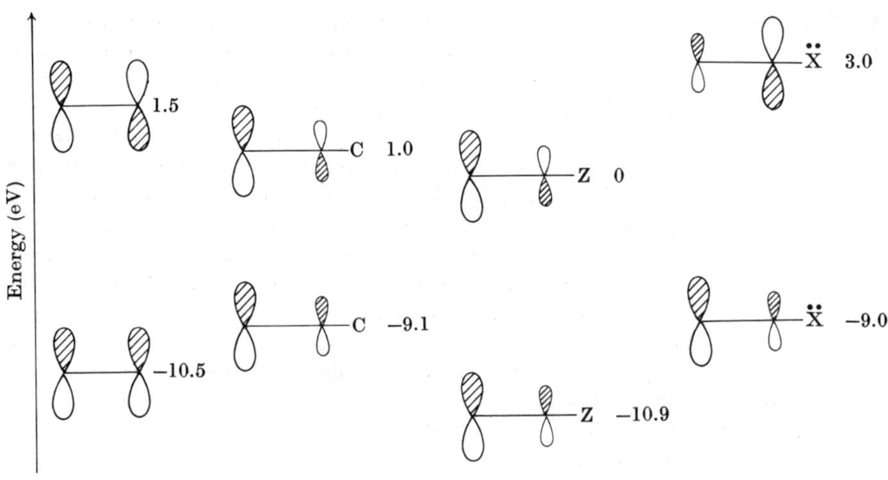

FIG. 3. C = conjugating substituent $CH=CH_2$, Ph etc.; Z = CHO, CN etc.; X = R, OR, NR_2 etc.

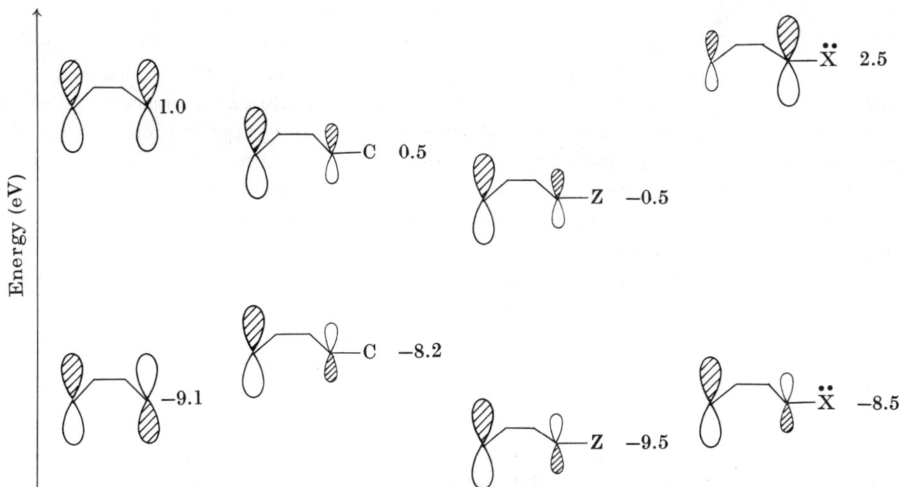

FIG. 4. For key see Figure 3.

For the normal type of Diels–Alder reaction between an electron-rich diene and an electron-deficient alkene the dominant interaction is between diene HOMO and dienophile LVMO, and 3,4-disubstituted cyclohexene formation is predicted and observed.

The LVMO of acrolein (**1**) is lowered in energy, and the coefficients are modified as in (**2**), by co-ordination of a proton or Lewis acid on the carbonyl-oxygen atom. The lower energy should result in higher reactivity, the greater disparity of orbital coefficients at C-3 and C-2 in enhanced regioselectivity, and the large C-1 coefficient in greater *endo*-selectivity (greater secondary interaction) for the complexed acrolein, all in line with experimental observations.[8]

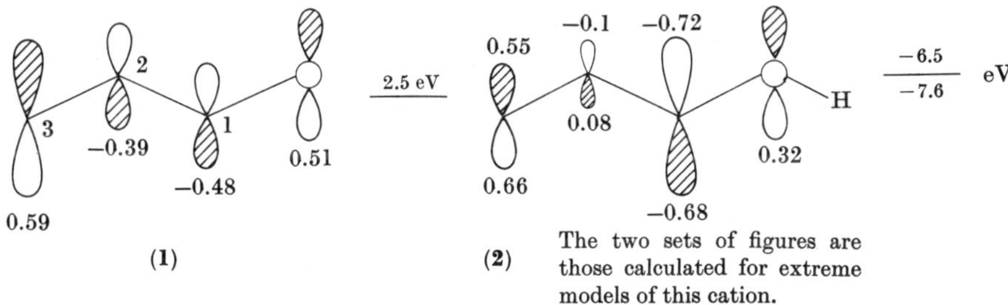

The two sets of figures are those calculated for extreme models of this cation.

Application of the above approach to non-polar 2+2-cycloadditions, assuming them to be concerted, leads in many cases to a predicted head-to-tail regioselectivity, whereas head-to-head selectivity is observed experimentally. This possibly provides a criterion of the stepwise character for such reactions. Similar disagreement between prediction and experiment also occurs for non-polar, but not for polar, Diels–Alder reactions.[12]

PMO theory has been used to explain perispecificity in cross-conjugated systems such as isobenzofulvenes.[13] A general discussion of the perturbation method of treating chemical reactions includes a brief section on cycloaddition.[14]

The concept of steric attraction, or repulsion, depending on the number of electrons involved has been discussed[15] and may well have relevance to certain secondary interactions in cycloadditions. The validity of estimating the energy of diradical intermediates from activation energies for other reactions assumed to pass through the same intermediate has been questioned, since *trans*-1,2-divinylcyclobutane rearranges to vinylcyclohexene with the stereochemistry expected for a concerted 1,3-sigmatropic migration.[16] Further discussion on the nature of 1,3 and 1,4 "diradical" species, and the question of whether these are species with energy minima as suggested by thermochemical analysis or with no minima (twixtylic) as indicated by quantum calculations, has appeared.[17] The problems involved in defining concertedness have been discussed.[18]

Direct thermal population of triplet states has been observed in the highly exothermic rearrangement of Dewar-benzene to benzene.[19]

A valuable review on polar cycloadditions[20] and reviews of cycloaddition,[21] addition reactions in general,[22] and stereochemistry of concerted reactions[23] have appeared.

2+4-Cycloaddition

The application of PMO theory to account for the rates, regiospecificity and effect of Lewis acid catalysts on Diels–Alder reactions has been discussed above. The cycloaddition of several fulvenes to a variety of dienes gives only 4+2-adducts with fulvene as diene component. This perispecificity and the regioselectivity observed are compatible with the above FMO model where the dominant interaction is between fulvene HOMO and diene LVMO.[24]

The effect of neighbouring non-conjugated double bonds on the addition of various dienophiles to *cis*-hexahydro-, *cis*-tetrahydro- and *cis*-dihydro-naphthalene is best explained in terms of steric rather than secondary orbital interactions.[25] *endo*-Selectivity is observed in the addition of cyclopentene, *cis*-but-2-ene and cycloheptene to 2-benzopyran-3-one and its 1-methyl derivative (**3**; $R^1 = R^2 = H$ and $R^1 = Me$, $R^2 = H$) and in the addition of cyclopentene to the diene (**4**), situations where steric and other effects are not expected to obscure attractive diene–*endo*-alkyl group interactions.[26]

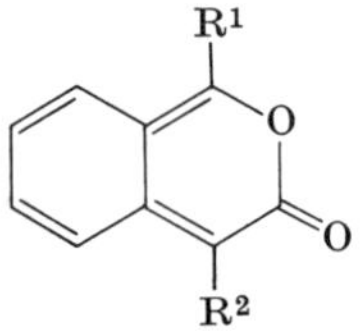

(**3**) 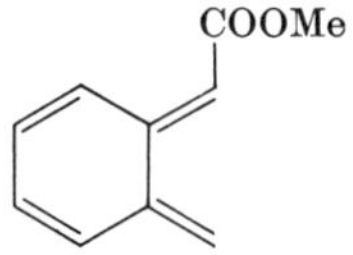 (**4**)

When R^1 and R^2 are replaced by phenyl groups, *exo*-addition predominates because the non-planar phenyl groups impede such *endo*-stabilizing interactions.[27] The relation between *exo*:*endo* ratio and the difference in activation volume for the corresponding transition states have been determined for the addition of various dienophiles to cyclopentadiene.[28] It has been suggested that the observed *syn-anti*-orientations in Diels–Alder reactions of cyclic dienes can be explained only if both secondary orbital interactions and van der Waals–London interactions are taken into account.[29] The fluoroalkyl substituent determines the adduct stereochemistry for the reaction of a series of fluorinated *trans*-crotonic acid derivatives with cyclopentadiene.[30]

Thermochemical data for the Diels–Alder reaction of cyclopentadiene and maleic anhydride have been obtained and discussed.[31] Differences in regiospecificity have been observed, but not explained, with the *cis–trans* dienophilic pair citracononitrile and mesocononitrile.[32] Formation of the racemic adduct between diethyl maleate and oxidatively liberated cyclobutadiene from an optically active tricarbonyl(cyclobutadiene)iron indicates that free cyclobutadiene is involved.[33] Significantly the thermal 2+4-adduct from cycloheptatriene and anthracene is not observed in the photochemical addition which gives 6+4- and 4+4-adducts, possibly via a diradical.[34] A 5-halo-substituent enhances the reactivity of cyclopentadiene as both diene and dienophile in its dimerization; spiroconjugation and a combination of spiroconjugation and negative hyperconjugation, respectively, is suggested to account for this; the effect of halogen is, however, dependent on the character of the dienophile since the halogen deactivates the diene in its reactions with electrophilic dienophiles.[35]

The high reactivity of octamethylnaphthalene, compared with naphthalene and hexamethylbenzene, towards dienophiles (and electrophiles) is attributed to severe *peri*-interactions which distort the π-system from planarity.[36]

The decrease in $\Delta G^{\ddagger}$ with increasing size of the *N*-alkyl substituent in the intramolecular Diels–Alder reaction of pentadienylacrylamides can be rationalized in terms of the effect of the substituent on the conformation of the substrate.[37] The intramolecular Diels–Alder reaction of 5-alkenylcyclohexa-1,3-dienes (**5**; $n = 3$) proceeds with the same regiospecificity as is observed for lower homologues (**5**; $n = 0$, 1, 2),

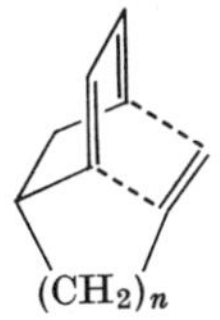

$(CH_2)_n$

(5)

probably mainly for steric reasons.[38] The opposite regiospecificity can be observed under the influence of appropriate substituent effects.[39] An intramolecular Diels–Alder reaction surprisingly accompanies dibromocyclopropanation of the tetraene (**6**) to give (**7**); the greater stability of adduct (**7**) than (**8**) may be due to trishomoconjugation.[40] The potential use of intramolecular Diels–Alder reactions in synthesis has been further underlined.[41]

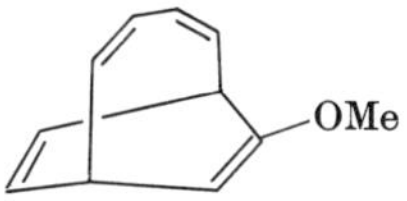

OMe

(6)

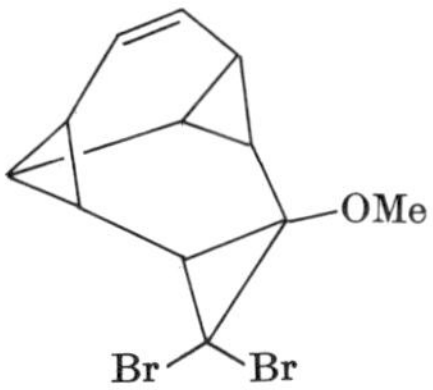

OMe Br Br

(7)

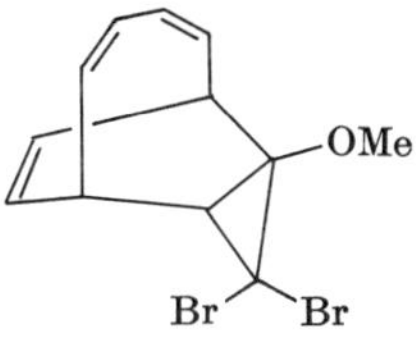

OMe Br Br

(8)

The reactions of steroidal dienes with tetracyanoethylene (TCNE) have been clarified.[42] Cycloaddition of vinylketene thioacetals to dienophiles provides a route to cyclohexanones that avoids the possible complication of 2+2-addition which exists for vinylketenes; the thio-substituents activate the diene systems.[43] The more reactive and easily handled 2-phenyl- and 2-thio-1,3-dioxol-4-ene provide attractive alternatives to acetylene in Diels–Alder reactions because the adducts undergo ready retro-1,3-dipolar cycloaddition either with base or thermally.[44] Although its adducts are more easily converted into cyclic hydrazines, 1,3,4-thiazoline-2,5-dione is a less reactive dienophile than 4-phenyltriazoline-2,5-dione.[45] The acid-catalysed $4\pi + 2\pi$-dimerization of ethyl

1,2-diazepine-1-carboxylate involves the first example of an enamine as dienophile.[46] A synthetically useful example of the rare Lewis acid-catalysed Diels–Alder reaction with inverse electron demand has been reported[47] (Scheme 1). The expected inertness of perchlorobutadienes towards maleic anhydride has been verified.[48] A product analysis of

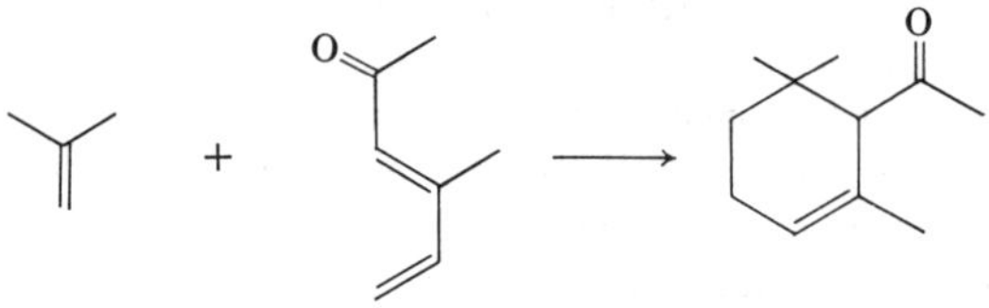

SCHEME 1.

the Diels–Alder reaction of various silyl- and germyl-cyclopentadienes with trichloro-(ethynyl)germane has been reported.[49] Relative reactivity,[50] kinetic,[51] and regio-[52] and stereo-chemical[53] studies for a variety of carbocyclic Diels–Alder reactions have been reported and stereoselectivity in Diels–Alder reactions has been reviewed.[54]

The acid-catalysed 2+4-cycloaddition of iminourethanes to cyclohexa-1,3-diene apparently involves a stepwise process in which acid-complexed imine is attacked by the diene in a cyclic transition state involving the *E*-iminourethane. This is deduced from an analysis of the *endo*:*exo* preference of the urethane substituent. The stepwise nature is suggested by formation of products arising from diversion of the intermediate cation (Scheme 2).[55]

$$RCH(NHCOOEt)_2 \xrightarrow{H^+} RCH{=}NCOOEt + NH_2COOEt$$

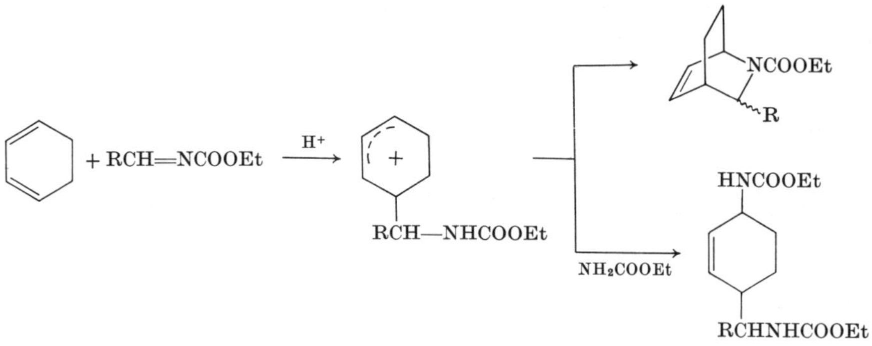

+ RCH=NCOOEt $\xrightarrow{H^+}$ RCH—NHCOOEt

NCOOEt R

NH_2COOEt

HNCOOEt

RCHNHCOOEt

SCHEME 2.

The Diels–Alder reaction of 3-*tert*-butyl-1-isopropyl-1,2-diazabuta-1,3-diene with electron-deficient dienophiles has been reported[56] and a two-stage mechanism has been suggested for the Diels–Alder reaction of 5-ethoxyoxazoles with diethyl azodicarboxylate.[57]

Several attempts to observe homo-Diels–Alder reactions between vinylcyclopropanes and various dienophiles have been described.[58–60] In most cases alternative modes of reaction resulted, but an example was found in the reaction of the homodiene (**9**) with TCNE;[58] and in the reaction of vinylcyclopropanes with chlorosulphonyl isocyanate, the products can be explained in terms of electrophilic attack by the reagent to give a

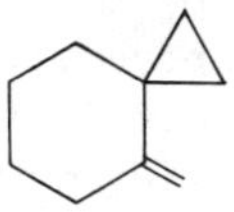

(9)

zwitterion which collapses to a homo-Diels–Alder adduct or to an open-chain product depending on the substituents in the homodiene.[60]

Polar 1,4-cycloaddition has been reviewed.[20,61] One article[20] gives a comprehensive discussion of the structures of different 1,4-dipolar species and of the mechanism of their cycloaddition. A new example has been found[62] of stereoselectivity in the polar 1,4-cycloaddition of 2-alkyl-3-methylisoquinolinium salts; such results are explicable in terms of electrostatic repulsion between discharging and developing positive charge in the transition state, and support the non-synchronous nature of these polar cycloadditions. Further linear correlations have been reported in the cycloadditions of 9-substituted acridizinium salts.[63] The stereo- and regio-specific formation of 5,6-dihydro-4*H* 1,3-thiazines from a thioamide, an aldehyde and an olefin has been rationalized by polar 1,4-cycloaddition[64] (Scheme 3).

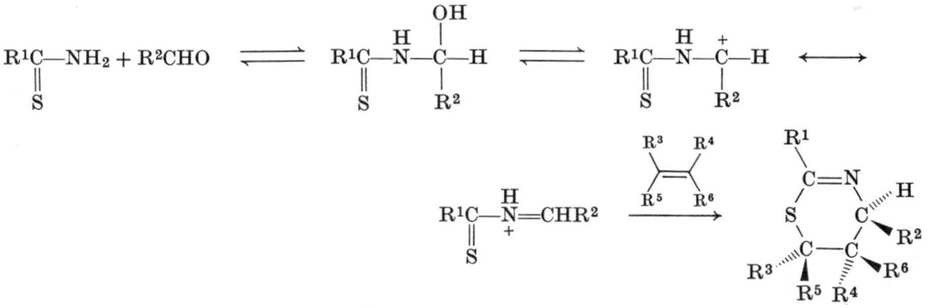

$R^1C(=S)\text{—}NH_2 + R^2CHO \rightleftharpoons R^1C(=S)\text{—}NH\text{—}CH(OH)R^2 \rightleftharpoons R^1C(=S)\text{—}NH\text{—}\overset{+}{C}HR^2 \longleftrightarrow R^1C(=S)\text{—}\overset{+}{N}H\text{=}CHR^2 \xrightarrow{R^3R^4C=CR^5R^6}$ 1,3-thiazine (R¹, R², R³, R⁴, R⁵, R⁶, H)

SCHEME 3.

2+3-Cycloaddition

A review of 1,3-dipolar cycloaddition to alkynes has appeared[65] and a review of azomethine ylides, azomethine imines and iminophosphoranes in synthesis contains a section on cycloaddition.[66] Perturbation FMO theory has thrown considerable light on 1,3-dipolar cycloaddition.[10,11] Reversal of the normal regiospecificity for the addition of nitrones to highly electron-deficient dipolarophiles, as predicted by the FMO model, has been observed.[67]

^{14}C Isotope effects for the addition of *N*,α-diphenylnitrone to styrene when labelled nitrone and α- and β-labelled styrene are used, show that the reaction is concerted.[68] Huisgen and his co-workers have reported extensive studies of the 1,3-dipolar addition of nitrile oxides[69] including fulminic acid itself.[70] Relative reactivity[71] and regioselectivity[72] data were discussed in detail in terms of the FMO models developed by Sustmann[73] and Houk.[10] Other workers have studied the regioselectivity of addition of nitrile oxides to α,β-unsaturated ketones[74] and to cyclopentadiene,[75] and substituent effects in the addition of nitrile oxides to styrenes have been interpreted in terms of a one-step mechanism with some asymmetry in the transition state.[76]

Further examples of 1,3-anionic cycloaddition[77] and the reverse reaction[78] have been reported. The first example of 1,3-dipolar cycloaddition of an azoxy-compound has been claimed for the reaction of benzocinnoline *N*-oxide (**10**) with dimethyl acetylenedicarboxylate, which parallels that previously reported for related *N*-iminobenzocinnolinium ylides. New regiochemical evidence confirms a cycloaddition, electrocyclic ring-opening sequence for the latter reaction and rules out the sort of alternative pathways that must be involved in the formally similar cycloadditions of cyclic nitrones with acetylenic esters.[79]

A 1,3-dipolar intermediate is involved in the thermal rearrangement of a *syn*-2-azatricyclo[4.1.0.0^{3,5}]heptane ester (**11**) to methyl 2,3-dihydroazepine-1-carboxylate (the *anti*-isomer, for which formation of the dipole is disallowed, rearranges much more

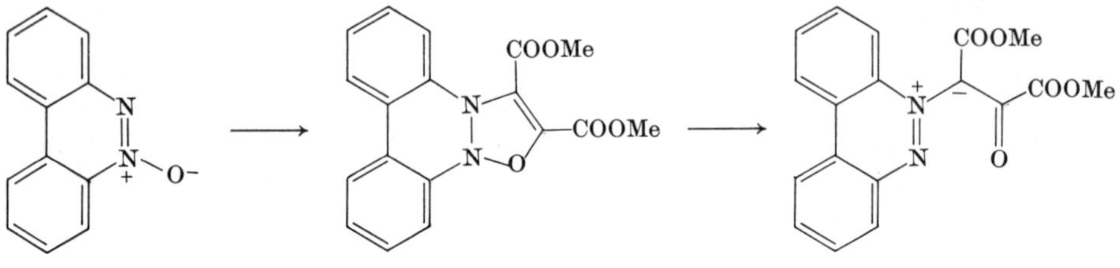

(**10**)

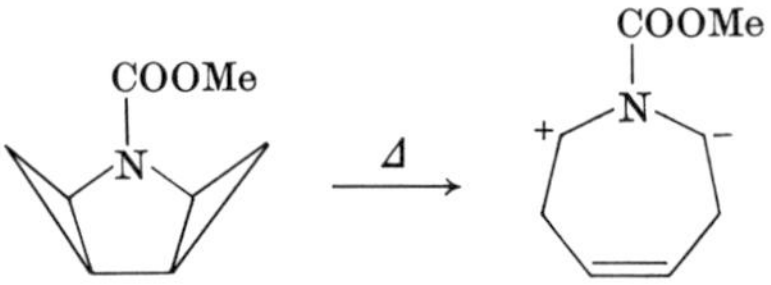

(**11**)

slowly) and in its addition to *N*-phenylmaleimide which is not therefore a bis-homo-Diels–Alder reaction. Similar behaviour probably explains the difference between cycloadditions of heteroquadricyclanes and of quadricyclanes.[80]

Anhydro-3-hydroxy-2,4,6-triphenylpyrylium hydroxide (**12**) resembles the corresponding pyridinium derivative in its 1,3-dipolar activity towards a variety of dipolarophiles.[81] The previously reported anomalous formation of (**13**) from the pyridinium betaine (**14**) and benzyne has now been shown to involve reaction of benzyne with the isomer (**15**) to which (**14**) readily rearranges.[82]

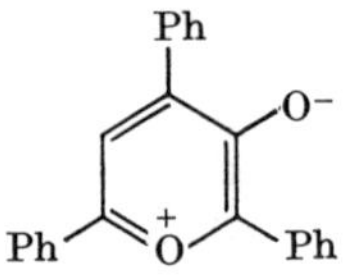

(**12**) 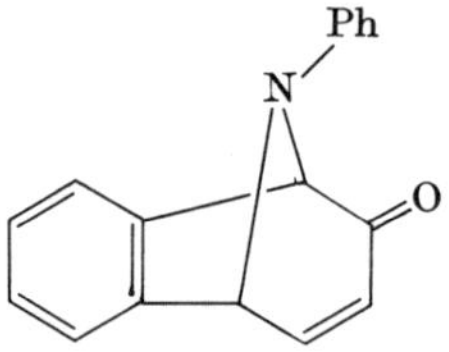 (**13**) 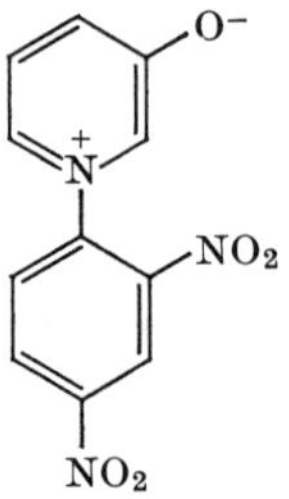 (**14**) 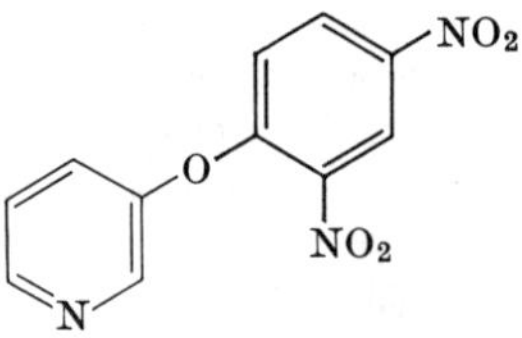 (**15**)

The stereochemistry of thiocarbonyl ylides is preserved in their addition to the carbonyl group of diphenylketene.[83] A non-linear correlation of rate with substituent effects for the addition of diazomethane to arylacetylenes suggests different mechanisms for electron-withdrawing and -releasing substituents; the ρ-value for the former is of the order expected for nucleophilic attack on the acetylene.[84] Kinetic evidence for the intermediacy of benzonitrile sulphide in the reaction of 5-phenyl-1,3,4-oxathiazol-2-one with dimethyl acetylenedicarboxylate to give dimethyl isothiazole-4,5-dicarboxylate has appeared.[85] Kinetic studies have also been reported for the cycloaddition of diphenyldiazomethane to organophosphorus compounds of type (**16**)[86] and for the ozonolysis of phenylethylenes[87] (see also Chapter 13). Parameters for the latter reaction are consistent with 1,3-dipolar cycloaddition as the initial step.

The cycloaddition of alkyl azides to the C=N of aryl, acyl, alkoxycarbonyl and sulphonyl isocyanates and of aryl azides to sulphonyl isocyanates has been studied. The reaction is reversible for alkyl azides and sulphonyl isocyanates and kinetic evidence supports a concerted mechanism with charge imbalance in the transition state.[88] Addition of alkyl azides to sulphonyl isothiocyanates occurs at the C=S bond and heating of the adducts gives carbodi-imides by a cycloreversion in which the novel external stabilized dipole (**17**) can be intercepted with enamines and ynamines.[89]

$CH_2{=}CR{-}P(O)X_2$

(**16**)

$PhCH_2N{=}C{=}NSO_2Ar$

$PhCH_2{-}\bar{N}{-}C(=NSO_2Ar){-}S^+$ (**17**)

$Et_2N{-}C{\equiv}CMe$

Et_2N Me $PhCH_2N$ S C NSO_2Ar

Factors affecting the approach of nitrones[90] and nitronic esters[91] to activated olefins have been determined. Formation of Δ^3-1,3,4-thiadiazoline 1-oxides from sulphines and 2-diazopropane is both regio- and stereo-specific (kinetic control).[92] With sterically hindered sulphines, 2+3-addition is replaced by non-stereospecific episulphide formation possibly via a zwitterion formed by nucleophilic attack by the diazo-carbon on the sulphine-sulphur.[93]

Both azomethine ylide and thiocarbonyl ylide reactivity has been observed for the thieno[3,4-*c*]pyrrole (**18**).[94] Normal 1,3-dipolar cycloadditions are observed for the nitrile ylides produced by photochemical cleavage of arylazirines, and photodimerization of arylazirines can be rationalized by 1,3-dipolar additions of these nitrile ylides to the arylazirines.[95] The high dipolarophilic reactivity of azirines has been demonstrated by other workers.[96] 1,3-Dipolar addition of benzonitrile *p*-nitrobenzylide to 3-phenyl-2*H*-azirines gives 1,3-diazabicyclo[3.1.0]hex-3-enes and secondary products derived therefrom.[97] The azlactone 2,4-diphenyl-Δ^2-oxazoline, however undergoes an ene reaction, via its enol tautomer with azirines.[97] *endo*-2,4,6-Triphenyl-1,3-diazabicyclo[3.1.0]hex-3-ene (**19**) undergoes both thermal and photo-induced conrotatory opening to give an interceptable azomethine ylide; the implications of the photo-induced conrotatory opening were discussed.[98]

The reactions of aldo-nitrones with carbon disulphide can be rationalized by initial dipolar cycloaddition, but primary adducts have not yet been detected.[99] Dibenzo-[1,3a,6a]triazapentalenes (**20**) behave as cyclic azomethine imines with acetylenic esters and tolane (particularly in the presence of an acid catalyst); a 2:1 adduct (**21**) is obtained with an excess of dimethyl acetylenedicarboxylate.[100] Examples of 1,3-dipolar activity of keto-carbenes,[101] and of carbene activity of benzonitrile oxide in cycloadditions,[102] have been reported. Attention has been drawn to the use of dibenzoylacetylene as a

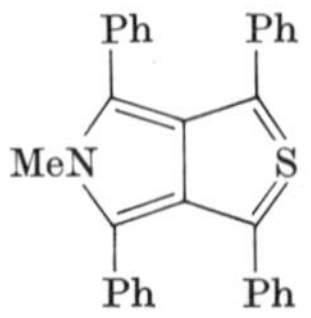

(**18**) 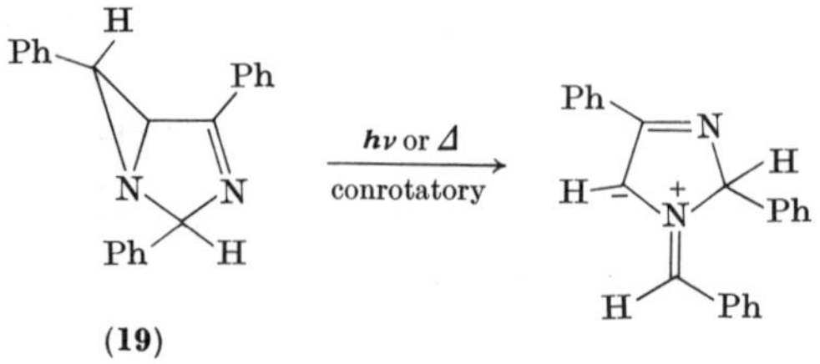 (**19**)

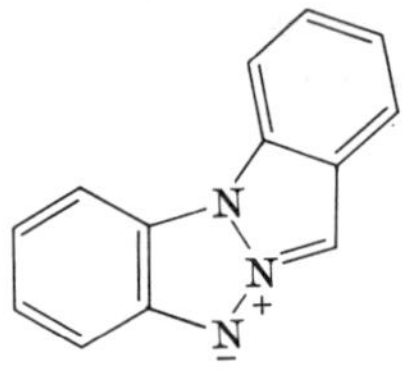

(**20**) 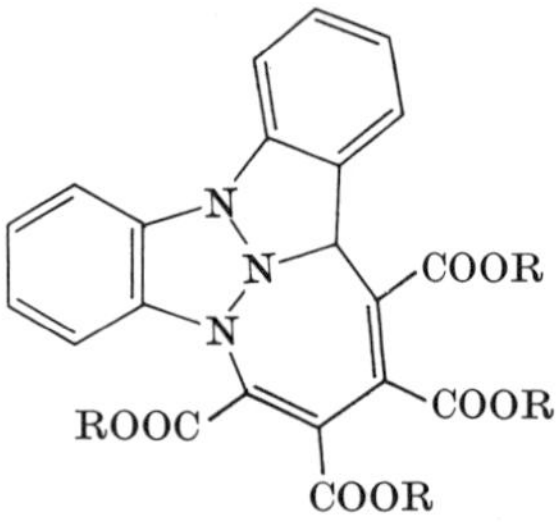 (**21**)

useful precursor to *o*-dibenzoyl heterocycles via Michael or cyclo-additions.[103] The first thione 5-imides have been prepared; they undergo apparently non-concerted 3+2-cycloaddition to enamines.[104] Nitrile ylides are formed by photolysis of 1-azetines (**22**) and in the absence of dipolarophiles give a new type of nitrile ylide dimer.[105]

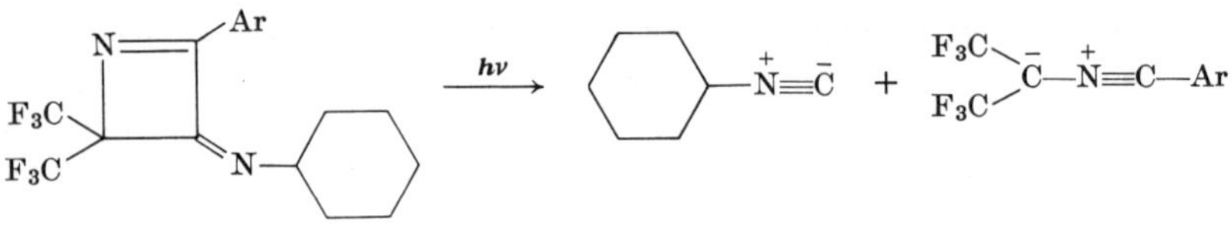

(**22**)

2+2-Cycloaddition

Bartlett and his co-workers have continued their studies of those olefins for which concerted Diels–Alder addition to dienes and diradical formation leading largely to vinylcyclobutane are finely balanced[106, 107] by using photosensitized reactions as models for reactions proceeding entirely through diradicals, the proportion of 1,4-adducts arising via diradicals has been shown to be low. Regioselectivity observed in cyclo-additions with trifluoroethylene and vinylidene fluoride indicates that radical accommodation is $CH_2 > CHF > CF_2$, which is not the order observed in free-radical attack on trifluoroethylene; initial electron-availability seems to be the important factor for the exothermic radical attack, and radical stabilization is important in the biradical formation.[107] In the addition of 1,2-dichloro-1,2-difluoroethylenes to *trans*-cyclo-octene, loss of configuration is observed for the former but not the latter;[108] the addition thus involves a diradical intermediate in which transannular interaction hinders loss of configuration of the cyclic portion and is not a concerted $\pi^2s + \pi^2a$ cycloaddition. Cycloaddition, presumably via a diradical, of ethylene to acrylonitrile has been achieved at high pressure and 300–345°.[109]

In a series of communications, Huisgen and Steiner[110] have reported detailed mechanistic studies of the 2+2-addition of TCNE to enol ethers. They conclude that the addition proceeds through a reversibly formed zwitterion for which the rates of dissociation and closure are comparable. Ring-closure of the zwitterion is approximately five times faster than internal rotation, thus accounting for the high stereo selectivity; this is in contrast to the behaviour of diradicals, where rotation is more than ten times faster than cyclization which is in turn approximately four times faster than dissociation. Details of the thermal 2+2-additions of cyclopropylethylenes to TCNE via zwitterionic intermediates have appeared.[111]

The mechanism shown in Scheme 4 has been proposed for the zeolite-catalysed cyclodimerization of cyclopropenes; optimum size of the zeolite cavities prevents polymerization, and the nature of active sites on the zeolite influences the rate.[112]

Inagaki and Fukui[113] have suggested that singlet oxygen and benzyne, and possibly other donor–acceptor pairs, interact through a transition state such as (**23**) which differs from the $\pi^2s + \pi^2a$ transition state considered by Hoffmann.

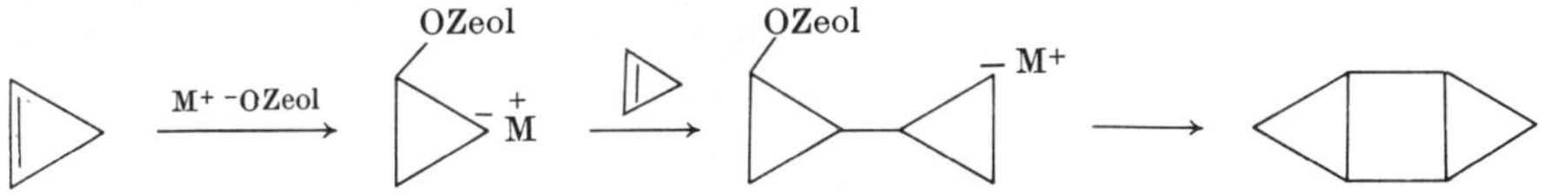

OZeol OZeol

M^+ $^-$OZeol $\overset{+}{M}$ M^+

Scheme 4.

HOMO

LVMO

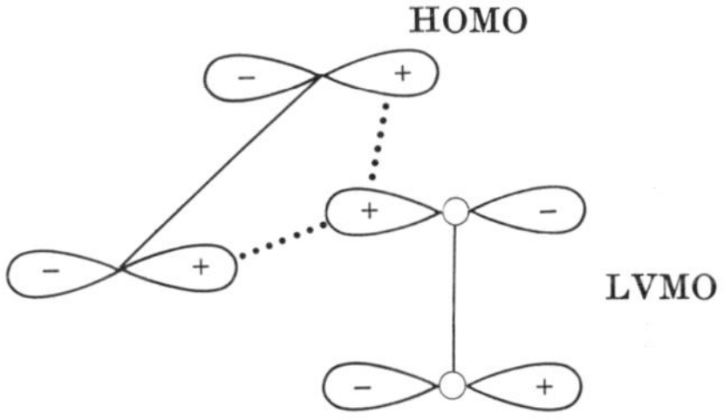

(**23**)

Several papers support the role of tetramethylene-ethane diradical intermediates in allene dimerization. Such species produced by alternative routes collapse to give similar product distributions to those found in corresponding allene cycloadditions.[114] The cycloaddition of *tert*-butylcyanoketene to optically active penta-2,3-diene gives a product distribution consistent with a stepwise reaction via a doubly allyl-stabilized intermediate zwitterion.[115] *cis*-Stereoselectivity in the dimerization of unsymmetrical keto-ketenes suggests a concerted $\pi^2s + \pi^2a$ process.[116] No clear mechanistic decision is made for the addition of diphenyl- and phenylchloro-ketenes to 1,1-dimethylallene, reactions which both occur preferentially at the unsubstituted double bond.[117] The triethylamine-catalysed dimerization of *tert*-butylcyanoketene gives the β-lactone dimer which is further decarboxylated to give 1,3-di-*tert*-butyl-1,3-dicyanoallene.[118]

Further examples of cyclobutenone formation from ketenes and ethoxyacetylene[119] and several examples of the addition of dichloroketene to cyclic olefins[120] have appeared, and the addition of ketenes to 1*H*-1,2-diazepines[121] has been studied. Vinylketene imines undergo 4+2-addition to electron-deficient olefins and acetylenes. Electron-donation from such species to the cumulene π^*-orbital which is necessary for 2+2-cycloaddition is unlikely.[122]

Activation parameters for the two modes of cleavage of 3-ethoxy-2,2-dimethylcyclobutan-1-one are interpreted in terms of a highly polar retro $\pi^2s + \pi^2a$ addition,[123] but similar studies for 2-chlorocyclobutanone indicate that the two modes involve different pathways.[124] Cycloadditions of enamines to electron-deficient allenes[125] and of isocyanates to ketene imines[126] have been reported; isolation of 2:1-adducts in the latter case implies a stepwise mechanism via a zwitterion. Butane sultone formation from sulphur trioxide and *cis*- and *trans*-butenes is stereospecific.[127]

Other Cycloadditions

A review of the scope and mechanism of cycloaddition of allyl cations has appeared,[128] and a convenient method for the generation of the 2-methoxyllayl cation for cycloaddition has been described.[129] Further studies of the $\pi^4s + \pi^2s$ additions of oxyallyls, obtained from α,α'-dibromo-ketones, to dienes have been reported.[130]

Stereochemically there are four six-electron and four eight-electron allowed modes for the formal $\sigma^2 + \pi^2 + \pi^2$ cycloaddition of alkenylidenecyclopropanes (**24**) to 4-phenyltriazoline-3,5-dione. Experimental data fit one of the eight-electron modes best. The high reactivity of alkenylidenecyclopropanes compared with methylenecyclopropanes in this type of reaction is attributed to interaction between the cyclopropane σ- and remote allene π-bonds.[131] Another example of this type of addition to a methylenecyclopropane has, however, appeared.[132] In contrast to the concerted reaction with 4-phenyltriazoline-3,5-dione the reaction of alkenylidenecyclopropanes with chlorosulphonyl isothiocyanate involves electrophilic attack at either C-4 or C-5, to give cyclopropane-cleaved products or β-lactams, respectively; the point of attack is governed by substituents in the cyclopropane ring, in particular by their ability to stabilize the "homoallylic" cation (**25**) produced by electrophilic attack at C-5.[133]

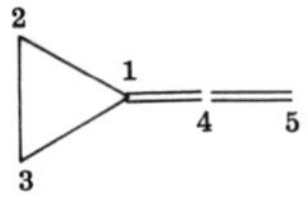

(**24**) 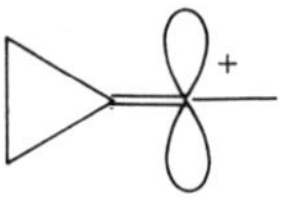 (**25**)

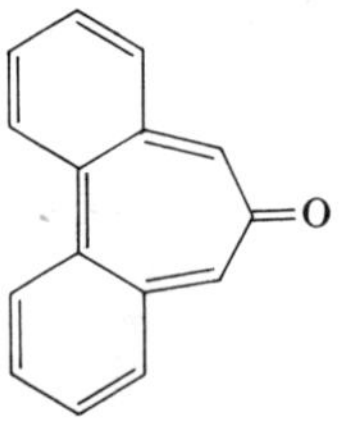

(**26**) 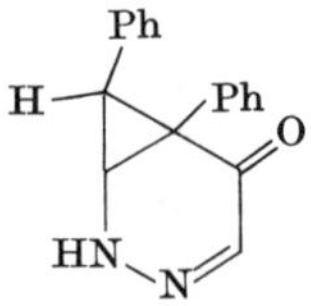 (**27**) 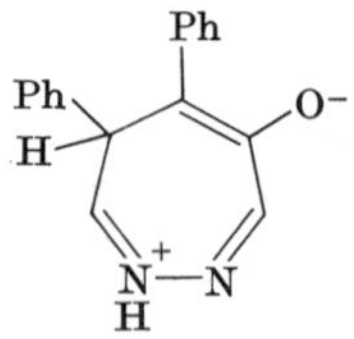 (**28**)

Steric effects have been recognised in 6+4-cycloadditions;[134] the dibenzotropone (**26**) has been reported as a transient but highly reactive triene in 6+4-cycloaddition[135] and it has been suggested that 2,3-diazabicyclo[4.1.0]hepten-5-one (**27**) dimerizes by way of its valence isomer (**28**) in a $\pi^4s + \pi^6s$ process.[136] Both 6+4- and 4+2-adducts are observed in the reaction of isobenzofuran with tropones,[137] and products from the addition of tropylium ion to cycloheptatriene can be rationalized as formed by rearrangement of the carbonium ions produced by 6+4- and 2+4-cycloaddition.[138]

Examples of $\pi^8s + \pi^2s$ addition have been reported for sulphenes to tropones,[139] maleic anhydride to tropthione,[140] and *N*-(*p*-methoxyphenyl)maleimide to methylenetriphenyl-2-troponylphosphoranes.[141] Adducts (**29**) from the first reaction lose SO_2, when heated, to give styrene derivatives by a process that formally involves a homocheletropic fragmentation.[139] With TCNE, tropthione gives 8,8-dicyanoheptafulvenes, possibly by 2+2-addition.[142]

A diradical intermediate is thought likely in the non-stereospecific reaction of dispiro[2.2.2.2]deca-4,9-diene (**30**) with butadienes to give [8]-paracyclophanes.[143] A formally allowed $\pi^4s + \pi^2a + \pi^2s + \pi^2a$ cycloaddition occurs between TCNE and [3,3]-paracyclophadiyne (**31**).[144] The thermal dimerization of cyclo-octatetraene has been analysed in the light of the recognition that the isomerization of cyclo-octatetraene

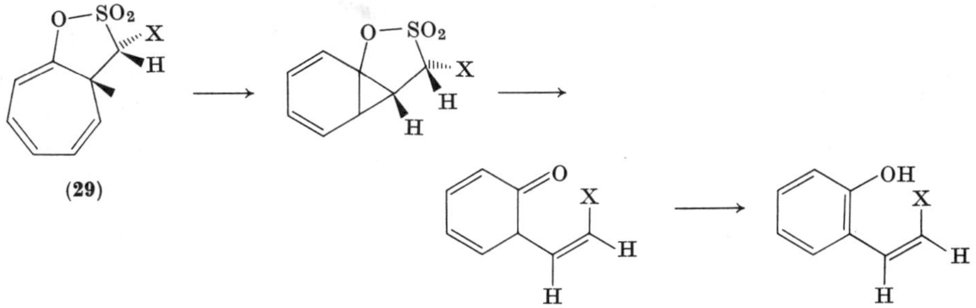

(**29**)

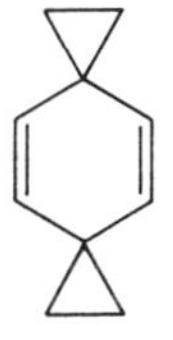

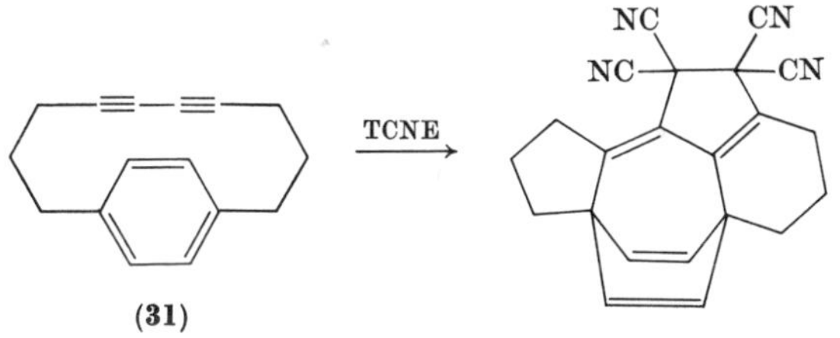

(**30**) (**31**)

to semibullvalene is allowed.[145] Conclusions from orbital-symmetry analysis for the additions of tetrafluorodisilylene are similar to those for ethylene.[146] The addition of SO (from thi-irene oxide) to dienes is remarkably stereospecific in view of the strong evidence that it involves triplet SO.[147] The cyclobutadiene (**32**) has been suggested as an intermediate in the formation of the 1,2-dibenzoylacetylene dimer[148] (**33**).

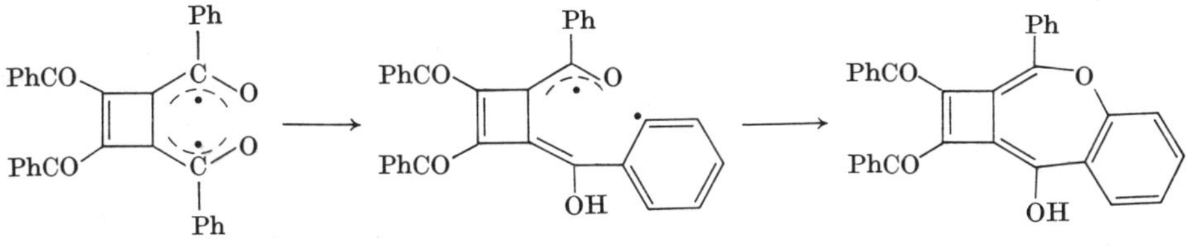

(**32**) (**33**)

Several examples of formal dipolar (not mesoionic) addition where the dipole is derived from ring-opening of a heterocycle have appeared.[149] Steric effects play an important rôle in the mode of closure of zwitterionic intermediates in the formation of 2:1-adducts from diphenylketene and imines and enamines.[150] A possible rationalization for the formation of the diphenylketene–1,1-diphenylethylene cycloadduct has been given.[151] The reaction of 8-oxoheptafulvenes with 2-methoxytropones involves 2+2-addition, 1,7-oxygen migration, 3,3-sigmatropic rearrangement and elimination of methanol.[152] Both 2+2- and 2+4-additions of hexafluorothioacetone have been observed.[153] The mode of addition of benzyne to cyclic olefins (2+2, 2+4 and ene reaction) is governed by the geometry of the diene segment.[154] The effect of catalytic amounts of Ag^+ on these reactions has also been investigated, and it has been explained in terms of a model different from that previously considered for such reactions.[155]

Full details of the addition of TCNE and chlorosulphonyl isocyanate to *cis*-bicyclo[6.1.0]nona-2,4,6-trienes have been published,[156] the additions are best rationalized as formation of a *trans*-1,3-bishomotropylium ion intermediate by electrophilic attack on the less stable folded triene conformer. Additions of the isocyanate to other systems that could lead to bicycloaromatic and antibicycloaromatic intermediate zwitterions have also been described.[157]

Cyclic aminimides are formed from isocyanates, such as phenyl isothiocyanate, and N^2-(dimethylamino)-N',N'-dimethylformamidine (Scheme 5, path a). With certain aryl isocyanates bearing electron-withdrawing groups, perhydro-*s*-triazinediones are

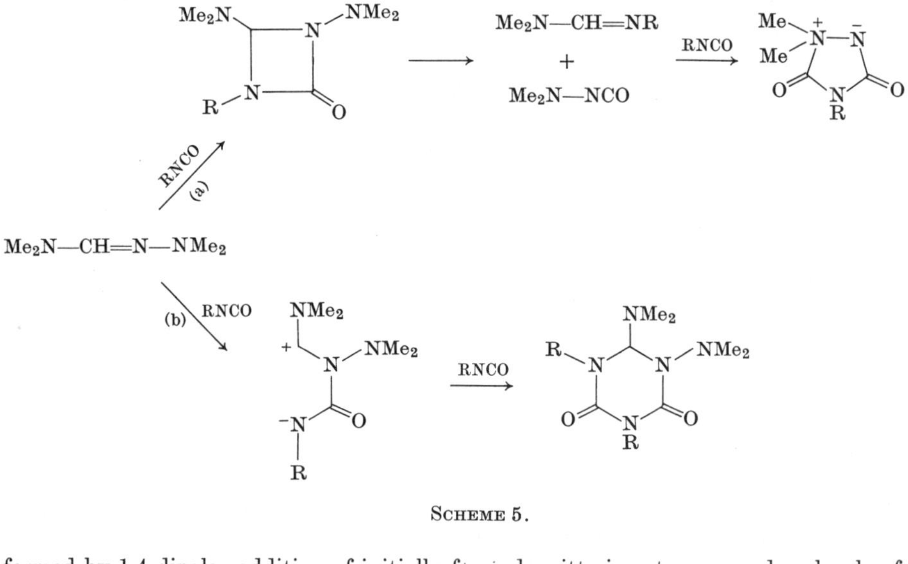

SCHEME 5.

formed by 1,4-dipolar addition of initially formed zwitterions to a second molecule of isocyanate (path b).[158] Kinetic studies indicate that two concurrent mechanisms are involved in the reaction of 1-*p*-tosylazocyclohexene with maleic anhydride.[159] Addition of tetrachlorobenzyne to 1,2,3,4-tetramethoxynaphthalene gives 1,2,3,4-tetrachlorodibenzo-1,2,5,8-tetramethoxysemibullvalene (formally $\pi^2s + \pi^2a + \pi^2a$ addition) in addition to the expected 1,4-adduct.[160] 2+1-Addition of hexafluoroacetone azine to phosphites and to hexamethylphosphorous triamide have been reported,[161] and also criss-cross adducts from benzoyl isocyanates and benzaldazines (thiobenzoyl isocyanate gives $4\pi + 2\pi$ adducts with aldazine as 2π-component).[162] Formation of 1,2-oxathian-4-one 2-oxides from ketene imines and ketene in anhydrous sulphur dioxide involves a ketene–SO_2 adduct.[163] Stereochemical studies indicate that the non-catalytic addition of D_2 to cyclopentadiene and transfer of hydrogen from cyclopentene to cyclopentadiene are concerted six-electron and ten-electron suprafacial processes, respectively.[164]

1,3-Addition, probably through a dipolar intermediate, occurs in the additions of TCNE to tricarbonyl(cycloheptatriene)iron[165] and tricarbonyl(cyclo-octatetraene)-iron;[166] chlorosulphonyl isocyanate adds 1,4 to the latter.[166] Cycloaddition of uni-particulate electrophiles to the iron–allenyl complex (**34**) involves intermediate cationic iron–acetylene complexes.[167] A σ-bonded metallocyclic intermediate has been isolated in a metal-catalysed forbidden 2+2-dimerization of norbornadiene,[168] and the underlying importance of conservation of co-ordinate bonding in transition-metal catalysis of allowed and forbidden reactions has been further underlined.[169]

$$h^5\text{—}C_5H_5Fe(CO)_2\text{—}CH{=}C{=}CH_2 \;(\mathbf{34}) \xrightarrow{X=Y} C_5H_5\overset{+}{Fe}(CO)_2\text{—}(HC{\equiv}C\text{—}CH_2\text{—}X\text{—}Y^-) \longrightarrow C_5H_5\text{—}Fe(CO)_2\text{—}(\text{ring } C{=}C\text{—}X\text{—}Y)$$

References

1 N. D. Epiotis, *J. Am. Chem. Soc.*, **95**, 1191 (1973).
2 *Org. Reaction Mech.*, **1972**, 170.
3 N. D. Epiotis, *J. Am. Chem. Soc.*, **95**, 1200 (1973).
4 N. D. Epiotis, *J. Am. Chem. Soc.*, **95**, 1206 (1973).
5 N. D. Epiotis, *J. Am. Chem. Soc.*, **95**, 1214 (1973).
6 K. N. Houk, *J. Am. Chem. Soc.*, **95**, 4092 (1973).
7 P. V. Alston, R. M. Ottenbrite and D. D. Shillady, *J. Org. Chem.*, **38**, 4075 (1973); see also O. Eisenstein and N. T. Anh, *Bull. Soc. Chim. France*, **1973**, 2721, 2723.
8 K. N. Houk and R. W. Strozier, *J. Am. Chem. Soc.*, **95**, 4094 (1973).
9 N. T. Anh and J. Seyden-Penne, *Tetrahedron*, **29**, 3259 (1973).
10 K. N. Houk, *J. Am. Chem. Soc.*, **94**, 8953 (1972); K. N. Houk , J. Sims, R. E. Duke, R. W. Strozier and J. K. George, *ibid.*, **95**, 7287 (1973); K. N. Houk, J. Sims, C. R. Watts and L. J. Kuskus, *ibid.*, **95**, 7301 (1973).
11 J. Bastide, N. El Ghandour and O. Henri-Rousseau, *Bull. Soc. Chim. France*, **1973**, 2290; J. Bastide and O. Henri-Rousseau, *ibid.*, p. 2294.
12 N. D. Epiotis, *J. Am. Chem. Soc.*, **95**, 5624 (1973).
13 M. N. Padden-Row, P. L. Watson and R. N. Warrener, *Tetrahedron Letters*, **1973**, 1033.
14 R. F. Hudson, *Angew. Chem. Internat. Ed.*, **12**, 36 (1973).
15 R. Hoffmann, C. C. Levin and R. A. Moss, *J. Am. Chem. Soc.*, **95**, 629 (1973); see also N. D. Epiotis and W. Cherry, *Chem. Comm.*, **1973**, 278.
16 J. A. Berson and P. B. Dervan, *J. Am. Chem. Soc.*, **95**, 267 (1973).
17 L. M. Stephenson, T. A. Gibson and J. I. Brauman, *J. Am. Chem. Soc.*, **95**, 2849 (1973).
18 A. H. Andrist, *J. Org. Chem.*, **38**, 1772 (1973).
19 P. Lechtken, R. Breslow, A. H. Schmidt and N. J. Turro, *J. Am. Chem. Soc.*, **95**, 3025 (1973).
20 R. R. Schmidt, *Angew. Chem. Internat. Ed.*, **12**, 212 (1973).
21 Y. Takeuchi, *Yuki Crosei Kagaku Kyokai Shi*, **1972**, 71; *Chem. Abs.*, **79**, 4473 (1973).
22 D. I. Davies, *MTP* (*Med. Tech. Publ. Co.*) *Int. Rev. Sci.*; *Org. Chem.*, *Ser.* 1, **10**, 49 (1973).
23 J. Mathieu, *Bull. Soc. Chim. France*, **1973**, 807.
24 K. N. Houk and L. J. Luskus, *J. Org. Chem.*, **38**, 3836 (1973).
25 B. M. Jacobson, *J. Am. Chem. Soc.*, **95**, 2579 (1973).
26 D. W. Jones and G. Kneen, *Chem. Comm.*, **1973**, 420; see *Org. Reaction Mechanisms*, **1972**, 172.
27 D. W. Jones and R. L. Wife, *Chem. Comm.*, **1973**, 421.
28 K. Seguchi, A. Sera and K. Maruyama, *Tetrahedron Letters*, **1973**, 1585.

[29] N. T. Anh, *Tetrahedron*, **29**, 3227 (1973).
[30] E. T. McBee, M. J. Keogh, R. P. Levek and E. P. Wesseler, *J. Org. Chem.*, **38**, 632 (1973).
[31] F. E. Rogers and S. W. Quan, *J. Phys. Chem.*, **77**, 828 (1973).
[32] W. R. Vaughan and D. R. Simonson, *J. Org. Chem.*, **38**, 566 (1973).
[33] E. K. G. Schmidt, *Angew. Chem. Internat. Ed.*, **12**, 777 (1973); R. H. Grubbs and R. A. Grey, *J. Am. Chem. Soc.*, **95**, 5765 (1973).
[34] T. Sasaki, K. Kanematsu and K. Hayakawa, *J. Am. Chem. Soc.*, **95**, 5632 (1973).
[35] R. Breslow, J. M. Hoffman and C. Perchonock, *Tetrahedron Letters*, **1973**, 3723.
[36] A. Oku, Y. Ohnishi and F. Mashio, *J. Org. Chem.*, **37**, 4264 (1972).
[37] H. W. Gschwend, A. O. Lee and H.-P. Meier, *J. Org. Chem.*, **38**, 2169 (1973).
[38] A. Krantz and C. Y. Lin, *J. Am. Chem. Soc.*, **95**, 5662 (1973).
[39] H. Greuter, G. Frater and H. Schmid, *Helv. Chim. Acta*, **55**, 526 (1972).
[40] M. J. Goldstein and S. A. Kline, *Tetrahedron Letters*, **1973**, 1085 1089.
[41] E. J. Corey and R. L. Danheiser, *Tetrahedron Letters*, **1973**, 4477; H. W. Gschwend, *Helv. Chim Acta*, **56**, 1763 (1973).
[42] A. L. Andrews, R. C. Fort and P. W. LeQuesne, *J. Org. Chem.*, **38**, 237 (1973).
[43] F. A. Carey and A. S. Court, *J. Org. Chem.*, **37**, 4474 (1972).
[44] W. K. Anderson and R. H. Dewey, *J. Am. Chem. Soc.*, **95**, 7161 (1973).
[45] E. J. Corey and B. B. Snider, *J. Org. Chem.*, **38**, 3632 (1973).
[46] B. Willig and J. Streith, *Tetrahedron Letters*, **1973**, 4167.
[47] R. C. Cookson and R. M. Tuddenham, *Chem. Comm.*, **1973**, 742.
[48] Z. Ordelt, *Coll. Czech. Chem. Comm.*, **38**, 1930 (1973).
[49] A. Laporterie, J. Dubac and P. Mazerolles, *J. Organometal. Chem.*, **46**, C3 (1972).
[50] A. I. Konovalov, B. N. Solomonov and A. N. Ustyugov, *Dokl. Akad. Nauk SSSR*, **211**, 102 (1973) *Chem. Abs.*, **79**, 91426 (1973); A. I. Konovalov, Ya. D. Samuilov, E. A. Berdnikov and V. V Plemenkov, *Dokl. Akad. Nauk SSSR*, **208**, 862 (1973); *Chem. Abs.*, **78**, 158594 (1973).
[51] A. I. Konovalov, G. I. Kamasheva and M. P. Loskutov, *Zh. Org. Khim.*, **8**, 2088 (1972); *Chem Abs.*; **78**, 57565 (1973); G. V. Golovkin, N. F. Kononov, and A. F. Plate, *Khim. Atsetilena Tr Vses Konf. 3rd*, **1968**, 270; *Chem. Abs.*, **79**, 4550 (1973); J. S. McConoghy and R. N. Moore, *Am Chem. Soc., Div. Petrol., Chem. Prepr.* **16**, B88 (1971); *Chem. Abs.*, **78**, 28738 (1973).
[52] V. F. Kucherov, S. Ya. Metylaeva and B. A. Rudenko, *Izv. Akad. Nauk SSSR, Ser. Khim.*, **1972** 2126; *Chem. Abs.*, **78**, 28821 (1973); S. Ya. Metylaeva, B. A. Rudenko and V. F. Kucherov, *Izv Akad. Nauk SSSR, Ser. Khim.*, **1973**, 1275; *Chem. Abs.* **79**, 91274 (1973).
[53] M. S. Salakhov, M. M. Guseinov, R. S. Salakhova, E. M. Treivus and F. G. Akhundova, *Azerb Khim. Zh.*, **1971**, 76; *Chem. Abs.*, **78**, 57603 (1973).
[54] Y. Kobuke, *Yuki Gosei Kagaku Kyokai Shi.*, **30**, 992 (1972); *Chem. Abs.*, **78**, 146896 (1973).
[55] G. R. Krow, R. Rodebaugh, R. Carmosin, W. Figures, H. Pannella, G. DeVicaris and M. Grippi *J. Am. Chem. Soc.*, **95**, 5273 (1973); G. R. Krow, R. Rodebaugh, M. Grippi, G. DeVicaris, C. Hyndman and J. Marakowski, *J. Org. Chem.*, **38**, 3094, (1973); see also G. R. Krow, R. Rodebaugh J. Marakowski and K. C. Ramey, *Tetrahedron Letters*, **1973**, 1899.
[56] K. N. Zelenim, V A. Nikitin and N. M. Anodina, *Khim. Geterotsikil. Soedin.*, **1973**, 124; *Chem Abs.*; **78**, 110399 (1973).
[57] N. Sh. Padyukova and V. L. Florent'ev, *Khim. Geterotskil. Soedin.*, **1973**, 600; *Chem. Abs.* **79** **52483** (1973).
[58] S. Sarel, A. Felzenstein and J. Yovell, *Chem. Comm.*, **1973**, 859.
[59] V. Useieli and S. Sarel, *J. Org. Chem.*, **38**, 1703 (1973).
[60] D. J. Pasto and A. F. T. Chen, *Tetrahedron Letters*, **1973**, 713.
[61] W. Schramm and G. Rembarz, *Wiss. Z. Univ. Rostock, Math.-Naturwiss. Reihe*, **21**, 151 (1973).
[62] C. K. Bradsher, F. H. Day, A. T. McPhail and P.-S. Wong, *Chem. Comm.*, **1973**, 156.
[63] T. G. Wallis, N. A. Porter and C. K. Bradsher, *J. Org. Chem.*, **38**, 2917 (1973).
[64] L. Abis and C. Giordano, *J.C.S. Perkin I*, **1973**, 771.
[65] J. Bastide, J. Hamelin, F. Texier and Y. Vo Quang, *Bull. Soc. Chim. France*, **1973**, 2555, 2871.
[66] C. G. Stuckwisch, *Synthesis*, **1973**, 469.
[67] J. Sims and K. N. Houk, *J. Am. Chem. Soc.*, **95**, 5798 (1973).
[68] B. M. Benjamin and C. J. Collins, *J. Am. Chem. Soc.*, **95**, 6145 (1973).
[69] K. Bast, M. Christl, R. Huisgen, W. Mack and R. Sustmann, *Chem. Ber.*, **106**, 3258 (1973); M Christl, R. Huisgen and R. Sustmann, *ibid.*, p. 3275.
[70] R. Huisgen and M. Christl, *Chem. Ber.*, **106**, 3291 (1973).
[71] K. Bast, M. Christl, R. Huisgen and W. Mack, *Chem. Ber.*, **106**, 3312. (1973).

[72] M. Christl and R. Huisgen, *Chem. Ber.*, **106**, 3345 (1973).
[73] *Org. Reaction Mech.*, **1972**, 176; **1971**, 168.
[74] G. Bianchi, C. de Micheli, R. Gandolfi, P. Grünanger, P. V. Finzi and O. V. de Pava, *J.C.S. Perkin I*, **1973**, 1148.
[75] G. Bailo, P. Caramella, G. Cellerino, A. G. Invernizzi and P. Grünanger, *Gazz. Chim. Ital.*, **103**, 47 (1973).
[76] A. Dondoni and G. Barbaro, *J.C.S. Perkin II*, **1973**, 1769.
[77] T. Kauffmann and R. Eidenschink, *Angew. Chem. Internat. Ed.*, **12**, 568 (1973); T. Kauffmann, A. Busch, K. Habersaat and E. Köppelmann, *ibid.*, p. 569.
[78] T. Kauffmann, A. Busch, K. Habersaat and B. Scheerer, *Tetrahedron Letters*, **1973**, 4047.
[79] S. R. Challand, C. W. Rees and R. C. Storr, *Chem. Comm.*, **1973**, 837.
[80] S. R. Tanny and F. W. Fowler, *J. Am. Chem. Soc.*, **95**, 7320 (1973).
[81] K. T. Potts, A. J. Elliott and M. Sorm, *J. Org. Chem.*, **37**, 3838 (1972).
[82] N. Dennis, B. Ibrahim, A. R. Katritzky and Y. Takeuchi, *Chem. Comm.*, **1973**, 292.
[83] R. M. Kellogg, *J. Org. Chem.*, **38**, 844 (1973).
[84] E. Stephan, L. VoQuang and Y. VoQuang, *Bull. Chim. Soc. France*, **1972**, 4781; **1973**, 2795; E. Stephan, L. VoQuang, Y. VoQuang and P. Cadiot, *Tetrahedron Letters*, **1973**, 245.
[85] R. K. Howe and J. E. Franz, *Chem. Comm.*, **1973**, 524.
[86] A. N. Pudovik, L. A. Stabrovskaya, G. I. Evstaf'ev, A. B. Remizov and R. D. Gareev, *Zh. Obshch. Khim.*, **42**, 1862 (1972); *Chem. Abs.*, **78**, 3490 (1973).
[87] H. Henry, M. Zador and S. Fliszár, *Can. J. Chem.* **51**, 3398 (1973).
[88] J.-M. Vandensavel, G. Smets and G. L'Abbé, *J. Org. Chem.* **38**, 675 (1973).
[89] E. V. Loock, J.-M. Vandensavel, G. L'Abbé and G. Smets, *J. Org. Chem.*, **38**, 2917 (1973).
[90] M. Joucla, J. Hamelin and R. Carriér, *Bull. Chim. Soc. France*, **1973**, 3116.
[91] R. Grée, F. Tonnard and R. Carriér, *Tetrahedron Letters*, **1973**, 453.
[92] B. F. Bononi, G. Maccagnani, L. Thijs and B. Zwanenburg, *Tetrahedron Letters*, **1973**, 3569.
[93] L. Thijs, A. Wagenaar, E. M. M. van Rens and B. Zwanenburg, *Tetrahedron Letters*, **1973**, 3589.
[94] K. T. Potts and D. McKeough, *J. Am. Chem. Soc.*, **95**, 2749 (1973).
[95] A. Padwa, M. Dharan, J. Smolanoff and S. I. Wetmore, *J. Am. Chem. Soc.*, **95**, 1945, 1954 (1973).
[96] M. Matsumoto and K. Maruyama, *Chem. Letters* (*Tokyo*), **1973**, 759.
[97] N. S. Narasimham, H. Heimgartner, H.-J. Hansen and H. Schmid, *Helv. Chim. Acta*, **56**, 1351 (1973).
[98] A. Padwa and E. Glazer, *J. Org. Chem.*, **38**, 284 (1973).
[99] D. St. C. Black and K. G. Watson, *Austral. J. Chem.* **26**, 2177 (1973).
[100] O. Tsuge and H. Samura, *Tetrahedron Letters*, **1973**, 597.
[101] L. T. Scott and W. D. Cotton, *J. Am. Chem. Soc.* **95**, 5416 (1973); B. M. Trost and P. L. Kinson, *Tetrahedron Letters*, **1973**, 2675.
[102] G. Lo Vecchio, G. Grassi, F. Risitano and F. Foti, *Tetrahedron Letters*, **1973**, 3777.
[103] K. T. Potts and A. J. Elliott, *J. Org. Chem.*, **38**, 1769 (1973).
[104] E. M. Burgess and J. R. Penton *J. Am. Chem. Soc.*, **95**, 279 (1973).
[105] K. Burger, W. Thenn and E. Muller, *Angew. Chem. Internat. Ed.*, **12**, 155 (1973).
[106] P. D. Bartlett, B. M. Jacobson and L. E. Walker, *J. Am. Chem. Soc.*, **95**, 146 (1973); B. M. Jacobson and P. D. Bartlett, *J. Org. Chem.*, **38**, 1030 (1973); see also D. Kaufmann and A. de Meijere, *Angew. Chem. Internat. Ed.*, **12**, 158 (1973).
[107] L. E. Walker and P. D. Bartlett, *J. Am. Chem. Soc.*, **95**, 150 (1973).
[108] R. Wheland and P. D. Bartlett, *J. Am. Chem. Soc.*, **95**, 4003 (1973).
[109] H. K. Hall, C. D. Smith and D. E. Plorde, *J. Org. Chem.*, **38**, 2084 (1973).
[110] R. Huisgen and G. Steiner, *J. Am. Chem. Soc.*, **95**, 5054, 5055, 5056 (1973); *Tetrahedron Letters*, **1973**, 3763, 3769; see also F. K. Fleischmann and H. Kelm, *Tetrahedron Letters*, **1973**, 3773.
[111] S. Nishida, I. Moritani and T. Teraji, *J. Org. Chem.*, **38**, 1878 (1973).
[112] A. J. Schipperijn and J. Lukas, *Rec. Trav. chim.*, **92**, 572 (1973).
[113] S. Inagaki and K. Fukui, *Bull. Chem. Soc. Japan*, **46**, 2240 (1973).
[114] T. Beetz and R. M. Kellogg, *J. Am. Chem. Soc.*, **95**, 7925 (1973); W. Grimme and H. J. Rother, *Angew. Chem. Internat. Ed.*, **12**, 505 (1973); W. R. Roth, M. Heiber and G. Erker, *ibid.*, p. 504; W. R. Roth and G. Erker, *ibid.*, p. 503.
[115] W. G. Duncan, W. Weyler and H. W. Moore, *Tetrahedron Letters*, **1973**, 4391.
[116] E. V. Dehmlow, *Tetrahedron Letters*, **1973**, 2573.
[117] P. R. Brook, J. M. Harrison and K. Hunt, *Chem. Comm.*, **1973**, 733.
[118] H. W. Moore and W. G. Duncan, *J. Org. Chem.*, **38**, 156 (1973).

[119] H. H. Wasserman, J. U. Piper and E. V. Dehmlow, *J. Org. Chem.* **38**, 145 (1973).
[120] P. W. Jeffs and G. Molina, *Chem. Comm.*, **1973**, 3.
[121] J. P. Luttringer and J. Streith, *Tetrahedron Letters*, **1973**, 4163.
[122] E. Sonveaux and L. Ghosez, *J. Am. Chem. Soc.*, **95**, 5417 (1973).
[123] K. W. Egger, *J. Am. Chem. Soc.*, **95**, 1745 (1973).
[124] J. Metcalfe and E. K. C. Lee, *J. Am. Chem. Soc.*, **95**, 4316 (1973).
[125] R. Gompper and D. Lach, *Angew. Chem. Internat. Ed.*, **12**, 567 (1973).
[126] N. U. Din, J. Riegl and L. Skattebøl, *Chem. Comm.*, **1973**, 271.
[127] M. Nagayama, O. Okumura, S. Noda and A. Mori, *Chem. Comm.*, **1973**, 841.
[128] H. M. R. Hoffmann, *Angew. Chem. Internat. Ed.*, **12**, 819 (1973).
[129] A. E. Hill, G. Greenwood and H. M. R. Hoffmann, *J. Am. Chem. Soc.*, **95**, 1338 (1973).
[130] S. Itô, H. Ohtani and S. Amiya, *Tetrahedron Letters*, **1973**, 1737; R. Noyori, Y. Baba, S. Makino and H. Takaya, *ibid.*, p. 1741.
[131] D. J. Pasto, A. F. T. Chen and G. Binsch, *J. Am. Chem. Soc.*, **95**, 1553 (1973).
[132] R. Noyori, N. Hayashi and M. Katô, *Tetrahedron Letters*, **1973**, 2983.
[133] D. J. Pasto, A. F. T. Chen, G. Ciurdaru and L. A. Paquette, *J. Org. Chem.*, **38**, 1015 (1973); see also R. Gompper and D. Lach, *Tetrahedron Letters*, **1973**, 2683, 2687.
[134] H. R. Pfaendler and H. Tanida, *Helv. Chim. Acta*, **56**, 543, 545 (1973).
[135] C. E. Hudson and N. L. Bauld, *J. Am. Chem. Soc.*, **95**, 3822 (1973).
[136] A. Nabeya, K. Kurita and J. A. Moore, *J. Org. Chem.*, **38**, 2954 (1973).
[137] H. Takeshita, Y. Wada, A. Mori and T. Hatsui, *Chem. Letters* (*Tokyo*), **1973**, 335.
[138] S. Itô, A. Mori, I. Saito, K. Sakan, H. Ishiyama and K. Sasaki, *Tetrahedron Letters*, **1973**, 2737.
[139] W. E. Truce and C.-I. M. Lin, *J. Am. Chem. Soc.*, **95**, 4426 (1973).
[140] T. Machiguchi, M. Hoshino, S. Ebine and Y. Kitahara, *Chem. Comm.*, **1973**, 196.
[141] I. Kawamoto, Y. Sugimura, N. Soma and Y. Kishida, *Chem. Letters* (*Tokyo*), **1972**, 931.
[142] T. Machiguchi, K. Okuma, M. Hoshino and Y. Kitahara, *Tetrahedron Letters*, **1973**, 2011.
[143] T. Tsuji and S. Nishida, *J. Am. Chem. Soc.*, **95**, 7519 (1973).
[144] T. Kaneda, T. Ogawa and S. Misumi, *Tetrahedron Letters*, **1973**, 3373.
[145] H. Iwamura and K. Morio, *Bull. Chem. Soc. Japan*, **45**, 3599 (1972); H. Iwamura, *Tetrahedron Letters*, **1973**, 369.
[146] C. S. Liu and J. C. Thompson, *Tetrahedron Letters*, **1973**, 1581.
[147] P. Chao and D. M. Lemal, *J. Am. Chem. Soc.*, **95**, 920, 922 (1973).
[148] M. Tsutsui, Y.-C. Hrung, J. N. Francis and K. Harasawa, *Chem. Letters* (*Tokyo*), **1973**, 557.
[149] R. Gompper and J. Stetter, *Tetrahedron Letters*, **1973**, 233.
[150] A. Hassner, M. J. Haddadin and A. B. Levy, *Tetrahedron Letters*, **1973**, 1015; M. J. Haddadin and A. Hassner, *J. Org. Chem.* **38**, 2650 (1973).
[151] J. E. Baldwin and D. S. Johnson, *J. Org. Chem.*, **38**, 2147 (1973).
[152] N. Morita, T. Asao, N. Iwagame and Y. Kitahara, *Chem. Letters* (*Tokyo*), **1973**, 67.
[153] T. Kitazume and N. Ishikawa, *Chem. Letters* (*Tokyo*), **1973**, 267.
[154] P. Crews and J. Beard, *J. Org. Chem.* **38**, 522 (1973).
[155] P. Crews and J. Beard, *J. Org. Chem.* **38**, 529 (1973).
[156] L. A. Paquette, M. J. Broadhurst, L. K. Read and J. Clardy, *J. Am. Chem. Soc.*, **95**, 4639 (1973); L. A. Paquette, M. J. Broadhurst, C. Lee and J. Clardy, *ibid.*, p. 4647.
[157] L. A. Paquette and M. J. Broadhurst, *J. Org. Chem.*, **38**, 1886, 1893 (1973).
[158] K. Seckinger, *Helv. Chim. Acta*, **56**, 2061 (1973).
[159] P. de Maria, F. Gasparrini, L. Caglioti and M. Ghedini, *J.C.S. Perkin II*, **1973**, 1922.
[160] F. Serratosa and P. Sota, *Tetrahedron Letters*, **1973**, 821.
[161] K. Burger, J. Fehn and W. Thenn, *Angew. Chem. Internat. Ed.*, **12**, 502 (1973).
[162] D. Tsuge and S. Kanemasa, *Bull. Chem. Soc. Japan*, **45**, 3591 (1972).
[163] J. M. Bohen and M. M. Joullié, *J. Org. Chem.*, **38**, 2652 (1973).
[164] F. A. L. Anet and F. Leyendecker, *J. Am. Chem. Soc.*, **95**, 156 (1973).
[165] M. Green, S. Heathcock and D. C. Wood, *J.C.S. Dalton*, **1973**, 1564.
[166] L. A. Paquette, S. V. Ley, M. J. Broadhurst, D. Truesdell, J. Fayos and J. Clardy, *Tetrahedron Letters*, **1973**, 2943.
[167] S. Raghu and M. Rosenblum, *J. Am. Chem. Soc.*, **95**, 3060 (1973).
[168] A. R. Fraser, P. H. Bird, S. A. Bezman, J. R. Shapley, R. White and J. A. Osborne, *J. Am. Chem. Soc.*, **95**, 597 (1973).
[169] F. D. Mango, *Tetrahedron Letters*, **1973**, 1509.

CHAPTER 13

Molecular Rearrangements

F. L. Scott

Pennwalt Chemical Corporation, Pharmaceuticals Division, P.O. Box 1710, Rochester, N.Y. 14603, U.S.A.

Aromatic Rearrangements

Benzene Derivatives

In the rearrangement of *o*-nitrotoluene (**1**) to anthranilic acid (**2**), careful ^{18}O studies show incorporation of one oxygen atom from the solvent water during the reaction (100°, 9 hr., in the dark, KOH solution) and are consistent with anthranil (**3**) as an intermediate.[1]

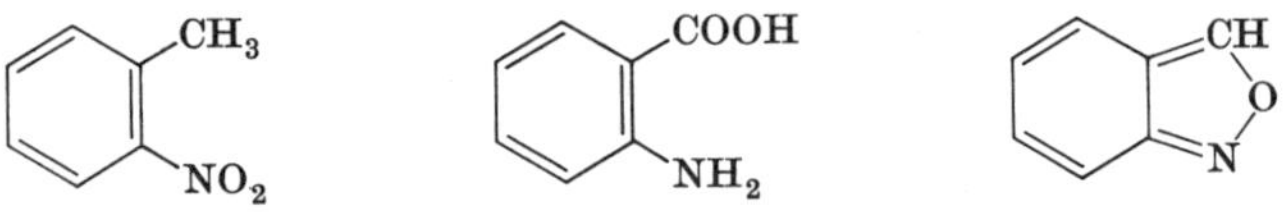

(**1**) (**2**) (**3**)

Several studies have been reported of rearrangements of *N*-alkylanilines. In one, the product mixture formed on thermolysis (at ca. 315° alone) of *N*-benzyl-*N*-methylaniline has been shown to arise from radical reactions,[2] whereas in the other[3] the rather unlikely suggestion is made that a key intermediate in the metal halide-catalysed rearrangement of *N*-alkylanilines is the species formed by metal halide "addition" to the carbon *ortho* to the amino-group.

An extension of the Smiles rearrangement has been described[4] which involves displacement of an aromatic amide group by an amine nitrogen, e.g. the amine-induced conversion of (**4**) into (**5**). As expected, the reaction is facilitated by activation of the

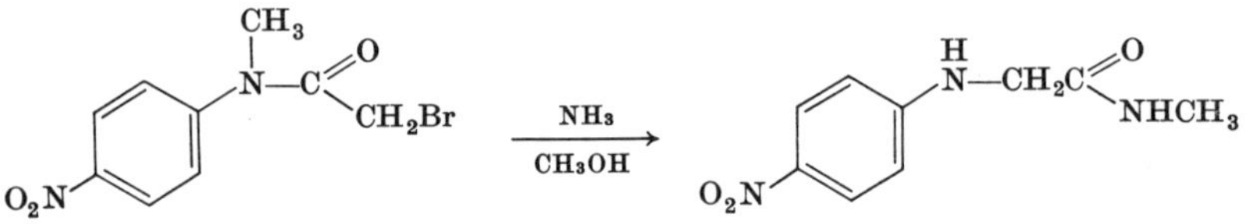

$\xrightarrow[CH_3OH]{NH_3}$

(**4**) (**5**)

aromatic ring by electron-withdrawing groups in the *o*- and *p*-positions. With sterically hindered amines as reagents [the conditions being the usual ones: room temperature, solvents MeOH or $(Me_2N)_3PO$] no rearrangement occurred.

For a general discussion of thermally induced side-chain to ring migration see ref. 5.

Miscellaneous hydrocarbon migrations have been described. For example, pyrolyses (at ca. 800°) of methyl *p*-tolylacetate, and *p*-ethyltoluene proceed via *p*-tolylcarbene (**6**) which forms both benzocyclobutene (**7**) and styrene (**8**).[5a] In the deuteriation of aromatic

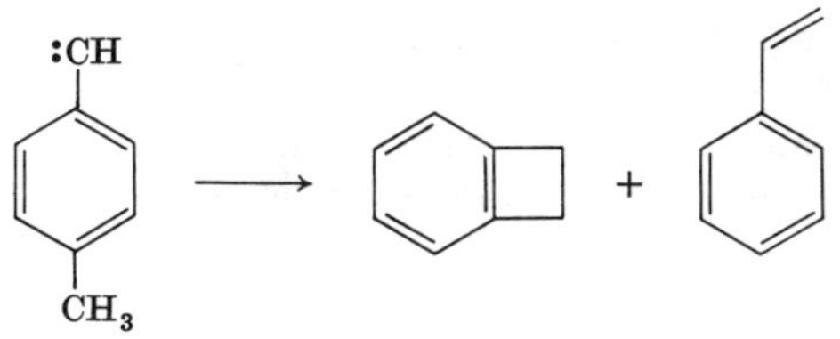

(6) (7) (8)

hydrocarbons with organoaluminium dihalides and traces of water as co-catalysts,[6] isomerization and intermolecular alkylations can interfere with deuteriation.

Photolysis of 2,4,6-tri-*tert*-butylnitrosobenzene (**9**) yields the rearrangement products (**10**) and (**11**) by radical processes.[7]

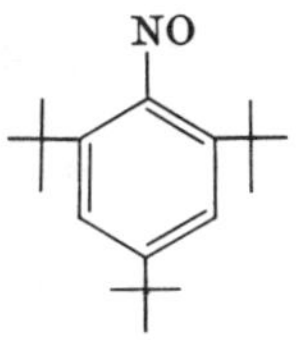

(9) 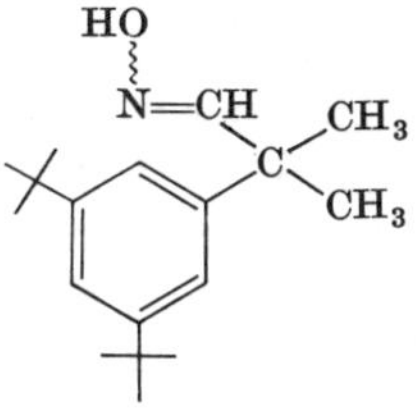 (10) 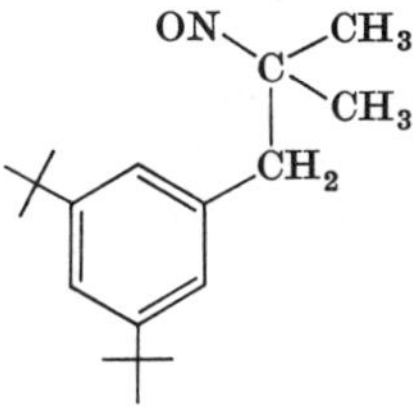 (11)

Several interesting quinone rearrangements have been studied. Ultraviolet irradiation of (**12**) in benzene gave among other products the fused dioxocyclopentanocyclopentene (**13**) [although in unspectacular (1%) yield!].[8] In one of a continuing series of papers, the acid-catalysed decomposition of 2,5-diazido-1,4-benzoquinones (**14**) provides a convenient route[9] to 3-azido-4-Cl-cyanoalkylidene)but-2-enolides (**15**). Depending on the nature of R^1 and R^2, the nitrile and azide groups in (**15**) may remain free or interact to

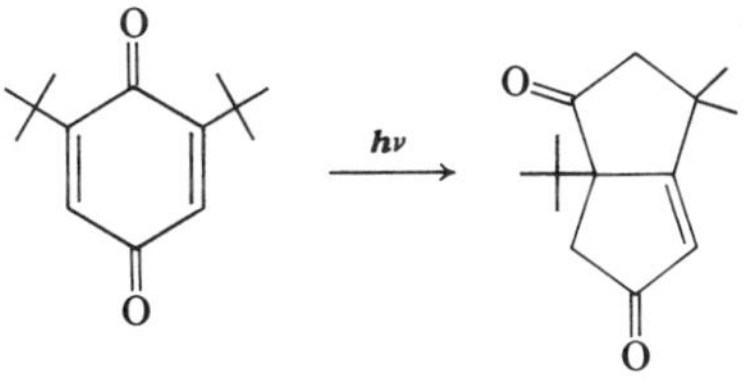

(12) (13) 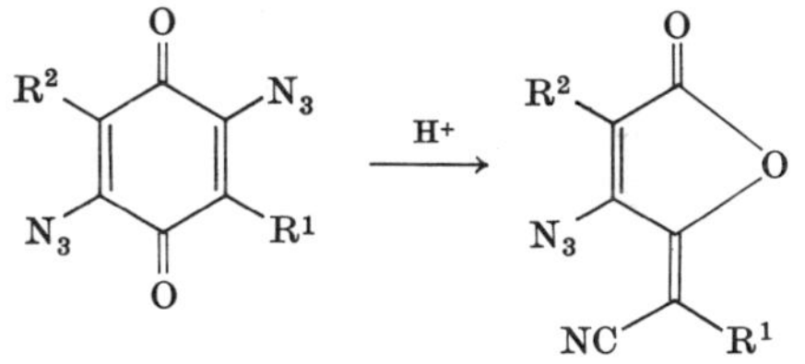 (14) (15)

form tetrazoles. While the emphasis in that report was largely synthetic, the same group in another paper[10] described mechanistic studies of the thermolysis (ca. 80–100°) of the monoazidoquinones (**16**) wherein the reaction took a different course and yielded

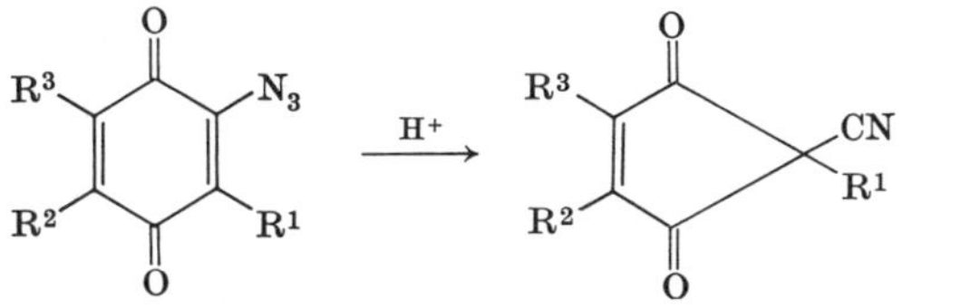

(16) (17) 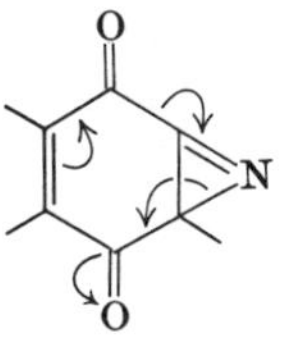 (18) 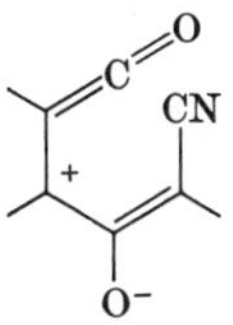 (19)

the 2-cyanocyclopent-4-ene-1,3-diones (**17**). The mechanistic pathway involved the sequence (**16**) → (**18**) → (**19**) → (**17**).

More complete details have now been published on the rearrangements of arenesulphenanilides (**20**) to *o*- and *p*-aminodiphenyl sulphides (**21**). With a wide variation

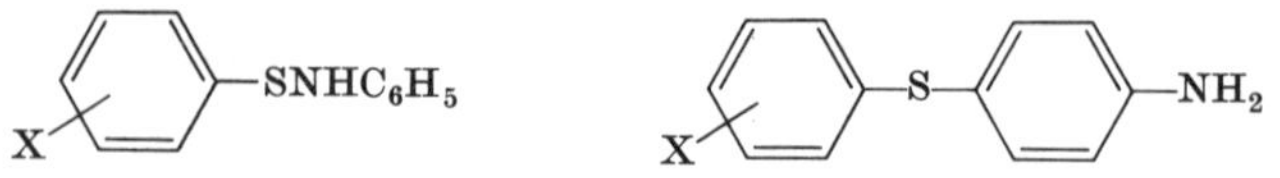

X–C_6H_4–$SNHC_6H_5$

(**20**)

X–C_6H_4–S–C_6H_4–NH_2

(**21**)

in X (4-Me, H, 4-Cl, 4-Br, 4-NO_2 or 3-NO_2) the generality of the process (at 190° in aniline) was confirmed.[11] Electron-withdrawing groups favoured rearrangement as did the presence of aniline hydrochloride. The reaction had to compete with disproportionation (to disulphides and azobenzenes) of the substrates (**20**) and the balance between rearrangement and disproportionation was very sensitive to environmental factors. The acid-catalysed process was examined further and evidence was accumulated to suggest (with reservations) that the rearrangement process was intramolecular.[12]

Competitive behaviour is also encountered in the Fischer–Hepp rearrangement of *N*-nitrosoamines. By adding *N*-methylaniline to the reaction mixture of *N*-methyl-*N*-nitrosoaniline in hydrochloric acid, k_{obs} was decreased to a limiting value equivalent to the rate coefficient for the rearrangement process alone.[13] There was a linear dependence of this limiting rate (k_2) on H_0 (slope −1.2 between 3M- and 6.5M-acid and a ring isotope effect $(k_2)_H:(k_2)_D$ of 2.4. The mechanism suggested for the rearrangement involved the σ-complex and π-complex pathway shown in Scheme 1.

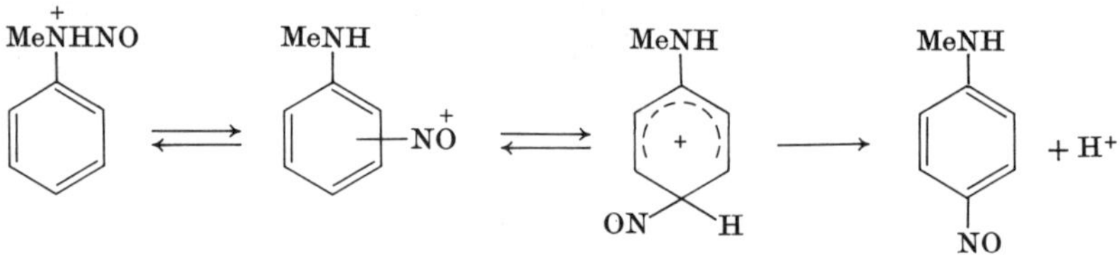

$$PhN^{+}H(Me)NO \rightleftharpoons [PhNHMe \cdot NO^{+}]_{\pi} \rightleftharpoons \sigma\text{-complex (ON, H)} \longrightarrow p\text{-ON}C_6H_4NHMe + H^{+}$$

SCHEME 1.

Continuing interest has been displayed in the Wallach rearrangement. The azoxybenzene (**22**) undergoes a photocatalysed rearrangement to the *o*-hydroxyazobenzene (**23**), the well-known photo-Wallach reaction. It has now been shown[14] that azoxybenzene undergoes a rapid protonation in its first excited state *before* the rearrangement that occurs in strongly acidic solutions, so that under these conditions the photochemical process is acid-catalysed. In an attempt to distinguish between ionic and free-radical mechanisms for this photoconversion of (**22**) into (**23**), Bunce and his colleagues[15] examined substituent effect in one ring (of the CHXY type), solvent variation and kinetic deuterium isotope effect, and concluded that the intramolecular process outlined in Scheme 2 was the one observed. The mode and role of the oxygen attack to form (**24**) is still uncertain. Buncel[16] in his continuing study of the reaction under discussion has examined the behaviour of the blocked system hexamethylazoxybenzene (**25**). Rearrangement of (**25**) did occur in moderately concentrated sulphuric acid and displayed

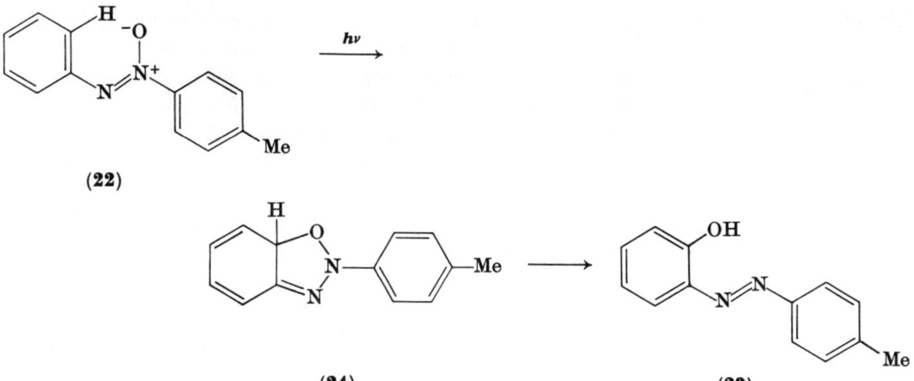

(22)

$h\nu$

(24) (23)

SCHEME 2.

an unusual course, yielding the alcohol (**26**). No methyl migration was encountered. The data could be rationalized by means of the dication (**27**) proposed previously.

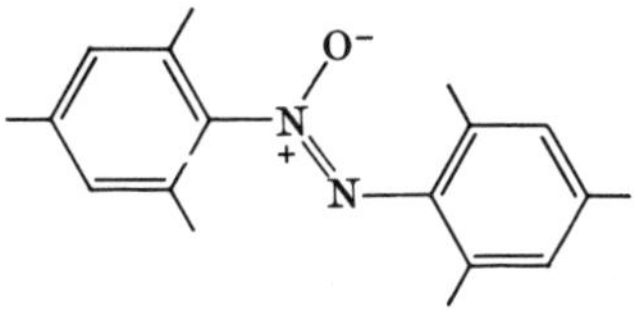

(25) 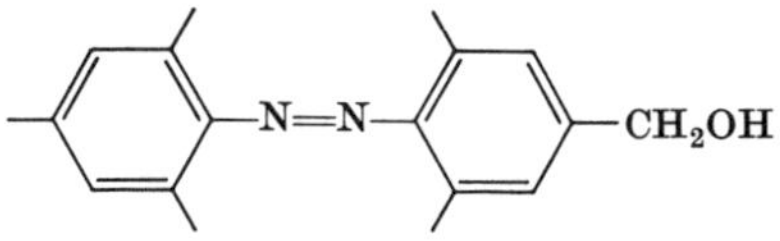 (26)

Interest is also being maintained in the classic benzidine rearrangement. Banthorpe and O'Sullivan[17] reported their studies on *N*-alkylated hydrazoarenes. The acid-catalysed rearrangement (which involves a high degree of semidine formation) of (**28**) was of the first order in acid and has a solvent isotope effect k_{D_2O}/k_{H_2O} of 2.3. Reaction was

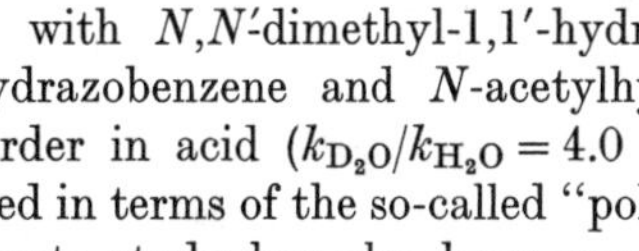

(27) 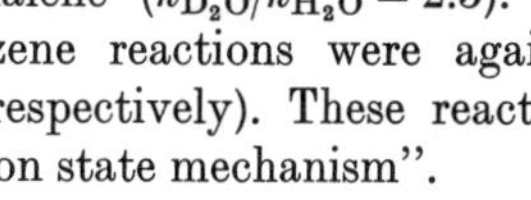 (28)

intramolecular and did not involve radicals. First-order reactions in acid were also observed with *N*,*N*′-dimethyl-1,1′-hydrazonaphthalene ($k_{D_2O}/k_{H_2O} = 2.3$). With *N*-methylhydrazobenzene and *N*-acetylhydrazobenzene reactions were again of the second order in acid ($k_{D_2O}/k_{H_2O} = 4.0$ and 3.1, respectively). These reactions were interpreted in terms of the so-called "polar transition state mechanism".

A separate study has also been reported with the *N*-acetyl compound (**29**).[18] In concentrated mineral acid (6.0*F*-$HClO_4$), it rearranges to (**30**) in nearly quantitative

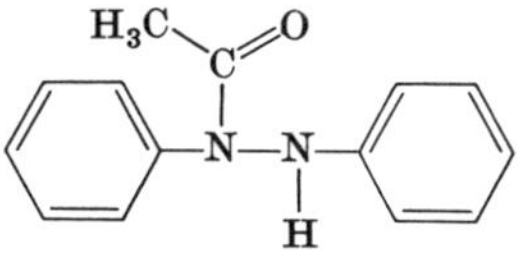

(29) 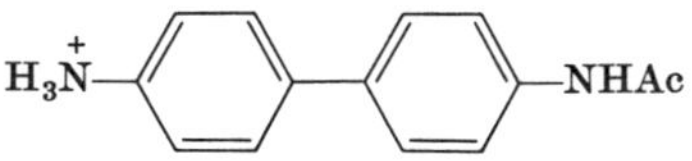 (30)

yield by an intramolecular process (established by appropriate perdeuterium-labelling). A plot of log k_{app} against H_0 gives a linear correlation with a slope of 0.833. The solvent isotope effect was $k_{H_2O}/k_{D_2O} = 1.27$.* A small substrate isotope effect was observed on the rate of rearrangement of the ring-perdeuteriated material ($k_H/k_D = 1.07$). Isotope studies also ruled out protonation at sites other than nitrogen. The electron-withdrawing effect of the *N*-acetyl group is regarded as usurping the catalytic role played by a second proton.

Despite the fact that tetrazenes are a ready source of radicals, and that photolysis of (**31**) can lead to benzidine rearrangement products, the initial products of such photolyses are the corresponding *N,N'*-dimethylhydrazoarenes and there is no direct evidence for radical involvement.[19] Rearrangement of the diazocine (**32**) (HCl, MeOH, 0°, few minutes) yielded the spiro-compound (**33**) which is the first isolated example of a stable *o*-semidine rearrangement intermediate.[20]

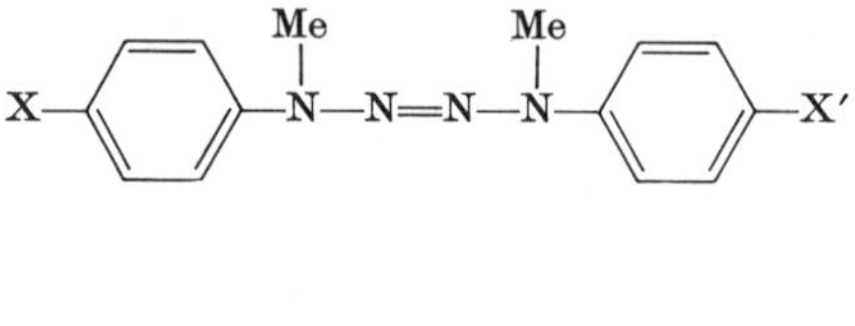

(**31**)

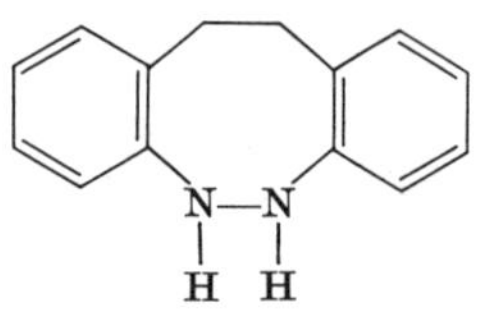

(**32**)

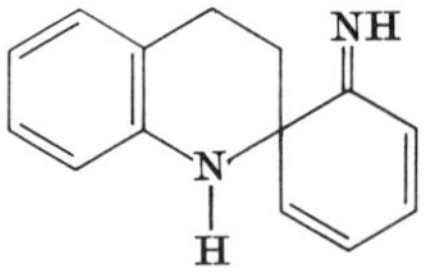

(**33**)

The Fries rearrangement of 2,5-dimethylphenyl propionate (**34**) with $AlCl_3/TiCl_3$ in nitromethane at 20° yielded the *p*-propionyl compound, the *o*-propionyl derivative (the first recorded example of this kind with such systems) and a methyl-migration product.[21] The rearrangements of *tert*-butylated phenols involves both loss of *tert*-butyl and its retention with normal Fries rearrangement.[22,23] Isomerization of phenyl benzoates in tetrachloroethane with trifluoromethanesulphonic acid as catalyst (24 hr., 170°), and of *o*-hydroxyaryl ketones under the same conditions, reveals[24] the Fries rearrangement as reversible, with thermodynamic control operating in product formation. The 1,8-shift of the acyl group in (**35**) which forms (**36**) in polyphosphoric acid[25] follows an inter-

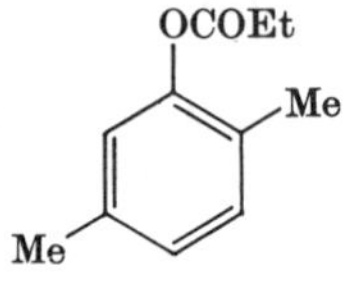

(**34**)

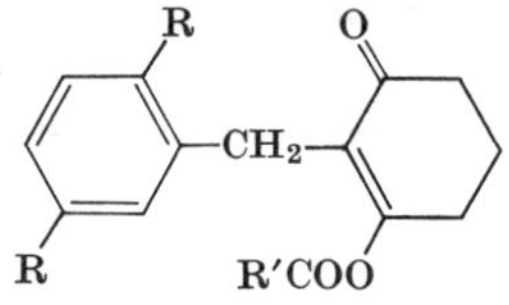

(**35**)

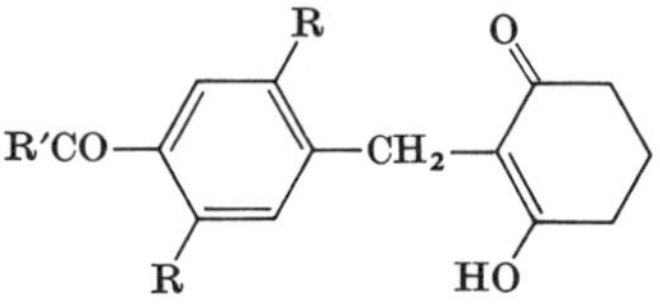

(**36**)

* Contrast the data cited from ref. 17.

molecular path and is of the Fries type. Kalmus and Hercules published an interesting discussion of the photo-Fries rearrangement of phenyl acetate.[25a]

Condensation of phenols with α,β-unsaturated carboxylic acids in polyphosphoric acid yields first acylation, followed by Fries rearrangement, and then cyclization to form substituted chromanones.[25b]

Some rearrangements of hydrazonyl ethers of type (**37**) have been described which can be classified as Chapman or Smiles types depending upon the mode of reaction. The thermal rearrangements[26] of both (**37**) and (**39**) to yield (**38**) display radical character-

$Ar^1{-}C(OAr^3){=}NNHAr^2$ (**37**) $Ar^1{-}CONHNAr^2Ar^3$ (**38**) $Ar^1{-}C(OAr^2){=}NNHAr^3$ (**39**)

istics (being catalysed by initiators and certain oxidizing agents). The aryl migrations were intramolecular (no cross-overs) and the substitution pattern of the migrating group was retained. Electron-withdrawing substituents retarded and electron-donating substituents enhanced reactivity, irrespective of whether the substituent was in Ar^1, Ar^2 or Ar^3 (ρ for Ar^2 was -2.07). A radical of the hydrazyl type was suggested as being the key intermediate in the rearrangement process. A formally similar process has also been described that occurs under base-catalysed conditions (triethylamine in boiling ethanol.)[27] Here again the reactions were intramolecular and a transition state of type (**40**) is involved. In more forcing conditions, an additional cyclization (**41** → **42**) was also realized.

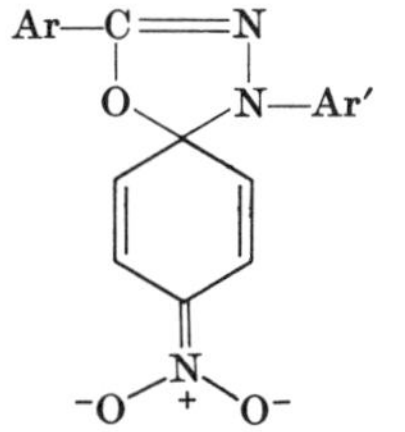

(**40**) 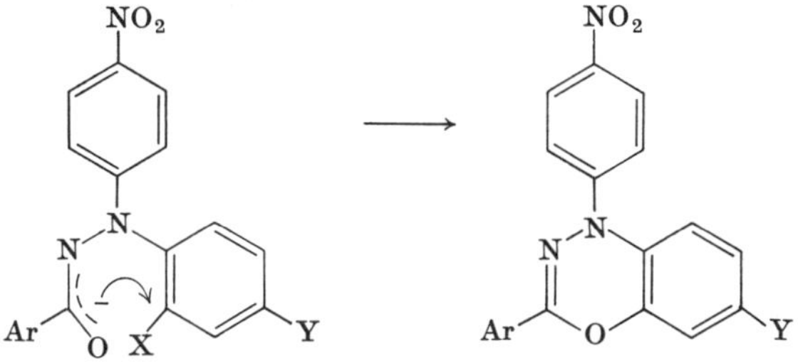 (**41**) (**42**)

The rearrangement of oxanilic hydrazonates similar to (**37**) has also been reported.[28] Under mass-spectral conditions, both Chapman rearrangements (of benzoylhydrazines[28a] and of benzimidates[28b]) and reverse Chapman reactions have been observed.

Butyl-lithium-catalysed Smiles rearrangements of both *o*-methyl-[29] and trifluoromethyl-diaryl sulphones[30] have been described.

An interesting synthesis of disubstituted acetylenes which involves the novel rearrangement (**43** → **44**) has been described (without mechanistic studies).[31]

$$ArCH_2SCCl_2Ar \rightarrow ArC{\equiv}CAr$$

(**43**) (**44**)

In the acid-catalysed cyclization of α-(*o*-arylthiophenyl)-substituted alcohols (**45**) to thioxanthen derivatives (**48**), *ipso*-attack occurs to form the spiro-species (**46**) [which undergoes a 1,2-sulphur shift and thereby effects rearrangement of R^2 (*vis-à-vis* the sulphur atom)] as well as *ortho*-attack (**47**) which leads to unrearranged products.[32]

(45) → (46)

↓ ↓

(47) → (48)

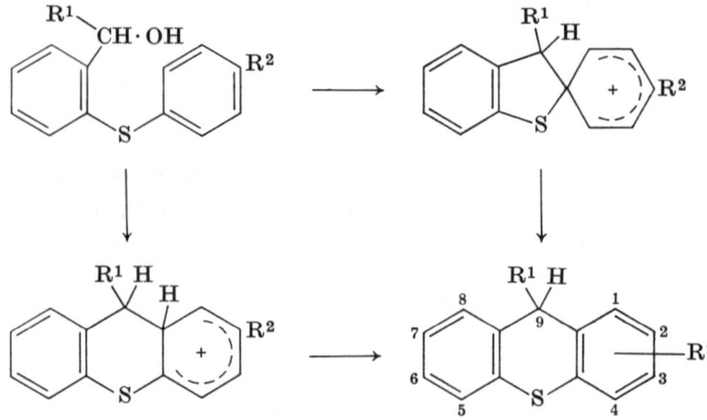

Heterocyclic Derivatives

The nature of the products obtained by the reactions of arenesulphonyl azides with tetrahydrocarbazoles is markedly affected by substituents at positions 1 and 9 (Scheme 3). With $R^1 = R^2 = H$, nitrogen is lost from the adduct (**49**), this being followed by nitrogen bridging and rearrangement to yield (**50**). When $R^1 = H$ and $R^2 = Me$, C—C bond migration occurs to give (**51**). When $R^1 = R^2 = Me$ the sequence (**49** → **52** → **53**) is followed.[33]

(50)

(49) (51)

(52) (53)

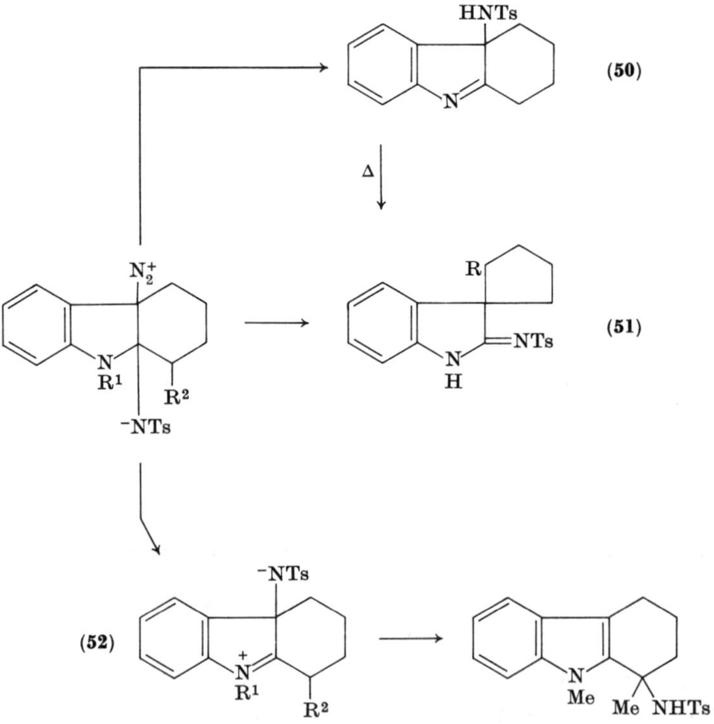

Scheme 3.

Photochemical excitation of the substituted 7-azanorbornadienes (**54**) yields the 3-azaquadricyclanes (**55**).[34] These in turn can cleave to the azepines (**56**), the balance between the various processes being sensitive to the *N*- and *C*-substituents.

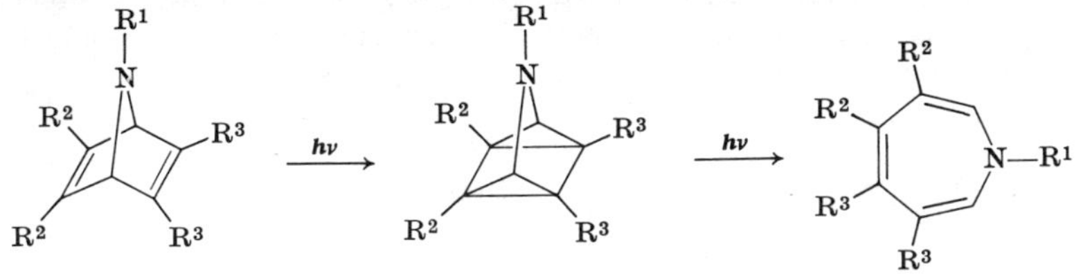

(**54**) (**55**) (**56**)

The reaction[35] of appropriately substituted 4-alkylpyridines (**57**) with sulphenyl halides may proceed by a pathway analogous to that observed in the reactions of pyridine *N*-oxides and acetic anhydride to yield side chain-substituted products (**58**)

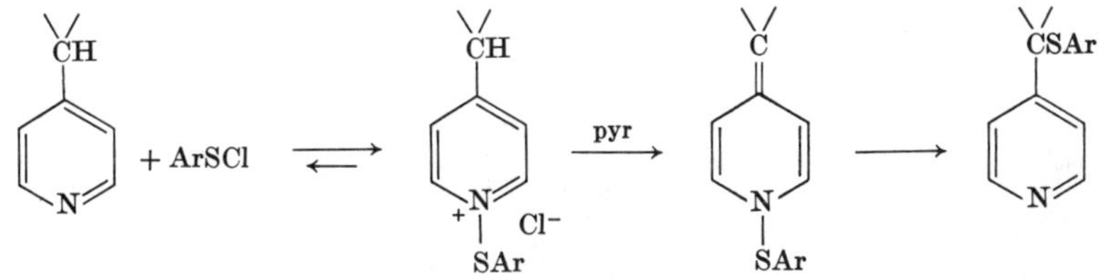

(**57**) (**58**)

SCHEME 4.

(Scheme 4). It has been suggested[36] that β-alkylation of pyridine (and quinoline) 1-oxides with acetylenes (e.g. PhC≡CCN) proceeds through a cyclic adduct (**59**) which may open in various ways leading ultimately to the product (**60**).

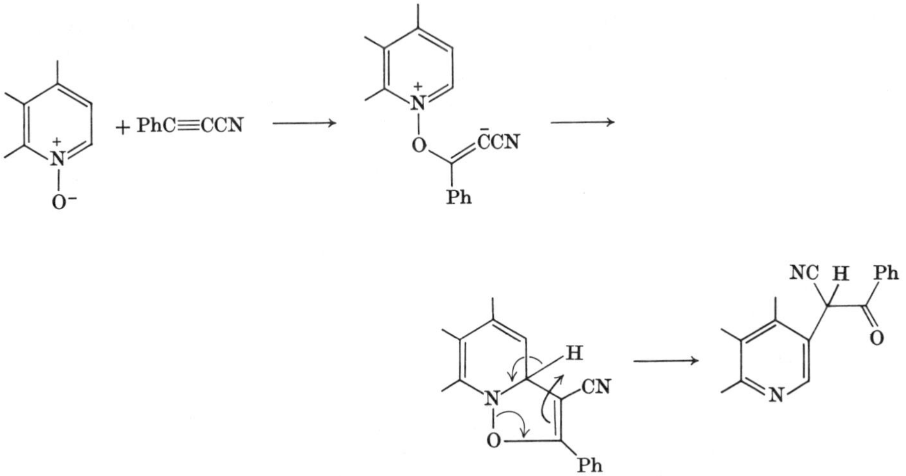

(**59**) (**60**)

The reaction of 3,3-disubstituted 3*H*-indoles with acyl chlorides in pyridine has been reported to yield 1:1 adducts (and derivatives thereof) as well as dimeric products.[37]

The acid-catalysed rearrangement of the optically active disubstituted 1,2-dihydroisoquinoline (**61**) gives an optically active product (**62**).[38] The reaction is intermolecular, and to explain the rearrangement a bimolecular exchange reaction is proposed in the course of which two molecules exchange their benzyl groups.

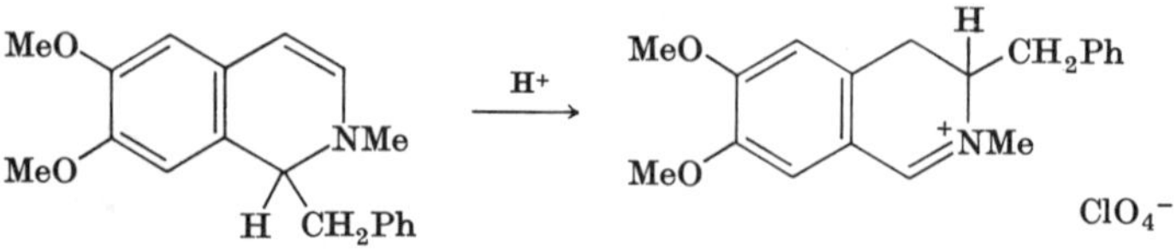

(**61**) (**62**)

In the photo-oxidation of phenyl-substituted isobenzofurans (**63**) to the ketones (**65**), the intermediate ozonide (**64**) is postulated; its cleavage would explain the formation of the various products isolated.[39]

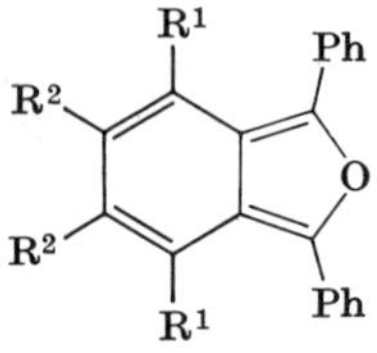

(**63**) 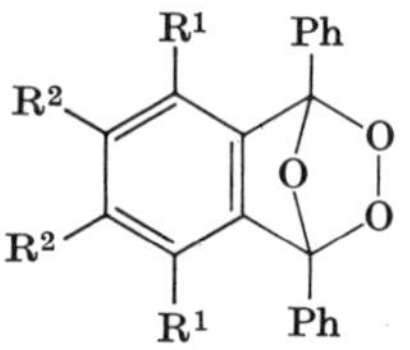 (**64**) 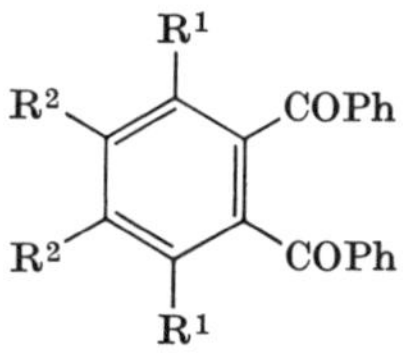 (**65**)

The acid-catalysed rearrangement of epoxycycloalkanols corresponds to cyclic ether formation involving intramolecular hydroxyl group displacement of the epoxide-oxygen with inversion of configuration.[40]

In an interesting study, Berson and his collaborators have examined[41] the requirements for concertedness in the cleavage of 3,4-diazabicyclo[4.2.0]alk-3-enes (**66**). The stereospecificities observed in the cleavages (**66** → **67**) and (**68** → **69**) point firmly to the decomposition of (**66**) and (**68**) as concerted cycloreversions.

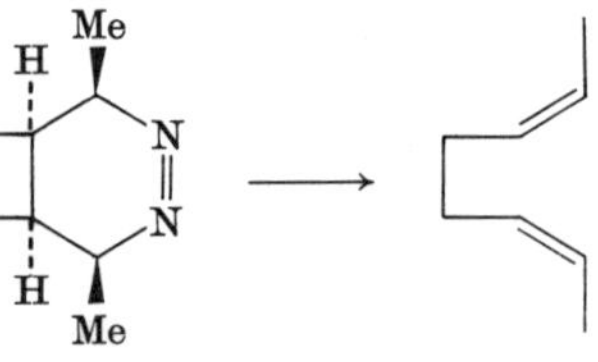

(**66**) (**67**) 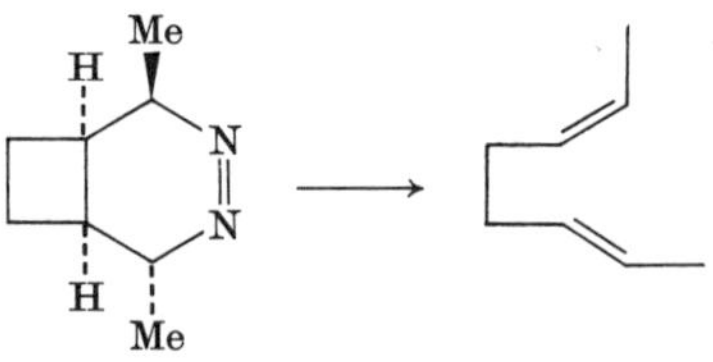 (**68**) (**69**)

The thermally induced (500–600°) rearrangement of 1-substituted imidazoles to 2-substituted isomers has been studied.[42] Migration of a methoxy-group (under very mild conditions; CCl_4, 30°) around the periphery of a pyrazole ring have been recently reported[43] (**70** → **71** + **72** + **73**) but without mechanistic data.

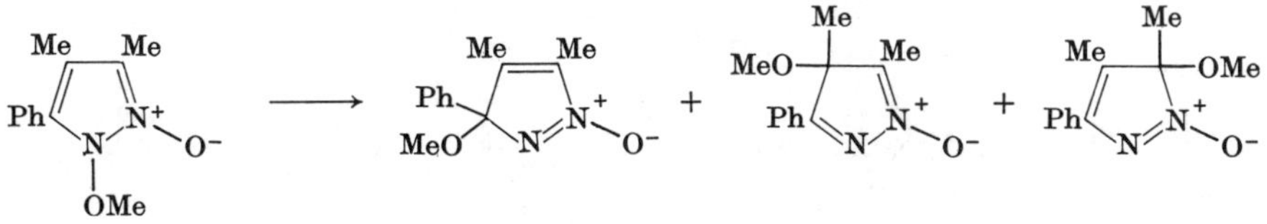

(70) (71) (72) (73)

The $AlCl_3$-catalysed rearrangement[44] of the benzoxazin-1-one (**74**) to the corresponding *N*-arylphthalimide (**77**) (in nitrobenzene) corresponds to the sequence: Beckmann rearrangement to yield the zwitterions (**75**) and (**76**), followed by internal trapping of the acylium ion to yield (**77**). Interestingly, when the solvent used to effect the rearrangement is an activated aromatic hydrocarbon Ar′H, then (**76**) can be trapped in a different way to yield the anilides (**78**).

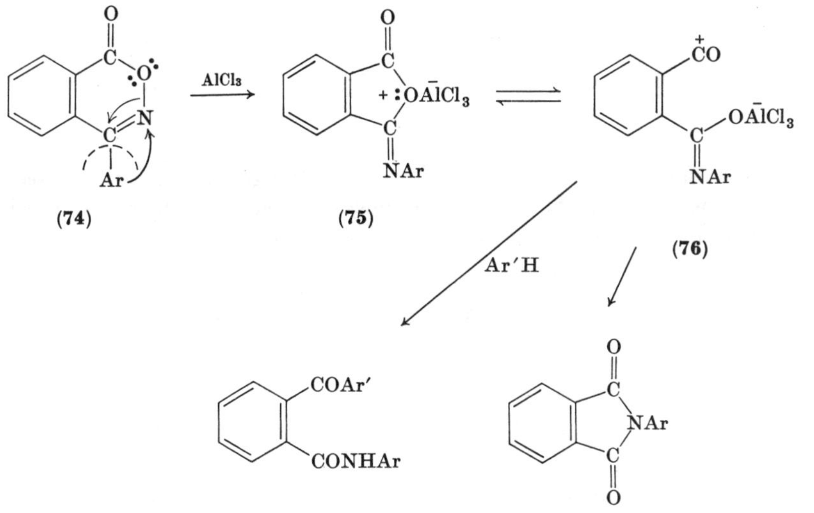

(74) (75) (76) (78) (77)

Reaction of the dihydro-1,2-diazepin-4-ol (**79**) with acylating agents[45] in the presence of weak bases can give, via bridging, the bicyclo-system (**80**), or the fused-ring (**81**) or a ring-contracted product (**82**). Compound (**81**) can also rearrange to give either

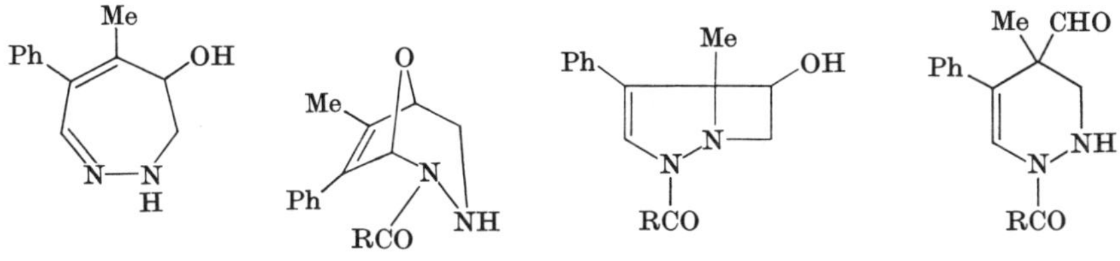

(79) (80) (81) (82)

(**80**) or (**82**). An acyldiazepinium cation-acyl betaine (**83**) is suggested as an intermediate in these interesting ring interconversions. The same group has reported rearrangement

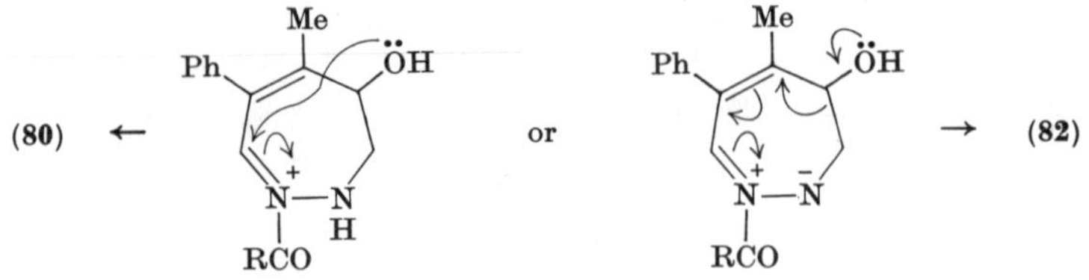

(83)

of (**84**) to (**85**), a reaction which occurs in high yields by the action of heat, acid or base.[46]

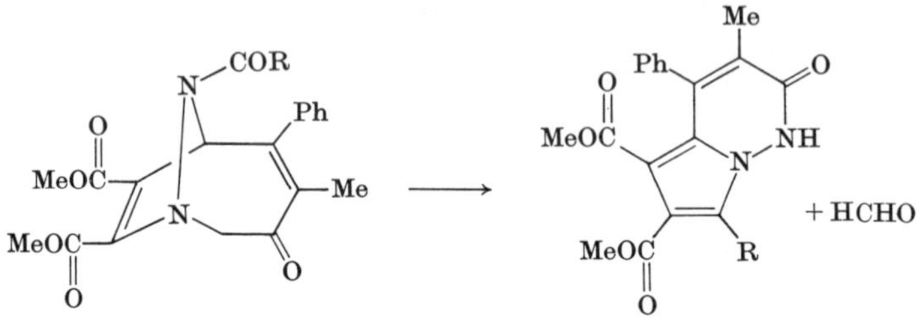

(**84**) (**85**)

The substituted thiopenam (**86**) undergoes rearrangement with ring expansion in trifluoroacetic acid to yield the 1,4-thiazepine (**87**); such reactions have considerable synthetic utility in designing novel bicyclic β-lactam systems.[47]

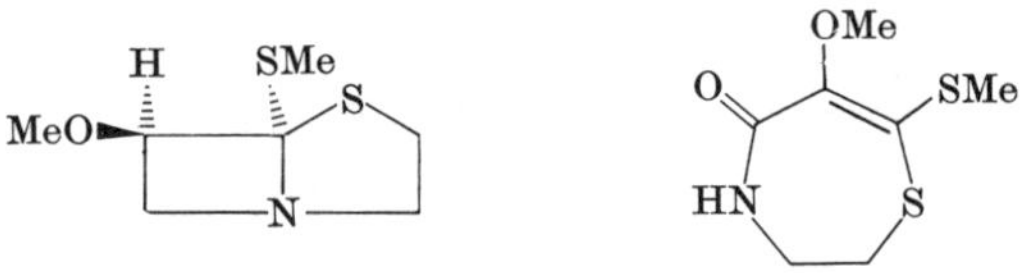

(**86**) (**87**)

Some interesting ring-interconversions have been reported also for 1,4-thiazine systems. Thus compound (**88**) is smoothly isomerized in boiling toluene to (**89**)—the new carbon—sulphur bond being formed with retention of configuration.[48] When the aziridine ring in (**88**) is opened with hydriodic acid, and the iodo-compound cyclized thermally to yield (**89**) the new carbon–sulphur bond is this time formed with inversion of configuration. The isomerization of (**88**) to (**89**) involves a 1,3-sulphur shift, the stereochemical data being most reasonably accounted for in terms of a concerted reorganization involving the thermodynamically unfavourable conformer (**90**) of the aziridine. The aziridine (**91**) may be an intermediate in the process which then represents an example of a [4.4]-dyotropic shift.[49]

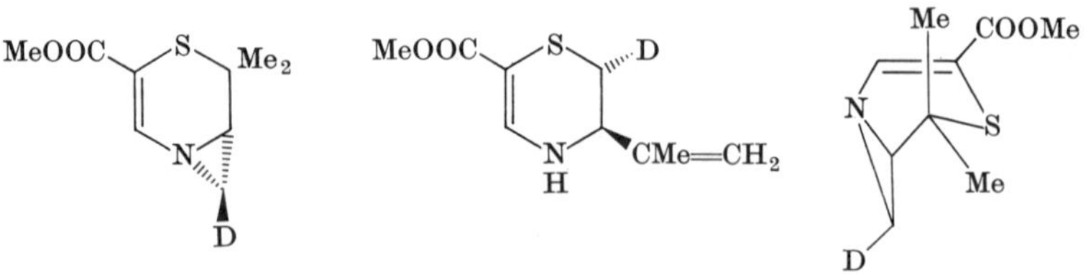

(**88**) (**89**) (**90**)

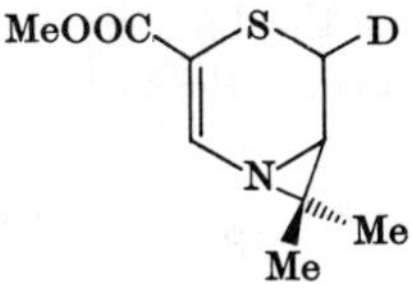

(91)

A Dimroth rearrangement has been observed with imidazo[1,2-*a*]pyridines activated by a nitro-group at position 6 or 8 in aqueous basic media, the reaction being (**92** → **93**).[50] Intermediates of two kinds were detected spectroscopically in the studies, those

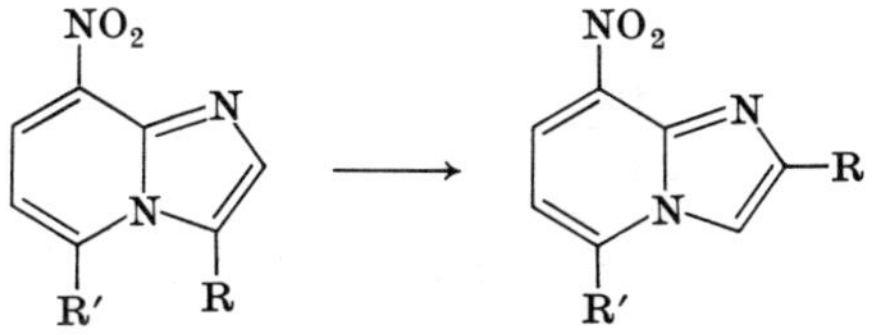

(92) (93)

corresponding to hydroxide attack at the 5-position forming highly delocalized anions whose rings could be opened whilst the imidazole unit remained intact. Other Dimroth rearrangements (for recent reviews see ref. 51) reported involve (**94**) → (**95**),[52] a new

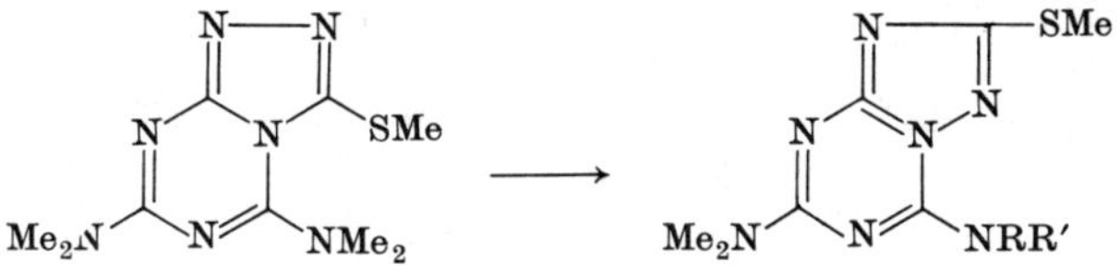

(94) (95)

type of reversible Dimroth rearrangement as in (**96** → **97** → **98** → **99**) (Scheme 5);[53] catalysis of azapurine Dimroth reorganizations by methylammonium salts has also been described.[54]

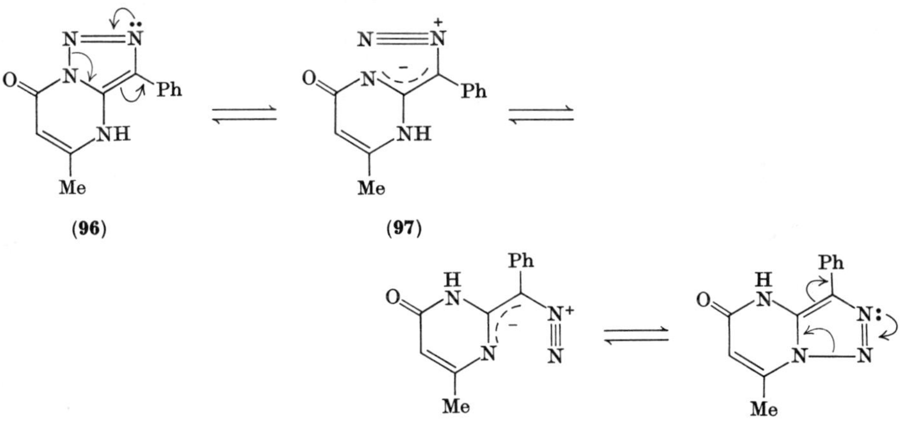

(96) (97)

(98) (99)

SCHEME 5.

The intermolecular character of the Lander rearrangement[55] of (**100**) to (**101**) has been studied recently.[56,57] A formally related reaction which constitutes a simple and

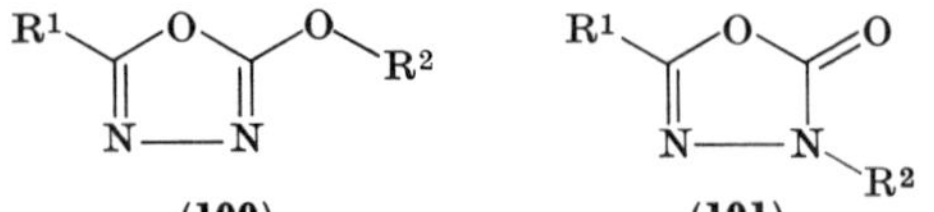

(**100**) (**101**)

effective way of synthesizing certain imidazoles is the base-catalysed interconversion of certain 1,2,4-oxadiazoles (**102**) to (**103**).[58]

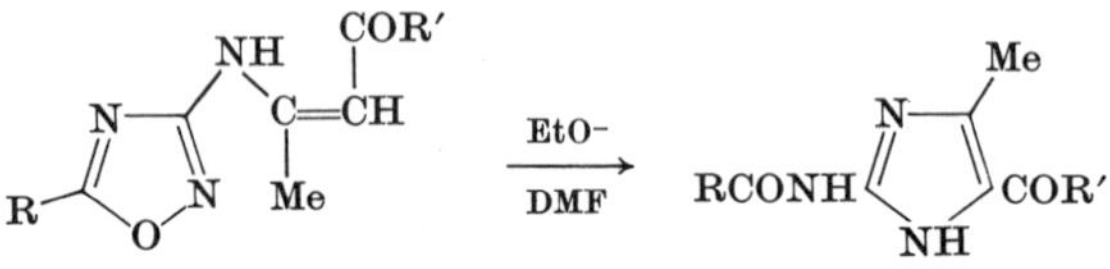

(**102**) (**103**)

The novel rearrangement of a substituted pyrazolo[1,5-*a*]pyrimidine (**104**) to a pyrazolo[3,4-*d*]pyrimidine (**105**) by treatment with hydrogen peroxide and sodium hydroxide has been observed.[59]

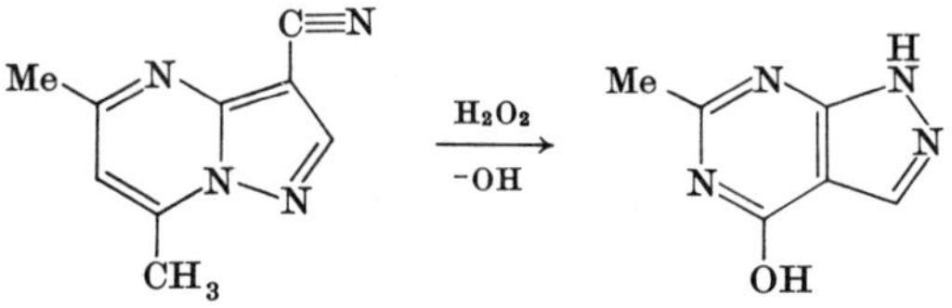

(**104**) (**105**)

Several N → O and N → S migrations (within the heterocyclic series) have been observed. Thus the (bistriazinylamino)phenols (**106**) are rearranged by base[60] to yield (**106a**) and finally (**107**). Similar reactions occur with an *o*-thiophenoxide ion.

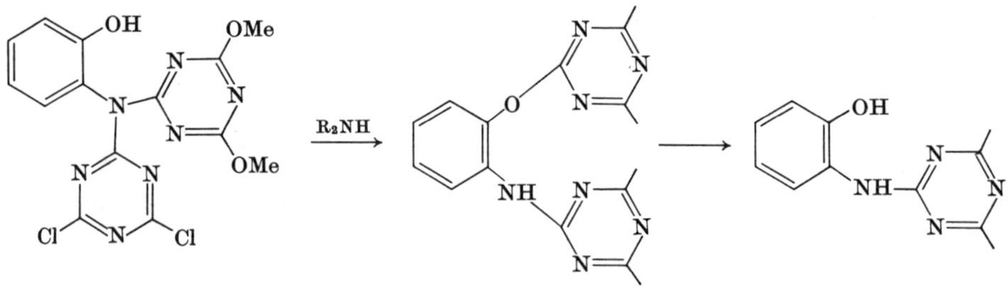

(**106**) (**106a**) (**107**)

By controlling the reaction conditions, the rearrangement can be intercepted at (**106a**). Under different conditions, the dihydrophthalazin-4-ones (**108**) readily undergo conversion into the *N*-substituted products (**109**).[61]

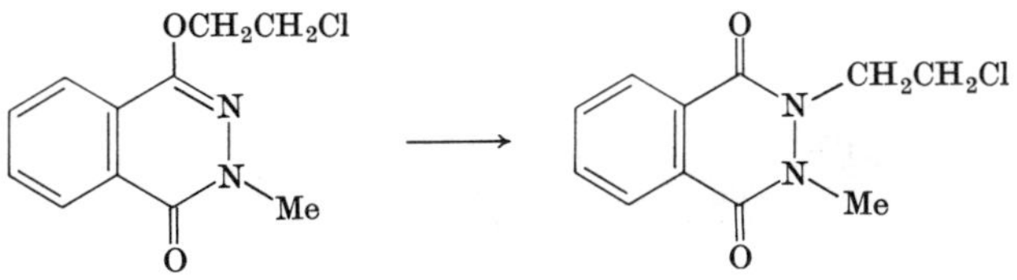

(**108**) (**109**)

A new 1,3-migration reaction has been reported in the smooth rearrangement of thiadiazoline 1-oxides such as (**110**) to thiadiazoles (**111**), probably by elimination (of $PhSO_2OH$) and readdition thereof.[62]

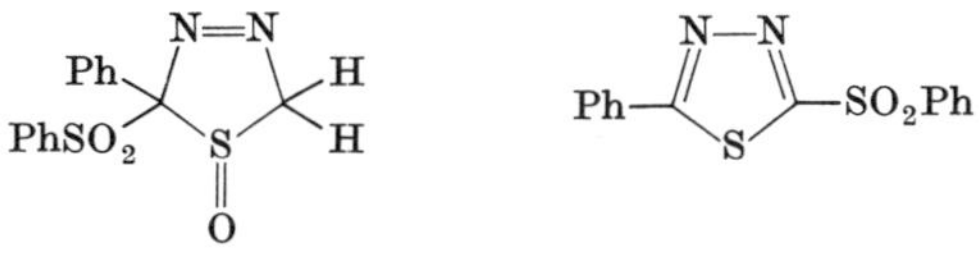

(**110**) (**111**)

In the reactions of 4-phenyl-1,2,4-triazoline-3,5-dione with vinyl esters, kinetic evidence points to the existence of an intermediary 1,4-dipole which then rearranges to products,[62a] as shown in Scheme 6.

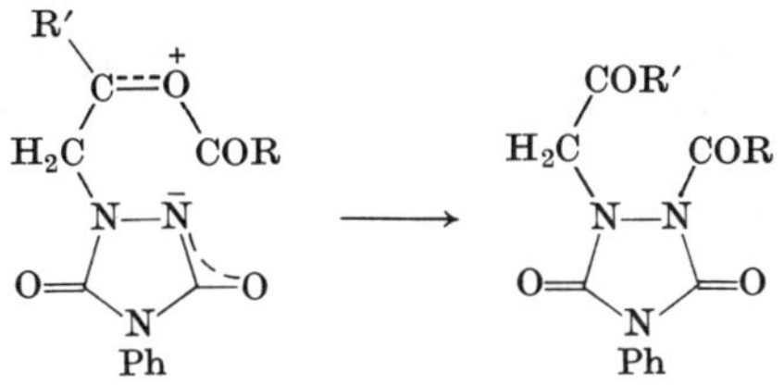

SCHEME 6.

Cyclohexadiene Derivatives

Kinetic studies have been reported on the rearrangements of a series of alkyl-substituted cyclohexa-2,5-dienones,[63,64] in both sulphuric and perchloric acid at 25°. Increasing acidity increased k_{obs} (all rates following the amide acidity function H_A). The mechanism corresponds to unimolecular alkyl migration in the dienone monocation and is an A-1 process (Bunnett–Olsen $\phi < 0$; faster $HClO_4$ than in H_2SO_4 of the same H_A). The same group has used the acid-catalysed rearrangements of 4,4-disubstituted cyclohexadienones to determine the migratory aptitude of methyl and ethyl groups (Et/Me = 25/1).[65]

In charge-induced sigmatropic reactions the charge serves only to accelerate appreciably thermal orbital-symmetry-allowed reactions. In charge-controlled sigmatropic reactions the charge determines the course of the transformations according to the Woodward–Hoffmann rules. In acylating systems, allylcyclohexadienones undergo charge-induced and charge-controlled reactions simultaneously.[66] Acid-catalysed rearrangement of methyl-substituted 2-propargylcyclohexadien-1-ones (**112**) in acetic anhydride involves the formation of acetoxybenzenium ions (**113**) which then undergo charge-controlled [3,4]- and [1,2]-sigmatropic shifts.[67]

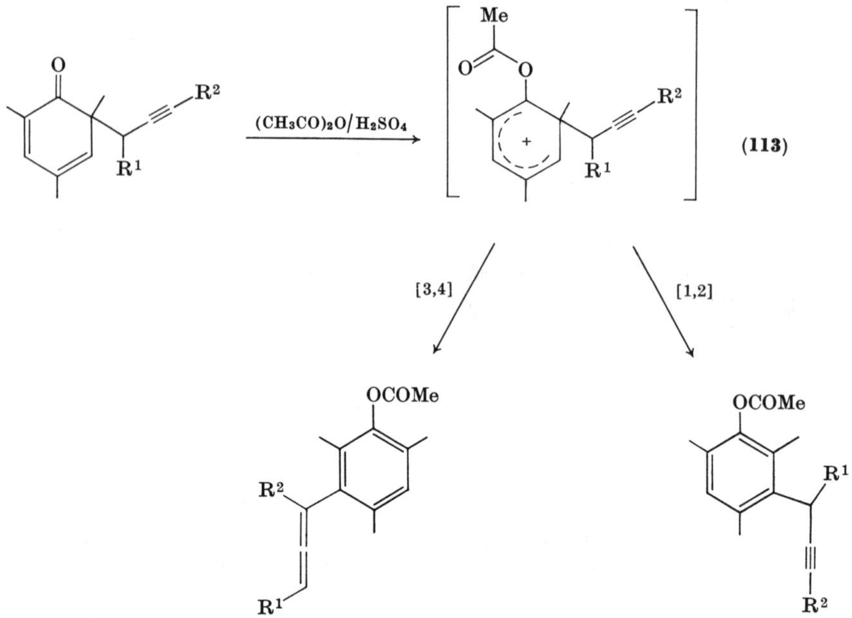

(112) $\xrightarrow{(CH_3CO)_2O/H_2SO_4}$ (113)

[3,4] [1,2]

OCOMe R^2 R^1

OCOMe R^1 R^2

The rearrangements of some substituted spirocyclopropanecyclohexadienones have been the subject of some controversy but a recent paper clears up the difficulty.[68] In the synthesis of the key compounds, several aromatic rearrrangements went previously undetected (Scheme 7). Finally, treatment of certain steroidal phenols with HF–SbF_5 causes effective dearomatization to dienones.[69]

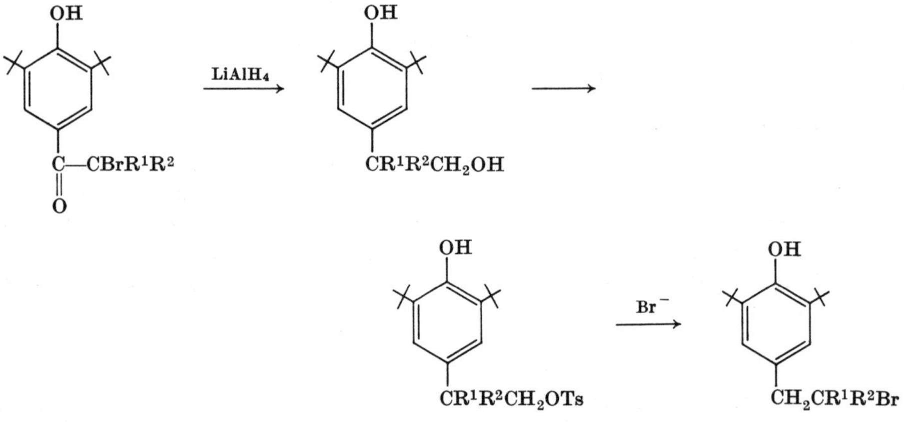

OH C—$CBrR^1R^2$ O $\xrightarrow{LiAlH_4}$ OH $CR^1R^2CH_2OH$ $\longrightarrow$

OH $CR^1R^2CH_2OTs$ $\xrightarrow{Br^-}$ OH $CH_2CR^1R^2Br$

SCHEME 7.

The kinetics of the dienol–benzene rearrangement[70] indicate involvement of a cyclohexadienyl cation which may isomerize before aromatizing. Cyclohexadienyl cations have also been invoked to explain the reaction path and stereochemistry involved in the skeletal rearrangements of furanoeremophilanes (**114**).[71] Cyclohexadienyl zwitterions are the key intermediates in the rearrangements involved in the cycloadditions of tetracyanoethylene to cyclohexa-1,4-dienes.[72] Finally, the phosphorus analogue of the semibenzene-benzene aromatizations has been realized in the conversion of (**115**) into (**116**); cross-over studies show the reaction to be intermolecular.[73]

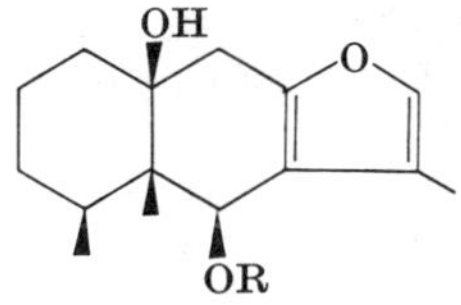

(**114**) 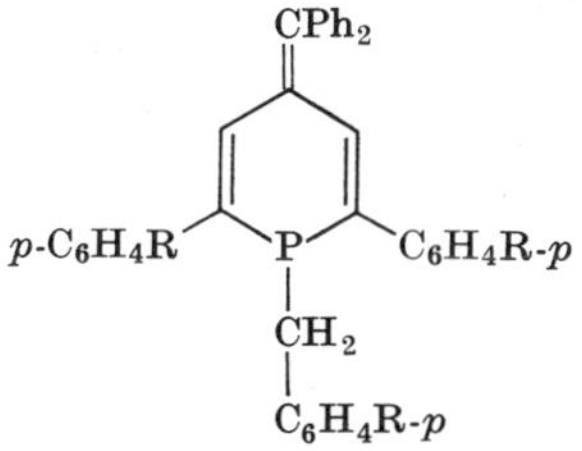 (**115**) 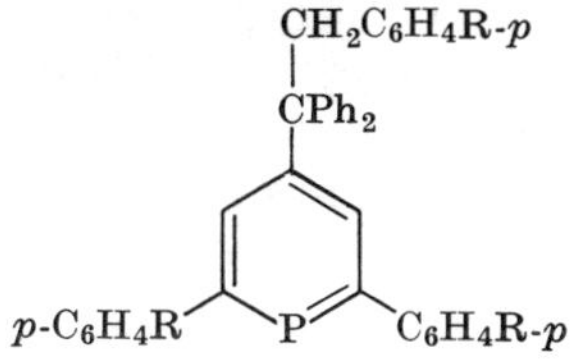 (**116**)

An oxygen walk has been observed[74] (via kinetic studies) as a complementary pathway to the NIH shift in the aromatization of indane 8,9-oxide (**117**), and indeed in this case accounts for 70% of the reaction. Only 30% of (**119**) is produced via (**118**). An oxygen

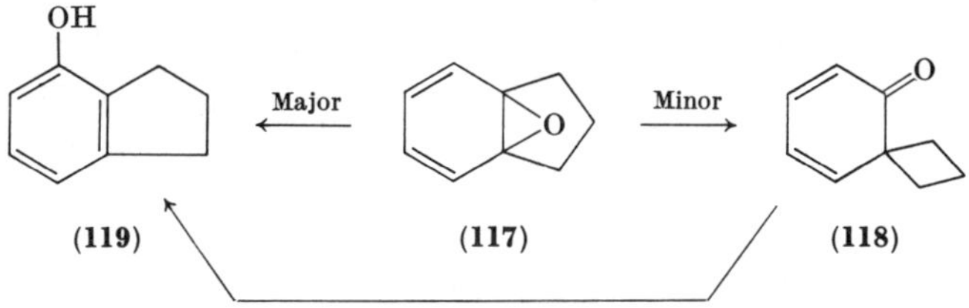

Major Minor

(**119**) (**117**) (**118**)

walk was also observed in the isomerization of (**120**) to (**121**), this migration constituting a suprafacial [1,5]-sigmatropic (and symmetry-allowed) shift.[75] Finally, an arene dioxide antibiotic, the first natural product containing such a group, has been isolated[76] and its ready valence isomerization to a 1,4-dioxocin has been noted.

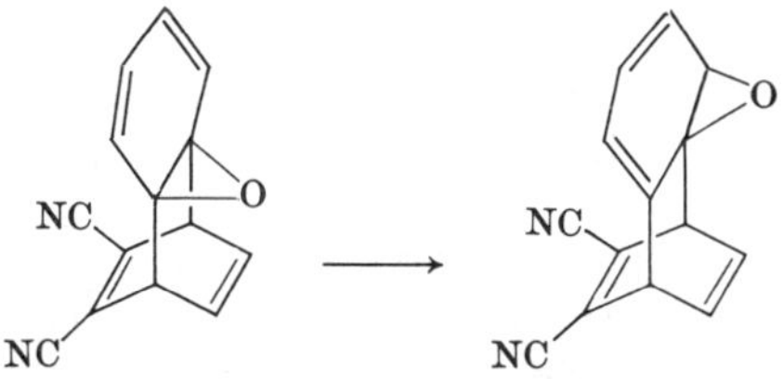

(**120**) (**121**)

Sigmatropic Rearrangements

Pertinent reviews include those of orbital-symmetry-forbidden reactions,[77] orbital-symmetry-disallowed energetically concerted reactions,[78] and the molecular-orbital counterpart of electron-pushing.[79] Some other general discussions have dealt with subjacent orbital control[80] (an electronic factor that may favour concertedness in Woodward–Hoffmann "forbidden" reactions), electron-repulsion in pericyclic transition states,[81] the effect of configuration-interaction on the stereoselectivity of sigmatropic reactions,[82] and sigmatropic reactions in compounds containing small rings.[83]

[3,3]-*Migrations*

Claisen and Related Rearrangements. Rearrangement of allyl aryl ethers to the corresponding 2-allylphenols (in trifluoroacetic acid) has been shown to be of the charge-induced [3,3]-type.[84] The thermal rearrangements (230–260°) of some halogen-substituted propargyl ethers, e.g. (**122**), involve[85] some fascinating reactions (Scheme 8 inter alia). The mechanism of the process involves a preliminary [3,3]-sigmatropic shift followed by radical reactions as outlined in the Scheme.

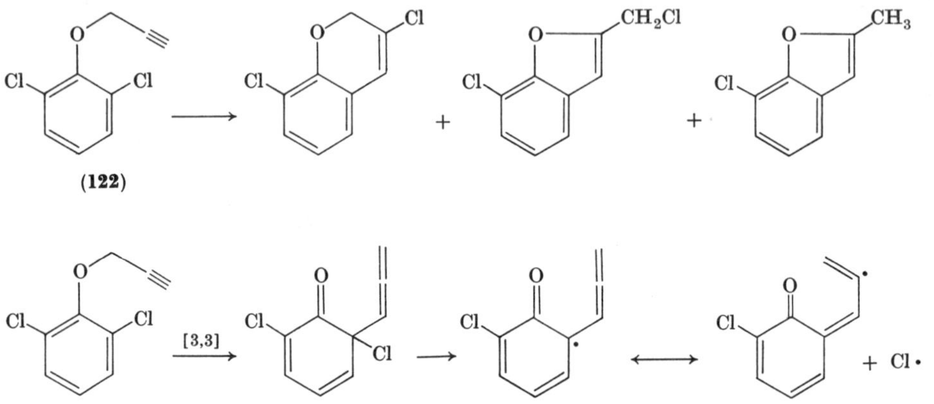

(**122**)

[3,3]

+ Cl•

SCHEME 8.

The kinetics of rearrangement of allyloxynaphthaquinones (**123**) have been found[86] to be relatively insensitive to variation of solvent (rates, at 90°, in $EtOCH_2CH_2OH$, DMSO and decalin are 1.97, 1.55 and 1.03, respectively). The thermodynamic parameters are, however, different for the reactions in these three media, $\Delta H^{\ddagger}$ (kcal mol^{-1}) being 20.52, 27.15 and 26.27, with $\Delta S^{\ddagger}$ (cal mol^{-1}) –24.0, –8.6 and 8.6, respectively. These and other data suggest that the small charge separation in the transition states of such Claisen rearrangements ($O^{\delta-}$ and $Ar^{\delta+}$) is affected by hydrogen-bonding, with consequent decrease of $\Delta H^{\ddagger}$ in protic media but with increase in $\Delta S^{\ddagger}$, the overall effects compensating. An equation for correlating the activation energies for such rearrangements with the electronic structures of the compounds undergoing change has been developed.[87]

Rearrangement of cinnamyl phenyl ether yields both *ortho-* and *para-*products, the former predominating in basic media, the latter in acidic or neutral solvents. Allyl phenyl ether, on the other hand, gives *ortho-*products in solvents of all kinds. The rates of rearrangement of the cinnamyl phenyl ether showed no discontinuity on changing from nematic to isotropic liquids, but were strongly reduced when the substrate was clathrated.[88]

The thermal and acid-catalysed rearrangements of allylic cyclohexadienones and phenol allyl ethers, and the geometries of the activated complexes of the [3,3]-sigmatropic processes involved have been analysed.[89]

The synthetic utility of successive Claisen and Cope rearrangements has been employed to prepare, e.g., 1,5-dienes from allylic alcohols.[90,91] Similarly, the Claisen rearrangements of prop-2-ynyl vinyl ethers have been used to produce antibiotic lactols.[92]

An interesting study has been made[93] of the site specificity of [3,3]-sigmatropic rearrangements of 3-allyl- (**124**) and 3-prop-2-ynyl-3*H*-indoles (**125**), the order of

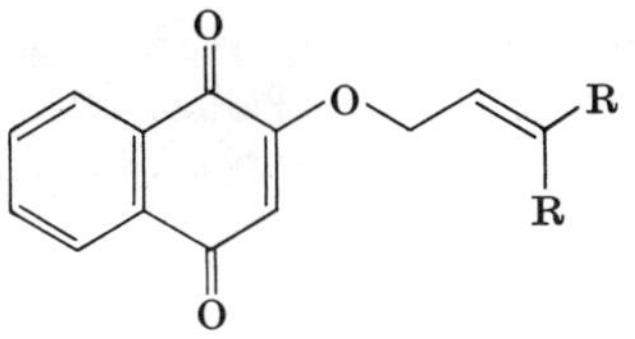

(**123**) 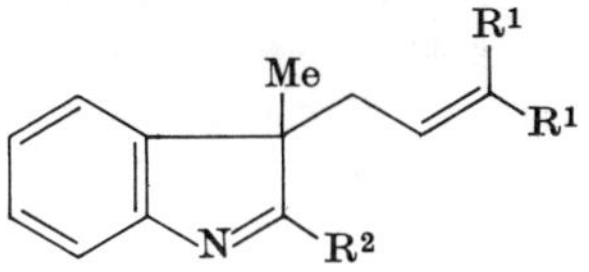 (**124**) 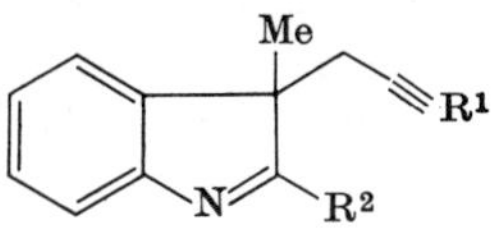 (**125**)

preference of migration terminus being C(2)-methyl (via a prior enamine tautomerism) > N ≫ C(4) in the aryl ring. A related investigation involved the pyrolysis (450–470°) of some *N*- and 3-allylindoles which are mutually interconverted under the reaction conditions, no N-to-2 sigmatropic shift being observed.[94] In some cases Plancher–Brunner rearrangements may interfere with Cope processes.[93b]

The thermal rearrangement (to the *o*-position) of *N*-allyl-*N*-tosylaniline proceeds smoothly in *N*,*N*-dibutylaniline in the presence of small quantities of triphenylphosphine.[95] Charge-induced aromatic amino-Claisen rearrangements have been observed with *N*-allylanilines in the presence of zinc chloride, or when the amine tetraphenylborates are heated in $(Me_2N)_3PO$.[96] With propargylamines, reactions are more complex and follow the pattern outlined in Scheme 9.[97]

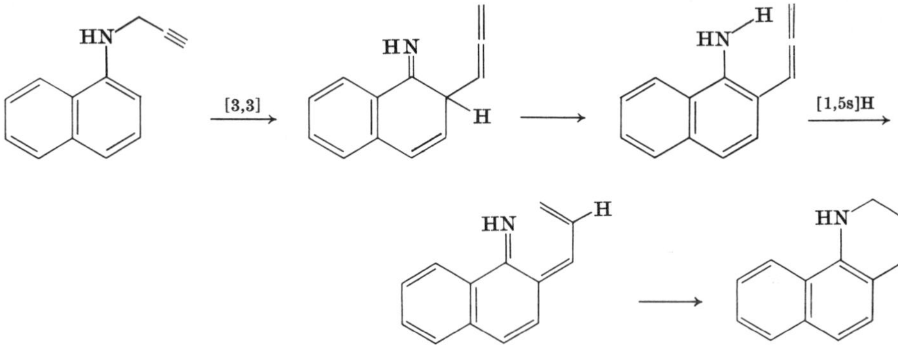

Scheme 9.

The thermal rearrangement of the allyl ether (**126**) gave the expected Claisen product (**127**) as well as the isomer (**128**). The formation of (**128**) involves a [1,5]-shift of the

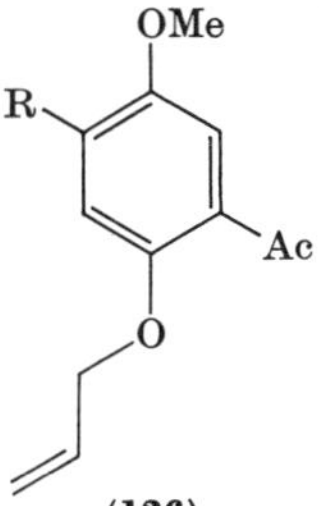

(**126**) 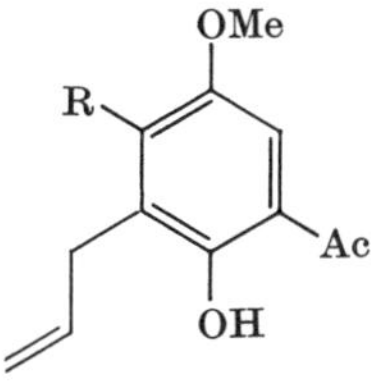 (**127**) 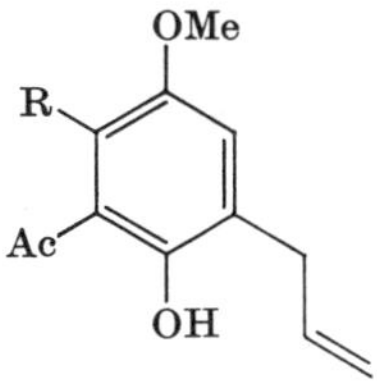 (**128**)

acetyl substituent. This type of reaction is without precedent as an accompaniment of the Claisen rearrangement.[98] CIDNP evidence has been obtained for singlet and triplet radical-pair encounters in the photo-Claisen rearrangement.[99]

Cope Rearrangements. An interesting interplay of steric effects has been observed in the rearrangement of *cis*-bicyclo[6.1.0]nona-2,6-dienes (**129**) to *cis*-bicyclo[5.2.0]nona-2,5-dienes (**130**).[100] The ease of rearrangement follows the order (for the 9-substituents),

R^1 R^2 ⇌ R^1 R^2

(129) **(130)**

anti-methyl > unsubstituted > *syn*-methyl > *gem*-dimethyl. The mechanism requires coiling of the ring before its stereospecific rearrangement, and the coiled conformation is susceptible to the steric effects of 9-alkyl-substitution. When the cyclopropyl unit ($>CR_2$) in (**129**) is replaced by an oxa- or aza-bridge, the same type of behaviour is observed, the relative rates of (irreversible) rearrangement to the appropriate (**130**) (replacing $>CR^2$ by either $>CH_2$, $>O$ or $>NR$) are in the order the aza-compound > carbocyclic > oxa-substituted.[101] In a related study,[102] the rapid reversible Cope rearrangements between the *sym*-oxabicyclo[5.1.0]octa-2,5-dienes (**131**, **132**) have been measured and confirm the point established earlier[101] that the activation energies for the Cope rearrangements of *cis*-divinyl-substituted three-membered rings lie closer together than formerly assumed.

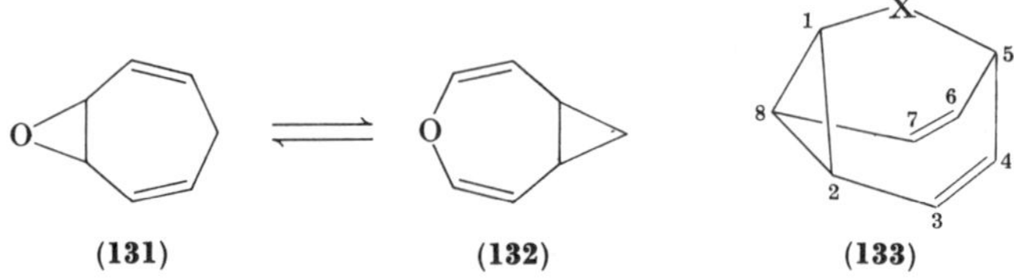

(131) **(132)** **(133)**

Examination of the kinetics of rearrangement of hexa-1,5-diene and various of its phenyl derivatives (2-, 3- and 2,5-) points to availability of a duality of mechanisms for Cope rearrangements, namely, biradical reactions and the more usually assumed pericyclic process.[103]

The free energies of activation for Cope rearrangements in a number of 9-substituted[104] barbaralanes (tricyclononadienes) (**133**) have been measured by both coalescence temperature and complete line-shape methods for the variation in X of CH_2, $C{=}CH_2$, $C{=}O$ and $^{+}C{-}OH$, $\Delta G^{\ddagger}$ (kcal mol^{-1}) values are 7.8, 9.5, 10.5 and >13.8, respectively. The most important factor in determining the relative rates is the electronic perturbation of the bishomobenzene transition state by the group X; differential strain factors within the series are negligible.

Rate constants have been determined[105] for the forward and reverse rearrangements of the Cope-related pair *trans,trans*-cyclodeca-1,5-diene (A) and *trans*-1,2-divinylcyclohexane (B); ratio of forward (A → B)/reverse (B → A) rate constants being $K_{200°} = 2 \times 10^4$. The ring strain of A relative to B is of the order of 12 kcal mol^{-1}. A series of *cis,trans*-cyclodeca-1,5-dienes underwent normal Cope arrangements, but when a furan ring was fused in the system, the rearrangement processes were strongly inhibited.[106] Efforts to prepare isomers of *trans*-cyclonona-1,2,6-triene yielded the Cope rearrangement product, whereas the corresponding *cis*-triene has a half-life of over 30 minutes at 150°,[107] (*cis*-2-Phenylvinyl)cyclopropane undergoes a slow Cope reaction at 100°, the product being trapped via anion formation and electrocyclic ring opening.[108] Pyrolysis (350°) of cyclohept-4-enylmethyl acetate (C) gave methylenecyclohept-4-ene along with its Cope rearrangement isomer (a substituted methylenecyclopentane)[109]; these results are in contrast to the thermal behaviour of (C) in flow systems.

The Cope rearrangement of vinylbicycloalkenes (**134**) constitutes a general method of preparing *cis*-annelated rings (**135**), with functionality capable of further synthetic exploitation.[110]

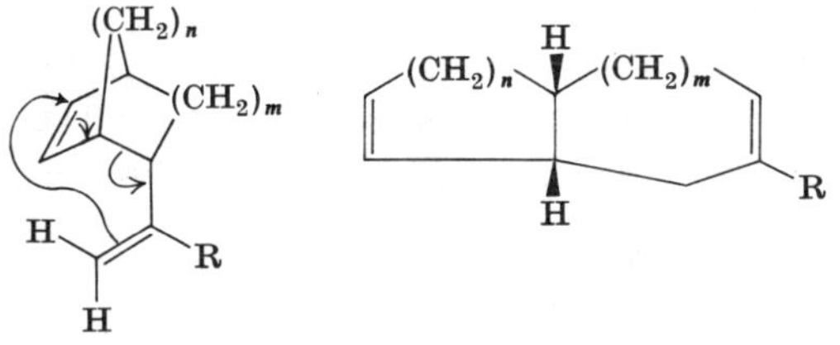

(**134**) (**135**)

Gas-phase thermolysis (>500°) of *cis*- and *trans*-propenylsilanes involves their interconversion. This process indicates the enhanced ability of silicon to participate in pericyclic transition states and is strongly suggestive of Si=C bonding.[111] Thermolysis (300–310°) of *cis*- and *trans*-1-(trimethylsilyloxy)-1-vinylcyclodec-3-ene results in 1,3-shift ring expansion (to cyclododec-5-en-1-one) with the silyloxy-Cope rearrangement playing a minor part; ring-strain effects have been evaluated therein. The evidence available suggests that concerted processes are involved but does not rule out diradical pathways.[112]

Other [3,3]-*Processes.* Thermolysis of dimethylenetricyclo[3.2.1.0]oct-3-enes generates allenyl-1-methylenecyclohexadienes *in situ*. These undergo [3,3]-sigmatropic shifts to form arylbutadienes, a process illustrative of the concerted thermal semibenzene → benzene rearrangements.[113] Certain electron-rich olefins have been found [114] to rearrange by two competing pathways, a thermally allowed [3,3]-sigmatropic process and a formally forbidden [1,3]-sigmatropic pathway.

The ketone bridge in the adducts (**136**) of substituted cyclopentadienones with azo-esters sharply increases the lability of the adducts towards thermal isomerization. A reversible [3,3]-sigmatropic rearrangement to the 1,3,4-oxadiazine derivatives (**137**) is observed along with a slower (and perhaps irreversible) [1,3]-sigmatropic rearrangement to the diazetidines (**138**), the stable thermal end products.[115]

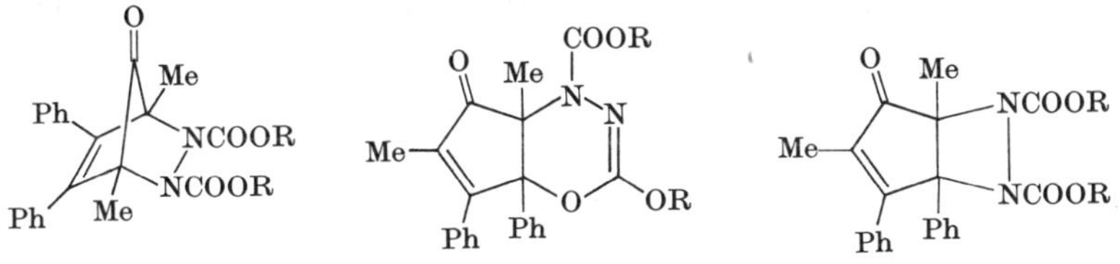

(**136**) (**137**) (**138**)

Pyrolysis of *N*-allylhydrazones (300°) yields, via a synchronous [3,3]-sigmatropic rearrangement, first allylated diazenes and then allylated hydrocarbons.[116] Unusually easy [3,3]-sigmatropic shifts are described for *N,N'*-diarylidene-ethylenediamine systems, the aza-Cope process (Scheme 10) taking place extremely readily.[117]

SCHEME 10.

[2,3]-*Migrations*

Thermolysis of oxime *O*-allyl ethers (180°, 10 hr.) leads via [2,3]-sigmatropic shifts to substituted nitrones.[118] Such shifts have been very neatly used for the stereoselective formation of asymmetry at quaternary carbon[119] (Scheme 11). When 3-chloroallyl sulphides and *N*-allylanilines are oxidized with *m*-chlorobenzoic acid, the resultant

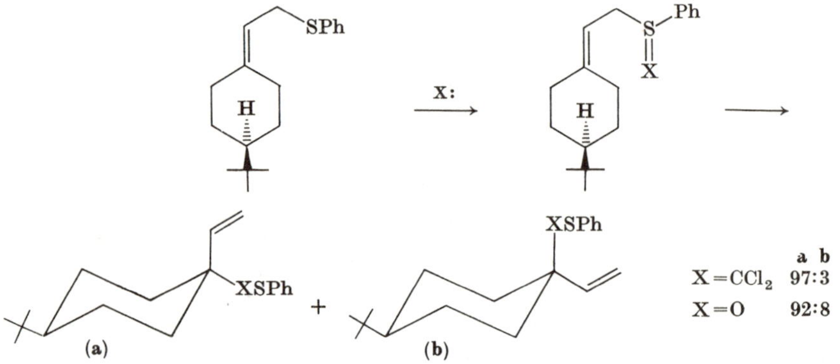

	a	b
$X{=}CCl_2$	97:3	
X=O	92:8	

SCHEME 11.

S- and *N*-oxides rapidly undergo [2,3]-sigmatropic isomerization to labile products that decompose internally with selective formation of conjugated enones.[120] The [2,3]-sigmatropic rearrangement of a chiral amine oxide was effected thermally (–20°, 24 days) with essentially complete transfer of chirality to trigonal carbon.[121] A similar procedure was applied (via Meisenheimer type [2,3]-sigmatropic rearrangements) to the conversion of ionones into damascones. A primary allylic alcohol was converted into the corresponding *N,N*-dimethylamine oxide; this was rearranged thermally ([2,3]-shift) to give a hydroxylamine which in turn was reductively cleaved to give the rearranged secondary or tertiary alcohol.[122]

The stereochemistry of the Polonovski reaction has been studied by using N[4]-oxides of 1,4-benzodiazepines (**139**) as substrates, with the chiral reagent camphoryl chloride.

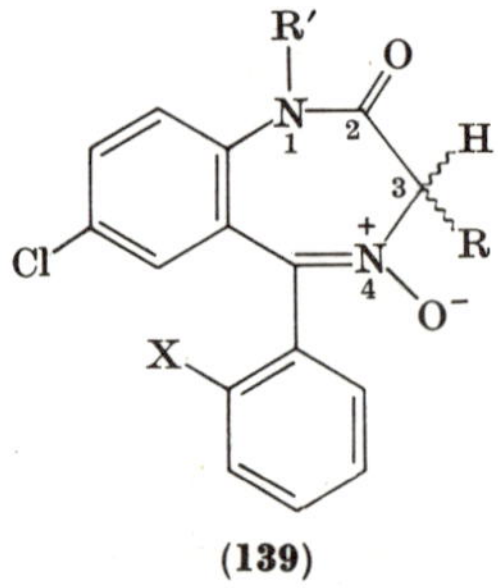

(**139**)

The mechanism involves deprotonation at C-3 followed by [2,3]-sigmatropic shift in the ammonium ylide produced. The carbanion involved displayed slight selectivity in its position of attack by the acyl unit.[123]

The *para*-allylic anilinium ylides (**140**) rearrange on heating via a [2,3]-shift (to **141**) followed by a [3,3]-migration to give (**142**) (R being H or Ph).[124] The corresponding *ortho*-ylides (**143**) yield both the ethers (**144**) (which are derived by a [1,4]-sigmatropic rearrangement)[125] and substituted allylic phenols (derived from consecutive [2,3]- and

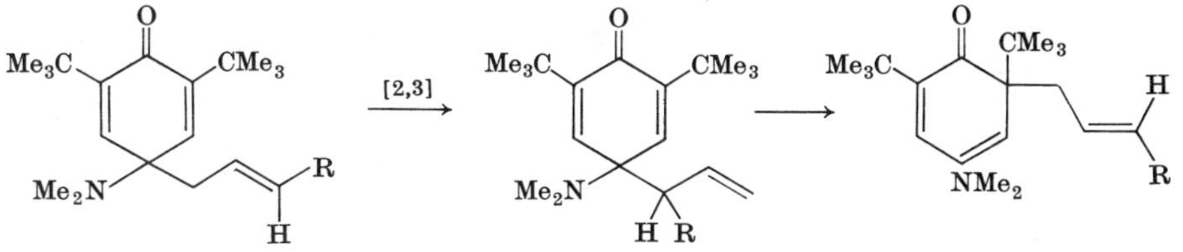

(**140**) (**141**) (**142**)

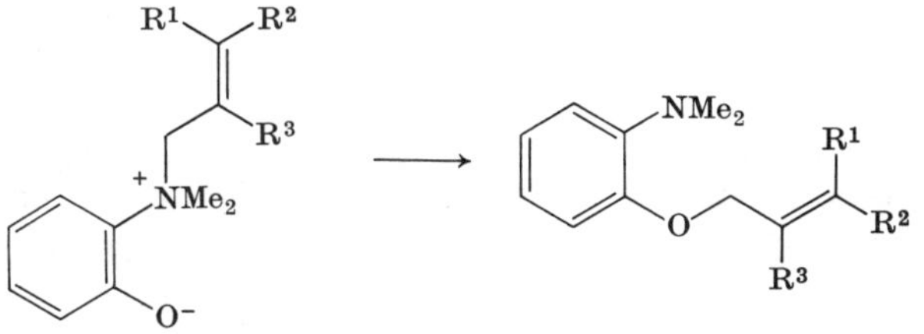

(**143**) (**144**)

[3,3]-sigmatropic rearrangements). While the [1,4]-shift has been demonstrated to be concerted, the possibility of competing radical-pair processes is revealed by CIDNP data.[126]

The relative rates of the [2,3]-sigmatropic rearrangement of the six-membered cyclic allylic ylides (**145**; X = NR or S) to (**146**) are consistent[127] with a concerted reaction

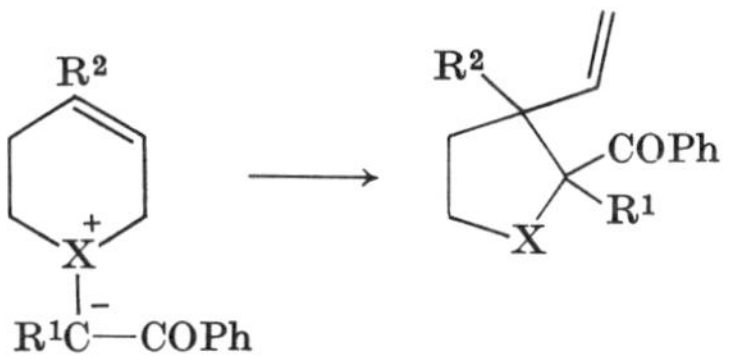

(**145**) (**146**)

mechanism. The analogous five-membered ylides (X = NR) rearrange by a [1,2]-sigmatropic shift. With acetylenic ammonium ylides, the rearrangements involve non-concerted processes rather than concerted [2,3]-sigmatropic shifts.[128] Acetylenic sulphonium ylides and allenic sulphonium ylides undergo smooth [2,3]-sigmatropic rearrangements to give allenes and conjugated dienes, respectively.[129]

Rapid [2,3]-shifts have been observed with allyl diselenides and are faster than for the sulphur analogues. The same effect is observed in the [1,3]-sigmatropic rearrangements of allyl selenides.[130]

The rates of several [2,3]-sigmatropic allyl sulphoxide–sulphenate rearrangements in steroidal systems have been shown to be influenced by chirality at sulphur. The rearrangements were clearly suprafacial with respect to the allyl system.[131]

[1,3]-*Migrations*

Calculations have been made by the MINDO/CI procedure for the 1,3-sigmatropic hydrogen shift in propene (calc. $E^{\ddagger}$ 49.2 kcal mol^{-1}) and for the suprafacial and antarafacial [1,5]-sigmatropic hydrogen shifts in *cis*-piperylene.[132] A mechanistic analysis of the four pathways (s = suprafacial, a = antarafacial, r = retention of configuration in the migrating carbon, i = inversion of configuration of the migrating carbon) si, sr, ar, ai has been made for the sigmatropic rearrangements of *trans*-1,2-*trans,trans*- and *trans*-1,2-*cis,trans*-dipropenylcyclobutane.[133] The major processes involved namely, si and sr, behave as concerted reactions with si allowed but only slightly preferred over sr which is forbidden. Both antarafacial reactions proved to be relatively slow. Such forbidden [1,3]-migrations (of the sr type) have been realized in the bicyclo[4.2.0]octene series, the experimental facts fitting a concerted mechanism for the formally forbidden process as well as for the allowed one.[134]

A variety of degenerate silyl shifts from O to O and from N to N can be classified as [1,3]-, [1,2]- etc. sigmatropic shifts but the evidence available is equally consistent[135] with the reaction formulated as an intramolecular nucleophilic displacement with a cyclic transition slate.

In a fine study, the kinetics, stereochemistry and mechanisms of the sila-allylic and silapropynylic rearrangements have been determined.[136] In the former case, the course of the [1,3]-sigmatropic migration of silicon has been identified as a concerted process with every act of migration accompanied by inversion of the silicon configuration. This indicates a preference for utilizing a 3*p*-orbital in bridging the allylic structure with conservation of orbital symmetry. This is the first case of [1,3]- or [1,5]-migration of silicon in which such preference for 3*p*-orbital utilization has been identified. The kinetic characteristics of the silapropynylic rearrangement indicate a symmetrical concerted transition state similar to that in the sila-allylic rearrangement, again occurring with complete inversion of the silicon configuration.

By means of but-3-en-2-yl(dimethylamino)ethylborane (**147**) as substrate,[137] the normally extremely ready [1,3]-sigmatropic shift of boron in such systems was slowed down to measurable rates. Cross-over data revealed the intramolecularity of the rearrangement. A similar conclusion was reached in the case of the rearrangement of tris-(2-methylallyl)borane.[138]

The thermal rearrangement of the benzylated pyrazine (**148**) to (**149**) represents the

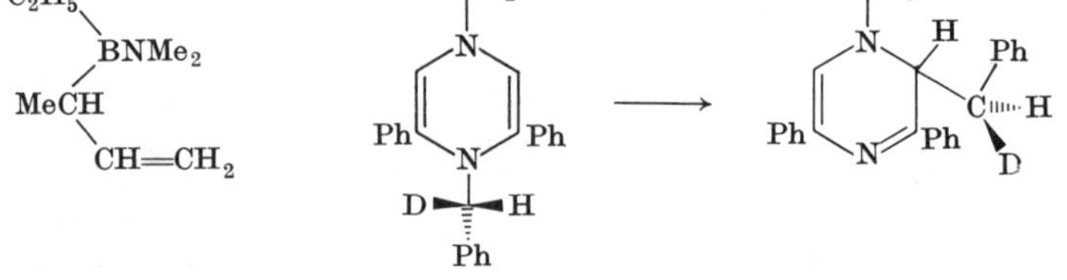

(**147**) (**148**) (**149**)

first clear-cut example of a [1,3]-sigmatropic shift with inversion involving nitrogen at the migration origin. The overall process showed minimal rate dependence on solvent, proceeded with ≥95% stereospecificity and displayed a small extra-cage free-radical component.[139] [1,3]-Sigmatropic C → N alkyl shifts have been reported in the cleavage reactions of some 1,4-diazocines.[140] These two approaches have been combined[141] to examine the stereochemistry of [1,3]-sigmatropic alkyl shifts from nitrogen to carbon and their reverse in certain pyrazine-based heterocyclic systems. The shifts involved were all of the allowed suprafacial [1,3]-types.

The base-induced conversion of *o*-dipropargylbenzene and 2,3-dipropargylnaphthalene to the corresponding allenes has been reported.[141a]

Other Sigmatropic Migrations

[1,5]-*Migrations.* An intramolecular [1,5]-shift of a formyl group has been detected in the thermolysis of methyl bicyclo[3.2.0]hept-2-en-7-ones.[142] From rate studies on the thermolysis of 1-methylcyclohexa-2,4-dienes, the formyl group undergoes [1,5]-sigmatropic shift faster than hydrogen by more than two orders of magnitude, whereas the methoxycarbonyl group is slower by a factor of about 70 and acetyl shows a migration aptitude comparable to that of hydrogen.[143] Thermolysis (150–190°, in decalin) of arylallenes causes their rearrangement via a [1,5]-hydrogen shift to yield *o*-quinodimethanes which then may cyclize to give dihydronaphthalenes and/or undergo [1,7]-sigmatropic H-shifts to give arylbutadienes. The primary isotope effect in the [1,5]-shift, i.e. k_H/k_D, is 3.45.[144] A kinetic study[145] of the rearrangement of 1,2,3,4,5-pentaphenylcyclopenta-2,4-dien-1-ol to 2,2,3,4,5-pentaphenylcyclopent-3-en-1-one shows the reaction is a [1,5]-sigmatropic rearrangement of the 1-phenyl group with $E^‡ = 36.1$ kcal mol^{-1} and $\Delta S^‡ = -7$ eu. The rates of hydrogen tautomerization of the 5-halocyclopentadienes (which are [1,5]-hydrogen shift reactions) reveal the striking fact that halogen-substitution slows such shifts, despite the fact that such substitution stabilizes the products.[146] A previously undetected intermolecular component has been observed in certain [1,5]-sigmatropic methyl shifts: pyrolysis of 1,5,5-trimethylcyclopentadiene (350°) revealed examples of radical, diradical and concerted rearrangements.[147]

The thermal rearrangements of *N*-nitropyrazoles to 3-nitropyrazoles involves [1,5]-sigmatropic shifts of the nitro-group.[148]

Miscellaneous. In the acid-catalysed rearrangements of 1,2-dihydro-1-hydroxy-2-methyl-2-(penta-2,4-dienyl)naphthalenes, [1,2]-sigmatropic shifts predominate.[149] The first examples of long-range concerted anionic rearrangements ([4,5]-sigmatropic rearrangements) have been reported;[150] they were realized in the rearrangements of allyl(pentadienyl)ammonium ylides (**150**) and of *N*-pentadienyl-2-oxyanilinium ylides (**151**), where [1,4]-sigmatropic shifts were also observed.[151] Photochemical rearrangements in 3,4-benzotropilidenes show that [1,3]-migrations may compete to the extent of ca. 10% with the more common [1,7]-shifts in these systems; this competition may result when the [1,3]-process does not involve disruption of the aromatic system.[152]

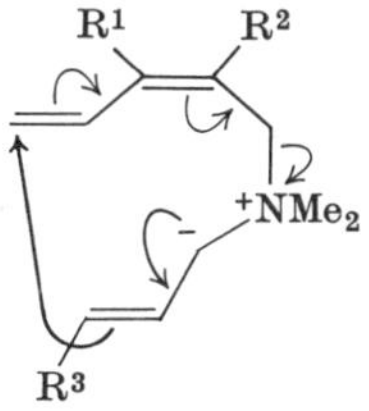

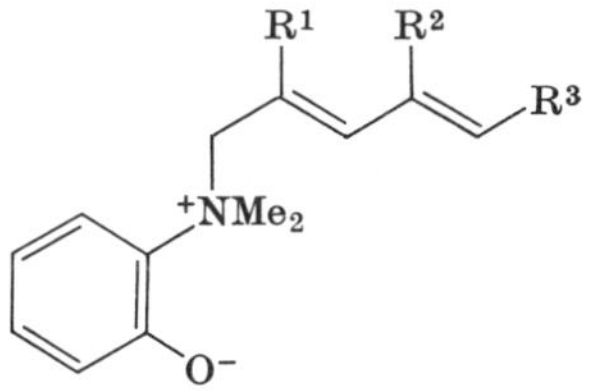

R¹ R² $^+NMe_2$ R³

(150)

R¹ R² R³ $^+NMe_2$ O⁻

(151)

Electrocyclic Reactions

Some relevant general discussions have been examination of the effects of substituents on the stereochemistry of electrocyclic reactions with the aid of perturbation theory at the one-electron level,[153] on the nature and definition of concertedness[154] and on the parity and stereochemistry of concerted reactions.[155]

Interconversion of the enolate ions (**152**) and (**153**) of eucarvone involves a hexatriene–cyclohexadiene disrotatory electrocyclization which has now been directly observed by NMR.[156] At 28°, the equilibrium mixture contains ca. 67% of (**153**).

Reaction of (**154**) with *tert*-butoxide in refluxing diglyme involves an electrocyclic (thermal conrotatory) cyclization to yield the 1,4-benzodiazepines (**155**) in a new synthesis of such ring systems.[157] The cyclization of *o*-diazoalkyl-styrenes and -stilbenes

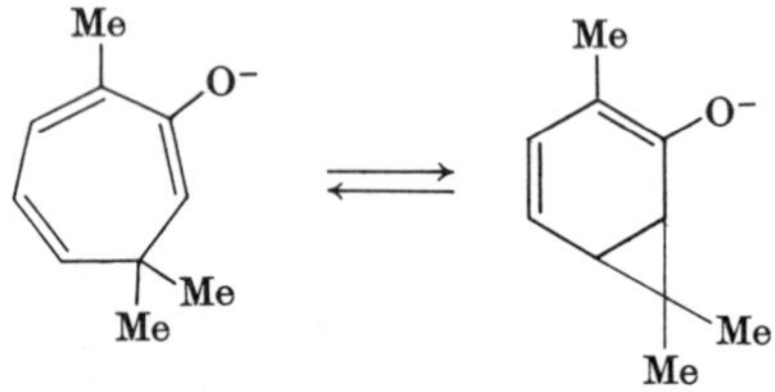

(**152**) (**153**)

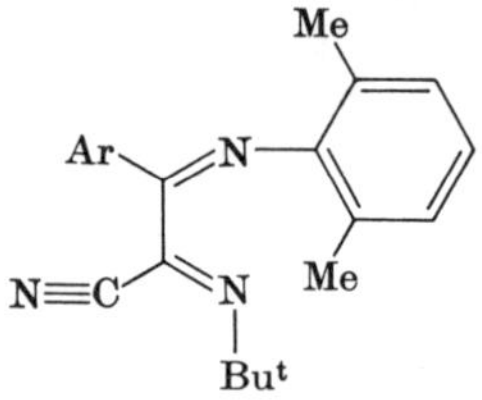

(**154**) 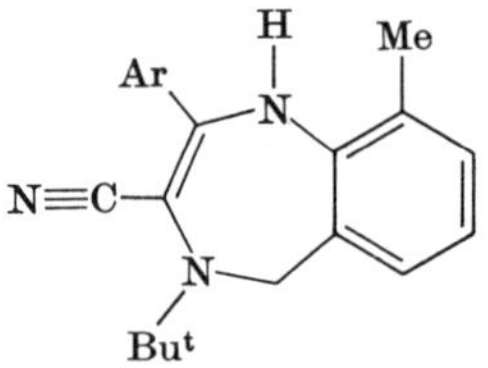 (**155**)

involves a novel 1,7-electrocyclic ring closure to give a substituted cyclohexadieno-2,3-benzodiazepine, which then aromatizes by a suprafacial [1,5]-sigmatropic hydrogen shift; this constitutes a potentially very useful route to such seven-membered heterocycles.[158]

Treatment of α,α-dibromodibenzylideneacetone with hydriodic acid and red phosphorus yields first a pentadienyl cation, which then undergoes a thermal ground-state concerted conrotatory [2 + 2]-electrocyclic reaction to yield a cyclopentenyl system.[159] A similar electrocyclization was involved in the acid-catalysed cyclizations of α,α-dimethyldibenzylideneacetone to cyclopentyl systems.[160]

Rates of thermal electrocyclization of a series of trienes and their geometrical interconversions have been examined, and structure–reactivity correlations have been made thereon.[161]

Anionic Rearrangements

[1,2]-*Migrations*

The reactions of benzophenone ketyls and alkyl iodides—reactions which result in Wittig rearrangements—display a clearly radical intermolecular component. A radical-pair mechanism is suggested for the intramolecular rearrangement as well.[162] The

Wittig rearrangement of *N,N*-dialkyl-2-(diarylmethoxy)acetamide (by use of NaH) yields *N,N*-dialkyl-3,3-diaryl-lactamides. When the diarylmethoxy-group was asymmetric, rearrangement resulted in diastereoisomeric reaction products.[163]

The first example of a 1,2-anionic rearrangement from oxygen to nitrogen was realized when bis(organosilyl)hydroxylamines were treated with alkyl-lithiums; this caused intramolecular migration of the organosilyl group and resulted in an equilibrium mixture (**156** ⇄ **157**), with the rearranged anion predominant.[164] A comparable radical rearrangement has also been observed.[165] Reaction of 1,2-diphenyl-1,1-bis(trimethylsilyl)-

$$R_3Si\bar{N}OSiR_3 \rightleftharpoons (R_3Si)_2NO^-$$

(**156**) (**157**)

ethane with butyl-lithium yielded a carbanionic 1,2-trialkylsilyl shift (5%) (**158**, +H⁺), but unexpectedly the major product (**159**) (90%) corresponded to a 1,4-trialkylsilyl shift.[166]

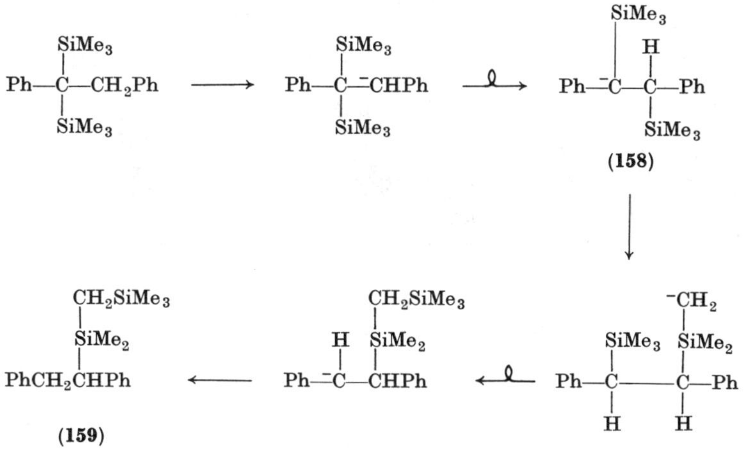

Ph—C(SiMe₃)₂—CH₂Ph ⟶ Ph—C(SiMe₃)₂—$\bar{C}$HPh ⟶ Ph—$\bar{C}$(SiMe₃)—CH(SiMe₃)—Ph (**158**)

↓

Ph—CH(SiMe₃)—CH(SiMe₂$\bar{C}$H₂)—Ph ⟶ Ph—$\bar{C}$H—CH(SiMe₂CH₂SiMe₃)Ph ⟶ PhCH₂CH(SiMe₂CH₂SiMe₃)Ph (**159**)

Treatment of ethyl 2,3-dioxobutyrate with aqueous base causes hydrolysis at pH <10, followed by carboxylate migration in a benzilic acid rearrangement (at pH >11.5 the intact ester group migrates first).[167]

Lithiation of (chloromethyl) phenyl sulphoxide, followed by reactions with aldehydes and work-up yielded[168] (phenylsulfinyl)methyl ketones by a reaction whose mechanism is still highly speculative, the overall process being outlined in Scheme 12.

$$PhSOCH_2Cl + BuLi \rightarrow PhSO\bar{C}HClLi^+ + RCHO + PhSOCH_2COR$$

SCHEME 12.

[1,3]- *and Other Migrations*

The base-catalysed prototropic propargylic rearrangement of hex-3-yne (with $NaNH_2$ in ethylenediamine) does not seem to involve an intramolecular 1,3-proton transfer (a so-called "conducted tour" mechanism[169]) as a major pathway; instead intermolecular proton abstraction and recapture seem to predominate.[170]

Base-catalysed [1,3]-shifts are also involved in the rearrangement of a bicyclo-octa-2,4-diene system to a tropylidene derivative,[171] and in the equilibration that occurs between cyclohex-3-enylmagnesium bromide and cyclopent-2-enylmethylmagnesium bromide on heating.[172] The driving force for ring contraction in the latter case (which occurs via a bicyclo[3.1.0]hexane intermediate) is ascribed to the inductively caused difference in stability between primary and secondary Grignard reagents (ca. 3.7 kcal mol^{-1}).

When *N*,*N*-dimethyl-5-trimethylsilythiophen-5-sulphonamide (**160**) was treated[173] with *n*-butyl-lithium–tetramethylene-ethylenediamine complex, rearrangement (involving a [2,3]-shift) occurred to yield (**161**) (Scheme 13).

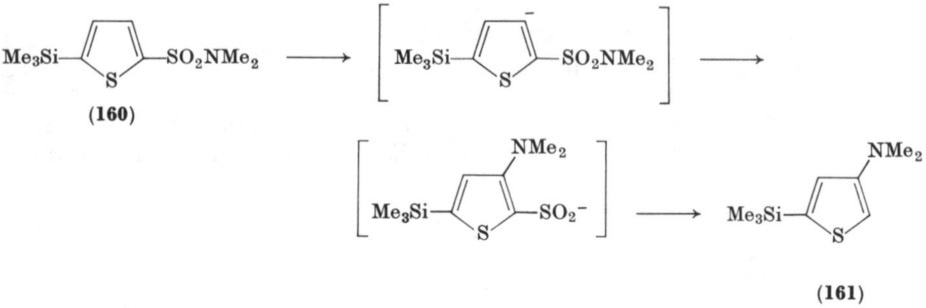

Me_3Si–(thiophene)–SO_2NMe_2 (**160**) ⟶ [Me_3Si–(thiophene$^-$)–SO_2NMe_2] ⟶ [Me_3Si–(thiophene, 3-NMe_2)–SO_2^-] ⟶ Me_3Si–(thiophene)–NMe_2 (**161**)

SCHEME 13.

Enolate-induced Rearrangements (*Favorskii Types*)

Reaction of α-halo-α′-aryl ketones (e.g. *cis*-2-chloro-6-phenylcyclohexanone) with NaOMe in MeOH (despite enhanced rates of enolate ion formation compared with the non-arylated substrate) does not yield a Favorskii rearrangement. (With k_{Br}/k_{Cl} for the α-halo-compound as ca. 4, proton-removal is largely rate-determining.) Reaction of these substrates with secondary amines did give Favorskii products (e.g. *trans*-2-phenylcyclopentanecarbonylpiperidide) in high yields. These results point to the intermediacy of enamines in the amine-induced Favorskii processes.[174]

The effect of methyl-substitution upon the Favorskii reaction of α-halo-ketones was probed[175] with substrates such as $PhCH_2COCMe_2X$ and $PhCHXCOCMe_2H$. With both, increasing concentrations of NaOMe in MeOH increased substantially the yield of Favorskii product. Deuterium-exchange studies showed no internal return to the enolate ion during the process.

While 2-bromo-5-methyl-5-phenylcyclohexanone gives higher yields of Favorskii product (compared with competitive pathways) on increasing the methoxide (in MeOH) concentration, 2-chloro(or bromo)-4-methyl-4-phenylcyclohexanone is much less subject to this concentration effect. 1,3-Diaxial effects favour the Favorskii process by inhibiting alternative reactions.[176]

Alkaline hydrolysis of 5,5,5-trichloropent-3-en-2-one ($Cl_3CCH{=}CHCOCH_3$) affords 5-chloropenta-2,4-dienoic acid by an interesting Favorskii type rearrangement. With methanolic sodium methoxide the methyl ester of both this acid and 5,5-dichloropent-4-enoic acid are obtained.[177]

Reaction of *N*-nitroso-*N*-propargylureas with 0.5N-NaOMe (in MeOH), despite its normal dissimilarity, has relevance to the Favorskii rearrangement. Reaction of an optically active substrate ($HC{\equiv}C{-}CHEt{-}N{=}N{-}O^-$) afforded the corresponding butyrate ($CH_3CHEtCOOMe$) with 88% inversion of configuration (12% racemization).[178] These data can be rationalized as in Scheme 14. The high stereospecificity of the overall

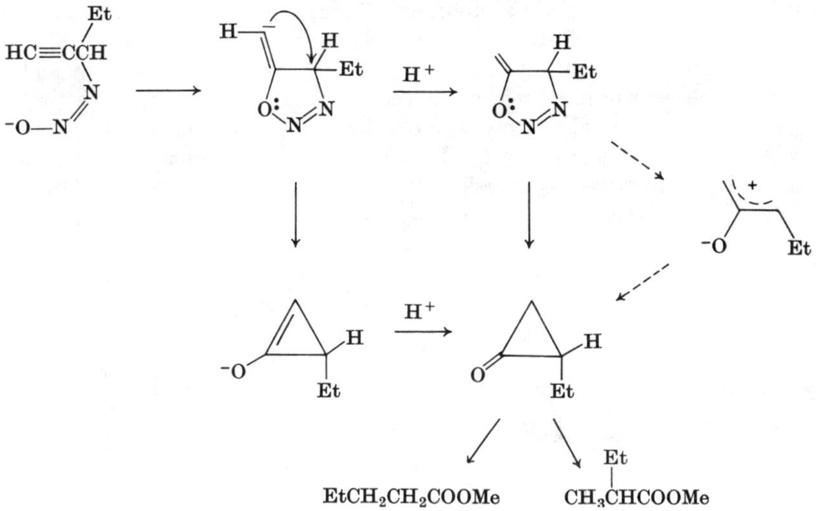

SCHEME 14.

process eliminates a planar oxyallyl cation as intermediate. The balance between internal cyclization of the intermediary propargyl diazotates ($R^1C{\equiv}C{-}CHR^2{-}N{=}N{-}O^-$) and other pathways of reaction has been explored for variation in R^1 and R^2.[178,179]

A reaction which, while clearly not a Favorskii rearrangement (and which is a mechanistic "waif", i.e., mechanism unknown) has a certain formal resemblance to it, is the reaction of *N*-methyl-α-bromoacetanilide ($PhNMeCOCH_2Br$) with EtMgBr containing a catalytic amount of $NiCl_2(PPh_3)_2$; this yields, amongst other materials, the rearranged product $(PhNMeCH_2CO)_2CH_2$.[180] It may turn out to be a simple aziridinone ring-formation followed by ring opening.

Cationic Rearrangements

General

Two distinct classes of base-catalysed carbonium ion rearrangements have been identified. *cis*-1,3,4,5-Tetramethylcyclopentenyl cation (**162**) rearranges in 84–100% H_2SO_4 to the *trans*-isomer (**163**); this in turn forms the 1,2,3,4-tetramethylcyclopentyl cation (**164**) (Scheme 15). The first reaction (**162** → **163**) displays a slope of log k against H_0 of

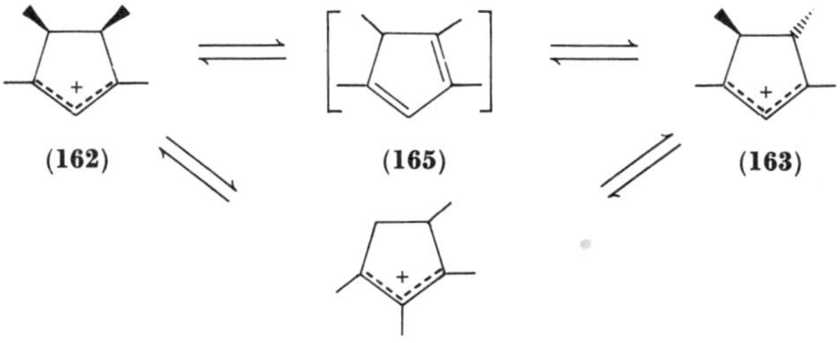

SCHEME 15.

0.94; it involves the diene (**165**) as an intermediate; it involves also base catalysis and is typical of the first type of base-catalysed carbonium ion rearrangements. The second type does not involve a diene intermediate, has a log k *vs.* H_0 slope of 0.45 and must involve a termolecular or intramolecular push-pull process.[181]

The use of deuterium-labelling as a probe for signal assignments and mechanistic studies by ^{13}C NMR spectroscopy has been reported.[182] In a complex rearrangement such as the D_2SO_4-catalysed conversion of humulene (**166**) into apollanol (**167**) where six different positions become partly deuteriated, the NMR tool permitted identification of all the labelled sites.

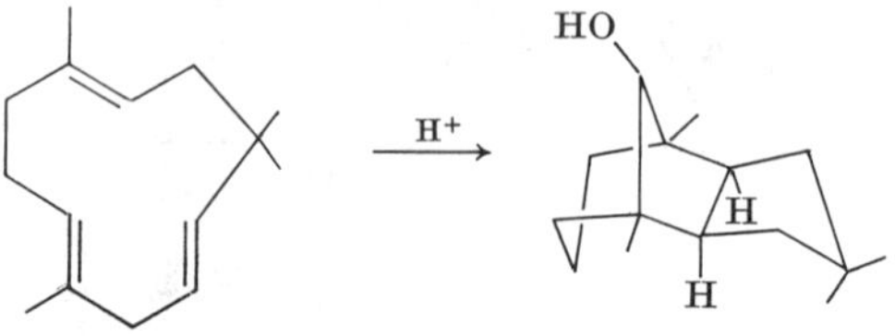

(**166**) (**167**)

Circumambulatory degenerate rearrangements in polyenyl cations have been examined. In the bicyclo[3.1.0]hex-3-enyl system the allowed suprafacial migration preserves the original *endo*- amd *exo*-nature of the 6-hydrogens at each step and has $\Delta F^{\ddagger} = 15$ kcal mol^{-1}. The next higher vinylogue (homotropylium ion) would require an interchange of 8-*exo*- and 8-*endo*-hydrogen atoms at each step. Examination of the behaviour of the [4-D]homotropylium ion indicates that circumambulatory reaction does not occur ($\Delta F^{\ddagger} \geqslant 26$–27 kcal mol^{-1}); instead there is conformational inversion via a planar classical cyclo-octatrienyl cation.[183] *Ab initio* molecular-orbital calculations help to interpret these results as due to loss of homoaromaticity in the homotropylium case in the transition state.[184]

γ-Radiolysis of cycloheptatriene vapour has been shown to yield the $C_7H_7^+$ ion largely as a rearranged benzyl type.[185] The predominant mechanism of the double McLafferty rearrangement of aliphatic ketones (and enolic ions) involves transfer of hydrogen to the carbon—carbon double bond.[186]

[1,2]-*Migrations*

Ab initio molecular-orbital calculations show that intramolecular [1,2]-shifts in simple organic radicals may be facilitated by protonation of the migrating group, intermediates of bridged and open forms of cation radicals being visualized.[187] This approach led to a mechanistic interpretation of the enzymic rearrangements of propane-1,2-diols to propanals.

The influence of topological effects upon the acid-catalysed rearrangements of alkylcyclohexyl ketones has been studied by using the DARC approach.[188] A series of [1,2]-shifts is involved[189] in the acid-catalysed ring contractions of various bicycloalkanes; for example bicyclo[7.1.0]decane yields derivatives of cyclo-octanes, cycloheptane and cyclohexane under the reaction conditions (50°, $ArSO_3H$, benzene).

The acid-catalysed rearrangement of the strained metacyclophane (**168**) has been proved to involve a series of [1,2]-migrations, the product being the naphthalene shown.[190] Rate constants and activation parameters have been measured for the degenerate rearrangements of 9-aryl-9,10-dimethylphenanthrenonium ions.[191]

Retention of configuration of the migrating carbon has been demonstrated in the

Wagner–Meerwein rearrangement caused by nitrosative deamination of the substrate (**169**).[192]

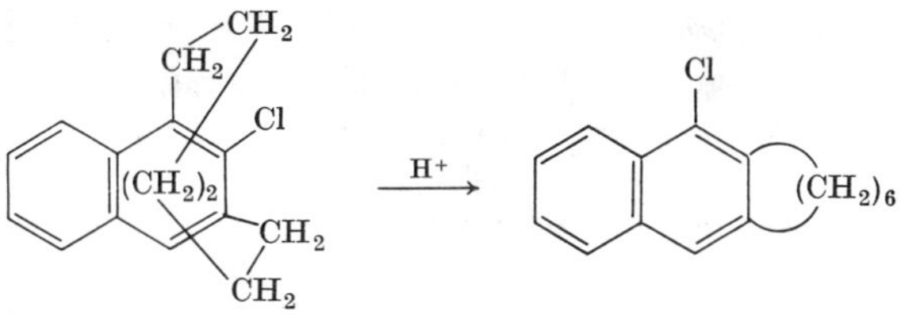

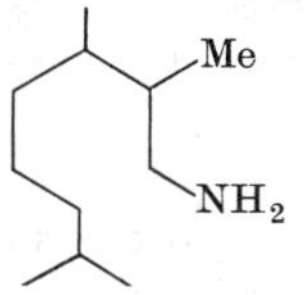

$$\xrightarrow{H^+}$$

(**168**) (**169**)

A pinacol-type mechanism is probably involved in a new ring-expansion reaction, which involves the decomposition of the magnesium salts of various halohydrins.[193] Treatment of either tetrahydroindane-*cis*- or *trans*-8,9-diol with concentrated sulphuric acid yielded a keto-spiran (**170**).[194] Reaction of 1,2-di-(*p*-methoxyphenyl)ethane-1,2-diol with refluxing 50% H_2SO_4 for 1 hour gave the expected rearrangement product (**171**) as well as *p*-$MeOC_6H_4CH_2COC_6H_4OMe$-*p*; under the same conditions for one minute, the diol yielded (**171**) and the isomeric epoxide (**172**). Heating (**171**) with Ac_2O reversed the pinacol–pinacolone rearrangement and gave the diacetate of starting glycol.[195]

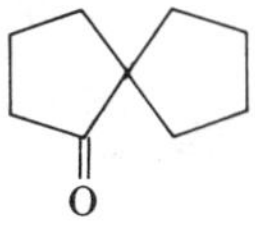

(*p*-$MeOC_6H_4)_2CHCHO$

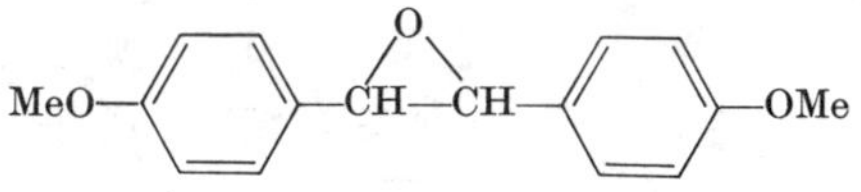

(**170**) (**171**) (**172**)

Reaction of [1-^{14}C]phenethyl iodide with either silver trifluoromethanesulphonate (triflate) or silver toluene-*p*-sulphonate in cyclohexane led to complete scrambling of the ^{14}C label over the C-1 and C-2 positions; with silver acetate, only 11% of scrambling occurred.[196]

The processes by which vinyl cations may be formed[197] in either solvolyses or electrophilic additions, though formally related, must be fundamentally very different. Some interesting work on electrophilic additions to alkenes and alkynes highlights this.[198] When the solvolyses of 1-methyl-2-phenyl-*cis*- and -*trans*-propenyl trifluoromethanesulphonate are compared, the differing rates of reaction and products (*trans* → ketone, *cis* → allene or adducts) and particularly the product stereochemistry all point to reaction of the *trans*-isomer by phenyl participation and a bridged vinyl cation (as an intimate ion-pair initially); the *cis*-ester ionizes preferentially to an open vinyl cation which may or may not then form a bridged spiroarene intermediate.[199] Deuterium isotope effects provide further evidence for a vinylidenephenonium ion in such reactions.[200] Additional evidence for aryl migration in vinylic cation systems and the stage at which this occurs is available from solvolysis studies with 2,2-diaryl-1-phenylvinyl bromides.[201]

Diazotization of 2-amino-1-hydroxy-1-phenylethylphosphonic acid, $NH_2CH_2C(C_6H_5)(OH)PO_3H_2$, yielded a carbonium ion that fragmented to inorganic phosphate (7%) but mostly rearranged by migration of the phospho-group to yield $C_6H_5COCH_2PO_3H$ in >80% yield.[202] In a more competitive situation, either a methyl group or a diphenylphosphinyl group (Ph_2PO) migrated to a carbonium ion centre. The cation-stabilizing ability of the group remaining at the migration site seemed to be of major importance in affecting the course of the rearrangement.[203] (Vinyl cations are also discussed in Chapter 9).

Rearrangement in Natural Product Systems

These studies are classified thus because, while their emphasis is synthetic rather than mechanistic, all the reactions concerned proceed through carbonium ions and all involve rearrangements.

Catalysis by BF_3–Et_2O causes a kinetically controlled, complete backbone rearrangement of the triterpenoid oxide, $3\alpha,4\alpha$-epoxyshionane.[204] Acid-catalysed cyclizations of α- and β-humulene yield a variety of azulene derivatives.[205] Protonation of *trans*-5,6-dihydro-5,6-dihydroxy-β-ionone yields a mixture of a 5-substituted 2-methylfuranone and an open-chain triketone.[206] The sesquiterpene longifolene rearranges in FSO_3H to stable cations which give good yields of some new C_{15} hexahydronaphthalenes on quenching; if the temperature of the reaction is controlled, the structure of the product can be determined.[207] Rearrangements of several abietic-type resin acids have been found to proceed through detectable (NMR) carbocations.[208] Thermodynamic control operates in the acid-catalysed rearrangements of A-noreuphenone.[209] Treatment of ketone (**173**) with protic (or Lewis) acids, yields the spiro-compound (**174**) by a 1,4-shift, whereas with acetyl methanesulphonate the tricyclic sesquiterpenoid cedrone (**175**) was obtained (Scheme 16).[210]

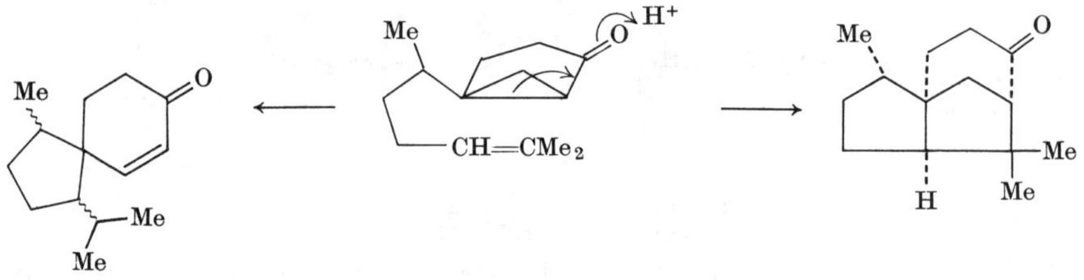

Me O Me Me Me (174) ← Me O H+ CH=CMe₂ (173) → Me O Me Me H (175)

SCHEME 16.

*Rearrangements in Polycyclic Systems**

Bicyclo[2.2.1]*heptanes.* A carbonium ion rearrangement of the Wagner–Meerwein type has been described that involves retention at the migration terminus.[211] The isomeric 7-*tert*-butylbicyclo[2.2.1]hept-2-en-7-ols afford isomeric dienes by Wagner–Meerwein rearrangement on treatment with thionyl chloride or thionyl bromide. Methyl migration in the non-classical 7-*tert*-butylbicyclo[2.2.1]hept-2-en-7-yl cation occurs with a substantial *syn*-preference.

By using ^{13}C-labelling at the 8-position, it has been found that racemization of camphene embodies four mechanisms, two in very minor amounts, namely tricyclene formation and 3,2-migration of *endo*-methyl, and two major (Wagner–Meerwein), processes, namely, 2,6-hydride shift and 3,2-migration of *exo*-methyl.[212]

The 10-isobornyl sultone (**176**) rearranges thermally to give, first, the *endo*-camphene sultone (**177**) and then the *exo*-camphene sultone (**178**). Mechanistic studies (optical, deuterium-labelling, etc.) indicate preferential operation of an *exo*-3,2-methyl shift in

* See also Chapter 8.

the formation of (**178**); several Wagner–Meerwein and 2,6-hydride shifts were also detected.[213]

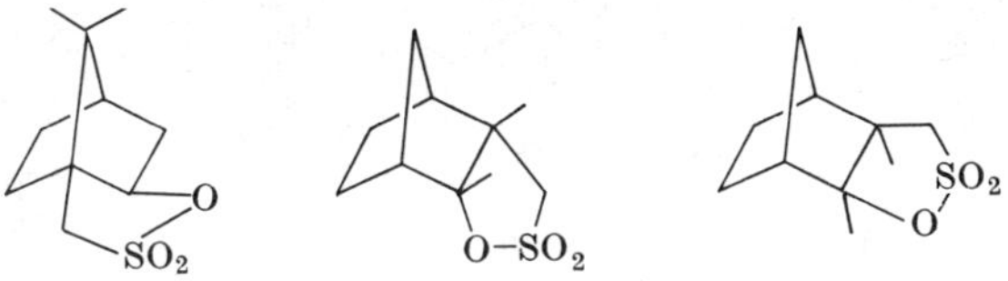

(**176**) (**177**) (**178**)

Under stable ion conditions, 2-methyl-*exo*-3-methyl-2-norbornyl cation was generated from a variety of appropriate precursors (olefins, alcohols and halides) at low temperatures.[214] The cation was characterized as a rapidly equilibrating carbonium ion by ^1H- and ^{13}C-NMR spectroscopy. When the solution was allowed to warm to –60°, the cation rearranged slowly to the isomeric 1,2-dimethyl-2-norbornyl cation.

The 1,2-diphenylnorbornyl cation was similarly observed in FSO_3H-SO_2ClF solution at –78° by means of ^1H- and ^{13}C-NMR spectroscopy.[215] It also behaved as a rapidly equilibrating cation, undergoing fast 1,2-Wagner–Meerwein shifts. The degree of σ-delocalization in the cation was very similar to that found for the 1,2-dimethyl analogue.

The long-standing controversy as to the nature of the 2-norbornyl cation has been resolved unequivocally in favour of the non-classical carbonium ion:[216] in a key paper, Olah and his collaborators examined the complete ^{13}C-Fourier transformation NMR spectrum of the 2-norbornyl cation at –70° and –150°, and also those of the 7-norbornenyl and 7-norbornadienyl cations. The CMR data clearly indicate that at low temperature all three ions are present as bridged carbonium ions. X-ray photoelectron (ESCA) spectra of the ions were also obtained and, since in these spectra the time scale of the measured ionization process is of the order of 10^{-16} sec, electron spectra can characterize individual ionic species regardless of possible Wagner–Meerwein rearrangements or hydrogen shifts. All the evidence attests to the existence of the 2-norbornyl cation as a non-classical bridged species. However, an opposite view of such spectroscopic-based conclusions has also been expressed.[217]

The conversion of camphor into nopinone has been carried out by rearrangement of 1-acetoxy-3,3-dimethylbicyclo[2.2.1]hept-*endo*-2-yl methanesulphonate which rearranges to nopinone in a *trans*-coplanar manner.[218]

The rearrangements of 2,3-dichloronorborn-2-ene epoxide to give three ketones involves 1,2-halogen shifts as well as Wagner–Meerwein rearrangements.[219] That rearrangement of 3-diazonorbornan-2-one on acid catalysis gives a bicyclo[3.1.1]heptane ketol has been confirmed.[220] Dehydration of 3α-deuteriotetrahydro-*exo*-bicyclopentadien-5-ol by acid suggests that the formation of 5,6-dihydro-*exo*-bicyclopentadiene proceeds through 2,3-dihydro-*exo*-dicyclopentadiene or the equilibrating ions formed from the latter by protonation followed by a 1,3-hydride shift, a 1,2-hydride shift and proton loss.[221] A common intermediate is indicated[222] in the solvolyses of (**179**), (**180**) and (**181**) in buffered 70% aqueous dioxan, inasmuch as product runs from those substrates lead to a common product mixture; substrates (**179**) and (**180**) react 7.2×10^3 and 7.7 times, respectively as fast as (**181**). When the bicyclo[4.1.0]heptanes (the caranes) were heated at 400–550° they underwent *cis–trans*-isomerization as well as opening of the cyclopropane ring.[223]

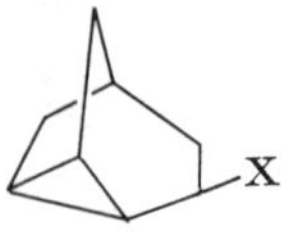

(179) 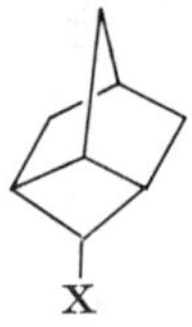 **(180)** 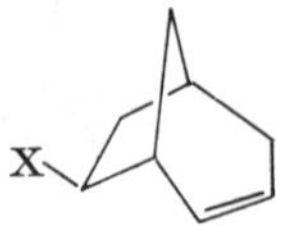 **(181)**

More Fused Systems. Passing alkyladamantanes over alumina at 200–250° results in extensive rearrangement of the tertiary adamantanyl cation, including 1,2- and 1,4-alkyl shifts, 2,4-hydride migrations, ring expansion and ring contraction.[244] The homoadamantan-4-yl cation readily rearranged to 2-methyladamantane and tricyclo-[5.3.1.0^{3,8}]undecane (**182**) in H_2SO_4–pentane mixtures.[225]

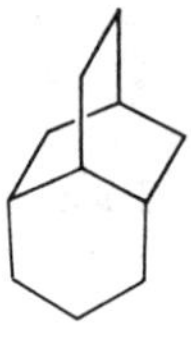

(182)

On suitable treatment trimethylsilyl ethers of 4,5-disubstituted homocubyl systems undergo extensive ring reorganization.[226] Reaction of the dibromide (**183**) with silver trifluoroacetate provides a convenient synthesis of homobullvalenone (**184**), which on treatment with BF_3 yields first (**185**) and then (**186**).[227] An intramolecular 1,3-hydride

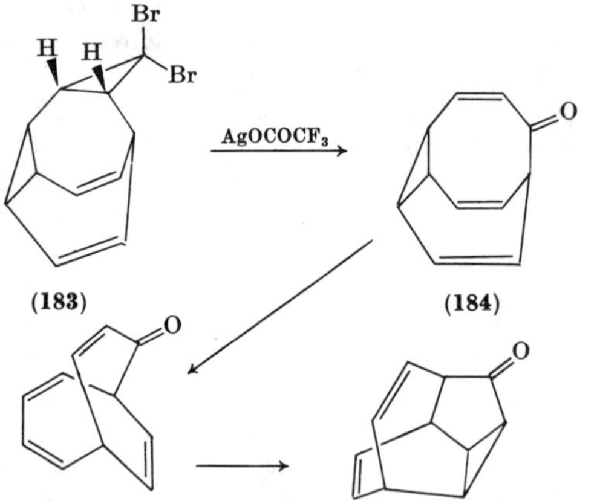

$AgOCOCF_3$

(183) **(184)**

(185) **(186)**

shift has been identified in the illustrated rearrangements of biadamantyl systems of type (**187**).[228]

Adamantane rearrangements have been examined by a molecular mechanics approach in a brief paper replete with references and interpretation.[229]

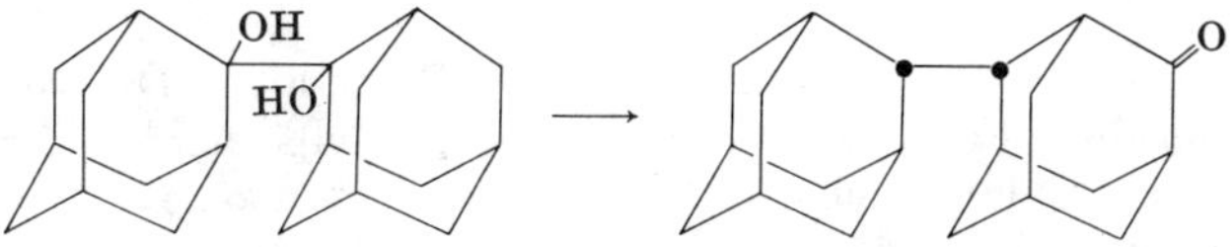

(187)

6-Methylenetricyclo[3.2.1.$0^{2,7}$]oct-3-en-8-ones [e.g. (**188**)] (compounds readily formed by Claisen rearrangements of 2,6-dialkyl-substituted phenylpropargyl ethers) on treatment with trifluoroacetic acid undergo extensive changes (including carbon monoxide extrusion by a linear cheletropic reaction) to afford polymethylated tropylium salts in a new and convenient synthesis.[230]

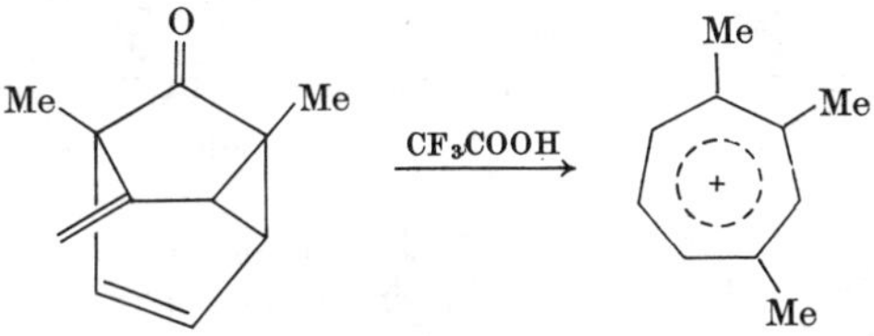

(188)

Miscellaneous

The rearrangement of the diazo-ketone, $ArOCMe_2COCHN_2$ yielding the compound $ArOCH_2COCMe{=}CH_2$ probably proceeds through an oxonium ion.[231] Treatment of the compounds (**189**) with Lewis acids in CS_2 solution yields the pyrylium salts (**190**).[232] Thermal decomposition of cyclopentylmethyl and cyclohexylmethyl dichloroborinates yields (via ion-pair intermediates) various halides including ring-expansion products.[233]

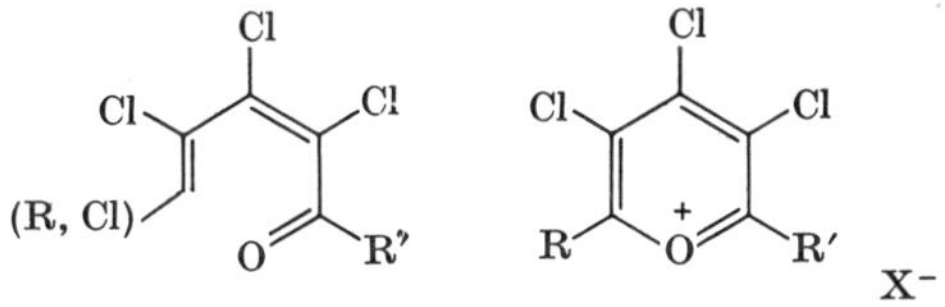

(189) (190)

Enol ethers react with arylsulphonyl azides to form unstable 2-triazolines, which lose nitrogen to yield cations that in turn undergo a Wagner–Meerwein shift, the overall effect being one of ring contraction.[234] A similar process is involved in the reaction of methylenecycloalkanes with arenesulphonyl azides except that here the overall effect is one of ring expansion.[235]

Metal-catalysed Rearrangements

Some general discussions relevant to this topic have been published.[236,237]

Acyclic Systems

Octacarbonyldicobalt-catalysed hydroformylation of 3-methyl-[3-D]-hex-1-ene, followed by oxidation and esterification, gives methyl 3-([1-D_1]ethyl)hexanoate; formation of this product involved a 1,2-migration of deuterium during the alkene hydroformylation.[238] Hydroformylation of the optically active form of the alkene yielded an ester with 70% overall retention of optical activity; the reaction sequence involves multiple isomerizations of the complexed alkene, these isomerizations preceding hydroformylation. Catalysis of the isomerization of 3-ethylpent-1-ene by iron carbonyl proceeds by a different mechanism, involving a π-allyl-metal hydride intermediate.[239] A mechanism of the latter type has also been demonstrated for the isomerization of alkenes catalysed by rhodium(III) hydrides[240] or by various palladium, platinum and rhodium complexes.[241]

Isomerization of allyl phenyl ethers by $PdCl_2(PhCN)_2$ or $RuCl_2(Ph_3P)_3$ in boiling benzene gives quantitatively phenyl propenyl ethers, with the *cis*-form predominant. Isomerization of allylphenols under similar conditions gives *trans*-(propenyl)-phenols.[242]

Different thermal and photochemical mechanisms apply to the decomposition of [σ-(but-3-enyl)-π-(cyclopentadienyl)triphenylphosphine]nickel (**191**) in benzene solution,[243] though both paths lead to the same product, *trans*-1,3-π-but-2-enyl-π-(cyclopentadienyl)nickel (**192**). When (**191**) labelled with deuterium in the 1- or 4-position of the butenyl group was used the C-1 and C-4 positions became equivalent on thermolysis; the photochemical reaction was regioselective and involved a predominant [1,3]-hydrogen shift.

Isomerization of *N*-allylamides by iron pentacarbonyl proceeds cleanly in ultraviolet light and is a useful synthetic procedure.[244] It involves intramolecular transfer of an allylic proton, but the subtle details of mechanism are not clear.

In a fine study,[245] the mechanism of the catalysis of the reversible change, propargyl ester $\rightleftharpoons$ allenyl ester (**193** $\rightleftharpoons$ **194**), by silver(I) ions has been reported, with optically

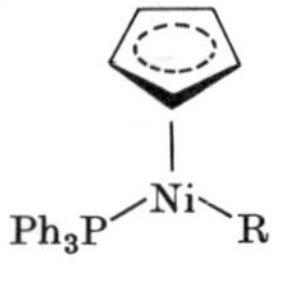

(**191**)  (**192**) 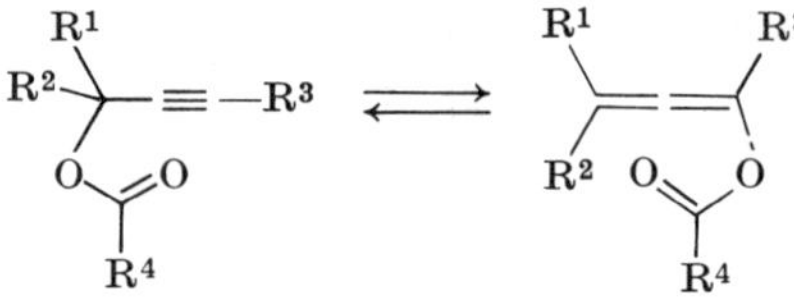 (**193**) (**194**)

active and diastereoisomeric esters as well as ^{14}C- and ^{18}O-labelling. The following were the main points: (1) The rate-determining step occurs in a silver(I) π-complex; (2) the shift of the carboxyl portion occurs intramolecularly and, as there is no ^{18}O-scrambling therein, ion-pairs can be excluded. In fact, the rearrangement occurs with inversion of the carboxyl portion, i.e., acyl-oxygen becomes alkyl-oxygen. These, together with various stereochemical results, show that the rearrangement is a [3,3]-sigmatropic reaction, of the charge-induced type.

NMR measurements show that the ligand-displacement reactions of (π-allyl)Pd(π-cyclopentadienyl) with Lewis bases (R_3P) proceed by a $\pi \rightarrow \sigma$ rearrangement of the allylic group as a primary step;[246] the final organic product is generally an allyl-substituted cyclopentadiene. A mechanism involving formation of intimate ion-pairs has been proposed for the rearrangements of platinum–chlorofluoro-olefin complexes.[247]

The palladium(II)-catalysed isomerization of *cis*- and *trans*-1-chloropropene has been studied. A possible mechanism involves rearrangement of a π-complex to a Pd(II)-σ-bonded carbonium ion which then undergoes rotation before reverting to a π-complex.[248]

The *cis–trans*-isomerizations of maleylacetone (**195**) ($\rightarrow$ **196**) and related compounds

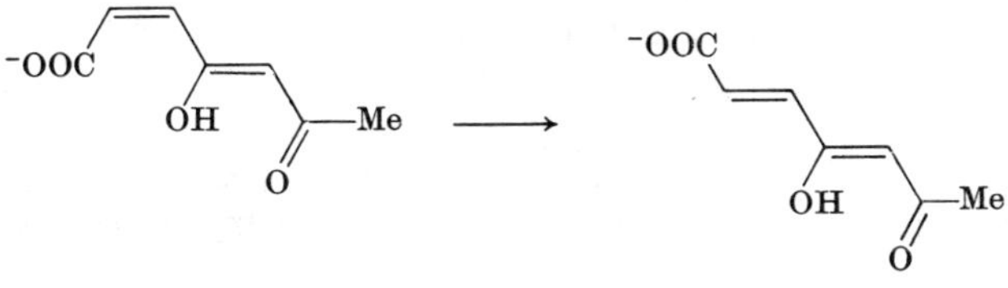

(**195**) (**196**)

are catalysed by silver ion.[249] The reaction does not involve radicals and proceeds without vinyl-proton exchange. Solvent isotope effects, pH-rate profiles, and the lack of general acid–base catalysis suggest that a π-complex between silver ion and the monoanion (**195**) is formed. This in turn loses a proton and cyclizes to a furanone, which readily undergoes rotation and then ring-opening to yield (**196**).

The ready cyclization of α,β-unsaturated ketoximes to isoxazoles in presence of palladium complexes of the type $(Ph_3P)_2PdCl_2$ as catalyst has been reported, but without mechanistic data.[250]

Cyclic Systems

Monocyclic Systems (where the rearranging system is a single ring or set of such rings). Methylenecyclopropane gives 1-methylene-2-vinylcyclopentane in the presence of nickel complexes as catalyst; butadiene was shown to be an intermediate in the reaction.[251]

The first successful direct proof of a Dewar-benzene intermediate in the thermal re-arrangement of a bicyclopropenyl compound has been reported.[252] Refluxing (**197**) in $(CH_2Cl)_2$ for 4 weeks afforded (**199**) and (**200**). Compound (**198**) was shown by NMR-spectroscopy to be an intermediate (Scheme 17). When $AgClO_4$ in benzene is used (50°, 2 hr), (**198**) and (**199**) are formed exclusively. A new mechanistic scheme is proposed

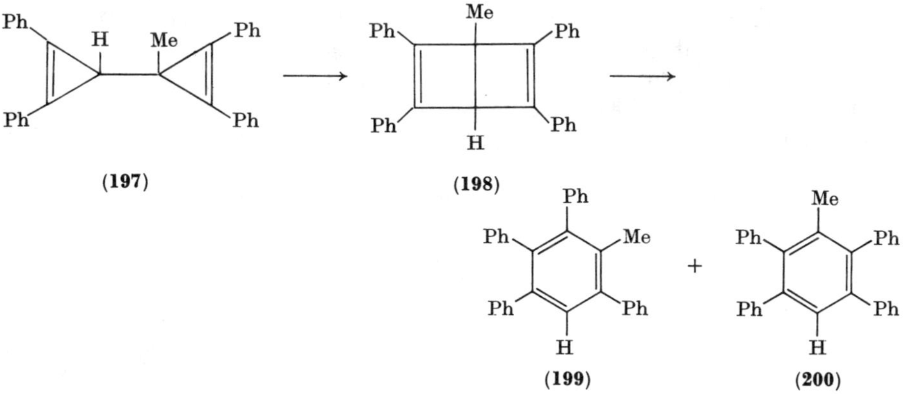

(**197**) (**198**) (**199**) (**200**)

Scheme 17.

for such processes which involves the following steps: (*a*) retrocarbene fission, catalysed by Ag(I) or heat; (*b*) ring expansion to a cyclobutenyl cation; (*c*) reaction of the inner ion-pair to give Dewar-benzenes. Consistent with this proposal is the fact that when the bridgehead-methyl group in (**197**) is replaced by a cyano-group, all rearrangements of the bipropenyl system are inhibited.[253] When the bridgehead-substituent is a methoxy-group, the rearrangements are greatly facilitated.

Vapour-phase pyrolysis of tricarbonyl[bis-1,2-(dideuteriomethylene)cyclobutane]iron at 170° resulted in hydrogen scrambling, by intermediacy of a σ-bonded iron with a π-allyl ligand.[254]

The effects of substrate geometry, transition-state geometry and reaction specificity have been examined for the allylic rearrangements **201** → **202** in which organoaluminium species are involved as intermediates.[255]

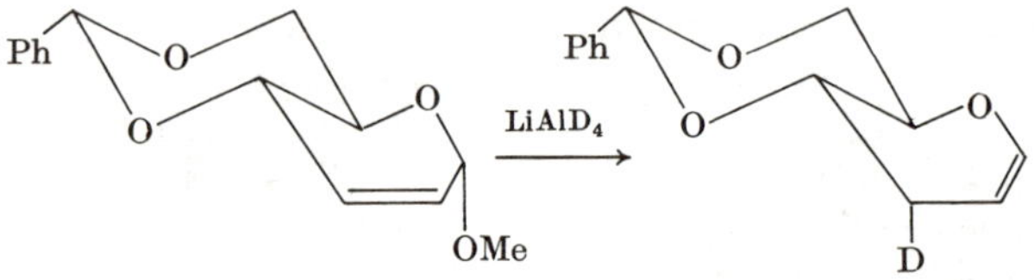

Ph O O O OMe —LiAlD$_4$→ Ph O O O D

(**201**) (**202**)

When any of a variety of substituted tricarbonyl(cyclohexadiene)iron complexes is heated in refluxing xylene for several hours, isomerization resulting from 1,5-hydrogen migration is observed.[256] From isotopic-labelling data and trapping of a co-ordinatively unsaturated diene intermediate, the *cis*-hydrogen shift does not appear to be concerted but is postulated as involving a metal hydride intermediate.

In the same way as silver ion catalyses the rearrangement of (**193**) so its presence catalyses the rearrangements of phenol propargyl ethers. Depending upon the aryl structure, the products may be allenyldienones, allenylphenols, 2*H*-chromenes or coumarones.[257]

The effects of additional ring fusion upon the cyclo-octatriene–bicyclo-octadiene equilibrium has been examined.[258] Fusion of a five-membered ring [as in (**203**)] stabilized the diene tautomer (**204**); at 58° the equilibrium constant, K, for (**204**)/(**203**) is ca. 33. Fusion of a six-membered ring as in (**205**) changed the isomer balance only slightly, K for (**206**)/(**205**) at 114° is ca. 1. With $Fe_2(CO)_9$, a mixture of (**203**) and (**204**) yields the $Fe(CO)_3$ complex of the diene form (**204**) alone. With $Mo(CO)_3(NCMe)_3$, the same starting mixture yields $Mo(CO)_3$ complexes of both forms (**203**) and (**204**). With a mixture of (**205**) and (**206**), $Fe_2(CO)_9$ yields mainly the $Fe(CO)_3$ complex of (**206**) with ca. 1% of the complex of (**205**). The other reagent, $Mo(CO)_3(NCMe)_3$, again yields $Mo(CO)_3$ complexes of both forms (**205**) and (**206**).

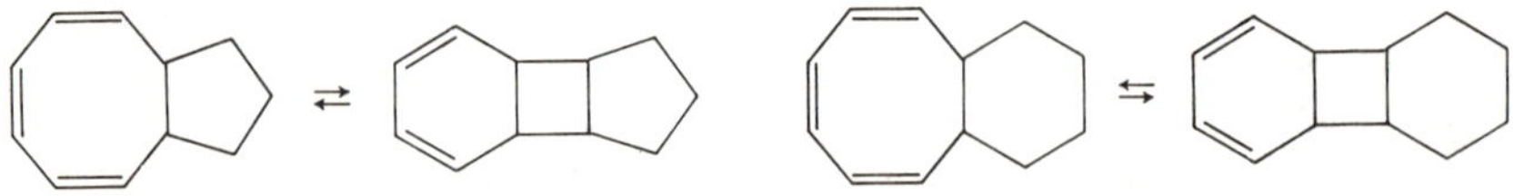

(**203**) ⇄ (**204**) (**205**) ⇆ (**206**)

Isomerization of *cis,cis*-cyclodeca-1,6-diene to *cis,trans*- and *cis,cis*-cyclodeca-1,5-diene was achieved by using $RhCl_3 \cdot 3H_2O$ in ethanolic solution.[259]

Reaction of *cis*-bicyclo[6.1.0]nonatriene either photochemically with $Fe(CO)_5$ or thermally with $Fe_2(CO)_9$ yields, amongst other products, tricarbonyl-(*cis*^{4}-cyclononatetraene)iron (**207**). At 101° (**207**) undergoes electrocyclic ring closure to the *cis*-8,9-dihydroindene isomer (**208**) ($\Delta F^{\ddagger} = 28.4$ kcal mol^{-1}). The ring closure of *cis*^{4}-cyclononatetraene is much more rapid ($t_{\frac{1}{2}} = 50$ min. at 23°; $\Delta F^{\ddagger} =$ ca. 23 kcal mol^{-1}. Low temperature (–120°) protonation of (**207**) yields the monocyclic cation (**209**).[260]

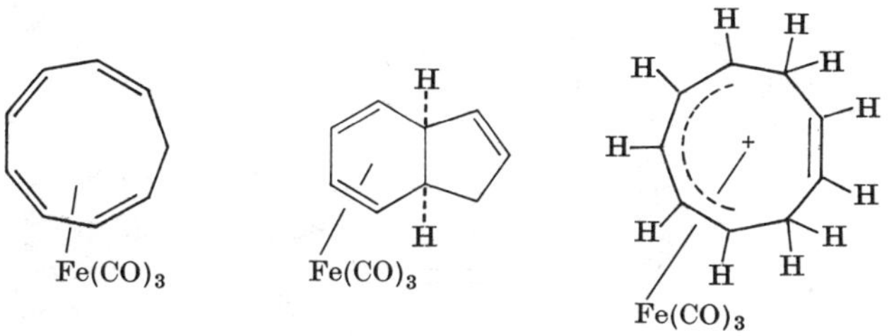

(**207**) (**208**) (**209**)

Polycyclic Systems. An interesting general discussion on the interactions of bicyclo-[1.1.0]butanes and transition-metal complexes has been published.[261]

The sensitivity to environmental factors of rearrangements of tricyclo[4.1.0.0^{2,7}]-heptane promoted by transition metals in methanol has been described. The product composition obtained by using $[Rh(CO)_2Cl]_2$ as catalyst can be determined by the rate of addition of the parent hydrocarbon to the methanolic solution of catalyst and/or base ($NaHCO_3$).[262] It is very sensitive to pH-variation.[263] Dauben's group[263] showed by infrared studies that (**210**) forms a dissociated methanolic carbene complex (**211**) which decomposes to yield diene (**212**), methanol and catalyst. When $RhCl_2(PhCN)_2$ is used as catalyst, a palladium π-allyl complex is formed that reacts with methanol to yield 2-methoxynorcarane (**213**). The use of mercuric salts and of oxymercuration reagents in ring-opening reactions with (**210**) has also been described.[264,265]

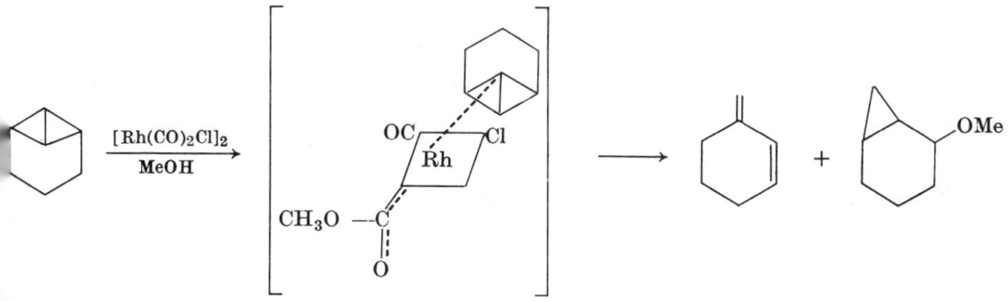

(**210**) (**211**) (**212**) (**213**)

1,1′-Bishomocubane (**214**) isomerizes in the presence of bis(cyclo-octa-1,5-diene)nickel to give a mixture of the dienes (**215**) and (**216**).[266]

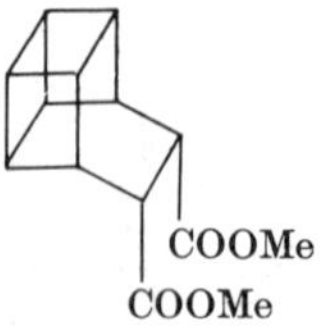

COOMe COOMe (**214**) COOMe COOMe (**215**) COOMe COOMe (**216**)

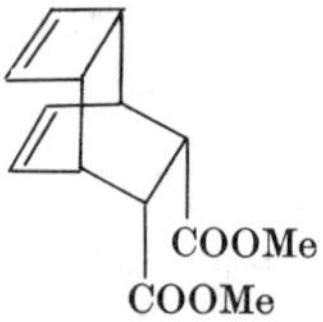

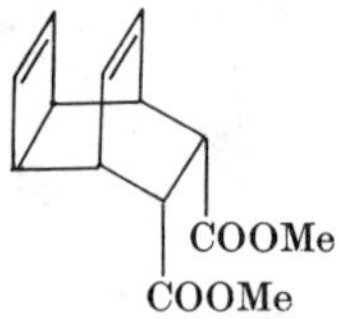

The first example of the isolation of a stable complex (**217**) between pentacarbonyltungsten and a tricyclic diene, semibullvalene, has been reported and used to determine the barrier height for its degenerate Cope rearrangement ($\Delta F^{\ddagger}$ = ca. 12 kcal mol^{-1}) (Scheme 18).[267]

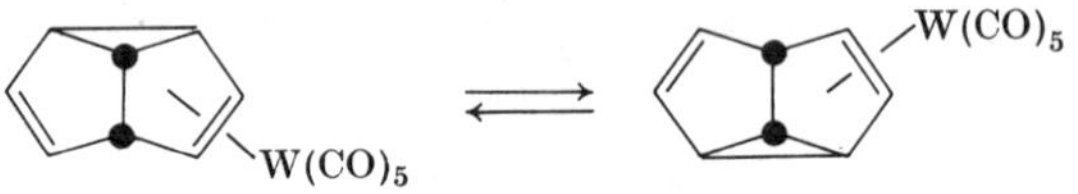

$W(CO)_5$ ⇌ $W(CO)_5$

(**217**)

SCHEME 18.

The first examples of 1,2-alkyl shifts during the hydrogenolysis of strained C_3-ring hydrocarbons in an excess of hydrogen over supported metals have been reported.[268] The involvement of a metallocarbonium ion (**218**) is suggested.

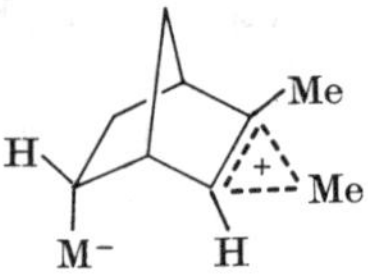

Me H Me M⁻ H

(**218**)

Reaction of (**219**) and (**220**) with $[Rh(CO)_2Cl]_2$ as catalyst in $CDCl_3$ yields the corresponding fulvenes (**221**).[269] Quadricyclanes such as (**222**) yield norbornadienes such as (**223**) as sole products.[270] From compounds such as (**219**) in methanol and the same catalyst or protons the products are cyclohexadienols or phenolic ethers (Scheme 19).[271] Similar results have been reported for oxanorbornadiene units fused in a chrysene system.[272]

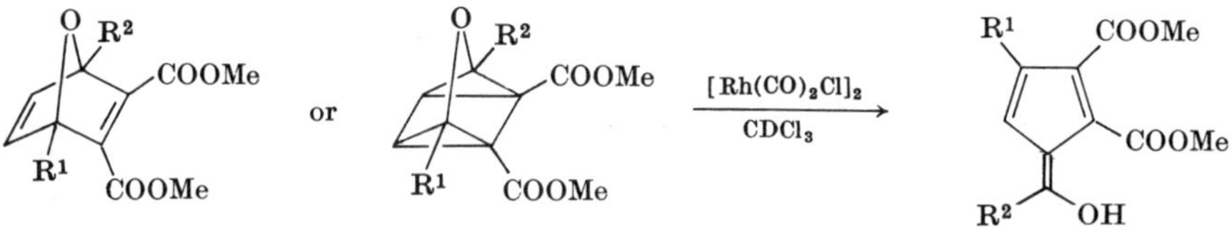

O R^2 COOMe R^1 COOMe (**219**) or O R^2 COOMe R^1 COOMe (**220**) $\xrightarrow[CDCl_3]{[Rh(CO)_2Cl]_2}$ R^1 COOMe COOMe R^2 OH (**221**)

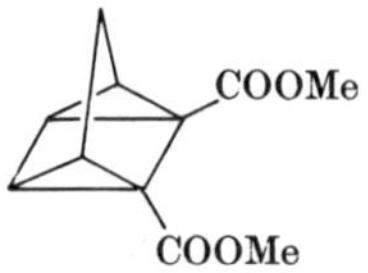

COOMe COOMe (**222**) COOMe COOMe (**223**) MeO COOMe COOMe (**224**)

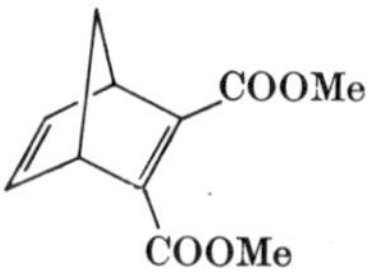

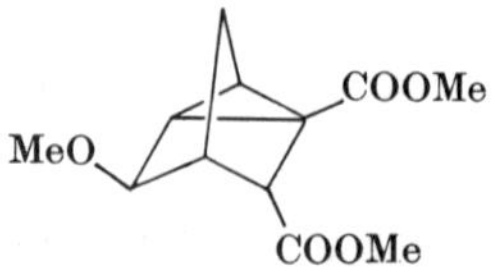

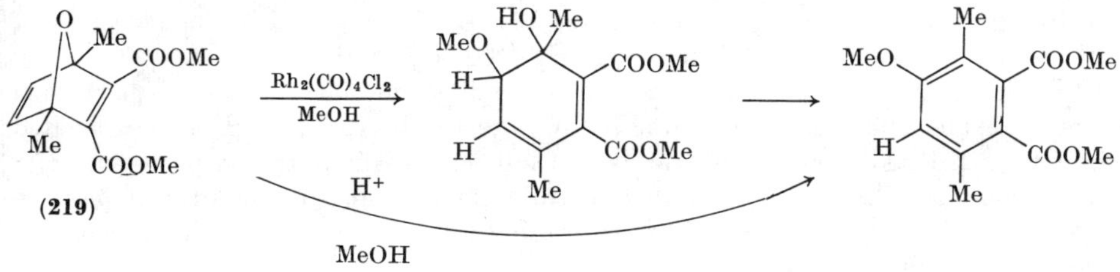

SCHEME 19.

With silver salts as catalysts, rupture of (**222**) to both diene (**223**) and nortricyclyl ethers (**224**) has been reported, an unusual side reaction being the reduction of substantial quantities of the silver to metal.[273]

Silver(I) ion also catalyses the isomerization of 1-substituted tricyclo[4.1.0.0^{2,7}]-heptanes (**225**) to a variety of products (Scheme 20).[274] With R = H, only the α-type

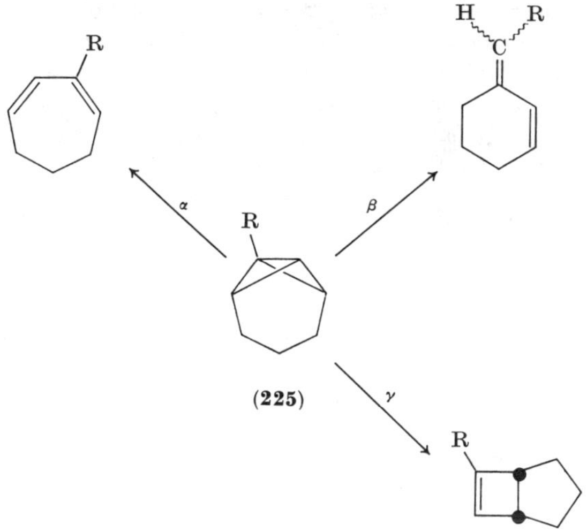

SCHEME 20.

of reaction is observed. With R = Me, types α (26%), β (29%) and γ (44%) occur. With R = COOMe, types α (97%), β (2.5%) and γ (0.5%) represents the product mixture. A mechanism involving edge attack followed by 1,2-carbon shifts and the involvement of an argentocarbonium ion appear to be involved. The reaction displays high stereoselectivity and moderate regioselectivity.

Rearrangements Involving Electron-deficient Heteroatoms

Hexamethylphosphoric triamide has been found to be a useful solvent *cum* reagent for the Beckmann rearrangement of a variety of ketoximes at 225–240°.[276]

The rearrangement of benzophenone oxime *O*-picryl ether in acetone has been studied in the presence of ^{18}O-labelled picric acid, both intramolecular and intermolecular shifts from N to C of the picryl group being detected. ^{18}O-Effects were also examined in the rearrangements of cyclohexanone oxime by $PhSO_2Cl$ in base, and in the hydrolysis of that oxime's *O*-toluene-*p*-sulphonate with $H_2S^{18}O_4$ or $K^{18}OH$.[277]

The rearrangements of cyclopentanone oxime and cyclohexanone oxime to 2-piperidone and 6-hexanolactam have been studied by infrared absorption with a decationated zeolite surface as heterogeneous catalyst. The reaction is initiated by proton-transfer from one of the hydroxy-groups in the zeolite lattice and is essentially complete at 120°.[278] This study has been extended to using a flow-system and a decationated Y zeolite containing palladium.[279]

Reaction of 3,4,5,6-tetrahydro-2*H*-azepin-7-yl hydrogen sulphate with cyclohexanone oxime–stannic chloride complex in ethylene chloride involves exchange of a sulphonyl group and ligand to give a 6-hexanolactam–stannic chloride complex and cyclohexanone oxime hydrogen sulphate, which in turn rearranges and regenerates the substituted azepinyl sulphate in the reaction system.[280]

The rearrangement of arenesulphonamides of arenehydroxamic acids, $ArSO_2NHC(=NOH)Ar'$, with PCl_5 or $SOCl_2$ leads to *N*-(arenesulphonyl)-*N*-'arylchloroformamidines, $ArSO_2N=CCl-NHAr'$). The starting oximes in the presence of aromatic sulphonyl chlorides and base undergo the Tiemann reaction.[281]

Treatment of 2-thiaisochroman-4-one oxime (**226**) with PCl_5, polyphosphoric acid (PPA) or toluene-*p*-sulphonyl chloride (in pyridine) led to normal Beckmann rearrangement yielding the benzothiazepin-2(1*H*)-one (**227**);[282] reaction with thionyl chloride led exclusively to fragmentation, the product being the chloromethyl sulphide (**228**).

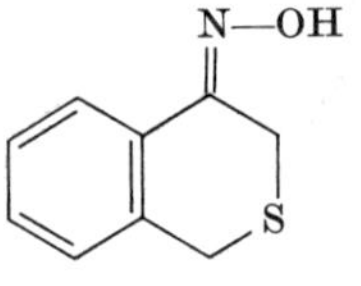

N—OH S H N O S CN CH_2SCH_2Cl

(**226**) 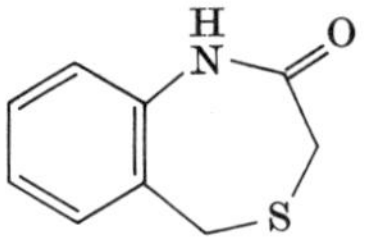 (**227**) 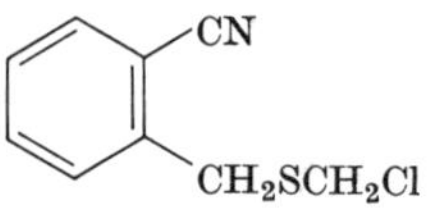 (**228**)

A similar result occurred with 3-thiacyclohexanone oxime. Both these fragmentation processes are ascribed to the involvement of thiacarbonium ions as reactive intermediates.

A novel unimolecular reaction leading to primary amine radical ions has been detected in the gas-phase decomposition of ions generated from substituted isoxazol-5(4*H*)-ones upon electron impact. The proposed mechanism implies aryl migration as the first step (Scheme 21) and this was unambiguously demonstrated by ^{15}N- and ^2H-labelling experi-

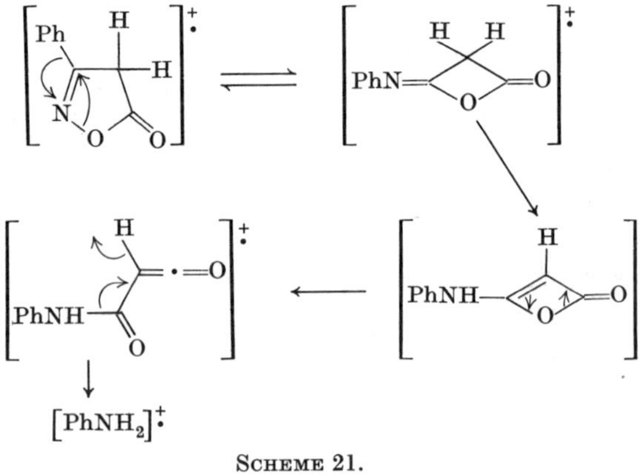

Ph H H N O O PhN= H H O =O PhNH H =O O PhNH H ·=O O $[PhNH_2]^{+\cdot}$

SCHEME 21.

ments as well as by metastable transition studies and exact mass measurements. The [1,2]-sigmatropic process is a fast reaction and the energetic and electronic factors involved suggest striking similarities to the Beckmann rearrangement in the condensed phase.[283]

Reaction of 5,6,7,8-tetrahydrocinnolin-5-one oximes (**229**) with PPA, H_2SO_4 or $POCl_3$–pyridine yield the corresponding 5-aminocinnolines (**230**) in good-to-excellent

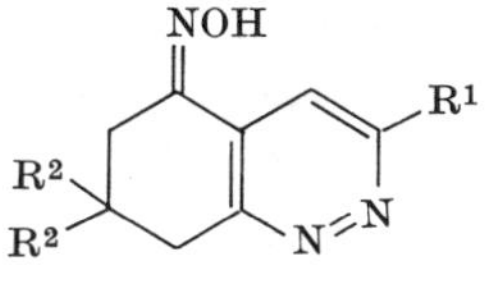

(**229**) 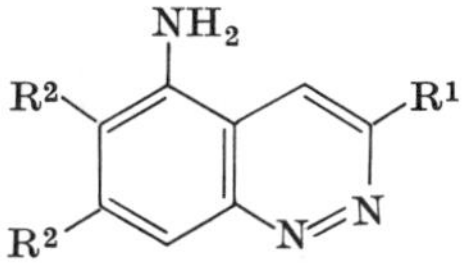 (**230**)

yields.[284] These reactions are new examples of the Semmler–Wolff rearrangement. Under Schmidt reaction conditions (NaN_3 and H_2SO_4), the free ketones (**229**; O in place of NOH) similarly take an abnormal course and yield amines (**230**).

Similar Semmler–Wolff aromatizations and abnormal Beckmann and Schmidt reactions were observed in the reactions of 3-alkyl-4,5,6,7-tetrahydro-1-phenylindazol-4-ones and their oximes (**231**) in PPA.[285] These oximes rearranged largely to the Semmler–Wolff products (**232**) and the abnormal (alkyl migration) Beckmann products (**233**) via

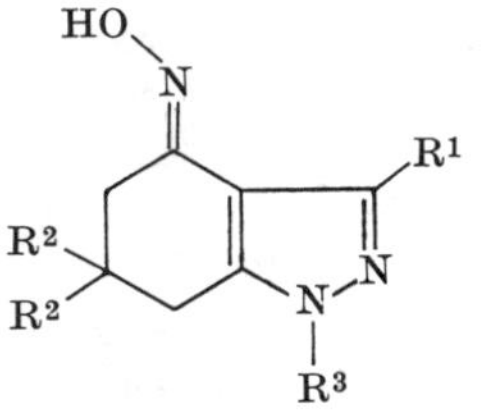

(**231**) 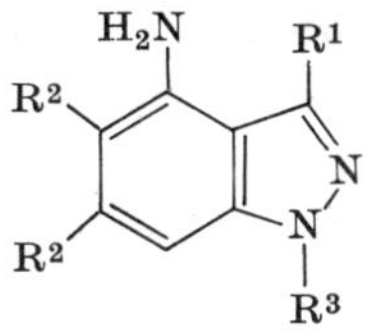 (**232**) 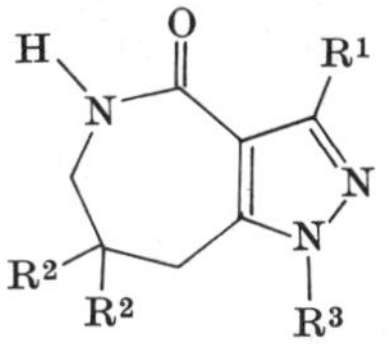 (**233**)

nitrenium ions. In the Schmidt reactions with the parent ketones, Semmler–Wolff products (ca. 18%), Beckmann alkyl migrations (70–100%) and aryl migrations (ca. 22%) were observed.

The Schmidt reaction has been applied to a series of oxa- and aza-benzocycloalkenones (**234**).[286] The pattern of reactions displayed is very dependent upon ring size. With five- and six-membered rings, alkyl migration occurs to between 80 and 100%, whereas with seven- and eight-membered-ring ketones aryl migration predominates (93–100%).

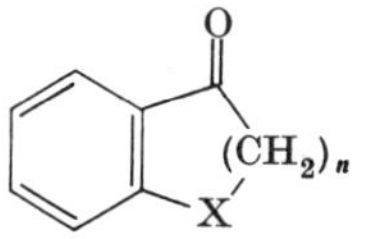

$n = 1,3$ or 4
X = O or NH

(**234**)

The true rate constants of the reactions of alkyl phenyl ketones with hydrazoic acid correlate with the substituent constants $\sigma^*(\rho = 2.28)$, showing that the limiting stage of the Schmidt reaction with ketones is apparently the addition of non-protonated hydrazoic acid to the protonated ketone.[287]

The ring expansion observed in the photochemical and thermal decompositions of 3-azidocyclohex-2-en-1-one to form a substituted azepinone by Curtius rearrangement did not occur when alkyl groups were substituted in the 2-position of the azide substrate; instead the reaction involved nitrene and azirine formation followed by ring opening (in MeOH) to yiel α-amino-dimethoxy-ketals.[288]

Photolysis of some substituted aryl azides in mesitylene–TFA (50:4) mixtures yielded the azepines (**235**) in the product mixture. Formation of such products represents a new type of reactivity for phenylnitrenes.[289]

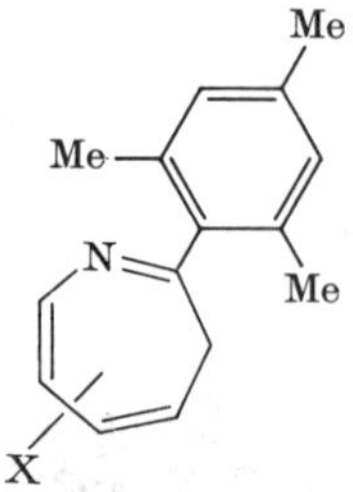

Me Me Me N X

(**235**)

A non-empirical LCAO–MO–SCF calculation shows that oxirene and formylcarbene, which are postulated intermediates in the Wolff rearrangement, have almost identical energies and that both are less stable than ketene by 70 kcal mol^{-1}.[290]

From the thermal Wolff rearrangement of diacyldiazomethanes ($ArCOC(N_2)COR$), products arising from the migration of both Ar and R groups have been isolated.[291] Relative migration aptitudes are mesityl 98, 2,4,6-$Br_3C_6H_2$ 8, Ph 3 and Me 2, suggesting that, while both steric and electronic factors are important, the steric factor is dominant when the two are in opposition. In a related study,[292] under thermal conditions *p*-$MeOC_6H_4$ > Me (3:1) in relative migratory aptitudes while H ≫ Ph. In an electron-impact-induced Wolff rearrangement H, Me and Ph had similar migratory aptitudes.[292]

A novel rearrangement of alkoxy-groups from carbon to nitrogen has been reported[293] (Scheme 22). Treatment of dialkyl *N*-(arenesulphonyloxyimido)carbonates (**236**) with alkoxides gave dialkyl (alkoxyimido)carbonates (**237** and **238**). The mechanism favoured for the process involves formation of an anion of type (**236A**) which then rearranges concertedly with loss of arenesulphonate ion.

$$(RO)_2C{=}N{-}OSO_2Ar \xrightleftharpoons{R'O^-} (RO)_2C{=}N{-}OR' + (R'O)(RO)C{=}NOR$$

(**236**) (**237**) (**238**)

$$(RO)_2C(OR')\bar{N}SO_2Ar$$

(**236A**)

SCHEME 22.

Reaction of sulphonhydrazides of type $(RCH_2)(R')NNHSO_2Ph$ with base to yield hydrazones R′NHN=CHR (the so-called diazene–hydrazone rearrangement) has been shown to involve the azo-compounds $R'N{=}NCH_2R$ as key intermediates,[294] these being derived from the rearrangement of an aminonitrene.

Nitrenium ions have been involved in a variety of rearrangements. The sequence outlined in Scheme 23, which involves as a key step cleavage of a hydroxylamine *O*-sulphonate (to generate a nitrenium species)[295] constitutes a good route for the conversion of cyclopropanones into β-lactams.[296]

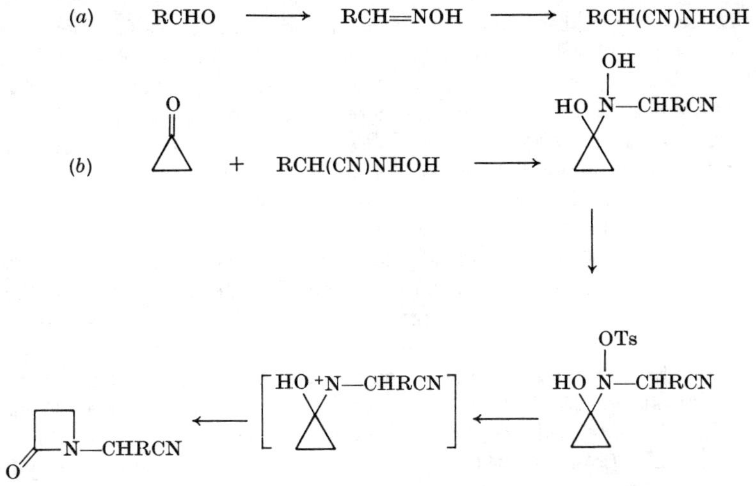

(*a*) RCHO ⟶ RCH=NOH ⟶ RCH(CN)NHOH

(*b*) cyclopropanone + RCH(CN)NHOH ⟶ HO N(OH)—CHRCN ⟶ HO N(OTs)—CHRCN ⟶ [HO ⁺N—CHRCN] ⟶ N—CHRCN

Scheme 23.

Reaction of 1-(*N*,*N*-dichloroamino)apocamphane (**239**) with aluminium chloride leads to the rearrangement product shown via a nitrenium ion.[297] On the other hand, reaction

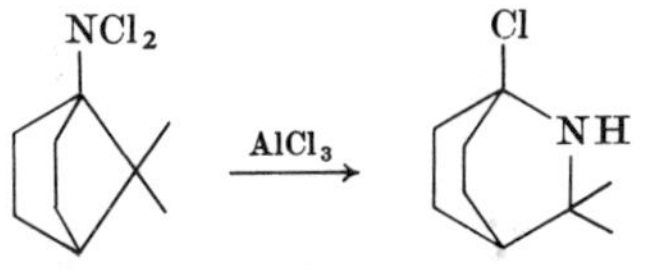

NCl_2 —$AlCl_3$⟶ Cl, NH

(**239**)

of a series of *N*,*N*-dichlorocarbinylamines (R_3CNCl_2) at low temperatures with aluminium chloride in methylene chloride does not appear to involve a nitrenium ion; instead the rearrangements observed appear to proceed by concerted processes. This conclusion is based on the low relative migration aptitudes, the lack of evidence for singlet–triplet conversion and the pronounced steric requirements in the rearrangement. The following relative migratory aptitudes were determined: Ph 18; *sec*-butyl 2.4; $PhCH_2$ 1.8; *n*-C_4H_9 1.0; H 0.09; and Me 0.05.[298]

N-Alkyl-*N'*-chloroamidines undergo rearrangement in the presence of bases such as NaOR, NaOH or Ag_2O to yield substituted ureas or carbodi-imides. The timing of halide loss and alkyl migration has been examined.[299]

Isomerizations

General

The MINDO/2 method has been used in a study of the structures of some carbenes and their intramolecular rearrangements. Methylcarbene is predicted to be strongly stabilized (by ca. 1 eV) by hyperconjugation and the rearrangements are predicted to involve migration of hydrogen gauche to the methine group.[300] The same technique has been applied to the rearrangements of singlet cyclopropylidene.[301]

A general and systematic approach to the topological analysis of isomerism and isomerization has been developed for compounds belonging to such systems as Ar_3ZX, Ar_2ZXY and Ar_2ZX (where X and Y represent ligands or pseudo-ligands and Z is a multivalent atom, e.g. C or P), wherein restricted rotation of aryl groups is displayed.[302]

The restricted rotation about the C—NMe_2 bond in a variety of *N*,*N*-diaryl-*N'*,*N'*-dimethylureas, their thio-analogues and their *S*-methylthiouronium salts has been examined kinetically. The process has a free energy of activation, $\Delta G^{\ddagger}_{Tc}$ (at the coalescence temperature) of between 9 and 17 kcal mol^{-1}.[303]

Conformational isomerism in some acyclic hydrazines has been studied by NMR.[304] The barriers to nitrogen *inversion* in 1,1-dibenzylhydrazines and tetrabenzylhydrazine are ($\Delta G^{\ddagger}$) 8.0 kcal mol^{-1} at −105°. Larger barriers found in hydrazines with *E*-related substituents are attributed to *hindered rotation* about the N—N bond. Finally, barriers to *rotation* about C—N bonds were observed in acetyl- and picryl-hydrazine.

The kinetics (at 200–250°) of the rearrangements depicted in Scheme 24 have been analysed. They provide evidence that at these temperatures (*R*)- and (*S*)-3-hydroxy-3-phenylbutan-2-ones (**240** and **241**) and their optically inactive isomer α-hydroxyisobutyrophenone (**242**) exist in a first-order, three-component, cyclic equilibrium. Rearrangement involves simultaneous O → O transfer of H and C → C transfer of Th or Me.[305]

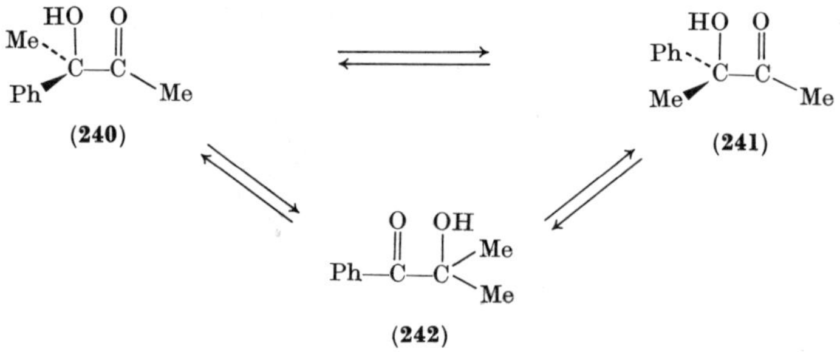

(**240**) (**241**) (**242**)

SCHEME 24.

The acetate-catalysed epimerization of 6,6,6-trichloro-5-hydroxy-4-methylhexan-3-one has been studied in glacial acetic acid. The reaction follows pseudo-first-order, reversible kinetics ($E^{\ddagger}$ 24.0 kcal mol^{-1}) and occurs by an enolization–ketonization pathway.[306]

A 4-pyridyl group attached to a carbon acid [as in *N*,*N*-dimethyl-1-(4-pyridyl)-ethylamine] introduces an isoinversion component into the base-catalysed racemization of the system with, for example, $k_e/k_\alpha \leqslant 0.8$. A conducted-tour mechanism for the isoinversion is proposed.[307]

The kinetics of epimerization of a series of dimethyl *cis*- and *trans*-cycloalkane-1,2-dicarboxylates have been measured. The *trans*-isomer is favoured with *K* (*trans*/*cis*) varying from 99 (cyclopropane) to 1.6 (1-methylcyclohexane). The rate effects ($5 > 7 \approx 4 > 6 > 3$) are explicable in terms of steric strain or its relief in a trigonal transition state and of relative ease of removal of the enolizable proton.[308]

The base-catalysed isomerization of the aromatic terpenoid taondiol (which involves epimerization at the 6a-methyl position) involves a free-radical mechanism.[309]

Tautomeric equilibria have been studied with substituted 4-(dimethylamino)azobenzene,[310] and with *N*-(acetoacetyl)-acetamide and -benzamide.[311]

Salt effects on the isomerization of triphenylmethyl isocyanide to the cyanide in acetonitrile or nitromethane have been studied. Added perchlorate ions raise, whereas added cyanide or halide ions depress, the rate.[312]

Cyclopropane and Cyclobutane Reorganizations

A modified CNDO approach has been applied to the isomerization of cyclopropanes. It was concluded that the actual mechanism operative will be determined by electronic factors, three alternative mechanisms being available depending upon the substituents present.[313]

Pyrolysis of [D_5]-labelled vinylmethylenecyclopropane (**243**) at 80° shows that the isomerization to [D_5]-3-methylenecyclopentene is accompanied by a competing degenerate methylenecyclopropane rearrangement that equilibrates (**243**) and (**244**). Both cyclopentenes (**245**) and (**246**) are formed on pyrolysis of (**243**) (Scheme 25). The

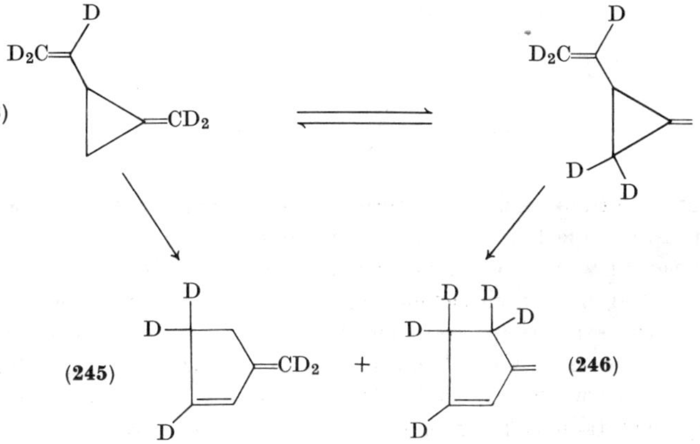

SCHEME 25.

distribution of label in the products points to the intervention of an orthogonal diradical as the major intermediate; there may be a minor concerted component, or alternatively the generation of diradicals that do not achieve geometrical symmetry.[314]

Experiments with the four racemic diastereoisomeric 1-cyano-2-ethylidene-3-methylcyclopropanes exclude concerted birotational mechanisms involving either the 45° parallel twisted allylic (Woodward–Hoffman, Moebius-controlled) or the 45° perpendicularly twisted allylic (Berson–Salem subjacent orbital-controlled) transition states from the set of structurally compatible, chirality-retaining, configuration-inverting mechanisms for the thermal rearrangements of methylenecyclopropanes.[315]

Thermal reorganization of dideuteriobiscyclopropylidene (**247**) at 214° proceeds according to Scheme 26 and involves two distinct isotope effects: (k_H/k_D) cleavage = 1.24, and (k'_H/k'_D) cyclization = 1.14. The cleavage isotope effect may be rationalized in terms of a steric secondary deuterium isotope effect upon the formation of a biradical intermediate. The isotope effect for cyclization is similar in magnitude to those observed in other cyclization processes. A concerted process cannot be ruled out for these transformations.[316]

The factors influencing the electronic configurations of diradical structures have been discussed in terms of perturbation theory and have been applied particularly to the intermediates in the thermal rearrangement of methylenecyclopropane derivatives.[317]

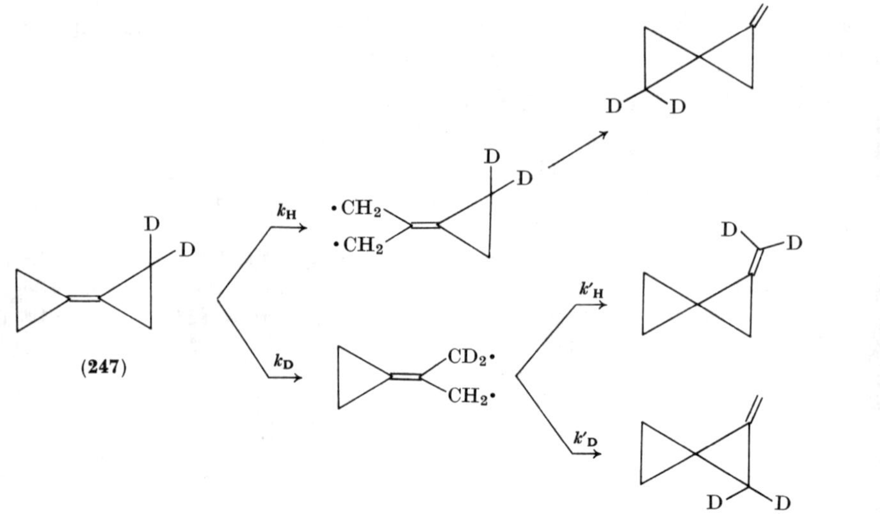

SCHEME 26.

The relative heats of formation of such intermediates have been calculated as a function of the angle of twist of a methylene group by a modified INDO method: in the favoured conformation the group was orthogonal to the plane of the remaining carbon atoms.[318]

Thermolysis of the *trans*-divinylcyclopropanes (**248**) and (**249**) yield the cycloheptadiene (**250**), the activation parameters being $\Delta H^{\ddagger}$ ca. 32.0 kcal mol^{-1}, $\Delta S^{\ddagger}$ ca. -1.0 eu. This rearrangement (photochemically realized) may be related to biosynthetic pathways for such compounds as (**250**).[319] The activation parameters of the *cis*-divinylcyclopropane–cycloheptadiene rearrangement are $\Delta H^{\ddagger} = 19.4$ kcal mol^{-1} and $\Delta S^{\ddagger} = -5.3$ eu.[320]

Rearrangement of 1-isopropylidene-2,2-dimethyl-2-(3-methylprop-1-enylidene)cyclopropane (**251**) at 80° does not yield the symmetrical material (**252**) but instead forms the cross-conjugated diene (**253**).[321]

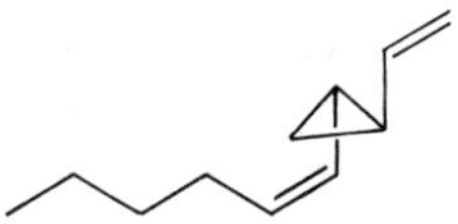

(**248**) 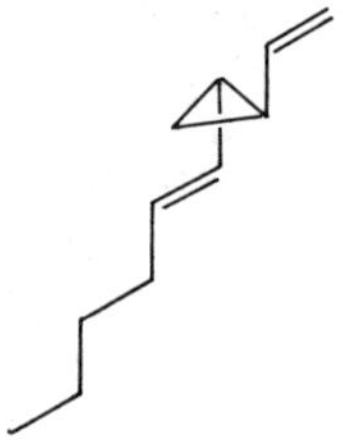 (**249**) 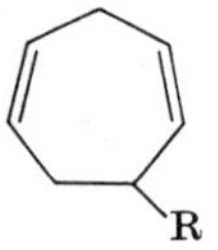 (**250**)

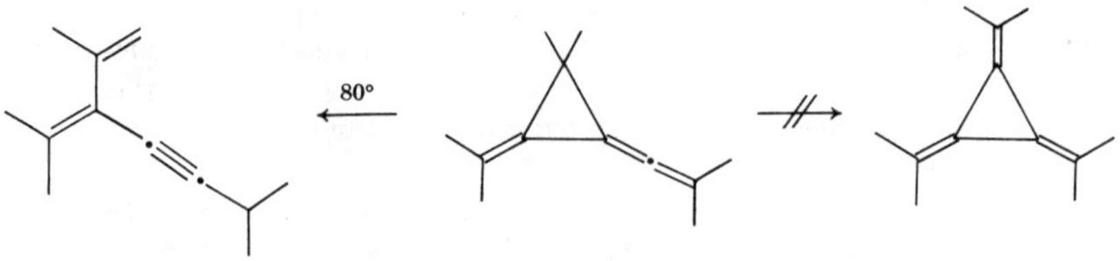

(**253**) (**251**) (**252**)

Reaction of 3,3-dimethyl-1,2-diphenylcyclopropene (**254**) with toluene-*p*-sulphonic acid in benzene yields the butadiene (**255**). Photolysis of (**254**) gives both (**255**) and its *trans*-diphenyl isomer. Vinylcarbene (**256**) mediates in the photochemical reaction.[322] Vinylcarbenes are similarly involved in the thermolysis (150–195°) of 1-*tert*-butyl-3,3-dimethylcyclopropene ($E^{\ddagger} = 29.8$ kcal mol^{-1}, $\Delta S^{\ddagger} = -17.8$ eu),[323] and in the racemization (160–190°, gas phase) of optically active 1,3-diethylcyclopropene; in the latter case, the cyclopropene must undergo ring opening, bond rotation and recyclization at a surprisingly rapid rate considering that the closed ring has 54 kcal mol^{-1} of strain energy; thermochemical analysis of the reaction suggests that ring cleavage and rotation occur simultaneously, leading directly to a vinylcarbene intermediate.[324]

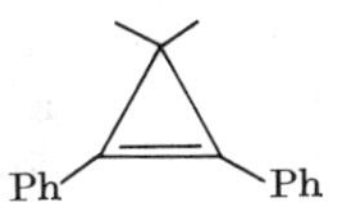

Ph Ph Ph Ph CH_2 H Ph Ph R

(**254**) 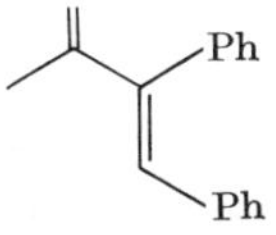 (**255**) 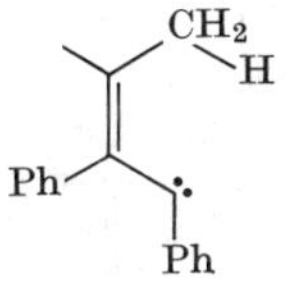 (**256**) 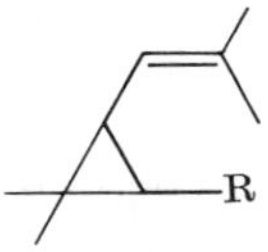 (**257**)

R = CH**:** or CMe**:**

Tri-*tert*-butylcyclopropenyl azide exhibits both solvent- and temperature-dependence in its NMR spectra as the result of chemical exchange of the azide function between three equivalent annular sites. The process proceeds by an ionization–recombination mechanism.[324a]

The *cis*- and *trans*-chrysanthemylcarbenes (**257**; R = CH) undergo thermal cleavage to yield fragmentation products, namely butadienes (70%), and ring-expansion to form cyclobutenes (30%). An added methyl substituent, as in (**257**; R = CMe), reverses these trends.[325]

The methylenecyclobutanone (**258**) undergoes ready rearrangement to the isomer (**259**) at 130° ($E^{\ddagger} = 33$ kcal mol^{-1}). A diradical or zwitterionic species (**260**) is suggested as intermediate. The rearrangement is solvent-insensitive, supporting the involvement of a diradical.[326]

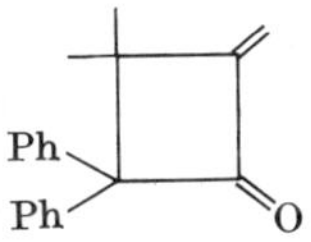

Ph Ph O Ph Ph O Ph O Ph

(**258**) 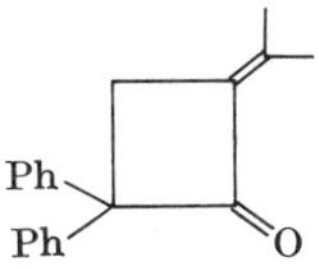 (**259**) 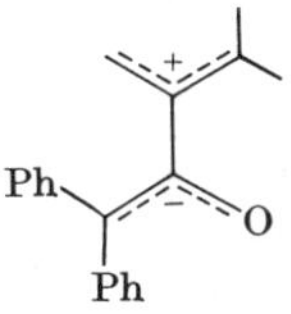 (**260**)

Pyrolysis of (1*R*,2*R*)-*trans*-1,2-divinylcyclobutane at 147° gives (*R*)-4-vinylcyclohexene. The rearrangement of the portion of starting material that did not undergo prior racemization gives 54% of (*R*)-(+)-substituted cyclohexene and 46% of (*S*)-(–)-product.[327] The stereochemistry of this rearrangement results from two competing stereospecific processes rather than from a small stereospecific component superimposed on a mechanism dominated by a stereorandom intermediate.*

* We have discussed this in more detail on p. 438.

The isomeric methylenecyclobutane systems (*E*)- and (*Z*)-1-ethylidene-2-methylcyclobutane and *cis*- and *trans*-1,3-dimethyl-2-methylenecyclobutane have been thermally equilibrated at 332°. The rate constants reveal some striking stereochemical effects. The latter pair rearrange preferentially to the *E*-isomer of 1-ethylidene-2-methylcyclobutane. The *Z*-isomer of this material isomerizes with predominant cleavage of the more substituted C-2 to C-3 bond; but the *E*-isomer does the opposite, the C-3 to C-4 bond breaking preferentially. Both the *E*- and the *Z*-isomer show net retention of geometrical configuration about the ethylidene unit during 1,3-carbon migration originating from C-2.[328]

Optically active (*Z*)-1-ethylidene-2-methylcyclobutane racemizes at 332° more slowly than (*Z*)-1-(1-deuterioethylidene)-2-methylcyclobutane equilibrates with (*Z*)-1-ethylidene-2-deuterio-2-methylcyclobutane; this result requires that at least some of the 1,3-carbon migration in the degenerate methylenecyclobutane isomerization (automerization) occurs with antarafacial allylic participation.[329] 1-(*Z*)-(1-Deuterioethylidene)-*trans*-3,4,4-trideuterio-2-methylcyclobutane (**261**) isomerizes degenerately at 332° to

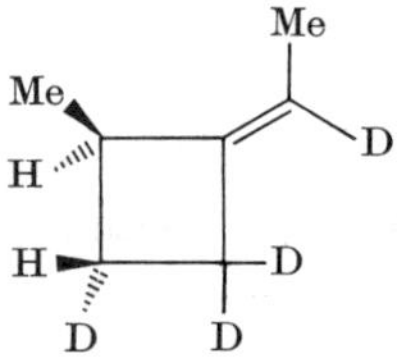

Me Me D H H D D D

(**261**)

give three other (*Z*)-1-ethylidene-2-methylcyclobutanes, one having the 3-deuterium label *cis* to the 2-methyl group, and the two *cis–trans*-isomers of 1-ethylidene-2,3,4,4-tetradeuterio-2-methylcyclobutane. The rate constants for these processes give a direct measure of the extent to which this methylenecyclobutane automerization by 1,3-carbon migration occurs with stereochemistry appropriate to an orbital-symmetry-allowed reaction, namely 77%. The quantitative estimate for the extent to which the 1,3-shift is antarafacial with respect to the allylic component is 65%.[330]

Polycyclic Rearrangements

The valence isomerization between cycloheptatrienes and norcaradienes has been reviewed.[331]

An intramolecular 1,5-hydrogen shift has been observed in 2-deuteriocyclohepta-1,3,5-trienes at 160°.[332]

Pyrolysis of a series of substituted cyclo-octatetraenes in a flow system at temperatures between 400° and 600° causes degenerate rearrangements [deuterium migration from C-3 and C-8 (e.g. to C-4 and C-7), and from C-5 and C-6] and alkyl migrations; involvement of bicyclo[4.2.0]octatrienes is implicated.[333]

On a theoretical basis, it has been shown by the MINDO method that thermal isomerization of cyclo-octatetraene to semibullvalene is an allowed process as a concerted cyclization in the ground state.[334]

Treatment of 2-allyl-4-hydroxycyclopent-2-enones with pyridine hydrochloride at 200° produces 2-propylcyclopent-2-ene-1,4-diones. The isomerization probably proceeds through intermediate 2-allylcyclopentane-1,4-diones.[335]

Pyrolyses of spirans of type (**262**) at 235° involves initial concerted conrotatory ring opening to yield the allylidenecyclopropane (**263**) which then at 440° undergoes homolysis to a singlet diradical leading ultimately to fulvenes of type (**264**).[336]

Equilibration of a series of methyl oxocycloalk-2-enecarboxylates (**265**) by means of 1,5-diazabicyclo[4.3.0]non-5-ene as a basic catalyst in benzene solution has been studied:

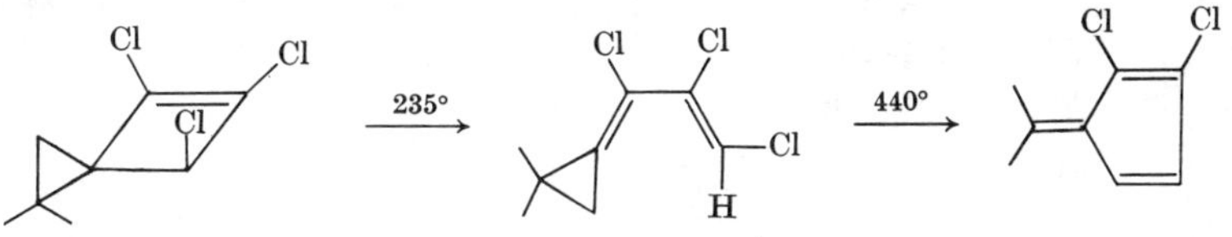

(**262**) (**263**) (**264**)

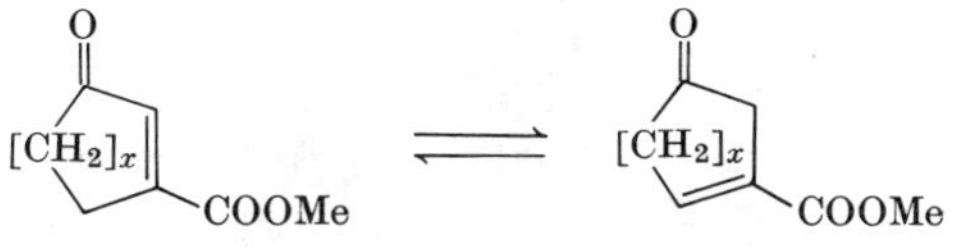

(A) (B)

(**265**)

with ring sizes 7, 8, 9 or 10, the percentages of isomer (B) present at equilibrium were 85, 97, >99 and > 99.[337]

Treatment of the 9-substituted 9-benzyl-*sym*-octahydrothioxanthenes (**266**) with HCl gave the isomerized products (**267**). Similarly the thiopyran (**268**) gave (**269**).[338]

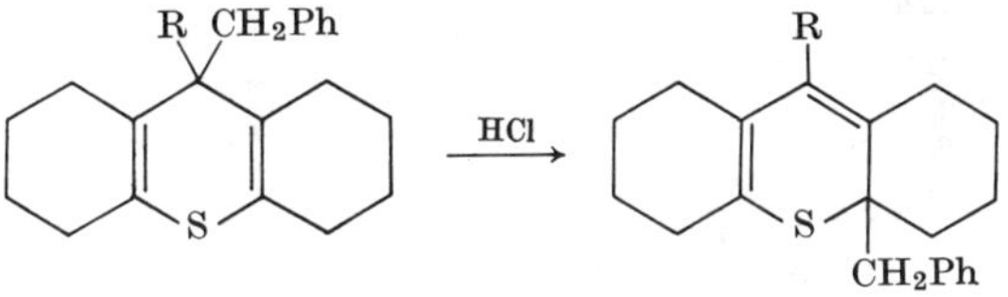

(**266**) (**267**)

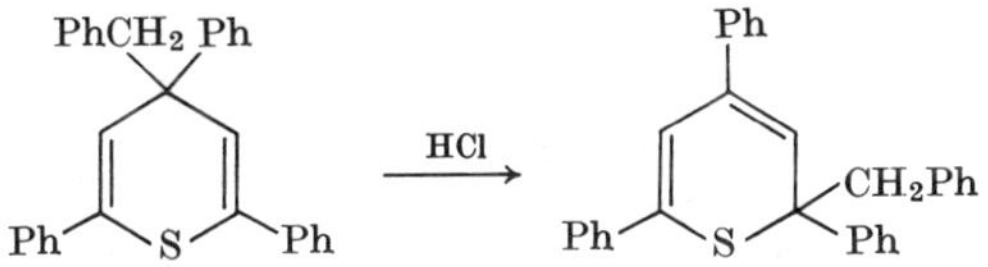

(**268**) (**269**)

The 4-hydroxypyrylium cations (**270**) in 96% H_2SO_4 undergo photoisomerizations to corresponding 2-hydroxypyrylium cations (**271**). These conversions involve the π,π^* excitation of (**270**) and bicyclic oxygen-bridged intermediates.[339]

Thermal rearrangement at 300° of bicyclo[2.1.0]pentane, dideuterio-labelled at position 5, involves C-1 to C-4 bond-cleavage and *C*(5)-H migration, to yield 2,3-dideuterio-cyclopentene.[340]

2-Methylbicyclo[2.1.0]pent-2-ene (A) is thermally isomerized to both 2-methylcyclopentadiene (B) and 1-methylcyclopentadiene (C). At 50° (gas phase), $k_{AB}/k_{AC} = 1.3$; in hexane (from 30 to 90°) $k_{AB}/k_{AC} = 12.2$. Such results may be explained by assuming that even in solution and to a constant extent over a 60° temperature range, collisional deactivation and further isomerization of vibrationally excited 2-methylcyclopentadiene are kinetically competitive.[341] A concurrent study[342] showed that gas-phase pyrolyses (50°) of 1- and 2-methylbicyclo[2.1.0]pent-2-ene leads to both (B) and (C), with (B) generally predominant; this work reiterates that cleavage of the central (1,4) bond in such bicyclo[2.1.0]pent-2-enes is the predominant isomerization path, hydrogen shifts occurring subsequently in the chemically activated cyclopentadienes produced.[342]

Pyrolysis (136–238°), both in the gas phase and in solution, of a variety of bicyclo-[2.2.0]hexanes (**272**) has been studied. An excellent group additivity effect has been found for the activation energies of ring opening [to the dienes (**273**)] for the following

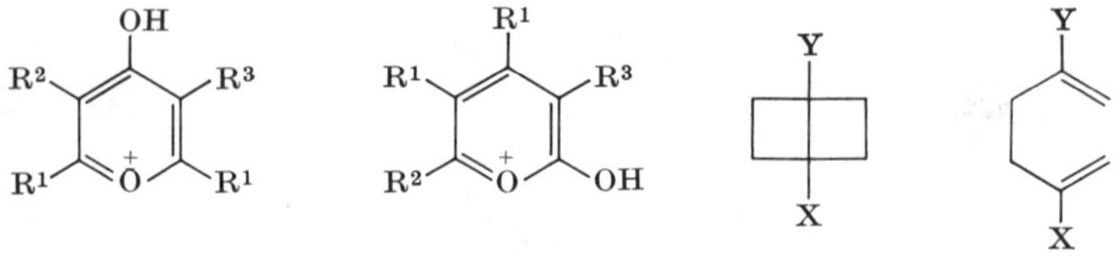

(**270**) (**271**) (**272**) (**273**)

sets of 1,4-substituted compounds: X,Y = H,H; H,Cl; and Cl,Cl; Cl,H; Cl,COOMe; COOMe, COOMe. This assumed that there is no 1,4-interaction in the transition complex.[343] With alkyl substitution in the compounds (**272**); X = Cl; Y = H, Me or Et) there is a small increase (1.2 kcal mol^{-1}) in activation energy when H is replaced by Me, and an additional small increase when it is replaced by Et. If the ring -opening involves a biradical intermediate, then these results are anomalous.[344]

Thermolysis of 6-methylenebicyclo[3.2.0]hept-2-ene (**274**) at 200° results in the interconversions outlined in Scheme 27. The involvement of a bisallyl radical is proposed as a key intermediate.[345]

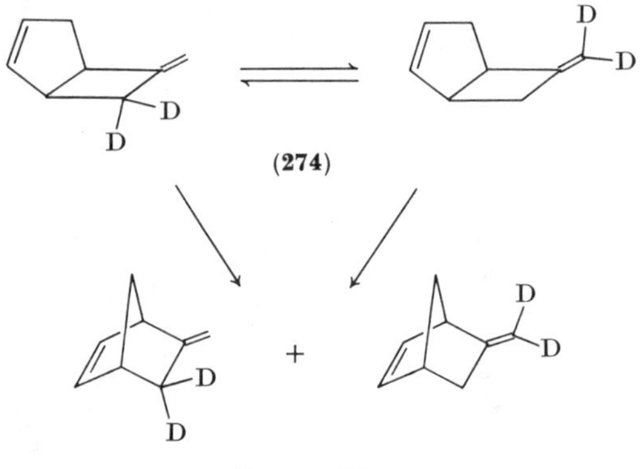

(**274**)

SCHEME 27.

Heating (see Scheme 28) neat *anti*-9-cyanobicyclo[6.1.0]nona-2,4,6-triene (**275**) at 139° yields the *syn*-counterpart (**276**), and a mixture of the *cis*-8,9-dihydroindenes (**277**)

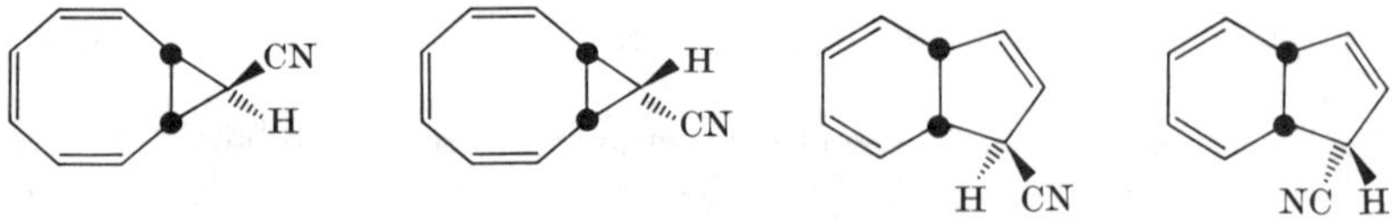

(**275**) (**276**) (**277A**) (**277B**)

SCHEME 28.

[$\Delta G^\ddagger$ for loss of (**275**) = 30.9 kcal mol^{-1}]. Isomer (**275**) is responsible for the rearrangement of the system to an 8,9-dihydroindene skeleton.[346]

The first example of a butadienylcyclopropane rearrangement has been observed in the thermolysis of labelled bicyclo[5.1.0]octa-2,4-diene (**278**) at >110°. This process is accompanied by a degenerate 1,5-H shift. In the two groups of four non-adjacent carbon atoms on the eight-membered ring periphery, the functions are exchanged by the sequence of these two reactions.[347]

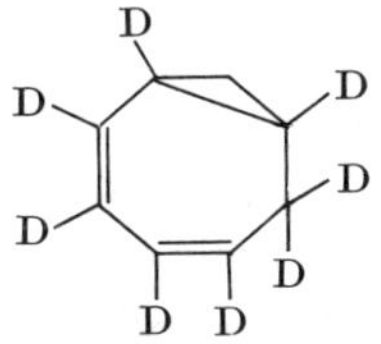

(**278**)

Homobullvalenone (**279**) rearranges at 85–105° via a vinylketene two-step rearrangement to its isomer (**280**).[348]

The first case in which a 2-substituted adamantane is the major constituent at equilibrium has been realized in the treatment of 1,1′-biadamantane (**281**) with $AlBr_3$ in cyclohexane at 60°, whereupon 25% of (**281**) and 70% of (**282**) were formed. In agreement with molecular-mechanics calculations, (**282**) is more stable than (**281**): **281** $\rightleftharpoons$ **282**, ΔG°_I (at 298°) = −0.51 kcal mol^{-1}, ΔH°_I = 1.11 kcal mol^{-1} and ΔS°_I = 5.4 cal deg^{-1} mol^{-1}.[349]

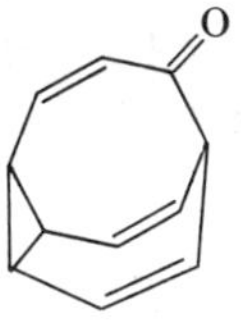

(**279**) 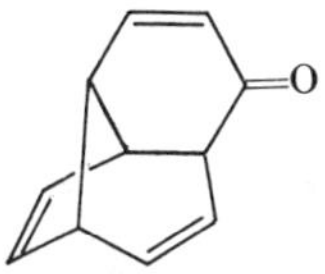 (**280**)

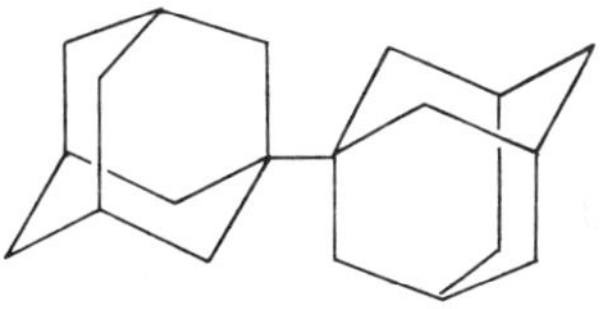

(**281**) 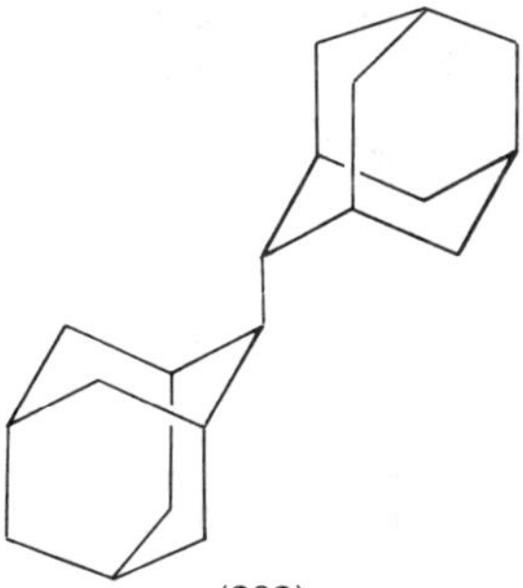 (**282**)

Thermal rearrangement of diademane (**283**) to the quinacene (**284**) occurs by concerted allowed cycloreversion ($E^\ddagger$ = 28.3 kcal mol^{-1}) and not by a diradical or dipolar route.[350]

A symmetry-allowed concerted mechanism has been proposed, based on all-valence electron-semiempirical SCF-MO theory for the thermal rearrangement of tricyclo-[3.3.0.0^{2,6}]octa-3,7-diene (**285**) to semibullvalene. The mechanism proposed involves

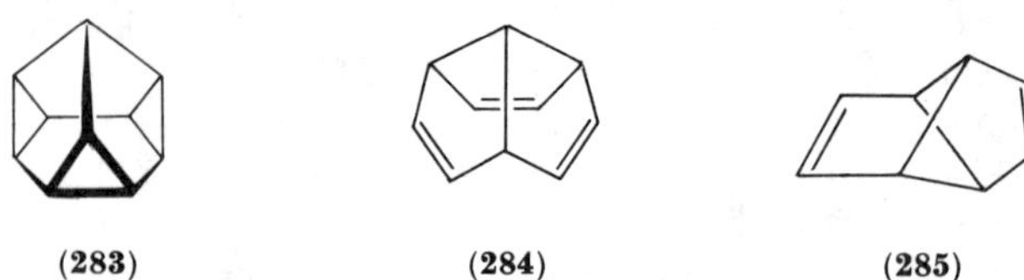

(283) (284) (285)

cleavage of an allylic bond by twisting the molecule, with C_2 symmetry maintained until the geometry of the transition state for the Cope rearrangement is obtained.[351] The same technique has been applied to the electronic structure and valence isomerization of *syn*- and *anti*-tricyclo[4.2.0.0^{2,5}]octa-3,7-dienes.[352]

Pyrolysis of barbaralane (tricyclo[3.3.1.0^{2,8}]nona-3,6-diene) at 300–500° gave 1- and 3-vinylcyclohepta-1,3,5-triene, *cis*-4a,7a-dihydroindene and allylbenzene.[353]

The first example of the existence of the "valene"-type valence isomer of non-alternant hydrocarbons has been reported in the synthesis of naphtho[1.8]tricyclo[4.1.0.0^{2,7}]-heptene (**286**). This compound readily rearranges in presence of $AgClO_4$ in benzene to pleiadiene (**287**).[354]

Refluxing a series of 5-methylenebicyclo[4.2.0]octa-2,7-dienes, e.g. (**288**), in xylene leads to mixtures of four compounds (**289a** and **b**) and (**290a** and **b**). This thermal isomerization takes place with moderate stereospecificity.[355]

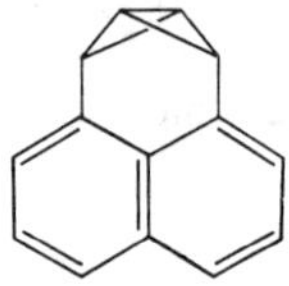

(286) 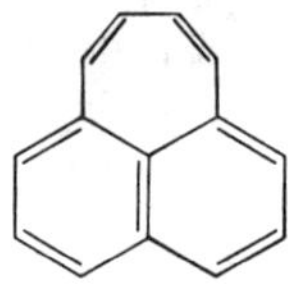 (287)

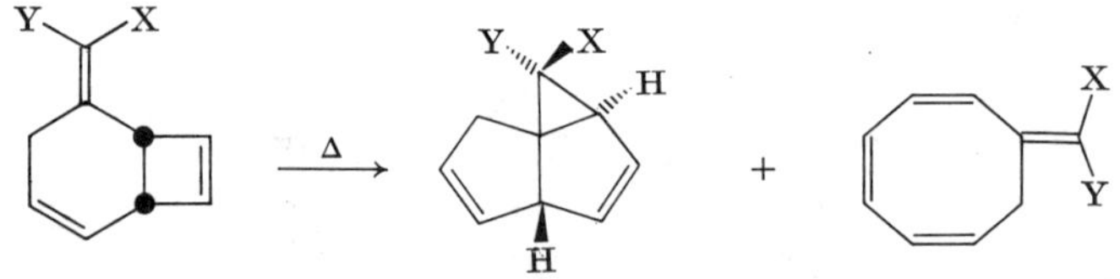

(288) (289) (290)

a, X = CN, Y = COOMe
b, X = COOMe, Y = CN

Pyrolysis of *exo*-tricyclo[3.2.1.0^{2,4}]octene (**291**) in the gas phase at 290° yields both bicyclo[3.2.1]octa-2,6-diene (**292**) and tetracyclo[3.3.0.0^{2,8}0^{4,6}]octane (**293**) (4:1). The *endo*-isomer of (**291**) gives (**293**) alone. While (**293**) appears to arise from a diradical intermediate, deuterium-labelling at C-3 establishes that (**292**) arises from a concerted process.[356]

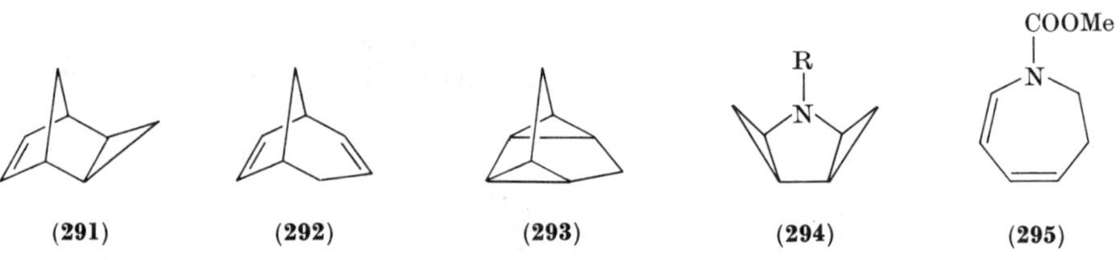

(291) (292) (293) (294) (295)

Heating methyl *syn*-2-azatricyclo[4.1.0.0^{3,5}]heptane-*N*-carboxylate (**294**) to 121° causes its rearrangement to the 2,3-dihydroazepine ester (**295**). The *anti*-isomer requires a temperature of 350° before it rearranges to (**295**). With deuterium-labelling in each of the apical positions of the cyclopropane rings in the *anti*-material, rearrangement produces an azepine with deuterium distributed statistically in positions 2, 3, 6 and 7, thus indicating that rapid 1,5-H shifts occur at 350°. Both (**294**) and its *anti*-isomer revert to a dipolar form, the *syn*-isomer proceeding through a smooth concerted $\pi^2 s + \pi^4 s$ cycloreversion, and the *anti*-isomer through a less energetically favoured $\pi^2 a + \pi^4 a$ process.[357]

Pyrolysis of *tert*-butyl *anti*-bicyclo[7.1.0]decane-2-percarboxylate yields a cyclopropylcarbinyl radical that does not give stereospecific ring expansion, both *cis*- and *trans*-cyclodecenes being formed.[358]

Reaction of 2′-cyclohexylcyclohexanespiro-3′-oxaziridine with iron(II) sulphate gave a considerable quantity of isomeric *N*,*N*′-dicyclohexyldecanedicarboxamides, some of these having a branched-chain structure produced by 1,5-radical rearrangements.[359]

Alkene Studies

CNDO/2 and INDO calculations have been applied to propenes of type MeCH═CHX where X is a first-row heteroatom containing at least one lone pair housed in a p_z atomic orbital. These studies show that bonding interactions of an attractive nature can occur between the methyl group and the heteroatom's p_z lone pair. This in turn is responsible for the lower energy of the *cis*- than of the *trans*-isomer and also for the lower rotational barrier of the *cis*- than of the *trans*-isomer.[360]

Equilibrium constants for reactions of the type *trans*-$XCH_2CH{=}CHY \rightarrow$ *trans*-$XCH{=}CHCH_2Y$ have been correlated by using an allowance for interaction of XCH_2 and Y (or X and CH_2Y) across the *trans*-vinylene group. Such interactions were calculated from σ_p and a proportionality constant τ_V. A double-bond stabilization parameter D_x was derived. Its magnitude, which increases with increasing tendency of the substituent to stabilize a carbon–carbon double bond to which it is attached, increases in the order MeO > Ph > F > C_nH_{2n+1} SMe ~ COOMe > NO_2 > CH_2OMe > CN > CH_2COOMe > Cl > SOMe > Br > SO_2Me.[361]

The *cis-trans*-isomerization of a series of halogen-substituted maleic esters has been studied.[362] Under certain conditions, alkoxycarbonyl (COOR) migration takes place and methylenemalonic acid derivatives are formed.[363]

Irradiation of the dithiolylidene ketones (**296**) in ethanolic solution converts them into the isomers (**297**). The reverse process occurs thermally, and its kinetics have

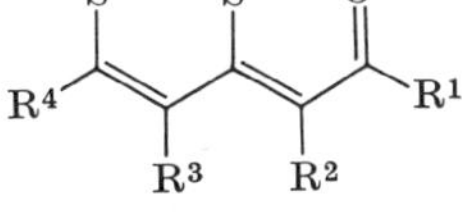

(**296**) 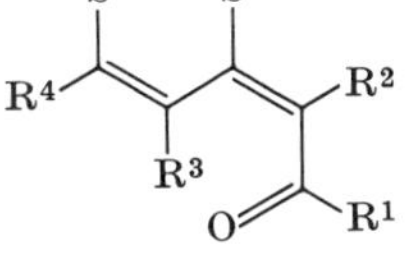 (**297**)

been studied. The rate constants, which varied over five powers of ten, were largely governed by steric factors.[364]

The mechanism of skeletal rearrangement of isomeric hexanes over platinum black in the presence of helium or hydrogen has been studied,[365] and the thermal rearrangements of acetylenes to allenes has been reviewed.[366] The kinetics of acid- and base-catalysed isomerization of 1,3-diphenylpropenes has been studied with the aid of ^{3}H- and ^{14}C-labelling; the Ingold B-S_E2, mechanism is not found to be operative therein.[367]

Isomerization of [3-D]but-1-ene over calcium oxide involves a 1,3-intramolecular hydrogen shift. The deuterium content in the product remained unchanged (relative to starting material), no D_2 was formed, and [1-D]but-1-ene was formed.[368]

The kinetics of positional isomerization of 2,3-dimethylbut-1-ene to 2,3-dimethylbut-2-ene catalysed by iodine atoms has been studied from 410 to 530°K. From the results, the stabilization energy of the dimethylbutenyl radical may be calculated as being from 16.7–18.5 kcal mol^{-1}.[369]

The reversible thermal isomerization of the positional isomers of several substituted silylindenes is sensitive to the nature of substituents on the silicon.[370]

Thermal *cis–trans*-isomerization (910°K, 3 mm, contact time per pass ca. 50 msec) of *cis,cis*-1,4-dideuteriobuta-1,3-diene shows that a portion proceeds through a diradical species; however, a correlated double isomerization is stereospecific as required for a cyclobutene pathway and is, in fact, the dominant mechanism.[371]

The gas-phase pyrolysis of *trans*-hexa-1,3,5-triene (at 588 and 640°K) has been studied. Reaction proceeds via *cis*-hexa-1,3,5-triene to yield cyclohexa-1,3-diene. Further unimolecular decomposition of this cyclohexa-1,3-diene (between 719 and 824°K) is symmetry-forbidden as a direct concerted process and proceeds through cyclohexa-1,4-diene.[372]

The prototropy of 2-methyl-1,3-diphenylpropenes under basic catalysis has been studied. The free energies of the carbanions generated from the substrate increase in the order *trans–trans* > *trans–cis*, indicating that 1,2-interactions in the carbanion are dominant, in contrast with related systems that do not possess the 2-methyl group).[373]

Thermal, photochemical and photosensitized decompositions of 1,2-diphenyldiazopropane yield (Z)- and (E)-1,2-diphenylpropenes amongst other products, *via* a carbene intermediate. The carbene rearrangements observed with 2,2-diphenyldiazopropane and with 1-(p-deuteriophenyl)-2-phenyldiazopropane all suggest the existence of a discrete intermediate in these migrations, this species being the so-called phantom singlet state.[374]

Action of deuterium fluoride on steroidal olefins such as (**298**) caused double-bond migrations by way of carbonium ion intermediates to yield products such as (**299**).[375] Thermal isomerization of steroidal boranes can provide a fast and simple synthesis of *cis*- and *trans*-hydroindanic steroidal systems.[376]

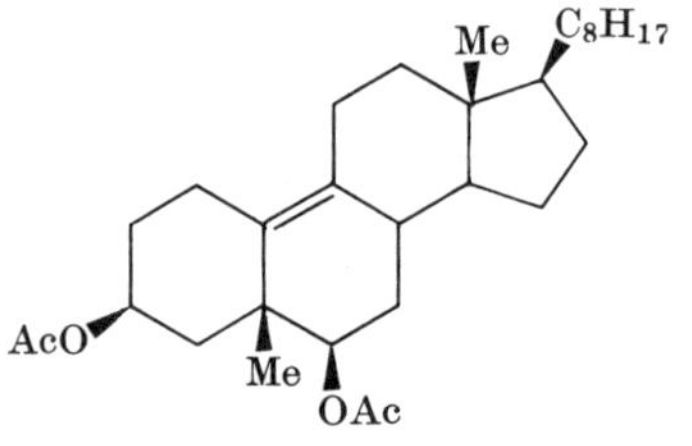

Me C_8H_{17} AcO Me OAc

(**298**)

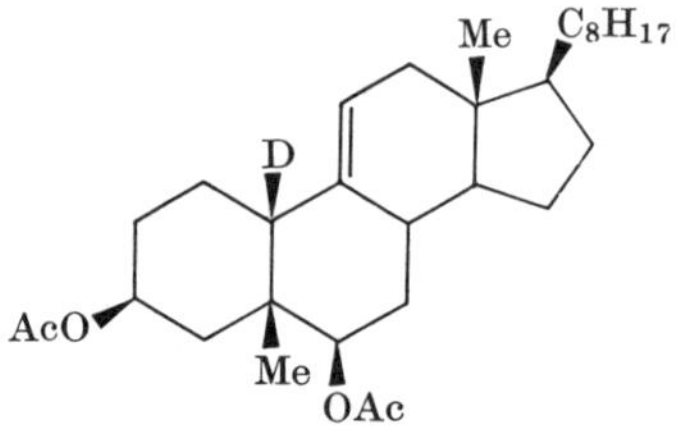

Me C_8H_{17} D AcO Me OAc

(**299**)

Imine Studies

There continues to be sustained interest in the geometric isomerization of imine derivatives, partly from the remarkable sensitivity of their *syn–anti*-isomerization rates to structural changes and partly because of the controversy concerning the mechanisms of such processes. This controversy hinges on whether the isomerization process involves a rotational pathway, an inversion route or a continuum between both,[377–379] and is reminiscent of the ongoing S_N1–S_N2 duality-or-merged debate.

A careful study in that regard has been reported by Herkstroeter.[380] He examined substituent and steric effects upon the rates of thermal *syn–anti*-isomerization of some pyrazolone (**300**) and benzoylacetanilide (**301**) azomethine dyes (on photochemical excitation. The rates were measured in a flash-photoelectric apparatus using degassed

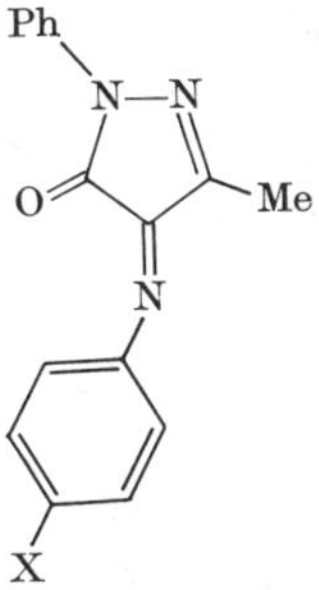

(**300**) 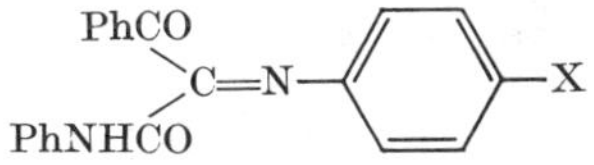 (**301**)

benzene solutions at room temperature, and k varied from ca. 10^{-1} to 10^3 sec^{-1}. For all the compounds studied, methyl groups attached adjacent to the azomethine bond increased the rate of isomerization. Changing the nature of the substituent X, from electron-donating to electron-withdrawing or *vice versa both* caused rate enhancement, i.e., the derived Hammett plots were V-shaped. With compounds (**300**), for example, ρ (electron-donating X groups) was –2.70 and for electron-attracting X groups was +1.88. Thus it appears that a duality of mechanisms operates: with electron-releasing groups in PhX, isomerization proceeds by rotation about the central azomethine bond, and with electron-attracting X groups, an *inversion* mechanism is the preferred path.

A transition state with considerable rotational character is also favoured for the *syn–anti*-isomerizations of aldehyde 4-pyridylhydrazones (ArCH=N–NHPy).[381] The rate constants at 25° in 70% acetic acid are of the order of 10^{-5} sec^{-1} and were determined by a rapid bromination technique. The ρ values for *C*-Ar substituent variation was –0.43, consistent with other work on semicarbazones (ρ = –0.65), guanylhydrazones (–0.36), 5-(1-benzyltetrazolyl)hydrazones (ρ = –0.69) and *p*-nitrophenylhydrazones (ρ = –0.29). With *C*-Ar = *p*-BrC_6H_4 $\Delta H^\ddagger = 18.7$ kcal mol^{-1}, $\Delta S^\ddagger = -17$ eu, and for *C*-Ar = *p*-ClC_6H_4 $\Delta H^\ddagger = 18.1$ kcal mol^{-1} and $\Delta S^\ddagger = -19$ eu, these hydrazone reorganizations being characterized by large negative entropies of activation.

In examining the *syn–anti*-isomerization of aromatic sulphines (**302**), the conclusion was also reached (not unequivocally) that the mechanism of isomerization involved rotation around the C=S bond. Because of ambiguities in the literature, the unambiguous

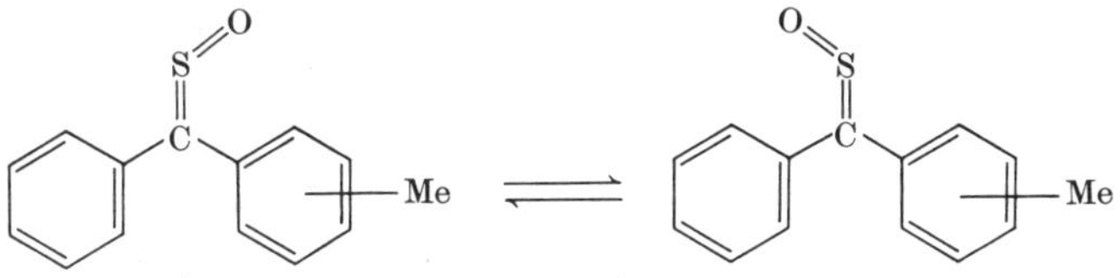

syn *anti*

(**302**)

configurational assignment of starting material was a prerequisite. This involved a variety of techniques including measurement of aromatic solvent-induced shift (ASIS) in the ^{1}H- and ^{13}C-NMR. The isomerizations were studied at temperatures between 110° and 180°, mostly in benzene, and the rate-constants at 110° were of the order of 10^{-5} sec.$^{-1}$ The entropies of activation were again large and negative, $\Delta S^\ddagger$ values from −21 to −29 eu being observed, whilst $\Delta H^\ddagger$ values ranged from 20 to 24 kcal mol^{-1}.[382]

syn- and *anti*-Isomers of *N*-arylmaleisoimides (**303**) and succinisoimides have been observed by using ^{1}H- and ^{13}C-NMR data. Solvent-variation and substituent effects have little effect on the *syn–anti*-equilibrium position. The rates of isomerization of two of the *N*-arylsuccinimides in CD_3CN at 60° were *N*-phenyl 45 sec^{-1} and *N*-*p*-methoxyphenyl 16 sec^{-1}. On the basis of somewhat slender evidence, it was concluded that the inversion process best represented the isomerization mechanism.[383]

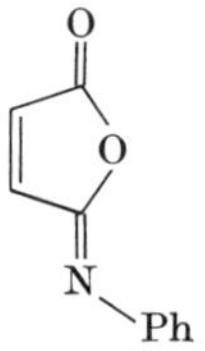

(**303**) *anti* (major)

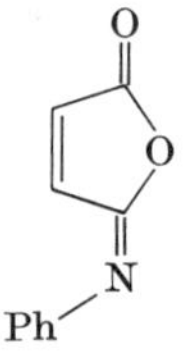

(**303**) *syn* (minor)

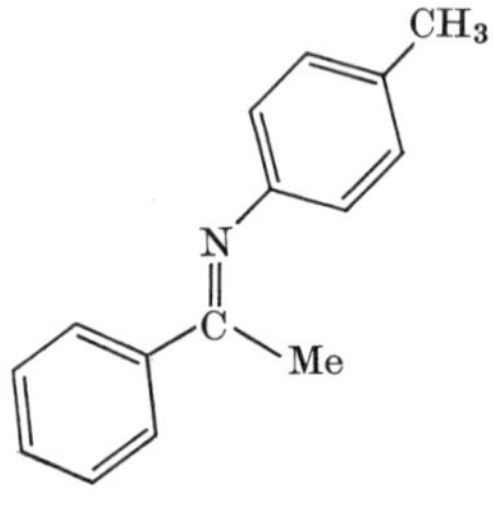

(**304**) Mc = ferrocenyl, ruthenocenyl or tricarbonyl(cyclobutadienyl)iron

Another interesting study involved the examination of the *syn–anti*-isomerization of *N*-(*p*-tolyl)imines of ferrocenyl, ruthenocenyl and [tricarbonyl(cyclobutadienyl)iron]-phenyl ketones (**304**).[384] The organometallic substituents were selected because of their demonstrated ability to stabilize an adjacent electron-deficient centre. The rates of isomerization of compounds (**304**) were studied by using the dynamic NMR technique in diphenyl ether or benzotrifluoride (from 50° to 90°). While reliable activation entropies could not be obtained (they ranged from −9 to +25 eu), more satisfactory $\Delta G^\ddagger$ values of the order of 18–19 kcal mol^{-1} were derived. Because these free energies of activation are very similar to those obtained with triarylimines of type $Ar_2C{=}NAr'$, wherein a lateral shift mechanism operates for the *syn–anti*-interconversions, the authors[384] conclude a similar mechanism applies to the geometric interconversions of compounds (**304**).

The first-order rate constants for thermal approach to equilibrium with the (*Z*)- and (*E*)-isomers of *N*-methyl-α-phenyl-α-*p*-tolylnitrone and of *N*-benzyl-α-phenyl-α-*p*-tolylnitrones have been measured in degassed *tert*-butyl alcohol solution between 130° and 160°. With compound (**305**) for example, $\Delta E^\ddagger$ was 33.6 kcal mol^{-1}, an energy barrier much higher than expected ($\Delta S^\ddagger$ was -4 ± 4 eu). The isomerizations are solvent-sensitive: changing from diethylcarbitol to *tert*-butyl alcohol at 135° reduces the rate for (**305**) by a factor of 35.[385]

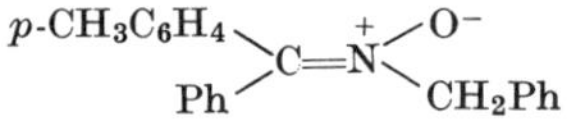

$$(p\text{-}CH_3C_6H_4)(Ph)C{=}\overset{+}{N}(O^-)CH_2Ph$$

(**305**)

In a cognate study,[386] the interconversion of sodium 4-methylbenzophenone (*Z*)-oximate into its (*E*)-isomer was studied; in an excess of methanolic $NaOCH_3$ in methanol, k_1 at 144° was 3×10^{-5} sec^{-1}. For this process, $E^\ddagger$ was 26 kcal mol^{-1} and $\Delta S^\ddagger$ – 20 eu. Thus, amongst diaryl ketimine derivatives, only *O*-alkyl oximes have appreciably greater configurational stability[387] than the oximate anions. By both extended Hückel calculations and CNDO/2 methods, the isomerization barriers (kcal mol^{-1}) were calculated for the acetone oximate anion system: CNDO/2 results were inversion 28 and rotation 106! The extended Hückel computations gave inversion 18 and rotation 25. These results certainly suggest that an inversion pathway is the more likely for this system.

Thermolysis of the (*Z*)- and (*E*)-isomers of *N*-benzyhydryl-α-phenyl-α-*p*-tolylnitrone as substrates leads to a variety of competing processes: (*a*) geometrical interconversion of the nitrones; (*b*) their decomposition leading to configurationally isomeric iminoxy-radicals; (*c*) their interconversion; and finally (*d*) radical recombination at both N and C centres. On disentangling these pathways, it was found that the iminoxy-radical underwent relatively very ready interconversions that were 10^5–10^6 times faster than those of the corresponding oximate anions.[388]

Proton NMR, IR, and UV spectroscopy establish that isatin β-thiosemicarbazones exist largely in the *syn*-form, stabilized by an intramolecular hydrogen bond.[389]

The NMR spectra of the (*E*)- and (*Z*)-isomers of acetaldehyde oxime and of dimethylamino- and (trimethylammonio)-acetaldehyde oxime have been reported. Increase in the steric load of the oxime, passage from an organic solvent to aqueous solutions and ionization of the oxime group each leads to a rise in the equilibrium *E*/*Z* ratio.[390]

Recently it was proposed[391] that an important factor in determining the (*E*)- to-(*Z*) isomer ratios in ketimines of type ArCMe=NMe (and also in diaryloximes and alkylarylhydrazones) is an *n*–π repulsive interaction between the lone pair of electrons on the sp^2-nitrogen atoms and the π-electrons of the aryl ring. This proposal is now discounted, the *n*–π effect being of minimal importance, at best, and the *E*-to-*Z* ratio data are reinterpreted in terms of the steric effects involved.[392]

Thermal isomerization (toluene, heat 140 h.) of *C*-(*o*-nitrostyryl)-*N*-phenylnitrone to *o*-nitrocinnamanilide has been reported.[393]

A stopped-flow technique has been used to determine the isomerization kinetics of *syn*-*p*-nitrobenzenediazotate. Between pH 10.06 and 12.60 the acidity of the medium played no effect, *k*(isomerization) at 20° being ca. 6×10^{-2} sec^{-1}. For the rearrangement $\Delta H^\ddagger$ was 14.2 kcal mol^{-1} and $\Delta S^\ddagger$ – 19.2 eu. It was concluded that the transition state in this isomerization is similar in structure to the final (*anti*)-product.[394]

Reaction of the (*E*)-isomers of *O*-methylbenzohydroximoyl chloride (**306**) with methoxide ion in 9:1 DMSO–CH_3OH leads to essentially complete inversion, the product being (**307**). The case is similar with the (*Z*)-isomer of (**306**) except that the basic catalyst

Cl–C(Ar)=N–OMe (Ar = 4-Y-C$_6$H$_4$)

(**306**)

MeO–C(Ar)=N–OMe (Ar = 4-Y-C$_6$H$_4$)

(**307**)

induces some *Z*-to-*E* conversion prior to displacement in that case. The rate of halide loss is not changed much when chloride is replaced by bromide. This is the first stereoselective inversion of configuration during nucleophilic substitution at sp^2-hybridized carbon to be reported. It proceeds by a *trans*-addition of methanol to (**306**) followed by *trans*-elimination of hydrogen chloride.[395]

Acyl Derivatives

Two groups[396,397] report the degenerate thermal acyl shifts (which the Russian group call acylotropic tautomerism) observed with 2-acyloxypent-2-en-4-ones (**308**). The kinetics of the rearrangement were studied by NMR spectroscopy, k being of the order of 10^3 sec^{-1} at 100°). When R was a *m*- or *p*-substituted phenyl group, the rates at 298°K gave $\rho = +0.81$ (using σ^+ constants). The rearrangements are intramolecular with $\Delta S^{\ddagger}$ values ranging from −17 to −36 eu and $\Delta H^{\ddagger}$ varying from 6 to 11 kcal mol^{-1}, and they proceed in accordance with the 1,5-sigmatropic shift outlined in Scheme 29.[396] Only the acylated *cis*-enol forms participate in the acyl shift. The second group's data[397] confirm the above.

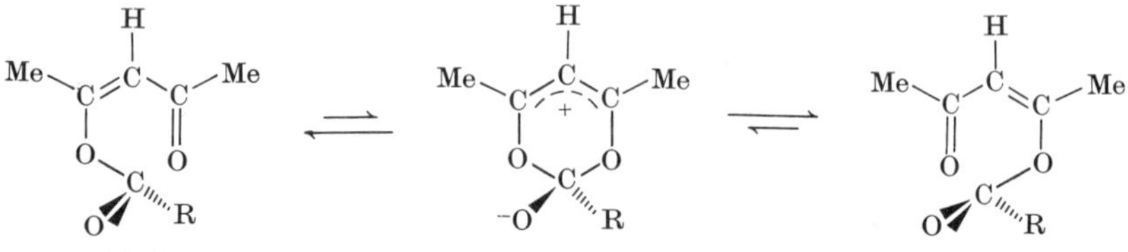

(**308**)

SCHEME 29.

Two alternative mechanisms have been observed in the rearrangements of some substituted acrylamides. The first of these is outlined in Scheme 30 where dithioacetic acid is added to dimethylaminopropynal and the disposition of the sulphur atoms in the product indicates the route adopted by the reaction, namely, formation of very labile primary adducts, followed by a [1,5]-sigmatropic shift and ring-opening.[398]

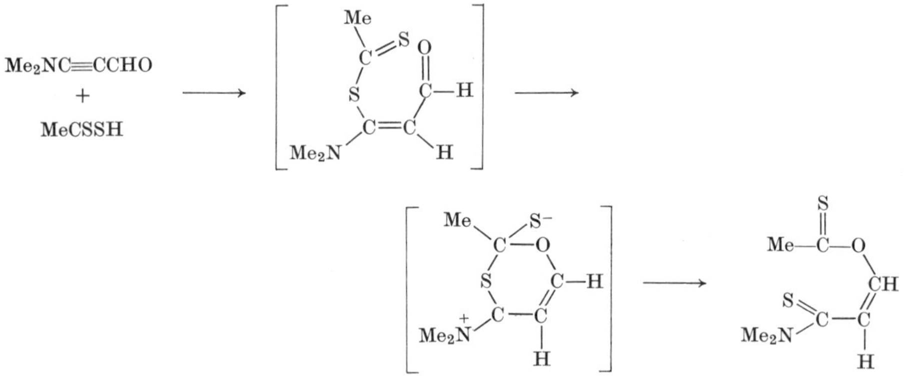

$Me_2NC{\equiv}CCHO$ + MeCSSH

SCHEME 30.

This mechanism is not operative when the protic materials added to the alkyne substrates are hydrohalogen acids, alcohols or amines. Thus, the acid-catalysed reaction of 3-chloro-3-(dimethylamino)propenal (labelled at the 1-position with ^{13}C) (**309**) yields 3-chloro-*N*,*N*-dimethylacrylamide containing the label exclusively at position 3. This reaction involves an ammonio-oxetene (**310**) as intermediate.[398,399] (Scheme 31).

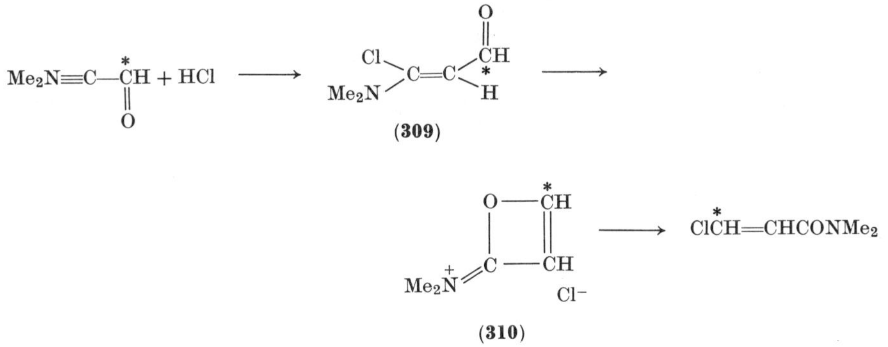

(**309**)

(**310**)

$ClCH=CHCONMe_2$

SCHEME 31.

Alk-2-ynyl alkanedithioates (**311**) rearrange at 130–140° in the presence of catalytic amounts of a tertiary amine to yield 1,3-dithiole derivatives (**312**), possibly via allenyl dithioates though these have not been isolated.[400] *S*-Alkyl xanthates of alk-2-enols

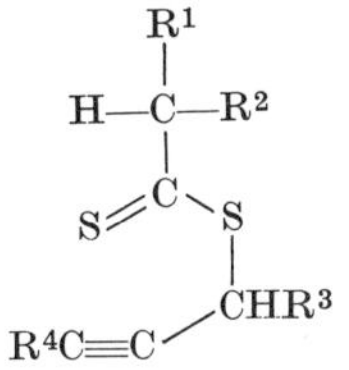

(**311**) 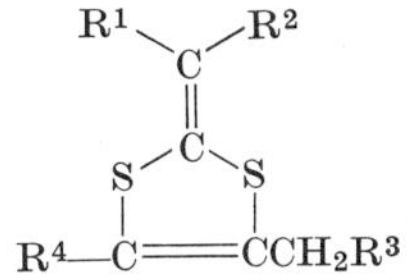 (**312**)

rearrange thermally to yield dithiocarbamates with an accompanying allylic shift. When a mixture of two xanthates having different *S*-alkyl and *O*-alkenyl groups was allowed to rearrange, no cross-over products were obtained. 3-Aryl-substituted allylic xanthates rearranged to the corresponding dithiocarbamates without an allylic shift.[401] Acid-catalysed rearrangement of a mixture of two thionocarbamates, namely, $Me_2N{-}C({=}S)OEt$ and $C_5H_{11}NC({=}S)OC_5H_{11}N$ gave four products, including those of cross-over, suggestive of an intermolecular rearrangement path.[402]

The thermal conversion of the oxime thionocarbamates (**313**) to thiolocarbamates (**314**) at 85° was examined for ^{13}C CIDNP effects. Clear-cut evidence for the formation of the radicals $Ar_2C{=}N\cdot$ (g ca. 2.003) and $Me_2NC({=}O)S\cdot$ (g 2.05) was obtained.[403]

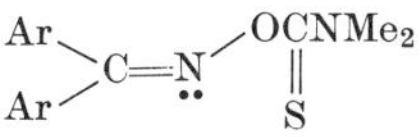

(**313**) 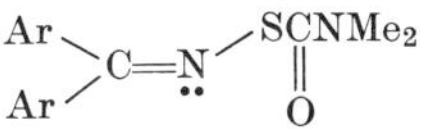 (**314**)

A neat way of modifying functionality in carbohydrate units exploits the principle outlined in Scheme 32.[404]

As part of an interesting approach to the asymmetric synthesis of chiral sulphoxides,

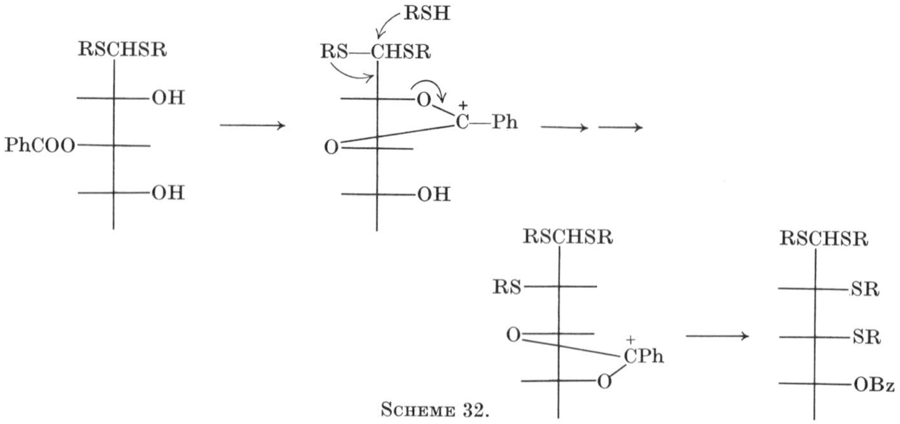

RSCHSR, PhCOO, OH, RSH, CPh, SR, OBz

SCHEME 32.

the stereochemistry of sulphinyl transfer in an *O*-sulphinylated ethanolamine (**315**) has been investigated. The rearrangement proceeds along two competitive paths, intra- and inter-molecular, the former path yielding a sulphinamide (**316**) with retention of configuration at sulphur (Scheme 33).[405]

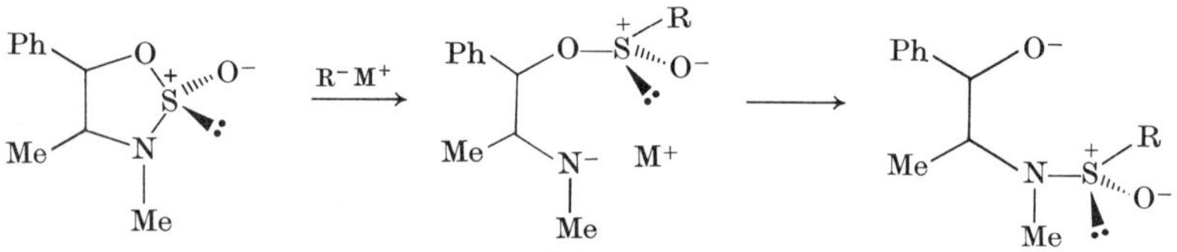

Ph, Me, R^-M^+, M^+, (**315**), (**316**)

SCHEME 33.

17β-Bromo- and 17β-chloro-16α,17α-oxoandrostan-3β-yl-acetate rearrange when heated above their m.p.'s to a mixture of 16α-halo- and 16β-halo-ketones, the overall effect being as outlined in Scheme 34. Ion-pairs are suggested as intermediates.[406]

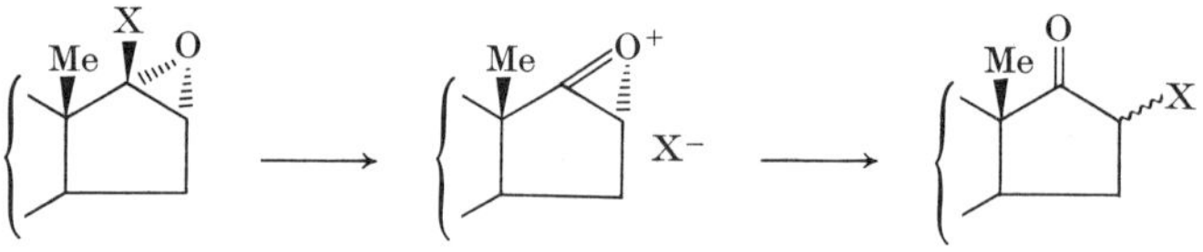

Me, X, O^+, X^-

SCHEME 34.

An N-to-N phenacyl migration (which is formally related to the Stevens rearrangement) is a key process in the ready transformation of 1,1-dimethyl-1-phenacylhydrazinium bromide (**317**) to 2-benzoyl-4-phenylimidazole (**318**) (Scheme 35).[407]

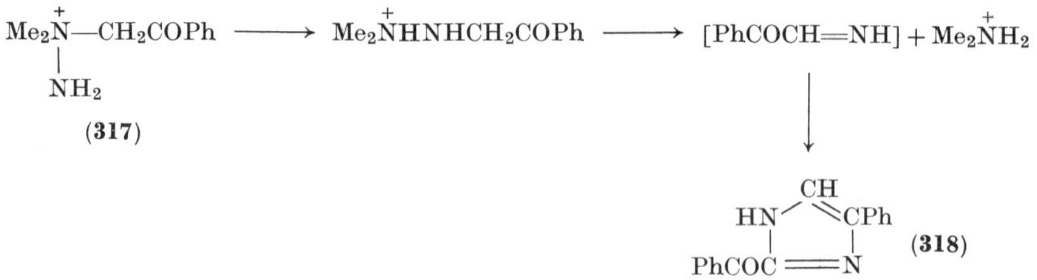

$Me_2\overset{+}{N}(NH_2)$—CH_2COPh (**317**) ⟶ $Me_2\overset{+}{N}HNHCH_2COPh$ ⟶ [PhCOCH=NH] + $Me_2\overset{+}{N}H_2$ ⟶ (**318**)

SCHEME 35.

Miscellaneous

The dynamic equilibrium between the *cis*- and *trans*-forms of the diene (**319**) and the cyclic compound (**320**) has been reported.[408]

$Me_2C{=}CHCH{=}C(COMe)COOEt$

(**319**)

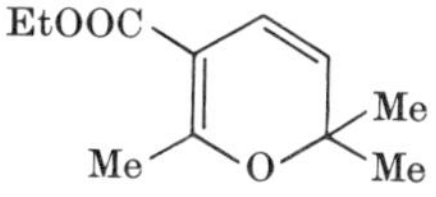

(**320**)

A Hammett treatment has been applied to the cyclization/ring-openings involved in the conversion of chalcones into flavanones (Scheme 36). A new set of constants σ_c, specific to the chalcone system, is proposed.[409] A study of the electron impact-induced

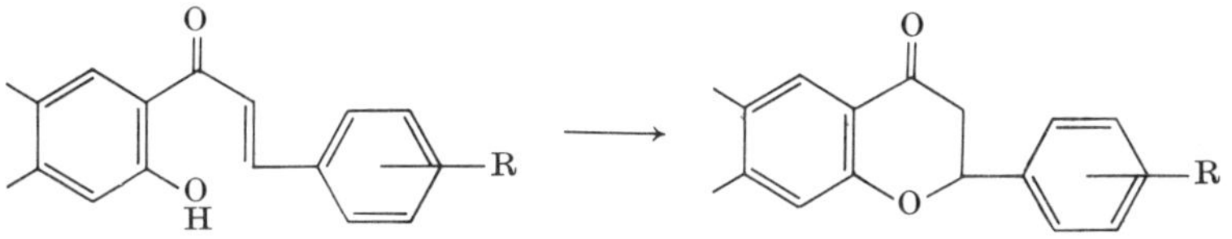

SCHEME 36.

isomerization of 2′-hydroxyacrylophenones and chroman-4-ones reveals that aryl-substituted isomers behave similarly under electron-impact and in solution whereas alkyl-substituted homologues differ in that regard.[410]

Reaction of 2-methyl-1,3-diphenylprop-2-en-1-one phenylhydrazone (**321**; $R^2 = Ph$, $R^1 = Me$) in acetic acid solution for 7 days at 60° gave a *trans*-pyrazoline (**323**), whereas

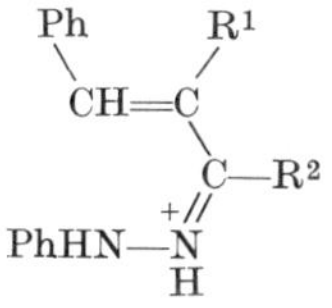

(**321**) 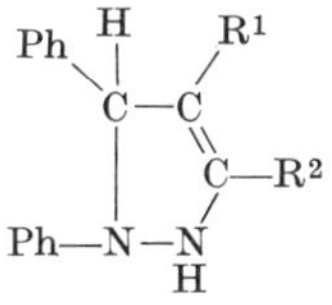 (**322**) 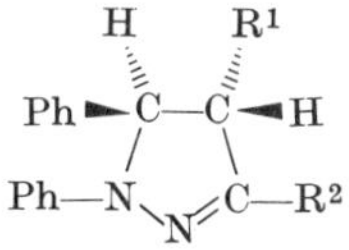 (**323**)

refluxing for 6 hours yielded a *cis–trans*-mixture (1:4), the reaction being under thermodynamic control. Under kinetic control, the product stereochemistry is determined by the step (**322**) to (**323**), an enamine–imine tautomeric shift.[411] The stereochemistry of the 2-pyrazolines obtained by reaction of hydrazine with 3-ethylpent-3-en-2-one has been clarified (after the incorrect literature assignment of structures to the starting ketone had been corrected).[412]

gem-α-Disubstituted succinaldehydic acids rapidly equilibrate with their hydroxy-lactone tautomers which are generally predominant. This behaviour may be exploited synthetically by appropriate selection of reagent.[413]

Refluxing *trans*-2-(2-aminobenzylidene)indan-1-ones in glacial acetic acid involves *trans–cis*-isomerization, followed by thermal cyclodehydration to yield 11*H*-indeno-[1,2-*b*]quinolines.[414]

Pyrolysis of the diphosphonic acid (**324**) yields both cyclodehydration (**325**) and cyclodephosphonation (**326**) products.[415]

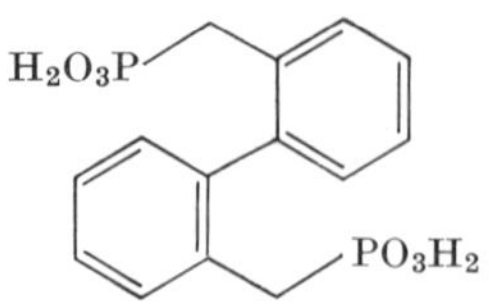

(**324**) 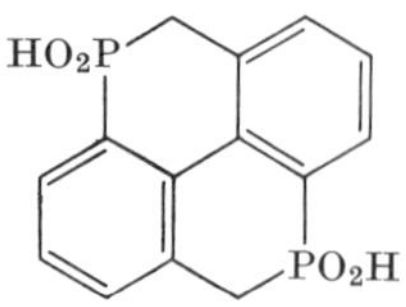 (**325**) 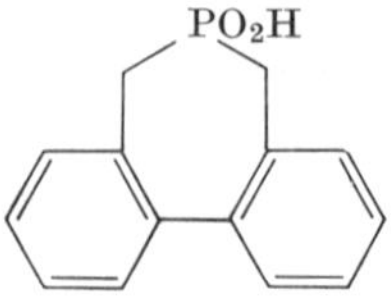 (**326**)

Heating ligularone (**327**) to 260–263° for 90 hr. gave a 50:50 mixture of (**327**) and isoligularone (**328**) by way of a diradical intermediate. The equilibrium mixture can be approached from either side.[416]

Mass-spectral data of alkylsulphinylamines (RNSO), which involve a novel McLafferty rearrangement, support the idea that the *trans*- are substantially ($\geqslant$10 kcal mol^{-1}) more stable than the *cis*-isomers.[417]

At elevated temperatures (80–90°), the α- and γ-nitrogen atoms of carbamoyl azides are exchanged by inversion of the azido-group (detected by ^{15}N-scrambling). This rearrangement involves reversible thermal dissociation to isocyanates and hydrazoic acid.[418]

The rearrangements of (**329**; X, Y = O, NMe) during preparation and thermolysis have been elucidated in terms of kinetically and thermodynamically controlled spirophosphorane ring cleavage and bis(phospholane) redistribution processes.[419]

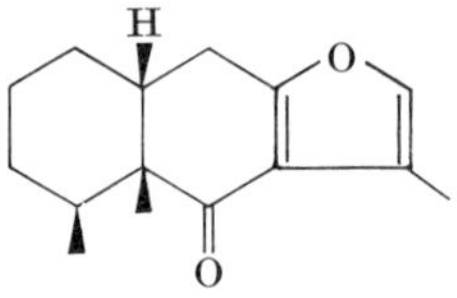

(**327**) 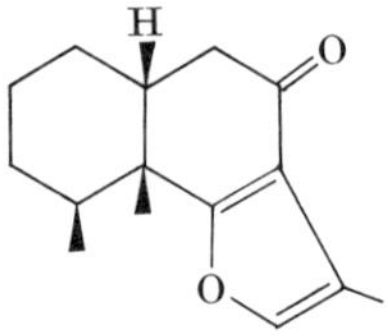 (**328**) 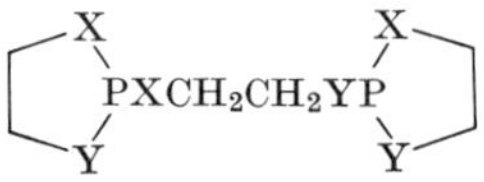 (**329**)

The thermal rearrangement of 4-carbonyl-substituted oxazoles gave a negative ρ for substituent variation in an aromatic ring in the 2-position (using σ^+ values in the Hammett plot) and appears to involve the sequence outlined in Scheme 37.[420]

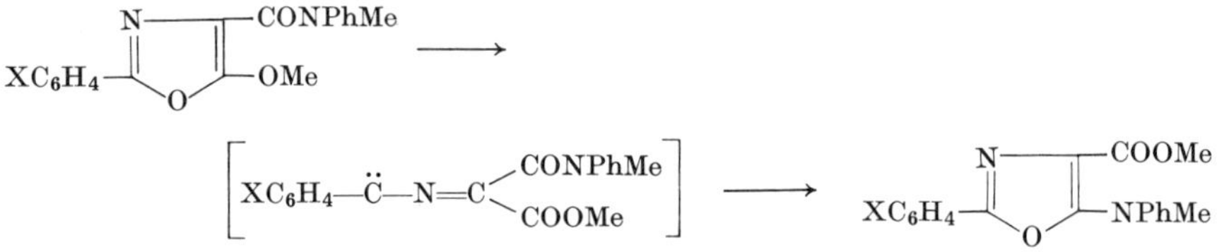

SCHEME 37.

Rearrangements Involving Ring-openings and Ring-closures

Three-membered Rings

The factors governing the closure of three-membered rings[421] and the energetics of neighbouring-group participation[422] have been reviewed.

The rates of cyclization, in dioxan–water and ethanol–water, of a series of *N*-ω-(haloalkyl)amines and of the corresponding arylsulphonamides in alcohol–alkoxide media have been studied. For the amines (25°), the relative ease of ring formation was 5- > 6- > 3- > 4-membered; for the sulphonamides the order was 5- > 3- > 6- > 4-membered. Bromide–chloride ratios (for three-membered ring formation) were 28:1 for amine cyclizations and 79:1 for the sulphonamides.[423]

In the cyclization of ω-haloalkyl sulphides the ease of ring-formation was 3- > 5-membered. The bromide–chloride ratios were 35:1 and 56:1 for three- and five-membered ring formation, respectively.[424]

An examination of the charge distribution in the transition state of the enolene rearrangement has been made by approaching the reversible rearrangement from both sides. Thus the pent-4-enophenones (**330**) rearrange at 217°, their rate data corresponding to a ρ value of –0.37 (with σ^+ constants). The dialkylacylcyclopropanes (**331**) ring-open at 125° to yield compounds (**330**), with a ρ value of +0.57.[425]

(**330**) (**331**)

Such thermal "ene reactions" of three-membered rings take place with a sigmatropic 1,5-hydrogen shift, concomitant ring cleavage and necessarily a *cis*-relation between the π-system and the alkyl group on the vicinal carbon atom. The thermal rearrangement (Scheme 38) of a set of *N*-*tert*-butylated aziridinyl ketones (**332**) differs dramatically from this concept in that the *trans*-isomer reacts faster than the *cis*-form.

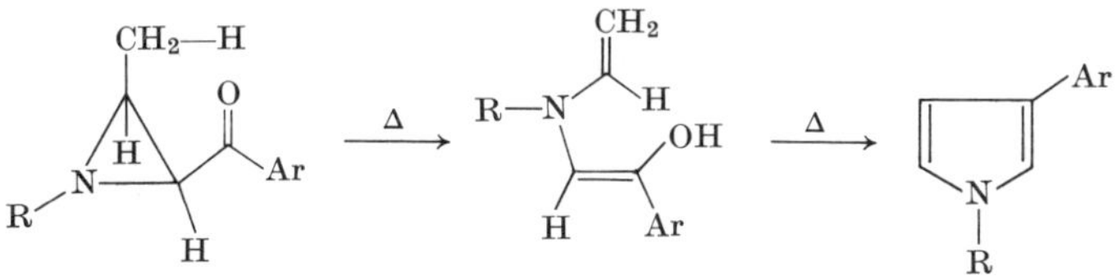

(**332**) (**333**)

SCHEME 38.

Thermolysis of *cis*-(**332**) gave (**333**) in methanol at 100°, with $k = 1.18 \times 10^{-4}$ sec^{-1}, $E_a = 28.6$ kcal mol^{-1} and $\Delta S^‡ = -2.3$ eu. The reaction kinetics with *trans*-(**332**) were complicated by the fact that it formed *cis*-(**332**) as well as the pyrrole (**333**) in methanol. However, the *trans*-(**332**)-to-(**333**) process proved to be 39 times faster than for the *cis*-isomer ($E_a = 26.4$ kcal mol^{-1}). Such aziridines readily undergo thermal cleavage to azomethine ylides by conrotation of the substituent groups. Conrotatory rotation in either direction for the *cis*-isomer causes steric interaction between one of the rotating groups and the adjacent *tert*-butyl group. Such an effect is not encountered with the *trans*-isomer. The reactive intermediate involved in the *trans–cis*-epimerization is a 3-enol form.[426]

Racemization of (+)-2,3-divinyloxirane occurs readily at 150° (log $k_1 = 11.3$ sec^{-1}) and involves a reversible conrotatory ring opening to form a carbonyl ylide ($R\bar{C}H-\overset{+}{O}=CHR$) ($\Delta S^‡ = -9$ eu). Such an intermediate can reclose to give an oxirane, a dihydrovinylfuran or a dihydro-oxepine. The last two processes are substantially slower than the racemization.[427]

Oxidation of *N*-aminophthalimide by lead tetra-acetate in the presence of various alkynes appears to yield 1*H*-azirines which then ring-open by an ion-pair mechanism and re-cyclize to yield the isolable 2*H*-azirines.[428]

Thermolysis of 2.3-diphenyl-2*H*-azirine at 250°, for 3 hr. yields a variety of substituted pyrroles, imidazoles, pyrazines and indoles. Their formation has been interpreted in terms of nitrene and carbene intermediates.[429]

Thermal isomerization (40–75°) of 3-methylpyrazole occurs readily [E_a = ca. 25.5 kcal mol^{-1}, k_1 (40°) = 6.2×10^{-5} sec^{-1}]. The data are consistent with isomerization as a two-step process but are not sufficient to indicate whether the key intermediate is diradical or ionic in nature.[430]

Pyrolysis of some spiro-α,β-epoxyketones at 260° resulted in formation of β-diketones derived by [1,2]-shift of a β-hydrogen or methyl group; in migratory aptitude, H > Me.[431]

The acid-catalysed ring opening of some spirocyclopropyl ketones having a phenyl substituent at the 2-position of the cyclopropane ring occurred with cleavage of the C-1 to C-2 bond whether or not this was the small-ring bond more overlapped with a *p*-orbital of the carbonyl group. The more important factor directing ring opening was the ability of the substituent concerned to delocalize a positive charge.[432]

Solvolysis of *cis*- and *trans*-3,4-epoxycyclopentyl toluene-*p*-sulphonate in buffered acetic acid does not involve neighbouring-group participation by the epoxy unit.[433]

The gas-phase thermal isomerization of pentachlorocyclopropane to 1,1,3,3,3-pentachloropropene has been studied (440–490K, 10–50 mm). The reaction is first-order (E_a ca. 40 kcal mol^{-1}) and takes place with migration of a chlorine atom (through a chlorine bridge) via a concerted transition-state.[434]

INDO and *ab initio* SCF calculations have been made on cyclopropanone and its possible isomerization products, oxyallyl or allene oxide. The *ab initio* results indicate that singlet oxyallyl is 83 kcal mol^{-1} less stable than cyclopropanone while allene oxide is 21 kcal mol^{-1} less stable. Although the conrotatory opening of cyclopropanone passes through a true transition state, the disrotatory opening is the pathway of lower energy and oxyallyl appears to be at, or near, the disrotatory energy maximum.[435]

Reaction of 1-methyl-2,2-diphenylcyclopropylamine with base leads very readily to ring opening to form 4,4-diphenylbutan-2-one as the sole product.

Contrary to earlier literature,[437] the isomerization of 2,3-dihydro-5-methylfuran to acetylcyclopropane has been found to be a reversible process with the equilibrium between 673 and 731K favour the ketone to an extent which increases with temperature.[438] The thermodynamic data indicate that any interaction between the carbonyl group and the cyclopropane ring has little overall effect on the stability of acetylcyclopropane.

There has been a study of the kinetics of the reversible isomerization of the dihydrofuran to acetylcyclopropane and of the irreversible conversion of the latter into *cis*- and *trans*-pent-3-en-2-one as well as into pent-4-en-2-one. The overall reactions can be rationalized by a diradical mechanism in which a stabilization energy of ca. 7 kcal mol^{-1} in the alkylacetonyl radical is operative.[439]

In an independent stereochemical and kinetic study, the thermal and photochemical conversion of a series of cyclopropyl ketones to isomeric furans and pentenes has been reported. The dihydrofurans were formed with predominant retention at the migrating carbon. A kinetic study of the rearrangement of 1-acetyl-1-methyl-2-phenylcyclopropane to 4,5-dihydro-2,3-dimethyl-5-phenylfuran at 255–288° provides an E_a of 48.1 kcal mol^{-1}. Again it was concluded that the thermal and the photochemical reaction occur by a non-concerted process through a common 1,3-diradical intermediate in which bond rotation and ring closure steps are competitive.[440]

Solution pyrolysis of either *cis*- or *trans*-1,2-diethynylthi-irane (**334**; *cis*) ([D_8]toluene,

100°) gave only desulphurized olefins, 90% retention of stereochemical configuration being observed. In a gas-phase nitrogen flow system, pyrolysis of (**334**; *trans*) still only leads to desulphurization but (**334**; *cis*) (250°) affords 5% of cyclobuta[*c*]thiophen (**335**). On storage the very reactive compound (**335**) dimerizes to yield (**336**).[441]

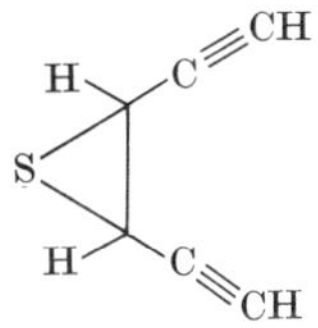

(**334**) 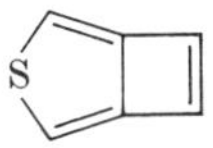 (**335**) 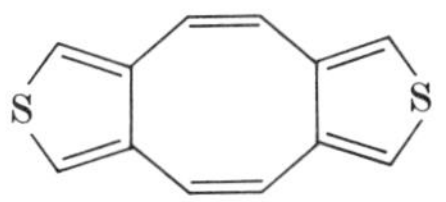 (**336**)

Four-membered Rings

Thermolyses of cyclobutenes containing electron-withdrawing substituents at C-1 and C-2 (CN, COOMe, $COOCH_2CH{=}CH_2$ or COCl) between 380° and 420° at 10^{-1} mm afford the corresponding 2,3-substituted buta-1,3-dienes.[442]

Five different acid-catalysed rearrangement patterns were observed with some aryl-substituted azetidin-2-ones. In concentrated sulphuric acid 1-arylazetidinones underwent ring cleavage with carbonium ion formation followed by ring-closure to yield substituted quinolin-2(1*H*)-ones. With BF_3 in toluene, the intermediary carbonium ions developed from the azetidinones attacked the solvent, yielding substituted propionamides. With BF_3 in chlorobenzene, the heterocycles yielded isomeric cinnamanilides. Finally, in the same conditions, more highly substituted azetidinones underwent more extensive fission, Schiff bases and substituted indene derivatives being isolated.[443]

When 2-alkylidenecyclobutanols were heated at 245° for 4 hr. or treated with 5% H_2SO_4 for 30 min at 100°, ring contraction to the corresponding cyclopropylcarbonyl compounds occurred. When 1,5-hydrogen transfer was possible, further reaction of these materials occurred leading to the isolation of homoallylic carbonyl derivatives.[444]

Cycloaddition of alkylhaloketenes with methylenecycloalkanes occurs in good yields to form spirocyclobutanones. These, upon reduction to alcohols followed by treatment, with base, undergo ring-contractions to spirocyclopropyl aldehydes and this constitutes an efficient synthetic route to such materials.[445]

Hydrolyses of *cis*- and *trans*-1-*tert*-butyl-2-methyl-3-azetidinyl toluene-*p*-sulphonate proceed with stereospecific retention of configuration in 60% aqueous acetone, the *cis*- reacting faster than the *trans*-form. Formation of an intermediary 1-azabicyclo-[1.1.0]butonium ion is the exclusive pathway by which the *cis*-material solvolyses and plays an important part in the solvolysis of the *trans*-substrate.[446]

Five-membered Rings

Reaction of *N*-chloro-1-methylcyclopentylamine with alkyl-lithiums results in alkyl migration, probably concurrent with chloride loss, to form the ring-expanded substituted piperidine, 2-methyl-1-azacyclohex-1-ene.[447]

N-Acyl derivatives of 2-amino-1-phenylpropan-1-ol readily cyclize under dehydrating conditions to form substituted phenyloxazolines. It has now been reported that, on treatment with phosphorus pentoxide in boiling decalin, these form substituted isoquinolines by ring expansion in good yields. It is suggested that such oxazolines are intermediates in the Pictet–Gams synthesis of isoquinolines.[448]

A new convenient synthesis of quinazolines based on the intramolecular rearrangement of an indole derivative has been reported. Treatment of 3-nitroso-2-phenylindole under Beckmann conditions ($POCl_3$ or TsCl in sulfolane at 200° for 1 hr.) gave 2-phenylquinazolin-4(3*H*)-one in 90% yield. By working under milder conditions the intermediacy of 2-benzaminobenzonitrile has been demonstrated.[449] The same principle and a variety of modifications have been used to induce ring-expansion in a 5-nitrosopyrrolopyrimidine to yield a fused pyrimido[4,5-*d*]pyrimidine system.[450]

An imidazole-to-pyrimidine ring expansion was encountered when 6-cyano-9-(2,3,5-tri-*O*-acetyl-β-D-ribofuranosyl)purine was treated with methanolic ammonia, this process involving a novel rearrangement of the purine ring (Scheme 39).[451]

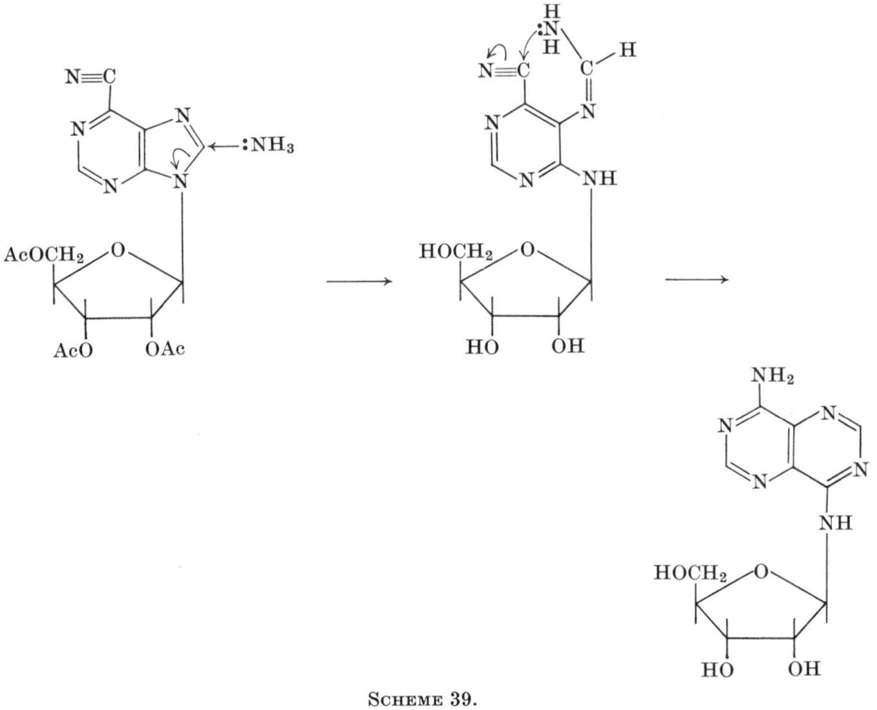

Scheme 39.

A new photoreaction of 2-alkylated indazoles in dilute acidic solution (pH 3–4) has been reported, ring expansion to two isomeric dihydroazepinones occurring. Thus, irradiation of 2,3-dimethylindazole in dilute sulphuric acid (pH 3.8) yields 7- and 3-acetyl-1,3-dihydro-2*H*-azepin-2-one; this probably involves the sequence N—N bond cleavage, imine hydrolysis, and nitrene insertion into the aromatic ring.[452]

Reaction of 2-acetylcyclopentanone with chloral in the presence of potassium carbonate probably formed (trichloroethylidene)cyclopentanone but this underwent a Favorskii type rearrangement (with ring contraction) to yield 2-(2-chlorovinyl)cyclobutene-carboxylic acid.[453]

Five-membered ring contractions have also been observed with some 1,2,3-triazole compounds. Thus vapour-phase pyrolysis (350–600°) of 4,5-disubstituted 1-phthalimido-1,2,3-triazoles gives 2*H*-azirines as the primary isolable products. These azirines undergo further reactions under the conditions of pyrolysis; if a methyl or ethyl 2-substituent is present, they cleave to nitriles, whereas with a phenyl 2-substituent they undergo ring expansion to indoles. 4-Methyl-5- and 6-methyl-4-phenyl-1-phthalimido-1,2,3-triazole give identical mixtures of azirines and their pyrolysis products, indicating that the products are formed through a common intermediate considered to be the antiaromatic 2-methyl-3-phenyl-1-phthalimido-1*H*-azirine.[454]

Six-membered Rings

An extensive series of carbene–carbene rearrangements has been described. For example, phenyl- and diphenyl-carbene undergo carbene–carbene rearrangements in the gas phase above 250° to give cycloheptatrienylidene and its 2-phenyl derivative, respectively, these aromatic carbenes being detected by formation of the dimers heptafulvalene and its 2,2′-diphenyl derivative (**337**), the first recorded stable heptafulvalene. Four-fold and five-fold carbene–carbene rearrangements have been detected in further studies. Whereas most of such rearrangements to date have been limited to the gas-phase at elevated temperatures, 4,5-benzocycloheptatrienylidene readily undergoes carbene–carbene rearrangement even at room temperature; the rearranged carbene shows properties chemically identical with those of β-naphthyldiazomethane.[455]

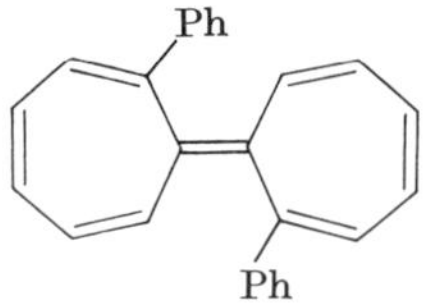

Ph

Ph

(**337**)

Whilst heating cyclopent-3-enylmagnesium bromide converts it into cyclopenta-2,4-dienylmagnesium bromide, the ring remaining intact, compounds (**338**) and (**340**) rearrange on such treatment and are interconverted with establishment of an equilibrium in which the primary Grignard reagent (**340**) is strongly favoured (Scheme 40). By deuterium-labelling it was shown that rearrangement occurs via a bicyclo[3.1.0]hexane derivative (**339**) which, however, could not be detected in the equilibrium. The driving

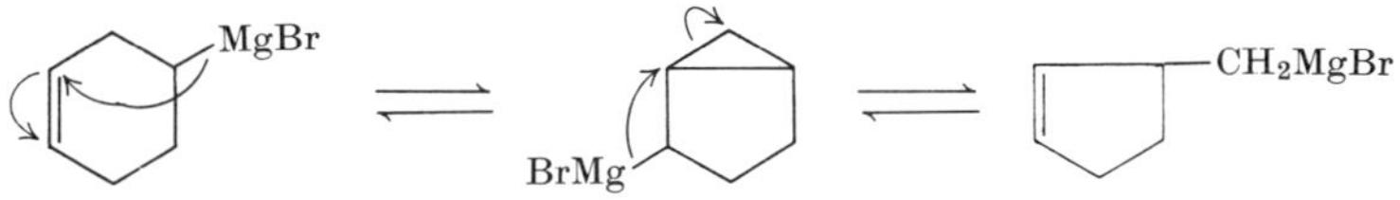

MgBr

BrMg

CH_2MgBr

(**338**) (**339**) (**340**)

SCHEME 40.

force for ring contraction is assigned to the inductively-caused difference in stability between primary and secondary Grignard reagents (ca. 3.7 kcal mol^{-1} in ΔH).[456] The overall effect of the 1-aryl substitution on this process is to decrease the rearrangement rate (perhaps for steric reasons); but, between the three arylated compounds used, the rate order for intramolecular addition within the substituted 4-phenylcyclohex-3-en-1-yl

Grignard reagents is m-$CF_3C_6H_4$ > Ph > p-MeC_6H_4 (approximate $\rho = +1.4$). The primary Grignard (cyclopentenylmethyl) reagent is again the favoured species present at equilibrium. A concerted four-centre addition mechanism is proposed for the rearrangement.[457]

The ring contraction observed on treating the deuteriated ketone (**341**) with base (a so-called homo-Favorskii reaction) yields compound (**342**) without scrambling of the label and does not, therefore, involve the symmetrical ketene (**343**) as an intermediate.[458]

A Favorskii-type rearrangement (**344** to **345**) appears to be involved in the ring contraction effected with 1,2,4-triazin-3-ones when they are treated with ethereal

(**341**) (**342**)

(**343**)

(**344**) (**345**)

chloramine at room temperature, yielding 1,2,3-triazoles. On treatment with hydroxylamine-O-sulphonic acid in base at 70°, the 1,2,4-triazin-3-ones appear to undergo initial N-amination at the 2-position, which is then followed by another mode of ring contraction (with nitrogen loss) affording imidazolin-2-ones in high yield.[459]

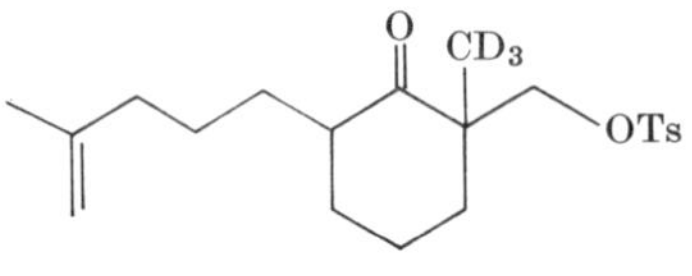

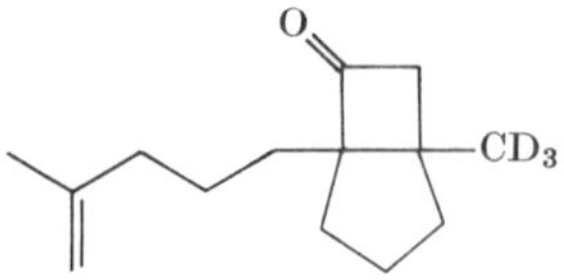

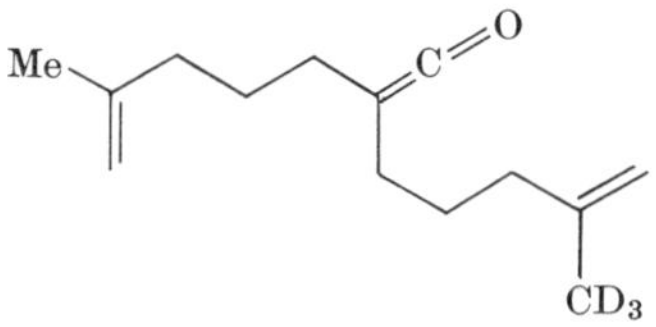

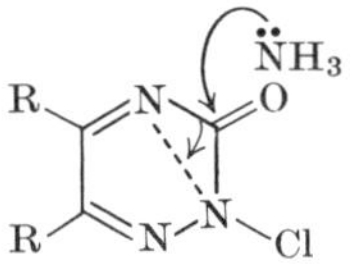

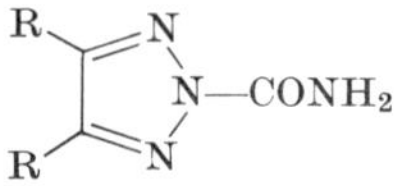

Meso-, racemic and optically active forms of 3,6-bis-[1-hydroxy-1-(4-methylphenyl)-ethyl]-1,2,4,5-tetrazines have been identified. By appropriate techniques these have been converted into the stereoisomers of the ring-contracted 3,5-disubstituted 1,2,4-triazoles and 2,5-disubstituted 1,3,4-oxadiazoles.[460]

The six-membered ring compounds (**346**) on heating at moderate temperatures undergo a thermal [2 + 4]-cycloreversion to yield a nitrile and the N-sulphonylurethane $R^1OOCN{=}SO_2$ which may then participate in subsequent cycloadditions.[461]

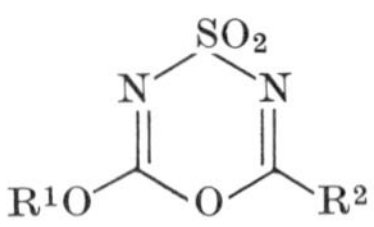

(**346**)

Other Ring Systems

The kinetics of the formation of lactones from ω-bromoalkanoate ions in the range of 7- to 12-membered rings have been investigated in 99% aqueous DMSO solutions at 50°. The reactivity sequence as far as ease of ring closure is in the order $7 \gg 8 \sim 9 < 10 < 11 < 12$. Ring-closure and ring-opening reactions were related to each other by a roughly inverted reactivity order, the eight- and nine-membered-ring lactones being much more reactive (towards hydrolysis) than the ten- and eleven-membered ones. The *cis–trans*-conformational effects appear to be an important contributing factor in determining the observed overall reactivity picture in lactone formation.[462]

Thermolysis of 1-methylbicyclo[3.2.0]hept-2-en-7-one between 489 and 565K involves two parallel first-order homogeneous isomerizations to yield 5-methylbicyclo[2.2.1]hept-5-en-2-one and 2-methylhepta-1,3,6-trien-1-one. No substituted cyclopentadiene was observed, although such compounds were formed when substituted bicyclo[3.2.0]hept-2-en-6-ones were used as substrates. Thus a change in the position of the carbonyl group from the 6- to the 7-position in the bicyclo[3.2.0]hept-2-one series brought about a change in mechanism of ring cleavage, from a polar-concerted (quasi-zwitterionic) pathway to one involving diradicals.[463]

Pyrolysis of *exo*-2,3-epithionorborn-5-ene (in benzonitrile at 160–170°, 19 hr.) gave 2-thiabicyclo[3.2.1]octa-3,6-diene. Heating the epithio-compound without solvent produced a polysulphide.[464]

Treatment of the fused-ring compound (**347**; R = H) with sodium methoxide in methanol resulted in its bishomoketonization with formation of the diketone (**348**). Using [D_1]methanol as solvent introduced two deuterons in an unusual stereospecific *exo–endo*-pathway.[465] Compound (**349**) was shown later to be the initial product of cleavage by methyl-lithium of (**347**; R = $SiMe_3$). In turn (**349**) readily formed (**350**), a reaction which is unusual since it involves not only skeletal rearrangement but also oxygen-to-oxygen migration of the trimethylsilyl group. The presence of the silyl group appeared to promote ring-contraction in this strained system.[466]

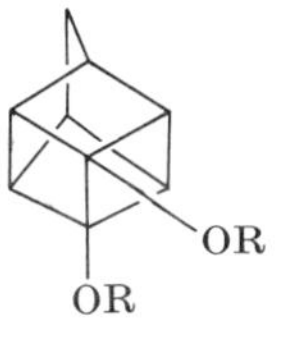

OR OR O O O $OSiMe_3$ O $OSiMe_3$

(**347**) 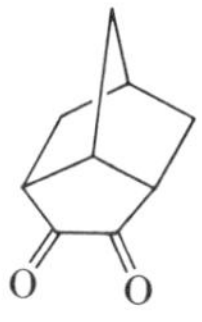 (**348**)  (**349**) 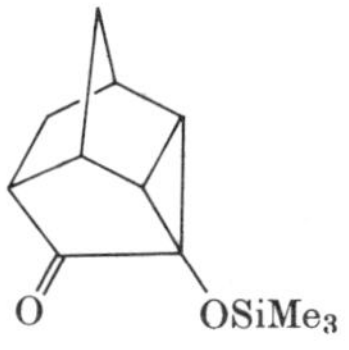 (**350**)

An asymmetric sigmatropic rearrangement has been reported. 2,3-Dibenzoyl-2,3-diazabicyclo[2.2.1]hept-5-ene (**351**) underwent asymmetric isomerization when catalysed by (+)-camphor-10-sulphonic acid in $CHCl_3$, the product (–)-*cis*-4-benzoyl-4,4a,5,7a-tetrahydro-2-phenylcyclopenta-1,3,4-oxadiazine (**352**) being formed in 1.34% enantiomeric excess. With benzene as solvent, the (+)-enantiomer was preferentially formed.[467]

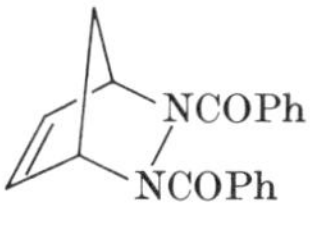

NCOPh NCOPh H O Ph N N H COPh

(**351**) 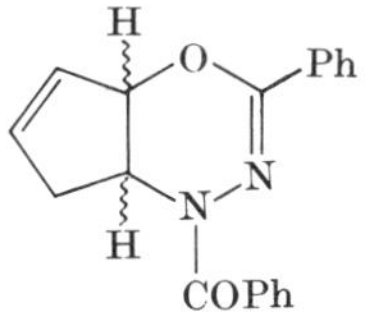 (**352**)

When 13,13-dihalo-1-methylbicyclo[10.1.0]tridecanes (**353**) (R = Cl or Br and (**354**) (R = Cl or Br) are heated in diglyme or quinoline, they form both elimination products (**355** and **356**) and rearrangement products (**357**), the last being formed by [1,3]-sigmatropic rearrangement.[468]

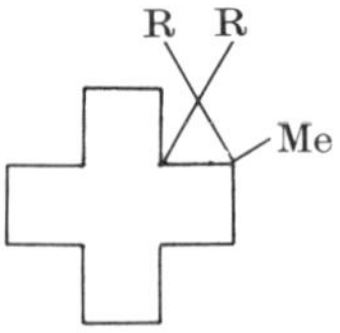

(**353**) 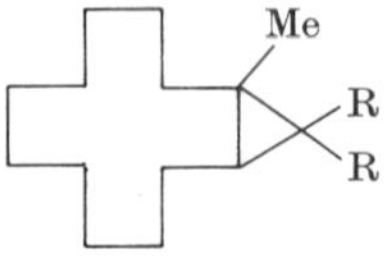 (**354**)

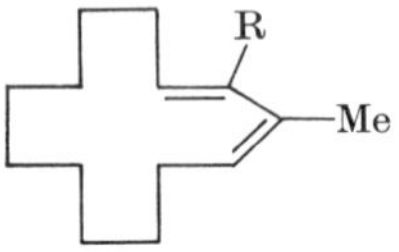

(**355**) 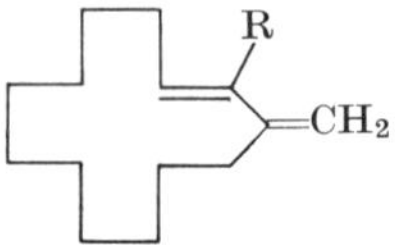 (**356**) 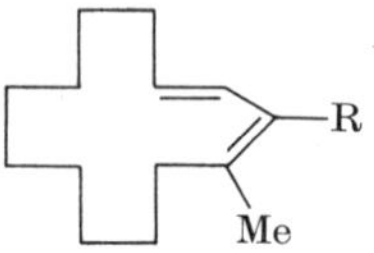 (**357**)

When 2,4,5,7-tetraphenyl-1,3-oxazepine was thermolysed, almost quantitative rearrangement to 3-benzoyl-2,4,5-triphenylpyrrole occurred. Acid hydrolysis of the oxazine, on the other hand, led to three different substituted pyrroles, the *C*-tetrasubstituted compound mentioned above, *N*-benzoyl-2,3,5-triphenylpyrrole and 2,3,5-triphenylpyrrole, as well as the ring-opened product, 1,2,4-triphenylbutane-1,4-dione.[469]

Addenda

Aromatic Rearrangements

A general text on molecular rearrangements has been published.[470]

The following isomerizations have been studied: dichlorobenzenes in $AlCl_3$–NaCl melts at 130–200°;[471] dihalobenzenes in HF–SbF_5;[472] chlorocyclo-octane in the presence of benzene and $AlCl_3$;[473] mesitylene over a variety of aluminosilicate, oxide and carbon catalysts;[474] and methylbenzothiophens at 350–400° with $ZnCl_2$–Al_2O_3 catalysts.[475]

Irradiation of 3-acetaminopyridine in ethanol is a route to the new 4-acetyl-3-aminopyridine.[476]

When 2-(acylamino)alkyl 2,4,6-triazinyl ethers are treated with base O-to-N triazinyl migration occurs.[60,477]

Thermal isomerization (in boiling nitrobenzene) of 6-(dimethylamino)-3,7-, -1,9- and -7,9-dimethylpurinium iodides lead to 6-(dimethylamino)-3,9-dimethylpurinium iodide.[478]

Aromatization and other rearrangements of spirodienones have been reviewed.[479]

Sigmatropic Rearrangements

The Claisen rearrangement of 1-benzyl-5-ethyl-1,2,3,6-tetrahydropyridin-3-ol is a key step in the total synthesis of the indole alkaloid (±)-tabersonine.[480]

Allyl aryl ethers that have no strongly electron-attracting substituent undergo a charge-induced [3,3]-sigmatropic rearrangement (BCl_3 in chlorobenzene at low temperatures) to give the corresponding *o*-allylphenols. The charge induction causes an increase in the reaction rate relative to the thermal Claisen rearrangement of ca 10^{10}. Under these conditions, allyl 2,6-dialkylaryl ethers give products that arise from a sequence of *ortho*-Claisen rearrangements followed by a [1,2]-, [3,3]- or [3,4]-shift of the allyl group.[481]

A synthesis of the coumarin dentatin (poncitrin) involves, as a key step, the Claisen rearrangement (180° in diethylaniline) of the 3,3-dimethylallyl ether (**358**).[482]

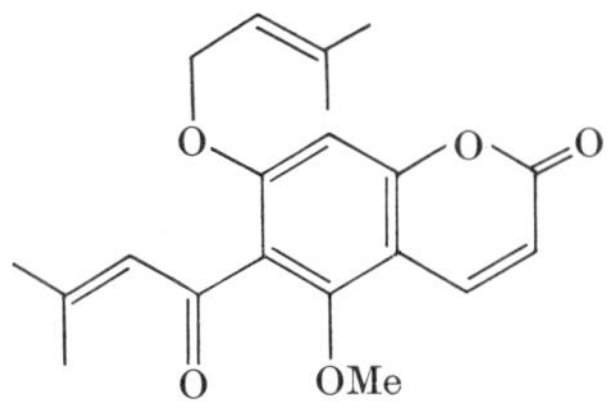

(**358**)

On thermolysis the trimethylsilyl derivative of *cis*-1-vinylcyclo-oct-3-en-1-ol undergoes predominant ring-contraction, to yield *cis*- and *trans*-1-(trimethylsilyloxy)-1,2-divinylcyclohexane. These in turn rearrange to yield *cis*- and *trans*-cyclodec-5-enones as well as 4-vinylcyclo-octanone (after working up). An equilibrium is established involving all six compounds. Though diradical pathways cannot be unequivocally ruled out, all the interconversions can be interpreted as a series of concerted reactions involving [1,3]- and [3,3]-sigmatropic rearrangements of starting material and of its primary rearrangement products. These rearrangements are of the silyloxy-Cope type. The starting enol underwent oxy-Cope rearrangements, its pyrolysis pattern corresponding largely to ring-expansion (again to yield the isomeric cyclodec-5-enones), these in turn rearranging to *cis*- and *trans*-bicyclo[5.3.0]decan-2-ones.[483]

The quaternary salts of *N*-(cyanomethyl)pyrrolidine with allylic halides, when treated with potassium *tert*-butoxide in liquid ammonia, yield ylides that undergo [2,3]-sigmatropic shift and yield $\beta\gamma$-unsaturated aldehydes.[484]

The novel allylic compounds CHCl=CRCH$(NHCOOR)_2$ undergo ready rearrangement to compounds ROOCNHCHClCR=CHNHCOOR.[485]

The [1,3]- and [1,5]-sigmatropic rearrangements of onium ylides have been reviewed.[486]

Electrocyclic Processes

The energetics of concerted electrocyclic reactions have been reviewed and a new general bond-order rule has been developed. This states that the electronically controlled electrocyclic reaction proceeds disrotatory if the generalized bond order between the two reacting centres is positive, and conrotatory if this bond order is negative in the reactant. An important consequence is the theoretical prediction of a new class of electrocyclic reactions: the concerted ring closure in the class of systems with zero or very small bond orders between the reacting centres is non-stereospecific and the product ratio can be shifted by introducing substituents.[487]

Thermolysis (100–180°) of the (*E,Z,E*)-, (*Z,Z,E*)- and (*Z,Z,Z*)-isomers of octa-2,4,6-triene results in substituted cyclohexadiene formation. The (*E,Z,E*)-form gives (132°, pentane) *cis*-5,6-dimethylcyclohexa-1,3-diene in a stereospecific disrotatory concerted cyclization ($\Delta H^\ddagger$ 28.6 kcal mol^{-1}, $\Delta S^\ddagger$ −7 eu). The remaining isomers (*Z,Z,E*) and (*Z,Z,Z*) are interconverted at 110° before cyclizing; the former undergoes electrocyclization at 178° to *trans*-5,6-dimethylcyclohexa-1,3-diene ($\Delta H^\ddagger$ 32 kcal mol^{-1},

$\Delta S^{\ddagger}$ −9 eu).[488] The thermolyses (between 125 and 195°) of a series of divinyl-, or dienyl-substituted cyclohexenes and aryl-substituted 1,3,5-trienes were studied. Kinetic data gave $\Delta H^{\ddagger}$ values between 25 and 35 kcal mol^{-1} and $\Delta S^{\ddagger}$ values from −1 to −15 eu.[489]

Anionic Rearrangements

7-Bromo-*cis*-bicyclo[4.3.0]nona-2,4-dien-8-ones [**359**; X = Br, Y = CO, R = R = H, Me or $(CH_2)_4$] react with potassium *tert*-butoxide in DMSO in short times at 25° to give bicyclo[4.2.1]nona-2,4,7-trien-9-ones [**361**; Y = CO, R = R = H, Me or $(CH_2)_4$]. The mechanism (Scheme 41) involves a Favorskii-type sequence, deprotonation, zwitterion formation (by π-assisted C-Br ionization), intramolecular capture of the zwitterion to form the bishomobenzene intermediate (**360**) which then undergoes retro-Diels–Alder reaction to give (**361**). These processes have been termed "bishomoconjugative rearrangements". The balance between the zwitterionic and alternative pathways when the bridging group Y is SO_2 (reaction being of the Ramberg–Bäcklund type) is also discussed.[490]

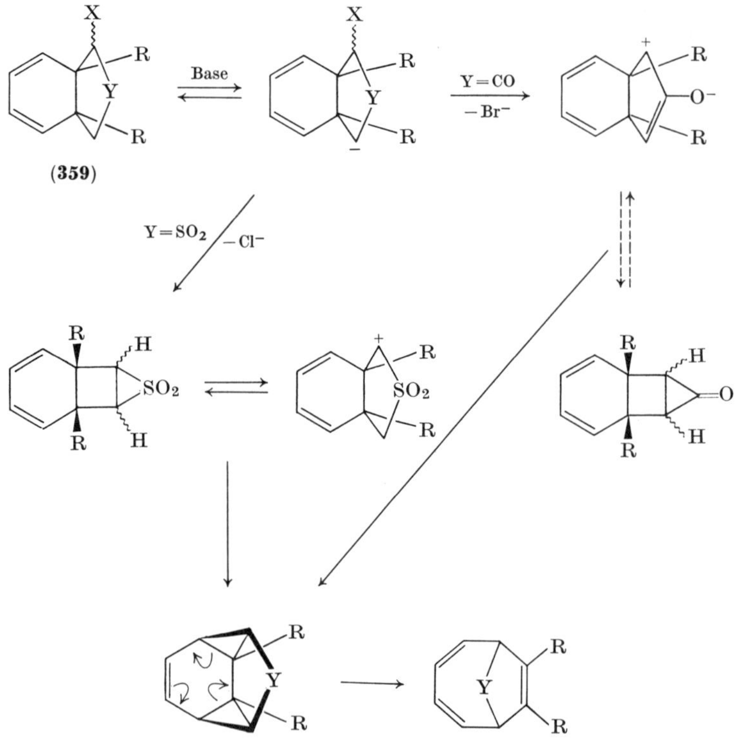

X, R, Y, R, (359), Base, X, R, Y, R, Y=CO, −Br⁻, R, O⁻, R, Y=SO₂, −Cl⁻, R, H, SO₂, R, H, R, SO₂, R, R, H, O, R, H, R, Y, R, R, Y, R, (360), (361)

SCHEME 41.

4-Aminobuta-1,2-dienyl selenides[$(C_2H_5)_2NCH_2CH{=}C{=}CHSeR$] undergo a Favorskii-type rearrangement when treated with sodium in liquid ammonia, to yield $(C_2H_5)_2NCH_2CH_2C{\equiv}CH$.[491]

Loss of thiols from $(RS)_2CHCH_2CH_2SR$ in the presence of potassium *tert*-butoxide produces 1,3-di(alkylthio)propenes together with 1,2-di(alkylthio)prop-1-enes.[492]

Reaction of *o*-$Bu^tSO_2C_6H_4C_6H_4Br$-*o* with butyl-lithium in THF (with C_3H_7SH added as a trap for butyl bromide liberated) gives (*o*-$Bu^tC_6H_4C_6H_4SO_2H$-*o*.[493]

Reaction of alkyl mesityl sulphones with butyl-lithium resulted in cyclization to thioxanthene 10,10-dioxides and Smiles rearrangements to substituted benzyl 4,6-dimethylbenzenesulphinic esters. Variation of the *p*-substituent influenced the rate of the Smiles reaction in the order H > Bu^t > Pr^i ~ Et > Me. In other ring positions, the effects of substituents were complicated because of their influence upon the cyclic and open-chain forms of the carbanion intermediates.[494]

Cationic Rearrangements

Deamination of *endo*- and *exo*-2-aminobicyclo[3.1.0]hexanes and solvolyses of the corresponding 2-chloro-compounds yielded very similar ratios of products (mixtures of *endo*- and *exo*-alcohols and of cyclohex-3-enol), indicating the intervention of a common 2-cation (which is regarded as non-classical with stabilization of the symmetrical homoallylic type).[495]

Electrophilic attack on hexamethyl-Dewar-benzene yielded various 5-substituted 1,2,3,4,5,6-hexamethylbicyclo[2.1.1]hexenyl cations. PMR and CMR data indicate that the positive charges therein reside mainly at C-2 and C-3.[496]

A marked temperature-dependence (between −70° and −40°) of the ^{13}C chemical shifts of a number of substituted tetramethylenehalonium ions, particularly 1,1-dimethyltetramethylenechloronium ion, has been observed. This is due to shifts in an equilibrium between the halonium ions and appropriate carbonium ions. The first observed ionization product when 1,6-dichlorohexane is dissolved in SbF_5–SO_2 solution is a 1-ethyltetramethylenehalonium ion.[497]

The bromination (or chlorination) of tetrafluorodihydrobenzobarrelene [5,6,7,8-tetrafluoro-1,4-dihydro-1,4-ethanonaphthalene] (**362**) leads to a mixture derived from

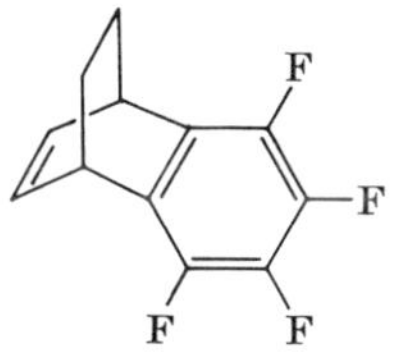

F F F F

(**362**)

equilibrating carbonium ions by skeletal rearrangements; the role therein of non-classical ions has been examined.[498] A similar comprehensive study has been made on the bromination of tetrafluorobenzobarrelene.[499] The skeletal rearrangements occurring during the reductive dehalogenations of derivatives of bicyclo[3.2.1]octadiene and tricyclo[3.2.1.0^{2,7}]octene have been described.[500] The roles of classical and non-classical carbonium ions in the skeletal rearrangements observed during the chlorination of 6,7-tetrafluorobenzobicyclo[3.2.1]octadienes have also been assessed.[501]

The isomeric mixtures of phenyl-alkanoic acids and alkylphenyl cyanides obtained by reaction of benzene with a substituted acrylic acid or acrylonitrile ($AlCl_3$ as catalyst) under both kinetic and thermodynamic control have been examined.[502]

The dehydration and rearrangement of the following alcoholic systems have been studied: 8-hydroxyoct-6-en-2-one and its homologues (products being dienes and dihydropyrans)[503] 1-(1-adamantanyl)-2-methylpropan-2-ol (fragmentation and elimination occurring);[504] and bicyclo[4.2.1]nonan-3- and -2-ol (bicyclo[3.3.1]non-2-enes and bicyclo[3.2.2]nonenes being isolated).[505]

The acid-catalysed cyclization of (**363**) to (**364** and **365**) (a highly efficient stereospecific olefinic cyclization) is a key step in a new approach to the total synthesis of oestrone.[506] The *p*:*o* ratio (**364**: **365**) in the products was very dependent upon the nature of the

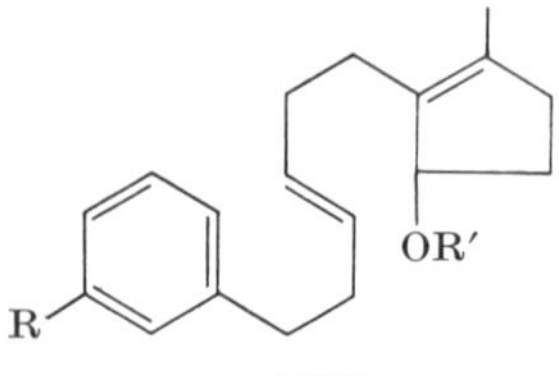

(**363**) 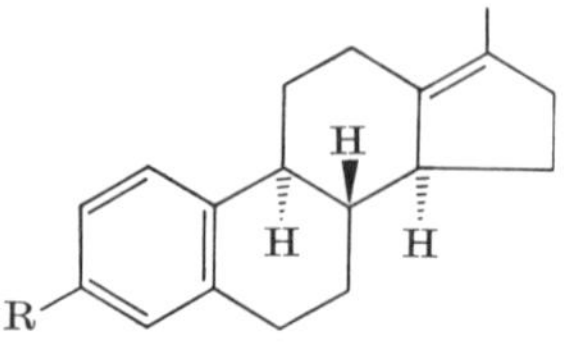 (**364**) 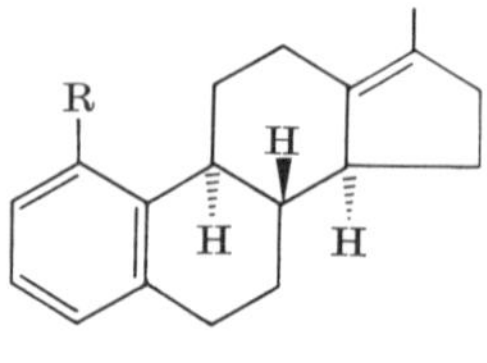 (**365**)

leaving group. This dependence together with the second-order anchimeric assistance in a two-step mechanism, the relatively low value (–1.4) of ρ for variation in R, and the absence of products derived from an intermediate bicyclic cation all suggest a mechanism involving the concerted formation of two rings.[507]

But-2-enyl phenyl sulphide at 300° in the presence of Al_2O_3 cyclizes to both substituted thiophens and thiochromans; it rearranges to both but-3-enyl and but-1-enyl phenyl sulphide and cleaves thiophenol.[508]

5-(*m*-Hydroxyphenyl)pent-1-ene, when heated neat at 225° for 1 hr., formed 5,6,7,8-tetrahydro-8-methyl-1-naphthol (50%) and -5-methyl-2-naphthol (50%), but in the gas phase, only the 8-methylated compound was formed.[509] The synthetic limits of analogous cyclizations have been outlined.[510]

Treatment of $ArCH_2CH{=}CH_2$ with thiocyanogen leads to normal and rearranged ($NCSCH_2CHArCH_2SCN$) products.[511] Similarly, when $ArCMe(OH)CMe{=}CH_2$ are used as substrates, the products are $MeCOCMeAr(CH_2SCN)$, kinetic data giving a ρ for Ar of –2.35. An intermediary episulphonium ion causes the Ar group to migrate.[512]

Product and rate studies of deamination and solvolysis of some *cis*-D-norsteroids involving both 16α- and 16β-substituents indicate that both series behave as if the 16-substituent is pseudoequatorial and ring C is a boat.[513]

Acid-catalysed rearrangement of the unsaturated alcohol (**366**) results in an abnormal backbone isomerization, with inversion of configuration occurring at the 14-position leading to a system containing a *trans*-C/D-ring junction.[514]

The effects of substituents at C-2 on the mechanism of stereomutation of allyl cations have been calculated. Relative energies were compared for planar allyl (A), perpendicular allyl (B) and cyclopropyl cations (C). The energies of (C) with Me or F as a single substituent were lower than those of (B), equivalently substituted; and the energies of (C) with an NH_2 or OH group were lower than those of either (A) or (B), similarly substituted.[515]

Adamantan-1-amine, on deamination ($NaNO_2$ in aqueous acetic acid), gave the corresponding 1-hydroxy-compound (60%). Adamantane-1,3-diamine gave adamantane, (4%), adamantan-1-ol (12%) and principally adamantane-1,3-diol (84%).[516]

Acid-catalysed rearrangement of triphenylallene gives 1,3-diphenylindene and mainly the dimer 1,3-diphenyl-2-R-indene [$R = -CHC_6H_5CH{=}C(C_6H_5)_2$]. The monomer does not react further with the parent allene in acid.[517]

In concentrated acid media, a series of 9-aryl-9,10-dimethylphenanthrenonium ions were generated and their PMR spectra were studied.[518] The kinetics of their degenerate rearrangement (automerization) were used to measure the relative migratory aptitude of the aryl groups ($\rho = -4.5$ at 50°).

Metal-catalysed Rearrangements

Heating 1,2,4,5,6,6-hexamethyl-3-methylenecyclohexa-1,4-diene and 2,3,4,4,5,6-hexamethylcyclohexa-2,5-dienone with pentacarbonyliron gave $Fe(CO)_3$ complexes of the isomeric cyclohexa-diene or -dienone, respectively, by a 1,5-hydride shift from the 1-methyl group to the *exo*-methylene group.[520]

The ester (−)-*trans*-$CH_3CH{=}CHCH(CH_3)CH(COOC_2H_5)_2$ was prepared from *cis*-pent-2-ene and sodium diethylmalonate in the presence of palladium(II) chloride and optically active ligands such as (+)-cyclohexyl-*o*-methoxyphenylmethylphosphine.[521]

Rearrangements to Electron-deficient Heteroatoms

Studies relevant to the Schmidt reaction have included examination of the basicity and acid-catalysed decomposition rates of hydrazoic acid,[522] and extension of the reaction to ketones (**367**). With X = O or NH, lactam formation took place with NH insertion between C=O and CH_2, with X = NTs, the reverse mode of insertion occurred.[523]

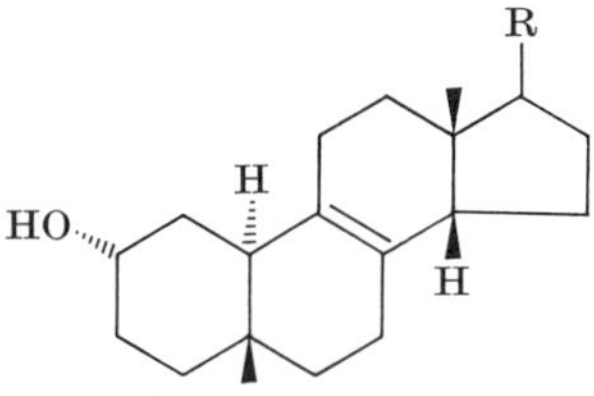

(**366**)

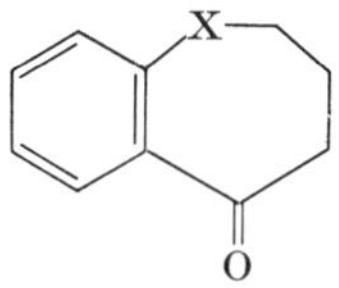

(**367**)

The rearrangement of the peresters $C_6H_5C(CH_3)_2OOC({=}O)R$ to the products $C_6H_5OC(CH_3)_2OC({=}O)R$ in glacial acetic acid (20–60°) gave $\rho^* = +1.1$ for variation in R.[524] The perester $C_6H_5C(CH_3)(C_2H_5)OOC({=}O)CH_3$ undergoes rearrangement in glacial acetic acid to yield $C_6H_5OC(CH_3)(C_2H_5)OC({=}O)CH_3$.[525]

Isomerizations

The structures and tautomerism of some unsymmetrically substituted furoxans has been reported.[526]

In a series of nitro-sugars of the type 1-deoxy-4-*O*-methyl-1-nitro*scyllo*inositol, the thermodynamic stabilities of the free nitro-compounds are in reverse order to those of their corresponding nitronate ions.[527]

Tautomeric equilibria have been studied with the following systems: 4-(hydroxybenzylidene)amines,[528] cyclohexanone enamine–imines,[529] 2-iminothiazolidin-4-ones,[530] 5-substituted rhodanines,[531] pyrazolin-5-ones, -thiones and -selones,[532] acetoacetic ester enamine–imines,[533] arylaminopropenethiones[534] and hydroxyarylmethyleneamines.[535]

1-(Dimethylvinylidene)-2-methyl-2-phenylcyclopropane undergoes rearrangement in solution at 130° or in the vapour phase at 350° to give 2-isopropylidene-1-methyl-3-methylene-1-phenylcyclopropane. Kinetics over the range 120–150° yield $E_a = 30.4$ kcal mol^{-1}, $\Delta S^{\ddagger} = -2.5$ eu, consistently with the intermediacy of an orthogonal diradical. At higher temperatures, both these cyclopropanes rearrange to 1,2,3-trisubstituted indene derivatives.[536]

Vinylmethylenecyclopropane rearranges in the gas phase or in solution to 3-methylenecyclopentene ($E_a = 26$ kcal mol^{-1}). The geometric isomers of vinylethylidenecyclopropane rearrange at different rates, yielding different mixtures of 3-ethylidenecyclopentene and 4-methyl-3-methylenecyclopentene. Both *cis*- and *trans*-1-methyl-3-methylene-2-vinylcyclopropane appear transiently in the reaction of the faster isomer. [1,3]-Sigmatropic shifts can alone account for all the products, although some [3,3]-shifts are not excluded. A concerted mechanism is favoured.[537]

The photochemical rearrangement of 2,2-dimethyl-1-phenylcyclopropane to 2-methyl-4-phenylbut-1-ene occurred with preferential hydrogen migration from the *trans*-methyl group.[538]

Orbital-symmetry-allowed conrotatory ring opening occurs with 1-cyano-2,3-diphenylcyclopropyl anions.[539] The rearrangements of 1-oxaspiro[2.2]pentanes (to cyclobutanones)[540] and of 2,3-dichloronorborn-2-ene epoxide (to three isomeric bicyclo[2.2.1]heptanes)[541] have been reported.

Treatment of bicyclo[5.1.0]octadienes with sodium methoxide in boiling methanol gave cycloheptatrienes, which underwent intramolecular hydrogen shifts at higher temperatures (200°).[542]

1-*cis*-Propenylspiro[2.6]nona-4,6,8-triene rearranges above 50° to the geometric isomers of 8-methylbicyclo[5.4.0]undeca-1,3,5,9-tetraene. 1-*trans*-Propenylspiro[2.6]-nona-4,6,8-triene rearranges at room temperature to the same pair of geometric isomers. Stepwise diradical processes are involved.[543]

The isomerization of bicyclo[4.1.0]heptane, bicyclo[5.1.0]octane, bicyclo[7.1.0]decane and *cis*- and *trans*-bicyclo[10.1.0]tridecane with toluene-*p*-sulphonic acid in benzene has been studied, olefin formation and ring-contraction reactions occurring.[544] 7-Methylene- and 7-methyl-bicyclo[4.1.0]hept-2-ene undergo hydrogenolysis of the cyclopropane unit during hydrogenation over Raney nickel.[545]

Biquadricyclanylidene (**368**) in the presence of catalytic amounts of palladium(II) chloride–norbornadiene complex isomerizes in solution to binorbornadienylidene (**370**) with norbornadienylidenequadricyclane (**369**) as an intermediate.[546]

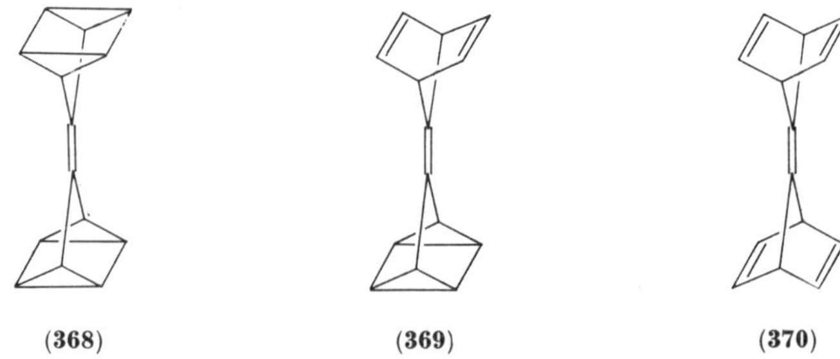

(**368**) (**369**) (**370**)

Thermolysis of all-*endo*-hexamethylbicyclo[2.2.0]hexane (137–177°) yields the all-*exo*-isomer and the ring-opened product *erythro*-3,4,5,6-tetramethylocta-(2*Z*,6*E*)-diene ($\Delta H^{\ddagger} = 140 \pm 7$ kJ mol^{-1}, $\Delta S^{\ddagger} = +1 \pm 15$ J mol^{-1} K^{-1}). Conversion of the C-1—C-4 σ-bond into a π-bond occurs, followed by conrotatory opening of the C-2—C-3 bond.[547]

Hex-1-en-4-yne isomerizes in base by parallel routes to hexa-1,3,4-triene and hex-2-en-4-yne.[548] Disproportionation of mixtures of alkenes in the presence of an aluminium–rhenium catalyst is accompanied by double-bond migration.[549] Isomerization of 6,10-dimethylundeca-4,5,9-trien-2-one on Al_2O_3 forms 3-*cis*-5-*trans*- and 3-*cis*-5-*cis*-isomers of 6,10-dimethylundeca-3,5,9-trien-2-one (pseudoionone).[550] The nitroalkene $(CH_3)_2CHC(NO_2){=}CH_2$ isomerizes (R_3N in THF) reversibly to a mixture of *cis*- and *trans*- $(CH_3)_2CHCH{=}CHNO_2$ and $(CH_3)_2C{=}CHCH_2NO_2$. The phenyl compound $C_6H_5C(NO_2){=}CH_2$ isomerizes irreversibly to $C_6H_5CH{=}CHNO_2$.[551] Rearrangement of $RSCH_2CH{=}CHR'$ to $RSCH{=}CH{-}CH_2R'$ (R being a nitrogen-heterocyclic group and R′ being H or C_6H_5), and *cis–trans*-isomerism in the products, have been followed by NMR spectroscopy.[552]

cis–trans-Isomerization of the following systems has been reported: 1,2-diphenylcyclopropane (acid-catalysed);[553] maleic and fumaric acid (thiourea-catalysed);[554] 1-phenyl-2-xylylcyclopropanes (acid-catalysed);[555] (−)-*cis*- and (+)-*trans*-carane (thermal)[556] and α-nitrocinnamic esters (thermal).[557]

Interconversion of the isomeric forms of isatin β-semicarbazone (*syn/anti*)[558] of phenyl *o*-benzoylbenzoate (open and cyclic forms),[559] and of 2-hydrazino- and 2-(methylhydrazino)-2,2-dimethylethanol (ring–chain),[560] have been studied. A kinetic study including medium effects, acidity and substituent variation, has been made of the base-catalysed isomerization of $RC{\equiv}CCHYCHR'(OH)$ to dihydrofurans.[561]

While the tautomerization of 1-chloro-1-nitroso-compounds (to chloro-oximes) has been detected,[562] 2-chloro-1-nitroso-compounds disproportionate rather than isomerize.[563]

Quaternary ammonium salts containing a penta-2,4-dienyl group in addition to a $\beta\gamma$-unsaturated group undergo thermal cyclization to substituted isoindoles and base-catalysed cyclization and ring-opening.[564]

Hydrogen shifts have been reported in oxidation of 5,5-dimethyl-2-heptanol by lead tetra-acetate[565] and in acid-catalysed rearrangement of *o*-alkenyldimethylbenzylamine oxides to aminophthalans and conjugated amino-ketones.[566]

Adamantanylcarbene rearrangements have been reported in the pyrolysis of adamantane-1-carboxaldehyde tosylhydrazone salts[567] and of adamantane-2-spiro-3′-diazirine.[568]

The equilibrium between $ArC[OGe(CH_3)_3]{=}CH_2$ and $(CH_3)_3GeCH_2COAr$ favours the *O*-germyl form with rising temperature and follows a ρ of 1.67.[569]

Rearrangements Involving Ring Openings and Ring Closures

Ring-opening reactions have been observed with (a) aziridines under electron-impact to yield species such as $RN{\equiv}CH$,[570] (b) 1-allyloxy-2,3-epoxypropane (where the direction of ring opening is a function of the reaction conditions),[571] (c) the oxides of *p*-substituted β,β-dimethylstyrenes, which undergo ring openings at various positions when they are treated with benzonitrile plus stannic chloride, to form 2-isoxazolines,[572] (d) thiophens and selenophens, which cleave with alkyl-lithiums to form substituted thio- and selenovinylacetylenes,[573] (e) substituted 1,3-dioxans, which cleave under γ-radiation to form esters,[574] and (f) 2,2-dialkyloxetanes which open to β-unsaturated primary alcohols in acid.[575]

The configuration of the breaking C—O bond determines the direction of skeletal rearrangements when 3-oxa-6,7-(tetrafluorobenzo)tricyclo[3.2.2.0$^{2,4\text{-}exo}$]nona-6,8-diene and related compounds react with HF or BF_3.[576] The reaction of 1-hydroxy-2,3-

diphenylcycloprop-2-ene-1-carboxylate with base causes ring expansion to a cyclobutene-3,4-dione derivative.[577]

The isomerization and hydrolysis of the anilides and methyl esters of $\alpha(\beta)$-chloro-$\beta(\alpha)$-methylthio- and $\alpha(\beta)$-chloro-$\beta(\alpha)$-phenylthio-isobutyric acid involve an episulphonium ion intermediate.[578]

The following ring interconversions have been studied: 2,5,5-triphenyl-2-oxazolin-4-one[579a] and -2-thiazolin-4-one[579b] into a benzo[*b*]thienyl system (with P_2S_5); pyrylium salts to pyridinium salts with sulphinylamines as reagents;[580] imidazolidones (hydantoins) to pyrimidinones (orotic acids);[581,582] thiophens and thiacyclohexanones to thiepin and thiocin compounds;[583] and ring-contractions of cycloheptatrienes in the presence of base.[584,585]

Photochemical Studies

While formally outside the scope of this Chapter, the following references are appended for general interest.

Aromatic and Heterocyclic Studies. Photochemical rearrangements of the following systems have been reported: *o*-nitrophenyl phenyl sulphoxide to 2-nitrosophenyl phenyl sulphone;[586] 4-phenyl-Δ^5-4-aza-3-oxo-steroids (6- to 8-membered ring-expansion occurring);[587] 7,7-dimethyl-2,5-diphenyl-4-diazanorcaradiene which undergoes a walk rearrangement to form 4,4-dimethyl-3,7-diphenyl-1,2-diazepine;[588] 3-hydroxyflavones to 3-aryl-3-hydroxy-1,2-indandiones;[589] *cis*-5-*tert*-butyl-2,2-dimethyl-3-phenyl-6-oxa-1-azabicyclo[3.1.0]hexane (forming pyrrolidones and acylazetidines);[590] and *N*-sulphonyliminopyridinium ylides (which yield either sulphonamide or 1,2-diazepines).[591]

Photochemical sigmatropic rearrangements have included observation of the enone-π-methane unit in [1.3]- and [3.3]-sigmatropic shifts,[592] and also [1,3]-sigmatropic vinyl rearrangements.[593]

The di-π-methane rearrangement has been studied in the following ways: a review;[594] its stereochemistry (using 1-phenyl-3-methyl-3-(1-*cis*-propenyl)cyclohex-1-ene);[595] its regiospecificity (using benzobicyclo[2.2.2]octadiene substrates);[596] its involvement in a norcaradiene–bisnorcaradiene rearrangement;[597] and its suppression in an allenic system.[598]

Other photochemical processes reported have included: the stereochemistry of the oxa-di-π-methane rearrangement;[599] isomerization of 1-hexyl (to 2-hexyl) radicals in the gas-phase;[600] conversion of 2,3-anthrabarrelene to 2,3-anthrasemibullvalene (by vinyl–vinyl bridging);[601] photoconversion of cycloheptatrienes and bicyclo[3.2.0]hepta-2,6-dienes,[602] and conversion of the latter into a tetracyclo[$3.2.0.0^{2,7}.0^{3,6}$]heptane;[603] photodimerization of some 1,2-diphenylcyclopropenes;[604] isomerization of cyclic α,β-unsaturated esters;[605] interaction of singlet oxygen and phenylcyclopropenes;[606] dimerization of some cyclopropene-carbonyl compounds;[607] and detection of the alkylideneoxophosphoranes $C_6H_5P(O){=}C(R)C_6H_5$ in the rearrangement of $(C_6H_5)_2P(O)C(R){=}N_2$.[608]

References

[1] P. Willadsen, B. Zerner and C. G. MacDonald, *J. Org. Chem.*, **38**, 3411 (1973).

[2] M. Z. A. Badr and M. M. Aly, *Can. J. Chem.*, **52**, 293 (1974).

[3] N. E. Lupes, *Rev. Rom. Chim.*, **17**, 1909 (1972); *Chem. Abs.*, **78**, 83628 (1973).

[4] N. W. Gilman, P. Levitan and L. H. Sternbach, *J. Org. Chem.*, **38**, 373 (1973).
[5] J. M. Patterson, N. F. Haidar, C.-Y. Shine and W. T. Smith, Jr., *J. Org. Chem.*, **38**, 3052 (1973).
[5a] W. J. Baron and M. R. De Camp, *Tetrahedron Letters*, **1973**, 4225.
[6] J. L. Garnett, M. A. Long, R. F. W. Vining and T. Mole, *Tetrahedron Letters*, **1973**, 4075.
[7] L. R. C. Barclay and I. T. McMaster, *Tetrahedron Letters*, **1973**, 4901.
[8] T. J. King, A. R. Forrester, M. M. Ogilvy and R. H. Thomson, *Chem. Comm.*, **1973**, 844, 956.
[9] W. Weyler, Jr., P. Germeraad and H. W. Moore, *J. Org. Chem.* **38**, 3865 (1973).
[10] W. Weyler, Jr., D. S. Pearce and H. W. Moore, *J. Am. Chem. Soc.*, **95**, 2603 (1973).
[11] F. A. Davis, E. R. Fretz and C. J. Horner, *J. Org. Chem.*, **39**, 690 (1973).
[12] F. A. Davis, C. J. Horner, E. R. Fretz and J. F. Stackhouse, *J. Org. Chem.*, **38**, 695 (1973).
[13] T. D. B. Morgan, D. L. H. Williams and J. A. Wilson, *J.C.S. Perkin II*, **1973**, 473.
[14] R. H. Squire and H. H. Jaffe, *J. Am. Chem. Soc.*, **95**, 8188 (1973).
[15] D. J. W. Goon, N. G. Murray, J. P. Schoch and W. J. Bunce, *Can. J. Chem.*, **51**, 3827 (1973).
[16] R. A. Cox and E. Buncel, *Can. J. Chem.*, **51**, 3143 (1973).
[17] D. V. Banthorpe and M. O'Sullivan, *J.C.S. Perkin II*, **1973**, 551.
[18] J. R. Cox, Jr., and M. F. Dunn, *J. Org. Chem.* **37**, 4415 (1972).
[19] V. J. Hill and H. J. Shine, *J. Am. Chem. Soc.*, **95**, 8102 (1973).
[20] W. W. Paudler, A. G. Zeiler and M. M. Goodman, *J. Heterocyclic Chem.*, **10**, 423 (1973).
[21] R. Martin and G. Coton, *Bull. Soc. chim. France*, **1973**, 1438.
[22] R. Martin and G. Coton, *Bull. Soc. chim. France*, **1973**, 1442.
[23] R. Martin, *Bull. Soc. chim. France*, **1973**, 3087.
[24] F. Effenberger, H. Klenk and P. L. Reiter, *Angew. Chem. Internat. Ed.*, **12**, 775 (1973).
[25] J. Correa, *Rev. Soc. Quim. Mex.*, **16**, 196 (1972); *Chem. Abs.* **78**, 3891 (1973).
[25a] C. E. Kalmus and D. M. Hercules, *J. Am. Chem. Soc.* **96**, 449 (1974).
[25b] C. Vandesande and M. Vandewalle, *Bull. Soc. Chim. Belges.*, **82**, 705 (1973).
[26] A. F. Hegarty, J. A. Kearney and F. L. Scott, *J.C.S., Perkin II*, **1973**, 1422.
[27] A. J. Elliott, M. S. Gibson, M. M. Kayser and G. A. Pawelchak, *Can. J. Chem.*, **51**, 4115 (1973).
[28] A. S. Shawali and M. K. Ahmad, *Bull. Chem. Soc. Japan*, **46**, 3625 (1973).
[28a] T. W. Bentley, R. A. W. Johnstone and A. F. Neville, *J.C.S. Perkin I*, **1973**, 449.
[28b] V. D. Goldberg, and M. M. Harris, *J.C.S. Perkin II*, **1973**, 1303.
[29] V. N. Drozd and O. I. Trifonova, *Zh. Org. Khim*, **8**, 2383 (1972); *Chem. Abs.*, **78**, 7116 (1973).
[30] V. N. Drozd and T. R. Saks, *Zh. Org. Khim.*, **9**, 2544 (1973).
[31] R. H. Mitchell, *Chem. Comm.*, **1973**, 955.
[32] G. Capozzi, G. Melloni and G. Modena, *J.C.S. Perkin I*, **1973**, 2250.
[33] A. S. Bailey, A. J. Buckley and J. F. Seager, *J.C.S. Perkin I*, **1973**, 1809.
[34] H. Prinzbach, G. Kaupp, R. Fuchs, M. Joyeux, R. Kitzing and J. Markert, *Chem. Ber.*, **106**, 3824 (1973).
[35] V. J. Traynelis and J. N. Rieck, *J. Org. Chem.*, **38**, 4334 (1973).
[36] R. A. Abramovitch and I. Shinkai, *Chem. Comm.*, **1973**, 569.
[37] K. Takayama, M. Isobe, K. Harano and T. Taguchi, *Tetrahedron Letters*, **1973**, 365.
[38] J. Knabe, R. Dörr, S. F. Dyke and R. G. Kinsman, *Tetrahedron Letters*, **1972**, 5373: see also J. Knabe and R. Dörr, *Arch. Pharm.*, **306**, 784 (1973).
[39] F. Nahavandi, F. Razmara and M. P. Stevens, *Tetrahedron Letters*, **1973**, 301.
[40] J. M. Coxon, M. P. Hartshorn and W. H. Swallow, *Austral. J. Chem.*, **26**, 2521 (1973).
[41] J. A. Berson, E. W. Petrillo, Jr., and P. Bickart, *J. Am. Chem. Soc.*, **96**, 636 (1974).
[42] C. G. Begg, M. R. Grimmett and P. D. Wethey, *Austral. J. Chem.*, **26**, 2435 (1973).
[43] F. T. Boyle and R. A. Y. Jones, *J.C.S. Perkin I.*, **1973**, 167.
[44] F. G. Boddar, A. F. M. Fahmy and N. F. Aly, *J.C.S. Perkin I*, **1973**, 2448.
[45] S. M. Rosen and J. A. Moore, *J. Org. Chem.*, **37**, 3770 (1972).
[46] J. A. Moore, R. C. Gearhart, O. S. Rothenberger, P. C. Thorstenson and R. H. Wood, *J. Org. Chem.*, **37**, 3774 (1972).
[47] A. K. Bose, J. L. Fahey and M. S. Manhas, *J. Heterocyclic Chem.*, **10**, 791 (1973).
[48] J. Kilchin and R. J. Stoodley, *J.C.S. Perkin I*, **1973**, 2460.
[49] M. T. Reetz, *Tetrahedron*, **29**, 2189 (1973).
[50] R. Jacquier, H. Lopez and G. Maury, *J. Heterocyclic Chem.*, **10**, 755 (1973).
[51] P. Guerret, R. Jacquier and G. Maury, *J. Heterocyclic Chem.*, **8**, 643 (1971); D. J. Brown in Mechanisms of Molecular Migrations" (B. S. Thyagarajan, Ed., Interscience-Wiley, New York 1968, Vol. 1, p. 209.
[52] A. B. DeMilo, J. E. Oliver and R. D. Gilardi, *J. Heterocyclic Chem.*, **10**, 231 (1973).
[53] D. R. Sutherland, G. Tennant and R. J. S. Vevero, *J.C.S. Perkin I*, **1973**, 943.

[54] A. Albert, *J.C.S. Perkin I*, **1973**, 2659.
[55] Review: J. W. Schulenberg and S. Archer, *Organic Reactions*, John Wiley, New York, 1968. Vol. 14, p. 3 *et seq.*
[56] M. Golbier and R. Milcent, *Tetrahedron Letters*, **1973**, 4465.
[57] M. Golbier and R. Milcent, *Bull. Soc. chim. France*, **1973**, 254.
[58] M. Ruccia, N. Visona and G. Cusmano, *Tetrahedron Letters*, **1972**, 4959.
[59] T. Novinson, R. K. Robins and D. E. O'Brien, *J. Heterocyclic Chem.*, **10**, 887 (1973).
[60] S. Tsunoda, T. Shiojima, Y. Hashida, S. Sekiguchi and K. Matsui, *Bull. Chem. Soc. Japan*, **46**, 3499 (1973).
[61] B. G. Pring and C. G. Swahn, *Acta Chem. Scand.*, **27**, 1891 (1973).
[62] B. Zwanenburg and A. Wagenaar, *Tetrahedron Letters*, **1973**, 5009.
[62a] K. B. Wagener and G. B. Butler, *J. Org. Chem.*, **31**, 3070 (1973).
[63] K. L. Cook and A. J. Waring, *J.C.S. Perkin II*, **1973**, 88.
[64] K. L. Cook and A. J. Waring, *J.C.S. Perkin II*, **1973**, 84.
[65] J. W. Pilkington and A. J. Waring, *Tetrahedron Letters*, **1973**, 4345.
[66] U. Widmer, J. Zsindely, H.-J. Hansen and H. Schmid, *Helv. Chim. Acta*, **56**, 75 (1973).
[67] U. Widmer, H.-J. Hansen and H. Schmid, *Helv. Chim. Acta*, **56**, 1895 (1973).
[68] L. H. Schwartz, R. V. Flor and V. P. Gullo, *J. Org. Chem.* **39**, 219 (1974).
[69] J. P. Gesson and J. C. Jacquey, *Tetrahedron Letters*, **29**, 3631 (1973).
[70] V. P. Vitullo and M. J. Cashen, *Tetrahedron Letters*, **1973**, 4823.
[71] M. Tada, Y. Tanahashi, Y. Moriyama and T. Takahashi, *Tetrahedron Letters*, **1972**, 5255.
[72] N. K. Hamer and M. E. Stubbs, *J.C.S. Perkin I*, **1972**, 2971.
[73] G. Märkl and D. E. Fisher, *Tetrahedron Letters*, **1973**, 223.
[74] P. Y. Bruice, G. J. Kasperek, T. C. Bruice, H. Yagi and D. M. Jerina, *J. Am. Chem. Soc.*, **95**, 1673 (1973).
[75] F.-G. Klärner and E. Vogel, *Angew. Chem. Internat. Ed.*, **12**, 840 (1973).
[76] D. B. Borders and J. E. Lancaster, *J. Org. Chem.*, **39**, 435 (1974).
[77] J. A. Berson, *Accounts Chem. Res.*, **5**, 406 (1972).
[78] J. E. Baldwin, A. H. Andrist and R. K. Pinschmidt, Jr., *Accounts Chem. Res.*, **5**, 402 (1972).
[79] H. E. Zimmerman, *Accounts Chem. Res.*, **5**, 393 (1972).
[80] J. A. Berson and L. Salem, *J. Am. Chem. Soc.*, **94**, 8917 (1972).
[81] W. T. Borden and L. Salem, *J. Am. Chem. Soc.*, **94**, 932 (1973).
[82] N. D. Epiotis, *J. Am. Chem. Soc.*, **95**, 1206 (1973).
[83] I. N. Domnin, *Sovrem. Prokl. Org. Khim.*, **1971**, 79; *Chem. Abs.*, **78**, 3362 (1973).
[84] U. Widmer, H.-J. Hansen and H. Schmid, *Helv. Chim. Acta*, **56**, 2644 (1973).
[85] N. Šorčević, J. Zsindely and H. Schmid, *Helv. Chim. Acta*, **56**, 1457 (1973).
[86] J. A. Miller and C. M. Scrimgeour, *J. C.S. Perkin II*, **1973**, 1137.
[87] S. Marcinkiewicz, *Bull Acad. Pol. Sci., Ser. Sci. Chim.*, **20**, 861 (1972); *Chem. Abs.*, **78**, 123717 (1973).
[88] M. J. S. Dewar and B. D. Nahlovsky, *J. Am. Chem. Soc.*, **96**, 460 (1974).
[89] A. Wunderli, J. Zsindely, H.-J. Hansen and H. Schmid, *Helv. Chim. Acta*, **56**, 989 (1973).
[90] B. Bowden, R. C. Cookson and H. A. Davis, *J.C.S. Perkin I*, **1973**, 2634.
[91] R. C. Cookson and N. W. Hughes, *J.S.C. Perkin I*, **1973**, 2738; see also R. C. Cookson and R. N. Rogers, *ibid.*, p. 2741.
[92] D. K. Black, Z. T. Fomum, P. D. Landor and S. R. Landor, *J.C.S. Perkin I*, **1973**, 1349.
[93] R. K. Bramely, J. Caldwell and R. Grigg, *J.C.S. Perkin I*, **1973**, 1913; (*b*) R. K. Bramely, J. Caldwell and R. Grigg, *Tetrahedron Letters*, **1973**, 3207.
[94] J. M. Patterson, A. Wu, C. S. Kook and W. T. Smith, Jr., *J. Org. Chem.*, **39**, 486 (1974).
[95] S. Jnada, S. Jkadu and M. Okazaki, *Chem. Letters*, **1973**, 1213.
[96] M. Schmid, H.-J. Hansen and H. Schmid, *Helv. Chim. Acta*, **56**, 105 (1973).
[97] H. Scheurer, J. Zsindely and H. Schmid, *Helv. Chim. Acta*, **56**, 478 (1973).
[98] C. P. Falshaw, S. A. Lane and W. D. Ollis, *Chem. Comm.*, **1973**, 491.
[99] W. Adam, H. Fischer, H.-J. Hansen, H. Heimgartner, H. Schmid and H.-R. Waespe, *Angew. Chem. Internat. Ed.*, **12**, 662 (1973).
[100] W. Grimme, *J. Amer. Chem. Soc.*, **95**, 2381 (1973).
[101] W. Grimme and K. Seel, *Angew. Chem. Internat. Ed.*, **12**, 507 (1973).
[102] H. Klein, W. Kursawa and W. Grimme, *Angew. Chem. Internat. Ed.*, **12**, 580 (1973).
[103] M. J. S. Dewar and L. E. Wade, *J. Am. Chem. Soc.*, **95**, 290 (1973); for some experimental studies along these lines, see W. R. Roth, M. Heiber and G. Erkev, *Angew. Chem. Internat. Ed.*, **12**, 504 (1973); W. R. Roth and G. Erkev, *ibid.*, p. 505; W. Grimme and H.-J. Rother, *ibid.*, p. 505.
[104] L. G. Greifenstein, J. B. Lambert, M. J. Broadhurst and L. A. Paquette, *J. Org. Chem.*, **38**, 1210 (1973).

[105] P. S. Wharton and D. W. Johnson, *J. Org. Chem.*, **38**, 4117 (1973).
[106] K. Takeda, I. Horibe and H. Minato, *J.C.S. Perkin I*, **1973**, 2212.
[107] J. A. Deyrup and M. Betkouski, *Tetrahedron Letters*, **1973**, 1131.
[108] E. N. Marvell and C. Lin, *Tetrahedron Letters*, **1973**, 2679.
[109] R. W. Thies and L. E. Schick, *J. Am. Chem. Soc.*, **96**, 456 (1974).
[110] G. Adames, R. Grigg and J. N. Grover, *Tetrahedron Letters*, **1974**, 363.
[111] J. Slutsky and H. Kwart, *J. Org. Chem.*, **38**, 3658 (1973).
[112] R. W. Thies and J. E. Billigmeier, *J. Am. Chem. Soc.*, **96**, 200 (1974).
[113] P. Gilgen, J. Zsindely and H. Schmid, *Helv. Chim. Acta*, **56**, 681 (1973).
[114] J. E. Baldwin and J. A. Walker, *J. Am. Chem. Soc.*, **96**, 596 (1974).
[115] D. Mackay, C. W. Pilger, and L. L. Wong. *J. Org. Chem.*, **38**, 2043 (1973).
[116] R. V. Stevens, E. E. McEntire, W. E. Barnett and E. Wenkert, *Chem. Comm.*, **1973**, 662.
[117] F. Vögtle and E. Goldschmitt, *Angew. Chem. Internat. Ed.*, **12**, 767 (1973).
[118] S. Ranganathan, D. Ranganathan, R. S. Sidha and A. K. Mehrotra, *Tetrahedron Letters*, **1973**, 3577.
[119] G. Andrews and D. A. Evans, *Tetrahedron Letters*, **1972**, 5121.
[120] P. T. Lansbury and J. E. Rhodes, *Chem. Comm.*, **1974**, 21.
[121] Y. Yamamoto, J. Oda and Y. Jnouye, *Chem. Comm.*, **1973**, 848.
[122] V. Rautenstrauch, *Helv. Chim. Acta*, **56**, 2492 (1973).
[123] V. Šunjić, F. Kajfež, D. Kolbah, H. Hofman and M. Štromar, *Tetrahedron Letters*, **1973**, 3209.
[124] S. Mageswaran, W. D. Ollis, I. O. Sutherland and Y. Thebtaranonth, *Chem. Comm.*, **1973**, 651.
[125] W. D. Ollis, I. O. Sutherland and Y. Thebtaranonth, *Chem. Comm.*, **1973**, 653.
[126] W. D. Ollis, I. O. Sutherland and Y. Thebtaranonth, *Chem. Comm.*, **1973**, 654.
[127] S. Mageswaran, W. D. Ollis and I. O. Sutherland, *Chem. Comm.*, **1973**, 656.
[128] W. D. Ollis, I. O. Sutherland and Y. Thebtaranonth, *Chem. Comm.*, **1973**, 657.
[129] P. A. Grieco, M. Meyers and R. S. Finkelhor, *J. Org. Chem.*, **39**, 119 (1974); see also P. A. Grieco, *Chem. Comm.*, **1972**, 702; D. A. Evans, G. C. Andrews, T. T. Fujimoto and D. Wells, *Tetrahedron Letters*, **1973**, 1389; P. A. Grieco and R. S. Finkelhor, *J. Org. Chem.*, **38**, 2245 (1973).
[130] K. B. Sharpless and R. F. Lauer, *J. Org. Chem.*, **37**, 3973 (1972).
[131] D. N. Jones, J. Blenkinsopp, A. C. F. Edmonds, E. Helmy and R. J. K. Taylor, *J.C.S. Perkin I*, **1973**, 2602.
[132] R. C. Bingham and M. J. S. Dewar, *J. Am. Chem. Soc.*, **94**, 9107 (1972).
[133] J. A. Berson and P. B. Dervan. *J. Am. Chem. Soc.*, **95**, 269 (1973).
[134] J. A. Berson and R. W. Holder, *J. Am. Chem. Soc.* **95**, 2037 (1973).
[135] H. J. Reich and D. A. Murcia, *J. Am. Chem. Soc.*, **95**, 3418 (1973).
[136] J. Slutsky and H. Kwart, *J. Am. Chem. Soc.*, **95**, 8678 (1973).
[137] K. G. Hancock and J. D. Kramer, *J. Am. Chem. Soc.*, **95**, 6463 (1973).
[138] V. S. Bogdanov, Yu. N. Bubnov, M. N. Bochkareva and B. M. Mikhailov, *Dokl. Akad. Nauk SSSR*, **201**, 605 (1971); *Chem. Abs.*, **78**, 28965 (1973).
[139] J. W. Lown and M. H. Akhtar, *Chem. Comm.*, **1973**, 511.
[140] J. W. Lown and M. H. Akhtar, *Tetrahedron Letters*, **1973**, 3727.
[141] J. W. Lown and M. H. Akhtar, *Tetrahedron Letters*, **1974**, 179.
[141a] C. M. Bowes, D. F. Montecalvo and F. Sondheimer, *Tetrahedron Letters*, **1973**, 3181.
[142] P. Schiess and P. Fünfschilling, *Tetrahedron Letters*, **1972**, 5191.
[143] P. Schiess and P. Fünfschilling, *Tetrahedron Letters*, **1972**, 5195.
[144] H. Heimgartner, J. Zsindely, H.-J. Hansen and H. Schmid, *Helv. Chim. Acta*, **56**, 2924 (1973).
[145] A. K. Youssef and M. A. Ogliaruso, *J. Org. Chem.*, **38**, 487 (1973).
[146] R. Breslow, J. M. Hoffman and C. Perchonock, *Tetrahedron Letters*, **1973**, 3723.
[147] M. R. Willcott III and I. M. Rathburn III, *J. Am. Chem. Soc.*, **96**, 948 (1974).
[148] J. W. A. M. Janssen, H. J. Koeners, C. G. Kruse and C. L. Habraken, *J. Org. Chem.*, **38**, 1777 (1973).
[149] H. Greuter, H.-J. Hansen and H. Schmid, *Helv. Chim. Acta*, **56**, 2479 (1973).
[150] T. Laird and W. D. Ollis, *Chem. Comm.*, **1973**, 658.
[151] W. D. Ollis, R. Somanathan and I. O. Sutherland, *Chem. Comm.*, **1973**, 661.
[152] K. A. Burdett, D. H. Yates and J. S. Swenton, *Tetrahedron Letters*, **1973**, 783.
[153] N. D. Epiotis, *J. Am. Chem. Soc.*, **95**, 1200, 1214 (1973).
[154] A. H. Andrist, *J. Org. Chem.*, **38**, 1772 (1973).
[155] J. Mathieu, *Bull Soc. chim. France*, **1973**, 807.
[156] A. J. Bellamy and W. Crilly, *Tetrahedron Letters*, **1973**, 1893.
[157] J. A. Deyrup and J. C. Gill, *Tetrahedron Letters*, **1973**, 4845.
[158] A. A. Reid, J. T. Sharp, H. R. Sood and P. B. Thorogood, *J. C.S. Perkin I*, **1973**, 2543.

[159] C. W. Shoppee and B. J. A. Cooke, *J.C.S. Perkin I*, **1973**, 2197.
[160] C. W. Shoppee and B. J. A. Cooke, *J.C.S. Perkin I*, **1973**, 1026.
[161] E. N. Marvell, G. Caple, C. Delphey, J. Platt, N. Polston and J. Tashiro, *Tetrahedron*, **29**, 3797 (1973); E. N. Marvell, G. Caple, B. Schatz and W. Pippin, *ibid.*, p. 3781.
[162] J. F. Garst and C. D. Smith, *J. Am. Chem. Soc.*, **95**, 6870 (1973).
[163] C. van der Stelt, W. J. Heus and A. Haasjes, *Red. Trav. chim.*, **92**, 492 (1973).
[164] R. West and P. Boudjouk, *J. Am. Chem. Soc.*, **95**, 3987 (1973).
[165] R. West and P. Boudjouk, *J. Am. Chem. Soc.*, **95**, 3983 (1973).
[166] J. J. Eisch and M.-R. Tsai, *J. Am. Chem. Soc.*, **95**, 4065 (1973).
[167] H. Rodé-Gowal and H. Dahn, *Helv. Chim. Acta*, **56**, 2070 (1973).
[168] I. Kuwajima and Y. Fukuda, *Tetrahedron Letters*, **1973**, 327.
[169] D. J. Cram, F. Willey, H. P. Fisher, H. M. Relles and D. A. Scott, *J. Am. Chem. Soc.*, **88**, 2759 (1966).
[170] J. H. Wotiz, P. M. Barelski and D. F. Koster, *J. Org. Chem.*, **38**, 489 (1973).
[171] B. Decocklereverend and M. Durand, *Compt. Rend., Ser. C*, **277**, 1247 (1973).
[172] A. Maercker and R. Geuss, *Chem. Ber.*, **106**, 775 (1973).
[173] D. W. Slocum and P. L. Gierer, *J. Org. Chem.*, **38**, 4189 (1973).
[174] F. G. Bordwell and J. Almy, *J. Org. Chem.*, **38**, 571 (1973).
[175] F. G. Bordwell and J. Almy, *J. Org. Chem.*, **38**, 575 (1973).
[176] F. G. Bordwell and J. G. Strong, *J. Org. Chem.*, **38**, 579 (1973).
[177] A. Takeda and S. Tsubai, *J. Org. Chem.*, **38**, 1709 (1973).
[178] W. Kirmse, A. Engelmann and J. Heese, *J. Am. Chem. Soc.*, **95**, 625 (1973); *Chem. Ber.*, **106**, 3073 (1973).
[179] W. Kirmse and A. Engelmann, *Chem. Ber.*, **106**, 3086 (1973).
[180] M. Mori, S. Nishimura and Y. Ban, *Tetrahedron Letters*, **1973**, 4951.
[181] N. W. K. Chiu and T. S. Sorensen, *Can. J. Chem.*, **51**, 2783 (1973).
[182] J. B. Stothers, C. T. Tan, A. Nickon, F. Huang, R. Sridhar, and R. Weglein, *J. Am. Chem. Soc.*, **94**, 8581 (1972).
[183] J. A. Berson and J. A. Jenkins, *J. Am. Chem. Soc.*, **94**, 8907 (1973).
[184] W. J. Hehre, *J. Am. Chem. Soc.*, **94**, 8909 (1972).
[185] S. Takamuku, K. Nakamura, K. Nagaoka and H. Sakurai, *Chem. Letters* (*Tokyo*), **1973**, 1303.
[186] G. Eadon, *J. Am. Chem. Soc.*, **94**, 8938 (1972).
[187] B. T. Golding and L. Radom, *Chem. Comm.*, **1973**, 939.
[188] R. Magne, E. Torreilles and L. Giral, *Bull. Soc. Chim. France*, **1973**, 2813.
[189] E. S. Balenkova and L. A. Karamysheva, *Zh. Org. Khim.*, **9**, 1667 (1973).
[190] W. E. Parham, D. E. Egberg and W. C. Montgomery, *J. Org. Chem.*, **38**, 1207 (1973).
[191] V. G. Shubin, D. V. Korchagina, B. G. Derendyoev, G. I. Borodkin and V. A. Koptyug, *Zh. Org. Khim.*, **9**, 1041 (1973); *Chem. Abs.*, **79**, 41683 (1973).
[192] T. Shono, K. Fujita and S. Kumai, *Tetradedron Letters*, **1973**, 3123.
[193] A. J. Sisti and M. Meyers, *J. Org. Chem.*, **38**, 4431 (1973).
[194] B. P. Mundy and R. D. Otzenberger, *J. Org. Chem.*, **38**, 2109 (1973).
[195] W. Tadros, A. B. Sakla, S. B. Awad and A. A. A. Helmy, *Helv. Chim. Acta*, **55**, 2808 (1972).
[196] C. C. Lee and D. Unger, *Can. J. Chem.*, **51**, 1494 (1973).
[197] Review: P. J. Stang, *Progr. Phys. Org. Chem.*, **10**, 276 (1973).
[198] K. Yates, G. H. Schmid, T. W. Regulski, D. G. Garratt, H.-W. Leung and R. McDonald, *J. Am. Chem. Soc.*, **95**, 160 (1973).
[199] P. J. Stang and T. E. Dueber, *J. Am. Chem. Soc.*, **95**, 2683 (1973).
[200] P. J. Stang and T. E. Dueber, *J. Am. Chem. Soc.*, **95**, 2686 (1973).
[201] Z. Rappoport and Y. Houminer, *J.C.S. Perkin II*, **1973**, 1506.
[202] G. Richtarski and P. Mastalerz, *Tetrahedron Letters*, **1973**, 4069.
[203] D. Howells and S. Warren, *J.C.S. Perkin II*, **1973**, 1645.
[204] S. Yamada, S. Yamada, Y. Moriyama, Y. Tanahashi and T. Takashi, *Tetrahedron Letters*, **1972**, 5043.
[205] Y. Naya and H. Yoshio, *Chem. Letters* (*Tokyo*), **1973**, 727.
[206] W. Skorianetz and G. Ohloff, *Helv. Chim. Acta*, **56**, 2025 (1973).
[207] D. G. Farnum, R. A. Maden and G. Mehta, *J. Am. Chem. Soc.*, **95**, 8692 (1973).
[208] G. Mehta, N. Pattnaik and S. K. Kapoor, *Tetrahedron Letters*, **1972**, 4947.
[209] J. L. Zundel, G. Wolff and G. Ourisson, *Bull. Soc. Chim. France*, **1973**, 3206.
[210] E. J. Corey, R. D. Balanson, *Tetrahedron Letters*, **1973**, 3153.
[211] F. R. S. Clark and J. Warkentin, *Can. J. Chem.*, **51**, 4090 (1973).
[212] C. W. David, B. W. Everling, R. J. Kilian, J. B. Stothers and W. R. Vaughan, *J. Am. Chem. Soc.*, **95**, 1265 (1973).

[213] D. R. Dimmel and W. Y. Yu,. *J. Org. Chem.*, **38**, 3778, 3782 (1973).
[214] G. A. Olah and G. Liang, *J. Am. Chem. Soc.*, **96**, 189 (1974).
[215] G. A. Olah and G. Liang, *J. Am. Chem. Soc.*, **96**, 195 (1974).
[216] G. A. Olah, G. Liang G. D. Mateescu and J. L. Riemenschneider, *J. Am. Chem. Soc.*, **95**, 8698 (1973).
[217] H. C. Brown and E. N. Peters, *J. Am. Chem. Soc.*, **95**, 2400 (1973).
[218] J. V. Paukstelis and B. W. Macharia, *Tetrahedron*, **29**, 1955 (1973).
[219] G. Nagendrappa and K. Griesbaum, *Chem. Ind.* (*London*) **1973**, 902.
[220] D. S. Selthi and P. Yates, *J. Am. Chem. Soc.*, **95**, 3820 (1973).
[221] M. Gates and J. L. Zabriskie, Jr., *J. Org. Chem.* **39**, 222 (1974).
[222] M. Geisel, C. A. Grob, W. Santi and W. Tschudi, *Helv. Chim. Acta*, **56**, 1055 (1973).
[223] I. I. Bardyshev, E. F. Buinova and B. G. Udarov, *Zh. Org. Khim.*, **9**, 1670 (1973).
[224] E. I. Bagrii, T. Yu. Frid and P. I. Sanin, *Neftekhimiya*, **12**, 797 (1972); *Chem. Abs.*, **78**, 71108 (1973).
[225] K. M. Majerski and Z. Majerski, *Tetrahedron Letters*, **1973**, 4915.
[226] R. D. Miller and D. L. Dolce, *Tetrahedron Letters*, **1973**, 5217.
[227] J. T. Groves and K. W. Ma, *Tetrahedron Letters*, **1973**, 5225.
[228] E. Boelema, J. H. Wieringa, H. Wynberg and J. Strating, *Tetrahedron Letters*, **1973**, 2377.
[229] E. M. Engler, M. Farcasiu, A. Sevin, J. M. Cense and P. von R. Schleyer, *J. Am. Chem. Soc.*, **95**, 5769 (1973).
[230] J. Peter-Katalinić, J. Zsindely and H. Schmid, *Helv. Chim. Acta*, **56**, 2796 (1973).
[231] N. A. Nelson, *J. Org. Chem.*, **38**, 3798 (1973).
[232] A. Roedig and H. A. Renk, *Chem. Ber.*, **106**, 3877 (1973).
[233] H. R. Hudson and P. A. Karam, *J.C.S. Perkin II*, **1973**, 1141.
[234] R. A. Wohl, *Tetrahedron Letters*, **1973**, 3111.
[235] R. A. Wohl, *J. Org. Chem.*, **38**, 3862 (1973).
[236] D. J. Cardin, B. Cetinkaya, M. J. Doyle and M. F. Lappert, *Chem. Soc. Rev.*, **2**, 99 (1973).
[237] P. S. Braterman and R. J. Cross, *Chem. Soc. Rev.*, **2**, 271 (1973).
[238] C. P. Casey and C. R. Cyr, *J. Am. Chem. Soc.*, **95**, 2240 (1973).
[239] C. P. Casey and C. R. Cyr, *J. Am. Chem. Soc.*, **95**, 2248 (1973).
[240] J. F. Nixon and B. Wilkins, *J. Organometal. Chem.*, **44**, C25 (1972).
[241] B. I. Cruikshank and N. R. Davies, *Austral. J. Chem.*, **26**, 1935 (1973).
[242] P. Golborn and F. Scheinmann, *J.C.S. Perkin I*, **1973**, 2870.
[243] J. M. Brown and K. Mertis, *J.C.S. Perkin II*, **1973**, 1993.
[244] A. J. Hubert, P. Moniotte, G. Goebbels, R. Warin and P. Teyssié, *J.C.S. Perkin II*, **1973**, 1954.
[245] H. Schlossarczyk, W. Sieber, M. Hesse, H.-J. Hansen and H. Schmid, *Helv. Chim. Acta*, **56**, 875 (1973).
[246] G. Parker and H. Werner, *Helv. Chim. Acta*, **56**, 2819 (1973).
[247] M. Green and G. J. Parker, *J.C.S. Dalton*, **1973**, 2099.
[248] P. M. Henry, *J. Org. Chem.*, **38**, 1140 (1973).
[249] R. A. Johnson and S. Seltzer, *J. Am. Chem. Soc.*, **95**, 5700 (1973).
[250] K. Maeda, T. Hosokawa, S.-I. Murahashi and I. Moritani, *Tetrahedron Letters*, **1973**, 5075.
[251] H. Takaya, N. Hayashi, T. Ishigami and R. Noyori, *Chem. Letters*, (*Tokyo*), **1973**, 813.
[252] R. Weiss and S. Andrae, *Angew. Chem. Internat. Ed.*, **12**, 150 (1973).
[253] R. Weiss and S. Andrae, *Angew. Chem. Internat. Ed.*, **12**, 152 (1973).
[254] J. J. Gajewski and C. N. Shih, *Tetrahedron Letters*, **1973**, 3959.
[255] S. Y.-K. Tam and B. Fraser-Reid, *Tetrahedron Letters*, **1973**, 4897.
[256] T. H. Whitesides and J. P. Neilen, *J. Am. Chem. Soc.*, **95**, 5811 (1973).
[257] U. Koch-Pomeranz, H.-J. Hansen and H. Schmid, *Helv. Chim. Acta*, **56**, 2981 (1973).
[258] F. A. Cotton and G. Deganello, *J. Am. Chem. Soc.*, **95**, 396 (1973).
[259] D. L. Schmitt and H. B. Jonassen, *J. Organometal. Chem.*, **49**, 469 (1973).
[260] E. J. Reardon, Jr., and M. Brookhart, *J. Am. Chem. Soc.*, **95**, 4311 (1973).
[261] R. Noyori, *Tetrahedron Letters*, **1973**, 1691.
[262] P. G. Gassman and R. R. Reitz, *J. Am. Chem. Soc.*, **95**, 3057 (1973).
[263] W. G. Dauben, A. J. Kielbania and K. N. Raymond, *J. Am. Chem. Soc.*, **95**, 7166 (1973).
[264] E. Müller, *Tetrahedron Letters*, **1973**, 1201.
[265] E. Müller, *Tetrahedron Letters*, **1973**, 1203.
[266] H. Takaya, M. Yamakawa and R. Noyori, *Chem. Letters* (*Tokyo*), **1973**, 781.
[267] R. M. Moriarty, C.-L. Yeh, E.-L. Yeh and K. C. Ramey, *J. Am. Chem. Soc.*, **94**, 9229 (1972).
[268] M. N. Akhtar, W. R. Jackson and J. J. Rooney, *J. Am. Chem. Soc.*, **96**, 276 (1974).
[269] A. Brieggink and H. Hogeveen, *Tetrahedron Letters*, **1972**, 4961.
[270] H. Hogeveen and B. J. Nusse, *Tetrahedron Letters*, **1973**, 3667.
[271] H. Hogeveen and T. B. Middelkoop, *Tetrahedron Letters*, **1973**, 3671.

[272] K. Reiff, U. Schumacher, G. Stuberrauch and W. Tochtermann, *Tetrahedron Letters*, **1973**, 1553.
[273] G. F. Koser, P. R. Pappas and S.-M. Yu, *Tetrahedron Letters*, **1973**, 4943.
[274] G. Zon and L. A. Paquette, *J. Am. Chem. Soc.*, **95**, 4456 (1973).
[276] R. S. Monson and B. M. Broline, *Can. J. Chem.*, **51**, 942 (1973).
[277] I. I. Kukhtenko, *Zh. Org. Khim.*, **7**, 333 (1971): *Chem. Abs.*, **78**, 57448 (1973).
[278] J. D. Butler and T. C. Poles, *J.C.S. Perkin II*, **1973**, 41.
[279] J. D. Butler and T. C. Poles, *J.C.S. Perkin II*, **1973**, 1262.
[280] M. Masaki, K. Fukui, M. Uchida, K. Yamamoto and I. Uchida, *Bull. Chem. Soc. Japan*, **46**, 3179 (1973).
[281] E. A. Abrazhanova and L. F. Pronskii, *Zh. Org. Khim*, **9**, 780 (1973).
[282] R. K. Hill and D. A. Cullison, *J. Am. Chem. Soc.*, **95**, 2923 (1973).
[283] G. Cum, P. D. Giannetto and N. Uccella, *J.C.S. Perkin II*, **1973**, 2038.
[284] K. Nagarajan and R. K. Shah, *Chem. Comm.*, **1973**, 926.
[285] A. J. Nunn and F. J. Rowell, *J.C.S. Perkin I*, **1973**, 2697.
[286] D. Misiti, V. Rimatori, and F. Gatta, *J. Heterocyclic Chem.*, **10**, 689 (1973).
[287] V. A. Ostrovskii, A. S. Enim and G. I. Koldobskii, *Zh. Org. Khim.*, **9**, 802 (1973).
[288] Y. Tamura, Y. Yoshimura, T. Nishimura, S. Kato and Y. Kita, *Tetrahedron Letters*, **1973**, 351.
[289] R. J. Sundberg and K. B. Sloan, *J. Org. Chem.*, **38**, 2052 (1973).
[290] A. C. Hopkinson, *J.C.S. Perkin II*, **1973**, 794.
[291] G. Heyes and G. Holt, *J.C.S. Perkin I*, **1973**, 1206.
[292] K.-P. Zeller, H. Meier and E. Müller, *Tetrahedron*, **28**, 5831 (1972).
[293] W. Lwowski and A. Milling, Jr., *J. Am. Chem. Soc.*, **95**, 609 (1973).
[294] B. V. Ioffe and L. A. Kartsova, *Tetrahedron Letters*, **1973**, 623.
[295] P. Gassman and G. Hartman, *J. Am. Chem. Soc.*, **95**, 449 (1973).
[296] H. H. Wasserman, E. A. Glazer and M. J. Hearn, *Tetrahedron Letters*, **1973**, 4855.
[297] R. D. Fisher, T. D. Bogard and P. Kovacic, *J. Am. Chem. Soc.*, **95**, 3646 (1973).
[298] R. E. White, M. B. Nazareno, M. R. Gleissner and P. Kovacic, *J. Org. Chem.*, **38**, 3902 (1973).
[299] T. Fuchigami, E. Ichikawa and K. Odo, *Bull. Chem. Soc. Japan*, **46**, 1765 (1973).
[300] N. Bodor and M. J. S. Dewar, *J. Am. Chem. Soc.*, **94**, 9103 (1972).
[301] N. Bodor, M. J. S. Dewar and Z. B. Maksic, *J. Am. Chem. Soc.*, **95**, 5245 (1973).
[302] D. Gust and K. Mislow, *J. Am. Chem. Soc.*, **95**, 1535 (1973).
[303] P. Hanson and D. A. R. Williams, *J.C.S. Perkin II*, **1973**, 2162.
[304] M. J. S. Dewar and W. B. Jennings, *J. Am. Chem. Soc.*, **95**, 1562 (1973).
[305] C. L. Stevens, F. E. Gleen and P. M. Pillai, *J. Am. Chem. Soc.*, **95**, 6301 (1973).
[306] E. Kiehlmann, F. Masaro and F. J. Slawson, *Can. J. Chem.* **51**, 3182 (1973).
[307] D. A. Jaeger, M. D. Broadhurst and D. J. Cram., *J. Am. Chem. Soc.*, **95**, 7525 (1973).
[308] D. S. Seigler and J. J. Bloomfield, *J. Org. Chem.*, **38**, 1375 (1973).
[309] A. G. González, M. A. Alvarez, J. Darias and J. D. Martin, *J.C.S. Perkin I*, **1973**, 2637.
[310] S. Yamamoto, *Bull. Soc. Chem. Japan*, **46**, 3139 (1973).
[311] O. A. Erastov, S. N. Ignat'eva and R. Kh. Saifullin, *Zh. Org. Khim*, **9**, 2016 (1973).
[312] T. Austad and J. Songstad, *Acta Chem. Scand.*, **26**, 3141 (1972).
[313] H. Kollmar, *J. Am. Chem. Soc.*, **95**, 966 (1973).
[314] J. C. Gilbert and D. P. Higley, *Tetrahedron Letters*, **1973**, 2075.
[315] W. von E. Doering and L. Birladeanu, *Tetrahedron*, **29**, 499 (1973).
[316] W. R. Dolbier, Jr., and J. H. Alonso, *J. Am. Chem. Soc.*, **95**, 4421 (1973).
[317] W. W. Schoeller, *Tetrahedron Letters*, **1973**, 2047.
[318] W. W. Schoeller, *Tetrahedron Letters*, **1973**, 2043.
[319] W. Pickenhagen, F. Näf, G. Ohloff, P. Müller and J.-C. Perlberger, *Helv. Chim. Acta*, **56**, 1868 (1973).
[320] J. M. Brown, B. T. Golding and J. J. Stofko, Jr., *Chem. Comm.*, **1973**, 319.
[321] G. Köbrich and B. Rösner, *Tetrahedron Letters*, **1973**, 2031.
[322] J. A. Pincock, R. Morchat and D. R. Arnold, *J. Am. Chem. Soc.*, **95**, 7536 (1973).
[323] R. D. Streeper and P. D. Gardner, *Tetrahedron Letters*, **1973**, 767.
[324] E. J. York, W. Dittmar, J. R. Stevenson and R. G. Bergman, *J. Am. Chem. Soc.*, **95**, 5680 (1973).
[324a] R. Curci, V. Lucchini, G. Modena, P. J. Kocienski and J. Ciabottani, *J. Org. Chem.*, **38**, 3149 (1973).
[325] T. Sasaki, S. Eguchi, M. Ohno and T. Umemura, *J. Org. Chem.*, **38**, 4095 (1973).
[326] P. R. Brook, J. M. Harrison and K. Hunt, *Chem. Comm.*, **1973**, 733.
[327] J. A. Berson and P. B. Dervan. *J. Am. Chem. Soc.*, **95**, 267 (1973).
[328] J. E. Baldwin and R. H. Fleming, *J. Am. Chem. Soc.*, **95**, 5249 (1973).
[329] J. E. Baldwin and R. H. Fleming, *J. Am. Chem. Soc.*, **95**, 5256 (1973).
[330] J. E. Baldwin and R. H. Fleming, *J. Am. Chem. Soc.*, **95**, 5261 (1973).

[331] T. Toda, *Yuki Crosei Kagaku Koyokai Shi*, **30**, 412 (1972); *Chem. Abs.*, **78**, 3350 (1973).
[332] V. A. Mironov, O. S. Chizhov, Ya. M. Kimel'fel'd and A. A. Akhrem, *Izv. Akad. Nauk. SSSR, Ser. Khim.*, **1972**, 2084; *Chem. Abs.*, **78**, 28984 (1973).
[333] L. A. Paquette, R. H. Meisinger and R. E. Wingard, Jr., *J. Am. Chem. Soc.*, **94**, 9224 (1972).
[334] H. Iwamura, *Tetrahedron Letters*, **1973**, 369.
[335] G. Pattenden and R. Storer, *Chem. Comm.*, **1973**, 875.
[336] M. F. Semmelhack and R. J. DeFranco, *J. Am. Chem. Soc.*, **94**, 8838 (1972).
[337] J. A. Hirsch and L. Y. Lin, *J.C.S. Perkin I*, **1973**, 1366.
[338] V. G. Kharchenko and A. A. Rassudova, *Zh. Org. Khim.*, **9**, 2177 (1973); *Chem. Abs.*, **80**, 36962 (1974).
[339] J. W. Pavlik and J. Kwong, *J. Am. Chem. Soc.*, **95**, 7914 (1973); J. W. Pavlik and E. L. Clennan, *ibid.*, p. 1697.
[340] J. E. Baldwin and G. D. Andrews, *J. Org. Chem.*, **38**, 1063 (1973).
[341] G. D. Andrews, M. Davalt and J. E. Baldwin, *J. Am. Chem. Soc.*, **95**, 5044 (1973).
[342] J. I. Braumann, W. E. Farneth and M. B. D'Amore, *J. Am. Chem. Soc.*, **95**, 5043 (1973).
[343] E. N. Cain and R. K. Solly, *J. Am. Chem. Soc.*, **95**, 4791 (1973).
[344] E. N. Cain and R. K. Solly, *J. Am. Chem. Soc.*, **95**, 7884 (1973).
[345] D. Hasselmann, *Tetrahedron Letters*, **1973**, 3739.
[346] A. G. Anastassiou and R. C. Griffith, *Tetrahedron Letters*, **1973**, 2379, 3067.
[347] W. Grimme and W. von E. Doering, *Chem. Ber.*, **106**, 1765 (1973).
[348] M. J. Goldstein and S. H. Dai, *J. Am. Chem. Soc.*, **95**, 933 (1973).
[349] J. Slutsky, E. M. Engler and P. von R. Schleyer, *Chem. Comm.*, **1973**, 685.
[350] A. de Meijere, D. Kaufmann and O. Schallner, *Tetrahedron Letters*, **1973**, 553.
[351] H. Iwamura and H. Kihara, *Chem. Letters* (*Tokyo*), **1973**, 71.
[352] H. Iwamura, H. Kihara, K. Morio and T. L. Kumi, *Bull. Chem. Soc. Japan*, **46**, 3248 (1973).
[353] H. Tsuruta, T. Kumugai and T. Mukai, *Chem. Letters* (*Tokyo*), **1972**, 981.
[354] I. Murata and K. Nakasuji, *Tetrahedron Letters*, **1973**, 47.
[355] K. Tomisawa and T. Mukai, *J. Am. Chem. Soc.*, **95**, 5405 (1973).
[356] R. B. Kinnel and P. K. Freeman, *Tetrahedron Letters*. **1973**, 4803.
[357] S. R. Tanny and F. W. Fowler, *J. Am. Chem. Soc.*, **95**, 7320 (1973).
[358] R. W. Thies and D. D. McRitchie, *J. Org. Chem.*, **38**, 112 (1973).
[359] E. G. E. Hawkins, *J.C.S. Perkin I*, **1973**, 2155.
[360] N. D. Epiotis, D. Bjorkquist and S. Sarkanen, *J. Am. Chem. Soc.*, **95**, 7558 (1973).
[361] J. Hine and N. W. Flackskam, *J. Am. Chem. Soc.*, **95**, 1179 (1973).
[362] J. L. Liard, B. Jasse and R. Poisson, *Bull. Soc. Chim. France*, **1973**, 3000.
[363] J. L. Liard, B. Jasse and R. Poisson, *Bull. Soc. Chim. France*, **1973**, 3006.
[364] C. Th. Pedersen and C. Lohse, *J.C.S. Perkin I*, **1973**, 2837.
[365] Z. Paal and P. Tetenyi, *J. Catal.*, **29**, 176 (1973); *Chem. Abs.* **78**, 158690 (1973).
[366] W. D. Huntsman, *Intra-Sci. Chem. Rep.*, **6**, 135 (1972); *Chem. Abs.*, **79**, 65281 (1973).
[367] J. M. Figuera, J. M. Gramboa and J. Santos, *An. Quím.*, **68**, 1201 (1972) *Chem. Abs.*, **78**, 83539 (1973).
[368] N. Tani, M. Makoto and Y. Yoneda, *Chem. Letters* (*Tokyo*), **1973**, 591; *Chem. Abs.*, **79**, 52609 (1973).
[369] A. S. Rodgers and M. C. R. Wu, *J. Am. Chem. Soc.*, **95**, 6913 (1973).
[370] P. E. Rakita and G. A. Taylor, *J. Organometal. Chem.*, **61**, 71 (1973).
[371] L. M. Stephenson, R. V. Gemmer and J. L. Brauman, *J. Am. Chem. Soc.*, **94**, 8620 (1972).
[372] S. W. Orchard and B. A. Thrush, *Chem. Comm.*, **1973**, 14.
[373] J. M. Gamboa, C. Saá and J. M. Figuera, *J.C.S. Perkin II*, **1973**, 2025.
[374] M. Pomerantz and T. H. Witherup, *J. Am. Chem. Soc.*, **95**, 5977 (1973).
[375] J. Barbier, C. Berrier, J. C. Jacquesy and R. Jacquesy, *Tetrahedron*, **29**, 1047 (1973).
[376] E. Mincione and F. Feliziani, *Chem. Comm.*, **1973**, 942.
[377] H. Kessler, P. F. Bley and D. Leibfritz, *Tetrahedron*, **27**, 1687 (1971); H. Kessler, *Angew. Chem. Internat. Ed.*, **9**, 219 (1970).
[378] M. Raban and E. Carlson, *J. Am. Chem. Soc.*, **93**, 685 (1971).
[379] J. M. Lehn, *Fortsch. Chem. Forsch.*, **15**, 311 (1970).
[380] W. G. Herkstroeter, *J. Am. Chem. Soc.*, **95**, 8686 (1973).
[381] A. F. Hegarty, P. J. Moroney and F. L. Scott, *J.C.S. Perkin II*, **1973**, 1466.
[382] B. F. Bonini, L. Lunazzi, G. Maccagnani and G. Mazzanti, *J.C.S. Perkin I*, **1973**, 2314.
[383] C. K. Sauers and H. M. Relles, *J. Am. Chem. Soc.*, **95**, 7731 (1973).
[384] R. Damrauer and T. E. Rutledge, *J. Org. Chem.*, **31**, 3330 (1973).

[385] T. S. Dobashi, M. H. Goodrow and E. J. Grubbs, *J. Org. Chem.*, **38**, 4440 (1973).
[386] E. J. Grubbs, D. R. Parker and W. D. Jones, *Tetrahedron Letters*, **1973**, 3279.
[387] D. Y. Curtin, E. J. Grubbs and C. G. McCarty, *J. Am. Chem. Soc.*, **88**, 2775 (1966); G. E. Hall, W. J. Middleton and J. D. Roberts, *ibid.*, **93**, 4778 (1971).
[388] T. S. Dobashi and E. J. Grubbs, *J. Am. Chem. Soc.*, **95**, 5070 (1973).
[389] A. B. Tomchin, I. S. Ioffe, Yu. V. Lepp and A. I. Kol'tsov, *Zh. Org. Khim.*, **9**, 1081 (1973).
[390] I. N. Somin and V. A. Gindin, *Zh. Org. Khim.*, **9**, 1993 (1973).
[391] J. Bjørgo, D. R. Boyd, C. G. Watson and W. B. Jennings, *Tetrahedron Letters*, **1972**, 1747.
[392] E. P. Kyba, *Tetrahedron Letters*, **1973**, 5117.
[393] B. C. Sharma, K. S. Bose and C. C. Patel, *Indian J. Chem.*, **11**, 1038 (1973).
[394] V. A. Ketlinskii and I. L. Bagal, *Zh. Org. Khim.*, **9**, 1915 (1973).
[395] J. E. Johnson, E. A. Nalley and C. Weidig, *J. Am. Chem. Soc.*, **95**, 2051 (1973).
[396] V. I. Minkin, L. P. Olekhnovich, Yu. A. Zhdanov, V. V. Kiselev, M. A. Voronov and Z. N. Budarina, *Zh. Org. Khim.*, **9**, 1319 (1973).
[397] A. Mannschreck and H. Dvorak, *Tetrahedron Letters*, **1973**, 547.
[398] A. Niederhauser, G. Bart, and N. Neuenschwander, *Helv. Chim. Acta*, **56**, 2427 (1973); see also A. Niederhauser and M. Neuenschwander, *ibid.*, pp. 1318, 1331.
[399] M. Neuenschwander and A. Niederhauser, *Chimia*, **27**, 379 (1973).
[400] J. Meijer, P. Vermeer, H. J. T. Bos and L. Brandsma, *Rec. Trav. Chim.*, **92**, 1067 (1973).
[401] K. Harano and T. Taguchi, *Chem. Pharm. Bull.* (*Japan*), **20**, 2348 (1972).
[402] Y. Kinoshita, M. Misaka, S. Kubota and H. Ishikawa, *Agr. Biol. Chem.*, **36**, 1975 (1972); *Chem. Abs.*, **78**, 57453 (1973).
[403] C. Brown, R. F. Hudson and A. J. Lawson, *J. Am. Chem. Soc.*, **95**, 6500 (1973).
[404] G. S. Bethell and R. J. Ferrier, *J.C.S. Perkin I*, **1973**, 1400.
[405] F. Wudl and T. B. K. Lee, *J. Am. Chem. Soc.*, **95**, 6349 (1973).
[406] P. Catsoulacos and A. Hassner, *Bull. Soc. Chim. France*, **1973**, 717.
[407] M. Koga and J.-P. Anselme, *Chem. Comm.*, **1973**, 53.
[408] Zh. A. Krasnaya, E. P. Prokov'ev, Sh. M. Zaripava and V. F. Kucherov, *Isv. Akad. Nauk SSSR, Ser. Khim.*, **1973**, 2356.
[409] A. Grouiller, P. Thomassery and H. Pacheco, *Bull. Soc. Chim. France*, **1973**, 3448.
[410] C. Vandesande and M. Vandewalle, *Bull. Soc. Chim. Belges.*, **82**, 775 (1973).
[411] H. Ferres, M. S. Hamdam and W. R. Jackson, *J.C.S. Perkin II*, **1973**, 936.
[412] J. Elguero and C. Marzin, *Bull. Soc. Chim. France*, **1973**, 3401.
[413] H. des Abbayes, C. Neveu and F. Salmon-Legagneur, *Bull. Soc. Chim. France*, **1973**, 2686.
[414] D. C. Lankin and H. Zimmer, *J. Heterocyclic Chem.*, **10**, 1035 (1973).
[415] C. N. Robinson and W. A. Pettit, *Tetrahedron Letters*, **1972**, 4977.
[416] M. Tada and T. Takahashi, *Tetrahedron Letters*, **1973**, 3999.
[417] J. R. Grunwell and A. Kochan, *J. Org. Chem.*, **38**, 1610 (1973).
[418] R. Galland and A. Heesing, *Chem. Ber.*, **106**, 2580 (7973).
[419] R. Burgada, H. Germa and M. Willson, *Tetrahedron*, **29**, 727 (1973).
[420] M. J. S. Dewar, P. A. Spanninger and I. J. Turchi, *Chem. Comm.*, **1973**, 925.
[421] C. J. M. Stirling, *J. Chem. Educ.*, **50**, 844 (1973).
[422] M. I. Page, *Chem. Soc. Rev.*, **2**, 295 (1973).
[423] R. Bird, A. C. Knipe and C. J. M. Stirling, *J.C.S. Perkin II*, **1973**, 1215.
[424] R. Bird and C. J. M. Stirling, *J.C.S. Perkin II*, **1973**, 1221.
[425] J. M. Watson, J. L. Irvine and R. M. Roberts, *J. Am. Chem. Soc.*, **95**, 3348 (1973).
[426] A. Padwa, D. Dean, A. Mazzu and E. Vega, *J. Am. Chem. Soc.*, **95**, 7168 (1973).
[427] R. J. Crawford, V. Vukov and H. Tokunaga, *Can. J. Chem.*, **51**, 3718 (1973).
[428] D. J. Anderson, T. L. Gilchrist, G. E. Gymer and C. W. Rees, *J.C.S. Perkin I*, **1973**, 550.
[429] J. H. Bowie and B. Nussey, *J.C.S. Perkin I*, **1973**, 1693.
[430] M. T. H. Liu and K. Toriyama, *Can. J. Chem.*, **51**, 2393 (1973).
[431] P. G. Khazinie and E. Lee-Ruff, *Can. J. Chem.*, **51**, 3173 (1973).
[432] P. Bennett, J. A. Donnelly, D. C. Meaney and P. O'Boyle, *J.C.S. Perkin I*, **1972**, 2982.
[433] J. M. Hornback, *J. Org. Chem.*, **38**, 4122 (1973).
[434] J. C. Ferrero, J. T. Cosa and E. H. Staricco, *J.C.S. Perkin II*, **1972**, 2382.
[435] A. Liberles, A. Greenberg and A. Lesk, *J. Am. Chem. Soc.*, **94**, 8685 (1972).
[436] H. M. Walborsky and P. E. Ronman, *J. Org. Chem.*, **38**, 4213 (1973).
[437] D. M. A. Armitage and C. L. Wilson, *J. Am. Chem. Soc.*, **81**, 2437 (1959).
[438] A. T. Cocks and K. W. Egger, *J.C.S. Perkin II*, **1973**, 197.

[439] A. T. Cocks and K. W. Egger, *J.C.S. Perkin II*, **1973**, 199.
[440] D. E. McGreer and J. W. McKinley, *Can. J. Chem.*, **51**, 1487 (1973).
[441] K. P. C. Vollhardt and R. G. Bergman, *J. Am. Chem. Soc.*, **95**, 7538 (1973).
[442] D. Belluš, K. von Bredow, H. Sauter and C. D. Weis, *Helv. Chim. Acta*, **56**, 3004 (1973).
[443] P. G. Bird and W. J. Irwin, *J.C.S. Perkin I*, **1973**, 2664.
[444] J. P. Barnier, J. M. Denis, J. R. Salaün and J. M. Conia, *Chem. Comm.*, **1973**, 103.
[445] W. T. Brady and A. D. Patel, *J. Org. Chem.*, **38**, 4106 (1973).
[446] R. H. Higgins and N. H. Cromwell, *J. Am. Chem. Soc.*, **95**, 120 (1973).
[447] L. W. Haynes and R. A. Amos, *J. Heterocyclic Chem.*, **10**, 875 (1973).
[448] A. O. Fitton, J. R. Frost, M. M. Zakaria and G. Andrew, *Chem. Comm.*, **1973**, 889.
[449] F. Yoneda, M. Higuchi and R. Nonaka, *Tetradehron Letters*, **1973**, 359.
[450] F. Yoneda and M. Higuchi, *Bull. Chem. Soc. Japan*, **46**, 3849 (1973).
[451] H. M. Berman, R. J. Rousseau, R. W. Mancuso, G. P. Kreishman and R. K. Robins, *Tetrahedron Letters*, **1973**, 3099.
[452] W. Heinzelmann and M. Märky, *Helv. Chim. Acta*, **56**, 1852 (1973).
[453] A. Takeda, S. Tsuboi, F. Sakai and M. Tanabe, *Tetrahedron Letters*, **1973**, 4961; A. Takeda and S. Tsuboi, *J. Org. Chem.*, **38**, 1709 (1973).
[454] T. L. Gilchrist, G. E. Gymer and C. W. Rees, *J.C.S. Perkin I*, **1973**, 555.
[455] W. M. Jones, R. C. Joines, J. A. Myers, T. Mitsuhashi, K. E. Krajca, E. E. Waali, T. L. Davis and A. B. Turner, *J. Am. Chem. Soc.*, **95**, 826 (1973).
[456] A. Maercker and R. Geuss, *Chem. Ber.*, **105**, 773 (1973).
[457] E. A. Hill and G. E.-M. Shih, *J. Am. Chem. Soc.*, **95**, 7764 (1973).
[458] S. Wolff and W. C. Agosta, *Chem. Comm.*, **1973**, 771.
[459] C. W. Rees and A. A. Sale, *J.C.S. Perkin I*, **1973**, 545.
[460] D. G. Neilson, S. Mahmood and K. M. Watson, *J.C.S. Perkin I*, **1973**, 335.
[461] E. M. Burgess and W. M. Williams, *J. Org. Chem.*, **38**, 1249 (1973).
[462] C. Galli, G. Illuminati and L. Mandolini, *J. Am. Chem. Soc.*, **95**, 8374 (1973).
[463] A. T. Cocks and K. W. Egger, *J.C.S. Perkin II*, **1973**, 835.
[464] T. Fujisawa and T. Kobori, *Chem. Comm.*, **1972**, 1298
[465] R. D. Miller and D. L. Dolce, *Tetrahedron Letters*, **1973**, 1151.
[466] R. D. Miller and D. L. Dolce, *Tetrahedron Letters*, **1973**, 5217.
[467] C. P. R. Jennison and D. Mackay, *Tetrahedron*, **29**, 1255 (1973).
[468] J. Casanova, G. Koukoua and B. Waegell, *Compt. Rend., C*, **275**, 507 (1972).
[469] C. L. Pedersen and O. Buchardt, *Acta Chem. Scand.*, **27**, 271 (1973).
[470] T. S. Stevens and W. E. Watts, *Selected Molecular Rearrangements*, Van Nostrand Reinhold, New York, 1973.
[471] Yu. G. Erykalov, V. G. Chirtulov and A. A. Spryskov, *Zh. Org. Khim.* **8**, 2089 (1973).
[472] Yu. G. Erykalov, A. P. Belukurova, I. S. Isaev and V. A. Koptyug, *Zh. Org. Khim.*, **9**, 343 (1973).
[473] V. M. Akhmedov, F. R. Alieva and M. A. Mardanov, *Zh. Org. Khim.*, **9**, 1653 (1973).
[474] D. É. Bodrina and A. P. Rudenko, *Zh. Org. Khim.*, **9**, 1465 (1973).
[475] A. N. Korepanov, T. A. Danilova and E. A. Viktorova, *Zh. Org. Khim.*, **9**, 641 (1973).
[476] J. T. Edward and L. Y.-S. Mo, *J. Heterocyclic Chem.*, **10**, 1047 (1973).
[477] T. Shiojima, Y. Hashida and K. Matsui, *Bull. Chem. Soc. Japan*, **46**, 3147 (1973).
[478] Kh. L. Muravich-Aleksandr, A. V. El'tsov and I. Él'-Sakka, *Zh. Org. Khim.*, **9**, 1288 (1973).
[479] R. S. Ward, *Chem. in Britain*, **9**, 444 (1973).
[480] F. E. Ziegler and G. B. Bennett, *J. Am. Chem. Soc.*, **95**, 7458 (1973).
[481] J. Borgulya, R. Madeja, P. Fahrni, H.-J. Hansen, H. Schmid and R. Barner, *Helv. Chim. Acta*, **56**, 14 (1973).
[482] D. Mowat and R. D. H. Murray, *Tetrahedron*, **29**, 2943 (1973).
[483] R. W. Thies, M. T. Wills, A. W. Chin, L. E. Schick and E. S. Walton, *J. Am. Chem. Soc.*, **95**, 5281 (1973).
[484] L. N. Mander and J. V. Turner, *J. Org. Chem.*, **38**, 2915 (1973).
[485] R. Ya. Popova, T. V. Protopopava and A. P. Skoldinov, *Zh. Org. Khim.*, **9**, 882 (1973).
[486] W. Ando, *Internat. J. Sulphur Chem., B*, **7**, 189 (1972).
[487] E. E. Weltin, *J. Am. Chem. Soc.*, **95**, 7650 (1973).
[488] E. N. Marvell, G. Caple, B. Schatz and W. Pippin, *Tetrahedron*, **29**, 3781 (1973).
[489] E. N. Marvell, G. Caple, C. Delphey, J. Platt, N. Polston and J. Tashiro, *Tetrahedron*, **29**, 3797 (1973); E. N. Marvell, *ibid.*, p. 3791.
[490] L. A. Paquette, R. H. Meisinger and R. E. Wingard, Jr., *J. Am. Chem. Soc.*, **95**, 2230 (1973).

491 M. L. Petrov, S. I. Radchenko, V. S. Kupin and A. A. Petrov., *Zh. Org. Khim.*, **9**, 663 (1973).

492 A. S. Atavin, A. I. Mikhaleva, V. A. Pestunovich, E. G. Chebotareva, V. I. Kaigorodova and B. A. Trofimov, *Izv. Akad. Nauk SSSR, Ser. Khim.*, **1973**, 2334; *Chem. Abs.*, **80**, 36660 (1974).

493 F. M. Stoyanovich, G. B. Chermanova and Ya. L. Gol'dfarb, *Izv. Akad. Nauk SSSR, Ser. Khim.*, **1973**, 2367; *Chem. Abs.*, **80**, 47590 (1974).

494 V. N. Drozd, O. I. Trifonova and V. V. Sergeichuk, *Zh. Org. Khim.*, **9**, 156 (1973).

495 A. S. Bloss, P. R. Brook and R. M. Ellam, *J.C.S. Perkin II*, **1973**, 2165.

496 H. Hogeveen and P. W. Kwant, *J. Am. Chem. Soc.*, **95**, 7315 (1973).

497 P. W. Henrichs and P. E. Peterson, *J. Am. Chem. Soc.*, **95**, 7449 (1973).

498 N. N. Povolotskaya, B. G. Derendyaev and V. A. Barkhash, *Zh. Org. Khim.*, **9**, 1878 (1973).

499 I. P. Lobanova, B. G. Derendyaev, M. I. Kollegova and V. A. Barkhash, *Zh. Org. Khim.*, **9**., 1883 (1973).

500 T. P. Lobanova, N. M. Slyn'ko, B. G. Derendyaev and V. A. Barkhash, *Zh. Org. Khim.*, **9**, 1893 (1973).

501 N. M. Slyn'ko, B. G. Derendyaev, M. I. Kollegova and V. A. Barkhash, *Zh. Org. Khim.*, **9**, 1901 (1973).

502 P. Four, *Bull. Soc. Chim. France*, **1973**, 3344.

503 F. Collonges and G. Descotes, *Bull. Soc. Chim. France*, **1973**, 3491.

504 F. N. Stepanov, L. I. Sidorova and N. L. Dovgani, *Zh. Org. Khim.*, **9**, 1545 (1973).

505 N. E. Belikova, M. Ordubadi, A. F. Platé, T. I. Pernk and É. T. Lippmaa, *Zh. Org. Khim.*, **9**, 1855 (1973).

506 P. A. Bartlett and W. S. Johnson, *J. Am. Chem. Soc.*, **95**, 7501 (1973).

507 P. A. Bartlett, J. I. Braumann, W. S. Johnson and R. A. Volkmann, *J. Am. Chem. Soc.*, **95**, 7502 (1973).

508 T. Abdin, T. A. Danilova and Ye. A. Viktorova, *Khim. Geterotsikl. Soedin.*, **1973**, 1337; *Chem. Abs.*, **80**, 47788 (1974).

509 C. Moreau and F. Rouessac, *Bull. Soc. Chim. France*, **1973**, 3427.

510 C. Moreau and F. Rouessac, *Bull. Soc. Chim. France*, **1973**, 3433.

511 V. R. Kartashov, E. V. Skorobogatova and I. V. Bodrikov, *Zh. Org. Khim.*, **9**, 214 (1973).

512 V. R. Kartashov, E. V. Skorobogatova and I. V. Bodrikov, *Zh. Org. Khim.* **9**, 535 (1973); *Chem. Abs.*, **78**, 147121 (1973).

513 J. Meinwald and A. J. Taggi, *J. Am. Chem. Soc.*, **95**, 7663 (1973).

514 A. Ambles, J.-C. Jacquesy and R. Jacquesy, *Bull. Soc. Chim. France*, **1973**, 2865.

515 L. Radom, J. A. Pople and P. von R. Schleyer, *J. Am. Chem. Soc.*, **95**, 8193 (1973).

516 I. Moiseev, V. P. Konovalova and S. S. Novikov, *Izv. Akad. Nauk SSSR, Ser Khim.*, **1973**, 2378; *Chem. Abs.*, **80**, 26819 (1974).

517 T. Greibrokk, *Acta Chem. Scand.*, **27**, 2252 (1973).

518 V. G. Shubin, D. V. Korchagina, G. I. Borodkin, B. G. Derendyaev and V. A. Koptyug, *Zh. Or. Khim.*, **9**, 1031 (1973).

519 V. G. Shubin, D. V. Korchagina, B. G. Derendyaev, G. I. Borodkin and V. A. Koptyug, *Zh. Org. Khim.*, **9**, 1041 (1973).

520 R. N. Berezina, E. P. Yablokova and V. G. Shubin, *Izv. Akad. Nauk SSSR, Ser. Khim.*, **1973**, 2273; *Chem. Abs.*, **80**, 37257 (1974).

521 B. M. Trost and T. J. Dietsch, *J. Am. Chem. Soc.*, **95**, 8200 (1973).

522 N. D. Agibalova, V. A. Ostrovskii, G. I. Koldobskii and A. S. Enin, *Zh. Org. Khim.*, **9**, 1580 (1973).

523 H. Kawamoto, T. Matsuo, S. Morosawa and A. Yokoo, *Bull. Chem. Soc. Japan*, **46**, 3898 (1973).

524 N. V. Yablokova, V. A. Yablokov and V. K. Mamushkin, *Zh. Org. Khim.*, **9**, 82 (1973).

525 V. A. Yablokov and S. A. Petrova, *Zh. Org. Khim.*, **9**, 211 (1973).

526 A. Gasco and A. J. Boulton, *J.C.S. Perkin II*, **1973**, 1613; J. Ackrell and A. J. Boulton, *J.C.S. Perkin I*, **1973**, 351; A. J. Boulton and S. S. Mathur, *J. Org. Chem.*, **38**, 1054 (1973).

527 J. Kovář and H. H. Baer, *Can J. Chem.*, **51**, 3373 (1973).

528 Yu. A. Bruk, *Zh. Org. Khim.*, **9**, 215 (1973).

529 A. De Savignac, M. Bon and A. Lattes, *Compt. Rend., C.*, **277**, 1367 (1973); *Chem. Abs.*, **80**, 70082 (1974).

530 S. M. Ramsh, K. A. V'yunov, A. I. Ginak and E. G. Sochilin, *Zh. Org. Khim.*, **9**, 412 (1973).

531 K. A. V'yunov, A. I. Ginak and E. G. Sochilin, *Zh. Org. Khim.* **9**, 817 (1973).

532 L. N. Kurkovskaya, N. N. Shapet'ko, I. Ya. Kvitko, Yu. N. Koshelev and E. M. Sof'ina, *Zh. Org. Khim.*, **9**, 821 (1973).

533 S. I. Yakimovich, V. A. Khrustalev and T. A. Favorskaya, *Zh. Org. Khim.*, **9**, 1382 (1973).

534 M. S. Korobov, L. E. Nivorozhkin and V. I. Minkin, *Zh. Org. Khim.*, **9**, 1717 (1973).

535 L. P. Olekhnovich, A. É. Lyubarskaya, M. I. Knyazhanskiv and V. I. Mivkin, *Zh. Org. Khim.*, **9**, 1724 (1973).
536 I. H. Sadlier and J. A. G. Stewart, *J.C.S. Perkin II*, **1973**, 278.
537 W. E. Billups, K. H. Leavell, E. S. Lewis and S. Vanderpool, *J. Am. Chem. Soc.*, **95**, 8096 (1973).
538 P. H. Mazzochi and R. S. Lustig, *J. Org. Chem.*, **38**, 4091 (1973).
539 M. Newcomb and W. T. Ford, *J. Am. Chem. Soc.*, **95**, 7186 (1973).
540 D. H. Aue, M. J. Mashishnek and D. F. Shelhamer, *Tetrahedron Letters*, **1973**, 4799.
541 G. Nagendrappa and K. Griesbaum, *Chem. Ind.* (*London*), **1973**, 902.
542 B. D.-Le Reverend and M. Durand, *Compt. Rend., C.*, **277**, 1247 (1973); *Chem. Abs.*, **80**, 70081 (1974).
543 E. E. Waali and W. M. Jones, *J. Am. Chem. Soc.*, **95**, 8114 (1973).
544 L. S. Balenkova and I. A. Karamysheva, *Zh. Org. Khim.*, **9**, 1667 (1973).
545 A. V. Tarakanova, E. M. Mil'vitskaya and A. F. Platé, *Zh. Org. Khim.*, **9**, 635 (1973).
546 H. Sauter, H.-G. Hörster and H. Prinzbach, *Angew. Chem. Internat. Ed.*, **12**, 991 (1973).
547 A. Sinnema, F. van Rantwijk, A. J. de Koning, A. M. van Wijk and H. van Bekkum, *Chem. Comm.*, **1973**, 364.
548 Ya. M. Slobodin and I. Z. Égenburg, *Zh. Org. Khim.*, **9**, 1791 (1973).
549 V. Sh. Fel'dblyum, T. I. Baranova, T. A. Tsailingol'd, N. V. Petrushanskaya, E. D. Tkachenko and G. A. Maksimenko, *Zh. Org. Khim.*, **9**, 874 (1973).
550 N. I. Zakharova, M. A. Miropol'skaya, T. M. Filippova, I. M. Kustanovich and G. I. Samokhvalvov, *Zh. Org. Khim.*, **9**, 512 (1973).
551 L. Leseticky, V. Fidler and M. Prochazka, *Coll. Czech. Chem. Comm.*, **38**, 459 (1973); *Chem. Abs.*, **78**, 147124 (1973).
552 M. Kocevar, B. Stanovnik and M. Tisler, *Croat. Chem. Acta*, **45**, 457 (1973); *Chem. Abs.*, **80**, 37043 (1974).
553 I. P. Stepanova, N. M. Loim, Z. N. Parnes and Yu. S. Shabarov, *Zh. Org. Khim.*, **9**, 521 (1973).
554 A. I. Konovalov, L. K. Konovalova and E. G. Kataev, *Zh. Org. Khim.*, **9**, 1830 (1973).
555 Yu. S. Shabarov, I. P. Stepanova, E.-L. Protasova and O. A. Subbotin, *Zh. Org. Khim.*, **9**, 960 (1973).
556 I. I. Bardyshev, É. F. Buinova and B. G. Udarov, *Zh. Org. Khim.*, **9**, 1670 (1973).
557 K. K. Babievskii, V. M. Belikov, A. I. Vinogradova and V. K. Latov, *Zh. Org. Khim.*, **9**, 1700 (1973).
558 A. B. Tomchin, I. S. Ioffe, V. V. Tret'yakova, Yu. V. Lepp and A. I. Kol'tsov, *Zh. Org. Khim.*, **9**, 1537 (1973).
559 V. V. Korshak, S. V. Vinogradova, S. N. Salazkin and A. A. Kul'kov, *Zh. Org. Khim.*, **9**, 640 (1973).
560 A. A. Potekhin and T. F. Barkova, *Zh. Org. Khim.*, **9**, 1180 (1973).
561 F. Mercier and R. Epsztein, *Bull. Soc. Chim. France*, **1973**, 3393.
562 A. D. Nikolaeva, V. S. Perekhod'ko and N. G. D'yachenko, *Zh. Org. Khim.*, **9**, 1624 (1973).
563 K. A. Ogloblin and D. M. Kunovskaya, *Zh. Org. Khim.*, **9**, 1547 (1973).
564 A. T. Babayan, K. Ts. Tagmazyan and G. O. Torosyan, *Zh. Org. Khim.*, **9**, 1156 (1973).
565 S. Milosavbjević, D. Jeremić and M. Lj. Mihailovic, *Tetrahedron*, **29**, 3547 (1973).
566 H. C. Lacey and K. L. Erickson, *Tetrahedron*, **29**, 4025 (1973).
567 M. Farcasiu, D. Farcasiu, M. Jones, Jr., and P. von R. Schleyer, *J. Am. Chem. Soc.*, **95**, 8207 (1973).
568 S. D. Isaev, A. G. Yurchenko, F. N. Stepanov, G. G. Kolyada, S. S. Novikov and N. F. Karpenko, *Zh. Org. Khim.*, **9**, 724 (1973).
569 A. N. Tvorogov, L. V. Goncharenko, I. Yu. Belavin, Yu. I. Baukor and I. F. Lutsenko, *Zh. Obshch. Khim.*, **43**, 441 (1973); *Chem. Abs.*, **78**, 14712 (1973).
570 R. G. Kostyanovskii, A. I. Ermakov, Kh. Khafizov and G. K. Kadorkina, *Izvst. Akad. Nauk SSSR, Ser. Khim.*, **1973**, 2646; *Chem. Abs.*, **80**, 70087 (1974).
571 B. F. Pishnamazzade and A. Kh. Mamishov, *Zh. Org. Khim.*, **9**, 1365 (1973).
572 V. N. Yandovskii and T. I. Temnikova, *Zh. Org. Khim.*, **9**, 1376 (1973).
573 S. Gronowitz and T. Frejd, *Acta Chem. Scand.*, **27**, 2242 (1973).
574 D. L. Rakhmankulov, D. A. Kaushanskii, S. S. Zlotskii, V. P. Nayanov and V. I. Isagulyants, *Zh. Org Khim.*, **9**, 631 (1973).
575 Yu. M. Portnyagin and T. M. Pavel', *Zh. Org. Khim.*, **9**, 890 (1973).
576 N. N. Povolotskaya, A. Yu. Spivak, G. D. Slyn'ko, M. I. Kollegova, B. G. Derendyaev, A. K. Petrov and V. A. Barkhash, *Zh. Org. Khim.*, **9**, 1869 (1973).
577 M. I. Komendantov, I. N. Domnin, R. M. Kenbaeva and T. N. Grigorova, *Zh. Org. Khim.*, **9**, 1420 (1973).
578 D. I. Greichute, Y. Y. Kulis and L. P. Rasteikene, *Zh. Org. Khim.*, **9**, 1837 (1973).
579a E. Koltai and K. Lempert, *Tetrahedron*, **29**, 2795 (1973).

579b E. Koltai, J. Nyitrai, K. Lempert, Gy. Horváth, A. Kalmán and Gy. Argay, *Tetrahedron*, **29**, 2783 (1973).
580 N. S. Zefirov, G. N. Dorofeenko and T. M. Pozdnykova, *Zh. Org. Khim.*, **9**, 387 (1973).
581 B. A. Ivin, G. V. Rutkovskii and E. G. Sochilin, *Zh. Org. Khim.*, **9**, 179 (1973).
582 B. A. Ivin, G. V. Rutkovskii, S. A. Andreev and E. G. Sochilin, *Zh. Org. Khim.*, **9**, 420 (1973).
583 D. N. Reinhoudt and C. G. Kouwenhoven, *Rec. Trav. Chim.*, **92**, 865 (1973); *Chem. Abs.*, **79**, 105058 (1973).
584 G. Biggi, A. J. DeHoog, F. Del Cima and F. Pietra, *J. Am. Chem. Soc.*, **95**, 7108 (1973).
585 G. Biggi, F. Del Cima and F. Pietra, *J. Am. Chem. Soc.*, **95**, 7101 (1973).
586 R. Tanikaga and A. Kaji, *Bull. Chem. Soc. Japan*, **46**, 3814 (1973).
587 R. P. Gandhi, M. Singh, Y. P. Sachdeva and S. M. Mukherji, *Tetrahedron Letters*, **1973**, 661.
588 H. E. Zimmerman and W. Eberbach, *J. Am. Chem. Soc.*, **95**, 3970 (1973).
589 T. Matsuura, T. Takemeto and R. Nakashima, *Tetrahedron*, **29**, 3337 (1973).
590 D. S. C. Black and K. G. Watson, *Austral. J. Chem.*, **26**, 2505 (1973).
591 R. A. Abramovitch and T. Takaya, *J. Org. Chem.*, **38**, 3311 (1973).
592 T. Sasaki, K. Kanematsu, K. Hayakawa and A. Kondo, *J. Org. Chem.*, **38**, 4100 (1973).
593 H. E. Zimmerman, D. W. Kurtz and L. M. Tolbert, *J. Am. Chem. Soc.*, **95**, 8210 (1973).
594 S. S. Hixson, P. S. Mariano and H. E. Zimmerman, *Chem. Rev.*, **73**, 531 (1973).
595 P. S. Mariano and J. Ko, *J. Am. Chem. Soc.*, **95**, 8670 (1973).
596 H. Hart and G. M. Love, *J. Am. Chem. Soc.*, **95**, 4592 (1973).
597 H. Dürr, H. Kober, I. Halberstadt, U. Neu, T. T. Coburn, T. Mitsuhashi and W. M. Jones, *J. Am. Chem. Soc.*, **95**, 3818 (1973).
598 D. C. Lankin, D. M. Chihal, G. W. Griffin and N. S. Bhacca, *Tetrahedron Letters*, **1973**, 4009.
599 J. I. Seeman and H. Ziffer, *Tetrahedron Letters*, **1973**, 4413.
600 K. W. Watkins, *J. Phys. Chem.*, **77**, 2938 (1973).
601 H. E. Zimmerman and D. R. Amick, *J. Am. Chem. Soc.* **95**, 3977 (1973).
602 A. R. Brember, A. A. Gorman and J. B. Sheridan, *Tetrahedron Letters*, **1973**, 475.
603 A. R. Brember, A. A. Gorman and J. B. Sheridan, *Tetrahedron Letters*, **1973**, 481.
604 G. D. DeBoer, D. H. Wadsworth and W. C. Perkins, *J. Am. Chem. Soc.*, **95**, 861 (1973).
605 S. Majeti and T. W. Gibson, *Tetrahedron Letters*, **1973**, 4889.
606 I. R. Polizer and G. W. Griffin, *Tetrahedron Letters*, **1973**, 4775.
607 M. I. Komendantov and I. N. Domnin, *Zh. Org. Khim.*, **9**, 939 (1973).
608 M. Regitz, H. Scherer, W. Illger and H. Eckes, *Angew. Chem. Internat. Ed.*, **12**, 1010 (1973).

Author Index 1973

Subject Index 1973

RETOUR LE: